T0329853

FOSSIL MAMMALS OF ASIA

FOSSIL MAMMALS OF ASIA

NEOGENE BIOSTRATIGRAPHY AND CHRONOLOGY

Edited by Xiaoming Wang, Lawrence J. Flynn, and Mikael Fortelius

Columbia University Press   *New York*

Supported by the Special Researches Program of Basic Science and Technology
(grant no. 2006FY120400), Ministry of Science and Technology, People's Republic of China.

Columbia University Press
Publishers Since 1893
New York   Chichester, West Sussex
cup.columbia.edu

Library of Congress Cataloging-in-Publication Data
Fossil mammals of Asia : Neogene biostratigraphy and chronology / edited by Xiaoming Wang,
Lawrence J. Flynn, and Mikael Fortelius.
p. cm.
Includes bibliographical references and index.
Summary: "This book is on the emergence of mammals in Asia, based largely on new fossil
finds throughout Asia and cutting-edge biostratigraphic and geochemical methods of dating
the fossils and their geological substrate"—Provided by publisher.
ISBN 978-0-231-15012- 5 (cloth)
1. Mammals, Fossil—Asia.   2. Paleontology—Neogene.   3. Paleontology—Asia.
4. Geology, Stratigraphic—Neogene.   5. Geology, Stratigraphic—Asia.   I. Wang,
Xiaoming, 1957– II. Flynn, Lawrence J. (Lawrence John), 1952–
III. Fortelius, Mikael, 1954–
QE881.F66 2012
569.095—dc23
2012008531

COVER IMAGE: About 15 Ma, in the early Middle Miocene of Central Asia, an agitated *Kubanochoerus*, with characteristically high facial horns, charges up a trail in a riparian habitat. The *Kubanochoerus* was a large, long-legged member of the pig family (Suidae), with individuals of some species exceeding 500 kg. It is charging past the fossorial rodent *Tachyoryctoides*, a fully subterranean extinct muroid unrelated to modern burrowing rodents. *Tachyoryctoides* was larger than the living Asiatic zokor, a mole rat of body mass typically 200 g or more. Both mammals are iconic for a large part of Asia, from Mongolia and China, eastward to the Aral Sea. They overlap in time, although *Kubanochoerus* characterizes Middle Miocene Tunggurrian-age faunas, while *Tachyoryctoides* is commonly found in Oligocene and Early Miocene assemblages. (Illustration by Mauricio Antón)

To Richard H. Tedford

1929–2011

We dedicate this book to Richard H. Tedford for the perceptive advancements that he stimulated in the field of evolutionary biology, as represented by the vertebrate fossil record of Asia. Dick's many passions included field documentation of Asian mammalian biostratigraphy, and he set a high standard for all to follow in championing multidisciplinary, international collaborations. His work inspires future generations of researchers.

CONTENTS

FOSSIL MAMMALS OF ASIA

Introduction

Toward a Continental Asian Biostratigraphic and Geochronologic Framework

XIAOMING WANG, LAWRENCE J. FLYNN, AND MIKAEL FORTELIUS

Strategically located between North America, Europe, and Africa, Asia is at the crossroads of intercontinental migrations of terrestrial mammals. Asia thus plays a crucial role in our understanding of mammalian evolution, zoogeography, and related questions about first appearances of immigrant mammals in surrounding continents and their roles as major markers of biochronology. As the largest continent, Asia is the locus of origination for many groups of mammals and/or a site of significant subsequent evolution. The temporal and spatial distributions of these mammals in Asia thus provide a vital link to related clades in surrounding continents (figure I.1; see figure I.3). Such a strategic role is particularly apparent during the Neogene (~23–2.6 Ma) when Asia was intermittently connected to Africa and North America, and widely connected to Europe. Asia also occupies the greatest range of climates and habitats, from tropics to arctic and from rainforests to desert zones, often boasting the most fossiliferous regions with fantastic exposures and producing some of the richest fossil mammal localities in the world. It is therefore no exaggeration that Asia is central to a global understanding of mammalian history.

Such importance and opportunity notwithstanding, with the exception of a few instances (such as northern Pakistan), Asian mammalian biostratigraphy lags behind that science in Europe and North America for historical reasons, and many unresolved issues become bottlenecks for a detailed understanding of mammalian evolution elsewhere. Despite a relatively late start, a tremendous surge has been seen in recent decades in indigenous re-

search, with international collaborations. Asian mammalian biostratigraphy is at a stage where local or regional frameworks are beginning to take shape, but there is no attempt at linking these regional syntheses to derive a continent-wide perspective. Asian vertebrate paleontologists are largely operating within the borders of their own countries, with infrequent communication across political boundaries. This is in contrast to situations in North America and Europe, where fluid exchange of information and ideas results in continuous refinement of continent-wide chronological schemes that are widely accepted among practitioners (e.g., Woodburne 1987; Steininger et al. 1996; Steininger 1999; Woodburne 2004).

During the last 30 years, an indigenous continental mammalian chronological system has been emerging, mostly based on existing, relatively well-studied faunas in China (Chiu et al. 1979; Li et al. 1984; Qiu 1989; Qiu and Qiu 1990, 1995; Tong et al. 1995; Qiu et al. 1999; Deng 2006). These compilations, however, suffer from some shortcomings. Foremost is constant looking to Europe as a reference for relative correlations. To a certain extent, this is inevitable as Asia and Europe constitute essentially a single continent during much of the Neogene and at any given time, the two "continents" share many faunal characteristics. However, this tendency of looking to the West for guidance also breeds a reluctance to build indigenous systems. As a result, discussions about chronology tend to make references to the European Neogene Mammal units (MN system), as if the latter's "stamp of approval" would somehow make a more

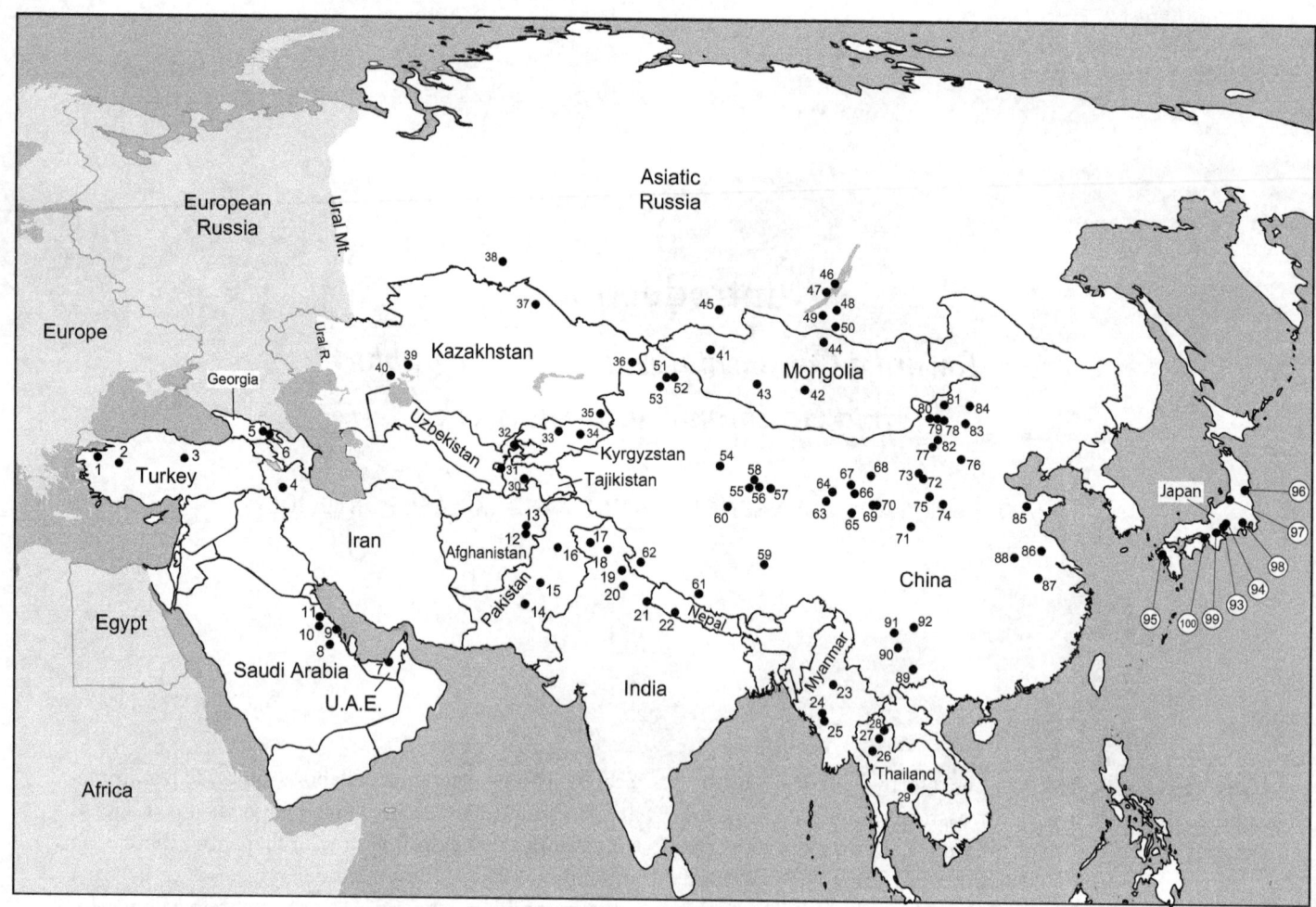

Figure I.1 Main Neogene vertebrate fossil-producing regions or localities in Asia discussed in this volume. A traditional definition of the continent of Asia is adopted here (areas without shade), even though such a definition is somewhat arbitrary and often does not represent natural boundaries of faunal provinces. The Aral Sea is currently much smaller than it is shown in this map.

Afghanistan: (*12*) Khurdkabul Basin; (*13*) Kabul Basin. **China**: (*51*) Botamoyin (XJ99005) section, Junggar Basin, Xinjiang Autonomous Region; (*52*) Chibaerwoyi section, Junggar Basin, Xinjiang Autonomous Region; (*53*) Dingshanyanchi section, Junggar Basin, Xinjiang Autonomous Region; (*54*) Xishuigou section, Tabenbuluk (Danghe) Basin, Gansu Province; (*55*) Olongbuluk section, Qaidam Basin, Qinghai Province; (*56*) Tuosu Nor section, Qaidam Basin, Qinghai Province; (*57*) Shengou section, Qaidam Basin, Qinghai Province; (*58*) Huaitoutala section, Qaidam Basin, Qinghai Province; (*59*) Bulong Basin, Tibetan Autonomous Region; (*60*) Kunlun Pass Basin, Qinghai Province; (*61*) Gyirong Basin, Tibetan Autonomous Region; (*62*) Zanda Basin, Tibetan Autonomous Region; (*63*) Guide Basin, Qinghai Province; (*64*) Xining Basin, Qinghai Province; (*65*) Linxia Basin, Gansu Province; (*66*) Zhangjiaping-Duitinggou section, Lanzhou Basin, Gansu Province; (*67*) Quantougou section, Lanzhou Basin, Gansu Province; (*68*) Tongxin Basin, Ningxia Autonomous Region; (*69*) Leijiahe, Lingtai area, Gansu Province; (*70*) Renjiagou, Lingtai area, Gansu Province; (*71*) Lantian area, Shaanxi Province; (*72*) Baode Basin, Shanxi Province; (*73*) Fugu area, Shaanxi Province; (*74*) Yushe Basin, Shanxi Province; (*75*) Jingle Basin, Shanxi Province; (*76*) Nihewan Basin, Hebei Province; (*77*) Damiao area, Inner Mongolia Autonomous Region; (*78*) Aoerban area, Inner Mongolia Autonomous Region; (*79*) Gashunyinadege area, Inner Mongolia Autonomous Region; (*80*) Tunggur Tableland, Inner Mongolia Autonomous Region; (*81*) Baogeda Ula area, Inner Mongolia Autonomous Region; (*82*) Jurh area, Inner Mongolia Autonomous Region; (*83*) Huade area, Inner Mongolia Autonomous Region; (*84*) Gaotege area, Inner Mongolia Autonomous Region; (*85*) Shanwang area, Shandong Province; (*86*) Sihong area, Jiangsu Province; (*87*) Nanjing area, Jiangsu Province; (*88*) Huainan area, Anhui Province; (*89*) Xiaolongtan Basin, Yunnan Province; (*90*) Lufeng Basin, Yunnan Province; (*91*) Yuanmou Basin, Yunnan Province; (*92*) Zhaotong (Chaotung) Basin, Yunnan Province. **Georgia**: (*5*) Bazaleti; (*6*) Eldari. **India**: (*17*) Ramnagar; (*18*) Nurpur; (*19*) Haritalyangar; (*20*) Chandigarh and Haripur Khol areas; (*21*) Kalagarh. **Iran**: (*4*) Maragheh. **Japan**: (*93*) Kani Basin, Gifu Prefecture; (*94*) Mizunami Basin, Gifu Prefecture; (*95*) Sasebo area, Nagasaki Prefecture; (*96*) Sendai area, Miyagi Prefecture; (*97*) Tochio area, Niigata Prefecture; (*98*) Aikawa area, Kanagawa Prefecture; (*99*) Iga-Omi Basin, Mie Prefecture; (*100*) Awaji Island, Hyogo Prefecture. **Kazakhstan**: (*35*) Aktau Mountain area; (*36*) Kalmakpay; (*37*) Pavlodar; (*39*) North Aral region; (*40*) northern Ustyurt region. **Kyrgyzstan**: (*33*) Ortok; (*34*) Djilgyndykoo and Akterek. **Mongolia**: (*41*) Altan-Teli and Hyargas Nor; (*42*) Valley of Lakes; (*43*) Kholobolchi Nor and Hung Kureh; (*44*) Shamar. **Myanmar**: (*23*) Chaungtha; (*24*) Yenangyaung; (*25*) Magway. **Nepal**: (*22*) Dang Valley. **Pakistan**: (*14*) Bugti; (*15*) Zinda Pir; (*16*) Potwar Plateau. **Russia**: (*38*) Novaya Stanitsa and Isakovka; (*45*) Tuva; (*46*) Tagay and Sarayskoe; (*47*) Aya Cave; (*48*) Tologoi 1; (*49*) Udunga; (*50*) Beregovaya. **Saudi Arabia**: (*8*) Al Jadidah; (*9*) Jabal Midra ash-Shamali; (*10*) Ad Dabtiyah; (*11*) As Sarrar. **Tajikistan**: (*30*) Daraispon; (*31*) Magian and Pedjikent. **Thailand**: (*26*) Li Mae Long Basin; (*27*) Mae Moh Basin; (*28*) Chiang Muan Basin; (*29*) Mun River Sand Pits. **Turkey**: (*1*) Paşalar; (*2*) Sinap; (*3*) central and western Anatolia. **United Arab Emirates**: (*7*) Al Gharbia. **Uzbekistan**: (*32*) Kairakkum.

reliable age determination. This is unfortunate because many Asian faunas are derived from basins with long and continuous sections, which, with careful magnetic calibrations, can offer chronological control superior to the long-distance correlations that the MN system ever can achieve.

This book is thus a coming-of-age attempt to synthesize the state of the art. By compiling mammal faunas from all major fossil-producing countries and regions in Asia, we hope to demonstrate that an Asian system can stand on its own, or at the very least be a starting point for further refinements that can ultimately build a major continental system in its own right. This book is the result of a collaborative effort by leading mammalian paleontologists of the world, who gathered in Beijing in 2009 and 2010 for two international conferences for the purpose of formulating an initial framework of Asian continental biostratigraphy (see following section). The complex nature of such a task, which often has to contend with incomplete information, makes it necessarily an interim solution intended to encourage additional research and further debate. A timely publication of this volume, however incomplete it may be in particular areas, stands to gain the most by laying down the principles and practices of mammalian biostratigraphy and geochronology from all regions and countries. Toward this goal, we are confident that a well-established mammalian biostratigraphic framework in Asia will contribute to a global picture of mammalian evolution in a refined chronological context.

BACKGROUND FOR BEIJING WORKSHOPS AND GENESIS OF THIS VOLUME

The idea of an Asian Neogene biostratigraphic meeting in Beijing with Asia-wide participation came up in late June 2007, while the senior author (X. W.) was in Beijing. The main impetus was the recognition that there is, thus far, no Asia-wide forum to discuss the feasibility of an Asian land mammal age system. As an emerging leader, China seems a natural place to take the initiative, as the country embarks on an unprecedented economic development with attendant renaissance in basic research. China also happens to straddle the mid-latitude desert zones that are often the best hunting grounds for vertebrate fossils in the world. Its long history of "dragon bone" hunting, going back hundreds of years in traditional medicine, gives it a head start in vertebrate paleontology.

Given these favorable conditions, a meeting proposal, with endorsements from Zhan-xiang Qiu, Zhu-ding Qiu, and Tao Deng, was submitted to the Institute of Verte-

brate Paleontology and Paleoanthropology (IVPP) in early July 2007. A symposium volume was also included in the proposal. However, organizational efforts did not begin in earnest until April 2008, when IVPP noted that such a meeting would be opportune as a celebration of its 80th anniversary. At this point, co-editors of this volume (LJF and MF) agreed to be involved in the meeting organization and editing of the symposium volume. The main challenge was to raise substantial funds to pay for participants who were otherwise unable to attend. Toward that end, we secured funding from the National Science Foundation (NSF, U.S.), National Natural Science Foundation (NSFC, China), and Society of Vertebrate Paleontology, as well as institutional support from the IVPP. In particular, we adopted the Critical Transitions workshop (a NSF–NSFC cofunded workshop series on the critical transitions in the history of life) as a unifying theme for international collaborations.

The "Neogene terrestrial mammalian biostratigraphy and chronology in Asia—a workshop and symposium toward the establishment of a continent-wide stratigraphic and chronologic framework" was convened at the IVPP over 3 days, June 8–10, 2009, followed by a 4-day post-conference field trip to the Linxia Basin in Gansu Province. More than 70 scholars and graduate students participated in the workshop, with representation from 19 countries, including Austria, China, Finland, France, Germany, Great Britain, Greece, India, Iran, Japan, Mongolia, Pakistan, Russia, Spain, Sweden, Thailand, Turkey, United Arab Emirates, and the United States.

It became apparent during the workshop that existing Chinese mammalian biostratigraphic divisions possess the best potential as the core of an Asian framework, as summarized by Woodburne: "The background of China's long and fundamental role in developing a chronologic system was clearly recognized in this regard, and the array of approaches to developing chronological systems portrayed at this conference provided the Chinese organizers with considerable examples to draw upon in furthering their goals" (unpublished report to the Society of Vertebrate Paleontology by M. O. Woodburne). As one of the chief architects of the Chinese system developed during the past 20 years, Zhan-xiang Qiu was tasked to form a working group for creating such a framework (Qiu et al., chapter 1, this volume). However, it was clear from the beginning, as well as in reviews of various drafts of manuscripts circulated during the workshop, that serious disagreements exist regarding conceptual issues as well as practical problems. Another forum would thus be necessary to give a full airing of the controversies. Toward that end, a second workshop was organized, again funded by the NSF-NSFC critical transitions theme. This second workshop was held at IVPP, March 8–9, 2010, and attended by a small group of key participants from the United States, Finland, and China.

This book, following a similar volume on North American mammals (Woodburne 2004), is the culmination of these efforts. It attempts to bring together state-of-the-art Asian biostratigraphy and geochronology with the widest representation possible.

SUMMARY OF WORKSHOP DISCUSSIONS AND RESOLUTIONS

One of the distinguishing features of the workshops is the open discussion about concepts and practices, as well as the diversity of opinions. Much reflection is given to practices elsewhere in the world. In particular, the European Neogene Mammal units (MN system) and North American Land Mammal Ages (NALMA) are closely scrutinized for strengths and weaknesses in the hope of building a better system. Many of the comments during the workshop are indicative of current sentiments regarding historic developments, and they are briefly summarized here as extractions from meeting minutes with original commentators cited in parentheses when appropriate.

There is general recognition that the European MN system, although very practical and widely used, has some serious limitations, mostly out of necessity rather than by design. By its own nature and often for lack of long stratigraphic sections with unambiguous superpositional relationships, the MN system is a formulation of biozonation that cannot distinguish diachrony even in cases of precise correlation, and the system would not be able to distinguish time differences in correlative faunas (L. Werdelin). Furthermore, correlation errors can be as much as two MN units above and below (M. Fortelius). Whenever possible, therefore, an Asian system should avoid the deficiencies in the MN zonation, which is undergoing revision to improve the basis of those units. For example, current work by Spanish colleagues is recalibrating MN units to base them on a true biostratigraphic framework (J. Agustí), but where this is done, the Iberian equivalents are often younger than those in mainland Europe.

Given the shortcomings of the MN system, the widely used chronostratigraphic stage ("golden spikes" and associated concepts) in the marine realm seems an attractive approach (M. Böhme). Furthermore, most of the marine Neogene stages have been ratified, and the All-China Stratigraphic Commission has been in full agreement with this approach and has attempted to set Chinese con-

tinental Neogene research in motion toward that goal (Z.-x. Qiu). However, there is strong opposition against a chronostratigraphic system by several participants, particularly those who champion an independent system as exemplified by the NALMA. The main problem with golden spikes is that, once nailed, they are no longer flexible, and an Asian system should be based on true biostratigraphy in multiple long sections that can be further refined and revised as new advancements come along (M. Woodburne; see further discussion in "Some Conceptual Issues").

Given the often messy developments of continental land mammal systems, some openly wonder if we should not simply do away with a land mammal age system and use numeric ages instead (F. Bibi). In fact, a biozonation has never been given a high priority in the Siwalik sequence (J. Barry), and people working in South Asia are generally content in talking about absolute ages rather than land mammal ages (L. Flynn). However, most seem to recognize that land mammal ages will always have a place in the formulation of a chronological system because the biological component can never be subjugated under isotopic dating or paleomagnetic dating (M. Woodburne).

Another issue of major concern is the spatial distribution of mammal fossils. Geography is of paramount importance for a super-continent as Eurasia that spans great longitudes and latitudes and crosses many climatic zones. In South Asia and Southeast Asia, roughly the modern Oriental Zoogeographic Province, mammals share much greater similarities during much of the Neogene, whereas the low latitude faunas in southern China and southeastern Asia are generally unlike those from mid-latitudes in north China and the rest of central Asia (R. Hanta; L. Flynn). Nonetheless, mid-latitude faunas can often be recognized along great longitudinal spans, such as the Pikermian chronofauna originally recognized from late Miocene localities in Greece, which have comparable equivalents in north China (M. Fortelius).

While conventional biostratigraphic approaches are perhaps best employed to generate regional stratifications, the rising field of computational ordering (seriation) of localities based on taxonomic presence/absence information (e.g., Alroy 1992, 1994, 1998, 2000; Fortelius et al. 2006; Puolamäki et al. 2006) may well offer more general systems. Ultimately based on the essentially irreversible evolution of lineages and communities (climate driven or intrinsic), computational approaches are especially attractive as a potential route toward a future, continent-wide mammal chronology. Indeed, a preliminary study by Alroy et al. (1998) already suggested that the biochronological signal was stronger for Western Eurasia as a whole than for its western and eastern parts separated at 20 degrees eastern longitude. Such a result reflects well-known phenomena of the fossil record: regional persistence of core lineages and the connectedness of coeval communities through long-range dispersal of species. For the operational detection of both of these, two requirements are critical: standardized taxonomy and presence of exceptionally well-sampled "Rosetta localities" (Alroy 1992). Therefore, a key priority for computational as well as for conventional approaches to biostratigraphy is coming to grips with problems of synonymy and regional taxonomic "dialects."

For the time being, it is clear that an Asian land mammal system faces some challenges common to all continents (fossil mammals are rare; sampling errors are high; diachrony is common) as well as unique challenges in Asia (uneven studies in different countries; lack of marine interface; shortage of datable volcanic rocks interbedded in sediments; high degree of zoogeographic differentiation; some degree of endemism). Recognizing these challenges, the workshop participants adopted the following resolutions by unanimous consent:

(1) an Asian chronologic system, independent from the European MN units, is needed; (2) such a system should be mainly based on biological events, associated with paleomagnetic and isotopic dates where available; (3) the existing Chinese system, imperfect as it is, can serve as a starting point that can evolve through time; (4) the benefit of such a system is a common framework in which hypotheses of biological events across the continent can be rigorously tested; (5) a committee headed by Zhan-xiang Qiu, Tao Deng, Zhuding Qiu, Chuan-kui Li, Zhao-qun Zhang, Ban-yue Wang, and Xiaoming Wang (additional expertise will be recruited as need arises) will work toward the above goal; and (6) additional subcommittees of relevant specialists to clean up taxonomies should be established.

SOME CONCEPTUAL ISSUES

Mammalian biostratigraphy has been and still is the primary means for Cenozoic terrestrial geochronology. Continental mammalian biostratigraphic frameworks are integral to related disciplines such as mammalian evolution, zoogeography, paleoecology, and paleoenvironment. Various chronologic frameworks have been established in all continents except Antarctica, but their qualities (precision and internal consistencies) vary greatly, with European and North American systems being the most

mature and those of other continents far less so. In developing an Asian land mammal system, much of the focus, both in workshop discussion and in subsequent manuscript development, has thus centered on the best practices in Europe and North America.

Although the Chinese land mammal age system has implicitly or explicitly adopted certain aspects of the European or North American practices, past iterations have mostly been concerned with articulations of the empirical evidence instead of an examination of the methodologies (e.g., Qiu 1989; Qiu and Qiu 1990; Qiu et al. 1999). An introspective assessment of current practices in the world thus represents a welcomed first step to construct a thoughtful system that is both methodologically defensible and practically useful.

From the beginning of the first workshop, it became clear that a European-style MN unit system has serious shortcomings because of its general lack of biostratigraphic underpinnings. The MN system, while widely practiced, offers little guidance as a model for Asia. Asia, like North America, possesses all the potential for developing a framework based on biostratigraphy in long stratigraphic sections. Nonetheless, the MN system is by far the most influential in Asian biochronologic developments due to the wide connections between the two continents and the large number of shared taxa in various ages. So pervasive are the MN units that it is not uncommon for Asian faunas to be directly compared to European ages/MN units or simply to be labeled with MN designation.

Unity with International Code vs. Regional Independence

One of the most controversial subjects during the two Beijing workshops is the desire to follow the *International Stratigraphic Guide* (*ISG*; Hedberg 1976; Salvador 1994). Intense debates center on the suitability of a chronostratigraphic system in continental settings with golden spikes (Global Stratotype Section and Point, or GSSP) nailed in a physical lithostratigraphic section. The debate is set against a background of recent trends in the Chinese stratigraphic community to adopt the *ISG* protocol, buoyed by the establishment of several Chinese GSSPs for the Mesozoic and Paleozoic eras (e.g., Yin et al. 2001; Chen et al. 2006). The All-China Stratigraphic Commission (2001) went as far as selecting many existing Chinese land mammal units as "stages" and briefly characterized each (within Neogene the following were included: Xiejian, Shanwangian, Tunggurian, Baodean, Gaozhuangian, and

Mazegouan). To push these efforts further, the commission distributed grants to the IVPP to flesh out Cenozoic stages in China, which resulted in some preliminary boundary selections, mostly coinciding with those endorsed by the *ISG* (e.g., Deng et al. 2003; Deng et al. 2004; Deng et al. 2006; Meng et al. 2006; Deng et al. 2007).

Whereas the GSSP standard promoted by the *ISG* is largely accepted in the marine stratigraphic community, it is far from certain how a continental system should proceed given its inherent problems in depositional gaps, rareness of fossils, patchiness in distribution, and insularity of paleoenvironments. While there is general agreement that such factors call for regionally limited chronological systems, commonly at the continental scale or smaller, opinions are deeply divided regarding how to construct such a system and whether such a system should be consistent with the *ISG* recommendations. A prominent example is the North American Land Mammal Age system, which enjoys wide acceptance among North American vertebrate paleontologists but is at variance from the recommendations of the *ISG*. Fundamental to the premise of the NALMA is the recognition that there is no inherent reason why events in land mammal evolution should coincide with those of marine organisms from half a world away. In fact, part of the initial impetus by the "Wood Committee" to establish the North American "provincial ages" is an attempt to avoid the "dangerous ambiguity, cumbersome circumlocution, or both" when trying to correlate to the European standard time scale (Wood et al. 1941:2).

Following the recommendations by the All-China Stratigraphic Commission (2001), Qiu et al. (chapter 1, this volume) propose a Chinese Regional Land Mammal Stage/Age system that they envision will ultimately transition to one fully consistent with the *ISG* standards. Chronostratigraphic boundaries of such a system are based on multiple criteria of lithostratigraphy, magnetostratigraphic reversals, and mammalian first appearances and faunal characterizations. In doing so, Qiu at al. point out that the NALMA also uses lithologic criteria, at least in the case of the lower boundary of the Arikareean. They further argue that land mammal ages cannot be equated to biochrons. In fact, in their opinion, biochrons have no place in a regional chronostratigraphic system. As a step further in making all land mammal stage/age systems conformable to the international standard, Qiu et al. propose that for those mammal ages whose lower boundaries are near the standard international boundaries of a higher rank, such as the Oligo-Miocene and Mio-Pliocene boundaries, the mammal age boundaries should coincide with the epoch boundaries.

Bringing their vast experience in the North American land mammal age system to bear, Woodburne, Tedford, and Lindsay (chapter 2, this volume) proposed a framework of an endemic North China mammalian biochronologic system as an evolving standard of temporal intervals that accounts for all of Neogene time without gaps or overlaps. They suggest that such a system represents informal biochronologic units, and until this system has been widely tested, formalized international chronostratigraphic standards should not be applied. Woodburne et al.'s premise is that a land mammal age system should always give fossil mammals prominent consideration. Methodologically, they strongly advocate for a single taxon definition of mammal age boundaries in order to minimize potential gaps and overlaps.

In a compromise approach, Meng et al. (chapter 3, this volume) used the Xiejian as an example to illustrate their single-criterion, single-taxon definition, but largely within the chronostratigraphic framework recommended by the *ISG*. As such, Meng et al.'s scheme allows future adjustments of boundary definition but it must be tied to a specific stratotype section. Their Xiejian example also explores the case where a stage/age in question roughly coincides with a major international boundary of higher rank (in this case, Oligo–Miocene boundary). They treat these two boundaries as strictly separate entities and place the Xiejian lower boundary 0.5 myr above the international Oligo–Miocene boundary.

This controversy pits chronostratigraphic boundary definition as a convention serving to standardize nomenclature against a more dynamic land mammal age scheme (as practiced by North American paleontologists), emphasizing empirical evidence and flexibility of shifting boundaries. To a certain extent, the former seems to signal a desire to move toward an internationally accepted, marine invertebrate norm, whereas the latter represents a more self-confident approach to a regional, continental mammal-based system divorced from *ISG* standards.

CURRENT STATE OF ASIAN BIOCHRONOLOGY

Primarily as a working hypothesis and overview aid, we compile a generalized chart to summarize the state of continental Neogene mammalian biostratigraphy and chronology, usually based on the most recent published updates in the respective regions, including those in this volume (figure I.2). Not intended as an original synthesis, these diagrammatic summaries provide a measure of consistency in presentation of existing stratigraphic frameworks and thus serve as a quick index for existing

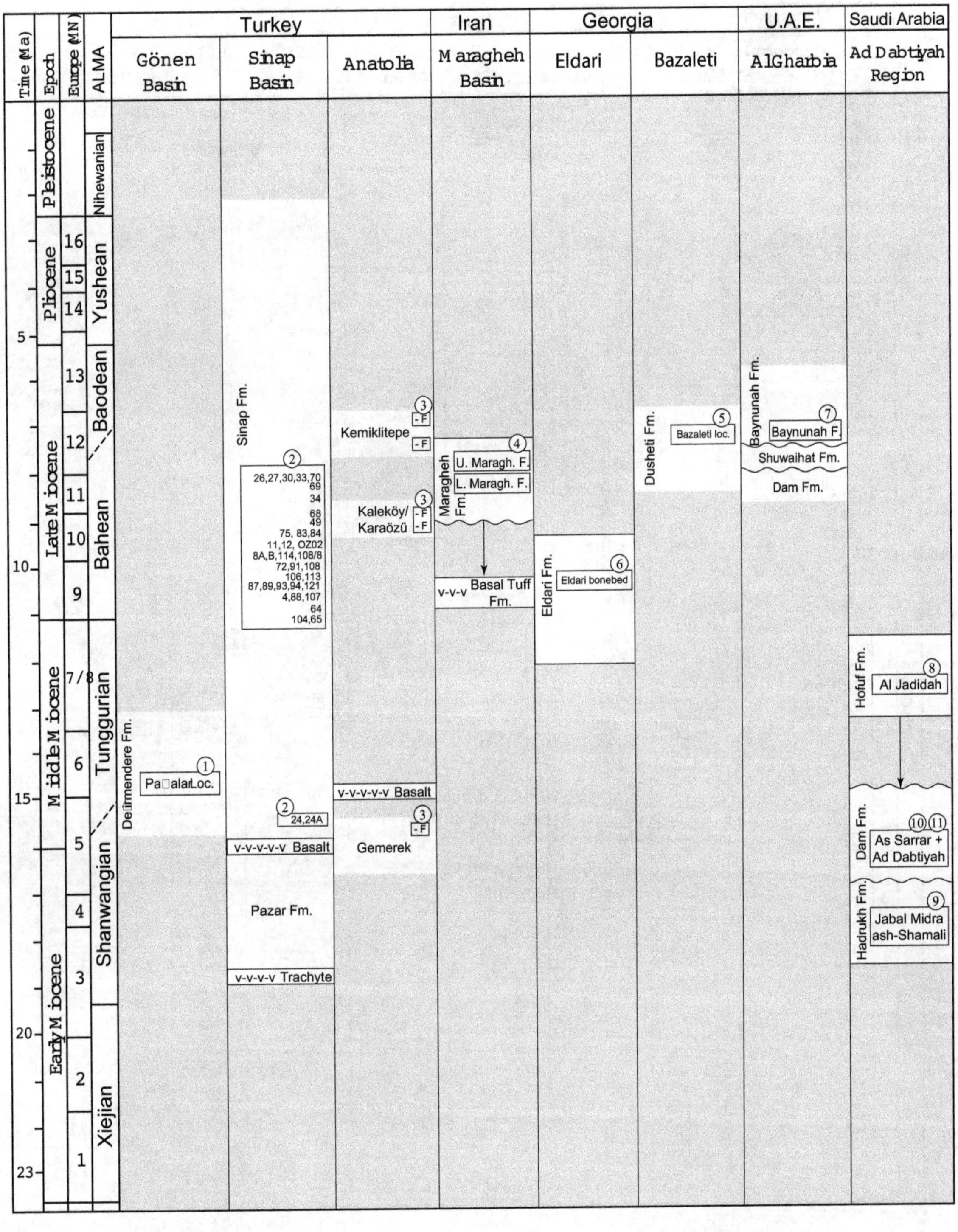

Figure I.2a Asian terrestrial Neogene vertebrate-producing strata, mammalian faunas, and faunistic complexes (abbreviated as "F" and "F C") or fossil-producing horizons (placed within a box), and their chronologic relationships. Solid lines above and below a block of strata indicate approximate duration of the strata (often constrained by magnetostratigraphy), and absence of such lines indicates uncertainty of the duration of sedimentation. We adopt the Neogene-Quaternary (Pliocene/Pleistocene) boundary at 2.6 Ma, as formally defined by the International Commission on Stratigraphy (Mascarelli 2009), and many of the faunas falling within the 1.8–2.6 Ma interval and formerly considered late Pliocene are not treated here. Locality numbers correspond to those in figure I.1. Major Neogene faunas and strata of western Asia: (*1*) Paşalar, Gönen Basin, Turkey (Andrews and Alpagut 1990); (*2*) Sinap, Turkey (Kappelman et al. 2003; numbers indicate select fossil localities); (*3*) central and western Anatolia, Turkey (Sen 1996; "- F" indicates fossil horizons); (*4*) Maragheh, Iran (Mirzaie Ataabadi et al., chapter 25, this volume); (*5*) Bazaleti, Georgia (Vekua and Lordkipanidze 2008; Vangengeim and Tesakov, chapter 23, this volume); (*6*) Eldari, Georgia (Vangengeim, Lungu, and Tesakov 2006; Vekua and Lordkipanidze 2008); (*7*) Al Gharbia, United Arab Emirates (Bibi et al., chapter 27, this volume); (*8*) Al Jadidah (Hofuf Formation), Saudi Arabia (Thomas 1983; Whybrow, McClure, and Elliott 1987; Whybrow and Clemens 1999; Flynn and Wessels, chapter 18, this volume); (*9*) Jabal Midra ash-Shamali (Hadrukh Formation), Saudi Arabia (Whybrow, McClure, and Elliott 1987; Whybrow and Clemens 1999; Flynn and Wessels, chapter 18, this volume); (*10*) Ad Dabtiyah (Dam Formation), Saudi Arabia (Whybrow, McClure, and Elliott 1987; Whybrow and Clemens 1999); (*11*) As Sarrar (Dam Formation), Saudi Arabia (Whybrow, McClure, and Elliott 1987; Whybrow and Clemens 1999; Flynn and Wessels, chapter 18, this volume).

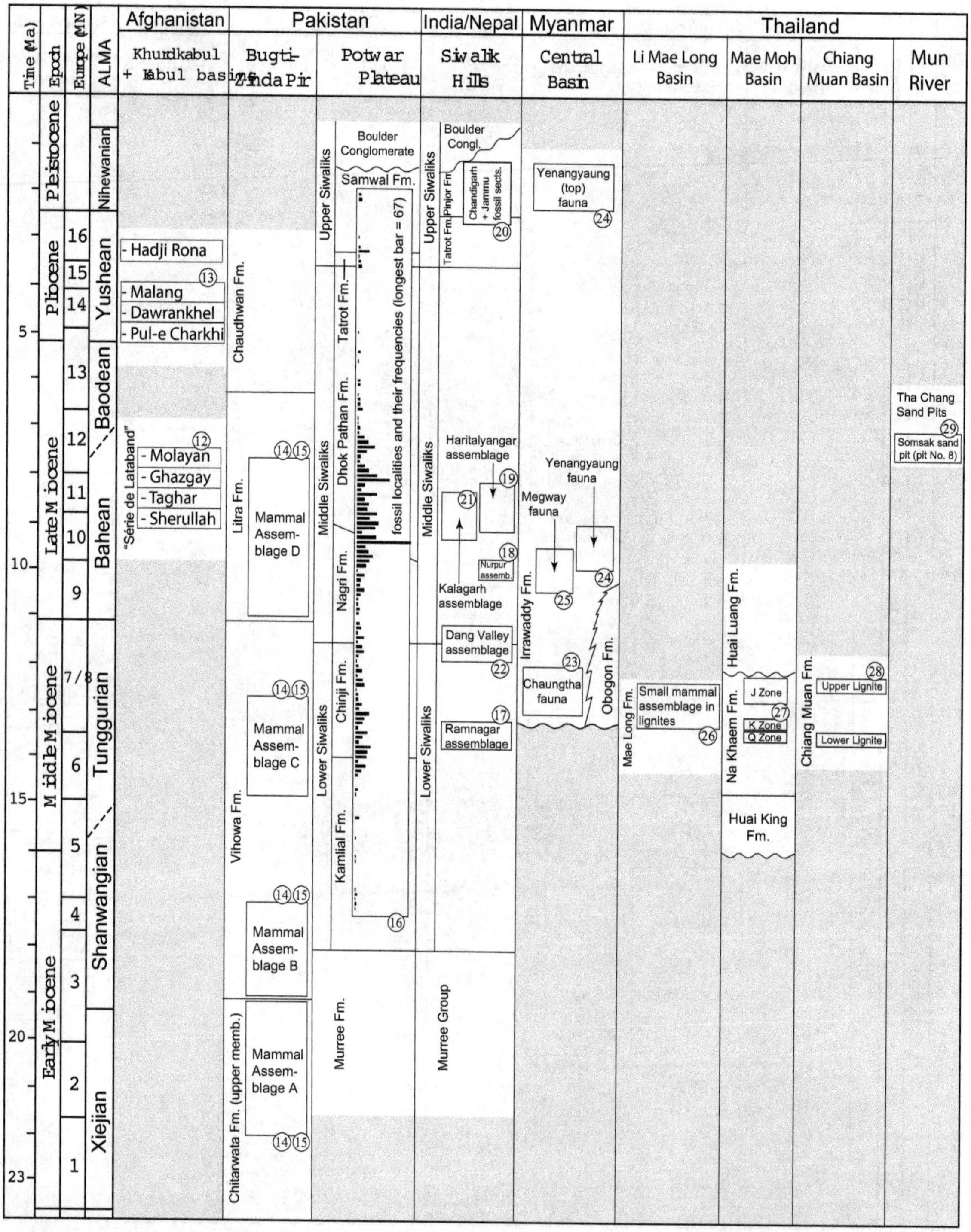

Figure I.2b Major Neogene faunas and strata of South and Southeast Asia: (*12*) Khurdkabul Basin, Afghanistan (Sen 2001); (*13*) Kabul Basin, Afghanistan (Brandy 1981; Sen 1983, 2001); (*14*) Bugti and (*15*) Zinda Pir, Pakistan (Antoine et al., chapter 16, this volume; Flynn et al., chapter 14, this volume); (*16*) Potwar Plateau (Siwaliks), Pakistan (Barry et al., chapter 15, this volume; Flynn et al., chapter 14, this volume); (*17*) Ramnagar, India (Patnaik, chapter 17, this volume); (*18*) Nurpur, India (Patnaik, chapter 17, this volume); (*19*) Haritalyangar, India (Patnaik, chapter 17, this volume); (*20*) Chandigarh (including Patiali Rao, Ghaggar, and Nadah sections) and Haripur Khol areas, India (Patnaik, chapter 17, this volume); (*21*) Kalagarh, India (Patnaik, chapter 17, this volume); (*22*) Dang Valley, Nepal (Patnaik, chapter 17, this volume); (*23*) Chaungtha, Myanmar (Chavasseau et al., chapter 19, this volume); (*24*) Yenangyaung, Myanmar (Chavasseau et al., chapter 19, this volume); (*25*) Magway, Myanmar (Chavasseau et al., chapter 19, this volume); (*26*) Li Mae Long Basin, Thailand (Mein and Ginsburg 1997; Ratanasthien 2002; Chaimanee et al. 2007); (*27*) Mae Moh Basin, Thailand (Chaimanee et al. 2007; Coster et al. 2010); (*28*) Chiang Muan Basin, Thailand (Coster et al. 2010); (*29*) Mun River Sand Pits, Thailand (Chaimanee et al. 2006; Hanta et al. 2008).

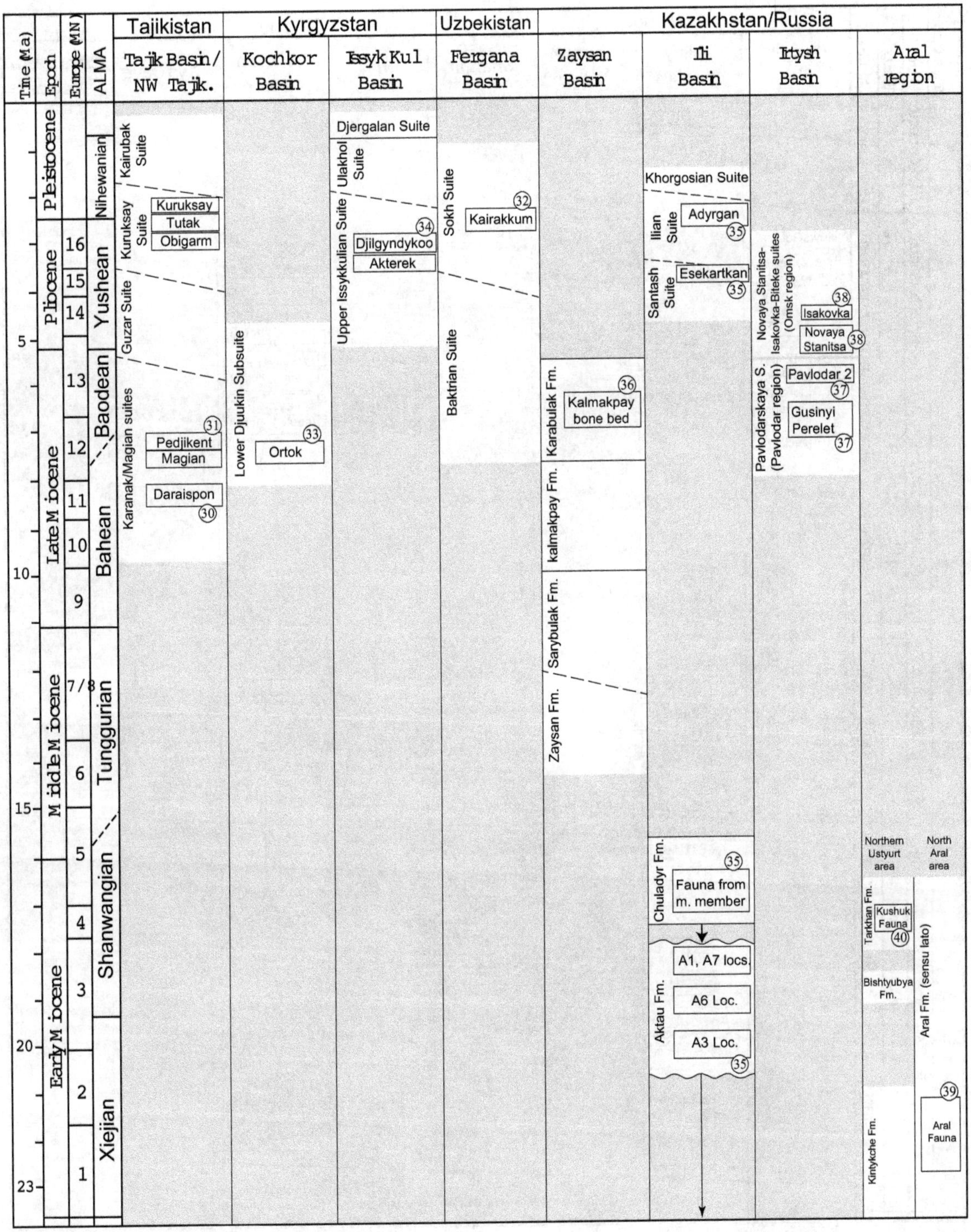

Figure I.2c Major Neogene faunas and strata of Central Asia: (*30*) Daraispon, Tajik Basin, Tajikistan (Sotnikova, Dodonov, and Pen'kov 1997; Vislobokova, Sotnikova, and Dodonov 2001); (*31*) Magian and Pedjikent, northwestern Tajikistan (Sotnikova, Dodonov, and Pen'kov 1997; Vislobokova, Sotnikova, and Dodonov 2001); (*32*) Kairakkum, Fergana Basin, Uzbekistan (Sotnikova, Dodonov, and Pen'kov 1997; Vislobokova, Sotnikova, and Dodonov 2001); (*33*) Ortok, Kochkor Basin, Kyrgyzstan (Sotnikova, Dodonov, and Pen'kov 1997; Vislobokova, Sotnikova, and Dodonov 2001); (*34*) Djilgyndykoo and Akterek, Issyk Kul Lake, Kyrgyzstan (Sotnikova, Dodonov, and Pen'kov 1997; Vislobokova, Sotnikova, and Dodonov 2001); (*35*) Aktau Mountain area (Kordikova and Mavrin 1996; Lucas et al. 1997; Kordikova, Heizmann, and Marvin 2000; stratigraphic nomenclature and faunal contents cannot be easily reconciled among cited authors), Esekartkan and Adyrgan (Sotnikova, Dodonov, and Pen'kov 1997; Vislobokova, Sotnikova, and Dodonov 2001), Ili Basin, Kazakhstan; (*36*) Kalmakpay, Zaysan Basin, Kazakhstan (Vangengeim et al. 1993; Sotnikova, Dodonov, and Pen'kov 1997; Vislobokova, Sotnikova, and Dodonov 2001; Lucas et al. 2009); (*37*) Pavlodar, Pavlodar region, Irtysh River, Kazakhstan (Gnibidenko 1990; Vislobokova, Sotnikova, and Dodonov 2001; Zykin, Zykina, and Zazhigin 2007); (*38*) Novaya Stanitsa and Isakovka, Omsk region, Irtysh River, Russia (Zykin and Zazhigin 2004; Zykin, Zykina, and Zazhigin 2007); (*39*) North Aral regions, Kazakhstan (Lopatin 2004); (*40*) northern Ustyurt region, Kazakhstan (Lopatin 2004).

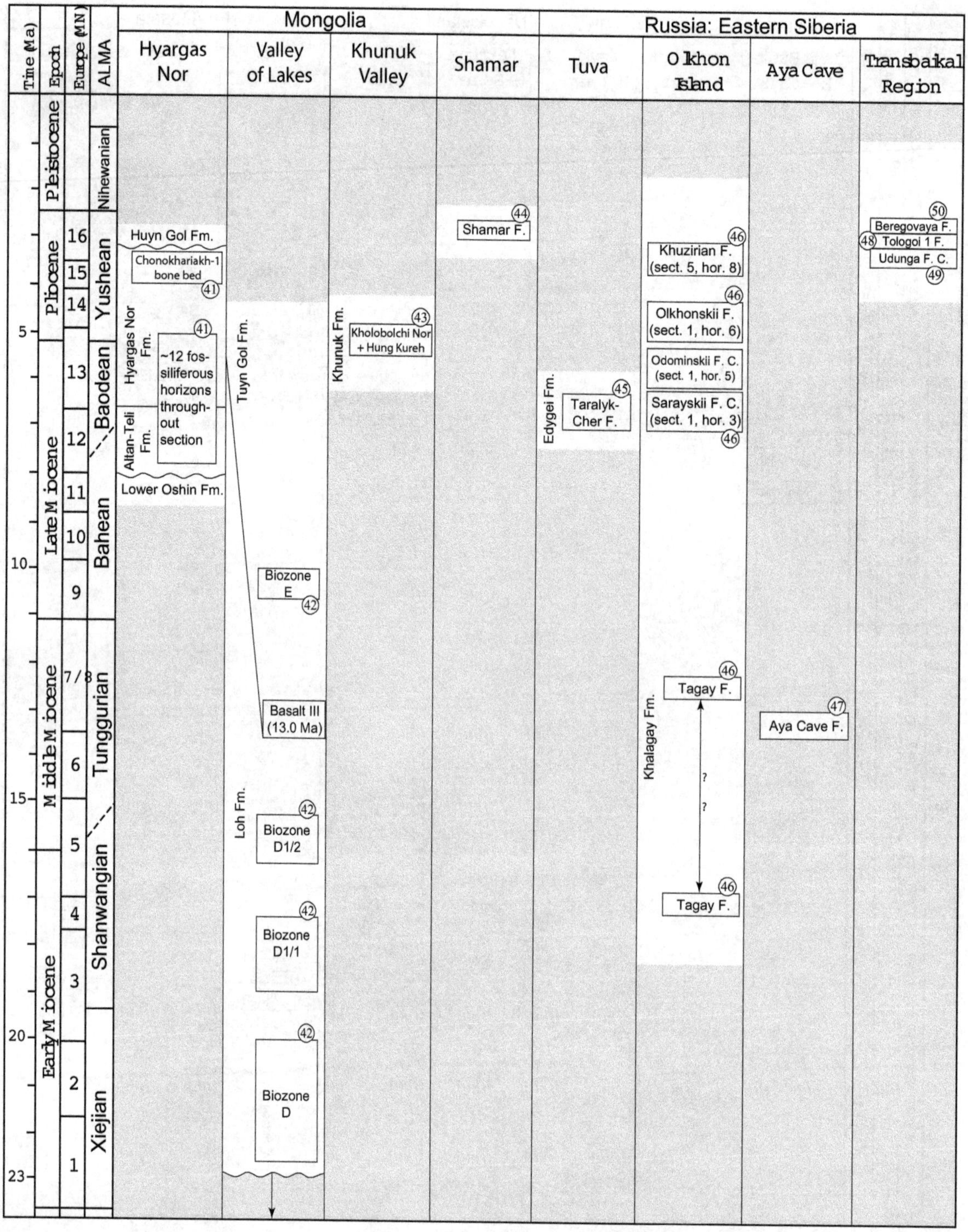

Figure I.2d Major Neogene faunas and strata of Mongolia and eastern Russia: (*41*) Altan-Teli and Hyargas (Khyargas, Khirgis) Nor, Mongolia (Pevzner et al. 1982; but see Tedford et al. 1991 for an alternative interpretation; Sotnikova 2006); (*42*) Valley of Lakes, Mongolia (Höck et al. 1999; Daxner-Höck et al., chapter 20, this volume); (*43*) Kholobolchi Nor and Hung Kureh, Mongolia (Flynn and Bernor 1987); (*44*) Shamar, Mongolia (Vislobokova, Sotnikova, and Dodonov 2001); (*45*) Taralyk-Cher, Tuva, Russia (Vislobokova 2009); (*46*) Tagay (Tagai) and Sarayskoe (Saray), Olkhon Island, Lake Baikal, Russia (arrows indicate widely divergent interpretations of the Tagay Fauna) (Daxner-Höck et al., chapter 22, this volume; Erbajeva and Alexeeva, chapter 21, this volume); (*47*) Aya Cave, western shore of Lake Baikal, Russia (Erbajeva and Filippov 1997; Sen and Erbajeva 2011); (*48*) Tologoi 1, (*49*) Udunga, and (*50*) Beregovaya of Transbaikal area, east of Lake Baikal, Russia (Erbajeva and Alexeeva, chapter 21, this volume).

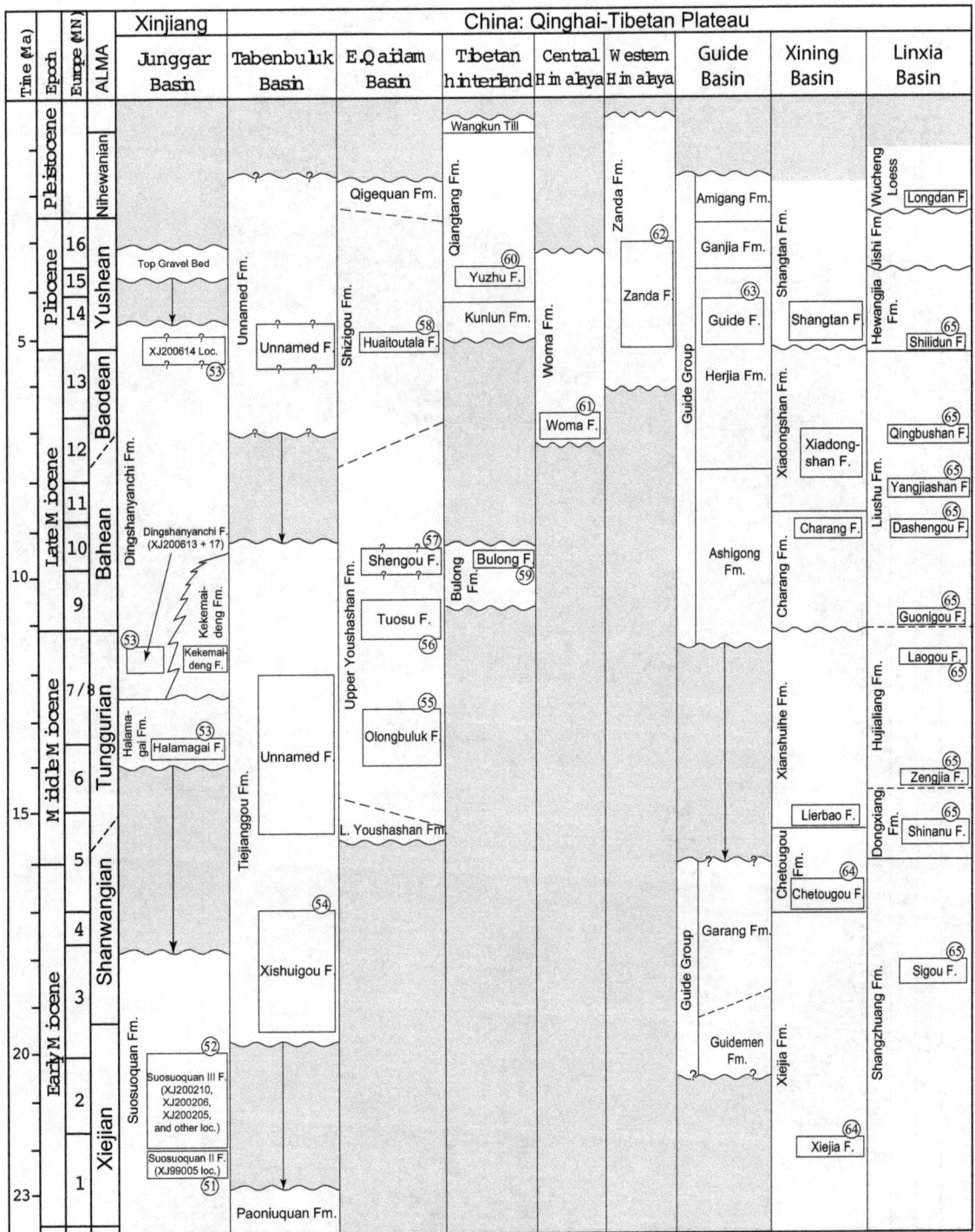

Figure I.2e Major Neogene faunas and strata of Xinjiang and the Tibetan Plateau: (*51*) Botamoyin (XJ99005) section, Junggar Basin, Xinjiang Autonomous Region (Meng et al. 2006; Meng et al., chapter 3, this volume); (*52*) Chibaerwoyi section, Junggar Basin, Xinjiang Autonomous Region (Meng et al. 2006; Meng et al., chapter 3, this volume); (*53*) Dingshanyanchi section, Junggar Basin, Xinjiang Autonomous Region (Meng et al. 2008); (*54*) Xishuigou Fauna, Tabenbuluk (Danghe) Basin, Gansu Province (Wang, Qiu, and Opdyke 2003; Wang et al., chapter 10, this volume); (*55*) Olongbuluk Fauna, Qaidam Basin, Qinghai Province (Wang et al. 2007; Wang et al. 2011; Wang et al., chapter 10, this volume); (*56*) Tuosu Fauna, Qaidam Basin, Qinghai Province (Wang et al. 2007; Wang et al. 2011; Wang et al., chapter 10, this volume); (*57*) Shengou Fauna, Qaidam Basin, Qinghai Province (Wang et al. 2007; Qiu and Li 2008; Wang et al., chapter 10, this volume); (*58*) Huaitoutala Fauna, Qaidam Basin, Qinghai Province (Wang et al. 2007; Wang et al. 2011; Wang et al., chapter 10, this volume); (*59*) Bulong (Biru) Fauna, Bulong Basin, Tibetan Autonomous Region (Huang et al. 1980; Zheng 1980; Wang et al., chapter 10, this volume); (*60*) Yuzhu Fauna, Kunlun Pass Basin, Qinghai Province (Song et al. 2005; Wang et al., chapter 10, this volume); (*61*) Woma Fauna, Gyirong Basin, Tibetan Autonomous Region (Huang et al. 1980; Yue et al. 2004; Wang et al., chapter 10, this volume); (*62*) Zanda Fauna, Zanda Basin, Tibetan Autonomous Region (Deng et al. 2011; Wang et al., chapter 10, this volume); (*63*) Guide Fauna, Guide Basin, Qinghai Province (Zheng, Wu, and Li 1985; Fang et al. 2005; Wang et al., chapter 10, this volume); (*64*) Xiejia and Chetougou faunas, Xining Basin, Qinghai Province (Li and Qiu 1980; Li, Qiu, and Wang 1981; Qiu et al., chapter 1, this volume); (*65*) Linxia Basin, Gansu Province (Deng et al., chapter 9, this volume; Qiu et al., chapter 1, this volume).

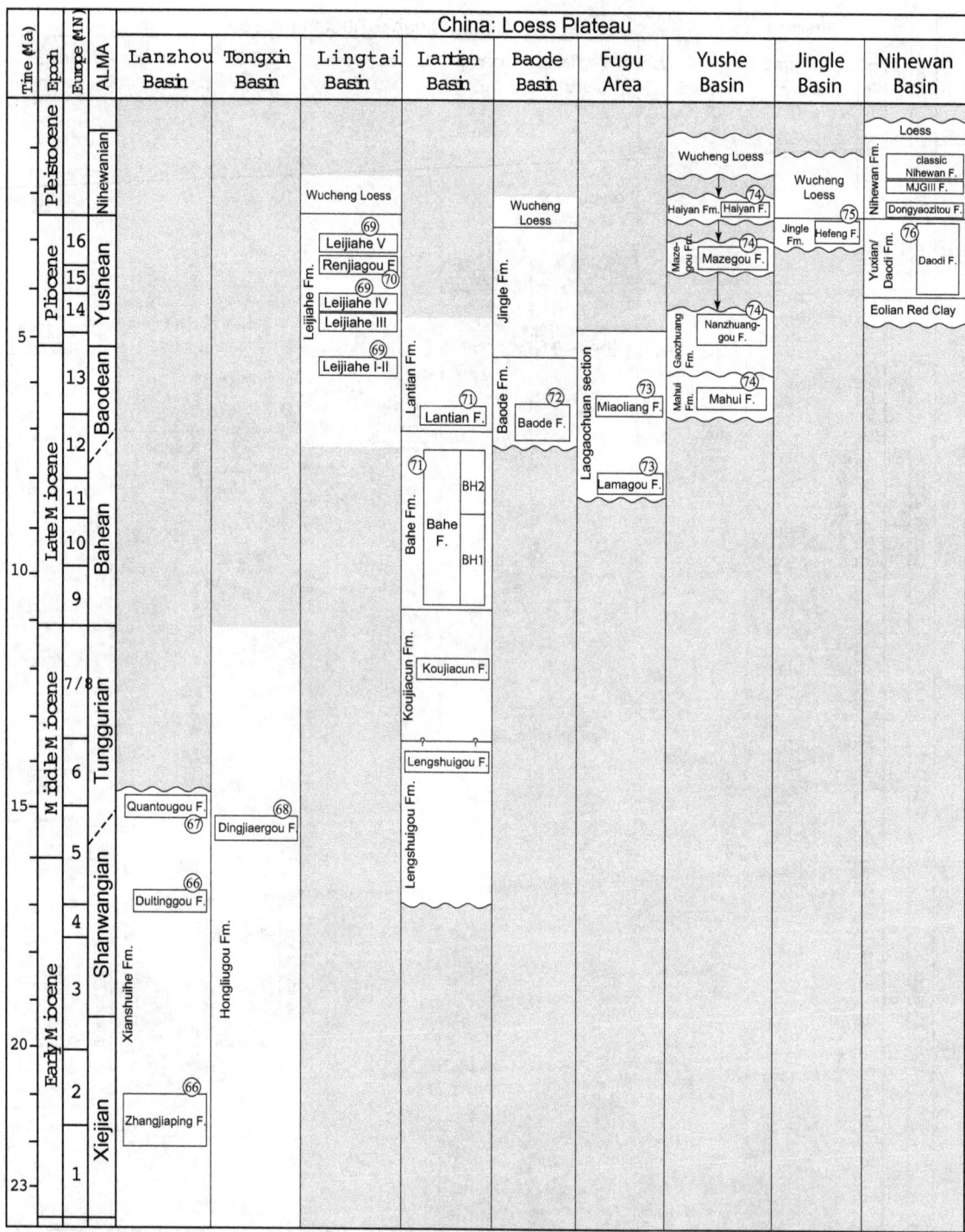

Figure I.2f Major Neogene faunas and strata of the Loess Plateau: (66) Zhangjiaping and Duitinggou faunas, Lanzhou Basin, Gansu Province (Qiu et al. 2001; Qiu et al., chapter 1, this volume); (67) Quantougou Fauna, Lanzhou Basin, Gansu Province (Qiu 2001; Qiu et al., chapter 1 this volume); (68) Dingjiaergou Fauna, Tongxin Basin, Ningxia Autonomous Region (Qiu et al., chapter 1, this volume); (69) Leijiahe biozones I–V, Lingtai, Gansu Province (Zheng and Zhang 2001; Qiu et al., chapter 1, this volume); (70) Renjiagou Fauna, Lingtai, Gansu Province (Zhang et al. 1999); (71) Bahe Fauna, Lantian Basin, Shaanxi Province (Zhang et al. 2002; Kaakinen and Lunkka 2003; Zhang et al., chapter 6, this volume); (72) Baode Fauna, Shanxi Province (Zhu et al. 2008; Kaakinen et al., chapter 7, this volume); (73) Laogaochuan section, Fugu area, Shaanxi Province (Xue, Zhang, and Yue 1995; Zhang et al. 1995; Xue, Zhang, and Yue 2006); (74) Mahui, Nanzhuanggou, and Mazegou faunas, Yushe Basin, Shanxi Province (Tedford et al. 1991; Flynn, Wu, and Downs 1997); (75) Hefeng Fauna, Jingle Basin, Shanxi Province (Chen 1994; Yue and Zhang 1998); (76) Daodi Fauna, Nihewan Basin, Hebei Province (Cai et al., chapter 8, this volume).

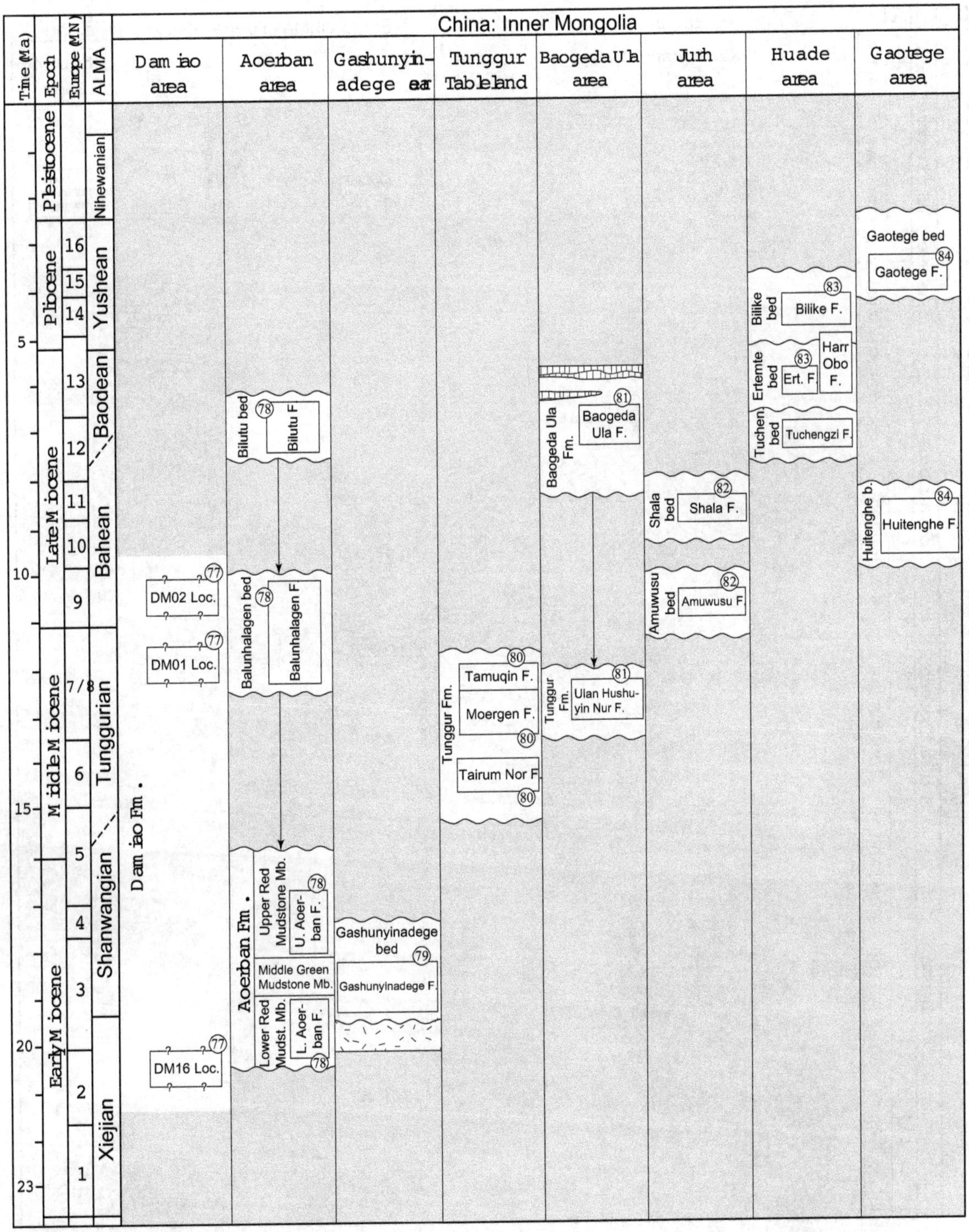

Figure I.2g Major Neogene faunas and strata of Inner Mongolia: (*77*) Damiao section, Siziwang Qi, Inner Mongolia (Zhang et al. 2011); (*78*) upper and lower Aoerban, Balunhalagen, and Bilutu faunas, Inner Mongolia Autonomous Region (Wang et al. 2009; Qiu, Wang, and Li, chapter 5, this volume); (*79*) Gashunyinadege Fauna, Inner Mongolia (Meng, Wang, and Bai 1996; Qiu, Wang, and Li, chapter 5, this volume); (*80*) Tairum Nor, Moergen, and Tamuqin faunas, Inner Mongolia (Qiu 1996; Wang, Qiu, and Opdyke 2003; Qiu, Wang, and Li, chapter 5, this volume); (*81*) Ulan Hushuyin Nur and Baogeda Ula faunas, Inner Mongolia (Qiu, Wang, and Li, chapter 5, this volume); (*82*) Shala and Amuwusu faunas, Inner Mongolia (Qiu, Wang, and Li, chapter 5, this volume); (*83*) Bilike, Ertemte, Harr Obo, and Tuchengzi faunas, Inner Mongolia (Fahlbusch, Qiu, and Storch 1983; Qiu and Storch 2000; Qiu, Wang, and Li, chapter 5, this volume); (*84*) Gaotege and Huitenghe faunas, Inner Mongolia (Li, Wang, and Qiu 2003; Xu et al. 2007; Qiu, Wang, and Li, chapter 5, this volume).

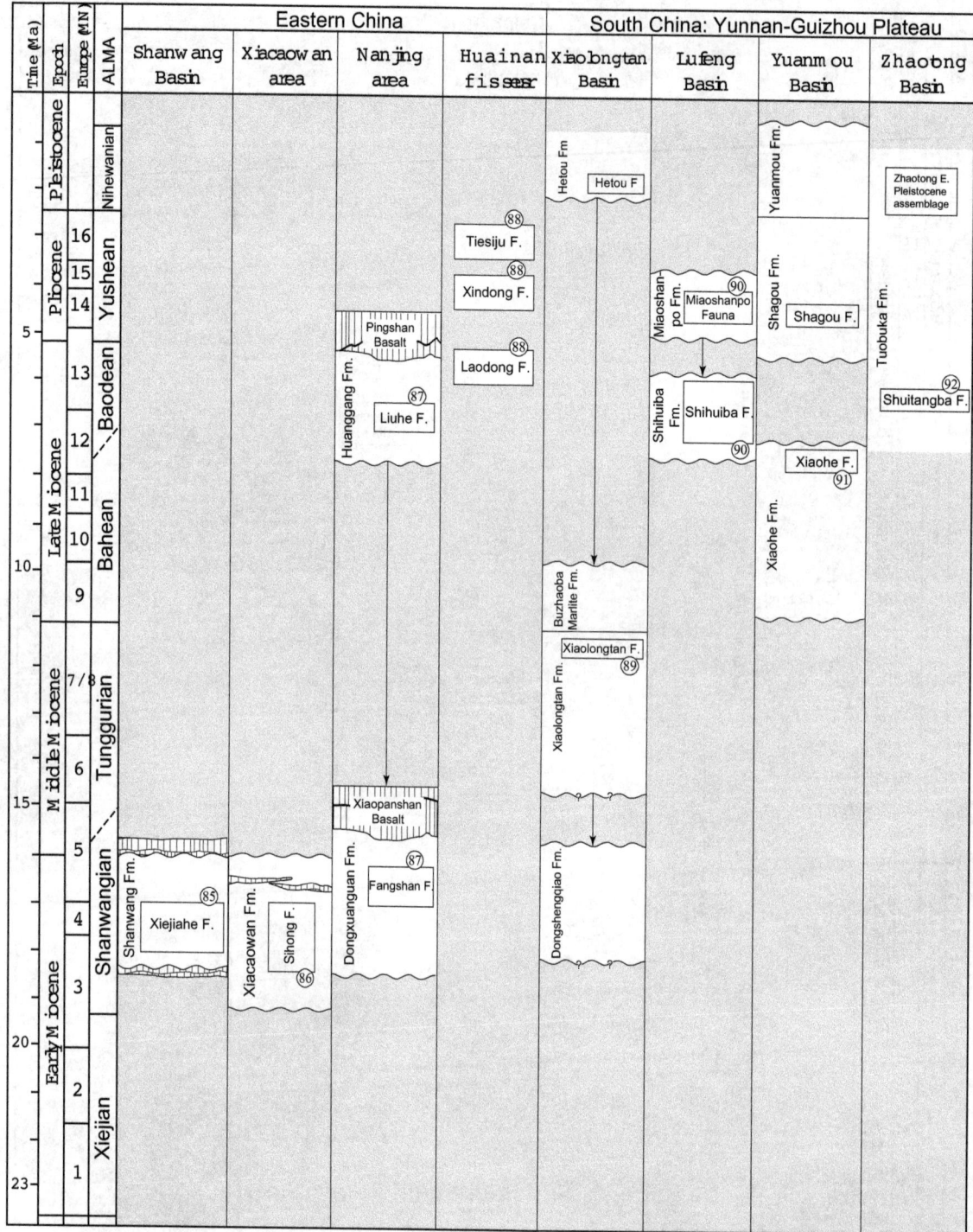

Figure I.2h Major Neogene faunas and strata of eastern China and Yunnan: (85) Xiejiahe Fauna, Shandong Province (Deng, Wang, and Yue 2008; Qiu and Qiu, chapter 4, this volume); (86) Xiacaowan Fauna, Jiangsu Province (Li et al. 1983; Qiu and Qiu, chapter 4, this volume); (87) Fangshan and Liuhe faunas, Jiangsu Province (Bi, Yu, and Qiu 1977; Qiu et al., chapter 1, this volume); (88) Laodong, Xindong, Tiesiju fissure faunas, Anhui Province (Jin, Kawamura, and Tatuno 1999; Jin 2004; Tomida and Jin 2009); (89) Xiaolongtan Fauna, Xiaolongtan Basin, Yunnan Province (Dong 2001; Dong and Qi, chapter 11, this volume); (90) Shihuiba and Miaoshanpo faunas, Lufeng Basin, Yunnan Province (Qi 1985; Chen 1986; Yue and Zhang 2006; Dong and Qi, chapter 11, this volume); (91) Xiaohe and Shagou faunas, Yuanmou Basin, Yunnan Province (Zhu et al. 2005; Dong and Qi, chapter 11, this volume); (92) Shihuiba Fauna, Zhaotong (Chaotung) Basin, Yunnan Province (Chow and Zhai 1962; Zhang et al. 1989; Denise Su, pers. comm.).

Japan

Time (Ma)	Epoch	Europe (MN)	ALMA	Kani Basin	Mizunami Basin	Sasebo area	Sendai area	Tochio area	Aikawa area	Iga-Omi Basin	Awaji Island
	Pleistocene		Nihewanian					Unnamed Fm.	Nakatsu Gr. (98) - F	Kobiwako Group - F	Osaka Group (100) - F
	Pliocene	16	Yushean					Shiroiwa Fm.		- F	
		15						(97) - F		- F	
5		14					Sendai Gr. Tatsunokuchi Fm. (96) - F	Ushigakubi Fm.		- F - F	
		13	Baodean				Kameoka Fm.	Araya Fm.		(99)	
	Late Miocene	12									
10		11	Bahean								
		10									
		9									
	Middle Miocene	7/8	Tunggurian								
15		6									
		5	Shanwangian			Nojima Group Minamitabira Fm.					
	Early Miocene	4		Mizunami Group Nakamu-Hiramaki Fm. - F - F Dota Loc. (93)	Mizunami Group Akeyo Fm. - F Hongo Fm. - F	Fukazuki Fm. (95) - F Oya Fm.					
		3			Toki Lignite-bearing Fm. (94) - F						
20		2	Xiejian	Hachiya Fm. - F							
23		1									

Figure I.2*i* Major Neogene faunas and strata of Japan: (*93*) Dota locality and other fossil sites (marked with an "- F"), Kani Basin (Tomida et al., chapter 12, this volume); (*94*) terrestrial vertebrate fossil sites (marked with an "- F"), Mizunami Basin (Tomida et al., chapter 12, this volume); (*95*) *Diatomys* locality ("- F"), Sasebo area (Tomida et al., chapter 12, this volume); (*96*) Sendai area (Tomida et al., chapter 12, this volume); (*97*) *Parailurus* locality ("- F"), Tochio area (Sasagawa et al. 2003; Nakagawa, Kawamura, and Taruno, chapter 13, this volume); (*98*) *Dolichopithecus* locality, Aikawa area (Nakagawa, Kawamura, and Taruno, chapter 13, this volume); (*99*) Iga-Omi Basin (Nakagawa, Kawamura, and Taruno, chapter 13, this volume); (*100*) Awaji Island, Hyogo Prefecture (Nakagawa, Kawamura, and Taruno, chapter 13, this volume).

works on fossil-producing basins. Efforts are made to preserve a sense of lithostratigraphic (formations, suites, etc.) and biostratigraphic relationships (fossil localities, faunas, faunistic complexes, etc.). Although we cite the original sources for individual columns, any errors or misinterpretations are entirely our own. Nor is this an exhaustive account of all Asian sites, although most of the well-known sites are included. Such an exercise invariably fails to capture the complexities and nuances of the regions being depicted, and readers are urged to examine the original sources (and citations within) for each locality or basin. In many basins, controversies exist for faunal interpretations, in some cases, with discrepancies of millions of years. In presenting individual stratigraphic columns, we did not attempt to analyze each regional scheme, although we did occasionally reinterpret magnetic correlations. The intention of this exercise is to put together, for the first time, all major fossil-mammal-producing regions in a series of charts, to draw attention to the different conceptual frameworks and different constructions of faunal relationships. We hope this will serve as a starting point to integrate various stratigraphic schemes.

It is also immediately clear that there is much unevenness in concepts and in practices. At the conceptual level, countries in the former Soviet Union (such as Georgia, Tajikistan, Uzbekistan, Kyrgyzstan, Kazakhstan, and Russia) or those influenced by the Soviet Union (Mongolia, and China to a lesser extent) have used stratigraphic schemes combining various notions of litho- or biochronology. The stratigraphic term "svita" (here translated as "suite"), is not only defined by lithologic characteristics (as in "formations" in western countries), but is also laden with a connotation of time, often indicated by fossil content. When western concepts of separation of litho- and biostratigraphy are applied to some of the areas, major discrepancies can occur that are difficult to reconcile, such as in the Aktau Mountain area (Kordikova and Mavrin 1996; Lucas et al. 1997) and in the Zaysan Basin (Sotnikova, Dodonov, and Pen'kov 1997; Lucas et al. 2000). As a result, our summaries for these countries are often not strictly comparable to those found elsewhere (see figure I.2c). These are areas that can benefit greatly by applications of consistent criteria to evaluate the stratigraphic schemes.

In practical correlations, the European MN system continues to exert influences in many areas. In some cases, such as countries in the former Soviet Union and western Asian countries, the MN units sometimes have been directly projected to the local strata. The European influence can also be felt as far as Southeast Asia, some-

times with disparate results, such as in the Li Mae Long Basin in northern Thailand (Ginsburg 1984; Mein and Ginsburg 1997; Chaimanee et al. 2007). This basin was considered to have a late Early Miocene small mammal fauna or even as earlier Miocene (MN 4), but ongoing work has benefitted from paleomagnetic data (Chaimanee et al. 2007) that, together with continued systematic studies, place the assemblage in the Middle Miocene. Examples like these highlight the perilous nature of long-distance correlation to a European system that is itself full of uncertainties and ambiguities and the importance of establishment of an indigenous biostratigraphy.

In stratigraphic resolution, existing Asian frameworks span a full spectrum of resolving power of continental biostratigraphy. At the finest scale, the Siwalik sequences in Pakistan, well constrained by 47 magnetic sections, boast consistent resolution of up to 200,000 years or less for 80% of the more than 1,000 fossil localities, and 100,000 years or less for 50% of the localities. Such a remarkable precision is pushing the resolving power in terrestrial sedimentation to the limits (Barry et al., chapter 15, this volume) and can rival the resolution of any basin of continental deposits in the world. At the other end of the spectrum, however, crude biochronologic characterizations, often without any independent calibrations, are still widely practiced, which is the norm in many Asian countries.

Although a true sense of biostratigraphy for individual taxa within reasonably fossiliferous spans is emerging for a number of basins (e.g., Sinap, Maragheh, Siwalik, Junggar Basin, Valley of Lakes, Qaidam Basin, Lingtai, Bahe, Yushe Basin), in the majority of regions in Asia, nominal "local faunas" or "faunas" are still widely used as a traditional way to communicate an aggregate of taxa often spanning a certain stratigraphic thickness representing a certain amount of time. More inclusive terms, such as chronofauna (or faunistic complex) can be useful concepts to construct ideas of larger biota that span greater geographic and temporal ranges.

From the perspective of geologic time represented in various Asian regions, our charts show that Early Miocene has the largest gaps in the fossil records of almost every region in Asia. This is especially true for the beginning of the Miocene (23–20 Ma), during which preciously few localities have any records and those that have some data are represented by mostly small mammals.

Another conspicuous gap in the Chinese coverage has somewhat unexpectedly turned out to be the early part of the Late Miocene, the temporal equivalent of the European Vallesian (11.2–9.5 Ma). The reason why this was not at first realized has to do in part with lack of stratigraphic control and in part with the monsoon-driven climate history of East Asia. Recent fieldwork has revealed that several key Chinese Late Miocene localities are significantly younger than previously assumed (Zhang et al., chapter 6, this volume; Kaakinen et al., chapter 7, this volume), leaving the Vallesian time equivalent of China remarkably poorly sampled. Recent rediscovery and correlation of Bohlin's *Tsaidamotherium* locality (now called Quanshuiliang section) in the Qaidam Basin reveal that much of the Quanshuiliang section corresponds to the magnetically calibrated Tuosu Fauna of the early Bahean (~11–10 Ma; Wang et al. 2011). However, while rich in large mammals, much of the fauna represents an endemic Tibetan Plateau assemblage that is not easily related to faunas elsewhere. It has also recently become clear that the general trend of climate change in China during the Late Miocene is the opposite of the global trend of increasing aridity seen in Europe and North America and that faunal evolution in China accordingly follows a different path (e.g., the reappearance in the record and survival of the anchitherine horse *Sinohippus* far into the Late Miocene). Evidence from multiple sources now shows that China instead became gradually more humid during this interval, most probably as a result of a strengthened summer monsoon (Fortelius et al. 2002; Passey et al. 2009). The Chinese mammal fauna of the early, dry part of the late Miocene is characterized by low diversity and endemism, and it is only the more humid part of the late Miocene, from about 8 Ma onward, that sees the proper, pancontinental "Hipparion fauna" established in China (Fortelius and Zhang 2006; Mirzaie Ataabadi et al., chapter 29, this volume). In this perspective, the potential for establishing a long stratigraphic sequence from Lingtai takes on a special importance, and it will therefore be of considerable interest and importance to see whether future fieldwork will verify the tentative Vallesian correlations suggested by Deng et al. (chapter 9, this volume).

ZOOGEOGRAPHIC COMPLEXITY

As the largest continent on Earth, Asia defies easy categorization. With vast latitudinal, longitudinal, and altitudinal spans, as well as the attendant climatic zonations, zoogeographic differentiations are profound (figure I.3). Indeed, in Alfred Russel Wallace's (1876:map 1) classic map of zoogeographic provinces, the boundary between Palearctic and Oriental provinces was drawn within Asia, mostly along the southern slopes of the Himalaya Range and its lateral extensions. In other words, northern Asia and Europe are zoogeographically more similar to each other than either is to southern Asia. Plate tectonics along the India-Asia collision zone and its resulting uplift of the Himalayan Range and Tibetan Plateau are thus critical factors imposing a first-order organization of the Asian continent. This pattern of modern zoogeographic division can be traced back in deep time at least to the early Neogene, if not earlier, based on fossil mammals (Qiu and Li 2003, 2005; Flynn 2008), consistent with the early attainment of a high Tibet (e.g., Rowley and Currie 2006; Quade et al. 2011). Climatic differentiations are similarly recorded by Neogene mammal records (Fortelius et al. 2002; Fortelius et al. 2003; Fortelius et al. 2006; Liu et al. 2009). Given such complexity in geography and climate, questions naturally arose during the workshops as to the feasibility of devising an Asia-wide land mammal age system that can work across major zoogeographic boundaries. If Europe and East Asia within the Palearctic Province are to have a separate chronologic system, shouldn't South Asia in the Oriental Province have its own?

The East–West Divide Between Europe and Asia

The Eurasian continent spans the entire eastern hemisphere and beyond. Since the disappearance of the epicontinental Turgai Sea by about early Oligocene, Europe and Asia have been a single connected landmass. Despite this continuity during the Neogene, however, distant faunas from the extreme ends in western Europe and eastern Asia show marked Early Miocene differences, although there is a tendency for increased similarity through time (Mirzaie Ataabadi et al., chapter 29, this volume). A climatic gradient is likely, since sheer distance alone probably cannot fully account for such faunal differences.

Diamond (1997) advanced the thesis that organismic (including human) migrations are easily achieved along the east–west axis because Earth's atmospheric variances are often organized latitudinally; that is, organisms can readily adapt to habitats of similar climatic zones of similar latitudes. This is in contrast to the north–south axis, which entails the crossing of climatic zones. By this argument, western Europe and eastern China, both of similar latitudes, should share more faunal similarities despite their vast geographic distance. Existence of distinct faunas from Europe and eastern Asia thus indicates climatic differentiations (wetter Europe vs. drier central and eastern Asia) or distinct environmental barriers, such as deserts in central Asia and the Tibetan Plateau. Faunal distinction through much of the Neogene (few species in common) is the strongest rationale for an Asian land

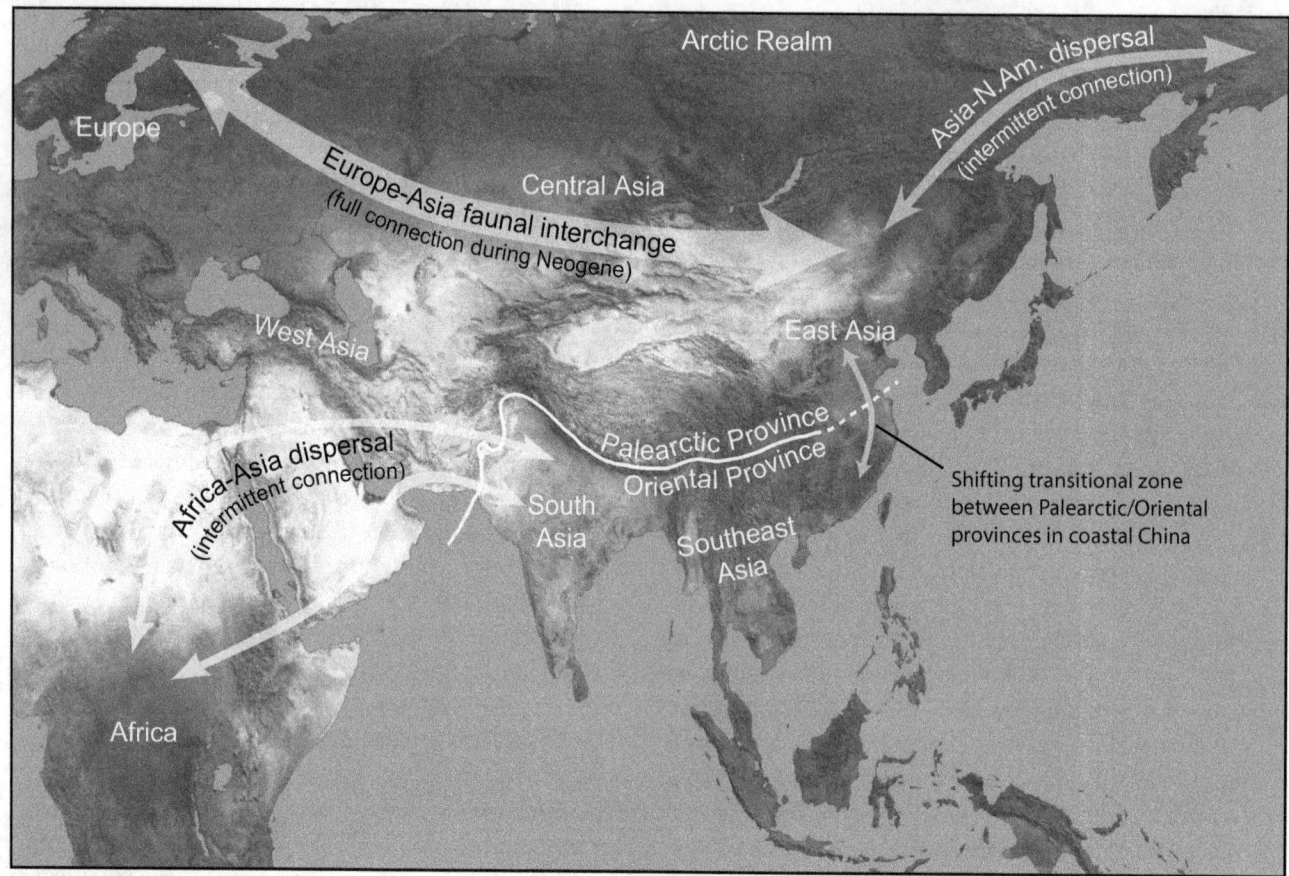

Figure I.3 Schematics of inter- and intracontinental faunal interchanges and dispersals centered around Asia. Europe–East Asia faunal interchange entails largely the same latitudes in the east–west direction; the main barrier is the Tibetan Plateau and adjacent arid regions of Central Asia. Except for mammals adapted to Arctic regions, Asia–North America dispersals include a large "vertical axis" component along longitude, and mammals must cross different climate zones in order to reach to the other side. Thin air and high mountains present formidable barriers along the southern slopes of the Himalaya, which form a sharp zoogeographic boundary; to the east along the east coast of China, however, the boundary becomes fuzzy and a transitional zone shifts through time along with climate changes. Africa–Asia connection is intermittent during the Neogene. Gray tones in continents roughly reflect the amount of vegetation: white or light gray indicates desert or dry environments and darker gray indicates more vegetation coverage. Width of arrows is suggestive of magnitude of terrestrial dispersals.

mammal age system independent of the European MN units. This is in contrast to North America, which has a much narrower longitudinal span, and its paleofaunas have even narrower distributions within the western half of North America (eastern North America is poorly fossiliferous). As a result, faunal differences between Pacific coastal states and the Great Plains are small enough to be subsumed within a single NALMA system.

Despite east–west faunal differentiations, however, broad faunal similarities can be recognized in much of western and Central Asia at select time periods. For example, the notion of a Pikermian paleobiome recognizes a wide swath of Eurasia during the late Miocene that is dominated by dry climate, increasingly open environments, and seasonally adapted mammals (Bernor et al.

1996). Such a widespread biome of long duration has been termed the Pikermian chronofauna (Eronen et al. 2009), which lends support for Asian land mammal ages spanning at least northern Asia. As demonstrated by Mirzaie Ataabadi et al. (chapter 29, this volume), such a concept may also be applicable in parts of Eurasia in the Middle Miocene, as represented by the Tunggurian chronofauna, although the evolving nature of this chronofauna from an earlier appearance in western Europe and migrating east to eastern China near the latest Middle Miocene implies diachrony. Such diachrony has obvious implications for correlation, a case in point being the carnivore genus *Dinocrocuta*, which in Europe and western Asia is a good indicator of early Late Miocene age and has been used to support a Vallesian correlation of Bahean age localities in China. Recent studies suggest, however, that *Dinocrocuta* has a primarily (or even exclusively?) Turolian age range in China, with the best-dated records so far clustering around 8 Ma (Zhang et al., chapter 6, this volume).

The North–South Divide Between North and South Asia

A first-order zoogeographic division between the Palearctic Province to the north and Oriental Province to the south was long recognized to be the result of Earth's surface processes (Wallace 1876). Such a clear distinction is rooted in the following two interrelated processes: erection of a formidable geographic barrier in Tibet-Himalaya and its lateral extensions, and formation of summer monsoons in South and East Asia and winter westerlies and northwesterlies in northern China and central Asia. This factor, coupled with major west–east river systems, distinguishes much of China. A Palearctic/Oriental-style provinciality can be recognized since the early to middle Miocene based on small-mammal records in eastern China (Qiu and Li 2003, 2005), large mammals from the northern rim of the Tibetan Plateau (Qiu et al. 2001), and small mammals from South Asia (Flynn 2008). Furthermore, in contrast to increasing faunal homogeneity between east and west Eurasia during the Neogene, the north–south faunal division became progressively more clearly delineated through time as Tibet was being uplifted and its climatic effects became more pronounced. The above process thus presents the biggest obstacle in the establishment of a truly Asia-wide land mammal age system.

Intermittent Connections Between Africa and South Asia

Faunal exchanges between Africa and South Asia, either by direct migration through the Arabian Peninsula or by indirect routes of western Europe (across the Strait of Gibraltar), are evidenced by records from the Siwaliks and equivalent deposits of Dera Bugti and Sulaiman areas (Barry et al. 1991; Flynn et al. 1995; Antoine et al., chapter 16, this volume). Being in similar latitudes and warm climates, the main control of Africa–South Asia dispersal was by intermittent land corridors. It is thus not surprising that South Asia often has the largest number of African elements outside of Africa, and an Ethiopian-Oriental connection seems to be recognizable (Flynn

and Wessels, chapter 18, this volume), featuring occasional dispersals in both directions, notably among rodents and primates.

Connection of North America and Asia

Since the early Miocene, immigrants to North America from Asia seem to suggest a closed Bering Strait for much of the time (Woodburne and Swisher 1995). The Bering Land Bridge undoubtedly acts as a filter that allows faunas in the Arctic realm to pass freely but severely limits those from middle or lower latitudes. Because of this limited faunal exchange, correlations of Asian and North American land mammal ages, which are entirely based on mid-latitude faunas, are not easy, and the NALMA did not have much influence on the developments of the Asian mammal system.

Contributions of Asia–North America faunal exchange are often asymmetrical; a large number of immigrant events have been recorded in North America, but far fewer mammals made it to Asia. Tedford et al. (2004:fig. 6.3) counted 88 allochthonous genera of Old World origin during the Miocene Epoch (Arikareean through Hemphillian); many of these became significant components of local communities in North America. With the exception of horses (*Anchitherium, Hipparion, Equus*), camels (*Paracamelus*), dogs (*Eucyon, Nyctereutes, Vulpes*), and several small mammals (the rabbit *Alilepus*, squirrels like *Marmota*, beavers, and birch mice [see Kimura, chapter 30, this volume]), mammals that dispersed from North America did not exert a corresponding presence in Asia. Although such a discrepancy potentially may be accounted for by sampling effects, at least in the Pliocene (Flynn et al. 1991), it is also possible that a larger Eurasian continent presented a more competitive environment for North American newcomers. One striking example is an early Pliocene Arctic North American fauna that shares close similarity with contemporaneous faunas from North China (Tedford and Harington 2003).

Embedded within the overall balance of exchange favoring entry of Asian forms into North America is a striking, apparently climate-driven exception. The dispersal of Eurasian ungulates into North America was discontinuous, greatly declining during the later Miocene. Between 15 and 5 million years ago only four ungulate genera of Eurasian origin are known from North America: *Pseudoceras, Neotragoceros, Platybelodon*, and *Tapirus* (Tedford et al. 2004). In contrast to the successful dispersal of horses and camels in the opposite direction during this interval, none of the new arrivals diversified after dispersal, and, with the exception of *Tapirus*, all had a short duration in the fossil record. Eronen et al. (in press) attribute this imbalance to the fact that during this time North America was significantly more arid than Eurasia, creating a situation where North American ungulates were literally pre-adapted to the conditions yet to appear in Asia, while Asian ungulates correspondingly lagged behind the environmental conditions already in place in North America. That the result is not due to sampling error is testified by continued successful dispersal of Eurasian carnivores into North America during the same interval, and by the fact that ungulate dispersal into North America resumed when the climatic imbalance disappeared in the Plio-Pleistocene.

CRITICAL TRANSITIONS

This book is the result of collaborative efforts in Beijing and a follow-up Los Angeles workshop, which are part of a Sino-U.S. collaborative research agenda on critical transitions in the history of life. The goal is to address critical transitions in geologic history that profoundly affect biological and environmental evolution on global scales. Once again, Asia, by its unique geographic position and geologic history, has much to offer in our understanding of global environmental changes. Mammal distributions in space (zoogeography) and time (biostratigraphy and geochronology) are two key components in any attempt to formulate ideas about paleoenvironmental change. In many ways, mammal biostratigraphy by itself offers evidences of critical transitions. In that sense, we hope this volume will provide the initial dataset and encouragement to stimulate further research on the various critical transitions.

Looming large among Asian Cenozoic geologic events is the rise of the Himalayan and Tibetan highlands and effects on the initiation of Indian and East Asian monsoon climates. Without doubt, Himalaya-Tibet, as an imposing physical entity in central Asia, is a first-order climate maker. Much debate, however, is centered on the timing and process of the coupling of mountain uplift and climate change and their feedback on erosion and weathering (e.g., Molnar 2005). From a paleontological perspective, mammals as a biological component and a chronological marker have much to offer in this debate.

The emergence of Himalaya-Tibet and the ensuing zoogeographic division of Palearctic and Oriental provinces affects mammal distributions in two ways. The rising Himalaya coupled with drastic changes in climatic zonation form an effective barrier for all but high-flying

birds. Once again, fossil mammals offer direct evidence for this profound change. Furthermore, as consumers of vegetation, mammalian ungulates are also invaluable for assessing plant compositions. Isotope ratios of dental enamels, hypsodonty indices, microwear, and mesowear have become critical means to deduce plant coverage, paleotemperature, and precipitation. As the field matures, such "ecometrics" (Eronen et al. 2010) are likely to become welded into an increasingly quantitative paleoenvironmental framework that can be used in conjunction with paleoclimate modeling to constrain and refine reconstructions of past conditions and processes (Eronen et al. 2009).

ACKNOWLEDGMENTS

It goes without saying that a volume such as this is not possible without the contributions from all authors—we express our gratitude to all who took the time to undertake this worthy project. The Institute of Vertebrate Paleontology and Paleoanthropology provided logistic support in the two Beijing workshops, and we thank the numerous graduate students from the IVPP for their help and participation. We thank our publisher Patrick Fitzgerald, senior manuscript editor Irene Pavitt, assistant editor Bridget Flannery-McCoy, copyeditors Karen Victoria Brown and Richard Camp, production editor Edward Wade, designer Milenda Lee, and indexer Maria Coughlin for their tireless work to ensure that this book will come to fruition. We also appreciate the valuable comments and suggestions by two anonymous volume reviewers. We are grateful to Alexey Tesakov for his review of a late draft of this chapter and for providing important Russian literature on several localities. This book and its companion workshops in Beijing and Los Angeles benefited from financial support from the Sedimentary Geology and Paleobiology program of the National Science Foundation (U.S.) and its counterpart in the Chinese National Natural Science Foundation. In connection to these financial assistances, we would like to acknowledge Raymond L. Bernor, H. Richard Lane, and Lisa Boush, whose sustained support are keys to our success in putting together the largest gathering of mammalian paleontologists working on Asian continental biostratigraphy. The Society of Vertebrate Paleontology made it possible for three young scholars to attend. Finally, but certainly not least emphatically, we are greatly indebted to Zhan-xiang Qiu, who not only produced the key summary chapter on Chinese land mammal ages/stages but was more than generous in his financial support of the production of this volume through his Special Researches Program of Basic

Science and Technology grant (No. 2006FY120400) from the Ministry of Science and Technology.

REFERENCES

All-China Stratigraphic Commission. 2001. *Regional Stratigraphic Guide of China and Its Explanation.* Revised ed. Beijing: Geological Publishing House.

Alroy, J. 1992. Conjunction among taxonomic distributions and the Miocene mammalian biochronology of the Great Plains. *Paleobiology* 18:326–343.

Alroy, J. 1994. Appearance event ordination: A new biochronologic method. *Paleobiology* 20:191–207.

Alroy, J. 1998. Diachrony of mammalian appearance events: Implications for biochronology. *Geology* 26:23–26.

Alroy, J. 2000. New methods for quantifying macroevolutionary patterns and processes. *Paleobiology* 26:707–733.

Alroy, J., R. L. Bernor, M. Fortelius, L. Werdelin. 1998. The MN System: Regional or Continental? *Mitteilungen der Bayerischen Staatssammllung für Paläontologie und historische Geologie* 38: 243–258.

Andrews, P. and B. Alpagut. 1990. Description of the fossiliferous units at Paşalar, Turkey. *Journal of Human Evolution* 19:343–361.

Barry, J. C., M. E. Morgan, A. J. Winkler, L. J. Flynn, E. H. Lindsay, L. L. Jacobs, and D. Pilbeam. 1991. Faunal interchange and Miocene terrestrial vertebrates of southern Asia. *Paleobiology* 17:231–245.

Bernor, R. L., N. Solounias, C. C. Swisher III, and J. A. Van Couvering. 1996. The correlation of three classical "Pikermian" mammal faunas—Maragheh, Samos, and Pikermi—with the European MN unit system. In *The Evolution of Western Eurasian Neogene Mammal Faunas,* ed. R. L. Bernor, V. Fahlbusch, and H.-W. Mittmann, pp. 137–154. New York: Columbia University Press.

Bi, Z.-g., Z.-j. Yu, and Z.-x. Qiu. 1977. First discovery of mammal remains from upper Tertiary deposits near Nanking. *Vertebrata PalAsiatica* 15:126–138.

Brandy, L. D. 1981. Rongeurs muroïdés du Néogène supérieur d'Afghanistan: Évolution, biogéographie, corrélations. *Palaeovertebrata* 25:133–149.

Chaimanee, Y., C. Yamee, B. Marandat, and J.-J. Jaeger. 2007. First middle Miocene rodents from the Mae Moh Basin (Thailand): Biochronological and paleoenvironmental implications. *Bulletin of Carnegie Museum of Natural History* 39:157–163.

Chaimanee, Y., C. Yamee, P. Tian, K. Khaowiset, B. Marandat, P. Tafforeau, C. Nemoz, and J.-J. Jaeger. 2006. *Khoratpithecus piriyai,* a Late Miocene hominoid of Thailand. *American Journal of Physical Anthropology* 131:311–323.

Chen, W.-y. 1986. Preliminary studies of sedimental environment and taphonomy in the hominoid fossil site of Lufeng. *Acta Anthropologica Sinica* 5:89–100.

Chen, X.-f. 1994. Stratigraphy and large mammals of the "Jinglean" Stage, Shanxi, China. *Quaternary Science* 1994:339–352.

Chen, X., J.-y. Rong, J.-x. Fan, R.-b. Zhan, C. E. Mitchell, D. A. T. Harper, M. J. Melchin, P.-a. Peng, S. C. Finney, and X.-f. Wang. 2006. The Global Boundary Stratotype Section and Point (GSSP) for the base of the Hirnantian Stage (the uppermost of the Ordovician System). *Episodes* 29:183–196.

Chiu, C.-s., C.-k. Li, and C.-t. Chiu. 1979. A preliminary review of the mammalian localities and faunas. *Annales Géologiques des Pays Helléniques,* non-ser. vol. 1:263–272.

Chow, M.-c. and R.-j. Zhai. 1962. Early Pleistocene mammals of Chaotung, Yunnan, with notes on some Chinese stegodonts. *Vertebrata PalAsiatica* 6:138–147.

Coster, P., M. Benammi, Y. Chaimanee, C. Yamee, O. Chavasseau, E.-G. Emonet, and J.-J. Jaeger. 2010. A complete magnetic-polarity stratigraphy of the Miocene continental deposits of Mae Moh Basin, northern Thailand, and a reassessment of the age of hominoid-bearing localities in northern Thailand. *Geological Society of America Bulletin* 122:1180–1191.

Deng, T. 2006. Chinese Neogene mammal biochronology. *Vertebrata PalAsiatica* 44:143–163.

Deng, T., S. Hou, and H. Wang. 2007. The Tunggurian Stage of the continental Miocene in China. *Acta Geologica Sinica* 81:709–721.

Deng, T., W.-m. Wang, and L.-p. Yue. 2003. Recent advances of the establishment of the Shanwang stage in the Chinese Neogene. *Vertebrata PalAsiatica* 41:314–323.

Deng, T., W.-m. Wang, and L.-p. Yue. 2006. The Xiejian stage of the continental Miocene series in China. *Journal of Stratigraphy* 30:315–322.

Deng, T., W.-m. Wang, and L.-p. Yue. 2008. Report on research of the Shanwangian and Baodean stages in the continental Neogene series of China. In *Reports on Establishing the Major Stages in China,* ed. Z.-j. Wang and Z.-g. Huang, pp. 13–31. Beijing: Geological Publishing House.

Deng, T., W.-m. Wang, L.-p. Yue, and Y.-x. Zhang. 2004. New advances in the establishment of the Neogene Baode Stage. *Journal of Stratigraphy* 28:41–47.

Deng, T., X.-m. Wang, M. Fortelius, Q. Li, Y. Wang, Z. J. Tseng, G. T. Takeuchi, J. E. Saylor, L. K. Säilä, and G.-p. Xie. 2011. Out of Tibet: Pliocene woolly rhino suggests high-plateau origin of Ice Age megaherbivores. *Science* 333:1285–1288.

Diamond, J. 1997. *Guns, Germs, and Steel: The Fates of Human Societies.* New York: Norton.

Dong, W. 2001. Upper Cenozoic stratigraphy and paleoenvironment of Xiaolongtan Basin, Kaiyuan, Yunnan Province. In *Proceedings of the Eighth Annual Meetings of Chinese Society of Vertebrate Paleontology,* ed. T. Deng and Y.-q. Wang, pp. 91–100. Beijing: China Ocean Press.

Erbajeva, M. A. and A. G. Filippov. 1997. Miocene small mammalian faunas of the Baikalian region. In *Actes du Congrès BiochroM'97,* ed. J.-P. Aguilar, S. Legendre, and J. Michaux, pp. 249–259. Mémoires et Travaux E.P.H.E., Institut de Montpellier, 21.

Eronen, J. T., M. Fortelius, A. Micheels, F. T. Portmann, K. Puolamäki and C. M. Janis. In press. Neogene aridification of the Northern Hemisphere. *Geology.*

Eronen, J. T., M. Mirzaie Ataabadi, A. Micheels, A. Karme, R. L. Bernor, and M. Fortelius. 2009. Distribution history and climatic controls of the Late Miocene Pikermian chronofauna. *Proceedings of the National Academy of Sciences* 106:11867–11871.

Eronen, J. T., P. D. Polly, M. Fred, J. Damuth, D. C. Frank, V. Mosbrugger, C. Scheidegger, N. C. Stenseth, and M. Fortelius. 2010. Ecometrics: The traits that bind the past and present together. *Integrative Zoology* 5:88–101.

Fahlbusch, V., Z.-d. Qiu, and G. Storch. 1983. Neogene mammalian faunas of Ertemte and Harr Obo in Nei Monggol, China. 1 Report on field work in 1980 and preliminary results. *Scientia Sinica*, ser. B, 26:205–224.

Fang, X.-m., M.-d. Yan, R. Van der Voo, D. K. Rea, C.-h. Song, J. M. Parés, J.-p. Gao, J.-s. Nie, and S. Dai. 2005. Late Cenozoic deformation and uplift of the NE Tibetan Plateau: Evidence from high-resolution magnetostratigraphy of the Guide Basin, Qinghai Province, China. *Geological Society of America Bulletin* 117:1208–1225.

Flynn, L. J. 2008. Paleobiogeographic affinity across southern Asia during the Miocene: Small mammals reflect the Oriental Realm. *Journal of Vertebrate Paleontology* 28:78A.

Flynn, L. J., J. C. Barry, M. E. Morgan, D. Pilbeam, L. L. Jacobs, and E. H. Lindsay. 1995. Neogene Siwalik mammalian lineages: Species longevities, rates of change, and modes of speciation. *Palaeogeography, Palaeoclimatology, Palaeoecology* 115:249–264.

Flynn, L. J. and R. L. Bernor. 1987. Late Tertiary mammals from the Mongolian People's Republic. *American Museum Novitates* 2872: 1–16.

Flynn, L. J., R. H. Tedford, and Z.-x. Qiu. 1991. Enrichment and stability in the Pliocene mammalian fauna of North China. *Paleobiology* 17:246–265.

Flynn, L. J., W.-y. Wu, and W. R. Downs. 1997. Dating vertebrate microfaunas in the late Neogene record of northern China. *Palaeogeography, Palaeoclimatology, Palaeoecology* 133:227–242.

Fortelius, M., J. Eronen, J. Jernvall, L.-p. Liu, D. Pushkina, J. Rinne, A. Tesakov, I. A. Vislobokova, Z.-q. Zhang, and L.-p. Zhou. 2002. Fossil mammals resolve regional patterns of Eurasian climate change over 20 million years. *Evolutionary Ecology Research* 4:1005–1016.

Fortelius, M., J. Eronen, L.-p. Liu, D. Pushkina, A. Tesakov, I. Vislobokova, and Z.-q. Zhang. 2003. Continental-scale hypsodonty patterns, climatic paleobiogoegraphy and dispersal of Eurasian Neogene large mammal herbivores. In *Distribution and Migration of Tertiary Mammals in Eurasia, A Volume in Honour of Hans de Bruijn*, ed. J. W. F. Reumer and W. Wessels, pp. 1–11. DEINSEA 10.

Fortelius, M., J. Eronen, L. Liu, D. Pushkina, A. Tesakov, I. Vislobokova, and Z.-q. Zhang. 2006. Late Miocene and Pliocene large land mammals and climatic changes in Eurasia. *Palaeogeography, Palaeoclimatology, Palaeoecology* 238:219–227.

Fortelius M., A. Gionis J. Jernvall and H. Mannila. 2006. Spectral ordering and biochronology of European fossil mammals. *Paleobiology* 32(2): 206–214.

Fortelius, M. and Z.-q. Zhang. 2006. An oasis in the desert? History of endemism and climate in the late Neogene of North China. *Palaeontographica*, part A 277:131–141.

Ginsburg, L. 1984. Les faunnes tertiaires du Nord de la Thaïlande. *Mémoires de la Société Géologique de France*, n.s., 47:67–69.

Gnibidenko, Z. N. 1990. Paleomagnetism and magnetic stratigraphy of Neogene deposits in Pre-Irtisch region. *Academy Sciences USSR (Siberian Branch), Geology Geophysics*:85–94.

Hanta, R., B. Ratanasthien, Y. Kunimatsu, H. Saegusa, H. Nakaya, S. Nagaoka, and P. Jintasakul. 2008. A new species of Bothriodontinae, *Merycopotamus thachangensis* (Cetartiodactyla, Anthracotheriidae) from the late Miocene of Nakhon Ratchasima, northeastern Thailand. *Journal of Vertebrate Paleontology* 28:1182–1188.

Hedberg, H. D. 1976. *International Stratigraphic Guide: A Guide to Stratigraphic Classification, Terminology, and Procedure*. First ed.

International Subcommission on Stratigraphic Classification of IUGS International Commission on Stratigraphy. New York: Wiley.

Höck, V., G. Daxner-Höck, H. P. Schmid, D. Badamgarav, W. Frank, G. Furtmüller, O. Montag, R. Barsbold, Y. Khand, and J. Sodov. 1999. Oligocene-Miocene sediments, fossils and basalts from the Valley of Lakes (Central Mongolia): An integrated study. *Mitteilungen der Österreichischen Geologischen Gesellschaft* 90: 83–125.

Huang, W.-b., H.-x. Ji, W.-y. Chen, C.-q. Hsu, and S.-h. Zheng. 1980. Pliocene stratum of Guizhong and Bulong Basin, Xizang. In *Paleontology of Tibet, Part 1*, ed. Qinghai-Tibetan Plateau Comprehensive Scientific Investigation Team of Chinese Academy of Sciences, pp. 4–17. Qinghai-Tibetan Plateau Scientific Investigation Series. Beijing: Science Press.

Jin, C.-z. 2004. Fossil leporids (Mammalia, Lagomorpha) from Huainan, Anhui, China. *Vertebrata PalAsiatica* 42:230–245.

Jin, C.-z., Y. Kawamura, and H. Taruno. 1999. Pliocene and early Pleistocene insectivore and rodent faunas from Dajushan, Qipanshan and Haimao in North China and the reconstruction of the faunal succession from the late Miocene to middle Pleistocene. *Osaka City University Journal of Geosciences* 42:1–19.

Kaakinen, A. and J. P. Lunkka. 2003. Sedimentation of the Late Miocene Bahe Formation and its implications for stable environments adjacent to Qinling mountains in Shaanxi, China. *Journal of Asian Earth Sciences* 22:67–78.

Kappelman, J., A. Duncan, M. Feseha, J. P. Lunkka, D. Ekart, F. McDowell, T. M. Ryan, and C. C. Swisher III. 2003. Chronology. In *Geology and Paleontology of the Miocene Sinap Formation, Turkey*, ed. M. Fortelius, J. Kappelman, S. Sen, and R. L. Bernor, pp. 41–66. New York: Columbia University Press.

Kordikova, E. G. and A. V. Mavrin. 1996. Stratigraphy and Oligocene-Miocene mammalian biochronology of the Aktau Mountains, Dzhungarian Alatau Range, Kazakhstan. *Palaeovertebrata* 25:141–174.

Kordikova, E. G., E. P. J. Heizmann, and A. V. Mavrin. 2000. Early Miocene Carnivora of Aktau Mountains, south eastern Kazakhstan. *Paläontologische Zeitschrift* 74:195–204.

Li, C.-k., Y.-p. Lin, Y.-m. Gu, L.-h. Hou, W.-y. Wu, and Z.-d. Qiu. 1983. The Aragonian vertebrate fauna of Xiacaowan, Jiangsu. 1. A brief introduction to the fossil localities and preliminary report on the new material. *Vertebrata PalAsiatica* 21:313–327.

Li, C.-k. and Z.-d. Qiu. 1980. Early Miocene mammalian fossils of Xining basin, Qinghai. *Vertebrata PalAsiatica* 18:198–214.

Li, C.-k., Z.-d. Qiu, and S.-j. Wang. 1981. Discussion on Miocene stratigraphy and mammals from Xining basin, Qinghai. *Vertebrata PalAsiatica* 19:313–320.

Li, C.-k., W.-y. Wu, and Z.-d. Qiu. 1984. Chinese Neogene: Subdivision and correlation. *Vertebrata PalAsiatica* 22:163–178 (in Chinese with English abstract).

Li, Q., X.-m. Wang, and Z.-d. Qiu. 2003. Pliocene mammalian fauna of Gaotege in Nei Mongol (Inner Mongolia), China. *Vertebrata PalAsiatica* 41:104–114.

Liu, L.-p., J. T. Eronen, and M. Fortelius. 2009. Significant mid-latitude aridity in the middle Miocene of East Asia. *Palaeogeography, Palaeoclimatology, Palaeoecology* 279:201–206.

Lopatin, A. V. 2004. The Early Miocene small mammals from the North Aral region (Kazakhstan) with special reference to their biostratigraphic significance. *Paleontological Journal* 38(suppl. 3): 217–323.

Lucas, S. G., B. U. Bayshashov, L. A. Tyutkova, A. K. Shamangara, and B. Z. Aubekerov. 1997. Mammalian biochronology of the Paleogene-Neogene boundary at Aktau Mountain, eastern Kazakhstan. *Paläontologische Zeitschrift* 71:305–314.

Lucas, S. G., R. J. Emry, B. U. Bayshashov, and L. A. Tyutkova. 2009. Cenozoic mammalian biostratigraphy and biochronology in the Zaysan Basin, Kazakstan. In *Papers on Geology, Vertebrate Paleontology, and Biostratigraphy in Honor of Michael O. Woodburne*, ed. L. B. Albright, III. *Museum of Northern Arizona Bulletin* 65:621–633.

Lucas, S. G., R. J. Emry, V. Chkhikvadze, B. U. Bayshashov, L. A. Tyutkova, F. A. Tleuberdina, and A. Zhamangara. 2000. Chapter 29, Upper Cretaceous-Cenozoic lacustrine deposits of the Zaysan Basin, eastern Kazakhstan. In *Lake Basins Through Space and Time*, ed. E. H. Gierlowski-Kordesch and K. R. Kelts, pp. 335–340. American Association of Petroleum Geologists Special Volume 46. Tulsa: American Association of Petroleum Geologists.

Mascarelli, A. L. 2009. Quaternary geologists win timescale vote. *Nature* 459:624.

Mein, P. and L. Ginsburg. 1997. Les mammifères du gisement Miocène inférieur de Li Mae Long, Thaïlande : Systématique, biostratigraphie et paléoenvironnement. *Geodiversitas* 19:783–844.

Meng, J., B.-y. Wang, and Z.-q. Bai. 1996. A new middle Tertiary mammalian locality from Sunitezuoqi, Nei Mongol. *Vertebrata PalAsiatica* 34:297–304.

Meng, J., J. Ye, W.-y. Wu, L.-p. Yue, and X.-j. Ni. 2006. A recommended boundary stratotype section for Xiejian Stage from northern Junggar Basin: Implications to related bio-chronostratigraphy and environmental changes. *Vertebrata PalAsiatica* 44:205–236.

Meng, J., J. Ye, W.-y. Wu, X.-j. Ni, and S.-d. Bi. 2008. The Neogene Dingshanyanchi Formation in northern Junggar Basin of Xinjiang and its stratigraphic implications. *Vertebrata PalAsiatica* 46:90–110.

Molnar, P. 2005. Mio-Pliocene growth of the Tibetan Plateau and evolution of East Asian climate. *Palaeontologia Electronica* 8:1–23.

Passey, B. H., L. K. Ayliffe, A. Kaakinen, Z. Zhang, J. T. Eronen, Y. Zhu, L. Zhou, T. E. Cerling, and M. Fortelius. 2009. Strengthened East Asian summer monsoons during a period of high-latitude warmth? Isotopic evidence from Mio-Pliocene fossil mammals and soil carbonates from northern China. *Earth and Planetary Science Letters* 277:443–452.

Pevzner, M. A., E. A. Vangengeim, V. I. Zhegallo, V. S. Zazhigin, and I. G. Liskun. 1982. Correlation of the upper Neogene sediments of entral Asia and Europe on the basis of paleomagnetic and biostratigraphic data. *International Geology Review* 25:1075–1085.

Puolamäki, K. M., Fortelius, and H. Mannila. 2006. Seriation in paleontological data using Markov Chain Monte Carlo methods. *PLoS Computational Biology* 2:e6.

Qi, G.-q. 1985. Stratigraphic summarization of *Ramapithecus* fossil locality, Lufeng, Yunnan. *Acta Anthropologica Sinica* 4:55–69.

Qiu, Z.-d. 1996. *Middle Miocene Micromammalian Fauna from Tungur, Nei Mongol*. Beijing: Academic Press.

Qiu, Z.-d. 2001. Cricetid rodents from the middle Miocene Quantougou Fauna of Lanzhou, Gansu. *Vertebrata PalAsiatica* 39:204–214.

Qiu, Z.-d. and C.-k. Li. 2003. Rodents from the Chinese Neogene: Biogeographic relationships with Europe and North America. In *Vertebrate Fossils and Their Context: Contributions in Honor of Richard H. Tedford*, ed. L. J. Flynn, pp. 586–602. *Bulletin of the American Museum of Natural History* 279.

Qiu, Z.-d. and C.-k. Li. 2005. Evolution of Chinese mammalian faunal regions and elevation of the Qinghai-Xizang (Tibet) Plateau. *Science in China*, ser. D, 48:1246–1258.

Qiu, Z.-d. and G. Storch. 2000. The early Pliocene micromammalian fauna of Bilike, Inner Mongolia, China (Mammalia: Lipotyphla, Chiroptera, Rodentia, Lagomorpha). *Senckenbergiana Lethaea* 80:173–229.

Qiu, Z.-d. and Q. Li. 2008. Late Miocene micromammals from the Qaidam Basin in the Qinghai-Xizang Plateau. *Vertebrata PalAsiatica* 46:284–306.

Qiu, Z.-x. 1989. The Chinese Neogene mammalian biochronology—its correlation with the European Neogene mammalian zonation. In *European Neogene Mammal Chronology*, ed. E. H. Lindsay, V. Fahlbusch, and P. Mein, pp. 527–556. New York: Plenum Press.

Qiu, Z.-x. and Z.-d. Qiu. 1990. The sequence and division of mammalian local faunas in the Neogene of China. *Journal of Stratigraphy* 14:241–260.

Qiu, Z.-x. and Z.-d. Qiu. 1995. Chronological sequence and subdivision of Chinese Neogene mammalian faunas. *Palaeogeography, Palaeoclimatology, Palaeoecology* 116:41–70.

Qiu, Z.-x., B.-y. Wang, Z.-d. Qiu, F. Heller, L.-p. Yue, G.-p. Xie, X.-m. Wang, and B. Engesser. 2001. Land mammal geochronology and magnetostratigraphy of mid-Tertiary deposits in the Lanzhou Basin, Gansu Province, China. *Eclogae Geologicae Helvetiae* 94:373–385.

Qiu, Z.-x., W.-y. Wu, and Z.-d. Qiu. 1999. Miocene mammal faunal sequences of China: Palaeozoogeography and Eurasian relationships. In *The Miocene Land Mammals of Europe*, ed. G. E. Rössner and K. Heissig, pp. 443–472. Munich: Dr. Friedrich Pfeil.

Quade, J., D. O. Breecker, M. Daëron, and J. Eiler. 2011. The paleoaltimetry of Tibet: An isotopic perspective. *American Journal of Science* 311:77–115.

Ratanasthien, B. 2002. Problems of Neogene biostratigraphic correlation in Thailand and surrounding areas. *Revista Mexicana de Ciencias Geológicas* 19:235–241.

Rowley, D. B. and B. S. Currie. 2006. Palaeo-altimetry of the late Eocene to Miocene, Lunpola Basin, central Tibet. *Nature* 439:677–681.

Salvador, A. 1994. *International Stratigraphic Guide—A Guide to Stratigraphic Classification, Terminology, and Procedure*. Second Edition. In *International Subcommission on Stratigraphic Classification of IUGS International Commission on Stratigraphy*, ed. A. Salvador. Boulder: Geological Society of America.

Sasagawa, I., K. Takahashi, T. Sakumoto, H. Nagamori, H. Yabe, and I. Kobayashi. 2003. Discovery of the extinct red panda "*Parailurus*" (Mammalia, Carnivora) in Japan. *Journal of Vertebrate Paleontology* 23:895–900.

Sen, S. 1983. Rongeurs et Lagomorphes du gisement Pliocène de Pul-e Charki, bassin de Kabul, Afghanistan. *Bulletin du Muséum National d'Histoire Naturelle, Paris*, 5th ser., 5C:33–74.

Sen, S. 1996. Present state of magnetostratigraphic studies in the continental Neogene of Europe and Anatolia. In *The Evolution of Western Eurasian Neogene Mammal Faunas*, ed. R. L. Bernor, V. Fahlbusch, and H.-W. Mittmann, pp. 57–63. New York: Columbia University Press.

Sen, S. 2001. Rodents and insectivores from the upper Miocene of Molayan, Afghanistan. *Palaeontology* 44:913–932.

Sen, S. and M. A. Erbajeva. 2011. A new species of *Gobicricetodon* Qiu 1996 (Mammalia, Rodentia, Cricetidae) from the Middle Miocene Aya Cave, Lake Baikal. *Vertebrata PalAsiatica* 49:257–274.

Song, C.-h., D.-l. Gao, X.-m. Fang, Z.-j. Cui, J.-j. Li, S.-l. Yang, H.-b. Jin, D. W. Burbank, and J. L. Kirschvink. 2005. Late Cenozoic high-resolution magnetostratigraphy in the Kunlun Pass Basin and its implications for the uplift of the northern Tibetan Plateau. *Chinese Science Bulletin* 50:1912–1922.

Sotnikova, M. V. 2006. A new canid *Nurocyon chonokhariensis* gen. et sp. nov. (Canini, Canidae, Mammalia) from the Pliocene of Mongolia. *Courier Forschungsinstitut Senckenberg* 256:11–21.

Sotnikova, M. V., A. E. Dodonov, and A. V. Pen'kov. 1997. Upper Cenozoic bio-magnetic stratigraphy of Central Asian mammalian localities. *Palaeogeography, Palaeoclimatology, Palaeoecology* 133:243–258.

Steininger, F. F. 1999. Chronostratigraphy, geochronology and biochronology of the Miocene "European Land Mammal Mega-Zone" (ELMMZ) and the Miocene "Mammal-Zones (MN-Zones"). In *The Miocene Land Mammals of Europe*, ed. G. E. Rössner and K. Heissig, pp. 9–24. Munich: Dr. Friedrich Pfeil.

Steininger, F. F., W. A. Berggren, D. V. Kent, R. L. Bernor, S. Sen, and J. Agustí. 1996. Circum-Mediterranean Neogene (Miocene and Pliocene) marine-continental chronologic correlations of European mammal units. In *The Evolution of Western Eurasian Neogene Mammal Faunas*, ed. R. L. Bernor, V. Fahlbusch, and H.-W. Mittmann, pp. 7–46. New York: Columbia University Press.

Tedford, R. H., L. B. Albright III, A. D. Barnosky, I. Ferrusquía-Villafranca, R. M. Hunt Jr., J. E. Storer, C. C. Swisher III, M. R. Voorhies, S. D. Webb, and D. P. Whistler. 2004. Mammalian biochronology of the Arikareean through Hemphillian interval (Late Oligocene through Early Pliocene Epochs). In *Late Cretaceous and Cenozoic Mammals of North America: Biostratigraphy and Geochronology*, ed. M. O. Woodburne, pp. 169–231. New York: Columbia University Press.

Tedford, R. H., L. J. Flynn, Z.-x. Qiu, N. O. Opdyke, and W. R. Downs. 1991. Yushe Basin, China: paleomagnetically calibrated mammalian biostratigraphic standard for the late Neogene of eastern Asia. *Journal of Vertebrate Paleontology* 11:519–526.

Tedford, R. H. and C. R. Harington. 2003. An Arctic mammal fauna from the Early Pliocene of North America. *Nature* 425:388–390.

Thomas, H. 1983. Les Bovidae (Artiodactyla, Mammalia) du Miocène moyen de la formation Hofuf (Province de Hasa, Arabie Saoudite). *Palaeovertebrata* 13:157–206.

Tomida, Y. and C.-z. Jin. 2009. Two new species of *Pliopentalagus* (Leporidae, Lagomorpha) from the Pliocene of Anhui Province, China, with a revision of *Pl. huainanensis*. *Vertebrata PalAsiatica* 47:53–71.

Tong, Y.-s., S.-h. Zheng, and Z.-d. Qiu. 1995. Cenozoic mammal ages of China. *Vertebrata PalAsiatica* 33:290–314.

Vangengeim, E. A., A. N. Lungu, and A. S. Tesakov. 2006. Age of the Vallesian lower boundary (continental Miocene of Europe). *Stratigraphy and Geological Correlation* 14:655–667.

Vangengeim, E. A., I. A. Vislobokova, A. Y. Godina, E. L. Dmitrieva, V. I. Zhegalla, M. V. Sotnikova, and P. A. Tleuberdina. 1993. On the age of mammalian fauna from the Karabulak Formation of the Kalmakpai River (Zaisan Depression, eastern Kazakhstan). *Stratigraphy and Geological Correlation* 1:37–44.

Vekua, A. and D. Lordkipanidze. 2008. The history of vertebrate fauna in eastern Georgia. *Bulletin of the Georgian National Academy of Sciences* 2:149–155.

Vislobokova, I. A. 2009. A new species of Megacerini (Cervidae, Artiodactyla) from the Late Miocene of Taralyk-Cher, Tuva (Russia), and remarks on the relationships of the group. *Geobios* 42(3):397–410.

Vislobokova, I. A., M. V. Sotnikova, and A. E. Dodonov. 2001. Late Miocene-Pliocene mammalian faunas of Russia and neighbouring countries. *Bullettino della Società Paleontologica Italiana* 40:307–313.

Wallace, A. R. 1876. *The Geographical Distribution of Animals. With a Study of the Relations of Living and Extinct Faunas as Elucidating the Past Changes of the Earth's Surface*. Vol. 1. New York: Harper.

Wang, X., Z.-d. Qiu, Q. Li, Y. Tomida, Y. Kimura, Z. J. Tseng, and H.-j. Wang. 2009. A new early to late Miocene fossiliferous region in central Nei Mongol: Lithostratigraphy and biostratigraphy in Aoerban strata. *Vertebrata PalAsiatica* 47:111–134.

Wang, X., Z.-d. Qiu, Q. Li, B.-y. Wang, Z.-x. Qiu, W. R. Downs, G.-p. Xie, J.-y. Xie, T. Deng, G. T. Takeuchi, Z. J. Tseng, M.-m. Chang, J. Liu, Y. Wang, D. Biasatti, Z. Sun, X. Fang, and Q. Meng. 2007. Vertebrate paleontology, biostratigraphy, geochronology, and paleoenvironment of Qaidam Basin in northern Tibetan Plateau. *Palaeogeography, Palaeoclimatology, Palaeoecology* 254:363–385.

Wang, X., Z.-d. Qiu, and N. O. Opdyke. 2003. Litho-, bio-, and magnetostratigraphy and paleoenvironment of Tunggur Formation (Middle Miocene) in central Inner Mongolia, China. *American Museum Novitates* 3411:1–31.

Wang, X., B.-y. Wang, Z.-x. Qiu, G.-p. Xie, J.-y. Xie, W. R. Downs, Z.-d. Qiu, and T. Deng. 2003. Danghe area (western Gansu, China) biostratigraphy and implications for depositional history and tectonics of northern Tibetan Plateau. *Earth and Planetary Science Letters* 208:253–269.

Wang, X., G.-p. Xie, Q. Li, Z.-d. Qiu, Z. J. Tseng, G. T. Takeuchi, B.-y. Wang, M. Fortelius, A. Rosenström-Fortelius, H. Wahlquist, W. R. Downs, C.-f. Zhang, and Y. Wang. 2011. Early explorations of Qaidam Basin (Tibetan Plateau) by Birger Bohlin—reconciling classic vertebrate fossil localities with modern biostratigraphy. *Vertebrata PalAsiatica* 49:285–310.

Whybrow, P. J. and D. Clemens. 1999. Arabian Tertiary fauna, flora, and localities. In *Fossil Vertebrates of Arabia, with Emphasis on the Late Miocene Faunas, Geology, and Palaeoenvironments of the Emirate of Abu Dhabi, United Arab Emirates*, ed. P. J. Whybrow and A. Hill, pp. 460–473. New Haven: Yale University Press.

Whybrow, P. J., H. A. McClure, and G. F. Elliott. 1987. Miocene stratigraphy, geology and flora (Algae) of eastern Saudi Arabia and the Ad Dabtiyah vertebrate locality. *Bulletin of the British Museum (Natural History) Geology* 41:371–381.

Wood, H. E., R. W. Chaney, J. Clark, E. H. Colbert, G. L. Jepsen, J. B. Reeside Jr., and C. Stock. 1941. Nomenclature and correlation of the North American continental Tertiary. *Bulletin of the Geological Society of America* 52:1–48.

Woodburne, M. O. 1987. *Cenozoic Mammals of North America: Geochronology and Biostratigraphy*. Berkeley: University of California Press.

Woodburne, M. O. 2004. *Late Cretaceous and Cenozoic Mammals of North America: Biostratigraphy and Geochronology*. New York: Columbia University Press.

Woodburne, M. O. and C. C. Swisher III. 1995. Land mammal high-resolution geochronology, intercontinental overland dispersals, sea level, climate, and vicariance. In *Geochronology, Time Scales, and Global Stratigraphic Correlation: A Unified Framework for a Historical Geology*, ed. W. A. Berggren, D. V. Kent, M. P. Aubry, and J. A. Hardenbol, pp. 335–364. Society of Economic Paleontology and Mineralogy Special Publications 54. Tulsa: Society for Sedimentary Geology.

Xu, Y.-l., Y.-b. Tong, Q. Li, Z.-m. Sun, J.-l. Pei, and Z.-y. Yang. 2007. Magnetostratigraphic dating on the Pliocene mammalian fauna of the Gaotege section, central Inner Mongolia. *Geological Review* 53:250–260.

Xue, X.-x., Y.-x. Zhang, and L.-p. Yue. 1995. Discovery and chronological division of the *Hipparion* fauna in Laogaochuan Village, Fugu County, Shaanxi. *Chinese Science Bulletin* 40:926–929.

Xue, X.-x., Y.-x. Zhang, and L.-p. Yue. 2006. Paleoenvironments indicated by the fossil mammalian assemblages from red clay-loess sequence in the Chinese Loess Plateau since 8.0 Ma B.P. *Science in China*, ser. D, 49:518–530.

Yin, H.-f., K.-x. Zhang, J.-n. Tong, Z.-y. Yang, and S.-b. Wu. 2001. The Global Stratotype Section and Point (GSSP) of the Permian-Triassic Boundary. *Episodes* 24:102–114.

Yue, L.-p., T. Deng, R. Zhang, Z.-q. Zhang, F. Heller, J.-q. Wang, and L.-r. Yang. 2004. Paleomagnetic chronology and record of Himalayan movements in the Longgugou section of Gyirong-Oma Basin in Xizang (Tibet). *Chinese Journal of Geophysics* 47:1135–1142.

Yue, L.-p. and Y.-x. Zhang. 1998. *Hipparion* fauna and magnetostratigraphy in Hefeng, Jingle, Shanxi Province. *Vertebrata PalAsiatica* 36:76–80.

Yue, L.-p. and Y.-x. Zhang. 2006. Paleomagnetic dating of *Lufengpithecus lufengensis* fossil strata. In *Lufengpithecus hudienensis Site*, ed. G.-q. Qi and W. Dong, pp. 252–255. Beijing: Science Press.

Zhang, Y.-p., Y.-z. You, H.-x. Ji, and S.-y. Ting. 1989. The Cenozoic deposits of the Yunnan region. *Professional Papers on Stratigraphy and Paleontology* 7:1–21.

Zhang, Y.-x., D.-h. Sun, Z.-s. An, and X.-x. Xue. 1999. Mammalian fossils from late Pliocene (lower MN 16) of Lingtai, Gansu Province. *Vertebrata PalAsiatica* 37:190–199.

Zhang, Y.-x., X.-x. Xue, and L.-p. Yue. 1995. Age and division of Neogene "red bed" of Laogaochuan, Fugu County, Shaanxi. *Journal of Stratigraphy* 19:214–219.

Zhang, Z.-q., A. W. Gentry, A. Kaakinen, L.-p. Liu, J. P. Lunkka, Z.-d. Qiu, S. Sen, R. S. Scott, L. Werdelin, S.-h. Zheng, and M. Fortelius. 2002. Land mammal faunal sequence of the late Miocene of China: New evidence from Lantian, Shaanxi Province. *Vertebrata PalAsiatica* 40:165–176.

Zhang, Z.-q., L.-h. Wang, A. Kaakinen, L.-p. Liu, and M. Fortelius. 2011. Miocene mammalian faunal succession from Damiao, central Nei Mongol and the environmental changes. *Quaternary Sciences* 31:608–613.

Zheng, S.-h. 1980. The *Hipparion* fauna of Bulong Basin, Biru, Xizang. In *Paleontology of Tibet, Part 1*, ed. Qinghai-Tibetan Plateau Comprehensive Scientific Investigation Team of Chinese Academy of Sciences, 33–47. Qinghai-Tibetan Plateau Scientific Investigation Series. Beijing: Science Press.

Zheng, S.-h., W.-y. Wu, and Y. Li. 1985. Late Cenozoic mammalian faunas of Guide and Gonghe basins, Qinghai Province. *Vertebrata PalAsiatica* 23:89–134.

Zheng, S.-h. and Z.-q. Zhang. 2001. Late Miocene-early Pleistocene biostratigraphy of the Leijiahe area, Lingtai, Gansu. *Vertebrata PalAsiatica* 39:215–228.

Zhu, R.-x., Q.-s. Liu, H.-t. Yao, Z.-t. Guo, C.-l. Deng, Y.-x. Pan, L.-q. Lü, Z.-g. Chang, and F. Gao. 2005. Magnetostratigraphic dating of hominoid-bearing sediments at Zhupeng, Yuanmou Basin, southwestern China. *Earth and Planetary Science Letters* 236:559–568.

Zhu, Y.-m., L.-p. Zhou, D.-w. Mo, A. Kaakinen, Z.-q. Zhang, and M. Fortelius. 2008. A new magnetostratigraphic framework for late Neogene *Hipparion* red clay in the eastern Loess Plateau of China. *Palaeogeography, Palaeoclimatology, Palaeoecology* 268:47–57.

Zykin, V. S. and V. S. Zazhigin. 2004. A new biostratigraphic level of Pliocene in West Siberia and the age of the lower-middle Miocene stratotype of the Beshcheul horizon. *Doklady Earth Sciences* 398:904–907.

Zykin, V. S., V. S. Zykina, and V. S. Zazhigin. 2007. Issues in separating and correlating Pliocene and Quaternary sediments of southwestern Siberia. *Archaeology, Ethnology & Anthropology of Eurasia* 2(30):24–40.

ADDENDUM

The editors of this volume count among our highly esteemed colleagues all of the contributors herein. Well into production of this book, we were deeply moved by the passing of one of our authors, Eleonora Vangengeim (1930–2012). Eleonora was a key figure in vertebrate paleontology at the Geological Institute of the Russian Academy of Sciences. Her expertise led and inspired a generation of paleontologists throughout the world. Her focus was biostratigraphy and the evolution of Neogene mammalian complexes, emphasizing the geological setting of the vertebrate remains upon which we piece together the terrestrial biotic history of Asia. We benefit from the rich legacy of her work. Thank you, Eleonora.

Part I

East Asia

Chapter 1

Neogene Land Mammal Stages/Ages of China

Toward the Goal to Establish an Asian Land Mammal Stage/Age Scheme

ZHAN-XIANG QIU, ZHU-DING QIU, TAO DENG, CHUAN-KUI LI, ZHAO-QUN ZHANG,

BAN-YUE WANG, AND XIAOMING WANG

Led mainly by European and North American geologists, the domain of stratigraphy entered into a state of rapid development after World War II. Foremost among these developments were the discoveries, improvements, and widespread uses of new dating methods (age determination by isotopes, magnetostratigraphy, geochemistry, sequence stratigraphy, and tuning of astronomical cycles), which greatly increased the accuracy and precision of age estimates. Also instrumental in this rapid development was the publication of the *International Stratigraphic Guide* (ISG; Hedberg 1976; Salvador 1994) and the *Revised Guidelines for the Establishment of Global Chronostratigraphic Standards* (Remane et al. 1996), which clarified the basic principles and standardized terminologies and procedures. A direct reflection of these improvements is the establishment of the Global Standard Stratotype-Section and Point (GSSP) of the marine stages that is the foundation of the global standard geologic time scale. From 1972 to the present, GSSPs for about 60 of the 100 stages in the Phanerozoic Eonothem have been ratified and codified. This process is embodied in the publication of *A Geologic Time Scale 2004* (Gradstein, Ogg, and Smith 2004).

Work on Neogene chronostratigraphy stands as one of the highlights of these developments. Of the eight Neogene stages (up to Piacenzian) in the Standard Global Chronostratigraphic (Geochronologic) Scale (SGCS), lower boundaries of six GSSPs have already been nailed, the boundary for another (Langhian) is all but settled, and only the lower boundary for the Burdigalian stage is still controversial. This progress especially benefits from studies, in the postwar period, of deep-sea drill cores and contained microfossils.

However, the fact remains that the importance of terrestrial stratigraphy and mammalian fossils did not gain sufficient recognition, nor did it gain an adequate expression in the *International Stratigraphic Guides*. Due to extremely large facies variation in terrestrial deposits and strong endemism, low abundance, and incompleteness of mammal fossil records, terrestrial stratigraphy differs greatly from invertebrate-based marine stratigraphy in methodology and working procedure. For lack of a uniform international standard, every continent, or even country, has established its own terrestrial stratigraphic system.

During the last three decades, progress has also been made in Chinese Neogene terrestrial stratigraphic studies. Most of the classic regions have been revisited, such as the Yushe and Baode areas in Shanxi, the Lantian area in Shaanxi, the Tunggur area in Inner Mongolia, and so on. New discoveries are made in well-exposed fossiliferous regions, such as Tongxin in Ningxia, eastern Nei Mongol (Inner Mongolia), the northern Junggar Basin in Xinjiang, the Linxia Basin in Gansu, the Qaidam Basin in Qinghai, and so on. Magnetostratigraphic work was also undertaken in several classic regions. Great gaps, however, still exist between China and its European and North American counterparts in terms of accumulation of fossils as well as such basic tasks as documentation of fossil occurrences and their biostratigraphic contexts.

This chapter is an attempt to reappraise the existing stages/ages in the Chinese Neogene in relation to the currently widely adopted approaches in terrestrial stratigraphy (Neogene Mammal unit [MN] and North American Land Mammal Age [NALMA]) and from the point of view of the *International Stratigraphic Guides*. We examine the principles, methods, and working procedures used for the establishment of Chinese stages/ages in the past. We propose a new Neogene chronostratigraphic framework that we consider more consistent with the reality of the state of research and conditions in China. This will provide a foundation for the establishment of a formal Chinese Regional Land Mammal Stage/Age system. Given that China possesses well-developed Neogene terrestrial strata that are richly endowed with fossil mammals, such a system should play a role in the establishment of a Centro-East Asian Land Mammal Stage/Age (CEALMS/A) scheme (Neogene System) in the future.

The numerical age data consistently used in this chapter are those based on the orbitally tuned calibrations for the Neogene, ATNTS2004 (Gradstein, Ogg, and Smith 2004).

For most of the definitions of stratigraphic terminology, we follow Woodburne (2004a:XI–XIV). Here, FHA (first historical appearance of Walsh) = first appearance datum (FAD), and LHA (last historical appearance of Walsh) = last appearance datum (LAD). For further explanations or alternative views, the readers are referred to articles by Walsh (1998, 2000).

REMARKS ON PRINCIPLES, METHODS, AND PROCEDURES

Different Approaches and Practices in Neogene Land Mammal Chronology and Stratigraphy

NALMA Scheme

Tedford (1970) systematically summarized in great detail the evolution of the North American terrestrial stratigraphic system based on fossil mammals. From the beginning of the last century, Osborn and Matthew (1909) organized the North American Cenozoic terrestrial strata into a series of "life zones" on the basis of lithological units containing representative mammalian genera. Although the system was limited by the knowledge then available and had no clearly defined boundaries, Tedford (1970:697) considered these "life zones" as "biostratigraphic units." On the other hand, Matthew (1924) devised a series of "faunal zones" based on stage of evolution

of the horses. Tedford noted that conceptually Matthew's series is a temporal arrangement entirely different from a chronostratigraphic system, but did not explicitly refer it directly to a bio- or geochronologic system.

Toward the end of the 1930s, the temporal sequencing of the terrestrial Tertiary deposits of North America and the mammal faunas contained therein had become evident and widely accepted. However, its progression was "seriously hampered by the confusing use of identical terms for rock units and for the time units" (Wood et al. 1941:3). The purpose and goal of the Wood Committee, as mandated by the Vertebrate Section of the Paleontological Society, was very clear. It was "to present a provincial time scale for the North American Tertiary," a system of "purely temporal significance" that will "cover all of Tertiary time," and to propose "a standardized terminology of purely temporal significance" (Wood et al. 1941:1, 6). More than 300 significant rock and faunal units were analyzed. As a result, 18 mutually exclusive "North American Provincial Ages" (NAPA) were formally proposed. As stated by Wood et al. (1941:1), these NAPAs were established "based on North American mammal-bearing units, . . . defined in terms of precisely analyzed faunas and the related stratigraphy." The fauna of each NAPA can be classified into four kinds: index fossils (confined within the NAPA), first appearances (at any point within the NAPA), last appearances (at any point within the NAPA), and characteristic fossils (common but not confined to the NAPA). In magnitude of time span, the NAPA was roughly equivalent to the Age in the European marine chronostratigraphy. As a replacement, Savage (1962) recommended the usage of "North American Land Mammal Age," or NALMA, which is now unanimously accepted.

The strong conceptual intention to build a pure time scale based on stage-in-evolution of mammals, conflicted with ambiguous and sometimes mutually contradictory methodologies applied and diverged from the ideals of the *ISG*, which hampered the further development of the NALMA time scale. Commenting on Wood committee's NALMA, Tedford (1970) pointed out that the conceptual underpinning of the Wood Committee proposal was divorced from their practices and that the two were self-contradictory. In practice, more than half of these "ages" were based on the time spans of specific rock units (Tedford thus called these ages "geochrons"). Even the names of these "ages" were taken from the rock units—for example, "the Arikareean Age equals the temporal span of the Arikaree Group." Thus defined, these "ages" could not possibly cover the entire time span of the Tertiary Period, because of numerous gaps and hiatuses between them. These "ages" were neither biostratigraphic units, because of lack of detailed biostratigraphic analysis, nor geochronologic "ages," which should be derived from established chronostratigraphic stages. Finally, Tedford came to the conclusion that these ages should be biochronologic units, "biochrons," based on evolutionary stages of fossil mammals. Tedford also clearly noticed the fact that, although the NALMA "are useful in depicting gross region-wide faunal change" (Telford 1970:696), it is "impossible to define the exact temporal limits or boundaries of these units without arbitrary selection of criteria" (Telford 1970:701). For such criteria, first appearances of taxa, especially of exotic forms, were listed.

As a leading land mammal stratigrapher, M. O. Woodburne has published since the 1970s a series of papers with the aim of clarifying and remedying the ambiguities and misconceptions of the original NALMA scale (Woodburne 1977). We owe much to him in developing the modern role of mammal biochronology in defining biostratigraphy and chronostratigraphy within the general domain of stratigraphy.

In early years, some mammal paleontologists also attempted to establish formal stages in accordance with the international stratigraphic guidelines, as exemplified by Savage, who proposed the Cerrotejonian and Montediablan Stages in 1955. More recently, the Wasatchian, Clarkforkian, Whitneyan, and Orellan also have been called stages.

In spite of the shortcomings in its original version, the NALMA time scale has proved vital and useful in practice. The majority of North American mammal paleontologists show great sympathy for it and retain this scheme, refining it where possible. Since the middle of the past century, persisting efforts have been made by several generations of paleontologists to enhance the quality of the NALMA scheme. As a result, a carefully refined and comprehensive framework for the Neogene has been accomplished by groups of authoritative paleontologists in 1987, and updated in 2004 (Tedford et al. 1987; Tedford et al. 2004).

We take the Arikareean NALMA as an example to illustrate the principles and methodology adopted by Tedford and colleagues for the Neogene. The type of the Arikareean Age originally designated by Wood et al. (1941:11) was the "Arikaree Group of western Nebraska, Agate being the most typical locality, with the limits as defined by Schultz (1938), but including the Rosebud." Originally, the Arikaree Group included, in ascending order, the Gering, Monroe Creek, and Harrison formations and was tentatively correlated with the Aquitanian Stage and the lower part of the Burdigalian Stage of the Miocene Epoch in the European time scale. Such a definition and correlation remained acceptable until the mid-1970s. Martin (1974)

published a short article about some rodent fossils including two specimens of the earliest appearance of *Plesiosminthus geringensis* from UNSM Mo-19, Durnal locality of the Gering Formation, eastern Wildcat Ridge of western Nebraska. Since the early 1980s, intensive stratigraphic and paleontologic investigations have continued (Swisher 1982; Tedford et al. 1985; Tedford et al. 1996; etc.). In their 1987 version of the North American Neogene faunal succession and biochronology, Tedford and colleagues defined the early Arikareean LMA by the earliest appearance of *Ocajila*, talpine moles, *Plesiosminthus*, and *Allomys*. Based on the fission-track and radiometric dating of the ashes of the Helvas Canyon Member of the Gering Formation then available (27.6–28.8 Ma), the age of the base of the Gering Formation was defined as ~29 Ma. In 1996, Tedford and colleagues agreed with the opinion of Swinehart et al. (1985) that the upper part of the Whitney Member should be separated from its lower part and be called "Brown Siltstone Beds." Further, Martin's *Plesiosminthus* specimens were assigned by Swisher to the upper half of the "Brown Siltstone Beds" (Swisher's Unit A). The other first appearance taxa, *Ocajila*, the talpine (*Scalopoides*), and *Allomys* (recognized as *Alwoodia*) were proven to be from higher levels, the Gering Formation and "Monroe Creek." New isotopic and paleomagnetic dating (Tedford et al. 1996:fig. 9) showed that *Plesiosminthus* appeared in chron 11n (~29.5 Ma), whereas the base of the "Brown Siltstone Beds," was located in the upper part of chron 11r (slightly older than 30 Ma). Instead of using the first appearance of *Plesiosminthus* to define the base of the Arikareean Age, as recommended by Woodburne's principle of single-taxon boundary definition, Tedford and colleagues chose the base of the "Brown Siltstone Beds" as the base of the Arikareean Age and put the base of the Arikareean Age at 30 Ma. Such a boundary definition of the Arikareean Age has been followed since then (Tedford et al. 2004; Albright et al. 2008). It is worth noting that Tedford and colleagues applied the same principle to use a lithologic rather than biologic criterion in boundary definition of a biochronologic unit, which had been criticized by Tedford himself in 1970.

MN Units in Europe

Prior to World War II, owing to the discreteness and generally loose constraint in age determination of the mammal faunas, mammal paleontologists had to attach themselves to the system of stratigraphic division based on marine deposits in the temporal ordering of mammal faunas. The geochronologic units above the age status (Epoch, Period, Era) were always used, whereas marine chronostratigraphic/geochronologic units, stage/age, were occasionally applied (e.g., Burdigalian, Vindobonian, etc.). During that period, the only proposed stage/age based on terrestrial deposits containing fossil mammals was that of Villafranchian, created by Lorenzo Pareto in 1865.

After World War II, the trend to establish terrestrial stages/ages was pursued chiefly by established mammal paleontologists. Several such stages/ages were proposed during the 1950s and 1960s—for example, the Vallesian (Crusafont-Pairó 1950, 1951), the Csarnotanum and Ruscinium (Kretzoi 1962), and the Turolian (Crusafont-Pairó 1965). Concise accounts of this history can be found in Lindsay and Tedford (1990), de Bruijn et al. (1992), and Lindsay (1997).

The difficulty in applying the marine stages to terrestrial sediments with mammal fossils had been particularly acutely felt in the practical work of paleontologists studying the Neogene and Quaternary mammal faunas. The unusual agreement reached at the Eighteenth International Geological Congress in 1948 about the equivalence between the marine Calabrian and the terrestrial Villafranchian stages may serve as a good example to demonstrate the futility of such an approach. The mid-1970s marked a critical period for mammal paleontologists to develop their own system of Neogene subdivision. In the early 1970s, while preparing a report on the Neogene land mammal subdivision for the Regional Committee of the Mediterranean Neogene Stratigraphy (RCMNS) Congress to be held in 1975, P. Mein analyzed altogether 204 mammal sites (132 from western European countries, 61 from eastern Europe, the former Soviet Union, and Turkey, and 11 from Africa) and arranged them into 16 MN "zones." Mein's chart was presented at various international colloquia (Montpellier and Madrid 1974; the Munich Symposium, April 1975; the Sixth RCMNS Congress in Bratislava, September 1975; and the Round Table at Madrid, September–October 1976). Mein's MN "zone" system aroused heated debates during and after these meetings (Azzaroli 1977; Lindsay 1990; Lindsay and Tedford 1990; Fahlbusch 1991; de Bruijn et al. 1992; Agustí and Moyà-Solà 1991; Agustí 1999; Agustí et al. 2001; etc.).

At first, the general opinions were rather negative, as summarized by Fahlbusch in 1976 and 1991. Sharp criticisms were centered on the following two points:

1. Mein's "zone" was primarily based on "stage of [mammal] evolution," without referring to any particular rock body. Therefore, it was something "outside the [international stratigraphic] rules." It was neither a formal zone (biozone) in biostratigraphic unit terms nor a unit in chronostratigraphy/geochronology.

2. Mein did not provide age determination of his "zones," nor was it conceptually possible to fill the entire Neogene time span without gaps or overlaps based on these "zones" defined by Mein. These criticisms seem to have hit on the vital points of Mein's subdivision system.

Since the mid-1970s, the European mammal paleontologists have been divided into two major groups. One group held the opinion that the MN system should be maintained based on the principle of "stability and continuity" (Fahlbusch 1991:161), despite theoretical ambiguity and methodological shortcomings. Mein himself, Fahlbusch, de Bruijn, and most of the French paleontologists were of this group. Mein seldom directly talked about the nature of his MN "zones" nor answered the above criticisms. He implicitly considered his "zones" biostratigraphic units while submitting his chart to the Sixth RCMNS Congress (Mein 1975), but later he admitted that his "zones" had chronostratigraphic value (Mein 1981:83). Regardless of these disagreements, Mein kept improving his MN "zones" (Mein 1979, 1981, 1990). The last version of Mein's chart with tables of mammalian genera was published in a report of the RCMNS working group on fossil mammals authored by seven members of the group, including Mein himself (de Bruijn et al. 1992). A series of changes were made in the 1992 version as follows. The MN system was formally acknowledged as biochronological subdivision. Accordingly, the troublesome term "zone" was substituted by the neutral "unit." The "reference faunas" or "reference localities" were particularly emphasized and singled out in the chart. Similarly, the "stage-of-evolution" was also emphasized, as opposed to the first and last appearances in characterizing the units. All the subdivisions within the "units" were suppressed, and MN 7 and MN 8 were joined together. In addition, it was "strongly advised to continue to use the Agenian, Orleanian, Astaracian, Vallesian, Turolian and Ruscinian . . . , regardless whether or not these terms indicate a stage or age or both" (de Bruijn et al. 1992:68).

The other group consists mainly of the mammal paleontologists and stratigraphers working in the Iberian Peninsula and the North Alpine Foreland Molasse area, where terrestrial deposits with rich mammal fossils are widely developed. Unsatisfied with Mein's MN system, paleontologists and stratigraphers from Spain and their colleagues tended to "redefine" Mein's MN units by building up real biostratigraphic biozones based on rock bodies with mammal fossils. During the next 15 years, extensive projects were undertaken in the major terrestrial basins in Spain (Ebro, Calatayud-Daroca, Vallès-Penedès, Duero, Teruel, Madrid, and Guadix-Baza). As a result, 24 such biozones were established and named primarily by small mammal genera and/or species. These biozones were grouped into formal geochronologic units (Early, Middle, Late Miocene, and Pliocene). Mein's MN units were considered to play "the role of hypothesis of bioevent succession" (Agustí and Moyà-Solà 1991:105). The 24 biozones were correlated to the MN units as well.

The rapid development and steady improvement of paleomagnetic techniques in the 1980s gave birth to the concept of the Integrated Magnetobiostratigraphic Time Scale (IMBTS), first suggested by Berggren et al. in 1985. The application of this concept to biostratigraphic work in Spain and the Alpine Molasse made it possible to define the boundaries of the biozones established in rock bodies in these areas. It is even more important that, through careful correlation between the local biozones and the MN units, time constraint of the MN units has become in many cases feasible and reliable. Thus, the MN units 1–17 were systematically revised in Spain (e.g., Agustí et al. 2001), based on the first appearances of selected small and large mammal taxa, and finely calibrated paleomagnetic ages. Whether this version of the MN system will stand up to further testing is a matter for the future.

In retrospect, when submitted, the MN system was rather flawed in concept and methodology, causing much confusion and criticism. However, it was based on tremendous numbers of discrete faunas, the data of which accumulated over more than 150 years of research and study, although the stratigraphic constraint of these faunas remained very poor in most cases. A biochronologic scheme probably fits this particular situation well, and the IBMTS concept and methodology help greatly to revive Mein's MN system.

Practice of Chinese Neogene Terrestrial Stratigraphy

Systematic studies of Chinese Neogene terrestrial stratigraphy began in the 1980s, mostly carried out by mammal paleontologists. In 1984, Chuan-kui Li, Wen-yu Wu, and Zhu-ding Qiu, following the "mammal age" approach by Fahlbusch (1976), first divided the Chinese terrestrial Neogene into seven "ages" (Xiejian, Shanwangian, Tunggurian, Bahean, Baodean, Jinglean, and Youhean), exactly corresponding to the seven European land mammal ages (Agenian, Orleanian, Astaracian, Vallesian, Turolian, Ruscinian, and Villanyian). As in Europe, Li et al. did not

follow the *International Stratigraphic Guide* (Hedberg 1976) or the *Stratigraphic Guide of China and Its Explanation* (All-China Stratigraphic Commission 1981). Therefore, their "ages" were not founded on the basis of preexisting "stages." They listed several mammal assemblages for each of the "ages," but did not explicitly specify which one is the more representative. The upper and lower boundaries of Li et al.'s "ages" follow completely the corresponding "ages" from Europe. If the nominal localities for the "ages" are used as the representative faunas, a large difference in time will result. For example, the lower limit of the Shanwang Formation probably is no older than 18 Ma, but the lower boundary of the Shanwangian "Age," if aligned with that of the European Orleanian, would be ~20 Ma. Yong-sheng Tong, Zheng, and Qiu (1995) essentially adopted the system of Li et al. Zhan-xiang Qiu (1990), conscious of the difference in concepts between the above "ages" and those of the *ISG*, as well as in the procedures and standards in establishing ages, proposed informal "mammal units" as a temporary substitute for the "ages" that are not embodied in geological entities. Zhan-xiang Qiu and colleagues (1999) followed up with a system of "Neogene mammal unit and superunit." Three superunits containing 11 units were proposed: *Aprotodon*-tataromyids (NMU 1–3), *Platybelodon-Alloptox* (NMU 4–7), and *Hipparion*-siphneines (NMU 8–11). Deng (2006), in his summary paper, used seven ages and 13 NMUs.

On the other hand, as *ISG* became more widely adopted in China, stratigraphers also actively pushed for the establishment of regional chronostratigraphic stages for Chinese Cenozoic terrestrial strata. The Amendment Group for Chinese Stratigraphic Guide was founded in 1992 and was charged to establish a Chinese chronostratigraphic scale as one of its tasks. Encouraged by the breakthrough of the first GSSP in Chinese marine strata (formal acceptance of the GSSP for the Ordovician Darriwilian Stage), the Amendment Group decided in 1999 to "attempt China's own series of stage names (including post-Permian continental stages)." During the Third Conference of Chinese Stratigraphy in 2000 and in the "Chinese Regional Stratigraphic Table" in the appendix of the revised edition of *Stratigraphic Guide of China and Its Explanation* (All-China Stratigraphic Commission 2001), six Neogene stages were listed: Xiejian, Shanwangian, Tunggurian, Baodean, Gaozhuangian, and Mazegouan. However, because these stages were not established strictly according to the requirements in the *Stratigraphic Guide of China and Its Explanation*, a program of "Studies for the Establishment of Major

Stages in Chinese Stratigraphy" was launched in 2001. Deng Tao was charged to lead the efforts in completing works on three stages—the Xiejian, Shanwangian, and Baodean—by 2005, a task that has been concluded (Deng, Wang, and Yue 2008a, 2008b).

Reconsideration of Some Issues in Neogene Land Mammal Stratigraphy

Terrestrial Mammal Fossils vs. Marine Microfossils

The historical development of the terrestrial mammal subdivision in North America, Europe, and China as reviewed in the preceding paragraphs clearly shows that the way of treatment for the subdivision of historical events in terrestrial mammal evolution has differed considerably, conceptually and methodologically, from that of the stratigraphers working on marine invertebrates and general stratigraphers who constitute the majority in the stratigraphic domain of the world's earth science. The dual developments by mammal and marine invertebrate paleontologists seem partly rooted in the intrinsic nature of the fossils occurrences themselves. Tedford (1970:698) attributed "the abandonment" of classical biostratigraphy "partly . . . to the intermittent nature of most vertebrate fossil occurrences." Azzaroli (1977:25) explicitly stated that "the concept of biozone . . . is not applicable to mammalian fossils." Prothero (1995:305) expressed the same notion, while Walsh (1998) made an analysis in greater detail regarding the differences of the terrestrial mammals from the pelagic marine invertebrate and marine microfossils.

The terrestrial mammal fossils differ from marine invertebrate fossils, especially pelagic microfossils, in several important ways. The first distinctive nature of terrestrial mammal fossils is their low absolute abundance of individuals compared to marine invertebrate occurrences in general. This in turn results in the low average rate of "finds" of mammal fossils. Rough estimates provided by Walsh (1998:174) show that "the average rate of 'finds' of fossils of a given mammal taxon is often orders of magnitude less than that for marine organisms." In fact, "the finds per meter (f/m) might range from 0.001 to 1 for a mammal taxon, whereas the f/m might be from thousands to millions for a typical marine microfossil taxon." The second is the localized distribution of the nonmarine deposits where the mammal fossils are usually embedded. The lacustrine and fluvial deposits, plus various overbank components, are laterally restricted (compared to

marine deposits), frequently isolated from each other, with rapid changes in lithofacies and numerous depositional hiatuses, and often lack widespread marker beds (with the exceptions of eolian sediments and volcanic ashes). The third is the intermittent nature of the mammal fossils themselves. In most instances, fossil mammals appear to be distributed sporadically, "tending to concentrate in few particular beds or even in pockets, separated from one another by beds devoid of mammalian remains" (Azzaroli 1977:25) or "in isolated pockets or quarries without stratigraphic superposition" (Prothero 1995:305). As a rule cited from Murphy (1994) by Walsh, "Any organism living in a subaerial environment is less likely to be preserved intact in sediments than an organism in or above marine deposits" (Walsh 1998:174). The fourth is the stronger geographic differentiation of mammal faunas. It is easy to understand that "the dispersal of many pelagic marine organisms may be assumed to be rapid over large oceanic areas." In contrast, "the epicontinental seas and mountain ranges, and vegetational changes caused by physiography and latitudial position" (Walsh 1998:174), all may serve as segregation factors for mammal dispersal. As a consequence, precise division and large-scale correlation are relatively difficult when biostratigraphy is based on terrestrial fossil mammals alone.

These aspects of land mammal fossils as tools of stratigraphic subdivision, especially for units of global significance, are negatives. This is why general stratigraphers prefer basing the global chronostratigraphic units on marine sediments. However, these negative aspects cannot alter the essence of mammal fossils as remains of organic bodies. In fact, such characters also exist for invertebrate and plant fossils in various combinations and degrees. In both editions of the ISG, the rarity of the organic fossils has been fully recognized. Salvador (1994:54) noted: "Fossils usually constitute only a minor, disseminated, fractional part of a rock stratum. Even within fossiliferous sequences, fossils are rarely found in every bed or formation, nor are they found everywhere along a bed or formation. There are barren spaces or intervals in all stratigraphic sequences." Salvador also listed "metamorphism, the vagaries of fossil preservation, the time required for migration, [and] accidents of collection" as factors influencing the incompleteness in preservation of fossils in general.

Apparently, the differences between the terrestrial mammal fossils and the marine invertebrate fossils have been overestimated by some mammal paleontologists. Most of the differences seem to be differences in degree, rather than in essence. Most of these shortcomings are being remedied through intensive in situ collecting, as evidenced by the tremendous augmentation of mammal fossils in North America in the postwar years.

On the other hand, as noted by Salvador (1994:54), "because of the irreversibility of organic evolution, fossils are particularly valuable in time-correlation of strata and in placing strata in their proper relative chronologic position." We may add here also three positive aspects of the mammal fossils. The first is the rapid evolutionary rate of mammals, compared with invertebrates and plants. The second is the better quality of the mammal fossils in general, enabling easy identification of the fossils at generic or specific levels, even when represented by fragmentary pieces and in small numbers. The third is the ever-increasing usage of micro-mammal fossils in vertebrate paleontology since the sieving technique was implemented in fossil collecting in the 1940s. In abundance and spatial distribution, the micro-mammal fossils can roughly match most of the marine invertebrate fossils.

LMA: Its Present State and Future Development

ARE LMAS BIOCHRONS IN THE SENSE OF THE *ISG*?

Most North American mammal paleontologists emphasize the temporal connotation of the LMAs and recognize them as biochrons under the category biochronology. The term of biochron was first proposed by Williams (1901:579–580), originally defined as "the endurance of organic [individual, species, genus, etc.] characters," which was interpreted by Salvador (1994:109) as "the absolute duration of a fauna or component parts of it." However, neither biochron nor biochronology has been formally acknowledged in the *ISG*. In the first version of the *ISG* (Hedberg 1976:48), biochron was only mentioned in passing: "The total time represented by a biozone may be referred to simply as its time or time-value, or biochron." In the second edition of the *ISG* (Salvador 1994:105, 109), these two terms were dropped, reserving them only in the appended glossary. Therein the term *biochron* is retained in larger type (those in more common use), whereas *biochronology* in smaller type (for terms that have received limited or no acceptance). Biochron is explained in two ways. In addition to Williams's original usage, it is also defined as "the total time represented by a biozone [biostratigraphic unit]." Biochronology is here defined as "the relative dating of geologic events based on biostratigraphic methods or evidence." Both terms are here narrowly interpreted as time equivalents of biozone and biostratigraphy. These *ISG* explanations were also

adopted by Woodburne (2004a:XI, 2004b:table 1.3). In this sense, it is difficult to squeeze an LMA into a biochron, even though the latter is established based on a preexistent biozone.

1. Biochron as defined in the *ISG* is generally much shorter in time duration than an LMA, which approximates a formal geochronologic age in magnitude. As it now stands, NALMA durations vary from 3.5 Ma to 12 Ma, even larger than for stage/ages, which are 2–10 Ma. It seems rather paradoxical that, on the one hand, the NALMAs are called biochrons or chrons (of biochronology), and, on the other hand, whenever the biozones (equivalents of chrons) are built, they are much smaller in magnitude, often being a very small part of the NALMAs.

2. The faunal content of an LMA is generally great and contains much more evolutionary information than that of a biochron in the sense of the *ISG*. The faunal content of an LMA usually cannot be properly dealt with as that of a biochron, unless the LMA is further subdivided. Various types of biozones (range zone, lineage zone, and even the assemblage zone) are difficult to apply to define the LMAs. The characterization of LMAs using four kinds of mammal fossils, proposed by Wood et al. in 1941, seems to fit quite well the goal of correlating the sections with various combinations of taxa.

3. The currently used principles to define the lower boundary of LMAs include not only biological criteria (first appearance datum [FAD] of taxon or taxa) but also nonbiological ones (lithologic markers, paleomagnetic reversals, and so on), as exemplified by the Arikareean (see previous discussion). Walsh discussed the phenomenon of mixed usage of multiple criteria in boundary definition in a specific section of his paper. Walsh (2000:763) separated the biostratigraphic units in two types: eubiostratigraphic units for the commonly used biostratigraphic units, and quasibiostratigraphic units for the units "whose boundaries are defined by other means." As examples, he listed "paleontologically distinct lithozone" (defined by lithologic contacts, marker beds, or erosion surfaces), "paleontologically distinct metrizone" (in terms of metric levels in a measured section), and "paleontologically distinct 'fuzzy' zone." Although we are not in favor of his sophisticated classification, we agree with him that the multiple-criterion approach is quite common in stratigraphic practice. In fact, the multiple-criterion approach has been clearly endorsed by the *ISG* as one of the diagnostic features for boundary definition in chronostratigraphy.

IS IT POSSIBLE TO ESTABLISH BIOCHRONOLOGY AS AN INDEPENDENT CATEGORY IN THE *ISG*?

As in the case for biochron, North American mammal paleontologists lay much emphasis on the role of biochronology. Woodburne's viewpoint is probably the most representative. According to Woodburne (2006:233): "The operation begins with biostratigraphy, is carried forward by biochronology, and ultimately can lead to chronostratigraphy. From this standpoint, biochronology deserves formalized recognition in stratigraphic codes and guides." Walsh (2000:164) holds a slightly different view that Periods, Epochs, and Ages are biochrons and that biochronology is a hierarchic system.

It seems to us that the community of stratigraphers and of geologists at large can hardly tolerate the existence of more than one category of purely temporal significance within the SGCS. It is understandable that the community of stratigraphers favors geochronology rather than biochronology, and several points support this.

1. By definition, the category of temporal significance to be used in stratigraphy must be based on rock bodies, not on evolutionary stages of organisms. We must remember that the whole point of all stratigraphic endeavors is to work with rock bodies and nothing else. In its strict sense of the term itself, biochronology seems to be attributed to the domain of biology.

2. If LMA, as a biochronologic unit, is accepted as a formal unit in the SGCS, an inevitable consequence would be increasing claims from paleontologists working on other groups of organism (e.g., ammonite-based, mollusk-based, foraminifera-based biochrons, etc.; see also Walsh 1998:163).

3. Conceptually, the boundaries of any biochronologic unit (biochron) are diachronous, since appearance and disappearance of any organism theoretically cannot be globally, even regionally, isochronous. This point has been clearly illustrated in the *ISG* (Hedberg 1976:79, fig. 12; Salvador 1994:83, fig. 13). Chronostratigraphic units, as defined in the *ISG*, are bounded by isochronous surfaces (Hedberg 1976:67) or horizons (Salvador 1994:78), using "distinct marker horizons, such as biozone boundaries or magnetic polarity reversals, that can be readily recognized and widely traced as time-significant horizons" (Salvador 1994:79). Such an isochrony is further guaranteed by a final ratification of a given GSSP.

4. Biochronology, were it allowed to exist as a special category in the *ISG*, is doomed to be a rankless system, as it is based on rankless biostratigraphic biozones.

This shortcoming renders it impossible to satisfy the basic requirement to encompass all classes of rocks—sedimentary, igneous, and metamorphic—and organize them into a ranked system. This could be fulfilled only by geochronology, which is defined in the *ISG*, since it is based on a ranked system of chronostratigraphy. The concept of biochronology can be useful, but only within the limits of the rankless biozone, and biochron can only be the time equivalent of the biozone.

NECESSITY AND POSSIBILITY OF REDEFINING LMA AS FORMAL REGIONAL STAGE/AGE

If the appeal to establish an independent category for biochronology is possibly rather unrealistic, the proposal to accept the duly modified LMA as a special type of geochronologic age might be welcomed.

First, compared with Wood et al.'s original LMA, the current NALMA scheme has been greatly refined and radically changed in many aspects. A large body of biostratigraphic work on the Neogene has been intensively carried out. With the rapid development of radioisotopic dating and paleomagnetic reversal stratigraphy since the 1970s, much effort has also been centered on the high-resolution boundary definition of all the NALMAs. As the biostratigraphy and boundary definition constitute the two primary bases in establishing chronostratigraphic units, the nature of the NALMA changed greatly in the direction of chronostratigraphy. As Woodburne (2006:240) noticed, "With the addition of a boundary definition and a biostratigraphic base, a NALMA will possess all data required to form the basis of a chronostratigraphic stage." In practice, Woodburne (1996:551) proposed that "boundaries of both biostratigraphic and chronostratigraphic units should be defined on single taxa." Woodburne (1996:547) used the FAD of *Hippotherium* in Europe and the Mediterranean area as "nearly isochronous within the limits of available resolution." This accords, at least in part, with the basic requirement in establishing chronostratigraphic units, as exemplified by stage as set in the *ISG* (Salvador 1994:79), that the boundaries should be "in principle isochronous." Walsh (1998) holds slightly different opinions as to the synchroneity of FHA (see following discussion). At any rate, most North American mammal paleontologists have intended to find lower boundaries as isochronous as possible for NALMAs, following the same core principle (isochroneity) in boundary definition of chronostratigraphic units as set in the *ISG*.

Second, the NALMAs approximate stage/age in magnitude (the lowest rank of chronostratigraphic/geochronologic unit), rather than any kind of zones/chrons (unique biostratigraphic/biochronologic unit).

Third, there is a practical need for establishing formal chronostratigraphic units in terrestrial deposits based on land mammals for the Cenozoic Erathem, which should be expressed in stratigraphic charts of global use. This need is acutely felt in particular by mammal paleontologists and stratigraphers working in areas where terrestrial deposits overwhelmingly predominate marine sediments or where the latter are lacking completely. Without these formal terrestrial units, the Cenozoic, at least the Tertiary System, would be left blank. Even in North America, where Cenozoic marine sediments are relatively developed, the call for redefining the LMA as formal chronostratigraphic and geochronologic units has never ceased (e.g., Clarkforkian, Wasatchian, Cerrotejonian, and Montediablan). The attitude of mainstream North American mammal paleontologists remains ambiguous. Tedford (1970:696) admitted that "a biostratigraphically founded sequence of chronostratigraphic units seems the best approach" and suggested a five-step procedure to reach the final goal to build local stages. Woodburne (2004b:14) stated, "When this ["a chronozone with both boundaries defined"] is accomplished, the entire package of strata can be combined into a chronostratigraphic unit of regional scale." Nevertheless, no leading mammal paleontologist has taken a resolute step forward to that goal.

Fourth, the attitude of the ISG toward the terrestrial deposits has changed. In an early stage of the stratigraphic endeavor, the marine deposits with invertebrate fossils played an almost exclusive role in stratigraphic work. The neglectful attitude toward terrestrial deposits and mammal fossils was manifested in the first version of the ISG (Hedberg 1976), wherein the role of terrestrial deposits was not mentioned at all. In the second edition of the ISG, when dealing with the establishment of the boundary stratotype of stage, it is claimed (Salvador 1994:79) that the boundary stratotype "should be within sequences of essentially continuous deposition, preferably marine (except in cases such as the stages based on mammalian faunas in regions of nonmarine Tertiary sequences or the Quaternary glacial stages)." The ISG thus permits the establishment of stages/ages based on fossil mammals in regional terrestrial settings. Unfortunately, the ISG did not fully appreciate the special importance of regional stages/ages in the Cenozoic Erathem and failed to state explicitly how such a system should be expressed in stratigraphic charts for global use.

WAYS TO REDEFINE LMA TO FORMAL REGIONAL AGE/STAGE

The principles and rules to be observed in establishing a stage as stated in the ISG (1994) can be summarized as follows:

1. A stage must be founded on rock bodies. In this particular case, there must be plenty of terrestrial deposits rich in mammal fossils.
2. A stage must be bounded by isochronous upper and lower boundaries. Emphasis should be placed on the lower boundary, since the upper boundary is defined by the lower boundary of succeeding stage.
3. A boundary should be associated with distinct marker horizons, such as biozone boundaries or magnetic polarity reversals that can be readily recognized and widely traced as time-significant horizons. At least a boundary stratotype should be selected and collectively and officially approved.
4. A series of requirements for the selection of a stage boundary stratotype (continuous deposition, rich fossils, less deformed, easy access, well studied, etc.) should be met.

In order to achieve the goal to transform the LMA to formal stage, these requirements must be fulfilled. For the first and fourth points, there is no problem for North America and Asia. In fact, most large areas possess terrestrial deposits rich in mammal fossils. France may be one of the few exceptions, wherein terrestrial deposits are less well developed compared to the rich mammal fossils accumulated through intensive collecting during the last two centuries. However, regarding the second and third points, two issues are specifically discussed later.

Defining lower boundary of LMA based on rules set in the ISG for stage

As clearly stated and illustrated in the ISG, the boundary definitions for biostratigraphic and chronostratigraphic units are conceptually different in theory. Take a taxon-range biozone-chronozone pair as an example (figure 1.1); the boundaries (irrespective of lower or upper) of such a biozone (biostratigraphy) in a given geographic area are inherently diachronous, as expressed by the lowest stratigraphic datums (LSDs) in all investigated sections of the area, while the lower boundary of the chronozone (chronostratigraphy) based on this biozone is an isochronous horizon fixed at some one point (conventionally the lowest LSD of all the LSDs, = FAD). Similarly, the biozone under discussion is limited in extent to the strata in which the fossil taxon occurs, while the chronozone includes all the strata of the same age as that represented by the total

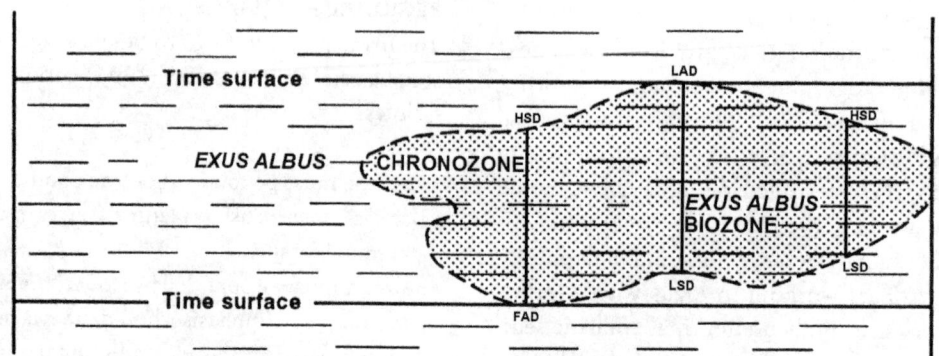

Figure 1.1 Relation between chronozone and biozone. Adapted from Salvador (1994:fig. 13). LSD = lowest stratigraphic datum; HSD = highest stratigraphic datum; FAD = first appearance datum; LAD = last appearance datum.

vertical range of the taxon, regardless of whether fossils of the taxon are present.

The majority of North American mammal paleontologists are, explicitly or implicitly, inclined to apply the same rules and procedures to define both the lower boundaries of the NALMAs and stages. Woodburne (1987a:15) stated that "lower boundaries of biozones [biostratigraphic], chronozones [chronostratigraphic], chrons [biochrons, biochronologic], stages [chronostratigraphic], or ages [geochronologic] based on such new occurrences [local introduction or allochthonous first occurrence] should be theoretically and operationally isochronous away from the stratotype." Walsh (1998:154) remains doubtful about the utility of Woodburne's isochroneity of FAD in both NALMAs and formal chronostratigraphic units. He agrees with Hedberg's view that "regardless of whether or not biological breaks were historically synchronous world-wide, our record of them world-wide in the Earth's strata can rarely, if ever, be so." Walsh (1998:154–155) made a compromise: he admitted that a part of his FHAs (first historical appearance, = FAD) may be essentially synchronous and called synchronous first historical appearances (SFHAs), while the other FHAs are significantly diachronous and called diachronous first historical appearances (DFHAs). Walsh (1998:163) urges North American mammal paleontologists to "accept the diachrony and evolving nature of our local and regional biostratigraphic and biochronologic units [NALMAs] and use them for what they were devised for: local and regional correlations of rocks."

The way to achieve isochroneity in lower boundary definition for North American mammal paleontologists is slightly different from that set in the *ISG*. According to the *ISG* (Salvador 1994:79), "Both [upper and lower] boundaries should be associated with distinct marker horizons." As later further elaborated (Gradstein, Ogg, and Smith 2004:23–24), there are two kinds of markers: a single primary marker that defines the lower boundary itself, and multiple secondary correlation markers that function as helpers to locate the lower boundary (lithologic, magnetic reversals, and so on). Gradstein, Ogg, and Smith (2004) further explain: "Most primary markers for GSSPs have been biostratigraphic events, but some have utilized other global stratigraphic episodes (e.g., iridium spike . . . carbon isotopic anomaly . . . C6Cn.2n . . . Milankovitch cycle)." Finding a primary marker at the lower boundary of a biozone based on marine microfossils is probably not a very difficult task. However, to do the same based on land mammal fossils in terrestrial deposits is probably not very easy, if not impossible. As discussed earlier, large mammal fossils inherently are extremely few in numbers of individuals and specimens, and the terrestrial deposits are extremely variable both vertically and horizontally, compared with microfossils in marine sediments. Except for some rare cases, as exemplified by Walsh's SFHA or possibly Woodburne's "*Hippotherium* Datum," many a FAD of a land mammal taxon or taxa in terrestrial deposits can hardly be a marker horizon at the same time. It may happen that the next time one visits the section where the LSD (or FAD) of a given taxon is established, no specimen of that taxon is found in monotonous, markerless deposits. In other words, in many cases, if we stick to the biologic criterion rigorously, it may not be possible to fix the point or horizon in a stratotype section. In such cases, the multiple-criterion rule as set in the *ISG* may be of great help. In terrestrial deposits, the magnetic reversal data can be particularly helpful. Being globally isochronous (≈1000 yr for an excursion between and/or within polarity reversals) and frequent phenomena, paleomagnetic reversals provide invaluable marker horizons for chronostratigraphic use. In view of the theoretical possibility to find in the future new specimens of a FAD taxon in deposits lower than the existing FAD level, it is reasonable to choose a magnetic reversal immediately underneath and nearest to the existing FAD horizon as the proxy marker.

Land Mammal Stage/Ages (LMS/As) as regional chronostratigraphic (geochronologic) units of global significance

In the *ISG* (Salvador 1994:78), the stage is considered "the smallest unit in the standard chronostratigraphic hierarchy that can be recognized at a global scale." Now, all the stages/ages of the entire Neogene System/Period, except for the base of the Burdigalian, have been formally defined in the ratified GSSP marine sections. However, this finely constructed system seems of little practical use in Neogene stratigraphic work, especially in the interior areas where marine sediments are practically lacking, as in China and many countries of central Asia. In fact, since the Jurassic and particularly during the Cenozoic, terrestrial strata increasingly take up a greater proportion of the land surface. Neogene terrestrial sediments far exceed marine strata in continental settings, especially so in North America and Asia.

Wide regional differentiation is one of the inherent characters of mammals. From the beginning, the Wood Committee (1941) emphasized that their LMAs should be provincial, not global. North American Neogene mammalian composition and evolution are different from those in Europe and Asia, and these differences are apparent from the characters of the NALMAs themselves. In Europe and Asia, such differentiations are also quite pronounced. During the early Neogene in China, faunas retain distinct Asian Oligocene characteristics for an extended period of time (approximately 3.5 myr). Major Miocene representative components appeared relatively late (around 19.5 Ma), and before the late Miocene, faunal compositions and characteristic members (represented by *Alloptox*, *Platybelodon*, and *Kubanochoerus*) are visibly different from those in Europe. The Asian *Hipparion* faunas also lack the kind of drastic transition from the forested type (Vallesian) to the grassland type (Turolian) as in Europe. Only during the Pliocene do faunas from China and Europe become closer to each other. It is thus apparent that mammal characteristics and evolution can only justify a regional stratigraphic system in continental settings.

Tremendous amounts of work undertaken by mammal paleontologists of the world since the 1950s prove clearly that a regional stratigraphic zonation scheme based on land mammals is not only indispensable for the countries or areas lacking marine sediments but also important for a better understanding of Cenozoic geology and the evolutionary history of mammals as a whole. Although the Neogene mammal faunas of each continent of the Old and New World are quite different in composition and at specific or generic levels, significant migrations between these continents occurred frequently. A good number of genera of proboscideans, equids, and carnivores are almost cosmopolitan in spatial distribution. Modified as suggested, the LMAs can be considered regional chronostratigraphic (geochronologic) units of global significance. Lack of representation of the terrestrial Neogene stratigraphic zonation in the global geologic time scale is irrational. Terrestrial Neogene stratigraphic zonations should appear in stratigraphic charts of global use.

Nature of European MN Units

Mein's MN unit conforms with Williams's original "biochron" in concept (see previous discussion). MN units are time units of biochronology in a pure sense, not the equivalent of biozones as later proposed in the *ISG*. Each MN unit is composed of a number of faunas closely similar in content and/or stage in evolution, without superposition control or defined boundaries. As such, the MN system cannot cover the whole time span without gaps, and it is not clear whether the units in fact overlap with each other. However, this system proves very useful for mammal paleontologists, especially when dealing with newly found mammal faunas in the same paleogeographic area. However, it is hardly possible to modify this system into biostratigraphic, chronostratigraphic, or geochronologic units without biostratigraphic work based on new sections with similar faunas. Such new work is being undertaken in Spain and Portugal.

Strategy and Procedures in Establishing the Regional Land Mammal Stage/Age in China

Primary Goal and Perspective

The comparative review of the two major systems (LMA and MN) in the previous section demonstrated that the LMA scheme shows clear advantages over the MN system. In addition to its wide acceptance in North America, the LMA scheme has also been gradually accepted in Europe (ELMA since Sen 1997), Asia (ALMA first for Paleogene since Meng and McKenna 1998), and South America (SALMA; see Flynn and Swisher 1995)—a trend of global acceptance seems in place. The LMA scheme is not only clearer in concept, constructed based on rich rock and time information, but also provides the possibility of self-modifying into chronostratigraphic units. When the first scheme of the Chinese Neogene stratigraphic subdivision based on land mammal faunas in terrestrial deposits was attempted by Li and colleagues in 1984, it was only possible to subdivide the Chinese Neogene into a few larger units of approximately stage/age magnitude, since the mammal paleontological information then available permitted only characterizing the bulk aggregate "chapters" of the mammal evolution in China. Attempts have also been made to apply the MN zonation system to the work in China, but in most cases the results were unsatisfactory, especially when the boundary definition was involved. We would like to apply the concept and methodology of the LMA scheme, as modified earlier, to establish a regional scheme of terrestrial chronostratigraphic units based on mammal fossils. If theoretically well founded and methodologically reasonably appropriate, such a scheme can well be accepted by mammal paleontologists and stratigraphers here in China and abroad. As it stands, similar geological setting and evolutionary history of mammals during the Neogene Epoch can also be observed in adjacent countries and areas, like Mongolia, the Asian part of Russia, Kazakhstan, Kyrgyzstan, Tajikistan, Korea, Japan, Vietnam, Thailand, and Myanmar. It is our desire that, in the near future, when work on mammal biostratigraphy and paleobiogeography in some of these areas is sufficiently known, a regional CEALMS/A scheme will emerge.

Procedure for Establishing Regional LMS/A

STAGE PRECEDES AGE

According to the *ISG*, stages are always established before ages, and ages are automatically derived from stages. In the case of LMA, however, such a process is reversed due to the special circumstances in historical development of the LMA (see previous discussion). In North America, where the NALMA has been universally accepted, the easiest way is to derive the stage name from the preexisting LMA, when the boundaries of the latter have been well defined and the biostratigraphic work of the whole LMA

has been adequately pursued and accomplished. Otherwise, recreation of new stages according to the *ISG* rules, followed by automatic award of age status, would cause much confusion. As it stands now in China, the LMA concept has so far not been solidified, and the Neogene subdivision in light of the modified LMA system is still in its initial stage. This special situation lends itself to starting directly from establishment of the chronostratigraphic unit. In so doing, the LMA can be changed to LMS/A.

FINDING AND DEFINING LOWER BOUNDARY AS FIRST PRIORITY

Base defines boundary

The priority of lower boundary over the upper one has been clearly set in the *ISG* (Salvador 1994:90): "The definition of a chronostratigraphic unit should place emphasis in the selection of the boundary-stratotype of its lower boundary; its upper boundary is defined as the lower boundary of the succeeding unit." If the two previously mentioned boundaries are defined, the whole time span (age) during which the given stage formed is settled, even when the rock and the faunas bracketed by the two boundaries are inadequately known. In searching for the lower boundary, the continuous rock sequence of paleontologically "critical time span" covering both the stage under discussion and the subjacent stage should be highlighted. This point is particularly important because in areas like China where the biostratigraphy is far from mature, with limited work done within the "critical time span," the pursuit of establishing stages hopefully can still be achieved. To this extent, the lower boundary definition is crucial in stage/age establishment for biostratigraphically less well-studied countries and areas.

Single-taxon definition

Woodburne (1977:233) suggested that "boundaries between chronostratigraphic units should be defined on the first appearance of a single taxon." In practice, this principle can sometimes be difficult to apply. In many situations, such as when fossils are preserved in a lens or pocket, few or no fossils may be present above and below the horizon in question, or when small fossils are screened where multiple species co-occur within a single point, it is difficult to choose a particular species to represent all first appearances. Furthermore, Woodburne's principle is a pure biostratigraphic concept, whereas in establishing stages in a chronostratigraphic system we should strive to follow the multiple-criterion rules in the *ISG* in seeking boundary stratotypes. Although we can use the first appearances of fossil mammals as boundaries, sometimes it is not easy to find a distinct marker horizon at the point of the first appearance, necessitating the downward search for such marker horizons, such as lithologic boundaries, characteristic magnetic reversals, or geochemical changes (such as the well-known iridium excursion).

Magnetostratigraphy

The rules and methods in lower boundary definition are amply demonstrated in the *ISG* (Salvador 1994:25–30, 78–80), and they should be carefully followed. In the *ISG* (Salvador 1994:79), magnetic polarity reversal was listed as one of the two major distinct markers (the other being the biozone) in boundary definition. Magnetostratigraphy plays an even more important and indispensable role in establishing LMS/As in China for a number of reasons. First, there are few volcanic ash layers feasible for $^{40}Ar/^{39}Ar$ dating in Neogene terrestrial deposits on the mainland of China. Second, there are few fossils of other organic groups that can play a similarly important role in relative age determination as mammal fossils do and could serve as a competing marker in boundary definition. Third, there are many Cenozoic basins containing thick terrestrial sequences suitable for carrying on magnetostratigraphic work in China. Finally, the particularly useful nature of magentostratigraphy is that magnetochrons are globally isochronous. Therefore, integrated biomagnetostratigraphic work is one of the indispensable prerequisites for choosing a Regional Stratotype Section and Point (RSSP).

Prior consideration of boundaries of ranking units higher than stage

There is one particular case in LMA boundary definition that has not been discussed in the *ISG*. In the case where the lower boundary of a stage to be established is very close to a formally ratified unit boundary of higher rank (Series, System, or Erathem), what shall we do with these two closely situated boundaries? We would like to recommend an auxiliary rule to solve this problem. Global standard lower boundaries of higher ranking units (series, system, etc.) should be considered prior to those of the regional stages. If the boundary of a stage being established is very close (say, <1 myr in time duration) to a ratified boundary of unit of higher rank, and no other better markers exist nearby, the higher-rank unit boundary can be considered as the lower boundary of the stage to be established. If such a suggestion can be adopted in a future revision of the *ISG*, it should not be difficult to look for boundary stratotypes for the establishment of

stages in terrestrial mammal-bearing strata. An argument for such a decision is practicality and convenience. After all, the chronostratigraphic (geochronologic) units higher than stage in Cenozoic Erathem have already been ratified by the International Commission on Stratigraphy (ICS) and unanimously accepted in the realm of earth science.

Chronostratigraphic boundary definition is in essence conventional, serving the purpose of standardization. A GSSP ("golden spike") should be collectively agreed and submitted, and eventually be formally ratified by the ICS. Since the LMS/A is a regional chronostratigraphic unit, its stratotype can be called an RSSP ("silver spike"). Candidate RSSPs can be proposed by stratigraphic organizations of any countries of the region concerned. Formal proposal of an RSSP should be agreed by a consensus of representative mammal paleontologists and stratigraphers of the region concerned in some forum (meeting or correspondence) and then submitted to the ICS.

As exemplified by Gradstein, Ogg, and Smith (2004:23) for the Lochkovian Stage of the Devonian System, "once the golden spike has been agreed, the discovery, say, of *Monograptus uniformis* below the GSSP does not require a redefinition of its position, but simply an acknowledgment that the initial level chosen was not in fact at the lowest occurrence of the particular graptolite." The same would hold true for the RSSP.

Unit Stratotype

Due to the relative rarity of fossil mammals, it is often difficult to find a single section that contains the desired temporal and mammalian contents that are completely representative of a LMS/A in a single body of rocks. Often, it is necessary that several sections or localities be used for such a purpose. Therefore, as unit stratotypes, LMS/As can have several sections representing different time periods or stages-of-evolution of the stage under discussion within the zoogeographic region concerned. Of these, the most representative section can be called the main Unit Stratotype, and the rest may be called auxiliary unit stratotypes.

DIVISION OF CHINESE NEOGENE LMS/A

Zoogeographic Division of Chinese Neogene Mammals

Based on geologic structures and depositional settings, Yun-tong Li (1984) and Jia-jian Zheng et al. (1999) divided

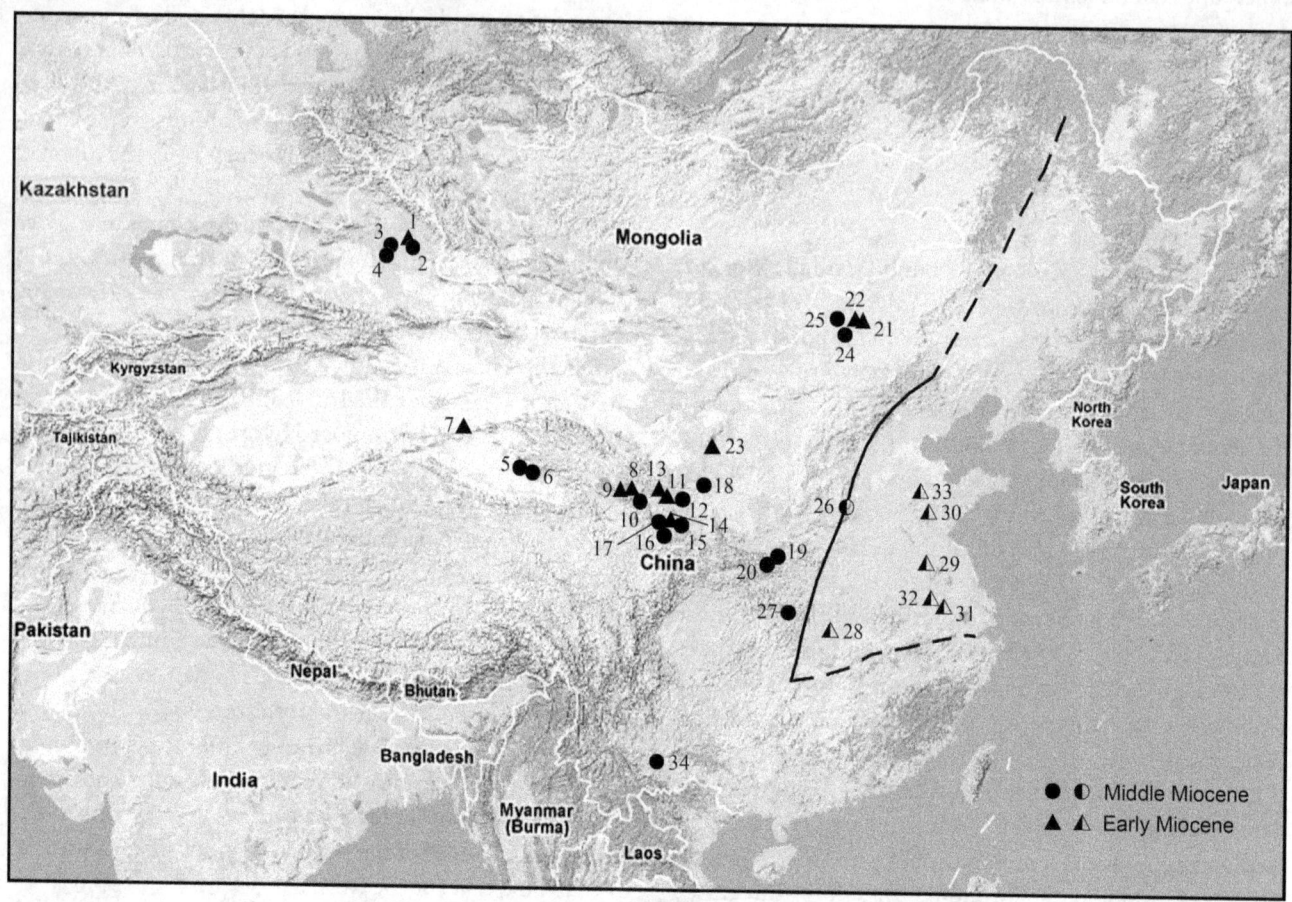

Figure 1.2 Paleozoogeographic subdivisions during Early–Middle Miocene of China mainland based on mammalian fossils: (*1*) Suosuoquan; (*2*) Halamagai; (*3*) Kekemaideng; (*4*) Dingshanyanchi; (*5*) Olongbuluk; (*6*) Tuosu Nor; (*7*) Xishuigou; (*8*) Xiejia; (*9*) Chetougou; (*10*) Lierbao; (*11*) Zhangjiaping; (*12*) Duitinggou; (*13*) Quantougou; (*14*) Sigou; (*15*) Shinanu; (*16*) Zengjia; (*17*) Laogou; (*18*) Dingjiaergou; (*19*) Lengshuigou; (*20*) Koujiacun; (*21*) Aoerban; (*22*) Gashunyinadege; (*23*) Urtu; (*24*) Tairun Nor; (*25*) Moergen; (*26*) Jiulongkou; (*27*) Erlanggang; (*28*) Xiaodian; (*29*) Sihong; (*30*) Xiejiahe; (*31*) Fangshan; (*32*) Puzhen; (*33*) Jiyang; (*34*) Xiaolongtan.

the Chinese Tertiary into 13 paleozoogeographic subareas. One key division separating the North-Northeast China area is the Helan-Liupan Mountains and Qinling Mountains–Huai River (present-day boundary between Palearctic and Oriental zoogeographic provinces). In the recently published "Stratigraphic Lexicon of China—General Introduction" by Yu-qi Cheng (2009:312, figs. 4–6) the Chinese Neogene was divided into four "superareas." The western boundary of the Northeast-North China super-area is the same, but its southern boundary is shifted to the south bank of Chang Jiang (Yangtze River). Mainly based on biostratigraphic characteristics, Yi-yong Zhang, Lan, and Yang (2000), on the other hand, devised a seven-area system. Of these, the North China Area borders to the west along the Taihang Mountains,

whereas its southern border is shifted to the center of Shandong Peninsula (Zhang, Lan, and Yang 2000:350, fig. 18-1). These subdivisions generally do not reflect known data of mammal fossils.

From the standpoint of fossil mammals, a prominent boundary line stretching in the northeast-southwest direction is visible during the Early Miocene—that is, along the eastern foothills of Taihang Mountains, then across the Yellow River and reaching to Wudang Mountains and the eastern slopes of the Shennongjia Mountains (figure 1.2). Lying west of this boundary line is the vast **North-West China Area** (including Tibet). Early Miocene mammal faunas of this area, such as Xiejia, Zhangjiaping, Suosuoquan II-III, Xishuigou, Urtu, and so on, contain many drier-climate adapted fossil mammals like ctenodactylids, tachyoryctoidids, dipodids, ochotonids, and primitive bovids. To the east of this boundary line is the territory including the present-day North China Plain and floodplains of the Yellow and Huai rivers, which can be called the **North China Coastal Area**. The most representative Early Miocene faunas of this sub-area are those of Xiejiahe and Sihong, yielding fossil vertebrates adapted to a warm and wet climate of temperate and subtropical zones, as crocodiles, turtles, chiropterans, castorids, suids, cervoids, and the like (unfortunately, so far no fossils are found in this region that are equivalent in age to the earliest Miocene Xiejian). The north end of the North China Coastal Area can extend to northeastern China as far as Shenyang. *Ansomys shantungensis* is one of the most characteristic rodents of the Xiejiahe fauna (Z.-d. Qiu 1987). A tooth of *Ansomys shantun-*

gensis was recovered from the drilling core in the top part of the Dongying Formation (attributed to "Oligocene–early Miocene") of the Jiyang Depression. The Jiyang Depression extends northward to the capital city of the Liaoning Province, Shenyang. The southern-most locality of the North China Coastal Area is Fangshan, situated south of the Yangtze River. It contains *Anchitherium* and some deer (Zhou and Hu 1954), as well as *Spanocricetodon* (Li 1977). We thus tentatively place the southern boundary of the North Asian Coastal Area just south of the Yangtze River. South of Yangtze River, other than the Fangshan locality, no Neogene fossil mammals have been recorded.

The Jiulongkou fauna in Cixian, Hebei Province, is located to the east of the Taihang Mountains, but most of its members apparently came from the paleo-Taihang Mountain slopes. This fauna contains a few Shanwang faunal

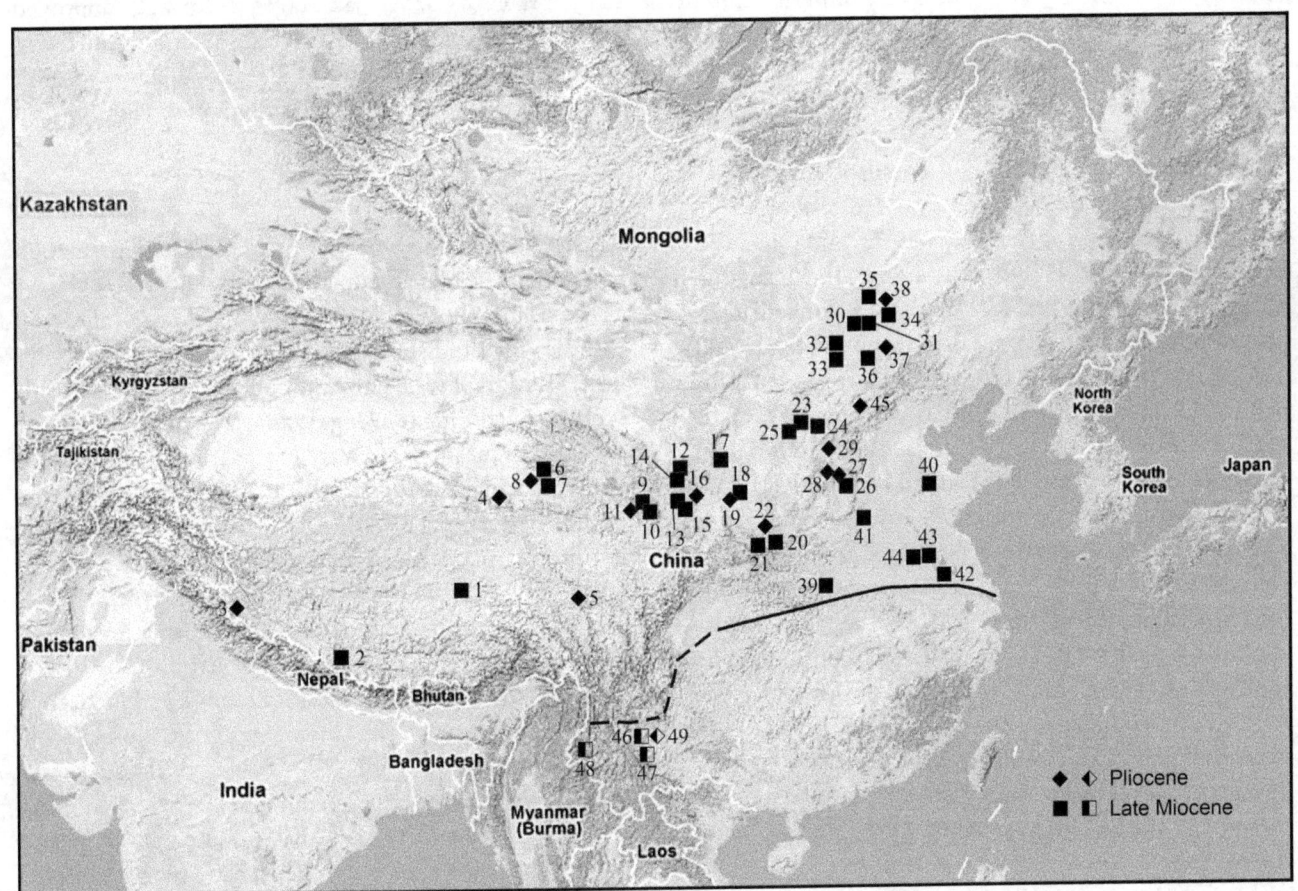

Figure 1.3 Paleozoogeographic subdivisions during Late Miocene–Pliocene of China mainland based on mammalian fossils. (*1*) Bulong; (*2*) Woma; (*3*) Zanda; (*4*) Yuzhu; (*5*) Dege; (*6*) Tuosu Nor; (*7*) Shengou; (*8*) Huaitoutala; (*9*) Charang; (*10*) Xiadongshan; (*11*) Shangtan; (*12*) Guonigou; (*13*) Dashengou; (*14*) Yangjiashan; (*15*) Qingbushan; (*16*) Shilidun; (*17*) Ganhegou; (*18*) Leijiahe; (*19*) Renjiagou; (*20*) Bahe; (*21*) Lantian; (*22*) Youhe; (*23*) Lamagou; (*24*) Baode; (*25*) Miaoliang; (*26*) Jiayucun; (*27*) Nanzhuanggou; (*28*) Mazegou; (*29*) Hefeng; (*30*) Balunhalagen; (*31*) Bilutu; (*32*) Amuwusu; (*33*) Shala; (*34*) Huitenghe; (*35*) Baogeda Ula; (*36*) Ertemte; (*37*) Bilike; (*38*) Gaotege; (*39*) Duodaoshi; (*40*) Balouhe; (*41*) Luwangfen; (*42*) Liuhe; (*43*) Laodong; (*44*) Huainan; (*45*) Daodi; (*46*) Xiaohe; (*47*) Shihuiba; (*48*) Baoshan; (*49*) Shagou.

components, such as crocodiles, chalicotheres (?*Anisodon* sp.), and *Sinomeryx* sp., but it also includes members from the western region, such as *Percrocuta*, chilotheres, and particularly large numbers of high-crowned bovids (*Turcocerus*), suggesting a drier climate. To the east of the southern end of the north–south boundary line is the Xiaodian locality, about 20 km south of Zhongxiang, Hubei Province, which produced *Phoberocyon youngi*, some cervids (Chen 1981), and suids (originally described as *Macaca* by Gu 1980). This fauna was originally considered Pliocene or slightly earlier, but it is likely to be similar to that of Xiejiahe. The north–south stretching boundary line may extend further southwestward along the eastern borders of the Hengduanshan Mountains.

Possibly during the Middle Miocene and definitely from the Late Miocene on, the importance of the north–south stretching boundary line became diminished. The major biogeographic boundary turned gradually latitudinal in direction (figure 1.3). The southern limit of Song, Li, and Zheng's (1983) "Inner Continental Forest-Grassland and Grassland" region extends southward to the Yangtze River. This is in agreement with the distribution pattern of the "North China *Hipparion* Fauna," as demontrated by the finding of such *Hipparion* faunas in Shandong Province (Jiang and Wu 1978), in the Liuhe area just north of Yangtze River (Bi, Yu, and Qiu 1977), and in Duodaoshi, Hubei (Yan 1978). The Hengduan Shan Mountain Ranges probably still belonged to the northern *Hipparion* realm because members of the Wangbuding Fauna (Pliocene) of Dege County in Ganzi Prefecture, Sichuan (Zong et al. 1996), are almost indistinguishable from the northern faunas. On the other hand, Neogene mammal faunas in Yunnan Province (Kaiyuan, Lufeng, Baoshan, Yuanmou, etc.) show clear tendency toward the modern oriental biogeographic province.

Judging by the Early Miocene mammal faunas recently found in Japan, Burma and Thailand, the Japan Islands, and the Indo-China Peninsula will eventually merge with the North China Coastal Area into the even larger **East Asian Subprovince**. On the other hand, based on the Miocene faunas so far available, Mongolia, the Asian part of Russia, Kazakhstan, Kyrgyzstan, and Tajikistan will merge with the North-West China Area into the **Central Asian Subprovince**. The above two subprovinces together form the **Centro-East Asian Province**.

Neogene Chinese Land Mammal Stage/Age

To minimize new names, we adopt most of the Miocene chronostratigraphic stage names of Li, Wu, and Qiu

(1984) as land mammal stage/age names in this chapter. To avoid confusion, some faunal names that are identical to land mammal ages are renamed based on alternative geographic names. Our suggested division of the Chinese Neogene land mammal ages is summarized in figure 1.4.

Xiejian LMS/A

NAME DERIVATION
The stage name Xiejian is based on the Xiejia Formation, which was coined by the Petroleum Survey Team under Qinghai Geological Survey. In 1978, a small local fauna was found from the basal part of the type Xiejia Formation and was described as the Xiejia Fauna by Li and Qiu (1980). Li, Wu, and Qiu (1984) first used the Xiejian Age and/or Stage casually and alternately: Age in Chinese text, Stage in English abstract, and Land Mammal Stage in their table 1. The Xiejian Stage was formally approved at the Third Congress of All-China Stratigraphic Commission held in 2000, and as a regional stage/age unit published by the All-China Stratigraphic Commission in 2001.

UNIT STRATOTYPE
Conventionally, the type section of the Xiejia Formation where the Xiejia local fauna was found has been viewed as the unit stratotype. At the request of the program of "Studies for the Establishment of Major Stages in Chinese Stratigraphy," Deng and colleagues restudied this section. According to Deng, Wang, and Yue (2008b), the section in question is located in a small gully (North Gully), north of the Xiejia village, 13 km southeast of the capital city, Xining, of Qinghai Province. The Cenozoic sediments, which are about 251.5 m thick, overlay in angular unconformity the Cretaceous red beds. The Cenozoic sediments are composed of three formations (from bottom to top): the Mahalagou Formation (81.6 m, Paleogene), the Xiejia Formation (112.5 m, Early Miocene), and the lower part of the Chetougou Formation (57.4 m, Early–Middle Miocene). The Xiejia local fauna was found only from a small lens-like pocket (3 m × 0.5 m), which was believed to lie in the middle part of layer 3 (>30 m above the base of the Xiejia Formation as counted from Deng et al.'s fig. 6). The entire lens was believed to be exhausted during the initial investigation by Li and Qiu in 1978, making it difficult to relocate the fossiliferous locality by subsequent workers, although our 2011 field work indicates that additional screenwashing may be feasible. The underlying Mahalagou Formation is barren. The mammal fossils of the overlying Chetougou For-

mations are sparse. Therefore, there is little possibility to choose the section in question as a candidate RSSP. On the other hand, the Xiejia local fauna is quite representative for this time span in northwestern China. The Xiejia local fauna contains about a dozen large and small mammals showing a distinctly Early Miocene character (see revised faunal list in appendix 1).

As for the numerical age of the Xiejia Formation and fauna, there are two different opinions based on different paleomagnetic data. Wu et al. (2006) made the first magnetostratigraphic sampling in the Xiejia type section. Of the 18 normal and 17 reversed intervals of the whole section, 7 normal (N5-N11) and 6 reversed (R5-R10) intervals correspond with the Xiejia Formation and are interpreted as from the top of C5Dn to the base of C6AAr.1n. The horizon of the Xiejia local fauna falls in R9, corresponding to C6Ar. Using the numerical dates of the ATNTS2004 (Gradstein et al. 2004), the Xiejia Formation is bracketed in 17.235–21.403 Ma (not 17.32–21.58 Ma as listed by Wu et al. 2006), and the Xiejia local fauna itself is around 21 Ma (C6Ar: 20.709–21.083). As to Wu et al.'s work (2006), two comments can be made. First, the position of the Xiejia local fauna in the type section was erroneously placed as more than 30 m above the base of the Xiejia Formation. A recently organized in situ inspection of the type section joined also by the discoverers of the fauna (Chuan-kui Li, Zhu-ding Qiu, and Shoubiao Xie, a villager who took part in the excavation in 1978) confirmed that the fossiliferous pocket had been found directly above the basal sandstones, about 10 m above the base of the Xiejia Formation. As a result, the site of the Xiejia local fauna site should not fall in R9, but in N10 in the Wu et al. paleomagnetic column, corresponding to C6AAn (21.083–21.159 Ma). Second, Wu et al.'s (2006) correlation was entirely based on the numbers of magnetic reversals without consideration of the relative lengths of the reversals so that the very short N7 was correlated with the C6n, which is the longest normal chron in the Early–Middle Miocene segment of the ATNTS2004. However, so far no other better matching interpretation could be made.

Dai et al. (2006) made a more extensive magnetostratigraphic study in the Xining Basin, with the Xiejia section chosen as the most representative of the basin. They concluded that the Xiejia local fauna site was bracketed between C7n and C7An—that is, corresponding with C7r (24.556–24.915 Ma in ATNTS 2004, not 25.183–25.496 Ma as in Dai et al. 2006). The age of the Xiejia Formation was estimated as 30–23 Ma, mainly Oligocene. This is in contradiction to paleontological evidence: the base of the Xiejia Formation should be younger

than 23 Ma based on the mammal fossils. It should be noted that the Xiejia local fauna site was even more erroneously located in the section. Here, it was placed only about 35 m below the base of the Chetougou Formation. The lower boundary of the Xiejia Formation seems to have been placed much lower than that defined by Li and Qiu (1980) and Wu et al. (2006), judging by the thickness of the Xiejia Formation (>160 m) given by Dai et al.

Neither of these paleomagnetic interpretations is fully satisfactory, so the question will remain open pending more detailed biomagnetostratigraphic work in the future.

REFERRED DEPOSITS AND FAUNAS

Suosuoquan-II Biozone (Junggar Basin, Xinjiang) This biozone was described in great detail by Meng et al. (2006), with a faunal list of 23 preliminarily identified taxa of micromammals. The faunal assemblage was found from locality XJ99005, about 10 km west of the main biomagnetostratigraphic section 02Tr. Most of the identified taxa are common with those of Xiejia at the generic level, but only a few specimens of *Parasminthus* are present. On the other hand, a large number of specimens of a more "modern" cricetid, *Democricetodon*, were found in this assemblage. By tracing in the field and based on biostratigraphic data, Meng et al. (2006) correlated the XJ99005 assemblage with the interval 52.75–63 m of the 02Tr section, which was interpreted as the interval from C6AAr.3r to C6Bn.1r—that is, **21.7–21.9 Ma.**

If this proves tenable, the Suosuoquan-II fauna would be slightly older than the Xiejia local fauna, if the Wu et al. (2006) age determination for the Xiejia fauna is preferred, which is around 21 Ma (see previous discussions). Paleontologically, the Suosuoquan-II fauna seems to be slightly younger than the Xiejia local fauna. We note that the Xiejia assemblage is characterized by the co-occurrence of *Yindirtemys, Tachyoryctoides, Parasminthus,* and *Sinolagomys,* implying its close affinities with the late Oligocene faunas, such as the Taben Buluk fauna of Gansu and the Suosuoquan-I or Tieersihabahe-I assemblages from Xinjiang (Li and Ting 1983; Meng et al. 2006). The Suosuoquan-II fauna, is characterized by the absence of archaic *Yindirtemys,* the presence of more "modern" *Plesiosminthus,* and the appearance of *Democricetodon* and some other commonly known early Miocene genera, which resemble the still-younger Lower Aoerban assemblage from Nei Mongol (Wang et al. 2009).

Suosuoquan-III Biozone (Junggar Basin, Xinjiang) Originally Meng et al. (2006) referred, without hesita-

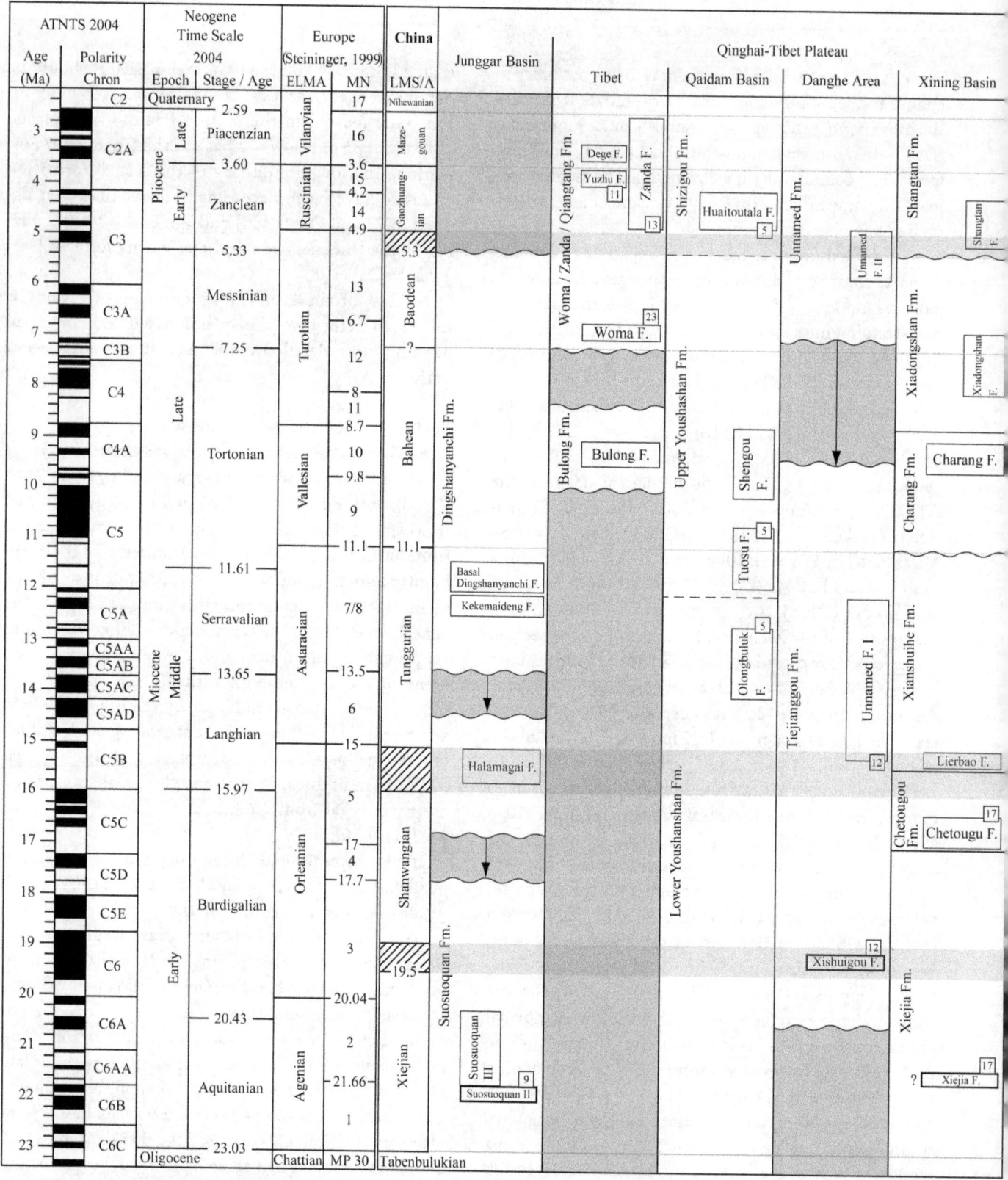

Figure 1.4 Division of the Chinese Neogene Land Mammal Ages and the distribution of mammalian faunas and strata. Reference numbers for age determinations: (*1*) Chen and Peng 1985; (*2*) Deng et al. 2008a; (*3*) Ding et al. 1999; (*4*) Fang et al. 2003; (*5*) Fang et al. 2007; (*6*) Flynn, Wu, and Downs 1997; (*7*) Kaakinen and Lunkka 2003; (*8*) Luo and Chen 1990; (*9*) Meng et al. 2006; (*10*) Qiu et al. 2001; (*11*) Song et al. 2005; (*12*) Sun, Zhu, and An 2005; (*13*) Wang et al. 2008; (*14*) Wang et al. 2009; (*15*) Wang et al. 2003; (*16*) Wei et al. 1993; (*17*) Wu et al. 2006; (*18*) Xu et al. 2007; (*19*) Xue, Zhang, and Yue 1995; (*20*) Yue and Zhang 1998; (*21*) Yue and Zhang 2006; (*22*) Yue et al. 2004b; (*23*) Yue et al. 2004a; (*24*) Zhu et al. 2005.

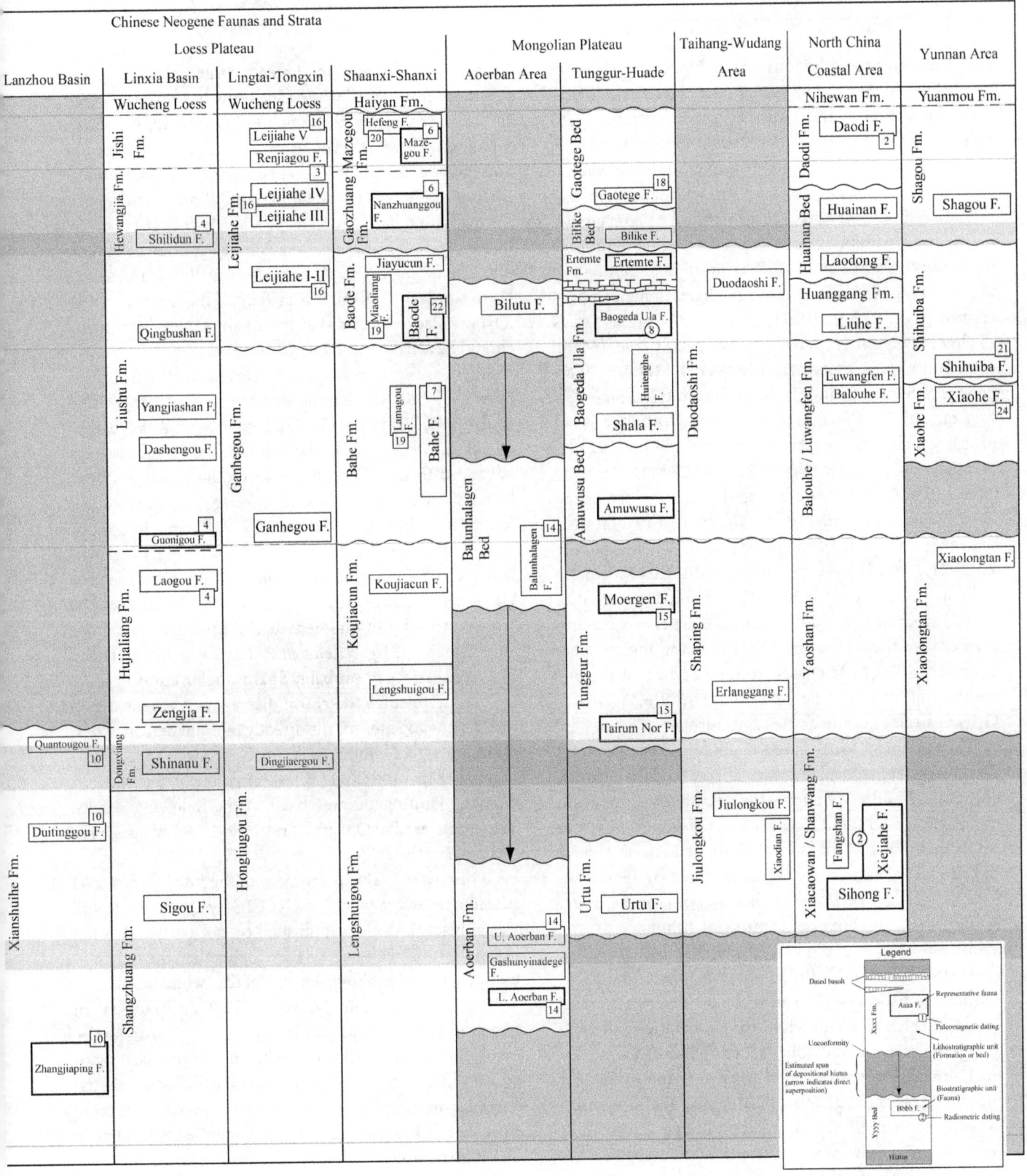

tion, the lower two horizons of this biozone, XJ200205 (at the level of 68 m) and XJ200206 (at 88 m) of the Tieersihabahe section, to the Xiejian Stage/Age. But the highest fossiliferous horizon, XJ200210 (at 105.5 m) was only tentatively included in the Suosuoquan-III and was referred to the Shanwangian Stage/Age with some degree of hesitation. Based on its faunal content and paleomagnetic dating, we are inclined to attribute all three fossiliferous levels to the Suosuoquan-III Biozone, and all of them to the Xiejian LMS/A. Paleontologically, although impoverished in content, XJ200210 contains the same taxa as the lower two horizons, even at the specific level (Meng et al. 2006:fig. 3). Elsewhere, Meng et al. (2006:230) argued that Suosuoquan-III was in the stratigraphic interval from 105.5 m (XJ200210) to 64 m (XJ200205, but not 68 m), which was correlated with the top of C6An.1r through the base of C6AAr.3r (21.16–21.7 Ma). Here, the cited age of C6An.1r was wrong. In fact, the age of the C6An.1r is 20.213–20.439 Ma. Therefore, the total time span of Suosuoquan-III should be 20.213–21.767, roughly **20.2–21.7 Ma**. This entire time span still falls within the Xiejian LMS/A, as defined in this chapter. Meng et al. (2006) separated the Suosuoquan-III into two parts simply because they drew the Xiejian–Shanwangian stage boundary at 20.43 Ma based on the notion that the former boundary should be equivalent with the Agenian–Orleanean stage boundary used in Europe.

Zhangjiaping local fauna (Lanzhou Basin, Gansu) The Zhangjiaping local fauna was discovered in the lower part of the Middle Member of the Xianshuihe Formation in Lanzhou Basin, Gansu Province. A section near the Miaozuizi gully, which is located about 1 km northwest of the Duitinggou village, was measured by Guang-pu Xie, Xiao-feng Chen, and Yi-zheng Li in January, 1991. A profile and a short description of lithology of this measured section are given in Chinese by Qiu et al. (1997:180–182, fig. 2). The Middle Member of the Xianshuihe Formation is ~225 m thick. Its lower part, composed of thick layers of white conglomeratic sandstone and gray-greenish sandstone (called "white sands" in the field) intercalated with red claystone, is the main part producing the Zhangjiaping local fauna. The fauna contains more than 30 taxa of small and large mammals, a recently revised list of which is provided in appendix 1. The large mammals, like *Turpanotherium*, *Aprotodon*, *Phyllotillon*, and *Hyaenodon*, are all advanced forms of the characteristic Oligocene groups in central Asia. A suspect piece of proboscidean tusk found by a local archaeologist was also mentioned (Qiu 1990:537; Qiu and Qiu 1995:45). However, the specimen has been lost so that the presence of proboscideans in this local fauna cannot

be accepted for certain. Later a large calcaneum was found from the basal white sandstone in another locality in the west wing of the syncline (Yanwagou, GL9508), which was once believed as proboscidean, but it may in fact belong to an unknown form of Rhinocerotidae. Therefore, there is no certain record of proboscideans in Zhangjiaping local fauna as previously thought. On the other hand, the presence of advanced species of *Tataromys*, *Yindirtemys*, and *Parasminthus* in this fauna shows clear similarity to the Xiejia fauna. The occurrence of *Sayimys* also indicates that the Zhangjiaping local fauna should be attributed to Early Miocene.

There are some ambiguities concerning the upper extension of the Zhangjiaping local fauna. According to Qiu et al. (1997, 2001), the GL 9303 was placed at the level of 67.5 m above the base of the "white sands." In addition to advanced forms of Oligocene micromammals as above listed, some earliest *Democricetodon* specimens were also found from the GL 9303. Thus, GL 9303 was thought by Qiu et al. as the latest phase of the Zhangjiaping local fauna. On the other hand, Flynn et al. (1999) placed the GL 9303 at a level about 120 m above the base of the "white sands" and claimed that a combination of *Democricetodon* + *Megacricetodon* + *Alloptox* had been found in it, considering it "probably Shanwangian equivalent."

Aiming to solve the above discrepancy, a field trip to the Duitinggou area for the directly relevant persons (Z.-x. Qiu, Z.-d. Qiu, Flynn, B.-y. Wang, G.-p. Xie, J. Ye, etc.) was organized in June 2011. Reexamination of the sections near the Duitinggou area has had the following results. The profile used in Qiu et al. (1997:fig. 2) was measured in 1991, when the micromammal site GL 9303 had not yet been discovered. The assignment of the fossiliferous level (the fourth "white sand") to GL 9303 by Qiu et al. in 1997 was a mistake. As is reconfirmed by our party, GL 9303 lies in the fifth "white sand," 10 m higher than the fourth— that is, about 77 m above the base of the "white sands."

The column used in Flynn et al. (1999:fig. 2) was taken from the section measured in the same Miaozuizi gully by Jie Ye et al. in 1993, along with the magnetostratigrapic sampling. However, the thickness of the "white sands" part measured by Ye's team was 110 m, about 40 m more than that of Xie's. The difference might be caused partly by different slopes the two teams took for their sections, and partly by different ways of measuring. As a quick field check by X. Wang using a simple algorithm based on telemetry tended to show the thickness of the part of the section in question might be only around 70 m, the thickness obtained by Ye's team seems overestimated. Since the magnetostratigraphic sampling is based on Ye et al.'s section, we refer to it in spite of its exaggerated thickness.

On the other hand, after a closer re-examination of the GL 9303 micromammals, Z.-d Qiu., Flynn, and B.-y. Wang provided the following preliminary list (at generic level): *Prodistylomys* sp., *Sinotamias*, *Plesiosminthus* sp., *Heterosminthus* spp., *Litodonomys* spp., *Sicista* sp., *Protalactaga* sp., *Cricetodon* sp., *Democricetodon* sp., *Desmatolagus* sp., and *Sinolagomys* sp. The trio also agreed that no *Megacricetodon* and *Alloptox* can be recognized with certainty. This faunal list shows clearly that GL 9303 is intermediate between the Zhangjiaping and Duitinggou local faunas. It could be equally viewed either as the later phase of the Zhangjiaping local fauna or the early phase of the Duitinggou one. However, based on the new interpretation of the magetostratigraphy (see following discussion), the former option is preferred here.

For the purpose of testing the feasibility of the rocks for future paleomagnetic work, a trial sampling at 1–5 m intervals from the Duitinggou section was conducted by Opdyke's team in 1993. As preliminarily interpreted by Opdyke et al. (1998), the Middle Member of the Xianshuihe Formation was correlated with the segment from the top of C6n to C6Cn.2n (Oligocene–Miocene boundary), or from ~19 Ma to 23.8 Ma (in BKSA 98), that is, from >18.748 to 23.03 Ma, according to ATNTS2004. Based primarily on the presumed presence of proboscidean fossils and an *Alloptox*+*Democricetodon*+*Megacricetodon* combination in the upper 100 m segment of the Duitinggou section, Flynn et al. (1999) reinterpreted the Middle Member as corresponding to the upper part of C5En (now 18.524–18.056 Ma) to above C5ADn (now 14.581–14.194 Ma). Therefore, the base of the Middle Member of the Xianshuihe Formation (the White Sand) was dated as "no more than about 19 Ma," and its top "in excess of 14 Ma" (Flynn et al. 1999:116). Qiu et al. (2001), based on the same data, reached essentially the same conclusion as Flynn et al. did, with slightly older age assignment for the top of the Middle Member. The Middle Member was correlated with the upper part of C6n (19.722–18.748 Ma) through C5Cn.1n (16.268–15.974 Ma)—that is, roughly from 19.3 Ma to ~16 Ma.

Neither of the above magnetostratigraphic interpretations can reconcile with our current knowledge of the Zhangjiaping local fauna as previously stated. The micromammal fossils show that the Zhangjiaping local fauna is very close to the Xiejia and Suosuoquan-II faunas. Its age should be close to them as well—that is, around 21 Ma. This leads us to think that the Zhangjiaping local fauna (GL 9303 included) is better to be correlated with the upper part of C6Ar (20.709–21.083 Ma) through the base of C6B.1n (21.936 Ma) or within C6Bn.2n (21.992–22.268 Ma), roughly 21–22 Ma (figure 1.5). This interpretation

differs from the previous ones (Opdyke et al. 1998, Flynn et al. 1999, and Qiu et al. 2001) in that the base of Zhangjiaping local fauna (GL 8801) is around 22 Ma rather than 23.8 Ma or around 19 Ma. The polarity reversal pattern of the upper part of the section, expressed by three rather long normal chrons separated by shorter reversed ones, seems closely compatible with C6n, C6An.1n and C6An.2n, as Opdyke et al. first proposed in 1998, but completely different from those suggested by Flynn et al. (1999) and Qiu et al. (2001).

Lower Aoerban Fauna (central Nei Mongol) The Aoerban Formation containing rich micro-mammal fossils of the Aoerban area in central Nei Mongol has been described in great detail by Wang et al. (2009), and summarized in Qiu, Wang, and Li, (chapter 5, this volume). Paleontologically, the Lower Aoerban Fauna indicates an age later than the Xiejia local fauna by the lack of more archaic taxa, such as *Parasminthus* and *Yindirtemys*, and the presence of more "modern" cricetid, *Democricetodon*. It may be slightly younger than the Suosuoquan-II–III and Zhanjiaping local faunas, which contain the three above listed genera. On the other hand, the Lower Aoerban Fauna is probably earlier than that of Sihong because of its lack of *Alloptox* and the proboscideans, both of which are present in the Sihong local fauna. Test studies of magnetostratigraphy indicate a normal interval between the two major screenwashing sites, IM0407 and IM0507 (Wang et al. 2009:fig. 3), which was interpreted as C6n (18.748–19.722 Ma) based on a few test samples. This interpretation seems inconsistent with the sedimentation rates. The normal chron C6n lasted about 1 myr, however, the magnetozone in the Aoerban Formation is less than 4 m of sediment, implying a sedimentation rate only about 4 mm/ky, five times slower than average sedimentation rates in Early Miocene deposits in North China (~20 mm/ky: 25 mm/ky in Tieersihabahe and Suosuoquan formations, 16.7 mm/ky in Qin'an section; see Meng et al. 2006:226). The Aoerban normal interval is more reasonably less than 0.2 myr in duration and more probably should be correlated with C6An.1n (20.04–20.213 Ma). The uppermost horizon of the Lower Aoerban Fauna (IM0407) lies about 2–3 m above the top of a short normal polarity interval, and the lowest horizon (IM0507) is about 7 m below the base of this normal interval. If our reinterpretation is more tenable than the previous one, the time duration of the Lower Aoerban Fauna should be roughly estimated as **20–20.3 Ma**.

Gaolanshan local fauna (Lanzhou Basin, Gansu) Described by Qiu and Gu (1988) from the north slope of the Gaolanshan hill in Lanzhou, the assemblage contains only five identified species (rodents revised

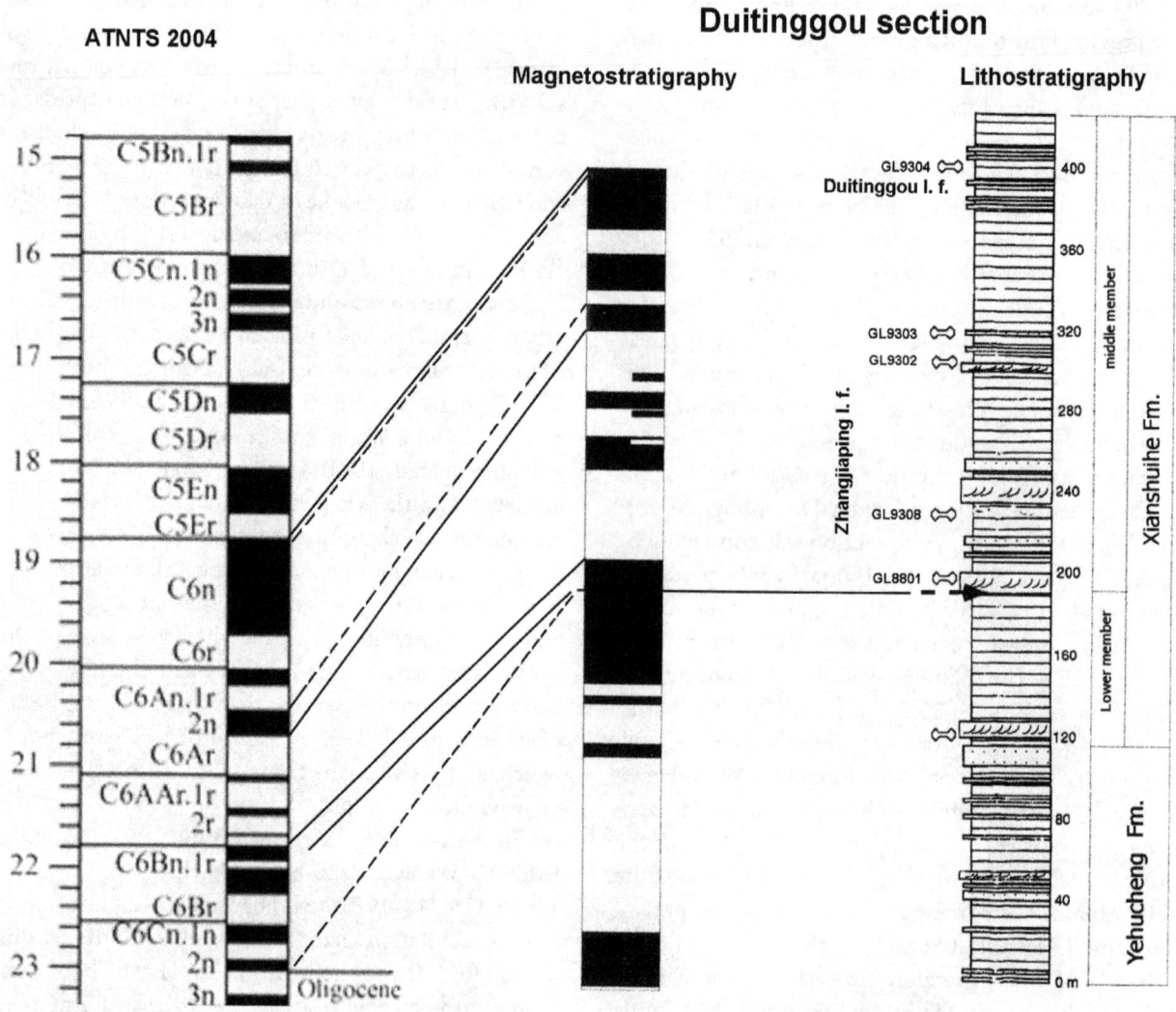

Figure 1.5 Magnetostratigraphic reinterpretation of the Xianshuihe Formation in the Lanzhou Basin. Solid line = interpretation in this chapter; dashed line = according to Opdyke et al. (1998).

by B.-y Wang): *Metexallerix gaolanshanensis, Yindirtemys suni, Y. grangeri, Tataromys sigmodon,* and *Tsaganomys altaicus.* The aggregate was first referred to the Xiejian Stage based mainly on the discovery of an advanced insectivore, *Metexallerix.* However, the presence of *Tataromys sigmodon* and *Tsaganomys altaicus* in this small fauna tends to show that the Gaolanshan aggregate is more probably of Oligocene age. As pointed out by Wang (2001:44), Tsaganomyinae are the "index fossils for the Asian Oligocene." Both *Tsaganomys altaicus* and *Tataromys sigmodon* have not been found in any other Miocene faunas, except the Gaolanshan local fauna.

Candidate RSSP As discussed, neither Xiejia nor Zhangjiaping can be chosen as RSSP. Recently Meng

et al. (2006) recommended the Tieersihabahe section in north rim of the Junggar Basin as the regional lower boundary stratotype section. The GSSP for the Oligocene–Miocene boundary, which coincides with the lower boundary of the Aquitanian Stage, was defined at the base of C6Cn.2n (23.03 Ma) and ratified in 1997. As did European mammal paleontologists, Meng and colleagues chose the Global Standard Oligocene–Miocene Series boundary as the lower boundary of the Xiejian Stage, without further explanation. This boundary does not coincide with the lithologic boundary between the underlying Tieersihabahe Formation and the Suosuoquan Formation but lies in the lower part of the Suosuoquan Formation, 7.25 m above its base. The boundary in question does not coincide with any boundaries of biostratigraphic zones either. Based on the XJ99005 section, where the largest samples of Suosuoquan-II fossils were collected, the lowest horizon of the Suosuoquan-II is located at the level of 52.75 m in the Tieersihabahe (02Tr) section—that is, **12.5 m above** the base of the lower boundary of the supposed Xiejian Stage. The uppermost horizon of the Suosuoquan-I, which is almost the same as the typical Oligocene Tieersihabahe assemblages, lies **3.75 m below** the supposed Xiejian Stage.

Meng et al.'s interpretation of the polarity reversals is not indubitable. Using the average sedimentation rate, which is around 25 mm/ky (Meng et al. 2006:226), to estimate the reversals against the thickness of deposits, it becomes evident that for the deposits of the Suosuoquan-II in the Tieersihabahe section, which is 10.25 m thick and interpreted as corresponding to the base of C6Bn.1r (21.992 Ma) through the top of C6AAr.3r (21.688 Ma), the sedimentation rate would be **33.7 mm/ky** (10.25 m/0.304 Ma). On the other hand, for the segment from 89.25 m to 127 m, interpreted as corresponding to the base of C6n (19.722 Ma) to the top of C6An.2n (20.439 Ma), the sedimentation rate would be as high as **52.65 mm/ky** (37.75 m/0.717 Ma). Such a great disparity in sedimentation rates seems unlikely. Either there are some undetected sedimentation gaps, or something is wrong with paleomagnetic interpretation. Another problem is the fact that the richest collection of the Suosuoquan-II zone was from a site (XJ99005) about 10 km west of the typical Tieersihabahe section (02Tr). Whether its correlation with the 02Tr section is tenable is to be verified in the future. The fact that the Suosuoquan-II assemblage does not contain any large mammal fossils is also unfavorable as a RSSP. This notwithstanding, the Tieersihabahe section satisfies the major requirements recommended in the *ISG* for establishment of stratotype section and point.

Meng et al.'s procedure does not conform with Woodburne's proposition (1996:551) that "boundaries of both biostratigraphic and chronostratigraphic units should be defined on single taxa," but it does not violate the rules in boundary definition of chronostratigraphic units as set in the *ISG* (see previous discussions). This is in agreement with our newly proposed auxiliary principle in stage boundary definition: prior consideration of boundaries of higher ranking units over that of stage.

Mammal fauna characterization The Xiejian chronofauna can be characterized by the retention of Oligocene holdovers and/or highly specialized Oligocene survivors. Almost all the families are those of Oligocene or even earlier age. Those that flourished particularly in the Oligocene, like tsaganomyids, ctenodactylids, tachyor-

yctoidids, hyaenodontids, and giant rhinoceroses, either disappeared or were drastically waning. The majority of genera are survivors of the Oligocene, with a few newcomer insectivores and rodents, especially Muroidea (the modern cricetids *Cricetodon* and *Democricetodon*), Gliridae, and Eomyidae. Taken as a whole, the Xiejian chronofauna can be viewed as an impoverished Oligocene fauna, composed mainly of advanced species of Oligocene taxa accompanied by a few Miocene components in the later half of the stage, prior to the immigration of proboscideans.

Index fossils: *Yindirtemys suni, Parasminthus xiningensis, P. huangshuiensis, Litodonomys lajeensis, Cricetodon youngi, Sinolagomys pachygnathus, Hyaenodon weilini, Aprotodon lanzhouensis, Phyllotillon huangheensis.*

First appearances within the stage: *Mioechinus, Quyania, Sayimys, Ansomys, Atlantoxerus, Sinotamias, Miodyromys, Microdyromys, Keramidomys, Ligerimys, Sicista, Cricetodon, Democricetodon, Protalactaga.*

Last appearance within the stage: *Tataromys, Yindirtemys, Parasminthus, Hyaenodon, Phyllotillon, Turpanotherium.*

Characteristic fossils: *Amphechinus, Sinolagomys, Prodistylomys, Tachyoryctoides, Atlantoxerus, Asianeomys, Aprotodon.*

Shanwangian LMS/A

NAME DERIVATION

The stage name Shanwangian is taken from the Shanwang Series, created by Young (1936). In 1960, the Shandong Geological Mapping Team renamed it as the Shanwang Formation in Chinese. Its first appearance in formal publication was a year later by Sun (1961), also in Chinese. As for the Xiejian Stage/Age, the Shanwangian Stage/Age was proposed by Li et al. in 1984. The name Shanwangian Stage was later adopted in the appendix of the revised edition of the *Stratigraphic Guide of China and Its Explanation* (All-China Stratigraphic Commission 2001). A detailed revision of this stage/age was made by Deng, Wang, and Yue (2008a).

UNIT STRATOTYPE

The Jiaoyanshan hill, which is situated 1 km east of Shanwang village and 1.5 km southwest of Xiejiahe village, is about 16 km ENE of Linqu county seat, Shandong Province. The west slopes of Jiaoyanshan have long been implicitly viewed as stratotypical for the stage. It was formally proposed as [unit]-stratotype of the Shanwangian Stage by Deng, Wang, and Yue (2008a). The Shanwang Formation consists of a sequence of ~100 m of fluviolacustrine sediments with richly fossiliferous diatomite in the midsection and two layers of basalt in the top part of the section. The topmost layer of grayish yellow sandstones and gravels (mainly seen near the Xiaoyaoshan west of the Jiaoyanshan) had been separately named as the Yaoshan Formation, but both Yan, Qiu, and Meng (1983) and Deng, Wang, and Yue (2008a) thought that fossils from both formations are very similar and a separate formation is not warranted. Underneath the Shanwang Formation, with an unconformable contact, lies a thick lava flow, of which only its top is exposed in this area and nearby.

The Shanwang Formation is well known for its richness in fossils of great varieties of organic groups (plants, insects, fishes, various groups of other vertebrates) and particularly for its exquisite state of preservation of mammal skeletons. This fauna has been customarily called Shanwang fauna. Since the major excavation sites of the mammal fossils are nearer to Xiejiahe village, and for the sake of avoiding confusion with the stage/age name, the fauna is renamed as the Xiejiahe Fauna. During the long history of studies since Young's (1937) first description of Shanwang fossils, lists of fossil mammals from the Xiejiahe Fauna have been augmented and emended several times (Yan, Qiu, and Meng 1983; Deng, Wang, and Yue 2008a). A recently revised mammal faunal list of the Xiejiahe Fauna is given by Qiu and Qiu (chapter 4, this volume) and in appendix 1. Based on the overall stage of evolution for mammals, faunas from the lower part of the Shanwang Formation (diatomite plus underlying sandstones and gravels) are comparable to those of the European MN 4. Only *Anisodon grande* is earlier than its European counterpart, which may be explained by the view that Asia was the center of diversification for chalicotheres. The mammals from the upper part of the Shanwang Formation (sandstones and conglomerates in Xiaoyaoshan) are probably equivalent to those of the lower half of MN 5 in Europe. From the standpoint of evolutionary stages of the fossils, the Shanwang Formation seems to fall in the range of 18–16 Ma or slightly younger.

Multiple dates for the Shanwang Formation are available. Deng, Wang, and Yue (2008a) evaluated the various dates and settled on a date of **18.05 ± 0.55 Ma** for the topmost age of the underlying basalt (the Niushan Formation), which should be close to the age of the lower boundary of the Shanwang Formation. The dating of the uppermost basalt of the Shanwang Formation varies from 7.86 Ma to 13.72 Ma and is not relevant for the upper boundary of the Shanwang Formation. As indicated in a summary of the volcanic activities of China by Liu (1999:32), the Early Miocene phase of volcanism in this area is estimated to be 21.48–15.77 Ma, this should also be the approximate age of the Shanwang Formation. Although the boundaries of the Shanwangian Stage/Age are still debatable, the Jiaoyanshan section is undoubtedly the best representative section for the Shanwangian Stage/Age and can serve as the main unit stratotype.

REFERRED DEPOSITS AND FAUNAS

Jijiazhuang fauna (Shandong) Recently, a large number of complete mammal skeletons (more than 600) were collected from the diatomite quarries in Jijiazhuang village of Changle County by the Tianyu Museum in Pingyi County, Shandong. The quarries lie only about 6 km east of Xiejiahe. The fossils from the Jijiazhuang quarries are essentially similar to those from Xiejiahe, with the same kind of preservation and the same major groups of animals. As the antlers of *Ligeromeryx* (=*Lagomeryx*) and the "ossicones" of *Sinomeryx* show greater complexity, the Jijiazhuang Fauna may be slightly more advanced, indicating a somewhat later age for the Jijiazhuang assemblage.

Sihong fauna (Jiangsu) Together with the Xiejiahe fauna, the Sihong fauna is currently revised by Qiu and Qiu (chapter 4, this volume). There is no doubt that the two faunas belong to the Shanwangian LMS/A. The two faunas differ in state of preservation and faunal composition. The preservation of the Xiejiahe specimens is much better than the Sihong ones, mostly complete skeletons in the former, but mostly isolated teeth in the latter. The Xiejiahe fauna is so far known to contain only about 23 species, mainly large mammals, while the Sihong fauna contains more than 40 taxa, most of which are micromammals. Almost 30 large mammals were initially listed from Sihong, but after a reexamination, only 10 remain. Within carnivorans, *Semigenetta huaiheensis* is closest in size and stage of evolution to the European *S. elegans* from the Wintershof-West (MN 3), but it is obviously smaller than the latter species. The sizes of p4 and m1 in *Pseudaelurus* cf. *lorteti* are between *P. transitorius* from Wintershof-West and later species (Qiu and Gu 1986). *Dorcatherium* first occurs in MN 4 of Europe. However, *Dorcatherium orientale* from Sihong is smaller and more primitive than all known European forms (Qiu and Gu 1991). In addition, there is a very small *Stephanocemas*, whose antler is only about half that of *S. thomsoni* from Tunggur and has six tines (as many as nine in those from Tunggur)—obviously much more primitive than the Tunggur form.

In general composition the Xiejiahe and Sihong faunas are very similar, but the Sihong Fauna is likely to be slightly older. Among the micromammals, *Ansomys* and *Plesiosciurus* form Sihong show more primitive characters than those of Xiejiahe, while of the large mammals the more advanced deer, *Ligeromeryx*, *Heterocemas*, and

Sinomeryx appeared in the Xiejiahe Fauna, while in Sihong only a very primitive and small-size *Stephanocemas* is present.

Xishuigou local fauna (Gansu) B.-y. Wang et al. (2003) recently found the small Xishuigou local fauna, in Danghe area, western Gansu. The small local fauna was found in the lower part (layers 1–6; 659 m thick) of the Tiejianggou Formation in the Xishuigou section. The lowermost layer 1 (14 m thick) produced the earliest Chinese shovel-tusked proboscidean, *Platybelodon dangheensis* (Wang and Qiu 2002b), and a primitive leptarctine *Kinometaxia guangpui* (Wang, Qiu, and Wang 2004), while in the basal part of the layer 3 (DH199909, 199911; ~250 m above the base) *Amphimoschus* cf. *artenensis* and *Turcocerus* sp. were found. "*Kansupithecus*" described by Bohlin (1946) and *Heterosminthus intermedius* collected from DH199903 in Tiejianggou section (Wang 2003) were also believed to have been found from this level. All these taxa are definitely Miocene forms. Platybelodonts are common in Middle Miocene deposits in northwestern China. However, *P. dangheensis* is the most primitive of the group. *Kinometaxia* is also the earliest member of the leptarctines in the world. In North America, *Schultzogale* was found in the earliest Hemingfordian (~19 Ma), slightly later than *Kinometaxia*. *Amphimoschus* only occurs in the European MN 3–4. The Xishuigou local fauna so far does not share anything with the faunas of the Xiejian Stage/Age. The magnetostratigraphy of the lower half (1300 m, up to the F1 fault) of the Xishuigou section (2723 m) offers a good match to the GPTS. The magnetostratigraphy indicates a normal interval for the layers 1–2 correlated to C6n (18.748–19.772 Ma). Regrettably, below DH199910 the Xishuigou section is truncated by a major thrust fault and as a result not all of C6n is present in the strata. The lower boundary of the section is now roughly considered as 19.5 Ma, and the lowest fossiliferous level (DH199910) may only be slightly younger than 19.5 Ma.

Gashunyinadege fauna (central Nei Mongol) This fauna of central Nei Mongol was first described by Meng, Wang, and Bai (1996) and reviewed by Qiu, Wang, and Li (chapter 5, this volume). Judging by its faunal composition (47 taxa; for faunal list, see Qiu, Wang, and Li, chapter 5, this volume), the Gashunyinadege Fauna should be intermediate between the Lower and Upper Aoerban Fauna in age, with the first appearance of *Alloptox* and *Megacricetodon*.

Upper Aoerban fauna (central Nei Mongol) This fauna seems rather like an impoverished Lower Aoerban Fauna, with fewer Oligocene "holdovers," as exemplified by the lack of Ctenodactylidae and the reduction of Aplo-

dontidae and Zapodidae species. However, it is certainly more advanced than the Lower Aoerban Fauna in appearance of a series of more advanced forms, like *Megacricetodon*, *Alloptox*, *Ligeromeryx*/*Lagomeryx*, and some indeterminable proboscidean specimens. All these later forms are also found in the Sihong fauna. Wang et al. (2009) correlated this fauna with the European MN 5, lower boundary which was ~17 Ma (Steininger et al. 1996). This may be too young. Since we reinterpreted the short normal polarity interval in the Lower Aoerban fauna as corresponding with C6An.1n rather than C6n as previously proposed, the age of the Upper Aoerban fauna should also be shifted downward accordingly. Judging by the fact that the sedimentation of the Aoerban Formation is basically continuous, and the base of the 17-m-thick Upper Red Mudstone Member containing the Upper Aoerban fauna lies only 7 m above the Lower Red Mudstone Member containing the Lower Aoerban fauna, the age of the upper member should not be much younger than the lower member, which is estimated as 20–20.3 Ma (see previous discussions).

Urtu local fauna (western Nei Mongol) This small fauna was found in 1988 and its faunal list was published by Wang and Wang (1990). The locality lies about 50 km northwest of the Alxa Zuoqi in western Nei Mongol. The Urtu Formation is composed of interbedded layers of red and yellow siltstone and silty sandstone, with a total thickness of 137 m. The fossils are mainly in the lower part (~55 m), concentrated in the interval, 31–35 m above the base. Major taxa include *Amphechinus minimus*, *Sinolagomys* cf. *ulungurensis*, *Distylomys qianlishanensis*, *Prodistylomys xinjiangensis*, *Tachyoryctoides* sp., *Megacricetodon* sp., *Gomphotherium* sp., and some Perissodactyla and Artiodactyla fossils. No paleomagnetic sampling has been attempted. This local fauna is generally correlated with the Duitinggou one.

Duitinggou local fauna (Lanzhou Basin, Gansu) This local fauna is primarily represented by the specimens found from GL9304, a locality at the top (layer 18) of the Middle Member of the Xianshuihe Formation in the Duitinggou section of the Lanzhou Basin, Gansu. The fossils are still under study by Flynn. A preliminary faunal list was given by Flynn et al. (1999): ?*Metexallerix* sp.,cf. *Sinolagomys*, *Alloptox minor*, cf. *Bellatona forsythmajori*, cf. *Sinotamias primitivus*, *Altantoxerus orientalis*, *Heterosminthus* sp. nov., *Protalactaga grabaui*, *Democricetodon*, *Megacricetodon*, *Prodistylomys* sp., and *Stephanocemas*. Flynn et al. (1999:112) considered it as "early Tungurian." As stated earlier, Flynn et al. (1999) tended to subdivide the top 120 m of the Middle Member of the Xianshuihe Formation of the Duitinggou section to two parts, represented by the lower GL 9303, containing a faunal assemblage of Shanwangian Age, and the upper GL 9304, correlatable with the basal part of the Tunggurian Age. Accordingly, the ages of the two localities were determined by Flynn et al. (1999) as 15.5 Ma and 14 Ma, respectively, corresponding to the segment from the base of C5Bn.2n to above the C5ADn.

A joint reexamination of the GL 9304 micromammal material by Z.-d. Qiu, Flynn, and B.-y. Wang recognized the following forms (identified to the genus level): *Mioechinus*? sp., *Metexallerix* sp., *Atlantoxerus* sp., *Heterosminthus* cf. *erbajevae*, *Litodonomys* spp., *Protalactaga* sp., *Democricetodon* sp., *Megacricetodon* sp., and *Alloptox* sp. Of them, the Oligocene holdovers like *Metexallerix* and *Litodonomys* have never been found from faunas of the Tunggurian Age. It is agreed that the GL9304 assemblage is better to be considered as the later phase of the Shanwangian fauna.

With the GL 9303 assemblage having been assigned to the Zhangjiaping local fauna (see previous discussion), the transition between the Zhangjiaping and Duitinggou local faunas should occur, in theory, somewhere within the deposits bracketed by the two localities. Without the help of additional paleontological data, magnetostratigraphy may help in this regard. The polarity reversal pattern of the section covering the segment between the GL9303 and GL9304 is characterized by three rather long normal chrons separated by shorter reversed ones. Consistent with the paleontology, this distinctive pattern could easily be correlated with C6n, C6An.1n, and C6An.2n (18.748–20.709 Ma). GL 9303 is located just around the base of the C6An.1n, and should be slightly younger than 20.709 Ma (see figure 1.5).

Chetougou local fauna (Qinghai) This small assemblage is a composite one, found from four small sites in the Xining Basin, Qinghai Province. These sites were assigned to the same Chetougou Formation, based mainly on lithologic characters. Stratigraphically, the Chetougou Formation is bracketed between the Xiejia and "Xianshuihe" formations. The fossils were identified (Qiu, Li, and Wang 1981) as *Heterosminthus* (originally *Protalactaga tungurensis*), *Megacricetodon sinensis*, ?*Eumyarion* sp., and some rhino and deer specimens. No paleomagnetic sampling specifically for Chetougou Formation was attempted. However, while studying the Xiejia Formation, Wu et al. (2006) continued their paleomagnetic sampling into the Chetougou Formation. The fossiliferous level was thought to be from the very base of the Chetougou Formation (Li and Qiu 1980:fig. 1). This level was correlated with the base of C5Cr (17.235 Ma). This inter-

pretation seems reasonable, since the polarity reversal pattern of this part of the section matches the GPTS well. This would mean that the age of the Chetougou local fauna may be around 17 Ma.

Shinanu local fauna (Linxia Basin, Gansu) Cao et al. (1990) (China University of Geosciences) found two fossiliferous layers in the ~50 m brownish-yellow sandstones and brownish-red mudstones. At the bottom of the section *Alloptox minor*, *A. chinghaiensis*, *Atlantoxerus* sp., *Megacricetodon sinensis*, *Heterosminthus tungurensis*, *Sayimys* cf. *S. obliquidens*, *Pseudaelurus guangheensis*, *Dorcatherium* sp., etc., were found. In the same locality, we have collected half an upper premolar of *Beliajevina tongxinensis* and a horncore and lower jaw with p4-m3 of *Turcocerus* cf. *T. noverca*. Stratigraphically, it should belong to the Dongxiang Formation (Deng et al. 2004). The assemblage shows characters transitional between the Shanwangian and Tunggurian LMS/A.

LOWER BOUNDARY DEFINITION

Almost from the very beginning since 1984, the lower boundary of the Shanwangian Stage/Age has been assumed to roughly correspond with those of Orleanean and/or Burdigalian (Li, Wu, and Qiu 1984). Deng (2006) drew this boundary at **20 Ma** according to Steininger (1999) age estimation for the Orleanian Mammal Faunal Unit, but later Deng, Wang, and Yue (2008a) shifted it to the base of C6r—that is, **20.5 Ma**. Meng et al. (2006) directly accepted the lower boundary of the Burdigalian Stage set in GTS2004—that is, **20.43 Ma**—as that of the Shanwangian Stage/Age.

The land mammal faunas of the Shanwangian Age clearly differ from those of the Xiejian in being basically Miocene in faunal composition, with only very limited Oligocene "holdovers," like some zapodids (*Plesiosminthus*, *Heterosminthus*, *Litodonomys*) and ochotonids (*Desmatolagus*, *Sinolagomys*). The transition and turnover from the Xiejian to the Shanwangian ages appears to be different from those in Europe. During this "critical time interval" a number of faunal events occurred.

1. The appearance of the proboscideans is unquestionably a very important event. Eurasian proboscideans are widely known to have emigrated from Africa. The earliest dated occurrence of proboscideans in China, so far known, is *Platybelodon dangheensis* from the Xishuigou section, paleomagnetically dated as 19.5 Ma. The Shovel-tusked elephants are the most characteristic middle Miocene animals in Asia, widely occurring in north and northwestern China (Tongxin in Ningxia, Linxia in Gansu,

Junggar in Xinjiang, etc.) and became extinct by the end of the Tunggurian Age. Other early Shanwangian records of proboscideans are from Urtu, Upper Aoerban, and Sihong faunas. None of them has been paleomagnetically dated. Paleontologically, all contain *Megacricetodon* with some Oligocene relic forms, like *Tachyoryctoides*. Compared with other faunas of Shanwangian Age, the Urtu local fauna could be closer to Gashunyinadege additionally by the presence of *Prodistylomys* and *Distylomys*. According to our new reinterpretation (see previous discussion), the Upper Aoerban Fauna is probably around 19 Ma. The Gashunyinadege Fauna could be slightly older than 19 Ma, and the Sihong fauna is generally considered slightly older than the Xiejiahe fauna, the lowest horizon of which was estimated around 18 Ma (Deng, Wang, and Yue 2008a). Therefore, we estimate the Sihong fauna at no more than 19 Ma. As a result of the previous discussion, the Danghe *Platybelodon* may be the earliest record of proboscideans so far known in China.

In Europe, proboscideans first occurred in MN3b, around 19–17.5 Ma, and this was named Proboscidean Event or Proboscidean Datum. From a recent report by Antoine et al. (2009), a proboscidean tusk fragment can be dated to the late Oligocene (before 24.6 Ma) in South Asia (Pakistan). However, specimens identifiable to *Gomphotherium* or deinotheres occur only in the top of Chitarwata Formation and lower Vihowa Formation (~20 Ma) and thrived thereafter.

2. The first appearances of *Megacricetodon* and *Alloptox* may also serve as indicators of the faunal turnover. Both forms were found from the Gashunyinadege, Upper Aoerban, and Duitinggou local faunas. The age of Duitinggou is still problematic (see previous discussion). Excepting for Duitinggou, all the other localitites yielding these two forms, including Urtu and Sihong (containing only *Megacricetodon*), are roughly estimated as around 19 Ma. Fossils of these animals are easily found in the field and each of these can serve as markers for defining the lower boundary of the Shanwangian LMS/A.

None of these candidates meet all requirements for being chosen as a lower boundary RSSP of the Shanwangian LMS/A. The Xishuigou section provides us with a good temporal constraint, but its base is separated from the underlying deposits by a hiatus, and the section as a whole is tectonically strongly deformed. The Aoerban section comprises the critical time interval covering both the Xiejian and Shanwangian LMS/A. This section has more potential for being chosen as a candidate RSSP. The Aoerban area may deserve more extensive fossil collecting and biomagnetostratigraphic investigation in the future.

MAMMAL FAUNA CHARACTERIZATION

The Shanwangian chronofauna can be viewed as the initial stage of overall faunal modernization. The sciurids and the myomorph rodents became predominant in the rodent fauna; eomyids, glirids, and zapodids entered their flourishing period; and the "modern cricetids" noticeably diversified. Many new genera appeared, although they represent Oligocene families. In the carnivorous guild, the Carnivora almost totally replaced the Creodonta (with the exception of the unique creodont skeleton recently found in Jijiazhuang). The ruminants flourished and the first proboscideans immigrated into the China mainland.

Index fossils: North-West China Area: *Kinometaxia*, *Platybelodon dangheensis*, *Amphimoschus*; North China Coastal Area: *Youngofiber*, *Diatomys*, *Primus*, *Neocometes*, *Alloeumyarion*, *Spanocricetodon*, *Phoberocyon*, *Ballusia*, *Semigenetta*, *Dionysopithecus*, *Platodontopithecus*, *Plesiotapirus*, *Hyotherium*, *Sihongotherium*, and *Ligeromeryx*.

First appearances within the stage: North-West China Area: *Alloptox*, *Leptodontomys*, *Protalactaga*, *Megacricetodon*, *Platybelodon*, and *Turcocerus*; North China Coastal Area: *Diatomys*, *Apeomys*, *Amphicyon*, *Pseudaelurus*, *Stegolophodon*, *Anchitherium*, *Anisodon*, *Plesiaceratherium*, *Dorcatherium*, and *Sinomeryx*.

Last appearance within the stage: North-West China Area: *Amphechinus*, *Metexallerix*, *Prodistylomys*, *Ligerimys*, *Asianeomys*, *Plesiosminthus*, *Litodonomys*, *Cricetodon*; North China Coastal Area: *Diaceratherium*.

Characteristic fossils: North-West China Area: *Mioechinus*, *Distylomys*, *Sayimys*, *Ansomys*, *Plesiosminthus*, *Atlantoxerus*, *Platybelodon*; North China Coastal Area: *Stegolophodon*, *Plesiaceratherium*, *Sinomeryx*.

Tunggurian LMA/S

NAME DERIVATION

The Tunggurian Age was also established by Li, Wu, and Qiu (1984), mainly based on the mammal faunas from the Tunggur Formation. The Tunggur Formation was named by Spock (1929). Later, the term Tunggurian Stage first appeared in the appendix of the revised edition of the *Stratigraphic Guide of China and Its Explanation* (All-China Stratigraphic Commission 2001), although it was not established by proper procedures.

UNIT STRATOTYPES

Wang, Qiu, and Opdyke (2003) systematically reviewed the Tunggur Formation and explored its litho-, bio-, and magnetostratigraphy and its paleoenvironments in great detail. Numerous sections are exposed along the rim of the Tunggur Tableland. Wang, Qiu, and Opdyke (2003) described in detail 12 sections worked on by the American Museum Third Central Asiatic Expedition, Sino-Soviet Paleontologic Expedition, and Institute of Vertebrate Paleontology and Paleoanthropology. In reality, the whole Tunggur Tableland can be considered as the type area of the Tunggurian LMS, since almost every section is known to produce the same fossil mammal assemblage traditionally called *Platybelodon* Fauna.

We owe the finding of Spock's original "type section" of the Tunggur Formation much to X. Wang's thorough search of the archives of the American Museum of Natural History (AMNH), and it should be nowadays called Mandelin Chabu (originally North Camp 1928). This section was investigated by an Institute of Vertebrate Paleontology and Paleoanthropology (IVPP) field team in 1986–1987 (probably a few hundred meters farther east of the type section). According to that field work (Qiu, Yan, et al. 1988), the exposed sediments of this site were only 26.6 m thick. Fossils collected from this site were *Anchitheriomys tungurensis* (originally *Amblycastor tungurensis*), *Bellatona forsythmajori*, ?*Melodon* sp., *Platybelodon grangeri*, Chalicotheriidae gen. et sp. indet., *Stephanocemas thomsoni*, *Dicrocerus* sp., and *Turcocerus* ?*noverca* (originally *Oioceros* ?*noverca*). Wang, Qiu, and Opdyke (2003) added two rhinos: *Acerorhinus zernovi* and *Hispanotherium tungurense*, described by Cerdeño in 1996 (from Gur Tung Khara Usu). Compared with 77 taxa of the Tunggurian Chronofauna from the Tunggur Tableland area, these 10 taxa from 26.6 m of sediments seem too few to adequately represent the whole Tunggurian LMS/A. Lithologically and paleontologically, the Mandelin Chabu section may represent only a small portion of the top part of the Tunggur Formation. Nevertheless, Mandelin Chabu section should be retained as the unit holostratotype of the Tunggurian LMS.

Apparently unsatisfied with Spock's type section, Deng, Hou, and Wang (2007) proposed the Tairum Nor section in the southern rim of the Tunggur Tableland (Roadmark 346) as a new stratotype. They also defined the lower boundary of the Tunggurian Stage there. Thus, according to Deng, Hou, and Wang (2007) the Tairum Nor section serves both as a unit stratotype and a lower boundary stratotype. As indicated by Wang, Qiu, and Opdyke (2003), the southern escarpment of the Tunggur Tableland (Tairum Nor area) was investigated by members of the American Museum Third Central Asiatic Expedition, and the type of *Platybelodon grangeri* was unearthed from a site of this escarpment. Deng et al.'s designation of the Tairum Nor section as the lower boundary stratotype of the Tunggurian LMS seems untenable, since it is based on the notion that the lower boundary of the Tunggurian Stage should coincide with that of the European Astaracian and therefore defined it at the lowest level of the occurrence of the Tunggurian Chronofauna in the Tairum Nor section. This seems in contradiction to the available paleontologic data, which show that the Tunggurian Chronofauna extends deep into deposits of older age (see following discussion). As a result, we would like to consider the Tairum Nor section as the **unit parastratotype** of the Tunggurian LMS.

The Tunggurian Chronofauna was subdivided into two faunas (the Tairum Nor and Moergen) by Wang, Qiu, and Opdyke (2003), but three faunas (including the Tamuqin at the top) by Qiu, Wang, and Li, (chapter 5, this volume). The Tairum Nor fauna occurs in the Lower Beds (reddish-brown or dark red, capped with yellow sandstone) of the Tunggur Formation, mainly exposed in the lower parts of the Aletexire and Tairum Nor sections. The fauna in question differs from the overlying faunas in having more archaic forms, like *Tachyoryctoides*, *Distylomys*, *Leptarctus*, and *Sthenictis*, which have not been found from the younger faunas, and more primitive forms of *Stephanocemas thomsoni*. The Moergen fauna is found from the main part of the Upper Beds of the Tunggur Formation, widely developed on the whole tableland but particularly well exposed along its north and west edges. This fauna is composed of the bulk of the large mammals collected in the past, with the exception of fossils found from the channel sandstones in the Tairum Nor section, and all the micromammal fossils from the Moergen section (more than 20 taxa). In reality, this fauna should be the Tunggur fauna in its strict and traditional sense. Revised faunal list of this fauna can be found in papers by Wang, Qiu, and Opdyke (2003) and Qiu, Wang, and Li (chapter 5, this volume). The Tamuqin local fauna collected at the top of the Moergen section, consists of only 13 forms identified at generic and/or specific levels. A "newcomer," *Steneofiber*, and some more advanced species of *Gobicricetodon* (*G. robustus*) and *Plesiodipus* (*P. progressus*), which have not been found in underlying faunas, appeared in this local fauna. Wang, Qiu, and Opdyke (2003:19) thought it was unnecessary to separate it as a new fauna.

Only two of the Tunggur Tableland sections were chosen for paleomagnetic study: Moergen and Tairum Nor. Wang, Qiu, and Opdyke (2003) originally interpreted the lowest normal magnetozone of the Tairum Nor section (the lowest fossiliferous level) as corresponding to C5Ar.2n (base: 12.878 Ma) and the upper third of the overlying reversed magnetozone (the upper boundary of the Tairum Nor fauna) as corresponding to C5Ar.1r (12.415–12.73 Ma; base of upper 1/3: 12.625 Ma). The top of the Moergen section lies roughly at the middle of C5r.3r (11.614–12.014 Ma; middle: 11.814 Ma). Therefore, the Tunggur Fauna in the type area spans roughly 11.8–12.6 Ma. Later, based mainly on faunal correlation, Qiu, Wang, and Li (2006) thought that the Tairum Nor Fauna should be 14–15 Ma old. Deng, Hou, and Wang (2007) offered another alternative interpretation for the Tairum Nor section: by aligning the second reversed magnetozone (R2) with C5Bn.1r (14.877–15.032 Ma). According to their figure 4, the base of C5Bn.1r (15.032 Ma) was correlated to the lower boundary of the Tairum Nor fauna, and hence the lower boundary of the Tunggurian Stage was placed at 15 Ma. However, according to Wang, Qiu, and Opdyke (2003:fig. 10), the lowest occurrence of the Tairum Nor fauna is lower, at the base of the N1 normal interval, which should correspond with C5Bn.2n (15.032–15.16 Ma), not where Deng et al. place it. Thus, the lower boundary of the Tunggurian Stage should be **15.16 Ma**, not 15.032 Ma. Taken as a whole, Deng et al.'s reinterpretation of an earlier Tairum Nor fauna seems more consistent with the paleontologic data. Nonetheless Qiu, Wang, and Li (chapter 5, this volume) questioned Deng et al.'s reinterpretation but pointed out that there is presently no completely satisfactory solution and that additional work is needed. It should be remembered that 15.16 Ma is only the lower boundary of the Tunggurian Chronofauna in the Tunggur type area, the earliest appearance of Tunggurian Chronofauna, and therefore the Tunggurian LMS might be much earlier as discussed in more detail in the next paragraphs. While accepting Deng et al.'s opinion for the Tairum Nor section, Qiu, Wang, and Li (chapter 5, this volume) maintained their own magnetostratigraphic interpretation for the Moergen section. As a result, the total time span for the Tunggur Formation was estimated as from ~**15.16 to 11.8 Ma**. However, this would imply a depositional hiatus of more than 1 myr between the Tairum Nor and the Moergen sections (Qiu, Wang, and Li, chapter 5, this volume, fig. 5.3). Further magnetostratigraphic work should be attempted in order to solve the dates of the boundaries of the Tunggur Formation.

REFERRED DEPOSITS AND FAUNAS

Qiu (1990) separated his Mammal Unit III (corresponding Tunggurian Stage/Age) into three faunal levels (from bottom to top): Tongxin, Lengshuigou, and Tunggur.

The Tongxin fauna included the faunas from the Koujiacun Formation (Shaanxi), Jiulongkou local fauna (Hebei), and fossil aggregate from the "Hsienshuiho" (=Xianshuihe) Formation (Gansu). The Tunggur fauna included faunas from the Halamagai Formation (Xinjiang), the Erlanggang locality of the Shaping Formation (Hubei), and Lingyanshan localities (Jiangsu). Except for Quantougou and Halamagai, no modern field collecting or magnetostratigraphic work has been undertaken at these localities.

Balunhalagen fauna (central Nei Mongol) This fauna (Qiu, Wang, and Li, chapter 5, this volume) consists of 35 forms, mainly common elements of the Tunggurian Chronofauna, with only a few more advanced genera, like *Prosiphneus* and *Brachyscirtetes*. In general, this fauna is very close to the Basal Dingshanyanchi fauna in the Junggar Basin (see following discussion), and it is here tentatively assigned to the latest stage of the Tunggurian LMA, with its top possibly extending into the Bahean LMS/A.

Halamagai fauna (Junggar Basin, Xinjiang) According to Ye, Wu, and Meng (2001), the Halamagai Formation consists of gray-greenish silty mudstone layers interbedded with grayish sandstones and conglomerates. Its type section thickness is about 50 m, with the basal conglomerates and the sandstones within the lower 20 m being rich in fossils. Of the 44 taxa identified at the genus or species level for the composite faunal list, 28 have been described and published (see appendix 1). Among the large mammals there are a number of archaic taxa that have not been discovered in Moergen and Tairum Nor faunas, like *Nimravus*, *Pseudaelurus*, *Gomphotherium* cf. *G. shensiensis*, and *Eotragus*, or taxa showing more primitive characters, although they are present in the Moergen and Tairum Nor faunas as well, like *Tungurictis spocki*, *Anchitherium gobiense*, *Stephanocems thomsoni*, and maybe *Lagomeryx*. It is to be noted that there are a few taxa representing the forerunners of some groups characterizing Late Miocene "*Hipparion*" faunas, like *Proticititherium*, *Thalassictis*, *Simocyon*, and *Chilotherium* (if its identification is tenable). The absence of *Bellatona*, which is an important member of the Moergen fauna, and the presence of *Tachyoryctoides*, which is present in the Tairum Nor fauna, but not the Moergen one, indicate that the Halamagai fauna is better correlated with Tairum Nor, or with Dingjiaergou.

Kekemaideng local fauna (Junggar Basin, Xinjiang) The Kekemaideng Formation is a sequence formed by coarse-grained sandstones and conglomerates, overlying the Halamagai Formation disconformably. It is only about 14 m thick in its type section on the south bank of the Ulunggur River (96DL). There are only a few forms found and briefly mentioned (Ye 1989; Ye, Wu, and Meng 2001): *Platybelodon* sp., *Chilotherium* sp., *Brachypotherium* sp., *Kubanochoerus* sp., *Dicrocerus grangeri*, and *Turcocerus kekemaidengensis*.

Basal Dingshanyanchi fauna (Junggar Basin, Xinjiang) This faunal assemblage, recently studied by Wu, Meng, and Ye (2009), records 26 mammal taxa. Of them, 19 were identified at and below the generic level and described in detail, and 13 are shared with the Moergen fauna.

Quantougou local fauna (Lanzhou Basin, Gansu) The locality was found by Andersson from Chuan Tou Kou (Quantougou) of Ping Fan Hsien (now Yongdeng County), Gansu (Young 1927:23; Hopwood 1935:20). It is now renamed as the Quantougou local fauna. Recent study of the section shows that the Xianshuihe (originally Hsienshuiho) Formation was named for a long sequence of sediments, maximum thickness of which is about 880 m (Qiu et al. 2001). It is composed of three members spanning from Oligocene through Middle Miocene. The Quantougou local fauna was found from the top of Upper Member of the Xianshuihe Formation in the Xiajie section of the Dahonggou area. Z.-d. Qiu (2001:304) listed 11 micromammal taxa as follows: *Mioechinus? gobiensis*, *Microdyromys wuae*, *Heterosminthus orientalis*, *Protalactaga grabaui*, *P. major*, *Plesiodipus leei*, *Megacricetodon sinensis*, *Ganocricetodon cheni*, *Paracricetulus schaubi*, *Mellalomys gansus*, *Myocricetodon plebius*, and Ochotonidae indet. Among large mammals there were *Kubanochoerus gigas* and *Gomphotherium wimani*. A majority of the above taxa are commonly shared by the Tairum Nor and Moergen faunas, except *Protalactaga grabaui*, which is present in Moergen but lacking in Tairum Nor. Paleomagnetic sampling at the Dahonggou and Xiajie sections indicates that the Xiajie section should correspond to the basal parts of C5Bn to C5Cn (15.16–16.721 Ma) and the Quantougou local fauna should be slightly older than 15 Ma.

Local fauna from "Hsienshuiho" Formation (Qinghai) A small group of mammal fossils found from the "Hsienshuiho" Formation in the Xiejia section, southeast of Xining, Qinghai, was described by Qiu, Li, and Wang (1981) and their stratigraphic position was summarized by Li, Qui, and Wang (1981). The faunal list consists of eight taxa: *Alloptox chinhaiensis*, *Plesiodipus leei*, *Gomphotherium connexus*, *G. wimani*, *Kubanochoerus minheensis*, *Stephanocemas chinghaiensis* (Young 1964:334), *Micromeryx* sp., and *Turcocerus* (?) *noverca*. No paleomagnetic sampling has been attempted for this formation. This assortment of fossils is possibly a mixed one. *Kubanochoerus minheensis*

and *Gomphotherium connexus* seem more primitive than those found in the Dingjiaergou fauna.

Hujialiang Formation (Linxia Basin, Gansu) The Hujialiang Formation was erected by Deng (2004a). Its lithology in the type section is mainly grayish-yellow gravelly sandstones, 20–30 m thick, yielding large amount of complete skulls and mandibles of shovel-tusked elephants, while at Laogou locality it is represented by a 6 m conglomerate yielding numerous isolated teeth of a great variety of carnivores, proboscideans, perissodactyls, and artiodactyls. So far, only two taxa were described from Laogou: *Hispanotherium matritense* (Deng 2003) and *Alicornops laogouense* (Deng 2004b). The latter is very close to *Acerorhinus zernowi* described by Cerdeño (1996) from Tung Gur in both morphology and size. As a whole, there is no problem that the fauna from the Hujialiang Formation in Gansu is very close to that of Dingjiaergou in Ningxia. Paleomagnetic work in the Linxia Basin began in the 1990s (Li 1995), but, due to errors in mammal identification and stratigraphic correlation, results are difficult to interpret (Deng et al. 2004). Results from renewed magnetostratigraphic work are used in this paper (see Guonigou fauna).

Dingjiaergou local fauna (Tongxin Basin, Ningxia) Dingjiaergou, a small village 20 km northeast of the Tongxin county seat, became well known for its production of rich shovel-tusked elephant fossils in the 1980s. The fossiliferous area extends from Dingjiaergou south and eastward, covering a total area of about 200 km². According to Huo et al. (1989:207–209), all the mammal fossils were from the Hongliugou Formation, which is composed of interbedded yellow sandstones and red-brown silty claystones. The Hongliugou Formation near Dingjiaergou is about 70 m thick. Mammal fossils were found from a large number of localities (no less than 20) in lower, middle, and upper parts of the formation. Guan (1988) and later with other colleagues, enumerated a long list of taxa without paying much attention to their exact stratigraphic position, or with highly confusing data when biostratigraphy was mentioned. Since most of the large mammal fossils were purchased from local people, stratigraphic position of specific mammal taxa can hardly be ascertained. What is more or less certain is that most of the skeletons and complete skulls and mandibles of *Platybelodon tongxinensis* and *Kubanochoerus gigas* were excavated from the upper part of the Hongliugou Formation near Dingjiaergou. *Platybelodon tongxinensis* was considered to be more primitive than *grangeri* from the Tunggur region in having narrower mandibular symphysis and lower tusks and thinner cement cover on the cheek teeth. *Kubanochoerus* was first studied by Qiu, Ye, and Huo (1988), identified as *K. lantianensis*, and later placed in *K. gigas* by Guan and Van der Made (1993). *Kubanochoerus gigas* is only slightly different from the type species, *K. robustus*, from the Belametchetskaya fauna (Georgia). As for the elasmotheres, Guan (1988) first described *Caementodon tongxinensis* from the Gujiazhuang, Dingjiaergou, and Huangjiashui localities. Later, Guan (1993) added another taxon, *Huaqingtherium qiui*. About their stratigraphic position, Guan (1993:200) stated that "*Huaqingtherium* came from the upper fossiliferous zone and *Caementodon* from lower one." While describing the teeth of *Hispanotherium matritense* from Laogou (Hezheng), Deng (2003) transferred the two Tongxin taxa to the same species, *H. matritense*, and considered the latter more primitive than the Tunggurian *H. tungurense*. Having checked the morphology and measurements of Guan's specimens again, we are inclined to retain the smaller species, *Caementodon tongxinensis*, as separate. It could probably belong to the genus *Beliajevina*, the type species of which is *B. caucasica*, an important component of the Belametchetskaya fauna. Among the carnivores, *Tongxinictis primordialis* (Qiu, Ye, and Cao 1988) may be also from the upper part of the Hongliugou Formation. It is certainly more primitive than the Tunggurian *Percrocuta tungurensis*. Skulls and mandibles of *Sansanosmilus* and *Gobicyon* were also found, but mainly in the south part of the area (Huangjiashui and Gunziling, 10–12 km south of Dingjiaergou). *Sansanosmilus* of the Tunggur area is represented only by an upper carnassial (originally identified as *Machairodus* (?) sp.; Colbert 1939:79), which is longer and more advanced in morphology than the Tongxin one. The Tongxin specimens of *Gobicyon* are smaller in size than those of the Moergen Fauna of the Tunggur area.

This faunal composition seems to indicate that the Dingjiaergou local fauna found from the upper part of the Hongliugou Formation is closest to the Belamechetskaya Fauna of the Caucasus. The latter is sandwiched between marine Chokrakian sediments and is correlated to the middle Badenian Stage in the middle Paratethys, with age around 15 Ma (Steininger 1999). The fossils found from the lower part of the Hongliugou Foramtion are likely to be even earlier in age.

LOWER BOUNDARY DEFINITION

Since the discovery of the co-occurrence of two giant-size animals, *Platybelodon* and *Kubanochoerus*, in the Tongxin Basin during the 1980s, it has become more and more clear that there must have been two distinctive phases in mammal evolutionary history after the Xiejian Age and before the advent of the famous "*Hipparion*" faunas in China. The early stage is represented by the faunas of the Shanwangian Age as previously defined. There were a few large animals, like *Gomphotherium*, *Plesiaceratherium*, and possibly *Anisodon*. *Gomphotherium* and *Anisodon* are represented by very poor material. On the contrary, a large number of complete skeletons, skulls, and jaws of *Plesiaceratherium* were unearthed. *Plesiaceratherium* is the only dominant Shanwangian large-sized animal. As we now know, the later (Tunggurian) stage is characterized by a group of large- or giant-sized animals such as *Platybelodon grangeri*, *P. tongxinensis*, *Hispanotherium*, *Acerorhinus* (*Alicornops*), *Kubanochoerus*, and *Macrotherium*. The available data show that some local faunas with the same closely related genera may represent earlier levels of evolution from lower strata, as shown by the fossils recently found from Ningxia, Gansu, and Qinghai. Among them the Tongxin area may have the best potential to find the "critical time span" around the Shanwangian-Tunggurian ages. Unfortunately, no elementary biostratigraphic (especially for micromammals) or magnetostratigraphic work has been systematically attempted.

Based primarily on the assumed identity between the lower boundaries of the Tunggurian with the European Astaracian, Deng (2006) put the lower boundary of the Tunggurian Stage at 15 Ma (base of C5Bn.1r). As a RSSP, the Tairum Nor section is deficient in its lack of any representative mammals of Shanwangian Age below the supposed boundary. Furthermore, their definition did not incorporate the data from the Tongxin and Linxia areas, which may be earlier in age.

The GTS2004 (Gradstein, Ogg, and Smith 2004) recommends the top of C5Cn.1n as the boundary for early Miocene and middle Miocene (15.97 Ma). The European Astaracian Land Mammal Age is defined as MN 6–8. According to Steininger (1999), MN 5 is 17–15 Ma and MN 6 is 15–13.6 Ma. Thus, the European Early–Middle Miocene boundary (15.97 Ma) occurs within MN 5. How should this boundary be treated in China? From the paleontological perspective of Tongxin and Linxia areas, the lower boundary of Tunggurian Stage likely should be extended downward, and it could conveniently coincide with the global standard Early–Middle Miocene boundary. As a compromise, we would suggest to put it temporarily between 15 and 16 Ma, pending further investigation, especially in the Linxia and Tongxin basins.

MAMMAL FAUNA CHARACTERIZATION

The Tunggurian Chronofauna can be viewed as a further-developed Shanwangian Chronofauna. *Mioechinus* replaced the archaic forms of erinaceids. New ochotonid, *Bellatona*, appeared, while eomyids, glirids, and zapodids became uncommon. Dipodids appeared and *Democricetodon-Megacricetodon* became highly diversified. Of the large mammals, *Platybelodon* (*grangeri*, *tongxinensis*, or *danovi*), *Kubanochoerus*, and *Hispanotherium* constitute the most characteristic trio of large-size mammals. In combination with the trio, the ruminants began to diversify, and percrocutas, ictitheres, nimravines, chilotheres, and listriodonts made their first appearances as well.

Index fossils: *Schizogalerix, Anchitheriomys, Steneofiber tunggurensis, Plesiodipus, Gobicricetodon flynni, Bellatona, Gobicyon, Leptarctus, Percrocuta, Tungurictis, Sansanosmilus, Lartetotherium, Hispanotherium, Kubanochoerus, Bunolistriodon,* and *Listriodon.*

First appearances within the stage: *Sinotamias, Gobicricetodon, Protictitherium, Thalassictis, Metailurus, ?Chilotherium, Palaeotragus,* and *Euprox.*

Last appearance within the stage: *Distylomys, Sayimys, Tachyoryctoides, Heterosminthus, Megacricetodon, Sinolagomys, Alloptox, Amphicyon, Hemicyon, Pliopithecus, Gomphotherium,* and *Anisodon.*

Characteristic fossils: *Heterosminthus, Democricetodon, Megacricetodon, Alloptox, Bellatona, Platybelodon, Anchitherium, Dicrocerus, Lagomeryx, Stephanocemas,* and *Turcocerus.*

Bahean LMS/A

NAME DERIVATION

The age name Bahean was created by Li et al. (1984) on the basis of the Bahe Formation in the Lantian area, Shaanxi Province, and was correlated to the European Vallesian. The rationale behind this was as follows:

1. Although mammals from the Bahe Formation are still a "*Hipparion* fauna," it contains a number of forms absent from the typical *Hipparion* fauna of Baode, such as *Dinocrocuta, Shaanxispira,* and "*Hippotherium*" grade species like "*Hipparion*" *weihoense.*
2. Lithologically, the variegated fluviolacustrine deposits of the Bahe Formation are different from the "*Hipparion* red clay," and it disconformably underlies the Lantian Formation ("*Hipparion* red clay").

Regarding the differences in faunal composition and ecology as insufficient, Qiu and Qiu (1990) recommended abolishing this age. The revised edition of the *Stratigraphic Guide of China and Its Explanation* (All-China Stratigraphic Commission 2001) also accepted the latter recommendation and did not use the Bahean Stage. Since 1997, a team initially headed by Z.-d. Qiu and Mikael Fortelius, then by S.-h. Zheng and Z.-q. Zhang, system-

atically explored the Neogene of the Lantian area again and resurrected the Bahean Age based on newly acquired data (Zhang et al. 2002). However, even to the present time, the Bahean Stage/Age is still not widely accepted (Qiu, Wang, and Li 2006; Wang et al. 2009).

UNIT STRATOTYPE

Liu, Ding, and Gao (1960) named the Bahe Formation. They described in detail the Shuijiazui section in Xiehu Zhen (town) of Lantian County, on the southern bank of Bahe River, where they discovered *Hipparion* and hyaenid fossils. Explorations in 1963–1965 by the Cenozoic Laboratory of the IVPP obtained an additional large quantity of fossils in the Shuijiazui section (Zhang et al. 1978). Liu, Li, and Zhai (1978) then made a detailed inventory of these fossils. The field team conducted lithostratigraphic studies (Kaakinen and Lunkka 2003) and systematically reviewed mammalian paleontology in this area (Zhang et al., chapter 6, this volume). The Shuijiazui section undoubtedly can serve as the unit stratotype of the Bahean Stage.

REFERRED DEPOSITS AND FAUNAS

Upper Youshashan Formation (Qaidam Basin, Qinghai) As summarized by Wang et al. (chapter 10, this volume), the Upper Youshashan Formation contains two major Upper Miocene mammal [local] faunas: the Tuosu and Shengou mammal faunas (abbreviated as TMF and SMF). The composite faunal list of TMF is as follows: *Ictitherium, Eomellivora, Chalicotherium brevirostris, Hipparion teilhardi,* Sivatheriinae indet., *Dicrocerus, Euprox* sp., *Olonbulukia tsaidamensis, Qurliqnoria cheni, Tossunnoria pseudibex, Tsaidamotherium hedini, Protoryx* sp., and *Tetralophodon.* The composite faunal list of SMF is *Ictitherium, Adcrocuta eximia, Plesiogulo, Promephitis parvus, Acerorhinus tsaidamensis, Dicerorhinus ringstromi, Hipparion* cf. *H. chiai, H. weihoense, H. teilhardi, Euprox* sp., *Gazella* sp., and *Amebelodon.* Compared with the faunal list composed by Bohlin, it seems clear that only *Stephanocemas* and *Lagomeryx* have been excluded from these two faunas, while several endemic bovids (e.g., *Olonbulukia*) are restricted to the TMF. Based on the biomagnetostratigraphic study of the Huaitoutala section (about 10 km west of Tuosu Lake) carried out by Fang et al. (2007), the strata bearing the TMF (1300–1900 m) correlate with the base of C5r.3r (12.014 Ma) to the lower two-thirds of C5n.2n (9.987–11.04 Ma). Such a correlation was slightly adjusted by Wang et al. (2011), and the time span of TMF is approximately 11.5–10.3 Ma. The paleomagnetic section (Fang et al. 2007:fig. 8; Wang et al., chapter 10, this volume, fig. 10.3) shows that only

CD9823 (with *Chalicotherium brevirostris*) is above the base of C5n.2n, all other localities (with *Dicrocerus, Euprox, ?Qurliqnoria*, etc.) are below C5n.2n and thus lower than the possible lower boundary of the Bahe LMS/A (see following discussion). It is to be noted here in particular that the majority of the *Hipparion* specimens described by Bohlin (1937) came from the Quanshuiliang section, where a magnetostratigraphic study is being undertaken. On the other hand, the *Hipparion* materials recently discovered in this area and described by Deng and Wang (2004) came from other areas (Tuosu Nor, Naoge, and Shengou), although new collections from the Quanshuiliang section are currently undescribed. Much of the current biostratigraphy scheme was based on correlation by means of marker beds and thickness measurements traced to the magnetostratigraphy of the Huaitoutala section carried out by Fang (Wang et al. 2007; Wang et. al., chapter 10, this volume), although new magnetic sections in Quanshuiliang, Naoge, and Shengou will eventually improve on our knowledge. Otherwise, the majority of the published *Hipparion* specimens from Shengou and Naoge sections are to be referred to SMF (Wang et al. 2007:fig. 11). Only the CD0205 sample in the Naoge section is from the uppermost part of the TMF. It is possible that the single astragalus of *H. weihoense* from CD0205, along with some unpublished teeth from the Quanshuiliang section, represents the earliest record of *Hipparion* in the Qaidam Basin. Its age should be only slightly older than 11 Ma. If this proves true in the future, the TMF could belong largely to the Tunggurian LMS/A, rather than to the Bahean LMS/A.

The SMF is Bahean LMS/A without doubt, judging by its faunal composition and its stratigraphic position in the Upper Youshashan Formation.

Liushu Formation (Linxia Basin) According to Deng et al. (chapter 9, this volume), the Liushu Formation contains four mammal assemblages (from bottom to top): the Guonigou, Dashengou, Yangjialing, and Qingbushan faunas. The first three have been attributed to the Bahean LMS/A, while the last (Qingbushan) is assigned to the Baodean LMS/A. The presence of *Dinocrocuta* and *Shaanxispira*, which are particularly characteristic of the Bahean LMS/A, are absent in Baodean LMS/A, and a number of forms evidently more primitive than directly related genera found only from the Baodean LMS/A, like *Hezhengia* (versus *Plesiaddax*) and *Parelasmotherium* (versus *Sinotherium*) in the Guonigou and Dashengou faunas corroborate such an assignment. However, paleontologically, neither the Yangjiashan nor the Qingbushan fauna are distinctive enough for referral to the Bahean and Baodean LMS/A, respectively. Both the Yangjiashan

and Qingbushan faunas lack *Dinocrocuta* and *Hezhengia*, but no particularly characteristic Baodean taxa are present either. Furthermore, no sharp difference in lithology can be seen between the sediments bearing these two faunas, in contrast to the Lantian stratigraphy. Also, there are no tenable magnetostratigraphic interpretations to support such an assignment.

Lamagou local fauna (Fugu, Shaanxi) This local fauna was first named by Xue, Zhang, and Yue (1995). Of this local fauna, only the specimens of *Dinocrocuta gigantea* were described (Zhang and Xue 1996). The other taxa preliminarily identified at and below generic level are *Ictitherium wongi, Homotherium* sp., *Plesiogulo* cf. *P. brachygnathus, Platybelodon* sp., *Hipparion chiai, H.* cf. *H. forstenae, Chilotherium harbereri, Acerorhinus hezhengensis, Sinotherium lagrelii, Palaeotragus* cf. *P. decipiens, Samotherium* sp., *Gazella gaudryi, Miotragocerus* sp., and *Plesiaddax* cf. *P. minor*. In addition, X. Wang. (1997) described a perfectly preserved skull in association with its mandible of *Simocyon primigenius* from the Lamagou local fauna. Most of the above taxa are commonly shared by both the Bahean and Baodean LMS/A. The presence of a good sample of *Dinocrocuta*, and *Platybelodon*, so far the only known specimen of this Tunggurian "holdover" in "Hipparion" faunas, seems to favor an older (Bahean) assignment. However, the possible occurrence of *Sinotherium* and *Plesiaddax* tends to argue to the contrary—that is, to the Baodean LMS/A. A preliminary paleomagnetic study was carried out by Xue, Zhang, and Yue (1995). According to their interpretation (Xue, Zhang, and Yue 1995:fig. 1), the normal reversal of the fossiliferous level was correlated with the lower part of "Epoch 7" —that is, 7–8 Ma. According to the currently used ATNTS2004, this should be somewhere in C4n.2n (7.695–8.108 Ma). If the paleomagnetic data are tenable to some degree, the fossiliferous level of the Lamagou local fauna should be around 8 Ma, therefore belonging to the Bahean LMS/A.

Amuwusu local fauna (central Nei Mongol) This local fauna was found in a locality called Amuwusu to the west of Jurh township. The sediments are fluvial sandstones and overbank mudstones. The fauna is composed of 34 forms, predominantly micromammals. As summarized by Qiu, Wang, and Li (chapter 5, this volume), the lack of typical Tunggurian forms, like *Plesiodipus, Megacricetodon, Alloptox*, and *Bellatona*, but the appearance of advanced forms, like *Castor, Paralactaga, Prosiphneus*, and *Sinozapus*, show clearly that the small mammal local fauna should be included in the Bahean LMS/A.

Shala fauna (central Nei Mongol) Like Amuwusu, this fauna consists of 34 forms. According to Qiu, Wang, and Li (chapter 5, this volume), it is younger than Amu-wusu in having more advanced *Dipus, Kowalskia,* and *Sinocricetus,* as well as the derived genus *Microtoscoptes.*

Huitenghe local fauna (central Nei Mongol) is also roughly contemporary with Shala. It is correlated to the European MN 12, and it is considered Bahean LMS/A by Qiu, Wang, and Li (chapter 5, this volume).

LOWER BOUNDARY DEFINITION

For Neogene terrestrial strata of the Palearctic Province, the lower boundary of the Late Miocene presents a very special situation. This boundary has long been fixed at the first appearance of three-toed horse (*Hippotherium* or *Hipparion*). With the improvement of dating techniques in recently years, this boundary gradually converged to around 11 Ma. Agustí et al.'s (1997) studies on the strato-type of the European Vallesian suggested that the earliest *Hippotherium/Hipparion* occurs at the Creu de Conill-20 locality in Can Guitart 1 section of Montagut region. As-sociated with the horses are *Hispanomys dispectus, Mio-tragocerus pannoniae, Ursavus* sp., and others. The fossilif-erous layer lies within a very short normal magnetozone correlated to C5r.1n (11.118–11.154 Ma) by Agustí et al., who thus concluded that the lower boundary of the Val-lesian was 11.1 Ma. Sen (1997) summarized data from Europe, western Asia, and southern Asia and concluded that, in almost every locality with early occurrences of *Hipparion* horses (Höwenegg, Sinap Tepe, Siwaliks, etc.), they lie within a long normal interval correlated to C5n.2n (9.987–11.04 Ma).

In GTS2004, the Middle Miocene–Late Miocene boundary is placed at 11.61 Ma—that is, at the base of marine Tortonian Stage. The GSSP of this boundary is located in the middle of the 76th sapropelic cycle at the Monte dei Corvi section in Ancona of northern Italy and coincides approximately with the last common occur-rence (LCO) of the calcareous nanofossil *Discoaster ku-gleri* and planktonic foraminifera *Globigerinoides subqua-dratus.* This is placed at the base of C5r.2n magnetochron (11.614 Ma) in ATNTS2004. As such, the marine and terrestrial boundaries differ by ~0.5 myr.

The base of the Bahe Formation in the unit stratotype section is not exposed in most areas along the south bank of Bahe. At the western end of Bailuyuan platform and western slope of Lishan mountains, the Bahe Formation lies unconformably on top of the Koujiacun Formation (Zhang et al. 1978). Only preliminary paleomagnetic re-sults are available for the Bahe Formation (Kaakinen 2005). As summarized by Zhang et al. (chapter 6, this volume), the earliest *Hipparion* is found in the locality

L 50, which is within the magnetochron C5n.2n (9.987–11.04 Ma) and considered 10.21 Ma.

Earlier appearances of Chinese *Hipparion* fossils so far known are from the Qaidam Basin in Qinghai and Linxia Basin in Gansu.

The earliest *Hipparion* record in the Linxia Basin is from the Guonigou locality in Nalesi of Dongxiang County. Based on unpublished magnetostratigraphy (X.-m Fang., pers. comm.), the fossiliferous deposits fall at the base of C5n.2n (11.04 Ma), or within C5r.1r (11.04–11.118 Ma). Associated with this *Hipparion* record are *Dinocro-cuta, Machairodus, Tetralophodon, Parelasmotherium,* and *Shaanxispira* (Deng et al., chapter 9, this volume). *Prodei-notherium sinensis* from the Bantu locality may be from this level as well. Unfortunately, at the Guonigou locality the Liushu Formation disconformably overlies the Hu-jialiang Formation so that some gaps may exist. This ren-ders the Guonigou section unsuitable for being chosen as a candidate for the lower boundary stratotype of the Ba-hean Stage. On the other hand, the Qaidam Basin may hold more promise of yielding the earliest *Hipparion* in China and could be a proper candidate for the RSSP in the future (see previous discussion). At any rate, the FAD of *Hipparion* in China must be around 11 Ma or slightly earlier.

MAMMAL FAUNA CHARACTERIZATION

The Bahean Chronofauna can be viewed as a chrono-fauna transitional from the typical Tunggurian *Platybe-lodon* Chronofauna to the typical Baodean *Hipparion* Chronofauna. The immigration of *Hippotherium* (or *Hipparion*) marks the beginning of the age. A series of advanced forms appeared, such as *Dinocrocuta, Parelas-motherium,* and *Shaanxispira,* having evolved from Early–Middle Miocene taxa, while typical Middle Mio-cene forms like *Hispanotherium* and *Kubanochoerus* van-ished. *Ochotona* and *Ochotonoma* replaced all the old Middle Miocene ochotonids. Myomorph rodents be-came predominant, with appearance of the cricetines, siphneines, and murines.

Index fossils: *Myocricetodon lantianensis, Nannocricetus primitivus, Progonomys, Huerzelimys, Prosiphneus qiui, Di-nocrocuta, Tetralophodon, Diceros, Parelasmotherium, Irano-therium, Shaanxispira, Hezhengia,* and *Miotragocerus.*

First appearances within the stage: *Castor, Lopho-cricetus, Sinozapus, Paralatactaga, Dipus, Brachyscirtetes, Nannocricetus, Kowalskia, Sinocricetus, Microtoscoptes, Abudhabia, Prosiphneus, Pararhizomys, Ochotona, Ochoto-noma, Indarctos, Agriotherium, Sinictis, Parataxidea, Melodon, Eomellivora, Promephitis, Simocyon, Adcrocuta,* ictitheres, *Machairodus, Prodeinotherium, Sinohippus, Hipparion, ?Sinotherium, Tapirus, Chleuastochoerus, Schansitherium, Honanotherium, Cervavitus, Gazella, Plesiaddax,* and *Sinotragus.*

Last appearance within the stage: *Mioechinus, Desmatolagus, Ansomys, Keramidomys, Heterosminthus, Protalactaga, Democricetodon, Myocricetodon, Gobicricetodon, Chalicotherium, Ursavus,* and *Euprox.*

Characteristic fossils: *Castor+Steneofiber, Protalactaga, Paralactaga, Plesiodipus, Prosiphneus, Indarctos, Promephitis,* and chilotheres.

Baodean LMS/A

NAME DERIVATION

While studying the *Hipparion* fauna of Baode area, Zdansky (1923) informally called the fossiliferous deposits "*Hipparionlehm*" or "roter Lehm." Later, Teilhard de Chardin (1929:16) described it as "red earth with the fossil remains of *Hipparion richthofeni* of Pontian age." Since then until the 1950s, it was casually called "*Hipparion richthofeni* red clay," "red *Hipparion* clay," or "red Pontian clay." Years later, Pei, Zhou, and Zheng (1963) started to use the Baode as a stage name in a report submitted to the All-China Stratigraphic Congress. The Baodean Stage is the earliest Neogene named stage in China. Li, Wu, and Qiu (1984) divided the Chinese strata containing *Hipparion* faunas into two stages, the Bahean (lower) and Baodean (upper), and their Baodean corresponded to the European Turolian Stage/Age (MN 11–13). Qiu and Qiu (1990) rejected the Bahean and used a single Baodean (sensu lato) to cover both. In the appendix of the revised edition of the *Stratigraphic Guide of China and Its Explanation* (All-China Stratigraphic Commission 2001), the Baodean was used in this broad sense. Recently (e.g., Qiu, Wang, and Li 2006), it was still being used this way. On the other hand, Zhang et al. (2002) and Deng (2006) resurrected the Bahean Age.

UNIT STRATOTYPE

Although no stratotype section was formally designated, the general intention of the Chinese paleontologists was clear; that is, Zdansky's "*Hipparionlehm*" of the Baode area in Shanxi (centered around Daijiagou and Jijiagou) should be the stratotype. Deng, Wang, and Yue (2008a) systematically studied the Baode *Hipparion* red clay. They proposed to select a best-exposed section among the localities worked by Zdansky in the Jijiagou valley as the lectostratotype. Such a section was fixed at a small gully on the south slope of the Jijiagou valley, situated

south of Jijiagou village (39°00'10.5"N, 110°09'48.5"E). The *Hipparion* red clay was divided into 13 beds with a total thickness of 60 m (including 15 m of basal conglomerates); of these, the lower ~45 m belongs to the Baode Formation and the upper ~15 m belongs to the Jingle Formation. There are two fossiliferous layers: the lower one is layer 9, which is the main bone bed, 15.5–17.5 m above the base of the red clay; the upper one is layer 11, just below the boundary between the Baode and Jingle formations. Paleomagnetic sampling was carried out as well. According to Yue et al. (2004a) and Deng, Wang, and Yue (2008a), layer 9 is correlatable with C3Ar (6.733–7.14 Ma) and C3Bn (7.14–7.212 Ma), while layer 11 falls just below C3n.4n (5.235 Ma) and is near the lower boundary of the Jingle Formation. The base of the red clay is interpreted as representing C4n.2n (7.695–8.108 Ma). Thus, the time span of the Baode Formation in the type section is estimated as 8–5.2 Ma.

Zhu et al. (2008) conducted a more extensive exploration of the *Hipparion* red clay in the Baode area. They chose three sections for paleomagnetic studies: the Tanyugou section (upper 30 m of the Baode Formation, 40.9 m of the Jingle Formation), the Yangjiagou-I section (upper 36 m of the Baode Formation, lower 7 m of the Jingle Formation), and the Yangjiagou-II section (upper 56 m of the Baode Formation). Three mammal fossil bone beds can only be determined in the Yangjiagou-II section: at the 26–28.6 m, 43.2–44.6 m, and 58.5–59.9 m levels (Zhu et al. 2008:fig. 3). Paleomagnetically they are correlated with the top part of C3An.2n, slightly below the middle of C3Ar, and the upper part of C3Br.1r (or the top of C3Br.2r), respectively (Zhu et al. 2008:fig. 6). Their paleomagnetic age scale, translated into ATNTS2004, yields C3An.2n = 6.436–6.733 Ma, C3Ar = 6.733–7.14 Ma, C3Br.1r = 7.212–7.251 Ma, and C3Br.2r = 7.285–7.454 Ma. The estimated age of the upper bone bed is about the same as Zhu et al. (2008) calculated, 6.43–6.54 Ma; the middle bone bed would be slightly older, around 6.937 Ma instead of 6.83–6.86 Ma; the lower bone bed would be 7.2 Ma or 7.3 Ma instead of 7.15–7.18 Ma, respectively. As a result, the age range of the fossiliferous layers of the Yangjiagou-II section is estimated as **6.4 Ma to 7.2 or 7.3 Ma**.

Kaakinen et al. (chapter 7, this volume) carried out several field seasons in the Baode area. They report the discovery of the original location of Zdansky's Loc. 30 in Daijiagou. Based on a stratigraphic survey of the fossiliferous area, they also reverse the correlations of localities 30 and 49 compared with Yue et al. (2004b) and Deng, Wang, and Yue (2008a), whose section was in Jijiagou.

Paleomagnetic sampling carried out through the famous Zdansky Loc. 30. suggests that the time span of the section ranged form 7.2 Ma to 5.5 Ma, with Loc. 30 itself at about 5.7 Ma. The latter age assignment is slightly beyond our expectation. It seems too young if the Loc. 30 fauna is compared with the Ertemte one, which is quite different from the former but has been generally considered the latest stage of the Baodean Chronofauna.

The Jijiagou section described by Deng, Wang, and Yue (2008a) is situated in the most fossiliferous area intensively studied by Zdansky in 1920s, and it has a long and well-exposed section covering the major part of the Baode Formation (55 m thick) superposed by Jingle Formation and Wucheng loess. The section is also feasible for reliable paleomagnetic study. We agree with Deng, Wang, and Yue (2008a) to recommend the Jijiagou section as a candidate unit stratotype for the Baodean LMS.

REFERRED DEPOSITS AND FAUNAS

Woma local fauna (Gyirong Basin, Tibet) This small fauna described by Ji, Xu, and Huang (1980) consists of 10 forms identified at and below generic level (including *Hipparion guizhongense*, *H.* sp., *Chilotherium xizangense*, *Palaeotragus microdon*, *Metacervulus capreolinus*, *Gazella gaudryi*, *Gazella* sp., *Hyaena* [*Adcrocuta?*] sp., and *Ochotona guizhongensis*). The local fauna is not characteristic enough for assignment either to the Bahean or to the Baodean LMS/A. However, a recent paleomagnetic study carried out by Yue et al. (2004a) indicated that it may belong to the Baodean LMS/A. They correlated the total sequence of the Woma Formation to the top of C3Br.2n (7.454 Ma) to C2An.2r (3.207–3.33 Ma), and the fossiliferous level to the base of C3Br.1r (7.251 Ma).

Miaoliang local fauna (Fugu, Shaanxi) As in the case of Lamagou, the Miaoliang local fauna was also created by Xue, Zhang, and Yue (1995). The taxa preliminarily identified at and below the generic level were listed as follows: *Adcrocuta eximia*, *Hipparion platyodus*, *Honanotherium* sp., *Chleuastochoerus stehlini*, *Eostyloceras blainvillei*, *Muntiacus* cf. *M. lacustris*, *Cervavitus novorossiae*, *C. demissus*, and *Procapreolus latifrons*. All five cervid taxa have been found also from the Yushe Group. On the other hand, *Adcrocuta*, *Hipparion platyodus*, and *Chleuastochoerus* are common forms of the Baodean LMS/A. Xue, Zhang, and Yue (1995) provided us with a paleomagnetic age for the Miaoliang local fauna as from ~7 Ma to 5.2 Ma. This seems in general accordance with the Baodean LMS/A.

Baogeda Ula local fauna (central Nei Mongol) This local fauna was briefly summarized with a complete faunal list by Qiu, Wang, and Li, (chapter 5, this volume).

The formation is about 70 m thick and composed mainly of fluviolacustrine deposits. The small fauna is composed of 26 forms (23 identified at and below generic level), mainly of micromammals. It was characterized by the first appearance of a series of "newcomers," in central Nei Mongol, like *Dipoides*, *Hansdebruijina*, and *Alilepus*, and assigned to the Baodean LMS/A. The immigrations of murids (*Hansdebruijina*) and the leporid *Alilepus* were considered important in central Nei Mongol.

Ertemte fauna (central Nei Mongol) This is by far the best-known latest Miocene mammal fauna of China. A summary of this fauna and its stratigraphic position is given by Qiu, Wang, and Li (chapter 5, this volume).

LOWER BOUNDARY DEFINITION

By *ISG* requirements, neither the Jijiagou nor the Yangjiagou section recommended by Deng, Wang, and Yue (2008a) and Zhu et al. (2008) is suitable as a lower boundary stratotype. Neither contains fossils of Bahean age below, and the terrestrial deposits rest unconformably on top of Carboniferous limestones. Direct contact of the Bahe and Lantian (carbonate nodule rich red clay, lithologically equivalent to Baode) formations is present in the Lantian area. According to Kaakinen and Lunkka (2003), the boundary between Bahe Formation and Lantian Formation has an age of 7.3 Ma. From data in Zhang et al. (chapter 6, this volume), the lower boundary of the Lantian Formation is 7.0 Ma in the Liujiaopo section, a few kilometers east of the main section at Shuijiazui. A similar section covering both Bahean and Baodean LMS/A is that of Laogaochuan (Fugu), where the yellow sandy clay yielding the Lamagou (Bahean) local fauna is overlapped by carbonate nodule rich red clay with the Miaoliang (Baodean) local fauna. The contact between them lies slightly lower than the "Epoch 6," which is here roughly interpreted as C3An (6.033–6.733 Ma). That means the lower boundary of the carbonate nodule rich red clay lies at about 6.8 Ma, just as in the Shuijiazui section in the Lantian area.

In the Baode area, the age of the lowest occurrence of the carbonate nodule rich red clay has been estimated differently: 8 Ma (Deng et al. 2004), 7.23 Ma (Zhu et al. 2008), and 7.2 Ma (Kaakinen et al., chapter 7, this volume). It cannot be excluded that the lower boundary of the "*Hipparion* red clay" is slightly diachronous.

On the other hand, based on systematic stratigraphic and paleomagnetic study of four sections in different parts of the Loess Plateau (Lingtai, Bajiazui, Zhaojiachuan, and Duanjiapo), An et al. (2000) came to the conclusion that the "red clay" was of eolian origin, consisting of seven pairs of interbedded paleosols (RS1–7) and red-dish pedogenic loess (RL1–7). Its sedimentation started at about 7.2 Ma in the vast territory of the Loess Plateau of North China, and this is also the onset of the East Asian summer monsoon.

These data tend to show that, if any lithological marker is needed for the lower boundary stratotype of the Bao-dean LMS, the lower limit of the "red clay" may well serve as a candidate. It could not be only accident that the Messinian Stage in the Standard Global Chronostrati-graphic Scale, based on similar climatologic and litho-logic grounds, has its GSSP fixed at 7.24 Ma.

However, from the point of view of mammal paleon-tology, the boundary is by no means so clear-cut. In the Shuijiazui section of the Lantian area, the segment repre-senting the interval from 7.7 Ma to 6.8 Ma is very poorly represented by fossils (Zhang et al., chapter 6, this vol-ume). The last appearance of Bahean forms, *Dinocrocuta gigantea* and *Shaanxispira baheensis*, in the Shuijiazui sec-tion occurred at 8.03 Ma. The upper part of the lower beds yielding the Lamagou local fauna in the Fugu area is practically barren for mammal fossils.

Comparing to the European MN system, the lower boundary of the Baodean LMS is close to that of MN 13. Steininger et al. (1996) placed this boundary at 7.1 Ma. This is mainly constrained by the top of the Samos Main Bone Bed (MN 12), radioisotopically dated as 7.1–7.3 Ma.

MAMMAL FAUNA CHARACTERIZATION

The Baodean Chronofauna can be briefly characterized as the climax in development of the Chinese "*Hipparion*" faunas, with highly diversified and densely populated grazers like hipparionines, gazelles, and bovids and their predators like ictitheres and hyenas. Archaic insectivores were replaced by new genera and species, and the leporids reappeared after a long hiatus in Asia. The myomorphs (zapodines, dipodids, cricetines, and murines) highly di-versified, reaching an unprecedented height in taxonomic variety and abundance. The last survivors of the Eomy-idae and Aplodontidae vanished at this time.

Index fossils: *Paranourosorex, Pseudaplodon, Rhinocer-odon, Hansdebruijnia, Alilepus, Leecyaena, Ancylotherium,* and *Urmiatherium.*

First appearances within the stage: *Erinaceus, Para-soriculus, Paenelimnoecus, Dipoides, Myomimus, Paralopho-cricetus, Pseudomeriones, Apodemus, Micromys, Huaxiamys, Orientalomys, Plesiogulo, Enhydriodon, Anancus, Sinomast-odon, Mammut, Stegodon, Shansirhinus,* and *Dihoplus.*

Last appearance within the stage: *Miodyromys, Mi-crodyromys, Leptodontomys, Democricetodon, Microtoscop-*

tes, Abudhabia, Prosiphneus, Indarctos, Parataxidea, Melo-don, Eomellivora, Adcrocuta, Machairodus, Sinohippus, Cervavitus, Plesiaddax, and *Sinotragus.*

Characteristic fossils: *Paralophocricetus, Sinocrice-todon, Anatolomys, Microtodon, Micromys, Prosiphneus, Ad-crocuta,* ictitheres, *Metailurus,* hipparionines of middle size, chilotheres, *Sinotherium, Plesiaddax,* and *Chleuastochoerus.*

Gaozhuangian LMS/A

NAME DERIVATION

Before 1990, the name Jinglean was widely used for the terrestrial sediments between the classical "*Hipparion* red clay" and the Quaternary. Qiu and Qiu (1990) began to recommend substituting the name Yushean for the Jing-lean. Their main objection was that the Jingle Formation, where the Jingle fauna is produced, is very thin, less than 10 m at the Xiaohong'ao section in Jingle county, and the chance to find more fossils is slight. In contrast, the Yushe Basin possesses a fairly continuous sedimentary sequence covering the time span from the Late Miocene through Early Quaternary. The Pliocene mammal-producing strata are up to 500 m thick and richly fossiliferous in both large and small mammals. At the Third Conference of Chinese Stratigraphy held in 2000, it was formally recommended to separate the Yushean Stage/Age into two: the Gaozhuan-gian and Mazegouan stage/ages (All-China Stratigraphic Commission 2001).

UNIT STRATOTYPE

Recent research on the Gaozhuang Formation and its mammal fossils are in the process of being monographed elsewhere, but several published summaries are available (Flynn, Tedford, and Qiu 1991; Tedford et al. 1991; Flynn et al. 1995; Flynn 1997; Flynn, Wu, and Downs 1997). As the nominee for the Gaozhuangian LMS/A, the Yushe Ba-sin is naturally to be chosen as its unit stratotype. Figure 1.5 is a simplified litho- and magnetostratigraphic chart as a basis for the establishment of the Gaozhuangian, as well as of the Mazegouan LMS/A.

As can be seen in figure 1.6, the correlation between the polarity reversal intervals and the formations of the Yushe Group remain the same as previously suggested. The Mahui Formation roughly corresponds to C3An (former chron 5) and the Gaozhuang Formation (sepa-rated into the Taoyang and Nanzhuanggou members) to C2Ar and C3 (former Gilbert). Adapted to the timing of ATNTS2004, the Mahui Formation (base not exposed) extends from 6.7 Ma to 5.8 Ma, the Taoyang Member from 5.8 Ma to 5.2 Ma, and the Nanzhuanggou Mem-ber from 4.9 Ma to 4.2 Ma with its top truncated. The

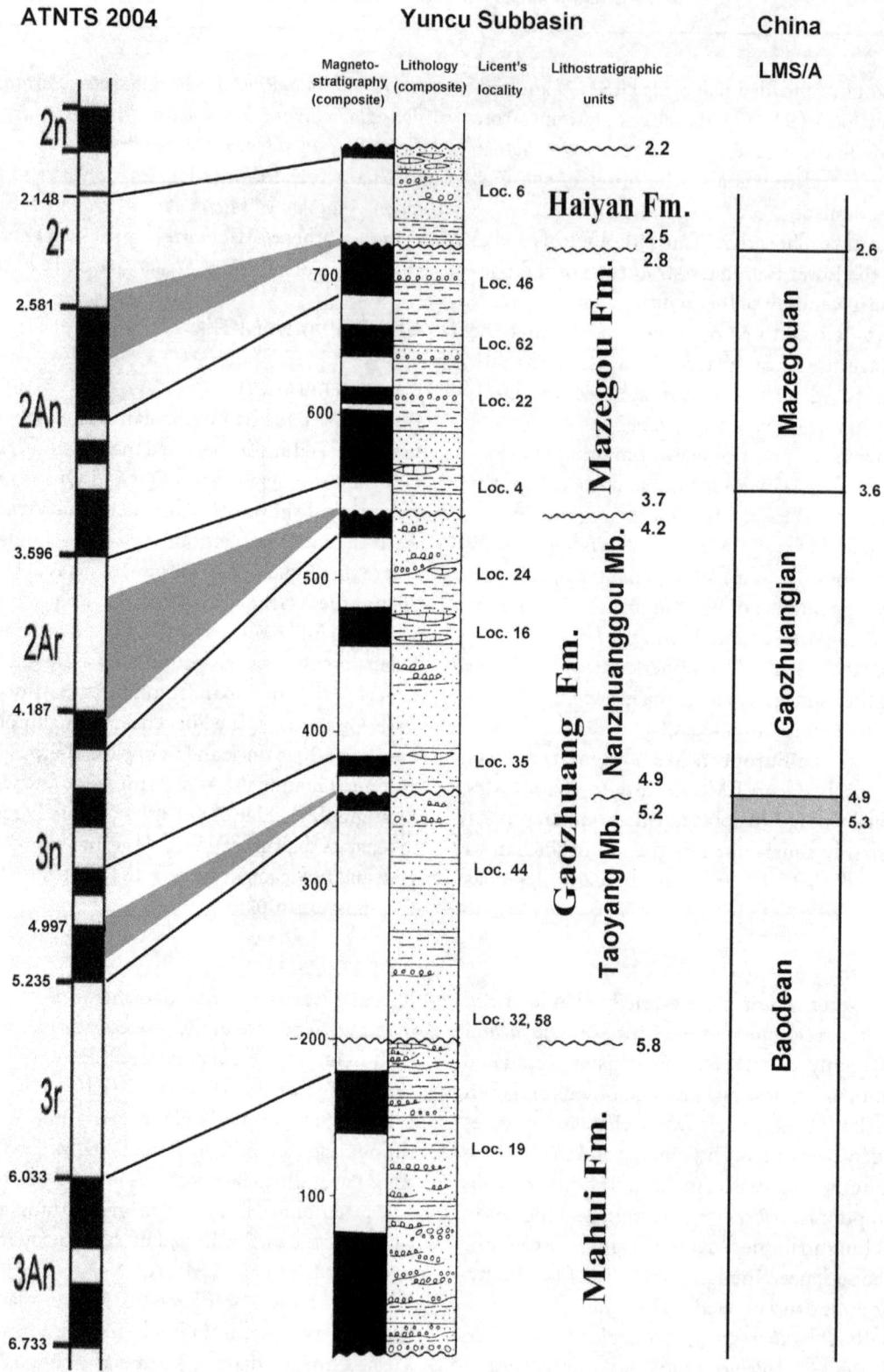

Figure 1.6 Lithological divisions and magnetostratigraphic interpretations of the Yushe Group in the Yuncu Subbasin of the Yushe Basin, with major Licent localities plotted (ages have been corrected based on ATNTS2004). Loc. 4: Zhaozhuangcun; Loc. 6: Haiyan; Loc. 16: Gaozhuang; Loc. 19: Lintou; Loc. 22: Malancun; Loc. 24: Nanzhuanggou; Loc. 32: Dongzhuang; Loc. 35: Wangjiagou; Loc. 44: Taoyang; Loc. 46: Qingyangping; Loc. 58: Niliuhe; Loc. 62: Zhangwagou. Gray shaded triangles represent postulated time spans of sedimentation gaps; gray shaded rectangle represents fuzzy boundary area between LMS/As (see text).

geochronological Mio–Pliocene boundary (5.33 Ma) now passes through the upper part of the Taoyang Member.

As stated by Flynn, Tedford, and Qiu (1991:250), while the whole section of the Yushe Group is generally well fossiliferous throughout, the Taoyang Member of the Gaozhuang Formation is poorly represented by fossils of both large and small mammals.

Licent's Loc. 32, Dongzhuang village, lies in the basal part of the Taoyang Member, and the fossils collected from this locality may really belong to the basal part of the Taoyang Member. Therein, *Anancus sinensis* (Tobien, Chen, and Li 1988:115), *Mammut shanxiensis* and *Sinomastodon intermedius* (G.-f. Chen, in preparation), and *Stegodon zdanskyi* and *Gazella gaudryi* (Teilhard de Chardin and Trassaert 1937, 1938) were described or revised. No records of other large mammals from this locality can be ascertained. According to Flynn, Wu, and Downs (1997), from the basal part of the Taoyang Member only the following micromammal forms were found: *Ochotona lagreli*, *Sinocastor anderssoni*, *Dipoides* sp., *Rhizomys shansius*, *Pseudomeriones abbreviatus*, and *Huaxiamys primitivus*. In addition, in the basal part of the "Gaozhuang Formation" in the Tancun Subbasin *Yanshuella primaeva*, *Alilepus annectens*, *Pliopetaurista rugosa*, *Neocricetodon grangeri*, *Prosiphneus murinus*, *Apodemus orientalis*, *Micromys chalceus*, and *Chardinomys* sp. were added. Among them only *Alilepus annectens*, *Rhizomys shansius*, *Micromys chalceus*, and *Chardinomys* sp. are the "newcomers" in the Yushe area. The total assemblage was considered by Flynn, Wu, and Downs (1997) very close to that of the latest Miocene Ertemte. Qiu, Wang, and Li (chapter 5, this volume) pointed out that the earliest occurrence of both *Chardinomys* and *Huaxiamys* was in the basal Pliocene Bilike local fauna in Nei Mongol, and these genera also occur in Biozone II of Leijiahe (see "Leijiahe Formation").

The upper part of the Taoyang Member is poorly represented by mammal fossils. Licent's Loc. 44, Taoyang, lies in this part. *Sinomastodon intermedius* (G.-f. Chen, in preparation), *Martes zdanskyi* (Teilhard de Chardin and Leroy 1945), *Hyaenictitherium hyaenoides* (Z.-x. Qiu, in preparation), and *Paracervulus bidens* (Teilhard de Chardin and Trassaert 1937) were recorded from this locality. Just recently, Deng et al. (2010) reported some mammal fossils found from the top part of the section near Taoyang village: *Hipparion platyodus*, *Proboscidipparion pater*, and *Gazella gaozhuangensis*.

The Nanzhuanggou Member is very richly fossiliferous. Most of the characteristic large and small mammals are present here. Licent's Loc. 35, Wangjiagou, and Loc. 58, Niliuhe, are located in the basal part of the Nan-

zhuanggou Member, where *Dipoides majori*, *Rhizomys shansius* (Teilhard de Chardin 1942), *Stegodon zdanskyi* (Teilhard de Chardin and Trassaert 1937), *Nyctereutes* sp. (unstudied), *Pliocrocuta pyrenaica orientalis* (Z.-x. Qiu 1987), a broken skull of *Agriotherium intermedium* (Z.-x. Qiu and Tedford, in preparation), and *Muntiacus lacustris* (under study by Wei Dong) were found.

REFERRED DEPOSITS AND FAUNAS

Some recently found Tibetan local faunas Wang et al. (chapter 10, this volume) made a summary account of the discovery of a series of Pliocene mammal localities during their recent explorations. The Zanda Basin is the most important one of their Pliocene mammal study areas. More than one fauna can potentially be recognized from the more than 800-m-thick sequence of fluviolacustrine sediments that span latest Miocene to early Pleistocene. However, with the exception of a new woolly rhino, *Coelodonta thibetana* (Deng et al. 2011), the rest of the Zanda mammalian assemblage is still largely undescribed. As an interim solution, a composite Zanda fauna is tentatively recognized: Soricidae indet., *Nyctereutes* cf. *N. tingi*, *Vulpes* sp., *Panthera* (*Uncia*) sp., *Meles* sp., *Chasmaporthetes* sp., *Hipparion zandaense*, *Coelodonta thibetana*, *Cervavitus* sp., ?*Pseudois* sp., *Antilospira/Spirocerus* sp., *Qurliqnoria* sp., Gomphotheriidae indet., *Aepyosciurus* sp., *Nannocricetus* sp., Cricetidae gen. et sp. nov., *Prosiphneus* cf. *P. eriksoni*, *Mimomys* (*Aratomys*) *bilikeensis*, *Apodemus* sp., *Trischizolagus* cf. *T. mirificus*, *Trischizolagus* cf. *T. dumitrescuae*, and as many as four species of *Ochotona*. Paleomagnetically, these mammals span from C3n.4n (4.997–5.235 Ma) and C2An.1r (3.032–3.116 Ma), thus belonging to the Gaozhuangian and earliest Mazegouan LMS/A.

Yuzhu (Kunlun Pass Basin) and Huitoutala (Qaidam) local faunas contain also some characteirstic Pliocene forms, like *Mimomys*, *Prosiphneus*, *Orientalomys*, and *Chasmaporthetes*. Magnetostratigraphy, as reinterpreted by Wang et al. (chapter 10, this volume), shows that the deposits yielding the Yuzhu aggregate are to be correlated to C2Ar (3.596–4.187 Ma).

Shilidun local fauna (Linxia Basin, Gansu) A slab of "bone-breccia," about 45 m × 1.5 m × 1 m in size, was unearthed from the Shilidun section of Guanghe county. The slab forms the base of "red clay," which is correlated with the Hewangjia Formation of Pliocene age. According to Deng et al. (chapter 9, this volume), this local fauna consists of at least 22 taxa preliminarily identified at and below generic level. Of them, so far only one taxon, *Hystrix gansuensis*, was described (Wang and Qiu 2002a). The majority of these taxa, such as *Parataxidea*, ictitheres,

Adcrocuta, Palaeotragus, and *Cervavitus,* are typical of the Baodean Age, but the co-occurrence of *Chasmaporthetes, Hipparion (Proboscidipparion) pater,* and *Hesperotherium* gave a strong hint that the fauna might be Pliocene, or at least straddle the Miocene–Pliocene boundary.

Bilike local fauna (central Nei Mongol) The small local fauna was collected from a single quarry of some hundred square meters in the middle part of a ~10-m-thick section near the village Bilike. It consists of 50 micromammal forms identified at and below the generic level, plus a few remains of proboscideans, *Hipparion,* and deer (Qiu and Storch 2000). It is characterized by the predominance of arvicolines murines (*Aratomys* occupies 31% of the specimens) and the high diversity and abundance of insectivores. Of these, 25 genera and 21 species are shared with Ertemte. It differs from the Ertemte local fauna by the appearance of *Aratomys, Chardinomys, Huaxiamys, Mimomys (Aratomys),* and *Trischizolagus.* No paleomagnetic sampling has been attempted. Judging by its faunal composition, it is younger than Ertemte but older than the Nanzhuanggou local fauna.

Gaotege local fauna (central Nei Mongol) According to Li, Wang, and Qui (2003), the local fauna was collected from the lower 20 m of variegated mudstone layers of a 60 m section. In addition to some fragmentary proboscidean material, *Hipparion* and *Gazella,* the fauna consists of 46 forms identified at and below generic level. Its micromammal composition is particularly close to that of Bilike in sharing about 50% of the same or similar species. Of the carnivores, *Chasmaporthetes, Nyctereutes,* and ?*Eucyon* are characteristic elements of the Nanzhuanggou local fauna. Magnetostratigraphic study of the fossiliferous part of the section (27.65 m) has been carried out recently (Xu et al. 2007). Xu's party detected two short normal events, interpreted as C3n.1n (Cochiti) and C3n.2n (Nunivak). These two normal chrons bracket all the fossil localities except one (DB03-1), which lies ~2 m above the upper normal chron. According to ATNTS2004, the age of the lower limit of the fossiliferous part of the section should be slightly younger than 4.493Ma, and the upper limit, slightly younger than 4.187 Ma. Therefore, the total time span of the Gaotege local fauna should be roughly 4.48–4.1 Ma, and the most richly fossiliferous level falls at about 4.3 Ma.

Leijiahe Formation (Gansu) The small area east of Leijiahe village in the southeastern border area of Gansu is one of the most classical areas producing rich Pliocene micromammal fossils in North China. Repeated collecting activities using wet-sieving techniques have been carried out since 1970s. Its faunal analyses and stratigraphic

position were summarized by Zheng and Zhang (2001). The micromammals so far found from the Leijiahe Formation amount to about 80 forms, most of which are identified at specific and generic levels. The whole section is subdivided into six biozones. Biozones I and II are correlated to the lower part of the Gaozhuang Formation (Taoyang Member), and Biozones III and IV to the upper part of the Gaozhuang Formation (Nanzhuanggou + Culiugou members). Biozones III and IV, yielding 45 forms, are much more fossiliferous than the Biozones I and II. The appearance of *Pliosiphneus lyratus, Dipus, Paralactaga* cf. *P. anderssoni, Huaxiamys primitivus, Apodemus qiui,* and *Chardinomys yushensis* marks the beginning of the Biozone III.

Magnetostratigraphically, Biozone II was roughly correlated to the interval of C3r (6.033–5.235 Ma), which was only slightly older than that of the Taoyang Member (5.9–5.2 Ma. Biozones III and IV (=III in Loc. 93001 section) were correlated to the interval from the base of C2An.3n (3.596 Ma) to the base of C3n.3n (4.896 Ma)— that is, roughly 4.9–3.6 Ma.

LOWER BOUNDARY DEFINITION

There are several options for the lower boundary definition of the Gaozhuangian Stage. As indicated earlier, from the basal part of the Taoyang Member (~5.9 Ma) *Alilepus annectens, Rhizomys shansius, Micromys chalceus, Chardinomys* sp., *Anancus sinensis, Mammut shanxiensis, Stegodon zdanskyi,* and *Gazella gaudryi* were found. Of them, only *Chardinomys* and *Anancus sinensis* are first-appearing forms in North China. The other listed forms have been known from Ertemte as well. Lithologically, the lower limit of the Gaozhuang Formation falls in the lower part of the chron C3r, roughly at 5.9 Ma. Theoretically, this can be chosen as the lower boundary of the stage. However, the *ISG* strongly recommends avoiding boundaries based on unconformities. The next fossiliferous level is that in the upper part of the Taoyang Member, Loc. 44, Taoyang (~5.2–5.3 Ma). From this level *Sinomastodon intermedius, Martes zdanskyi, Hyaenictitherium hyaenoids* and *Paracervulus bidens* were recorded, but practically no micromammal fossils were found. Of these four forms, *Martes zdanskyi* and *Paracervulus bidens* probably represent the first appearances of these two genera in North China. The next faunal turnover occurred in the interval 4.9–4.6 Ma. *Chardina truncatus* and *Chardinomys yusheensis* appeared at the very base of this turnover interval (4.9 Ma), followed by the appearance of a number of other taxa, such as *Neocricetodon* cf. *N. grangeri, Allocricetus* sp., *Cricetinus* sp., *Germanomys* sp., *Apode-*

mus qiui, *Huaxiamys downsi*, and *Micromys tedfordi*. None of these forms have been found in the Ertemte fauna. It should be noted that the lower limit of the Biozone III of the Leijiahe area has the same dating. This seems a strong support for choosing 4.9 Ma as the lower boundary of the Gaozhuangian Stage. If the principle "prior consideration of boundaries of higher ranking units over that of stage" (see previous discussions) is acknowledged, the Miocene–Pliocene lower boundary GSSP (5.33 Ma) can also be adopted. In this case, the Miocene–Pliocene boundary would fall somewhere in the upper part of the Taoyang Member. Of the three options, the latter two are better than the first one, since fossils found from the lower part of the Taoyang Member resemble those of Ertemte, which is generally considered as latest Miocene in age. We would temporarily draw a fuzzy interval from 5.33 Ma through 4.9 Ma to represent the lower boundary of the Gaozhuangian Stage.

Taking the richness of fossils, the lithologic continuity, and the other requirements proposed by the *ISG*, any of the above three areas—Yushe, Leijiahe, and Shilidun—may serve as a candidate of RSSP pending further investigation.

MAMMAL FAUNA CHARACTERIZATION

The Gaozhuangian Chronofauna can be briefly characterized as the first renewed fauna posterior to the Miocene "red clay *Hipparion*" fauna, with the extinction of some genera of Late Miocene origin but the appearance of a series of new genera and advanced species in major mammal groups, especially siphneines, murids, arvicolids, hyaenids, hipparionines, and bovids. All taxa are members of living mammal families, but only a small proportion of genera persist to the present day. The immigration events of the canids and camelids from North America into the China mainland mark the beginning of the stage/age.

Index fossils: *Sulimskia, Chardinomys yusheensis, Pliosiphneus lyratus,* and *Huabeitragus yusheensis.*

First appearances within the stage: *Desmana, Lunanosorex, Chardinomys, Allorattus, Chardina, Mesosiphneus, Mimomys (Aratomys), Trischizolagus, Nyctereutes, Eucyon, Ursus, Pliocrocuta, Chasmaporthetes, Hipparion (Cremohipparion), Hipparion (Plesiohipparion), Proboscidipparion, Coelodonta, Sus, Paracamelus, Muntiacus, Paracervulus, Axis, Antilospira,* and *Lyrocerus.*

Last appearance within the stage: *Quyania, Parasoriculus, Paenelimnoecus, Nannocricetus, Sinocricetus, Myomimus, Lophocricetus, Paralophocricetus, Sinozapus, Pararhizomys, Plesiogulo, Simocyon,* ictitheres, *Hip-*

parion (*Hipparion*), chilotheres, *Chleuastochoerus,* and *Dorcadoryx.*

Characteristic fossils: *Dipoides, Dipus, Pseudomeriones, Chardinomys, Huaxiamys, Mimomys (Aratomys), Mesosiphneus, Trischizolagus, Agriotherium, Eucyon, Nyctereutes, Anancus, Sinomastodon, Stegodon, Shansirhinus, Hipparion (Plesiohipparion), Hipparion (Cremohipparion), Proboscidipparion pater, Paracamelus,* and *Lyrocerus.*

Mazegouan LMS/A

NAME DERIVATION

The Mazegouan LMS/A is based on the Mazegou Formation of the Yushe Basin. Together with the Gaozhuangian Stage, it was proposed and accepted at the Third Conference of Chinese Stratigraphy held in 2000 (see previous discussion).

UNIT STRATOTYPE

The Mazegou Formation of the Yuncu Subbasin is chosen as the unit stratotype of the Mazegouan LMS/A. The formation is composed of violet mudstone interbedded with yellow conglomerates and cross-bedded sandstone layers, with its total thickness of ~200 m. It is over- and underlain by the Haiyan and Gaozhaung formations, respectively. There are minor depositional hiatuses between these three formations. Based on magnetostratigraphic study (see figure 1.6), the hiatus between Mazegou and Gaozhuang formations may occupy the interval of the upper part of C3n.1n through the upper part of C2r, lasting approximately 0.5 myr (4.2–3.7 Ma). The hiatus between Mazegou and Haiyan formations may occupy the interval from the lower part of C2An.1n to the lower part of C2r.2r, lasting approximately 0.5 myr (2.9-2.4 Ma). This results in the age for the Mazegou Formation being 3.7–2.9 Ma.

The Mazegou Formation is particularly fossiliferous for large mammals. Many of Licent's and Frick's famous localities like Zhangwagou, Baihai, Zhaozhuang, and so on are located in the areas where only Mazegou Formation is exposed. Good samples of micromammals have been gathered from this area as well. Preliminary faunal list is given in faunal characterization.

REFERRED DEPOSITS AND FAUNAS

Leijiahe Formation (Gansu) As far as the micromammals are concerned, Biozone V closely corresponds to the Mazegou fauna. It contains 48 forms identified at and below generic level. A large number of rodents and lagomorphs are commonly shared with those from the Mazegou Formation. They are *Ochotona*

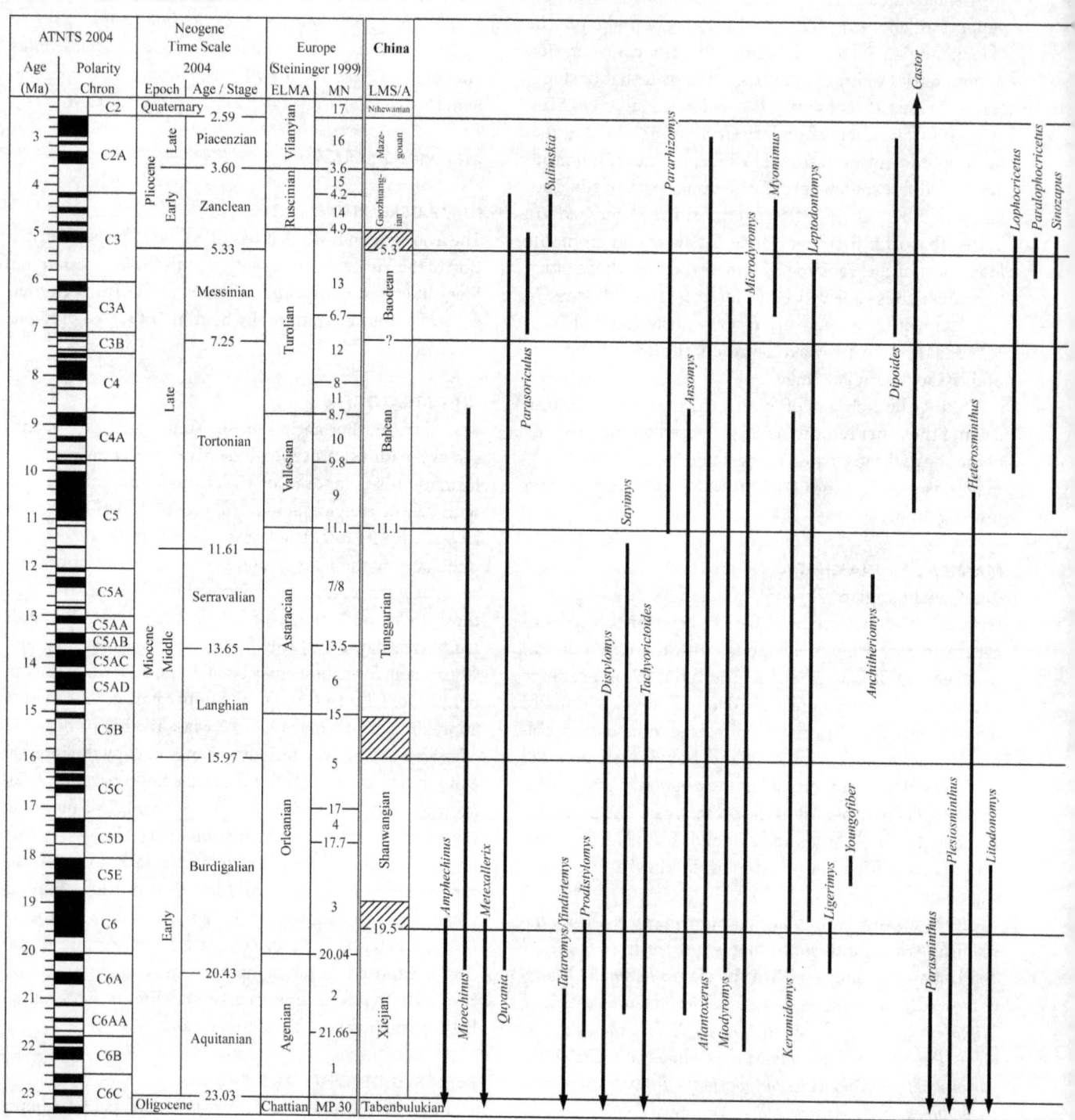

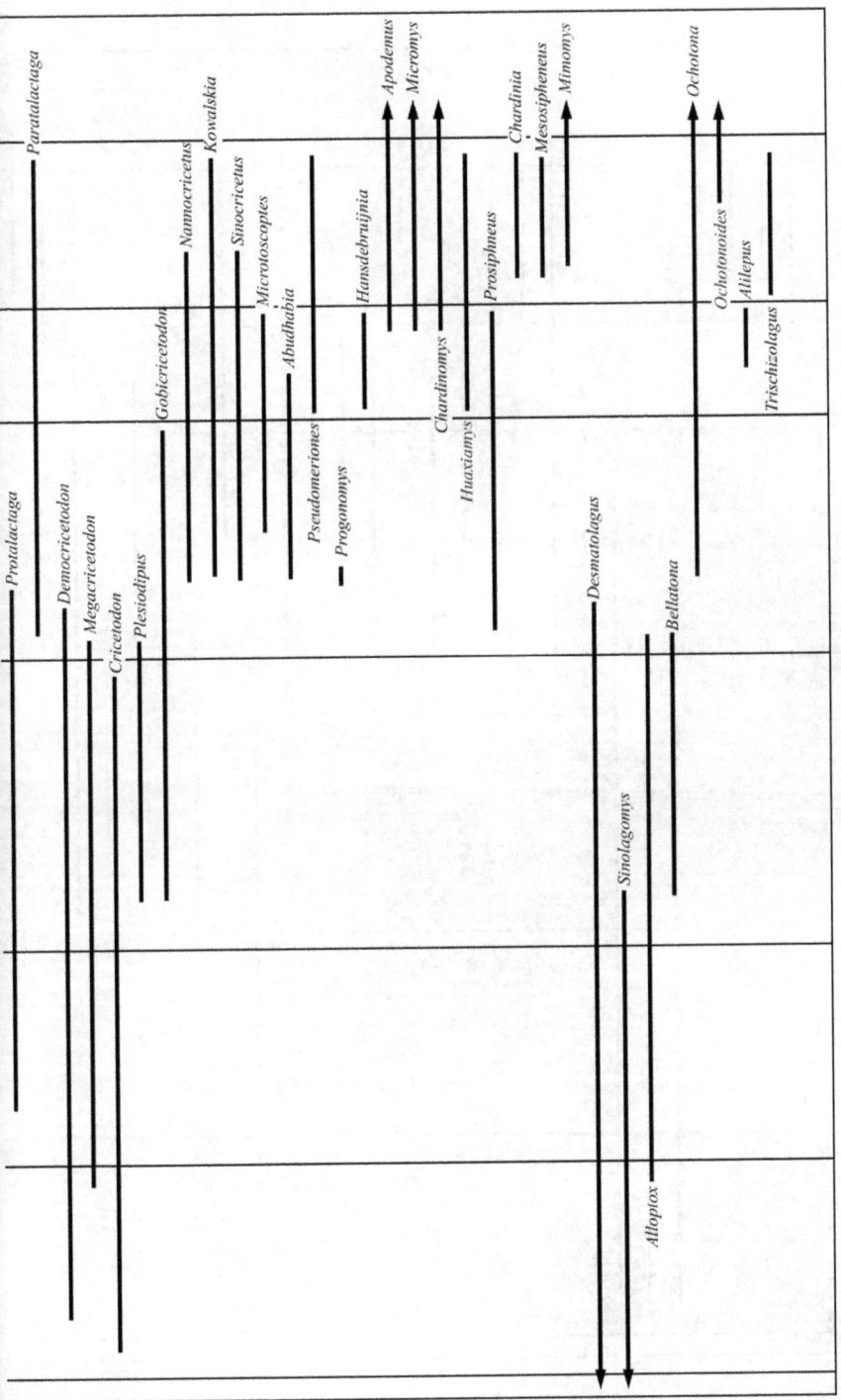

Figure 1.7a Historic ranges of major Chinese Neogene mammalian genera (Insectivora and Rodentia). *Amphechinus*: Olig., Gashunyinadege; *Mioechinus*: Lower Aoerban, Shala; *Metexallerix*: Olig., Gashunyinadege; *Quyania*: Lower Aoerban, Gaotege; *Parasoriculus*: Baogeda Ula, Bilike; *Sulimskia*: Bilike, Gaotege; *Tataromys/Yindirtemys*: Olig., Zhangjiaping; *Prodistylomys*: Suosuoquan-III, Gashunyinadege; *Distylomys*: Olig., Tairum Nor; *Sayimys*: Zhangjiaping, Basal Dingshanyanchi; *Tachyoryctoides*: Olig., Tairum Nor; *Pararhizomys*: Guonigou, Gaotege; *Ansomys*: Zhangjiaping, Shala; *Atlantoxerus*: Lower Aoerban, Leijiahe-V; *Miodyromys*: Lower Aoerban, Bilutu; *Microdyromys*: Suosuoquan-II, Bilutu; *Myomimus*: Bilutu, Nanzhuanggou; *Keramidomys*: Lower Aoerban, Huitenghe (Shala); *Leptodontomys*: Gashunyinadege, Ertemte; *Ligerimys*: Lower Aoerban, Gashunyinadege; *Youngofiber*: Sihong; *Anchitheriomys*: Moergen; *Dipoides*: Baogeda Ula, Mazegou; *Castor*: Amuwusu, Quat.; *Parasminthus*: Olig., Zhangjiaping; *Plesiosminthus*: Olig., Sigou; *Heterosminthus*: Olig., Amuwusu; *Litodonomys*: Olig., Sigou; *Lophocricetus*: Bahe (10 Ma), Bilike; *Paralophocricetus*: Ertemte, Bilike; *Sinozapus*: Amuwusu, Bilike; *Protalactaga*: Late Zhangjiaping (GL 9303), Bahe (10.1 Ma); *Paratalactaga*: Amuwusu, Leijiahe-V; *Democricetodon*: Suosuoquan-II, Amuwusu; *Megacricetodon*: Gashunyinadege, Balunhalagen; *Cricetodon*: Xiejia, Basal Dingshanyanchi; *Plesiodipus*: Tairum Nor, Balunhalagen; *Gobicricetodon*: Tairum Nor, Huitenghe; *Nannocricetus*: Bahe (10 Ma), Leijiahe-III; *Kowalskia*: Bahe (9.8 Ma), Leijiahe-V; *Sinocricetus*: Bahe (9.9 Ma), Leijiahe-III; *Microtoscoptes*: Shala, Ertemte; *Abudhabia*: Bahe (9.9 Ma), Baogeda Ula; *Pseudomeriones*: Baogeda Ula, Leijiahe-V; *Progonomys*: Bahe (10-9.7 Ma); *Hansdebruijnia*: Baogeda Ula, Ertemte; *Apodemus*: Ertemte, Quat.; *Micromys*: Ertemte, Quat.; *Chardinomys*: Taoyang, Quat.; *Huaxiamys*: Mahui, Leijiahe-V; *Prosiphneus*: Amuwusu, Ertemte; *Chardina*: Nanzhuanggou, Leijiahe-V; *Mesosiphneus*: Nanzhuanggou, Mazegou; *Mimomys*: Zanda, Quat.

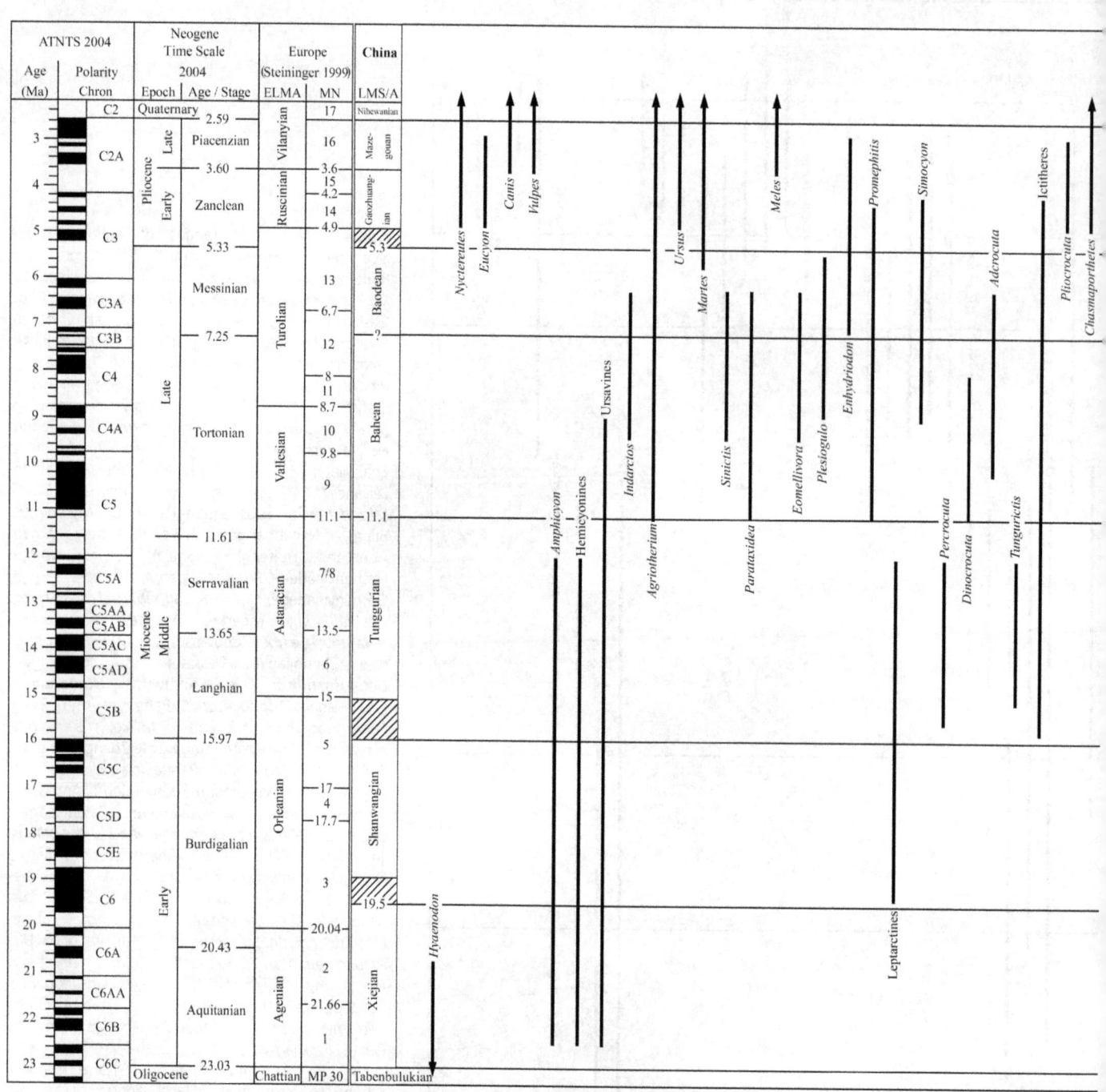

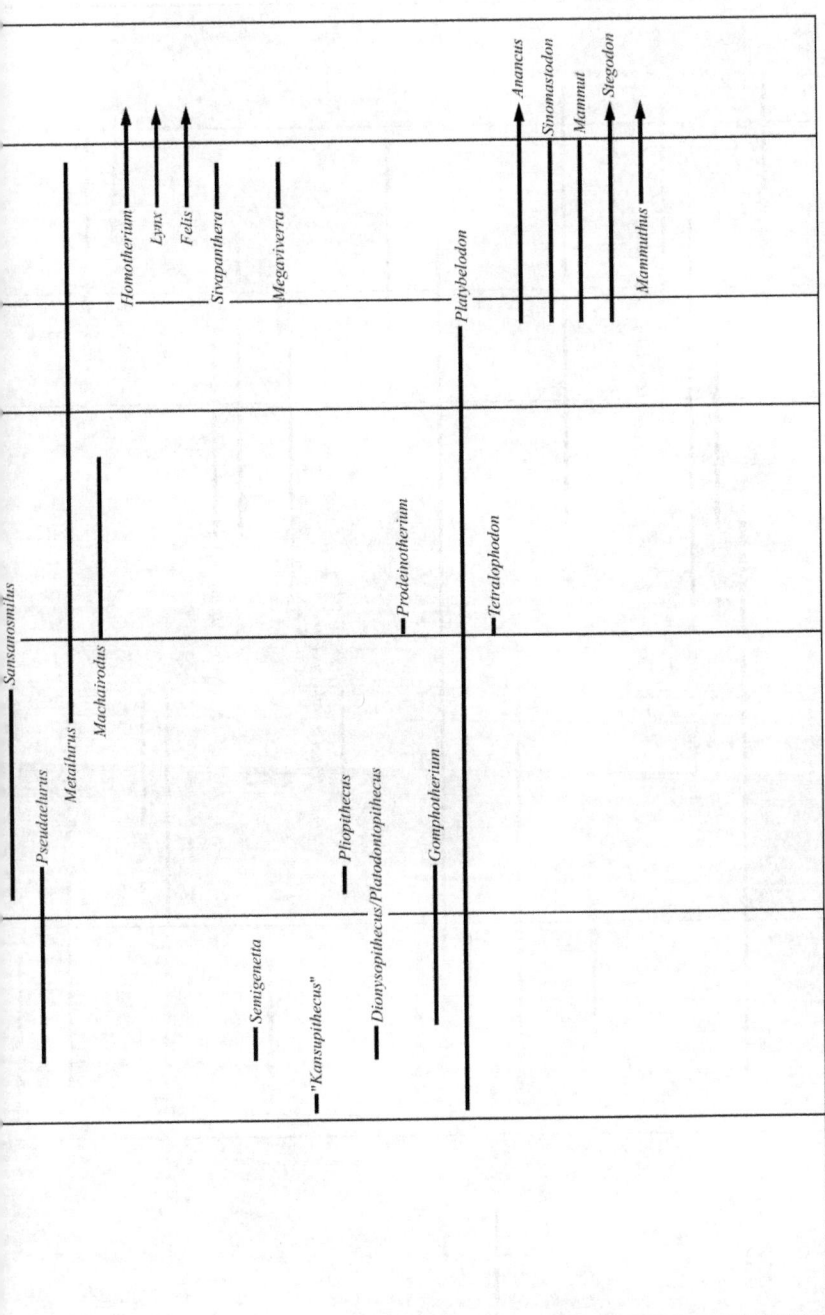

Figure 1.7b Historic ranges of major Chinese Neogene mammalian genera (Lagomorpha, Carnivora, Primates, and Proboscidea). *Desmatolagus*: Olig., Amuwusu; *Sinolagomys*: Olig., Halamagai; *Alloptox*: Gashunyinadege, Balunhalagen; *Bellatona*: Tairum Nor, Balunhalagen; *Ochotona*: Bahe (9.9 Ma), Quat.; *Ochotonoides*: Mazegou, Quat.; *Alilepus*: Baogeda Ula, Ertemte; *Trischizolagus*: Bilike, Leijiahe-V; *Hyaenodon*: Olig., Zhangjiaping; *Nyctereutes*: Nanzhuanggou, Quat.; *Eucyon*: Nanzhuanggou, Mazegou; *Canis*: Mazegou, Quat.; *Vulpes*: Mazegou, Quat.; *Amphicyon*: Xiejiahe, Moergen; Hemicyonines: Xiejiahe, Moergen; Ursavines: Xiejiahe, Liushu (mid); *Indarctos*: Liushu (mid), Mahui; *Agriotherium*: ?Liushu, Quat.; *Ursus*: Nanzhuanggou, Quat.; *Martes*: Taoyang, Quat.; *Sinictis*: Liushu (mid), Baode; *Parataxidea*: Liushu (lower), Baode; *Meles*: Mazegou, Quat.; *Eomellivora*: Liushu (mid), Baode; *Plesiogulo*: Lamagou, Taoyang; *Enhydriodon*: Baode, Mazegou; *Promephitis*: Liushu (lower), Huainan; Leptarctines: Xishuigou, Moergen; *Simocyon*: Lamagou, ?Nanzhuanggou; Percrocutas: Dingjiaergou, Moergen; *Dinocrocuta*: Liushu, Lamagou; *Adcrocuta*: ?Bahe (9.9 Ma), Mahui; *Tungurictis*: Tairum Nor, Moergen; Ictitheres: Halamagai, Nanzhuanggou; *Pliocrocuta*: Nanzhuanggou, Mazegou; *Chasmaporthetes*: Nanzhuanggou, Quat.; *Sansanosmilus*: Dingjiaergou, Moergen; *Pseudaelurus*: Sihong, Halamagai; *Metailurus*: Moergen, Mazegou; *Machairodus*: Guonigou, Baode; *Homotherium*: Mazegou, Quat.; *Lynx*: Mazegou, Quat.; *Felis*: Mazegou, Quat.; *Sivapanthera*: Mazegou; *Semigenetta*: Sihong; *Megaviverra*: Mazegou; "*Kansupithecus*": Xishuigou; *Pliopithecus*: Dingjiaergou, Halamagai; *Dionysopithecus/Platodontopithecus*: Sihong; *Prodeinotherium*: Guonigou; *Gomphotherium*: Xiejiahe, Halamagai; *Platybelodon*: Xishuigou, Miaoliang; *Tetralophodon*: Guonigou - ?; *Anancus*: Dongzhuang, Quat.; *Sinomastodon*: Dongzhuang, Mazegou; *Mammut*: Dongzhuang, Mazegou; *Stegodon*: Dongzhuang, Quat.; *Mammuthus*: Mazegou, Quat.

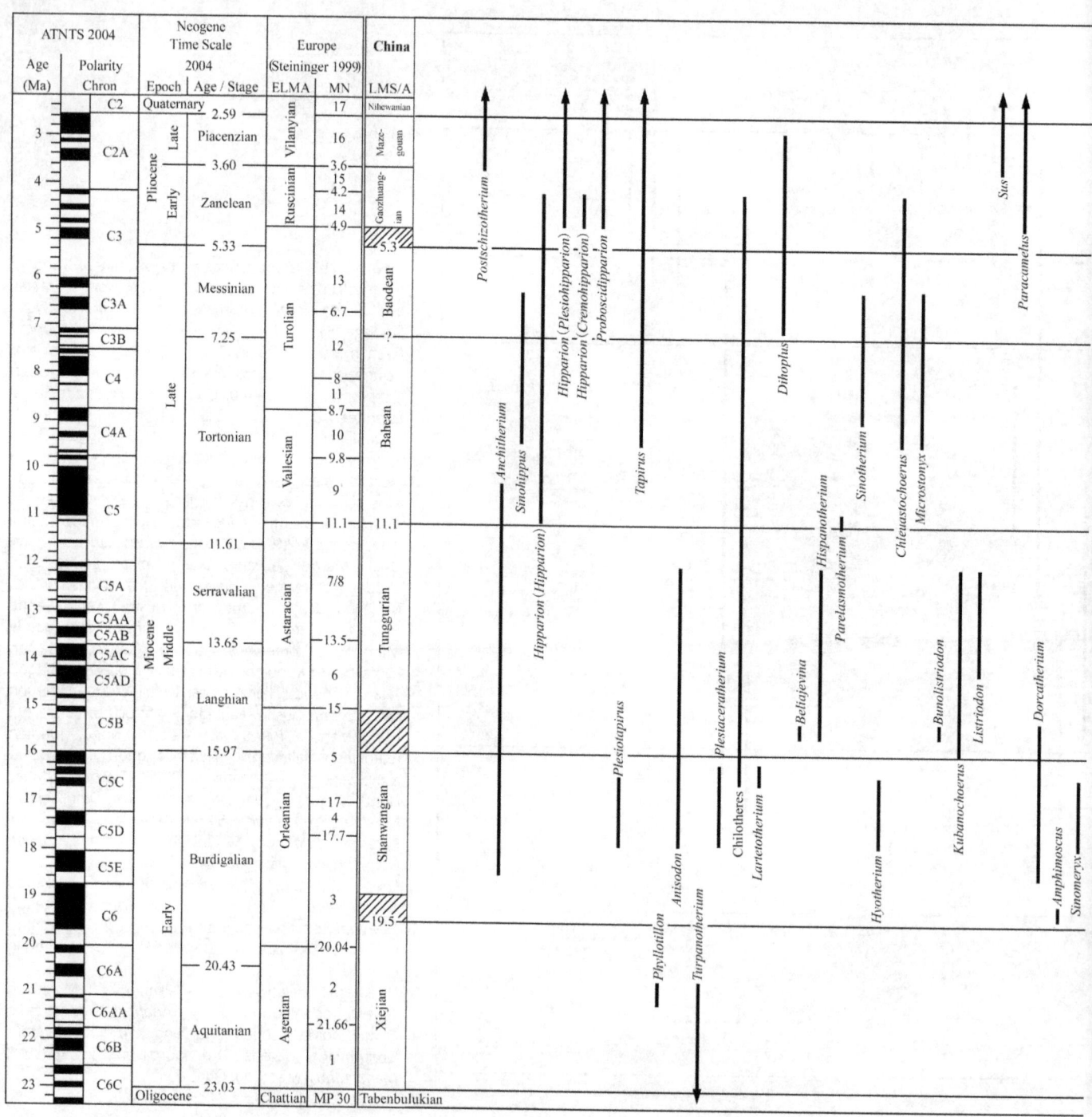

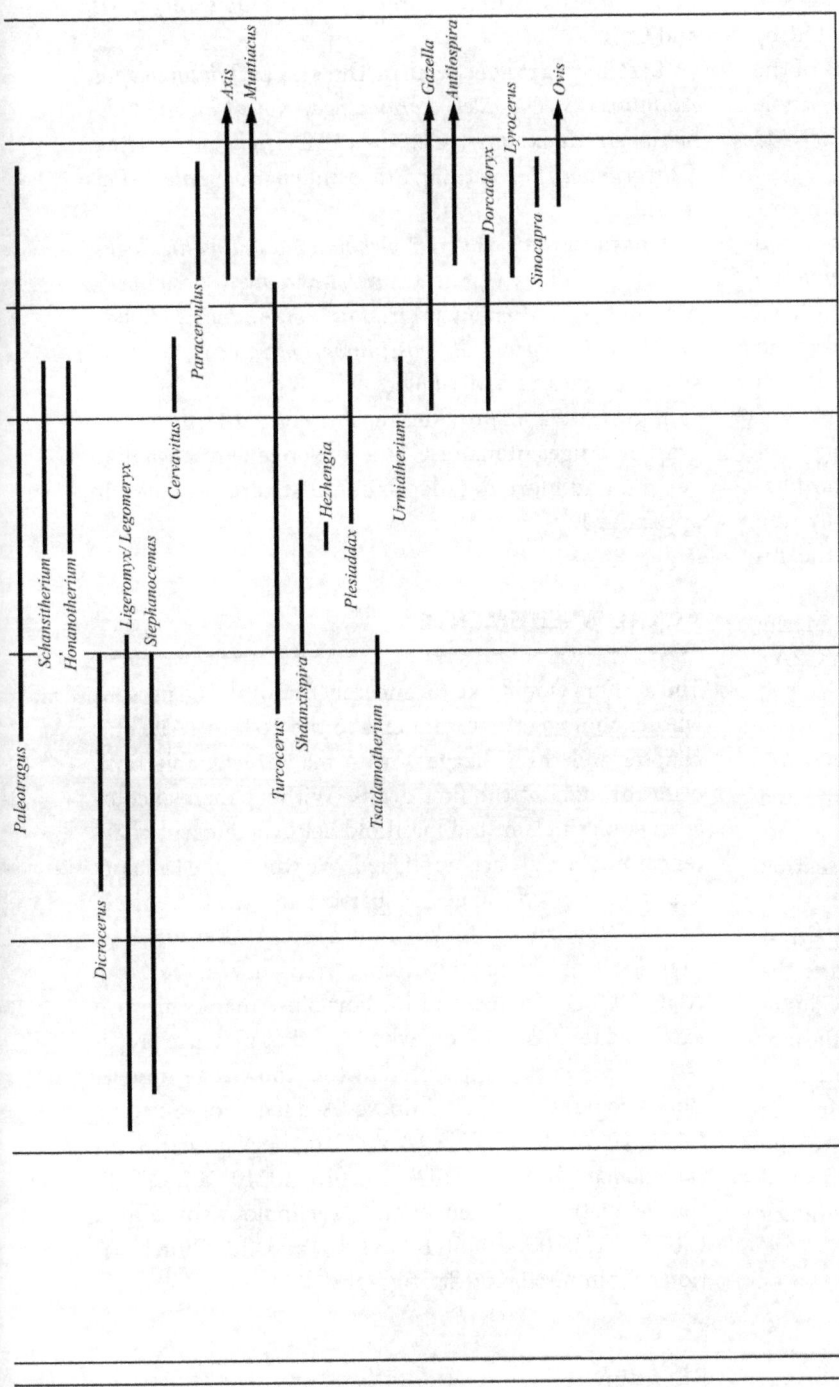

Figure 1.7c Historic ranges of major Chinese Neogene mammalian genera (Hyracoidea, Perissodactyla and Artiodactyla). *Postschizotherium*: Mazegou, Quat.; *Anchitherium*: Sihong, Amuwusu; *Sinohippus*: Liushu (mid?), Baode; *Hipparion* (*Hipparion*): Guonigou, Nanzhuanggou; *Hipparion* (*Plesiohipparion*): Nanzhuanggou, Quat. *Hipparion* (*Cremohipparion*): Nanzhuanggou; *Proboscidipparion*: Nanzhuanggou, Quat.; *Plesiotapirus*: Xiejiahe; *Tapirus*: Liushu (mid?), Quat.; *Phyllotillon*: Zhangjiaping; *Anisodon*: Xiejiehe, Moergen; *Turpanotherium*: Olig., Zhangjiaping; *Plesiaceratherium*: Xiejiehe, Cixian; Chilotheres: Cixian, Nanzhuanggou; *Lartetotherium*: Cixian; *Dihoplus*: Baode, Mazegou; *Beliajevina*: Dingjiaergou; *Hispanotherium*: Dingjiaergou, Moergen; *Parelasmotherium*: Guonigou; *Sinotherium*: Lamagou, Baode; *Hyotherium*: Xiejiahe; *Chleuastochoerus*: Liushu (mid), Nanzhuanggou; *Microstonyx*: Liushu (mid), Baode; *Bunolistriodon*: Dingjiaergou; *Kubanochoerus*: "Hsienshuiho" (Minhe), Moergen; *Listriodon*: Lengshuigou, Moergen; *Sus*: Mazegou, Quat.; *Paracamelus*: Nanzhuanggou, Quat.; *Dorcatherium*: Sihong, Shinanu; *Amphimoschus*: Xishuigou; *Sinomeryx*: Xiejiahe; *Palaeotragus*: Moergen, ?Mazegou; *Schansitherium*: Liushu (mid), Baode; *Honanotherium*: Liushu (mid), Mahui; *Dicrocerus*: Tairum Nor, Tuosu (lower); *Ligeromeryx/Lagomeryx*: Upper Aoerban, Balunhalagen; *Stephanocemas*: Sihong - Tuosu (lower); *Cervavitus*: Mahui, Miaoliang; *Paracervulus*: Nanzhuanggou, Mazegou; *Axis*: Nanzhuanggou, Quat.; *Muntiacus*: Nanzhuanggou, Quat.; *Turcocerus*: Xishuigou, Kekemaideng; *Shaanxispira*: Guonigou, Bahe (8.18 Ma); *Hezhengia*: Dashenggou, Liushu (mid); *Plesiaddax*: Lamagou, Baode; *Tsaidamotherium*: Tuosu (lower), Liushu (lower?); *Urmiatherium*: Baode; *Gazella*: Mahui, Quat.; *Antilospira*: Zanda, Quat.; *Dorcadoryx*: Mahui, Nanzhuanggou; *Lyrocerus*: Nanzhuanggou, Mazegou; *Sinocapra*: Mazegou; *Ovis*: Mazegou, Quat.

lagreli, Ochotonoides complicidens, Dipus fraudator, Cricetinus, Allocricetus bursae, A. ehiki, Cricetulus, Mesosiphneus paratingi, Cromeromys, Chardinomys louisi, and *Micromys tedfordi.* Of these taxa, *Ochotonoides* and *Mesosiphneus paratingi* are the most important indicators of the Mazegouan Age. The time interval represented by the Leijiahe Biozone V is slightly longer than that of the Mazegou Formation, estimated as covering the whole time span of the Gauss Chron—that is, 3.6–2.59 Ma (Zheng and Zhang 2001).

Daodi local fauna (Nihewan, Hebei) The sediments yielding this local fauna are fluviatile sandstones and mudstones, ~30 m, sandwiched between the "*Hipparion* red clay" and the classical "Nihewan beds." The local fauna contains 16 taxa of micromammals identified at and below generic level. The forms commonly shared with those of Mazegouan Age are *Ochotona lagreli, Hypolagus schreuderi, Dipus fraudator, Mesosiphneus paratingi, Allocricetus, Germanomys, Cromeromys irtyshensis, Chardinomys louisi,* and *Micromys.* Magnetostratigraphically, the fossiliferous layers were thought to correspond to the upper half of the Gauss Chron—that is, around 3 Ma.

Candidate RSSP The Mazegou Formation in the Yuncu Subbasin extends downward into the chron C2Ar, at ~3.7 Ma, or even lower. However, the basal part of the Mazegou Formation is poorly exposed in the Yuncu Subbasin. The lowest fossiliferous layers fall in chron C2An.3n—that is, no earlier than 3.6 Ma. This is in good agreement with the data obtained from the Leijiahe section. We thus recommend choosing the Leijiahe section as a candidate RSSP for the Mazegouan LMS/A.

The International Commission on Stratigraphy (Mascarelli 2009) has recently elected to formally define the base of the Quaternary at the C2 and C2A (Gauss–Matuyama) boundary (2.59 Ma). The Mazegou Formation ranges from 3.7 Ma to 2.9 Ma, with its top truncated by the overlying Haiyan Formation, the lowest limit of which is about 2.4 Ma (see figure 1.6). In this case, the Mazegouan LMS/A would be the last Neogene terrestrial stage/age. However, the Neogene–Quaternary boundary RSSP should be sought elsewhere.

MAMMAL FAUNA CHARACTERIZATION

The Mazegouan Chronofauna is the last Neogene land mammal fauna prior to the *Equus* immigration. It is composed of a large number of survivors of the Gaozhuangian Chronofauna and direct ancestors of living forms.

Index fossils: *Mesosiphneus paratingi, Youngia omegodon, Cromeromys irtyshensis, Megaviverra,* and *Sinocapra.*

First appearances within the stage: *Ochotonoides, Hypolagus schreuderi, Allocricetus, Germanomys, Vulpes, Canis, Meles, Pliocrocuta perrieri, Chasmaporthetes ossifragus progressus, Homotherium, Lynx, Felis, Sivapanthera, Mammuthus, Postschizotherium, Sus, Pseudois, Megalovis,* and *Ovis.*

Last appearances within the stage: *Trischizolagus, Atlantoxerus, Dipoides, Paralactaga, Kowalskia, Pseudomeriones, Huaxiamys, Chardina, Mesosiphneus, Eucyon, Enhydriodon, Pliocrocuta, Proboscidipparion pater, ?Palaeotragus,* and *Lyrocerus.*

Characteristic fossils: *Ochotonoides, Trischizolagus, Dipus, Chardinomys, Huaxiamys, Micromys, Pliosiphneus, Mesosiphneus, Mimomys, Agriotherium, Anancus, Hipparion (Plesiohipparion), Proboscidipparion pater, Sus, Antilospira, Pseudois,* and *Megalovis.*

Figure 1.7a–c provides a summary of the stratigraphic ranges of major Chinese Neogene mammalian genera, and more detailed faunal lists are provided in appendix 1.

ACKNOWLEDGMENTS

The authors would like to sincerely thank all the persons who encouraged us to undertake the task to write this chapter and the colleagues who made invaluable criticism throughout the first drafts. Without their encouragement, criticism, and manifold help, the present chapter cannot be duly accomplished. We would like to thank the following colleagues in particular: R. H. Tedford, M. O. Woodburne, E. H. Lindsay, R. L. Bernor, L. J. Flynn, H. de Bruijn, Jin Meng, Wen-yu Wu, Jie Ye, J. Agustí, G. D. Koufos, and M. Fortelius. Thanks are also extended to Su-kuan Hou, who helped us in figure preparation. This work is supported by the Knowledge Innovation Program of the Chinese Academy of Sciences (KZCX-YW-Q09, KZCX2-YW-120), the National Natural Science Foundation of China (40730210, 40232023), the Ministry of Science and Technology of China (2006FY120300, 2006CB806400), and the China National Commission on Stratigraphy.

REFERENCES

Agustí, J. 1999. A critical reevaluation of the Miocene mammal units in Western Europe: Dispersal events and problems of correlation. In *The Evolution of Neogene Terrestrial Ecosystems in Europe*, ed. J. Agustí, L. Rook, and P. Andrews, pp. 84–112. Cambridge: Cambridge University Press.

Agustí, J., L. Cabrera, M. Garcés, W. Krijgsman, O. Oms, and J. M. Parés. 2001. A calibrated mammal scale for the Neogene of western Europe: State of the art. *Earth-Science Reviews* 52:247–260.

Agustí, J., L. Cabrera, M. Garcés, and J. M. Parés. 1997. The Vallesian mammal succession in the Vallès-Penedès basin (northeast Spain): Paleomagnetic calibration and correlation with global events. *Palaeogeography, Palaeoclimatology, Palaeoecology* 133:149–180.

Agustí, J. and S. Moyà-Solà. 1991. Spanish Neogene mammal succession and its bearing on continental biochronology. *Newsletters on Stratigraphy* 25:91–114.

Albright III, L. B., T. J. Fremd, M. O. Woodburne, C. C. Swisher III, B. J. MacFadden, G. R. Scott. 2008. Revised chronostratigraphy and biostratigraphy of the John Day Formation (Turtle Cove and Kimberly members), Oregon, with implications for updated calibration of the Arikareean North American Land Mammal Age. *Journal of Geology* 116:213–237.

All-China Stratigraphic Commission. 1981. *Stratigraphic Guide of China and Its Explanation*. Beijing: Science Press.

All-China Stratigraphic Commission. 2001. *Stratigraphic Guide of China and Its Explanation*. Rev. ed. Beijing: Geological Publishing House.

An, Z.-s., D.-h. Sun, M.-y. Chen, Y.-b. Sun, L. Li, and B.-q. Chen. 2000. Red Clay sequences in Chinese loess plateau and recorded paleoclimate events of the late Tertiary. *Quaternary Sciences* 20:435–446.

Antoine, P.-O., J.-L. Welcomme, L. Marivaux, I. Baloch, M. Benammi, and P. Tassy. 2009. First record of Paleogene Elephantoidea (Mammalia, Proboscidea) from the Bugti Hills of Pakistan. *Journal of Vertebrate Paleontology* 23:977–980.

Azzaroli, A. 1977. Mammal units versus biozones. *Trabajos sobre Neogeno-Cuaternario* 7:25–27.

Berggren, W. A., D. V. Kent, and J. A. van Couvering. 1985. Neogene geochronology and chronostratigraphy. In *The Chronology of the Geological Record*, ed. N. J. Snelling, 211–260. Geological Society of London Memoir 10. London: Geological Society of London.

Bi, S.-d., W.-Y. Wu, J. Ye, and J. Meng. 1999. Erinaceidae from the Middle Miocene of northern Junggar Basin, Xinjiang Uygur Autonomous Region, China. In *Proceedings of the 7th Annual Meeting o fthe Chinese Society of Vertebrate Paleontology*, ed. Y.-q Wang. and T. Deng, 157–165. Beijing: China Ocean Press.

Bi, Z.-g., Z.-j. Yu, and Z.-x. Qiu. 1977. First discovery of mammal remains from upper Tertiary deposits near Nanking. *Vertebrata PalAsiatica* 15:126–138.

Bohlin, B. 1937. Eine Tertiäre säugetier-fauna aus Tsaidam. *Sino-Swedish Expedition Publication (Palaeontologia Sinica, ser. C 14)* 1:3–111.

Bohlin, B. 1946. The fossil mammals from the Tertiary deposit of Taben-buluk, Western Kansu, Part II: Simplicidentata, Carnivora, Artiodactyla, Perissodactyla, and Primates. *Sino-Swedish Expedition Publication (Palaeontologia Sinica, n.s. C 8B)* 28:1–259.

Cao, Z.-x., H.-j. Du, Q.-q. Zhao, and J. Cheng. 1990. Discovery of the middle Miocene fossil mammals in Guanghe District, Gansu and their stratigraphic significance. *Geoscience* 4:16–29.

Cerdeño, E.1996. Rhinocerotidae from the Middle Miocene of the Tung-gur Formation, Inner Mongolia (China). *American Museum Novitates* 3184:1–43.

Chen, D.-g. and Z.-c. Peng. 1985. K-Ar ages and Pb-Sr isotopic characteristics of Cenozoic volcanic rocks in Shandong, China. *Geochimica* 4:293–303.

Chen, G.-f. 1981. A new species of *Amphicyon* from the Pliocene of Zhongxiang, Hubei. *Vertebrata PalAsiatica* 19:21–25.

Chen, G.-f. 1988a. Remarks on the *Oioceros* species (Bovidae, Artiodactyla, Mammalia) from the Neogene of China. *Vertebrata PalAsiatica* 26:157–172.

Cheng, Y.-q. 2009. *Stratigraphic Lexicon of China—General Introduction*. Beijing: Geological Publishing House.

Colbert, E. H. 1939. Carnivora of the Tung Gur Formation of Mongolia. *Bulletin of American Museum of Natural History* 76:47–81.

Crusafont-Pairó, M. 1950. La cuestión del llamado Meótico español. *Arrahona* 1:3–9.

Crusafont-Pairó, M. 1951. El sistema miocénico en la depresión española del Vallés-Penedés. *International. Geological. Congress "Reports of XVIII Session, Great Britain 1948"* 11:33–43.

Crusafont-Pairó, M. 1965. Observations au travail de M. Freudenthal et P. Y. Sondaar sur les nouveaux gisements à *Hipparion* d'Espagne. *Koninklijke Nederlanse Akademie van Wetenschappen*, Proceedings, ser. B, 68, 3:121–126.

Dai S., X.-m. Fang, G. Dupont-Nivet, C.-h. Song, J.-p. Gao, W. Krijgsman, C. Langereis, and W.-l. Zhang. 2006. Magnetostratigraphy of Cenozoic sediments from the Xining Basin: Tectonic implications for the northeastern Tibetan Plateau. *Journal of Geophysical Research* 111, B11102, doi:10.1029/2005JB004187, 2006: pp. 1–19.

de Bruijn H., R. Daams, G. Daxner-Höck, V. Fahlbusch, L. Ginsburg, P. Mein, and J. Morales. 1992. Report of the RCMNS working group on fossil mammals, Reisensburg 1990. *Newsletters on Stratigraphy* 26:65–118.

Delson, E., J. Guan, and T. Harrison. 1990. A new species of *Pliopithecus* from the Middle Miocene of China and the generic taxonomy of the Pliopithecidae. *American Journal of Physical Anthropology* 81:214.

Deng, T. 2003. New material of *Hispanotherium matritense* (Rhinocerotidae, Perissodatyla) from Laogou of Hezheng County (Gansu, China), with special reference to the Chinese Middle Miocene elasmotheres. *Geobios* 36:141–150.

Deng, T. 2004a. Establishment of the Middle Miocene Hujialiang Formation in the Linxia Basin of Gansu and its features. *Journal of Stratigraphy* 28:307–312.

Deng, T. 2004b. A new species of the rhinoceros *Alicornops* from the Middle Miocene of the Linxia Basin, Gaunsu, China. *Palaeontology* 47:1427–1439.

Deng, T. 2006. Chinese Neogene mammal biochronology. *Vertebrata PalAsiatica* 44:143–163.

Deng, T., S.-k. Hou, and H. Wang. 2007. The Tunggurian Stage of the continental Miocene in China. *Acta Geologica Sinica* 81:709–721.

Deng, T., S.-k. Hou, T.-m. Wang, and Y.-q. Mu. 2010. The Gaozhuangian Stage of the Continental Pliocene Series in China. *Journal of Stratigraphy* 34:225–240.

Deng, T. and X.-m. Wang. 2004. Late Miocene *Hipparion* (Equidae, Mammalia) of Eastern Qaidam Basin in Qinghai, China. *Vertebrata PalAsiatica* 42:316–333.

Deng, T., X.-m. Wang, M. Fortelius, Q. Li, Y. Wang, Z. J. Tseng, G. T. Takeuchi, J. E. Saylor, L. K. Säilä, and G.-p. Xie. 2011. Out of Tibet: Pliocene woolly rhino suggests high-plateau origin of Ice Age megaherbivores. *Science* 333:1285–1288.

Deng, T., X.-m. Wang, X.-j. Ni, and L.-p. Liu. 2004. Sequence of the Cenozoic mammalian faunas of the Linxia Basin in Gansu, China. *Acta Geologica Sinica* 78:8–14.

Deng, T., W.-m. Wang, and L.-p. Yue. 2008a. Report on research of the Shanwangian and Baodean Stages in the continental Neogene Series of China. In *Reports on Establishing the Major Stages in China*, ed. Z.-j. Wang and Z.-g Huang, pp. 13–31. Beijing: Geological Publishing House.

Deng, T., W.-m. Wang, and L.-p. Yue. 2008b. Report on research of the Xiejian Stage in the continental Neogene Series of China. In *Reports on Establishing the Major Stages in China*, ed. Z.-j. Wang and Z.-g. Huang, pp. 32–40. Beijing: Geological Publishing House.

Ding, Z.-l., S.-f. Xiong, J.-m. Sun, S.-l., Yang, Z.-y. Gu, and T.-s. Liu. 1999. Pedostratigraphy and paleomagnetism of a ~7.0 Ma eolian loess-red clay sequence at Lingtai, Loess Plateau, north-central China and the implications for paleomonsoon evolution. *Palaeogeography, Palaeoclimatology, Palaeoecology* 152:49–66.

Fahlbusch, V. 1976. Report on the International Symposium on Mammalian Biostratigraphy of the European Tertiary. *Newsletter of Stratigraphy* 5:160–167.

Fahlbusch, V. 1991. The meaning of MN zonation: considerations for a subdivision of the European continental Tertiary using mammals. *Newsletters on Stratigraphy* 24:159–173.

Fang, X.-m., C. Garzione, R. Van der Voo, J.-j. Li, and M.-j. Fan. 2003. Flexural subsidence by 29 Ma on the NE edge of Tibet from the magnetostratigraphy of Linxia Basin, China. *Earth and Planetary Science Letters* 210:545–560.

Fang, X.-m., W. Zhang, Q. Meng, J. Gao, X. Wang, J. King, C. Song, S. Dai, and Y. Miao. 2007. High-resolution magnetostratigraphy of the Neogene Huaitoutala section in the eastern Qaidam Basin on the NE Tibetan Plateau, Qinghai Province, China and its implication on tectonic uplift of the NE Tibetan Plateau. *Earth and Planetary Science Letters* 258:293–306.

Flynn, J. J. and C. C. Swisher III. 1995. Cenozoic South American land mammal ages: Correlation to global chronologies. In *Geochronology, Time Scales, and Global Stratigraphic Correlation: A Unified Framework for a Historical Geology*, ed. W. A. Berggren, D. V. Kent, M. P. Aubry, and J. A. Hardenbol, pp. 317–333. Society of Economic Paleontologists and Mineralogists Special Publications 54. Tulsa: Society for Sedimentary Geology.

Flynn, L. J. 1997. Late Neogene mammalian events in North China. In *Actes du Congrès BiochroM'97*, ed. J.-P. Aguilar, S. Legendre, and J. Michaux, pp. 183–192. Mémoires et Travaux E.P.H.E., Institut de Montpellier, 21.

Flynn, L. J., W. Downs, N. Opdyke, K.-n. Huang, E. Lindsay, J. Ye, G.-p. Xie, and X.-m. Wang. 1999. Recent advances in the small mammal biostratigraphy and magnetostratigraphy of Lanzhou Basin. *Chinese Science Bulletin* 44 (Supplement):105–118.

Flynn, L. J., Z.-x. Qiu, N. O. Opdyke, and R. H. Tedford. 1995. Ages of key fossil assemblages in the late Neogene terrestrial record of northern China. In *Geochronology, Time Scales, and Global Stratigraphic Correlation: A Unified Framework for a Historical Geology*, ed. W. A. Berggren, D. V. Kent, M.-P. Aubrey, and J. A. Hardenbol, pp. 365–373. Society of Economic Paleontology and Mineralogists Special Publications 54. Tulsa: Society for Sedimentary Geology.

Flynn, L. J., R. H. Tedford, and Z.-x. Qiu. 1991. Enrichment and stability in the Pliocene mammalian fauna of North China. *Paleobiology* 17:246–265.

Flynn, L. J., W.-y. Wu, W. R. Downs III. 1997. Dating vertebrate microfaunas in the late Neogene record of Northern China. *Palaeogeography, Palaeoclimatology, Palaeoecology* 133:227–242.

Gradstein, F. M., J. G. Ogg, and A. G. Smith. 2004. *A Geologic Time Scale 2004*. Cambridge: Cambridge University Press.

Gu, Y.-m. 1980. A Pliocene macaque's tooth from Hubei. *Vertebrata PalAsiatica* 18:324–326.

Guan, J. 1988. The Miocene strata and mammals from Tongxin, Ningxia and Guanghe, Gansu. *Memoir of Beijing Natural History Museum* 42:1–21.

Guan, J. 1991. The character analysis and phylogeny discussion on the shovel-tusked mastodonts. *Memoir of Beijing Natural History Museum* 50:1–21.

Guan, J. 1993. Primitive elasmotherines from the middle Miocene, Ningxia (Northwestern China). *Memoir of Beijing Natural History Museum* 53:200–207.

Guan, J. and J. Van der Made. 1993. Fossil Suidae from Dingjiaergou near Tongxin, China. *Memoir of Beijing Natural History Museum* 53:150–199.

Hedberg, H. D., ed. 1976. *International Stratigraphic Guide—A Guide to Stratigraphic Classification, Terminology, and Procedure*. First Edition. International Subcommission on Stratigraphic Classification of IUGS International Commission on Stratigraphy. New York: Wiley.

Hopwood, A. T. 1935. Fossils Proboscidea from China. *Palaeontologia Sinica* C, 9(3):1–108.

Huo, F.-c., X.-s. Pan, G.-l. You, S.-c. Zheng, H.-r. Liao, T.-x. Zhou, G.-d. Zhang, G.-x. Xu, Z.-c. Liu, and Z.-x. Gui. 1989. *Introduction to Geology of Ningxia*. Beijing: Science Press (in Chinese).

Ji, H.-x., Q.-q. Xu, and W.-b. Huang. 1980. The *Hipparion* fauna from Guizhong Basin, Xizang. In *Paleontology of Xizang*, Book 1, *The Comprehensive Scientific Expedition to the Qinghai-Xizang Plateau*, ed. The Chinese Academy of Sciences, pp. 18–32. Beijing: Science Press.

Jiang, Z.-x. and W.-y. Wu. 1978. The first discovery of Pliocene vertebrate fossils and the subdivision of the Late Cenozoic deposits in central part of Shantung. *Vertebrata PalAsiatica* 16:193–200.

Kaakinen, A. 2005. A long terrestrial sequence in Lantian: A window into the late Neogene palaeoenviornnments of northern China. Ph.D. diss., University of Helsinki.

Kaakinen, A. and J. P. Lunkka. 2003. Sedimentation of the Late Miocene Bahe Formation and its implications for stable environments adjacent to Qinling mountains in Shaanxi, China. *Journal of Asian Earth Sciences* 22:67–78.

Kretzoi, M. 1962. Fauna und Faunenhorizont von Csarnóta. *Annual Report of the Hungarian Geological Institute of 1959:*340–395.

Li, C.-k. 1977. A new Miocene cricetodont rodent of Fangshan, Nanking. *Vertebrata PalAsiatica* 15:67–75.

Li, C.-k. and Z.-d. Qiu. 1980. Early Miocene mammalian fossils of Xining basin, Qinghai. *Vertebrata PalAsiatica* 18:198–214.

Li, C.-k., Z.-d. Qiu, and S.-j. Wang. 1981. Discussion on Miocene stratigraphy and mammals from Xining basin, Qinghai. *Vertebrata PalAsiatica* 19:313–320.

Li, C.-k. and S.-y. Ting. 1983. The Paleogene mammals of China. *Bulletin of Carnegie Museum of Natural History*, 21:1–98.

Li, C.-k., W.-y. Wu, and Z.-d. Qiu. 1984. Chinese Neogene: Subdivision and correlation. *Vertebrata PalAsiatica* 22:163–178 (in Chinese with English abstract).

Li, J.-j. 1995. *Uplift of Qinghai-Xizang (Tibet) Plateau and Global Change*. Lanzhou: Lanzhou University Press.

Li, Q., X.-m. Wang, and Z.-d. Qiu. 2003. Pliocene mammalian fauna of Gaotege in Nei Mongol (Inner Mongolia), China. *Vertebrata PalAsiatica* 41:104–114.

Li, Y.-t. 1984. The Tertiary System of China. *Stratigraphy of China* 13:1–362.

Lindsay, E. H. 1990. The setting. In *European Neogene Mammal Chronology*, ed. E. H. Lindsay, V. Fahlbusch, and P. Mein, pp. 1–14. New York: Plenum Press.

Lindsay, E. 1997. Eurasian mammal biochronology: An overview. *Palaeogeography, Palaeoclimatology, Palaeoecology* 133:117–128.

Lindsay, E. H. and R. H. Tedford. 1990. Development and application of land mammal ages in North America and Europe: A comparison. In *European Neogene Mammal Chronology*, ed. E. H. Lindsay, V. Fahlbusch, and P. Mein, pp. 601–624. New York: Plenum Press.

Liu J.-q. 1999. *Volcanos of China*. Beijing: Science Press. 1–219 (in Chinese).

Liu, T.-s., M.-l. Ding, and F.-q. Gao. 1960. Section of Cenozoic deposits of Xi'an-Lantian, Shaaxi. *Scientia Geologica Sinica* 1960:199–208.

Liu, T.-s., C.-k. Li, and R.-j. Zhai. 1978. Fossil vertebrates from the Pliocene of Lantian County, Shaanxi Province. *Professional Papers of Stratigraphy and Palaeontology* 7:149–200.

Luo, X.-q. and Q.-t. Chen. 1990. Preliminary study on geochronology for Cenozoic basalts from Inner Mongolia. *Acta Petrologica et Mineralogica* 9:37–46.

Martin, L. D. 1974. New rodents from the Lower Miocene Gering Formation of western Nebraska. *Occasional Papers of the Museum of Natural History, the University of Kansas* 32:1–12.

Mascarelli, A. L. 2009. Quaternary geologists win timescale vote. *Nature* 459:624.

Matthew, W. D. 1924. Correlation of the Tertiary formations of the Great Plains. *Geological Society of America Bulletin* 35:743–754.

Mein, P. 1975. Résultats du Groupe de Travail des Vertébrés. In *Report on Activity of the RCMNS Working Groups*, ed. J. Senes, pp. 78–81. Bratislava: SAV.

Mein, P. 1979. Rapport d'activité de groupe de travail vértebrés mise a jour de la biostratigraphie du Néogène basée sur les mammifères. *Annales géologiques des Pays Helléniques*, non-series vol. III:83–88.

Mein, P. 1981. Mammal zonation: Introduction. *Annales géologiques des Pays Helléniques*, non-series vol. IV:83–88.

Mein, P. 1990. Updating of MN Zones. In *European Neogene Mammal Chronology*, ed. E. H. Lindsay, V. Fahlbusch, and P. Mein, pp. 73–89. New York: Plenum Press.

Meng, J. and M. C. McKenna, 1998. Faunal turnovers of Paleogene mammals from the Mongolian Plateau. *Nature* 395:364–367.

Meng, J., B.-y. Wang, and Z.-q. Bai. 1996. A new middle Tertiary mammalian locality from Sunitezuoqi, Nei Mongol. *Vertebrata PalAsiatica* 34:297–304.

Meng, J., J. Ye, W.-y. Wu, L.-p. Yue, and X.-j. Ni. 2006. A recommended boundary stratotype section for Xiejian Stage from northern Junggar Basin: Implications to related bio-chronostratigraphy and environmental changes. *Vertebrata PalAsiatica* 44:205–236.

Murphy, M. A. 1994. Fossils as a basis for chronostratigraphic interpretation. *Neues Jahrbuch für Geologie und Paläontologie Abhandlungen* 192:255–271.

Opdyke, N. D., L. J. Flynn, E. H. Lindsay, and Z.-x. Qiu. 1998. The magnetic stratigraphy of the Yehucheng and Xianshuihe Formations of Oligocene-Miocene age near Lanzhou city, China. *Paleontology Working Meeting Lanzhou*, p. 51.

Osborn, H. F. and W. D. Matthew. 1909. Cenozoic mammal horizons of western North America. *United States Geological Survey Bulletin* 361:1–138.

Pei, W.-z., M.-z. Zhou, and J.-j. Zheng. 1963. Cenozoic Erathem of China. In *Selected Scientific Reports at All-China Stratigraphic Congress*, ed. All-China Stratigraphic Commission, pp. 1–31. Beijing: Science Press.

Prothero, D. R. 1995. Geochronology and magnetostratigraphy of Paleogene North American land mammal "ages": An update. In *Geochronology, Time Scales, and Global Stratigraphic Correlation: A Unified Framework for a Historical Geology*, ed. W. A. Berggren, D. V. Kent, M. P. Aubry, and J. A. Hardenbol, pp. 305–315. Society of Economic Paleontologists and Mineralogists Special Publications 54. Tulsa: Society for Sedimentary Geology.

Qiu, Z.-d. 1987. The Aragonian vertebrate fauna of Xiacaowan, Jiangsu.—7. Aplodontidae (Rodentia, Mammalia). *Vertebrata PalAsiatica* 25:283–296.

Qiu, Z.-d. 2001. Glirid and gerbillid rodents from the Middle Miocene Quantougou Fauna of Lanzhou, Gansu. *Vertebrata PalAsiatica* 39:299–306.

Qiu, Z.-d., C.-k. Li, and S.-j. Wang. 1981. Miocene mammalian fossils from Xining Basin, Qinghai. *Vertebrata PalAsiatica* 19:156–173.

Qiu, Z.-d. and G. Storch. 2000. The early Pliocene micromammalian fauna of Bilike, Inner Mongolia, China (Mammalia: Lipotyphla, Chiroptera, Rodentia, Lagomorpha). *Senckenbergiana Lethaea* 80:173–229.

Qiu, Z.-d., X.-m. Wang, and Q. Li. 2006. Faunal succession and biochronology of the Miocene through Pliocene in Nei Mongol (Inner Mongolia). *Vertebrata PalAsiatica* 44:164–181.

Qiu, Z.-x. 1987. Die Hyaeniden aus dem Ruscinium und Villafranchium Chinas. *Münchner Geowissenschaftliche Abhandlungen* A, 9:1–108.

Qiu, Z.-x. 1990. The Chinese Neogene mammalian biochronology—its correlation with the European Neogene mammalian zonation. In *European Neogene Mammal Chronology*, ed. E. H. Lindsay, V. Fahlbusch, and P. Mein, pp. 527–556. New York: Plenum Press.

Qiu, Z.-x. and Y.-m. Gu. 1986. The Aragonian vertebrate fauna of Xiacaowan, Jiangsu—3. Two carnivores: *Semigenetta* and *Pseudaelurus*. *Vertebrata PalAsiatica* 24:20–31.

Qiu, Z.-x. and Y.-m. Gu. 1991. The Aragonian vertebrate fauna of Xiacaowan, Jiangsu—8. *Dorcatherium* (Tragulidae, Artiodactyla). *Vertebrata PalAsiatica* 29:21–37.

Qiu, Z.-x. and Z.-g. Gu. 1988. A new locality yielding mid-Tertiary mammals near Lanzhou, Gansu. *Vertebrata PalAsiatica* 26: 198–213.

Qiu, Z.-x. and Z.-d. Qiu. 1990. The sequence and division of mammalian local faunas in the Neogene of China. *Journal of Stratigraphy* 14:241–260.

Qiu, Z.-x. and Z.-d. Qiu. 1995. Chronological sequence and subdivision of Chinese Neogene mammalian faunas. *Palaeogeography, Palaeoclimatology, Palaeoecology* 116:41–70.

Qiu, Z.-x. and B.-y. Wang. 2007. Paracerathere fossils of China. *Palaeontologia Sinica*, n.s. C, 29:1–396.

Qiu, Z.-x., B.-y. Wang, Z.-d. Qiu, G.-p. Xie, J.-y. Xie, and X.-m. Wang. 1997. New advances in studies of the Xianshuihe Formation of Lanzhou Basin, Gansu. In *Evidence for Evolution—Essays in Honor of Prof. Chungchien Young on the Hundredth Anniversary of His Birth*, ed. Y.-S. Tong, Y.-y. Zhang, W.-y. Wu, J.-l. Li, and L.-q. Shi, pp. 177–192. Beijing: China Ocean Press (in Chinese).

Qiu, Z.-x., B.-y. Wang, Z.-d. Qiu, F. Heller, L.-p. Yue, G.-p. Xie, X.-m. Wang, and B. Engesser. 2001. Land mammal geochronology and magnetostratigraphy of mid-Tertiary deposits in the Lanzhou Basin, Gansu Province, China. *Eclogae Geologicae Helvetiae* 94: 373–385.

Qiu, Z.-x., B.-y. Wang, and J.-y. Xie. 1998. Mid-Tertiary chalicothere (Perissodactyla) fossils from Lanzhou, Gansu, China. *Vertebrata PalAsiatica* 36:297–318.

Qiu, Z.-x., W.-y. Wu, and Z.-d. Qiu. 1999. Miocene mammal faunal sequences of China: Palaeozoogeography and Eurasian relationships. In *The Miocene Land Mammals of Europe*, ed. G. E. Rössner and K. Heissig, pp. 443–472. Munich: Dr. Friedrich Pfeil.

Qiu, Z.-x. and J.-y. Xie. 1997. A new species of *Aprotodon* (Perissodactyla, Rhinocerotidae) from Lanzhou Basin, Gansu, China. *Vertebrata PalAsiatica* 35:250–267.

Qiu, Z.-x., D.-f. Yan, G.-f. Chen, and Z.-d. Qiu. 1988. Preliminary report on the field work in 1986 at Tung-gur, Nei Mongol. *Chinese Science Bulletin* 33:399–404.

Qiu, Z.-x., J. Ye, and J.-x. Cao. 1988. A new species of *Percrocuta* from Tongxin, Ningxia. *Vertebrata PalAsiatica* 26:116–127.

Qiu, Z.-x., J. Ye, and F.-c. Huo. 1988. Description of a *Kubanochoerus* skull from Tongxin, Ningxia. *Vertebrata PalAsiatica* 26:1–19.

Remane, J., M. G. Basset, J. W. Cowie, K. H. Gohrbandt, H. R. Lane, O. Michelsen, and N. Wang. 1996. Revised guidelines for the establishment of global chronostratigraphic standards by the International Commission on Stratigraphy (ICS). *Episodes* 19:77–81.

Salvador, A. 1994. *International Stratigraphic Guide—A Guide to Stratigraphic Classification, Terminology, and Procedure*. 2nd Edition. In *International Subcommission on Stratigraphic Classification of IUGS International Commission on Stratigraphy*, ed. A. Salvador. Boulder: Geological Society of America.

Savage, D. E. 1955. Nonmarine lower Pliocene sediments in California: A geochronostratigraphic classification. *University of California Publication of Geological Science* 31:1–26.

Savage, D. E. 1962. Cenozoic geochronology of the fossil mammals of the western hemisphere. *Revista del Museo Argentino de Ciencias Naturales "Bernardino Rivadavia" y Instituto Nacional de Investigación de las Ciencias Naturales, Ciencias Zoológicas* 8:53–67.

Schultz, C. B. 1938. The Miocene of western Nebraska. *American Journal of Science* 35:441–444.

Sen, S. 1997. Magnetostratigraphic calibration of the European Neogene mammal chronology. *Palaeogeography, Palaeoclimatology, Palaeoecology* 133:181–204.

Song, C.-h., D.-l. Gao, X.-m. Fang, Z.-j. Cui, J.-j. Li, S.-l. Yang, H.-b. Jin, D. W. Burbank, and J. L. Kirschvink. 2005. Late Cenozoic high-resolution magnetostratigraphy in the Kunlun Pass Basin and its implications for the uplift of the northern Tibetan Plateau. *Chinese Science Bulletin* 50:1912–1922.

Song, Z.-c., H.-m. Li, and Y.-h. Zheng. 1983. Miocene floristic regions of China. In *Paleobiogeographic Provinces of China*, ed. Editorial Committee of Fundamental Theory of Palenotological Book Series, pp. 178–184. Beijing: Science Press.

Spock, L. E. 1929. Pliocene beds of the Iren Gobi. *American Museum Novitates* 394:1–8.

Steininger, F. F. 1999. Chronostratigraphy, geochronology and biochronology of the Miocene "European Land Mammal Mega-Zone (ELMMZ)" and the Miocene "Mammal-Zones (MN-Zones)." In *The Miocene Land Mammals of Europe*, ed. G. E. Rössner and K. Heissig, pp. 9–24. Munich: Dr. Friedrich Pfeil.

Steininger, F., W. A. Berggren, D. V. Kent, R. L. Bernor, S. Sen, and J. Agustí. 1996. Circum-Mediterranean Neogene (Miocene and Pliocene) marine-continental chronologic correlations of European mammal units. In *The Evolution of Western Eurasian Neogene Mammal Faunas*, ed. R. L. Bernor, V. Fahlbusch, and H.-W. Mittmann, pp. 7–46. New York: Columbia University Press.

Sun, A.-l. 1961. Note on fossils snakes from Shanwang. Shantung. *Vertebrata PalAsiatica* 1961:306–312.

Sun, J.-m., R.-x. Zhu, and Z.-s. An. 2005. Tectonic uplift in the northern Tibetan Plateau since 13.7 Ma ago inferred from molasse deposits along the Altyn Tagh Fault. *Earth and Planetary Science Letters* 235:641–653.

Swinehart, J. B., V. L. Souders, H. M. DeGraw, and R. F. Diffendal. 1985. Cenozoic paleogeography of western Nebraska. In *Cenozoic Paleogeography of West-Central United States. Rocky Mountain Section, Society of Economic Paleontologists and Mineralogists*, ed. R. M. Flores and S. S. Kaplan, pp. 204–229. Denver: Rocky Mountain Paleogeography Symposium 3.

Swisher III, C. C. 1982. Stratigraphy and biostratigraphy of the eastern portion of the Wildcat Ridge, western Nebraska. M.S. thesis, University of Nebraska, Lincoln.

Tedford, R. H. 1970. Principles and practices of mammalian geochronology in North America. *Proceedings of the North American Paleontological Convention* Part F:666–703.

Tedford, R. H., L. B. Albright III, A. D. Barnosky, I. Ferrusquía-Villafranca, R. M. Hunt Jr., J. E. Storer, C. C. Swisher III, M. R. Voorhies, S. D. Webb, and D. P. Whistler. 2004. Mammalian biochronology of the Arikareean through Hemphillian interval (Late Oligocene through Early Pliocene Epochs). In *Late Cretaceous and Cenozoic Mammals of North America: Biostratigraphy and Geochronology*, ed. M. O. Woodburne, pp. 169–231. New York: Columbia University Press.

Tedford, R. H., L. J. Flynn, Z.-x. Qiu, N. O. Opdyke, and W. R. Downs. 1991. Yushe Basin, China; paleomagnetically calibrated mammalian biostratigraphic standard for the late Neogene of eastern Asia. *Journal of Vertebrate Paleontology* 11:519–526.

Tedford, R. H., T. Galusha, M. F. Skinner, B. E. Taylor, R. W. Fields, J. R. Macdonald, J. M. Rensberger, S. D. Webb, and D. P. Whistler. 1987. Faunal succession and biochronology of the Arikareean through Hemphillian interval (Late Oligocene through earliest Miocene epochs) in North America. In *Cenozoic Mammals of North America: Geochronology and Biostratigraphy*, ed. M. O. Woodburne, pp. 153–210. Berkeley: University of California Press.

Tedford, R. H., J. B. Swineheart, R. M. Hunt Jr., and M. R. Voorhies. 1985. Uppermost White River and lowermost Arikaree rocks and fuanas, White River Valley, northwestern Nebraska and their correlation with South Dakota. *Dakoterra* 2:335–352.

Tedford, R. H., J. B. Swineheart, C. C. Swisher III, D. R. Prothero, S. A. King, and T. E. Tierney. 1996. The Whitneyan-Arikareean transition in the High Plains. In *The Terrestrial Eocene-Oligocene Transition in North America*, ed. D. R. Prothero and R. J. Emry, pp. 312–334. New York: Cambridge University Press.

Teilhard de Chardin, P. 1929. The times of the Loess and early man in China. *Contributions of the Department of Geography and Geology, Yenching University Peiping* 27:14–16.

Teilhard de Chardin, P. 1942. New rodents of the Pliocene and Lower Pleistocene of North China. *Publications de Institut de Géobiologie* 9:1–101.

Teilhard de Chardin, P. and P. Leroy 1945. Les Mustélidés de Chine. *Publications de Institut de Géobiologie* 11:1–56.

Teilhard de Chardin, P. and M. Trassaert 1937. The Proboscideans of South-Eastern Shansi (Yushê basin). *Palaeontologia Sinica*, ser. C 13(1):1–85.

Teilhard de Chardin, P. and M. Trassaert 1938. Cavicornia of South-Eastern Shansi. *Palaeontologia Sinica*, n.s. C 6:1–107.

Tobien, H., G.-f. Chen, and Y.-q. Li 1988. Mastodonts (Proboscidea, Mammalia) from Late Neogene and Early Pleistocene of the People's Republic of China. Part 2. *Mainzer Geowissenschaftliche Mitteilungen* 17:95–220.

Tong, Y.-s., S.-h. Zheng, and Z.-d. Qiu. 1995. Cenozoic mammal ages of China. *Vertebrata PalAsiatica* 33:290–314.

Walsh, S. L. 1998. Fossil datum and paleobiological event terms, paleontostratigraphy, chronostratigraphy, and the definition of land mammal "age" boundaries. *Journal of Vertebrate Paleontology* 18: 150–179.

Walsh, S. L. 2000. Notes on geochronologic and chronostratigraphic units. *Geological Society of America Bulletin* 113:704–713.

Wang, B.-y. 1997. The mid-Tertiary Ctenodactylidae (Rodentia, Mammalia) of eastern and central Asia. *Bulletin of the American Museum of Natural History* 234:1–88.

Wang, B.-y. 2001. On Tsaganomyidae (Rodentia, Mammalia) of Asia. *American Museum Novitates* 3317:1–50.

Wang, B.-y. 2003. Dipodidae (Rodentia, Mammalia) from the mid-Tertiary deposits in Danghe area, Gansu, China. *Vertebrata PalAsiatica* 41:89–103.

Wang, B.-y. and Z.-x. Qiu. 2002a. A porcupine from Late Miocene of Linxia Basin, Gauns, China. *Vertebrata PalAsiatica* 40:23–33.

Wang, B.-y. and Z.-x. Qiu. 2002b. A new species of *Platybelodon* (Gomphotheriidae, Proboscidea, Mammalia) from early Miocene of the Danghe area, Gansu, China. *Vertebrata PalAsiatica* 40:291–299.

Wang, B.-y., Z.-x. Qiu, X. Wang, G.-p. Xie, J.-y. Xie, W. R. Downs, Z.-d. Qiu, and T. Deng. 2003. Cenozoic stratigraphy in Danghe area, Gansu Province, and uplift of Tibetan Plateau. *Vertebrata PalAsiatica* 41:66–75.

Wang, B.-y. and P.-y. Wang. 1990. Discovery of Early Miocene mammal fauna from Urtu area, Alxa Zuoqi, Nei Mongol. *Chinese Science Bulletin* 35:607–611.

Wang, S., W. Zhang, X. Fang, S. Dai, and O. Kempf. 2008. Magnetostratigraphy of the Zanda basin in southwest Tibet Plateau and its tectonic implications. *Chinese Science Bulletin* 53:1393–1400.

Wang, X. 1997. New cranial material of *Simocyon* from China, and its implications for phylogenetic relationship to the red panda (*Ailurus*). *Journal of Vertebrate Paleontology* 17:184–198.

Wang, X. 2004. New materials of *Tunguristis* (Hyaenidae, Carnivora) from Tunggur Formation, Nei Mongol. *Vertebrata PalAsiatica* 42:144–153.

Wang, X., Z.-d. Qiu, Q. Li, Y. Tomida, Y. Kimura, Z. J. Tseng, and H.-j. Wang. 2009. A new early to late Miocene fossiliferous region in central Nei Mongol: Lithostratigraphy and biostratigraphy in Aoerban strata. *Vertebrata PalAsiatica* 47:111–134.

Wang, X., Z.-d. Qiu, Q. Li, B.-y. Wang, Z.-x. Qiu, W. R. Downs, G.-p. Xie, J.-y. Xie, T. Deng, G. T. Takeuchi, Z. J. Tseng, M.-m. Chang, J. Liu, Y. Wang, D. Biasatti, Z. Sun, X. Fang, and Q. Meng. 2007. Vertebrate paleontology, biostratigraphy, geochronology, and paleoenvironment of Qaidam Basin in northern Tibetan Plateau. *Palaeogeography, Palaeoclimatology, Palaeoecology* 254: 363–385.

Wang, X., Z.-d. Qiu, and N. O. Opdyke. 2003. Litho-, bio-, and magnetostratigraphy and paleoenvironment of Tunggur Formation (Middle Miocene) in central Inner Mongolia, China. *American Museum Novitates* 3411:1–31.

Wang, X., Z.-x. Qiu, and B.-y. Wang. 2004. A new leptarctine (Carnivora: Mustelidae) from the early Miocene of the northern Tibetan Plateau and implications of the phylogeny and zoogeography of basal mustelids. *Zoological Journal of the Linnean Society* 142:405–421.

Wang, X., Z.-x. Qiu, and B.-y. Wang. 2005. Hyaenodonts and Carnivorans from the Early Oligocene to Early Miocene of Xianshuihe Formation, Lanzhou Basin, Gansu Province, China. *Palaeontologia Electronica* 8:1–14.

Wang, X., B.-y. Wang, Z.-x. Qiu, G.-p. Xie, J.-y. Xie, W. R. Downs, Z.-d. Qiu, and T. Deng. 2003. Danghe area (western Gansu, China) biostratigraphy and implications for depositional history and tectonics of northern Tibetan Plateau. *Earth and Planetary Science Letters* 208:253–269.

Wang, X., G.-p. Xie, Q. Li, Z.-d. Qiu, Z. J. Tseng, G. T. Takeuchi, B.-y. Wang, M. Fortelius, A. Rosenström-Fortelius, H. Wahlquist, W. R. Downs, C.-f. Zhang, and Y. Wang. 2011. Early explorations of Qaidam Basin (Tibetan Plateau) by Birger Bohlin—reconciling classic vertebrate fossil localities with modern biostratigraphy. *Vertebrata PalAsiatica* 49(3):285–310.

Wei, L.-y., M.-y. Chen, H.-m. Zhao, J.-m. Sun, and X. Lu. 1993. Magnetostratigraphic study on the late Miocene-Pliocene lacustrine sediments near Leijiahe. In *Monograph of the Meeting in Honor of Prof. Yuan Fuli on the Hundredth Anniversary of His Birth*. Beijing: Seismological Press, pp. 63–67.

Williams, H. S. 1901. The discrimination of time-values in geology. *Journal of Geology* 9:570–585.

Wood, H. E., R. W. Chaney, J. Clark, E. H. Colbert, G. L. Jepsen, J. B. Reeside Jr., and C. Stock. 1941. Nomenclature and correlation of the North American continental Tertiary. *Bulletin of the Geological Society of America* 52:1–48.

Woodburne, M. O. 1977. Definition and characterization in mammalian chronostratigraphy. *Journal of Paleontology* 51:220–234.

Woodburne, M. O. 1987a. Principles, classification, and recommendations. In *Cenozoic Mammals of North America: Geochronology and Biostratigraphy*, ed. M. O. Woodburne, pp. 9–17. Berkeley: University of California Press.

Woodburne, M. O. 1987b. Mammal ages, stages, and zones. In *Cenozoic Mammals of North America: Geochronology and Biostratigraphy*, ed. M. O. Woodburne, pp. 18–23. Berkeley: University of California Press.

Woodburne, M. O. 1996. Precision and resolution in mammalian chronostratigraphy: Principles, practices, examples. *Journal of Vertebrate Plaeontology* 16:531–555.

Woodburne, M. O. 2004a. Definitions. In *Late Cretaceous and Cenozoic Mammals of North America: Biostratigraphy and Geochronology*, ed. M. O. Woodburne, pp. xi–xv. New York: Columbia University Press.

Woodburne, M. O. 2004b. Principles and procedures. In *Late Cretaceous and Cenozoic Mammals of North America: Biostratigraphy and Geochronology*, ed. M. O. Woodburne, pp. 1–20. New York: Columbia University Press.

Woodburne, M. O. 2006. Mammal ages. *Stratigraphy* 3:229–261.

Wu, L.-c., L.-p. Yue, J.-q. Wang, F. Heller, and T. Deng. 2006. Magnetostratigraphy of stratotype section of the Neogene Xiejian Stage. *Journal of Stratigraphy* 30:50–53.

Wu, W.-y. 1988. The first discovery of Middle Miocene rodents from the Northern Junggar Basin, China. *Vertebrata PalAsiatica* 26:250–264.

Wu, W.-y., J. Meng, and J. Ye, 2003. The discovery of *Pliopithecus* from Northern Junggar Basin, Xinjiang. *Vertebrata PalAsiatica* 41:76–86.

Wu, W.-y., J. Meng, and J. Ye, et al. 2009. The Miocene mammals from Dingshanyanchi Formation of North Junggar Basin, Xinjiang. *Vertebrata PalAsiatica* 47:208–233.

Wu, W.-y., J. Ye, and B.-c. Zhu. 1991. On *Alloptox* (Lagomorpha, Ochotonidae) from the middle Miocene of Tongxin, Ningxia Hui Autonomous Region, China. *Vertebrata PalAsiatica* 29:204–229.

Xu, Y.-l., Y.-b. Tong, Q. Li, Z.-m. Sun, J.-l. Pei, and Z.-y. Yang. 2007. Magnetostratigraphic dating on the Pliocene mammalian fauna of the Gaotege section, central Inner Mongolia. *Geological Review* 53:250–260.

Xue, X.-x., Y.-x. Zhang, and L.-p. Yue. 1995. Discovery and chronological division of the *Hipparion* fauna in Laogaochuan Village, Fugu County, Shaanxi. *Chinese Science Bulletin* 40:926–929.

Yan, D.-f. 1978. On the geological age of Duodaoshi Formation, Jingxiang Region, Hubei. *Vertebrata PalAsiatica* 16:30–32.

Yan, D.-f., Z.-d. Qiu, and Z.-y. Meng. 1983. Miocene stratigraphy and mammals of Shanwang, Shandong. *Vertebrata PalAsiatica* 21:210–222.

Ye, J. 1989. Middle Miocene artiodactyls from the Northern Junggar Basin. *Vertebrata PalAsiatica* 27:37–52.

Ye, J., Z.-x. Qiu, and G.-d. Zhang. 1992. *Bunolistriodon intermedius* (Suidae, Artiodactyla) from Tongxin, Ningxia. *Vertebrata PalAsiatica* 30:135–145.

Ye, J., W.-y. W, and J. Meng. 2001a. Tertiary stratigraphy in the Ulungur River Area of Northern Junggar Basin of Xinjiang. *Journal of Stratigraphy* 25:193–200.

Ye, J., W.-y. Wu, and J. Meng, 2001b. The age of the Tertiary strata and mammal faunas in Ulungur River Area of Xinjiang. *Journal of Stratigraphy* 25:283–286.

Ye, J., W.-y. Wu, and J. Meng, 2005. *Anchitherium* from the Middle Miocene Halamagai Formation of Northern Junggar Basin, Xinjiang. *Vertebrata PalAsiatica* 43:100–109.

Young, C.-c. 1927. Fossil Nagetiere aus Nord-Chinas. *Palaeontologia Sinica*, ser. C 5(3):1–82.

Young, C.-c. 1936. On the Cenozoic geology of Itu, Changlo and Linchü districts (Shantung). *Bulletin of the Geological Society of China* 15:171–188.

Young, C.-c. 1937. On a Miocene mammalian fauna from Shantung. *Bulletin of the Geological Society of China* 17:209–245.

Young, C.-c. 1964. On a new *Lagomeryx* from Lantian, Shensi. *Vertebrata PalAsiatica* 8:329–340.

Yue, L.-p., T. Deng, R. Zhang, Z.-q. Zhang, F. Heller, J.-q. Wang, and L.-r. Yang. 2004a. Paleomagnetic chronology and record of Himalayan movements in the Longgugou section of Gyirong-Oma Basin in Xizang (Tibet). *Chinese Journal of Geophysics* 47:1135–1142.

Yue, L.-p., T. Deng, Y.-x. Zhang, J.-q. Wang, R. Zhang, L.-r. Yang, and F. Heller. 2004b. Magnetostratigraphy of stratotype section of the Baode Stage. *Journal of Stratigraphy* 28:48–63.

Yue, L.-p. and Y.-x. Zhang. 1998. *Hipparion* fauna and magnetostratigraphy in Hefeng, Jingle, Shanxi Province. *Vertebrata PalAsiatica* 36:76–80.

Yue, L.-p. and Y.-x. Zhang. 2006. Paleomagnetic dating of *Lufengpithecus lufengensis* fossil strata. In *Lufengpithecus hudienensis Site*, ed. G.-q. Qi and W. Dong, pp. 252–255. Beijing: Science Press.

Zdansky, O. 1923. Fundorte der *Hipparion* fauna um Pao-Te-Hsien in N. W. Shanxi. *Bulletin of the Geological Survey of China* 5:83–92.

Zhang, Y.-p., W.-b. Huang, Y.-j. Tang, H.-x. Ji, Y.-z. You, Y.-s. Tong, S.-y. Ting, X.-s. Huang, and J.-j. Zheng. 1978. Cenozoic Erathem of the Lantian area, Shaanxi. *Institute of Vertebrate Paleontology and Paleoanthropology Memoir* A 14:1–64.

Zhang, Y.-x. and X.-x. Xue. 1996. New materials of Dinocrocuta gigantea found in Fugu County, Shaanxi Province. *Vertebrata PalAsiatica* 34:18–26.

Zhang, Y.-y., X. Lan, and H.-r. Yang. 2000. Biostratigraphy of Paleogene and Neogene of China (Chapter 18). In *Stratigraphical Studies in China (1979–1999)*, ed. Chinese Academy of Sciences Nanjing Institute of Geology and Paleontology, pp. 347–372. Hefei: University of Science and Technology of China Press.

Zhang, Z.-q., A. W. Gentry, A. Kaakinen, L.-p. Liu, J. P. Lunkka, Z.-d. Qiu, S. Sen, R. S. Scott, and L. Werdelin. 2002. Land mammal faunal sequence of the late Miocene of China: New evidence from Lantian, Shaanxi Province. *Vertebrata PalAsiatica* 40:165–176.

Zhang, Z.-q., A. W. Gentry, A. Kaakinen, L.-p. Liu, J. P. Lunkka, Z.-d. Qiu, R. S. Scott, L. Werdelin, S. Zheng, and M. Fortelius. 2002. Land mammal faunal sequence of the late Miocene of China: New evidence from Lantian, Shaanxi Province. *Vertebrata PalAsiatica* 40:165–176.

Zheng, J.-j., X.-x. He, S.-w. Liu, Z.-j. Li, X.-s. Huang, G.-f. Chen, and Z.-d. Qiu. 1999. *Stratigraphical Lexicon of China—Tertiary System*. Beijing: Geological Publishing House.

Zheng, S.-h. and Z.-q. Zhang. 2001. Late Miocene–Early Pleistocene biostratigraphy of the Leijiahe area, Lingtai, Gansu. *Vertebrata PalAsiatica* 39:215–228.

Zhou, M.-z. and C.-k. Hu. 1954. The occurrence of *Anchitherium aurelianense* at Fangshan, Nanking. *Acta Palaeontologica Sinica* 4:525–533.

Zhu, R.-x., Q.-s. Liu, H.-t. Yao, Z.-t. Guo, C.-l. Deng, Y.-x. Pan, L.-q. Lu, Z.-g. Chang, and F. Gao. 2005. Magnetostratigraphic dating of hominoid-bearing sediments at Zhupeng, Yuanmou Basin, southwestern China. *Earth and Planetary Science Letters* 236:559–568.

Zhu, Y., L. Zhou, D. Mo, A. Kaakinen, Z. Zhang, and M. Fortelius. 2008. A new magnetostratigraphic framework for late Neogene *Hipparion* Red Clay in the eastern Loess Plateau of China. *Palaeogeography, Palaeoclimatology, Palaeoecology* 268:47–57.

Zong, G.-f., W.-y. Chen, X.-s. Huang, and Q.-q. Xu. 1996. *Cenozoic Mammals and Environment of Hengduan Mountains Region*. Beijing: Ocean Press.

APPENDIX 1

Representative Mammalian Faunal Lists

Detailed faunal lists for various Neogene Chinese Land Mammal Ages are available from Li, Wu, and Qiu (1984) and Qiu and Qiu (1995). Updated lists from some localities are also available from various authors in this volume. These include Meng et al. (chapter 3) for Ulungu He localities in Xinjiang and the Xiejia Fauna; Qiu and Qiu (chapter 4) for the Xiejiahe and Sihong faunas; Qiu, Wang, and Li, (chapter 5) for Inner Mongolian localities; Zhang et al. (chapter 6) for the Bahe and Lantian formations in Lantian area; and Deng et al. (chapter 9) for the Linxia Basin in Gansu. This appendix only lists those that are not present in the above chapters or those that have been revised. Main literature cited in parentheses "()", "[]" indicates original identifications, and "*" indicates revised identifications in this chapter.

XIEJIA LOCAL FAUNA

Sinolagomys pachygnathus
*Cricetodon youngi** [*Eucricetodon youngi*]
*Parasminthus xiningensis** [*Plesiosminthus xiningensis*]
*Parasminthus huangshuiensis** [*Plesiosminthus huangshuiensis*]
*Litodonomys lajeensis** [*Plesiosminthus lajeensis*]
Yindirtemys suni [*Tataromys suni*] (B.-y. Wang 1997)
Yindirtemys xiningensis [*Tataromys* sp.] (B.-y. Wang 1997)
Tachyoryctoides kokonorensis
*Turpanotherium elegans** [*Brachypotherium* sp.]
Elasmotheriinae gen. et sp. indet.* [*Brachypotherium* sp.]

Sinopalaeoceros xiejiaensis [*Oioceros* (?) *xiejiaensis*] (G.-f. Chen 1988a)

ZHANGJIAPING LOCAL FAUNA
(mainly based on Z.-x. Qiu et al. 2001, augmented by L. Flynn)

RODENTIA* (REVISED BY B.-Y. WANG)
Tataromys plicidens
Tataromys sp.
Yindirtemys grangeri
Y. deflexus
Y. ambiguus
Bounomys sp.
Prodistylomys sp. (only from GL 9303)
Sayimys sp.
Tachyoryctoides cf. *kokonorensis*
Ansomys sp.
Atlantoxerus sp.
Sinotamias sp. (only from GL 9303)
Anomoemys sp.
Parasminthus asiae-orientalis
P. tangingoli
Heterosminthus orientalis
Protalactaga sp.
Democricetodon sp. (only from GL 9303)
Cricetodon? sp.
LAGOMORPHA
Desmatolagus pusillus
Sinolagomys kansuensis

S. ulunguensis
S. pachygnathus

CARNIVORA

Hyaenodon weilini (Wang, Qiu, and Wang 2005)
Ictiocyon cf. *I. socialis* (Wang, Qiu, and Wang 2005)

PERISSODACTYLA

Aprotodon lanzhouensis (Qiu and Xie 1997)
Phyllotillon huangheensis (Qiu et al. 1998)
Turpanotherium elegans (Qiu and Wang 2007)

XISHUIGOU LOCAL FAUNA

(Tiejianggou-I and Dh199909-11 in
B.-y. Wang et al. 2003)

Kinometaxia guangpui (Wang, Qiu, and Wang 2004)
Platybelodon dangheensis (Wang and Qiu 2002b)
Amphimoschus cf. *A. artenensis*
Turcocerus sp.

SIGOU LOCAL FAUNA (Guanghe County, LX 200051; identified by Z.-d. Qiu and B.-y. Wang, May 20/2009)

Mioechinus sp.
Sayimys sp.
Atlantoxerus sp.
Plesiosminthus sp.
Heterosminthus sp.
Litodonomys sp.
Protalactaga sp.
?*Democricetodon* sp.
Sinolagomys ulunguensis

DINGJIAERGOU FAUNA (Qiu and Qiu 1995)

RODENTIA

Sayimys sp.
Tachyoryctoides sp.
Atlantoxerus sp.
Steneofiber sp.
Leptodontomys sp.
Prodryomys sp.
Heterosminthus sp.
Protalactaga grabaui
Paralactaga sp.
Megacricetodon sp.
Democricetodon sp.

LAGOMORPHA

Alloptox gobiensis (Wu, Ye, and Zhu 1991)

PRIMATES

Pliopithecus zhanxiangi (Delson, Guan, and Harrison 1990)

CARNIVORA

Gobicyon sp.
Hemicyon sp.
Tongxinictis primordialis
Sansanosmilus sp.

Proboscidea

Platybelodon danovi [*P. tongxinensis*, Guan 1991]

PERISSODACTYLA

*Hispanotherium matritense**
*Beliajevina tongxinensis**

ARTIODACTYLA

Kubanochoerus gigas (Guan and Van der Made 1993)
Bunolistriodon intermedius (Ye, Qiu, and Zhang 1992)
Stephanocemas sp. *Eotragus* sp.
Turococerus sp.

HALAMAGAI LOCAL FAUNA (Ye et al. 2001a and b; Bi et al. 1999; Ye, Wu, and Meng 2005, Wu, Meng, and Ye 2003)

INSECTIVORA

**Schizogalerix duolebulejinensis*
**Mioechinus*? aff. *M. gobiensis*

RODENTIA

Sayimys sp.
Tachyoryctoides sp.
**Sinomylagaulus halamagaiensis*
Eutamias sp.
Miodyromys sp.
Microdyromys sp.
**Atlantoxerus giganteus*
**A. junggarensis*
Paleaeosciurus sp.
Steneofiber depereti
**Anchitheriomys tungurensis*
Protalactaga grabaui
Democricetodon sp.
Megacricetodon sp.

LAGOMORPHA

**Plicalagus junggarensis*
Alloptox gobiensis

PRIMATES

**Pliopithecus bii*
**Amphicyon ulungurensis*
**Nimravus* sp.
**Pseudaelurus cuspidatus*
**Tunguristis spocki*

*Protictitherium intermedium
*P. sp.
*Thalassictis chinjiensis
*Gobicyon sp
*Oligobunis? sp.
*Alopecocyon goeriachensis
*Simocyon sp.
Proboscidea
*Gomphotherium cf. G. shensiensis
*G. sp.
*Zygolophodon junggarensis
*Z. sp.
*Platybelodon sp.
PERISSODACTYLA
*?Chilotherium sp.
Aceratherium sp.
*Anchitherium gobiense
ARTIODACTYLA
*Lagomeryx sp.
*Stephanocemas aff. S. thomsoni
*Eotragus halamagaiensis
Micromeryx, sp.
Palaeomeryx sp.

NANZHUANGGOU FAUNA

INSECTIVORA
Soriculus praecursus
Yanshuella primaeva
Desmana kowalskae
RODENTIA
Eutamias cf. ertemtensis
Tamiasciurus sp.
Pliopetaurista rugosa
Myomimus sp.
Castor anderssoni
?Eucastor youngi
Dipoides majori
Hystrix sp.
Dipus fraudator
Allocricetus sp.
Neocricetodon grangeri
Prosiphneus lyratus
P. antiquus
Mesosiphneus praetingi
Chardina truncates
Germanomys sp.
Mimomys sp.
Apodemus qiui
M. tedfordi

aff. Karnimata hipparionum
Chardinomys yusheensis
Huaxiamys downsi
Allorattus sp.
LAGOMORPHA
Ochotona lagreli
Trischizolagus sp.
Hypolagus sp.
CARNIVORA
Nyctereutes tingi
N. sinensis
Eucyon davisi
E. zhoui
Agriotherium intermedium
?Ursus sp.
?Helarctos sinomalayanus
Plesiogulo major
Martes sp.
Simocyon sp.
?Hyaenictitherium wongii
Chasmaportetes kani
Pliocrocuta orientalis
Metailurus major
Proboscidea
Anancus sinensis
Mammut shanxiensis
Sinomastodon intermedius
Stegodon zdanskyi
PERISSODACTYLA
?Hipparion (Hipparion) platyodus
H. (Plesiohipparion) houfenense
H. (Cremohipparion) licenti
Proboscidipparion pater
Shansirhinus cornutus
Dihoplus megarhinus
ARTIODACTYLA
Chleuastochoerus stehlini
Sus sp.
Paracamelus gigas
?Paracervulus bidens
?P. brevis
?P. attenuatus
Muntiacus lacustris
Axis shansius
Gazella gaozhuangensis
?G. yusheensis
G. niheensis
Dorcadoryx triquetricornis
Huabeitragus yusheensis
Lyrocerus satan
Sinoreas cornucopia

MAZEGOU FAUNA

INSECTIVORA
cf. *Erinaceus* sp.
cf. *Blarinoides* sp. nov.
Beremendia pohaiensis
Yanshuella primaeva
Scaptochirus sp.

RODENTIA
Pliopetaurista rugosa
Eucastor youngi
Dipoides majori
Hystrix sp.
Dipus fraudator
Rhizomys shansius
?*Cricetulus* sp.
Mesosiphneus paratingi
Germanomys sp.
Mimomys sp.
Apodemus zhangwagouensis
Micromys tedfordi
Chardinomys nihowanicus

LAGOMORPHA
Ochotonoides complicidens
Ochotona sp.
?*Trischizolagus* sp.
Hypolagus sp.

CARNIVORA
Vulpes chikushanensis
Nyctereutes tingi
N. sinensis
Eucyon davisi
E. zhoui
Canis chihliensis
Agriotherium sp.
Ursus sp.
?*Parameles suillus*
Enhydriodon aonychoides
Promephitis sp.
?*Pannonictis* sp.
Vormela prisca
Martes zdanskyi
?*Meles chiai*
Chasmaporthetes kani
C. cf. *ossifragus*
Pliocrocuta orientalis
?*P. perrieri*
Crocuta honanensis
Megaviverra sp.
Homotherium hengduanshanense
Metailurus major
M. parvulus
Lynx shansius
Felis peii
Panthera sp.
Sivapanthera sp.

Proboscidea
Anancus sinensis
Sinomastodon intermedius
?*Mammuthus meridionalis*

Hyracoidea
Postschizotherium intermedium

PERISSODACTYLA
Hipparion (*Plesiohipparion*) *huangheense*
?*Proboscidipparion pater*
P. sinense
Shansirhinus cornutus
Dihoplus sp.

ARTIODACTYLA
Sus sp.
Paracamelus gigas
?*Palaeotragus* sp.
Paracervulus bidens
Muntiacus lacustris
Axis shansius
Gazella yusheensis
G. sinensis
Antilospira licenti
Lyrocerus satan
Sinocapra minor
Ovis yushenesis
Megalovis sp.
?*Bison palaeosinensis*

APPENDIX 2

Stratigraphic Ranges of Some Mammalian Taxa

INSECTIVORA

Amphechinus: Olig., Gashunyinadege

Mioechinus: Lower Aoerban, Shala

Metexallerix: Olig., Gashunyinadege

Quyania: Lower Aoerban, Gaotege

Parasoriculus: Baogeda Ula, Bilike

Sulimskia: Bilike, Gaotege

RODENTIA

Ctenodactylidae

Tataromys/Yindirtemys: Olig., Zhangjiaping

Prodistylomys: Suosuoquan-III, Gashunyinadege

Distylomys: Olig., Tairum Nor

Sayimys: Zhangjiaping, Basal Dingshanyanchi

Tachyoryctoididae

Tachyoryctoides: Olig., Tairum Nor

Pararhizomys: Guonigou, Gaotege

Aplodontidae

Ansomys: Zhangjiaping, Shala

Sciuridae

Atlantoxerus: Lower Aoerban, Leijiahe-V

Gliridae

Miodyromys: Lower Aoerban, Bilutu

Microdyromys: Suosuoquan-II, Bilutu

Myomimus: Bilutu, Nanzhuanggou

Eomyidae

Keramidomys: Lower Aoerban, Huitenghe (Shala)

Leptodontomys: Gashunyinadege, Ertemte

Ligerimys: Lower Aoerban, Gashunyinadege

Castoridae

Youngofiber: Sihong

Anchitheriomys: Moergen

Dipoides: Baogeda Ula, Mazegou

Castor: Amuwusu, Quat.

Zapodidae

Parasminthus: Olig., Zhangjiaping

Plesiosminthus: Olig., Sigou

Heterosminthus: Olig., Amuwusu

Litodonomys: Olig., Sigou

Lophocricetus: Bahe (10 Ma), Bilike

Paralophocricetus: Ertemte, Bilike

Sinozapus: Amuwusu, Bilike

Dipodidae

Protalactaga: Late Zhangjiaping (GL 9303), Bahe (10.1 Ma)

Paralactaga: Amuwusu, Leijiahe-V

Cricetidae

Democricetodon: Suosuoquan-II, Amuwusu

Megacricetodon: Gashunyinadege, Balunhalagen

Cricetodon: Xiejia, Basal Dingshanyanchi

Plesiodipus: Tairum Nor, Balunhalagen

Gobicricetodon: Tairum Nor, Huitenghe

Nannocricetus: Bahe (10 Ma), Leijiahe-III

Kowalskia: Bahe (9.8 Ma), Leijiahe-V

Sinocricetus: Bahe (9.9 Ma), Leijiahe-III

Microtoscoptes: Shala, Ertemte

Gerbillidae

Abudhabia: Bahe (9.9 Ma), Baogeda Ula

Pseudomeriones: Baogeda Ula, Leijiahe-V

Muridae

Progonomys: Bahe (10–9.7 Ma)

Hansdebruijnia: Baogeda Ula, Ertemte
Apodemus: Ertemte, Quat.
Micromys: Ertemte, Quat.
Chardinomys: Taoyang, Quat.
Huaxiamys: Mahui, Leijiahe-V
Siphneidae
   Prosiphneus: Amuwusu, Ertemte
   Chardina: Nanzhuanggou, Leijiahe-V
   Mesosiphneus: Nanzhuanggou, Mazegou
Arvicolidae
   Mimomys: Zanda, Quat.
LAGOMORPHA
   Desmatolagus: Olig., Amuwusu
   Sinolagomys: Olig., Halamagai
   Alloptox: Gashunyinadege, Balunhalagen
   Bellatona: Tairum Nor, Balunhalagen
   Ochotona: Bahe (9.9 Ma), Quat.
   Ochotonoides: Mazegou, Quat.
   Alilepus: Baogeda Ula, Ertemte
   Trischizolagus: Bilike, Leijiahe-V
CREODONTA
   Hyaenodon: Olig., Zhangjiaping
CARNIVORA
  Canidae
   Nyctereutes: Nanzhuanggou, Quat.
   Eucyon: Nanzhuanggou, Mazegou
   Canis: Mazegou, Quat.
   Vulpes: Mazegou, Quat.
  Amphicyonidae
   Amphicyon: Xiejiahe, Moergen
  Ursidae
   Hemicyonines: Xiejiahe, Moergen
   Ursavines: Xiejiahe, Liushu (mid)
   Indarctos: Liushu (mid), Mahui
   Agriotherium: ?Liushu, Quat.
   Ursus: Nanzhuanggou, Quat.
  Mustelidae
   Martes: Taoyang, Quat.
   Sinictis: Liushu (mid), Baode
   Parataxidea: Liushu (lower), Baode
   Meles: Mazegou, Quat.
   Eomellivora: Liushu (mid), Baode
   Plesiogulo: Lamagou, Taoyang
   Enhydriodon: Baode, Mazegou
   Promephitis: Liushu (lower), Huainan
   Leptarctines: Xishuigou, Moergen
  Procyonidae
   Simocyon: Lamagou, ?Nanzhuanggou
  Hyaenidae
   Percrocuta: Dingjiaergou, Moergen
   Dinocrocuta: Liushu, Lamagou

Adcrocuta: ?Bahe (9.9 Ma), Mahui
Tungurictis: Tairum Nor, Moergen
Ictitheres: Halamagai, Nanzhuanggou
Pliocrocuta: Nanzhuanggou, Mazegou
Chasmaporthetes: Nanzhuanggou, Quat.
Nimravidae
   Sansanosmilus: Dingjiaergou, Moergen
Felidae
   Pseudaelurus: Sihong, Halamagai
   Metailurus: Moergen, Mazegou
   Machairodus: Guonigou, Baode
   Homotherium: Mazegou, Quat.
   Lynx: Mazegou, Quat.
   Felis: Mazegou, Quat.
   Sivapanthera: Mazegou
Viverridae
   Semigenetta: Sihong
   Megaviverra: Mazegou
PRIMATES
   "*Kansupithecus*": Xishuigou
   Pliopithecus: Dingjiaergou, Halamagai
   Dionysopithecus/Platodontopithecus: Sihong
Proboscidea
   Prodeinotherium: Guonigou
   Gomphotherium: Xiejiahe, Halamagai
   Platybelodon: Xishuigou, Miaoliang
   Tetralophodon: Guonigou, ?
   Anancus: Dongzhuang, Quat.
   Sinomastodon: Dongzhuang, Mazegou
   Mammut: Dongzhuang, Mazegou
   Stegodon: Dongzhuang, Quat.
   Mammuthus: Mazegou, Quat.
Hyracoidea
   Postschizotherium: Mazegou, Quat.
PERISSODACTYLA
  Equidae
   Anchitherium: Sihong, Amuwusu
   Sinohippus: Liushu (mid?), Baode
   Hipparion (*Hipparion*): Guonigou, Nanzhuanggou
   Hipparion (*Plesiohipparion*): Nanzhuanggou, Quat.
   Hipparion (*Cremohipparion*): Nanzhuanggou
   Proboscidipparion : Nanzhuanggou, Quat.
  Tapiridae
   Plesiotapirus: Xiejiahe
   Tapirus: Liushu (mid?), Quat.
  Chalicotheriidae
   Phyllotillon: Zhangjiaping
   Anisodon: Xiejiehe, Moergen
  Rhinocerotidae
   Turpanotherium: Olig., Zhangjiaping
   Plesiaceratherium: Xiejiehe, Cixian

Chilotheres: Cixian, Nanzhuanggou
Lartetotherium: Cixian
Dihoplus: Baode, Mazegou
Beliajevina: Dingjiaergou
Hispanotherium: Dingjiaergou, Moergen
Parelasmotherium: Guonigou
Sinotherium: Lamagou, Baode

ARTIODACTYLA
  Suiformes
    Hyotherium: Xiejiahe
    Chleuastochoerus: Liushu (mid), Nanzhuanggou
    Microstonyx: Liushu (mid), Baode
    Bunolistriodon: Dingjiaergou
    Kubanochoerus: "Hsienshuiho" (Minhe), Moergen
    Listriodon: Lengshuigou, Moergen
    Sus: Mazegou, Quat.
  Camelidae
    Paracamelus: Nanzhuanggou, Quat.
  Ruminantia
    Dorcatherium: Sihong, Shinanu
    Amphimoschus: Xishuigou
    Sinomeryx: Xiejiahe
    Palaeotragus: Moergen, ?Mazegou

Schansitherium: Liushu (mid), Baode
Honanotherium: Liushu (mid), Mahui
Dicrocerus: Tairum Nor, Tuosu (lower)
Ligeromeryx/Lagomeryx: Upper Aoerban, Balunhalagen
Stephanocemas: Sihong, Tuosu (lower)
Cervavitus: Mahui, Miaoliang
Paracervulus: Nanzhuanggou, Mazegou
Axis: Nanzhuanggou, Quat.
Muntiacus: Nanzhuanggou, Quat.
Turcocerus: Xishuigou, Kekemaideng
Shaanxispira: Guonigou, Bahe (8.18 Ma)
Hezhengia: Dashenggou, Liushu (mid)
Plesiaddax: Lamagou, Baode
Tsaidamotherium: Tuosu (lower), Liushu (lower?)
Urmiatherium: Baode
Gazella: Mahui, Quat.
Antilospira: Zanda, Quat.
Dorcadoryx: Mahui, Nanzhuanggou
Lyrocerus: Nanzhuanggou, Mazegou
Sinocapra: Mazegou
Ovis: Mazegou, Quat.

Chapter 2

North China Neogene Biochronology
A Chinese Standard

MICHAEL O. WOODBURNE, RICHARD H. TEDFORD, AND EVERETT H. LINDSAY

It is readily apparent that the Neogene record of fossil mammals in China is both temporally extensive and representative as well as geographically diverse. The following discussion presents the thesis that Neogene chronologic analysis in China will be best served when an endemic standard is developed that is independent of other biochronologic systems, such as MN zones in Europe or NALMAs in North America. We propose that North China has the best stratigraphic record for this purpose and that such a biochronologic system should be developed first within a given biogeographic region, and then extended outward as justified, to integrate with other biochronologies. We also propose that it is of fundamental importance to begin by establishing independent lithostratigraphic and biostratigraphic frameworks and to build biochronologies therefrom. Here, we focus on the record in North China, defined as the region co-extensive with the past and present distribution of myospalacine and certain microtine rodents. This region surrounds Beijing and continues westward to include the Yellow River, the Tibetan Plateau, and adjacent Kazakhstan, and northward to include Mongolia and adjacent parts of Russia, as indicated in figure 2.1. Our focus recognizes the substantial and increasing role played by this region in documenting the evolutionary development of Chinese mammal faunas. We also appreciate that the strength of this record is increasingly enhanced by the application of magnetostratigraphy to the rock sequences

Richard H. Tedford is deceased.

that contain the fossils, thereby providing an independent means of chronologic analysis and correlation (figure 2.2). These factors combine to establish the North China succession as an evolving standard to which local fossil mammal sequences can be aligned within China and correlated to those in other regions. The development of this succession also will highlight the evolutionary history of Chinese Neogene mammals both endemically and with respect to dispersals to and from adjacent regions.

In this chapter, we present a methodology designed to accomplish the establishment of the North China mammalian record as the nucleus of a regional Chinese Neogene biochronology. This is followed by local examples having stratigraphic successions of proven or potential utility in this regard. We then offer a revised Chinese Neogene mammal biochronology, concluding with examples that might lead toward the development of a fully defined chronostratigraphic system in the future.

Definitions and Abbreviations

ASSEMBLAGE ZONE "A stratum or body of strata characterized by a distinctive assemblage or association of three or more fossil taxa that, taken together, distinguishes it in biostratigraphic character from adjacent strata" (Salvador 1994:62–63).

BIOCHRONOLOGIC UNIT An interval of geologic time based on biological data, usually a genus or species, alone or in association with others (after Salvador 1994:109).

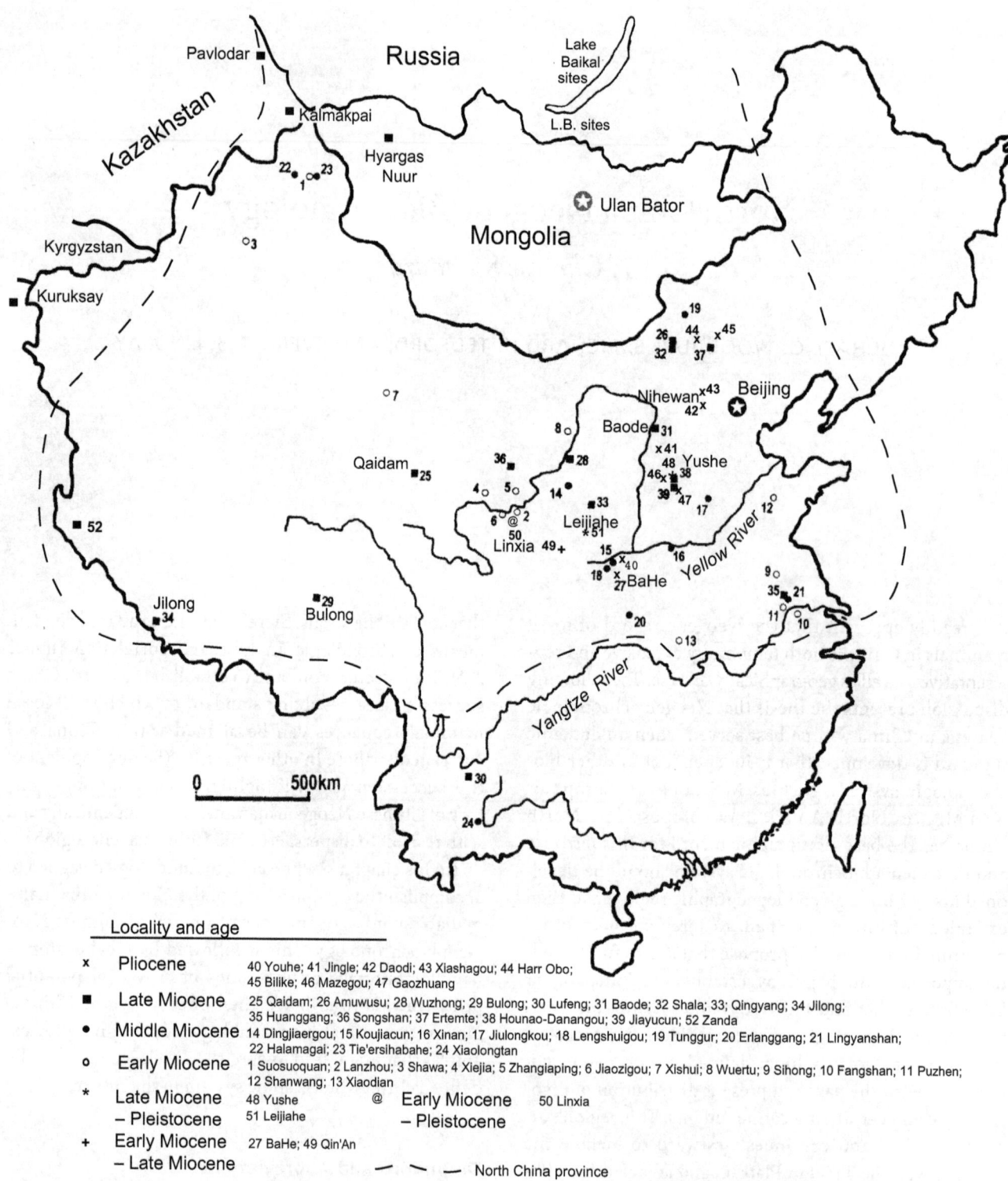

Figure 2.1 Map of the Chinese Neogene mammal localities. After Qiu, Wu, and Qiu (1999), with Pliocene sites added from Qiu and Qiu (1995).

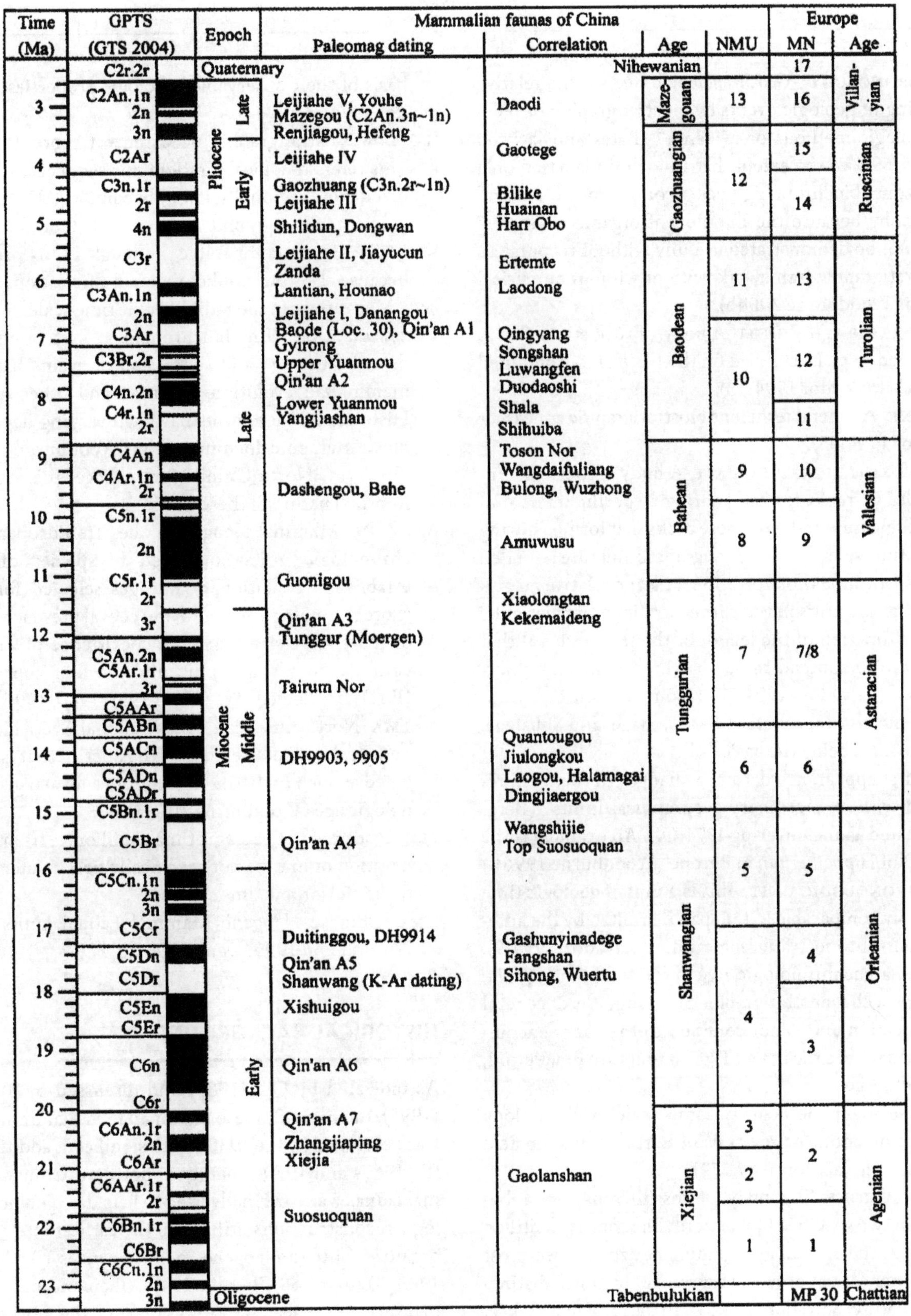

Mammalian faunas of China / Europe correlation chart

Time (Ma)	GPTS (GTS 2004)	Epoch	Paleomag dating	Correlation	Age	NMU	MN	Age
	C2r.2r	Quaternary			Nihewanian		17	
3	C2An.1n / 2n / 3n	Pliocene — Late	Leijiahe V, Youhe / Mazegou (C2An.3n~1n) / Renjiagou, Hefeng	Daodi	Mazegouan	13	16	Villanyian
4	C2Ar	Pliocene — Early	Leijiahe IV	Gaotege	Gaozhuangian	12	15	Ruscinian
	C3n.1r / 2r		Gaozhuang (C3n.2r~1n) / Leijiahe III	Bilike / Huainan			14	
5	4n		Shilidun, Dongwan	Harr Obo				
	C3r		Leijiahe II, Jiayucun / Zanda	Ertemte		11	13	Turolian
6	C3An.1n / 2n		Lantian, Hounao	Laodong	Baodean			
7	C3Ar	Miocene — Late	Leijiahe I, Danangou / Baode (Loc. 30), Qin'an A1 / Gyirong	Qingyang / Songshan			12	
	C3Br.2r		Upper Yuanmou / Qin'an A2	Luwangfen / Duodaoshi		10		
8	C4n.2n / C4r.1n / 2r		Lower Yuanmou / Yangjiashan	Shala / Shihuiba			11	
9	C4An / C4Ar.1n / 2r		Dashengou, Bahe	Toson Nor / Wangdaifuliang / Bulong, Wuzhong	Bahean	9	10	Vallesian
10	C5n.1r / 2n			Amuwusu		8	9	
11	C5r.1r / 2r / 3r		Guonigou	Xiaolongtan / Kekemaideng				
12	C5An.2n / C5Ar.1n / 3r		Qin'an A3 / Tunggur (Moergen)		Tunggurian	7	7/8	Astaracian
13	C5AAr / C5ABn	Miocene — Middle	Tairum Nor					
14	C5ACn / C5ADn / C5ADr		DH9903, 9905	Quantougou / Jiulongkou / Laogou, Halamagai / Dingjiaergou		6	6	
15	C5Bn.1r			Wangshijie / Top Suosuoquan				
16	C5Br / C5Cn.1n / 2n / 3n		Qin'an A4		Shanwangian	5	5	Orleanian
17	C5Cr / C5Dn / C5Dr		Duitinggou, DH9914 / Qin'an A5 / Shanwang (K-Ar dating)	Gashunyinadege / Fangshan / Sihong, Wuertu			4	
18	C5En / C5Er	Miocene — Early	Xishuigou			4	3	
19	C6n		Qin'an A6					
20	C6r / C6An.1r / 2n		Qin'an A7					
21	C6Ar / C6AAr.1r / 2r		Zhangjiaping / Xiejia	Gaolanshan	Xiejian	3	2	Agenian
						2		
22	C6Bn.1r / C6Br / C6Cn.1n		Suosuoquan			1	1	
23	2n / 3n	Oligocene		Tabenbulukian			MP 30	Chattian

Figure 2.2 Subdivision and correlation of Chinese Neogene mammalian localities. After Deng (2006:fig. 2).

BIOCHRONOLOGY *"Geochronology* based on the relative dating of geologic events by biostratigraphic or paleontologic methods or evidence" (Bates and Jackson 1987:69). To the extent that a biochron is based on a biozone, biochronology has a connection to biostratigraphy because the duration of organic characters cannot be demonstrated usefully without recourse to a stratigraphic framework, with or without numerical data (Woodburne 2004b).

BIOSTRATIGRAPHIC UNIT A body of rock strata that is defined or characterized on the basis of its contained fossils (Salvador 1994:53).

BIOZONE A general term for a biostratigraphic zone (Salvador 1994:55).

CHRONOSTRATIGRAPHIC UNIT A body of rocks that includes all rocks formed "during a specific interval of geologic time and only those rocks that formed during that time span. Chronostratigraphic units are bounded by isochronous horizons. The rank and relative magnitude of the units in the chronostratigraphic hierarchy are a function of the length of the time interval that their rocks subtend, rather than of their physical thickness" (Salvador 1994:77). Common chronostratigraphic units are System, Series, Stage, and Substage, with Stage being the basic unit.

FAD First appearance datum, a change "in the fossil record with extraordinary geographical limits" (Berggren and Van Couvering 1974:IX). An example is the FAD of *Hippotherium* in Europe (Woodburne 1996).

GEOCHRONOLOGIC UNIT This is a unit of geologic time exactly comparable to the span delimited by the antecedent chronostratigraphic unit (Salvador 1994:9). Thus, a geochronologic Age is the temporal equivalent of a chronostratigraphic Stage, but it is of general application and is not confined to the rocks encompassed within the Stage. Typical units are Era, Period, Epoch, Age.

GPTS Geomagnetic Polarity Time Scale. A chronology based on counting reversals of Earth's magnetic field (Bates and Jackson 1987:272).

INTERVAL-ZONE The body of fossiliferous strata between two specified biohorizons. The zone is identified only on the basis of its bounding horizons (lowermost occurrence, uppermost occurrence, or other distinctive feature; Salvador 1994:59). In the present instance, we utilize Interval-zones composed of the LO (lowest stratigraphic occurrence) of two taxa, one LO for each boundary, and choose a characterizing assemblage from those taxa found within the zone so defined.

LITHOSTRATIGRAPHIC UNIT Bodies of rock, bedded or unbedded, that are defined and characterized on the basis of their observable lithologic properties (Salvador 1994:31).

LO Lowest stratigraphic occurrence (Aubry 1997:17). This refers to a local stratigraphic sequence. If recognized more regionally, it may become a FAD. The LSD (Lowest Stratigraphic Datum) of Opdyke et al. (1977:334) is comparable, although LO is preferred because "Datum" implies regional correlation.

MA Megannum in the radioisotopic time scale.

MAMMAL AGE The biochronologic system used to describe the age and succession of events based on mammalian evolution in China and other regions. Historically, these units have had varying degrees of biostratigraphic documentation (Woodburne 2004a). The term "North China Mammal age" preserves the informal nature of these units.

MN ZONE Mammal Neogene Zones are European biochronologic units composed of species of well-established evolutionary lineages selected for their short chronological record, the coeval presence of two or more large-sized taxa the coexistence of which is confined to a single unit, and the first appearance (FAD) of selected taxa (Mein 1975, 1980, 1996).

NALMA North American Land Mammal Age, a biochronological system used in North America that is designated as such to distinguish it from a formal geochronologic age (Woodburne 2004a:xiv).

M.Y. A segment of geologic time a million years either in duration or in time not specifically tied to intervals of the radioisotopic time scale.

NMU Chinese Neogene Mammal Faunal Units (Qiu, Wu, and Qiu 1999; Deng 2006:fig. 2).

HISTORICAL BACKGROUND

As indicated by Qiu (1989) and discussed by Tedford (1995), the basic Chinese mammalian faunal framework was developed by the 1930s, with significant additions in the 1960s and 1970s, mainly in northern China. Mammal faunas were originally used to help date geomorphologic reconstructions utilized in the earliest attempts to organize China's Neogene rock sequences (Andersson 1923; Teilhard de Chardin 1941). Then and now, fossil mammals have been important in being the most widely applicable means by which the age of China's continental Cenozoic deposits can be established. In contrast to other regions, volcanic activity has been relatively sparse during the Cenozoic Era in China, so that radioisotopic means for independently assessing the age of the strata are not widely applicable in China.

The Neogene mammal faunal sequence was originally proposed by Chiu (=Qiu), Li, and Qiu (1979), and most recently discussed by Li, Wu, and Qiu (1984), Qiu (1989), Qiu and Qiu (1995), Tedford (1995), and Qiu, Wu, and Qiu (1999). In these discussions it is apparent that the fossil mammal record in China is abundant and widely distributed, including many taxa that are endemic to Asia, and the relative sequences of which can be proposed on phylogenetic bases, as well as by direct superposition.

Thus, Chiu (=Qiu), Li, and Qiu (1979) noted the endemic nature of the Chinese faunas and used wide-ranging taxa shared with European successions to derive correlations to conventional time scales, with varying degrees of success. Whereas in many cases faunal similarity can be found with European units, other Chinese faunas show affinity to those of southwestern Asia (Siwaliks), which underscores the varied biogeographic affinities of the Chinese faunal record. Similar discussions, with a growing wealth of detailed information, are supplied by Qiu and Qiu (1995) and Qiu, Wu, and Qiu (1999), but the basic endemism of Chinese faunas still is apparent. This is discussed in Tedford (1995), who identified the intervals for which a relatively large number of taxa are shared with Europe (up to 50%; MN 6–7, MN 11, MN 13–14, and MN 16), versus those intervals in which taxa are mostly distinct (less than 30% shared taxa; MN 1–2, MN 3–5), for example.

Tedford (1995) noted the growing application of magnetostratigraphic methods to fossil-bearing successions in China, which enhanced the means by which they may be compared with mammal-bearing units in other regions, and in that context demonstrated the integration of biostratigraphic and magnetostratigraphic data in the Yushe Basin of eastern China. Tedford (1995) concluded that the goal of developing a more completely documented Neogene fossil mammal succession in China is a topic for future work.

This goal was addressed by publications such as Qiu, Wu, and Qiu (1999), utilizing the then known lengthy fossiliferous records in northern and northwest China, to emphasize their affinity with those of Kazakhstan and thereby to correlate more securely with mammal records of the Paratethys of Europe, along with increased information regarding late Miocene faunas of Inner Mongolia. Qiu, Wu, and Qiu (1999) also elaborated the mammal faunal unit system begun by Qiu (1989; see also Qiu and Qiu 1995).

As originally conceived by Qiu (1989:table 2) the six intervals of Chinese mammal units were numbered I–VI (rather than named). They were named Mammal Units as informal categories that might become the bases for Chinese mammal ages. Emphasis clearly was on the fossil biota. As examples of biogeographic influence, mammal unit I was noted as having affinities with the Bugti fauna of Pakistan; elements of mammal unit II showed close relationships with European taxa; mammal unit III consisted of mostly endemic forms; mammal unit IV was heralded by the appearance of the horse, *Hipparion*, and its nearly contemporaneous European and African records; mammal unit V, in contrast, reflected a North American affinity and mammal unit VI was based on the North American (*Equus*) and African elephant (*Archidiskodon*) immigrants.

Subsequently, Qiu and Qiu (1995) named six Chinese Mammal Ages, modified from those proposed by Li, Wu, and Qiu (1984), which mostly corresponded to the informal mammal units of Qiu (1989). Deng (2006:fig. 2) added two more, designated all of them as Chinese Neogene Mammal Faunal Units (NMU) and increased their number from 11 to 13 (see figure 2.2 [China NMU column]). Deng (2006) also took the very important step of integrating the mammal associations with the magnetic polarity record so as to provide an independent means of correlation with the global geologic time scale.

TOWARD A NORTH CHINA NEOGENE TIME SCALE

Methods and Principles

As discussed in Woodburne (2004b), development of a biochronologic system, such as a sequence of mammal ages, is intended to result in a sequence of temporal intervals that accounts for all of, in this case, Neogene time without gaps or overlaps. It is critical to preserve the basic integrity of the system at the very beginning. This means that the process of developing a biochronology begins with the demonstration of an empirically objective lithostratigraphy, the description and recognition of which depends solely on lithologic criteria without regard for origin or age (table 2.1). The following is paraphrased from Woodburne (1977, 1996, 2004a, 2004b, 2006).

Lithostratigraphic units. The purpose of a lithostratigraphic unit is to demonstrate a physical framework at a level of detail pertinent to the study at hand and to maintain that framework independently of any other consideration. As indicated in table 2.1, the Formation is the fundamental lithostratigraphic unit, defined on lithologic criteria without regard for age or geologic origin (see "Definitions and Abbreviations").

Table 2.1

Summary of Stratigraphic Categories and Concepts

Lithostratigrapic[a]	Biostratigraphic[b]	Chronostratigraphic[c]	Geochronologic[d]
Formation	Various kinds of biozones	Chronozone	Chron
Member			
Bed			
Horizon			

[a]Physical, descriptive unit based on lithological characteristics without regard to age or source of deposition. A formation may be lithologically heterogeneous or homogeneous. A member usually is more lithologically homogeneous and may be interpreted as to lithogenesis. A bed commonly is of limited thickness and lithologically homogeneous. A horizon is of very limited (conceptually zero) thickness but is a traceable marker.

[b]Fundamentally a physical unit, descriptive of the fossils in their stratigraphic context. Procedurally independent of other stratigraphic units, biostratigraphic units can be developed for their biochronological significance and transformed into the paleontological basis for recognizing chronostratigraphic units. These are commonly various kinds of biozones, such as taxon-range zones, concurrent-range zones, or lineage zones that are based on the distribution of taxa without regard for abundance.

[c]This is a physical unit in that it is the rock deposited during an interval of geologic time. It is a conceptual unit in the sense that the means by which it is recognized (most commonly fossils) are presumed to have a unique temporal component. Once a chronostratigraphic (time-rock) unit is created a corresponding geochronologic (geologic time) unit of equal rank also is created. Chronostratigraphic units are the fundamental means for building a time-rock record that accounts for all of geologic time, without gaps or overlaps. Their development requires the utilization of a type section, and potentially reference sections depending on where unit boundaries are best demonstrated. The chronozone is the basic unit.

[d]This is a conceptual and intangible unit that represents an interval of geologic time. It is not a stratigraphic unit, although it may correspond to the time span of such a unit. It is thus possible to speak of events that transpired during, e.g., the Ypresian age without reference to a specific section of strata. The chron is the basic unit.

Biostratigraphic units. The next step in the process requires the demonstration of the biostratigraphy, with the documentation of the stratigraphic succession of fossils within the lithostratigraphic sequence forming an empirical framework parallel to that of lithostratigraphy. A biostratigraphic unit is a physical, descriptive unit, the only interpretive aspect of which is the identification of the fossil taxonomy. A species may be recognized on the basis of a single specimen in the rock record, or on a stated level of the abundance of the relevant morphology. The precision of a boundary based on the first stratigraphic or last stratigraphic occurrence of a given species will be affected by the method of recognition. The biostratigraphy should be as refined as pertinent to the anticipated resolution of the eventual biochronologic correlation.

Biochronology. The next step in the formalized process, which moves from an empirical biostratigraphy ultimately to a chronostratigraphy, requires the interpretation that the biostratigraphic sequence developed in one place and recognized in another is of temporal significance. This interpretive step is known as biochronology. Biochronology is rooted in the irreversibility of organic evolution and documented as to its place in time and space by the stratigraphic framework in which it is shown. Once placed in an empirical stratigraphic sequence, the biotic properties can be evaluated for their relationship to any other kind of potentially significant temporal information, such as magnetostratigraphic or radioisotopic data. It is important to emphasize that a magnetostratigraphic chronology is developed in the same manner, from a well-documented stratigraphic sequence with data tied to precise lithologic units irrespective of fossil content. Only after the magnetic properties are identified (in the laboratory) can they be evaluated for independent chronologic information, such as the sequence of reversals. A similar corollary applies to radioisotopic data in that the relevant samples are obtained and analyzed independently of the biostratigraphic data. The ultimate goal is to arrive at a regional biostratigraphy and independent chronologic documentation that enable the original biostratigraphic record to be integrated with other temporal systems.

Neither North American (NALMAs; Woodburne, 2006) nor Chinese (this chapter) mammal biochronologic units are part of a fully formalized process. Both examples enjoy stratigraphic documentation that usually is capable of demonstrating their relative age relationships (in contrast to many MN units; e.g., Steininger 1999). Both examples are capable of providing important means of correlating terrestrial (and some marine) deposits, and both are enhanced by independent (radioisotopic, stable isotope, and magnetochronologic) chronologic methods. On the one hand, the Chinese biochronological units are extremely valuable in correlation; on the other, they are not yet at a point where they can form the basis of chronostratigraphic Stages. It is a thesis of the present chapter that this second step, development of a stable chronostratigraphic system in terrestrial deposits, is a goal for the future and that the North China succession holds great promise in that regard. When more formalized documentation is provided and thoroughly tested, the Chinese biochronologic units will support moving to the next level, that of chronostratigraphy.

Chronostratigraphic units. These are physical units in that they are composed of the rocks that formed during a specified interval of geologic time (see table 2.1). They are interpreted as being of temporal significance. Examples shown herein employ Interval Zones to develop a biostratigraphic zonation with succinctly defined boundaries that would lack gaps or overlaps with respect to sub- and superjacent biozones. The two name-bearing taxa in the interval zone (such as the *Cupidinimus avawatzensis–Hipparion forcei* Interval Zone) each represent a LO (see "Definitions and Abbreviations") and provide the definition of the beginning of the zone, followed by the definition of the base of the next succeeding zone. The taxa that occur within the stratigraphic interval, so defined, can provide the characterizing group of taxa by which the zone may also be recognized from place to place. Whereas the North China mammal ages offered herein clearly are informal biochronologic units, the point of the present demonstration is that the North China record also is at this moment capable of supporting detailed formalized biostratigraphic zonations. Until all pertinent data have been compiled on a more widespread basis, and the sequence of biochronologic events is rigorously tested, it is not appropriate to designate Chinese mammal units as chronostratigraphic Stages.

Geochronologic units. As indicated in table 2.1, these arise as soon as a chronostratigraphic unit has been formally established. They are interpretive units considered to represent an interval of time coeval with that of the antecedent chronostratigraphic category (X Stage automatically defines the X Age).

The basic premise of the previous discussion is that any time scale based on organic evolution needs to be founded on a biostratigraphic framework. Philosophically, stratigraphic sequences need to be identified that contain all of past time, without overlap, in as clear and unambiguous a manner as possible. The problem is that most of the rock or stratigraphic record consists of hiatuses interspersed between pulses of deposition. Unfortunately, we rarely (if ever) can identify all hiatuses in the rock record, and so far we have no secure method to recognize when all of the hiatuses of a particular stratigraphic section have been eliminated. The problem of "insignificant hiatuses" has never even been addressed. In order to document a complete temporal record, we need to compile many similar records of the same stratigraphic-temporal interval to reduce the chance of missing some significant segment of that temporal record. In the early days of biostratigraphy, Albert Oppel realized that the presence of fossils with overlapping stratigraphic ranges in multiple sections is a reliable means of eliminating significant gaps, or hiatuses, in those stratigraphic sequences; hence, the development of concurrent-range zones. With current technology, we have additional means of identifying where significant hiatuses in the stratigraphic record occur; however, the recognition of fossils with overlapping ranges remains a primary objective in biostratigraphy for confidence in the removal of hiatuses. Currently, the key to developing reliable chronostratigraphy is to produce multiple stratigraphic sequences with numerous taxa having overlapping stratigraphic ranges. This is not always possible, and always requires much work and time, but it is essential for the development of a reliable chronostratigraphy. In general, the more densely sampled stratigraphic sections should produce the more complete sequences, with fewer hiatuses. Here is where magnetostratigraphy may be useful. If the thickness of a correlated magnetozone is larger in one section than in another section, we usually assume there are more hiatuses in the section with the thinner magnetozone. In short, many stratigraphic sections of the same temporal interval are required to develop a complete representation for the entire Neogene, and means must be employed to facilitate clear and unambiguous correlations between sections that are considered temporally equivalent. This is most important when a boundary between two chronostratigraphic units is selected; if that boundary is placed in a stratigraphic section where there is a significant hiatus, it is certain to cause problems (and confusion) with later correlation. This is the prime reason that boundaries should be selected only after we have gained confidence that most stratigraphic gaps have been identified and/or eliminated. Our rallying cry should be "Let there be no boundaries placed between chronostratigraphic intervals without complete knowledge of the fauna in those intervals." Whereas independent means, such as radioisotopic and magnetic polarity chronologies, are invaluable in such demonstrations, internal consistency requires that the biostratigraphies be correlated, first, by biochronology. This requires, in turn, that the boundaries between the biochronologic units must be defined, but only after we are confident that all (or the most important) stratigraphic gaps have been filled.

Some Lingering Problems

Chinese Mammal Units Are Characterized but (Mostly) Not Defined

Qiu (1989) illustrated the use of representative faunal units to help organize the understanding of faunal evolution in China. An example, the Lanzhou Fauna, was considered as the oldest Neogene mammalian unit in Mammal Unit I and part of the Xiejian Stage of Li, Wu, and Qiu (1984). Qiu (1989) provided the Lanzhou Fauna with a characteristic assemblage of taxa, gave a stratigraphic context, and suggested correlation to other regional faunas, based on both taxonomic as well as stage-of-evolution considerations. The Neogene age of the faunas of Unit I is derived from an interpretation that its faunas are more derived than those conventionally assigned to the Oligocene.

The chronologically next-younger Xiejia Fauna exhibits the role played by intercontinental dispersal. The characterizing taxa for the Xiejian include a large component of rodents, as well as a few carnivores and ungulates. Among these, *Plesiosminthus*, *Eucricetodon*, and *Brachypotherium* are considered as represented in Europe, as well as in China. Whereas subsequent taxonomic revisions have altered the generic nomenclature of some of these forms, elements of the Xiejian fauna still share sufficiently close phyletic relationships with European taxa so as to promote a correlation on that basis and thereby to integrate portions of the Chinese time scale with that of Europe. Still, neither the Xiejian fauna, the Xiejian Stage, nor Mammal Unit I was given a taxonomic definition (sensu Woodburne 1977, 1996, 2004b).

The stratigraphic first appearance of the equid genus, *Hipparion*, "marks" (=defines) the beginning of Mammal Unit IV of Qiu (1989), who indicates that the Qaidam (Tsaidam) fauna, Qinghai, is considered as the oldest such fauna in China. This general framework was followed by Qiu and Qiu (1995) except that the then equivalent Baodean Age faunal unit was "characterized" by the "coexistence of elements of the *Anchitherium* fauna and the earliest *Hipparion* faunas" (51). Qiu, Wu, and Qiu (1999) briefly discuss the "*Hipparion* datum" event, but do not use it to define NMU 8.

Mammal Unit IV (NMU 8) was once defined biochronologically, but that practice has not been continued, although Deng (2006) noted that MN 9 in Europe (to which NUM 8 is correlated) is "defined by FAD of *Hipparion*." Deng (2006) considered the Guonigou fauna (see figure 2.2 [Paleomag column]) with an early species

of *Hipparion* (*H. dongxiangense*) in China, as 11.1 Ma in age based on paleomagnetic data, but he did not explicitly define NMU 8 on the FAD of *Hipparion*. The use of immigrant taxa to "characterize" Mammal Unit VI by Qiu (1989) has been noted earlier.

In summary, a strong and viable succession of mammalian faunas and biochrons forms the basis of the Chinese Neogene mammalian time scale (see figure 2.2 [China Faunal Age column]). At the same time, the question of boundaries between these units has not been addressed. The magnetostratigraphic calibration of the NMU units by Deng (2006) is admirable and forms a useful basis for correlation both within and beyond Asia. Still, the tempo and mode of mammalian evolution in Asia proceed at their own pace. The results of that evolution likely will be expressed at points in Neogene time that do not conform to those seen in other zoogeographic provinces, such as Europe. Study of mammalian evolution in Asia needs to be reflected by a time scale pertinent to that evolution, and the remainder of this report will be addressed to that goal.

Separation of Lithostratigraphic, Biostratigraphic, and Chronostratigraphic Unit Names

As described by Salvador (1994) and illustrated in Woodburne (2004a), it is very important to separate the development of these three stratigraphic units from one another and to maintain separate names for the intervals that result. This is especially critical when the goal is to devise units of temporal correlation with respect to different inter- or intrabasinal sequences. As documented below (Yushe Basin), a biostratigraphic array of taxa commonly is developed without regard for lithostratigraphic (formational) boundaries. Using the lithostratigraphic name of one unit to apply to the other can only result in confusion. If the intervals based on biostratigraphy become sufficiently replicated regionally as to justify using them to propose a biocorrelation, then the boundaries of those units will be based on biological, rather than lithological, criteria. The two sets of units (lithological and biochronological) thus deserve to be recognized by different names so as to preserve their separate identities as well as criteria for recognition. As also discussed previously, a chronostratigraphic unit (such as a Stage) represents the strata deposited during the time span being represented, and the recognition of that time span can be based on paleontological as well as other (radioisotopic, magnetochronologic) criteria. Once established, a Stage (or other chronostratigraphic unit) is the precursor for the equivalent geochronologic unit (Age), and in that context the two units have the same basic name. Other examples of name repetition are, for instance, the Yushe Group for a lithostratigraphic unit, Yushean for the biochron, and Yushe Stage for a chronostratigraphic unit.

Definition of Biochrons

At present, mammal ages are units of biocorrelation not formally recognized in stratigraphic codes or guides (e.g., Salvador 1994), but nevertheless mammal ages have been shown to have great utility in assessing the age of mammal-bearing units, as abundantly illustrated by Chiu (=Qiu), Li, Chiu (1979), Li, Wu, and Qiu (1984), Qiu (1989), Qiu and Qiu (1995), Tedford (1995), Qiu, Wu, and Qiu (1999), and Deng (2006).

Woodburne (1977, 1996, 2006) has proposed that single-taxon definitions are the only appropriate means by which mammalian biochrons can be unambiguously defined and correlated and that multiple-taxon "definitions" are actually characterizations. Both procedures have a role to play in developing mammalian biochronological units (best developed when proceeding from a biostratigraphic demonstration).

Single-taxon definitions may inherit sources of imprecision, such as a statistical versus a first-record definition of the FAD of a taxon when encountered in a stratigraphic interval that reflects a speciation event, or a question of precision when faced with an immigration event relative to the time of the origin of the immigrant in its source area. But a characterization is composed of a given number of single-taxon occurrences, each of which is subject to the same set of potential errors as indicated earlier, in addition to the demonstrably much more important consideration that each taxon is produced by its own set of evolutionary processes. Unless forced extrinsically, or made to appear synchronous by geologic processes, it is very unlikely that any two taxa will have the same sojourn through time and result in an actually synchronous joint first occurrence, either temporally or geologically.

Nevertheless, single-taxon definitions are very important for unambiguous establishment and recognition of unit boundaries. Multiple-taxon characterizations are extremely important for correlating units from place to place. Whereas it may not always be possible to find the defining taxon in various local sections, the presence of the unit in question can be determined by finding elements of the characterizing taxa. Although it may be more common to find a specimen of a characterizing taxon than a defining taxon, this should not deter nomination of boundaries on the basis of single taxa.

Gaps in Sequences

Figure 2.3 is based on Deng (2006:fig. 2). It differs from figure 2.2 (also based on Deng 2006:fig. 2) in including only faunas calibrated by paleomagnetic means and in not using solid lines for unit boundaries, because they are all effectively undefined in any case. The intent is to show where there may be gaps or overlaps in the succession of these Chinese Neogene mammalian faunas based on paleomagnetic calibrations, on the one hand, but also to illustrate, on the other, those sections that are stratigraphically lengthy and targets for further work in constructing a Neogene time scale for China that has no gaps or overlaps and can account for all of Neogene time.

Stages and Ages in the Chinese Neogene Time Scale

Perhaps reflecting European conventions at times, Li, Wu, and Qiu (1984) and various subsequent authors have utilized the chronostratigraphic category of Stage in reference to Chinese mammalian faunal units. In not being constructed as "bodies of rock formed during a specified interval of geologic time" (Salvador 1994), the Chinese mammalian units do not so qualify. Similarly, a geochronologic Age ("the units of time during which chronostratigraphic units were formed" [Salvador 1994]) is formally dependent on the codification of an antecedent chronostratigraphic Stage. Hence, the use of North American Land Mammal Ages to differentiate them from geochronologic Ages (Savage 1955). In this context, it is clearly appropriate to designate Asian faunal units as NMUs (e.g., Deng 2006), and it would be prudent to discontinue the use of Age (upper-case A) when referring to Chinese faunas.

APPROACHES TO A NORTH CHINA STANDARD GEOCHRONOLOGY FOR THE NEOGENE

Regional Sequences with Paleomagnetic Data

The following illustrates examples drawn from the literature that contribute to achieving the current goal (see figure 2.3). In most cases, the examples are chosen on the basis of their representing chronologically long and stratigraphically detailed frameworks, aided by paleomagnetic data, within which mammalian evolution in North China can be documented and developed as a collective standard for Neogene faunal demonstration and correlation.

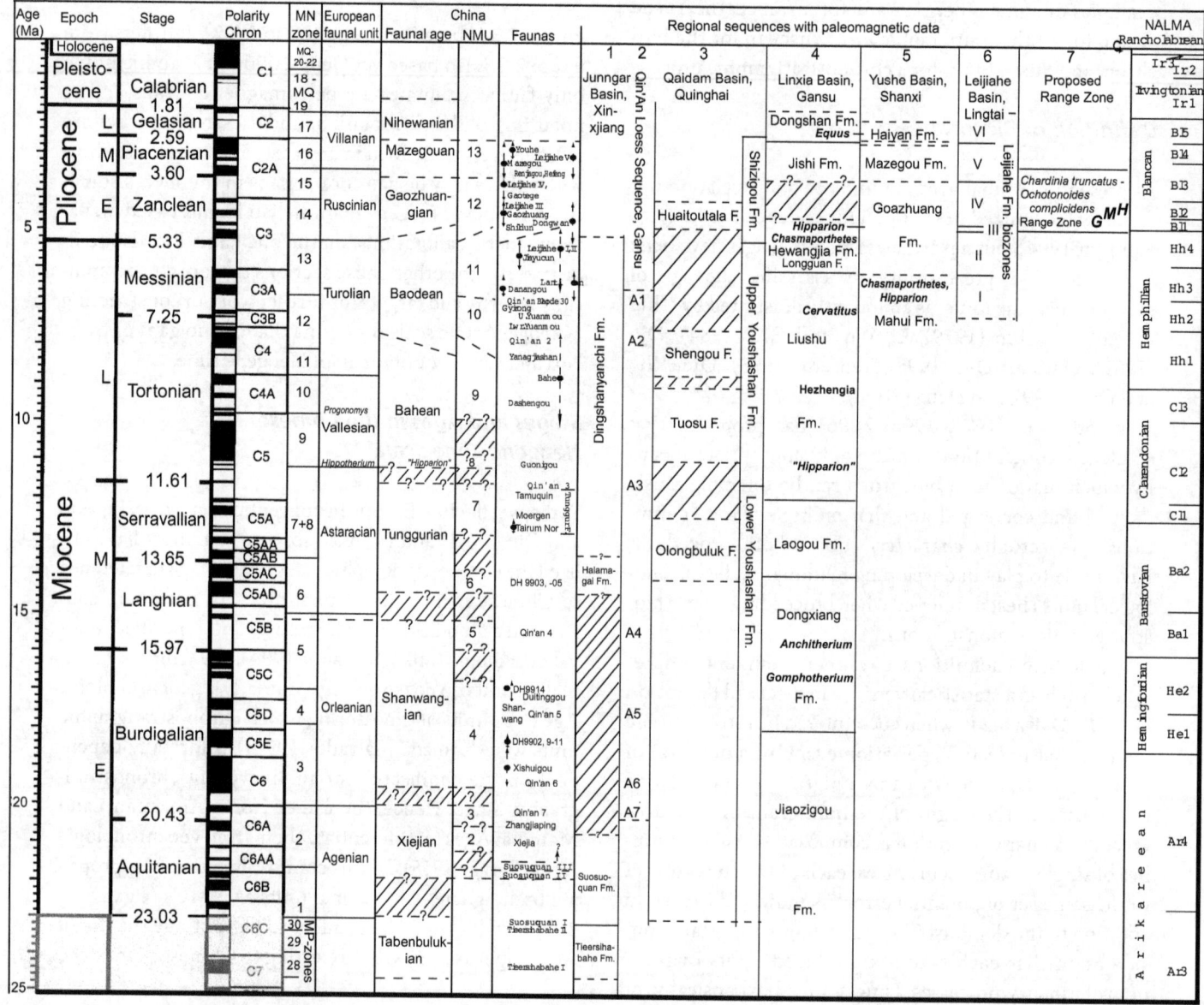

Figure 2.3 Biochronology of Asian Neogene and late Oligocene mammal-bearing units. See appendix for documentation of Neogene MN zonation. Columns for China Faunal Age, NMU, and Faunas are discussed in the text. Fauna ages are based on magnetostratigraphic information directly associated with each unit. Where there is an age range, this is indicated by arrows, with the unit in question identified by the solid black circle. Columns 1–7 regarding individual districts are discussed in the text.

The Xiejian Stage

As summarized in column 1 of figure 2.4, Meng et al. (2006) have made a valuable advance in the documentation of the Xiejian "Stage." Meng et al. (2006) describe a 300-m-thick stratigraphic sequence of nonmarine strata referred to the Tieersihabahe, Suosuoquan, and Halamagai formations in the Junggar Basin of Xinjiang Province (see figures 2.1 [1] and 2.3 [Junggar Basin column]), build-

ing on the work of Ye, Meng, and Wu (2003). Meng et al. (2006) record mammalian fossils from a number of sites in these formations (see figure 2.4). Those from the Tieersihabahe Formation are grouped as the Tieersihabahe Mammal Assemblage Zones I and II. TMAZ I is found in the 9–9.5 m interval above the base of the Tieersihabahe Formation in their main measured section (see figure 2.4). Meng et al. (2006) cite paleontological criteria for correlating zones T-I and T-II to the Tabenbulukian fau-

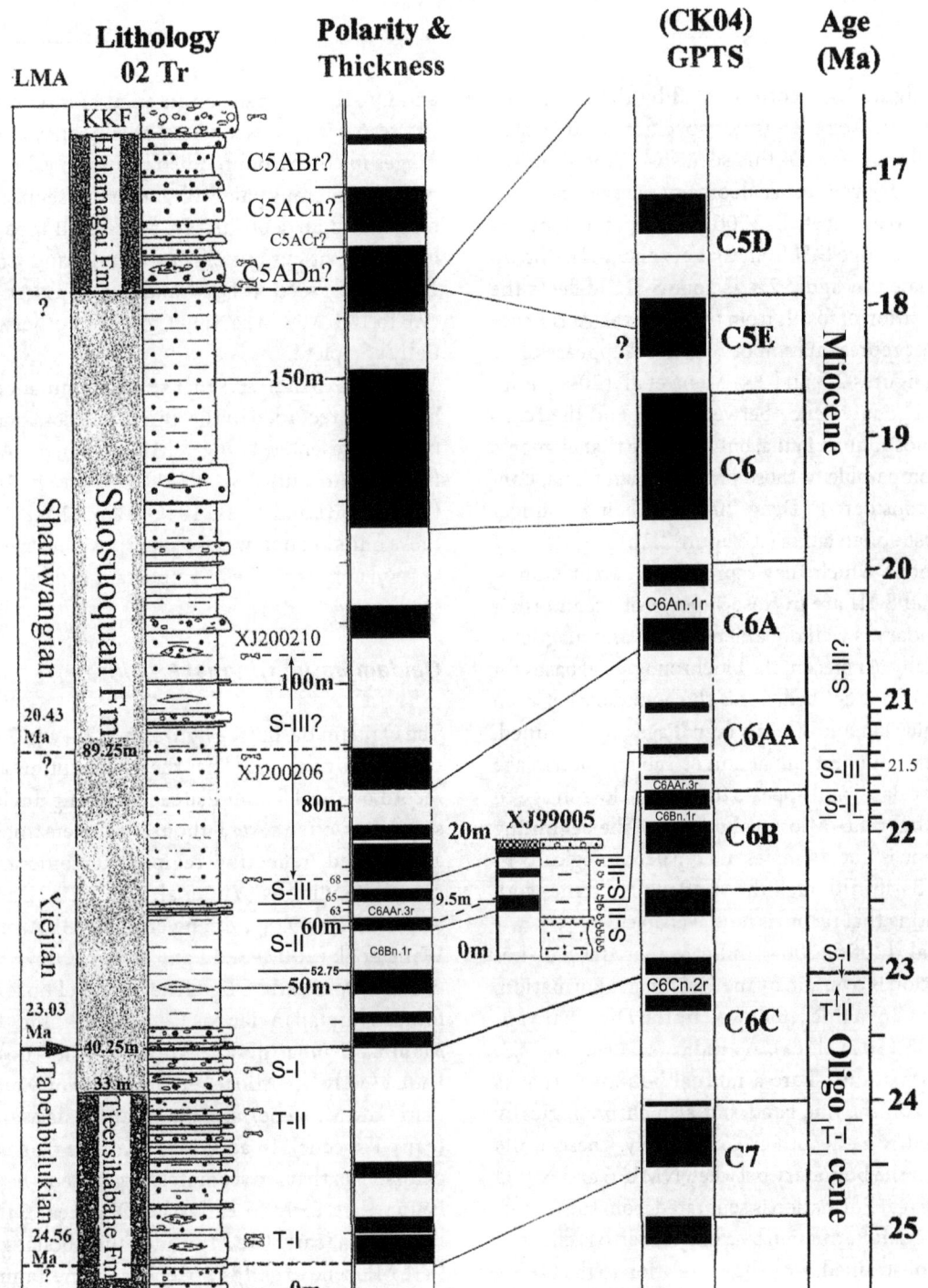

Figure 2.4 Biostratigraphy, magnetostratigraphy, and chronology of the Xiejian Stage boundary stratotype, Junggar Basin, Xinjiang, China. After Meng et al. (2006:fig. 4). T-1, S-III are Mammal Assemblage Zones of the Tieersihabahe and Suosuoquan formations, respectively. XJ200201, XJ99005, etc., are fossil localities, with bone symbols showing stratigraphic position (symbols truncated in XJ99005 due to space limitations). The XJ99005 section contains the main sites of S-II and S-III. These units are correlated to the Polarity and Thickness column based on paleomagnetic patterns aided by paleontological data. Ti-1 to S-III? are aligned in the Age (Ma) column based on paleomagnetic patterns (Meng et al., 2006). Magnetic polarity time scale (CK 04) GPTS is after Luterbacher et al. (2004).

nal age (see figure 2.4), corroborated by the paleomagnetic assessment. Zone S-I is not specifically correlated to the Tabenbulukian, but this seems warranted on the basis of its age derived from paleomagnetic criteria.

According to Meng et al. (2006), 1 m separates the upper boundary of zone S-II from that of zone S-III. The interval designated in figure 2.4 as zone S-III? reflects the sparse distribution of fossils from that interval. At the moment, the best representation of zone S-III appears to be as shown on figures 2.3 and 2.4. Meng et al. (2006) indicate a "rough" equivalency between S-III and the "conventional" Suosuoquan fauna, but also note that elements of S-II are comparable to those of the Xiejian fauna, conventionally considered (Deng 2006) as being younger than the Suosuoquan fauna (see figure 2.2).

The extent to which they represent different faunas, zones S-II and S-III are in close superposition and their mutual boundary is well documented bio- and magnetostratigraphically. At present the biochronological bases for these zones are not defined, nor is the boundary between the Tabenbulukian and Xiejian faunal ages well defined. Zone S-I is the youngest faunal unit of Tabenbulukian age in the Junggar Basin and appears to be very close in age to that of the Oligocene–Miocene boundary. The beginning of the Neogene is not yet represented paleontologically in the Junggar Basin, although the stratigraphic framework for documenting that record is now available.

Meng et al. (2006, 2008) indicate that the Suosuoquan Formation is overlain by the Halamagai Formation, which is about 26 m thick, followed by the Dingshanyanchi Formation (50 m thick). As indicated in figure 2.3, these two formations share a mutual boundary that is conformable, on the one hand, but also chronologically unconstrained, on the other. Collectively, these units might contain the boundary between NMU 6 and NMU 7. The Halamagai Formation is separated from the underlying Suosuoquan Formation by a gap that also is chronologically unconstrained, with the correlation to the GPTS shown on figure 2.4 being suggested only, based on biochronology. The Dingshanyanchi Formation may be as young as NMU 12 based on the presence of taxa such as *Hipparion (Plesiohipparion) houfenense* (Meng et al. 2008).

Qin'An Region, Gansu Province

Figure 2.3 (Regional sequences, column 2) indicates the presence of other stratigraphically long sequences amenable to documentation of the Neogene succession in China. Guo et al. (2002) discuss a long sequence of loess in the Qin'An region (see figure 2.1 [49]). The succession is about 250 m thick, occurs in two main sequences, QA-I

and QA-II, and spans an interval of time from about 22 Ma to 6 Ma. The sequence of siltstones and claystones ranges in color from reddish brown to yellow-brown, includes numerous paleosols, and represents a suite of primarily aeolian deposits. The succession apparently shows hiatuses that cut out chrons C5n.1n and C5n.1r (9.8 to 10.0 Ma) in section QA-1 and chrons C4Ar.1r to C5n.1r (9.0 to 10.0 Ma) in section QA-II, but otherwise is essentially complete.

Figure 2.3 indicates that seven mammal-bearing levels have been recorded in the Qin'An succession, with NMU units represented being NMU 3 (Qin'An A7), NMU 4 (Qin'An A6 and A5), NMU 5 (Qin'An A4), NMU 7 (Qin'An A3), and NMU 10 (Qin'An A2 and A1). Whereas these units are not in close superposition, the Qin'An succession appears to be an important candidate for further study.

Qaidam Basin, Qinghai Province

The Qaidam Basin (see figures 2.1 [25] and 2.3 [Regional sequences, column 3]) has the most continuous Cenozoic record on the Tibetan Plateau, and long fluvio-lacustrine sequences, often several thousand meters thick, are folded and faulted, reflecting regional tectonic and environmental conditions (Yin et al. 2007; Yin, Dang, Wang et al. 2008; Yin, Dang, Zhang et al. 2008). As discussed by Wang et al. (2007) and Fang et al. (2007), the sequence extends from early Oligocene to early Pliocene age, with four mammalian faunas recognized: the Olongbuluk Mammal Fauna (middle Miocene), the Tuosu Mammal Fauna (early late Miocene), the Shengou Mammal Fauna (early late Miocene), and the Huaitoutala Mammal Fauna (early Pliocene). In addition, there are faunas comprised of fossil fish that are separately organized: the Lulehe Fish Fauna (late early to early late Oligocene), the Shengou Fish Fauna (early late Miocene), the Eboliang Fish Fauna (late Miocene to early Pliocene), and the Yahu Fish Fauna (early Pliocene). These faunas are correlated to a magnetic polarity zonation (figures 2.5 and 2.6).

Collectively, the Qaidam Basin succession reaches 15,000 m in thickness, is distributed over an area of 137,000 km^2, and reflects collision of the Indian and Asian tectonic plates. In spite of tectonic influences, the Cenozoic deposits remain relatively undeformed. The Cenozoic fluvio-lacustrine sediments range in age from the late Eocene to the present, with a marked increase in accumulation rate in the early Pliocene.

The *Olongbuluk Mammal Fauna* occurs in the upper part of the Youshashan Formation (figures 2.5–2.7) and is represented by *Hispanotherium matritense, Acerorhinus*

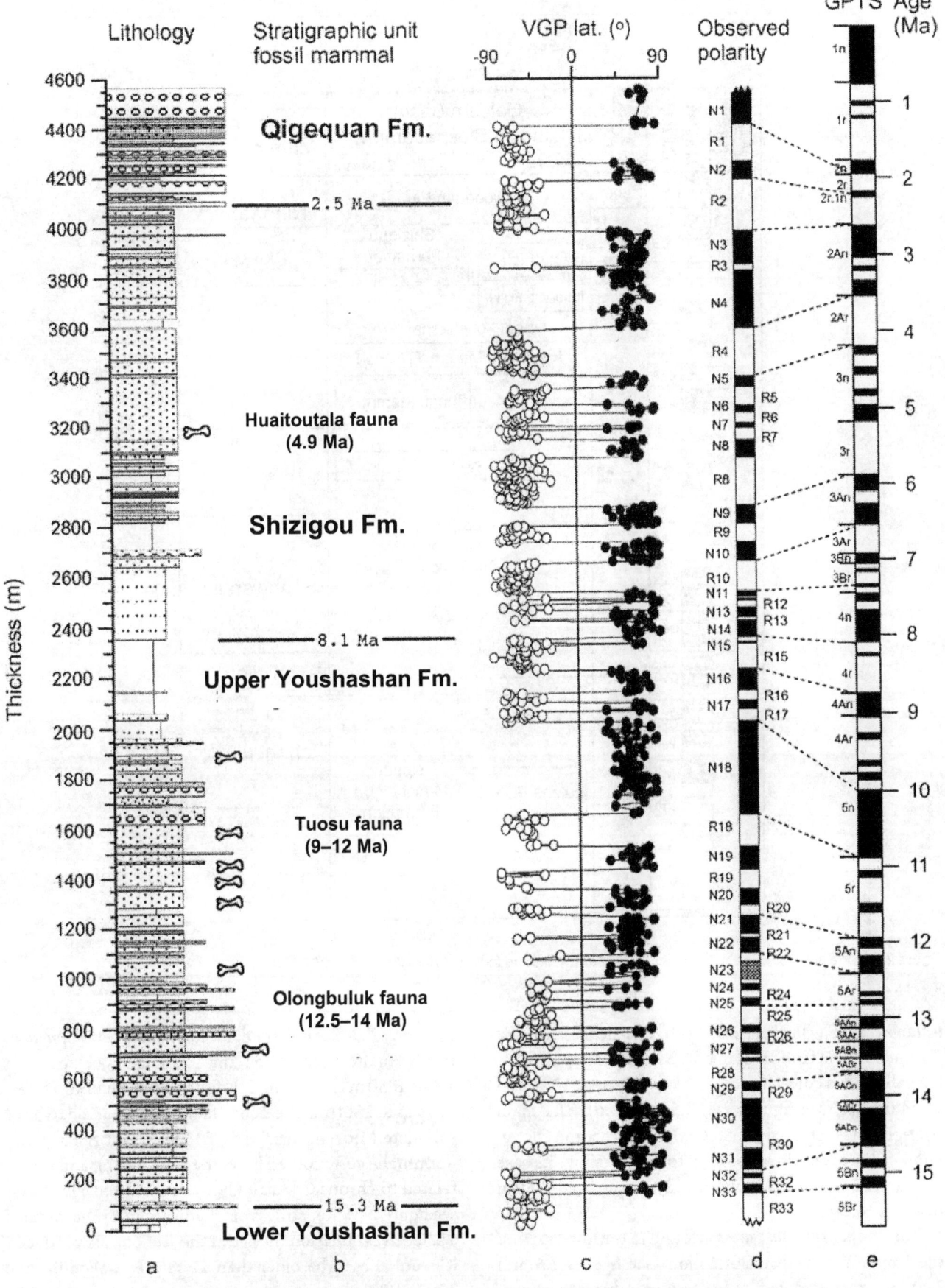

Figure 2.5 Magnetic polarity stratigraphy of mammal-bearing units in the Qaidam Basin. After Fang et al. (2007:fig. 8). Ma ranges of faunas follow Wang et al. (2007). Bone symbols show the stratigraphic position of fossil localities.

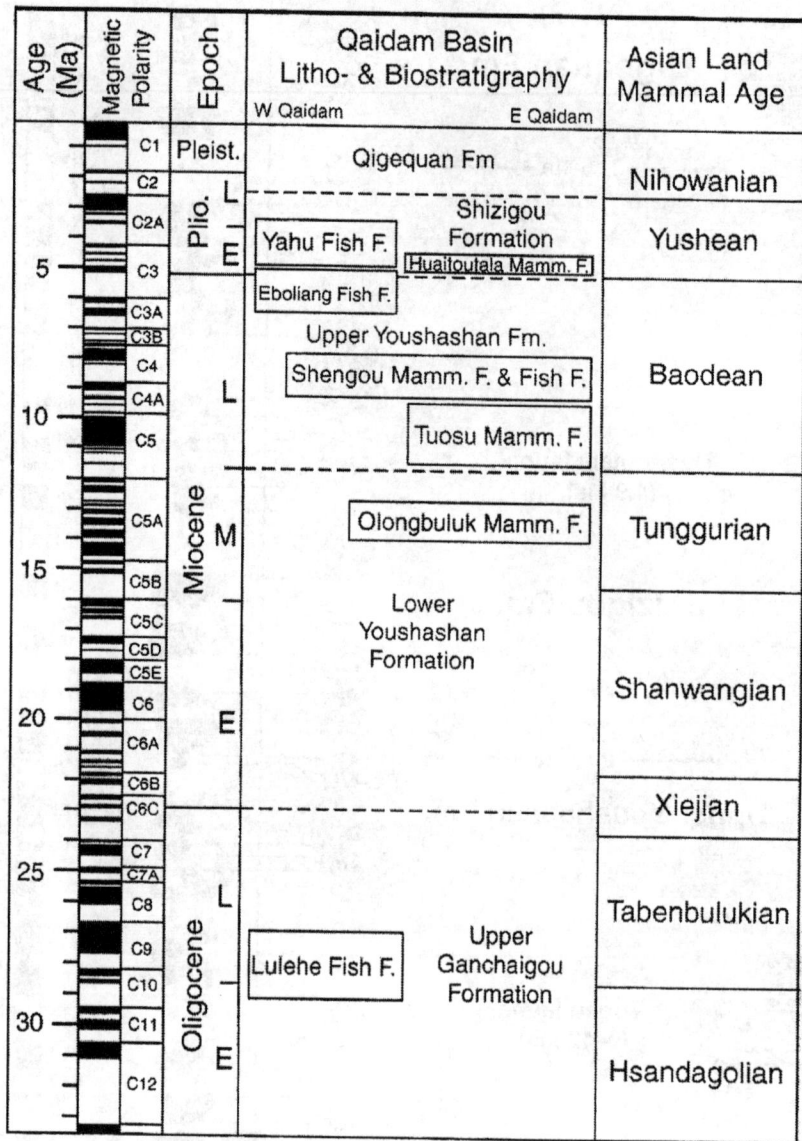

Figure 2.6 Correlation of Qaidam Basin mammal- and fish-bearing faunas. After Wang et al. (2007:fig. 12).

tsaidamensis, Lagomeryx tsaidamensis, Stephanocemas, and *?Dicroceros.* The fauna is of middle Miocene age paleontologically and is correlated to C5Ar to C5Cn, or 12.42 Ma to 14.09 Ma (see figures 2.3, 2.5, and 2.6). The main stratigraphic section (Huaitoutala) shows that the Olongbuluk fauna occurs in beds 700 m to 1350 m above its base (see figure 2.5). There is an unfossiliferous gap from 1350 m to about 1850 m.

The *Tuosu Mammal Fauna* is found in the lower part of the Upper Youshashan Formation (see figures 2.5 and 2.6). It is characterized by *Ictitherium, Adcrocuta eximia, Chalicotherium brevirostris, Hipparion teilhardi, Dicroceros, Euprox, Olonbulukia tsaidamensis, Qurliqnoria, Tos-*

sunnoria, Tsaidamotherium, Protoryx, and *Tetralophodon.* In the Huaitoutala section, the Tuosu fauna occurs from about 1850 m to 2200 m. It is followed by an unfossiliferous gap to 2300 m. The Tuosu fauna is considered to be of early late Miocene age (early Baodean East Asian Land Mammal Age as coined by Wang et al. 2007), and is correlated to chrons C4Ar to C5r (about 9.00 to 11.00 Ma; see figure 2.3). Regardless of whether or not *H. teilhardi* is a primitive hipparion the age of the section that produced it is unlikely to be older than 11 Ma (e.g., Woodburne, 2007, 2009).

The *Shengou Mammal Fauna* is obtained from the Upper Youshashan Formation (see figures 2.3, 2.6, and 2.7).

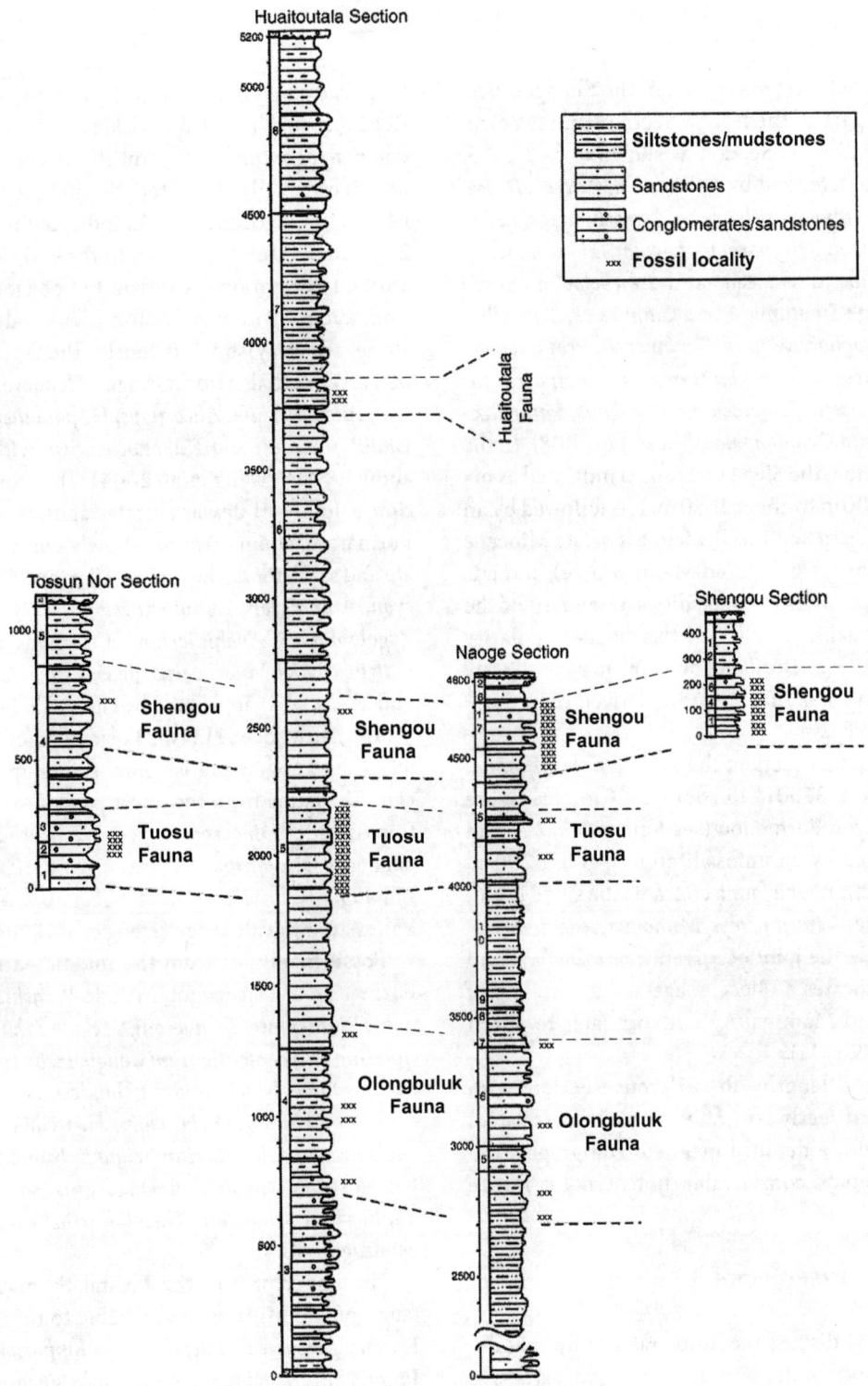

Figure 2.7 Stratigraphic sections of mammal-bearing units in the Qaidam Basin. After Wang et al. (2007:fig. 11).

It is found mainly in the lower part of the Shengou Section and upper part of the Naoge Section, and is correlated to the Huaitoutala Section as shown in figure 2.5. The fauna is characterized by *Ictitherium, Adcrocuta eximia, Plesiogulo, Promephitis parvus, Acerorhinus tsaidamensis, D. ringstromi, Hipparion* cf. *chiai, H. weihoense, H. teilhardi, Euprox, Gazella,* and *Amebelodon.* Small mammals include *Sinotamias, Sciurotamias* cf. *S. pusillus, Pliopetaurista, Lophocricetus* cf. *L. xianensis, Protalactaga, Stylodipus?, Myocricetodon lantianensis, Nannocricetus primitivus, Sinocricetus, Huerzelerimys exiguus, Pararhizomys, Ochotona,* and *Ochotonoma* (Qiu and Li 2008). In the Huaitoutala section, the Shengou fauna is indicated as occurring from 2300 m to about 2650 m. It is followed by an unfossiliferous gap to 3700 m. The fauna is of late Miocene age (Baodean North China Land Mammal Age), and it is correlated to the faunas from localities 12 and 19 of the Lantian Basin (Kaakinen, 2005), with a magnetic polarity age of 9.90 to 9.95 Ma. The Shengou fauna may be slightly younger in having taxa that are more derived than those from the Lantian Basin.

In the Huaitoutala section, the *Huaitoutala Mammal Fauna* occurs from 3700 m to about 2850 m (see figure 2.5) in the Shizigou Formation (see figures 2.3, 2.5, and 2.6). It is followed by an unfossiliferous gap to 5200 m, the top of the section. The fauna contains the small mammals *Orientalomys/Chardinomys, Mimomys, Pseudomeriones,* and a soricid. The joint occurrence of *Mimomys* and *Chardinomys* indicates a Pliocene age and a correlate of the Yushean Land Mammal Age. It correlates to chron C3n.3n (4.80–4.90 Ma).

In providing a lengthy fossiliferous section with which faunas are directly associated or can be correlated, in association with a detailed magnetostratigraphy, the Qaidam Basin holds considerable potential for future refinement.

Linxia Basin, Gansu Province

Deng et al. (2004) discuss the lithostratigraphy, magnetostratigraphy, and faunal correlations of an extensive succession of nonmarine strata in the Linxia Basin, Gansu Province (see figures 2.1 [50] and 2.3 [Regional sequences, column 4]). Fang et al. (2003) developed a detailed magnetostratigraphy for this succession, but this was not employed by Deng et al. (2004) and this is followed here. As indicated in figure 2.8, the composite stratigraphic section is about 500 m thick and is comprised of a variety of nonmarine sediments. The Neogene sequence begins with the Jiaozigou Formation, about 110 m of brownish-yellow sandstone and brownish-red mudstone that con-

tain *Tsaganomys, Dzungariotherium orgosense, Allacerops, Ronzotherium, Aprotodon, Schizotherium,* and *Paraentelodon macrognathus.* The giant rhinoceros, *Dzungariotherium,* is especially characteristic, and the fauna appears to be of late Oligocene age. As indicated in figures 2.3 and 2.8, it may extend upward into the early Miocene. The Jiaozigou Formation is overlain by the Dongxiang Formation, about 80 m of reddish-brown sandstone and mudstone, and grayish white marls. The fauna appears to be of early medial Miocene age (*Hemicyon, Gomphotherium, Anchitherium, Alicornops, Hispanotherium matritense, Chalicotherium,* and *Kubanochoerus*), with a basal age of about 18 Ma (Deng et al. 2004). The Dongxiang Formation is followed upward stratigraphically by the Laogou Formation, about 40 m of yellowish-gray fine conglomerate and sandstone. The fauna (*Alloptox, Pliopithecus, Hemicyon, Amphicyon, Gomphotherium, Platybelodon grangeri, Zygolophodon, Anchitherium, Alicornops, Hispanotherium matritense, Kubanochoerus gigas, Listriodon, Palaeotragus,* and *Turcocerus*) appears to be late medial Miocene in age.

The late Miocene Liushu Formation follows stratigraphically the Laogou and is composed of up to 100 m of red clay. The fauna from the lower part of the Liushu Formation includes *Dinocrocuta gigantea, Machairodus, Tetralophodon, "Hipparion" dongxiangense, Parelasmotherium simplum, P. linxiaense,* and *Shaanxispira,* apparently of Vallesian-equivalent age (Deng et al., 2004).

Fossil mammals from the middle part of the Liushu Formation at Dashengou include *Pararhizomys hipparionum, Promephitis, P. hootoni, Melodon majori, Sinictis, Ictitherium, Hyaenictitherium wongii, H. hyaenoides, Dinocrocuta gigantea, Machairodus palanderi, Felis, Tetralophodon exoletuus, Hipparion, H. chiai, H. weihoense, Acerorhinus hezhengensis, Chilotherium wimani, Iranotherium morgani, Chleuastochoerus stehlini, Dicrocerus, Samotherium, Honanotherium schlosseri, Gazella, Hezhengia bohlini,* and *Miotragocerus.*

The upper part of the Liushu Formation contains a fauna of late Miocene comparable to that of Baode, with *Hystrix gansuensis, Pararhizomys hipparionum, Simocyon, Promephitis, P. hootoni, Parataxidea sinensis, Plesiogulo, Ictitherium, Hyaenictitherium wongi, H. hyaenoides, Adcrocuta variabilis, Machairodus palanderi, Metailurus minor, Felis, Hipparion coelophyes, H. dermatorhinum, Acerorhinus hezhengensis, Chilotherium wimani, Dicerorhinus ringstromi, Ancylotherium, Chleuastochoerus stehlini, Microstonyx major, Metacervulus, Cervavitus novorossiae, Honanotherium schlosseri, Palaeotragus microdon, Miotragocerus, Sinotragus, Protoryx,* and *Gazella.*

The next superposed unit is the Lower Pliocene Hewangjia Formation, about 50 m thick when present and

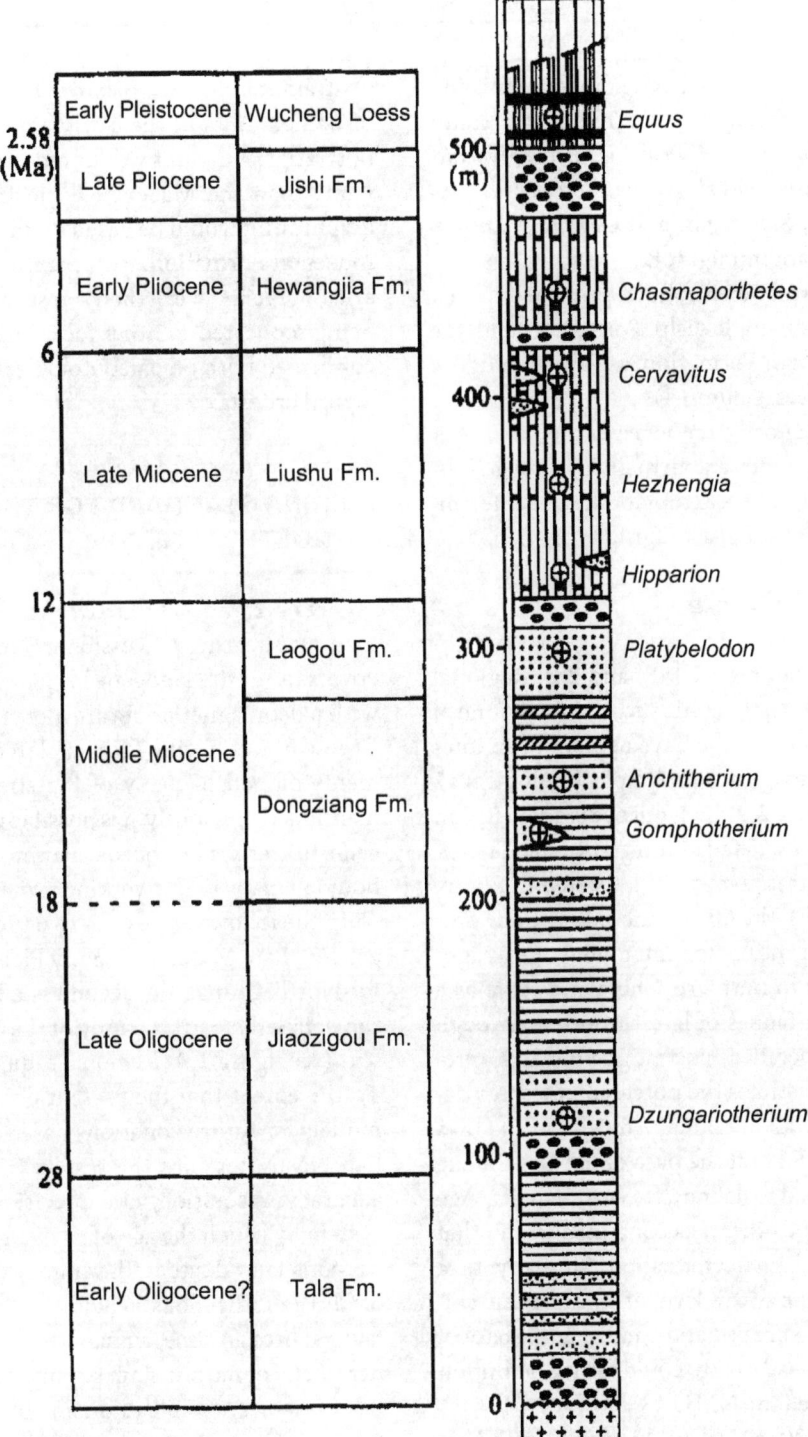

Figure 2.8 Correlation of Linxia Basin lithostratigraphy and magnetostratigraphy showing the location of formation units and fossil sites. After Fang et al. (2004:fig. 3).

composed of red clay and a basal conglomerate. In addition to *Hipparion*, the fauna from the red clay contains *Hystrix gansuensis*, *Promephitis*, *Chasmaporthetes*, *Hyaenictitherium wongi*, *Shansirhinus ringstromi*, *Cervavitus novorossiae*, *Palaeotragus*, *Sinotragus*, and *Gazella*. *Chasmaporthetes* is an important immigrant.

The Hewangjia Formation is followed by the basically unfossiliferous conglomeratic Jishi Formation and the *Equus*-bearing Dongshan Formation, as shown in figure 2.3 (Regional sequences, column 4).

In summary, the Cenozoic sequence in the Linxia Basin has the potential to represent virtually the entire Neogene with only minor hiatuses except for the unconformities that separate the Hewangjia and Jishi formations.

Yushe Basin, Shanxi Province

The Yushe Basin (see figures 2.1 [48] and 2.3 [Regional sequences, column 5]) stratigraphic succession was one of the first in China to support the development of an integrated litho-, bio-, and magnetostratigraphy (Tedford et al. 1991). As shown in figure 2.9, the sequence is about 800 m thick and demonstrates a detailed integration of magnetochronology and biostratigraphy. As indicated in Flynn, Wu, and Downs (1997), the taxonomic associations provide well-documented means for improving age assignments of other faunas in northern China and serve as a reference standard for faunas of latest Miocene through Pleistocene age. The detailed biostratigraphic array presented in figure 2.9 illustrates the potential power for defining mammal age units in China. Any one of the taxa that have an apparent FAD at the base of the Gaozhuang Formation could be used to define the Gaozhuangian Age (and NMU 12), with the other taxa contained within that unit contributing to its characterization. Similarly, taxa with a FAD at about the 650 m level of stratigraphic column, within the Mazegou Formation and near the base of the upper half of chron C2An.3n, could form a definition for the Mazegouan Age (and NMU 13).

Taxa at about the 800 m level, within the Haiyan Formation, could be utilized as a definition for Nihewanian Age (see figure 2.3 [Regional sequences, column 5]). In both of the latter two examples, it is clear that the biochronological definition would place the lower boundary of the unit within, not at, the lithostratigraphic base of its formational context. The Mazegou example shows the potential confusion that could result when the lithostratigraphic and biochronologic unit have the same name. If the boundary definitions were replicated in other places, thus demonstrating the utility of the unit so defined, it would be possible to defend the inauguration of a chrono-

nostratigraphic unit. In order to preserve the autonomy of named units in their original context, the new chronostratigraphic unit would require a distinctly different name than the Mazegou Formation. The chronostratigraphic unit would be based on the biostratigraphy of the Mazegou Formation, with the isochroneity of its boundaries agreeing with their biostratigraphical definitions being replicated in other locations and their correlation confirmed both on paleontological and magnetochronological criteria.

A CHINA STANDARD FOR THE LATE NEOGENE OF NORTHEASTERN ASIA

Figures 2.2 and 2.3 illustrate the fact that the Neogene mammal-bearing succession in North China reasonably covers the entire Neogene Period. Those units associated with paleomagnetic data are increasingly well developed. Even without having fully explored the potential refinements offered in many of the stratigraphic successions mentioned previously, it is possible to propose a Late Miocene to Early Pleistocene mammal age succession embodying many of the working principles proposed herein.

In the following we offer two proposals. The first is regarding Late Miocene to Early Pleistocene mammal ages for North China. The second is a biozonation based on the detailed biostratigraphy of the Yushe and Lingtai basins (see figure 2.3 [Regional sequences, columns 5–7]). To the extent that the present examples can be used as models for future zonations based on Chinese mammalian biostratigraphy, the results could revolutionize the accuracy, resolution, and precision of the chronologic system by which the age of its Neogene sedimentary successions is evaluated. The framework would reflect first of all the indigenous evolution of Chinese mammalian faunas through time, augmented by immigrant taxa from peripheral or more distant regions. In order to focus on the basic importance and autonomy of its mammalian evolution, the Chinese framework will be best served by having its own intrinsic unit nomenclature and its own intrinsic bases of definition, characterization, recognition, and correlation.

Proposed Late Miocene to Early Pleistocene Mammal Ages for North China

The following proposals are intended to aid in the definition and characterization of Chinese mammal ages that would apply to a part of China, north of the Yangtze River

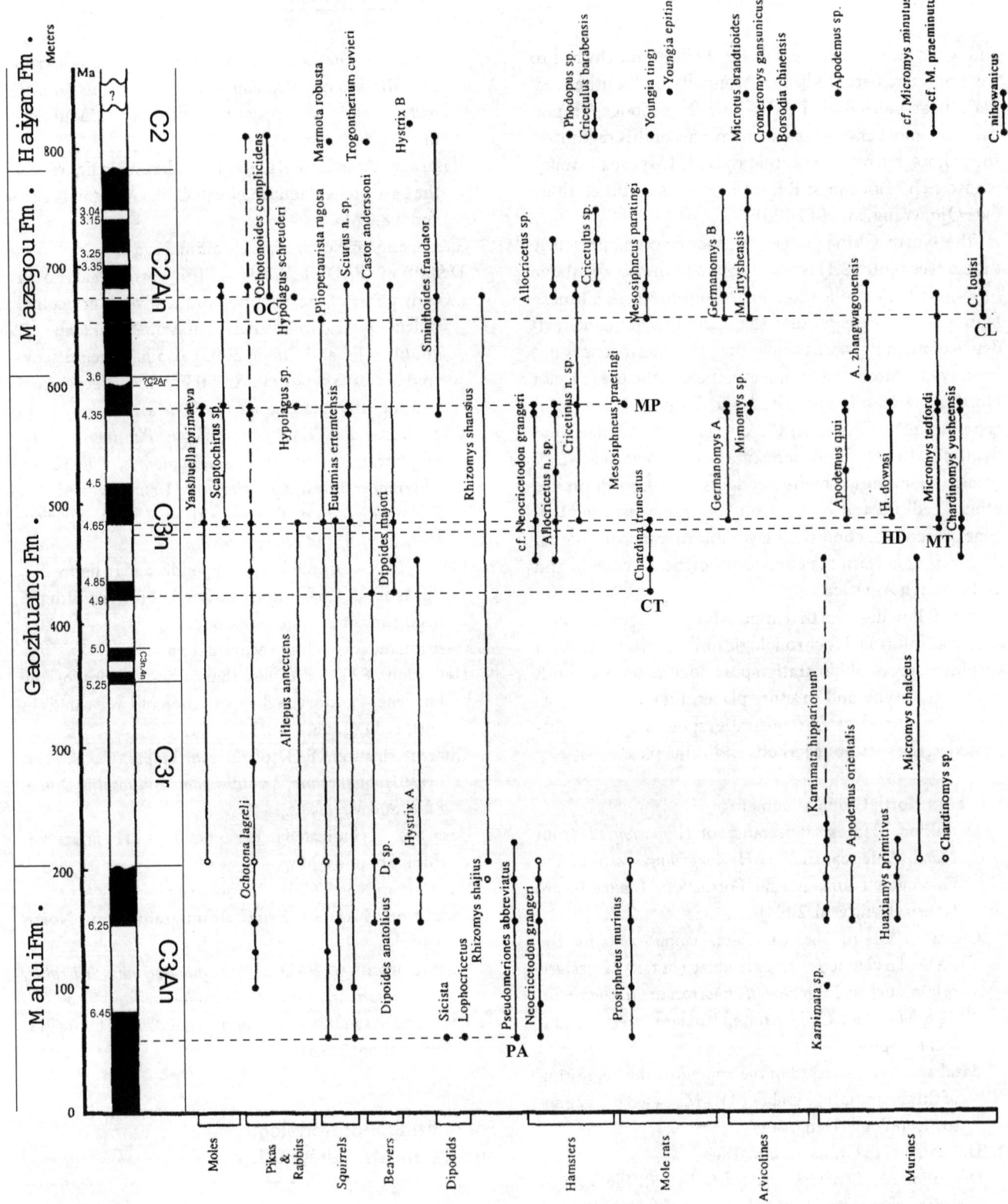

Figure 2.9 Biostratigraphy and magnetic polarity chronology of the Yushe Basin succession. After Flynn, Wu, and Downs (1997), with magnetic polarity chrons added based on their text and the ages they assigned to intervals; upgraded to Lourens et al. (2004). FADs of species shown in table 2.2 and figure 2.11. PA = *Pseudomeriones abbreviatus*; CT = *Chardina truncatus*; MT = *Micromys tedfordi*; HD = *Huaxiamys downsi*; MP = *Mesosiphneus praetingi*; CL = *Chardinomys louisi*; OC = *Ochotonoides complicidens*. The biostratigraphic extent of the *Chardina truncatus–Ochotonoides complicidens* Range Zone is indicated.

and westward across the Tibetan Plateau, northward to eastern Kazakhstan, adjacent Mongolia, and southern Siberia near Lake Baikal (see figure 2.1). Zoogeographically, the most characteristic mammals of this region are the zokors, burrowing muroid rodents (Myospalacinae), whose rich Miocene to Recent history is confined there (see Qiu, Wang, and Li 2006).

The North China zoogeographic region as presently drawn (see figure 2.1) is widely open to the north (above 55°N), with an implication that Neogene North Chinese faunas likely were present there. At the moment, evidence of mammal faunas older than the Pleistocene is not preserved in arctic Siberia. Nevertheless, the presence of North American *Hipparion s.l.* and Canidae in the Miocene of North China, on the one hand, and Asiatic endemics in the North American Miocene record at 45°N (such as *Parailurus*, meline badgers, and *Ursus*), on the other (Tedford and Harrington 2003), suggest the Neogene presence of conditions favorable to transcontinental dispersal of certain taxa between northeastern Asia and arctic North America.

The following North China Mammal ages are presented as informal biochronologic units at this time. With varying degrees of biostratigraphic documentation, both in the stratotype and in other places, they could be formally developed as chronostratigraphic stages as discussed in the section "Methods and Principles."

1. Bahean North China Mammal age

    Definition. Earliest appearance of *Hipparion s.l.* from North America, such as *H. dongxiangense* from the Guonigou fauna, Liushu Formation, Linxia Basin, Gansu (Deng et al. 2004).

    Characterization. The Guonigou fauna contains the FAD of a characteristic assemblage of taxa of western origin, such as *Progonomys, Dinocrocuta, Machairodus, Tetralophodon, Chilotherium, Parelasmotherium*, and *Shaanxispira*.

    Basal age. As discussed in the appendix, the beginning of this mammal age is about 11.1 Ma based on magnetochronology (Deng, 2006).

2. Baodean North China Mammal age

    Definition. FAD of *Ochotona* (cf. *O. lagreli*). The ochotonids early achieved a nearly world-wide distribution, but by the later Miocene they inhabited only Eurasia and North America.

    Characterization. This includes elements of the diverse assemblages of the "*Hipparion* red clays" of Shanxi, Shaanxi, Henan, and Gansu Chinese provinces that contain several species of this equid along with the FADs of *Prosiphneus, Alilepus, Castor, Indarctos, Ictither-*

ium, Adcrocuta, Gazella, Dorcadoryx, Chleuastochoerus, and *Microstonyx. Stegodon* and murid rodents (*Karnimata, Huaxiamys*) appear late in the age (Mahui Formation, Yushe Basin).

    Basal age. As indicated in the appendix and in figure 2.3, the base of the Baodean North China Mammal age is about 8.3 Ma.

3. Gaozhuangian North China Mammal age

    Definition. FAD of *Mimomys*. This corresponds to an early phase of the evolution of the diverse arvicoline rodents typical of northern Eurasia. *Promimomys* (*P. asiaticus*; Jin and Zhang 2005) also has been discovered in a cave deposit in Anhui Province, determined to be of early Pliocene age (ca 5.0 Ma).

    Characterization. FAD of *Hypolagus, Pliosiphneus, Mesosiphneus, Chardinomys, Germanomys, Pliohyaena, Chasmaporthetes, Agriotherium, Ursus, Nyctereutes, Proboscidipparion, Plesiohipparion, Cremohipparion, Paracamelus*, and *Sinomastodon*.

    Basal age. As indicated in the appendix and in figure 2.3, this is about 5.5 Ma and nearly synchronous with the beginning of the Pliocene Epoch.

4. Mazegouan North China Mammal age

    Definition. FAD of *Canis*, the earliest record of the Pliocene–Pleistocene diversification of large canids in northern Eurasia.

    Characterization. FAD of *Eucyon, Vulpes, Canis, Felis, Lynx, Homotherium, Archidiskodon, Antilospira, Dama, Rusa*, and *Megalovis.*

    Basal age. As indicated in the appendix and in figure 2.3, this is about 4.0 Ma.

5. Nihewanian North China Mammal age

    Definition. FAD of *Equus*, immigrant from North America.

    Characterization. FAD of *Youngia, Borsodia, Microtus, Trogontherium, Marmota, Eucladoceros*, and *Bison.*

    Basal age. As indicated in the appendix and in figure 2.3, this is about 2.6 Ma.

Formalizing Biochronologic and Chronostratigraphic Units

The first attempt to formalize biochronologic units was made by Wood et al. (1941), wherein NALMAs were defined. Subsequent research and critique have modified aspects of the stratigraphic and taxonomic content of the NALMAs, but the basic framework has withstood the test of almost 70 years. Tedford (1970) emphasized the critical importance of fossils, rather than lithostratigraphy, to characterize, recognize, and discuss the mammalian framework, and Woodburne (2006) followed the example of Savage (1955) in emphasizing the link between biostratigraphy and chronostratigraphy, with biochronologic units needing a clear definition, as well. We suggest that Chinese Neogene mammal chronology (as well as the North American record) can be strengthened by the following procedural steps that apply to both biochronological and chronostratigraphic units. A goal of the present exercise is to move beyond focusing on faunal units when developing a correlation of the Chinese Neogene fossil record.

Fundamental to this operation is the documentation of a stratigraphically long and productive fossiliferous sequence and identification of the locally and potentially regionally important biochronologic events contained therein. Such events include the evolutionary development of a new species or genus, or its introduction via dispersal from an outlying district. The exercise most likely would begin in a well-studied sequence that had been shown to have a well-defined and well-described biostratigraphy for taxa with a strong potential for regional correlation. Demonstrating that potential would involve replication of the targeted biostratigraphic pattern in other, closely located, and then more regionally widespread, biostratigraphic sections. These sections should contain a sufficient number of similar and/or identical fossils in a pattern that is replicated from place to place so as to be considered correlative. The proposed correlation could be tested and aided by independent radioisotopic or magnetostratigraphic chronologic data.

Definition. This aspect of boundary identification is discussed in prior as well as subsequent parts of this article. Once its boundaries have been determined, a new biochronologic unit can be defined and named. In the following section, we utilize the chronostratigraphic interval zone as the zonal unit of choice, such as the *Chardina truncatus–Ochotonoides complicidens* Interval Zone. Even though biochronologic units are not formalized in current stratigraphic codes or guides (e.g., Salvador 1994), the biochronologic equivalent could be the *Chardina truncatus–Ochotonoides complicidens* Interval Biozone, meaning that its base is defined by the LO of *C. truncatus* and its top by the base of the next overlying zone, based on the LO of *O. complicidens.*

Characterization. Once defined, a biochronologic or chronostratigraphic unit is characterized by the occurrence of a suite of taxa that on the whole is distinctive of that unit relative to those above or below. This characterization might be unique, but experience indicates that it is likely that some taxa also will be found in other

units. Thus, on average, the greater the number of fossil taxa identified, the better characterized the unit may become.

Zonation. As mentioned earlier, the purpose of the present exercise is to move beyond focusing on faunal units when developing a correlation of the Chinese Neogene fossil record, and a similar goal has been expressed relative to NALMAs (e.g., Woodburne 2006). Of course faunal correlations are valuable, and it is well recognized that faunal units of NALMAs as well as Chinese NMUs are a primary basis for correlation. But at least some intervals assigned to NALMAs have gone beyond that scale of correlation and use actual biozonations for portraying, assessing, and discussing time intervals that are no longer "faunas" (e.g., Lindsay 1972; Gingerich and Clyde 2001; Secord et al. 2006).

It is for this reason that developing biostratigraphies and biochronological zonations is such an important next step, to advance beyond the historical and classical faunas developed since the nineteenth century. Thus, in a very well-developed succession, the traditional "faunas" would be variously "thick" snapshots of the biochronology (depending on how many meters of section embrace a given fauna). But within that context, the thicker and more complete biochronology, and its correlation to other biozones, would potentially illuminate a greater number of useful temporal markers than would individual faunal units, and would also provide data upon which to assess the fidelity of the sequence, so that the inevitable gaps or overlaps that result from the faunal snapshot method would be minimized or eliminated. Other independent chronologic tools, (e.g., radioisotopic, magnetostratigraphic, stable isotopes) also should be utilized, to strengthen and constrain temporal placements. However, correlation of the biozones must always be based on biologic criteria.

Formalization. Once the preceding elements of this exercise have been accomplished and a biozonation has been shown to be sufficiently replicated from place to place so as to have a demonstrated validity within a given region, such as all or part of northern China, the zonal record justifies formalization. That usually entails designation of one or more type sections, and especially ones that contain the unit boundaries, along with a statement of the defining and characterizing criteria, providing a name for the unit and showing its relationship to sub- and superjacent zones as well as its lateral extent. Whereas the faunal unit basis of Chinese NMU correlations has provided an internally coherent and useful chronologic system, we believe that this system can be improved by incorporating the suggestions offered here.

One further suggestion is to establish a Chinese Stratigraphic Commission whose responsibility should include making recommendations for stratigraphic and chronologic definitions and application of terminology.

A Potential Future Late Neogene Chinese Chronostratigraphy

The following example is intended to illustrate the development of a chronostratigraphic zonation that might be developed for a Pliocene mammal-bearing succession in North China. As indicated earlier, there is a distinct progression from biostratigraphy to biochronology to chronostratigraphy and then to geochronology. The Interval Zone is the biostratigraphic unit here utilized, and it now seems possible to suggest that a fully developed Chinese Neogene chronostratigraphy could be developed in the future. The examples cited here could represent some of the chronostratigraphic units that might result from such a process.

The operation begins with the development of a biostratigraphy based on the Neogene formations of the Yushe Basin (Flynn, Wu, and Dowms 1997), where it is considered possible to establish taxonomic successions in a sequence where there are few hiatuses or overlaps, at least within formational units (see figure 2.9). The faunal sequence found in the Leijiahe Formation, Lingtai (Zheng and Zhang 2001) is chosen as the target for comparison in light of its well-developed succession (figures 2.10 and 2.11).

Zheng and Zhang (2001) integrated the biostratigraphy and magnetostratigraphy of the Leijiahe Formation, Leijiahe, Lingtai (see figures 2.1 [51] and 2.3 [Regional sequences, column 6]). The main stratigraphic section in the Leijiahe Formation is about 60 m thick and ranges in age from about 7.2 Ma to 1.8 Ma based on associated magnetochronology (see figure 2.10). The regional biozones are added from Zheng and Zhang (2001:fig. 2). Figure 2.11 shows the biostratigraphic ranges of seven taxa taken from Zheng and Zhang (2001:fig. 3). Note that the boundary between biozones III and IV is taken at 4.8 Ma, based on figure 2.10, rather than 4.25 Ma as shown in Zheng and Zhang (2001:fig. 3).

The biostratigraphy of Zheng and Zhang (2001) indicates that the stratigraphic first occurrences of *Chardina truncatus*, *Ochotonoides complicidens*, and *Chardinomys louisi* are closely comparable in age in the Yushe and Leijiahe areas, based on their correlation to the GPTS (see figures 2.9 and 2.11 and table 2.2). As seen in figure 2.1, these regions are separated by about 500 km.

For the purpose of this example, the Yushe bio- and magnetostratigraphy is taken as the standard upon which to base late Miocene to Pleistocene biocorrelations to other sites in northern China, as well as other parts of Asia. A new *Chardina truncates–Ochotonoides complicidens* Interval Zone could be inaugurated, with its boundaries defined on the stratigraphic first occurrence of the named taxa, which each represent a LO (see figure 2.3 [Regional sequences, column 7]). The zone could be characterized on the joint occurrence of *Chardina truncatus*, *Huaxiamys downsi*, *Mesosiphneus praetingi*, *Chardinomys yusheensis*, and *Dipoides majori* (known from Yushe), and *Allocricetus bursae*, *Cricetinus mesolophidus*, and *Chardinomys yusheensis* (from Lingtai; Zheng and Zhang 2001).

Note that the taxa chosen for this proposed chronostratigraphic zonation are based on their presence in the two regions. Other taxa could be used, with different stratigraphic results, but at the moment the named taxa are usefully shared between the two regions in question.

Lithostratigraphically, the *Chardina truncatus–Ochotonoides complicidens* Interval Zone occurs in the upper 200 m of the Gaozhuang Formation and the lower 80 m of the Mazegou Formation in the Yushe Basin (see figure 2.3 [Regional sequences, columns 5 and 7]). It would straddle the potential boundary between the Gaozhuangian and Mazegouan faunal ages (see figure 2.3). The *Chardina truncatus–Ochotonoides complicidens* Interval Zone would range in age from about 4.9 Ma to 3.4 Ma and would thus have a duration of about 1.5 m.y. It would effectively differ from all Chinese NMU units and European MN units in having an explicitly defined biostratigraphically based lower (as well as upper) boundary. Identifying the interval zone in other Chinese districts would depend primarily on the taxa that provide its biochronological definition and characterization. Recognition of the span of time involved could be based on magnetochronology or any other independent chronologic system.

The biogeographic strength of the *Chardina truncatus–Ochotonoides complicidens* Interval Zone would stem from its being based on mammalian evolution in China. Its major utility would be to help order the chronology of mammalian evolution in China. In having its own association with the magnetic polarity time scale, the *Chardina truncatus–Ochotonoides complicidens* Interval Zone could be compared with other mammalian chronologic systems, such as the European MN zonation or the North American Land Mammal Ages. More importantly, it would be possible to discuss the mammal chronology of China independently from that of other districts. In having

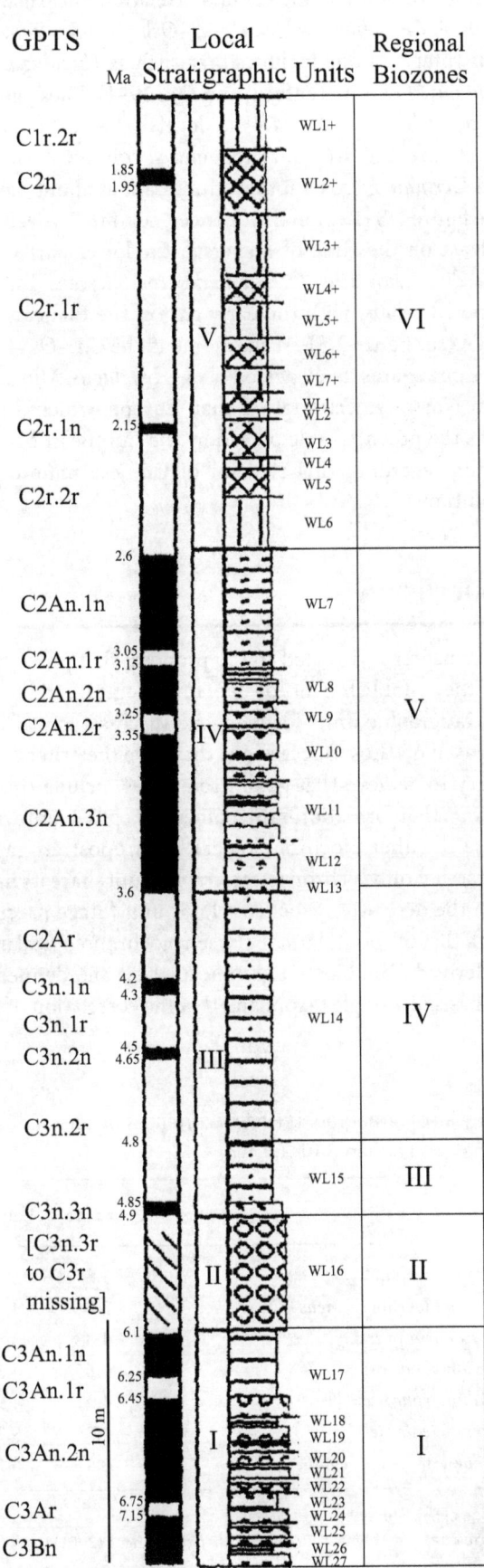

Figure 2.10 Lithostratigraphy and magnetostratigraphy of the Leiji-ahe Formation, Lingtai. After Zheng and Zhang (2001:fig. 1), with magnetic chrons added based on the ages supplied for intervals indicated therein; upgraded to Lourens et al. (2004), with numerical ages taken therefrom. Regional biozones added from Zheng and Zhang (2001:fig. 2).

distinctly defined biological boundaries, the *Chardina truncatus–Ochotonoides complicidens* Interval Zone would be more completely established than virtually all European MN zonations (Steininger et al. 1996; Steininger 1999; van Dam et al. 2001) and many North American Land Mammal Ages, many of which lack a detailed biostratigraphic demonstration (e.g., Woodburne 2006).

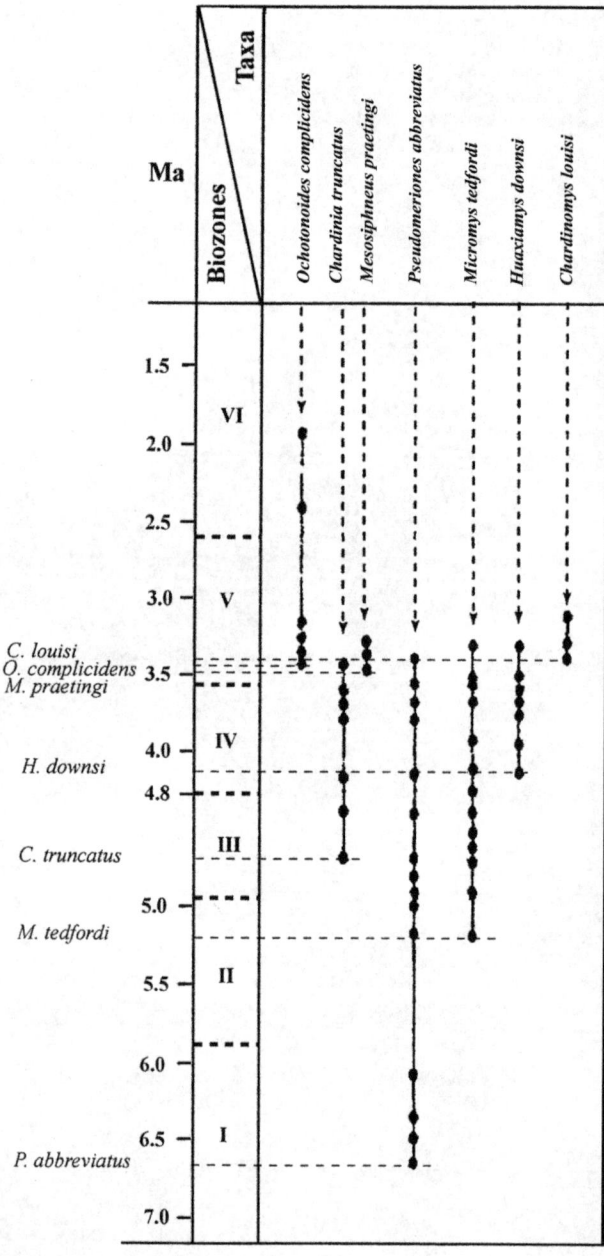

Figure 2.11 Biostratigraphy of selected elements of Leijiahe biozones. After Zheng and Zhang (2001:fig. 3). Note that the boundary between biozones III and IV is taken at 4.8 Ma rather than 4.25 Ma as in Zheng and Zhang (2001), based on the magnetic polarity chronology shown in figure 2.10.

Regarding intercontinental correlation, the Yushe record of the *Chardina truncatus–Ochotonoides complicidens* Interval Zone includes taxa such as *Hypolagus* and *Mimomys* (Flynn, Tedford, and Qiu 1991). These genera occur at about the 4.3–4.4 m.y. level in the Yushe section (see figure 2.3 [Regional sequences, column 7, *H, M*]), and *Germanomys* is another immigrant at about 4.6 Ma (see figure 2.3 [Regional sequences, column 7, *G*]). Based at least on the time of dispersal, the lower part of the *Chardina truncatus–Ochotonoides complicidens* Interval Zone correlates with the early part of the Blancan NALMA (see figure 2.3). At the moment, the *C.t.–O.c.* Interval Zone shares its dispersed taxa (*Hypolagus, Mimomys*) with North America rather than Europe, which diminishes the potential role played by the European MN zonation regarding this element of Chinese mammalian evolution.

CONCLUSION

This chapter is directed to the proposal that procedures for the establishment of biochronological and chronostratigraphic units for Neogene successions in China be based on those successions that have the criteria necessary to achieve that goal. Procedures include the following: that formation-rank names be kept separate from those of either biochronological or chronostratigraphic units; that only a chronostratigraphic unit share its name with the derivative geochronologic unit (Stage precedes Age); that the boundaries between biochronological units be derived from biostratigraphic analysis and defined on the basis of a single taxon; and that the correlation of bio-

Table 2.2

Age (Ma) of Stratigraphic First Occurrences of Selected Taxa, Yushe and Lantian Basins, China

	Yushe	Leijiahe
Chardinomys louisi	3.5	3.4
Ochotonoides complicidens	3.4	3.45
Mesosiphneus praetingi	4.35	3.5
Huaxiamys downsi	4.65	4.35
Chardina truncatus	4.9	4.9
Micromys tedfordi	4.7	5.2
Pseudomeriones abbreviatus	6.5	6.75

NOTE: See figures 2.9 and 2.11.
SOURCE: After Flynn et al. (1997); Zheng and Zhang (2001).

chronological units can be based on its characteristic suite of taxa, whether or not the defining taxon is present. The preceding sections have illustrated some of the districts in China that appear to be amenable to achieving these goals.

At the same time, our understanding of the paleozoogeographic and evolutionary patterns across Holarctica will benefit directly from an improved biochronology. At present, the zoogeographic limits of the late Neogene North China region (see figure 2.1) are fairly well defined in its southern extent relative to surrounding regions. In contrast, we know little of the biotic history of northeastern Asia and its relationships with arctic North America. The Asian record reaches only a short distance north of the latitude of Lake Baikal (55°N), but recent discoveries in the high arctic of North America (Ellesmere and Devon Islands above 75°N) in late Miocene (Tedford and Harrington 2003) and early Miocene (Rybczynski, Dawson, and Tedford 2009) time give a glimpse of the compositions of such faunas. In addition, immigrant clades of eastern Asian and North American origin at mid-latitude sites (40°N) reflect dispersal episodes from clades endemic to one or the other continents. This signals not only favorable conditions for Beringian immigration but, by inference, the nature of the mammal groups populating the intervening arctic terrains. The presently limited faunal lists for these Miocene North American high-latitude sites are surprisingly dominated by Eurasiatic clades of both small and large mammals, some of which also appear in mid-continental North American faunas at that time (e.g., early Hemingfordian; Tedford et al. 2004). For example, the early Hemingfordian cohort of taxa is dominated by small- to large-size insectivorous to carnivorous taxa, such as *Plesiosorex, Antesorex, Potamotherium, Amphictis, Amphicyon, Craterogale, Edaphocyon, Leptarctus, Miomustela, Phoberogale,* and *Ursavus*. Still, a variety of herbivorous taxa also occur, demonstrating the additional ecological breadth of the dispersal corridor (*Aletomeryx, Brachypotherium, Oreolagus, Desmatolagus, Euroxenomys, Floridaceras*). The early Hemingfordian was the ecologically most diverse and taxonomically most numerous of the dispersals to North America during the Neogene, but nearly 16 such episodes (Tedford et al. 2004:fig. 6.3; Bell et al. 2004) contributed a variety of groups to that record. An eastern Asian affinity for such dispersals is displayed by numerous taxa (mostly carnivores and rodents) throughout the Neogene, among which can be listed *Mimomys, Ophiomys, Microtus, Ogmodontomys, Pliopotamys, Bison, Mammuthus, Lepus, Smilodon, Mustela, Lutra, Propliophenacomys, Trigonictis, Castor, Plesi-*

ogulo, Pliozapus, Enhydritherium, Kansasimys, Megasminthus, Pliogale, Plithocyon, Mionictis, and *Sthenictis*. The North American presence of taxa such as these underscores the importance of developing a North China mammal chronology to help describe the time and place of their origin.

This history is becoming recognized in terms of changing zoogeography across the Northern Hemisphere as we search for faunal generalities that can be used to describe land mammal associations typifying given time spans. A well-developed North China biochronologic system will provide a stable chronologic framework from which to illustrate faunal interrelationships with other areas and enhance our understanding of paleozoogeographic and evolutionary patterns across Holarctica in the Neogene.

REFERENCES

Albright, L. B., T. J. Fremd, M. O. Woodburne, C. C. Swisher III, B. J. MacFadden, and G. R. Scott. 2008. Revised chronostratigraphy and biostratigraphy of the John Day Formation (Turtle Cove and Kimberly members), Oregon, with implications for updated calibration of the Arikareean North American Land Mammal Age. *Journal of Geology* 116:213–237.

Andersson, J. G. 1923. Essays on the Cenozoic of northern China. *Memoirs of the Geological Survey of China*, ser. A 3:152p.

Andersson, K. and L. Werdelin. 2005. Carnivora from the late Miocene of Lantian, China. *Vertebrata PalAsiatica* 10:256–271.

Aubry, M. P. 1997. Interpreting the (marine) stratigraphic record. In *Actes du Congrès BiochroM'97*, ed. J.-P. Aguilar, S. Legendre, and J. Michaux, pp. 15–32. Mémoires et Travaux E.P.H.E., Institut de Montpellier, 21.

Bates, R. L., and J. A. Jackson, eds. 1987. *Glossary of Geology*. Alexandria, Va.: American Geological Institute.

Bell, C. J., E. L. Lundelius, Jr., A. D. Barnosky, R. W. Graham, E. H. Lindsay, D. R. Ruez, Jr., H. A. Semken, Jr., S. D. Webb, R. J. Zakrzewski, and M. O. Woodburne. 2004. The Blancan, Irvingtonian, and Rancholabrean mammal ages. In *Late Cretaceous and Cenozoic Mammals of North America: Biostratigraphy and Geochronology*, ed. M. O. Woodburne, pp. 232–314. New York: Columbia University Press.

Berggren, W. A., F. J. Hilgen, and C. G. Langereis. 1995. Late Neogene (Pliocene-Pleistocene) chronology: New perspectives in high-resolution stratigraphy. *Geological Society of America Bulletin* 107:1272–1287.

Berggren, W. A., D. V. Kent, C. C. Swisher III, and M. P. Aubry. 1995. A revised Cenozoic geochronology and chronostratigraphy. In *Geochronology, Time Scales, and Global Stratigraphic Correlation: A Unified Framework for a Historical Geology*, ed. W. A. Berggren, D. V. Kent, M. P. Aubry, and J. A. Hardenbol, pp. 129–212. Society of Economic Paleontology and Mineralogists Special Publications 54. Tulsa: Society for Sedimentary Geology.

Berggren, W. A. and J. Van Couvering. 1974. The late Neogene: Biostratigraphy, geochronology and paleoclimatology of the last 15 million years in marine and continental sequences. *Palaeogeography, Palaeoclimatology, Palaeoecology* 16:1–216.

Chen, G.-f. and Z.-q. Zhang. 2004. *Lantiantragus* gen. nov. (Urmiatheriinae, Bovidae, Artiodactyla) from the Bahe Formation, Lantian, China. *Vertebrata PalAsiatica* 42(3):205–215.

Chiu, C.-s., C.-k. Li, and C.-t. Chiu. 1979. The Chinese Neogene: A preliminary review of the mammalian localities and faunas. *Annales Géologiques des Pays Helléniques*, non-series vol. 1: 263–272.

Deng, T. 2006. Chinese Neogene mammal biochronology. *Vertebrata PalAsiatica* 44(2):143–163.

Deng, T., X. Wang, X.-j. Ni, and L. Liu. 2004. Sequence of Cenozoic mammalian faunas of the Linxia Basin, in Gansu, China. *Acta Geologica Sinica* 78(1):8–14.

Fang, X.-m, C. Garzione, R. van der Voo, J.-j. Li, and M. Fan. 2003. Flexural subsidence by 29 Ma on the NE edge of Tibet from the magnetostratigraphy of the Linxia Basin, China. *Earth and Planetary Science Letters* 210:545–560.

Fang, X-m., W. Zhang, Q. Meng, J. Gao, X. Wang, J. King, C. Song, S. Dai, and Y. Miao. 2007. High-resolution magnetostratigraphy of the Neogene Huaitoutala section in the eastern Qaidam Basin on the NE Tibetan Plateau, Qinghai Province, China and its implications on tectonic uplift of the NE Tibetan Plateau. *Palaeogeography, Palaeoclimatology, Palaeoecology* 258: 293–306.

Flynn, L. J., R. H. Tedford, and Z.-x. Qiu. 1991. Enrichment and stability in the Pliocene mammalian fauna of China. *Paleobiology* 17(3):246–265.

Flynn, L. J., W. Wu, and W. R. Downs III. 1997. Dating vertebrate microfaunas in the Neogene record of northern China. *Palaeogeography, Palaeoclimatology, Palaeoecology* 133:227–242.

Gingerich, P. D. and Clyde, W. C. 2001. Overview of mammalian biostratigraphy in the Paleocene-Eocene Fort Union and Willwood formations of the Bighorn and Clark's Fork basins. In *Paleocene-Eocene Stratigraphy and Biotic Change in the Bighorn and Clarks Fork Basins, Wyoming*, ed. P. E. Gingerich, pp. 1–14. University of Michigan Papers on Paleontology 33.

Guo, Z.-t., W. F. Ruddiman, Q.-z. Hao, H.-b. Wu, Y.-s. Qiao, R.-x. Zhu, S. Peng, J.-j. Wei, B.-y. Yuan, and T.-s. Liu. 2002. Onset of Asian desertification by 22 Myr ago inferred from loess deposits in China. *Nature* 416:159–163.

Harzhauser, M., G. Daxner-Höck, and W. E. Piller. 2004. An integrated stratigraphy of the Pannonian (late Miocene) in the Vienna Basin. *Austrian Journal of Earth Sciences* 95/96:6–19.

Höck, V., G. Daxner-Höck, H. P. Schmid, D. Badamgarav, W. Frank, G. Furtmüller, O. Montag, R. Barsbold, Y. Khand, and J. Sodov. 1999. Oligocene-Miocene sediments, fossils, and basalts from the Valley of Lakes (central Mongolia)—an integrated study. *Mitteilungen Österreich Geologisches Gesellschaft* 90:83–125.

Jin, C.-z. and Y.-q. Zhang. 2005. First discovery of *Promimomys* (Arvicolidae) in East Asia. *Chinese Science Bulletin* 50:327–332.

Kaakinen, A. 2005. A long terrestrial sequence in Lantian: A window into the late Neogene palaeoenvironments of northern China. Ph.D. diss., University of Helsinki.

Kaakinen, A. and Lunkka, J. P. 2003. Sedimentation of the Late Miocene Bahe Formation and its implications for stable environments adjacent to Qinling mountains in Shanxi, China. *Journal of Asian Earth Sciences* 22:67–78.

Li, C.-k., W.-y. Wu, and Z.-d. Qiu. 1984. Chinese Neogene: Subdivision and correlation. *Vertebrata PalAsiatica* 22(3):163–178.

Li, Q., X.-m. Wang, and Z.-d. Qiu. 2003. Pliocene mammalian fauna of Gaotege in Nei Mongol (Inner Mongolia), China. *Vertebrata PalAsiatica* 41(2):1–114.

Lindsay, E. H. 1972. Small mammals from the Barstow Formation, California. *University of California Publications in Geological Sciences* 93:1–104.

Liu, T.-s., C.-k. Li, and R.-j. Zhai. 1978. Pliocene vertebrates of Lantien, Shensi. In *Tertiary Mammalian Fossils of the Lantien District, Shensi*. Part II. Chinese Academy Geological Sciences, *Professional Papers on Stratigraphy and Paleontology* 7:149–199.

Lourens, L., F. Hilgen, N. J. Shackleton, J. Laskar, and D. Wilson. 2004. The Neogene Period. In *A Geologic Time Scale 2004*, ed. F. Gradstein, J. Ogg, and A. Smith, pp. 409–440. Cambridge: Cambridge University Press.

Luterbacher, H. P., J. R. Ali, H. Brinkhuis, F. M. Gradstein, J. J. Hooker, S. Monechi, J. G. Ogg, J. Powell, U. Röhl, A. Sanfilippo, and B. Schmitz. 2004. The Paleogene Period. In *A Geologic Time Scale 2004*, ed. F. Gradstein, J. Ogg, and A. Smith, pp. 384–408. Cambridge: Cambridge University Press.

Mein, P. 1975. Resultats du Groupe de Travail des Vertebres. In *Report on Activity of the RCMNS Working Groups (1971–1975)*, Bratislava, 78–81.

Mein, P. 1980. Updating of MN Zones. In *European Neogene Mammal Chonology*, ed. E. H. Lindsay, V. Falbusch, and P. Mein, pp. 73–90, NATO ASI Series (A) 180. New York: Plenum Press.

Mein P. 1996. European Miocene mammal biochronology. In *Land Mammals of Europe*, ed. G. E. Rössnere and K. Heissig, pp. 25–38. Munich: Dr. Friedrich Pfeil.

Meng, J., J. Ye, W.-y. Wu, X.-j. Ni, and S.-d. Bi. 2008. The Neogene Dingshanyanchi Formation in northern Junggar Basin of Xinjiang and its stratigraphic implications. *Vertebrata PalAsiatica* 46(2):90–110.

Meng, J., J. Ye, W.-y. Wu, L.-p. Yue, and X.-j. Ni. 2006. A recommended boundary stratotype section for Xiejian Stage from northern Junggar Basin: Implications to related bio-chronostratigraphy and environmental changes. *Vertebrata PalAsiatica* 44(3):205–236.

O'Connor, J., D. R. Prothero, X. Wang, Q. Li, and Z. Qiu. 2008. Magnetic stratigraphy of the lower Pliocene Gaotege beds, Inner Mongolia. In *Neogene Mammals*, ed. S. G. Lucas, G. S. Morgan, J. A. Speilman, and D. R. Prothero, pp. 431–436. *New Mexico Museum of Natural History and Science Bulletin* 44.

Opdyke, N. D., E. H. Lindsay, N. M. Johnson, and T. Downs. 1977. The paleomagnetism and magnetic polarity stratigraphy of the mammal-bearing sections of Anza Borrego State Park, California. *Quaternary Research* 7:316–329.

Qiu, Z. 1989. The Chinese Neogene mammalian biochronology—its correlation with the European Neogene mammalian zonation. In *European Neogene Mammal Chronology*, ed. E. H. Lindsay, V. Fahlbusch, and P. Mein, pp. 527–556. NATO ASI series. New York: Plenum Press.

Qiu, Z.-d. and Q. Li. 2008. Late Miocene micromammals from the Qaidam Basin in the Qinghai-Xizang Plateau. *Vertebrata PalAsiatica* 46(4):284–306.

Qiu, Z.-d., X.-m. Wang, and Q. Li. 2006. Faunal succession and biochronology of the Miocene through Pliocene in Nei Mongol (Inner Mongolia). *Vertebrata PalAsiatica* 44(2):164–181.

Qiu, Z.-x. and Z.-d. Qiu. 1995. Chronological sequence and subdivision of Chinese Neogene mammalian faunas. *Palaeogeography, Palaeoclimatology, Palaeoecology* 116:41–70.

Qiu, Z.-x., W.-y. Wu, and Z.-d. Qiu. 1999. Miocene mammal faunal sequence of China: Palaeozoogeography and Eurasian relationships. In *The Miocene Land Mammals of Europe*, ed. G. Rössner and K. Heissig, pp. 443–455. Munich: Dr. Friedrich Pfeil.

Qiu, Z. and Xie, J. 1998. Notes on *Parelasmotherium* and *Hipparion* fossils from Wangji, Dongxziang, Gansu. *Vertebrata PalAsiatica* 36(1):13–23.

Rybszynski, N., M. R. Dawson, and R. H. Tedford. 2009. A semi-aquatic mammalian carnivore from the Miocene epoch and the origin of Pinnipedia. *Nature* 458:1021–1024.

Salvador, A., ed. 1994. *International Stratigraphic Guide*. 2nd ed. Boulder: International Union of Geological Sciences and the Geological Society of America.

Savage, D. E. 1955. Nonmarine lower Pliocene sediments in California: A geochronologic-stratigraphic classification. *University of California Publications in Geological Sciences* 31:1–26.

Secord, R., P. D. Gingerich, M. Elliot Smith, W. Clyde, P. Wilf, and B. S. Singer. 2006. Geochronology and mammalian biostratigraphy of middle and upper Paleocene continental strata, Bighorn Basin, Wyoming. *American Journal of Science* 306:211–245.

Steininger, F. F. 1999. Chronostratigraphy, Geochronology and Biochronology of the Miocene "European Land Mammal Mega-Zones (ELMMZ)" and the Miocene "Mammal-Zones (MN-Zones)." In *The Miocene Land Mammals of Europe*, ed. G. E. Rössner and K. Hessig, pp. 9–24. Munich: Dr. Friedrich Pfeil.

Steininger, F. F., W. A. Berggren, D. V. Kent, R. L. Bernor, S. Sen, and J. Agusti. 1996. Circum-Mediterranean Neogene (Miocene and Pliocene) marine–continental chronologic correlations of European mammal units. In *The Evolution of Western Eurasian Neogene Mammal Faunas*, ed. R. L. Bernor, V. Fahlbusch, and H.-W. Mittman, pp. 7–46. New York: Columbia University Press.

Tedford, R. H. 1970. Principles and practices of mammalian geochronology in North America. *Proceedings of the North American Paleontological Convention, Part F*: 66–703.

Tedford, R. H. 1995. Neogene mammalian biostratigraphy in China: Past, present and future. *Vertebrata PalAsiatica* 10:272–289.

Tedford, R. H., L. B. Albright III, A. D. Barnosky, I. Ferrusquía-Villafranca, R. M. Hunt, Jr., J. E. Storer, C. C. Swisher III, M. R. Voorhies, S. D. Webb, and D. P. Whistler. 2004. Mammalian biochronology of the Arikareean through Hemphillian interval (Late Oligocene through Early Pliocene Epochs). In *Late Cretaceous and Cenozoic Mammals of North America*, ed. M. O. Woodburne, pp. 169–231. New York: Columbia University Press.

Tedford, R. H., L. J. Flynn, Z. Qiu, N. D. Opdyke, and W. R. Downs. 1991. Yushe Basin, China; paleomagnetically calibrated mammalian biostratigraphic standard for the late Neogene of eastern China. *Journal of Vertebrate Paleontology* 11(4): 519–526.

Tedford, R. H. and C. R. Harrington. 2003. An Arctic mammal fauna from the early Pliocene of North America. *Nature* 425:388–390.

Teilhard de Chardin, P. 1941. *Early Man in China*. Pékin (Beijing): Institute de Géo-Biologie.

Tong, Y.-s., S.-h. Zheng, and Z.-d. Qiu. 1995. Cenozoic mammal ages of China. *Vertebrata PalAsiatica* 33(4):290–314.

van Dam, J. A., L. Alcalá, A. Alonso Zara, J. P. Calvo, M. Garcés, and W. Krijgsman. 2001. The upper Miocene mammal record from the Teruel–Alfambra region (Spain). The MN system and continental stage/age concepts discussed. *Journal of Vertebrate Paleontology* 21(2):367–385.

Wang. X. 1998. Carnivora from middle Miocene on northern Junggar Basin, Xinjiang Autonomous Region, China. *Vertebrata PalAsiatica* 36(3):218–243.

Wang, X., Z. Qiu, Q. Li, B. Wang, Z. Qiu, W. R. Downs, G. Xie, J. Xie, T. Deng, G. T. Takeuchi, Z. J. Tseng, M. Chang, J. Liu, Y. Wang, D. Biasatti, Z. Sun, X. Fang, and Q. Meng. 2007. Vertebrate Paleontology, biostratigraphy, geochronology, and paleoenvironment of Qaidam Basin in northern Tibetan Plateau. *Palaeogeography, Palaeoclimatology, Palaeoecology* 254:363–385.

Wang, X., Z. Qiu, and N. D. Opdyke. 2003. Litho-, bio-, and magnetostratigraphy and paleoenvironment of the Tunggur Formation (middle Miocene) in central Inner Mongolia, China. *American Museum Novitates* 3411:1–31.

Wood II, H. E., R. W. Chaney, J. Clark, E. H. Colbert, G. L. Jepsen, J. B. Reeside, Jr., and C. Stock. 1941. Nomenclature and correlation of the North American continental Tertiary. *Geological Society of America Bulletin* 52:1–48.

Woodburne, M.O. 1977. Definition and characterization in mammalian chronostratigraphy. *Journal of Paleontology* 51(2): 220–234.

Woodburne, M. O. 1996. Precision and resolution in mammalian chronostratigraphy: Principles, practices, examples. *Journal of Vertebrate Paleontology* 16(3):531–555.

Woodburne, M. O. 2004a. Definitions. In *Late Cretaceous and Cenozoic Mammals of North America: Biostratigraphy and Geochronology*, ed. M.O. Woodburne, pp. xi–xv. New York: Columbia University Press.

Woodburne, M. O. 2004b. Principles and procedures. In *Late Cretaceous and Cenozoic Mammals of North America: Biostratigraphy and Geochronology*, ed. M. O. Woodburne, pp. 1–22. New York: Columbia University Press.

Woodburne, M. O. 2006. Mammal Ages. *Stratigraphy* 3(4):229–261.

Woodburne, M. O. 2007. Phyletic Diversification of the *Cormohipparion occidentale* complex (Mammalia; Perissodactyla, Equidae), late Miocene, North America, and the origin of the Old World *Hippotherium* Datum. *American Museum of Natural History Bulletin* 454.

Woodburne, M. O. 2009. The early Vallesian vertebrates of Atzelsdorf (late Miocene, Austria. 9. *Hippotherium* (Mammalia, Equidae). *Annales Naturhistorishes Museum Wien* 111A:575–593.

Yin, A., Y. Dang, L.-c. Wang, W.-m. Jiang, S.-p. Zhou, X. Chen, G. E. Gehrels, and M. W. McRivette. 2008. Cenozoic tectonic evolution of Qaidam basin and its surrounding regions (Part 1): The southern Qilian Shan-Nan Shan thrust belt and northern Qaidam basin. *Geological Society of America Bulletin* 120(7–8):813–846.

Yin, A., Y. Dang, M. Zhang, X. Chen, and M. W. McRivette. 2008. Cenozoic tectonic evolution of the Qaidam basin and its surrounding regions (Part 3): Structural geology, sedimentation, and regional tectonic reconstruction. *Geological Society of America Bulletin* 120(7–8):847–876.

Yin, A., Y. Dang, M. Zhang, M. W. McRivette, W. P. Burgess, and X. Chen. 2007. Cenozoic tectonic evolution of Qaidam Basin and its surrounding regions (part 2): Wedge tectonics in southern Qaidam Basin and the eastern Kunlun Range. *Geological Society of America Special Paper* 433:369–390.

Ye, J., J. Meng, and W. Wu. 2003. Oligocene/Miocene beds and faunas from Tieersihabahe in the northern Junggar Basin of Xinjiang. *American Museum Bulletin* 279:568–585.

Zhang, Z.-q., A. W. Gentry, A. Kaakinen, L.-p. Liu, J. P. Lunkka, Z.-d. Qiu, S. Sen, R. S. Scott, L. Werdelin, S.-h. Zheng, and M. Fortelius. 2002. Land mammal faunal sequence of the late Miocene of China: New evidence from Lantian, Shaanxi Province. *Vertebrata PalAsiatica* 40:165–176.

Zheng, S.-h. and Z.-q. Zhang. 2001. Late Miocene–Early Pleistocene biostratigraphy of the Leijiahe area, Lingtai, Gansu. *Vertebrata PalAsiatica* 39(3):215–228.

APPENDIX

This appendix is an explanation of the faunal units and their chronologic relationships summarized in figure 2.3.

AGE, EPOCH, STAGE, AND POLARITY CHRON

The Oligocene is after Luterbacher et al. (2004:fig. 21.4). The Miocene–Holocene follows Lourens et al. (2004).

NEOGENE MN ZONE, EUROPEAN FAUNAL UNIT CHRONOLOGY, AND PARATETHYS STAGES

These are based on Steininger (1999) and Steininger et al. (1996), but updated on the basis of their magnetostratigraphic correlations as portrayed in Lourens et al. (2004). The Paleogene MP-zones and European Faunal Unit chronology is from Luterbacher et al. (2004).

1. Base of MN1 and the Agenian European Land Mammal Mega-Zone (ELMMZ)

Steininger (1999) correlates the base of the Agenian and MN1 with the base of chron C6Cn.2n and the base of the Aquitanian Stage. Lourens et al. (2004) correlate the base of that magnetic chron as 23.03 Ma (0.77 Ma younger than the time scale of Berggren, Hilgen, and Langereis (1995).

2. Base of MN2

This is the base of chron 6CAAr.2r according to Steininger (1999; see Berggren, Hilgen, and Langereis 1995:fig. 5) or 21.65 Ma in Lourens et al. (2004).

3. Base of MN 3 and base of the Orleanian ELMMZ

This is the base of chron C6r in Steininger (1999) or the base of chron C6 in Lourens et al. (2004), or 20.04 Ma.

4. Base of MN 4

This is prorated by Steininger (1999) as within chron C5Dr and at 18 Ma. In the Lourens et al. (2004) time scale, the equivalent position is about 17.9 Ma.

5. Base of MN 5

This is located by Steininger (1999) as within chron C5Cr, or at 17.0 Ma. The equivalent position within chron C5C in the Lourens et al. (2004) time scale also is at 17.0 Ma.

6. Base of MN 6 and base of Astaracian ELMMZ

Steininger (1999) locates this at the base of chron C5Bn.1r, at 15.0 Ma. In Lourens et al. (2004), the equivalent position also is at 15.0 Ma.

7. Base of MN 7 and 8

This is taken at the base of chron C5ABn in Steininger (1999), at 13.5 Ma. In Lourens et al. (2004), the equivalent position is about 13.55 Ma.

8. Base of MN 9 and base of Vallesian ELMMZ

Steininger (1999) uses the base of chron C5r.1n as the base of the MN 9 at 11.1 Ma (Berggren, Hilgen, and Langereis 1995:fig. 5). Steininger (1999) correlates that boundary with Pannonian Zone C of the Central Paratethys. Harzhauser, Daxner-Höck, and Piller (2004) indicate that the base of Pannonian Zone C, with the earliest *Hippotherium*-bearing sites (Atzelsdorf, Gaiselberg, Mariathal), correlates with the upper part of chron C5r.2r at 11.2 Ma. This is followed here.

9. Base of MN 10

Steininger (1999) calibrates this as the base of chron C4Ar.3r, at 9.74 Ma. Lourens et al. (2004) show the equivalent position at about 9.8 Ma (base of chron C4A).

10. Base of MN 11 and base of Turolian ELMMZ

This is at the base of chron C4r.2r and 8.67 Ma in Steininger (1999; see Berggren, Hilgen, and Langereis 1995:fig. 5). The equivalent position in Lourens et al. (2004) is the base of C4 at 8.8 Ma.

11. Base of MN 12

Steininger (1999) shows this at the base of chron C4n.2n or at 8.06 Ma (Berggren, Hilgen, and Langereis 1995:fig. 5). Lourens (2004) show the equivalent position at about 8.15 Ma.

12. Base of MN 13

Steininger (1999) indicates that this is at the base of chron C3An.2n, or 6.55 Ma (Berggren, Hilgen, and Langereis 1995:fig. 5). Lourens et al. (2004) indicate the equivalent position at 6.7 Ma.

13. Base of MN 14

Steininger (1999) shows this as the base of chron C3n.3n, Sidufjal event, or at 4.9 Ma in Berggren, Hilgen, and Langereis 1995:fig. 6). Lourens et al. (2004) indicate the equivalent position at 4.9 Ma.

14. Base of MN 15

Steininger et al. (1996) show this as at the very top of chron C3n.1n, at 4.15 Ma in Berggren, Hilgen, and Langereis (1995:fig. 6). Lourens et al. (2004) show the equivalent position at 4.2 Ma.

15. Base of MN 16

Steininger et al. (1996) indicate that this is within chron C2An.3n, at 3.4 Ma. Lourens et al. (2004) show the equivalent position at 3.4 Ma.

16. Base of MN 17

Steininger et al (1996) show this at the base of chron C2r.2r, or 2.6 Ma. Lourens et al. (2004) show the equivalent position at the base of chron C2, at 2.6 Ma.

17. Top of MN 17

Steininger et al. (1996) show this at the base of chron C2n, or 1.95 Ma. Lourens et al. (2004) have the equivalent position at about 1.95 Ma.

CHINA

The Chinese units are from Deng (2006), except that the boundaries between the Chinese Faunal ages and Neogene Mammal Units (NMU) are correlated to the basic chronological system based on the paleomagnetic interpretations provided therein. The purpose of this is to base the age and potential boundaries of the Chinese units on the paleomag-

netic information, rather than relying only upon mammal correlations. The following comments are given in that regard, following Deng (2006) and other authors.

Xiejian Age Faunas (NMU 1–3)

The Xiejian age was based on the Xiejian fauna by Li, Wu, and Qiu (1984), which, with other faunas, was correlated with MN 1–2. Qiu and Qiu (1995) included MN 3 in the Xiejian Age (Deng, 2006).

1. The base of the Xiejian age and the base of NMU 1. Qiu, Wu, and Qiu (1999) indicate that the Suosuoquan fauna, Xinjiang, is the oldest Neogene mammal fauna in China and correlative with MN 1. Deng (2006) indicates that MN 1 is coeval with NMU 1. Meng et al. (2006) describe the mammalian biostratigraphy and magnetostratigraphy of the Tieersihabahe and Suosuoquan formations of the Junggar basin (see figure 2.1 [1]). The stratigraphically lowest Neogene mammal assemblage is known as the Suosuoquan Mammal Assemblage Zone (SMAZ) II, correlated to an interval in the Suosuoquan Formation 19.75 to 30 m above its contact with the Tieersihabahe Formation (see figure 2.4). SMAZ II is correlated to chron C6Bn.1n and to the base of chron C6AAr.3r, or from 21.9 to 21.7 Ma (Meng et al. 2006). The position of the Suosuoquan Mammal Assemblage Zone II is shown at this level in figure 2.3. Meng et al. (2006) note that SMAZ II is "comparable" to the Xiejia fauna of China (see figure 2.1 [4]) and the fauna of Biozone D in the Valley of Lakes, Mongolia (Höck et al., 1999).

In this context, Suosuoquan Mammal Assemblage Zone I occurs stratigraphically 16.25 m below SMAZ II and contains a fauna considered to be of late Oligocene age. SMAZ I occurs at the top of a polarity zone correlated as chron C6Cn.2r, or about 23.03 Ma in age, so there remains a temporal gap of about 1.3 Ma between the Suosuoquan fauna (SMAZ II) and the beginning of the Neogene (see figure 2.3). Meng et al. (2006) correlate SMAZ I with the Tabenbulukian fauna (see figure 2.4), also of latest Oligocene age (Qiu, Wu, and Qiu 1999), so the likely lower boundary of the Xiejian fauna remains as shown in figure 2.3. Meng et al. (2006) also correlate the immediately older Tieersihabahe II and I Mammal Assemblage Zones to the Tabenbulukian (see figure 2.3).

2. Age of the Xiejian fauna. Meng et al. (2006) indicate that the Suosuoquan Mammal Assemblage Zone III occurs immediately above SMAZ II and correlates with chron C6AAr.3r to C6An.1r, or from 21.7 to 20.45 Ma (Lourens et al. 2004), but the upper limit of this range accommodates sites with sparse fossils from 50 and 72.5 m above the base of the Suosuoquan Formation. The sites with the best fau-

nal representatives of SMAZ III occur in beds (XJ99005) correlated as being equivalent to strata 28 to 30 m above the base of the formation (see figure 2.4). Those beds, in turn, are correlated to the base of chron C6Bn.1n and to the base of chron C6AAr.1n, or from 21.8 to 21.6 Ma (Luterbacher et al. 2004). This appears to be the best age for SMAZ III (see figure 2.3). Meng et al. (2006) correlate the "conventional" Suosuoquan fauna about with SMAZ III.

In addition, Deng (2006) considers the Xiejian fauna, of the Xiejia Formation, Quinghai (see figure 2.1 [4]), to correlate with chron C6Ar, at about 21 Ma, effectively coeval with the potential upper extent of SMAZ III (see figures 2.2 and 2.3).

In figure 2.3, the Xiejian fauna is correlated as shown, with the best elements of SMAZ III being older and directly following SMAZ II, as well as recognizing the "rough equivalency" of SMAZ III with the "conventional" Suosuoquan fauna as stated by Meng et al. (2006). If SMAZ II represents NMU 1 (as the oldest Neogene fauna in China), this leaves a potential gap of about 0.7 Ma between the calibrated mammal faunas of NMU 1 and NMU 2, provided the latter is best represented by the Xiejian fauna. It is additionally important that the succession in the Suosuoquan Formation described by Meng et al. (2006) has the potential to determine the age of the boundary between NMU 1 and NMU 2, and that further collecting in the Tieersihabahe and Suosuoquan formations has the potential to fill in the earliest Neogene gap in the faunal record in China.

3. NMU 3 is represented by the Qin'an 7 and Zhangjiaping faunas, Gansu, correlated as being 20.2 and 20.5 Ma old, respectively (see figure 2.2). Although shown in succession in Deng (2006; see figure 2.2), the boundary between NMU 2 and NMU 3 is not documented biochronologically or paleomagnetically. It is important that the Qin'An succession summarized by Guo et al. (2002) has the potential for establishing a paleomagnetically calibrated faunal sequence that ranges from NMU 3 through NMU 10 (see figure 2.3). On the other hand, this succession apparently does not contain the potential to address the boundary between NMU 2 and NMU 3. Qiu, Wang, and Li (2006:table 1) show faunas of Xiejian and Shanwangian age (as well as others). The arrangement of the Suosuoquan faunas differs from that of Meng et al. (2006), regarding which no comments are offered by Qiu, Wang, and Li (2006).

Shanwangian Age Faunas (NMU 4–5)

The Shanwangian age was coined by Li, Wu, and Qiu (1984) and discussed by Tong, Zheng, and Qiu (1995), who appar-

ently correlated it with NMU 4–5, as used by Deng (2006; see figure 2.2).

1. Base of the Shanwangian age and NMU 4. The two oldest faunas cited by Deng (2006) of the Shanwangian age and NMU 4 are Xishiugou and Qin'an 6, from Gansu. As noted by Deng (2006), the Xishiugou fauna correlates with chrons C6n to C6En, or an age of 19.5 to 18.5 Ma. The Qin'an 6 fauna correlates with an age of 19.4 Ma. In that the youngest calibrated Xiejian fauna (Qin'an 7) is 20.2 Ma, the Xiejian and Shanwangian ages are separated by a gap of about 0.7 m.y. (see figure 2.3). Meng et al. (2008) suggest that the base of the Shanwangian correlates to an age of 20.43 Ma, apparently using the sparse fossils of SAZ III? at locality XJ200206 (see figure 2.4) that have a "Shanwangian affinity" for this purpose. Until this biocorrelation can be made more secure, the arrangement in figure 2.3 follows Deng (2006).

2. Top of NMU 4. The fauna from DH9914, Danghe, is the youngest calibrated fauna attributed to NMU 4 by Deng (2006); correlated with C5Cr and an age of 17.3 to 16.7 Ma (see figures 2.2 and 2.3).

3. Base of NMU 5. The only calibrated fauna attributed to NMU 5 by Deng (2006) is that of Qin'an 4, from chron C5Br and an age of 15.5 Ma. Faunas attributed to NMU 4 and NMU 5 bridge a gap of about 1.2 m.y. (see figure 2.3).

4. Top of NMU 5 and Shanwangian Age. This is the same as in (3). The Shanwangian Age has a calibrated top of about 15.5 Ma.

Tunggurian Age Faunas (NMU 6–7)

The Tunggurian Age was based on the Tung Gur fauna (Li, Wu, and Qiu 1984) and recently discussed by Deng (2006) and Qiu, Wang, and Li (2006). As indicated in figures 2.2 and 2.3, the Tunggurian Age includes faunas correlated with NMU 6 and NMU 7. As discussed below, the Tunggur fauna (as distinct from the Tunggurian Age) correlates with NMU 7.

1. Base of the Tunggurian age and NMU 6. Gansu-area faunas DH9903 and DH 9905 are cited by Deng (2006) as the oldest (and sole) calibrated units of NMU 6 (as well as Tunggurian) age. In being correlated to chron C5ACr, these faunas are 14.2 Ma old. This suggests that the presently calibrated base of the Tunggurian age is separated by a gap of as much as 1.3 m.y. from the uppermost Shanwangian (see figure 2.3).

2. Base of NMU 7. Wang, Qiu, and Opdyke (2003) described the litho-, bio-, and magnetostratigraphy of the Tunggur Formation, Inner Mongolia, and recognized three faunal intervals. The stratigraphically lowest of these, the

Tairum Nor fauna, was calibrated at 12.6 to 12.9 Ma in the Lourens et al. (2004) time scale (base of chron C5Ar.2n to near the top of chron C5Ar.1r), and this is suggested in figure 2.3 as representing the lower part of NMU 7. In this context, the lower boundary of NMU 7 is separated from the upper boundary of NMU 6 by a gap of about 1.3 m.y.

3. The Tunggur Fauna. Qiu, Wang, and Li (2006) use the term, Tunggur fauna (=*Platybelodon* fauna), to refer to all of the faunas from the Tunggur Formation (see figure 2.1 [19]), effectively comparable to the Tunggurian chronofauna as used by Wang (1998). This begins with the Tairum Nor fauna, which occurs stratigraphically directly below the Moergen fauna. The Moergen fauna is represented by locality MOII in the Moergen stratigraphic section (Wang, Qiu, and Opdyke 2003) and is correlated with the base of chron C5An.2n (12.5 Ma). The upper fauna from the Tunggur Formation in this section, locality MOV, produced the Tamuqin fauna (see figure 2.3), correlated within the lower part of chron C5r.3r, or11.9 Ma. The Tunggur mammalian fauna appears to range in age from 12.9 to 11.9 Ma in the Luterbacher et al. (2004) time scale.

4. Top of Tunggurian age and NMU 7. According to Deng (2006), the Qin'An A3 fauna is of NMU 7 age. It appears to fall at the base of chron C5r.3r, or about 11.5 Ma (see figure 2.3), and it appears to be the youngest fauna of NMU 7 age. If so, the top of NMU 7 (and also the top of the Tunggurian age) would post-date the top of the Tunggur fauna by about 0.4 m.y.

Bahean Age Faunas (NMU 8–9)

Li, Wu, and Qiu (1984) named this age from the Bahe fauna (see figure 2.1 [27]) and considered it to be the temporal equivalent to the Vallesian mammal age of Europe (see figures 2.2 and 2.3), heralded in both regions by the FAD of *Hipparion*. Qiu, Wang, and Li (2006) combine the Bahean and Baodean ages into a single unit, but Deng (2006) separates the two, as followed here. The succession from Lantian (=Bahe; see figure 2.1 [27]) is fundamental to discussions of the Bahe and Baode faunas and their equivalents. As discussed by Zhang et al. (2002), the Bahe Formation is about 300 m thick, with a basically conglomeratic lower 50 m, but this is followed upward by a succession of finer-grained units bearing fossils. Kaakinen (2005) revised the thickness to about 280 m, indicated a correlation to chrons C5n.2n and C4Ar (Kaakinen, 2005:fig 4), or an age of 11.0 to 7.0 Ma in the Lourens et al. (2004) time scale, with the fossil-bearing part of the formation ranging in age from 10.3 to 7.7 Ma (see figure 2.3). The Lantian Formation "red clay" is about 50 m thick, directly overlies (possibly grada-

tionally within a meter), and apparently extends in age to about the base of chron C2An.3n (about 3.6 Ma), although the Lantian fossil mammals only extend from 6.8 to 6.0 Ma. Liu, Li, and Zhai (1978) developed the first major biostratigraphic discussion of the Bahe Formation and described the horses *Hipparion weihoense* and *H. chiai* from the upper half of the Bahe deposits, with further discussions of the sediments and fauna seen in Zhang et al. (2002), Kaakinen and Lunkka (2003), Chen and Zhang (2004), and Andersson and Werdelin (2005).

1. Base of the Bahean age and NMU 8. As discussed in Deng (2006), the NMU 8 is represented by the Guonigou fauna, Gansu, calibrated as being within chron C5r.1r, at an age of about 11.1 Ma. The presence of *Hipparion dongxiangense* in this fauna is compatible with a Vallesian-equivalent age, but the morphology of the type material (Qiu and Xie 1998) shows a distinctly simpler enamel pattern than typical of *Hippotherium primigenium* from Vallesian sites in Europe (Woodburne 2007). In any case, the age of the Guonigou fauna suggests a gap of about 0.9 m.y. between the base of the Bahean age and the top of the Tunggurian age (see figure 2.3). If the top of the Tunggurian is defined by the base of the Bahean, then there would be no gap between them.

2. Base of NMU 9. Deng (2006) indicates that the Bahe fauna, Shaanxi, ranges in age from 9.9 to 7.7 Ma and that the Dasengou fauna, Gansu, correlates with chron C4Ar.2r, or an age of about 9.5 Ma. These data indicate that the base of NMU 9 is separated from the calibrated top of NMU 8 by a gap of about 1.2 m.y. As discussed in Deng (2006), the horses *Hipparion weihoense* and *H. chiai* occur in the Dasengou fauna. Whereas these species generally have been attributed as representing a Vallesian-equivalency for the faunas in which they are found, the current situation suggests that they are not of earliest Vallesian age. This is congruent with the fact that two, differently derived, taxa of hipparions co-occur in this fauna (and the other faunas) according to the interpretations of M. O. Woodburne. On this basis, alone, it would be unlikely that such faunas would be of earliest Vallesian age.

3. Top of the Bahean Age and NMU 9. As earlier, the Bahe fauna ranges upward to about 7.7 Ma. As discussed in the following section, this range overlaps the age of the earliest faunas considered applicable to the Baodean Age and NMU 10 (see figure 2.1).

Baodean Age Faunas (NMU 10–11)

Deng (2006) places the Baodean as equivalent to the Turolian European mammal age and to contain NMU 10 and 11 faunas. Qiu, Wang, and Li (2006) note that Li, Wu, and Qiu (1984) defined the Baodean Age on the *Hipparion* faunas of northern China, especially the Baode of Shanxi (see figure 2.1 [31]), Wudu of Gansu, Ertemte of Nei Mongol (see figure 2.1 [37]), and the hominoid fauna of Lufeng, Yunnan (see figure 2.1 [30]). The Amuwusu fauna (see figure 2.1 [26]), considered by Qiu, Wang, and Li (2006) as an early element of the Baodean, is treated as early Bahean by Deng (2006; see figure 2.2).

1. Base of the Badoean age and NMU 10. Deng (2006) records an age of 8.3 Ma for the Yanagjiashan fauna, Gansu, correlated with chron C4r.1n. Deng (2006) also notes that the Xiaohe and Lielao faunas, Yunnan, are contemporaneous and correlated with chron C4r.1r, or 8.2 to 8.1 Ma in age. The Qin'an 2 fauna, Gansu (see figure 2.1 [49]), correlates with chron C4n.2n, or an age of 8.15 to 7.7 Ma. These three faunas overlap in age with the upper limit of the NMU 9 Bahe fauna, and thus with that of the upper part of the Bahean Age.

2. Top of NMU 10. Deng (2006) indicates that the Qin'an 1 and Baode 30 localities (see figure 2.1 [49, 31]) represent the upper part of NMU 10. The Baode 30 and Qin'an 1 sites correlate with chron C3Ar and an age of 7.15 to 6.7 Ma (see figure 2.1).

3. Base of NMU 11. Deng (2006) notes that the Lantian fauna, Shaanxi, ranges in age from 6.8 to 6.0 Ma and thus slightly overlaps the upper age of the localities assigned to NMU 10 by about 0.1 m.y. (see figure 2.1).

4. Top of NMU 11. According to Deng (2006), the upper limit of NMU 11 is represented by the Leijiahe I and II faunas, Gansu, with a duration of 6.6 to 4.9 Ma.

Gaozhuangian Age Faunas (NMU 12)

This follows Deng (2006). The Gaozhuangian age is based on the Gaozhuangian fauna, Yushe Basin, Shanxi. Qiu, Wang, and Li (2006) use Yushean Age for the equivalent unit and note that it, also, was based in part on the Gaozhuangian fauna.

1. Base of the Gaozhuangian age and NMU 12. Deng (2006) indicates that the Dongwan fauna, Gansu, ranges in

age from 5.3 to 4.2 Ma as the oldest ranging faunal unit of MNU 12 age. Consequently, the base of NMU 12 and the Gaozhuangian age overlaps the upper limit of NMU 11 and the Baodean Age by about 0.4 m.y. Were it not for the apparent discrepancy following from the age of NMU 12 based on the Dongwan fauna, the Baodean, Gaozhuangian, and Mazegouan ages (and associated NMU units) would be directly superpositional and coherent based on the Leijiahe faunas of Lingtai, Gansu, region.

2. Top of the Gaozhuangian age and NMU 12. Deng (2006) notes that the Leijiahe IV fauna, Gansu, ranges in age from 4.2 to3.6 Ma, and Qiu, Wang, and Li (2006) suggest a similar age for the Gaotege fauna (see figures 2.2 and 2.3). O'Connor et al. (2008) interpret a comparable age for the Gaotege fauna (4.6–3.6 Ma) in correlating it to chrons C2Ar–C3n.2n. The fauna is reviewed by Li, Wang, and Qiu (2003).

Mazegouan Age (NMU 13)

This age is based on the Mazegou fauna, Shanxi (see figure 2.1 [46]), correlated to chron C2An.3n to C2An.1n, or from 3.6 to 2.6 Ma, comparable to the Leijiahe V fauna, Gansu. The Renjiagou, Gansu, and Hefeng, Shanxi, faunas range from 3.5 to 3.4 Ma. The Youhe fauna (see figure 2.1 [40]) ranges in age from 3.15 to 2.6 Ma.

1. Extent of Mazegouan age (MNU 13). This age and the Leijiahe V MNU 13 fauna appear to follow directly the MNU 12 Leijiahe IV fauna, so the Mazegouan Age directly supercedes the Gaozhuangian Age. See the earlier discussion of the Yushe Basin sequence, however, regarding a distinction between the Mazegouan Age and the Mazegou Formation.

North American Land Mammal Ages (NALMA)

This follows Woodburne (2004b:figs. 8.8 and 8.9) with the subdivisions within the Arikareean NALMA revised according to Albright et al. (2008).

Chapter 3

A Single-Point Base Definition of the Xiejian Age as an Exemplar for Refining Chinese Land Mammal Ages

JIN MENG, JIE YE, WEN-YU WU, XI-JUN NI, AND SHUN-DONG BI

Definitions and Abbreviations

CLMA Chinese land mammal ages. The CLMA are used here as an informal biochronological system so that when spelled out in the text, the "land mammal age" is written in lower case. A CLMA is a time span based on the biochronologic succession of mammalian evolution that is recognized by taxa with distinctive morphological features.

FAD First appearance datum, wherein the first stratigraphic occurrence of a taxon is considered to have been synchronous over a specified geographic region (Woodburne 1996).

GSSP The Global Stratotype Section and Point (Gradstein, Ogg, and Smith 2004).

LO Lowest stratigraphic occurrence (Aubry 1995). The stratigraphic (descriptive) first occurrence of a taxon (Woodburne 2006).

LOS Lowest stratigraphic occurrence in the stratotype. The LOS is a point specified in the stratotype section that is marked by the lowest occurrence of a chosen species in the section. The LOS may be younger than the LO.

LSD Lowest Stratigraphic Datum (Opdyke et al. 1977). As originally stated this is effectively the same as LO, but Walsh (1998) stressed reserving the term "datum" for temporal, rather than stratigraphic, connotations, and this is followed by Woodburne (2006). As a temporal connotation, the phrase may be termed as "oldest" instead of "lowest."

LDS Lowest stratigraphic datum in the stratotype. Similar to the relationship of the LSD to LO, the LDS is the temporal connotation of the LOS. The LDS can be understood as a time plane that contains LOS.

The workshop "Neogene Terrestrial Mammalian Biostratigraphy and Chronology in Asia" was held June 8–14, 2009, in Beijing, with a goal "toward the establishment of a continent-wide stratigraphic and chronologic framework." The vigorous discussions at the symposium illustrated diverse opinions on Asian stratigraphy and in particular how to build or refine a biochronologic system based on fossil mammals from terrestrial sediments. Discussions focused on both empirical and conceptual issues of the Asian Neogene, with implications applicable to similar issues for the Asian Paleogene. Unlike Europe and North America, where similar discussions have been in place for decades and have generated numerous publications, we have begun to engage a substantial discussion on land mammal ages of Asia. For those who work on continental rocks and faunas of the Cenozoic, it seems universally agreed that a biochronological system based on mammals is a useful and powerful tool with which we can investigate regional geohistorical events, even though the current system is far from satisfactory.

During the last few decades, a considerable effort has been made to establish Neogene Chinese land mammal ages, which have come to be widely used as the Asian system (Li, Wu, and Qiu 1984; Qiu and Qiu 1990, 1995; Tong, Zheng, and Qiu 1995; Qiu, Wu, and Qiu 1999; Deng

2006). Discussions about the conceptual aspects of biochronologic analysis are relatively rare and usually brief when such a discussion does occur. For instance, in their work on the Cenozoic mammal ages of China, Tong, Zheng, and Qiu (1995:292–293, original in Chinese) stated that "the establishment of a land mammal age is generally based on evolutionary replacement of mammals at the family and/or generic level," and there was no further discussion why that should be the case. Nearly every land mammal age consists primarily of a composite fauna, in which species are grouped together based on their similar evolutionary stages but are not associated with detailed biostratigraphic data (Tong, Zheng, and Qiu 1995; Ting 1998). It is also true that none of the ages have well-defined boundaries. Although first appearances of taxa have been used to delimit the ages (e.g., Ting 1998), these definitions are made outside biostratigraphic context. One reason for a lack of boundary definition of the Chinese mammal ages, as suggested by Tong, Zheng, and Qiu (1995), is that paleomagnetic and radiometric data were sparse in Chinese Cenozoic sequences, which would make the discussion difficult. Therefore, boundaries of these ages are either abstract or rely on other data, such as polarity chrons. The problems are not unique to Chinese land mammal ages (CLMA), and they also exist among North American land mammal ages (NALMA) after decades of discussions and debates about the inadequacy of the mammal ages without detailed biostratigraphic information (Savage 1962, 1977; Tedford 1970; Woodburne 1977, 1996, 2004, 2006; Lindsay 2003; Lindsay and Tedford 1990; Walsh 2005a, 2005b).

The CLMA, however, are more or less a hybrid of the NALMA and the European MN zonations. Partly because of different patterns of paleogeographical distributions and exchanges of mammals during the Paleogene and Neogene, the Paleogene CLMAs were more influenced by the NALMAs (Tong et al. 1995; Wang et al. 1998; Ting 1998), whereas the Neogene CLMA were more influenced by the European MN zonations (Li, Wu, and Qiu 1984; Tong et al. 1995; Qiu and Qiu 1999; Deng 2006; Qui, Wang, and Li 2006). In both, the CLMA were established primarily on the basis of intercontinental correlations instead of regional biostratigraphy. The Xiejian age, for instance, was established by directly correlating with European MNs.

The NALMA was originally proposed as a provincial time scale for the North American continental Tertiary, and by using such a provincial time scale controversies caused by correlations with European standard marine units could be avoided (Wood et al. 1941; Woodburne 2006). The European MN zonations were proposed

based on evolutionary stages of European mammals that are often collected from isolated localities with relatively poor stratigraphic context (Mein 1975, 1989, 1999; Fahlbusch 1991). However different, the NALMA and the MN zones are based on their respective regional paleontology. Although the NALMA and MN zones have been adopted as biochronologic units, or intended to be so, this has not been universally recognized. For instance, Steininger (1999:11) considered the MN zones as biostratigraphic units and even proposed to convert the European mammal ages into biomegazones.

As Tedford (1995) pointed out, the European MN zonation is not always suitable for stratigraphic division and correlation in East Asia. Compared to the Asian record, the MN zones have less stratigraphical control, which makes it difficult to transform them into biochronological units under the ISG. Meng et al. (2006:207, original in Chinese) therefore suggested: "[T]he best method is first to establish a Chinese biochronologic time scale that reflects regional biological events. Because rock sequences and faunas with clearly observable superpositional relationships have been known from many areas in China, there is unique advantage to establish the biochronologic/biochronostratigraphic system for the Chinese terrestrial Tertiary." This system must be based on detailed stratigraphic work, and this is the context in which we presented our study of the Junggar Basin, with the focus on the lower boundary of the Xiejian (Meng et al. 2006; Meng et al. 2008). Wherever applicable, correlations of the Chinese land mammal ages (CLMA) with the European MNs or NALMAs are unquestionably helpful, but the basis of the CLMA should be the regional (Chinese or Asian) biostratigraphy, and boundaries of the CLMA should be defined based on regional mammalian evolutionary events, which do not have to correlate perfectly with those of the European and North American units.

In principle, the mammal age system differs from, and should be independent of, other chronologic systems because it relies exclusively on events of mammalian evolution as the boundary markers, such as the evolutionary first occurrence of a taxon, as well as characteristic content, such as the characteristic faunas. If we employ other means as part of the definitions of the CLMA, such as magnetic chrons as we did for defining the lower boundary of the Xiejian (Deng 2006; Meng et al. 2006), we mix two different criteria, and the resulting CLMA are no longer a biochronologic system based solely on mammals. The critical work for building the CLMA is how to define the lower boundaries of the ages. Currently, nearly all boundaries of the Chinese (Asian) mammal ages are defined by either arbitrary correlations with marine chronostratigraphic units or by polarity chrons. If we regard mammal ages strictly as a biochronologic system, then the boundaries should be defined by mammalian events, such as evolutionary first occurrences of species.

To participate in the long overdue discussion in Asia, we presented our work, both empirical and conceptual, from Xinjiang, northwest China, at the symposium. In this chapter, we use the Xiejian age and its lower boundary as an example to address some general problems for CLMA. We first introduce the Xiejian concept as a mammal age of the Chinese terrestrial Neogene and discuss its current problems by contrasting the stratigraphic data and studies from the type Xiejia locality with those from the Junggar Basin, Xinjiang. Then, we present a base definition of the Xiejian age using the stratigraphic data from the Junggar Basin to illustrate how we establish and/or improve the CLMA. We think a boundary definition based on a thorough stratigraphic framework is perhaps the most important aspect of the CLMA. With the proposed definition, we will extend our discussions to some more general conceptual issues surrounding the mammal ages.

We understand that "formal" or "official" regional Stages/Ages have been proposed based on the CLMA (e.g., NCSC 2002), but most of them were poorly or inadequately formulated and defined. Our definition and discussion will focus on the CLMA as an informal biochronologic system on the basis of the best stratigraphic data we have. We do not intend to convert the definition of the Xiejian age into a formal definition of a chronostratigraphic unit in this chapter, for at least two reasons. First, the related biostratigraphy relevant to the Xiejian in China, even that in the type Xiejia locality, as we will show, needs systematic and more detailed documentation. Second, the base of the Shanwangian age, the successive unit younger than the Xiejian, is so poorly defined that the time span of the Xiejian age cannot be determined with any confidence. However, we recognize that a local biochronological boundary as defined in this chapter may have the potential to be converted into a formal chronostratigraphic/chronologic boundary in the future, and our discussion provides an example on how and what we should do when defining such a boundary in the CLMA system.

THE XIEJIAN CHINESE MAMMAL AGE

The Type Section of Xiejia

The Xiejia section is located near the Xiejia village, Huangzhong County, Qinghai Province. Following the initial usage of the Qinghai geological survey in 1978, Li and Qiu (1980) formally named the Xiejia Formation and fauna, considering the age of the fauna early Miocene. Li, Qiu, and Wang (1981) discussed the Xiejia fauna and the lithology of the formation in more detail and recognized that the Xiejia fauna shares most genera with the Late Oligocene Taben-buluk fauna but totally differs from the latter at the species level. Li, Wu, and Qiu (1984: table 1) initially proposed the Xiejian as a mammal age and correlated it to the European MN 1–4. The Xiejian age has since been widely used as an early Miocene biochronostratigraphic unit in Asia (Tong et al. 1995; Höck et al. 1999; Qui, Wu, and Qiu 1999). In their proposal of Cenozoic mammal ages of China, Tong et al. (1995) correlated the Xiejian with the European Agenian and North American Arikareean, respectively, and stated that "the Xiejian age is characterized by retention of holdover or highly specialized Oligocene survivors" (Tong, Zheng, and Qiu 1995:310). More recently, Deng (2006) correlated Xiejian with the European MN 1 and MN 2 (Agenian), and Deng, Wang, and Yue (2006) provided an updated location map and description of the section.

As in other Neogene CLMA, recognition of the Xiejian age is primarily based on correlations with European MN zonations (Li and Ting 1983; Russell and Zhai 1987; Tong, Zheng, and Qiu 1995). Based on Steininger, Bernor, and Fahlbusch (1990), some workers (Lucas 1997; Lucas, Kordikova, and Emry 1998) thought that the Oligocene-Miocene boundary should be within the Xiejian. With the recalibration of the MN zones, in which the lower boundary of MN 1 is correlated to the base of Aquitanian (Steininger et al. 1997; Steininger 1999), the lower boundary of the Xiejian was correlated to the base of MN 1 and Aquitanian (NCSC 2002) and more specifically with the base of polarity chron C6Cn.2n (Deng, Wang, and Yue 2006; Meng et al. 2006) at 23.03 Ma—that is, the beginning of the Miocene (Shackleton et al. 2000; Lourens et al. 2004; Ogg and Smith 2004).

In the introduction to the Chinese Regional Chronostratigraphy (Geochronology) (NCSC 2002), an official document on Chinese stratigraphy, the Xiejian Stage was defined with five features: (1) the Xiejia section as the type section; (2) representative mammals including *Sinolagomys pachygnathus*, *Atlantoxerus*, *Eucricetodon youngi*, *Parasminthus xiningensis*, *Tataromys suni*, and *Tachyoryctoides kokonorensis*; (3) a brief description of lithology of the type section; (4) correlation with formations of similar age in other areas of China; and (5) a rough age correlation to the Aquitanian.

Work intended to consolidate the NCSC (2002) proposal of the Xiejian Stage was conducted by Deng,

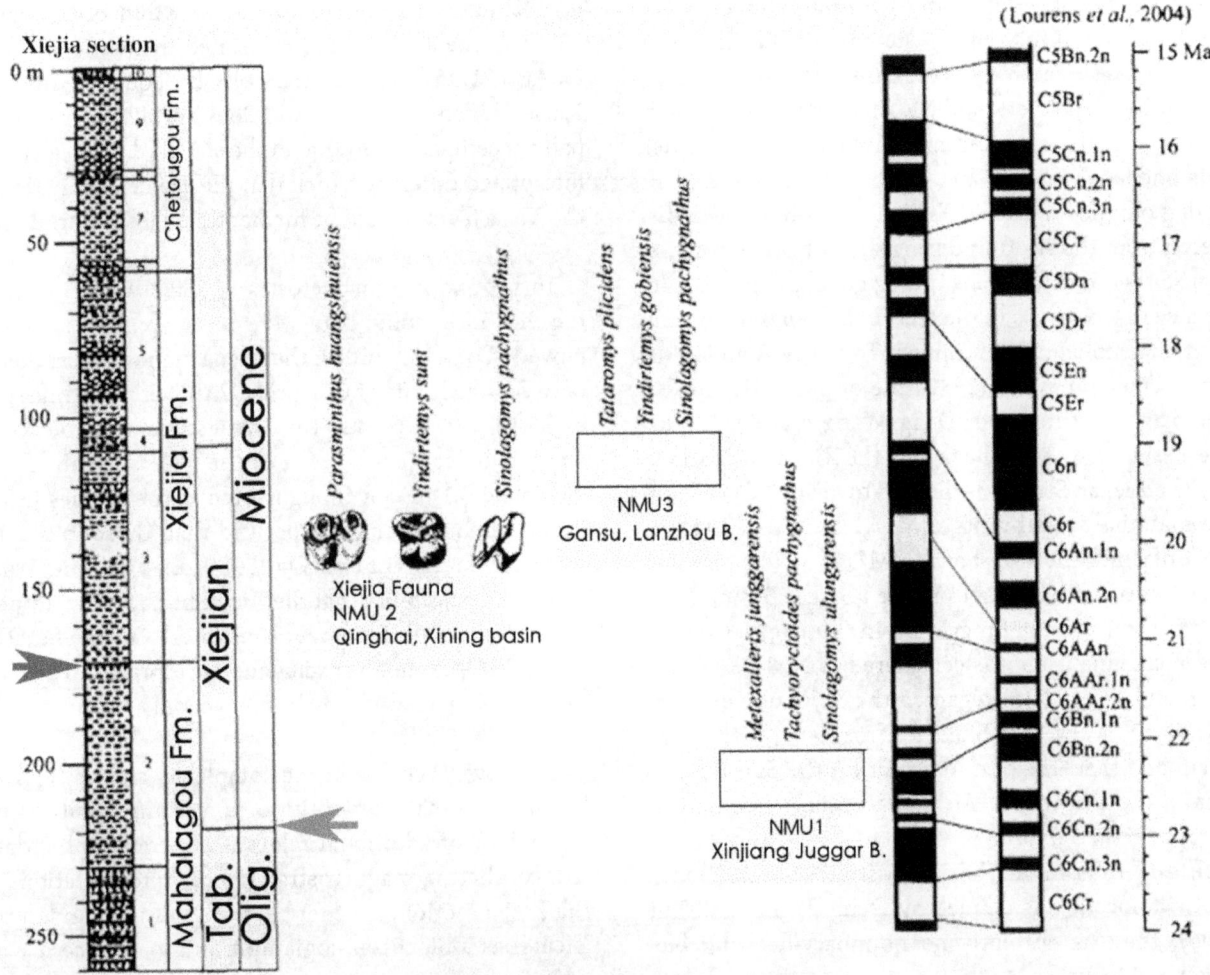

Figure 3.1 The stratigraphy of the Xiejia section, modified from Deng et al. (2006). The white diamond indicates the estimated position where the Xiejia fossils were collected. The dark gray arrow to the left points to the base of the Xiejia Formation and the light gray arrow to the right points to the base of the Xiejian Stage. The Xiejia fauna would project to about 21 Ma on the standard polarity time scale.

Wang, and Yue (2006:322). The authors specified the Xiejia section as the boundary stratotype for the Xiejian Stage and made the following conclusion:

The Xiejian Stage can be correlated with the marine Aquitanian Stage in the International Stratigraphic Chart and the lower boundary of both stages is defined at the base of the paleomagnetic Chron C6Cn.2n with an age of 23.0 Ma. At the Xiejia section paleomagnetic data indicate that the lower boundary of the Xiejian Stage is located within the successive deposits of reddish brown massive mudstone in the upper part of the Mahalagou Formation about 48 m from the base of the overlying Xiejia Formation. Biostratigraphic analyses suggest that the Xiejian Stage includes three Neogene mammal faunal units, NMU1 to NMU3. The Xiejia fauna discovered from the Xiejia section is re-

garded as NMU2, and the Suosuoquan fauna from the northern Junggar Basin, Xinjiang, and the Zhangjiaping fauna from the Lanzhou Basin, Gansu correspond to NMU1 and NMU3, respectively.

Several problems exist in this definition (figure 3.1).

1. Although the Xiejian is intended to be a biochronologic unit, its lower boundary is defined by a paleomagnetic chron, not by a mammalian evolutionary event.

2. "Local or regional chronostratigraphic units should adhere to the same rules established for the units of the Standard Global Chronostratigraphic Scale" (Salvador 1994:87). Although the Xiejian Stage was proposed as a Chinese regional stage, it does not meet the requirements for a chronostratigraphic unit as suggested in the

ISG. For instance, there is no specific point identified in the section that is recognizable with multiple methods, particularly fossil mammals. The base of the Xiejian is often defined subjectively as the base of the Miocene.

3. In the Xiejia section, the fossils were found in literally one spot in the section, from a 3-m-long and 0.5-m-thick sandstone lens (Li, Qiu, and Wang 1981). After the lens was dug out, no additional Xiejian fossils have been discovered from the stratum during the last three decades. Because the sandstone lens is long gone, the precise occurrence of the fossils in the section becomes uncertain. Based on lithological descriptions by Li and Qiu (1980) and Li, Qiu, and Wang (1981), the original Xiejia fossil assemblage is estimated by Deng, Wang, and Yue (2006) to be about 80 m above the base of the Xiejia Formation.

4. The Xiejian Stage was thought to include three Neogene mammal faunal units, NMU 1 to NMU 3 (Deng 2006), of which the lowest one (NMU 1) was represented by the Suosuoquan fauna from the Junggar Basin, Xinjiang (see discussion in Junggar Basin, Xinjiang), whereas the Xiejia fauna itself was considered as NMU 2. Deng, Wang, and Yue (2006) regarded the Suosuoquan fauna of Xinjiang as the best biological marker to define the Xiejian and that "the best indicator for the Suosuoquan fauna is the insectivore *Metexallerix junggarensis*," the first appearance datum of the genus in China (Deng 2006:148). Thus, in the "boundary stratotype" section at the Xiejia locality, as defined by Deng, Wang, and Yue (2006), there is currently no boundary-defining biomarker and fauna.

5. Because the Xiejia fauna shares most genera with the Late Oligocene Tabenbuluk fauna (Li, Wu, and Qiu 1984), it is thought to be older than Suosuoquan and could even be latest Oligocene (Zhuding Qiu, personal communication). Among the fossils collected from the Xiejia locality, only about 10 taxa can be identified to generic and/or species level (Deng, Wang, and Yue 2006). The sample size of the fauna seems not large enough to tell how distinctive its evolutionary stage is from that of the Tabenbuluk fauna, nor does it show its precise evolutionary (temporal) relationships with the Oligocene Tieersihabahe I or II or any of the Suosuoquan I, II, or even III series of assemblages (plate 3.1). In other words, faunal correlation alone cannot ensure its precise temporal position when correlated to the Junggar assemblages.

6. The age determination of the Xiejia fauna by magnetic polarity chrons is questionable and controversial. Wu et al. (2006) employed an interpolation method to interpret the magnetic sequence. They correlated the Xiejia polarity sequence with chrons C5Bn.2n–C7c.1n and estimated a time span of 9.21 myr (15.01–24.22 Ma)

for the Xiejia section. Then, the authors identified the stratigraphic position of the Xiejia fauna within bed 3, depth 137 m in the section, and correlated the fauna to chron C6Ar, ~21 Ma. Because the polarity sequence from the Xiejia section is complex and does not display a distinct pattern comparable to that of the standard scale, it can be interpreted differently such that the age determination of the Xiejia fauna should be further tested (see figure 3.1).

In yet another magnetostratigraphic work, in which the biostratigraphic data were more appropriately employed (Dai et al. 2006), the Xiejia fauna was bracketed between the limits of C7n and C7An, with the time span of 25.183–25.496 Ma. This age estimate of the Xiejia fauna is older than the age of the Tieersihabahe fauna (see plate 3.1). Even though some of the species in the Xiejia fauna appear more derived than those in the Tabenbuluk fauna (Li and Qiu 1980; Li, Qiu, and Wang 1981), the possibility that the Xiejia fauna is actually part of the Tabenbulukian age, thus latest Oligocene (Dai et al. 2006), cannot be excluded. If this proves to be true, then the Xiejia fauna may be considered a subunit of the Tabenbulukian.

Moreover, because of the complexity stemming from the poor faunal composition, uncertain fossil occurrence, lack of additional relevant biostratigraphic data, and conflicting magnetostratigraphic interpretations in the Xiejia section, another possibility is to abandon the Xiejian as a biochronologic unit and to replace it with the Suosuoquanian as the land mammal age representing the early Miocene of China. This possibility is worth further deliberation in the course of refining the CLMA.

Junggar Basin, Xinjiang

We have provided a study on the Xiejian age from the Junggar Basin, Xinjiang (Meng et al. 2006), and related regional geology and paleontology has been presented in several works (e.g., Ye, Wu, and Meng 2001a, 2001b; Ye, Meng, and Wu 2003; Meng et al. 2008; Sun et al. 2010). The study emphasized selection of the lower boundary stratotype of the Xiejian from an area outside the type locality. Our intention is to stimulate discussions on how to improve the CLMA and to define the lower boundary of mammal ages. We also want to show that the stratotype for a CLMA does not have to be the original nominal section. Discussions of this kind have been common among paleontologists and stratigraphers in Europe and North America since at least the 1940s (Wood et al. 1941), but they have been limited in Asia.

In northern Xinjiang, the Suosuoquan Formation consists primarily of reddish silt that is homogenous in lithology over a large area (Meng et al. 2006, 2008) and is shown to be aeolian in origin (Sun et al. 2010). At the Tieersihabahe section at the east side of the Chibaerwoyi escarpment, the Suosuoquan Formation is continuous with the underlying Tieersihabahe Formation and is separated by a hiatus from the overlying Halamagai Formation. The Tieersihabahe and Suosuoquan formations span the Oligocene-Miocene transition and are fossiliferous. Our surface collecting and screenwashing during the last decade have uncovered many fossil levels in the area. Most of these fossils have been precisely tied to the strata in the section, with GPS logs and meter numbers, as well as to the magnetic polarity sequence (Meng et al. 2006, 2008; Wu et al. 2006; see plate 3.1). The stratigraphy of the area provides a unique opportunity to address the Xiejian CLMA.

In early studies, the age determination of the fossil assemblage from the region was controversial. Based on faunal correlation, the Suosuoquan fauna was considered early or middle Miocene (Peng 1975; Tong et al. 1990), late Oligocene (Tong et al. 1987, 1990), ranging from late Oligocene to early middle Miocene (Wu et al. 1998), or from earliest Miocene to middle Miocene (Ye et al. 2000). With the magnetic polarity sequence as an independent chronological reference, we were able to overcome some problems in dating the fossil assemblage and provided better age estimates for the biozones (Meng et al. 2006). Biostratigraphic and magnetic polarity evidence show that the Paleogene–Neogene transition lies in the basal part of the Suosuoquan Formation (see plate 3.1). Because of its continuity, rich fossils, and paleomagnetic data, the Tieersihabahe section proves to be a potential boundary stratotype section for the base of the Xiejian, as we recommended in Meng et al. (2006).

In our work (Meng et al. 2006), we recognized two fossil assemblages in the Tieersihabahe Formation and three in the lower part of the Suosuoquan Formation and regarded them as assemblage zones. We stated (2006:211–212, original in Chinese): "These biozones are fossil assemblages with unique compositions and distributed in specific ranges of the rock bodies. Each biozone has its temporal dimension, but is not here regarded as a biochronologic unit. With detailed biostratigraphic study in the future, these biozones may serve as bases for biochronologic units." Our intention at the time was to document the biostratigraphic data at the finest level possible and to entertain discussions on boundary definition of the Xiejian, which we thought could also be applicable to other boundaries for the CLMA. However, we still followed the NCSC's proposal (2002) and placed the lower boundary of the Xiejian at the point correlative to the base of chron C6Cn.2n in the Tieersihabahe section (see plate 3.1). This, again, is not a boundary defined by biomarker, even though we specified the distance between the lowest occurrences of taxa in the S-II assemblage zone and the base of chron C6Cn.2n. We also concluded that the early Miocene S-II assemblage zone contains many relic genera from the Oligocene. This conclusion was not based on faunal correlation, but on observed superpositional relationships of the faunas that were calibrated using the polarity time scale (Meng et al. 2006).

In a follow-up study (Meng et al. 2008), we pointed out again that the base of the Xiejian age as currently defined (Deng, Wang, and Yue 2006; Meng et al. 2006) was defined by a magnetic polarity chron, not on biological evidence. We suggested that more discussions are needed to nail the lower boundary stratotype of the Xiejian age (Meng et al. 2008). We also stated that there is no need to equate the divisions and boundaries of the CLMA with those of European MN zones, nor is it necessary, if possible at all, to correlate them exactly with those of the marine chronostratigraphic stages. Following Tedford (1995), we suggested that the best way to accomplish a meaningful CLMA system is to establish a regional biochronological system rather than relying on correlations with European MN zones or with North American land mammal ages; it is better to base this system on regional fossil mammals that are stratigraphically well documented and chronologically calibrated to the best degree possible.

We understood that the establishment of a biochronologic unit depends on its biostratigraphic data. The work we have conducted in northern Xinjiang is still ongoing and is far from satisfactory for recognition as a stratotype, but it provides a case study contrasting with the work at the Xiejia type locality. The Tieersihabahe section has several advantages over the Xiejia section:

1. Multiple levels of fossils, both Late Oligocene and Early Miocene, have observable superpostional relationships and are stratigraphically well documented. They provide us with a higher degree of confidence in their relative temporal sequence and evolutionary trends. The fossils have been collected by screenwashing from various levels and show a relatively dense distribution within the rock sequence. These biostratigraphic data furnish a robust and accurate framework for a biochronologic unit.

2. The S-II fossil assemblage presumably correlative to the Xiejian fauna from the Tieersihabahe section is richer in species than that of the Xiejia fauna and therefore provides a greater power for correlation.

3. The lowest occurrence of the Xiejian fauna in the Tieersihabahe section is closer to the conceptual base of the Xiejian age so that it can provide a boundary definition that is more consistent with the current usage of the Xiejian.

4. The paleomagnetic polarity sequence is more consistent with the standard polarity time scale, perhaps because the fine-grained and evenly distributed aeolian silts are more suitable for magnetostratigraphy. The polarity sequence is integrated with multiple levels of fossils so that its calibration and correlation are more reliable.

In terms of the biostratigraphic record, there are currently two weak areas in the Tieersihabahe section:

1. There are about 16 m of strata between the S-I and S-II in the section that are poorly sampled. This interval contains the C6Cn.2n—thus the base of the Miocene. It is uncertain whether the S-II assemblage would extend to a lower position. Future collecting may narrow or fill the gap. But even so, the section stands now as the best known for addressing the Xiejian age.

2. Although all fossils listed for the Tieersihabahe section (Meng et al. 2006; see plate 3.1) are from the same escarpment that is continuously exposed, those of S-II assemblage are from site XJ99005, several kilometers away from the Tieersihabahe transect where the paleomagnetic samples and other fossils were collected. This is a common problem for mammal fossil records anywhere, in that it is difficult to have a continuous fossil record from a simple transect in the outcrop of the type locality. Lateral tracing of the beds, fossils collected at various sites between the Tieersihabahe transect and XJ99005, and redundant paleomagnetic information from site XJ99005 help to key the S-II assemblage in the Tieersihabahe section. Because the S-III taxa coexist at XJ99005 and in the Tieersihabahe transect (Meng et al. 2006: fig. 3; see plate 3.1), it is highly likely that S-II fossils would occur in the Tieersihabahe transect. Hopefully, future investigations will recover S-II fossils at or near the Tieersihabahe transect to test the current placement of the S-II assemblage in the section.

A PROPOSAL FOR A SINGLE-POINT BASE DEFINITION OF THE XIEJIAN AGE

Here we entertain a possible base definition for the Xiejian age, assuming the Xiejian will remain a valid unit. Our focus is on the base definition of the unit, but we also address other aspects related to the age. We understand that the decision to formally accept such a definition needs agreement of experts in the field of geochronology. This definition is presented as an example of how the CLMA should and could be defined. In the exercise, we follow the general format of the proposal for the Global Stratotype Section and Point (GSSP; Salvador 1994) in the hope that a land mammal age so defined can be easily converted into a formal Stage/Age when the regional stratigraphic data permit in the future. Although we follow the general guides for establishing a boundary stratotype for the Xiejian, we do not adopt the "Golden Spike" concept in our definition, because we think that in principle a boundary defined by a biological event changes along with our knowledge of the biostratigraphy in question. The boundary should be flexible to allow for improvement.

Name. Xiejian.

Rank of the boundary. Xiejian age, a basic unit of Chinese land mammal ages. Subunits may be contained in the age (see *Subdivision*).

Locality. Tieersihabahe at the Chibaerwoyi escarpment, northwest of the Dure village in the northern rim of the Junggar Basin, Xinjiang, China (Meng et al. 2006:fig. 1).

Boundary stratotype. The Tieersihabahe section is at the Tieersihabahe locality, a transect on the west side of the Chibaerwoyi escarpment, starting at N46°40.422', E88°28.259' and ending at N46°39.434', E88°28.745 (Ye, Meng, and Wu 2003).

Nominal stratotype. The name-bearing section for the Xiejian is at the Xiejia village, Huangzhong County, Qinghai Province.

Base definition. Specimens of *Democricetodon sui.* (Maridet et al. 2011) have been collected from a series of levels in the site XJ99005. Its lowest known stratigraphic occurrence can be estimated at 54 m above the base of the Tieersihabahe section (see plate 3.1), 21 m above the base of the Suosuoquan Formation and 13.75 m above the Oligocene–Miocene boundary. This point is the lowest stratigraphic occurrence of *Democricetodon sui.* in the boundary stratotype (LOS of *Democricetodon sui.*) and is chosen as the defining point for the base of the Xiejian. Future investigation may discover even a lower occurrence of the species in the section. When that happens, the lower boundary of Xiejian should be shifted accordingly.

Basal age. The lowest known occurrence of *Democricetodon sui.* in the section coincides with the top of C6Cn.1n or the basal part of C6Br (see plate 3.1), which is calibrated roughly as 22.56 Ma following the magnetic time scale of the Neogene (Ogg and Smith 2004).

Lithological marker. The Suosuoquan Formation is primarily reddish aeolian dust that is widely distributed in northern Xinjiang (Ye, Meng, and Wu 2003; Sun et al. 2010). The lithology is distinctive from the underlying Tieersihabahe Formation. Although the defined boundary of the Xiejian does not coincide with the lithological boundary, their relative positions help to locate the former (see plate 3.1).

Mineralogy and isotope. The rare-earth element distributions and the trace element compositions of the Suosuoquan deposits are nearly identical to those of the Loess Plateau aeolian dust (Sun et al. 2010). However, isotopic data of ^{143}Nd/^{144}Nd vs. ^{87}Sr/^{86}Sr and ^{87}Sr/^{86}Sr vs. 1/[Sr]ppm of the Junggar aeolian deposits are distinctive from those of the Loess Plateau, indicating that the Junggar aeolian dust must have come from a different source area. The particle sizes of the Junggar Tertiary aeolian dust are much finer and better sorted than those of the Loess Plateau, suggesting a longer transport distance and more even mixing.

Magnetostratigraphy. The magnetic polarity sequence of the Tieersihabahe section is reported elsewhere (Wu et al. 2006; Meng et al. 2006; Sun et al. 2010; see plate 3.1). This sequence is well calibrated and consistent with multiple fossil assemblages. The known lowest occurrence of *Democricetodon sui.* coincides with the top of C6Cn.1n. This polarity sequence is of good, if not better, quality compared to that of the Lemme-Carrosio section for the Neogene base GSSP (Steininger et al. 1997).

Time span. The time span of the Xiejian age depends on what is the base definition of the Shanwangian age, which is the successive unit younger than Xiejian. This boundary is currently not resolved (see the following discussion). It was correlated to the base of C6An.1r, with the age of 20.43 (Deng et al. 2004; Meng et al. 2006; see plate 3.1). More recently, Deng (2006) correlated the base of the Shanwangian to the base of C6r, with the age of 20 Ma (Ogg and Smith 2004; Lourens et al. 2004). For further discussion, see Qiu and Qiu (chapter 4, this volume) and Qiu, Wang, and Li (chapter 5, this volume).

Rock interval. Given the tentative basal age of the Shanwangian (see following discussion), the upper limit of the Xiejian sediment is estimated at 113 m in the Tieersihabahe section using the magnetic polarity scale. This rock interval ranges from 54 m to 108 m or 113 m in the section, depending on the choice of the basal age of the Shanwangian; either will contain the biostratigraphic content that we currently divided into the S-II and S-III assemblage zones (see plate 3.1).

Biological characterization. We divide the biological characterization of the Xiejian into two parts: those from the stratotype and those from other regions:

1. The rock interval of the Xiejian in the Tieersihabahe stratotype contains two fossil assemblages. The lower assemblage, Suosuoquan assemblage II, includes *Amphechinus* cf. *A. minimus, Amphechinus* cf. *A. bohlini, Sinolagomys* cf. *S. kansuensis, Atlantoxerus* sp. nov., Sciuridae gen. et sp. indet.1, Sciuridae gen. et sp. indet.2, *Parasminthus* cf. *P. tangingoli, Bohlinosminthus parvulus, Plesiosminthus* sp.1 (large), *Plesiosminthus* sp.2 (small), *Litodonomys* sp. 1–2, *Heterosminthus* cf. *H. firmus, Heterosminthus* cf. *H. nanus, Democricetodon sui., Tachyoryctoides* sp. nov., *Asianeomys fahlbuschi, Asianeomys engesseri, Asianeomys* sp., *Microdyromys* sp., and *Palaeogale* sp. The upper assemblage, Suosuoquan assemblage III, includes *Amphechinus bohlini, A.* cf. *A. minimus, Metexallerix junggarensis, Sinolagomys ulungurensis, Atlantoxerus* sp. nov., Dipodidae gen. et sp. nov., *Litodonomys* sp. 1–3, *Heterosminthus* aff. *H. firmus, Heterosminthus* cf. *H. nanus, Protalactaga* sp., *Cricetodon* sp. nov., *Democricetodon sui., Tachyoryctoides* spp. nov. 1–2, *Prodistylomys xinjiangensis, Prodistylomys lii, Palaeogale* cf. *P. sectoria, Aprotodon* sp., and Cervidae gen. et sp. indet. The stratigraphic distributions of the species are illustrated in plate 3.1 and in Meng et al. (2006:fig. 3). The identification of the taxa has been slightly changed since Meng et al. (2006), but the stratigraphic distributions of all the taxa remain the same. Although two assemblage zones are currently recognized, the S-III zone was tentatively grouped (Meng et al. 2006) and may be further divided, that is, those in the light blue area to the right (in the interval from 85 m to 107 m in the section) in plate 3.1 may form another biozone.

2. Taxa from the Xiejia locality include *Sinolagomys pachygnathus, Atlantoxerus* sp., *Eucricetodon youngi, Parasminthus xiningensis, P. huangshuiensis, P. lajeensis, Yindirtemys suni, Tachyoryctoides kokonorensis, Sinopalaeoceros xiejiaensis,* and *Diceratherium* sp. (Li, Wu, and Qiu 1984; Deng, Wang, and Yue 2006). Species from other localities may be included upon correlation.

Subdivision. There are potentially three subdivisions within the Xiejian (Deng 2006). However, we consider that the subdivision of the Xiejian is premature (see the following discussion), and we do not propose subunits here.

Preceding age. The base of the Xiejian age defines the top of the Tabenbulukian age. In the Tieersihabahe section, Tabenbulukian assemblages exist as T-I, T-II, and S-I assemblages (Meng et al. 2006). The superpositional relationship confirms the age relationships of the Tabenbulukian and Xiejian that were established for faunas originally known from two geographic regions (Bohlin 1942, 1946; Li and Qiu 1980; Li, Wu, and Qiu 1984).

Geographic extent. The geographic extent of the Xiejian as a CLMA is by definition within China, but it should be applicable elsewhere in Asia as far as faunal correlation can be extended.

Depositional environment and paleoclimatological interpretations. As primarily aeolian dust, the Suosuoquan deposits are widely spread in several isolated basins at different altitudes and rest on rocks of different ages in northern Xinjiang Province (Ye, Meng, and Wu 2003; Meng et al. 2006). The dusts were considered transported by the westerlies, instead of the northwest winter monsoon that had been responsible for the aeolian deposits on the Loess Plateau (Sun et al. 2010). The wind pattern is similar to the present wind system in China, characterized by the monsoon climate prevailing in eastern China and the westerlies prevailing in northwestern China (Sun et al. 2010). This suggests that the wind regime and climate similar to the modern pattern in central Asia have developed at least since the end of the Oligocene about 24 Ma—that is, starting the deposition of the Suosuoquan Formation. Regression of the Tethys Sea exposed the land area of Kazakhstan that subsequently experienced erosion and aridification in the early Oligocene (Akhmetyev et al. 2005) and potentially provided dust sources. Clift, Blusztajn, and Nguyen (2006) revealed that initial surface uplift in eastern Tibet occurred no later than 24 Ma, and climatic modeling suggests that Asian climate is affected significantly by the uplift of the Tibetan Plateau (Kutzbach et al. 1989; An et al. 2001; Broccoli and Manabe 1992). The rising Tibetan Plateau could have had a profound effect on the mid-latitude climate in the Asian interior. Recovered fossils for the time span are dominated by small mammals, particularly rodents and lagomorphs, in terms of species and specimen numbers.

COMPARISON

Before furthering discussion on the features of the definition, we think it helpful to compare the proposed definition of the Xiejian with the definition of the Global Stratotype Section and Point (GSSP). We choose two GSSPs to compare: one for the Permian–Triassic boundary (PTB, the base of Triassic as well as Mesozoic) and the other for the base of Neogene.

The Permian–Triassic Boundary (PTB) Stratotype

We start here with the "Yinkengian Stage," the lowest stage of Chinese marine Triassic, correlative to the Induan. The stratotype of this regional stage was defined at the type locality in Guichi, Anhui Province (NCSC 2002). Because the GSSP for the PTB was later chosen at the Meishan D section, Changxing, Zhejiang Province, the stratotype of the Chinese "Yinkengian Stage" was redefined at the Meishan D section (NCSC 2002). This shows that the stratotype of a regional stage does not have to be the nominal section.

Within the Meishan D section, there exists a conodont lineage consisting of *Hindeodus latidentatus, H. parvus,* and *Isarcicella isarcica* through the strata of the latest Permian to the earliest Triassic. The first occurrence of *H. parvus* is approved as the PTB marker. Two ash beds, one 13 cm below and the other 8 cm above the PTB, were dated as 251.23 Ma and 251.2 Ma, respectively, so that the PTB age has been set at 251 Ma (Yin et al. 2001; Gradstein, Ogg, and Smith 2004). This example shows that the boundary of a major global chronostratigraphic unit is defined by a biological event, the evolutionary first occurrence of a single species, *Hindeodus parvus,* instead of directly by the ash beds that give numerical ages. The radiometric dates are used to calibrate the biological event that defines the boundary of a major chronostratigraphic unit.

The traditional index fossil of the PTB was the ammonoid *Otoceras,* but the conodont *Hindeodus parvus* was later proposed to substitute *Otoceras* as the boundary marker (Yin et al. 1986; Yin et al. 2001). This indicates that the defining species for a boundary can be changed, provided compelling evidence becomes available.

The GSSP for the Base of Neogene

This GSSP is defined at the 35m level from the top of the Lemme-Carrosio section, Italy. This point coincides with the base of C6Cn.2n, and as Steininger et al. (1997:26–27) wrote: "This unambiguous reversal pattern and the associated excellent biostratigraphic markers have been among the main reasons for defining the base of the Neogene at metre 35 of the Lemme-Carrosio section." Unlike the case of PTB, there is no known biological event coinciding with the defining point, but the precise stratigraphic distances of various fossil species to the point are known from the stratotype.

This example shows that the boundary of a global chronostratigraphic unit can be defined by any means. In this case, it is basically defined by a polarity chron with the help of other properties.

Comments

Salvador (1994:87) stated: "Local or regional chronostratigraphic units should adhere to the same rules established for the units of the Standard Global Chronostratigraphic Scale." We followed this suggestion when defining the Xiejian age, although it is not yet a formal regional chronostratigraphic unit. We hope a well-constructed definition of a land mammal age can be more smoothly converted into a formal Stage/Age in the future.

The comparisons show that our definition of the Xiejian age is comparable to those of GSSPs in all aspects. There is no question that the same approach defining the GSSPs can be applied to definition of a mammal age, as we attempted in the previous discussion. Mammal teeth from a mammal species, a perfect analogue of conodont elements from a conodont animal, can be used as biological markers to define the boundaries of mammal ages. The differences are that in mammal ages, mammal species are the only biomarkers used to define the boundary. In addition, mammal species are primarily from terrestrial deposits and have a geographically more restricted distribution. In general, however, the quality of the definition of a mammal age can be as good as, if not better than, that of a GSSP. For instance, the critical feature that nails down the GSSP for the base of Neogene at the Lemme-Carrosio section is its magnetic polarity data, which in our view does not seem any better than the magnetic sequence from the Tieersihabahe section.

DISCUSSION

In this section, we discuss in more detail conceptual aspects of some features included in the base definition of the Xiejian age.

Boundary Stratotype

The proposed boundary stratotype is a different section from that of the Xiejian namesake. Whether the Tieersihabahe section will be eventually chosen as the stratotype for the Xiejian is a matter of agreement. Currently, we think it is much better in all aspects than the Xiejia section, although future work may improve the quality of the Xiejia section. This, however, is a minor issue. A more

important subject here is that a boundary stratotype should be specified when we define a mammal age. As stated in Salvador (1994:31): "All stratigraphic units are composed of *rock* and thus have 'rock character.'" We agree, and we think this should be applied to the biochronological units, as we did here. Logically, we must have a set of criteria on how to select a boundary stratotype to serve as the definition of a biochronologic unit, such as the land mammal age.

Requirements for the selection of boundary stratotypes for mammal ages can follow those for the global chronostratigraphic units (Salvador 1994). The following are requirements we think are essential, derived from Salvador (1994) with modifications necessary for application in mammal ages:

1. The stratotypes must be selected in terrestrial sections that are essentially continuous, well exposed, structurally least deformed, and with an ample thickness of strata below, above, and laterally. For strata above the boundary, it is preferable that the section contains sediments spanning the entire interval of the unit.
2. The fossil content in the stratotype section should be abundant, distinctive, well preserved, and representing at least a fauna (assemblage) that is as geographically widely distributed and as taxonomically diverse as possible. It is preferable that multiple levels of fossils through the interval of the unit are present so that the evolutionary stage can be empirically demonstrated.
3. There should be a well-studied mammal species, whose lowest occurrence can be used to provide the boundary-defining point in the stratotype, and that point can be calibrated by a dating method independent of evolution, such as radiometric dating, magnetic polarity time scale, or a combination of various methods.
4. The section should contain as many markers or other features favorable for long-distance correlation as possible, such as various fossils, magnetic polarity reversals, numerically datable beds, reliable widespread lithofacies (such as volcanic ash and perhaps aeolian sediments), and climatic and geochemical features.

As presented earlier, we think the Tieersihabahe section roughly meets these criteria.

Nominal Stratotype

The Xiejian is an example to illustrate the difference between a nominal (name-bearing) stratotype and a boundary stratotype. As a biochronologic unit, the name

Xiejian is from the locality name and the Xiejia section, not the Xiejia Formation, which is part of the Xiejia section (Deng, Wang, and Yue 2006). In fact, the base of the Xiejian sediments extends to the lower lithological unit, the Mahalagou Formation, in the Xiejia section (see figure 3.1). Deng, Wang, and Yue (2006) did designate the Xiejia section as the boundary stratotype for the Xiejian Stage. Given the existing problems we outlined earlier, this designation is premature to say the least. Future investigation at the Xiejia section may or may not improve its stratigraphic data, but regardless, the section remains as the nominal stratotype that serves to fix the geographic name of the Xiejian for the purpose of nomenclatural stability and priority (Walsh 2001, 2005a).

The Single-Point Base Definition

The general base definition for a biochronologic unit, the mammal age in this case, may be phrased as *a specific point representing the lowest stratigraphic occurrence of one mammal species in the stratotype section*. Several further observations should be addressed.

1. There are several species from the Tieersihabahe section that may be chosen to define the boundary (see plate 3.1). We prefer to use *Democricetodon sui* (Maridet et al. 2011) because this species has a denser record and occurs at site XJ99005 as well as in the transect where the paleomagnetic data were collected. The stratigraphic data help pinpoint the lowest known occurrence of the species in the Tieersihabahe section. In addition, this species probably represents the first occurrence of the genus, which has a continental distribution. For instance, a similar species has been reported from the D zone from the Valley of Lakes, Mongolia (Höck et al. 1999). In contrast, we do not think *Metexallerix junggarensis* is a good boundary marker, because it occurs in a younger stratum and is currently known only in one instance at a site between the Tieersihabahe transect and XJ99005.

As we mentioned previously, there is a 16 m interval of poorly sampled strata between the known first occurrence of *Democricetodon sui* and the S-I zone (see plate 3.1). Therefore, it is possible that future collecting may extend the occurrence of *Democricetodon sui* into even lower strata. When that happens, the base definition can be adjusted accordingly, if the same species is still used to define the boundary. However, this change is within the same stratotype, which differs from the situation caused by possible discovery of an earlier occurrence of the defining species in other places if the conventional bound-

ary definition is followed (see the following discussion). If its taxonomy proves to be controversial or another species turns out to be represented by better biostratigraphic records, *Democricetodon sui* may be substituted by the other species.

2. We have chosen a rodent, *Democricetodon sui*, as the defining species. Should we use small mammals to define the boundary? Our answer is yes. Large mammals may have the advantage to disperse rapidly over geographically wider areas than small mammals, but small mammals also disperse widely. One of the examples is the Late Paleocene *Alagomys*, one of smallest basal Glires, found in Asia and North America (Dashzeveg 1990; Dawson and Beard 1996; Meng et al. 1994; Meng et al. 2007). However, as we will discuss in more detail, geographic distribution is not a required feature for a boundary-defining species. Phylogenetic and stratigraphic data are more critical for boundary definition.

While disapproving the arguments of Lucas (1993) on a land vertebrate age system, Woodburne (2006: 232) argued that "to conflate a chronological system based on fossil mammals with one based on other land vertebrates (e.g., Sullivan and Lucas 2004) is counterproductive to the potential resolving power of both of them, and may mask their independent evolutionary patterns." We agree with this logic, but to extend the argument in defining the land mammal ages, should we use, if we can, a more exclusive group of mammals, such as rodents, in order to keep the maximum resolving power and independent evolutionary pattern of the group?

Although it may not be practical to restrict the boundary-defining taxa for CLMA to certain groups, we expect more species of Glires (rodents and lagomorphs) being used for the purpose for the following reasons:

- Glires have the most extensive geological distributions in the Asian Cenozoic record, from at least the middle Paleocene onward.
- Many groups of Glires are either endemic or originated in Asia, and they probably have the most complete evolutionary history compared to other groups of mammals.
- Glires are diverse, so that they provide a greater potential for choices of defining species and for correlation. With the screenwashing technique, lagomorphs and rodents have been commonly found in Tertiary strata in which large mammals are sparse. Moreover, the screenwashing method has recovered a much denser record of these small mammals than any other mammals, as in the case of the Tieersihabahe section.

- Their relatively larger numbers of specimens, compared to most large mammals, make quantitative study of these groups feasible.
- They are morphologically distinctive and the diverse morphologies can be used to accomplish a fine division of biozones, and therefore subunits of the mammal age.

3. In previous work, it is a common practice that a general definition is given to a mammal age as a unit or a time span. For instance, Ting (1998) defined the Gashatan Land Mammal Age as "the time between the first appearance of the order Rodentia, represented by *Tribosphenomys*, and the first appearance of the order Perissodactyla, represented by *Orientolophus*." Here, we emphasize the "base definition of an age" instead of "definition of an age." The base definition is more consistent with the preference of the ISG for boundary stratotype definition (Salvador 1994), instead of for section stratotype definition (Hedberg 1976), of a chronostratigraphic unit. Once the lower boundaries of two successive ages are defined, the time span of the older age is known.

4. Previous works, such as the one by Ting (1998) on the Gashatan Land Mammal Age, often use the earliest occurrence or first appearance of a species to define the boundary of a biochronologic unit without any stratigraphic context. Similarly, Woodburne, Tedford, and Lindsay (chapter 2, this volume) define the Baodean North China Mammal Age as FAD of *Ochotona* (*O*. cf. *lagreli*). Such a definition is abstract, lacks the rock character, and is unstable in that the earliest occurrence of the species could be shifting from place to place, depending on which record is regarded as the earliest. We do not think the definition without the stratigraphic context is consistent with the definition for a chronostratigraphic/ chronologic unit, such as a GSSP. The definitions of biochronologic units are recorded in the rocks, not written in the text. What we are doing is simply to describe the material definition in words to our best knowledge.

5. In some studies, a stratotype section (Gunnell et al. 2009) or nominal section (Tong, Zheng, and Qiu 1995; Deng 2006) is specified as part of the definition for a biochronologic unit, but the specific defining point is not identified in the rock sequence. In our definition, we tie the lowest occurrence of the defining taxon to a specific point in the section that meets the basic requirements for a boundary stratotype (see plate 3.1). The way we define the boundary is consistent with the following statement in the ISG (Salvador 1994:29): "A boundary-stratotype should be based on a single point in a designated sequence of rock strata, serving to indicate the position of the boundary horizon at one place. (Lateral extension of the boundary horizon in any direction from this point is accomplished by stratigraphic correlation.)." In other words, what we really need in defining the base of a mammal age is to identify the best point we can find in the best section known (i.e., the stratotype).

Although the defined point may be indicated using a physical and artificial marker, as we did, for convenience of recognition in the field, we contest the practice of defining the boundaries of mammal ages by fixed, permanent "Golden Spikes" driven into the boundary stratotype sections, because the boundaries based on biological events should be flexible and subject to change with new biostratigraphic discovery and revised taxonomy of the defining taxon.

6. For a biochronological unit based on mammals, such as the Xiejian age, the best method to make the single-point definition of the boundary stratotype is Woodburne's (1977, 1987, 1996, 2006) proposal of using the stratigraphic first occurrence of a single species. Woodburne (2006:239–240) advocated that boundaries between biostratigraphic units be placed on the basis of immigration events or at evolutionary first occurrences, and he considered that use of either criterion had an operational aspect of determining precision in correlation: "Whereas the genesis of an evolutionary first occurrence begins, at least in theory, with superior stratigraphic control in a local rock section, an immigration event is, by definition, distant from its site of origin. Nevertheless, a taxonomic event of endemic origin also involves a consideration of dispersal in more distant locations. In either case, therefore, and regardless of whether the origin of a new taxon can be documented, its FAD is comprised of a number of LOs at either an inter- or intracontinental perspective."

We agree that the stratigraphic first (lowest, as we use here) occurrence of a single species should be used to define the base of a biochronological unit because this is the only way to avoid any gap or overlap in time and is consistent with the definition of a chronostratigraphic unit. We would argue, however, that the evolutionary first occurrence, such as the case of *Hindeodus parvus* for the PTB (Yin et al. 2001), is better than an immigration event as the criterion for boundary definition because it provides a theoretically oldest occurrence for the boundary-defining species. Any immigrant occurrences of the same species into other geographic areas will be either synchronous with or younger but will never be older than the evolutionary first occurrence. Therefore, an immigrant species is likely not the oldest record compared to the range of the same species found in other areas. In terms of stability of definition, it is always better to use the evolutionary first occurrence as the marker because the possibility that the boundary will be shifted downward, pending new discoveries, is smaller.

An ideal boundary-defining event is the evolutionary first appearance of a species that has a wide geographic distribution. But in reality, there are various difficulties in determining the evolutionary first occurrence of a species (Woodburne 1996; Lindsay 2003; Walsh 2005b), and it is not easy to find such an ideal species in the fossil mammal record of any age. For instance, in the Xiejian case, the phylogenetic relationship of *Democricetodon* is unclear, such that we have to use the lowest occurrence of the taxon in the Tieersihabahe section, which in our understanding is the best stratigraphic datum for the base definition of the Xiejian. We hope the need for precise definition of biochronologic units will encourage detailed phylogenetic studies of these mammals. For improving the CLMA, it is a practical way to use the LOS of a species as the biomarker to define the lower boundary.

7. Following the single-point boundary definition, we note that the arguments for synchronous, diachronous, and precision of biological events are logically unnecessary in the definition of the boundary. Walsh (1998) proposed the "Synchronous First/Last Historical Appearance." Woodburne (2006:239–240) stated that "regardless of whether the origin of a new taxon can be documented, its FAD is comprised of a number of LOs at either an inter- or intracontinental perspective." These statements or concepts show that the authors are assuming a FAD that can be recognized by or consist of multiple physical points in space. We think this recognition is difficult, if possible at all.

For boundary definition, a single point in a designated sequence of rock is necessary and sufficient to indicate the position of the boundary horizon (Salvador 1994). We will never be certain whether multiple physical points in space are in a synchronous time plane, but we are certain that a time plane must contain the point represented by the biological event, that is, the LOS. The time plane that contains the LOS is the LDS. Whether other points in space, or occurrences of a species, are also on the same LDS is a matter of correlation, in which precision depends on various qualities of data and methods used.

A geographically broadly distributed species is traditionally preferred as a boundary-defining species (e.g., Lindsay 2003). Logically, if the single-point definition for the boundary horizon is accepted, the geographic extent is not a necessary feature for the defining taxon. A geographically broadly distributed species is preferable for correlation but not necessarily required for definition. Correlation is difficult to accomplish by occurrences of the boundary-defining species alone even if it is widespread geographically. Instead, correlation must depend on the fauna and all means associated with the defining species.

8. We want to emphasize that selection of the boundary-defining taxon should be part of the selection of a boundary stratotype. This is logical, for it makes no sense that a boundary-defining species is chosen from a nonstratotype section, or a stratotype does not contain that species. As required for selection of a stratotype, the boundary-defining species must be associated with a rich fauna that can be used for correlation. This binding relationship between the boundary-defining species and the stratotype reveals an interesting situation. For instance, if the same *Democricetodon sui* was found from a section X in a different area and was demonstrated to be older than the currently known earliest occurrence in the stratotype, then based on the traditional definition, the base defining point of the Xiejian ought to be moved to section X. But what if section X does not meet the criteria for a stratotype? This further boils down to a conceptual question: Shall we use "the earliest known occurrence (evolutionary first appearance) of a species" or "the earliest occurrence of the species in the stratotype" as the definition if we know the latter is slightly younger than the former? We prefer the latter because the base definition of a biochronologic unit should contain as much stratigraphic information as possible for age determination and correlation. This implies that the boundary definition will be adjusted if new biostratigraphic data are obtained in the stratotype. If new biostratigraphic data are known from other sections, the definition will not be changed, but the correlation will.

Given these points, we use the LOS instead of LO as the descriptive term for a boundary definition, whether or not the occurrence in the stratotype is the actual or known earliest occurrence of the defining species and regardless of its geographic distribution. We use the LDS as the temporal connotation of the LOS.

Basal Age

The basal age of the Xiejian, marked by the first occurrence of *Democricetodon sui* (see plate 3.1), is obtained independently using the magnetic polarity time scale; it is not determined by faunal correlation or mammalian evolution. A critical issue here is that the basal age of the Xiejian does not coincide with, but is actually younger than, the basal age of the Miocene. The base of the Miocene has been determined elsewhere (Steininger 1999;

Lourens et al. 2004; Ogg and Smith 2004). In the Tieersihabahe section, the point correlative to the base of the Miocene can be identified by magnetochronology (bottom arrow in plate 3.1), but that point is not currently marked by any mammalian event. It is natural that the basal age and duration of a mammal age, which is intended as a regional system, would differ from global chronostratigraphic units. Future discovery of *Democricetodon sui* in lower strata may make the basal age of the Xiejian older. If the lower boundary of the Xiejian happens to correlate with the base of the Miocene, it is fine; if it does not, then the top of the Tabenbulukian would extend into the Miocene.

Time Span

The CLMA younger than the Xiejian is the Shanwangian. Once we know the base definition of the Shanwangian, we know the time span of the Xiejian. Unfortunately, the base of the Shanwangian is currently uncertain. In its type locality, a basalt underlying the Shanwang Formation gave a date indicating "greater than 16.78 Ma" for the base of the Shanwangian (NCSC 2002). The base of the Shanwangian has been correlated to the base of MN 3, with the age of 20.5 Ma (Deng, Wang, and Yue 2003) or to the base of C6An.1r, with the age of 20.43 (Deng et al. 2004). More recently, following the correlation of the base of MN 3 with that of C6r (Steininger 1999; Lourens et al. 2004), Deng (2006) reconsidered the basal age of the Shanwangian as 20 Ma. For further discussion, see Qiu et al. (chapter 1, this volume). Given the base definition of the Xiejian proposed in this paper, the longest age span of the Xiejian would be from 22.56 Ma to 20 Ma.

Given this time span, we can locate the upper bound of the sequence that belongs to the Xiejian in our section and then include all fossils within the sequence as characteristic members of the Xiejian age. These fossils, listed in the biological characterization of the boundary definition, will serve as the nucleus for characterization and correlation of the Xiejian age.

Biological Characterization

In our proposed definition, we divide the biological characterization of the Xiejian age into two parts, the local fauna from the stratotype and that from other regions. We think it is necessary to distinguish these categories because the biostratigraphy of the stratotype is based on direct observation and serves as the core data that characterize the age as well as provide the basis for correlation. Data from other regions included in the biological characterization of the Xiejian are based on correlation or interpretation that they are coeval. Even though we note the potential multiple correlations of the Xiejia fauna with the Late Oligocene Tieersihabahe assemblages or S-I, we still follow the convention and list the Xiejia fauna as part of the biological characterization in the definition. Given the estimated time span of the Xiejian and the estimated upper boundary of the Xiejian sediments in the Tieersihabahe section, we consider that the Suosuoquan assemblages II and III both belong to the Xiejian (see plate 3.1), which differ from the Late Oligocene Tieersihabahe fauna (Tabenbulukian) mainly at the species level.

Subdivision

Deng, Wang, and Yue (2006) considered that the Xiejian Stage includes three Neogene mammal faunal units, NMU 1 to NMU 3, represented by the Suosuoquan fauna from Xinjiang, the Xiejia fauna from Qinghai, and the Zhangjiaping fauna from Lanzhou Basin, Gansu, respectively (see figure 3.1). The NMU 1 is considered equivalent to MN 1, and NMU 2–3 are correlated with the MN 2. This temporal arrangement of the three units was probably based on the relative evolutionary stages of mammals, but the division seems premature at its current status. For instance, the Suosuoquan fauna used by Deng, Wang, and Yue (2006) is largely based on the fauna collected in 1982. Our comparison shows that the Xiejia fauna is more correlative to S-II but is more primitive than S-III. However, as we noted (Meng et al. 2006), if the polarity scale of the Xiejia section (Wu et al. 2006) is correct, then the age estimate of roughly 21 Ma for the Xiejia fauna would be 1 m.y. younger than S-II and even slightly younger than most of S-III. This indicates that the current subdivision of the age is not well supported.

From the Tieersihabahe section, it seems possible that the Xiejian be divided into two subunits on the basis of S-II and S-III assemblage zones, but we think further investigation is needed to test this.

Preceding Age

The base of the Xiejian age defines the top of the Tabenbulukian age. In the Tieersihabahe section, Tabenbulukian assemblages exist as T-I, T-II, and S-I zones (Meng et al. 2006). The superpositional relationship confirms the age relationships of the Tabenbulukian and Xiejian that were established on the basis of faunas originally known from two geographic regions. The Xiejian fauna is similar to the Tabenbulukian fauna at the generic level but different at the species level. These differences are useful to differentiate the two ages as well as to reflect the evolutionary changes of mammals during the Paleogene-Neogene transition.

Geographic Extent

The Xiejian was proposed as a Chinese mammal age, but it has been used as a biochronologic unit in Asia. The potential selection of the Tieersihabahe section as the boundary stratotype for the Xiejian will make the Xiejian as a regional age useful for central Asia and will be applicable as far as the biological evidence extends.

CONCLUSION

We briefly review works surrounding the Xiejian Chinese land mammal age, with focus on the stratigraphic data from the Xiejia locality and Junggar Basin. We point out the possibilities that the Xiejia fauna may represent a subunit of the Late Oligocene Tabenbulukian and that a new biochronologic unit, the Suosuoquanian, may be used to replace the Xiejian. Nonetheless, our presentation focuses on the procedure and format of how a land mammal age as a biochronologic unit should be defined.

Assuming the Xiejian remains as a valid and improvable land mammal age of China, we propose a single-point base definition for it, following the general format for defining a GSSP. We demonstrate that with detailed stratigraphic data, a land mammal age can be as rigorously constructed, defined, characterized, and correlated as defining a chronostratigraphic unit in the standard of the ISG. The key component of the CLMA is a thorough biostratigraphic framework. A CLMA so defined can therefore serve as the basis of a formal chronostratigraphic/geochronologic unit. Although the stratigraphic data from the Junggar Basin need improvement, the current results show that a robust CLMA system is hopeful. Such a system better reflects the regional evolutionary history of mammals and provides a well-defined and more accurate biochronology for investigations of geohistory in the terrestrial record of the region. In addition to the necessary fieldwork for more thorough stratigraphic data, we note that conceptual aspects of the CLMA need more attention, and we hope our proposal of the single-point base definition of the Xiejian age will inspire more discussions on how to reconstruct and improve the CLMA.

ACKNOWLEDGMENTS

We are grateful to Drs. M. O. Woodburne, R. H. Tedford, Z.-x. Qiu, E. Lindsay, Z.-d. Qiu, L. Flynn, M. Fortelius, R. Bernor, T. Deng, and X.-m. Wang for extensive discussion on the subject presented in the paper. Comments by Drs. M. O. Woodburne, R. Tedford, L. Flynn, X.-m. Wang. and G. Daxner-Höck greatly improved the manuscript. We thank the National Natural Science Foundation of China (NSFC 40872032), Major Basic Research Projects of MST of China (2006CB806400), NSF, IVPP, and Society of Vertebrate Paleontology, Chinese Academy of Sciences for financial support of the symposium; and Z.-x. Qiu and X.-m. Wang for the invitation of our presentation at the symposium.

REFERENCES

Akhmetyev, M. A., A. E. Dodoniv, M. V. Somikova, I. I. Spasskaya, K. V. Kremenetsky, and V. A. Klimanov. 2005. Kazakhstan and Central Asia (plains and foothills). In *Cenozoic Climatic and Environmental Changes in Russia*, ed. A. A. Velichko and V. P. Nechaev, pp. 139–161. Geological Society of America Special Paper 382.

An, Z.-s., J. E. Kutzbach, W. L. Prell, and S. C. Porter. 2001, Evolution of Asian monsoons and phased uplift of the Himalaya-Tibetan Plateau since late Miocene times. *Nature* 411:62–66.

Aubry, M. P. 1995. From chronology to stratigraphy: Interpreting the stratigraphic record. In *Geochronology, Time Scales, and Global Stratigraphic Correlation: A Unified Framework for a Historical Geology*, ed. W. A. Berggren, D. V. Kent, M. P. Aubry, and J. Hardenbol, pp. 213–274. Society of Economic Paleontologists and Mineralogists Special Publication 54. Tulsa: Society for Sedimentary Geology.

Bohlin, B. 1942. The fossil mammals from the Tertiary deposit of Taben-buluk, Western Kansu. Part I: Insectivora and Lagomorpha. *Palaeonologia Sinica NSC*, 8a:1–113.

Bohlin, B. 1946. The fossil mammals from the Tertiary deposit of Taben-buluk, Western Kansu. Part II: Simplicidentata, Carnivora, Artiodactyla, Perissodactyla, and Primates. *Palaeonologia. Sinica NSC*, 8b:1–259.

Broccoli, A. J. and S. Manabe. 1992, The effects of orography on mid-latitude Northern Hemisphere dry climates. *Journal of Climate* 5:1181–1201.

Clift, P. D., J. Blusztajn, and D. A. Nguyen. 2006. Largescale drainage capture and surface uplift in eastern Tibet before 24 Ma. *Geophysical Research Letters* 33:L19403, doi:10.1029/2006GL027772.

Dai, S., X.-m. Fang, G. Dupont-Nivet, C.-h. Song, J.-p. Gao, W. Krijgsman, C. Langereis, and W.-l. Zhang. 2006. Magnetostratigraphy of Cenozoic sediments from the Xining Basin: Tectonic implications for the northeastern Tibetan Plateau. *Journal of Geophysical Research* 111: B11102, doi:10.1029/2005JB004187.

Dashzeveg, D. 1990. New trends in adaptive radiation of early Tertiary rodents (Rodentia, Mammalia). *Acta Zoologica Cracoviensia* 33:37–44.

Dawson, M. R. and C. K. Beard. 1996. New Late Paleocene rodents (Mammalia) from Big Multi Quarry, Washakie Basin, Wyoming. *Palaeovertebrata* 25:301–321.

Deng, T. 2006. Chinese Neogene mammal biochronology. *Vertebrata PalAsiatica* 44:143–163.

Deng, T., X.-m. Wang, X.-j. Ni, L.-p. Liu, and Z. Liang. 2004. Cenozoic stratigraphic sequence of the Linxia Basin in Gansu, China and its evidence from mammal fossils. *Vertebrata PalAsiatica* 42:46–66 (in Chinese with English abstract).

Deng, T., W.-m. Wang, and L.-p. Yue. 2003. Recent advances of the establishment of the Shanwang Stage in the Chinese Neogene. *Vertebrata PalAsiatica* 41:314–323.

Deng. T., W.-m. Wang, and L.-p. Yue. 2006. The Xiejian Stage of the continental Miocene Series in China. *Journal of Stratigaphy* 30(4):315–322 (in Chinese with English abstract).

Fahlbusch, V. 1991. The meaning of MN-zonation: Considerations for a subdivision of the European continental Tertiary using mammals. *Newsletters on Stratigraphy* 24(3):157–173.

Gradstein, F, J. Ogg, and A. Smith. 2004. *A Geological Time Scale.* Cambridge: Cambridge University Press.

Gunnell, G. F., P. C. Murphey, R. K. Stucky, K. E. B. Townsend, P. T. Robinson, J.-P. Zonneveld, and W. S. Bartels. 2009. Biochronological zonation of the Bridgerian and Uintan North American Land Mammal Ages. In *Papers on Geology, Vertebrate Paleontology, and Biostratigraphy in Honor of Michael O. Woodburne,* ed. L. Barry Albright III, pp. 297–330. *Museum of Northern Arizona Bulletin* 65.

Hedberg, H. C., ed. 1976. *International Stratigraphic Guide.* New York: Wiley.

Höck, V., G. Daxner-Höck, H. P. Schmid, D. Badamgarav, W. Frank, G. Furtmüller, O. Montag, R. Barsbold, Y. Khand, and J. Sodov. 1999. Oligocene-Miocene sediments, fossils, and basalts from the Valley of Lakes (Central Mongolia)—an integrated study. *Mitteilungen der Österreichischen Geologischen Gesellschaft* 90:83–125.

Kutzbach, J. E., P. J. Guetter, W. F. Ruddiman, and W. L. Prell. 1989. Sensitivity of climate to Late Cenozoic uplift in Southern Asia and the American West: Numerical experiments. *Journal of Geophysical Research* 94:18393–18407.

Li, C.-k. and Z.-d. Qiu. 1980. Early Miocene mammalian fossils of Xining Basin, Qinghai. *Vertebrata PalAsiatica* 18:198–214 (in Chinese with English summary).

Li, C.-k., Z.-d. Qiu, and S.-j. Wang. 1981. Discussion on Miocene stratigraphy and mammals from Xining Basin, Qinghai. *Vertebrata PalAsiatica* 19:313–320.

Li, C-k. and S.-y. Ting 1983. The Paleogene mammals of China. *Bulletin of Carnegie Museum of Natural History* 21:1–98.

Li, C.-k., W.-y. Wu, and Z.-d. Qiu. 1984. Chinese Neogene: Subdivision and correlation. *Vertebrata PalAsiatica* 22(3):163–178 (in Chinese with English summary).

Lindsay, E. H. 2003. Chronostratigraphy, biochronology, datum events, land mammal ages, stage of evolution, and appearance event ordination. *Bulletin of the American Museum of Natural History* 279:212–230.

Lindsay, E. H. and R. H. Tedford. 1990. Development and application of land mammalages in North America and Europe, a comparison. In *European Neogene Mammal Chronology.* NATO Advanced Science Institute Series 180, ed. E. H. Lindsay, V. Fahlbusch, and P. Mein, pp. 601–624. New York: Plenum Press.

Lourens, L., F. Hilgen, N. J. Shackleton, J. Laskar, and D. Wilson. 2004. The Neogene Period. In *A Geologic Time Scale 2004,* ed. F. Gradstein, J. Ogg, and A. Smith, pp. 409–440. Cambridge: Cambridge University Press.

Lucas, S. G. 1993. Vertebrate biochronology of the Jurassic-Cretaceous boundary, North American western interior. *Modern Geology* 18:371–390.

Lucas, S. G. 1997. Holarctic fossil mammals and boundaries of Paleogene Series boundaries. In *Proceedings of the 30th International Geological Congress, Vol. 11: Stratigraphy,* ed. N.-w., Wang and J. Remane, pp. 189–199. Utrecht: Brill.

Lucas, S. G., E. G. Kordikova, and R. J. Emry. 1998. Oligocene stratigraphy, sequence stratigraphy, and mammalian biochronology north of the Aral Sea, Western Kazakstan. *Bulletin of Carnegie Museum Natural History* 34:313–348.

Luterbacher, H. P., J. R. Ali, H. Brinkhuis, F. M. Gradstein, J. J. Hooker, S. Monechi, J. G. Ogg, J. Powell, U. Röhl, A. Sanfilippo, and B. Schmitz. 2004. The Paleogene Period. In *A Geologic Time Scale 2004,* ed. F. Gradstein, J. Ogg, and A. Smith, pp. 384–408. Cambridge: Cambridge University Press.

Maridet, O., W.-y. Wu, J. Ye, S.-d. Bi, X.-j. Ni, and. J. Meng. 2011. Earliest occurence of *Democricetodon* in China, in the Early Miocene of the Junggar Basin (Xinjiang) and comparison with the genus *Spanocricetodon. Vertebrata PalAsiatica* 49(4):393–405.

Mein, P. 1975. Résultats du Groupe de Travail des Vertébrés. In *Report on Activity of the RCMNS Working Groups (1971–1975),* Bratislava, pp. 78–81.

Mein, P. 1989. Updating of MN Zones. In *European Neogene Mammal Chonology,* ed. E. H. Lindsay, V. Falbusch, and P. Mein, pp. 73–90. NATO ASI Series (A) 180. New York: Plenum Press.

Mein, P. 1999. European Miocene mammal biochronology. In *The Miocene land Mammals of Europe,* ed. G. E. Rössner and K. Heissig, pp. 25–38. Munich: Dr. Friedrich Pfeil.

Meng, J., X.-j. Ni, C.-k Li, K. C. Beard, D. L. Gebo, Y.-q. Wang, and H.-j. Wang. 2007. New material of Alagomyidae (Mammalia, Glires) from the Late Paleocene Subeng locality, Inner Mongolia. *American Museum Novitates* 3570:1–30.

Meng, J., A. R. Wyss, M. R. Dawson, and R.-j. Zhai. 1994. Primitive fossil rodent from Inner Mongolia and its implications for mammalian phylogeny. *Nature* 370:134–136.

Meng, J., J. Ye, W.-y. Wu, X.-j. Ni, and S.-d. Bi. 2008. The Neogene Dingshanyanchi Formation in northern Junggar Basin of Xinjiang and its stratigraphic implications. *Vertebrate PalAsiatica* 46(2):90–110.

Meng, J., J. Ye, W.-y. Wu, L.-p. Yue, and X.-j. Ni. 2006. A recommended boundary stratotype section for Xiejian Stage from northern Junggar Basin: Implications to related bio-chronostratigraphy and environmental changes. *Vertebrata PalAsiatica* 44(3): 205–236.

National Committee of Stratigraphy, China (NCSC). 2002. *Introduction to the Table of Chinese Regional Chronostratigraphy (Geochronology).* Beijing: Geological Publishing House (in Chinese).

Ogg, J. G. and A. G. Smith. 2004. The geomagnetic polarity time scale. In *A Geological Time Scale 2004,* ed. F. Gradstein, J. Ogg, and A. Smith, pp. 63–86. Cambridge: Cambridge University Press.

Opdyke, N. D., E. H. Lindsay, N. M. Johnson, and T. Downs. 1977. The paleomagnetism and magnetic polarity stratigraphy of the mammal-bearing sections of Anza Borrego State Park, California. *Quaternary Research* 7:316–329.

Peng, S.-l. 1975. Cenozoic vertebrate localities and horizon of Dzungaria Basin, Sinkiang. *Vertebrata PalAsiatica* 13:185–189 (in Chinese).

Qiu, Z.-x. and Z.-d. Qiu. 1990. Neogene local mammalian faunas: Succession and ages. *Jounal of Stratigraphy* 14:241–260 (in Chinese).

Qiu, Z.-x. and Z.-d. Qiu. 1995. Chronological sequence and subdivision of Chinese Neogene mammalian faunas. *Palaeogeography, Palaeoclimatology, Palaeoecology* 116:41–70.

Qiu, Z.-d., X.-m. Wang, and Q. Li. 2006. Faunal succession and biochronology of the Miocene through Pliocene in Nei Mongol (Inner Mongolia). *Vertebrata PalAsiatica* 44(2):164–181.

Qiu, Z.-x., W.-y. Wu, and Z.-d. Qiu. 1999. Miocene mammal faunal sequence of China: Palaeozoogeography and Eurasian relationships. In *The Miocene Land Mammals of Europe*, ed. G. Rössner and K. Heissig, pp. 443–455. Munich: Dr. Friedrich Pfeil.

Russell, D. E. and R.-j. Zhai. 1987. The Palaeogene of Asia: Mammals and stratigraphy. *Mémoires du Muséum National d'Histoire Naturelle*, ser. C, *Sciences de la Terre* 52:1–488.

Salvador, A., ed. 1994. *International Stratigraphic Guide*. 2nd ed. Boulder: International Union of Geological Sciences and the Geological Society of America.

Savage, D. E. 1962. Cenozoic geochronology of the fossil mammals of the western hemisphere. *Revista del Museo Argentino de Ciencias Naturales "Bernardino Rivadavia" y Instituto Nacional de Investigación de las Ciencias Naturales, Ciencias Zoológicas,* 8(4):53–67.

Savage, D. E. 1977. Aspects of vertebrate paleontological stratigraphy and geochronology. In *Concepts and Methods of Biostratigraphy*, ed. E. G. Kauffmann and J. E. Hazel, pp. 427–442. Stroudsburgh: Dowden, Hutchinson and Ross.

Shackleton, N. J., M. A. Hall, I. Raffi, L. Tauxe, and J. Zachos. 2000. Astronomical calibration age for the Oligocene-Miocene boundary. *Geology* 28:447–450.

Steininger, F. F. 1999. Chronostratigraphy, geochronology and biochronology of the Miocene "European Land Mammal Mega-Zones (ELMMZ)" and the Miocene "Mammal-Zones (MN-Zones)." In *The Miocene Land Mammals of Europe*, ed. G. E. Rössner and K. Hessig, pp. 9–24. Munich: Dr. Friedrich Pfeil.

Steininger, F. F., M. P. Aubry, W. A. Berggren, M. Biolzi, A. M. Borsetti, J. E. Cartlidge, F. Cati, R. Corfield, R. Gelati, S. Iaccarino, C. Napoleone, F. Ottner, F. Rögl, R. Roetzel, S. Spezzaferri, F. Tateo, G. Villa, and D. Zevenboom. 1997. The Global Stratotype Section and Point (GSSP) for the base of the Neogene. IGCP Project 329. *Episodes* 20, N.1:23–28.

Steininger, F. F., R. L. Bernor, and V. Fahlbusch. 1990. European Neogene marine/continental chronologic correlations. In *European Neogene Mammal Chronology*, ed. E. H. Lindsay, V. Fahlbusch V, and P. Mein, pp. 15–24. New York: Plenum Press.

Sullivan, R. M. and S. G. Lucas. 2004. The Kirtlandian, a new land-vertebrate "age" for the Late Cretaceous of western North America. In *Geology of the Zuni Plateau*, ed. S. G. Lucas, S. C. Semken,

W. R. Berglof, and D. S. Ulmer-Scholle, 369–378. New Mexico Geological Society Fifty-fourth Annual Field Conference. Albuquerque: New Mexico Geological Society.

Sun, J.-m., J. Ye, W.-y. Wu, X.-j. Ni, S.-d. Bi, Z.-q. Zhang, W.-m. Liu, and J. Meng. 2010. Late Oligocene-Miocene mid-latitude aridification and wind patterns in the Asian interior. *Geology* 38: 515–518.

Tedford, R. H. 1970. Principles and practices of mammalian geochronology in North America. *Proceedings of the North American Paleontological Convention*, Part F:66–703.

Tedford, R. H. 1995. Neogene mammalian biostratigraphy in China: Past, present and future. *Vertebrata PalAsiatica* 10:272–289.

Ting, S.-y. 1998. Paleocene and early Eocene land mammal ages of Asia. In *Dawn of the Age of Mammals in Asia*, ed. K. Beard, and M. R. Dawson, pp. 124–147. *Bulletin of Carnegie Museum of Natural History* 34.

Tong, Y.-s., T. Qi, J. Ye, J. Meng, and D.-f. Yan. 1987. Tertiary beds and fossils from north Junggur Basin, Xing Jiang Province. In *Fossil Vertebrates and Stratigraphy of Xingjiang*, pp. 1–61. Beijing: Academia Sinica.

Tong, Y.-s., T. Qi, J. Ye, J. Meng, and D.-f. Yan. 1990. Tertiary stratigraphy of the north of Junggar Basin, Xinjiang. *Vertebrata PalAsiatica* 28:59–70 (in Chinese with English summary).

Tong, Y.-s., S.-h. Zheng, and Z.-d. Qiu. 1995. Cenozoic mammal ages of China. *Vertebrata PalAsiatica* 33(4):290–314.

Walsh, S. L. 1998. Fossil datum terms, paleobiological event terms, paleontostratigraphy, chronostratigraphy, and the definition of land mammal "age" boundaries. *Journal of Vertebrate Paleontology* 18:150–179.

Walsh, S. L. 2001. Notes on geochronologic and chronostratigraphic units. *Geological Society of America Bulletin* 113 (6):704–713.

Walsh, S. L. 2005a. The role of stratotypes in stratigraphy. Part 2. The debate between Kleinpell and Hedberg, and a proposal for the codification of biochronologic units. *Earth-Science Reviews* 70:47–73.

Walsh, S. L. 2005b. The role of stratotypes in stratigraphy. Part 3. The Wood Committee, the Berkeley school of North American mammalian stratigraphy paleontology, and the status of provincial golden spikes. *Earth-Science Reviews* 70:75–101.

Wang, X.-m., J. Ye, J. Meng, W.-y. Wu, L.-p. Liu, and S.-d. Bi. 1998. Canivora from middle Miocene of Northern Junggar Basin, Xinjiang Autonomous Region, China. *Vertebrata PalAsiatica* 36(3):218–243 (in Chinese with English summary).

Wood II, H. E., R. W. Chaney, J. Clark, E. H. Colbert, G. L. Jepsen, J. B. Reeside, Jr., and C. Stock. 1941. Nomenclature and correlation of the North American continental Tertiary. *Geological Society of America Bulletin* 52:1–48.

Woodburne, M. O. 1977. Definition and characterization in mammalian chronostratigraphy. *Journal of Paleontology* 51(2):220–234.

Woodburne, M. O. 1987. Principles, classification, and recommendations. In *Cenozoic Mammals of North America: Geochronology and Biostratigraphy*, ed. M. O. Woodburne, pp. 9–17. Berkeley: University of California Press.

Woodburne, M. O. 1996. Precision and resolution in mammalian chronostratigraphy: Principles, practices, examples. *Journal of Vertebrate Paleontology* 16(3):531–555.

Woodburne, M. O. 2004. Principles and Procedures. In *Late Cretaceous and Cenozoic Mammals of North America: Biostratigraphy and Geochronology*, ed. M. O. Woodburne, pp. 1–22. New York: Columbia University Press.

Woodburne, M. O. 2006. Mammal Ages. *Stratigraphy* 3(4):229–261.

Wu, L.-c., L.-p. Yue, J.-q. Wang, F. Heller, and T. Deng. 2006. Magnetostratigraphy of stratotype section of the Neogene Xiejian Stage. *Journal of Stratigraphy* 30:50–53.

Wu, W.-y., J. Ye, J. Meng, X.-m. Wang, L.-p. Liu, S.-d. Bi, and W. Dong. 1998. Progress of the study of Tertiary biostratigraphy in north Junggar Basin. *Vertebrata PalAsiatica* 36:24–31 (in Chinese with English summary).

Ye, J., J. Meng, and W.-y. Wu. 2003. Oligocene/Miocene beds and faunas from Tieersihabahe in the northern Junggar Basin of Xinjiang. *American Museum of Natural History Bulletin* 279:568–585.

Ye, J., W.-y. Wu, and J. Meng. 2001a. On the age of Tertiary rock uints and the contained mammalian faunas in Ulungur River area of

Xinjiang. *Journal of Stratigraphy* 25:283–287 (in Chinese with English summary).

Ye, J., W.-y. Wu, and J. Meng. 2001b. Tertiray stratigraphy in the Ulungur River area of the northern Junggar Basin of Xinjiang. *Journal of Stratigraphy* 25:193–200 (in Chinese with English summary).

Ye, J., W.-y. Wu, J. Meng, S.-d. Bi, and S.-y. Wu. 2000. New results in the study of Tertiary biostratigraphy in the Ulungur River region of Xinjiang, China. *Vertebrata PalAsiatica* 38:192–202.

Yin, H.-f., F.-q. Yang, K.-x. Zhang, and W.-p. Yang. 1986. A proposal to the biostratigraphic criterion of Permian/Triassic boundary. *Memorie della Societa Geologica Italiana* 34:329–344.

Yin, H.-f., K.-x. Zhang, J.-n. Tong, Z.-y. Yang, and S.-b. Wu. 2001. The Global Stratotype Section and Point (GSSP) of the Permian-Triassic Boundary. *Episodes* 24 (2):102–114.

Chapter 4

Early Miocene Xiejiahe and Sihong Fossil Localities and Their Faunas, Eastern China

ZHU-DING QIU AND ZHAN-XIANG QIU

A BRIEF INTRODUCTION TO THE LOCALITIES

The Shanwang Basin, located about 20 km east of the Linqu County seat in Shandong Province, is a classic Miocene locality in China (figure 4.1). Its abundant fossil animals and plants, as well as its diatomite mineral resources in the basin sediments, have attracted numerous scholars. Early studies can be traced to descriptions of fossil vertebrates by Matsumoto (1926). In the 1930s, C. C. Young's works on the biostratigraphy of the basin resulted in the establishment of "Shanwang System" and "Shanwang Fauna," then thought to be middle or late Miocene in age (Young 1936, 1937). Several scholars subsequently investigated the basin structure, depositional environment, and fossil flora (Juan 1937; Teilhard de Chardin 1939; Hu and Chaney 1940). Wars and political instability resulted in a hiatus of more than 10 years, and research resumed in the 1950s. In 1960, a Shandong geological survey team renamed the "Shanwang System" as "Shanwang Formation," which has been followed in subsequent publications (Sun 1961; Pei, Zhou, and Zheng 1963) and was adopted in the Chinese Stratigraphical Lexicon in 1999. During the Third National Stratigraphic Committee conference, "Shanwang Stage" was proposed as a chronostratigraphic unit (National Stratigraphic Committee 2001). In the research program "Establishment of Major Chinese Stratigraphic Stages" launched in 2001, the Shanwang Fauna was designated as the representative fauna for the Shanwangian Chinese Mammal Age, and Shanwang Formation was designated as the type section of Shanwangian Stage (Li et al. 1984; Qiu and Qiu 1995; Tong, Zheng, and Qiu 1995; Qiu et al. 1999; Deng, Wang, and Yue 2008). Given the existence of the terms Shanwangian age and Shanwangian stage, we propose a new name—Xiejiahe Fauna, based on the nearby village name—for Shanwang Fauna, in order to avoid confusion with the nomenclature of ages (stages) and to allow room for naming new faunas in surrounding regions.

Sihong is located approximately 180 km northwest of Nanjing in Jiangsu Province. Although there are more than ten known fossil localities, Miocene fossils are concentrated in Xiacaowan, Songlinzhuang, Shuanggou, Zhengji, and Qizui localities (see figure. 4.1). Fossils from these localities apparently are contemporaneous, and thus all included in the Sihong Fauna. Young (1955) was the first to report vertebrate fossils from Xiacaowan localities. He called the fossiliferous layer the "Xiacaowan System" and considered it middle Pleistocene. Subsequent discoveries suggested that it should be Miocene in age (Chow and Wang 1964; Chow and Li 1978). Lin (1980) renamed the Xiacaowan System as Xiacaowan Formation, which was formally adopted by the Chinese Stratigraphical Lexicon in 1999. Due to the close similarities in composition and character of Xiacaowan and Xiejiahe faunas, these two have been commonly treated as coeval faunas. In most Chinese studies on land mammal age relationships and attempts to establish stages, fossils from Xiacaowan are always treated as an important reference fauna for Shanwangian age.

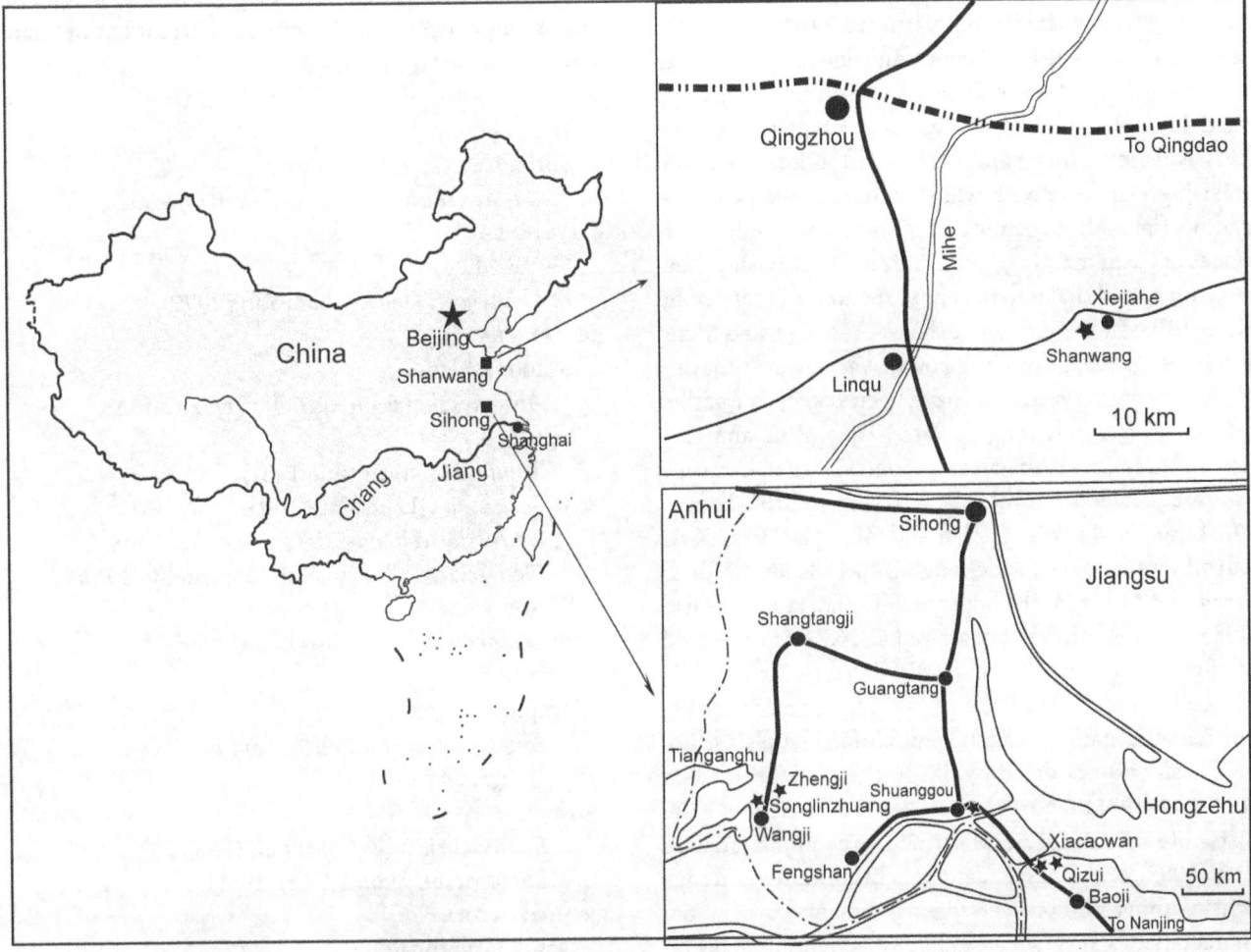

Figure 4.1 Map of the Xiejiahe and Sihong fossil localities.

In recent decades, as work in the Shanwang Basin and the Sihong area progressed, it has become increasingly clear that these two regions have tremendous potential in biostratigraphy. Abundant fossils, particularly terrestrial mammals that are of special chronological significance, include large and medium-sized taxa as well as small mammals. Coupled with datable basalts that helped to constrain age relationships, a substantial amount of multidisciplinary research has already been accomplished in these two regions. Optimal conditions such as these are rare in China.

Fossil mammals produced from these two regions have always attracted great attention, but reports in the last 30 years on new discoveries and taxonomic revisions have been scattered in a number of journals. Given the importance of these two faunas in the Chinese land mammal age system and in the establishment of chronostratigraphic stages, we propose a new faunal list based on our

reevaluation of existing fossil mammal collections, and we attempt to elaborate on the faunal characteristics.

XIEJIAHE FAUNA

Geologic Background and Faunal Composition

The Shanwang Basin is a graben controlled by the Great Luzhong (central Shandong) Fault. A sequence of diatomite clay containing tuffaceous clasts was deposited during the Miocene, accompanied by multiple basaltic eruptions. Miocene sediments are 100 m in thickness and best exposed in the southern margin of the basin near the village of Xiejiahe. Only about 0.3 km² of exposures are available. Young (1936, 1937) initially placed the fossil-producing sediments above the "Qingshan System" (presently Niushan Basalt) in his "Shanwang System" and assigned it an age of middle or late Miocene. In 1960, while changing the name to Shanwang Formation, the Shandong Geological Survey restricted it to the diatomaceous clays between the upper and lower basalts. The Shanwang Formation is mainly exposed in the Jiaoyan Shan (mountain) area and consists of brownish-yellow tuffaceous sandstones and gravels in the lower section, grayish-green to grayish-white diatomite shales in the middle (laterally becoming grayish-yellow sandy mudstones at basin margins), grayish-green to grayish-yellow mudstones and shales in the upper section, and topped by andesite basalt and carbonate shales (Yan, Qiu, and Meng 1983; Deng, Wang, and Yue 2008). Various basalt layers in the Shanwang sequence have been dated, but earlier results yielded wide-ranging dates. Fang, Zhu, and He's (1980) study on the magnetic property of the Shanwang basalts concluded that volcanic activities occurred around 27 Ma to 17 Ma. Wang, Zhu, and Zhang's (1981) K-Ar dates, however, gave a wide-ranging 44.1 Ma for the lower basalt and 24 Ma for the upper basalt. Zhu, Hu, and Zhao (1985), on the other hand, arrived at 16.79 Ma and 9.66 Ma for the same basalts. Jin (1985) dated (K-Ar method) the lower basalt at 18.87 ±0.49 Ma, a figure more consistent with the mammal chronology. Unfortunately, the upper basalt was not dated by Jin. In the same year, Chen and Peng (1985) reported a date for the basalt above the diatomite at 18.05 ± 0.55 Ma, which, in combination with Jin's (1985) lower basalt date, brackets the fossiliferous strata within a million years. Based on mammalian faunal characteristics, magnetostratigraphy, and preliminary radioisotopic analysis, the Shanwang Formation is likely to be around 18 Ma (Deng, Wang, and Yue 2008). In addition to exposures in the Shanwang Basin, the Shanwang Formation is also exposed in other areas in the Linqu region, its characteristic diatomaceous shales being present in nearby Baojiahe, Qingshan, and Dachegou.

Within the Shanwang Basin, the main fossiliferous layers are in the middle diatomite shales. By our incomplete tally, the fossil layers are known to produce 11 species of fishes, 4 species of amphibians, 2 species of reptiles, 5 species of birds, 23 species of mammals, 272 species of insects, 155 species of macroscopic plants, and 100 species or varieties of diatoms (Yeh 1980, 1981; Yan, Qiu, and Meng 1983; Gao 1986; Li and Wang 1987; Zhang 1989; Zhou 1992; Tao, Sun, and Yang 1999; Hou et al. 2000; Liu, Fortelius, and Pickford 2002; Deng, Wang, and Yue 2008).

Recovery of Xiejiahe mammal fossils benefits from diatomite mining, because new materials and new species are continuously uncovered. New descriptions and revisions are frequently made (Li 1974; Xie 1979, 1982; Qiu 1981; Qiu, Yan, Jia, and Sun 1985, 1986; Qiu, Yan, Jia,

and Wang 1985; Qiu and Sun 1988; Qiu, Yan, and Sun 1991; Qiu, Fortelius, and Pickford 2002; Storch and Qiu 2004; Qiu and Yan 2005), and our current list of Shanwang mammals is as follows:

INSECTIVORA
Soricidae
Lusorex taishanensis Storch and Qiu 2004
CHIROPTERA
Vespertilionidae
Shanwangia unexpectuta Yang 1977
RODENTIA
Aplodontidae
Ansomys shanwangensis Qiu and Sun 1988
Sciuridae
Tamiops asiaticus (Qiu 1981)
Sciurus lii Qiu and Yan 2005
Oriensciurus linquensis Qiu and Yan 2005
Plesiosciurus aff. *P. sinensis* Qiu and Lin 1986
Diatomyidae
Diatomys shantungensis Li 1974
CARNIVORA
Amphicyonidae
Amphicyon confucianus Young 1937
Ysengrinia sp.
Ursidae
Ballusia orientalis (Qiu et al. 1985)
Phoberocyon youngi (Chen 1981)
PROBOSCIDEA
Gomphotheriidae
Gomphotherium sp.
PERISSODACTYLA
Chalicotheriidae
Anisodon sp.
Rhinocerotidae
Plesiaceratherium gracile Young 1937
Diaceratherium sp.
Tapiridae
Plesiotapirus yagii (Matsumoto 1921)
ARTIODACTYLA
Suidae
Sinapriculus linquensis Liu, Fortelius, and Pickford 2002
Hyotherium shanwangense Liu, Fortelius, and Pickford 2002
Bunolistriodon? penisulus (Chang 1974)
Palaeomerycidae
Sinomeryx tricornis (Qiu et al. 1985)
Cervidae
Ligeromeryx colberti (Young 1937)
Heterocemas simpsoni Young 1937

Discussion

Small mammals are relatively underrepresented because rocks from the fossil-producing layers are not suitable for screening techniques. The eight known species are obviously not a full reflection of true faunal composition. For example, rodents that are common elsewhere during this period, such as cricetids, eomyids, and glirids, are so far absent. Large mammals are represented by 25 species, most of them preserved as articulated skeletons.

The insectivore *Lusorex* and bat *Shanwangia* are so far not found outside Shanwang. *Lusorex*, a member of Heterosoricinae, shares many morphological similarities with *Wilsonsorex* from the Hemingfordian of North America, and they may be sister taxa in the early Miocene faunal exchange of North America and Asia (Storch and Qiu 2004). Rodents from the Xiejiahe Fauna are widely distributed, such as *Ansomys* present in Sihong and central Inner Mongolia in China and in the Miocene of Kazakhstan, but also in late Oligocene to middle Miocene strata of North America (Qiu 1987; Wang et al. 2009; Lopatin 1997; Hopkins 2004; Kelly and Korth 2005; Korth 2007). *Tamiops*, a modern genus of sciurids in forested habitats of the Oriental zoogeographic province, occurs in the late Miocene Shihuiba and Xiaohe faunas of Yunnan (Qiu and Ni 2006), but the species from Xiejiahe is obviously more primitive than the Yunnan forms. *Sciurus*, *Plesiosciurus*, and *Diatomys* are all found in the Sihong Fauna. Furthermore, *Diatomys* is known in the early Miocene of Pakistan and Japan (Flynn et al. 1998; Tomida et al., chapter 12, this volume).

The true condition of *Amphicyon confucianus*, known by only a right mandible with p3 and m1 and an isolated m3, is still unclear due to a lack of better materials. The m1 is similar to those of the European *A. giganteus* and *A. major*, but is slightly longer and more slender than the former and overall closer to the latter. The Xiejiahe specimen differs from the latter in its larger premolar, apparently a primitive character. *A. major* occurs in MN 4–8 in Europe, common in MN6. Therefore, *A. confucianus* suggests an age earlier than MN 6.

After a careful comparison, we now assign to *Ysengrinia* sp. a specimen (P4-M3 and m1-m2) formerly identified as belonging to Thaumastocyoninae. Ginsburg (1966) established *Ysengrinia* based on *Pseudocyon gerandianus* Viret (1929). *Ysengrinia* is mainly diagnosed by its "pear-like" M1, a character identical to that in the Xiejiahe specimen. Additional characters, such as the very small M2 and m2 and high m1 protoconid, are also consistent with features of *Ysengrinia*. On the other hand, the

M1 in *Thaumastocyon* has a posteriorly skewed triangular outline along its lingual border, and the crown height difference between its protoconid and paraconid is also relatively low. In Europe, *Thaumastocyon* occurs in MN5–9, whereas *Ysengrinia* is restricted to MN 2–4.

When Qiu, Yan, Jai, and Wang (1985) described *Ursavus orientalis*, they considered its size and morphology to be closest to the European *U. elmensis*. Later, Ginsburg and Morales (1998) erected a new genus, *Ballusia*, based on *U. elmensis*. Therefore, the Xiejiahe form should also be emended as *Ballusia orientalis*. This species is overall slightly smaller than *B. elmensis* but is morphologically somewhat derived, as is mainly seen in its longer M1 and M2 relative to the P4. It is probably slightly later than MN 3.

During European MN 3–6, no tapir was present. The Xiejiahe fossil tapir can only be compared to fossils from Japan. The Japanese *Plesiotapirus yagii* is from the Hiramaki Formation. According to Tomida and Setoguchi (1994), the age of the Hiramaki Formation is 17–18 Ma, or equivalent to late MN 4.

Plesiaceratherium gracile is very close to the European *P. fahlbuschi* in both size and morphology, and this species pair may be considered to be ecological substitutes in their respective regions. The latter occurs in MN 5.

Two additional perissodactyls are recent discoveries. One is a chalicothere, consisting of a metatarsal III and phalanx I, as well as a phalanx I from the forelimb. In both size and shape, they are consistent with the European *Anisodon grande*. The latter species is from MN5–7 in Europe. Another perissodactyl is *Diaceratherium*, consisting of a complete skeleton housed in the Shandong Natural History Museum. This skeleton is similar to *D. aurelianense* in the European MN4 and is more derived than *D. aginense* from MN2. Yan and Heissig (1986) pointed out that some teeth described as *Plesiaceratherium gracile* by C. C. Young may belong to the genus *Brachypotherium*, but these also represent *Diaceratherium*.

Liu et al. (2002) have systematically revised fossil suids from Xiejiahe, and they concluded that the suids are equivalent to the European MN 4.

Astibia, Morales, and S. Moyà-Solà (1998) recently erected a new genus, *Sinomeryx*, for *Palaeomeryx tricornis*. Additional materials suggest that the originally described vertically oriented occipital appendages were not correctly reconstructed, due to postmortem lateral compression on the original specimen. To the contrary, this appendage is actually flattened and fork-like, extending posteriorly. Existing evidence suggests that *Palaeomeryx* had supraorbital "ossicones," but proof for possession of an occipital appendage is still lacking. Based on what we know, "ossicones" and occipital appendages in these deer (*Ampelomeryx*) first appeared in the European MN 4. By MN 5, *Triceromeryx* and *Tauromeryx* also appeared. The antlers in these three deer are different. Specimens from Xiejiahe have very short and thick "ossicones," which are thickened at the base and are anteroposteriorly elongated. These characters differ from those of other genera. The occipital appendage of the Xiejiahe species is closer to that in *Tauromeryx* and has a shallow division toward the posterior end with dorsoventrally compressed cross-section.

Azanza and Ginsburg's (1997) recent revision of lagomerycids concluded that large lagomerycids represented by *Lagomeryx praestans* should all be included in their new genus *Ligeromeryx*. Several species have been named for Xiejiahe lagomerycids based on the number and size of tines in antlers and the main stem. However, European materials indicate extreme variability in antler morphology of this group. Accordingly, we recognize only two lagomerycids from Xiejiahe: *Ligeromeryx colberti* and *Heterocemas simpsoni*. The latter is not seen in Europe. The former is closest to *L. praestans* from MN 3 of Europe but is more derived.

Although small mammals from Xiejiahe are not very diverse, judging by the ranges and distributions and morphological characteristics in *Lusorex, Ansomys, Tamiops, Sciurus, Orensciurus*, and *Diatomys*, the small mammal fauna should be later than Oligocene and earlier than late Miocene. Large mammals from Xiejiahe are comparable to assemblages of European MN4 and thus indicate an early Miocene age.

Rodent composition from Xiejiahe suggests similarity to that of the Sihong Fauna, sharing all genera except *Tamiops* and *Orensciurus*. However, only one large mammal is shared with the Sihong Fauna: *Plesiaceratherium*. Small mammals from Xiejiahe have many endemic forms, showing almost no zoogeographic relationship to contemporaneous European faunas. In contrast, with the exception of two genera, *Plesiotapirus* and *Sinapriculus*, all other Xiejiahe large mammals also occur in the early to middle Miocene strata of Europe, which constitutes more than 85% of the large mammals. Furthermore, *Amphicyon, Ysengrinia, Phoberocyon*, and *Gomphotherium* are also present in the early(?) to middle Miocene of North America.

In recent years, Tianyu Museum in Pingyi, Shandong Province, has amassed a large quantity of articulated skeletons of mammals from the Jijiazhuang area in neighboring Changle County. Preliminary observations suggest that its composition is almost identical to that of Xiejiahe, but some taxa are represented by more com-

plete materials. New forms are also present, such as multiple skeletons of *Dorcatherium*. Furthermore, a skeleton of a giant Creodonta represents the first of its kind in East Asia. Judging from its cranial morphology, this creodont is different from all known genera, including *Hyainailouros* (Europe) and *Megistotherium* (Africa) of similar age and size. It is likely a highly specialized descendant of *Pterodon*. In addition, there is a large number of *Sinomeryx tricornis* (previously *Palaeomeryx tricornis*) skeletons. Significantly, some skulls are well preserved, and their "bony antlers" are far more complex than was originally described: very long posterior antler with divergent distal ends, the antler anterodorsal to the orbit large and robust with many tubercles on the anterior surface. This indicates that skeletons from Changle County are more derived than those in Xiejiahe, thus suggesting a slightly younger age for the former.

SIHONG FAUNA

Geologic Background and Faunal Composition

The Xiacaowan Formation consists mainly of a sequence of fluviolacustrine detrital sediments with exposures scattered along either bank of Huai River to the west of Lake Hongze Hu. It is sandwiched between the underlying sandstones and mudstones of the Oligocene Fengshan and Sanduo formations and the overlying Quaternary Douchong Formation (Lin 1980). The lower part of the formation includes grayish-green, grayish-white, and purplish-red mudstones and sandy mudstones, occasionally containing gravelly sandstones and, south of the Huai River, multiple layers of basalts (no radiometric date is available). The upper part includes grayish yellow, grayish green, and purplish red mudstones and gravelly or muddy sandstones with marls locally. The thickness varies greatly, ranging from 8 m to 93 m.

Fossils are mainly produced in the upper part of the Xiacaowan Formation. Vertebrate fossils are mainly found in Songlinzhuang, Shuanggou, and Zhengji localities, including 11 species of fishes, 4 species of amphibians, 2 species of reptiles, 5 species of birds, and at least 35 species of mammals (Li et al. 1983; Hou 1984, 1987). Besides vertebrates, other fossils include plants, pollen and spores, ostracods, charophytes, bivalves, and gastropods. Since the description of *Dionysopithecus shuangouensis* from Songlinzhuang by Chuankuei Li (1978), the Sihong fossil mammal localities have attracted the interests of Chinese paleontologists, resulting in several excavations and small mammal screenings in this region in the 1980s. In 1983, Chuankuei Li and others made preliminary reports on the vertebrate faunas, followed by revisions and additions, as well as further descriptions and research during subsequent years (Gu and Lin 1983; Li et al. 1983; Qiu 1987, 2010; Qiu and Lin 1986; Qiu and Gu 1986, 1991; Gu 1989; Wu 1986, 1995; Qiu and Qiu 1995; Storch and Qiu 2002). The following list represents currently known fossil mammals (asterisk denotes unconfirmed records):

DIDELPHIMORPHIA
Didelphidae
Sinoperadectes clandestinus Storch and Qiu 2002
INSECTIVORA
Erinaceidae
Lanthanotherium sp.
Soricidae
Crocidosorex sp.
CHIROPTERA
Vespertilionidae
Myotis sp.
Vespertilionidae gen. et sp. indet.
RODENTIA
Ctenodactylidae
Sayimys sp.
Tachyoryctoididae
Tachyoryctoides sp. (previously Rhizomyidae gen. et sp. indet.)
Aplodontidae
Ansomys orientalis Qiu 1987
Sciuridae
Eutamias sihongensis Qiu and Lin 1986
Sciurus lii Qiu and Yan 2005
Plesiosciurus sinensis Qiu and Lin 1986
Shuanggouia lui Qiu and Lin 1986
Parapetaurista tenurugosa Qiu and Lin 1986
Sciurinae gen. et sp. indet.
Gliridae
Microdyromys orientalis Wu 1986
Eomyidae
Apeomys sp.
Castoridae
Youngofiber sinensis (Young 1955)
Diatomyidae
Diatomys cf. *D. shantungensis* Li 1974
Cricetidae
Alloeumyarion sihongensis Qiu 2010
Cricetodon wanhei Qiu 2010
Democricetodon suensis Qiu 2010
Primus pusillus Qiu 2010
Megacricetodon sinensis Qiu and Li 1981

Platacanthomyidae
Neocometes sp.
LAGOMORPHA
Ochotonidae
Alloptox sihongensis Wu 1995
CARNIVORA
Amphicyonidae
*Amphicyon sp.
Ursidae
*Ursidae gen. et sp. indet.
Mustelidae
*Proputorius sp.
*Mustela sp.
Hyaenidae
*Proctititherium sp.
Viverridae
Semigenetta huaiheensis Qiu and Gu 1986
Felidae
Pseudaelurus (*Schizailurus*) cf. *P. lorteti* Gaillard 1899
PROBOSCIDEA
Elephantidae
Stegolophodon hueiheensis (Chow 1959)
CETACEA
Delphinidae
Delphinus sp.
PERISSODACTYLA
Equidae
Anchitherium sp.
Rhinocerotidae
Plesiaceratherium gracile (previously *Brachypotherium pugnator*, see Chow and Wang 1964)
ARTIODACTYLA
Suidae
Suidae gen. et sp. indet.
Tayassuidae
*Pecarichoerus sp.
Tragulidae
Dorcatherium orientale Qiu and Gu 1991
Giraffidae
*Palaeotragus sp.
Cervidae
*Micromeryx sp.
*Dicrocerus sp.
*Amphimoschus sp.
Stephanocemas sp.
*Lagomeryx sp.
PRIMATES
Pliopithecidae
Dionysopithecus shuangouensis Li 1978
Platodontopithecus jianhuaiensis Gu and Lin 1983

There are 25 species of small mammals in the Sihong Fauna. In addition to the exceedingly rare Asian marsupial, common late Cenozoic insectivores, bats, rodents, and lagomorphs are present. These include several families common in the late Paleogene and early Neogene. The composition seems to be that of a typical Eurasian fauna of that time. Among the small mammals, sciurids and cricetids are dominant groups. "Modern cricetids" are already diverse, suggesting that the fauna has entered a relatively flourishing period for this group. Aplodontids, castorids, and glirids are modestly represented, whereas marsupials, eomyids, and platacanthomyids are rather scarce.

Large mammals from the Sihong region are mostly represented by isolated teeth. Almost 30 species were listed in the initial publications. We were either unable to locate the original specimens or unable to confirm the earlier identifications for a large number of these, all marked with an "*" in the previous list. As a result, only the following 11 taxa are presently confirmed: *Dionysopithecus shuangouensis, Platodontopithecus jianhuaiensis, Semigenetta huaiheensis, Pseudaelurus* cf. *P. lorteti, Stegolophodon hueiheensis, Delphinus* sp., *Anchitherium* sp., *Plesiaceratherium gracile,* Suidae gen. et sp. indet., *Dorcatherium orientale,* and *Stephanocemas* sp. Of these, four have not been formally published: *Delphinus, Anchitherium,* Suidae gen. et sp. indet., and *Stephanocemas. Delphinus* is represented by isolated teeth, and its specific status cannot be determined. *Anchitherium* also consists of only a single incisor and a few cheek tooth fragments, all small and difficult to further identify. A few fragmentary teeth are the material basis for Suidae gen. et sp. indet.; its generic identity is even more difficult to determine. *Stephanocemas* consists of a relatively complete shed antler plus a few antler fragments. Its size is very small, only about half that of *S. thomsoni* from the Tunggur Formation of Inner Mongolia, and there are only six tines, as compared to as many as nine in the Tunggur form—apparently far more primitive than the latter.

Faunal Comparison

Although Xiejiahe small mammals are still very limited, of the six known rodents, four—*Ansomys, Sciurus, Plesiosciurus,* and *Diatomys*—are shared with Sihong, suggesting a close relationship for these two faunas. Shared large mammals between Xiejiahe and Sihong, on the other hand, are relatively few, positively confirmed so far only by *Plesiaceratherium gracile.* The primates, anchitheres, and crown-antlered deers from the Sihong Fauna are completely absent in the Xiejiahe Fauna. On the other hand, the Xiejiahe tapirs, chalicotheres, palaeomerycids, and lagomerycids are all absent from the Sihong Fauna. The recently discovered Changle Fauna (see previous discussion) contains common Sihong elements, such as *Dorcatherium* and *Semigenetta* (both under study), and implies additional common elements yet to be discovered. Not far from the Sihong area, small quantities of mammal fossils have also been found in Fangshan and Puzhen in Jiangsu Province. The Fangshan locality is known to produce *Anchitherium,* which also occurs at Sihong, whereas the Puzhen locality shares several genera, such as *Plesiaceratherium, Hyotherium, Stephanocemas,* and *Dicrocerus,* with either Sihong or Xijiahe faunas. Both of these localities have overall mammal compositions similar to those from Sihong and Xiejiahe, suggesting similar ages (Li, Wu, and Qiu 1984).

Several genera in the Sihong Fauna, such as *Tachyoryctoides, Ansomys, Eutamias, Cricetodon, Microdyromys, Democricetodon, Megacricetodon,* and *Alloptox,* are known from Aoerban and Gashunyinadege faunas in the early Miocene of central Inner Mongolia, as well as in the middle Miocene Moergen Fauna, making these assemblages comparable (Qiu, Wang, and Li, chapter 5, this volume). Of these eight genera, the Lower Aoerban Fauna shares only five with the Sihong Fauna. The former seems earlier than Sihong because its "modern cricetids" are not very diverse, its lagomorphs have more distinctly Oligocene characteristics, and the genus *Alloptox* is yet to appear. Gashunyinadege and Upper Aoerban faunas, on the other hand, share six and seven genera with the Sihong Fauna, respectively; "modern cricetids," such as *Democricetodon* and *Megacricetodon,* are common in these faunas, and *Alloptox* also occurs. Although the Moergen Fauna also shares six genera with the Sihong Fauna and has abundant *Democricetodon* and *Megacricetodon,* archaic Oligocene families (Ctenodactylidae and Tachyoryctoididae) had completely disappeared from Moergen. Furthermore, its rich cricetid group includes advanced genera (e.g., *Plesiodipus* and *Gobicricetodon*), and its typically later families (Dipodidae) suggest a more modern character. Our comparison, therefore, indicates that the Gashunyinadege and Sihong faunas are closest in age because they share identical archaic families and possess cricetids at a comparably advanced stage of evolution (Qiu 2010). However, we do recognize that the Gashunyinadege and Sihong faunas differ in several respects. As in other early to middle Miocene faunas from Inner Mongolia, Gashunyinadege has many kinds of erinaceines, zapodids, and dipodids, as well as a high diversity of ochotonids, and its sciurids are dominated by ground squirrels. These differences are due to ecology and environmental variation in these faunas. Inner Mongolian small mammal faunas were adapted to a relatively drier environment similar to modern temperate open grasslands.

Some small mammal genera from Sihong (e.g., *Sayimys*, *Democricetodon*, *Megacricetodon*, and *Alloptox*) frequently appear in Miocene localities in northwestern China, such as Zhangjiaping and Quantougou in Gansu, Xiejia and Lierpu in Qinghai, and Tieersihabahe in Xinjiang (Tieersihabahe and Suosuoquan formations; Qiu 2001; Qiu, Li, and Wang 1981; Ye, Wu, and Meng 2001; Meng et al. 2006). However, these four genera occur together only in the Tieersihabahe localities in the early middle Miocene Halamagai Formation, and they appear singly in other localities. Furthermore, *Megacricetodon* and *Alloptox* in the Halamagai Formation seem to be more derived than their counterparts from the Sihong Fauna (Wu 1995; Qiu 2010). Other localities share too few elements with Sihong to permit a sense of relationship. Although the Suosuoquan II assemblage has *Democricetodon* and *Microdyromys* in common with the Sihong Fauna, its many archaic elements, low diversity of cricetids, and lack of *Alloptox* all suggest an earlier age than the Sihong Fauna. Similarly, despite the presence of *Tachyoryctoides* and *Cricetodon* in the Xiejia and Zhangjiaping faunas, the species there are distinctly more primitive than those from Sihong. Furthermore, the two faunas show no sign of *Democricetodon*, *Microdyromys*, or *Alloptox*, undoubtedly pointing to earlier ages than the Sihong Fauna. Finally, Quantougou and Lierpu from Gansu and Qinghai have advanced genera like *Plesiodipus* and *Protalactaga* that are closer in age to Tunggur faunas in Inner Mongolia and likely later than the Sihong Fauna.

The Sihong Fauna also shares some genera with the D1 horizon of the Valley of Lakes regions in Mongolia and the Murree and Kamlial faunas of Pakistan, including *Democricetodon*, *Megacricetodon*, *Sayimys*, *Ansomys*, and *Microdyromys* from the Valley of Lakes, *Primus* and *Sayimys* from Murree, and *Democricetodon* and *Megacricetodon* from Kamlial (Daxner-Höck and Badamgarav 2006; Flynn et al. 1998; Lindsay 1988; de Bruijn, Hussain, and Leinders 1981). *Primus* is not seen in the Kamlial Fauna, and its *Democricetodon* and *Megacricetodon* are already relatively advanced. Although *Democricetodon* and *Megacricetodon* are not present in the Murree Fauna, its "modern cricetids" are relatively abundant. It thus seems that the Sihong Fauna is closer to the Murree

Fauna. Additionally, two of the genera from the Sihong Fauna, *Youngofiber* and *Diatomys*, are present in the early Miocene of Japan (Tomida et al., chapter 12, this volume).

Biochronology

The Sihong Fauna is generally regarded as early or middle Miocene, belonging to the Chinese Shanwangian Land Mammal Age and correlative to the European Orleanian Land Mammal Age (Li, Wu, and Qiu 1984; Tong, Zheng, and Qiu 1995; Qiu and Qiu 1995; Qiu, Wu, and Qiu 1999). Of the 14 families of small mammals in the Sihong Fauna, all originated before Miocene except for Platacanthomyidae (restricted to Miocene strata). However, 18 genera (more than 80% of known genera from Sihong) are present in the Miocene or later in both the New and Old World, such as *Lanthanotherium*, *Eutamias*, *Sciurus*, *Democricetodon*, and *Megacricetodon*. On the other hand, the Sihong Fauna also contains Ctenodactylidae and Tachyoryctoididae that survived to, or nearly to, the early middle Miocene of northern Asia, as well as Aplodontidae and Eomyidae that disappeared in China only by the end of the late Miocene. Therefore, the Sihong Fauna must be earlier than the Pliocene. In Europe, the appearance of *Cricetodon*, *Democricetodon*, and *Megacricetodon* is a characteristic of the early to middle Miocene. The majority of known genera from Sihong are also early to middle Miocene taxa in Europe or Asia. *Crocidosorex* and *Apeomys* are only found in the early Miocene of Europe (McKenna and Bell 1997; Engesser 1999). Similarly, *Youngofiber*, *Diatomys*, and *Primus* are restricted to the early Miocene in Asia. The stage of evolution in *Cricetodon wanhei* from Sihong is similar to *C. aliveriensis* and *C. meini*, MN 4/5 of Europe and western Asia. Sihong species of *Democricetodon*, *Megacricetodon*, and *Alloptox* possess relatively primitive morphologies when compared to those from the middle Miocene Tunggur faunas in Inner Mongolia (Wu 1995; Qiu 2010). Presence of *Primus* suggests the Sihong cricetid assemblage is earlier than the Chinji Fauna from the Indian subcontinent and close to the Murree Fauna. Consequently, the age of the Sihong small mammal fauna is closest to the Xiejiahe Fauna from Shandong and the Gashunyinadege Fauna of Inner Mongolia and correlates to early Miocene Shanwangian Chinese Land Mammal Age, approximately equivalent to the European Orleanian or MN 3–4. Because of the primitive characters in *Ansomys* and *Plesiosciurus* when comparing species with the Xiejiahe Fauna, we cannot rule out the possibility that the Sihong Fauna is slightly earlier than

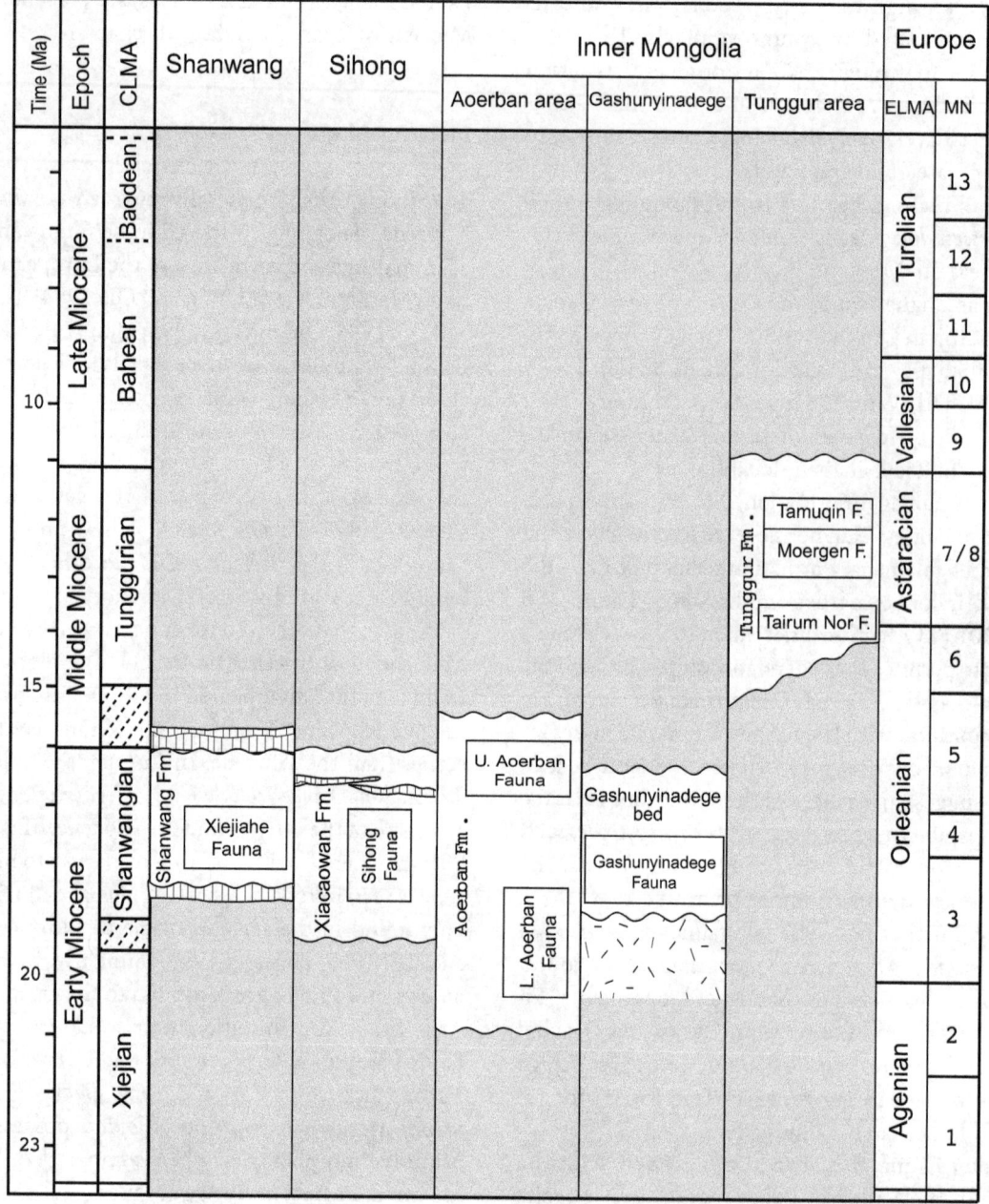

Figure 4.2 Positions of the Xiejiahe and Sihong faunas in the Chinese Miocene mammal sequence. Vertical hatching indicates basalts.

the Xiejiahe Fauna. Nevertheless, the fundamental similarities of these two faunas show that they represent the same land mammal age (figure 4.2).

Zoogeographic Relationship

Of the 23 genera of small mammals from Sihong, only 5 are known to be restricted to China: *Sinoperadectes, Ple-*

siosciurus, Shuanggouia, Parapetaurista, and *Alloeumyarion.* An additional five genera are also endemic to Asia: *Sayimys, Diatomys, Tachyoryctoides, Primus,* and *Alloptox.* The rest are widely distributed in Miocene strata of Europe or North America, including at least 10 genera in the early and middle Miocene of Europe, accounting for 43% of known genera. Among our confirmed nine genera of large mammals, six are also present in the Miocene of Europe: *Semigenetta, Pseudaelurus, Anchitherium, Plesi-*

Table 4.1

Comparisons of Distributions for Selected Genera from the Sihong Fauna

	South Asia	West Asia	Central Asia	Europe	North Africa	North America
Lanthanotherium				*		*
Crocidosorex				*		
Myotis				*	*	*
Sayimys	*					
Tachyoryctoides			*			
Ansomys			*			*
Eutamias	*	*	*	*	*	*
Microdyromys				*	*	
Apeomys				*		
Cricetodon		*		*		
Democricetodon	*	*	*	*	*	
Primus	*					
Megacricetodon	*	*		*		
Neocometes				*		
Alloptox		*				
Semigenetta				*		
Pseudaelurus				*	*	*
Anchitherium				*		*
Plesiaceratherium				*		
Dorcatherium	*			*	*	
Stephanocemas				*		

aceratherium, *Dorcatherium*, and *Stephanocemas*. This indicates a close zoogeographic relationship between East Asia and Europe with widespread biotic interchange and dispersal (table 4.1). There are also genera shared with South Asia, central Asia, western Asia, North Africa, and North America, but these numbers are fewer than with Europe. Environmental variability possibly accounts for these differences. In addition, although fossils from the Japanese Miocene are still rare, three genera are already shared with the Sihong Fauna: *Youngofiber*, *Diatomys*, and *Alloptox*. Genera in common with North America are rather limited. With the exception of the widespread bats and chipmunk *Eutamias*, only four genera seem to have crossed Beringia: *Lanthanotherium*, *Ansomys*, *Pseudaelurus*, and *Anchitherium*, indicating restricted biological dispersal between Asia and the New World.

Within the modern zoogeographic system, China straddles the Palearctic and Oriental provinces. Lacking an oceanic isolation mechanism as in the cases of all other zoogeographic boundaries, there is a mixed transitional zone between these two provinces. Sihong fossil localities are exactly in this zone, and the Sihong Fauna also seems to suggest the coexistence of elements from both provinces, a phenomenon observed in other modern transitional regions. Within the Sihong Fauna, modern families or subfamilies from the Oriental Province include Galericinae (*Lanthanotherium*), Diatomyidae, Platacanthomyidae (*Neocometes*), and Tragulidae, and at the same time, characteristic families from the Palearctic Province include Castoridae and Ochotonidae plus the largely Palearctic Cricetidae. Among extinct families, Asian Tapiridae are restricted to the Oriental Province at the present time, the Tachyoryctoididae are characteristic of the Palearctic Province, and Ctenodactylidae are Asian-African. In all, the Sihong Fauna seems to indicate influence of both the Oriental and Palearctic provinces during the early Miocene.

Ecology and Environment

From the standpoint of modern biological distribution, the presence of chipmunks (*Eutamias*), gymnures (*Lanthanotherium*), pigs, and crown-antlered deer (*Stephanocemas*) indicates a bushy habitat. Tree squirrels (*Sciurus, Plesiosci-urus,* and *Shuanggouia*), however, indicate continuous forests, and *Parapetaurista*, in particular, would require relatively tall trees. Cricetids and anchithere horses, however, may indicate more open environments, and ochotonids (*Alloptox*) may also share a similar preference. Most castorids require abundant water, and the coexistence of fishes and cetaceans further suggests the existence of a large body of water. Clearly, the Sihong Fauna implies a mosaic of ecological environments rather than a single habitat. Furthermore, Sihong galericines, diatomyids, and platacanthomyids, plus all genera within Sciuridae, all point to a relatively warm and humid climate during the early Miocene. The Xiejiahe Fauna also hints at a climate similar to that of Sihong, but its physical environment may be somewhat less complex, consisting of predominantly lakeshore forests, indicated also by the depositional environment. Paleoenvironments for the Sihong and Xiejiahe faunas are substantially different from those in central Inner Mongolia.

CONCLUSION

Most of the fossil mammals from the Shanwang Basin and Sihong region appear in the early and middle Miocene of Europe, suggesting widespread mammalian dispersals between Europe and Asia during the early Neogene and providing a foundation for correlation between faunas of these two continents.

Xiejiahe Fauna shares some rodent genera with Sihong Fauna, indicating affinity of these faunas and similar age. Within these two faunas, most of the families of small mammals originated before the Miocene, but most of the genera appear only in the Miocene. Overall composition of large mammals is similar to that of early to middle Miocene faunas of Europe, but the assemblages contain many archaic families and primitive genera and species. They are early Miocene in age, more or less equivalent to the European Orleanian Land Mammal Age, or MN 3–4.

The Sihong region is located along the transitional zone of the present-day Oriental and Palearctic provinces. Among modern families of small mammals in the Sihong Fauna, members of both the Palearctic and Oriental provinces are present, indicating that the composition of the Sihong Fauna shows the transitional characteristics of the two modern zoogeographic provinces.

Both Xiejiahe Fauna and Sihong Fauna contain some members that are indicative of a warm and humid climate, probably representing low-altitude environments closely controlled by an oceanic climate and quite different from those in the Miocene of the Mongolian and Xinjiang plateaus.

ACKNOWLEDGMENTS

We are grateful to Chuankui Li for his valuable suggestions and for reading an early draft of this chapter. This research is funded by the Knowledge Innovation Program of the Chinese Academy of Sciences (No. KZCX2-YW-120) and the Chinese National Natural Science Foundation (Nos. 40730210). We thank Xiaoming Wang for translating this paper into English.

REFERENCES

Astibia, H., J. Morales, and S. Moyà-Solà. 1998. *Tauromeryx,* a new genus of Palaeomerycidae (Artiodactyla, Mammalia) from the Miocene of Tarazona de Aragón (Ebro Basin, Aragón, Spain). *Bulletin de la Société géologique de France* 169(4):471–477.

Azanza, B. and L. Ginsburg. 1997. A revision of the large lagomerycid artiodactyls of Europe. *Palaeontology* 40(2):461–485.

Chang Y.-p. 1974. Miocene suids from Kaiyuan, Yunnan and Linchu, Shantung. *Vertebrata PalAsiatica* 12(2):117–123.

Chen, G.-f. 1981. A New species of *Amphicyon* from the Pliocene of Zhongxiang. Hebei. *Vertebrata PalAsiatic* 19(1):21–25.

Chen, D. and Z. Peng, 1985. K-Ar ages and Pb, Sr isotopic characteristics of Cenozoic volcanic rocks in Shandong, China. *Geochimica* 4:293–303.

Chow, M.-c. 1959. New species of fossil Proboscidea from South China. *Acta Palaeontogica Sinica* FZX(4):251–258.

Chow, M.-c. and C.-k. Li. 1978. A correction of the age of the Hsiatsaohwan Formation and its mammalian fauna. *Journal Stratigraphy* 2(2):122–130.

Chow, M.-c. and B.-y. Wang. 1964. Fossil vertebrates from the Miocene of northern Kiangsu. *Vertebrata PalAsiatica* 8(4):341–351.

Daxner-Höck, G. and D. Badamgarav. 2006. Geological and stratigraphic setting. *Annalen Naturhistorischen Museums Wien* 108A:1–24.

de Bruijn, H., S.-T. Hussain, and J.-M. Leinders. 1981. Fossil rodents from the Murree formation near Banda Daud Shah, Kohat, Pakistan. *Proceeding Koninklijke Nederlandse Akademie van Wetenschappen B* 84(1):71–99.

Deng, T., W.-m. Wang, and L.-p. Yue. 2008. Report on Shanwangian and Baodean stages in Chinese terrestrial Neogene. In *Establishment of Stages in Major Chinese Stratigraphic Sections (2001–2005),* ed. Z.-j. Wang and Z.-g. Huang, pp. 12–31. Chinese Geological Survey Report No. 2007001. Beijing: Geological Press.

Engesser, B. 1999. Family Eomyidae. In *The Miocene Land Mammals of Europe,* ed. G. E. Rossner and K. Heissig, pp. 319–335. Munich: Dr. Friedrich Pfeil.

Fang, D.-j., X.-y. Zhu, and L.-z. He. 1980. Paleomagnetic research on the Cenozoic basalt in Linqu Region, Shandong and its stratigraphic significance. *Journal of Zhejiang University* 1:49–57.

Flynn, L. J., W. Downs, M. E. Morgan, J. C. Barry, and D. Pilbeam. 1998. High Miocene species richness in the Siwaliks of Pakistan. In *Advances in Vertebrate Paleontology and Geochronology*, ed. Y. Tomida, L. J. Flynn, and L.L. Jacobs, pp. 167–180. Tokyo: National Science Museum Monographs, no. 14.

Galliard, C. 1899. Mammifères Miocènes nouveaux ou peu connus de la Grive-St. Alban (Isère). *Archives du Muséum d'Histoire Naturelle* (Lyon) 7:1–79.

Gao, K.-q. 1986. A new spadefoot toad from the Miocene of Linqu, Shandong with a restudy of *Bufo linquensis* Young 1977. *Vertebrata PalAsiatica* 24(1):63–74.

Ginsburg, L. 1966. L' *"Amphicyon"* ambiguous des Phosphorites du Quercy. *Bulletin du Muséum National d'Histoire Naturelle* (Paris) 37(4):724–730.

Ginsburg, L. and J. Morales. 1998. Les Hemicyoninae (Ursidae, Carnivora, Mammalia) et les formes apparentées du Miocènes inférieur et moyen d'Europe occidentale. *Annales de Paléontologie* 84(1):71–123.

Gu, Y.-m. 1989. *Platodontopithecus jianghuainesis*. In *Early Humankind in China*, ed. R. K. Wu, X. Z. Wu, and S .S. Zhang, pp. 267–269. Beijing: Science Press,.

Gu, Y.-m. and Y.-p. Lin. 1983. First discovery of *Dryopithecus* in East China. *Acta Anthropologica Sinica* 2:305–314.

Hopkins, S. S. B. 2004. Phylogeny and biogeography of the genus *Ansomys* Qiu, 1987 (Mammalia: Rodentia: Aplodontidae) and description of a new species from the Barstovian (mid-Miocene) of Montana. *Journal of Paleontology* 78:731–740.

Hou, L.-h. 1984. The Aragonian vertebrate fauna of Xiacaowan, Jiangsu – 2. Aegypinae (Falconiformes, Aves). *Vertebrata PalAsiatica* 22(1):14–19.

Hou, L.-h. 1987. The Aragonian vertebrate fauna of Xiacaowan, Jiangsu – 6. Aves. *Vertebrata PalAsiatica* 25(1):57–68.

Hou, L.-h., Z.-h. Zhou, F.-c. Zhang, and J.-d. Li. 2000. A new vulture from the Miocene of Shandong, eastern China. *Vertebrata PalAsiatica* 38(2):104–110.

Hu, H.-h. and R. W. Chaney. 1940. A Miocene flora from Shantung Province, China. *Carnegie Institution of Washington Publication* 507:1–147.

Jin, L.-y. 1985. K-Ar ages of Cenozoic volcanic rocks in the middle segment of the Tancheng-Lujiang fault zone and stages of related volcanic activity. *Geological Review* 31(4):309–315.

Juan, V. C. 1937. Diatomaceous earth in Shanwang, Linchu Shantung. *Bulletin of the Geological Society of China* 17(2):183–192.

Kelly, T. S. and W. W. Korth. 2005. A new species of *Ansomys* from the Late Hemingfordian (Early Miocene) of northwestern Nevada. *Paludicola* 5:85–91.

Korth, W. W. 2007. A new species of *Ansomys* (Rodentia, Aplodontidae) from the Late Oligocene (latest Whitneyan–earliest Arikareean) of South Dakota. *Journal of Vertebrate Paleontology* 27(3):740–743.

Li, C.-k. 1974. A probable geomyid rodent from Middle Miocene of Linchu, Shantung. *Vertebrata PalAsiatica* 12(1):43–53.

Li, C.-k., Y.-p. Lin, Y.-m. Gu, L.-h. Hou, W.-y. Wu, and Z.-d. Qiu. 1983. The Aragonian vertebrate fauna of Xiacaowan, Jiangsu. *Vertebrata PalAsiatica* 21(4):313–327.

Li, C.-k., W.-y. Wu, and Z.-d. Qiu. 1984. Chinese Neogene: Subdivision and correlation. *Vertebrata PalAsiatica* 22(3):163–178.

Li, J.-l. and B.-z. Wang. 1987. A new species of *Alligator* from Shanwang, Shandong. *Vertebrata PalAsiatica* 25(3):199–207.

Lin, S.-l. 1980. Discussion on the division and its age of Miocene series of the Lower Huai River. *Journal of Stratigraphy* 4(2):136–144.

Lindsay, E. H. 1988. Cricetid rodents from Siwalik deposits near Chinji Village. Part I: Megacricetodontinae, Myocricetodontinae and Dendromurinae. *Palaeovertebrata* 18(2):95–154.

Liu, L.-p., M. Fortelius, and M. Pickford. 2002. New fossil Suidae from Shanwang, Shandong, China. *Journal of Vertebrate Paleontology* 22(1):152–163.

Lopatin, A. V. 1997. The first find of *Ansomys* (Aplodontidae, Rodentia, Mammalia) in the Miocene of Kazakhstan. *Palaeontological Journal* 31:667–670.

Matsumoto, H. 1921. Descriptions of some new fossil mammals of Kani District, Province of Mino. *Science Reports Tohoko Imperial University*, 2nd ser. 5(3):76–91.

Matsumoto, H. 1926. On a new fossil race of bighorn sheep from Shantung, China. *Science Reports of the Tohoku imperial University*, 2nd ser. (Geology) 10:39–41.

McKenna, M. C. and S. K. Bell. 1997. *Classification of Mammals: Above the Species Level*. New York: Columbia University Press.

Meng, J., J. Ye, W.-y. Wu, L.-p. Yue, and X.-j. Ni. 2006. A recommended boundary stratotype section for Xiejian Stage from northern Junggar Basin: Implications to related bio-chronostratigraphy and environmental changes. *Vertebrata PalAsiatica* 44:205–236.

National Stratigraphic Committee. 2001. *A Guide to Chinese Stratigraphy and an Explanation to the Guide*. Rev. ed. Beijing: Geological Press.

Pei, W.-z., M.-c. Zhou, and J.-j. Zheng. 1963. *Cenozoic of China*. Beijing: Science Press.

Qiu, Z.-d. 1981. A new sciuroptere from Middle Miocene of Linqu, Shandong. *Vertebrata PalAsiatica* 19(3):228–238.

Qiu, Z.-d. 1987. The Aragonian vertebrate fauna of Xiacaowan, Jiangsu—7. Aplodontidae (Rodentia, Mammalia). *Vertebrata PalAsiatica* 25(4):283–296.

Qiu, Z.-d. 2001. Cricetid rodents from the middle Miocene Quantougou Fauna of Lanzhou, Gansu. *Vertebrata PalAsiatica* 39(3):204–214.

Qiu, Z.-d. 2010. Cricetid rodents from the Early Miocene Xiacaowan Formation, Sihong, Jiangsu. *Vertebrata PalAsiatica* 48(1):28–47.

Qiu, Z.-d., and C.-k. Li. 1981. Miocene mammalian fossils from Xining Basin, Qinghai. *Vertebrata PalAsiatica* 19(2):156–173.

Qiu, Z.-d., C.-k. Li, and S.-j Wang. 1981. Miocene mammalian fossils from Xining Basin, Qinghai. *Vertebrata PalAsiatica* 19(2):169–182.

Qiu, Z.-d. and Y.-p. Lin. 1986. The Aragonian vertebrate fauna of Xiacaowan, Jiangsu—5. Sciuridae (Rodentia, Mammalia). *Vertebrata PalAsiatica* 24(3):191–205.

Qiu, Z.-d. and X.-j. Ni. 2006. Small mammals. In *Lufengpithecus hudienensis Site*, ed. G.-q. Qi and W. Dong, pp. 113–130. Beijing: Science Press.

Qiu, Z.-d. and B. Sun. 1988. New fossil micromammals from Shanwang, Shandong. *Vertebrata PalAsiatica* 26(1):50–58.

Qiu, Z.-d. and C.-l. Yan. 2005. New sciurids from the Miocene Shanwang Formation, Linqu, Shandong. *Vertebrata PalAsiatica* 43(3):194–207.

Qiu, Z.-x. and Y.-m. Gu. 1986. The Aragonian vertebrate fauna of Xiacaowan, Jiangsu—3. Two carnivores: *Semigenetta* and *Pseudaelurus*. *Vertebrata PalAsiatica* 24(1):20–31.

Qiu, Z.-x. and Y.-m. Gu. 1991. The Aragonian vertebrate fauna of Xia-caowan, Jiangsu—8. *Dorcatherium* (Tragulidae, Artiodactyla). *Vertebrata PalAsiatica* 29(1):21–37.

Qiu, Z.-x. and Z.-d. Qiu. 1995. Chronological sequence and subdivision of Chinese Neogene mammalian faunas. *Palaeogeography Palaeoclimatology Palaeoecology* 116:41–70.

Qiu, Z.-x., W.-y. Wu, and Z.-d. Qiu. 1999. Miocene mammal faunal sequence of China—paleozoogeography and Eurasian relationships. In *The Miocene Land Mammals of Europe*, ed. G. E. Rossner and K. Heissig, pp. 443–455. Munich: Dr. Friedrich Pfeil.

Qiu, Z.-x., D.-f. Yan, H. Jia, and B. Sun. 1985. Preliminary observations on the newly found skeletons of *Palaeomeryx* from Shanwang, Shandong. *Vertebrata PalAsiatica* 23(2):173–195.

Qiu, Z.-x., D.-f. Yan, H. Jia, and B. Sun. 1986. The large-sized ursid fossils from Shanwang, Shandong. *Vertebrata PalAsiatica* 24(3):182–194.

Qiu, Z.-x., D.-f. Yan, H. Jia, and B.-z. Wang. 1985. Dentition of *Ursavus* skeleton from Shanwang, Shandong Province. *Vertebrata PalAsiatica* 23(4):264–275.

Qiu, Z.-x., D.-f. Yan, and B. Sun. 1991. A new genus of Tapiridae from Shanwang, Shandong. *Vertebrata PalAsiatica* 29(2):119–135.

Storch, G. and Z.-d. Qiu. 2002. First Neogene marsupial from China. *Journal of Vertebrate Paleontology* 22(1):179–181.

Storch, G. and Z.-d. Qiu. 2004. First complete heterosoricine shrew: A new genus and species from the Miocene of China. *Acta Palaeontologica Polonica* 49(3):346–358.

Sun, A. L. 1961. Note on fossil snakes from Shanwang, Shantung. *Vertebrata PalAsiatica* 5(4):306–312.

Tao, J.-r., B. Sun, and H. Yang. 1999. The flora of macro plants in Shanwang. In *Fossil Plants of Shanwang*, ed. B. Sun, pp. 13–89. Jinan: Shandong Science and Technology Publishing House.

Teilhard de Chardin, P. 1939. The Miocene cervids from Shantung. *Bulletin of the Geological Society of China* 19(2):269–278.

Tong, Y.-h., S.-h. Zheng, and Z.-d. Qiu. 1995. Cenozoic mammal ages of China. *Vertebrata PalAsiatica* 33(4):290–314.

Tomida, Y. and T. Setoguchi. 1994. Tertiary rodents from Japan. In *Rodent and Lagomorph Families of Asian Origins and Diversification*, ed. Y. Tomida, C.-k. Li, and T. Setoguchi, pp. 185–195. National Science Museum Monograph 8. Tokyo: National Science Museum.

Viret, J. 1929. Les faunes de mammifères de L'Oligocène supérieur de la Limagne bourbonnaise. *Annales de l'Université de Lyon*, n.s. 1, 47:1–327.

Wang, H.-f., B.-q. Zhu, and Q.-f. Zhang. 1981. K-Ar dating of Cenozoic basalt in Linqu Region, Shandong. *Geochemistry* 4:321–328.

Wang, X.-m., Z.-d. Qiu, Q. Li, Y. Tomida, Y. Kimura, Z. J. Tseng, and H. J., Wang. 2009. A new Early to Late Miocene fossiliferous region in central Nei Mongol: Lithostratigraphy and biostratigraphy in Aoerban Strata. *Vertebrata PalAsiatica* 47(2):111–134.

Wu, W.-y. 1986. The Aragonian vertebrate fauna of Xiacaowan, Jiangsu—4. Gliridae (Rodentia, Mammalia). *Vertebrata PalAsiatica* 24(1):32–42.

Wu, W.-y. 1995. The Aragonian vertebrate fauna of Xiacaowan, Jiangsu—9. Ochotonidae (Lagomorpha, Mammalia). *Vertebrata PalAsiatica* 33(1):48–60.

Xie, W.-m. 1979. First discovery of the *Palaeotapirus* in China. *Vertebrata PalAsiatica* 17(2):146–148.

Xie, W.-m. 1982. New discovery on aceratherine rhinoceros from Shanwang in Linchu, Shandong. *Vertebrata PalAsiatica* 20(2):133–137.

Yan, D.-f. and K. Heissig. 1986. Revision and autopodial morphology of the Chinese-European rhinocerotid genus *Plesiaceratherium* Young, 1937. *Abhandlungen der Bayerische Staatssammlung für Paläontogie und historische geology* 14:81–110.

Yan, D.-f., Z.-d. Qiu, and Z.-y. Meng. 1983. Miocene stratigraphy and mammals of Shanwang, Shandong. *Vertebrata PalAsiatica* 21(3):210–222.

Yang (Young), J.-j. 1977. On some Salienia and Chiroptera from Shanwang, Linqu, Shandong. *Vertebrata PalAsiatica* 15(1):76–80.

Ye, J., W.-y. Wu, and J. Meng. 2001. The Age of Tertiary strata and mammal faunas in Ulungur River area of Xinjiang. *Journal of Stratigraphy* 25(3):193–287.

Yeh, H.-k. 1980. Fossil birds from Linqu, Shandong. *Vertebrata PalAsiatica* 18(2):116–125.

Yeh, H.-k. 1981. Third note on fossil bird from Linqu, Shandong. *Vertebrata PalAsiatica* 19(2):149–155.

Young, C. C. 1936. On the Cenozoic geology of Itu, Changlo and Kinchu districts (Shantung). *Bulletin of the Geological Society of China* 15(2):171–188.

Young, C. C. 1937. On a Miocene mammalian fauna from Shantung. *Bulletin of the Geological Society of China* 17(2):209–244.

Young, C. C. 1955. On a new *Trogontherium* from Hsiatsaohwan, Shihhunghsien and with notes on the mammalian remains from Chi-Tsu, Wuhohsien, Anhwei. *Acta Palaeontologica Sinica* 3(1):55–66.

Zhang J.-f. 1989. *Fossil insects from Shanwang, Shandong, China.* Jinan: Shandong Science and Technology Publishing House.

Zhou, J.-j. 1992. A new cobitid from the Middle Miocene of Shanwang, Shandong. *Vertebrata PalAsiatica* 30(1):71–76.

Zhu, M., G.-h. Hu, and D.-z. Zhao. 1985. Potassium-Argon dating of Neogene basalt in Shanwang area, Shandong Province. *Petrological Research* 5:47–59.

Chapter 5

Neogene Faunal Succession and Biochronology of Central Nei Mongol (Inner Mongolia)

ZHU-DING QIU, XIAOMING WANG, AND QIANG LI

The Nei Mongol Autonomous Region (or Chinese Inner Mongolia) contains some of the richest vertebrate fossil localities in the world. Its Neogene mammals are among the earliest in Asia described by pioneering explorers in the beginning of the twentieth century. Not surprisingly, vertebrate paleontology in Nei Mongol played some of the key roles in the establishment of a continental biochronological record in Asia. Such a role will continue to exert lasting influence in an Asiatic chronological framework in the foreseeable future.

In a tectonically stable region, terrestrial sediments in Nei Mongol are often scattered in small basins that produce extraordinarily rich fossil mammals. Most of the basins have been only minimally disturbed since their deposition, and sediments are exposed by erosional escarpments that are often patchy in distribution and frequently partially exposed. Such a geological setting is conducive to discoveries of an array of richly fossiliferous localities that are limited in distribution and stratigraphic thickness. Further compounded by a scarcity of volcanic intertonguing in the sediments, vertebrate paleontologists are forced to confront vast riches of fossils that often lack direct superpositional relationships.

The fossil localities are centered mostly in a triangular area enclosed by three major cities of the region: Xilinhot to the northeast, Erenhot (Erlian) to the northwest, and Zhangjiakou (Kalgan) to the south (figure 5.1). Investigations of fossil mammals in this area can be dated back to the early part of last century, when explorations were made by the Swedish geologist J. G. Andersson (1923) at

Ertemte, the French paleontologist P. Teilhard de Chardin (1926a, 1926b) at Gaotege, and the American geologists and paleontologists from the American Museum of National History at Tunggur (Andrews 1932). These pioneering works not only initiated biostratigraphic studies in this area but also rendered these places as important classic Neogene localities in China. Furthermore, the collections from Tunggur, representing the richest of the Chinese middle Miocene faunas, became the basis for the Tunggurian Land Mammal Age in East Asia. After a long cessation of fieldwork, investigation and excavations in this region resumed in the mid-1980s. Particularly in the last decade or so, quite a number of new localities with fossiliferous deposits of different ages were found, and abundant small mammal assemblages were recovered from these localities, thanks to the screenwashing technique, as well as some remains of larger- and medium-size mammals (Meng, Wang, and Bai 1996; Qiu 1996; Qiu and Wang 1999; Qiu and Storch 2000; Wang et al. 2009).

Preliminary studies have shown that fossil small mammals found in this area contain genera exceeding 80% of the mammal diversity known in the Neogene deposits of equivalent ages in northern China, and these localities produced a faunal sequence spanning six of the seven Neogene Chinese Land Mammal Stage/Ages (LMS/A), from Xiejian through Gaozhuangian. It is evident that central Nei Mongol is an important region in biostratigraphic and biochronologic study, because these Neogene assemblages, with their highly diverse

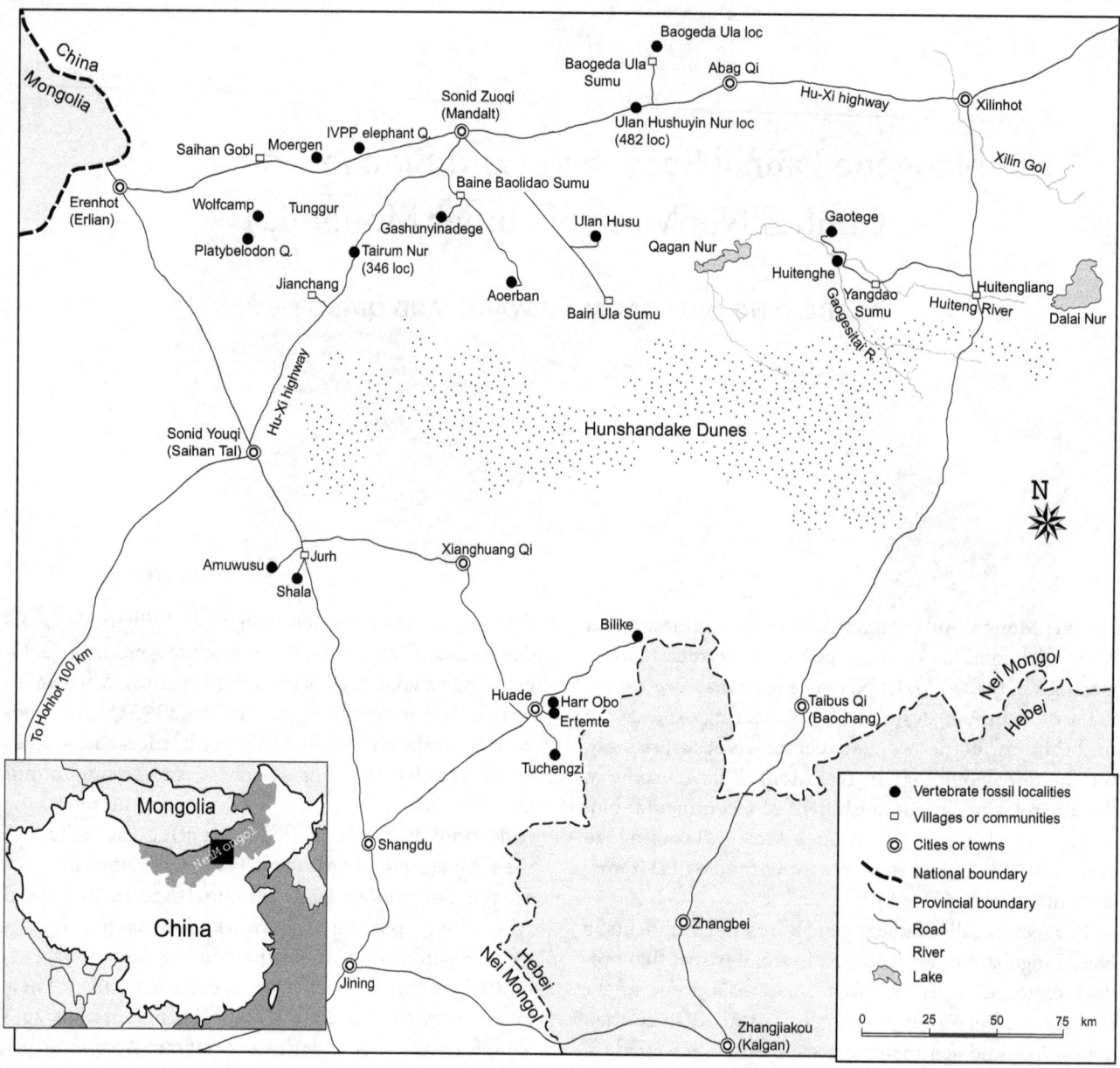

Figure 5.1 Map of central Nei Mongol and vertebrate fossil localities.

and abundant materials, some in demonstrable superpositional relationships, represent most of the Miocene and Pliocene Epochs. Furthermore, these assemblages are from a limited geographic area with little topographic differentiation, show stable community structure, and exhibit overall gradual changes in generic composition through geologic time. It is likely that they represent a single biogeographic region with a relatively stable ecosystem. Such a favorable circumstance is conducive to minimizing ecologic or zoogeographic complications in tracing faunal succession and testing biochronological

significance. Thus, in-depth studies of these assemblages will help not only to recognize the faunal succession in central Nei Mongol but also to understand the definition and characterization of Asian Neogene biochronology. One limitation is that the isolated localities, often scattered over long distances, usually have short fossil-bearing sections that do not allow easy independent age control. This poses a difficult problem in placing the biochronology in a time scale, and our understanding of the faunal successions must necessarily rely on studies of phylogenetic relationship and faunal change.

The main framework of Neogene mammal faunal succession in central Nei Mongol has been largely established by faunal seriation (Qiu and Wang 1999; Qiu, Wang, and Li 2006). The purpose of this work is to review this succession based on current studies, to update their biochronologic definition and characterization through the analysis of assemblages, and to integrate the faunal evidence with isotopic ages and magnetostratigraphy.

The term "Fauna" used in this chapter is restricted as an essentially contemporaneous assemblage derived from a limited geographic area, and almost corresponds to the "Local Fauna" used by some American scientists (Tedford et al. 1987). The "earliest and latest appearances" refer to the local lowest and highest occurrences of certain taxa or immigrants to this region. The updated faunal lists are presented as an appendix in this chapter, and for faunal descriptions the reader is referred to work by relevant authors.

English spellings of Mongolian geographic names follow those in *Atlas of the People's Republic of China* (first English edition, 1989), jointly published by the Foreign Languages Press and the China Cartographic Publishing House. Such standardization necessitates a few changes in spellings from usages in previous publications.

FAUNAL SUCCESSION

Aoerban Area

Aoerban is located about 60 km southeast of Sonid Zuoqi (see figure 5.1). Fossiliferous Neogene sediments containing Early Miocene mammalian fossils are exposed in a 3- × 2-km area of this region. The mammal faunal succession and the stratigraphic sequences in this locality were initially discussed by Wang et al. (2009). The Aoerban Formation is divided into three members: the Lower Red Mudstone Member, the Middle Green Mudstone Member, and the Upper Red Mudstone Member. The richly fossiliferous deposits in the lower member contain the Lower Aoerban Fauna, consisting of 44 taxa of mammals belonging to families of Oligocene occurrence but with more than half the genera known in the Miocene of Europe or Asia—for example, *Proscapanus, Atlantoxerus, Keramidomys, Prodryomys,* and *Democricetodon.* Most genera in this fauna are commonly known in older strata of Late Oligocene or Early Miocene age (e.g., *Amphechinus, Metexallerix, Prodistylomys/Distylomys, Tachyoryctoides, Ligerimys, Plesiosminthus, Sinolagomys, Palaeogale*), but elements such as *Protalactaga, Plesiodipus, Megacricetodon, Alloptox,* and *Bellatona,* typical of nearby middle

Miocene localities, are absent. The fauna shares a majority of genera of micromammals with assemblages of Early Miocene age in Asia, such as Xiejia of Qinghai, Zhangjiaping of Gansu, Suosuoquan of Xinjiang, and Sihong of Jiangsu in China, and the D horizon of the Valley of Lakes in Mongolia (Li and Qiu 1980; Qiu and Qiu 1995; Qiu et al. 1997; Meng et al. 2006; Daxner-Höck and Badamgarav 2007; Qiu and Qiu, chapter 4, this volume). However, it is different from the Xiejia Fauna, the Zhangjiaping Fauna, and the Suosuoquan Mammal Assemblage Zone I (S-I zone) in its absence of the relatively archaic genera *Parasminthus* and *Yindirtemys* and presence of a more "modern" cricetid, *Democricetodon.* The fauna is also distinguishable from the Sihong Fauna by its less diverse and specialized modern cricetids, as well as its absence of *Alloptox* and Proboscidea. It is essentially comparable to the Suosuoquan Mammal Assemblages Zone II and Zone III (S-II and S-III zones) and the assemblage from the D horizon of the Valley of Lakes in sharing most of the taxa at the generic level. The Lower Aoerban Fauna has at least 14 genera in common with European Orleanian faunas. It is worth mentioning that *Democricetodon* sp. from Aoerban is comparable in size and morphology to some primitive species like *D. franconicus* of Europe. Its eomyids *Pentabuneomys* and *Ligerimys* are restricted to MN 3–4 in Europe (Engesser 1999). Thus, the age of the Lower Aoerban Fauna is likely Early Miocene in the Xiejian Chinese Land Mammal stage/age, or equivalent to MN 2–3 (European mammal unit).

Conformably overlying the Middle Green Mudstone Member are red mudstones producing the Upper Aoerban Fauna. The fauna, composed of 25 species, is less diverse than, but it shares almost all families with the Lower Aoerban Fauna. Distinct differences of the fauna at the family level from the Lower Aoerban Fauna are the complete disappearance of Ctenodactylidae, the noticeable decline of Aplodontidae and Zapodidae, and the occurrence of Proboscidea. On the generic level, it is characterized by the lack of a large number of genera present in the Lower Aoerban Fauna, such as *Amphechinus, Tachyoryctoides, Asianeomys, Litodonomys,* and *Sinolagomys,* and the appearance of genera absent in the Lower Aoerban Fauna,—for example, *Megacricetodon, Cricetodon, Alloptox,* and *Ligeromeryx/Lagomeryx.* The Upper Aoerban Fauna shares more genera with the Sihong Fauna (e.g., *Ansomys, Eutamias, Democricetodon, Megacricetodon, Cricetodon, Alloptox,* and *Ligeromeryx/Lagomeryx*) than with the Lower Aoerban Fauna, suggesting closer age to the former. The presence of a proboscidean in this fauna currently represents the earliest appearance of the group in central Nei Mongol, although this record is probably later than those from Tabenbuluk (Wang and Qiu 2002; Wang, Qiu, and Opdyke 2003). The age of the Upper Aoerban Fauna probably belongs to late Early Miocene—that is, late Shanwangian LMA/S, or roughly equivalent to MN 4. The age estimates also appear to be borne out by a very primitive *Ligeromeryx/Lagomeryx* and the latest appearance of *Palaeogale* in China and the lack of typical Middle Miocene elements in the Upper Aoerban Fauna, such as *Plesiodipus* and *Bellatona*, which are commonly seen in Tunggurian assemblages.

Disconformably or unconformably overlying the Upper Red Mudstone are orange-red sandstones and siltstones of the Balunhalagen bed. The coarser-grained clastics of fluviatile deposits at the base of the bed, often within conglomeratic lenses, are packed with remains of mammals representing the Balunhalagen Fauna. The fauna, consisting of 35 taxa of almost exclusively small mammals, shows further decrease of the Oligocene families and some archaic genera. It lacks not only Ctenodactylidae, *Amphechinus*, *Asianeomys*, *Ligerimys*, and *Litodonomys*, which occurred in the Lower Aoerban Fauna, but also Tachyoryctoididae and *Plesiosminthus* known from the Upper Aoerban Fauna. The Balunhalagen Fauna is characterized by the dominance of genera commonly known in Tunggurian faunas of the Middle Miocene, such as *Keramidomys*, *Leptodontomys*, *Miodyromys*, *Microdyromys*, *Heterosminthus*, *Protalactaga*, *Democricetodon*, *Megacricetodon*, *Plesiodipus*, *Gobicricetodon*, *Alloptox*, and *Bellatona*, and by the occurrence of *Prosiphneus*, *Brachyscirtetes*, and *Ochotona?*, which are absent from Tunggurian faunas but abundant in faunas of Late Miocene age (see the following discussion). The small mammal fauna from Balunhalagen is generally comparable to assemblages from the lower part of the Dingshanyanchi Formation in the Junggar Basin of the Xinjiang Autonomous Region or to the A4 small mammal assemblage in the Qin'an section of Gansu Province (Meng et al. 2008; Guo et al. 2002). However, the Balunhalagen Fauna is probably younger than these latter assemblages, which resemble the Moergen Fauna or Tamuqin Fauna of Tunggurian age (see following discussion). Additionally, the Balunhalagen Fauna has in common nine genera with European faunas of Astaracian age, the Steinheim Fauna of Germany and the Anwil Fauna of Switzerland, indicating a comparable age to European MN 7–8. The presence of the Late Miocene elements in the Balunhalagen Fauna, however, seems to indicate the fauna has a later age, probably late Middle Miocene to early Late Miocene—that is, straddling Late Tunggurian/earliest Bahean of LMS/A.

On the top rim of the Aoerban profile is the Bilutu bed in disconformable contact with the underlying Balun-

halagen bed. The channel fillings at the base of the Bilutu bed yield rich remains of mammals. The Bilutu assemblage, composed of 53 taxa of small mammals, contains a mixture of taxa of varying age relationships due to the reworking of sediments from below. Among the 43 genera, except certain long-lived genera such as *Mioechinus*, *Yanshuella*, *Quyania*, *Ansomys*, *Eutamias*, and *Sicista*, the assemblage includes elements typical or common in the Middle Miocene, such as *Heterosminthus*, *Protalactaga*, *Democricetodon*, *Megacricetodon*, *Cricetodon*, *Plesiodipus*, and *Gobicricetodon*, and members that typically appear in the Late Miocene, *Erinaceus*, *Prospermophilus*, *Myomimus*, *Lophocricetus*, *Paralactaga*, *Dipus*, *Kowalskia*, *Sinocricetus*, *Nannocricetus*, *Microtoscoptes*, *Microtodon*, *Anatolomys*, *Pseudomeriones*, *Hansdebruijnia*, *Micromys*, and *Ochotona*, for example. If the principle of biasing correlation toward the more advanced elements in the fauna is followed, then equivalence with assemblages from Baogeda Ula or Ertemte of central Nei Mongol is indicated. Judging from the composition and evolutionary stage of the muroids, the Bilutu Fauna appears to be bounded in time by the Baogeda Ula and Ertemte faunas (see following discussion), probably slightly earlier than the Ertemte Fauna and belonging in the middle to late Late Miocene—that is, Baodean LMS/A or roughly equivalent to the MN 12–13 units of Europe (Wang et al. 2009).

Gashunyinadege Locality

This locality is at the Gashunyinadege valley about 35 km southwest of Sonid Zuoqi (see figure 5.1). The sedimentary basin fill lies in angular unconformity over a granite basement and consists of residual reddish and yellowish sandy clays exposed along several gullies in a 1- × 2.5-km area of the valley, probably contemporary with the Aoerban Formation. Scattered fossil remains in the middle part of the section yield the Gashunyinadege Fauna, most of which are of small mammals. Biostratigraphy of the site was first introduced by Meng, Wang, and Bai (1996), and subsequent work has added important data (Qiu, Wang, and Li 2006).

The assemblage contains 47 taxa of mammals and is dominated by the families Erinaceidae, Ctenodactylidae, Aplodontidae, and Ochotonidae. Some Oligocene or Early Miocene genera, such as *Amphechinus*, *Prodistylomys/Distylomys*, *Tachyoryctoides*, *Microdyromys*, *Plesiosminthus*, *Desmatolagus*, and *Sinolagomys*, are rather common in the fauna, and typical Middle Miocene genera, such as *Plesiodipus*, *Protalactaga*, and *Bellatona*, are lacking. The presence of numerous "relict" forms from the Oligocene and the absence of genera typical of the Middle Miocene seem to suggest an Early Miocene age. The Gashunyinadege Fauna is evidently younger than the Early Miocene Xiejia Fauna and the Suosuoquan I Assemblage in its lack of *Parasminthus* and *Yindirtemys*, and the presence of "modern" cricetids. It mostly resembles the Lower Aoerban Fauna in sharing the same families and most of the same genera, but it differs from the latter in the abundant cricetids and the presence of *Leptodontomys*, *Megacricetodon*, and *Alloptox*, indicative of its more advanced status. The Gashunyinadege Fauna is also similar to the Upper Aoerban Fauna in sharing *Yanshuella*, *Aralomys*, *Ansomys*, *Keramidomys*, *Miodyromys*, *Plesiosminthus*, *Democricetodon*, *Megacricetodon*, *Alloptox*, and the like, and in the dominance of diverse Eomyidae, Gliridae, and Zapodidae. However, it suggests an earlier age than the Upper Aoerban Fauna in retaining some families and genera known from earlier faunas (e.g., Ctenodactylidae, *Tachyoryctoides*, *Asianeomys*, *Litodonomys*, and *Sinolagomys*, which are absent from the Upper Aoerban Fauna). The Gashunyinadege Fauna seems to fill the gap between the assemblages from the Lower Red Mudstone Member and the Upper Red Mudstone Member of the Aoerban Formation. It has at least eight genera (*Amphechinus*, *Metexallerix*, *Distylomys*, *Heterosminthus*, *Litodonomys*, *Democricetodon*, *Atlantoxerus*, and *Sinolagomys*) in common with the Suosuoquan III Assemblage (Meng et al. 2006) and can be roughly compared to the assemblage from the upper part of Suosuoquan Formation in the Ulungu area of the Junggar Basin, Xinjiang, in the joint occurrence of *Atlantoxerus*, *Megacricetodon*, and *Alloptox*, as well as in lacking *Platybelodon* and *Stephanocemas* commonly known in the middle Miocene (Ye, Wu, and Meng 2001).

It has fewer diverse cricetids than does Sihong, but it shares *Tachyoryctoides*, *Ansomys*, *Eutamias*, *Microdyromys*, *Democricetodon*, *Megacricetodon*, *Alloptox*, and a large *Amphicyon* with the Shanwangian Fauna (Qiu and Qiu 1995). In all, the age of the Gashunyinadege Fauna probably belongs to the late Early Miocene—that is, Shanwangian LMS/A, or roughly equivalent to MN3 of European land mammal zonations.

Tunggur Tableland

The Tunggur Formation is exposed along the edge of the Tunggur Tableland in the northwest part of this region. The formation, composed dominantly of fluvial deposits in the northern exposures, but less so in the southern ex-

posures, has a maximum thickness of less than 80 m in individual sections and can be divided into two sedimentary units. The upper unit is light colored with sandstones and mudstones, while the lower unit is mainly red or lavender mudstones with a channel sandstone in the middle. Exposures of both units produce mammalian remains. The pioneering exploration made by the Central Asiatic Expedition from the American Museum of Natural History in the 1920s and 1930s resulted in the discovery of the classic "Tunggur Fauna," or the *Platybelodon* fauna (Andrews 1932; Osborn 1929; Osborn and Granger 1931, 1932; Colbert 1934, 1936a, 1936b, 1939a, 1939b, 1940; Wood 1936; Dawson 1961). Since then, significant works have added important data to the fauna (Li 1963; Zhai 1964; Qiu et al. 1988; Qiu 1996; Qiu and Wang 1999; Wang, Qiu, and Opdyke 2003).

The Tunggur Formation produced up to 77 taxa of mammals, representing the most diverse and abundant middle Miocene fauna in China (Wang, Qiu, and Opdyke 2003). The small mammal fauna is characterized by the latest occurrence of Ctenodactylidae and Tachyoryctoididae, the decline of Eomyidae and Zapodidae, the diversity and abundance of Cricetidae, Gomphotheriidae, and Artiodactyla, and the appearance of Dipodidae. At the generic level, the absence of some commonly known Early Miocene taxa, such as *Amphechinus*, *Asianeomys*, *Plesiosminthus*, and *Sinolagomys*, and the presence of newcomers, such as *Anchitheriomys*, *Plesiodipus*, *Gobicricetodon*, *Bellatona*, and *Platybelodon*, are distinctive. In a broad sense, the "*Platybelodon* fauna" is known from more than 10 sites of the Tunggur Formation exposed along the northern, western, and southern edges of the platform. The genera *Platybelodon*, *Mioechinus*, *Heterosminthus*, *Plesiodipus*, *Gobicricetodon*, *Megacricetodon*, *Alloptox*, and *Bellatona* are discovered throughout the formation and typify the Tunggurian age. Nevertheless, the assemblages in the Tunggur Formation appear to represent a chronofauna of still unknown duration. However, biostratigraphic work (mainly that of small mammals) in the last 20 years seems to have allowed us to recognize the following three superposed faunal units in the Tunggur Formation (Qiu, Wang, and Li 2006).

Represented by the assemblages from the Aletexire locality and the lower part of Tairum Nur or Roadmark 346 sections (the lower unit of the Tunggur Formation), the lower faunal unit is the Tairum Nur Fauna with small mammals *Tachyoryctoides* and *Distylomys*, and immigrants *Leptarctus* and *Sthenictis*, larger mammals not seen elsewhere in the formation. In addition, *Atlantoxerus*, *Heterosminthus*, *Gobicricetodon*, *Tungurictis*, and *Stephanocemas* in the fauna demonstrate primitive morphologies

(Wang, Qiu, and Opdyke 2003). Both *Tachyoryctoides* and *Distylomys* occur frequently in Oligocene and early Miocene faunas of North China, but are not known in the later assemblages from the tableland. Although they are present in the Gashunyinadege Fauna, the absence of *Gobicricetodon*, *Plesiodipus*, *Bellatona*, and *Platybelodon* in that fauna suggests its older age. An equivalent of the Tairum Nur Fauna occurs in the Dingjiaergou fauna in Tongxin, Ningxia Hui Autonomous Region; these two faunas share *Tachyoryctoides*, *Atlantoxerus*, *Heterosminthus*, *Megacricetodon*, *Alloptox*, *Platybelodon*, *Percrocuta*, *Sansanosmilus*, and *Stephanocemas* (Qiu and Qiu 1995). The Tairum Nur Fauna also shows close affinities with the Halamagai Fauna from the Halamagai Formation of Xinjiang in having co-occurrence of genera such as *Tachyoryctoides*, *Alloptox*, *Tungurictis* (*Protictitherium*), and *Platybelodon* (Wang et al. 1998; Ye, Wu, and Meng 2001).

The middle fauna is the Moergen Fauna, containing the assemblages from the lower part of the upper unit of the Tunggur Formation, mainly including collections from classic localities, such as Wolf Camp, the *Platybelodon* Quarry, and also localities mostly excavated in the last 20 years, such as Moergen (MOII), Huerguolajin, Aoershun Chaba, and the upper red beds of Tairum Nur section at Roadmark 346. The fauna is composed of at least 69 taxa of mammals, forming the basis of the faunal characterization of the middle Miocene and typifying the Tunggurian LMS/A. Among the 51 recognized genera, many are commonly known in the middle Miocene of the Holarctic Region, such as *Desmanella*, *Anchitheriomys*, "*Monosaulax*," *Keramidomys*, *Leptodontomys*, *Microdyromys*, *Miodyromys*, *Heterosminthus*, *Protalactaga*, *Plesiodipus*, *Gobicricetodon*, *Democricetodon*, *Megacricetodon*, *Alloptox*, *Bellatona*, *Platybelodon*, *Anchitherium*, *Listriodon*, *Stephanocemas*, *Lagomeryx*, *Micromeryx*, *Dicrocerus*, and *Turcocerus*. Of these, *Heterosminthus*, *Plesiodipus*, *Democricetodon*, *Megacricetodon*, *Alloptox*, *Bellatona*, *Platybelodon*, *Anchitherium*, *Lagomeryx*, *Stephanocemas*, *Micromeryx*, *Dicrocerus*, and *Turcocerus* dominate the fauna. The Moergen Fauna includes *Heterosminthus orientalis*, *Protalactaga major*, *P. grabaui*, *Democricetodon lindsayi*, *D. tongi*, *Megacricetodon sinensis*, *Plesiodipus leei*, *Miodyromys* sp., *Keramidomys fahlbuschi*, and *Alloptox gobiensis*, indicating correlation with the Quantougou Fauna in Gansu and the assemblage from the basal beds of the Dingshanyanchi Formation in Xinjiang (Qiu 2000, 2001; Meng et al. 2008).

Thirty-seven genera in this fauna, amounting to nearly two-thirds of the total, occur in European Miocene faunas, and 14 genera show affinities with North Africa, but only 6 with southern Asia (the Siwalik Chinji assem-

blages). About one-third of the total are congeneric with North American relatives, of which 9 genera are also known from Europe. The magnetostratigraphy of the fossiliferous strata at Moergen was correlated to the chrons C5An.1n through part of C5Ar.3r (Wang, Qiu, and Opdyke 2003), with a revised age range of 12.0–13.0 Ma (Lourens et al. 2004).

The upper fauna, represented by the assemblage from the top of the Moergen section (MoV of Qiu 1996), is the Tamuqin Fauna with *Steneofiber* and some derived species of the genera known from the Moergen Fauna, such as *Gobicricetodon robustus* and *Plesiodipus progressus*. Magnetic stratigraphy shows that the two horizons, producing the Moergen and the Tamuqin faunas, respectively, are approximately 0.5 m.y. apart (Wang, Qiu, and Opdyke 2003).

Thus, the three faunas from the Tunggur Formation belong to the Middle Miocene in age, as does the Tunggurian LMS/A. Faunal correlation suggests that the Moergen and Tamuqin faunas might be equivalent to MN 7+8 (Qiu 1996). The older Tairum Nur Fauna may correlate to MN 6.

Baogeda Ula Area

Two formations, the Tunggur Formation and the Baogeda Ula (Baogedawula; Mongolian for "Sacred Mountain") Formation are exposed about 25 km west of the town of Abag Qi in the north of this region. Exposures of red mudstones of possibly the easternmost extension of the Tunggur Formation at Ulan Hushuyin Nur locality (Road Mark 482 of Hohhot-Xilinhot Highway) yielded nine taxa of small mammals. The assignment to the Tunggurian Age, close to the Moergen Fauna, is supported by the presence of *Atlantoxerus*, *Heterosminthus*, *Protalactaga*, *Democricetodon*, *Megacricetodon*, *Alloptox*, and *Bellatona*. A grayish gravel bed unconformably on top of the red mudstones contains a *Hipparion* fauna and would represent the base of another stratigraphic unit.

The type section of the Baogeda Ula Formation near the village of Baogeda Ula Sumu is capped by more than one layer of basalts. The capping basalts have yielded dates ranging from 14.57 Ma to 3.85 Ma (Luo and Chen 1990). Sample locations of these dates, however, are not specific enough to be sure of which basalt corresponds to those that cap the vertebrate fossil localities, although we (Qiu, Wang, and Li 2006:fig. 2) singled out a date (7.11 ± 0.48 Ma) for sample B48 of Luo and Chen (1990:table 1) to be a possible candidate. The formation, composed of fluviolacustrine deposits with intercalated basalts and a maximum thickness of about 70 m, is mainly exposed along the southern and the western edges of Qagan Ula hill (Chagan Ula Yixile, Mongolian for white hill lawn).

Exposures along the western edges of the tableland produce the Baogeda Ula Fauna composed of mammalian remains from different fossiliferous horizons and sites. Remains of *Hipparion* were first reported from the grayish-white sandy mudstones in the middle part of the section by a geological team from the Bureau of Geology and Mineral Resources of Nei Mongol Autonomous Region (1991). More materials of various fossil mammals, especially those of small mammals, have been added to the fauna in the last decade (Qiu and Wang 1999; Qiu, Wang, and Li 2006). Except for the hyaenids studied by Tseng and Wang (2007) and murids by Storch and Ni (2002), no other specimens have been described in detail.

The Baogeda Ula assemblage, composed of 26 mammals, shares *Lophocricetus, Paralactaga, Dipus, Kowalskia, Sinocricetus, Microtoscoptes, Prosiphneus,* and *Ochotona* with the Shala Fauna (see the following discussion). However, tentative identifications of the taxa demonstrate that the Baogeda Ula Fauna is younger than the Shala Fauna in having some newcomers, Muridae, Leporidae, and the genera *Parasoriculus, Dipoides, Pararhizomys,* and *Abudhabia,* and in lacking some archaic elements—for example, *Ansomys* (Aplodontidae), *Miodyromys* and *Microdyromys* (Gliridae), and *Pentabuneomys* (Eomyidae). In addition, *Paralactaga, Microtoscoptes,* and *Ochotona* in the fauna show more advanced stages than those from Shala (see the following section). It is thus likely that the Baogeda Ula Fauna is younger than the Shala Fauna. About half of the small mammals in the fauna are congeneric with the Bilutu Fauna, suggesting a partial correlation of the Baogeda Ula Formation to the Bilutu beds of Aoerban. Both radiometric and faunal evidence seem to restrict the Baogeda Ula Fauna to the Baodean (late Miocene), and a correlation to middle Turolian (MN12) can be suggested by the presence of the primitive *Hansdebruijnia* and the earliest occurrence of *Microtoscoptes* (Mein 1999; Storch and Ni 2002). A faunal succession contained within the thick deposits of the Baogeda Ula bed deserves further attention to sort out its relatively scattered fossil sites known to produce both large and small mammals and their relationship with interfingering datable basalts.

Jurh Area

To the west of Jurh (Zhurihe) are situated two sites, Amuwusu and Shala, where fossiliferous Neogene sedi-

ments are developed with a thickness of less than 10 m each. The fluvial sandstones and overbank mudstones in these localities preserve fossil mammal remains, and taxa from the Amuwusu and Shala faunas are both Late Miocene in age. Most taxa of the faunas are undescribed, but an updated preliminary faunal list of the assemblages has been given by Qiu, Wang, and Li (2006).

The Amuwusu Fauna, composed of 34 taxa of mammals, includes 60% of the identifiable genera known in the Tunggur Fauna. However, it lacks some genera dominant in the Tunggur Fauna, such as *Plesiodipus, Megacricetodon, Alloptox, Bellatona,* and *Platybelodon,* and records newcomers that frequently occur in late Miocene and Pliocene faunas—for example, *Castor, Paralactaga, Prosiphneus, Sinozapus,* and *Ictitherium.* This assemblage shows primitive elements and may represent an early appearance of these groups in central Nei Mongol. The fauna is more or less comparable with assemblages from Balunhalagen in Aoerban area (see previous section) and the lower part of Bahe, Shaanxi, especially in the joint occurrence of *Eutamias, Protalactaga,* and *Prosiphneus.* The Amuwusu Fauna is, however, older than the Bahe Fauna because the latter contains *Lophocricetus, Nannocricetus, Progonomys,* and *Hipparion* that are typical genera of the Late Miocene, but it is probably younger than the Balunhalagen Fauna because it lacks more genera that are commonly known in the Middle Miocene—for example, *Megacricetodon, Plesiodipus,* and *Alloptox.* The Amuwusu Fauna suggests an early Late Miocene age, that is, Early Bahean LMA/S, or equivalent to MN 9.

The Shala Fauna contains 33 taxa of mammals. Among the 27 recognizable genera, 11 are shared either with faunas in the Tunggur Formation or the Amuwusu Fauna, and 20 with latest Miocene or Pliocene faunas, such as the Ertemte Fauna and the Bilike Fauna (see the following section). The Shala Fauna is characterized by the disappearance of many elements of Early and Middle Miocene origin that survived in the Amuwusu Fauna, such as "*Monosaulax,*" *Heterosminthus, Protalactaga, Democricetodon,* and *Gobicricetodon,* by the occurrence of quite a number of newcomers that are commonly known in late Miocene and Pliocene faunas, such as *Prospermophilus, Dipus,* and *Kowalskia,* and by the dominance of *Lophocricetus, Nannocricetus, Sinocricetus, Prosiphneus,* and *Ochotona.* It shares three genera, *Paralactaga, Lophocricetus,* and *Nannocricetus,* with the Bahe Fauna, Shaanxi, and the Shengou Fauna, Qinghai, but seems to be slightly younger than the latter two faunas in having some advanced genera, such as *Dipus, Kowalskia, Sinocricetus,* and *Microtoscoptes,* and in lacking some archaic taxa, such as *Protalactaga* and *Myocricetodon.* It is evidently older than the late Late Miocene faunas (such as the Baogeda Ula and Ertemte faunas; see following section) for its absence of the family Muridae. Therefore, the age of the Shala Fauna is considered to be middle Late Miocene—that is, Late Bahean LMA/S, or roughly equivalent to the European MN 11.

Huade Area

The richly fossiliferous Neogene sediments in the Huade area in the southern part of this region have been known since the early part of last century (Andersson 1923). Investigations of the scattered outcrops of the fluviolacustrine deposits in the 1980s have resulted in the recognition of the following four assemblages of different ages.

The earliest assemblage of this area was collected by the Sino-Soviet Expedition in 1959 from a quarry at Tuchengzi, about 12 km southeast of the town of Huade (Chow and Rozhdestvensky 1960). Two small mammals, *Prosiphneus* and *Ochotona*, can be added to the fauna, in addition to the large mammals reported by Zhai (1962) and Qiu (1979)—for example, *Sinohippus, Hipparion, Chilotherium, Aceratherium, Moschus, Cervocerus, Dorcadoryx, Plesiaddax, Palaeotragus,* and *Samotherium.* This suite of taxa indicates correlation of the Tuchengzi Fauna with other Baodean *Hipparion* faunas of northern China. The *Prosiphneus* shows more primitive features than *P. eriksoni* from Ertemte, and the presence of *Sinohippus, Plesiaddax, Palaeotragus,* and *Samotherium* seems to suggest a relatively early Late Miocene age for the fauna, possibly early Baodean, or equivalent to MN 12 of Europe.

The Ertemte assemblage was discovered and first excavated by Andersson and his Chinese collectors in 1919–1920 from the Ertemte site located 4 km southeast of Huade town (Andersson 1923) and described by Schlosser (1924). Excavations and screenwashing in a thin layer of lacustrine deposits by paleontologists from China and Germany in 1980 have greatly enhanced the fauna both in diversity and abundance (Fahlbusch, Qiu, and Storch 1983). The materials collected have been mostly described in the last 20 years (Storch and Qiu 1983; Qiu 1985, 1987, 1991, 2003; Wu 1985, 1991; Fahlbusch 1987, 1992; Storch 1987, 1995; Fahlbusch and Moser 2004). The Ertemte Fauna, composed of at least 66 species of both small and large mammals, represents the most diverse and abundant Late Miocene fauna in China. The big assemblage of small mammals is characterized by the high diversity of insectivores and myomorph rodents, the maximum generic diversity of ground squirrels and lophate cricetids, the dominance of *Lopho-*

cricetus, Paralophocricetus, Microtodon, Prosiphneus, and *Ochotona,* and the common occurrence of *Yanshuella* and *Sorex* among insectivores; *Prospermophilus, Sicista, Lophocricetus, Sinocricetus, Nannocricetus, Kowalskia, Microtodon, Prosiphneus,* and *Micromys* among rodents; and the lagomorph *Ochotona.* The Ertemte Fauna is close to the Baogeda Ula Fauna in age, but it is definitely younger than the latter in its lack of *Rhinocerodon* and *Abudhabia* and in the presence of *Paenepetenyia, Pseudaplodon, Myomimus, Lophocricetus grabaui, Paralophocricetus, Microtodon, Apodemus, Orientalomys,* and *Micromys.* The fauna was found to be closely comparable with assemblages from the Mahui Formation and the lower part of Gaozhuang Formation of Yushe, Shanxi (Flynn, Tedford, and Qiu 1991). It also shares quite a number of genera with the Late Miocene–Early Pliocene assemblages from Lingtai, Gansu (Zhang and Zheng 2000; Zheng and Zhang 2001). Its lack of *Chardinomys, Huaxiamys, Allorattus, Mimomys,* and *Trischizolagus* and the presence of *Hansdebruijnia* (*Occitanomys*) and *Micromys chalceus* suggests a relationship to Biozone I of the Xiaoshigou section rather than Biozones II or III of the Wenwanggou section in Lingtai. The Ertemte Fauna has been regarded as late Baodean, latest Miocene, based on the studies of fossil small mammals, or equivalent to MN 13 (Fahlbusch 1987, 1992; Fahlbusch and Moser 2004; Qiu 1985, 1987, 1991, 2003; Storch 1987, 1995; Storch and Qiu 1983; Wu 1985, 1991). Such an age assignment is also supported by large mammals—for example, *Promephitis alexejewi,* an ictithere hyaenid, and *Hipparion mongolicum* (Qiu, Wang, and Li 2006). About half of the genera of mammals occur in European Turolian faunas.

The site yielding the Harr Obo Fauna is located 3 km north of Ertemte. A handful of remains of perissodactyls and artiodactyls were recovered from this locality (Andersson 1923; Schlosser 1924), and small mammals were collected by the Chinese-German joint project from a pile of matrix dug out from an irrigation well of 7 m depth (Fahlbusch, Qiu, and Storch 1983). The Harr Obo Fauna, composed of 46 taxa of mammals, has almost the same genera as the Ertemte Fauna in all the more frequently occurring elements of small mammals. Significant differences of the two assemblages are the presence of *Trischizolagus* and *Rhagapodemus* in Harr Obo, and slightly more derived features seen in some taxa, such as *Ochotona* and *Brachyscirtetes* (Storch 1987; Qiu 1987, 2003). Thus, the Harr Obo Fauna is probably slightly younger than the Ertemte Fauna and would be early Gaozhuangian (Early Pliocene).

The Bilike site, about 50 km northeast of Huade, produces the youngest assemblage in the southern region of this study, including a large quantity of remains of small mammals and some fragments of *Hipparion*, proboscideans, and cervids. The Bilike Fauna, composed of 49 species of micromammals, is taxonomically the most diverse Pliocene micromammalian fauna in China (Qiu and Storch 2000). It shows strong similarities to the Ertemte Fauna in its dominance of myomorph rodents and the high diversity of insectivores, but distinct compositional differences from the latter can be detected. Among insectivores, *Desmana*, *Petenyia*, *Lunanosorex*, *Parasoriculus*, and *Sulimskia* replaced *Zelceina*, *Alloblarinella*, *Paenepetenyia*, *Cokia*, and *Paranourosorex*; among rodents, archaic genera, such as *Pseudaplodon*, *Leptodontomys*, *Microtoscoptes*, and *Hansdebruijnia*, disappeared, and advanced ones, such as *Mimomys* (*Aratomys*), *Chardinomys*, *Huaxiamys*, and *Allorattus*, appeared; and among lagomorphs, *Alilepus* was completely replaced by *Trischizolagus*. In addition, some commonly known members from Ertemte—for example, *Prospermophilus*, *Lophocricetus*, *Paralophocricetus*, *Anatolomys*, and *Microtodon*—are quite rare or absent in Bilike. A striking faunal change from Ertemte to Bilike is the disappearance of two families, Aplodontidae and Eomyidae, and the appearance of the typically Pliocene Arvicolidae. Dominant genera in this assemblage are *Sicista*, *Mimomys* (*Aratomys*), *Prosiphneus*, and *Ochotona*. *Mimomys* is as abundant in Bilike as the genera *Lophocricetus* and *Microtodon* in Ertemte. The Bilike Fauna generally resembles Biozone II of the Xiaoshigou section and Biozone III of the Wenwanggou section at Leijiahe, Gansu (Zhang and Zheng 2000; Zheng and Zhang 2001), especially in their joint occurrence of *Chardinomys*, *Huaxiamys*, *Allorattus*, and *Trischizolagus* that are otherwise unknown in other assemblages of the Huade area. The fauna is closely comparable to assemblages from the Gaozhuang Formation in Yushe, but it is probably older than those from the upper part of the formation. The Bilike Fauna is considered to be Gaozhuangian age (Pliocene), equivalent to early Ruscinian, or MN 14 (Qiu and Storch 2000).

Gaotege Area

The Gaotege area is situated about 80 km southeast of Abag Qi (Banner) in the eastern part of the study region. Scattered Neogene sediments are exposed along the Gaogesitai River (also known as the Bayin River; see figure 5.1), the Huiteng River, and their drainage regions. Two fossiliferous localities, Huitenghe and Gaotege, preserve locally abundant fossil mammal remains. Investigations of this area were initiated in 1920s (Teilhard de

Chardin 1926a, 1926b), and important taxonomic additions were made from the two localities (Li, Wang, and Qiu 2003).

The Huitenghe locality is about 5 km southwest of Bayan Ula (a volcanic cone) at the junction where the Huiteng River (Chiton-gol of Teilhard de Chardin 1926a) joins the Gaogesitai River. Red beds, cut in places by a layer of basalts, contain remains of large and small mammals. Teilhard de Chardin (1926b) called this "basin du Chiton-gol" and briefly described a few large mammals (*Martes anderssoni*, a hyaenid, *Chilotherium* sp., *Hipparion* sp., the then-new, *Moschus primaevus*, two other artiodactyls, and a proboscidean). The assemblage is now composed of 20 small mammal taxa, dominated by *Ochotona* and *Lophocricetus*. It generally resembles the Shala Fauna, especially in sharing 11 of 19 genera with the latter and in its lack of murids. Although the Huitenghe Fauna includes some members that are present in the Amuwusu Fauna, it is younger than the latter because of the appearance of some advanced elements—for example, *Paranourosorex*, *Prospermophilus*, *Kowalskia*, and derived species of *Prosiphneus* and *Ochotona*. However, this assemblage suggests that the fossiliferous deposits possibly have a larger range in age than those of Shala, because it includes some primitive elements, such as *Gobicricetodon*, *Democricetodon*, and *Desmatolagus* that are otherwise frequently present in Middle Miocene faunas, and because of a lack of advanced dipodids but presence of derived *Lophocricetus*. Thus, the Huitenghe Fauna is tentatively assigned to late Bahean or early Baodean LMA/S (middle Late Miocene), or roughly equivalent to the MN 10–11.

The Gaotege exposure is an isolated hill located about 8 km north of Teilhard's "Chiton-gol" site. On the southern face of the hill, exposures of light-colored deposits yield five fossiliferous layers (layers 2, 3, 4, and the bottom and the top of layer 5; Li, Wang, and Qiu 2003; Li 2010) at the middle part of the section. Teilhard de Chardin and Licent made a small collection from what they called "Gouochtock oula" in 1924, and a preliminary report on new material of the Gaotege Fauna was given by Li, Wang, and Qiu (2003). Continuing study of the collections has progressed to the extent that it is now possible to recognize some of the details of faunal successions that are contained within the fluviolacustrine sediments exceeding 70 m in thickness. The Lower Gaotege Fauna from localities DB00-4-7, DB02-1-6, and DB03-1 is composed of at least 51 taxa, dominated by small mammals. The fauna is similar to the Bilike Fauna in community structure and composition of small mammals, sharing nearly 90% of the genera and about half of the species. It can be differentiated from the Bilike Fauna by absence of the family Gliridae and some Late Miocene–originated genera such as *Lophocricetus*, *Kowalskia*, and *Orientalomys*, and the occurrence of *Sinocricetus* sp. nov. In addition, the Lower Gaotege Fauna shows a marked drop in insectivore and zapodid diversity, a rise in dipodids and murids, a dominance of *Mimomys*, *Prosiphneus*, *Chardinomys*, and *Ochotona*, and a common occurrence of *Dipus*, *Nannocricetus*, *Micromys*, *Chardinomys*, and *Huaxiamys*. It is likely that the fauna is slightly younger than the Bilike Fauna but older than the late Mazegouan faunas like the Daodi, Mazegou, and Jingle faunas in North China. The Lower Gaotege Fauna seems to belong in the Gaozhuangian, equivalent to lower MN 15.

Compared to our previous summaries (Li, Wang, and Qiu 2003; Qiu, Wang, and Li 2006), it is now possible to recognize a superposed fauna, the Upper Gaotege Fauna, which is represented by the assemblage from site DB03-2 of the Gaodege section, about 15 m above the Lower Gaotege Fauna. The upper Gaotege assemblage contains essentially the same taxa as the lower one, but it differs in the absence of Castoridae (*Castor* and *Dipoides*) that are quite common in the lower assemblage, the occurrence of *Mimomys* cf. *M. orientalis*, and in the taxonomic difference of *Hipparion* (*H. insperatum* in the lower fauna and *H. huangheense* in the upper fauna). This upper assemblage is comparable to the Daodi, Mazegou, and Jingle faunas in stage of evolution for a few mammals, but it shows more primitive status in other rodents. Thus, the Upper Gaotege Fauna is still considered to be Gaozhuangian age, representing the youngest fauna in central Nei Mongol.

BIOCHRONOLOGY

The Chinese Neogene biochronologic framework is developed by seriating mammal faunas based mainly on the stage of evolution of mammals and faunal correlation. A set of Chinese Neogene Land Mammal Stages/Ages (LMS/A) has been established and modified in the last 30 years (Chiu, Li, and Chiu 1979; Li, Wu, and Qiu 1984; Qiu and Qiu 1995; Tong, Zheng, and Qiu 1995; Qiu, Wu, and Qiu 1999). Currently, five Miocene biochronologic units (Xiejian, Shanwangian, Tunggurian, Bahean, Baodean) and two in the Pliocene (Gaozhuangian and Mazegouian) are recognized as the basis of the Chinese Land Mammal Stage/Age framework (Qiu et al., chapter 1, this volume).

Of the seven Chinese Neogene Land Mammal Stages/ Ages, six are demonstrably present in central Nei Mongol (figure 5.2; table 5.1). The following characterizations of these stages/ages are mainly based on first appearance (or occurrence) and last appearance data (FAD and LAD) of taxa pertinent to the Neogene sediments from central Nei Mongol. In some cases, the first appearances are of local or regional significance only, and their appearances elsewhere in China are discussed in Qiu et al. (chapter 1, this volume).

Xiejian

Li, Wu, and Qiu (1984) first used the Xiejian age based on taxa from the Xiejia Formation in Xining Basin, Qinghai. Subsequently, Qiu and Qiu (1995) assigned the Zhangjiaping assemblage from Lanzhou Basin, Gansu, as a reference fauna of the age. Meng, Wang, and Bai (2006) correlated the Suosuoquan mammal assemblages from Junggar Basin, Xinjiang, with the Xiejia Fauna. The Xiejian faunas are characterized by retention of holdover or highly specialized Oligocene survivors. Ctenodactylidae, Tachyoryctoididae, and Zapodidae in Rodentia and archaic genera in Ochotonidae and Erinaceidae are dominant in small mammals. At the same time, some genera of muroids that are absent in the late Oligocene made their first appearances. The Xiejian is recognized by the first appearance of *Atlantoxerus*, *Microdyromys*, *Cricetodon*, and *Democricetodon* and the last appearance of *Yindirtemys* and *Parasminthus*. The Lower Aoerban Fauna shows its Xiejian characteristics in its dominant presence of the above taxa and in having the characteristic taxa of this age— *Amphechinus*, *Tachyoryctoides*, *Prodistylomys*, *Atlantoxerus*, *Asianeomys*, and *Sinolagomys*. The absence of *Yindirtemys* and *Parasminthus* and the presence of *Democricetodon* and *Microdyromys* seem to imply that the Lower Aoerban Fauna is younger than the Xiejia Fauna, the Zhangjiaping Fauna, and the assemblage from S-I zone. It is even slightly younger than the assemblages from S-II zone for its absence of *Parasminthus* and the presence of *Sicista* and other newcomers, indicative of the Lower Aoerban Fauna representing the latest one in the Xiejian age.

Shanwangian

This mammal age was proposed by Li, Wu, and Qiu (1984) based on the classic Shanwang Fauna and the Sihong Fauna. Tong, Zheng, and Qiu (1995) defined the Shanwangian as the initiation of great diversification of myomorph rodents, the complete replacement of archaic carnivores by modern ones, the further decline of perissodactyls, and the flourishing of ruminants. The

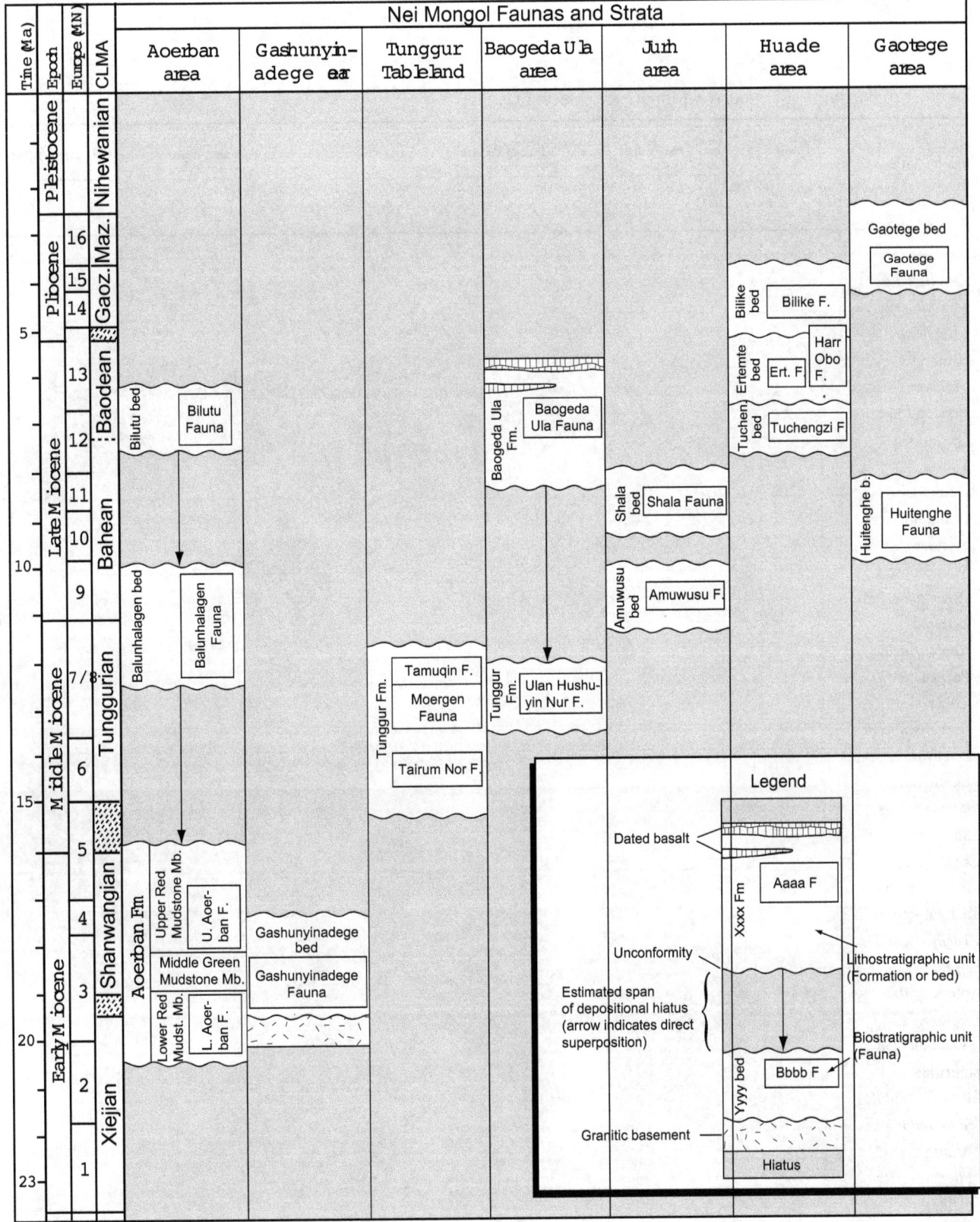

Figure 5.2 Correlation chart showing chronologic sequences of rock units and contained fossil assemblages in central Nei Mongol.

Table 5.1

Mammalian Biochron of Central Nei Mongol for Selected Genera

Taxa	Xiejian	Shanwangian		Tunggurian			Bahean		Baodean		Gaozhuangian	
	L	E	L	E	M	L	E	L	E	L	E	L
Erinaceidae												
Amphechinus	*	+										
Metexallerix	*	+										
Mioechinus	*							+				
Erinaceus										*		
Talpidae												
Proscapanus	*					+						
Quyania	*										+	
Yanshuella	*										+	
Desmanella	*					+						
Desmana											*	
Soricidae												
Mongolosorex	*					+						
Paranourosorex								*		+		
Zelceina									*	+		
Alloblarinella									*	+		
Paenepetenyia									*	+		
Cokia									*	+		
Paenelimnoecus										*	+	
Petenyia											*	
Lunanosorex											*	
Parasoriculus											*	
Sulimskia											*	
Ctenodactylidae												
Pro/Distylomys	*			+								
Tachyoryctoididae												
Tachyoryctoides	*			+								
Aralomys	*		+									
Aplodontidae												
Ansomys	*							+				
Pseudaplodon									*	+		
Sciuridae												
Prospermophilus							*				+	
Sinotamias					*					+		
Atlantoxerus	*						+					
Pliopetaurista							*			+		
Gliridae												
Prodryomys	*	+										
Miodyromys	*							+				
Microdyromys	*							+				
Myomimus										*	+	
Eomyidae												
Keramidomys	*							+				
Leptodontomys		*								+		

Taxa	Xiejian L	Shanwangian E	Shanwangian L	Tunggurian E	Tunggurian M	Tunggurian L	Bahean E	Bahean L	Baodean E	Baodean L	Gaozhuangian E	Gaozhuangian L
Asianeomys	*	+										
Pentabuneomys	*							+				
Ligerimys	*	+										
Castoridae												
Anchitheriomys				*+								
"Monosaulax"				*			+					
Steneofiber					*+							
Castor							*					
Dipoides									*		+	
Zapodidae												
Plesiosminthus	*		+									
Heterosminthus	*						+					
Litodonomys	*	+										
Sicista	*											
Lophocricetus								*			+	
Paralophocricetus										*	+	
Caridiocranius								*				
Sinozapus							*				+	
Eozapus										*		
Dipodidae												
Protalactaga					*		+					
Paralactaga							*				+	
Brachyscirtetes							*				+	
Dipus								*				
Cricetidae												
Eucricetodon?	*		+									
Democricetodon	*						+					
Megacricetodon		*				+						
Cricetodon			*				+					
Plesiodipus				*			+					
Gobicricetodon				*				+				
Nannocricetus							*				+	
Kowalskia							*				+	
Sinocricetus							*				+	
Anatolomys							*				+	
Microtodon							*				+	
Microtoscoptes							*			+		
Rhinocerodon									*+			
Gerbillidae												
Abudhabia									*+			
Pseudomeriones										*	+	
Muridae												
Hansdebruijnia									*	+		
Apodemus										*		

(continued)

Table 5.1 (*continued*)

Taxa	Xiejian	Shanwangian		Tunggurian			Bahean		Baodean		Gaozhuangian	
	L	E	L	E	M	L	E	L	E	L	E	L
Micromys										*		
Orientalomys										*	+	
Chardinomys											*	
Huaxiamys											*	
Allorattus											*+	
Siphneidae												
Prosiphneus							*				+	
Chardina											*	
Arvicolidae												
Mimomys											*	
Incertae Familiae												
Pararhizomys									*		+	
Leporidae												
Alilepus									*	+		
Trischizolagus											*	
Ochotonidae												
Desmatolagus	*						+					
Sinolagomys	*	+										
Alloptox		*				+						
Bellatona				*		+						
Ochotona							*					
Palaeogalidae												
Palaeogale	*	+										
Barbourofelidae												
Sansanosmilus					*+							
Hyaenidae												
Tungurictis				*	+							
Percrocuta					*+							
Hyaenictitherium												
Chasmaporthetes											*+	
Felidae												
Metailurus					*+							
Canidae												
Nyctereutes											*+	
Ursidae												
Plithocyon					*		+					
Amphicyonidae												
Amphicyon		*+										
Gobicyon					*+							
Mephitidae												
Promephitis										+		
Mustelidae												
Leptarctus				*+								
Sthenictis				*+								

Taxa	Xiejian	Shanwangian		Tunggurian			Bahean		Baodean		Gaozhuangian	
	L	E	L	E	M	L	E	L	E	L	E	L
Meles										*+		
Martes										*	+	
Equidae												
Anchitherium				*			+					
Hipparion									*		+	
Chalicotheriidae												
Chalicotherium				*+								
Rhinocerotidae												
Acerorhinus			*	+								
Hispanotherium				*+								
Cervidae												
Lagomeryx			*	+								
Stephanocemas			*	+								
Dicroceros			*	+								
Euprox							*+					
Suidae												
Listriodon				*+								
Kubanochoerus				*+								
Sus										*+		
Bovidae												
Turcocerus				*+								
Gazella												
Procapreolus										*+		
Gomphotheridae												
Platybelodon			*	+								

* = the earliest appearance; + = the latest appearance.

Gashunyinadege Fauna and the Upper Aoerban Fauna show this change for Rodentia. Taxonomic differences of Nei Mongol (various zapodids and ground squirrels) from Sihong and Shanwang assemblages in eastern China (*Diatomys* and tree squirrels) may reflect the East Asia biogeographic and ecological differences, whereas the co-occurrence of *Tachyoryctoides, Ansomys, Eutamias, Microdyromys, Democricetodon, Megacricetodon,* and *Alloptox* indicate that the Nei Mongol faunas are comparable to the Sihong Fauna, and they closely match Shanwangian LMA/S.

Eomyidae, Zapodidae, and Ochotonidae are rather diversified and flourished during Shanwangian in central Nei Mongol. Characteristic mammals include *Amphechinus, Prodistylomys/Distylomys, Tachyoryctoides, Aralomys, Ansomys, Prodryomys, Asianeomys, Ligerimys, Plesiosminthus, Litodonomys,* and *Sinolagomys*. The first records may include *Megacricetodon* and *Alloptox*. The last records include *Amphechinus, Metexallerix, Aralomys, Asianeomys, Ligerimys, Plesiosminthus, Litodonomys, Sinolagomys,* and a moderately advanced muroid (*Eucricetodon*?).

Immigrant taxa and faunal characters are used to mark two phases of the Shanwangian LMS/A. The early Shanwangian, represented by the Gashunyinadege Fauna, is defined by the first appearance of *Leptodontomys, Megacricetodon,* and *Alloptox* and the last appearance of *Amphechinus, Prodryomys, Ligerimys, Litodonomys,* and *Sinolagomys.* The late Shanwangian is recognized in the Upper Aoerban Fauna by its immigrants of *Oriensciurus*? and *Cricetodon* among small mammals and Proboscidea and *Lagomeryx* among large mammals, by the noticeable decline of Ctenodactylidae, Tachyoryctoididae, Aplodontidae, and

Zapodidae, by the loss of many early Shanwangian genera, such as *Tachyoryctoides*, *Asianeomys*, and *Palaeogale*, and by the last appearance of *Aralomys*, *Plesiosminthus*, and *Eucricetodon*?.

Tunggurian

Li, Wu, and Qiu (1984) based the Tunggurian Age on taxa from the Tunggur Tableland. Tong, Zheng, and Qiu (1995) presented a biologic characterization of the age in terms of the major radiation of proboscideans, the diversification of ruminants, the earliest appearance of the families Dipodidae and Hyaenidae, and the genera *Plesiodipus* and *Platybelodon*, as well as the last appearance of Ctenodactylidae and Tachyoryctoididae, *Megacricetodon*, *Alloptox*, *Hemicyon*, *Plesiaceratherium*, and so on. Qiu, Wang, and Li (2006) proposed an informal tripartite subdivision of the Tunggurian. The following may serve as a modified definition of the three subdivisions of the Tunggurian LMS/A.

The early Tunggurian is based on taxa from the lower unit of the Tunggur Formation at Tairum Nur (lower red bed and middle-channel sandstones) and Aletexire of the Tunggur Tableland. The earliest appearance of *Plesiodipus*, *Gobicricetodon*, *Bellatona*, *Leptarctus*, *Tungurictis*, *Sthenictis*, *Sansanosmilus*, *Stephanocemas*, *Platybelodon*, and *Dicroceros* and the latest appearance of Ctenodactylidae and Tachyoryctoididae characterize the early Tungurian in Nei Mongol. However, with the recent discovery of *Platybelodon* from the early Miocene strata of Tabenbuluk (Wang and Qiu 2002), the widespread appearances of *Platybelodon* in the middle Miocene of central Inner Mongolia, formerly a major indicator of the Tunggurian Age, may represent a local first appearance.

The middle Tunggurian is based on taxa from the lower part of the upper unit of the Tunggur Formation. It is defined by immigrant Castoridae from North America (*Anchitheriomys* and "*Monosaulax*"), the earliest appearance of Dipodidae (*Protalactaga*), *Plithocyon*, *Amphicyon*, *Gobicyon*, *Metailurus*, *Serridentinus*, *Zygolophodon*, *Anchitherium*, *Chalicotherium*, *Micromeryx*, *Dicroceros*, and *Turcocerus*, and the latest appearance of *Proscapanus*, *Desmanella*, *Mongolosorex*, *Listriodon*, *Kubanochoerus*, *Tungurictis*, *Percrocuta*, and *Sansanosmilus*. This period is characterized by the dominance of the Myomorpha (Rodentia) and the flourishing of Proboscidea and Artiodactyla. *Mioechinus*, "*Monosaulax*," *Heterosminthus*, *Plesiodipus*, *Democricetodon*, *Megacricetodon*, *Alloptox*, *Bellatona*, and *Platybelodon* are common at this time, and their joint

occurrence is considered characteristic for faunas in the Tunggur Formation.

The late Tunggurian is only represented by the Tamuqin Fauna from the top of the Moergen section, containing immigrant *Steneofiber* from Europe, and the earliest appearance of *Plesiodipus progressus* and *Gobicricetodon robustus*, and the last appearance of *Plesiodipus*, *Megacricetodon*, *Alloptox*, and *Bellatona*.

Bahean

Bahean was first proposed by Li, Wu, and Qiu (1984) based on taxa from the Bahe Formation at Lantian, Shaanxi, and discarded by Qiu and Qiu (1995) and Qiu, Wu, and Qiu (1999). Biostratigraphic work in central Nei Mongol agrees with the decision to resurrect this age (Qiu et al., chapter 1, and Zhang et al., chapter 6, this volume). The Bahean is established for taxa from the lower part of the Bahe Formation in Shaanxi and is represented locally by the Amuwusu beds at Jurh. The Amuwusu Fauna represents early Bahean LMS/A based on immigrant Meniscomyinae, the first appearance of Siphneidae (=Myospalacinae) and *Hipparion*, the genera *Castor*, *Sinozapus*, *Paralactaga*, *Brachyscirtetes*, and *Ictitherium*, and the last appearance of *Mioechinus*, "*Monosaulax*," *Heterosminthus*, *Democricetodon*, *Gobicricetodon*, *Desmatolagus*, *Plithocyon*, *Anchitherium*, and *Micromeryx*.

The lower Bahe Formation, representing middle Bahean age, contains the first appearance of Muridae, *Sciurotamias*, *Salpingotus*, *Cardiocranius*, *Lophocricetus*, *Nannocricetus*, *Kowalskia*, *Abudhabia*, and *Pararhizomys* and the last appearance of *Protalactaga* and *Myocricetodon*.

The earliest appearance in the Shala Fauna of *Paranourosorex*, *Prospermophilus*, *Pliopetaurista*, *Lophocricetus*, *Dipus*, *Cardiocranius*, *Sinocricetus*, *Anatolomys*, *Kowalskia*, *Microtoscoptes*, and *Ochotona* and the last occurrence of *Ansomys*, *Miodyromys*, *Microdyromys*, and *Keramidomys* characterize the late phase of Bahean.

Baodean

Li, Wu, and Qiu (1984) defined the Baodean Age based on *Hipparion* faunas primarily from some classic localities of northern China, including Baode of Shanxi, Qingyang and Wudu of Gansu, Ertemte of Nei Mongol, and the hominoid fauna of Lufeng, Yunnan. As for the definition and characterization of the Baodean, Tong, Zheng, and Qiu (1995) emphasized the further diversification of myomorph rodents, the dominance of Hyaenidae, Felidae, and Mustelidae among Carnivora, of *Hipparion* and *Chilotherium* (Perissodactyla), and the high abundance of Proboscidea and Artiodactyla. The Baogeda Ula, Bilutu, Tuchengzi, and Ertemte faunas from central Nei Mongol possess components comparable to other *Hipparion* faunas in North China and may represent part of this LMS/A. The Baodean can be divided into the following two intervals.

The early Baodean, represented by the Baogeda Ula Fauna, is recognized by the first occurrence of *Parasoriculus, Dipoides, Abudhabia, Pararhizomys, Hansdebruijnia,* and *Alilepus.* Immigration of murids, gerbillids, and the genus *Rhinocerodon* into central Nei Mongol and the first reappearance of leporids after a long hiatus in Asia during the Oligocene and most of the Miocene were important events in this region. The defining taxa for the late Baodean, based on the Ertemte Fauna, include the first appearance of *Erinaceus, Zelceina, Alloblarinella, Paenepetenyia, Cokia, Paenelimnoecus, Pseudaplodon, Myomimus, Paralophocricetus, Eozapus, Microtodon, Pseudomeriones, Apodemus, Micromys,* and *Orientalomys;* the dominance of *Yanshuella, Sorex, Lophocricetus, Sinocricetus, Microtodon, Prosiphneus, Micromys,* and *Ochotona;* and the latest appearance of Aplodontidae and Eomyidae, *Sinotamias, Microtoscoptes, Hansdebruijnia,* and *Alilepus.*

Gaozhuangian

This LMS/A is proposed based on the Gaozhuang Fauna from Yushe, Shanxi (Qiu et al., chapter 1, this volume). The Gaozhuangian faunas often contain primitive species of modern genera and are characterized by retention in mammal families in the Baodean LMS/A, except Aplodontidae and Eomyidae, by the first appearance of Arvicolidae and the immigrant Caninae and Camelidae from North America. Many genera of late Miocene origin became extinct or were gradually replaced by newcomers. All taxa are members of living mammal families, yet only a small proportion of genera persist to the present day.

The Bilike Fauna seems suitably referred to the Gaozhuangian LMS/A. Commonly known families in the Miocene, such as Aplodontidae and Eomyidae, have disappeared. Some genera originated in the late Miocene, such as *Paranourosorex, Pliopetaurista, Lophocricetus, Microtoscoptes, Microtodon, Hansdebruijnia,* and *Alilepus,* became completely extinct or declined significantly.

The Gaozhuangian Fauna is also defined by the earliest appearance of *Desmana, Petenyia, Lunanosorex, Sulimskia, Chardinomys, Huaxiamys, Allorattus, Mimomys* (*Aratomys*), and *Trischizolagus* and the latest appearance of, among others, *Parasoriculus, Lophocricetus, Paralophocricetus, Sinozapus, Anatolomys, Microtodon,* and *Orientalomys.* The Lower Gaotege Fauna, containing a few newcomers such as *Sinocricetus* sp. nov. and the last appearance of *Desmana* and *Lunanosorex,* demonstrates a close community structure and taxonomic composition with the Bilike Fauna but shows distinct decline of soricids and zapodids, as well as flourishing of high-crowned voles. We consider it to be a later Gaozhuangian age, but the possibility of the Upper Gaotege Fauna, including a few Mazegouan members, cannot be excluded due to its presence of more derived voles and species of *Hipparion* from the upper part of deposits.

MAGNETOCHRONOLOGY

Aoerban Formation

With a relatively thick section, the Aoerban Formation is well suited for magnetostratigraphic studies. As discussed previously, the Aoerban Formation also includes the local first appearance of proboscideans, an important immigration event in the definition of the beginning of the Shanwangian Age. Thus, an independent magnetic age determination of this event would surely help in calibrating the precise age. However, aside from a feasibility study by Liddicoat et al. (2007; also shown in Wang et al. 2009:fig. 3), in which only a few sites were sampled, nothing about its magnetostratigraphy has been published so far (an attempt by a group from the Nanjing University was not published; Yang Shenyü, pers. comm.).

Despite such a paucity of data, we (Wang et al. 2009:132) commented that sites 3–6 in Liddicoat et al. (2007:fig 5) appear to represent the long normal chron C6n (18.748–19.722), commonly associated with the first appearance of Proboscidea in Eurasia. However, on grounds of depositional rates, Qiu et al. (chapter 1, this volume) pointed out that such a short normal chron, only 4–5 m in maximum thickness, is unlikely to be C6n, which spans about 1 m.y. in duration. Instead, they proposed an alternative interpretation of C6An.1n (20.040–20.213 Ma) for this short interval. While this latter interpretation is preferable given available evidence, we continue to caution that all such interpretations must be treated as speculative until future studies of a much greater sampling density become available.

Tunggur Formation

Paleomagnetic studies of a 34 m Tairum Nur section yielded three normal and three reversed chrons that were tentatively correlated to C5An.2n through C5Ar.3r (12.2–13.0 Ma), and a 43 m Moergen section revealed four reversed and three normal chrons that were correlated to C5r.3r through part of C5Ar.2r (11.8–12.8 Ma; Wang, Qiu, and Opdyke 2003). In an effort to define a chronostratigraphic "Tunggurian Stage," Deng, Hou, and Wang (2007) realigned the Tairum Nur magnetostratigraphy to C5ADn through C5Br (14.2–16.0 Ma in Lourens et al. 2004), mostly based on a supposed correlation with the base of the European Astaracian Land Mammal Age. However, they made no attempt to see how this would affect the Moergen section. Instead, Deng, Hou, and Wang (2007:711) observed a 3–5° northward dip of the Tairum Nur strata, exact location of observation not given, and one we failed to replicate in repeated attempts. Presumably this putative dip was meant to demonstrate a superpositional relationship between the Tairum Nur and Moergen sections, although we would point out that a 3–5° northward dip at Tairum Nur translates to a 1800–3000 m vertical drop of the strata projected to the northern escarpments where Moergen section is located (35 km in horizontal distance from the Tairum Nur section), in contrast to a partial overlap in the northern and southern exposures in our earlier correlations (Wang, Qiu, and Opdyke 2003:fig. 10).

Unfortunately, little additional paleomagnetic investigation in the Moergen or adjacent strata has been made since our 2003 magnetostratigraphy, which contains several segments of missing magnetic sites due to sampling difficulties in coarse-grained sandstones. A few preliminary test samples were obtained at Aletexire section, about 16 km to the east of Moergen, but only a handful of sites were sampled, all of them normally oriented, and a meaningful correlation is not feasible at this time (Liddicoat et al. 2007). The situation is further compounded by the fact that we are still unable to physically trace the northern (Moergen) and southern (Tairum Nur) exposure, due to discontinuities between exposures, especially by a long gap between the *Platybelodon* Quarry and western end of the Tairum Nur exposures. Until an alternative means of correlation can be found, faunal comparison is still the only means of relative dating. Our research on small mammals increasingly suggests that the Tairum Nur Fauna may be earlier than we had concluded in our initial magnetic interpretation (Wang, Qiu, and Opdyke 2003; Qiu, Wang, and Li 2006), and Deng, Hou, and Wang's (2007) reinterpretation of our Tairum Nur magnetic correlation may be a viable alternative.

However, such an interpretation demands an adjustment to the correlation of the Moergen magnetostratigraphy, unless one is willing to accept a substantial gap between the Tairum Nur and Moergen sections—that is, either a long depositional hiatus within the Tunggur Formation or a large segment of the middle section being buried underground (as seems implicitly assumed in Deng, Hou, and Wang 2007). We arrived at a new compromise of correlating the lower Moergen section to magnetochrons much older than we had initially assumed (figure 5.3). Such a correlation is based on our sense that the small mammals from the lower part of the upper red mudstones at Tairum Nur (an assemblage recovered from a thin greenish mudstone lens a few meters above the channel sandstone at the Road Mark 356 locality) are overall comparable to Moergen small mammals (MoII); that is, we still maintain a small overlap of strata between the upper Tairum Nur and lower Moergen sections. This new correlation suggests that the Tunggur Formation spans at least chrons C5B to C5ABr (13.7–11.8 Ma), although additional work to sample chrons possibly missed in earlier studies may extend this range downward to as old as 15 Ma.

Although this new correlation satisfies some faunal considerations, it also demands a somewhat lower sedimentation rate for the Moergen section than the Tairum Nur section, despite the more coarse-grained sediments in the Moergen section. This result therefore must be tested in future magnetic investigations that sample multiple sections in the Tunggur Formation, as well as additional magnetostratigraphy in other middle Miocene basins, such as the Tongxin Basin in Ningxia Province and the Halamagai Formation in the Junggar Basin of Xinjiang Province (see preliminary result in Meng, Wang, and Bai 2006).

Gaotege

A 60 m magnetic section was collected in 2005 at ~1 m intervals, and the entire section showed reversed polarity (O'Connor et al. 2008). An independent magnetic study

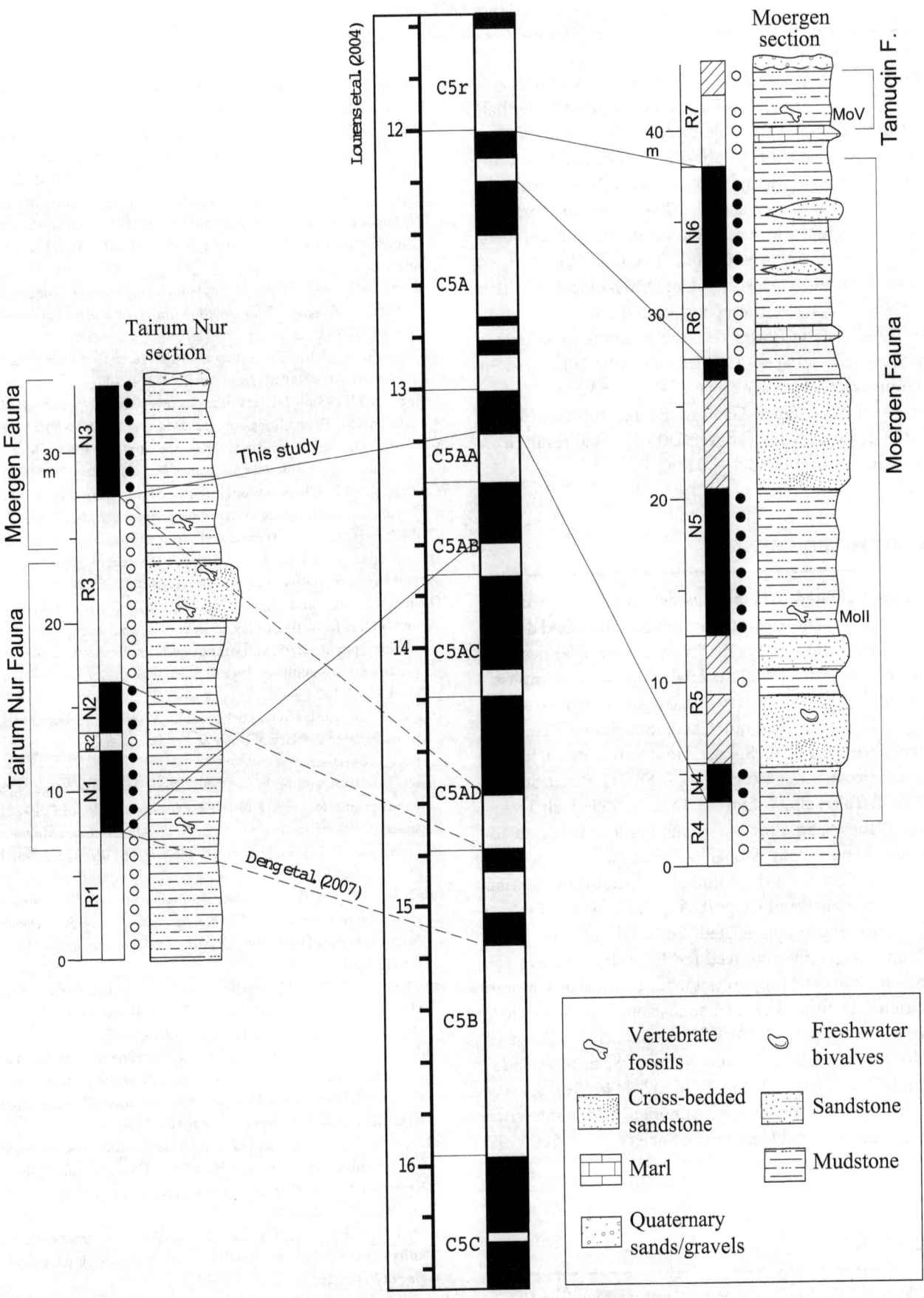

Figure 5.3 Reinterpretation of magnetochronology of the Tunggur Formation. Modified from Wang et al. (2003).

shortly afterward (but published earlier) sampled at a much higher density (~0.5 m intervals) in the lower half of the section and was able to detect two short normal chrons (Xu et al. 2007). Both studies correlated the Gaotege magnetostratigraphy to C2Ar–C3n.2n—that is, 3.59–4.63 Ma (Lourens et al. 2004). The two short normal chrons in the lower section essentially bracket fossil localities of the Lower Gaotege Fauna (DB00-4-7 and DB02-1-6), with the exception of DB03-1 locality, which is about 3 m above the upper normal chron (Xu et al. 2007:fig. 7). The Lower Gaotege Fauna is thus further constrained between C3n.1n (Cochiti) and C3n.3n (Nunivak), with an age range of 4.18–4.63 Ma. Interpolation of the Upper Gaotege Fauna, represented by DB03-2, which is 18 m above BD03-1, would result in an age estimate of about 3.8–4.0 Ma.

ACKNOWLEDGMENTS

We are indebted to numerous field personnel over the past quarter century for the accumulation of field data in the classic localities, the discovery of new localities, and collections of new fossil materials. This work is impossible without the enthusiastic participation by the following individuals: Shanqin Chen, Shaokun Chen, Tao Deng, Wenqing Feng, Sukuan Hou, Yuri Kimura, Joseph C. Liddicoat, Libo Pang, Qinqin Shi, Gerhard Storch, Gary T. Takeuchi, Yukimitsu Tomida, Zijie Jack Tseng, Hongjiang Wang, Ping Wang, and Weimin Wang, in addition to numerous Mongolian friends. We especially thank Deliger from the Xilinhot Cultural Relic Station for his guidance and support. A detailed review by Yuri Kimura is much appreciated. Editorial effort by Larry Flynn has greatly improved the English as well as the content. Financial support was made possible by Chinese National Natural Science Foundation (No. 40702004, 40911120091, 40730210), Chinese Academy of Sciences (No. KZCX2-YW-120), and National Science Foundation (U.S.) (EAR-0924142, 0716507, 0446699), Society of Vertebrate Paleontology summer field conference fund, and National Geographic Society (Nos. 6004–97 and 6771–00).

REFERENCES

Andersson, J. G. 1923. Essays on the Cenozoic of northern China. *Memoirs of the Geological Survey of China*, ser. A 3:1–152.

Andrews, R. C. 1932. The new conquest of Central Asia, a narrative of the explorations of the Central Asiatic Expeditions in Mongolia and China, Natural History of Central Asia. *American Museum of Natural History* 1:1–678.

Chiu, C.-s, C.-k. Li, and C.-t. Chiu. 1979. The Chinese Neogene: A preliminary review of the mammalian localities and faunas. *Annales Géologiques des Pays Helléniques*, non-ser. vol. 1:263–272.

Chow, M.-c. and A. K. Rozhdestvensky. 1960. Exploration in Inner Mongolia—a preliminary account of the 1959 field work of the Sino-Soviet Paleontological Expedition (SSPE). *Vertebrata Pal-Asiatica* 4(1):1–10.

Colbert, E. H. 1934. Chalicotheres from Mongolia and China in the American Museum. *Bulletin of the American Museum of Natural History* 67:353–387.

Colbert, E. H. 1936a. Palaeotragus in the Tung Gur Formation of Mongolia. *American Museum Novitates* 874:1–17.

Colbert, E. H. 1936b. Tertiary deer discovered by the American Museum Asiatic Expeditions. *American Museum Novitates* 856:1–21.

Colbert, E. H. 1939a. Carnivora of the Tung Gur Formation of Mongolia. *Bulletin of the American Museum of Natural History* 76:47–81.

Colbert, E. H. 1939b. A new anchitheriine horse from the Tung Gur Formation of Mongolia. *American Museum Novitates* 1019:1–9.

Colbert, E. H. 1940. Som cervid teeth from the Tung Gur Formation of Mongolia, and additional notes on the genera *Stephanocemas* and *Lagomeryx*. *American Museum Novitates* 1062:1–6.

Daxner-Höck, G. and D. Badamgarav. 2007. Oligocene-Miocene vertebrates from the Valley of Lakes (Central Mongolia): Morphology, phylogenetic and stratigraphic implications. 1. Geological and stratigraphic setting. *Annalen Naturhistorisches Museum Wien* 108 A:1–24.

Dawson, M. R. 1961. On two ochotonids (Mammalia, Lagomorpha) from the later Tertiary of Inner Mongolia. *American Museum Novitates* 2061:1–15.

Deng, T., S. Hou, and H. Wang. 2007. The Tunggurian Stage of the continental Miocene in China. *Acta Geologica Sinica* 81:709–721.

Engesser, B. 1999. Family Eomyidae. In *The Miocene Land Mammals of Europe*, ed. G. E. Rössner and K. Heissig, pp. 319–335. Munich: Dr. Friedrich Pfeil.

Fahlbusch, V. 1987. The Neogene mammalian faunas of Ertemte and Harr Obo in Inner Mongolia (Nei Mongol), China. 5. The genus Microtoscoptes (Rodentia: Cricetidae). *Senckenbergiana lethaea* 67:345–373.

Fahlbusch, V. 1992. The Neogene mammalian faunas of Ertemte and Harr Obo in Inner Mongolia (Nei Mongol), China. 10. *Eozapus* (Rodentia). *Senckenbergiana lethaea* 72:199–217.

Fahlbusch, V. and M. Moser. 2004. The Neogene mammalian faunas of Ertemte and Harr Obo in Inner Mongolia (Nei Mongol), China. 13. The genera *Microtodon* and *Anatolomys* (Rodentia: Cricetidae). *Senckenbergiana lethaea* 84:323–349.

Fahlbusch, V., Z.-d. Qiu, and G. Storch. 1983. The Neogene mammalian faunas of Ertemte and Harr Obo in Nei Mongol, China. 1. Report on field work in 1980 and preliminary results. *Scientia Sinica*, ser. B 26:205–224.

Flynn, L. J., R. H. Tedford, and Z.-x. Qiu. 1991. Enrichment and stability in the Pliocene mammalian fauna of North China. *Paleobiology* 17:246–265.

Guo, Z.-t., W. F. Ruddiman, Q.-z. Hao, H.-b. Wu, Y.-s. Qiao, R.-x. Zhu, S.-z. Peng, J.-j. Wei, B.-y. Yuan, and T.-s. Liu. 2002. Onset of Asian desertification by 22 Myr ago inferred from loess deposits in China. *Nature* 416:159–163.

Li, C.-k. 1963. A new species of *Monosaulax* from Tung Gur Miocene, Inner Mongolia. *Vertebrata PalAsiatica* 7:240–244.

Li, C.-k. and Z.-d. Qiu. 1980. Early Miocene mammalian fossils of Xining basin, Qinghai. *Vertebrata PalAsiatica* 18(3):198–214.

Li, C.-k., W.-y. Wu, and Z.-d. Qiu. 1984. Chinese Neogene: Subdivision and correlation. *Vertebrata PalAsiatica* 22(3):163–178.

Li, Q. 2010. Note on the cricetids from the Pliocene Gaotege locality, Nei Mongol. *Vertebrata PalAsiatica* 48(3):247–261.

Li, Q., X.-m. Wang, and Z.-d. Qiu. 2003. Pliocene mammalian fauna of Gaotege in Nei Mongol (Inner Mongolia), China. *Vertebrata PalAsiatica* 41(2):104–114.

Liddicoat, J. C., X.-m. Wang, Z.-d. Qiu, and Q. Li. 2007. Recent palaeomagnetic and magnetostratigraphic investigations on and around the Tunggur Tableland, central Nei Mengol (Inner Mongolia). *Vertebrata PalAsiatica* 45(2):110–117.

Lourens, L., F. Hilgren, N. J. Shackleton, J. Laskar, and J. Wilson. 2004. The Neogene Period. In *A Geologic Time Scale 2004*, ed. F. M. Gradstein, J. G. Ogg, and A. G. Smith, pp. 409–440. Cambridge: Cambridge University Press.

Luo, X.-q. and Q.-t. Chen. 1990. Preliminary study on geochronology for Cenozoic basalts from Inner Mongolia. *Acta Petrologica et Mineralogica* 9(1):37–46.

Mein, P. 1999. European Miocene mammal biochronology. In *The Miocene Land Mammals of Europe*, ed. G. E. Rössner and K. Heissig, pp. 25–38. Munich: Dr. Friedrich Pfeil.

Meng, J., B.-y. Wang, and Z.-q. Bai. 1996. A new middle Tertiary mammalian locality from Sunitezuoqi, Nei Mongol. *Vertebrata PalAsiatica* 34(4):297–304.

Meng, J., J. Ye, W.-y. Wu, X.-j. Ni, and S.-d. Bi. 2008. The Neogene Dingshanyanchi Formation in northern Junggar Basin of Xinjiang and its stratigraphic implications. *Vertebrata PalAsiatica* 46(2):90–110.

Meng, J., J. Ye, W.-y. Wu, L.-p. Yue, and X.-j. Ni. 2006. A recommended boundary stratotype section for Xiejian Stage from northern Junggar Basin: Implications to related biochronostratigraphy and environmental changes. *Vertebrata PalAsiatica* 44:205–236.

O'Connor, J., D. R. Prothero, X.-m. Wang, Q. Li, and Z.-d. Qiu. 2008. Magnetic stratigraphy of the lower Pliocene Gaotege beds, Inner Mongolia. In *Neogene Mammals*, ed. S. G. Lucas, G. S. Morgan, et al., pp. 431–436. *New Mexico Museum of Natural History and Science Bulletin* 44.

Osborn, H. F. 1929. The revival of central Asiatic life. *Natural History* 29:2–16.

Osborn, H. F. and W. Granger. 1931. The shovel-tuskers, Amebelodontinae, of Central Asia. *American Museum Novitates* 470:1–12.

Osborn, H. F. and W. Granger. 1932. *Platybelodon grangeri*, three growth stages, and a new serridentine from Mongolia. *American Museum Novitates* 537:1–13.

Qiu, Z.-d. 1979. Some mammalian fossils from the Pliocene of Inner Mongolia and Gansu (Kansu). *Vertebrata PalAsiatica* 17(3):221–235.

Qiu, Z.-d. 1985. The Neogene mammalian faunas of Ertemte and Harr Obo in Inner Mongolia (Nei Mongol), China. 3. Jumping mice. *Senckenbergiana lethaea* 66:39–67.

Qiu, Z.-d. 1987. The Neogene mammalian faunas of Ertemte and Harr Obo in Inner Mongolia (Nei Mongol), China. 6. Hares and

pikas—Lagomorpha: Leporidae and Ochotonidae. *Senckenbergiana lethaea* 67:375–399.

Qiu, Z.-d. 1991. The Neogene mammalian faunas of Ertemte and Harr Obo in Inner Mongolia (Nei Mongol), China. 8. Sciuridae (Rodentia). *Senckenbergiana lethaea* 71:223–255.

Qiu, Z.-d. 1996. *Middle Miocene Micromammalian Fauna from Tunggur, Nei Mongol*. Beijing: Science Press.

Qiu, Z.-d. 2000. Insectivore, dipodoidean and lagomorph from the middle Miocene Quantougou fauna of Lanzhou, Gansu. *Vertebrata PalAsiatica* 38(4):287–302.

Qiu, Z.-d. 2001. Cricetid rodents from the middle Miocene Quantougou Fauna of Lanzhou, Gansu. *Vertebrata PalAsiatica* 39(3):204–214.

Qiu, Z.-d. 2003. The Neogene mammalian faunas of Ertemte and Harr Obo in Inner Mongolia (Nei Mongol), China. 12. Jerboas—Rodentia: Dipodidae. *Senckenbergiana lethaea* 83:135–147.

Qiu, Z.-d. and G. Storch. 2000. The Early Pliocene micromammalian fauna of Bilike, Inner Mongolia, China (Mammalia: Lipotyphla, Chiroptera, Rodentia, Lagomorpha). *Senckenbergiana lethaea* 80(1):173–229.

Qiu, Z.-d. and X.-m. Wang. 1999. Small mammal faunas and their ages in Miocene of central Nei Mongol (Inner Mongolia). *Vertebrata PalAsiatica* 37(2):120–139.

Qiu, Z.-d., X.-m. Wang, and Q. Li 2006. Faunal succession and biochronology of the Miocene through Pliocene in Nei Mongol (Inner Mongolia). *Vertebrata PalAsiatica* 44(2):164–181.

Qiu, Z.-x. and Z.-d. Qiu. 1995. Chronological sequence and subdivision of Chinese Neogene mammalian faunas. *Palaeogeograph, Palaeoclimatology, Palaeoecology* 116:41–70.

Qiu, Z.-x., B.-y. Wang, Z.-d. Qiu, G.-p. Xie, J.-y. Xie, and X.-m. Wang. 1997. Recent advances in study of the Xianshuihe Formation in Lanzhou Basin. In *Evidence for Evolution—Essays in Honor of Prof. Chungchien Young on the Hundredth Anniversary of His Birth*, ed. Y.-s. Tong, Y.-y. Zhang, W.-y. Wu, J.-l. Li, and L.-q. Shi, pp. 177–192. Beijing: China Ocean Press.

Qiu, Z.-x., W.-y. Wu, and Z.-d. Qiu. 1999. Miocene mammal faunal sequence of China: Paleozoogeography and Eurasian relationships. In *The Miocene Land Mammals of Europe*, ed. G. E. Rössner and K. Heissig, pp. 443–455. Munich: Dr. Friedrich Pfeil.

Qiu, Z.-x., D.-f. Yan, G.-f. Chen, and Z.-d. Qiu. 1988. Preliminary report on the field work in 1986 at Tung-gur, Nei Mongol. *Chinese Science Bulletin (Kexue Tongbao)* 33:399–404.

Schlosser, M. 1924. Tertiary vertebrates from Mongolia. *Palaeontologia Sinica*, ser. C 1(1):1–119.

Storch, G. 1987. The Neogene mammalian faunas of Ertemte and Harr Obo in Inner Mongolia (Nei Mongol), China. 7. Muridae (Rodentia). *Senckenbergiana lethaea* 67:401–431.

Storch, G. 1995. The Neogene mammalian Faunas of Ertemte and Harr Obo in Inner Mongolia (Nei Mongol), China. 11. Soricidae (Insectivora). *Senckenbergiana lethaea* 75:221–251.

Storch, G. and X.-j. Ni. 2002. New late Miocene murids from China (Mammalia, Rodentia). *Geobios* 35:515–521.

Storch, G. and Z.-d. Qiu. 1983. The Neogene mammalian faunas of Ertemte and Harr Obo in Inner Mogolia (Nei Mongol), China. 2. Moles—Insectivora: Talpidae. *Senckenbergiana lethaea* 64:89–127.

Tedford, R. H., T. Galusha, M. F. Skinner, B. E. Taylor, R. W. Fields, J. R. Macdonald, J. M. Rensberger, S. D. Webb, and D. P. Whistler. 1987. Faunal succession and biochronology of the Arikareean through Hemphillian interval (Late Oligocene through earliest Pliocene epochs) in North America. In *Cenozoic Mammals of North America: Geochronology and Biostratigraphy*, ed. M. O. Woodburne, pp. 153–210. Berkeley: University of California Press.

Teilhard de Chardin, P. 1926a. Étude géologique sur la région du Dalai-Nor. *Mémoirs de la Société Géologique de France* 7:1–56.

Teilhard de Chardin, P. 1926b. Déscription de mammifères tertiaires de Chine et de Mongolie. *Annales de Paléontologie* 15:1–52.

Tong, Y.-s., S.-h. Zheng, and Z.-d. Qiu. 1995. Cenozoic mammal age of China. *Vertebrata PalAsiatica* 33(4):290–314.

Tseng, Z. J. and X.-m. Wang. 2007. The first record of the Late Miocene *Hyaenictitherium hyaenoides* Zansky (Carnivora: Hyaenidae) in Inner Mongolia and evolution of the genus. *Journal of Vertebrate Paleontology* 27(3):699–708.

Wang, B.-y. and Z.-x. Qiu. 2002. A new species of *Platybelodon* (Gomphotheriidae, Proboscidea, Mammalia) from early Miocene of the Danghe area, Gansu, China. *Vertebrata PalAsiatica* 40:291–299.

Wang, X.-m., Z.-d. Qiu, Q. Li, Y. Tomida, Y. Kimura, Z. J. Tseng, and H.-j. Wang. 2009. A new Early to Late Miocene fossiliferous region in central Nei Mongol: Lithostratigraphy and biostratigraphy in Aoerban strata. *Vertebrata PalAsiatica* 47(2):111–134.

Wang, X.-m., Z.-d. Qiu, and N. D. Opdyke 2003. Litho-, bio-, and magnetostratigraphy and paleoenvironment of Tunggur Formation (Middle Miocene) in central Inner Mongolia, China. *American Museum Novitates* 3411:1–31.

Wang, X.-m., B.-y. Wang, Z.-x. Qiu, G.-p. Xie, J.-y. Xie, W. R. Downs, Z.-d. Qiu, and T. Deng. 2003. Danghe area (western Gansu, China) biostratigraphy and implications for depositional history

and tectonics of northern Tibetan Plateau. *Earth and Planetary Science Letters* 208:253–269.

Wang, X.-m., J. Ye, J. Meng, W.-y. Wu, L.-p. Liu, and S.-d. Bi. 1998. Carnivora from middle Miocene of Northern Junggar Basin, Xinjiang Autonomous Region, China. *Vertebrata PalAsiatica* 36(3):218–243.

Wood, A. E. 1936. Two new rodents from the Miocene of Mongolia. *American Museum Novitates* 865:1–7.

Wu, W.-y. 1985. The Neogene mammalian faunas of Ertemte and Harr Obo in Inner Mongolia (Nei Mongol), China. 4. Dormice—Rodentia: Gliridae. *Senckenbergiana lethaea* 66:68–88.

Wu, W.-y. 1991. The Neogene mammalian faunas of Ertemte and Harr Obo in Inner Mongolia (Nei Mongol), China. 9. Hamsters: Cricetinae (Rodentia). *Senckenbergiana lethaea* 71:68–88.

Xu, Y.-l., Y.-b. Tong, Q. Li, Z.-m. Sun, J.-l. Pei, and Z.-y. Yang. 2007. Magnetostratigraphic dating on the Pliocene mammalian fauna of the Gaotege section, central Inner Mongolia. *Geological Review* 53(2):250–260.

Ye, J., W.-y. Wu, and J. Meng 2001. The Age of Tertiary strata and mammal faunas in Ulungur River area of Xinjiang. *Journal of Stratigraphy* 25(3):193–287.

Zhai, R.-j. 1962. On the generic character of *Hypohippus zitteli*. *Vertebrata PalAsiatica* 6(1):48–55.

Zhai, R.-j. 1964. *Leptarctus* and other Carnivora from the Tung Gur Formation, Inner Mongolia. *Vertebrata PalAsiatica* 8:18–32.

Zhang, Z.-q. and S.-h. Zheng. 2000. Late Miocene–Early Pliocene biostratigraphy of Loc. 93002 section, Lingtai, Gansu. *Vertebrata PalAsiatica* 38(4):282–294.

Zheng, S.-h. and Z.-q. Zhang. 2001. Late Miocene–Early Pleistocene biostratigraphy of the Leijiahe area, Lingtai, Gansu. *Vertebrata PalAsiatica* 39(3):215–228.

APPENDIX

Mammal Taxa in Major Nei Mongol Faunas

LOWER AOERBAN FAUNA

INSECTIVORA
  Erinaceidae
    Amphechinus sp.
    Mioechinus cf. *M. gobiensis*
    Metexallerix sp.
  Talpidae
    Proscapanus sp.
    Quyania sp.
    Yanshuella sp.
    Desmanella sp.
  Soricidae
    Mongolosorex sp.
    Sorex? sp. 1
    Sorex? sp. 2

RODENTIA
  Ctenodactylidae
    Prodistylomys/Distylomys sp.
  Tachyoryctoididae
    Tachyoryctoides sp. 1
    Tachyoryctoides sp. 2
    Aralomys sp. 1
  Aplodontidae
    Ansomys sp. 1
    cf. *Ansomys* sp. nov.
  Aplodontidae gen. et sp. nov.
  Sciuridae
    Eutamias cf. *E. ertemtensis*

    Eutamias sp.
    Atlantoxerus sp.
  Gliridae
    Prodryomys sp.
    Miodyromys sp.
    Microdyromys sp.
  Eomyidae
    Keramidomys sp. 1
    Asianeomys sp
    Pentabuneomys sp.
    Ligerimys sp.
  Zapodidae
    Plesiosminthus cf. *P. barsboldi*
    Plesiosminthus sp. nov.
    Heterosminthus nanus
    Heterosminthus sp.
    Litodonomys sp. nov.
    Sicista sp. nov.
  Zapodidae gen. et sp. nov. 1
  Zapodidae gen. et sp. nov. 2
  Cricetidae
    Eucricetodon? sp.
    Democricetodon sp. 1

LAGOMORPHA
  Ochotonidae
    Desmatolagus ? sp.
    Sinolagomys sp.
  Ochotonidae gen. et sp. nov.

CARNIVORA
  Palaeogalidae
    Palaeogale sp.
  Mustelidae indet.
PERISSODACTYLA
  Chalicotheriidae
  Schizotheriinae indet.
  Rhinocerotidae
  Rhinocerotidae indet.

UPPER AOERBAN FAUNA

INSECTIVORA
  Erinaceidae
    Mioechinus cf. *M. gobiensis*
  Talpidae
    Proscapanus sp.
    Quyania sp.
    Yanshuella sp.
  Soricidae
    Sorex? sp. 1
    Sorex? sp. 2
RODENTIA
  Tachyoryctoididae
    Aralomys sp. 2
  Aplodontidae
    Ansomys sp. 1
  Sciuridae
    Eutamias cf. *E. ertemtensis*
    Eutamias sp.
    Oriensciurus? sp.
    Atlantoxerus sp.
  Gliridae
    Miodyromys sp.
  Eomyidae
    Keramidomys sp. 1
    Pentabuneomys sp.
  Zapodidae
    Plesiosminthus sp. nov.
    Heterosminthus sp.
    Sicista sp. nov.
  Cricetidae
    Eucricetodon? sp.
    Democricetodon sp.2
    Megacricetodon sp.
    Cricetodon sp. 1
LAGOMORPHA
  Ochotonidae
    Desmatolagus? sp.
    Alloptox sp.

Proboscidea
  Proboscidea indet.
ARTIODACTYLA
  Cervidae
    Ligeromeryx/Lagomeryx sp.

GASHUNYINADEGE FAUNA

INSECTIVORA
  Erinaceidae
    Amphechinus sp.
    Mioechinus cf. *M. gobiensis*
    Metexallerix sp.
  Talpidae
    Proscapanus sp.
    Quyania sp.
    Yanshuella sp.
    Desmanella sp.
  Soricidae
    Sorex? sp. 1
    Sorex? sp. 2
    Mongolosorex sp.
  Heterosoricinae indet.
RODENTIA
  Ctenodactylidae
    Prodistylomys/Distylomys sp.
  Tachyoryctoididae
    Tachyoryctoides sp. 1
    Tachyoryctoides sp. 2
    Aralomys sp.
  Aplodontidae
    Ansomys sp.
    cf. *Ansomys* sp. nov.
  Aplodontidae gen. et sp. nov.
  Sciuridae
    Eutamias sp.
    Atlantoxerus sp.
  Gliridae
    Prodryomys sp.
    Miodyromys sp.
    Microdyromys sp.
  Eomyidae
    Keramidomys sp. 1
    Leptodontomys sp.
    Asianeomys sp
    Pentabuneomys sp.
    Ligerimys sp.
  Zapodidae
    Plesiosminthus cf. *P. barsboldi*
    Plesiosminthus sp. nov.

Heterosminthus nanus
Heterosminthus sp.
Litodonomys sp. nov.
Sicista sp. nov.
Zapodidae gen. et sp. nov. 1
Zapodidae gen. et sp. nov. 2
Cricetidae
    Eucricetodon? sp.
    Democricetodon sp. 1
    Democricetodon sp. 3
    Megacricetodon sp.
LAGOMORPHA
Ochotonidae
    Desmatolagus? sp.
    Sinolagomys sp.
    Alloptox sp.
Ochotonidae gen. et sp. nov.
CARNIVORA
Palaeogalidae
    Palaeogale sp.
Amphicyonidae
    Amphicyon sp.
Mustelidae indet.

TAIRUM NUR FAUNA

INSECTIVORA
Erinaceidae
    Mioechinus cf. *M. gobiensis*
RODENTIA
Ctenodactylidae
    Distylomys tedfordi
Tachyoryctoididae
    Tachyoryctoides sp.
Sciuridae
    Atlantoxerus cf. *A. orientalis*
Zapodidae
    Heterosminthus cf. *orientalis*
Cricetidae
    Gobicricetodon cf. *G. flynni*
    Plesiodipus leei
    Megacricetodon sp.
LAGOMORPHA
Ochotonidae
    Desmatolagus? moergenensis
    Alloptox gobiensis
    Bellatona forsythmajori

CARNIVORA
Mustelidae
    Leptarctus neimenguensis
    Aelurocyon? sp.
    Sthenictis neimonguensis
Hyaenidae
    Tungurictis spocki
PROBOSCIDEA
Gomphotheriidae
    Platybelodon grangeri
PERISSODACTYLA
Rhinocerotidae
    Acerorhinus zernowi
ARTIODACTYLA
Cervidae
    Stephanocemas thomsoni
    Dicrocerus sp.

MOERGEN FAUNA

INSECTIVORA
Erinaceidae
    Mioechinus gobiensis[†‡]
    Mioechinus sp.[†‡]
Erinaceidae indet.[†‡]
Talpidae
    Proscapanus sp.[†‡]
    Quyania sp.[†‡]
    Yanshuella sp.[†‡]
    Desmanella sp.[†‡]
Talpidae indet.[†‡]
Soricidae
    Mongolosorex qiui[†‡]
Soricinae indet.[†‡]
Soricidae indet.[†‡]
CHIROPTERA
CHIROPTERA INDET.[†‡]
RODENTIA
Sciuridae
    Eutamias aff. *E. ertemtensis*[†‡]
    Sinotamias primitivus[†]
    Atlantoxerus orientalis[†‡]
Gliridae
    Microdyromys wuae[†‡]
    Miodyromys sp.[†‡]

[†]MoII at Moergen loc and its laterally correlative localities.
[‡]Upper red bed of 346 loc.

Eomyidae
  Keramidomys fahlbuschi[†‡]
  Leptodontomys lii[†]
  Leptodontomys aff. *gansus*[†‡]
Castoridae
  Anchitheriomys tungurensis[†‡]
  "*Monosaulax*" *tungurensis*[†‡]
  Hystricops? sp.[†‡]
Zapodidae
  Heterosminthus orientalis[†‡]
Dipodidae
  Protalactaga grabaui[†‡]
  Protalactaga major[†‡]
Cricetidae
  Gobicricetodon flynni[†]
  Gobicricetodon sp.[†]
  Plesiodipus leei[†]
  Democricetodon lindsayi[†‡]
  Democricetodon tongi[†]
  Megacricetodon sinensis[†‡]
LAGOMORPHA
Ochotonidae
  Desmatolagus? moergenensis[†‡]
  Alloptox gobiensis[†‡]
  Bellatona forsythmajori[†‡]
Ursidae
  Plithocyon teilhardi[†]
  Pseudarctos sp.[†]
Amphicyonidae
  Amphicyon tairumensis[†]
  Gobicyon macrognathus[†]
Mustelidae
  Melodon? sp.[†]
  Mionictis sp.[†]
  Martes sp.[†]
Hyaenidae
  Tungurictis spocki[†]
  Percrocuta tungurensis[†]
Felidae
  Metailurus mongoliensis[†]
Nimravidae
  Sansanosmilus sp.[†]
PROBOSCIDEA
Gomphotheriidae
  Serridentinus gobiensis[†]
  Platybelodon grangeri[†]
Mammutidae
  Zygolophodon sp.[†]

PERISSODACTYLA
Equidae
  Anchitherium gobiense[†]
Chalicotheriidae
  Chalicotherium brevirostris[†]
Rhinocerotidae
  Acerorhinus zernowi[†]
  Hispanotherium tungurense[†]
Rhinocerotidae indet.
ARTIODACTYLA
Suidae
  Listriodon mongoliensis[†]
  Kubanochoerus sp.[†]
Giraffidae
  Palaeotragus tungurensis[†]
Cervidae
  Stephanocemas thomsoni[†]
  Lagomeryx triacuminatus[†]
  Dicrocerus grangeri[†]
  Micromeryx sp.[†]
Bovidae
  Turcocerus grangeri[†]
  Turcocerus noverca[†]

TAMUQIN FAUNA

INSECTIVORA
Erinaceidae
  Mioechinus gobiensis
Talpidae
Talpidae indet.
RODENTIA
Aplodontidae
  Ansomys sp.
Sciuridae
  Eutamias aff. *E. ertemtensis*
Castoridae
  Steneofiber sp.
Zapodidae
  Heterosminthus orientalis
Dipodidae
  Protalactaga grabaui
Cricetidae
  Gobicricetodon robustus
  Democricetodon lindsayi
  Democricetodon tongi
  Megacricetodon sinensis
  Rhinocerodon sp.

LAGOMORPHA
  Ochotonidae
    Alloptox gobiensis
    Bellatona forsythmajori

BALUNHALAGEN FAUNA

INSECTIVORA
  Erinaceidae
    Mioechinus cf. *M. gobiensis*
  Talpidae
    Proscapanus sp.
    Quyania sp.
    Yanshuella sp.
  Soricidae
  Heterosoricinae indet.
  Soricinae indet.
RODENTIA
  Aplodontidae
    Ansomys sp. 2
  Sciuridae
    Eutamias cf. *E. ertemtensis*
    Eutamias sp.
    Atlantoxerus sp.
  Sciuridae indet. 1
  Gliridae
    Miodyromys sp.
    Microdyromys sp.
  Eomyidae
    Keramidomys sp. 2
    Leptodontomys cf. *L. lii*
    Pentabuneomys sp.
  Zapodidae
    Heterosminthus orientalis
    Sicista sp.
  Zapodidae indet.
  Dipodidae
    Protalactaga grabaui
    Protalactaga sp.
    Brachyscirtetes sp.
  Cricetidae
    Eucricetodon? sp. 2
    Cricetodon sp. nov.
    Democricetodon lindsayi
    Megacricetodon sinensis
    Plesiodipus cf. *P. progressus*
    Gobicricetodon cf. *G. flynni*
    Gobicricetodon cf. *G. robustus*
  Siphneidae (=Myospalacinae)
    Prosiphneus sp.

LAGOMORPHA
  Ochotonidae
    Desmatolagus? sp.
    Alloptox sp.
    Bellatona sp.
    Ochotona sp.
ARTIODACTYLA
  Cervidae
    Ligeromeryx/Lagomeryx sp.

AMUWUSU FAUNA

INSECTIVORA
  Erinaceidae
    Mioechinus gobiensis
  Erinaceidae indet.
  Talpidae
    Yanshuella cf. *Y. primaeva*
  Soricidae
  Soricinae indet.
  Heterosoricinae indet.
RODENTIA
  Aplodontidae
    Ansomys sp.
  Meniscomyinae indet.
  Sciuridae
    Eutamias cf. *E. ertemtensis*
    cf. *Miopetaurista* sp.
  Gliridae
    Microdyromys sp. 1
  Eomyidae
    Keramidomys sp. 2
    Leptodontomys sp.
    Pentabuneomys sp.
    Asianeomys sp.
  Castoridae
    "*Monosaulax*" sp.
    Castor sp.
  Zapodidae
    Heterosminthus cf. *orientalis*
    Sinozapus sp.
  Dipodidae
    Protalactaga cf. *P. graubaui*
    Paralactaga sp.
  Cricetidae
    Democricetodon sp. 3
    Gobicricetodon sp. 1
    Gobicricetodon sp. 2

Cricetidae indet.
Siphneidae (=Myospalacinae)
  Prosiphneus qiui
LAGOMORPHA
Ochotonidae
  Desmatolagus? sp.
  Ochotona? sp.
CARNIVORA
Hyaenidae
  Ictitherium sp.
Ursidae
  Plithocyon cf. *P. teilhardi*
Proboscidea indet.
PERISSODACTYLA
Equidae
  Anchitherium sp.
Rhinocerotidae indet.
ARTIODACTYLA
Cervidae
  Micromeryx sp.
  Euprox sp.

HUITENGHE ([†]) AND SHALA ([‡]) FAUNAS

INSECTIVORA
Erinaceidae
  Mioechinus sp.[†]
Erinaceidae indet.[‡]
Talpidae
  Proscapanus sp.[†]
  Yanshuella primaeva[‡]
  Yanshuella sp.[†]
  cf. *Asthenoscapter* sp.[‡]
Soricidae
  Paranourosorex sp.[†‡]
  Sorex cf. *S. ertemteensis*[‡]
Soricinae indet.[†‡]
RODENTIA
Aplodontidae
  Ansomys sp.[†‡]
Sciuridae
  Eutamias cf. *ertemtensis*[†‡]
  Prospermophilus cf. *orientalis*[†‡]
  Sinotamias sp.[‡]
  Pliopetaurista sp.[‡]
Gliridae
  cf. *Prodryomys* sp.[‡]
  Miodyromys sp.[‡]
  Microdyromys wuae[†]

  Microdyromys sp. 1[‡]
  Microdyromys sp. 2[‡]
Eomyidae
  Keramidomys fahlbuschi[†]
  Leptodontomys sp.[†‡]
  Pentabuneomys sp.[‡]
Zapodidae
  Lophocricetus cf. *L. gansus*[†‡]
  Sicista sp.[‡]
  Sinozapus sp.[†‡]
Dipodidae
  Paralactaga sp. 1[‡]
  Paralactaga sp. 2[‡]
  Dipus sp. nov.[‡]
  Cardiocranius sp.[‡]
Cricetidae
  Gobicricetodon cf. *G. flynni*[†]
  Democricetodon lindsayi[†]
  Nannocricetus sp.[†]
  Kowalskia sp.[†‡]
  Sinocricetus sp.[‡]
  cf. *Sinocricetus* sp.[‡]
  Microtoscoptes sp.[‡]
  cf. *Microcricetus* sp.[‡]
  Anatolomys sp.[‡]
Cricetidae indet.[‡]
Siphneidae (=Myospalacinae)
  Prosiphneus sp.[†‡]
LAGOMORPHA
Ochotonidae
  Desmatolagus sp.[†]
  Ochotona cf. *O. lagreli*[‡]
  Ochotona sp.[†]

BAOGEDA ULA FAUNA

INSECTIVORA
Erinaceidae
Erinaceidae indet. 1
Erinaceidae indet. 2
Soricidae
  Parasoriculus sp.
RODENTIA
Gliridae
Gliridae indet.
Castoridae
  Dipoides sp.
Zapodidae
  Lophocricetus cf. *L. gansus*

Dipodidae
  Paralactaga sp. 1
  Paralactaga sp. 2
  Dipus sp. nov.
Cricetidae
  Kowalskia sp.
  Nannocricetus sp.
  Sinocricetus sp.
  cf. *Sinocricetus* sp.
  Microtoscoptes sp.
  Anatolomys sp.
  Rhinocerodon sp.
Gerbillidae
  Abudhabia sp.
Muridae
  Hansdebruijnia perpusilla
Siphneidae (=Myospalacinae)
  Prosiphneus sp.
FAMILY INDET.
  Pararhizomys hipparionum
LAGOMORPHA
Leporidae
  Alilepus sp.
Ochotonidae
  Ochotona cf. *O. lagreli*
CARNIVORA
Mephitidae
  Promephitis sp.
Mustelidae indet.
Hyaenidae
  Hyaenictitherium hyaenoides
PERISSODACTYLA
Equidae
  Hipparion sp.
Rhinocerotidae indet.
ARTIODACTYLA
Giraffidae indet.
Antilopinae
  Gazella sp. 1
  Gazella sp. 2

BILUTU FAUNA

INSECTIVORA
Erinaceidae
  Mioechinus cf. *M. gobiensis*
  Erinaceus sp.
Talpidae
  Proscapanus sp.

Quyania sp.
Yanshuella sp.
Soricidae
  *Heterosoricinae indet.
  Soricinae indet.
RODENTIA
Aplodontidae
  Ansomys sp. 2
Sciuridae
  Eutamias cf. *E. ertemtensis*
  Eutamias sp.
  Sciurus sp.
  Atlantoxerus sp.
  Prospermophilus sp.
Sciuridae indet. 2
Gliridae
  Miodyromys sp.
  Microdyromys sp.
  Myomimus sinensis
Eomyidae
  Keramidomys sp. 1
  Keramidomys sp. 2
  Pentabuneomys sp
  Leptodontomys sp.
Zapodidae
  Heterosminthus cf. *orientalis*
  Sicista sp.
  Lophocricetus grabaui
  Lophocricetus sp.
Dipodidae
  Protalactaga grabaui
  Paralactaga suni
  P. cf. *P. anderssoni*
  Dipus fraudator
Cricetidae
  *Eucricetodon? sp. 2
  Cricetodon sp. 2
  Democricetodon lindsayi
  Megacricetodon sinensis
  Plesiodipus leei
  Plesiodipus aff. *P. progressus*
  Plesiodipus sp.
  Gobicricetodon cf. *robustus*
  Kowalskia sp.
  Sinocricetus sp.
  Nannocricetus sp.
  Microtoscoptes sp.
  Microtodon sp.

* Taxon probably reworked from lower horizons.

    Anatolomys sp.
    Rhinocerodon sp.
  Cricetidae indet. 2
  Gerbillidae
    Pseudomeriones sp.
  Siphneidae (=Myospalacinae)
    Prosiphneus sp. nov.
  Muridae
    Hansdebruijnia pusilla
    Micromys chalceus
    "Karnimata" hipparionum
LAGOMORPHA
  Ochotonidae
    *Desmatolagus? sp.
    *Alloptox? sp.
    *Bellatona sp.
    Ochotona cf. *O. lagreli*

ERTEMTE([†]) AND HARR OBO([‡]) FAUNAS

INSECTIVORA
  Erinaceidae
    Erinaceus mongolicus[†]
  Talpidae
    Quyania chowi[†‡]
    Yanshuella primaeva[†‡]
  Soricidae
    Sorex minutoides[†‡]
    S. ertemteensis[†‡]
    S. pseudoalpinus[†‡]
    Zelceina kormosi[†‡]
    Alloblarinella sinica[†]
    Paenepetenyia zhudingi[†‡]
    Cokia kowalskae[†‡]
    Paranourosorex inexspectatus[†‡]
    Paenelimnoecus obtusus[†]
RODENTIA
  Aplodontidae
    Pseudaplodon asiaticus[†‡]
  Sciuridae
    Eutamias ertemtensis[†‡]
    Sciurus sp.[†‡]
    Prospermophilus orientalis[†‡]
    Sinotamias gravis[†‡]
    Petinomys auctor[†‡]
    Pliopetaurista rugosa[†‡]

  Gliridae
    Myomimus sinensis[†‡]
  Eomyidae
    Leptodontomys gansus[†‡]
  Castoridae
    Castor anderssoni[†]
    Dipoides majori[†‡]
  Zapodidae
    Lophocricetus grabaui[†‡]
    Paralophocricetus pusillus[†‡]
    Sicista sp.[†‡]
    Eozapus similis[†‡]
  Dipodidae
    Paralactaga suni[†‡]
    Brachyscirtetes wimani[†‡]
    Dipus fraudator[†‡]
  Cricetidae
    Sinocricetus zdanskyi[†‡]
    Nannocricetus mongolicus[†‡]
    Kowalskia similis[†‡]
    K. neimengensis[†‡]
    Anatolomys teilhardi[†‡]
    Microtodon atavus[†‡]
    Microtoscoptes praetermissus[†‡]
  Gerbillidae
    Pseudomeriones abbreviatus[†‡]
  Siphneidae (=Myospalacinae)
    Prosiphneus eriksoni[†‡]
  Muridae
    Apodemus orientalis[†‡]
    Orientalomys cf. *O. similis*[†‡]
    Hansdebruijnia pusilla[†‡]
    Karnimata? hipparionum[†‡] *Micromys chalceus*[†‡]
    Rhagapodemus sp.[‡]
LAGOMORPHA
  Leporidae
    Alilepus annectens[†‡]
    Trischizolagus sp.[‡]
  Ochotonidae
    Ochotona lagreli[†‡]
    Ochotona minor[†‡]
CARNIVORA
  Mustelidae
    Meles suillus[†]
    Promephitis alexejervi[†]
    Martes andersson[†]
    Martes sp.[†]
  Hyaenidae
    Ictitherium aff. *I. hipparionum*[†]
    Hyaena sp.[†]

Felidae
  Machairodus sp.[†]
  Felis sp[†]
Proboscidea
  Mastodon sp.[†]

PERISSODACTYLA
  Equidae
    Hipparion richthofeni[†]
  Rhinocerotidae
    Rhinoceros habereri[†]

ARTIODACTYLA
  Suidae
    Sus cf. *S. hyotherioides*[†]
    Sus sp.[†]
  Cervidae
    Alcicephalus sp.[†]
    Axis apecisus[†]
    "Paracervulus" brevi[†]
    Moschus grandaevus[†]
  Bovidae
    Procapreolus rutimeyer[†]
    P. latifrons[†]

BILIKE ([†]) AND GAOTEGE ([‡]) FAUNAS

INSECTIVORA
  Erinaceidae
    Erinaceus mongolicus[†]
    Erinaceus sp.[‡]
  Talpidae
    Quyania cf. *chowi*[†]
    Yanshuella primaeva[†‡]
    Desmana sp.[†‡]
  Soricidae
    Sorex minutoides[†]
    S. ertemteensis[†‡]
    S. pseudoalpinus[†‡]
    Sorex sp.[†‡]
    Petenyia katrinae[†‡]
    Lunanosorex cf. *lii*[†‡]
    Parasoriculus tongi[†]
    Sulimskia ziegleri[†‡]
    Paenelimnoecus obtusus[†]
CHIROPTERA
  Vespertilionidae
    Myotis cf. *annectans*[†]
    Myotis cf. *horsfieldii*[†]
    Myotis sp.[†]
    Vespertilio cf. *sinensis*[†]
    Murina sp.[†]

RODENTIA
  Sciuridae
    Eutamias ertemtensis[†‡]
    Sciurus yusheensis[†]
    Prospermophilus cf. *P. orientalis*[†‡]
  Gliridae
    Myomimus sinensis[†]
  Castoridae
    Castor anderssoni[†‡]
    Castor zdanskyi[†]
    Dipoides majori[†]
  Castoridae indet. 1[‡]
  Castoridae indet. 2[‡]
  Zapodidae
    Lophocricetus grabaui[†]
    Paralophocricetus pusillus[†]
    Sicista wangi[†]
    Sicista sp.[†‡]
    Eozapus similis[‡]
    Sinozapus volkeri[†‡]
  Dipodidae
    Paralactaga suni[†‡]
    Brachyscirtetes cf. *robustus*[†]
    Brachyscirtetes sp.[‡]
    Dipus fraudator[†]
    Dipus cf. *D. fraudator*[†]
  Cricetidae
    Sinocricetus progressus[†‡]
    Nannocricetus mongolicus[†‡]
    Kowalskia zhengi[†]
    Kowalskia cf. *similis*[†]
    Anatolomys cf. *teilhardi*[†]
    Microtodon cf. *atavus*[†]
  Cricetidae indet.[‡]
  Gerbillidae
    Pseudomeriones abbreviatus[†]
    P. cf. *P. abbreviatus*[‡]
    Pseudomeriones sp.[‡]
  Avicolidae
    Mimomys bilikeensis[†]
    Mimomys cf. *bilikeensis*[‡]
    Mimomys cf. *M. orientalis*[‡]
  Siphneidae (=Myospalacinae)
    Prosiphneus cf. *eriksoni*[†]
    Chardina sp.[‡]
    Chardina sp. nov.[‡]
  Muridae
    Apodemus lii[†‡]
    Apodemus sp.[†‡]
    Orientalomys sinensis[†]
    Micromys kozaniensis[†]

M. cf. M. kozaniensis[‡]
Chardinomys bilikeensis[†]
Chardinomys sp.[‡]
Huaxiamys downsi[‡]
Huaxiamys sp.[†]
Allorattus engesseri[†]
Incertae familiae
Pararhizomys sp.[‡]
LAGOMORPHA
Leporidae
Trischizolagus mirificus[†]
Trischizolagus sp.[‡]
Ochotonidae
Ochotona minor[†]
Ochotona sp.[‡]
CARNIVORA
Mustelidae
Lutra sp.[‡]
Pannonictis sp.[‡]
Mustela sp.[‡]

Hyaenidae
Chasmaporthetes sp.[‡]
Canidae
Nyctereutes sp. nov.[‡]
Eucyon? sp.[‡]
PROBOSCIDEA
Mastodon sp.[‡]
PERSSODACTYLA
Equidae
Hipparion insperatum[‡]
H. huangheense[‡]
Rhinocerotidae
Rhinoceros sp.[‡]
ARTIODACTYLA
Cervidae
Cervidae indet.[‡]
Giraffidae
Giraffidae indet.[‡]
Camelidae
Camelidae indet.[‡]

Chapter 6

Mammalian Biochronology of the Late Miocene Bahe Formation

ZHAO-QUN ZHANG, ANU KAAKINEN, LI-PING LIU, JUHA PEKKA LUNKKA, SEVKET SEN, WULF A. GOSE, ZHU-DING QIU, SHAO-HUA ZHENG, AND MIKAEL FORTELIUS

The use of fossils, especially of terrestrial mammals, as "dragon bones" in Chinese traditional medicine gave early paleontologists in China a head start, since many rich fossil localities were already known and readily accessible. More than two hundred species of Late Miocene terrestrial mammals have by now been recorded from China (Teilhard de Chardin and Leroy 1942; Fortelius 2009), representing almost every terrestrial mammal family of that age known from Eurasia. However, for lack of precise provenance and/or detailed stratigraphic work, for many years fossils were clumped into broad faunal aggregates, such as the "*Hipparion* fauna" (Schlosser 1903; Kurtén 1952).

The sedimentary sequence exposed along the Bahe River near Lantian in Shaanxi Province is one of the most complete available for subdivision of Late Miocene mammal biostratigraphy. Detailed stratigraphic correlation and paleontological work was first undertaken there during the years 1958–1965 (Liu, Ding, and Gao 1960; Zhang et al. 1978). Zhou (1978) named three "fossil zones," represented mainly by fossil assemblages from Koujiacun (locality 59S10), Shuijiazui (59S5, 59S6), and Jiulaopo (59S3), respectively. Based on apparent stratigraphic superposition and faunal composition in Lantian, Li, Wu, and Qiu (1984) proposed the Bahean (Late Miocene) and Baodean (latest Miocene) stages. The Bahean Stage was at that time tentatively correlated with the European Vallesian Stage. However, Qiu and Qiu (1995) eliminated the Bahean stage and extended the Baodean stage to represent the time of the Late Miocene *Hipparion* faunas in its entirety. The Baodean stage in this sense encompassed five successive faunas, including the Bahe fauna. Based on the faunal succession of Qiu and Qiu (1995), Qiu, Wu, and Qiu (1999) adopted the Neogene mammal faunal unit (NMU) system to further subdivide the Chinese Neogene faunas. The Shuijiazui fauna (formerly Bahe local fauna) from the middle and upper part of the Bahe Formation was named NMU 9 and correlated approximately with the later part of the Vallesian (Mein's MN 10).

In addition to the well-known biogeographic complications of long-distance biostratigraphic correlation, we must here draw attention to another difficulty in the sequencing and especially the correlation of Chinese mammal localities with the European MN system using the concepts of fauna and local fauna (Qiu 1990; Qiu, Wu, and Qiu 1999). Faunas, either from an isolated locality or composited into a local fauna from multiple localities or levels, represent only a time point rather than a time period, a fact that can easily cause confusion between sequence and time equivalence. For example, the isolated Ertemte fauna (a time point) was widely accepted to be of the latest Miocene and correlated with MN 13 (a time period) (Storch 1987; Qiu 1987; Wu 1991). The Baode fauna (a composited fauna from different levels and multiple localities), which was considered to be earlier, was accordingly thought to be correlated with MN 11–12 (Deng 2006). And since the Lantian fauna from the basal part of the Lantian Formation was correlated with the

Table 6.1

History of Subdivision and Correlation of Chinese Late Miocene

Epoch		MN	Li et al. (1984) Age	Qiu and Qiu (1995) Age	Local faunas	Qiu et al. (1999) Age	NMU (RF)	Deng (2006) Age	NMU (RF)	Zhang and Liu (2005) Age	Biozone
Late Miocene	Turolian	13			Ertemte		11	Baodean	11	Baodean	3
	Turolian	12	Baodean	Baodean/ Lufengian	Lufeng	Baodean	10	Baodean	10	Bahean	2 "BF"
	Turolian	11			Baode						
	Vallesian	10	Bahean (BF)		Bahe	9 (SJZ)		Bahean	9 (BF)		1
	Vallesian	9			Tsaidam	8			8		

NOTE: RF = Representative Fauna; BF = Bahe Fauna; SJZ = Shuijiazui fauna.

taxonomically similar Baode fauna, which was thought to be time-equivalent to the European Turolian stage, the apparently reasonable correlation of the underlying Bahe Fauna with the Vallesian was widely accepted (Li, Wu, and Qiu 1984; Deng 2006: table 6.1).

One of the purposes of our five field seasons during 1997–2001 in the Lantian area was to derive a chronology for faunas from different horizons and initiate a biochronologic division for the Late Miocene by efforts from multiple disciplines (e.g., sedimentology, magnetostratigraphy, biostratigraphy and isotopic geochemistry). Since then stratigraphic and sedimentologic studies of the sections have been reported by Kaakinen and Lunkka (2003), Andersson and Kaakinen (2004) and Kaakinen (2005). Systematic studies on most of the mammal groups have also been published (Qiu, Zheng, and Zhang 2004a, 2004b; Li and Zheng 2005; Zhang, Flynn, and Qiu 2005; Qiu, Zheng, and Zhang 2008; Zhang et al. 2008). We have also discussed the implications of the Lantian sequence for Chinese mammal biochronology and proposed three mammal biozones for the Late Miocene (Zhang and Liu 2005).

Here we synthesize the lithologic profile of the Bahe Formation, plot old localities and our new fossil localities in it, update the faunal list, calibrate the age of the main localities by paleomagnetic dating (Kaakinen 2005), and further discuss the biochronology. Finally, we will char-

acterize and clarify the concept of the Bahean age and tentatively propose its resurrection.

LITHOSTRATIGRAPHY AND DEPOSITIONAL ENVIRONMENTS

The existing stratigraphical scheme for the Lantian region was originally established on the basis of field observations undertaken during the 1960s (Zhang et al. 1978). Subsequent studies based on vertebrate paleontology and magnetostratigraphy (e.g., Qiu and Qiu 1995; Zheng, An, and Shaw 1992; Sun et al. 1997) have, in part, refined the scheme, mainly in the Neogene part of the succession. The current understanding of stratigraphy, age relationships, and depositional environments of the Bahe and Lantian formations are based on Zhang et al. (2002), Kaakinen and Lunkka (2003), Andersson and Kaakinen (2004), and Kaakinen (2005).

The sedimentological and stratigraphical work established a total of 30 vertical sections along the cliffs of the Bailuyuan Plateau, spanning more than 1 km in thickness. These sections serve as a basis for both the division of the sedimentary record into several lithofacies associations as well as interpretation of their depositional environments. The profiles were correlated lithostratigraphically and magnetostratigraphically.

The Bahe Formation has a maximum thickness of about 280 m in the study area. The dominant sedimentary facies are overbank deposits that crop out as massive, reddish-brown siltstones typically containing pedogenic features such as mottling, root traces, and carbonate nodule formation. The reddish-brown siltstones representing well-drained floodplain deposits are often interdigitated with poorly drained floodplain deposits that are grayish green in color. These fine-grained floodplain deposits are well exposed at all levels in the formation ranging up to 20 m in thickness. Apart from the floodplain sediments, the overbank deposits include crevasse splays and unconfined sheetflows that are organized in thin (commonly less than 1 m thick), fining-upward sequences. These occur at several levels within the overbank deposits. The floodplain deposits also comprise a few lacustrine marls that form laterally persistent ridges providing reliable guide horizons for correlation. The channel fills observed in the Bahe Formation are coarse grained and typically form broad sheets in the outcrops. The lack of lateral accretion features and their broad profiles correspond to frequent shifting of a river channel in the basin, indicative of a channel with relatively low sinuosity.

The Lantian Formation, consisting of a basal conglomerate of varying thickness and an upper unit of mainly structureless red-brown clayey siltstones and associated carbonate nodules, overlies the Bahe Formation with a contact relationship of a regional angular unconformity. The Lantian Formation, corresponding to late Neogene lithological units elsewhere in the Chinese Loess Plateau, has been informally referred to as "*Hipparion* Red Clay" by Zdansky (1923), and numerous studies have confirmed that these deposits are mainly of aeolian in origin (e.g., Sun et al. 1997; Sun et al. 1998; Ding et al. 1998; Guo et al. 2002; Lu, Vandenberghe, and An 2001). Our field observations show that the lower part of the "Red Clay" sequence in Lantian is at least partly of fluvial origin. However, the fluvial imprint is restricted to the lowermost part of the Lantian Formation, and except for the so-called boundary conglomerate, the fluvial deposits within the Red Clay sequence are thin and localized. The cessation of major fluvial channels indicates that the palaeo-Bahe streambed was deflected from this area during deposition of the bulk of the Lantian Formation. It seems likely that this major facies change between the formations was caused by a combination of regional tectonics and climate change that can all be considered as ultimately driven by the uplift of the Himalayas/Tibetan Plateau (i.e., the uplift of the Qinling mountains; Zhang et al. 1978), extensional tectonism in the Weihe graben (Bellier et al. 1988; Bellier et al. 1991), and intensification of the Asian monsoon system (Kutzbach, Prell, and Ruddiman 1993).

FOSSIL LOCALITIES

Liu, Ding, and Gao (1960) found seven localities from the Bahe and Lantian formations, 59S1–6 and 59S10. Excavations in the 1960s were mainly focused on these localities. Field locality numbers 63710 (=59S3), 63702.L1 (=59S5), and 63702.L4 (=59S6) were also used at this time (Liu, Li, and Zhai 1978). During our field seasons (1997–2001), we used a sequential numbering system ("Loc.N"). For the screenwashing sampling levels in the main section, the prefix MS was used to separate them from other localities.

By lithologic correlation, several of our localities turn out to be in the same level as the previously collected localities. Loc.30 and 32 are correlated with 59S6; Loc.5, 6, 10, 11, and 33 with 59S5; and Loc.42 with 59S3. On the basis of the lithology and composition of the fossil assemblage from 59S10, Zhang et al. (2002) suggested that 59S10 represents a slump from an upper level, correlated by them with 59S3. We accept this interpretation here.

MAGNETOSTRATIGRAPHY

A robust correlation of the Lantian sequence to the Neogene time scale is proposed on the basis of its characteristic pattern of magnetozones, supported by faunal records and lithostratigraphy.

As a result of a lithostratigraphic correlation between our magnetostratigraphic sections, known respectively as the Main, the Pyramid, the Locality 42, and the Liujiapo sections, it is possible to obtain firm tie points for the magnetostratigraphic correlation (figure 6.1). A characteristic limestone unit recognized in the Main Section at the 102 m level occurs in the basal part of the Pyramid Section at the 25 m level, as shown in figure 6.1. Hence, the normal polarity zone in the Main Section (at about the 97–115 m level; see figure 6.1) can be equated to the normal polarity at about the 21–39 m level in the Pyramid Section. In addition, the calcareous paleosol overlying Locality 6 has been used for correlation between the upper part of Pyramid and lower part of Locality 42 sections. This allows correlation of a long normal polarity zone in Locality 42 to the topmost normal magnetozone in the Pyramid Section. The Liujiapo and Locality 42 sections were correlated using a laterally extensive

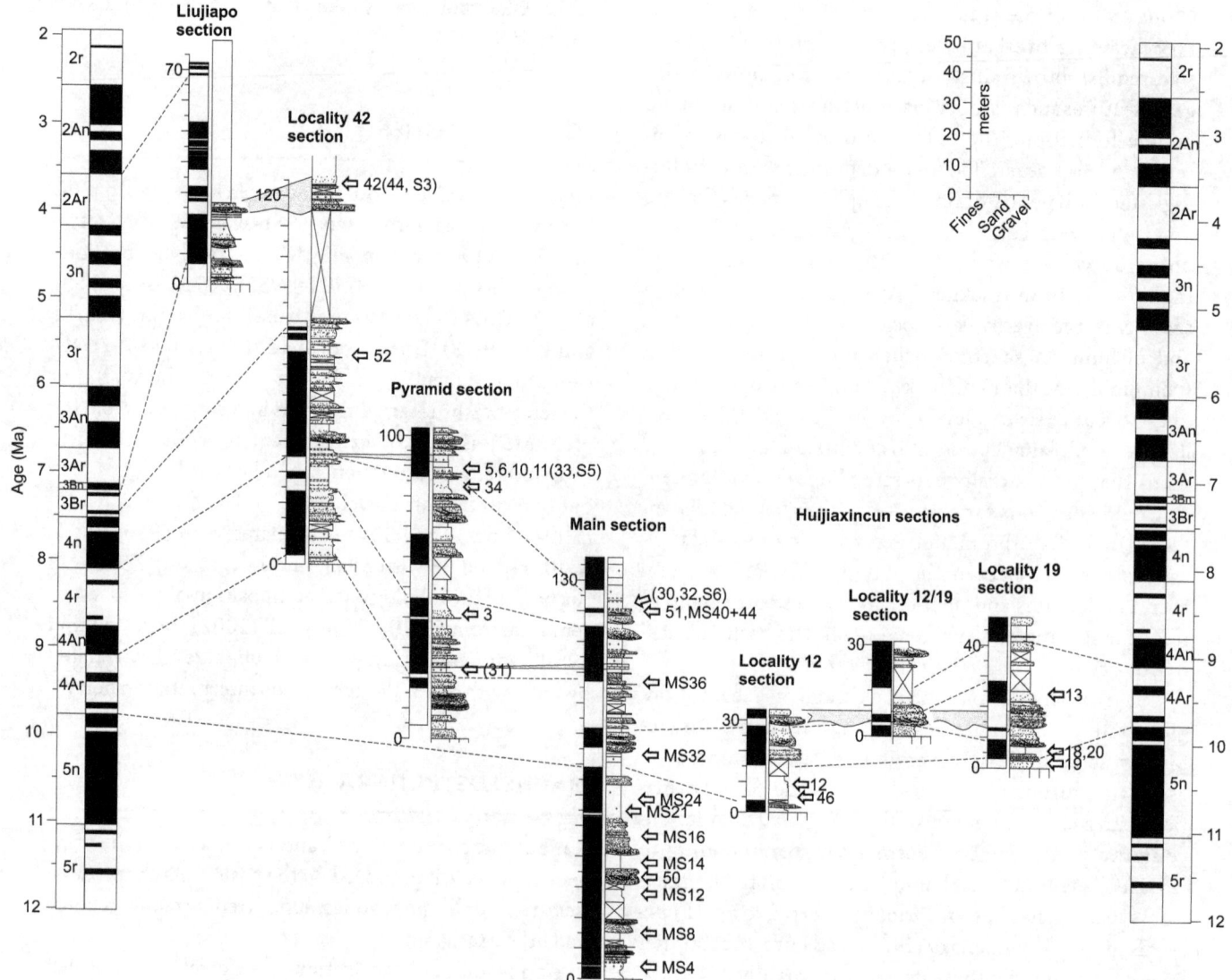

Figure 6.1 The main stratigraphic sections and magnetic polarity stratigraphies for the Lantian area. After Kaakinen and Lunkka (2003) and Kaakinen (2005). Lithostratigraphic correlations, based on the tracing of different marker horizons, are shown as gray lines, and dashed lines represent magnetostratigraphical correlations between the studied sections with the ATNTS 2004. Main fossil localities in these sections are also indicated; the localities in parentheses indicate fossil finds traced to the level presented in the log.

boundary conglomerate as a marker horizon. From this, it follows that the moderately long normal polarity of Locality 42 and the normal polarity zone underlying the marker conglomerate can be equated magnetostratigraphically. Although there is no lithostratigraphic tie between sections in the Huijiaxincun area and those closer to the town of Lantian, it is evident, on the basis of the mammal content, that localities in the Huijiaxincun area are significantly older than Locality 6 in the Pyramid Section. This implies that the polarity pattern obtained from the Huijiaxincun sections should be corre-

lated with similar patterns well below the topmost normal magnetozone in the Pyramid Section.

Correlation to the ATNTS 2004

The Lantian sequence provides a composite magnetostratigraphic section extending from chron 5n to chron 2An in the Astronomically Tuned Neogene Time Scale (ATNTS) of Lourens et al. (2004). The polarity sequences for the seven studied sections are long, and with the exception of the Huijiaxincun sections, allow independent correlations to the ATNTS. The distinctive long intervals of normal polarity in the main and Locality 42 sections were used as the principal anchor points for the ATNTS correlation and were correlated, respectively, with chrons C5n.2n and C4n.2n. The remaining magnetozones were correlated to the ATNTS based on the assumption that most chrons were detected as a result of the high sampling density (at intervals of 0.4–1.6 m in the Bahe Formation and ~0.4 m in the Lantian Formation). As a consequence, extrapolations based on rate of sedimentation date the basal and top ages to ca. 11 Ma and 3 Ma, respectively. The data provide no evidence for a major stratigraphic gap, as every polarity zone except C3Br.1r is accounted for.

BIOSTRATIGRAPHY

Sampling

Paleomagnetic dating shows that the Bahe Formation covers a time period from about 11 Ma to 7 Ma, with a total thickness of about 280 m. Most of the larger vertebrate fossils and some of the smaller ones were discovered by excavation and prospecting the surface of naturally eroded outcrops. In total, 52 fossil localities were discovered from the Bahe and Lantian formations (44 and 8 localities, respectively). Twenty-six of the fossil localities included mammalian material. Nine localities (Loc.6, 12, 30, 31, 34, 42, 44, 45, and 49) were excavated to some extent. Seven localities (Loc.3, 6, 12, 13, 19, 30 and 42) were screenwashed, and most of these localities also have remains of large species. In addition, due to the scarcity of micromammals in the Lantian sequence, 20 stratigraphic levels from the Main Section were also screenwashed, 10 of them (MS4, MS8, MS12, MS14, MS16, MS21, MS24, MS32, MS36, and MS40+44) producing small mammal fauna. Therefore, the resulting micromammal record does not represent a random sample, as in the sampling of large mammals, but is a result of deliberate effort of screenwashing of exceptionally large amounts of sediment (about 50 tons).

Mammalian fossil localities are widely scattered from the base upward to the top of the section. The distribution of fossil localities in the formation is, however, not even (figure 6.2b). The majority of mammal fossils occur in floodplain facies of the Bahe Formation. In the Lantian Formation, the fossil localities are scarce and mainly concentrated in the fluvial deposits of the lowermost Lantian Formation. Only very few specimens were associated with

the aeolian facies above. This scarcity is a real pattern because collection bias is unlikely after the extensive effort on prospecting the Lantian Formation (Kaakinen 2005).

The majority of the fossil localities occur between 10.8 Ma and 8.8 Ma, with a second concentration between 8.3 Ma and 7.8 Ma. Only a few bone fragments

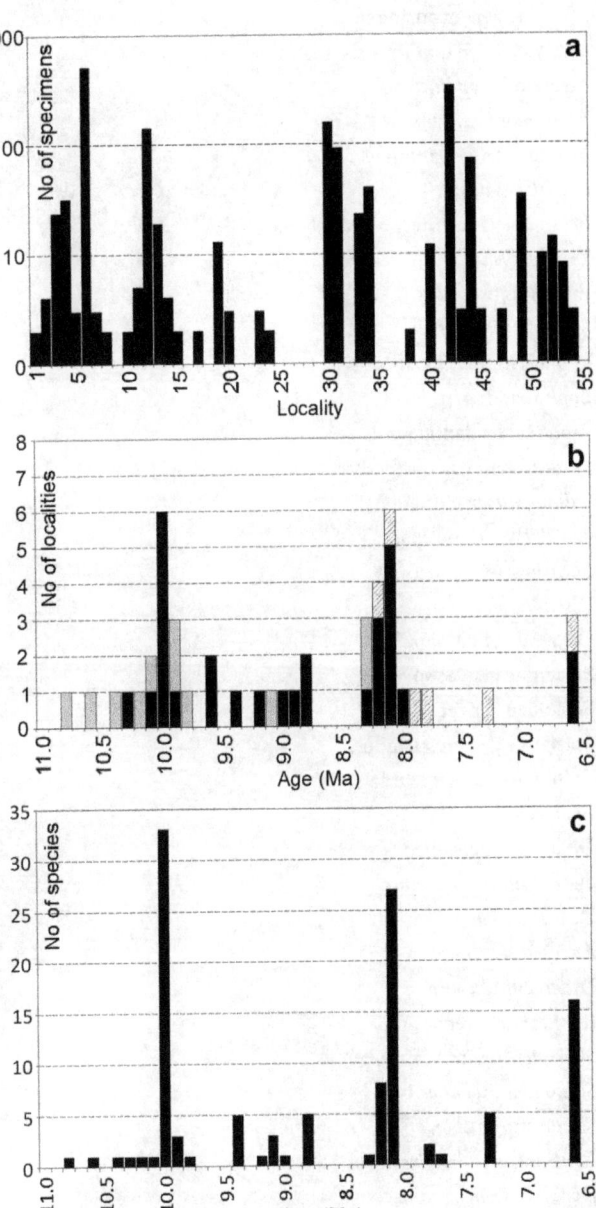

Figure 6.2 (a) Number of large mammal specimens collected from individual fossil localities. The old and MS localities are not included. Note a logarithmic scale on vertical axis. (b) Histogram showing the distribution of fossil localities in the Bahe Formation and the lowermost Lantian Formation per unit time. The gray color indicates MS localities; ruled pillars indicate old localities. (c) Histogram showing the distribution of different species in the Bahe Formation and lowermost Lantian Formation per unit time.

Table 6.2

Fossil Mammals from Lantian Localities with Their Age Estimation

Age (Ma)	10.79	10.62	10.39	10.29	10.23	10.08	10	9.98	9.95	9.92	9.9
Locality/Taxa	MS4	MS8	MS12	50	MS14	MS16	19	46	20/18/12	MS21	MS24
Progonomys sinensis							x		x		
Muridae gen. et sp. indet.								x		x	
Myocricetodon lantianensis							x		x		
Myocricetodon liui									x		
Abudhabia baheensis								x	x		
Nannocricetus primitivus	x							x	x	x	
Kowalskia indet.											x
Prosiphneus licenti											
Typhlomys sp.											
Eutamias lishanensis		x					x	x			
Sciurotamias pusillus							x		x		
Lophocricetus xianensis					x		x		x		
Lophocricetus sp.			x								
Protalactaga lantianensis					x		x		x		
Paralactaga sp.											
Salpingotus primitivus							x				
Cardiocranius pusillus							x				
Pararhizomys qinensis											
?Pseudaplodon sp.											
Ochotona cf. O. lagreli									x		
Soricinae gen. et sp. indet.							x		x		
Soricidae gen. et sp. indet.1											
Soricidae gen. et sp. indet.2								x			
Talpidae gen. et sp. indet.							x				
Erinaceus sp.											
Dinocrocuta gigantea											
cf. Metailurus parvulus											
cf. Metailurus major							x		x		
Adcrocuta eximia											
?Adcrocuta eximia							x				
Ictitherium viverrinum											
Hyaenictitherium cf. wongii											
Shaanxispira baheensis											
Shaanxispira chowi											
Lantiantragus longirostralis									x		
Dorcadoryx orientalis											
Gazella cf. lydekkeri											
Gazella sp.1											
Gazella sp.2									x		
Gazella paotehensis											
Protoryx indet.											
Schansitherium cf. tafeli											

9.75	9.37	9.23	9.14	8.95	8.84	8.79	8.28	8.21	8.15	8.07	7.91	7.84	7.71	7.26	6.6
MS32	13	35	MS36	31	3	38	MS40+44/51	32/30/S6	34	33/11/10/6/5/S5	S1	S2	52	S4	42/44/S3
	x					x									
			x				x			x					
										x					
x	x									x					
						x		x		x					
															x
															x
	x		x		x					x					
	x					x									
										x					
			x												
	x														
										x					
						x									
										x					
										x		x			
										x					
										x					
															x
															x
															x
										x					
							x								
										x					
										x					
				x						x					
												x	x	x	
															x
							x			x					
										x					

(continued)

Table 6.2

(continued)

Age (Ma)	10.79	10.62	10.39	10.29	10.23	10.08	10	9.98	9.95	9.92	9.9
Locality/Taxa	MS4	MS8	MS12	50	MS14	MS16	19	46	20/18/12	MS21	MS24
Palaeotragus cf. *decipiens*											
Palaeotragus microdon											
Metacervulus sp.											
Cervavitus novorossiae											
Procapreolus latifrons											
Microstonyx major											
Chleuastochoerus stehlini											
Hipparion chiai											
Hipparion weihoense											
Hipparion cf. *weihoense*											
Hipparion sp.1				x					x		
Hipparion sp.2											
Hipparion sp.3											
Hipparion cf. *hippidiodus*											
Hipparion sp.4											
Hipparion sp.5											
Hipparion sp.6											
Chalicotheriidae gen. et sp. indet.											
Stephanorhinus orientalis									x		
cf. *Brachypotherium* sp.											
cf. *Chilotherium* sp.									x		
Chilotherium "gracile"											
Chilotherium habereri											
Chilotherium indet.											
Acerorhinus paleosinense											
Acerorhinus indet.											
Tetralophodon exoletus											

were found from sediments between 7.7 Ma and 6.7 Ma (see figure 6.2b). As expected, numbers of specimens and taxa recorded from different horizons reflect sampling and taphonomy rather than original taxon richness (see figure 6.2a and c). For example, poor outcrop availability may explain the rarity of bone remains in the uppermost Bahe Formation. Nevertheless, the mammal fossil record from the Bahe Formation is quite adequate in a terrestrial biostratigraphic context. A total of 21 mammalian fossil horizons were identified in the Bahe Formation, with one main horizon from the base of the Lantian Formation. There are six horizons in the Bahe Formation and one in the Lantian Formation having five or more species (see figure 6.2c). The very

rich concentration at the Loc.6 level produced 27 species, while Loc.12 and Loc.19, from the lower part of the Bahe Formation, produced 12 and 16 species, respectively. In total, we found 53 species from the Bahe Formation, and 16 species from the base of the Lantian Formation (table 6.2).

Faunal Composition

Until now, fossil mammals discovered from the Bahe Formation include the following number of species: 17 rodents, 1 lagomorph, 5 insectivores, 3 carnivores, 12 artiodactyls, 14 perissodactyls, and 1 proboscidean.

9.75	9.37	9.23	9.14	8.95	8.84	8.79	8.28 MS40+44/51	8.21 32/30/S6	8.15	8.07 33/11/10/6/5/S5	7.91	7.84	7.71	7.26	6.6 42/44/S3
MS32	13	35	MS36	31	3	38			34		S1	S2	52	S4	
								x		x					
														x	x
										x					
															x
															x
															x
															x
							x	x		x	x			x	
							x	x		x					
														x	
							x	x		x					
										x					
															x
															x
							x								
										x					
										x					
															x
							x							x	
															x
								x						x	
										x					
								x		x					

Rodents and ungulates are the most diversified groups for the entire duration of the Bahean. In contrast to the absence of suids and extreme rarity of cervids in the Bahe Formation, suids are present and cervids are abundant in Loc.42 (basal Lantian Formation). The carnivoran fauna from the basal Lantian Formation is also remarkably different from that of the Bahe Formation, with *Ictitherium viverrinum* being dominant (Zhang 2005; Andersson and Werdelin 2005). Although about 10 tons of sediment from Loc.42 were screenwashed, only two species of rodents were found. The lack of small mammals from the Lantian Fauna does not, however, preclude the recognition of a major faunal turnover event between the Bahe Formation and the Lantian Formation (Zhang et al. 2002).

"Small" Mammals

Perhaps the most significant progress in the biostratigraphic aspect of our Lantian projects is the finding of small mammals from several different horizons. Among the small mammals, *Progonomys* plays a key role for biostratigraphy. Based on the intermediate evolutionary grade between *Progonomys cathalai* and *P. woelferi*, the new species *Progonomys sinensis* Qiu, Zheng, and Zhang (2004a) is believed to best correlate with the Late Vallesian (MN10) of the European mammal chronology. The earliest record of *Progonomys sinensis* is from Loc.19, dated to 10 Ma, and the last record from Loc.38, with an age estimate of 8.79 Ma. Another murid taxon referred to as Muridae gen. et sp. indet., with close affinities to *Progonomys sinensis*, occurs between 9.98 Ma to 8.07 Ma.

Three species of Gerbillidae were discovered from the Bahe Formation (Qiu, Zheng, and Zhang 2004b). *Myocricetodon lantianensis* was thought to be closely allied with *M. plebius* from Middle Miocene locality Quantougou, Gansu. It is distributed from the lower to the middle part of the Bahe Formation, dated from 10 Ma to 8.07 Ma. *Abudhabia baheensis* has a similarly long duration, dated from 9.98 Ma to 8.07 Ma. Its presence at Lantian currently represents the earliest known record of the genus. *Myocricetodon liui*, a larger form, has only been retrieved from Loc.12 (9.95 Ma).

Two cricetids have been found in the Bahe Formation, *Nannocricetus primitivus* and *Kowalskia* indet. (Zhang, Zheng, and Liu 2008). *Nannocricetus primitivus* is present in multiple levels in the Bahe Formation (dated from 10.79 Ma to 8.07 Ma), including specimens from Loc.6 and Loc.30, which display minor differences. *Kowalskia* is poorly documented, represented only by one m2, dated to about 9.9 Ma (Zhang, Zheng, and Liu [2008] mistakenly cited the age as about 8 Ma). This might be one of the earliest records of this genus except for some unpublished data from Europe mentioned by Kälin (1999).

Fossil dipodids are relatively rich in the Bahe Formation (Li and Zheng 2005). The advanced form *Protalactaga lantianensis* is the youngest known species of this genus. Its stratigraphic distribution is exactly the same as *Myocricetodon lantianensis* (10–8.07 Ma). *Paralactaga*, though only represented by a single tooth, occurred at about 9.14 Ma, some 1 myr earlier than the last occurrence of *Protalactaga lantianensis*, its stem ancestor. Two new species of Cardiocraniinae were recorded from the lower part of the formation. Two zapodids were also recorded from the lower part of the Bahe Formation, *Lophocricetus xianensis* and *Lophocricetus* sp. (Qiu, Zheng, and Zhang 2008). *Lophocricetus xianensis* is considered to be the most primitive species of the genus. The indeterminate species dated to 10.39 Ma also shows intermediate characters between the Early and Middle Miocene *Heterosminthus* and Late Miocene *Lophocricetus*. Although conventionally regarded as being evolutionarily conservative, sciurids from the Bahe Formation display distinctive characters. The new species *Sciurotamias pusillus*, ranging from 10 Ma to 8.79 Ma, has several primitive characters compared with the species from Yuanmou and Lufeng (Qiu, Zheng, and Zhang 2008).

There is also one species of *Pararhizomys*, *P. qinensis* from Loc.13 (dated to 9.37 Ma), which is more primitive than *P. hipparionum* from Fugu and Qin'an (Zhang,

Flynn, and Qiu 2005). While ochotonids are abundant in most of the later Miocene and Pliocene faunas, only a few specimens tentatively referred to *Ochotona* cf. *O. lagreli* were recovered from the basal part of the Bahe Formation (Qiu et al. 2003). The genus *Ochotona* is generally thought to have evolved from the Middle Miocene *Bellatona* (Dawson 1961; Qiu 1996). The record of the *Ochotona* from our Loc.12 (9.95 Ma) might be one of the earliest of the genus. Insectivores are sporadically recorded from different localities and need further study.

"Large" Mammals

One of the most widely cited taxa from the Bahe Formation is *Dinocrocuta gigantea*, which has long been cited as a characteristic species for the fauna and used for sequencing and correlation (Li, Wu, and Qiu 1984; Qiu, Xie, and Yan 1988; Qiu 1990; Qiu and Qiu 1995; Qiu, Wu, and Qiu 1999; Fejfar 1997; Deng 2006). However, Zhang and Xue (1996) questioned whether *D. gigantea* can be correlated exclusively to the Vallesian in China, because multiple co-occurrences of the species with *Hipparion* faunas of Turolian aspect are known. This argument is reinforced by the latest paleomagnetic age estimate for the Lamagou fauna at about 7.8 Ma (Xue, Zhang, and Yue 2006; Zhang 2005). The occurrence of *D. gigantea* in the Bahe Formation spans a very short time interval, from 8.07 Ma to 7.84 Ma, remarkably close to the age of Lamagou. It appears increasingly likely that this giant carnivore, while not surviving as long as other hyaenids, did extend its range into the Turolian. Specimens of ?*Metailurus parvulus*, ?*M. major*, and cf. *Adcrocuta eximia* have also been collected from the Bahe Formation (Andersson and Werdelin 2005), but the quality of the material is insufficient for refining the biostratigraphy further.

Fossil bovids are by far the most abundant large mammals in the Bahe Formation. From Loc.12 (dated to 9.95 Ma), *Lantiantragus longirostralis* (Chen and Zhang 2004) and a small *Gazella* were found. Two species of *Shaanxispira* including *S. chowi* described in 1978 (Liu, Li, and Zhai, 1978) and *S. baheensis* Zhang (2003) are recorded from a relatively short interval (8.21–8.07 Ma). The genus *Shaanxispira* is considered closely related to *Lantiantragus*, and both are assigned to the subfamily Urmiatheriinae (Zhang 2003; Chen and Zhang 2004). The well-documented *Urmiatherium* and *Plesiaddax* from the Baode *Hipparion* faunas are considered to be the last and most specialized taxa in this group.

Systematic study of a new species of *Dorcadoryx*, *D. orientalis* (Chen 2005), dated to 8.07 Ma, suggests that it is more primitive than other species referred to this genus, and it appears to represent the earliest record of the genus.

Many well-preserved specimens of *Gazella* cf. *G. lydekkeri* were collected from Loc.31 and other adjacent localities. This unexpected species is most probably an immigrant with a short range (8.95–8.07 Ma) and seems to be unrelated to any other later *Gazella* species from China.

There are three forms of giraffids. Except those referred to *Palaeotragus decipiens* and *P. microdon* (Liu, Li, and Zhai 1978), we also collected brachydont upper teeth assigned to *Schansitherium* cf. *S. tafeli* from Loc.6, where it coexisted with *Palaeotragus decipiens*. The small-size giraffe *Palaeotragus microdon* is recorded from the topmost level of the Bahe Formation and lasted into the base of the Lantian Formation. In contrast to the extremely rich deer material from the Lantian Formation, only a few specimens of *Metacervulus* sp. were found from Loc.33 in the Bahe Formation.

Fossil rhinoceros material from the Bahe Formation was first described by Liu, Li, and Zhai (1978). They recorded three taxa: *Chilotherium gracile* from localities 59S4 and 59S6, *Dicerorhinus orientalis* (=*Stephanorhinus orientalis*) and ?*Brachypotherium* sp. from locality 59S5. They did not find any rhinoceros from the Lantian Formation.

Restudy of the material housed in the IVPP collections essentially confirms these taxonomic attributions. Although *C. gracile* is usually regarded as a synonym of *C. habereri* (Heissig 1975), probably based on an unusually small individual, it should be noted that the Bahe material is indeed small and of a generally primitive appearance compared with the large sample of *C. habereri* from the Baode Formation described by Ringström (1924). We therefore retain the nomen *C. "gracile"* for this material, to underscore the fact that it appears to represent an earlier form than typical *C. habereri*. This small form also compares well with *C. kiliasi* from Pentalophos in Greece (MN 11; Geraads and Koufos 1990; Fortelius et al. 2003). In contrast, the *S. orientalis* material is typical of the species, very similar to Ringström's Baode sample as noted by Liu, Li, and Zhai (1978).

As noted by Zhang et al. (2002), the main biostratigraphic message that emerges from the rhinoceros material is unchanged. *Stephanorhinus orientalis* ranges have no detectable change (from Loc.12 to Loc.6). *Acerorhinus palaeosinensis* (or possibly an unrecognized taxon confused with it) has an even longer range (from Loc.12 to Loc.42). *Brachypotherium* is represented by a single individual and seems unique for the Late Miocene of China. Only *Chilotherium "gracile"* (from 59S4, 59S6) suggests a possible hint of a difference from the larger and more hypsodont *C. habereri* from Baode by its relatively small size and primi-

tive appearance. Unfortunately, the only specimen from Lantian confidently referable to *Chilotherium habereri* is from the (Miocene) Gongwangling locality (V 3142) of unknown lithostratigraphic position but correlated with the Lantian Formation by Liu, Li, and Zhai (1978).

Fossil equids are also very rich in the entire Bahe Formation (10.29–7.26 Ma). Although systematic study of this group is still pending, we tentatively recognized the two characteristic species, "*Hipparion*" *chiai* and "*Hipparion*" *weihoense* in our collection. These two species coexisted over most of the interval of 8.21–7.26 Ma.

It is quite peculiar that we found no specimens of proboscideans in the five field seasons. No new records can therefore be added to the old collection, which includes a single species, *Tetralophodon exoletus* from 59S5 and 59S6.

Biostratigraphy

As table 6.2 shows, fossil records from the Bahe Formation are far from ideal for high-resolution biostratigraphic subdivision. As is typically the case, large fossils occur sporadically, and their recovery largely depends on the sampling effort. Small mammals, on the other hand, are much better sampled thanks to the availability of screen-washing. Although the uppermost part of the section remains to be sampled, the combined fossil record from the Bahe and Lantian formations may nevertheless currently be the best basis for understanding Chinese Late Miocene mammal biostratigraphy.

The magnetic polarity succession is nearly complete in the Lantian sequence, indicating that the deposition was rather continuous throughout the sequence and that there are no marked hiatuses in the area (Kaakinen 2005). Lithologically, the Bahe Formation displays a series of similar sediment cycles from the base to the top, although there is a slight overall upwardly fining trend (Kaakinen and Lunkka 2003). Indications of moderately stable environments also come from oxygen and carbon isotope compositions showing relatively invariant values throughout the Bahe Formation (Kaakinen, Sonninen, and Lunkka 2006). Mammal fossil distribution also shows a remarkable lack of change during this time. Nevertheless, we can recognize two biostratigraphic units based on the duration of certain characteristic species and faunal assemblages.

Unit 1 (BH1) ranges from the base to Loc.38 level, about 11–8.79 Ma (figure 6.3). Mammalian assemblages from this interval are uniform. However, some primitive elements, such as *Myocricetodon liui*, *Lophocricetus xianensis*, and *Lantiantragus*, are present in the lower part, while some derived species or immigrants occur in the upper horizons—for example, the first records of *Paralactaga* and *Gazella* cf. G. *lydekkeri*. The characteristic taxon of this unit is *Progonomys sinensis*.

Unit 2 (BH2) covers the upper part of the formation (see figure 6.3). It includes almost all the old localities from the Bahe Formation collected in the 1960s. Small

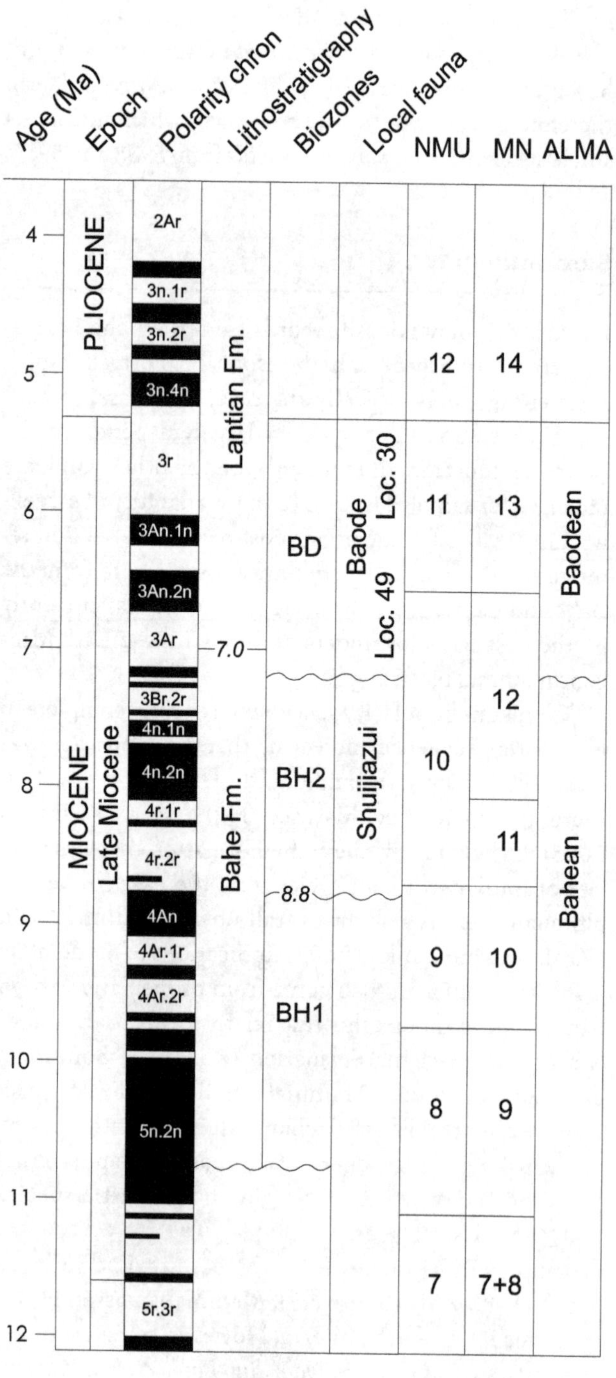

mammal assemblages differ slightly from the earlier unit. Large mammal assemblages are characterized by the co-occurrence of *Dinocrocuta gigantea*, *Shaanxispira*, *Hipparion chiai*, and *Hipparion weihoense*.

BAHEAN: CHINESE LATE MIOCENE LAND MAMMAL AGE

Taken as a whole, the Bahe fauna is obviously distinctive, separated from the Middle Miocene records from China and other areas by a significant faunal turnover event. Characteristic Middle Miocene large mammals, such as *Platybelodon*, *Anchitherium*, *Kubanochoerus*, *Turcocerus*, *Dicroceros*, *Stephanocemas*, and Amphicyonidae are not present in the Bahean fauna. Small mammals also show marked change. *Desmatolagus*, *Alloptox*, and *Bellatona* are lacking, and *Ochotona* is the only known lagomorph genus for the Bahean age. Ancient cricetids, such as *Democricetodon*, *Megacricetodon*, and *Gobicricetodon*, were replaced by cricetini, *Nannocricetus* and *Kowalskia*. The zapodid *Heterosminthus* was replaced by *Lophocricetus*. Castorids *Anchitheriomys* and "*Monosaulax*" did not survive the Middle/Late Miocene boundary. A few taxa, such as *Myocricetodon* and *Protalactaga*, did survive briefly into the Bahean fauna (table 6.3).

The first appearance of "*Hipparion*" in the Old World has been specified as defining the base of the Vallesian mammal age in Spain (Crusafont-Pairó 1951) and in recognizing the beginning of the Vallesian age in other Old World localities by subsequent workers (Woodburne et al. 1996). The earliest Old World hipparionine horse is *Hippotherium primigenium* from the Pannonian Basin, Vienna, with an estimated age of 11.3 Ma (based on stratigraphic relationships in the Pannonian Basin; Woodburne 2007). The earliest record of *Hipparion* from the Bahe Formation is from Loc.50, dated to 10.29 Ma by the paleomagnetic extrapolation, slightly earlier than the first

Figure 6.3 Biochronology of the Late Miocene Bahe Formation.

Table 6.3
Characteristic Taxa of the Suggested Bahean Age

First Occurrence	*Progonomys, Abudhabia, Nannocricetus, Kowalskia, Lophocricetus, Paralactaga, Sciurotamias, Ochotona, Lantiantragus, Shaanxispira, Dinocrocuta, Hipparion, Gazella*
Last Occurrence	*Myocricetodon, Progonomys, Protalactaga, Dinocrocuta, Lantiantragus, Shaanxispira*

occurrence of *Progonomys* (10 Ma). Considering the poor coverage and rare fossils from the base of the Bahe Formation, we tentatively propose the base of the Bahean age to be correlated with the base of the Vallesian for the time being, pending further exploration for earlier records of "*Hipparion*" from China.

As we have already discussed (Zhang et al. 2002), there is a remarkable faunal turnover near the Bahe/Lantian boundary. From the open and dry environments of the Bahean age, there occurred a change to more humid and closed conditions at the beginning of Baodean (Kaakinen, Sonninen, and Lunkka 2006; Passey et al. 2009), based in part on changes in stable isotopes preserved in soils and large mammal enamel. The latest paleomagnetic research in the Baode Red Clay showed the age of the lowest fossil level to be about 7.15–7.18 Ma (Zhu et al. 2008). This horizon is slightly younger than 59S4 (7.26 Ma; the highest fossil level from the Bahe Formation). Although the lithological boundary between the Bahe and Lantian Formations is slightly later in time than the beginning of red clay deposition in the Baode area, we provisionally suggest the Bahean/Baodean boundary at about 7.2 Ma (see figure 6.3). We discuss the details of the Baodean Land Mammal Age separately (Kaakinen et al., chapter 7, this volume).

DISCUSSION

Although they did not formally define stratigraphic units, Li, Wu, and Qiu (1984) used the Bahe fauna as the representative of the Bahean Age. Relocation of the localities and paleomagnetic correlation shows that the Bahe fauna of Li et al. was restricted to the Upper biozone (BH2) of the present study.

Contrary to the abundance of Baodean age faunas in China, there are few Bahean localities. Among them is the rich Lamagou fauna (Xue, Zhang, and Yue 1995), also known as the Wangdaifuliang fauna (Qiu, Wu, and Qiu 1999), from Fugu County, northernmost part of Shaanxi Province. Recent discussion of the Fugu paleomagnetic section suggested the Lamagou fauna is about 7.8 Ma in age (Xue, Zhang, and Yue 2006). Although the fossils have not been fully described, the Lamagou fauna appears similar in faunal composition to BH2, including the characteristic species *Dinocrocuta gigantea* and *Hipparion chiai*.

The Qaidam (Tsaidam) fauna described by Bohlin (1937) was considered to be early Late Miocene by having the earliest occurrence of *Hipparion* together with elements of *Anchitherium* fauna, such as *Stephanocemas*, *Lagomeryx*, and "*Dicrocerus*" (Qiu and Qiu 1995). Recent extensive field investigation in the Qaidam Basin showed that Bohlin's Qaidam fauna was actually from different horizons, mixing Middle Miocene and Late Miocene elements (Deng and Wang 2004; Wang et al. 2007). Wang et al. (2007) named four mammal faunas from the Qaidam Basin, the Olongbuluk mammal fauna, Tuosu fauna, Shengou mammal and fish fauna, and Pliocene Huaitoutala mammal fauna. The previously mentioned typical middle Miocene taxa were from the Olongbuluk fauna. The Tuosu fauna includes the specialized bovid taxa, *Tsaidamotherium*, *Olonbulukia*, *Qurliqnoria*, and *Tossunoria* as well as *Hipparion teilhardi*, *Chalicotherium brevirostris*, and *Adcrocuta eximia*. According to Wang et al. (2007), the Tuosu fauna covers the magnetochrons C4Ar–C5r (about 9.10–12.00 Ma). Biochronologic correlation with the Bahe fauna is difficult owing to the dearth of common taxa. Nevertheless, the specialized Urmiatheriini and Caprinae taxa from the Tuosu fauna do not favor the earliest Late Miocene age. The later Shengou fauna produced both large and small mammals, including the characteristic *Hipparion weihoense* and *Hipparion* cf. *H. chiai*. Its advanced *Lophocricetus* and murid *Huerzelerimys* prevent the correlation with the BH1 from Lantian, but it might match BH2.

From the Linxia Basin, Gansu Province, very rich Cenozoic faunas where recently discovered (Deng et al. 2004). The Late Miocene faunas from the Liushu Formation, the Guonigou, Dashengou, and Yangjiashan faunas, were correlated to MN 9–11, respectively, by referring to unpublished paleomagnetic data (Deng 2005). Detailed correlation needs further paleomagnetic and systematic study.

The Amuwusu fauna from Inner Mongolia was thought to be one of the representative fauna of earliest Late Miocene age, including both Middle Miocene elements and Late Miocene newcomers (Qiu and Qiu 1995; Qiu, Wang, and Li 2006). This fauna shares *Eutamias*, *Protalactaga*, and *Paralactaga* with the Bahe Fauna, however, it possesses no murids, *Ochotona*, nor *Lophocricetus*. Z.-x. Qiu (pers. comm.) did find some *Hipparion* material when he first discovered this locality, the only site where *Anchitherium* and *Hipparion* co-occur. Unfortunately this material was lost, and no new hipparionines were found in later investigations. If the occurrence of *Hipparion* is confirmed by future discovery, this fauna might be the earliest one so far known from the Bahean age.

CONCLUSION

The late Neogene record of mammalian fossils unearthed from the sedimentary sequence in Lantian is among the most complete in China. Our work has shown that the area is one of the few where the early late Miocene is recorded.

The sedimentary sequence consists of fluvial deposits of the Bahe Formation, extending from ca. 11 Ma to 7 Ma, overlain by the *Hipparion* Red Clay deposits of the Lantian Formation. Magnetic polarity stratigraphy demonstrates that the majority of fossil localities are dated to between 10.8 Ma and 6.6 Ma, with the most densely sampled intervals dating to 10.8–8.8 Ma and 8.3–7.8 Ma.

Sedimentological observations and isotope composition suggest moderately stable and dry conditions for the Bahe Formation, in accord with mammals adapted to relatively open and arid conditions. Fossil mammals reveal little evidence of faunal change in the Bahe Formation. After this strikingly stable interval a drastic change in sedimentological regime from fluviatile to mainly eolian occurs at the Bahe–Lantian Formation boundary. The change coincides with shifts in isotope values and a major faunal turnover event. The paleoecological context of the lowermost Lantian Formation localities represents closed habitats implying more humid environments than in the underlying Bahe Formation.

ACKNOWLEDGMENTS

We thank all participants in our Lantian project for their hard work and enjoyable cooperation. We also appreciate stimulating discussions with our colleagues, Prof. Zhan-xiang Qiu, Chuan-kui Li, Zhu-ding Qiu, and Tao Deng from the IVPP. Ray Bernor and Xiaoming Wang gave helpful and constructive comments. This work was supported by the Major Basic Research Projects (2006CB806400), NSFC (40711130639, 41072004, 40872018), Academy of Finland (44026, 34080), and by several exchange grants between the Chinese Academy of Sciences and Academy of Finland (MF, AK, and JPL) and Finnish Cultural Foundation (AK).

REFERENCES

Andersson, K. and A. Kaakinen. 2004. Floodplain processes in the shaping of fossil bone assemblages: An example from the Late Miocene, Bahe Formation, Lantian, China. *GFF* 126:279–287.

Andersson, K. and L. Werdelin. 2005. Carnivora from the Late Miocene of Lantian, China. *Vertebrata PalAsiatica* 43(4):256–271.

Bellier, O., J. L. Mercier, P. Vergély, C.-x. Long, and C.-z. Ning. 1988. Évolution sédimentaire et tectonique du graben cénozoïque de la Wei He (province du Shaanxi, Chine du Nord). *Bulletin de la Société Géologique de France* 4 (6):979–994.

Bellier, O., P. Vergély, J. L. Mercier, C.-z. Ning, N.-g. Deng, M.-c. Yi, and C.-x. Long. 1991. Analyse tectonique et sédimentaire dans les monts Li Shan (Province du Shaanxi—Chine du Nord): datation des regimes tectoniques extensifs dans le garben de la Wei He.). *Bulletin de la Société Géologique de France* 162:101–112.

Bohlin, B. 1937. Eine Tertiäre Säugetier-Fauna aus Tsaidam Sino-Swedisch Expedition. *Palaeontologica Sinica*, ser. C, 14(1):1–111.

Chen, G.-f. 2005. *Dorcadoryx* Teilhard et Trassaert 1938 (Bovidae, Artiodactyla) from the Bahe Formation of Lantian, Shaanxi Province, China. *Vertebrata PalAsiatica* 43(4): 272–282.

Chen, G.-f. and Z.-q. Zhang. 2004. *Lantiantragus* gen. nov. (Urmiatherinae, Bovidae, Artiodactyla) from the Bahe formation, Lantian, China, *Vertebrata PalAsiatica* 42(3):205–215 (in Chinese with English summary).

Crusafont-Pairó, M. 1951. El sistema Miocénico en la depression española del Vallés Penedés. *International Geological Congress, Report of XVIII Session, Great Britain* 11:33–43.

Dawson, M. 1961. On two ochotonids (Mammalia, Lagomorpha) from Late Tertiary of Inner Mongolia. *American Museum Novitates* 2061:1–15.

Deng, T. 2005. Character, age and ecology of the Hezheng Biota from northwestern China. *Acta Geologica Sinica* 79:739–750.

Deng, T. 2006. Chinese Neogene mammal biochronology. *Vertebrata PalAsiatica* 44(2):143–163.

Deng, T. and X.-m. Wang. 2004. Late Miocene *Hipparion* (Equidae, Mammalia) of eastern Qaidam Basin in Qinghai, China. *Vertebrata PalAsiatica* 42(4):316–333.

Deng, T., X.-m. Wang, X.-j. Ni, L.-p. Liu, and Z. Liang. 2004. Cenozoic stratigraphic sequence of the Linxia Basin in Gansu, China, and its evidence from mammal fossils. *Vertebrata PalAsiatica* 42(1):45–66.

Ding, Z.-l., J.-m. Sun, T.-s. Liu, R.-x. Zhu, S.-l. Yang, and B. Guo. 1998. Wind-blown origin of the Pliocene red clay formation in the central Loess Plateau, China. *Earth and Planetary Science Letters* 161:135–143.

Fejfar, O. 1997. Miocene Biochronology. In *Actes du Congrès BiochroM'97*, ed. J.-P. Aguilar, S. Legendre, and J. Michaux, pp. 795–802. Mémoires et Travaux E.P.H.E., Institut de Montpellier, 21.

Fortelius, M. 2009. Neogene of the Old World Database of Fossil Mammals (NOW), University of Helsinki. http://www.helsinki.fi/science/now/.

Fortelius, M., K. Heissig, G. Sarac, and S. Sen. 2003. Rhinocerotidae (Perissodactyla). In *Geology and Paleontology of the Miocene Sinap Formation, Turkey*, ed. M. Fortelius, J. W. Kappelman, S. Sen, and R. L. Bernor, pp. 282–307. New York: Columbia University Press.

Geraads, D. and G. Koufos. 1990. Upper Miocene Rhinocerotidae from Pentalophos-1, Macedonia, Greece. *Palaeontographica A* 210:151–168.

Guo, Z.-t., W. F. Ruddiman, Q.-z. Hao, H.-b. Wu, Y.-s. Qiao, R.-x. Zhu, S. Z. Peng, J. J. Wei, B. Y. Yuan, and D. S. Liu. 2002. Onset of Asian dersertification by 22 Myr ago inferred from loess deposits in China. *Nature* 416:159–163.

Heissig, K. 1975. Rhinocerotidae aus dem Jungtertiär Anatoliens. *Geologisches Jahrbuch (B)*:145–151.

Kaakinen, A. 2005. A long terrestrial sequence in Lantian: A window into the late Neogene palaeoenviornments of northern China. Ph.D. diss., University of Helsinki.

Kaakinen, A. and J. P. Lunkka. 2003. Sedimentation of the late Miocene Bahe formation and its implications for stable environments adjacent to Qinling mountains in Shaanxi, China. *Journal of Asian Earth Sciences* 22:67–78.

Kaakinen, A., E. Sonninen, and J. P. Lunkka. 2006. Stable isotope record in paleosol carbonates from the Chinese Loess Plateau: Implications for late Neogene paleoclimate and paleovegetation. *Palaeogeography, Palaeoclimatology, Palaeoecology* 237: 359–369.

Kälin, D. 1999. Tribe Cricetini. In *The Miocene Land Mammals of Europe*, ed G. E. Rössner and K. Hessig, pp. 373–387. Munich: Dr. Friedrich Pfeil.

Kurtén, B. 1952. The Chinese *Hipparion* fauna. *Commentationes Biologicae, Societas Scientiarum Fennica* 13:1–82.

Kutzbach, J. E., W. L. Prell, and W. F. Ruddiman. 1993. Sensitivity of Eurasian climate to surface uplift of the Tibetan Plateau. *Journal of Geology* 101:177–190.

Li, C.-k., W.-y. Wu, and Z.-d. Qiu. 1984. Chinese Neogene: Subdivision and correlation. *Vertebrata PalAsiatica* 22(3):163–178 (in Chinese with English summary).

Li, Q. and S.-h. Zheng. 2005. Note on four species of dipodids (Dipodidae, Rodentia) from the Late Miocene Bahe Formation, Lantian, Shaanxi. *Vertebrata PalAsiatica* 43(3):283–296 (in Chinese with English summary).

Liu, D.-s., M.-l. Ding, and F.-q. Gao. 1960. Cenozoic stratigraphic sections between Xi'an and Lantian. *Geology Science* 4:199–208 (in Chinese).

Liu, D.-s., C.-k. Li, and R.-j. Zhai. 1978. Pliocene vertebrates of Lantian, Shensi: Tertiary mammalian fauna of the Lantian District, Shensi. *Professional Papers of Stratigraphy and Paleontology* 7:149–200 (in Chinese).

Lourens, L., F. Hilgen, N. J. Shackleton, J. Laskar, and D. Wilson. 2004. The Neogene period. In *A Geologic Time Scale 2004*, ed. F. Gradstein, J. Ogg, and A. G. Smith, pp. 409–440. Cambridge: Cambridge University Press.

Lu, H., J. Vandenberghe, and Z.-s. An. 2001. Aeolian origin and palaeoclimatic implications of the "Red Clay" (north China) as evidenced by grain-size distribution. *Journal of Quaternary Science* 16:89–97.

Passey, B., L. Ayliffe, A. Kaakinen, Z.-q. Zhang, J. Eronen, Y.-m. Zhu, L.-p. Zhou, T. Cerling, and M. Fortelius. 2009. Strengthened East Asian summer monsoons during a period of high-latitude warmth? Isotopic evidence from Mio-Pliocene fossil mammals and soil carbonates from northern China. *Earth and Planetary Science Letters* 277:443–452.

Qiu, Z.-d. 1987. The Neogene mammalian faunas of Ertemte and Harr Obo in Inner Mongolia (Nei Mongol), China. 6. Hares and Pikas-Lagomorpha: Leporidae and Ochotonidae. *Senckenbergiana Lethaea* 67(5/6): 375–399.

Qiu, Z.-d. 1996. *Middle Miocene Small Mammalian Fauna from Tungur, Nei Mongol*. Beijing: Science Press.

Qiu, Z.-d., X.-m. Wang, and Q. Li. 2006. Faunal succession and biochronology of the Miocene through Pliocene in Nei Mongol (Inner Mongolia). *Vertebrata PalAsiatica* 44(2):164–181.

Qiu, Z.-d., S.-h. Zheng, S. Sen, and Z.-q. Zhang. 2003. Late Miocene micromammals from the Bahe Formation, Lantian, China. In *Distribution and Migration of Tertiary Mammals in Eurasia*, ed. J. W. F. Reumer and W. Wessels, pp. 443–453. DEINSEA 10.

Qiu, Z.-d., S.-h. Zheng, and Z.-q. Zhang. 2004a. Murids from the late Miocene Bahe formation, Lantian, Shaanxi. *Vertebrata PalAsiatica* 42(1):67–76.

Qiu, Z.-d., S.-h. Zheng, and Z.-q. Zhang. 2004b. Gerbillids from the late Miocene Bahe formation, Lantian, Shaanxi. *Vertebrata PalAsiatica* 42(2):193–204.

Qiu, Z.-d, S.-h. Zheng, and Z.-q. Zhang. 2008. Sciurids and zapodids from the Late Miocene Bahe Formation, Lantian, Shaanxi. *Vertebrata PalAsiatica* 46(2):111–123.

Qiu, Z.-x. 1990. The Chinese Neogene mammalian Biochronology—its correlation with the European Neogene Mammalian Zonation. In *European Neogene Mammal Chronology*, ed. E. L. Lindsay, V. Fahlbusch, and P. Mein, pp. 527–556. NATO ASI Series (A) 180.

Qiu, Z.-x. and Z.-d. Qiu. 1995. Chronological sequence and subdivision of Chinese Neogene mammalian faunas. *Palaeogeography, Palaeoclimatology, Palaeoecololgy* 116:41–70.

Qiu, Z.-x., W.-y., Wu, and Z.-d. Qiu. 1999. Miocene mammal faunal sequence of China: Paleozoogeography and Eurasian relationships. In *The Miocene Land Mammals of Europe*, ed. G. E. Rössener and K. Heissig, pp. 443–455. Munich: Dr. Friedrich Pfeil.

Qiu, Z.-x., J.-y. Xie, and D.-f. Yan. 1988. Discovery of the skull of *Dinocrocuta gigantea*. *Vertebrata PalAsiatica* 26:128–138 (in Chinese with English summary).

Ringström, T. 1924. Nashörner der Hipparion-Fauna Nord-Chinas. *Palaeontologica Sinica* C 1(4):1–156.

Schlosser, M. 1903. Die Fossilen Saugethiere Chinas nebst einer Odontographie der recenten Antilopen. *Abh Bayerischen Akademie der Wissenschaften* 22:1–221.

Storch, G. 1987. The Neogene mammalian faunas of Ertemte and Harr Obo in Inner Mongolia (Nei Mongol), China. 7. Muridae (Rodentia). *Senckenbergiana Lethaea* 67 (5/6):401–431.

Sun, D.-h., Z.-s. An, J. Shaw, J. Bloemendal, and Y.-b. Sun. 1998. Magnetostratigraphy and paleoclimatic significance of Late Tertiary aeolian sequences in the Chinese Loess Plateau. *Geophysical Journal International* 134:207–212.

Sun, D.-h., D.-s. Liu, M.-y. Chen, Z.-s. An, and J. Shaw. 1997. Magnetostratigraphy and palaeoclimate of Red Clay sequences from Chinese Loess Plateau. *Science in China*, ser. D, 40:337–343.

Teilhard de Chardin, P. and P. Leroy. 1942. *Chinese Fossil Mammals*. Peking: Institute de Géo-Biologie.

Wang, X.-m., Z.-d. Qiu, Q. Li, B.-y. Wang, Z.-x. Qiu, W. R. Downs, G.-p. Xie, J.-y. Xie, T. Deng, G. T. Takeuchi, Z. J. Tseng, M.-m. Chang, J. Liu, Y. Wang, D. Biasatti, Z.-c. Sun, X.-m. Fang, and Q.-q. Meng. 2007. Vertebrate paleontology, biostratigraphy, geochronology, and paleoenvironment of Qaidam Basin in northern Tibetan Plateau. *Palaeogeography, Palaeoclimatology, Palaeoecology* 254:363–385.

Woodburne, M. O. 2007. Phyletic diversification of the *Cormohipparion* occidentale complex (Mammalia; Perissodactyla, Equidae), Late Miocene, North America, and the origin of the old world *Hippotherium* datum. *Bulletin of the American Museum of Natural History* 306:1–138.

Woodburne, M. O., G. Theobald, R. L. Bernor, C. C. Swisher, H. König, and H. Tobien. 1996. Advances in the Geology and Stratigraphy at Höwenegg, Southwestern Germany. In *The Evolution of Western Eurasian Neogene Mammal Faunas*, ed. R. L. Bernor, V. Fahlbusch, and H-W. Mittman, pp. 106–123. New York: Columbia University Press.

Wu, W.-y. 1991. The Neogene mammalian faunas of Ertemte and Harr Obo in Inner Mongolia (Nei Mongol), China. 9. Hamsters: Cricetinae (Rodentia). *Senckenbergiana Lethaea* 71(3/4):257–305.

Xue, X.-x., Y.-x. Zhang, and L.-p. Yue. 1995. Discovery and chronological division of the *Hipparion* faunas in Laogaochuan Village, Fugu County, Shaanxi. *Chinese Science Bulletin* 40(11):926–929.

Xue, X.-x., Y.-x. Zhang, and L.-p. Yue. 2006. Paleoenvironments indicated by the fossil mammalian assemblages from red clay-loess sequence in the Chinese Loess Plateau since 8.0 Ma B.P. *Science in China*, ser. D 49(5):518–530.

Zdansky, O. 1923. Fundorte der *Hipparion*-fauna um Pao-Te-Hsien in NW Shansi. *Bulletin of the Geological Survey China* 5:69–82.

Zhang, Y.-p., W.-b. Huang, Y.-j. Tang, H.-x. Ji, Y.-z. You, Y.-s. Tong, S.-y. Ding, X.-s. Huang, and J.-j. Zheng. 1978. Cainozoic of the Lantian area, Shaanxi. *Memoirs of Institute of Vertebrate Paleontology and Paleoanthropology, Academia Sinica, Special Publication A* 14:1–64 (in Chinese).

Zhang, Y.-x and X.-x. Xue. 1996. New materials of *Dinocrocuta gigantea* found in Fugu County, Shaanxi Province. *Vertebrata PalAsiatica* 34(1):18–26

Zhang, Z.-q. 2003. On a new species of *Shaanxispira* (Bovidae, Artiodactyla) from Bahe Formation, Lantian, Shaanxi Province. *Vertebrata PalAsiatica* 41(3):230–239.

Zhang, Z.-q. 2005. New materials of *Dinocrocuta* (Percrocutidae, Carnivora) from Lantian, Shaanxi Province, China, and remarks on Chinese Late Miocene biochronology. *Geobios* 38: 685–689.

Zhang, Z.-q., L. J. Flynn, and Z.-d. Qiu. 2005. New materials of *Pararhizomys* from Northern China. *Paleontologica Electronica* 8.1. 5A.

Zhang, Z.-q., A. Gentry, A. Kaakinen, L.-p. Liu, J.-P. Lunkka, Z.-d. Qiu, S. Sen, R. Scott, L. Werdelin, S.-h. Zheng, and M. Fortelius. 2002. Land mammal faunal sequence of the late Miocene of China: New evidence from Lantian, Shaanxi Province. *Vertebrata PalAsiatica* 40(3):165–176.

Zhang, Z.-q. and L.-p. Liu. 2005. The Late Neogene Mammal Biochronology in the Loess Plateau, China. *Annales de Paléontologie* 91(3):257–266.

Zhang, Z.-q., S.-h. Zheng, and L.-p. Liu. 2008. Late Miocene cricetids from the Bahe Formation, Lantian, Shaanxi Province. *Vertebrata PalAsiatica* 46(4):307–316.

Zheng, H.-b., Z.-s. An, and J. Shaw. 1992. New contributions to Chinese Plio-Pleistocene magnetostratigraphy. *Physics of the Earth and Planetary Interiors* 70:146–153.

Zhou, M.-z. 1978. Tertiary mammalian faunas in the Lantian area, Shaanxi Province. *Professional Papers of Stratigraphy and Paleontology* 7:98–08 (in Chinese).

Zhu, Y.-m., L.-p. Zhou, D.-w. Mo, A. Kaakinen, Z.-q. Zhang, and M. Fortelius. 2008. A new magnetostratigraphic framework for late Neogene *Hipparion* Red Clay in the eastern Loess Plateau of China. *Palaeogeography, Palaeoclimatology, Palaeoecology* 268:47–57.

Chapter 7

Stratigraphy and Paleoecology of the Classical Dragon Bone Localities of Baode County, Shanxi Province

ANU KAAKINEN, BENJAMIN H. PASSEY, ZHAO-QUN ZHANG, LI-PING LIU, LAURI J. PESONEN, AND MIKAEL FORTELIUS

The study of Chinese Neogene mammals dates back to more than a hundred years ago (Koken 1885; Schlosser 1903). The first known scientific fossil collection in the Baode area was made in 1920 by Johan Gunnar Andersson (Qiu, Huang, and Guo 1987), the mining adviser of the Chinese government from 1914 to 1926. With the help of a considerable number of missionaries, J. G. Andersson tried to locate and identify the deposits that had yielded the rich fossils bought from drugstores and described by Schlosser (1903). In 1922, his cooperation with C. Wiman of the University of Uppsala, with financial support from the Swedish Parliament and private donors, led to the field exploration and systematic collection of fossil mammals from Baode and the adjacent area by a young Austrian paleontologist, Otto Zdansky (Andersson 1923). Zdansky (1923, English translation by Jokela et al. 2005) gave a detailed description of the fossil localities yielding *Hipparion* fauna in Baode County (Pao-Te-Shien) and labeled them on an accurate topographic map. His detailed survey in this area was unique in the early exploration of the *Hipparion* fauna anywhere. Extensive and systematic studies of these fossils were then published in the *Palaeontologia Sinica* (Ringström 1924; Bohlin 1926, 1935; Sefve 1927; Pearson 1928; Teilhard de Chardin and Piveteau 1930; Wiman 1930). These studies were the foundation of Chinese Neogene mammal work and gave the Baode *Hipparion* fauna its classic status in the field.

Later on, the Baode *Hipparion* fauna, comparable with the Paratethys Pontian faunas, was widely accepted as a standard fauna for Chinese Late Miocene. Pei, Zhou, and Zheng (1963) erected the Baodean stage based on the well-known fauna and the widespread red clay. Li, Wu, and Qiu (1984) clarified its relationships with the European Late Miocene, suggesting a correlation with the Turolian "age" (land mammal–based faunal unit). Qiu (1990), Qiu and Qiu (1990) and Tong, Zheng, and Qiu (1995) merged the Bahean and Baodean ages into a single Baodean age, correlated with both the Vallesian and Turolian. Although it has been adopted as a Chinese regional chronostratigraphic stage (National Stratigraphic Committee 2002), the content of the Baodean Stage is not well defined (Deng et al. 2004).

Zdansky reported 19 localities from the Baode area (including Baode, Fugu, and Hequ counties), 9 of them shown on his map from 1923 in the area recently studied by us. The most productive localities were 30, 43, 44, 49, 108, 109, 110, and 114, all from a limited area of about 10 km² in Baode County (Kurtén 1952). The fossils occurred mainly as rich pockets and were extracted from extensive tunnel networks, which according to Zdansky (1923) all lay at the same level, ca. 25 m above the base of the red clay. As Zdansky (1923) explained, and as further discussed by Kurtén (1952), all the "*Hipparion* Red Clay" localities of North China were thought to have produced roughly contemporary faunas. Schlosser (1903) introduced the idea that two faunal provinces were sampled in these localities: a northern "forest fauna" and a southern "steppe fauna." Kurtén (1952) developed this theme and showed that the localities could indeed be divided into sets by their faunal content and herbivore hypsodonty. He named them the gaudryi, dorcadoides, and mixed faunas, after the presence of their characteristic species of *Gazella*. Kurtén specifically noted that the faunas defined by taxonomic content showed a geographic separation, with the "mixed" faunas at the boundary between "gaudryi" faunas to the southeast and "dorcadoides" faunas to the northwest. Some later work also regarded the localities as essentially contemporary and regarded the differences between faunas as ecological, due to their geographic disposition (Kurtén 1985; Qiu, Huang, and Guo 1987).

Beginning with the first magnetostratigraphic study of the red clay sequence at Laogaochuan in Fugu County (Xue, Zhang, and Yue 1995), an alternative interpretation has been introduced. According to this view, the taxonomic and ecological differences between the faunas are due not to geography but to time. Xue, Zhang, and Yue (1995) recognized two different faunas: the earlier Lamagou fauna, similar in its characteristics to the steppe/dorcadoides faunas, and the later Miaoliang fauna, similar to the forest/gaudryi faunas. They therefore concluded that the differences observed by earlier workers and attributed by them to spatial distribution were in fact due to temporal trends. Based on magnetostratigraphy in the Fugu area in the west side of the Yellow River, Xue, Zhang, and Yue (1995) suggested that the age of the Lamagou fauna was about 7–8 Ma, the Miaoliang fauna about 5.2–7 Ma. Their study appears to have been the first to suggest that the climatic trend within the Late Miocene of China was toward more humid conditions, opposite to the global trend of mid-latitude aridification. Xue and colleagues have followed up and refined these studies in later papers (e.g., Xue, Zhang, and Yue 2006) supporting the view that the ecological differences are due to age and not to geographic disposition.

There are thus two seemingly competing hypotheses, both well supported by data. This, together with the classic nature of the Baode localities, prompted us to undertake the work reported here. From 2001 to 2005, we spent four field seasons in the Baode area, focusing on the relocation of Zdansky's localities and on a litho- and biostratigraphic study of their context. We were able to find most of the old localities according to Zdansky's topographic map. In this chapter, we try to synthesize information on lithostratigraphy, magnetostratigraphy, biostratigraphy, and barometric measurements to relocate the old localities and to build a framework for resolving the time-space conflict in the understanding of the Baodean faunas.

GEOLOGICAL SETTING

The Baode area in the northeastern Loess Plateau lies on the Shanxi Highlands tectonomorphic province (Xu and Ma 1992; Xu, Ma, and Deng 1993) between the Yellow River to the west and the Lüliang Shan tilted uplift to the east (figure 7.1). The area shows considerable relief with a maze of deep ravines and gullies carved into the plateau. Mass wasting and gully erosion have typically reached the pre-Cenozoic basement that consists of sedimentary rocks deposited during the Carboniferous and Permian. The late Neogene strata are exposed in all the gullies and rest on the basement with a slight angular unconformity. The first lithostratigraphic classification for the Neogene strata in Baode area was established by Zdansky (1923), who created the term "*Hipparion* red clay" to encompass the widely distributed fossiliferous red clay deposits underlying the Quaternary loess cover. This pioneering work has served as a basis for all later paleontological and stratigraphical studies conducted in the area. Later work by Teilhard de Chardin and Young (1930, 1931) separated it into two units, the Baode and Jingle Formations. The former refers to a town beside Yellow River some 10 km west of the study area. The first study that provided chronology for the Baode Formation in the Baode area based on paleomagnetic reversal stratigraphy was that of Yue et al. (2004). They suggested a basal age of 8.0 Ma for the Baode Formation. Recent paleomagnetic study by Zhu et al. (2008) dates the basal age for the studied Baode Formation to 7.23 Ma and places the boundary between the Baode and Jingle formations at 5.23 Ma. The Baode Formation is typically up to ca. 60 m thick in the study area, although the composite thickness has been reported to be 100 m (Zhu et al. 2008). Zhu et al. (2008) extend the Baode Formation to cover the underly-

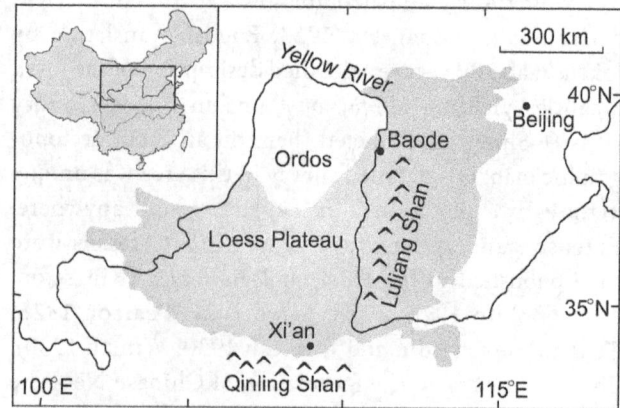

Figure 7.1 Map of the Chinese Loess Plateau region. Baode is situated in the northeastern Loess Plateau, on the eastern side of the Yellow River.

ing Luzigou Formation, as opposed to the original definition by Zdansky (1923), and the lithology in this lower part of the Baode Formation is typically dominated by conglomerates of varying thickness. The upper portion, representing the bulk of the Baode Formation, consists of reddish clays and silts with alternating layers of calcretes. Coarser lithologies are sparse in this upper part. The Jingle Formation overlies Baode Formation with a conformable contact. It consists of clays and silts with distinctive red coloration and carbonate nodule rich horizons, whereas sand and gravel lithologies are virtually absent. The distribution of the Jingle Formation is more restricted; in our study area only the lower part of this formation is exposed with a thickness of less than 10 m, and even more often, it is

completely lacking from the Neogene sequence. The thickest and most complete outcrop of Jingle Formation is described from Tanyugou, around 10 km south of the present study area (Zhu et al. 2008). In the literature the Baode and Jingle formations are commonly referred to as the late Neogene "Red Clay" deposits of the Chinese Loess Plateau. These "Red Clay" deposits have, in recent years, been widely accepted as being primarily eolian in origin and providing a valid terrestrial record for East Asian monsoon variability (e.g., Ding, Sun, Liu et al. 1998; Guo et al. 2001; Guo et al. 2002; Lu, Vandenberghe, and An 2001; Miao et al. 2004).

The rich fossil mammal fauna from Baode is collected from the Baode Formation, in particular from the bone

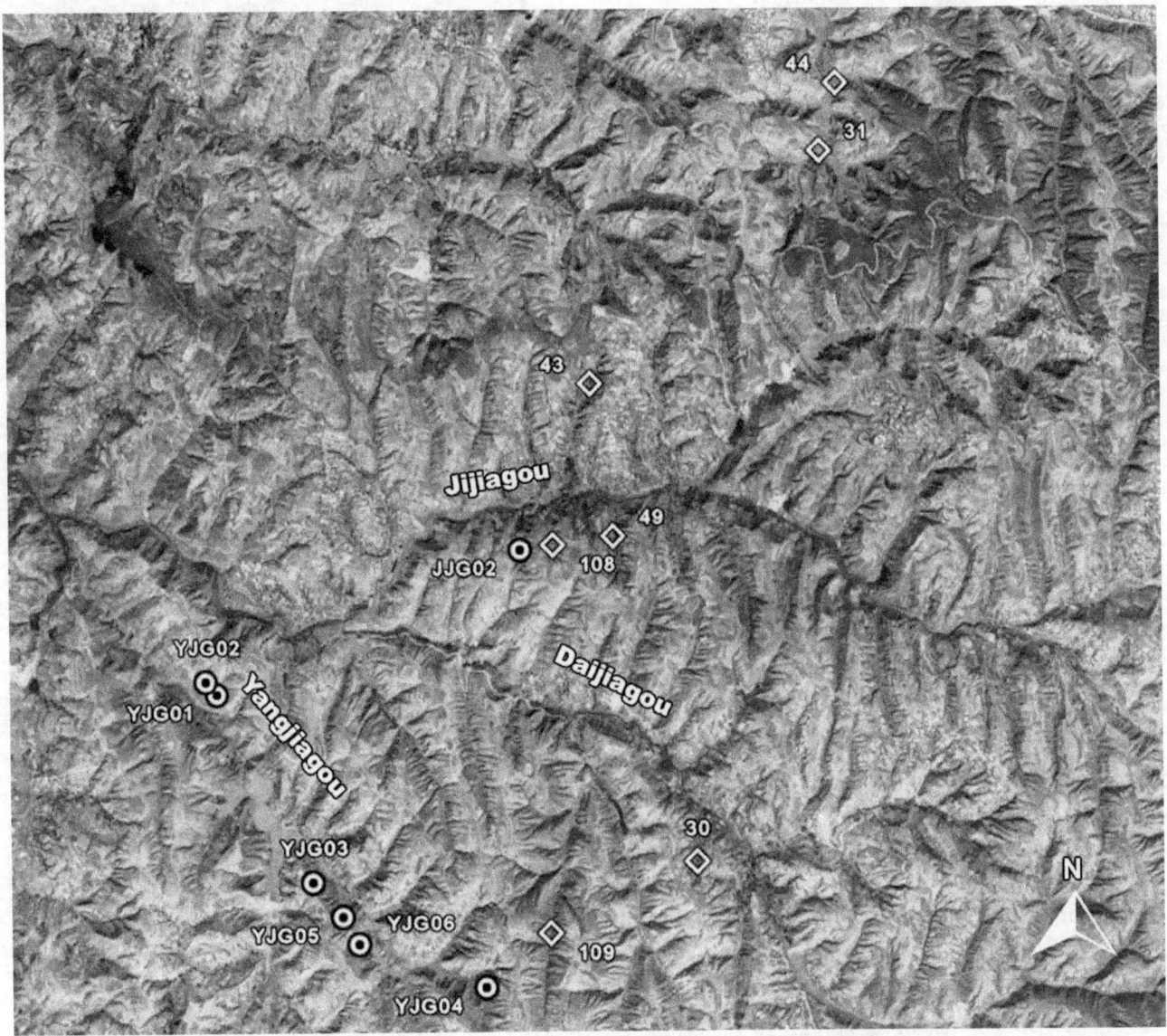

Figure 7.2 Satellite image of the study area with the main gullies Jijiagou, Daijiagou, and Yangjiagou indicated. The old Zdansky localities are labeled with a diamond; the new localities with an open circle.

mines of Jijiagou and adjacent gullies (figure 7.2). The paleontological content of the Jingle Formation here is scarce and presents little of biostratigraphic significance.

VERTEBRATE PALEONTOLOGY

Paleontological data for this study come from two sources: large mammal fossils are from the old collections, whereas small mammals were collected during our field seasons mainly by screenwashing the spoil heaps of both old and new tunnel entrances. The local "dragon bone" mining is still active; hence, the fossil localities are exposed by the tunnels, much as described by Zdansky (1923; translation by Jokela et al. [2005]).

Large Mammals

Zdansky recorded a large number of localities in Baode; among them, 30, 108, and 49 are the richest and best-known localities. The updated large mammalian fossil lists from the three classic localities are given here.

Locality 30

Indarctos sinensis, Mustela palaeosinensis, Plesiogulo brachygnathus, Parataxidea sinensis, Hyaenictitherium wongii, H. hyaenoides, Adcrocuta eximia, Machairodus palanderi, Machairodus horribilis, Machairodus sp., *Metailurus major, Metailurus parvulus,* "*Hipparion*" *dermatorhinum,* "*Hipparion*" *forstenae,* "*Hipparion*" *coelophyes,* "*H.*" *platyodus,* "*H.*" sp., *Chilotherium anderssoni, Chilotherium* sp., *Sinotherium lagrelii, Chleuastochoerus stehlini, Cervavitus novorossiae,* Cervidae gen. et sp. indet., *Palaeotragus microdon, P.* sp., *Samotherium* sp. I et II, *Urmiatherium intermedium, Plesiaddax deperiti, Miotragocerus spectabilis, Gazella* ?*paotehensis, G. dorcadoides, G.* sp., ?*Dorcadoryx lagrelii,* ?*Dorcadoryx anderssoni, Sinotragus wimani,* and Bovidae gen. et sp. indet. (Kurtén 1952; Gentry 1971; Werdelin and Solounias 1989; Bernor, Qiu, and Hayek 1990; Vislobokova 1990; Andersson and Werdelin 2005; Chen 2005; Qiu, Zheng, and Zhang 2008).

Locality 49

Amphicyon indet. (probably a milk tooth of an elephant; Zhan-xiang Qiu, pers. comm.), *Proputorius minimus, Plesiogulo brachygnathus, Lutra aonychoides, Melodon majori,* ?*Melodon incertum, Promephitis* cf. *maeotica, Hyaenictitherium wongii, H. hyaenoides, Ictitherium viverrinum, Adcro-*

cuta eximia, ?*Lycyaena dubia, Machairodus palanderi, Megantereon* sp., *Metailurus parvulus, Felis* sp., "*Mammut borsoni,*" "*Mammut* sp.," "*Hipparion*" *plocodus,* "*H.*" *platyodus,* "*H.*" *hippidiodus,* "*H.*" *coelophyes,* "*H.*" sp., *Stephanorhinus orientalis, Chilotherium* sp., *Acerorhinus paleosinense, Chleuastochoerus stehlini, Propotamochoerus hyotherioides,* Suidae gen. et sp. indet., *Cervavitus novorossiae, Procapreolus latifrons,* Cervidae gen. et sp. indet., *Palaeotragus microdon, P.* sp., *Honanotherium schlosseri, Urmiatherium intermedium, Gazella gaudryi, G.* sp. aff. *gaudryi, G.* ?*paotehensis, G. dorcadoides, G.* sp., ?*Dorcadoryx lagrelii, Palaeoryx sinensis,* and Bovidae gen. et sp. indet. (Kurtén 1952; Heissig 1975; Tobien, Chen, and Li 1988; Werdelin and Solounias 1989; Bernor, Qiu, and Hayek 1990; Vislobokova 1990; Deng 2000; Fortelius et al. 2003; Andersson and Werdelin 2005; Chen 2005).

Locality 108

Sinictis dolicognathus, Mustela palaeosinensis, Plesiogulo brachygnathus, Hyaenictitherium wongii, H. hyaenoides, Adcrocuta eximia, Machairodus palanderi, "*Hipparion*" *ptychodus,* "*Hipparion*" *forstenae,* "*H*". sp., *Chilotherium* sp., Cervidae gen. et sp. indet., *Palaeotragus microdon, P.* cf. *coelophrys, P.* sp., *Urmiatherium intermedium, Gazella* ?*paotehensis, G. dorcadoides,* ?*Dorcadoryx lagrelii,* and Bovidae gen. et sp. indet. (Kurtén 1952; Werdelin and Solounias 1989; Bernor, Qiu, and Hayek 1990; Andersson and Werdelin 2005; Chen 2005).

Small Mammals

Compared to the enormous richness of large mammals, small mammals are very rare from the Baode collections. Young (1927) described some beavers from locality 30 and 108 respectively, which were later on revised by Xu (1994) to *Dipoides anatolicus* (locality 30), and *Castor anderssoni, C. zdanskyi* (locality 108).

We made a major effort to recover small mammals during our fieldwork and managed to find small mammals from three localities. Except for the two jerboas, *Paralactaga suni* and *Dipus fraudator* from locality 30 described by Liu et al. (2008), the study of small mammals is still in progress. We provide a preliminary list here:

Locality 30

Orientalomys sp., *Hansdebruijnia* sp., *Pseudomeriones* cf. *latidens, Nannocricetus mongolicus, Dipus* sp. (*Sminthoides*), *Paralactaga* sp., and *Eutamias* sp.

YJG01

Pseudomeriones sp. nov., *Prosiphneus* sp., *Nannocricetus mongolicus*, *Hansdebruijnia* sp., *Dipus* sp., *Ochotona* indet., Leporidae gen. et sp. indet., and Sciuridae gen. et sp. indet.

JJG02

Pseudomeriones indet., *Nannocricetus mongolicus.*, and *Hansdebruijnia* indet.

SEDIMENTARY CHARACTERISTICS AND LITHOSTRATIGRAPHY

Detailed sedimentological sections through the Baode Formation were measured at two old localities, 30 and 49, which we were able to place in magnetic sections. Supplementary observations were collected at locality 108. The locality 49 and 108 sections are exposed in adjacent erosional ravines in Jijiagou (see figures 7.2 and 7.3), whereas the locality 30 section is situated around 2.5 km to SSE in Daijiagou (see figures 7.2 and 7.4). Although direct lithostratigraphic correlation between the two areas in Jijiagou and Daijiagou is hampered by facies changes and lack of distinct marker beds, all three sections have sediments with similar characteristics and they also represent well the general sedimentary succession observed in the area. The strata in the area are relatively undisturbed, the beds are nearly horizontal or dip gently (5°–10°) mainly toward NNE-ENE. There are local variations in the strike and dip, some of which may represent local paleotopography and others may be caused by faults obscured by modern soil and vegetation.

Figure 7.3 Satellite image of the locality 49 area. Symbols as in figure 7.2.

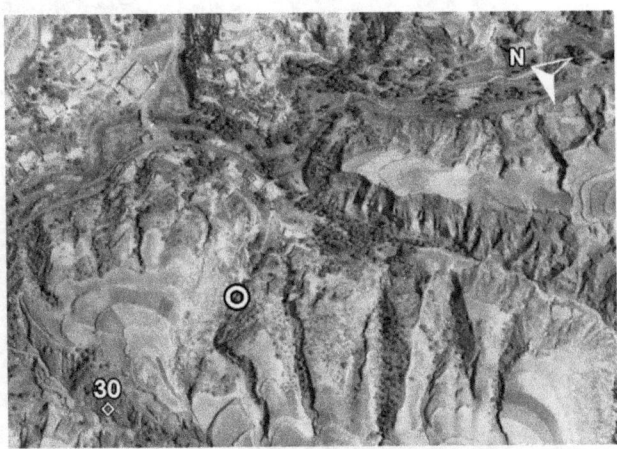

Figure 7.4 Satellite image of the locality 30 area. The collapsed tunnel opening at corresponding level is indicated with an open circle. The sedimentological observations and paleomagnetic study were done on the opposite wall.

In the Baode area, the late Miocene deposits rest upon a Paleozoic basement. Below the unconformity, the basement is often strongly weathered. The lower boundary of the Baode Formation is often marked by a basal conglomerate (such as in the locality 49 section; figure 7.5), but sometimes the base of the succession lacks such basal conglomerate (locality 30 section; figure 7.6). In Jijiagou, the thickness of this basal unit measures up to 10 m but is also highly variable due to irregularity of the lower contact, which displays a few meters of topographic relief in some outcrops. The basal unit consists of predominantly poorly sorted pebble cobble conglomerate with occasional boulders up to a meter across. The fabric is for the most part clast-supported. The surrounding matrix is yellowish to red-brown silt and fine sand, locally forming islands of matrix-supported texture. The structure is mainly massive or crudely stratified, although there are occasionally pebble interbeds and rare sand-silt bands within the conglomerate. The sedimentary characteristics of this unit suggest emplacement by high-energy sediment-bulked fluvial activity. The inverse grading of clasts along with large clast size in some conglomerate beds in the lowermost part of the sequence suggests they might represent viscous debris-flow deposits.

The sedimentary succession overlying the basal conglomerates is built up by numerous 1–3-m-thick sedimentary cycles typically beginning with red-brown clay or clayey silt coarsening up to yellowish-brown silt and capped by laterally continuous calcrete horizon. Subordinate conglomerate and sand beds are interbedded within these fine-grained deposits. The fine-grained facies is

Locality 49

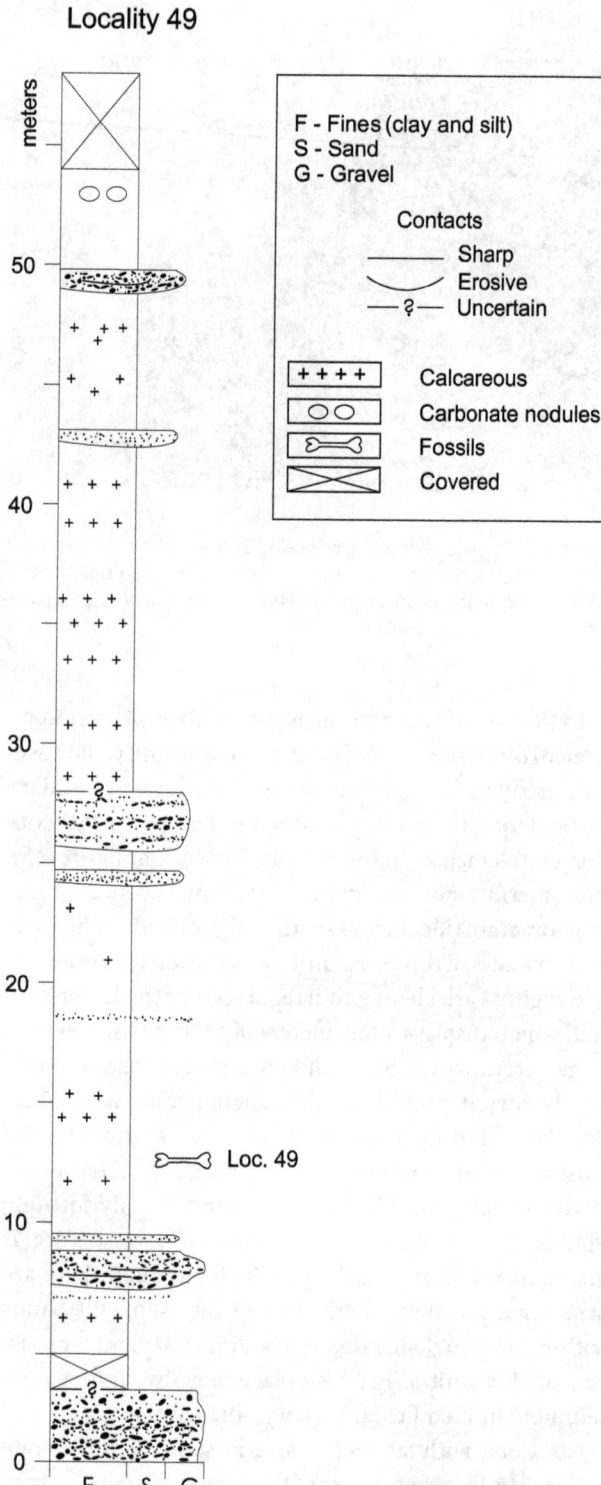

F - Fines (clay and silt)
S - Sand
G - Gravel

Contacts
——— Sharp
⌣ Erosive
— ? — Uncertain

+ + + + Calcareous
○○ Carbonate nodules
⊂⊃ Fossils
⊠ Covered

⊂⊃ Loc. 49

Figure 7.5 Sedimentary log of locality 49 section.

notably structureless and commonly contains pedogenic features such as mottling and ferromanganese coatings. The conglomerates cutting into the fine-grained facies are characteristically faintly stratified pebble to cobble conglomerates with calcareous sand matrix, or occur as horizontally stratified gravel-sand couplets. The individual units are up to 2–3 m thick. They have erosional bases in places marked with scours; the upper contact to the overlying siltstones is sharp. The clasts are subrounded and in some beds imbricated. The conglomerates occur typically as sheet-like complexes that may stretch laterally for more than 100 m. Thin (<1 m) units, however, of this facies are generally less than a few tens of metres in lateral extent. The suite of characteristics in these conglomerate units suggests that they probably resulted from low-sinuosity fluvial channels and unconfined sheet flow processes. Conglomerate clast imbrications and channel scour trends suggest flow from northerly and easterly directions. As a whole, the sedimentary sequence comprises an upward fining megacycle with higher sand and conglomerate content in its lower part.

In all studied sections, the old tunnels and spoil heaps described by Zdansky (1923) were relocated. In locality 49 (see figure 7.5), the tunnel level was established at the lowest part of the succession, around 12 m above the uppermost Paleozoic-Baode Formation contact, while in the adjacent gully, locality 108 was found from a higher stratigraphic level in the local section, around 25 m above the base. In locality 30 environs, the locality level was traced to the next gully over, where a collapsed tunnel opening was identified at the corresponding level. The sedimentological observations were done on the opposite wall, exposing an accessible stratigraphic section with a thickness of around 50 m. Locality 30 level was correlated to the topmost part of the logged section with Jacob's staff and Abney level (see figure 7.6).

MAGNETOSTRATIGRAPHY

Paleomagnetic reversal data were collected from the locality 30 section (see figure 7.6). The stratigraphic sampling interval averaged around 0.3 m between sites, but varied widely (from 0.1 m to 4.2 m) in order to obtain suitable lithologies—siltstones and mudstones. In the lower part, the conglomerates and sandy nature of the sediments did not allow thorough sampling. At each site, the surficial weathered material was removed to sample in sediments as fresh as possible. Altogether, 123 block samples were oriented in the field and cut into 2.5 cm cubes in the lab. Following sample preparation, paleo-

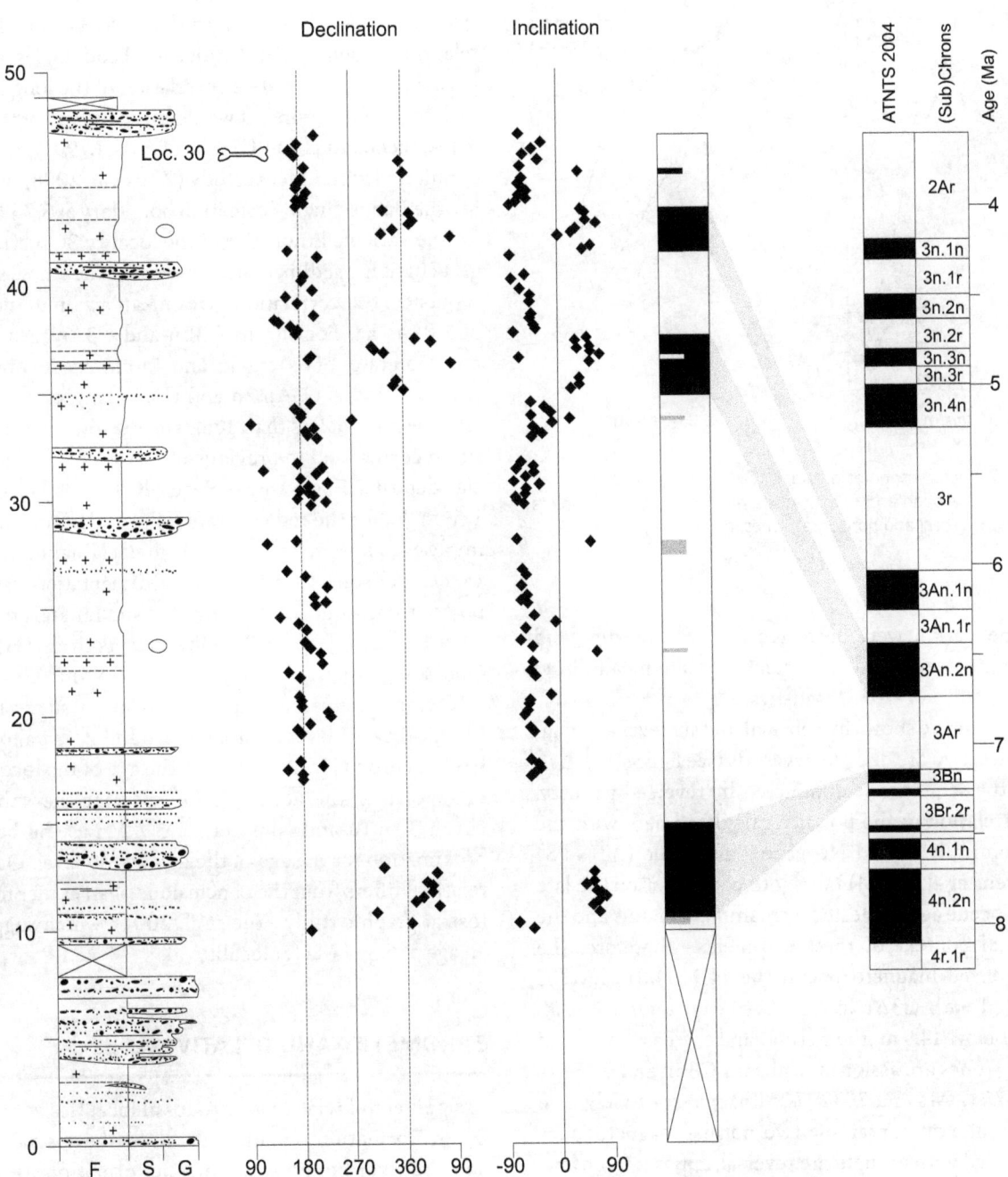

Figure 7.6 Magnetostratigraphy and lithological column of the locality 30 section and correlation of the observed magnetochrons with the geomagnetic reference timescale from Lourens et al. (2004). In the polarity column, black denotes normal polarity and white denotes reversed polarity. Short bars represent single site reversals; gray-shaded bars represent intermediate directions.

magnetic analysis in the laboratory included standard stepwise alternating field demagnetization and measurement of the direction and intensity of natural remanent magnetization decay by means of a 2G three-axis SQUID magnetometer at the Solid Earth Geophysics research laboratory of Helsinki University.

Typical demagnetization diagrams for normal and reversed samples are shown in figure 7.7. In some cases, the polarity was not possible to define either due to unstable behavior during the course of demagnetization or due to mixed polarities or possibly due to intermediate directions. The stable directions obtained after alternating field

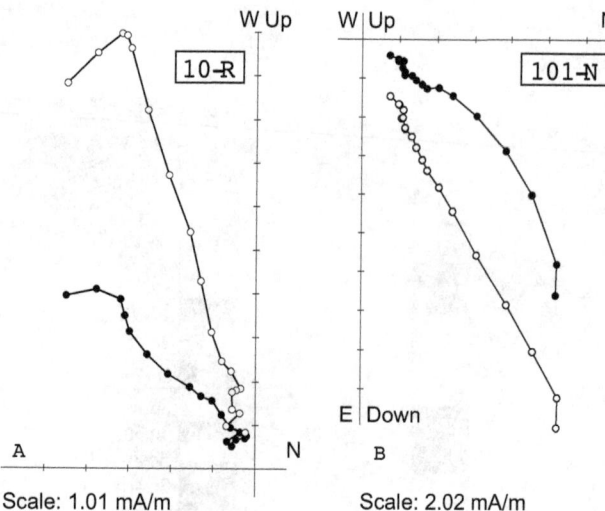

Figure 7.7 Vector endpoint diagrams (Zijderveld 1967) of selected samples, where open and solid symbols represent vector endpoints projected in vertical and horizontal planes, respectively.

demagnetization were corrected for bedding dips and these values were used to calculate the site-mean directions using Fisher (1953) statistics.

The sequence shows five normal and six reversed magnetozones; two are one-site reversals (see figure 7.6). As a whole, the sequence is dominated by reversed polarity. The correlation of the polarity calculated here with the Astronomically Tuned Neogene Time Scale (ATNTS) of Lourens et al. (2004) takes into consideration the late Late Miocene age suggested by mammal remains and the geological context of these sediments. Therefore, the long reversed magnetozone at the 14.9–34.8 m level is correlated with C3Ar (6.733–7.140 Ma; Lourens et al. 2004). Below 14.9 m, the normal and the basal reversed magnetozones are assigned to chrons C3Bn and C3Br.1r (7.140–7.212 Ma and 7.212–7.251 Ma), respectively. The 35–37.6 interval containing two normal magnetozones interbedded with a single site reversal, appears to be correlative with C3An.2n (6.436–6.733 Ma). The reversed magnetozone above matches chron C3An.1r (6.252–6.436 Ma). The interval of 41.7–43.4 consists of a normal magnetozone, correlative with chron C3An.1n (6.033–6.252 Ma). The section ends in the reversed magnetozone of C3r (5.235–6.033 Ma).

To conclude, our correlation is based on the assumption that the long reversed polarity interval is correlated with 3Ar and, consequently, delimiting the locality 30 magnetostratigraphy section between chrons C3Br.1r and C3r on Lourens et al.'s (2004) reference scale. The resulting extrapolations based on sedimentation rates

suggest the sampled portion of the locality 30 section spans ca. 7.2 Ma to no younger than 5.5 Ma. No other correlation involving Late Miocene chronologies appears reasonable: an alternative correlation of the long reversal to C3r is unlikely since it would place the top of the studied sequence to chron C3n.2r (4.631–4.799 Ma), an age conflicting with a recent study (Zhu et al. 2008) positioning the Baode-Jingle Formation boundary at 5.34 Ma.

The variable lithologies of the locality 30 section suggest changing sedimentation rates. For the whole studied sequence, the correlation gives mean accumulation rate of 2.8 cm/ka. For chrons C3Bn and C3Ar, mean accumulation rates of 6.0 cm/ka and 4.9 cm/ka are obtained, whereas chrons C3An.2n and C3An.1n give exceptionally low rates of less than 1cm/ka. The low rates of deposition correspond to previous observations made on red clay deposits. For example, Evans et al. (1991) calculated 1.5 cm/ka for the red clay in Baoji; results from Lantian imply even lower rates of ca. 1.4 cm/ka (Kaakinen 2005). In comparison, average linear sedimentation rates reported from sequences in Dongwan and Lingtai are somewhat higher, 2.1 and 3.0 cm/kyr, respectively (Hao and Guo 2004; Ding, Sun, Yang, and Liu 1998)

Our magnetostratigraphic correlation suggests that the locality 30 level is situated around 2.5 m above the base of chron C3r. We estimate the age of the locality by extrapolating the accumulation rate for the subjacent chron 3An.1n and consider ca. 5.7 Ma as the best approximation for the age of the locality 30 level. Our correlation differs from the conclusions of an earlier magnetostratigraphic study (Yue et al. 2004), which suggested an age of 6.5–7 Ma for locality 30.

BAROMETRY AND RELATIVE AGES

Assignment of relative ages to fossil localities within the Baode Formation is made difficult by the absence of reliable marker horizons within the formation itself, the undulating contact with underlying sedimentary rocks resulting from infilling of paleotopography, and the typically erosive contact with the overlying loess. The single reliable marker horizon relevant to these rocks appears to be the conformable contact between the Baode Formation and the overlying Jingle Formation (sensu Zhu et al. 2008). This contact has mostly been removed by erosion in the vicinity of localities 49, 30, and 108, but important exposures are preserved in Yangjiagou about 1.7 km due west of locality 30 (Zhu et al. 2008). In addition, the Baode-Jingle contact appears to be regionally extensive, with exposures at Tanyugou about 10 km south of locality 30 (Zhu et al. 2008), and likely correlative exposures at Shilou another ~150 km to the south (Xu et al. 2009).

We measured sections and utilized barometric pressure measurements to develop a local framework for relative elevations of Baode Formation contacts and fossil localities (table 7.1 and figure 7.8). Faults were discovered that influence the absolute elevations of the following localities: YJG-01, YJG-02, YJG-06, and YJG-05. The remaining localities appear to reside on relatively undeformed bedrock with little or no stratigraphic dip, and we believe that their relative heights can be loosely interpreted in terms of relative ages of the deposits.

Based on this barometric survey and the accompanying magnetostratigraphic information (this study; Zhu et al. 2008), we conclude that the oldest Baode fossil localities include localities 43, 44 (lower), 31, 49, and JJG-02; the youngest locality is locality 30; and intermediate-aged localities include 44 (upper), 108, and YJG-03 (see figure 7.8). Our magnetostratigraphic study places locality 30 at chron C3r, at an age of about 5.7 Ma. The basically similarly preserved thicknesses and sedimentary successions in Jijiagou (locality 49) and Daijiagou (locality 30) environs allow age estimates of 6.5 Ma and 7.0 Ma for locality 108 and 49 levels, respectively. In these cases, the age estimates are achieved by using the sediment accumulation rates calculated for locality 30 sec-

tion and the stratigraphic distance between the locality 30 level and the locality in question. However, as locality 108 and 49 levels cannot be physically traced to the paleomagnetic section, such correlation has a degree of uncertainty.

BIOCHRONOLOGY

In his paper on the localities of the *Hipparion* fauna of Baode County, Zdansky (1923) mentioned the similarity of the Baode *Hipparion* Clay fauna to that of Pikermi. Andersson (1923) further referred the *Hipparion* clays from Shanxi, Shaanxi, Henan, and Gansu to Lower Pliocene or Upper Miocene. One year earlier, Teilhard de Chardin (1922) explicitly correlated the *Hipparion* fauna from Qingyang to the Pontian, then regarded by European workers as Pliocene. Following Teilhard de Chardin and Young (1930), the Pontian *Hipparion* fauna from North China was then widely cited to be Pliocene until 1984, when Li, Wu, and Qiu emphasized the Baode *Hipparion* fauna to be comparable with the Late Miocene Turolian stage. Their subdivision of Late Miocene into Bahean and Baodean and the correlation with the European Vallesian and Turolian, respectively, was followed by Qiu, Huang, and Guo (1987). Shortly after, Qiu and Qiu (1990, 1995)

Table 7.1

Relative Heights of Baode Formation Contacts and Fossil Localities

Locality	Carboniferous-Baode Contact[a]	Fossil Locality Tunnel Level[b]	Baode-Jingle Contact[c]	Baode-Loess Contact[d]
Locality 30	1	54	Eroded	57
Locality 31	Buried	26[e]	56[f]	Not measured
Locality 43	16, 17	25	Eroded	Not measured
Locality 44	20	26, 40	Eroded	Not measured
Locality 49	0, 14	27	Eroded	Not measured
Locality 108	10	40	Eroded	Not measured
JJG 02	2, 18	24	Eroded	Not measured
YJG 03	Buried	32	Eroded	54
YJG Magneto. Section	Buried	—	61	70

NOTE: Units are in meters. Heights are relative to the lowest Carboniferous-Baode contact surface measured at locality 49 and were determined using barometric pressure differences and direct measurement of section. The estimated error is ±3 meters.

[a] Disconformable contact. Two values are given where significant paleotopography is observed at the contact surface.

[b] These include relocated fossil localities (localities 30, 31, 43, 44, 49, 108), and new localities (JJG 02, YJG 03). YJG magnetostratigraphic section is that of Zhu et al. (2008).

[c] Conformable contact. Removed by erosion at most localities.

[d] Disconformable contact typically showing significant paleotopography. This contact is present at all localities, though its elevation was not typically measured.

[e] Original tunnel buried and located below the given elevation.

[f] Sighted on nearby outcrop using hand level.

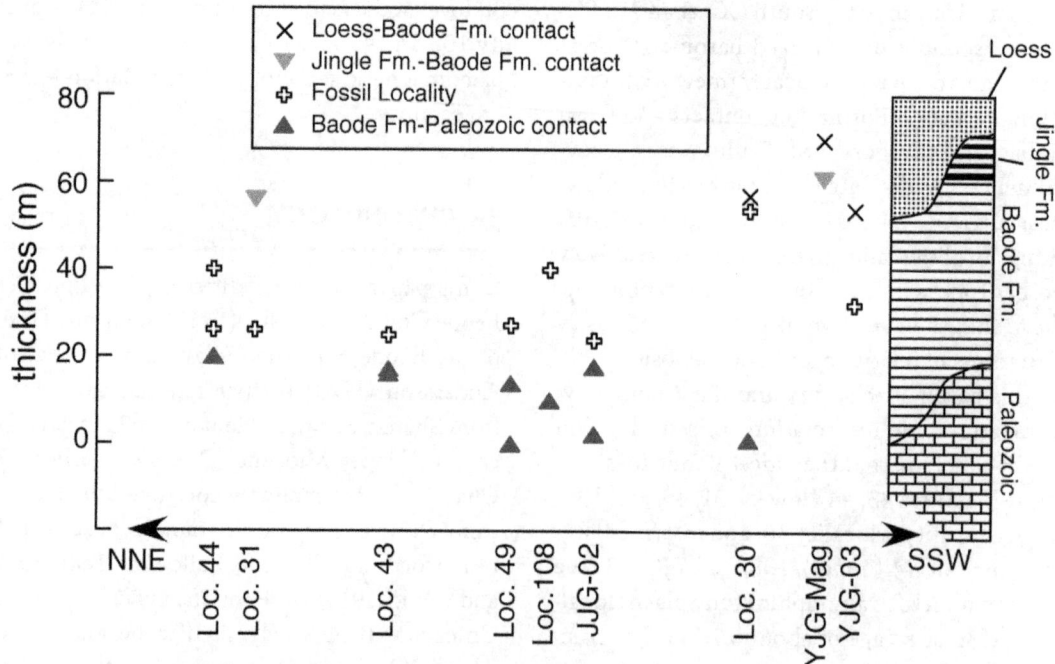

Figure 7.8 Relative elevations of the Baode Formation, upper and lower contacts, and fossil localities. The localities are arranged from north-northeast to south-southwest. The elevations were determined using a combination of ambient pressure-corrected barometry and direct measurement of section using a Jacob staff; estimated uncertainty is ±3 m. Variability in the elevations of upper and lower contacts is primarily due to the erosional topography before and after deposition of the Baode Formation, as depicted in the simplified stratigraphic column to the right.

and Tong, Zheng, and Qiu (1995) merged Bahean into Baodean. Qiu, Wu, and Qiu (1999) considered the Bahe Fauna as representative of their NMU 9 and listed the Baode fauna as representative of their NMU 10, while comparing it with the Pikermi, Samos, and Maragheh faunas, correlated with MN 11+12. Based on new findings from Lantian sections, we (Zhang et al. 2002; Zhang and Liu 2005; Zhang et al., chapter 6, this volume) suggested to resurrect the Bahean Age and considered the Baodean as representing only the later Late Miocene age.

This recent interpretation of the Baodean was reinforced by discovery of small mammals from some Baode localities. The Baode small mammal assemblages show significant difference from the Bahe Fauna, and confirm its younger age relative to the Bahean (Liu et al. 2008; Zhang et al., chapter 6, this volume). Two recent genera, *Dipus* and *Castor*, have their first appearances there. Ancient taxa such as *Myocricetodon* and *Protalactaga* disappeared, and *Dipoides* and *Pseudomeriones* dispersed into North China from North America and West Asia, respectively. The advanced murids *Hansdebruijnia* and *Orientalomys* replaced the primitive *Progonomys*, and *Nannocricetus mongolicus* replaced the primitive species *N. primitivus* from the Bahe Formation.

Updated biostratigraphy and previously published paleomagnetic dating (Zhu et al. 2008) of the sequences in the Baode area confine the faunas between 7.2 Ma and 5.3 Ma. This time interval is roughly correlated to MN 13 in Europe (Steininger et al. 1996; Steininger 1999).

PALEOECOLOGY

The paleoecological context of fossil herbivores from Baode has recently been studied using hypsodonty, mesowear, and stable isotopes (Fortelius et al. 2002; Passey 2007; Passey et al. 2007; Liu et al. 2008; Passey et al. 2009), and the following is a summary of the more salient results of these studies.

The late Miocene witnessed a global expansion of C_4 vegetation, both in terms of biomass, as gauged by stable isotopic studies of soil carbonates, and in importance in mammalian herbivore diets, as gauged by isotopic studies of fossil tooth enamel (Cerling et al. 1997). The stable isotope results from Baode fossil mammals clearly identify these faunas as part of the "C_4 world," either in the early stages of or post-dating the main expansion event (Passey et al. 2009). Mammal diets delineate along morphological lines, with lower-crowned animals consuming primarily C_3 vegetation, consistent with browsing diets, and with intermediate- and higher-crowned animals consuming some fraction of C_4 vegetation, consistent with mixed feeding or grazing diets (figure 7.9). These results support the Schlosser–Kurtén concept—largely based on hypsodonty—of the distinction between browsing, forest-dwelling "gaudryi faunas" and more open-country, mixed feeding or grazing "dorcadoides" faunas. Indeed, an isotopic study of *Gazella gaudryi* and *G. dorcadoides* revealed a distinct dietary (δ^{13}C) and

behavioral/physiological (δ^{18}O) difference between these species, in the same direction as expected from their contrasting tooth morphology: the higher-crowned gazelles (*G. dorcadoides*-type) consumed some C_4 vegetation, whereas the lower-crowned *G. gaudryi*-type gazelles consumed pure C_3 diets (Passey et al. 2007). The mesowear results (Fortelius et al., unpublished data) have been particularly enlightening in revealing that the putative "grazing taxa" (i.e., many of those associated with Kurtén's dorcadoides faunas) were actually rather more "mixed feeding" taxa, with very few individuals exhibiting the extreme wear and abrasion characteristic of modern grassland grazers (Fortelius and Solounias 2000; Solounias et al., chapter 31, this volume).

When the combined morphological and geochemical data are considered in a spatial and temporal framework, a picture emerges of habitats that were variable in time but on the whole more wooded than present-day habitats. A clear environmental gradient existed in the Loess Plateau during the late Miocene, with more wooded (and presumably humid) environments in the south and more open (drier) environments in the north, similar to the present-day ecological gradient maintained by the East Asian monsoons (Passey et al. 2009). In addition, northern China as a whole is notable as a region that became more humid during the late Miocene, which was a time of cooling and drying elsewhere (Fortelius et al. 2002). Intensification of the East Asian summer monsoon beginning in the late Miocene (or a weakening of the winter monsoon) is one mechanism that could account for this reversal (Fortelius et al. 2002; Passey et al. 2009).

The unusual convergence of morphological, stable isotopic, and taxonomic abundance data at localities 49 and 30 permits an investigation of community dynamics across a paleobiome boundary. Locality 49 resides low in the Baode Formation (this study) and is a "mixed locality" in the parlance of Kurtén (1952), preserving many low-crowned, presumably closed-habitat species, and fewer higher-crowned, presumably open habitat species. Locality 30, on the other hand, resides high in the Baode Formation and is a "dorcadoides locality" that is relatively enriched in higher-crowned, open habitat taxa. The open habitats are also supported by the small mammals; those in locality 30 show a high proportion of dipodids (Liu et al. 2008). Today, the family Dipodidae is restricted to the arid areas of the Palearctic Region (Ma et al. 1987; Wang 2003), and cohabiting herbivores also suggest dipodid habits in a distinctly arid context after 10 Ma (Liu et al. 2008).

Several taxa occur at both localities, and with the benefit of mesowear and stable isotope data, we can examine how different taxa adapted to different environments.

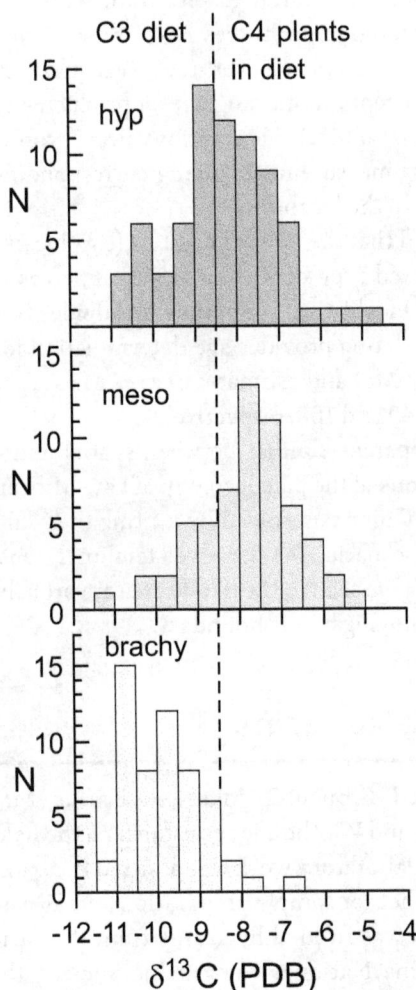

Figure 7.9 Histograms showing the distribution of carbon isotope values among brachydont, mesodont, and hypsodont Baodean herbivores, including cervids, giraffids, bovids, rhinocerotids, and equids. Dashed vertical line represents a −8.6‰ "cutoff" value, separating lower values consistent with pure C_3 diets and higher values consistent with some C_4 vegetation in diet. Carbon isotope ratios are statistically indistinguishable for hypsodont and mesodont taxa, but both are significantly different from brachydont taxa. Data from Passey (2007).

Plate 7.1 depicts the results of this analysis. In summary, we find that of five brachydont, C_3-feeding taxa, all common at locality 49, three are not present at locality 30 (*G. gaudryi*, *Honanotherium*, and *Chleuastochoerus*), and two are present but less common at locality 30 (*Cervavitus*, *G. paotehensis*). Two C_3-feeding species at locality 49 begin to consume C_4 vegetation at locality 30, but remain uncommon (*Palaeotragus*, *Tragoreas*). In contrast, three less common consumers of C_4 vegetation at locality 49 become dominant at locality 30 and continue consuming C_4 vegetation (*G. dorcadoides*, *Urmiatherium*, *Hipparion*). In addition, of three taxa examined here that were not present at locality 49 but that were present at locality 30, two consumed C_4 vegetation, and all were either hypsodont (*Chilotherium*, *Sinotherium*) or mesodont (*Samotherium*). Finally, mesowear patterns indicate more abrasive diets at locality 30 compared to locality 49 for four of five taxa studied.

At first sight, this temporal pattern appears to contradict the finding of a Late Miocene trend in North China from drier toward more humid conditions, established both locally (Xue, Zhang, and Yue 1995, 2006; Kaakinen 2005; Kaakinen et al. 2006; Zhang et al., chapter 6, this volume) and globally (Fortelius et al. 2002; Liu, Eronen, and Fortelius 2009). It is therefore important to emphasize that the change toward a more "steppe-like" assemblage takes place within the context of the Baodean sensu stricto, a time interval characterized as a whole as more humid than the preceding Bahean (Fortelius et al. 2002; Zhang et al., chapter 6, this volume). Furthermore, it is not possible to determine from the available data whether the change is a sustained one, or whether we are sampling opposite phases of ongoing cyclic change. In either case, the proximity to the boundary between two biomes (Kurtén 1952; Fortelius et al. 2002; Sun and Wang 2005) is the most likely reason for such apparently major changes and faunal mixing in the Baode area.

Thus, the apparent conflict between temporally and spatially based hypotheses to explain faunal differences in the Baodean interval is resolved by their fusion into a spatio-temporal scenario. Whether time or space is more important depends on the scale and the location of the data within the time–space continuum. For the entire Late Miocene, temporal change is the dominant factor, driving all of North China toward more humid conditions (Fortelius et al. 2002; Passey et al. 2009). During Baodean time, the spatial separation of two biomes is the main reason for changes. Finally, for the localities situated near the biome boundary in Baode and Fugu counties, the two factors interact strongly. A climatic change

that elsewhere would appear minor may be amplified in this area by an ecological boundary shifting across it. Even minor geographic distances may, under this scenario, have a locally pronounced ecological significance at any one time.

CONCLUSION

Although the fossils from Baode area have been known for more than a century, their stratigraphic position has remained unresolved. We have successfully relocated most of the localities collected by Zdansky, have investigated their sedimentological context, and have placed several of them in a stratigraphic framework.

Our stratigraphic studies have shown that the fossil localities represent at least three levels, in contrast to the earlier conception of a single level producing all faunas. The stratigraphical observations were confirmed by barometry measurements giving corresponding relative positions to the localities.

We find that the oldest localities (e.g., locality 49) are of the mixed type of Kurtén (1952), whereas the youngest (e.g., locality 30) are of the dorcadoides type. Paleomagnetic dating provides age determination for locality 30 (ca. 5.7 Ma) and estimates of ca. 7 Ma and 6.5 Ma for localities 49 and 108, respectively.

The apparent conflict between spatial and temporal explanations of the paleoecology of Late Miocene faunas of North China is resolved by uniting them in a spatio-temporal scenario. The perceived temporal trends of ecological change seen in the Baode area are probably amplified by a moving biome boundary.

ACKNOWLEDGMENTS

We thank J. Eronen, T. Jokela, A. Karme, M. Mirzaie Ataabadi, and W. Zhou for assistance in the field. Paleomagnetic laboratory work was assisted by S. M. Hoxha; we thank her for sample preparation and measurement. This work was funded by grants from the Academy of Finland, the National Geographic Society, the Major Basic Research Projects (2006CB806400), the National Science Foundation of China (40672010, 40711130639, 40872018), the Finnish Cultural Foundation, the Ella and Georg Ehrnrooth Foundation, and the Emil Aaltonen Foundation. We thank Larry Flynn, Nikos Solounias and Xiaoming Wang for their helpful and constructive reviews.

REFERENCES

Anderson, J. G. 1923. Essays on the Cenozoic of northern China. *Geological Survey of China Memoir*, ser. A, 3:1–152.

Andersson, K. and L. Werdelin. 2005. Carnivora from the Late Miocene of Lantian, China. *Vertebrata PalAsiatica* 43:256–271.

Bernor, R. L., Z.-x. Qiu, and L. Hayek. 1990. Systematic revision of Chinese *Hipparion* species described by Sefve, 1927. *American Museum Novitates* 2984:1–60.

Bohlin, B. 1926. Die Familie Giraffidae. *Palaeontologia Sinica*, ser. C, 4:1–179.

Bohlin, B. 1935. Cavicornier der Hipparion-Fauna Nord Chinas. *Palaeontologia Sinica*, ser. C, 9:1–166

Cerling, T. E., J. M. Harris, B. J. MacFadden, M. G. Leakey, J. Quade, V. Eisenmann, and J. R. Ehleringer. 1997. Global vegetation change through the Miocene/Pliocene boundary. *Nature* 389:153–158.

Chen, G.-f. 2005. Dorcadoryx Teilhard et Trassaert, 1938 (Bovidae, Artiodactyla) from the Bahe Formation of Lantian, Shaanxi Province, China. *Vertebrata PalAsiatica* 43:272–282.

Deng, T. 2000. A new species of *Acerorhinus* (Perissodactyla, Rhinocerotidae) from the Late Miocene in Fugu, Shaanxi, China. *Vertebrata PalAsiatica* 38:203–217.

Deng, T., W.-m. Wang, L.-p. Yue, and Y.-x. Zhang. 2004. New advances in the establishment of the Neogene Baode Stage. *Journal of Stratigraphy* 28:41–47 (in Chinese with English abstract).

Ding, Z.-l., J.-m. Sun, T.-s. Liu, R.-x. Zhu, S.-l. Yang, and B. Guo. 1998. Wind-blown origin of the Pliocene red clay formation in the Chinese Loess Plateau. *Earth and Planetary Science Letters* 161:135–143.

Ding, Z.-l., J.-m. Sun, S.-l. Yang, and T.-s. Liu. 1998. Preliminary magnetostratigraphy of a thick eolian red clay-loess sequence at Lingtai, the Chinese Loess Plateau. *Geophysical Research Letters* 25:1225–1228.

Evans, M. E., Y. Wang, N. Rutter, and Z.-l. Ding. 1991. Preliminary magnetostratigraphy of the Red Clay underlying the loess sequence at Baoji, China. *Geophysical Research Letters* 18:1409–1412.

Fisher, R. S. 1953. Dispersion on a sphere. *Proceedings of the Royal Society of London*, ser. A 217:295–305.

Fortelius, M., J. Eronen, J. Jernvall, L.-p. Liu, D. Pushkina, J. Rinne, A. Tesakov, I. Vislobokova, Z.-q. Zhang, L.-p. Zhou. 2002. Fossil mammals resolve regional patterns of Eurasian climate change over 20 million years. *Evolutionary Ecology Research* 4:1005–1016.

Fortelius, M., K. Heissig, G. Saraç, and S. Sen. 2003. Rhinocerotidae (Perissodactyla). In *Geology and Paleontology of the Miocene Sinap Formation, Turkey*, ed. M. Fortelius, J. Kappelman, S. Sen, and R. L. Brenor, pp. 282–307. New York: Columbia University Press.

Fortelius, M. and N. Solounias. 2000. Functional characterization of ungulate molars using the abrasion-attrition wear gradient: A new method for reconstructing paleodiets. *American Museum Novitates* 3301:1–36.

Gentry, A. W. 1971. The earliest goats and other antelopes from the Samos Hipparion fauna. *Bulletin of the British Museum, Geology* 20:229–296.

Guo, Z.-t., S.-t. Peng, Q.-z. Hao, P. E. Biscaye, and T.-s. Liu. 2001. Origin of the Miocene–Pliocene red earth formation at Xifeng in northern China and implications for paleoenvironments. *Palaeogeography, Palaeoclimatology, Palaeoecology* 170:11–26.

Guo, Z.-t., W. F. Ruddiman, Q.-z. Hao, H.-b. Wu, Y.-s. Qiao, R.-x. Zhu, S.-z. Peng, J.-j. Wei, B.-y. Yuan, and T.-s. Liu. 2002. Onset of Asian desertification by 22 Myr ago inferred from loess deposits in China. *Nature* 416:159–163.

Hao, Q.-z. and Z.-t. Guo. 2004. Magnetostratigraphy of a late Miocene–Pliocene loess–soil sequence in the western Loess Plateau in China. *Geophysical Research Letters* 31:L09209. doi:10.1029/2003GL019392.

Heissig, K. 1975. Rhinocerotidae aus dem Jungtertiär Anatoliens. *Geologisches Jahrbuch* ser. B 15:145–151.

Jokela, T., J. T. Eronen, A. Kaakinen, L.-p. Liu, B. Passey, Z.-q. Zhang, and M.-k. Fu. 2005. Translation of Otto Zdansky's "The Localities of the *Hipparion* Fauna in Baode County in N. W. Shanxi" (1923) from the German. *Paleontologia Electronica* 1(8):3A:10p.

Kaakinen, A. 2005. A long terrestrial sequence in Lantian: A window into the late Neogene palaeoenviornments of northern China. Ph.D. diss., University of Helsinki, Helsinki.

Kaakinen, A., E. Sonninen, and J. P. Lunkka. 2006. Stable isotope record in paleosol carbonates from the Chinese Loess Plateau: Implications for late Neogene paleoclimate and paleovegetation. *Palaeogeography, Palaeoclimatology, Palaeoecology* 237:359–369.

Koken, E. 1885. Über fossile Säugethiere aus China. *Paläontogische Abhandlungen* 3:31–114.

Kurtén, B. 1952. The Chinese *Hipparion* fauna. *Commentationes Biologicae, Societas Scientiarum Fennica* 13:1–82.

Kurtén, B. 1985. *Thalassictus wongii* (Mammalia, Hyaenidae) and related forms from China and Europe. *Bulletin of the Geological Institutions of the University of Uppsala* 11:79–90.

Li, C.-k., W.-y. Wu, and Z.-d. Qiu. 1984. Chinese Neogene: Subdivision and correlation. *Vertebrata PalAsiatica* 22:163–178 (in Chinese with English summary).

Liu, L.-p., J. T. Eronen, and M. Fortelius. 2009. Significant midlatitude aridity in the middle Miocene of East Asia. *Palaeogeography, Palaeoclimatology, Palaeoecology* 279:201–206.

Liu L.-p., Z.-q. Zhang, N. Cui, and M. Fortelius. 2008. The Dipodidae (Jerboas) from Loc.30 of Baode and their environmental significance. *Vertebrata PalAsiatica* 46:25–32.

Lourens, L., F. Hilgen, N. J. Shackleton, J. Laskar, and D. Wilson. 2004. The Neogene period. In *A Geologic Time Scale 2004*, ed. F. Gradstein, J. Ogg, and A. G. Smith, pp. 409–440. Cambridge: Cambridge University Press.

Lu, H.-y., J. Vandenberghe, and Z.-s. An. 2001. Aeolian origin and palaeoclimatic implications of the "Red Clay" (north China) as evidenced by grain-size distribution. *Journal of Quaternary Science* 16:89–97.

Ma, Y., F.-g. Wang, S.-k. Jin, and S.-h. Li. 1987. *Glires (Rodents and Lagomorphs) of Northern Xinjiang and Their Zoogeographical Distribution*. Beijing: Science Press (in Chinese).

Miao, X-d., Y.-b. Sun, H.-y. Lu, and J. A. Mason. 2004. Spatial pattern of grain size in the Late Pliocene "Red Clay" deposits (North China) indicates transport by low-level northerly winds. *Palaeogeogrphy, Palaeoclimatology, Palaeoecology* 206:149–155.

National Stratigraphic Committee. 2002. *Regional Chronostratigraphic (Geochronologic) Table of China*. Beijing: Geological Publishing House.

Passey, B. H. 2007. Stable isotope paleoecology: Methodological advances and applications to the Late Neogene environmental history of China. Ph.D. diss., University of Utah.

Passey, B. H., L. K. Ayliffe, A. Kaakinen, Z.-q. Zhang, J. T. Eronen, Y.-m. Zhu, L.-p. Zhou, T. E. Cerling, and M. Fortelius. 2009. Strengthened East Asian summer monsoons during a period of high-latitude warmth? Isotopic evidence from Mio-Pliocene fossil mammals and soil carbonates from northern China. *Earth and Planetary Science Letters* 277:443–452.

Passey, B. H., J. T. Eronen, M. Fortelius, and Z.-q. Zhang. 2007. Paleodiets and paleoenvironments of late Miocene gazelles from North China: Evidence from stable carbon isotopes. *Vertebrata PalAsiatica* 45:118–127.

Pearson, H. S. 1928. Chinese fossil Suidae. *Palaeontologia Sinica*, ser. C 5:1–75.

Pei, W.-z., M.-z. Zhou, and J.-j. Zheng. 1963. *Chinese Cenozoic Erathem*. Beijing: Science Press (in Chinese).

Qiu, Z.-d., S.-h. Zheng, and Z.-q. Zhang. 2008. Sciurids and zapodids from the Late Miocene Bahe Formation, Lantian, Shaanxi. *Vertebrata PalAsiatica* 46:111–123.

Qiu, Z.-x. 1990. The Chinese Neogene mammalian biochronology—its correlation with the European Neogene mammal zonation. In *European Neogene Mammal Chronology*, ed. E. H. Lindsay, V. Fahlbusch, and P. Mein, pp. 527–556. New York: Plenum Press.

Qiu, Z.-x., W.-l. Huang, and Z.-h. Guo. 1987. The Chinese Hipparionine fossils. *Palaeontologia Sinica*, n.s. C, 25:1–250.

Qiu, Z.-x. and Z.-d. Qiu. 1990. Neogene local mammalian faunas: Succession and ages. *Journal of Stratigraphy* 14:241–260 (in Chinese with English abstract).

Qiu, Z.-x. and Z.-d. Qiu. 1995. Chronological sequence and subdivision of Chinese Neogene mammalian faunas. *Palaeogeography, Palaeoclimatology, Palaeoecolology* 116:41–70.

Qiu, Z.-x., W.-y. Wu, and Z.-d. Qiu. 1999. Miocene mammal faunal sequence of China: Paleozoogeography and Eurasian relationships. In *The Miocene Land Mammals of Europe*, ed. G. E. Rössner and K. Heissig, pp. 443–455. Munich: Dr. Friedrich Pfeil.

Ringström, T. J. 1924. Nashörner der *Hipparion*-Fauna Nord-China. *Palaeontologia Sinica*, ser. C, 1:1–159.

Schlosser, M. 1903. Die fossilen Säugethiere Chinas nebst einer Odontographie der recenten Antilopen. *Abhandlungen der Bayerischen Akademie der Wissenschaften* 22:1–221.

Sefve, I. 1927. Die Hipparionen Nord-Chinas. *Palaeontologia Sinica*, Series C 4:1–94.

Steininger, F. F. 1999. Chronostratigraphy, geochronology and biochronology of the Miocene "European Land Mammal Mega-Zones (ELMMZ)" and the Miocene "Mammal-Zones (MN-Zones)." In *The Miocene Land Mammals of Europe*, ed. G. E. Rössner and K. Heissig, pp. 9–24. Munich: Dr. Friedrich Pfeil.

Steininger, F. F., W. A. Berggren, D. V. Kent, R. L. Bernor, S. Sen, and J. Agustí. 1996. Circum-Mediterranean Neogene (Miocene-Pliocene) marine-continental chronologic correlations of European mammal units. In *The Evolution of Western Eurasian Neogene Mammal Faunas*, ed. R. L. Bernor, V. Fahlbusch, and H.-W. Mittmann, pp. 7–46. New York: Columbia University Press.

Sun, X.-j., and P.-x. Wang. 2005. How old is the Asian monsoon system? Palaeobotanical records from China. *Palaeogeography, Palaeoclimatology, Palaeoecology* 222:181–222.

Teilhard de Chardin, P. 1922. Sur une faune de mammifères Pontiens provenant de la Chine septentrionale. *Comptes Rendus de l'Académie des Sciences Paris* 175: 979–981.

Teilhard de Chardin, P. and J. Piveteau. 1930. Les Mammifères fossiles de Nihowan (Chine). *Annales de Paléontologie* 19:1–134.

Teilhard de Chardin, P. and C. C. Young. 1930. Some correlations between the geology of China proper and geology of Mongolia. *Bulletin of the Geological Society of China* 9:119–125.

Teilhard de Chardin, P. and C. C. Young. 1931. Fossil mammals from the Late Cenozoic of Northern China. *Palaeontologia Sinica*, ser. C, 9:1–67.

Tobien H., G.-f. Chen, and Y.-q. Li. 1988. Mastodonts (Proboscidea, Mammalia) from the Late Neogene and Early Pleistocene of the People's Republic of China. Part 2. The genera *Tetralophodon, Anacus, Stegotetrabelodono, Zygolophodon, Mammut, Stegolophodon*; some generalities on the Chinese Mastodons. *Mainzner Geowissenschaften Mitteilungen* 17:95–220.

Tong, Y.-s., S.-h. Zheng, and Z.-d. Qiu. 1995. Cenozoic mammal ages of China. *Vertebrata PalAsiatica* 33:290–314.

Vislobokova, I. A. 1990. *The Fossil Deer of Eurasia*. Moscow: Nauka (in Russian).

Wang, Y.-x. 2003. *A Complete Checklist of Mammal Species and Subspecies in China—A Taxonomic and Geographic Reference*. Beijing: China Forestry Publishing House (in Chinese with English summary).

Werdelin, L. and N. Solounias. 1989. Studies of fossil hyaenids: The genus Adcrocuta Kretzoi and the interrelationships of some hyaenid taxa. *Zoological Journal of the Linnean Society* 98:363–386.

Wiman, C. 1930. Fossile Schildkroten aus China. *Palaeonologia Sinica*, ser. C, 6:5–53.

Xu, X.-f. 1994. Evolution of Chinese Castoridae. In *Rodent and Lagomorph Families of Asian Origins and Diversification*, ed. Y. Tomida, C.-k. Li, and T. Setoguchi. *National Science Museum Monographs, Tokyo* 8:77–97.

Xu, X.-w. and X.-y. Ma. 1992. Geodynamics of the Shanxi Rift system, China. *Tectonophysics* 208:325–340.

Xu, X.-w., X.-y. Ma, and Q.-d. Deng. 1993. Neotectonic activity along the Shanxi rift system, China. *Tectonophysics* 219:305–325.

Xu, Y., L.-p. Yue, J.-x. Li, L. Sun, B. Sun, J.-y. Zhang, J. Ma, and J.-q. Wang. 2009. An 11-Ma-old red clay sequence on the eastern Chinese Loess Plateau. *Palaeogeography, Palaeoclimatology, Palaeoecology* 284:383–391.

Xue, X.-x., Y.-x. Zhang, and L.-p. Yue. 1995. Discovery and chronological division of the *Hipparion* faunas in Laogaochuan Village, Fugu County, Shaanxi. *Chinese Science Bulletin* 40:926–929.

Xue, X.-x., Y.-x. Zhang, and L.-p. Yue. 2006. Paleoenvironments indicated by the fossil mammalian assemblages from red clay-loess sequence in the Chinese Loess Plateau since 8.0 Ma B.P. *Science in China*, ser. D, 49:518–530.

Young, C. C. 1927. Fossile Nagetiere aus Nord-China. *Palaentologia Sinica* 5:1–82.

Yue, L.-p., T. Deng, Y.-x. Zhang, J.-q. Wang, R. Zhang, L.-r. Yang, and F. Heller. 2004. Magnetostratigraphy of stratotype section of Baode Stage. *Journal of Stratigraphy* 28:48–63 (in Chinese with English abstract).

Zdansky, O. 1923. Fundorte der Hipparion-Fauna um Pao-Te-Hsien in NW-Shansi. *Bulletin of the Geological Survey of China* 5:69–81.

Zhang, Z.-q. and L.-p. Liu. 2005. The Late Neogene mammal bio-chronology in the Loess Plateau, China. *Annales de Paléontologie* 91:257–266.

Zhang, Z.-q., A. Gentry, A. Kaakinen, L.-p. Liu, J. P. Lunkka, Z.-d. Qiu, S. Sen, R. Scott, L. Werdelin, S.-h. Zheng, and M. Fortelius. 2002. Land mammal faunal sequence of the late Miocene of China: New evidence from Lantian Shaanxi Province. *Vertebrata PalAsiatica* 40:165–176.

Zhu, Y.-m., L.-p. Zhou, D.-w. Mo, A. Kaakinen, Z.-q. Zhang, and M. Fortelius. 2008. A new magnetostratigraphic framework for late Neogene *Hipparion* Red Clay in the eastern Loess Plateau of China. *Palaeogeography, Palaeoclimatology, Palaeoecology* 268:47–57.

Zijderveld, J. D. A., 1967. A. C. demagnetization of rocks: analysis of results. In *Methods in Palaeomagnetism*, ed. D. W. Collinson, K. M. Creer, and S. K. Runcorn, pp. 254–286. Amsterdam: Elsevier.

Chapter 8

Review of the Litho-, Bio-, and Chronostratigraphy in the Nihewan Basin, Hebei, China

BAO-QUAN CAI, SHAO-HUA ZHENG, JOSEPH C. LIDDICOAT, AND QIANG LI

Famous for yielding the classic early Pleistocene Nihowan fauna and a number of paleolithic relics, the Nihewan (=Nihowan) Basin is located about 140 km northwest of Beijing and about 55 km southwest of Xuanhua (=Hsuan Hua Fu) (figure 8.1). Physiographically, the Nihewan Basin in its restricted sense is similar to the Yangyuan Basin (Barbour, Licent, and Teilhard de Chardin 1926; Teilhard de Chardin and Piveteau 1930). However, in a broader sense, the Nihewan Basin once was interpreted as the combination of the Yangyuan and Yuxian basins (Wu, Sun, and Yuan 1980; Lin 1984) and included even the Datong, Zhuolu, Huailai, and Xuanhua basins (Wei 1978; Han 1982; Liu and Xia 1983). As now interpreted, the Nihewan Basin is a late Cenozoic intermontaine basin with dimensions of ~80 × 20 km and contains a thickness greater than 150 m of fluviolacustrine and eolian deposits along the banks of the Sanggan (=Sangkanho) and Huliu (=Hou-liu-ho) rivers; the basin is controlled tectonically by NE–SW faults.

Herein, we review lithostratigraphic, biostratigraphic, and chronostratigraphic studies of the Nihewan Basin since the 1920s when vertebrate fossils first were found. Cenozoic sediments constitute four general groups: early Pliocene red beds (Shixia Formation), late Pliocene reddish and dark-colored fluviolacustrine and palustrine sandy clay (Daodi Formation), Pleistocene gray-greenish or gray-yellowish fluviolacustrine sand, gravel, and clay (Nihewan Formation), and overlying loess. We determine how the Nihewan beds of Barbour, Licent, and Teilhard de Chardin (1926) match this stratigraphy. A section at Danangou serves as the stratotype of the Nihewan Formation for its well-developed Pleistocene strata with abundant fossil mammals. Based on the small mammals collected from twenty sections in the Nihewan basin, fossil assemblage-zones help correlate the classic Nihowan fauna of Teilhard de Chardin and Piveteau (1930).

Most Chinese Quaternary researchers are inclined to use the palaeomagnetic transition of the Gauss and Matuyama chrons (2.6 Ma) as the beginning of Pleistocene time (Liu et al. 2000; An and Ai 2005). In recent years, this viewpoint has become widely accepted (Kerr 2008, 2009). We agree with using 2.6 Ma as the boundary of Pliocene and Pleistocene.

Nihewan Basin Sensu Stricto

Barbour, Licent, and Teilhard de Chardin (1926) defined the range of Nihowan deposits as occupying the whole of the large basin at the junction of the Sanggan and the Hou-liu rivers. In the valley of the Sanggan River proper, the deposits abut the barrier of gneiss and Paleozoic sediments that lies athwart the gorge of the river in a southeast–northwest direction. To the west the deposits thin out, and on the north they are bounded by a range of crystalline and Sinian rocks and on the south by the mountain chain of the Hsiao Wutaishan. Teilhard de Chardin and Piveteau (1930:5), on the other hand, described it as follows: "Topographiquement, le bassin de Nihowan représente un large berceau, long de 20 kilomètres et large

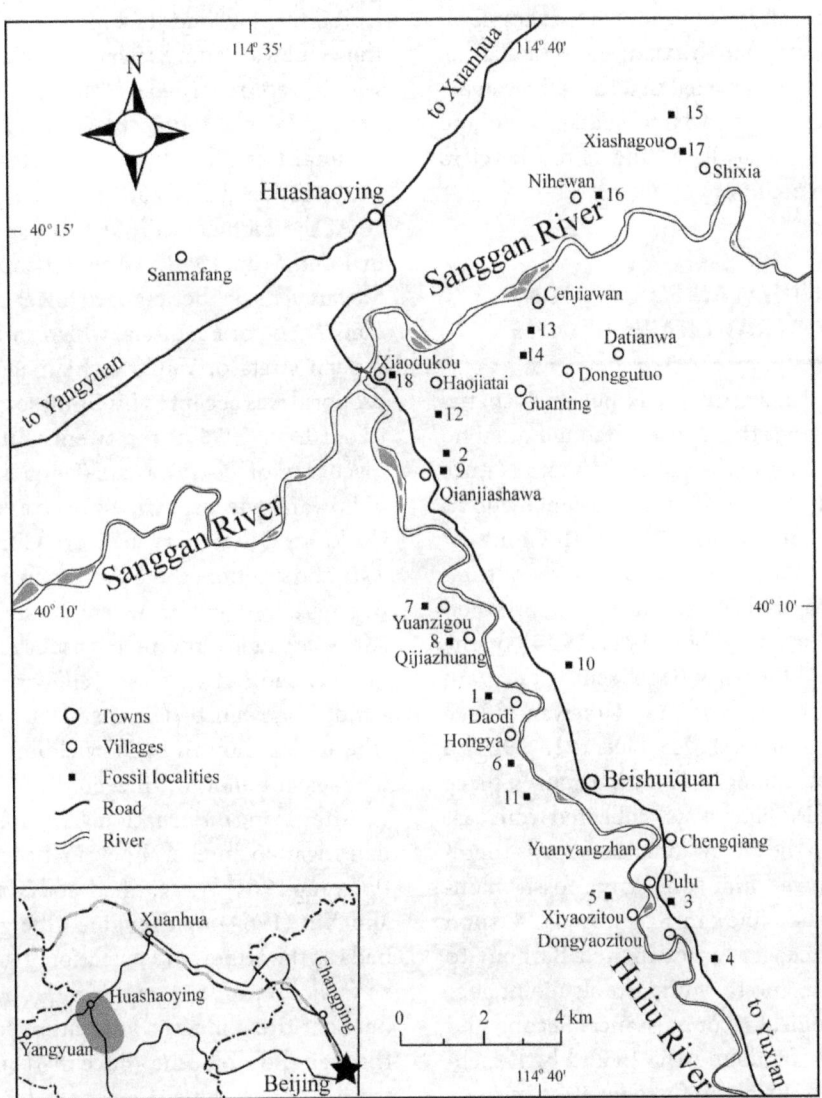

Figure 8.1 Map of the Plio-Pleistocene mammalian localities in the Nihewan Basin: (*1*) Laowogou; (*2*) Donggou; (*3*) Pulu (=Niutoushan); (*4*) Danangou; (*5*) Huabaogou; (*6*) Hongya Nangou; (*7*) Yuanzigou; (*8*) Hougou; (*9*) Xiaoshuigou; (*10*) Lianjiegou; (*11*) Jiangjungou; (*12*) Tai'ergou; (*13*) Majuangou; (*14*) Xiaochangliang; (*15*) Jinggou; (*16*) Baicaogou; (*17*) Dashuigou; (*18*) Haojiatai.

de 15 environ, bordé au Nord et au Sud par des chaînes cristallines et siniennes, et limité à l'Est par une croupe gneissique" (Topographically, Nihowan Basin is a broad trough 20 km long and about 15 wide, bounded on the north and south by Sinian crystalline rocks, and limited to the east by a ridge of gneiss).

Nihewan Basin Sensu Lato

Wei (1978) expanded the geographical range of the Nihewan deposits to the Xujiayao area and the Datong Basin.

Jia and Wei (1980), however, only used the term Datong Basin instead of the Yangyuan and Yuxian basins. Wu, Sun, and Yuan (1980) introduced the concept that the Nihewan strata were developed in fault-controlled basins in the Sanggan and Huliu rivers drainage area and that the outcrops narrowed in the north or northeastern parts of the basin. Han (1982) reported that "Nihewan beds" were widely developed from east Yangyuan and Yuxian to west Datong and Shuoxian with observed thickness of 300–1100 m (results of seismic survey and drilling, the numbers apparently including subsurface components). Liu and Xia (1983) also supported a broad concept of the

Nihewan deposits that includes a series of intermountain basins in the Sanggan River Valley—for example, in Datong, Tianzhen, Yangyuan, Yuxian, and Huailai basins. Finally, Lin (1984) mentioned that the Nihewan deposits were widely exposed in large spatial scale and thickness along the Sanggan River and Huliu River in Yangyuan and Yuxian counties.

NIHOWAN BEDS, NIHEWAN FORMATION, AND OTHER LITHOSTRATIGRAPHIC UNITS

The initial paleontolgical work in the Nihewan Basin began in 1921. To construct the "Musée Haong ho Pai ho de Tientsin" (predecessor of the present Tianjin Nature Museum), French paleontologist Père E. Licent made an appeal to the Catholic missionaries in North China for their attention to vertebrate fossils. In that year, in response to Licent's proposal, Père Vincent, the Mission Apostolique of Kalgan and abbot (1921–1924) of the Catholic Church in Nihewan village, sent a telegram about discovering fossil animals near Nihewan village and mailed a package of fossils to Licent. In 1923, a farmer brought British geologist G. B. Barbour "a piece of a silicified femur of *Rhinoceros* sp." collected from east of Yangyuan County in the Nihewan Basin. Shortly thereafter, Père Vicent showed him some large fossil mammals collected from the Nihewan Basin. In 1924, soon after receiving the package, Licent invited Barbour to visit the Nihewan area, and they undertook a joint geological and paleontological reconnaissance. Second and third visits were made by them in 1925 and by P. Teilhard de Chardin and E. Licent in October 1926, respectively. Their pioneering investigation focused on the sedimentary platform along the Sanggan River banks (Barbour 1924, 1925; Barbour, Licent, and Teilhard de Chardin 1926; Xie, Li, and Liu 2006:Appendix). G. B. Barbour (1924) first used Nihowan beds to represent the lower greenish and brownish freshwater deposits in the Nihewan Basin. The following year, he regarded the Nihowan beds as only belonging to the upper members of the "Nihowan series" (Barbour 1925). Barbour, Licent, and Teilhard de Chardin (1926) redefined the Nihowan beds as a sequence of greenish and whitish torrential and fluviolacustrine sediments interbedded within the top yellowish loess and the bottom reddish clay in the Nihewan Basin. They suggested that the Nihowan beds near Nihewan village could be subdivided into four zones: sand and gravel, intermediate sands and clay, upper sand and gravel, and white marls. They also emphasized that nearly all the fossil mammals were collected from a "middle zone of sands and clays" around the Nihewan and Xiashagou (=Hsia-sha-kou) villages, which were regarded as "the true Nihowan beds." This opinion was accepted by some of the later researchers (Teilhard et Chardin and Piveteau 1930; Black et al. 1933; Teilhard de Chardin 1941).

At the Eighteenth International Geological Congress in London in 1948, it was recommended that the Nihowan beds be the standard lower boundary strata of the Pleistocene of East Asia, which can be correlated to European strata of Villafranchian age (Young 1950). This proposal was accepted by Chinese geologists and paleontologists in 1954 at the twenty-fifth anniversary of the discovery of Peking Man (Zhou 1955). Thereafter, the Nihowan beds were widely used as the standard strata of the lower Quaternary in North China. However, lacking a strict definition, the Nihowan beds were used variously in generalized or narrow senses. The Nihowan beds, sensu lato, was usually interpreted as the fluviolacustrine deposits between red clay and yellowish loess. On the other hand, Nihowan beds, sensu stricto, was limited to the "the true Nihowan beds" yielding the classic Nihowan fauna near Nihewan village.

After a long interruption, Chinese researchers resumed investigation in the Nihewan Basin in the early 1960s (Ouyang 1964; Wang, Zhao, and Xin 1964). Wang, Zhao, and Xin (1964) first renamed the generalized Nihewan beds as the Nihewan Formation. Huang et al. (the Cenozoic Nihewan Stratigraphical Group 1974) first pointed out that the Nihewan Formation spanned the Pliocene through the Holocene. Since then, more than 10 new formation names came to be applied in the Nihewan Basin, as follows.

Xujiayao Formation

Based on the fossil mammals of the Xujiayao Paleolithic site, including *Bos primigenius* from west of Nihewan village and *Cervus elaphus* near a pump station of the Bajiao village. Wei (1978) selected the Xujiayou section as the stratotype of the upper Pleistocene Xujiayao Formation.

Hutouliang Formation and Que'ergou Formation

Based on his studies on ostracods from Hutouliang and Que'ergou, Yangyuan County, Huang (1980) proposed the Middle Pleistocene Hutouliang Formation and early Middle Pleistocene Que'ergou Formation.

Shixia Formation

Huang et al. (1974) found hipparionine fossils in the lower red bed in the Luanshigedagou in Hongya and called the red sequence the "*Hipparion* bed." Huang and Guo (1981), based on lithology and fossil mollusks, formally named the "*Hipparion* bed" as the Shixia Formation, considering it to crop out along the banks of Huliuhe and in the Shixia region. This formation is exposed at Luanshigedagou, Road Cut (Pulu), and Pump Station (Yangshuizhan) of Hongya village. Pang (2003) divided Huang and Guo's (1981) Shixia Formation into two formations: the lower Hongyacun Formation and middle and upper Shixia Formation on the typical section. He also regarded Wang's (1982) Yuxian Formation and Cai's (1987) Daodi Formation as junior synonyms of the Shixia Formation. Liu et al. (2009) regarded the red bed underlaying the Nihewan Formation only as the Shixia Formation.

Yuxian Formation and Huliuhe Formation

Based on the large mammals collected from the lower and upper red clay beds of the Huabaogou section, Wang (1982) proposed the Huliuhe Formation and Yuxian Formation, respectively. Zhang, Zheng, and Liu (2003) regarded the Huliuhe Formation as lithologically indistinguishable from the overlying Yuxian Formation and suggested abandoning the name Huliuhe Formation but retaining the Yuxian Formation. Pang (2003) thought Wang's Yuxian Formation to be synomymous with the "Eocene Yuxian Formation in the 1979 North China Regional Stratigraphic Table (Hebei Province volume)" and therefore invalid, but this was already mistaken because the original 1979 paper stated that it is Oligocene in age. At the same time, Pang considered the Huliuhe Formation to be synonymous with Shixia Formation. Liu et al. (2009) regarded both the Yuxian and Huliuhe formations as junior synonyms of the Shixia Formation.

Dongyaozitou Formation

Based on the fossil mammals collected from the lower part of a sandy-gravel cliff in the Danangou gully near Dongyaozitou village, Tang and Ji (1983) proposed the Dongyaozitou Formation. They chose the exposure described by Tang (1980) as the stratotype of this formation, which was divided into eight layers. Fossils occur in layers 1, 3, and 4, which are equal to layers 2, 3, and 4 described by Cai et al. (2004). Cai et al. (2004) consid-

ered that this formation partly belongs to the Nihewan Formation and suggested abandoning it.

Dahonggou Formation and Xiaodukou Formation

Chen et al. (1986:table 1) first mentioned Dahonggou and Xiaodukou formations. Later, Chen (1988) formally used these two names in dividing strata of the Nihewan Basin into four formations: the Pliocene Dahonggou Formation, lower Pleistocene Nihewan Formation, middle Pleistocene Xiaodukou Formation, and upper Pleistocene Xujiayao Formation. Yang et al. (1996) and Min, Chi, and Zhu (2000) continued to use the name Xiaodukou Formation. Liu et al. (2009) were of the opinion that the Dahonggou Formation is a junior synonym of the Shixia Formation.

Daodi Formation

Cai (1987) undertook screenwashing and collected abundant small mammal fossils at eight localities along the banks of the lower Huliu River in the Yangyuan and Yuxian basins, which occur in the the Daodi, Hongya Nangou, Qianjiashawa, Qijiazhuang, Yuanzigou, Jiangjungou, Pulu (=Niutoushan), and Beimajuan sections. The fossiliferous layers were usually a series of dark-colored clayey silt interbedded with silty clay, which were interbedded with gravels and calcareous nodules. Based on the similar small mammalian complexes from these eight localities, Du et al. (1988) proposed the Daodi Formation. Zhang, Zheng, and Liu (2003) regarded it as a junior synonym of the Yuxian Formation. Pang (2003), on the other hand, thought that it was a junior synonym of Huang and Guo's (1981) Shixia Formation. In recent years, we collected richer small mammals from many layers of Laowogou section, and restricted Daodi Formation to layer 1–9 (Cai et al. 2004; Li, Zheng, and Cai 2008). It seems that layer 1–2 of this section should belong to the Shixi Formation.

Xishuidi Formation and Dayuwan Formation

Liu (1989), mainly based on changes in ostracod assemblages in Pump Station (Yangshuizhan), Hutouliang, and other sections at Hongya, divided the late Cenozoic fluvio-lacustrine sediments in the Nihewan Basin into (from lower to upper) the Pliocene Shixia Formation, early Pleistocene Nihewan Formation, middle Pleistocene Hutouliang Formation, late Pleistocene Xishuidi Formation, and Holocene Dayuwan Formation.

Nangou Formation

Xia (1993) proposed this formation based on the lower parts of the Hongya Nangou section and the Pulu section, which are lithologically characterized by brownish or brown-yellowish clayey siltstone and gravels interbedded with lenses of gray-black clay. He regarded the overlying and underlying strata of this formation as the Nihowan beds and red clay, respectively.

Haojiatai Formation

Xia (1993) proposed the middle Pleistocene Haojiatai Formation for the upper strata of Nihowan beds. The upper part of the Haojiatai section was chosen as its stratotype, which is characterized by brown-yellowish clayey siltstone interbedded with gray-greenish silty clay and yellow-greenish fine sand and gravels. Zhang, Min, and Zhu (2003) redefined this formation based on the Tai'ergou section, and Min et al. (2006) divided the strata of the Tai'ergou section into the Yuxian Formation, Nihewan Formation, Haojiatai Formation, and Malan Formation (loess) from bottom to top.

Jing'erwa Formation

Based on the lithostratigraphic, biologic, and paleomagnetic features of a borehole near Jing'erwa village, Yangyuan County, Min, Chi, and Zhu (2000) suggested dividing the Quaternary lacustrine sediments of the Nihewan Basin into three formations: the lower Pleistocene Nihewan Formation, the lower-middle Pleistocene Xiaodukou Formation, and a new formation, the late Pleistocene Jing'erwa Formation.

Qianjiashawa Formation

Zhang, Min, and Zhu (2003) proposed the Qianjiashawa Formation based on the middle part of the Tai'ergou section, which was characterized by "the thick-bedded fluvial gravels, sands, fine sands, and silts with clayey silts." However, this name was abandoned in a subsequent paper for the same profile (Min et al. 2006).

Hongyacun Formation

Pang (2003) divided the late Cenozoic beds in the Nihewan Basin into (from lower to upper): the lower Pliocene Hongyacun Formation, upper Pliocene Shixia Formation, lower Pleistocene Nihewan Formation, middle Pleistocene

Hutouliang Formation, upper Pleistocene Xujiayao, and Malan formations. He reshuffled Huang and Guo's (1981) Shixia Formation, and renamed both Wang's (1982) Huliuhe Formation and Chen's (1988) Dahonggou Formation as the Hongyacun Formation. It is apparent that his Hongyacun Formation is a junior synonym of the Huliuhe and Dahonggou formations and should be abandoned.

CLASSIC NIHOWAN FAUNA AND OTHER FAUNAS IN NIHEWAN BASIN

According to the descriptions in Teilhard de Chardin and Piveteau (1930), the classic Nihowan fauna included 42 species among 31 genera. Most of the original material was collected from "the true Nihowan beds" around Nihowan and Xiashagou villages during the field seasons of Teilhard de Chardin and others in 1924–1926. Qiu (2000) reexamined the components of the Nihowan fauna and chose 20 species to compare with those of the European Villafranchian faunas at the species level. He concluded that the classic Nihowan fauna most closely resembled the early late Villafranchian Olivola fauna of Italy. In his opinion, the classic Nihowan fauna could be about 1.8 Ma in age, which is very close to the Pliocene/Pleistocene boundary at the Vrica stratotype section of the marine Calabrian stage (Qiu 2000).

Following a long hiatus after excavations by G. B. Barbour, E. Licent, and Teilhard de Chardin in 1924–1926, Huang et al. (1974) reported scattered fossil mammals found from the Luanshigedagou gully south of Hongya village, the Dashui gully southeast of Xiashagou, and the Haojiatai and Hutouliang profiles. Since then, scattered fossil mammals have been discovered in succession from different Nihewan sites, but they were not described in detail (Jia and Wei 1976; Jia, Wei, and Li 1979; Wei 1976, 1978, 1983). During the last 30 years, dozens of sites (see figure 8.1) and abundant fossils have been found (Tang 1980; Tang, You, and Li 1981; Zheng 1981; Wang 1982; Tang and Ji 1983; Li 1984; Qiu 1985; Tang, Li, and Chen 1995), especially small mammals by screenwashing (Cai 1987; Cai 1989; Zheng and Cai 1991; Cai and Qiu 1993; Cai and Li 2004; Cai et al. 2004; Erbajeva and Zheng 2005; Min et al. 2006; Zheng, Cai, and Li 2006; Cai, Zheng, and Li 2007, 2008; Li, Zheng, and Cai 2008).

Dongyaozitou Fauna

Tang (1980); Tang, You, and Li (1981); and Tang and Ji (1983) described this "Late Pliocene/Early Pleistocene

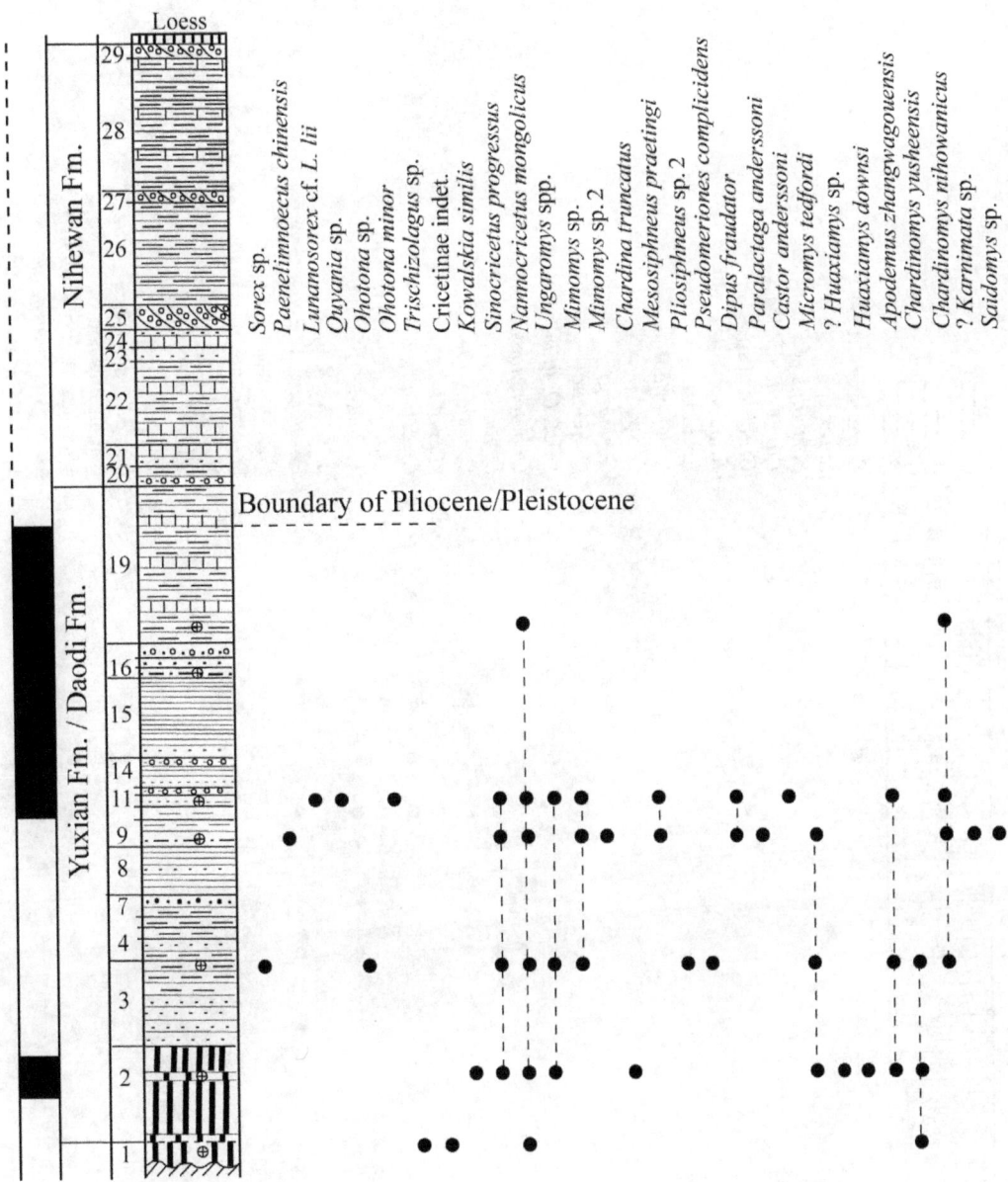

Figure 8.2 Small mammal stratigraphic ranges in the Laowogou section.

transitional" mammalian complex collected from the lower part of the sandy gravel cliff in Danangou gully, Dongyaozitou village, Yuxian County. The fossils included 12 species: ?*Dipoides* sp., *Nyctereutes* cf. *sinensis*, *Lynx variabilis*, *Zygolophodon* sp., *Coelodonta antiquitatis*, *Proboscidipparion sinense*, *Hipparion* cf. *houfenense*, *Paracamelus* sp., *Palaeotragus progressus*, *Antilospira yuxianensis*, *Gazella sinensis*, and ?*Axis* sp.

Daodi Fauna

This fauna is dominated by late Pliocene small mammals. The fossils were mainly collected from layers 3 through 13 (belonging to the Daodi Formation) of the Laowogou section (Du et al. 1988). The small mammals are listed in figure 8.2. The large mammals include *Proboscidipparion sinense*, ?*Chilotherium* sp., *Axis shansius*, *Cervus* sp., and *Antilospira* sp. (Cai 1987; Cai et al. 2004).

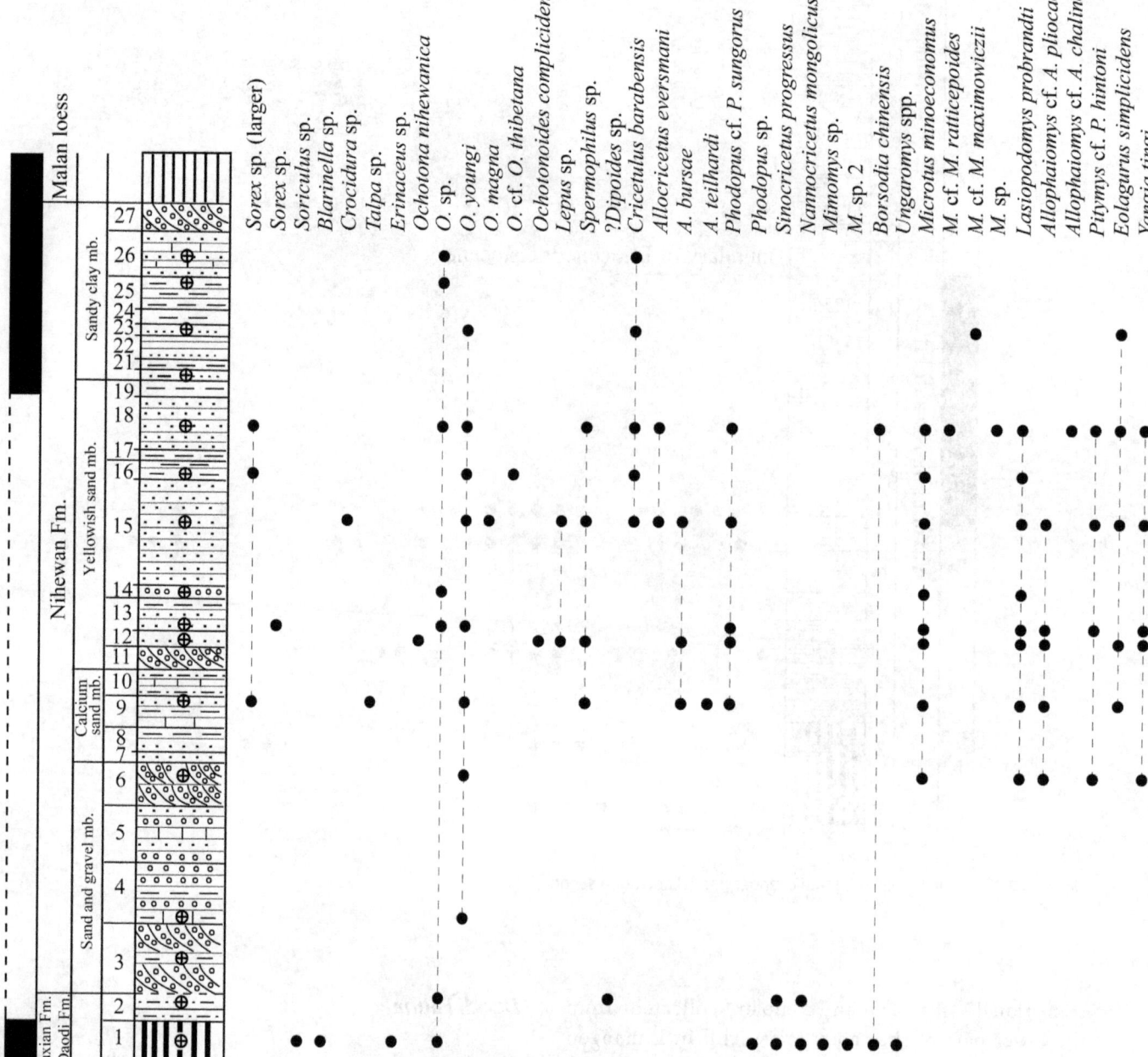

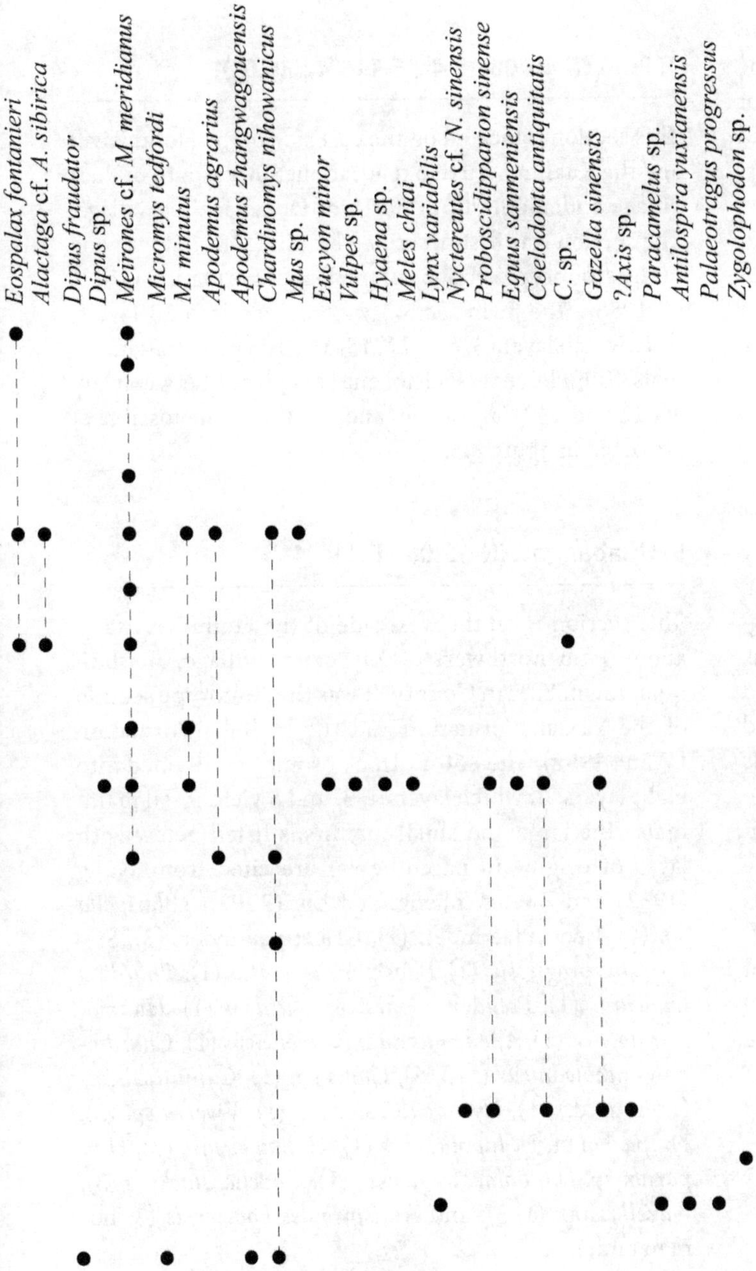

Figure 8.3 Distribution of the fossil mammals of the Plio-Pleistocene Danangou section.

FOSSILIFEROUS SECTIONS IN NIHEWAN BASIN

The numbers refer to the locations shown in figure 8.1.

1. Laowogou (N 40°08'59", E 114°39'31")

This site is located in the Laowogou gully about 750 m north of Daodi village, Yangyuan County. It was the most important stratotype of the Daodi Formation for Du et al. (1988). Cai (1987) described this section and has collected abundant small mammals during the past 20 years. The section is 128 m in thickness and divided into 29 layers. The fossils were reported from layers 1, 2, 3, 9, 11, and 19 (see figure 8.2). The Pliocene/Pleistocene boundary was postulated to be at the top of layer 19, based on fossil mammal assemblages (Cai et al. 2004).

2. Danangou (N 40°05'36", E 114°43'35")

Wang et al. (1975) first described this section in a study on ostracods and foraminifers. Tang (1980), Tang, You, and Li (1981), and Tang and Ji (1983) reported the Dongyaozitou mammal fauna collected from this profile. Zheng (1981) and Li (1984) described eight large and three small fossil mammals from this section. Zheng and Cai (1991) reported 15 species of fossil small mammals collected by Cai in 1986. On the basis of the measurement by Wang et al. (1975), the 97 m section was subdivided into four members and includes 27 layers. In the 2001 and 2002 field seasons, Cai and his colleagues undertook high-resolution stratigraphical and fossil collecting work (Cai et al. 2004). They found thousands of fossils from 16 layers (figure 8.3 [1, 2, 3, 4, 6, 9, 12, 13, 14, 15, 16, 18, 20, 23, 25, and 26]). The Pliocene/Pleistocene boundary was postulated to be located between layers 2 and 3.

3. Donggou (N 40°12'11.7", E 114°38'45.2")

This section is on the east side of the Huliu River and about 500 m northeast of Qianjiashawa village, Huashaoying town, Yangyuan County. It was once proposed as one of the stratotype sections of the Quaternary Nihewan standard strata (Yang et al. 1996). Yuan et al. (1996) measured a total thickness of 121 m and described the strata in detail, whereas Yang et al. (1996) described only the lower 56 m of the section, and Deng et al. (2008) in-

cluded only the upper 90 m of the section. The lower 56-m-thick profile was divided into 28 layers, of which layers 2, 4, 7, 11, 16, and 19 yield small mammals. The Pliocene/Pleistocene boundary was placed between layers 10 and 11 by biostratigraphy (Zheng, Cai, and Li 2006). The fossil taxa are listed in figure 8.4.

4. Pulu (N 40°06'13.4", E 114°42'36.6")

This section is located on the east side of the Huliu River and the Xuanhua-Yuxian road, about 500 m south of Pulu village and about 700 m north of Dongyaozitou village. The section was first described by Lin (1984). It was one of the stratotype sections of the Daodi Formation (Du et al. 1988). The 82 m section was divided into 23 layers, from which layers 3, 6, 9, 12, 15, and 16 yield small mammals. The Pliocene/Pleistocene was placed between layers 12 and 13 (Cai, Zheng, and Li 2007). The fossil taxa are listed in figure 8.5.

5. Huabaogou (N 40°06', E 114°41')

This section is on the west side of the Huliu River and about 1 km northwest of Xiyaozitou village, Beishuiquan town, Yuxian County. It was the stratotype section of the Yuxian Formation and the Huliuhe Formation (Wang 1982). The ~50-m-thick profile was divided into eight layers, of which layers 1, 3, and 5 yield fossil mammals. The large and small mammals listed below with layer of origin (in parentheses) are cited from Wang (1982) and Zhang, Zheng, and Liu (2003): ?*Blarinella* sp. (5), Leporidae indet. (1), Cricetinae indet. (3), *Sinocricetus progressus* (1), *Pliosiphneus lyratus* (1), *Chardina truncatus* (1), *Pseudomeriones complicidens* (1), *Huaxiamys downsi* (1), *Apodemus zhangwagouensis* (1), *Chardinomys nihowanicus* (1, 3, 5), *Canis* sp. (1), *C. multicuspus* (nom. nud.) (1), *Nyctereutes sinensis* (1), *Viverra* sp. (3), *Hipparion* cf. *H. hippidiodus* (1), *H. houfenense* (3), *Hipparion* sp. (1), *Palaeotragus* sp. (1), *Gazella blacki* (1, 3), *Gazella* spp. (1, 3), and *Antispiroides hopeiensis* (3; nomen nudum).

6. Hongya Nangou (N 40°08'07.3", E 114°39'57.1")

This section is located on the west side of the Huliu River and about 500 m south of Hongya village. This 25 m profile was one of the stratotype sections of the Daodi Formation and is divided into seven layers (Du et al. 1988),

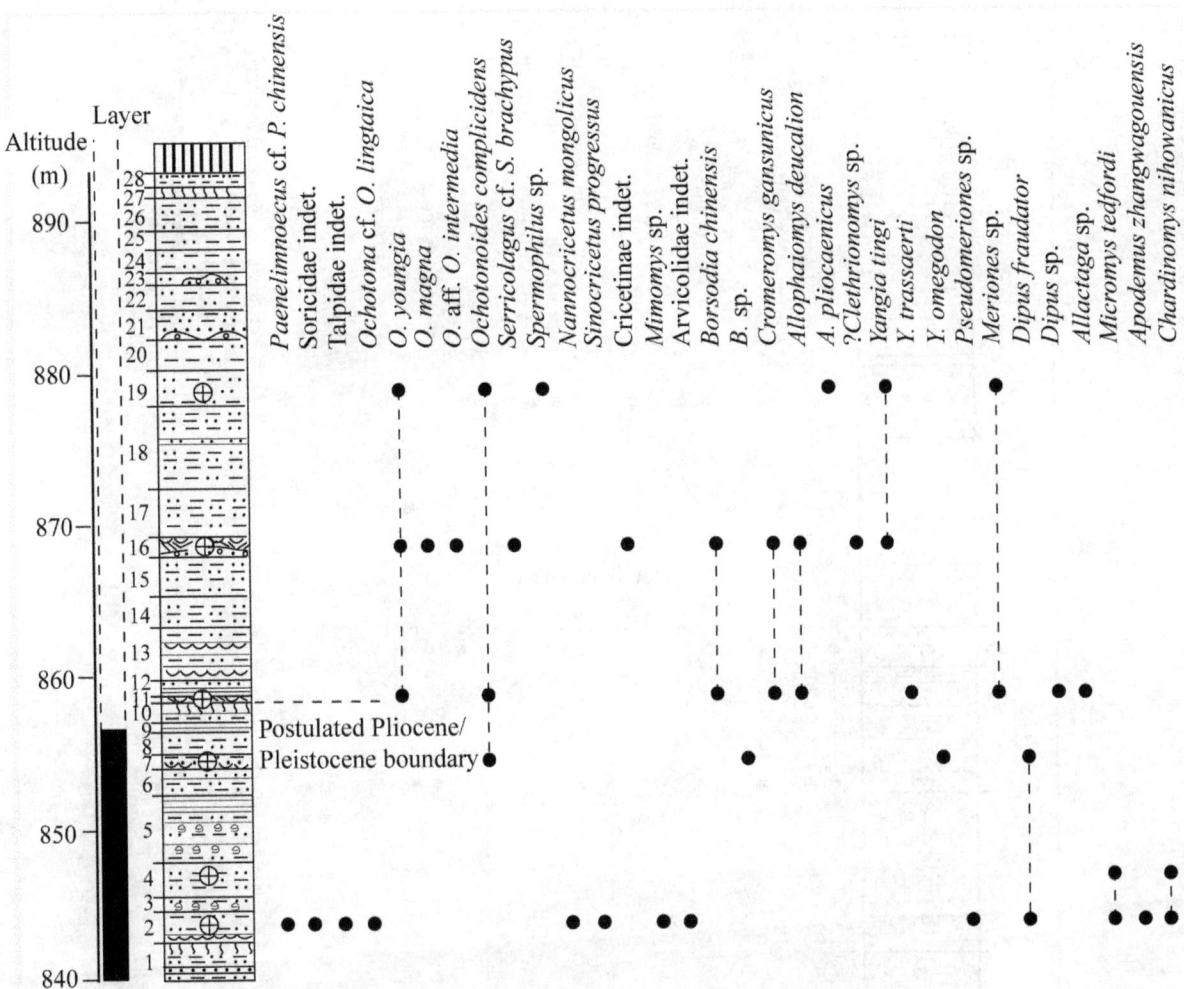

Figure 8.4 Small mammal stratigraphic ranges for the Donggou section.

from which layers 1 and 4 yield small mammals, including *Lunanosorex* cf. *L. lii* (1, 4), *Beremendia* sp. (4), *Sorex* sp. (small) (4), *Yanshuella* sp. (4), *Ochotona minor* (1), *O. erythrotis* (1), *O.* sp. (4), *Hypolagus schreuderi* (1, 4), *Nannocricetus mongolicus* (1, 4), *Sinocricetus progressus* (1, 4), ?*Kowalskia* sp. (4), *Mimomys* sp. 2 (1, 4), *Germanomys* sp. (1, 4), *Mesosiphneus paratingi* (1, 4), *Dipus fraudator* (1, 4), *Mus* sp. (1), *Micromys tedfordi* (1), *Huaxiamys downsi* (1), *Apodemus zhangwagouensis* (1), and *Chardinomys nihowanicus* (1, 4); see Li, Zheng, and Cai (2008; numbers in parentheses represent layers).

7. Yuanzigou (N 40°10′05.8″, E 114°38′28.9″)

This section is on the west bank of the Huliu River and about 300 m west of Yuanzigou village. This series of

palustrine sediments is 21 m in thickness and can be divided into six layers, from which layers 2 and 4 yield small mammals (Li, Zheng, and Cai 2008). These are *Ochotona* cf. *O. lagreli* (2), *Pliopentalagus nihewanensis* (2), *Sinocricetus progressus* (2, 4), *Ungaromys* (=*Germanomys*) sp. (2, 4), *Mimomys* sp. 2 (2, 4), *Apodemus zhangwagouensis* (4), and *Micromys tedfordi* (4). (Numbers in parentheses represent layers.)

8. Hougou (N 40°09′38.6″, E 114°38′53.1″)

This section is located on the west side of the Huliu River and about 300 m west of Qijiazhuang village, Yangyuan County. The 30-m-thick paludal deposit was one of the stratotype sections of the Daodi Formation and is divided into six layers (Du et al. 1988). Sixteen species of

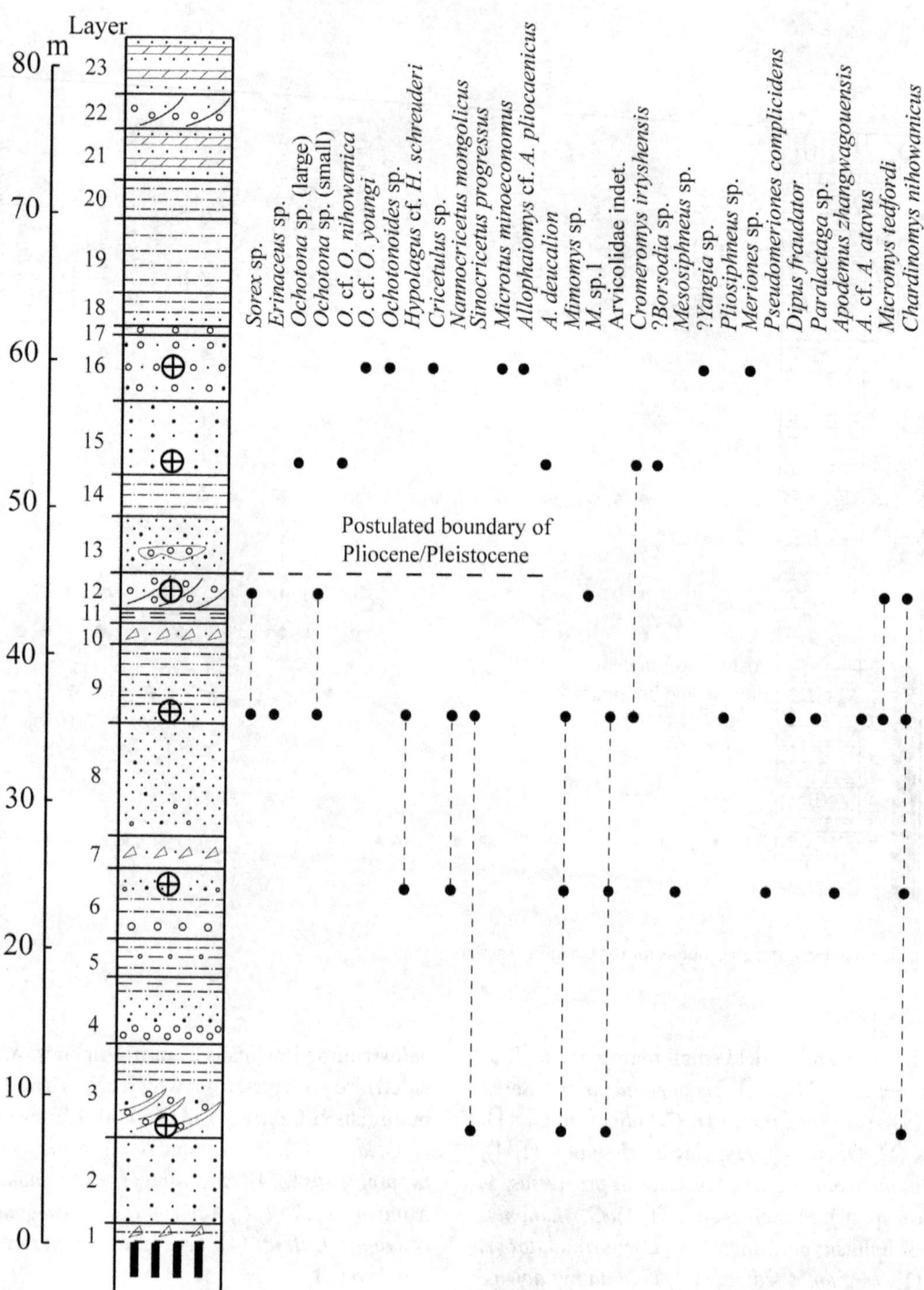

Figure 8.5 Small mammal stratigraphic ranges for the Pulu section.

small mammals found from layers 4 and 5 included *Lunanosorex* cf. *L. lii* (4, 5), ?*Beremendia* sp. (4), *Quyania* sp. (5), *Hypolagus schreuderi* (4, 5), *Pliopentalagus nihewanensis* (5), *Nannocricetus mongolicus* (5), *Sinocricetus progressus* (4), *Mimomys* sp. (4, 5), *M.* sp. 2 (4), *Ungaromys* spp.

(4, 5), *Mesosiphneus praetingi* (4, 5), *Dipus fraudator* (4), *Micromys tedfordi* (5), *Apodemus zhangwagouensis* (5), *Chardinomys nihowanicus* (4, 5), and *Karnimata* sp. (5); see Li, Zheng, and Cai (2008; numbers in parentheses represent layers).

9. Xiaoshuigou (N 40°11'43", E 114°38'40.6")

This section is located at the east side of the Huliu River and about 300 m east of Qianjiashawa village, Huashaoying town, Yangyuan County. It was one of the stratotype sections of the Daodi Formation (Du et al. 1988). This 18-m-thick profile is a series of palustrine sediments and can be divided into seven layers, from which layers 1, 2, and 4 produce *Hypolagus schreuderi* (4), *Nannocricetus mongolicus* (1, 4), *Sinocricetus progressus* (2, 4), *Mimomys* sp. (1, 2), *M.* spp. 1 and 2 (4), *Ungaromys* spp. (1, 2), *Mesosiphneus paratingi* (2, 4), *Dipus fraudator* (4), *Micromys tedfordi* (4), *Apodemus zhangwagouensis* (1, 4), and *Chardinomys nihowanicus* (1, 4); see Li, Zheng, and Cai (2008; numbers in parentheses represent layers).

10. Lianjiegou (N 40°09'18.3", E 114°40'38")

This section is on the east side of the Huliu River and about 1.5 km southeast of Beimajuan village, Beishuiquan town, Yuxian County. The profile is a series of 22-m-thick palustrine sediments and can be divided into 14 layers. Only layer 7 yields small mammals: *Ochotona* cf. *O. largreli*, *Pliopentalagus nihewanensis*, *Sinocricetus progressus*, *Mimomys* sp., *Mesosiphneus praetingi* and *Dipus fraudator* (Li, Zheng, and Cai 2008).

11. Jiangjungou (N 40°08', E 114°39')

This section (=southwest gully section in Du et al. 1988) is on the west side of the Huliu River and about 1.5 km west of Beishuiquan town, Yuxian County. This series of 18-m-thick palustrine sediments can be divided into seven layers. Cai collected fossil rodents by screenwashing in 1984. Seven species of small mammals were found from only the bottom of layer 1. They are *Ochotona minor?*, *Nannocricetus mongolicus*, *Mimomys* sp., *Ungaromys* spp., *Mesoiphneus paratingi*, *Dipus fraudator*, and *Chardinomys nihowanicus* (Li, Zheng, and Cai 2008).

12. Tai'ergou (N 40°12'52", E 114°38'22")

This section is exposed on the east bank of the Huliu River and south of Xiaodukou village, Huaoshaoying town. According to the data published by Zhang, Min, and Zhu (2003) and Min et al. (2006), this profile totals 151 m in thickness and is divided into 146 layers in 12 lithologic members. The Brunhes/Matuyama and Matuyama/Gauss boundaries were located at 33 m and 123 m depth, respectively. Small mammals were found only in the younger ninth (T2, 97th layer, ~68 m) and older third (T1, 29th layer, ~131 m) members. The fossils from the third member were reexamined by Li, Zheng, and Cai (2008) and include 15 species: *Sorex* sp., *Erinaceus* sp., *Ochotona* cf. *O. lingtaica*, *Hypolagus* cf. *H. schreuderi*, *Nannocricetus mongolicus*, *Sinocricetus progressus*, *Mimomys* sp., *Mesosiphneus paratingi*, *Dipus* sp., *Sicista* sp., *Mus* sp., *Micromys tedfordi*, *Apodemus zhangwagouensis*, *Chardinomys nihowanicus*, and *?Saidomys* sp. Judging from the fossil components and paleomagnetic results, this small mammal complex was considered to be greater than 2.6 Ma in age. The small mammals from T2 include *Ochotona youngi*, *Lepus* sp., *Spermophilus* sp., *Cricetulus barabensis*, *Microtus minoeconomus*, *Lasiopodomys probrandti*, *Allophaiomys* cf. *A. pliocaenicus*, *Pitymys* cf. *P. hintoni*, *?Eolagurus* sp., *Yangia tingi*, *?Eospalax* sp., and *Meriones* cf. *M. meridianus*. T2 is within the Jaramillo Normal Subchron (1.072–0.988 Ma; Gradstein, Ogg, and Smith 2004) and thus is estimated to be around 1.0 Ma.

13. Haojiatai (N 40°13'11.3", E 114°37'46.3")

Huang et al. (1974) first described this section, which is located about 50 m east of Xiaodukou village. Liu and Xia (1983), Lin (1984), Xia (1993), and Yang et al. (1996) also undertook lithostratigraphic work on this section. The more than 110-m-thick profile is a series of light-colored siltstone and silty clay and usually is interpreted as the center of the ancient Nihewan Lake. Due to its lacustrine facies, fossil mammals occur rarely in this section. Huang et al. (1974) reported *Equus sanmeniensis* from the bottom of a yellowish sand bed in the upper part of the section. In the 2005 field season, Cai and his colleagues only collected *Ochotona youngi* from the bottom of the section by screenwashing.

14. Majuangou (N 40°13'13", E 114°39'50.9")

This profile is on the south side of the Sanggan River and 1.2 km southwest of Cenjiawan village, Huashaoying town. The section is about 55 m thick and is divided into 21 layers, of which layers 1, 7, and 17 (=MJG III, MJG I and Banshan paleolithic sites, respectively) yield fossil mammals and artifacts; layer 4 produces only lithics (=MJG II). Twenty species of mammals collected from Majuangou III by Cai and his colleagues in 2001 and 2005 include *Sorex* sp., *Erinaceus* sp., *Ochotona youngi, Ochotonoides complicidens, Allactaga* sp., *Spermophilus* sp., *Cricetulus* sp., *Yangia tingi, ?Episiphneus* sp., *Micromys* sp., *Chardinomys nihowanicus, Cromeromys gansunicus, Borsodia chinensis, Allophaiomys deucalion,* Canidae gen. et sp. indet., *Canis* sp., *Mammuthus trogontherii, Hipparion* sp., Rhinocerotidae indet., and Cervidae indet. (Cai and Li 2004; Cai, Li, and Zheng 2008). Mammals from Majuangou I include *Ochotona* sp., *Allophaiomys* sp., *Cromeromys* sp., *Borsodia* sp., Proboscidea indet., Rhinocerotidae indet., and Cervidae indet. Banshan site (=layer 17) produced large mammals, including *Canis* sp., elephantids, *Equus* sp., rhinocerotids, and cervids (Wei 1990), and small mammals *Ochotona* sp., *Ochotonoides* sp., *Lepus* sp., *Allocricetus ehiki, ?Cricetulus* sp., *Borsodia chinensis,* and *Allophaiomys pliocaenicus* (Cai, Li, and Zheng 2008). According to the magnetic work of Zhu et al. (2004), the ages of MJG III, MJG II, MJG I, and Banshan were 1.66, 1.64, 1.55, and 1.32 Ma, respectively. However, Cai, Li, and Zheng (2008) considered the age of MJG III to be possibly older than 1.66 Ma based on the evolutionary stage of the arvicolids.

15. Xiaochangliang (N 40°13'50", E 114°39'44")

This section is located south of the Sanggan River and about 800 m northeast of Guanting village, Yangyuan County. The 97-m-thick profile was divided into 26 layers, from which the sixth layer yields fossil mammals, turtle, invertebrates, and stone artifacts (Tang, Li, and Chen 1995). Screenwashing was undertaken in this layer, and nine additional species of small mammals were collected (Zhang, Kawamura, and Cai 2008). The total reported fossil mammals included *Sorex* sp., *Ochotona youngi, Cricetulus barabensis, Allophaiomys deucalion* (should be *A. pliocaenicus*), *Borsodia chinensis,* Arvicolidae indet., *Yangia tingi, Micromys minutus, Apodemus agrarius, Chardinomys nihowanicus, Martes* sp., *Crocuta licenti, Palaeoloxodon* sp. (should be *Mammuthus trogontherii*), *Coelodonta antiquitatis* (should be *C. nihowanensis*), *Equus sanmeniensis, Proboscidipparion sinense, Hipparion* sp., *Cervus* sp., *Gazella* sp., and Bovinae indet. (Tang, Li, and Chen 1995; Zhang, Kawamura, and Cai 2008). The magnetic work of Zhu and colleagues (Zhu et al. 2001; Zhu et al. 2004) estimates the artifact and fossil layer at 1.36 Ma.

16. Dashuigou (N 40°16', E 114°42')

This section is located southeast of Xiashagou village. The 44-m-thick profile was divided into 15 layers and

consists mainly of yellow-greenish or gray-yellowish sand and gravel. Huang et al. (1974) found eight species of large mammals from the middle part of the section. The mammals include *Vulpes* sp., *Canis chihliensis, Machairodus nihowanensis, Rhinoceros sinensis, Equus sanmenensis, Cervus (Elaphurus) bifurcatus, C. (Rusa) elegans,* and *Gazella sinensis.*

17. Jinggou (N 40°16'44.2", E 114°42'37.8")

This section is located at the Jinggou gully, which is on the north side of the Sanggan River and about 700 m north of Xiashagou village, Huashaoying town, Yangyuan County. The main body of this section is a series of mottled sand and gravel beds, which can be roughly correlated to the sand and gravel member of the Danangou section. Only two species of small mammals, *Yangia trassaerti* and a hypsodont sciurid (Aepyosciurinae), were found from the section during the 2005 field season by Cai and his colleagues.

18. Baicaogou (N 40°15'30.9", E 114°41'25.1")

This section is on the north side of the Sanggan River and about 700 m east of Nihewan village. The section is mainly sandstone, siltstone, and gravel and roughly correlates to the sand and gravel member of Danangou section. In 2001, Cai and Li screenwashed the matrix from the lower part of the section and found only two species of small mammals, *Allactaga* cf. *A. sibirica* and *Borsodia chinensis.*

19. Xujiayao Paleolithic site (N 40°06', E 114°59')

This paleoanthropologic and paleolithic site is located between Xujiayao village, Yanggao County, Shanxi Province, and Houjiayao village, Yangyuan County, Hebei Province (outside the border of the map in figure 8.1). Vertebrate and human fossils and lithics were found from the gray-brownish clay (Jia and Wei 1976). Twenty-two species of fossil mammals coexisting with artifacts include *Crocidura* cf. *C. wongi, Scaptochirus moschatus, Lepus* cf. *L. wongi, Ochotona* cf. *O. thibetana, Eospalax fontanieri, Microtus brandtioides, Meriones* sp., *Rattus rattus, Canis lupus, Panthera* cf. *P. tigris, Equus przewalskyi, E. hemionus, Coelodonta antiquitatis, Sus* sp., *Megaloceros ordosianus, Cervus (P.) gray, Cervus (E.) elaphus, Spiroceros peii, S. hsuchiayaoca, Procapra przewalskii, Gazella subgutturosa,* and *Bos primigenius* (Jia, Wei, and Li 1979). The usual view of the age of Xujiayao is late Pleistocene and estimated to be about 0.1 Ma. The age is based on a combination of mammal fauna, magnetostratigraphy, and U-series dating (Jia, Wei, and Li 1979; Chen et al. 1984; Liu, Su, and Jin 1992). However, some researchers believe that the age of Xujiayao paleolithic site should be older than late Pleistocene (Løvlie et al. 2001; Fan, Su, and Løvlie 2002).

20. Hutouliang area (N 40°09', E 114°28')

Huang et al. (1974) first reported *Coelodonta* sp. from the upper part of Hutouliang profile, which was seated in Yangyuan County and about 20 km west of Haojiatai (outside the border of the map in figure 8.1). Gai and Wei (1977) listed 15 speices of fossil mammals from the "sandy loess" of nine paleolithic sites around Hutouliang village. One of us (SZ) reexamined and identified 15 species, including *Lasiopodomys brandti* (*Microtus brandtioides*), *Spermophilus daurica* (*Citellus citellus mongolicus*), *Eospalax fontanieri* (*Myospalax fontanieri*), *Tscherskia varians* (*Cricetulus varians*), *Canis lupus*, *Palaeoloxodon namadicus*, *Equus przewalskii*, *Equus hemionus*, *Coelodonta antiquitatis*, *Sus scrofa*, *Cervus* sp., *Procapra* ?*przewalskii*, *Gazella subgutturosa*, ?*Spiroceros* sp., and *Bos* sp. The age of these paleolithic sites was considered to be latest Pleistocene and the same age or later than the Upper Cave site in Zhoukoudian (Gai and Wei 1977; Xie, Li, and Liu 2006).

LITHOSTRATIGRAPHIC DIVISION AND CORRELATION IN THE NIHEWAN BASIN

The Cenozoic sediments in the Nihewan Basin have four components (figure 8.6), which from bottom upward are early Pliocene red clay and sandy clay (Shixia Formation), late Pliocene reddish and gray-black fluviolacustrine and palustrine sandy clay (Daodi Formation), Quaternary brown- or gray-yellowish fluviolacustrine sand, gravel, and clay (Nihewan Formation), and overlying Malan loess.

Red Beds–Shixia Formation

Red clay is usually considered as the underlying strata of the traditional Nihowan beds or Nihewan Formation. In the history of studies on the Nihewan Basin, there were seven lithologic units related to red clay or reddish sediments. These are named as the Shixia Formation by

Huang and Guo (1981), Yuxian Formation and Huliuhe Formation by Wang (1982), Dahonggou Formation by Chen (1988), Daodi Formation by Du et al. (1988), Nangou Formation by Xia (1993), and Hongyacun Formation by Pang (2003). Their type sections are all restricted to a small region on the west bank of the Huliu River. It is possible that the formation names are established multiple times. Of these, Liu et al. (2009) reconsidered the Yuxian and Huliuhe formations to be junior synonyms of the Shixia Formation. They also pointed out that the red beds mainly consist of brick red, dirty red, or brownish red clays, silts, and gravels, a sequence of flood and alluvial deposits. On the basis of Liu et al. (2009), we further treat the above seven formations as follows.

1. Huang and Guo's (1981) Shixia Formation is the first name for the lower red beds in Nihewan Basin. It is a valid name and should have priority, although it probably partially overlaps with the Daodi Formation on the Road Cut (Pulu) section.

2. The Yuxian Formation (Wang 1982), as stated by Pang (2003), has been defined in the North China Regional Stratigraphic Table (Hebei Province and Tianjin volume) (1979:23, 102) as a sequence of gray, grayish-black, yellowish-green, and purplish-red basalt containing green clays, its representative sections being in Da'aoshan and Luotuoling of Yangyuan and Zuopo of Yuxian. Therefore, Wang's Yuxian Formation is an invalid name in the Nihewan Basin and should be abandoned.

3. The Huliuhe Formation (Wang 1982) was established based on the Huabaogou section near the Xiyaozitou village, Yuxian County. The type section is very close (only about 3 km) to the Nangou section in Hongya, where Huang and Guo (1981) established their Shixia Formation. The two sections are lithologically traceable and produce similar mammals. The Huliuhe Formation thus should be a junior synonym of the Shixia Formation and must be abandoned.

4. The Dahonggou Formation, according to Chen et al. (1986) and Chen (1988), was established mainly for sections at Dahonggou of Shixia, Nangou, and Pump Station of Hongya. It is highly consistent in lithology, distribution, and reference sections with the Shixia Formation named by Huang and Guo (1981). It is thus a redundant name, a junior synonym to be abandoned.

5. The Daodi Formation (Du et al. 1988) is a sequence of deep-red fluvial and palustrine sediments and can be seen only along the banks of the Huliuhe. The Laowogou section near Daodi village was selected as the type for its very long section (Cai et al. 2004). Its lithology is roughly comparable to the "lower Pleistocene lower member of

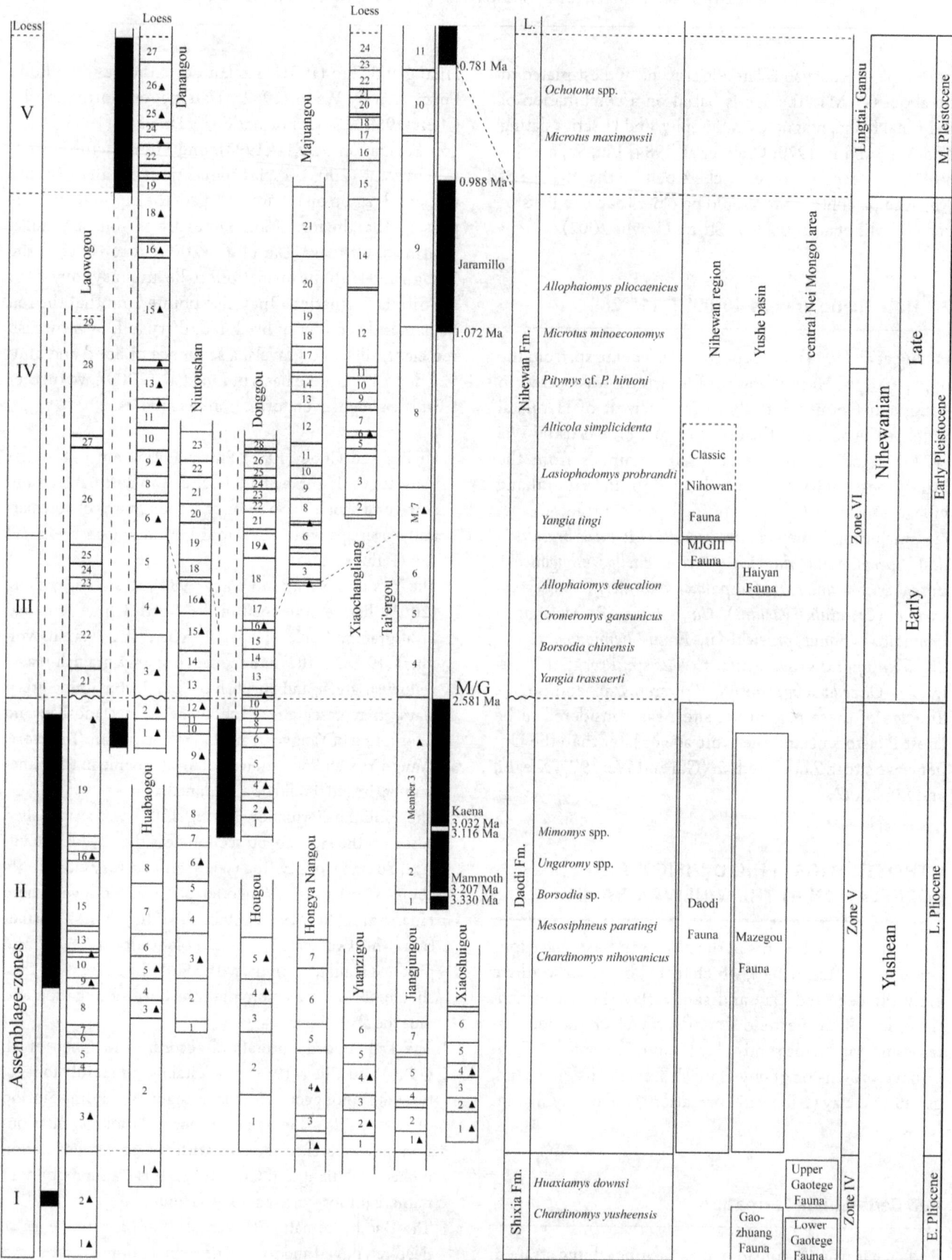

Figure 8.6 Biostratigraphic correlation among the Plio-Pleistocene sections in the Nihewan Basin. Solid black triangles represent the fossil layers. In the vertical columns, black represents normal polarity and white represents reverse polarity. Magnetic column of Tai'ergou section is cited from Min et al. (2006).

Nihewan Formation" at the type sections of the Shixia Formation (Huang and Guo 1981) in Nangou, and Pump Station in Hongya, and is disconformably overlain on top of the Shixia Formation. It represents an independent rock unit and is a valid name.

6. The Nangou Formation (Xia 1993) is very similar in lithology, distribution, and section selection to the Daodi Formation (Du et al. 1988). It is a redundant name, a junior synonym of the latter, and should be abandoned.

7. Pang's (2003) Hongyacun Formation was a reshuffling of Huang and Guo's (1981) Shixia Formation and was defined as the middle and lower members of the Shixia Formation. We regard such subdivision as inappropriate, similar to the establishment of two units based on lithologically inseparable red beds at the Huabaogou section (Wang 1982). We consider it an invalid name to be abandoned.

Thus, the Shixia Formation is the only valid name representing a sequence of late Cenozoic red sediments unconformably on top of basement rocks in the Nihewan Basin. Additional names, such as "Nanyulin Formation" by Yunsheng Han (1982) and "Jingle Formation" by the Shanxi Provincial Geologic and Mineral Bureau (1997: 304), have been used for Nihewan strata. Both of these latter names were established in Shanxi Province, far from the Nihewan Basin, and are unlikely to be traceable to the local strata.

Barbour, Licent, and Teilhard de Chardin (1926) first mentioned red clay near the Sanggan River Gorge (=Shixia) and considered it to be the weathering materials of the basement rocks with a speculated "Pontian" age (Teilhard de Chardin and Piveteau 1930). Huang et al. (1974) thought that red beds from Hongyacun and Shixia were lithologically different, but were unable to ascertain if the differences are due to age or different facies of the same age. Huang and Guo (1981:fig. 8) regard the red sediments from Luanshigedagou from south of Hongyacun and the lower red beds at Shixia as belonging to the same sequence. Liu et al. (2009), however, considered the Pliocene red beds in Nihewan to be of different origins, an opinion with which we agree.

Daodi Formation

A series of fluviopalustrine sediments underlying the Nihewan cross-bedded sands and overlying the red clays is widely exposed at Daodi, Hongya Nangou, Yuanzigou, Qijiazhuang, Jiangjungou, Qianjiashawa, Beimajuan, and Pulu along both sides of the Huliu River. The sediments are characterized by light-reddish, dark-grayish, gray-greenish silty clay and light-yellowish siltstone interbedded with dark-colored sandy clay. Based on the lithology, Du et al. (1988) chose the Laowugo section in the Daodi area as the type section of the Daodi Formation. They regarded the Daodi Formation as overlying the Yuxian Formation through a disconformity. Xia (1993) named the Nangou Formation for the same strata. It is obvious that the Nangou Formation is a junior synonym of the Daodi Formation. Zhang, Zheng, and Liu (2003) considered that there exist lihtofacies changes between the Yuxian Formation and the Daodi Formation and that they should be different contemporaneous facies. They regarded that the Yuxian Formation had priority over the Daodi Formation. After our analysis of the above seven named formations, these controversies can be simplified into a single question: Is the Daodi Formation synonymous with the Shixia Formation? The answer is no. The Shixia Formation named by Huang and Guo (1981:figs. 2, 3) is actually the same as Huang et al.'s (1974) "*Hipparion* red clay," as typified by sections along the banks of Huliuhe. Du et al.'s (1988) Daodi Formation also crops out in the same general area, but the two are lithologically distinguishable. In fact, the Daodi Formation is equivalent to the lower member of the Nihewan Formation of Huang and Guo (1981); that is, it is above the Shixia Formation. We further point out that the Daodi Formation is missing in the Shixia region and that the Nihewan Formation is directly above the Shixia Formation in an unconformable relationship.

Nihewan Formation

Among the 20 sections mentioned previously, the Danangou section is here recommended as the stratotype of the Nihewan Formation. It has a thick, developed sequence with abundant fossil mammals from many different layers. Moreover, the contact relationship of the strata with the overlying loess and underlying Daodi Formation can be easily recognized by its gravel beds, even though gravel beds are normally not recommended as boundaries of formational units. The Nihewan Formation can be interpreted as a large sedimentary cycle, litholgoically subdivided into four members that from top downward are as follows:

Member 4, sandy clay member (layers 19–27): fluviolacustrine sediments consisting of gray-greenish and light-brownish silty clay containing five small sedimentary cycles

Member 3, yellowish sand member (layers 11–18): fluvial sediments consisting of yellowish fine sandstone and siltstone

Member 2, calcium sand member (layers 7–10): fluviolacustrine sediments consisting mainly of light-brownish and gray-greenish calcium siltstone and silty clay interbedded with calcium slabs

Member 1, sand and gravel member (layers 3–6): fluviolacustrine sediments formed by light-yellowish and grayish calcium gravels with cross-bedded sandstones

Members 4 and 3 are continuous and are widely distributed on the east side of Huliu River, where they can be observed from south of Danangou to north of the Xiaodukou and Xiaochangliang areas. The lower part of member 1 is usually a gravel bed, which is well developed at the edge of the basin but changes rapidly to fine gravel and sandstone or silty clay in the center of the basin.

As reviewed in the introduction, it seems that most of the other lithologic units are partly or entirely local aspects of the Nihewan Formation and are not valid as formations. Thus, names such as the Xujiayao, Hutouliang, Que'ergou, Dongyaozitou, Xiaodukou, Haojiatai, Jing'erwa, and Qianjiashawa formations should be abandoned as formal rock units.

BIOSTRATIGRAPHIC SEQUENCE IN THE NIHEWAN BASIN

Based on the components of the mammal fossils from Laowogou, Danangou, and Donggou, especially on evolutionary stages of the small mammals, five assemblage zones can be recognized (see figure 8.6).

Huaxiamys downsi–Chardinomys yusheensis Assemblage Zone

The typical strata are the layers 1 and 2 of Laowogou and layer 1 of the Huabaogou section. This assemblage is characterized by retaining certain common Pliocene genera and speices, such as *Trischizolagus, Kowalskia, Chardinomys yusheensis, Huaxiamys downsi, Chardina truncatus, Pliosiphneus lyratus*, and *Hipparion* cf. *H. hippidiodus*.

It seems that this assemblage can be correlated to those of the upper Gaozhuang Formation of the Yushe Basin, Shanxi, and Zone IV of Lingtai, Gansu. The coexistence of *Chardinomys yusheensis, Huaxiamys downsi, Chardina truncatus*, and *Pliosiphneus lyratus* has the age range of 4.8–3.6 Ma in Yushe (Flynn, Wu, and Downs III

1997) and 4.8–3.3 Ma in Lingtai (Zheng and Zhang 2001). Thus, this assemblage zone seems be of early Pliocene age.

Mimomys–Ungaromys Assemblage Zone

This assemblage typically occurs in layers 3–19 of the Laowogou, layers 1–10 of the Donggou, and layers 1–2 of the Danangou sections. Compared to the older *Huaxiamys downsi–Chardinomys yusheensis* assemblage zone, the second assemblage lacks the early taxa, such as *Trischizolagus, Kowalskia*, and *Huaxiamys downsi*. The primitive *Chardinomys yusheensis, Chardina truncatus*, and *Pliosiphneus lyratus* are replaced by derived *Chardinomys nihowanicus, Mesosiphneus paratingi*, and *M. praetingi*. The most noticeable character of this assemblage is the blossoming of arvicolids, especially of *Mimomys* and *Ungaromys*. A primitive species of *Borsodia* first occurrs in this assemblage.

This assemblage can be correlated to those of the Mazegou Formation in the Yushe Basin and Zone V of Lingtai (Flynn, Wu, and Downs III 1997; Zheng and Zhang 2001). It also can roughly be compared to the late Pliocene Jingle fauna (Zhou 1988). This assemblage is the core of the Daodi fauna. According to the paleomagnetic interpretation of Li and Wang (1991), the Gauss–Matuyama boundary occurs 70 m above the bottom of the Laowogou section. This result is supported by additional paleomagnetic analysis by us. The lower part of Unit 19 records normal polarity, and the upper part of Unit 19 records reversed polarity. The change in polarity occurs between samples 19-5 (66.7 m, normal polarity) and 19-6 (69.7 m, reverse polarity), which is about 14 m above the streambed. According to our faunal correlation and paleomagneitc data, the *Mimomys–Ungaromys* assemblage zone would be Late Pliocene (~3.4–2.6 Ma).

Allophaiomys deucalion–Cromeromys gansunicus Assemblage Zone

Typical strata for this assemblage-zone are layers 11–16 of the Donggou and layers 3–5 of the Danangou sections. This assemblage is characterized by the absence of Pliocene genera and species common in the first and second assemblage zones. *Mimomys* and *Ungaromys* have been replaced by *Cromeromys gansunicus* and rootless *Allophaiomys deucalion*. *Borsodia* sp. has evolved into the more progressive *Borsodia chinensis*. The zokor *Yangia* occurrs instead of *Mesosiphneus*.

An evolutionary lineage within the genus *Allophaiomys* is known to occur in the early Pleistocene records: *A. terrae-rubrae—A. deucalion—A. pliocaenicus*. *A. terrae-rubrae* was found only in locality 93001 of Lingtai, Gansu, and locality 18 of Zoukoudian. Its age falls in the 2.58–2.14 Ma range (lower Matuyama reversed chron; Zheng and Zhang 2001). At present, *A. deucalion* is considered as late Villanyian to early Biharian in age at its type locality, Villany-5 in Hungary (Kowalski 2001). In China, *A. deucalion* is present in the layers WL$_5$+ to WL$_2$+ of Lingtai, Gansu, which was paleomagnetically dated about 2.0–1.8 Ma (Zheng and Zhang 2001). In addition, it occurred about 2.25–1.96 Ma at the Russian Tizdar locality (Pevzner, Tesakov, and Vangengeim 1998).

Borsodia chinensis is an important element of the few small mammals of the classic Nihewan fauna. It has been found from layers 11 and 16 of the Donggou section (Zheng, Cai, and Li 2006) and MJG III (Cai, Li, and Zheng 2008) and Xiaochangliang (Zhang, Kawamura, and Cai 2008) paleolithic sites in the Nihewan Basin. The paleomagnetic data of MJG III and Xiaochangliang indicate an age of 1.66 Ma and 1.36 Ma, respectively (Zhu et al. 2001, 2004). *Yangia trassaerti* is also found in the Haiyan Formation of Yushe Basin, whose age should be older than the Olduvai normal chron (Flynn, Wu, and Downs III 1997). *Cromeromys gansunicus* has a long temporal range, about 3.6–1.8 Ma (Zheng, Cai, and Li 2006), although Flynn, Wu, and Downs III (1997) applied the name *Cromeromys irtyshensis* for populations at the older end of this time range. Overall, our third assemblage probably has an age range of 2.6–1.8 Ma, equivalent to the European late Villanyian or MN 17.

Allophaiomys pliocaenicus–Microtus minoeconomus Assemblage Zone

The fossils are typically produced from layers 6–18 of Danangou, layer 19 of Donggou, and T2 of Tai'ergou sections. *Borsodia* and *Cromeromys* are extinct in this assemblage. The derived *Allophaiomys pliocaenicus* has replaced *A. deucalion*. Some extant genera make their first local appearance, such as insectivores *Crocidura* and *Soriculus*, and arvicolids *Pitymys*, *Alticola*, *Lasiopodomys*, and *Microtus*.

Allophaiomys pliocaenicus is a typical element in European early Biharian faunas. Its type locality, Betfia-II in Romania, has a younger age than the end of the Olduvai normal chron (Repenning 1992). In China, *A. pliocaeni-*

cus is known from the L15 layer of the loess-paleosol sequence in Gongwangling, Lantian, Shaanxi, with a paleomagnetic date of 1.263–1.240 Ma (Zheng and Li 1990; Ding et al. 2002).

Some primitive species of extant genera are present in this assemblage, such as *Microtus minoeconomus* and *Lasiopodomys probrandti* (Zheng and Cai 1991). Another species of *Microtus*, *M. ratticepoides*, is only produced from layer 18 of the Danangou section and is extinct in the European early Toringian age (Fejfar and Heinrich 1981).

The zokor *Yangia tingi* appears in the Wucheng section, Xixian, Shanxi, with paleomagnetic date of 1.9 Ma (Yue and Xue 1996), and is last recorded in the Chenjiawo paleoanthropological site near Lantian, with an age range of 0.71–0.684 Ma (Ding et al. 2002).

Cai et al. (2004) included taxa from layer 18 of the Danangou section into this assemblage. Our latest paleomagnetic results show that layer 18 is normal in polarity (Brunhes), whereas the polarity in layer 15 is reversed (Matuyama). We postulate that the Early–Middle Pleistocene boundary should fall between layers 15 and 18. This fourth assemblage possibly ranges in age from 1.8 Ma to 0.7 Ma, which would coincide with the European Biharian or late Villafranchian.

Microtus maximowiczii Assemblage Zone

This assemblage occurs at the top of the Nihewan Formation in layers 19 to 27 of the Danangou section. Only seven species were collected from this interval. Five species are survivors from the fourth assemblage, and only *Ochotona* spp. and *Microtus maximowiczii* are new elements. The precise composition of this assemblage is still uncertain.

AGE OF THE FAUNAS IN NIHEWAN BASIN

Classic Nihowan Fauna

The fossils of this fauna were collected from strata near Xiashagou and Nihewan villages. Some researchers are still doubtful about the identity and contemporaneity of the fossil-producing strata. In 2001, two of us (Cai and Li) visited Father Zhao Guoxian of the Catholic Church in Nihewan village, who was a bystander at Licent's excavation when he was 10 years old. In his memory, the fossils were mainly collected from the bottom of the Baicaogou section near Nihewan village (see figure 8.1 [16]) and the middle part of the Jinggou section near Xiasha-gou village (see figure 8.1 [15]).

During this visit, Cai and Li took a trial sample to wash from the bottom of Baicaogou and collected two lower jaws of *Borsodia chinensis*, which was a typical rodent of the classic Nihowan fauna. In 2005, the authors undertook screenwashing in the Jinggou section but failed to get any valuable fossils. In fact, the Nihewan Formation is only developed in a 30–50 m section in the Xiashagou-Nihewan area. It seems that the strata of "true Nihowan beds" should be correlated to the sand and gravel member of the Danangou section and would belong to the lower part of the Nihewan Formation (Cai et al. 2004). After Qiu's identification (Qiu 2000; Qiu, Deng, and Wang 2004) and reexamination by one of us (SZ), the classic Nihowan fauna should include:

SPECIES	LOCALITIES
Erinaceus amurensis cf. *E. a. dealbatus* Swinhoe 1870	?Xiashagou
Ochotonoides complicidens (Boule and Teilhard 1928)	Xiashagou
Yangia tingi (Young 1927)	Xiashagou
Borsodia chinensis (Kormos 1934)	Xiashagou
Allactaga sibirica cf. *A. s. annulata* (Milne-Edwards 1867)	Xiashagou
Canis chihliensis Zdansky 1924	Nihewan
Canis chihliensis var. *palmidens* Teilhard and Piveteau 1930	Nihewan
Vulpes chikushanensis(?) Young 1930 (pro *Vulpes* sp.)	Nihewan
Eucyon minor (Teilhard and Piveteau 1930) (pro *Canis chihliensis* var. *minor*)	Nihewan
Nyctereutes sinensis Schlosser 1903	Nihewan
Selenarctos thibetanus (Guvier 1823) (pro *U.* cf. *U. etruscus*)	Nihewan
Eirictis pachygnatha (Teilhard and Piveteau 1930)	Xiashagou
Meles chiai Teilhard 1940 (pro *M.* cf. *M. leucurus*)	Xiashagou
Lutra licenti Teilhard and Piveteau 1930	
Chasmaporthetes progressus (Qiu 1987) (pro *Hyaena* sp.)	
Crocuta honanensis (Zdansky 1924)	
Pachycrocuta licenti (Pei 1934)	Xiashagou, Nihewan
Homotherium crenatidens Fabrini 1890 (pro *Machairodus nihowanensis*, partim)	Xiashagou
Megantereon nihowanensis (Teilhard and Piveteau 1930) (pro *Machairodus nihowanensis*, partim)	Xiashagou
Lynx shansius Teilhard and Leroy 1945	
Felis sp.	
Sivapanthera cf. *S. pleistocaenica* (Zdansky 1925)	
Mammuthus trogontherii Pohlig 1885 (pro *Palaeoloxodon* cf. *P. namadicus*)	Nihewan
Postschizotherium chardini von Koenigswald 1932	Nihewan
Hipparion (*Proboschidipparion*) *sinense* Sefve 1927	Xiashagou
Equus sanmeniensis Teilhard and Piveteau 1930	Xiashagou, Nihewan
Coelodonta nihowanensis Kahlke 1969 (pro *Rhinoceros* cf. *tichorhinus*)	Xiashagou, Nihewan
Elasmotherium sp.	Nihewan
Sus cf. *S. lydekkeri* Zdansky 1928	
Paracamelus gigas (Schlosser 1903)	Nihewan
Muntiacus bohlini (Teilhard 1940)	
Eucladoceros boulei (Teilhard and Piveteau 1930)	Xiashagou
Cervus elegans (Teilhard and Piveteau 1930)	Xiashagou
Cervus (*Rusa*) sp.	
Elaphurus bifurcatus (Teilhard and Piveteau 1930)	
Spiroceros wongi Teilhard and Piveteau 1930	Xiashagou, Nihewan
Gazella sinensis Teilhard and Piveteau 1930	Xiashagou
Gazella cf. *G. subgutturosa* (Guldenstaedt 1780)	
Ovis shantungensis Matsumoto 1926	
Ovis or *Capra* sp.	
Bison palaeosinensis Teilhard and Piveteau 1930	Nihewan
Bovinae gen. et. sp. indet.	

Of only five species of small mammals described in the classic Nihowan fauna (Teilhard de Chardin and Piveteau 1930), *Ochotonoides complicidens*, *Allactaga* sp. and *Yangia tingi* belong to the *Allophaiomys pliocaenicus–Microtus minoeconomus* assemblage zone. *Borsodia chinensis* survived until the age of Xiaochangliang as a relic, about 1.36 Ma. Large mammals collected from the Danangou section described by Li (1984) include *Hyena* sp., *Canis chihliensis*,

Meles chiai, Vulpes sp., *Proboscidipparion sinense, Equus sameniensis, Coelodonta antiquitatis* (should be *C. nihowanensis*), and *Gazella sinensis*. These species are the common elements of the Nihowan fauna. Their horizons correspond to layer 17 of the Danangou section in this chapter, or within the range of the *Allophaiomys pliocaenicus–Microtus minoeconomus* assemblage zone. The fossil mammals collected from the Xiaochangliang Paleolithic site also show certain similarity to the classic Nihowan fauna (Tang, Li, and Chen 1995; Zhang, Kawamura, and Cai 2008). Qiu (2000) considered the age of Nihowan fauna closest to that of the European early late Villafranichan Olivola fauna, at about 1.8 Ma. We partially agree with Qiu's opinion that the classic Nihowan fauna is basically concentrated in a short period of time. We are inclined to regard the classic Nihowan fauna as belonging to our fourth—the *Allophaiomys pliocaenicus–Microtus minoeconomus*—assemblage zone (see figure 8.6).

Dongyaozitou Fauna

This fauna was collected from the second (first in Tang 1980), third, and fourth layers of the Danangou cliff (Tang 1980; Tang, You, and Li 1981). Eight of the species can be relocated to exact strata, but four species added later have no stratigraphic documentation (Tang and Ji 1983). *Hipparion* cf. *H. houfenense* and *Antilospira yuxianensis* collected from the second layer are common species in Pliocene faunas, whereas *Lynx variabilis* and *Paracamelus* sp. appeared earlier than the Nihowan fauna. The third layer only produced *Zygolophodon* sp. Large mammals from the fourth layer share some genera and species, such as *Coelodonta antiquitatis, Gazella sinensis*, and *Axis* sp., with those of the Nihowan fauna.

In fact, the fossil contents from the second and fourth layers are different. Futhermore, there exists a distinct disconformity between the second and overlying third and fourth layers. The fossils from the second layer should belong to our *Mimomys–Ungaromys* assemblage zone, while those from third and fourth layers correspond to our *Allophaiomys pliocaenicus–Microtus minoeconomus* assemblage zone. We consider the Dongyaozitou fauna to be a mixed-fossil assemblage. This is consistent with an unconformable stratigraphic relationship.

Daodi Fauna

The materials of this fauna were collected from the dark-colored fluviopalustrine strata of the lower parts of the Daodi, Hongya Nangou, Yuanzigou, Qijiazhuang, Jiangjungou, Qianjiashawa, Beimajuan, and Pulu sections (Cai 1987). This fauna consists only of small mammals, which account for 29 species (Li, Zheng, and Cai 2008). The Daodi fauna is characterized by numerous widespread Pliocene genera and species, such as *Sinocricetus progressus, Nannocricetus mongolicus, Mimomys* and *Ungaromys* spp., *Psudomeriones complicidens, Pliosiphineus* sp., *Chardinomys nihowanicus*, and *Apodemus zhangwagouensis*. It can be correlated to the Mazegou fauna of the Yushe Basin or biozone V of Lingtai, Gansu, which share many of these elements. The paleomagnetic dating of the Mazegou fauna and biozone V of Lingtai are both about 3.6–2.6 Ma (Flynn, Wu, and Downs III 1997; Zheng and Zhang 2001). The Daodi fauna also can be roughly correlated to the upper fauna of Gaotege, Nei Mongol, based on common presence of *Sinocricetus progressus, Nannocricetus mongolicus, Pseudomeriones complicidens*, and *Mimomys* with cementum. Magnetic dating places the upper Gaotege fauna at about 4.0–3.8 Ma (Qiu, Wang, and Li, chapter 5, this volume). Daodi assemblage would be younger than the upper Gaotege fauna in its first appearance of derived taxa, such as *Ungaromys, Mesosiphneus praetingi*, and *Chardinomys nihowanicus*. As mentioned previously, this fauna belongs to our *Mimomys–Ungaromys* assemblage zone and is latest Pliocene in age (see figure 8.6).

MJG III Fauna

This fauna contained 20 species of fossil mammals and is characterized by *Allophaiomys deucalion, Borsodia chinensis*, and *Cromeromys gansunicus* (Cai and Li 2004; Cai, Li, and Zheng 2008). The MJG III fauna should belong to our *Allophaiomys deucalion–Cromeromys gansunicus* assemblage zone (see figure 8.6). We further postulate that its age would be at least as old as the bottom of Olduvai normal chron—that is, older than 1.9 Ma (Cai, Li, and Zheng 2008)—which is somewhat older than the estimates by Zhu et al. (2004).

CHRONOSTRATIGRAPHY IN THE NIHEWAN BASIN

Top of the Nihewan Formation

The top strata of the Nihewan Formation are well developed on the east side of the Huliu River, especially in the Danangou section, but they are not well developed in the Nihewan and Xiashagou areas that yield classical Ni-howan fauna. The Nihewan Formation is unconformable, with the overlying Malan loess capped by a gravel bed, layer 27 of the Danangou section. This lithological contact is very pronounced in the Nihewan Basin. Judging from the normal polarity present in layer 18 of the Danangou section, the top of the Nihewan Formation of this section possibly has a middle Pleistocene or younger age. Xia, Zhang, and Chen (1993) dated the top strata of the Hutouliang section by ESR to 90 ka. Yuan et al. (1996), on the other hand, regarded the age of the topmost strata of the Nihewan Formation as about 0.97–0.13 Ma. Min et al. (2006) dated the topmost strata beneath Malan Loess at about 112 ka. Most recently, Zhao et al. (2010) dated the bottom of the loess-paleosol sequence in the upper part of the Haojiatai profile at about 128 ka by recuperated OSL of fine-grained quartz. It seems to us that the question remains whether or not the Nihewan Formation extends into the late Pleistocene.

Nihewanian and Pliocene/Pleistocene Boundary

Qiu and Qiu (1990, 1995) recommended using Yushean as the standard Pliocene mammalian unit in North China. The Yushean ranges from 5.3 Ma to 2.6 Ma in age. It seems that using 2.6 Ma as the boundary for the Yushean and Nihewanian biochrons has been widely accepted by most Chinese Neogene or Quaternary researchers. Yushean is subdivided as Gaozhuangian (roughly equal to European Ruscinian, or MN 14–15) and Mazegouan (roughly equal to early Villanyian, or MN 16) subages (Qiu et al., chapter 1, this volume). Deng (2006) also suggested a Mazegouan age for the Daodi fauna. Using such schemes, the first and second fossil assemblage zones in the Nihewan Basin thus should belong to Yushean Land Mammal Age.

According to our biostratigraphy, the Pliocene/Pleistocene boundary should also be the boundary of our assemblage zone II and III, which is located between second and third layers of Danangou, layers 19 and 20 of Laowogou, layers 12 and 13 of Pulu, layers 10 and 11 of Donggou, and members 3 and 4 of Tai'ergou sections (Cai and Li 2004; Min et al. 2006; Zheng, Cai, and Li 2006; Cai, Zheng, and Li 2007; see figure 8.6).

Tong, Zheng, and Qiu (1995) subdivided the Nihewanian into early, middle, and late subzones and regarded the age of the classic Nihowan fauna as middle Nihewanian. According to our lithologic and biostratigraphic data, we suggest that it is more reasonable to divide the Nihewanian into two parts: the third and fourth assemblage zones represent the early and late Nihewanian Land Mammal Age, respectively.

MAGNETOSTRATIGRAPHY

Initial study of magnetostratigraphy in the Nihewan Basin began in the early 1980s (Li and Wang 1982, 1991; Chen 1988), and as sampling and analytic laboratory equipment and techniques improved, data on the magnetic chronology were refined (Yuan et al. 1996; Zhu et al. 2001; Wang et al. 2004; Min et al. 2006; Deng et al. 2008). Of these investigations, Deng et al.'s (2008) was the most systematic in the sampling of fossiliferous sections. Deng et al. (2008) correlate the long normal zone in the lowest part of the Hongya and Huabaogou sections with that from the middle part of the Donggou section (their Donggou section is only the middle and upper parts of the entire Donggou exposure) and interpret it as the Olduvai Normal Subchron (1.945–1.778 Ma; Gradstein, Ogg, and Smith 2004). Such a correlation is inconsistent with the Pliocene mammal fossils from these two sections—that is, the *Huaxiamys downsi–Chardinomys yusheensis* assemblage zone and the *Mimomys–Ungaromys* assemblage zone. Therefore, we consider it a more probable interpretation to match the normal chron from the 75–110 m interval of the Hongya section and the bottom of the Huabaogou section.

To further test the accuracy of recognizing the Plio-Pleistocene boundary (the Gauss-Matuyama boundary) based on biochronological considerations, one of us (JCL) took paleomagnetic samples from the Donggou, Laowogou, and Danangou sections in 2005 and 2006. The results indicate a normal interval for layers 1–8 of the Donggou section, layer 1 of the Danangou section, and layers 2 and 9–19 of the Laowogou section. Mammals from these layers all belong to the *Mimomys–Ungaromys* assemblage zone, and we thus consider these normal samples to represent the Gauss Normal Chron, therefore older than 2.58 Ma. As for the normal interval above layer 18 in the Danangou section, it seems likely to be the Brunhes Normal Chron based on such advanced mammal forms as *Microtus maximowiczii*, although the overall composition of the fossil assemblage remains unclear.

CONCLUSIONS

Integrating diverse lithostratigraphic and biostratigraphic studies on the Nihewan Basin, we recognize the Nihowan beds named by Barbour (1924) as a series of fluviolacustrine sediments between late Pleistocene loess and early Pliocene "*Hipparion* red clay." The deposits can be considered as the late Pliocene Daodi Formation and the Pleistocene Nihewan Formation, respectively.

The Danangou section near Dongyaozitou village is recommended as the stratotype of the Nihewan Formation.

Based mainly on the small mammals from the Danangou, Pulu, Donggou, and Laowogou sections, five assemblage zones from oldest to youngest can be recognized: the *Huaxiamys downsi–Chardinomys yusheensis* assemblage zone, *Mimomys–Ungaromys* assemblage zone, *Allophaiomys deucalion–Cromeromys gansunicus* assemblage zone, *Allophaiomys pliocaenicus–Microtus minoeconomus* assemblage zone, and *Microtus maximowiczii* assemblage zone (only the last three occur in the Nihewan Formation).

The classic Nihowan fauna possibly is restricted to the *Allophaiomys pliocaenicus–Microtus minoeconomus* assemblage zone and is about 1.8 Ma in age.

ACKNOWLEDGMENTS

We express thanks for the great assistance of colleagues who participated in the Nihewan project, especially Profs. Zhu-ding Qiu and Zhao-qun Zhang, and Wei Zhou and Ying-qi Zhang from the IVPP, Xin-tian Wang, Zhi-gang Tan from Xianmen Universtiy, and Ye Jin from China University of Geosciences. Many thanks are also given to Xiao-ming Wang and Lawrence J. Flynn for their critique of the manuscript. The Nihewan project was supported by the Knowledge Innovation Program of the Chinese Academy of Sciences (grant no. KZCX2-YW-120), the Major Basic Research Projects (2006CB806400) of MST of China, the National Natural Science Foundation of China (grant nos. 40472012, 40702004, 40730210), NSF-RFBR grant no. 40911120091, and a Barnard College faculty research grant to JCL.

REFERENCES

An, Z.-s. and L. Ai. 2005. Imperfect geologic time scale pending future of the Quaternary. *Journal of Stratigraphy* 29(2):99–103 (in Chinese with English abstract).

Barbour, G. B. 1924. Preliminary observations in the Kalgan area. *Bulletin of the Geological Society of China* 3(2):167–168.

Barbour, G. B. 1925. The deposits of Sang Kan Ho valley. *Bulletin of the Geological Society of China* 4(1):53–55.

Barbour, G. B., E. Licent, and P. Teilhard de Chardin. 1926. Geological study of the deposits of the Sangkanho basin. *Bulletin of the Geological Society of China* 5(3–4):263–278.

Black, D., P. Teilhard de Chardin, C. C. Young, and W.-c. Pei. 1933. Fossil man in China. *Memoir of the Geological Society of China*, ser. A, p. 11.

Cai, B.-q. 1987. A preliminary report on the Late Pliocene micromammalian fauna from Yangyuan and Yuxian, Hebei. *Vertebrata PalAsiatica* 25(2):124–136 (in Chinese with English abstract).

Cai, B.-q. 1989. Fossil lagomorphs from the Late Pliocene of Yangyuan and Yuxian, Hebei. *Vertebrata PalAsiatica* 27(3):170–181 (in Chinese with English summary).

Cai, B.-q. and Q. Li. 2004. Human remains and the environment of Early Pleistocene in the Nihewan Basin. *Science in China*, ser. D, Earth Sciences 47(5):437–444.

Cai, B.-q., Q. Li and S.-h. Zheng. 2008. Fossil mammals from Majuangou section of Nihewan Basin, China and their age. *Acta Anthropologica Sinica* 27(2):129–142 (in Chinese with English summary).

Cai, B.-q. and Z.-d. Qiu. 1993. Murid rodents from the Late Pliocene of Yangyuan and Yuxian, Hebei. *Vertebrata PalAsiatica* 31(4):267–293 (in Chinese with English summary).

Cai, B.-q., Z.-q. Zhang, S.-h. Zheng, Z.-d. Qiu, Q. Li, and Q. Li. 2004. New advances in the stratigraphic study on representative sections in the Nihewan Basin, Hebei. *Professional Papers of Stratigraphy and Palaeontology* 28:267–285 (in Chinese with English abstract).

Cai, B.-q., S.-h. Zheng, and Q. Li. 2007. The Plio-Pleistocene small mammals from the Niutoushan section of the Yuxian Basin, China. *Vertebrata PalAsiatica* 45(3):232–245 (in Chinese with English summary).

Chen, M.-n. 1988. *Study on the Nihewan Beds*. Beijing: China Ocean Press (in Chinese with English abstract).

Chen, M.-n., Y.-s. Wang, S.-f. Wang, B.-x. Luo, Q. Wang, J. Yue, M.-m. Jiang, and S.-h. Ge. 1986. Study of Nihewan beds in basin of Yangyuan-Yuxian Hobei Province, China. *Bulletin of the Chinese Academy of Geological Sciences* 15:149–160 (in Chinese).

Chen, T.-m., S.-x. Yuan, S.-j. Gao, and J. Shi. 1984. The study on uranium-series dating of fossil bones and an absolute age sequence for the main Paleolithic site of North China. *Acta Antropologica Sinica* 3(3):259–269 (in Chinese).

Deng, C.-l., R.-x. Zhu, R. Zhang, H. Ao, and Y.-x. P. 2008. Timing of the Nihewan formation and faunas. *Quaternary Research* 69:77–90.

Deng, T. 2006. Chinese Neogene mammal biochronology. *Vertebrata PalAsiatica* 44(2):143–163.

Ding, Z.-l., E. Derbyshire, S.-l. Yang, Z.-w. Yu, S.-f. Xiong, and T.-s. Liu. 2002. Stacked 2.6-Ma grain-size record from the Chinese loess based on five sections and correlation with the deep-sea $\delta^{18}O$ record. *Paleoceanography* 17(3):501–521.

Du, H.-j., A.-d. Wang, Q.-q. Zhao, and B.-q. Cai. 1988. Daodi Formation—a new Pliocene stratigraphic unit in Nihewan district. *Earth Science-Journal of China University of Geosciences* 13(5):561–568 (in Chinese).

Erbajeva, M. and S.-h. Zheng. 2005. New data on Late Miocene-Pleistocene ochotonids (Ochotonidae) from North China. *Acta Zoologica Cracoviensia* 48A(1–2):93–117.

Fan X.-z., P. Su, and R. Løvlie. 2002. Magnetostratigraphic evidence for the age of the Xujiayao Paleolithic layer and the Xujiayao Formation. *Journal of Stratigraphy* 26(4):248–252 (in Chinese with English abstract).

Fejfar, O. and W. D. Heinrich. 1981. Zür biostratigraphischen Untergliederung des kontinentalen Quartärs in Europa anhand von Arvicoliden (Rodentia Mammalia). *Eclogae Geologicae Helvetiae* 74(3):997–1006.

Flynn, L. J., W.-y. Wu, and W. R. Downs III. 1997. Dating vertebrate microfaunas in the late Neogene record of northern China. *Palaeogeography Palaeoclimatology and Palaeoecology* 133:227–242.

Gai, P. and Q. Wei. 1977. Discovery of the Late Palaeolithic site at Hutouliang, Hebei. *Vertebrata PalAsiatica* 15(4):287–300 (in Chinese).

Gradstein, F. M., J. G. Ogg, and A. G. Smith. 2004. *A Geologic Time Scale 2004.* Cambridge: Cambridge University Press.

Han, Y.-s. 1982. On the "Nihewan bed." *Journal of Stratigraphy* 6(2):121–127 (in Chinese).

Huang, B.-r. 1980. Preliminary studies on Pleistocene ostracods from middle to lower reaches of Sanggan River. *Chinese Science Bulletin* 25(6):277–278 (in Chinese).

Huang, B.-y., and S.-y. Guo. 1981. Discussion on stratigraphic division, geological period and palaeography of the Nihewan according to the mollusca. *Bulletin Tianjin Institute Geology Mineral resources Research* 4:17–30 (in Chinese with English abstract).

Huang, W.-b., Y.-j. Tang, G.-f. Zong, Q.-q. Xu, Z.-d. Qiu, W.-l. Huang, F.-l. Li, and J.-y. Liu. 1974. Observation on the Later Cenozoic of Nihowan Basin. *Vertebrata PalAsiatica* 12(2):99–108 (in Chinese).

Jia, L.-p. and Q. Wei. 1976. A Palaeolithic site at Hsu-chia-yao in Yangkao County, Shanxi Province. *Acta Archaeologica Sinica* (2):97–114 (in Chinese with English summary).

Jia, L.-p., Q. Wei, and C.-r. Li. 1979. Report on the excavation of Hsuchiayao Man site in 1976. *Vertebrata PalAsiatica* 17(4):277–293 (in Chinese with English summary).

Jia, L.-p. and Q. Wei. 1980. Some animal fossils from the Holocene of N. China. *Vertebrata PalAsiatica* 18(4):327–333 (in Chinese with English abstract).

Kerr, R. A. 2008. A time war over the period we live in. *Science* 319:402–403.

Kerr, T. A. 2009. The Quaternary Period wins out in the end. *Science* 324:1249.

Kowalski, K. 2001. Pleistocene rodents of Europe. *Folia Quaternaria* 72:1–389.

Li, H.-m. and J.-d. Wang. 1982. Magnetostratigraphic study of several typical geologic sections in North China. In *Quaternary Geology and Environment of China*, ed. T.-S. Liu, pp. 33–38. Beijing: Ocean Press (in Chinese).

Li, H.-m. and J.-d. Wang. 1991. The latest advance in Quaternary magnetostratigraphy of China. In *Quaternary Geology and Environment in China*, ed. T.-s. Liu, pp. 158–167. Beijing: Science Press.

Li, Q., S.-h. Zheng, and B.-q. Cai. 2008. Pliocene biostratigraphic sequence in the Nihewan Basin, Hebei, China. *Vertebrata PalAsiatica* 46 (3):210–232 (in Chinese with English summary).

Li, Y. 1984. The Early Pleistocene mammalian fossils of Danangou, Yuxian, Hebei. *Vertebrata PalAsiatica* 22(1):60–68 (in Chinese with English abstract).

Lin, Y.-z. 1984. Some sections of the Nihewan Formation in Yangyuan and Yuxian, Hebei. *Journal of Stratigraphy* 8(2):152–160 (in Chinese).

Liu, C., P. Su, and Z.-x. Jin. 1992. Discovery of Blake episode in the Xujiayao Paleolithic site, Shanxi, China. *Scientia Geologica Sinica* 1(1):87–95.

Liu, H.-k. F.-g. Wang, J.-m. Xu, J. Zhang, H. Zhao, Y.-l. Zhang, F.-l. An, X. Yin, Y.-z. Huang, L.-m. Wang, S.-x. Fan, Y.-p. Ji, Z.-w. Bi, H.-l. Mao, H.-l. Ji, L.-j. Liu, R. Ma, and J.-z. Wang. 2009. Disscusion on several typical late Cenozoic stratigraphic units in North China. *Acta Geoscientica Sinica* 30(5):571–580 (in Chinese with English abstract).

Liu, Q.-s. 1989. Evolution of the fossil population of ostracodes and change of ancient environment in Nihewan. *Journal of Beijing Normal University* (Natural Science) 1:78–84 (in Chinese with English abstract).

Liu, T.-s., Y.-f. Shi, R.-j. Wang, Q.-h. Zhao, Z.-m. Jian, X.-r. Cheng, P.-x. Wang, S.-m. Wang, B.-y. Yuan, X.-z. Wu, Z.-x. Qiu, Q.-q. Xu, W.-b. Huang, W.-w. Huang, Z.-s. An, and H.-y. Lu. 2000. Table of Chinese Quaternary stratigraphic correlation remarked with climate change. *Quaternary Sciences* 20(2):108–128 (in Chinese with English abstract).

Liu, X.-q. and Z.-k. Xia. 1983. A suggestion on the division and correlation of the Nihewan Formation. *Marine Geology & Quaternary Geology* 3(1):75–85 (in Chinese with English abstract).

Løvlie, R., P. Su, X.-Z. Fan, Z.-J. Zhao, and C. Liu. 2001. A revised paleomagnetic age of the Nihewan Group at the Xujiayao Palaeolithic Site, China. *Quaternary Science Review* 20: 1341–1353.

Min, L.-r., Z.-q. Chi, and G.-x. Zhu. 2000. Division of Quaternary lacustrine beds in Jing'erwa borehole of the Yangyuan Basin. *Acta Geologica Sinica* 74(2):108–115 (in Chinese with English abstract).

Min, L.-r., Z.-h. Zhang, X.-s. Wang, S.-h. Zheng, and G.-x. Zhu. 2006. The basal boundary of the Nihewan Formation at the Tai'ergou section of Yangyuan, Hebei Province. *Journal of Stratigraphy* 30(2):103–108 (in Chinese with English abstract).

Ouyang, Q. 1964. Lower Pleistocene terraces of Yangyuan Basin lake shores. *Abstracts in Proceedings of the Second Academic Conference by Commission on Chinese Quaternary Research* (in Chinese).

Pang, Q.-q. 2003. Late Cenozoic stratigraphy, subdivision, and chronology of Nihewan Basin. In *Mineral Resources, Geologic Environments, and Economic Managements: A Volume Celebrating the 50th Anniversary of Shijiazhuang College of Economics*, ed. Y.-R. Lu and D.-H. Hao, pp. 17–31. Beijing: Geological Press.

Pevzner, M., A. Tesakov, and E. Vangengeim. 1998. The position of the Tizdar locality (Taman peninsula, Russia) in the magnetochronological scale. *Paludicola* 2(1):95–97

Qiu, Z.-d. 1985. A New Ochotonid from Nihewan Bed of Yuxian, Hebei. *Vertebrata PalAsiatica* 23(4):266–286 (in Chinese with English summary).

Qiu, Z.-x. 2000. Nihewan fauna and Q/N boundary in China. *Quaternary Sciences* 20(2):142–154 (in Chinese with English abstract).

Qiu, Z.-x., T. Deng, and B.-y. Wang. 2004. Early Pleistocene mammalian fauna from Longdan, Dongxiang, Gansu, China. *Palaeontlogica Sinica*, n.s. C 27:1–198 (in Chinese with English summary).

Qiu, Z.-x. and Z.-d. Qiu. 1990. Neogene local mammalian faunas: Succession and ages. *Journal of Stratigraphy* 14(4):241–260 (in Chinese).

Qiu, Z.-x and Z.-d. Qiu. 1995. Chronological sequence and subdivision of Chinese Neogene mammalian faunas. *Palaeogeography, Palaeoclimatology, and Palaeoecology* 116:41–70.

Repenning, C. A. 1992. *Allophaiomys* and the age of the Olyor Suite, Krestovka Sections, Yakutia. *US Geological Survey Bulletin* 2037:1–98.

Shanxi Provincial Geology and Mineral Bureau. 1997. *Research on National Stratigraphic Divisions—Lithostratigraphy of Shanxi Province*. Wuhan, China: University of Geoscience Press.

Tang, Y.-j. 1980. Note on a small collection of Early Pleistocene mammalian fossils from northern Hebei. *Vertebrata PalAsiatica* 18(4):314–323 (in Chinese with English abstract).

Tang, Y.-j. and H.-x. Ji. 1983. A Pliocene-Pleistocene transitional fauna from Yuxian, northern Hebei. *Vertebrata PalAsiatica* 21(3):245–254 (in Chinese with English abstract).

Tang, Y.-j., Y. Li, and W.-y. Chen. 1995. Mammalian fossils and the age of Xiaochangliang Paleolithic site of Yangyuan, Hebei. *Vertebrata PalAsiatica* 33(1):74–83 (in Chinese with English summary).

Tang Y.-j., Y.-z. You, and Y. Li, 1981. Some new fossil localities of Early Pleistocene from Yangyuan and Yuxian Basins, Northern Hopei. *Vertebrata PalAsiatica* 19(3):256–268 (in Chinese with English abstract).

Teilhard de Chardin, P. 1941. *Early Man in China*. Pékin (Beijing): Institut de Géo-biologie.

Teilhard de Chardin, P. and J. Piveteau. 1930. Les mammifères fossiles de Nihowan (Chine). *Annales de Paléontologie* 19:3–132.

Tong, Y.-s., S.-h. Zheng, and Z.-d. Qiu. 1995. Cenozoic mammal ages of China. *Vertebrata PalAsiatica* 33(4):290–314 (in Chinese with English summary).

Wang, A.-d. 1982. Discovery of the Pliocene mammalian faunas from the Nihewan district and their significance. *Chinese Science Bulletin* 27(4):227–229 (in Chinese).

Wang, K.-j., G.-m. Zhao, and Y.-x. Xin. 1964. Neotectonism in Nihewan region. *Science and Technology Information of North China Geology* 1:19–25.

Wang, P.-x., Q.-b. Min, J.-x. Lin, and Z.-t. Cui. 1975. Discovery of semi-saline feraminifera fauna in several Cenozoic basins in eastern China and their significance. *Professional Papers of Stratigraphy and Palaeontology* 2:1–36 (in Chinese).

Wang, X.-s., Z.-y. Yang, R. Løvlie, and L.-r. Min. 2004. High-resolution magnetic stratigraphy of fluvio-lacustrine succession in the Nihewan Basin, China. *Quaternary Science Reviews* 23:1187–1198

Wei, Q. 1976. Recent find of fossil *Palaeoloxodon namadicus* from Nihewan Beds, NW Hebei. *Vertebrata PalAsiatica* 14(1):53–58 (in Chinese with English summary).

Wei, Q. 1978. New discoveries of from Nihewan Beds and its stratigraphical significance. In *Collected Papers on Paleoanthropology*, ed. Institute of Vertebrate Paleontology and Paleoanthropology, pp. 136–150. Beijing: Science Press (in Chinese).

Wei, Q. 1983. A new *Megaloceros* from Nihowan Beds. *Vertebrata PalAsiatica* 21(1):87–95 (in Chinese with English summary).

Wei, Q. 1990. Banshan Paleolithic site from the lower Pleistocene in the Nihewan Basin in northern China. *Acta Anthropologica Sinica* 13(3):223–238 (in Chinese with English abstract).

Wu, Z.-r., J.-z. Sun, and B.-y. Yuan. 1980. Subdivision and recognition of the Nihewan strata. *Scientia Geologica Sinica* (1):87–95 (in Chinese with English abstract).

Xia, Z.-k. 1993. On the age and division of the "Nihowan beds." In *Proceedings in the Conference in Honor of the 100th Anniversary of Professor Yuan Fuli*, pp. 11–14. Beijing: Geology Press (in Chinese).

Xia, Z.-k., J. Zhang, and T.-m. Chen. 1993. Discovery of stramatolites from Nihewan strata and its significance. *Science in China* 23(8):874–879 (in Chinese).

Xie, F., J. Li, and L.-q. Liu. 2006. *Paleolithic Archeology in the Nihewan Basin*. Shijiazhuang: Huashan Literature and Arts Press (in Chinese).

Yang, Z.-g., H.-m. Lin, G.-w. Zhang, and S.-j. Wang. 1996. The Lower Pleistocene in Nihewan Basin. In *Quaternary Stratigraphy in China and Its International Correlation—A Summary by China National Working Group for International Geological Programme (IGCP), Project 296*, ed. Z.-G. Yang and H.-M. Lin, pp. 109–126. Beijing: Geological Publishing House (in Chinese).

Young, C.-C. 1950. The Plio-Pleistocene boundary in China. In *International Geological Congress "Report of the Eighteenth Session, Great Bratain, 1948," Part IX. Proceedings of Section H: The Pliocene-Pleistocene Boundary*, ed. K. P. Oakley, pp. 115–125. Stockport: Co-operative Wholesale Society.

Yuan, B.-y., R.-x. Zhu, W.-l. Tian, J.-x. Cui, R.-q. Li, Q. Wang, and F.-h. Yan. 1996. On the age, stratigraphy and comparision of the Nihewan Formation. *Science in China*, ser. D 26(1):67–75 (in Chinese).

Yue, L.-p. and X.-x. Xue. 1996. *Palaeomagnetism of Chinese Loess*. Beijing: Geological Publishing House (in Chinese with English summary).

Zhang, Y.-q., Y. Kawamura, and B.-q. Cai. 2008. Small mammal fauna of Early Pleistocene age from the Xiaochangliang site in the Nihewan Basin, Hebei, northern China. *Quaternary Research* 47(2):81–92.

Zhang, Z.-h, L.-r. Min, and G.-x. Zhu. 2003. Lithostratigraphic division of the Nihewan fluvial-lacustrine strata at the Tai'ergou section in Yangyuan, Hebei Province. *Geological Bulletin of China* 22 (6):379–383 (in Chinese with English abstract).

Zhang, Z.-q., S.-h. Zheng, and J.-b. Liu. 2003. Pliocene micromammalian biostratigraphy of Nihewan Basin, with comments on the stratigraphic division. *Vertebrata PalAsiatica* 41(4):306–313 (in Chinese with English summary).

Zhao, H., Y.-c. Lu, C.-m. Wang, J. Chen, J.-f. Liu, and H.-l. Mao. 2010. ReOSL dating of aeolian and fluvial sediments from Nihewan Basin, northern China and its environmental application. *Quaternary Geochronology* 5:159–163.

Zheng, S.-h. 1981. New discovered small mammals in the Nihowan Bed. *Vertebrata PalAsiatica* 19(4):348–358 (in Chinese with English summary).

Zheng, S.-h. and B.-q. Cai. 1991. Micromammlian fossils from Danangou of Yuxian, Hebei. In *Contribution to the XIII INQUA*, ed. Institute of Vertebrate Paleontology and Paleoanthropology, Academica Sinica, pp. 100–131. Beijing: Beijing Science & Technology Publishing House (in Chinese with English summary).

Zheng, S.-h., B.-q. Cai, and Q. Li. 2006. The Plio-Pleistocene small mammals from Donggou section of Nihewan Basin, Hebei,

China. *Vertebrata PalAsiatica* 44(4):320–331 (in Chinese with English summary).

Zheng, S.-h. and C.-k. Li. 1990. Comments on fossil arvicolids of China, In *International Symposium Evolution, Phylogeny and Biostratigraphy of Arvicolids (Rodentia, Mammalia)*, ed. O. Fejfar and W.-D. Heinrich, pp. 431–442. Prague: Geological Survery.

Zheng, S.-h. and Z.-q. Zhang. 2001. Late Miocene-Early Pleistocene biostratigraphy of the Leijiahe area, Lingtai, Gansu. *Vertebrata PalAsiatica* 39(3):215–228 (in Chinese with English summary).

Zhou, M.-z. 1955. Lives and natural environments of Chinese fossil man as seen from vertebrate fossils. In *Discoveries and Studies of Human Fossils from China— Symposium of 25th Anneversary of the Discovery of First Chinese Hominin Skull*, ed. M.-r. Guo, C.-c.

Young, W.-z. Pei, M.-z. Zhou, R.-k. Wu, L.-p. Jia, pp. 19–38. Beijing: Science Press (in Chinese).

Zhou, X.-y. 1988. The Pliocene micomammalian fauna from Jingle, Shanxi—a discussion of the age of Jingle Red Clay. *Vertebrata PalAsiatica* 26(3):181–197 (in Chinese with English summary).

Zhu, R.-x., K. A. Hoffman, R. Potts, C.-l. Deng, Y.-x. Pan, B. Guo, C.-d. Shi, Z.-t. Guo, B.-y. Yuan, Y.-m. Hou, and W.-w. Huang. 2001. Earliest presence of humans in northeast Asia. *Nature* 413:413–417.

Zhu, R.-x., R. Potts, F. Xie, K. A. Hoffman, C.-l. Deng, C.-d. Shi, Y.-x. Pan, H.-q. Wang, R.-p. Shi, Y.-c. Wang, G.-h. Shi, and N.-q. Wu. 2004. New evidence on the earliest human presence at high northern latitudes in northeast Asia. *Nature* 431:559–562.

Chapter 9

Late Cenozoic Biostratigraphy of the Linxia Basin, Northwestern China

TAO DENG, ZHAN-XIANG QIU, BAN-YUE WANG, XIAOMING WANG, AND SU-KUAN HOU

In the 1950s, the Geological Survey Team of Gansu Province found some *Hipparion* fauna fossils in the Neogene red deposits widely distributed in the Linxia Basin, including *Hipparion, Chilotherium, Palaeotragus,* and *Gazella*. This team also found *Equus, Lynx, Ochotona,* and *Myospalax* in Quaternary deposits of this area. Furthermore, local farmers in Hezheng County in the Linxia Basin had found some "dragon bones" during the 1950s. The local government reported to the Institute of Vertebrate Paleontology and Paleoanthropology (IVPP) of the Chinese Academy of Sciences about the farmers' discoveries. In 1962, one of us (Z.-x. Qiu) visited Hezheng and went to the village Yangdujia for further investigation.

In the 1980s, the scientific value of "dragon bones" from the Linxia Basin was finally recognized by vertebrate paleontologists. Since 1987, Qiu and his colleagues have studied and published several new fossil mammals, including *Acerorhinus hezhengensis* (Qiu, Xie, and Yan 1987), *Dinocrocuta gigantea* (Qiu, Xie, and Yan 1988), *Paraentelodon macrognathus* (Qiu, Xie, and Yan 1990), *Agriotherium inexpectans* (Qiu, Xie, and Yan 1991), and *Hipparion dongxiangense* (Qiu and Xie 1998). Contrary to original reports by the geological survey that the Cenozoic deposition started from the Pliocene, the discovery by Qiu et al. demonstrated the presence of not only a rich late Miocene *Hipparion* fauna but also a typical late Oligocene *Dzungariotherium* fauna. Separately, Guan (1988), Cao et al. (1990), and Guan and Zhang (1993) studied the middle Miocene mammalian fauna around Guanghe County in the Linxia Basin, and they reported two fossiliferous layers in the middle Miocene deposits, one that is higher, represented by *Platybelodon, Anchitherium, Hispanotherium,* and *Kubanochoerus*, and a lower one containing small mammals. In the 1990s, a research group of Lanzhou University developed a series of fruitful geological works in the Linxia Basin and also found many mammal fossils. However, these fossils were not associated with exact locality information, making them a mixture of different horizons (Gu et al. 1995a 1995b). Since the end of the 1980s, the Linxia Basin has become well known for its rich Late Cenozoic fossil mammals. Despite this promising start, however, most fossils have not been described or studied in detail, due to the huge numbers of specimens accumulated in a short time.

In 2000, the Hezheng County government invited us to study the fossil collection acquired by them, which was subsequently deposited in a newly established Hezheng Paleozoological Museum. In the Linxia Basin, fossiliferous areas occur throughout almost the whole of Hezheng County, the western half of Guanghe County, the southern half of Dongxiang County, Linxia City, and the eastern border area of Linxia and Jishishan counties, covering about 1300 km² (figure 9.1). The localities in the Linxia Basin (see appendix) are notable for abundant, relatively complete, well-preserved, and sometimes partially articulated bones of large mammals, which often occur in dense concentrations. Many new species of the late Oligocene *Dzungariotherium* fauna, the middle Miocene *Platybelodon* fauna, the late Miocene *Hipparion* fauna, and the early Pleistocene *Equus* fauna have been

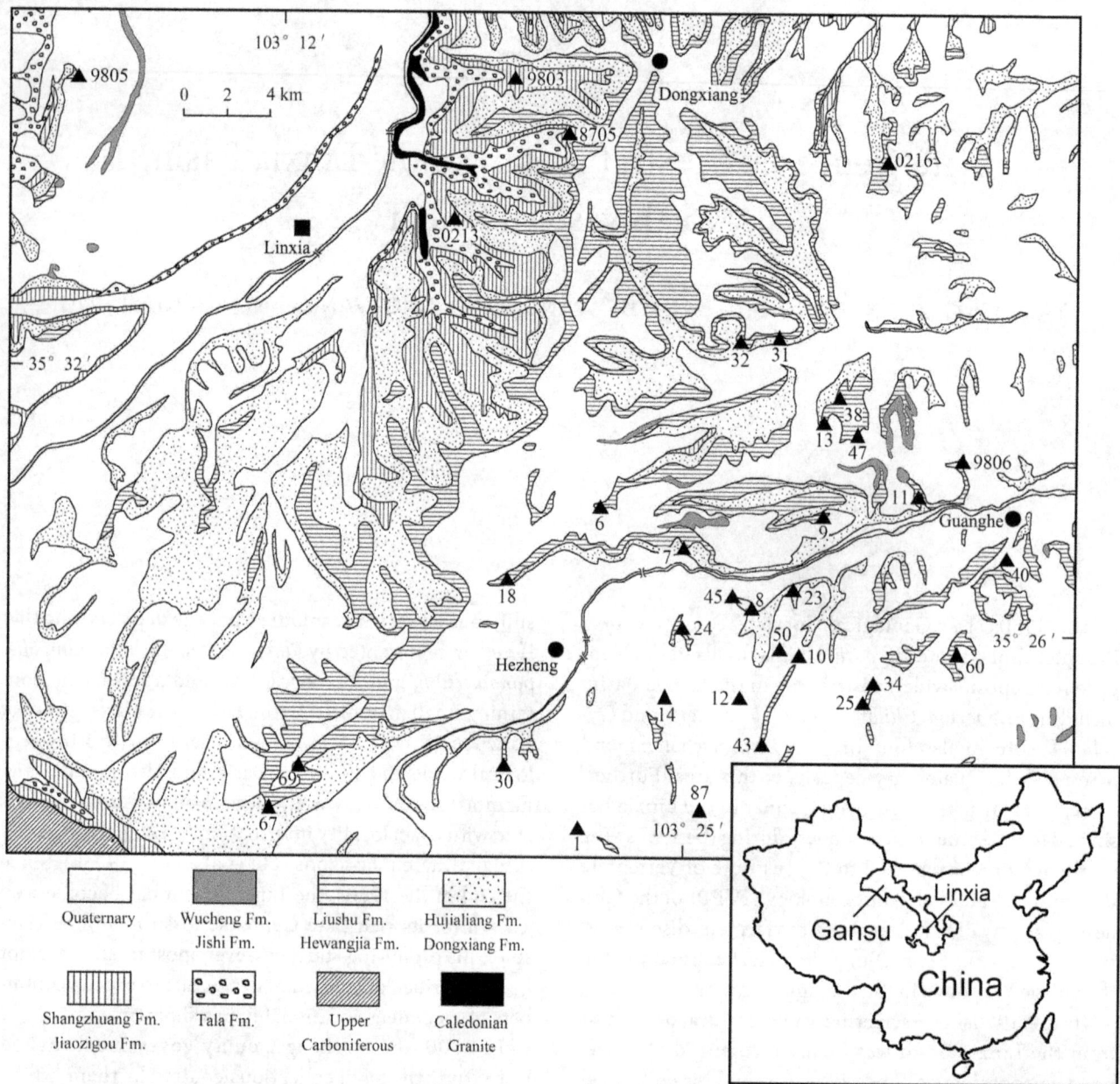

Figure 9.1 Geological map of the Linxia Basin, Gansu Province, China, and representative mammal fossil localities in this basin: GPM 8705 Ji-aozigou; LX 9803 Tala; LX 9805 Yinchuan; LX 9806 Shinanu; LX 0213 Yagou; LX 0216 Wangji; Loc. 1: Dashengou (LX 0011); Loc. 6: Yangjiashan (LX 0004); Loc. 7: Laogou (LX 0003); Loc. 8: Hujialiang (LX 0002); Loc. 9: Houshan (LX 0008); Loc. 10: Songshugou (LX 0030); Loc. 11: Sigou (LX 0007); Loc. 12: Longjiawan (LX 0022); Loc. 13: Shuanggongbei (LX 0009); Loc. 14: Zhongmajia (LX 0049); Loc. 18: Shanjiawan (LX 0025); Loc. 23: Dalanggou (LX 0001); Loc. 24: Shanzhuang (LX 0027); Loc. 25: Nanmiangou (LX 0020); Loc. 30: Heilinding (LX 0035); Loc. 31: Bantu (LX 0043); Loc. 32: Guonigou (LX 0042); Loc. 34: Yangwapuzi (LX 0018); Loc. 38: Longdan (LX 0010); Loc. 40: Shilidun (LX 0014); Loc. 43: Huaigou (LX 0029); Loc. 45: Zhujiachuan (LX 0032); Loc. 47: Shitougou (LX 0201); Loc. 50: Shilei (LX 0031); Loc. 60: Duikang (LX 0701); Loc. 67: Mida (LX 0601); Loc. 69: Wangjiashan (LX 0501); Loc. 87: Hualinqishe (LX 0703). Modified after Deng et al. 2004b.

described from the Linxia Basin since 2000, including rodents, lagomorphs, primates, carnivores, proboscide-ans, perissodactyls, and artiodactyls.

Besides abundant mammal fossils, the Linxia Basin also produces other Late Cenozoic fossils of different an-

imals and plants. Reptiles are represented mainly by a great number of specimens of *Testudo* from the late Mio-cene red clay deposits, which is an important member of the *Hipparion* fauna. The *Testudo* fossils generally are mass burials of complete shells, and a large number of individuals of different ages and sizes are concentrated together. This high concentration seems to indicate mass death and rapid burial to form spectacular turtle shell beds. Some bird fossils are also found from the late Miocene red clays, for example the earliest ostrich fossils in China (Hou et al. 2005; Wang 2007, 2008; Zhang et al. 2010). However, invertebrates are very sparse, and only *Cypridea* was found from the Pliocene sandstone and conglomerate lens. The Late Cenozoic deposits in the Linxia Basin do not produce large plant fossils and contain only rare pollens (Ma, Li, and Fang 1998).

Other geologists have also published many papers on various aspects of Cenozoic geology and paleoclimatology in the Linxia Basin, such as sedimentology (Wang et al. 1998; Wang and Fang 2000; Gong, Zhang, and Huang 2005; Song et al. 2005; Xu et al. 2008), chronology (Chen et al. 1996; Fang et al. 2003), tectonics (Zheng et al. 2006), geochemistry (Zhong et al. 1998; Song et al. 2007), and geophysics (Fang et al. 2007). Some of these studies, however, have misallocated fossil horizons, with consequent errors in interpretation of magnetostratigraphy (Li et al. 1995; Fang et al. 1997; Fang et al. 2003).

Fossiliferous rocks in China, especially northern and northwestern China, as well as Yunnan Province, record important Neogene faunas in different depositional contexts. In a few instances, including the Late Cenozoic strata of the Linxia Basin, deposition is essentially continuous through almost all of the Neogene strata. The Linxia sequence represents the most complete and successive section of the Late Cenozoic in China (figure 9.2), but considerable work remains to be done on geological background and stratigraphical correlations. Because most existing specimens were collected by local people (the "dragon bone" collectors), much basic contextual information is lacking, such as the exact localities and horizons. The majority of fossil collections from the Linxia Basin were acquired from fossil dealers and local collectors, who, in turn, may have purchased specimens from excavators. Most fossils were extracted from elaborate tunnels that can be hundreds of meters into the hills. Most tunnels, however, are horizontal and tend to follow a fossiliferous horizon, and their entrances tend to give an indication of stratigraphic relationships, although the interiors of the tunnels were not always examined by us. To mitigate the loss of contextual information, we engaged the largest fossil dealer in Hezheng area, Yongchang Zhao and his three sons, who have supplied the majority of fossils to the Hezheng Museum over the past decade, and devised a system of "locality numbers" usually referring to specific tunnel entrances. We then systemati-

cally and repeatedly checked their locality numbers on site, in some cases by verifying actual fossils in the spoil piles. The resulting database of localities (see appendix) includes records that are associated with some measure of stratigraphic context, but other questionable records are excluded in this study.

Our present field programs attempt to describe the abundant mammalian fossils from the Linxia Basin, to determine their exact localities and horizons to the extent possible, to update the faunal lists from specific localities, to establish lithological and biostratigraphic criteria for correlation, and to conduct paleomagnetic studies of key sections of the Linxia Basin. Despite the above ambitious goals, significant challenges remain both at the level of original data (uncertainties in purchased specimens passing through the hands of commercial collectors) and at various levels of interpretations (lithostratigraphic and magnetostratigraphic correlations). This reciprocal process of new discoveries and verification of old localities is expected to last for years in the future. This chapter thus represents an interim summary of these efforts.

GEOLOGICAL SETTING

The Linxia Basin is located at the triple junction of the northeastern Tibetan Plateau, the western Qinling Mountains, and the Loess Plateau, delineated by high-angle deep thrusts. The lateral extent of the Linxia Basin is marked by structural boundaries on the northern, western, and southern edges, but its eastern margin is poorly determined. The basin is filled with 700–2000 m of Late Cenozoic deposits, mainly red in color and dominated by lacustrine siltstones and mudstones punctuated by fluvial conglomerates or sandstones and other overbank deposits, and 30–200 m of Quaternary loess sediments. The Daxia and Tao rivers, two tributaries of the Yellow River, intermittently incise the whole Late Cenozoic strata, which provide good access for study (Li et al. 1995). To the west and south, the major basin-bounding faults within the Tibetan Plateau are the Leijishan and North Qinling faults, respectively. The Cenozoic deposits of the Linxia Basin begin in the late Eocene and lap over Cretaceous deposits in the Maxian Mountains to the north. Throughout the central part of the Linxia Basin, the oldest deposits were laid down unconformably on the granite of presumed Paleozoic age. To the southwest of the Linxia Basin, the Tibetan Plateau consists of Devonian-Permian terrestrial and marine deposits and Triassic submarine fan deposits, which were shed by the

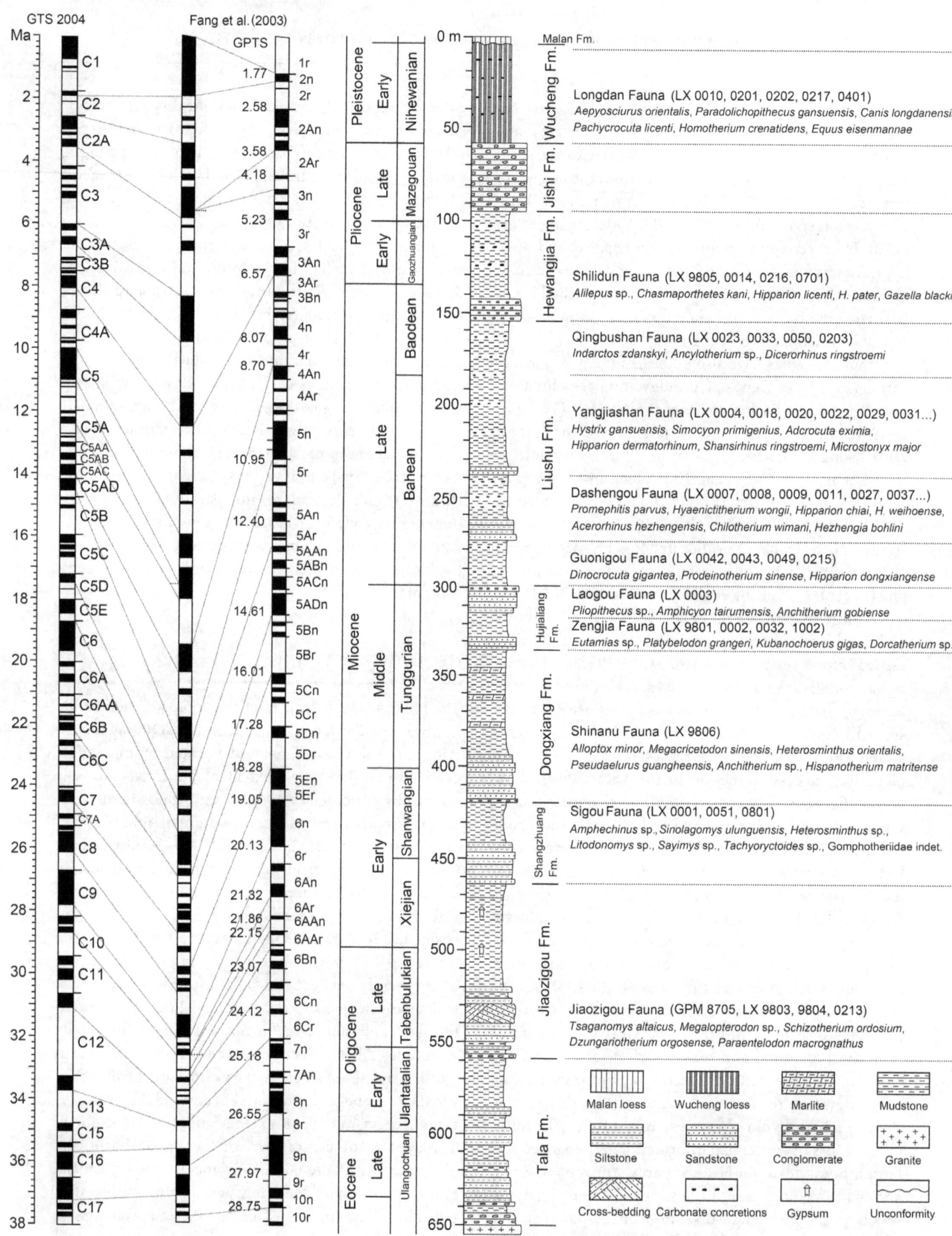

Figure 9.2 Composite stratigraphical section of the Cenozoic deposits of the Linxia Basin with mammalian faunas and new correlation to GTS 2004 (Gradstein, Ogg, and Smith 2004) for the magnetostratigraphy of Fang et al. (2003). Support for our reinterpretations of magnetic correlations is based largely on faunal assessments, although some are based on our own unpublished magnetic sections.

east–southeast striking Qinling mountain belt to the east of the plateau (Fang et al. 2003).

Prior to the 1960s, Cenozoic deposits of the Linxia Basin were simply called the Gansu Series. In 1965, the First Regional Geologic Survey Team of the Gansu Geological Bureau created a new name, Linxia Formation, for the whole Cenozoic sequence and subdivided it into four members (1 to 4), treating the whole sequence as Pliocene based on *Hipparion* fossils found in the uppermost part of the sequence. Qiu, Xie, and Yan (1990) created a new formation, Jiaozigou Formation, for the lower part of the section, corresponding to the members 1 and 2 of the Linxia Formation, thus restricting the Linxia Formation to the members 3 and 4. A giant rhino–entelodon–gomphothere assemblage was found at layer 5 of the newly established Jiaozigou Formation, and it was then considered early Miocene, leaving the Linxia Formation (sensu stricto) as middle to late Miocene. A research group of Lanzhou University further subdivided the Linxia sequence in the 1990s, nominating a series of new lithological units and dating them mainly by paleomagnetic means (Li et al. 1995; Fang et al. 1997). Deng et al. (2004a, 2004b) and Qiu, Wang, and Deng (2004b) revised the lithological sequence and adopted the following units: the Oligocene Tala and Jiaozigou formations; the Miocene Shangzhuang, Dongxiang, Hujialiang, and Liushu formations; the Pliocene Hewangjia and Jishi formations; and the early Pleistocene Wucheng Formation (see figure 9.2). These formations are summarized here, beginning with the oldest.

The Tala Formation consists of brownish-red sandstones and mudstones with some penecontemporaneous and secondary gypsum crystals. The Tala Formation is seen only in deep gullies along the eastern bank of the Daxia River in Dongxiang County, such as Maogou, Jiaozigou, and Yagou. The Jiaozigou Formation consists of brownish conglomeratic coarse sandstones with large cross-bedding and bearing the *Dzungariotherium* fauna in the lower part and brownish-red silty mudstones with secondary gypsum in the upper part. The Jiaozigou Formation has the same distribution as the Tala Formation. Both are exposed in the drainage of the Daxia River, not crossing the Nanyang Mountain, the water divide between the Daxia and Guangtong rivers.

The Shangzhuang Formation consists of yellowish-brown carbonate-cemented medium sandstone and brownish-red silty mudstones. This formation has a more widespread distribution, expanding to the Guangtong River drainage, but the lower sandstones are not exposed in this area. The Dongxiang Formation consists of brownish-red mudstones and siltstones intercalated

with a number of bluish-gray or grayish-white marly beds of 0.5–1 m thickness. The marls are well developed in exposures along the Daxia River drainage, especially in Dongxiang County, but less developed, or even absent, in the Guangtong River drainage. The Hujialiang Formation consists of sandstones and conglomerates with rich fossils of the *Platybelodon* fauna. It is distributed in the eastern part of the Linxia Basin, especially in Hezheng and Guanghe counties. The Liushu Formation consists of light yellowish-brown carbonate-cemented siltstones intercalated with a few thin beds of mudstones and marls, developing substantial mottles and big carbonate nodules and bearing the abundant fossils of the *Hipparion* fauna. Toward the southern Linxia Basin, more conglomerates are intercalated in the Liushu Formation.

The Hewangjia Formation consists of yellowish-brown calcareous mudstones encompassing big carbonate nodules, with basal conglomerates that are thicker in the western part of the Linxia Basin and thinner in the eastern half. The Jishi Formation consists of gray and partially carbonate-cemented coarse conglomerates (up to 20 cm clasts) with massive bedding or roughly imbricated structure and many intercalated lenses of siltstones and mudstones.

The Wucheng Formation consists of yellowish hard loess with layers of calcite concretions and contains the rich *Equus* fauna. It is distributed in the eastern part of the Linxia Basin and correlated to the siltstones of the Dongshan Formation in the weastern part.

REPRESENTATIVE FOSSIL LOCALITIES

The Late Cenozoic faunas of the Linxia Basin are represented by large, diverse samples from a number of fossiliferous horizons. The Linxia faunas come from many localities (see appendix) in the arid, highly seasonal country, and the fossil-rich localities were (and are today) far from being exhausted.

The Linxia Basin deposits are mostly flat-lying and their exposures are controlled by the down cutting of rivers. Since Qiu, Xie, and Yan (1987) reported the first mammalian fossil locality in Dashengou of Hezheng County, more than 100 sites have been found in the Linxia Basin. Deng et al. (2004b:fig. 1) noted 27 sites, including 3 Oligocene localities, 21 Miocene localities, 2 Pliocene localities, and 1 Pleistocene locality. Since then, we have added more localities, such as Qingbushan, Mida, Wangjiashan, Hualinqishe, and Duikang (see figure 9.1), and we present the detailed stratigraphic distributions for these (see figure 9.2). Some representative localities are introduced as follows.

Jiaozigou (Dongyuan, Dongxiang)

Jiaozigou is a ravine, some 10 km southwest of Dongxiang, and is situated in the northwest part of the Linxia Basin. In 1986, J. Y. Xie purchased some mammal fossils from a drugstore in Dongxiang, including teeth of giant rhino, a fragment of proboscidean tusk, and some teeth of a very specialized entelodont. Xie was told that the fossils were found at Jiaozigou. Later, Qiu and Xie explored this locality (Gansu Provincial Museum, Lanzhou, Gansu Province [GPM] 8705, 35°38'03.8"N, 103°19'22.0"E, 2060 m) and nominated the strata bearing giant rhinos and other taxa as the Jiaozigou Formation, with the exposures at Jiaozigou as its type section (Qiu, Xie, and Yan 1990). After several short visits to the fossil locality, they succeeded in collecting, in situ, more materials of giant rhino and entelodont. However, they failed to find any proboscidean fossils.

Qiu, Xie, and Yan (1990) published the Jiaozigou materials, including *Gomphotherium* sp., *Dzungariotherium orgosense*, Rhinocerotidae gen. et sp. indet., and *Paraentelodon macrognathus*. They considered the age to be early Miocene because of the presumed presence of proboscideans. Local farmers are still digging sands for construction at this site, and fossils of the giant rhino fauna are found frequently. After their own repeated failures to find proboscidean remains in the Jiaozigou locality, Qiu, Wang, and Deng (2004b) were inclined to view their earlier reported proboscidean tusk fragment as being mixed in from younger deposits. With the elimination of the proboscidean, the age of the Jiaozigou Fauna was then determined as late Oligocene. According to the present subdivision, the strata in this section (Qiu, Xie, and Yan 1990:fig. 2) from bottom to top include Tala, Jiaozigou, Shangzhuang, Dongxiang, Hujialiang, and Liushu formations, with a total thickness of 350 m. The Tala Formation (layers 1–4 of this section) does not present its basal part in this section, and its brownish-red mudstones interbed with brownish-yellow sandstones and conglomerates. The Jiaozigou Formation includes layer 5 and the lower part of layer 6. The upper part of layer 6 is the Shangzhuang Formation. The Dongxiang Formation includes layers 7–8. The Hujialiang Formation includes layer 9. Layer 10 represents the bottom of the Liushu Formation.

The Jiaozigou fossils are produced from layer 5 in the Jiaozigou Formation. Jiaozigou is the type locality of *Par-*

aentelodon macrognathus. This species is apparently closest to *P. intermedius* from the Benara Fauna (Gabunia 1964). The molars, especially the lower ones, of these two forms are so close that it is difficult to separate them. However, *P. macrognathus* is more progressive than *P. intermedius* (Qiu, Xie, and Yan 1990).

Yagou (Dongyuan, Dongxiang)

Yagou (IVPP vertebrate localities of Linxia Basin [LX] 0213, 35°35'47.9"N, 103°16'39.2"E, 1973 m) is named "Manshan" on the topographic map. The name "Yagou" comes from the local people's oral account. In July 2002, the Hezheng Paleozoological Museum (HPM; Hezheng County, Linxia Hui Autonomous Prefecture, Gansu Province) collected some mammalian fossils, including the same forms as the Jiaozigou Fauna, such as giant rhinos and entelodonts, which purportedly come from Yagou in Dongyuan Township, Dongxiang County. In fact, Yagou and Jiaozigou are situated in the same ravine system, and Maogou, Jiaozigou, and Yagou are closely spaced neighboring ravines. Subsequently, T. Deng and X.-j. Ni went to Yagou to investigate this locality, demonstrating that these fossils came from the first thick sandstones and conglomerates near the base of the section (Qiu, Wang, and Deng 2004b:fig. 1). Our detailed surveys indicate that fossils from Jiaozigou, Yagou, and Magou (Tala) came from the same horizon—that is, the basal sandstones and conglomerates of the Jiaozigou Formation. The exposed strata at Yagou are the Caledonian granite, and the Tala, Jiaozigou, Shangzhuang, and Dongxiang formations from bottom to top. Our visual traverse and marker-bed tracing confirmed that the sections of the three ravines are fully comparable.

Qiu, Wang, and Deng (2004a) reported two species of giant rhinos that jointly appeared at Jiaozigou—that is, *Dzungariotherium orgosense* and a new species *Paracerather-ium yagouense*. However, Qiu and Wang (2007) indicated that the small size and the high-crowned cheek teeth of *P. yagouense* are features most characteristic of their new genus *Turpanotherium*, thus tentatively transferring the species to *Turpanotherium*. Qiu, Wang, and Deng (2004b) also described other forms from Yagou, including *Tsaganomys altaicus*, *Megalopterodon* sp., *Schizotherium ordosium*, Hyracodontidae gen. et sp. indet., *Ardynia* sp., *A. altidentata*, *Aprotodon lanzhouensis*, and *Paraentelodon* cf. *macrognathus*. Taken as a whole, all 10 forms from Yagou show clearly that the fauna is late Oligocene in age (table 9.1; figure 9.3).

Table 9.1

Mammals from the Jiaozigou Formation (Late Oligocene) of the Linxia Basin

Rodentia	Hyracodontidae	Rhinocerotidae
Tsaganomyidae	Hyracodontidae gen. et sp. indet.	*Ronzotherium* sp.
Tsaganomys altaicus	*Ardynia altidentata*	*Aprotodon lanzhouensis*
Creodonta	*Ardynia* sp.	Artiodactyla
Hyaenodontidae	*Allacerops* sp.	Entenodontidae
Megalopterodon sp.	Paraceratheriidae	*Paraentelodon macrognathus*
Perissodactyla	*Dzungariotherium orgosense*	
Chalicotheriidae	*Paraceratherium yagouense*	
Schizotherium ordosium		

Dalanggou (Maijiaxiang, Guanghe)

Dalanggou (LX 0001; 35°26'50.7"N, 103°27'28.5"E; 2060 m) is located in a gully on the south bank of the Guangtong River, where the Shangzhuang, Dongxiang, Hujialiang, and Liushu formations were exposed and covered by the late Pleistocene Malan Loess. Fossils from this locality were collected from the conglomerate lens in the red mudstones at the upper part of the Shangzhuang Formation, mainly including several skulls of *Choerolophodon guangheensis* and other gomphotheres (Wang and Deng 2011). Deng (2006b) described an isolated lower tooth of *Aprotodon* sp. A mandible of *Turcocerus* has also been prepared (table 9.2; see figure 9.3). The size and characters of the rhinocerotid lower cheek teeth from Dalanggou are close to those of *Aprotodon lanzhouensis* from the adjacent Lanzhou Basin (Qiu and Xie 1997). In the Lanzhou Basin, *A. lanzhouensis* was found in the early Miocene white sandstone of the Xianshuihe Formation. *A. lanzhouensis* is also found from the late Oligocene yellow sandstones of the Jiaozigou Formation in the Linxia Basin, but these fossils are markedly smaller than those from Lanzhou and Dalanggou, and the size difference was considered to reflect sexual dimorphism (Qiu, Wang, and Deng 2004b).

Shinanu (Shinanu, Guanghe)

The strata at Shinanu (or Wangshijie in Deng et al. 2004b; LX 9806; 35°29'51.1"N, 103°32'59.8"E; 2263 m) include the Shangzhuang, Dongxiang, Hujialiang, Liushu, and Jishi formations. Cao et al. (1990) reported fossils discovered from Shinanu, consisting of at least 16 species, belonging to six orders. These fossils were collected from two layers: the upper (fossil-bearing bed II)

from the Hujialiang Formation and the lower (fossil-bearing bed I) from the basal sandstones of the Dongxiang Formation. Compared to other Neogene mammal faunas, especially on the basis of *Megacricetodon* and other micromammals, the fossils in the lower layers show close affinities to the early Astaracian elements of Europe, and they are much closer to the mammals found at Danshuilu in the Xining Basin, Qinghai, an early Tunggurian mammalian locality. Therefore, an early Tunggurian age is assumed for the lower layer. Mammals in the upper layers demonstrate apparently advanced characteristics, and they include representatives of late Tunggurian age. The mammalian fossils from the lower layers are Talpidae indet., *Alloptox minor*, *A. chinghaiensis*, *Atlantoxerus* sp., *Sayimys* cf. *obliquidens*, *Megacricetodon sinensis*, *Heterosminthus orientalis*, *Pseudaelurus guangheensis*, *Dorcatherium* sp., and *Moschus* sp., and those from upper layers include *Alloptox* sp., *A. guangheensis*, and *Pseudaelurus* sp. (table 9.3; see figure 9.3).

Deng et al. (2004b) also discovered fossils of the *Anchitherium* fauna from the basal sandstones of the Dongxiang Formation at Shinanu (Wangshijie), including *Hemicyon* sp., *Gomphotherium* sp., *Anchitherium* sp., *Hispanotherium matritense*, *Alicornops* sp., and *Chalicotherium* sp. The sizes of *Anchitherium* and *Hispanotherium* from Shinanu are obviously smaller than those from the Tunggur Formation in Inner Mongolia, consistent with early Tunggurian age.

Hujialiang (Maijiaxiang, Guanghe)

Hujialiang (LX 0002; 35°26'23.1"N, 103°26'37.6"E; 2150 m) is the stratotype of the Hujialiang Formation and bears abundant *Platybelodon* fossils, including many complete skulls and mandibles from juveniles to old

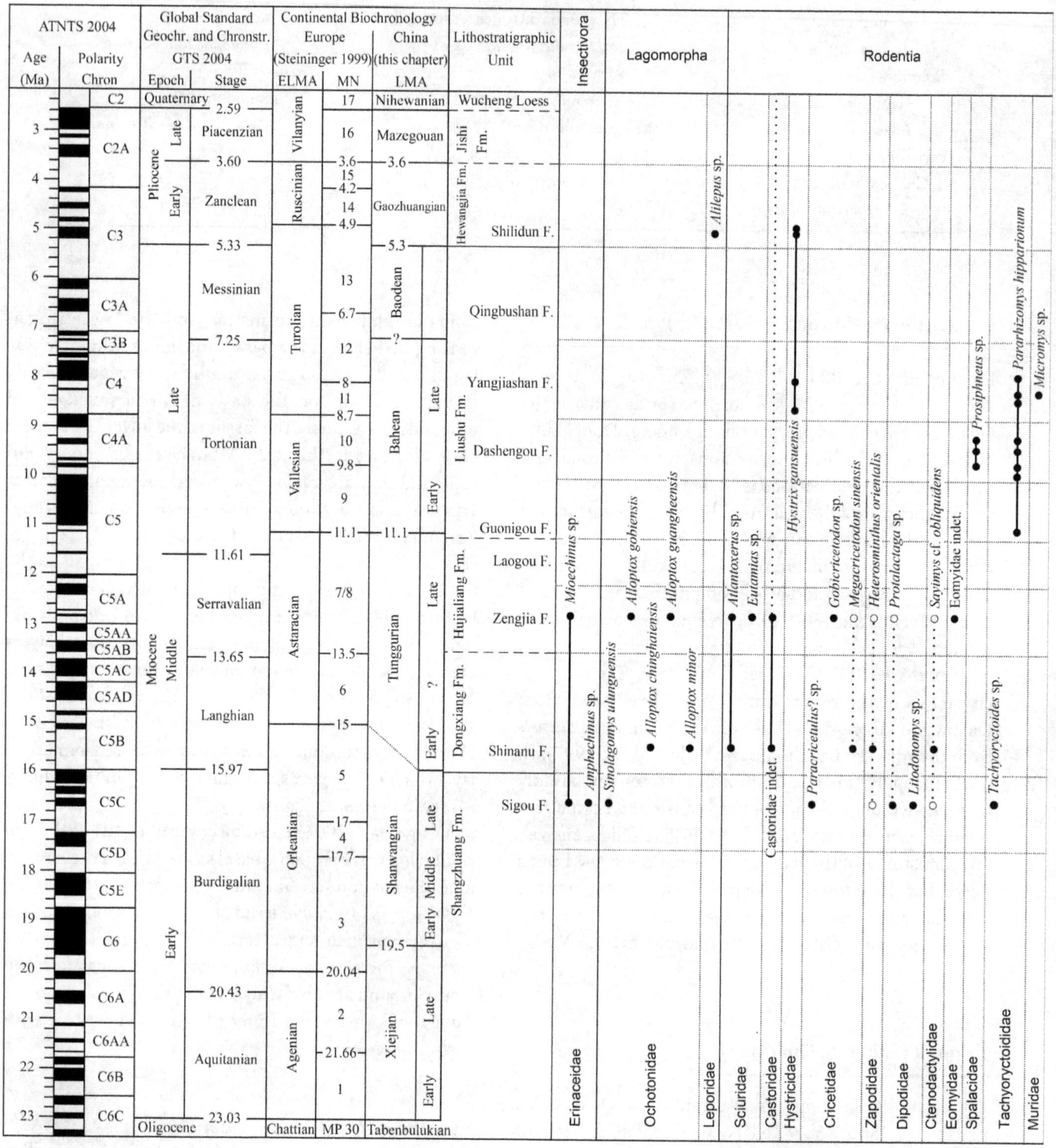

ATNTS 2004
Age (Ma)
Polarity Chron
Global Standard Geochr. and Chronstr. GTS 2004
Epoch
Stage
Continental Biochronology
Europe (Steininger 1999)
ELMA
MN
China (this chapter)
LMA
Lithostratigraphic Unit
Insectivora
Lagomorpha
Rodentia
C2
C2A
C3
C3A
C3B
C4
C4A
C5
C5A
C5AA
C5AB
C5AC
C5AD
C5B
C5C
C5D
C5E
C6
C6A
C6AA
C6B
C6C
Quaternary
Pliocene
Late
Early
Miocene
Middle
Oligocene
2.59
Piacenzian
3.60
Zanclean
5.33
Messinian
7.25
Tortonian
11.61
Serravalian
13.65
Langhian
15.97
Burdigalian
20.43
Aquitanian
23.03
Villanyian
Ruscinian
Turolian
Vallesian
Astaracian
Orleanian
Agenian
Chattian
MP 30
Nihewanian
Mazegouan
Gaozhuangian
Baodean
Bahean
Tunggurian
Shanwangian
Xiejian
Tabenbulukian
Wucheng Loess
Jishi Fm.
Hewangjia Fm.
Shilidun F.
Liushu Fm.
Qingbushan F.
Yangjiashan F.
Dashengou F.
Guonigou F.
Hujialiang Fm.
Laogou F.
Zengjia F.
Dongxiang Fm.
Shinanu F.
Shangzhuang Fm.
Sigou F.
Mioechinus sp.
Amphechinus sp.
Sinolagomys ulunguensis
Alloptox gobiensis
Alloptox chinghaiensis
Alloptox gianghensis
Alloptox minor
Alilepus sp.
Atlantoxerus sp.
Eutamias sp.
Castoridae indet.
Hystrix gansuensis
Paracricetulus? sp.
Gobicricetodon sp.
Megacricetodon sinensis
Heterosminthus orientalis
Protalactaga sp.
Litodonomys sp.
Sayimys cf. obliquidens
Eomyidae indet.
Tachyoryctoides sp.
Prosiphneus sp.
Parahizomys hipparionum
Micromys sp.
Erinaceidae
Ochotonidae
Leporidae
Sciuridae
Castoridae
Hystricidae
Cricetidae
Zapodidae
Dipodidae
Ctenodactylidae
Eomyidae
Spalacidae
Tachyoryctoididae
Muridae

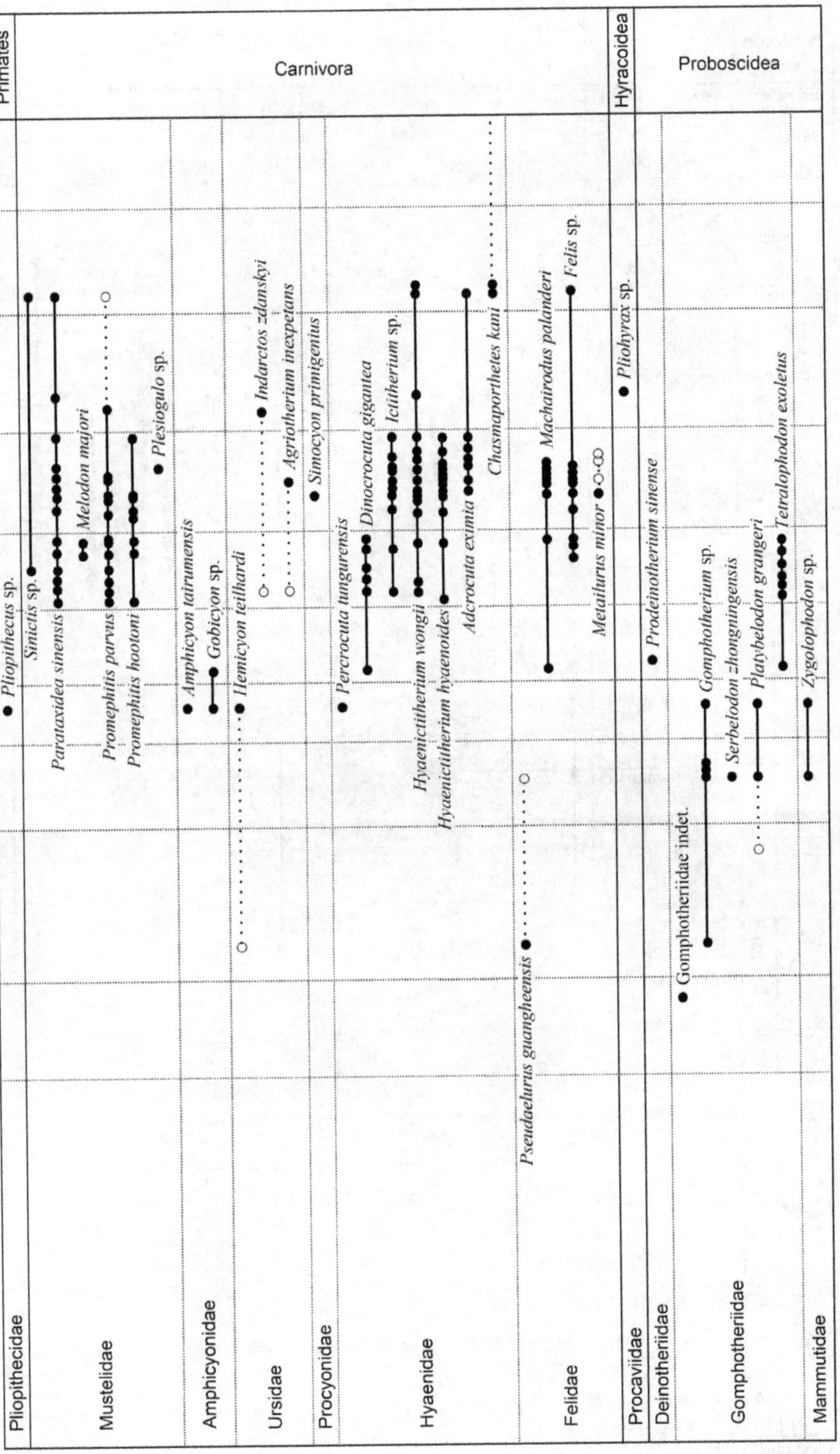

Figure 9.3 Biostratigraphical range of Neogene mammalian fossils from the Linxia Basin: (*A*) Insectivora, Lagomorpha, Rodentia, Primates, Canivora, and Proboscidea; (*B*) Perissodactyla and Artiodactyla. Solid circles represent the labeled taxa, and open circles represent uncertain records of the labeled species. Solid lines connect the same taxa, and dotted lines connect different species of the same genus.

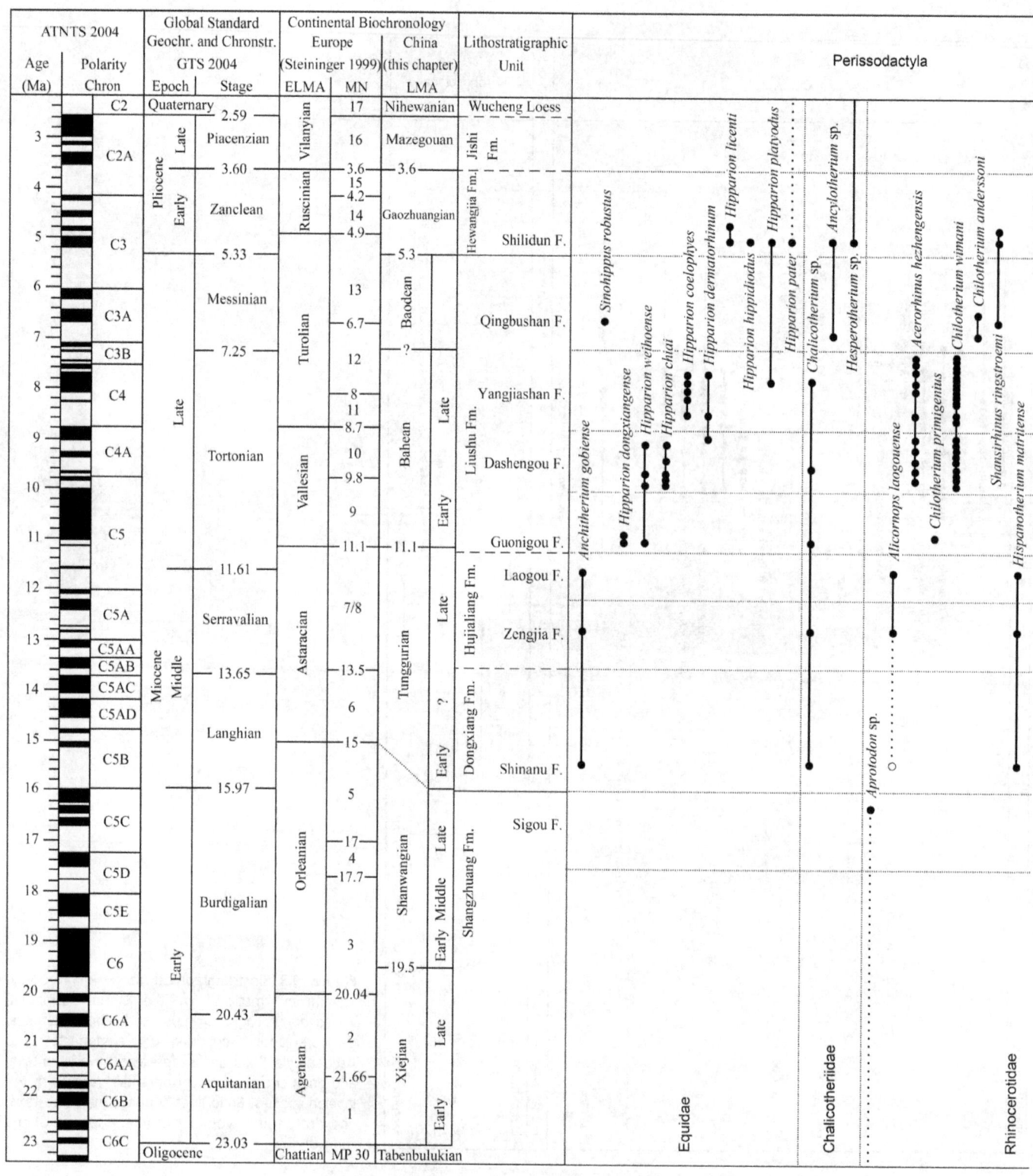

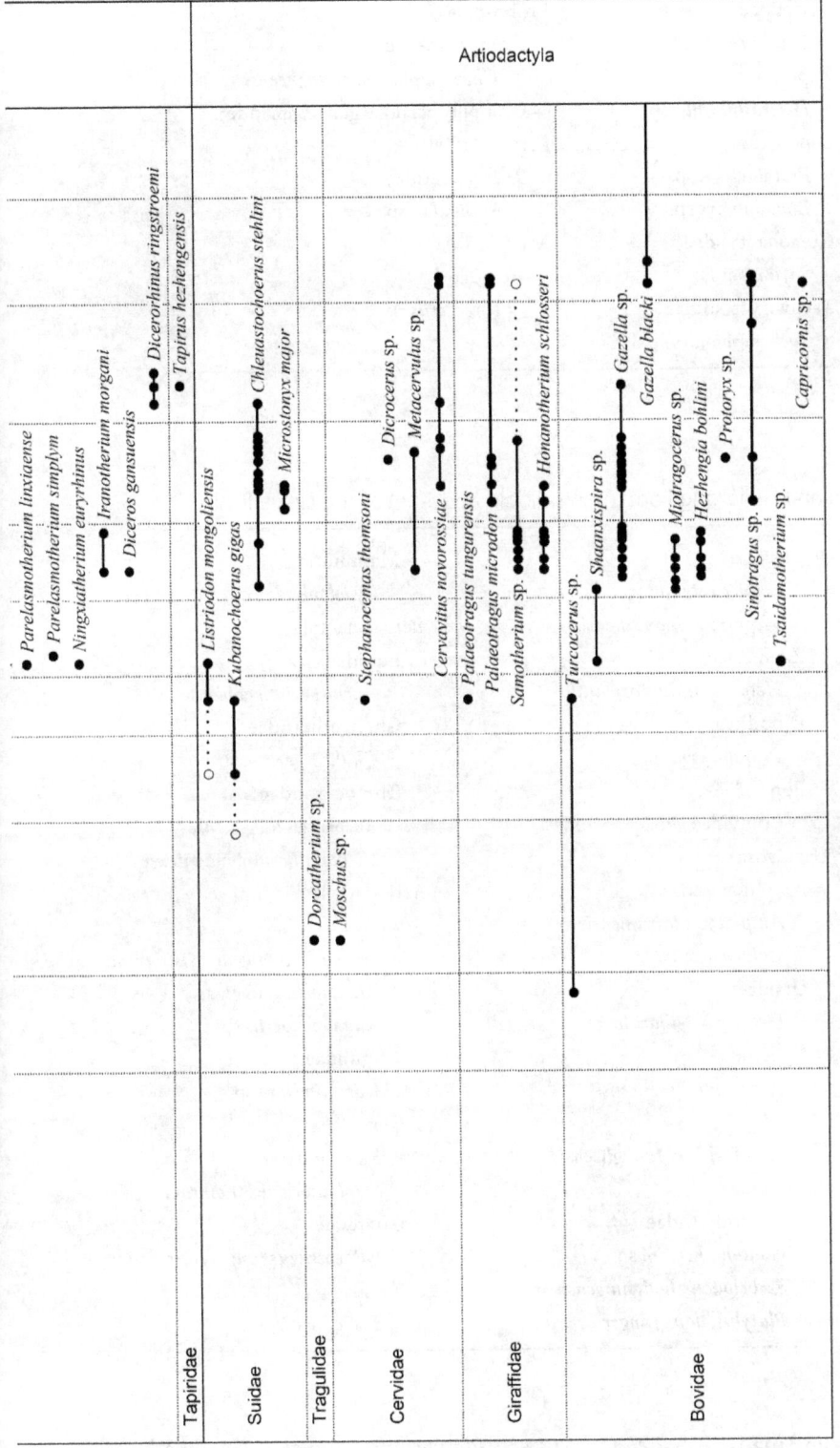

Figure 9.3 (continued)

Table 9.2

Mammals from the Shangzhuang Formation (Early Miocene) of the Linxia Basin

Insectivora	Cricetidae	Proboscidea
Erinaceidae	*Paracricetulus*? sp.	Gomphotheriidae
Mioechinus sp.	Zapodidae	*Choerolophodon guangheensis*
Amphechinus sp.	*Heterosminthus* sp.	Gomphotheriidae gen. et sp. indet.
Lagomorpha	Dipodidae	Perissodactyla
Ochotonidae	*Protalactaga* sp.	Rhinocerotidae
Sinolagomys ulunguensis	*Litodonomys* sp.	*Aprotodon* sp.
Rodentia	Ctenodactylidae	Artiodactyla
Sciuridae	*Sayimys* sp.	Bovidae
Atlantoxerus sp.	Tachyoryctoididae	*Turcocerus* sp.
	Tachyoryctoides sp.	

Table 9.3

Mammals from the Dongxiang Formation and Hujialiang Formation (Middle Miocene) of the Linxia Basin

Insectivora	Cricetidae	Mammutidae
Erinaceidae	*Gobicricetodon* sp.	*Zygolophodon* sp.
Mioechinus sp.	*Megacricetodon sinensis*	Perissodactyla
Talpidae	Zapodidae	Equidae
Talpidae gen. et sp. indet.	*Heterosminthus orientalis*	*Anchitherium gobiense*
Primates	Dipodidae	Chalicotheriidae
Pliopithecidae	*Protalactaga*? sp.	*Chalicotherium* sp.
Pliopithecus sp.	Eomyidae	Rhinocerotidae
Lagomorpha	Eomyidae gen. et sp. indet.	*Alicornops laogouense*
Ochotonidae	Carnivora	*Hispanotherium matritense*
Alloptox gobiensis	Amphicyonidae	Artiodactyla
Alloptox chinghaiensis	*Amphicyon tairumensis*	Suidae
Alloptox guangheensis	*Gobicyon* sp.	*Listriodon mongoliensis*
Alloptox minor	Ursidae	*Kubanochoerus gigas*
Ochotonidae gen. et sp. indet.	*Hemicyon teilhardi*	*Kubanochoerus* sp.
Rodentia	Hyaenidae	Tragulidae
Sciuridae	*Percrocuta tungurensis*	*Dorcatherium* sp.
Eutamias sp.	Felidae	Cervidae
Atlantoxerus sp.	*Pseudaelurus guangheensis*	*Moschus* sp.
Castoridae	Proboscidea	*Stephanocemas thomsoni*
Castoridae gen. et sp. indet.	Gomphotheriidae	Giraffidae
Ctenodactylidae	*Gomphotherium* sp.	*Palaeotragus tungurensis*
Sayimys cf. *obliquidens*	*Serbelodon zhongningensis*	Bovidae
Sayimys sp.	*Platybelodon granger*	*Turcocerus* sp.

adults. Guan (1988) reported a middle Miocene mammalian fauna from Maijiaxiang, which may have been collected from the Hujialiang locality according to our interviews with the local farmers. The Hujialiang assemblage includes Castoridae gen. et sp. indet., *Alloptox* sp., *Pseudaelurus* sp., *Platybelodon grangeri*, *Gomphotherium* sp., *Serbelodon zhongningensis*, *Chalicotherium* sp., *Anchitherium gobiense*, *Hispanotherium matritense*, *Listriodon* sp., *Kubanochoerus* sp., and Cervidae indet. (see table 9.3 and figure 9.3).

Laogou (Sanhe, Hezheng)

Laogou is a small gully on the north bank of Guangtong River in Hezheng County (see figure 9.1) where Neogene deposits are well exposed. This locality (LX 0003; 35°28'05.3"N, 103°24'50.5"E; 2200 m) is 7.5 km northeast of the county seat, near the village Yangdujia. The strata at Laogou include the Shangzhuang, Dongxiang, Hujialiang, Liushu and Jishi formations and the capping Quaternary deposits. The Jishi and Quaternary conglomerates form three terraces at Laogou (Deng et al. 2004b:fig. 3). Local farmers have dug sands for construction for many years, and there are several deep tunnels over 300 m long. This locality produced the most abundant middle Miocene fossils in the Linxia Basin from fine conglomerates and sandstones of the Hujialiang Formation. Lithological sequences at Laogou were described by Deng (2003) and Deng et al. (2004b:fig.3). The Hujialiang Formation consists of grayish-yellow gravelly sandstone, which is generally unconsolidated, although occasionally it forms hard sheets. The sands are mostly medium grained and contain many fine gravels. The thickness of the Hujialiang Formation is 6 m at Laogou. The overlying Liushu Formation is light yellowish-brown carbonate-cemented siltstones (also referred to as red clays), and the underlying Dongxiang Formation is alternating thin-bedded light brown and light orange mudstones with several grayish-white marly bands.

Deng (2003) described isolated teeth of *Hispanotherium matritense* from Laogou, and Deng (2004a) established a new acerathere species, *Alicornops laogouense*, based on specimens from Laogou. Judging by the numbers of specimens recovered, proboscideans were dominant and rhinocerotids were secondary in the Laogou fauna. The known fossils include *Pliopithecus* sp., *Amphicyon tairumensis*, *Gobicyon* sp., *Hemicyon teilhardi*, *Percrocuta tungurensis*, *Gomphotherium* sp., *Platybelodon grangeri*, *Anchitherium gobiense*, *Alicornops laogouense*, *Hispanotherium matritense*, *Listriodon mongoliensis*, *Kubanochoerus gigas*, and *Palaeotragus tungurensis* (see table 9.3 and figure 9.3). *H. matritense* from Laogou is more primitive than *H. tungurense* of the Tunggur Formation in Inner Mongolia. *A. laogouense* is much larger than that of *A. simorrense* from Simorre and Villefranche d'Astarac (Deng 2004a).

Wangji (Wangji, Dongxiang)

Wangji (LX 0215; 35°37'24.5"N, 103°30'30.1"E; 2350 m) is situated in a small gully, Maqigou, near Wangji Township. J.-y. Xie purchased some isolated teeth from a Chinese drugstore in Dongxiang in 1988. The fossils were reportedly unearthed from Wangji. The teeth belong to only two forms: an elasmothere rhino, smaller and more primitive than the well-known *Sinotherium*, and a new *Hipparion* of very small size. Qiu and Xie (1998) reported four upper cheek teeth of a new species, *Hipparion dongxiangense*, and three isolated upper molars of *Parelasmotherium simplum* from Wangji (table 9.4), but their stratigraphic position could not be precisely established. Deng et al. (2004b) found that the Wangji fossiliferous bed is located at the base of the very thick red clays of the Liushu Formation, 5–10 m above the underlying sandstones and conglomerates of the Hujialiang Formation.

Although the *Hipparion* material from Wangji is too poor to draw any conclusion as to true affinity, it is safe to

Table 9.4

Mammals from the Lower Part of the Liushu Formation (Early Late Miocene) of the Linxia Basin

Rodentia	Proboscidea	Rhinocerotidae
Tachyoryctoididae	Deinotheriidae	*Chilotherium primigenius*
Pararhizomys hipparionum	*Prodeinotherium sinense*	*Parelasmotherium simply*
Pararhizomyinae gen. et. sp. nov.	Gomphotheriidae	*Parelasmotherium linxiaense*
Carnivora	*Tetralophodon exoletus*	*Ningxiatherium euryrhinus*
Amphicyonidae	Perissodactyla	Artiodactyla
Gobicyon sp.	Equidae	Suidae
Hyaenidae	*Hipparion dongxiangense*	*Listriodon mongoliensis*
Dinocrocuta gigantea	*Hipparion weihoense*	Bovidae
Felidae	Chalicotheriidae	*Shaanxispira* sp.
Machairodus palanderi	*Chalicotherium* sp.	*Tsaidamotherium* sp.

say that these small teeth are readily distinguishable from all the hipparionine species so far known from China and adjacent areas. The size of *H. dongxiangense* is smaller than that of *H. parvum*, the smallest known species of *Hipparion* in China, and its characteristic structures of the hypocone and the hypocone groove frequently occur among middle Miocene hipparionines from North America but are infrequent among later hipparionines. As a result, *H. dongxiangense* seems to indicate a relatively early age among known Chinese hipparionines (Qiu and Xie 1998).

Guonigou (Nalesi, Dongxiang)

Guonigou (LX 0042; 35°33'01.1"N, 103°26'15.6"E; 2234 m) is midway along the highway between Dongxiang and Sanjiaji, Guanghe. The exposed strata at Guonigou include the Dongxiang, Hujialiang, Liushu, and Jishi formations, capped by thick late Pleistocene loess. In the brownish-red mudstones of the Dongxiang Formation, several characteristic grayish-white marly bands are very distinct. Rocks of the Hujialiang Formation are gray and yellow gravelly sandstone, from which fossils of the *Platybelodon* fauna are produced. The Liushu Formation consists of blocky and muddy siltstones, and the fossiliferous bed is near its base, with rich fossils of the *Hipparion* fauna. An unconformity exists between the Liushu Formation and the Jishi Conglomerate.

Deng (2001c) established a new elasmothere, *Parelasmotherium linxiaense*, based on isolated teeth from Guonigou. The accompanying mammalian fossils include *Dinocrocuta gigantea*, *Machairodus* sp., *Tetralophodon* sp., *Hipparion dongxiangense*, and *Shaanxispira* sp. (see table 9.4 and figure 9.3). Judging from the slightly more advanced characters of *P. simplum* from Wangji compared with those in *P. linxiaense*, the Guonigou fossil bed may be somewhat older than the Wangji locality. Deng (2007) described a complete skull of *P. linxiaense* from Guonigou. Deng (2008) then described a new, relatively large elasmothere, *Ningxiatherium euryrhinus*, based on a complete skull from Guonigou.

Bantu (Nalesi, Dongxiang)

Bantu (LX 0043; 35°33'08.4"N, 103°27'40.7"E; 2228 m) is very close to Guonigou, separated by only 2.5 km. The strata at Bantu include the Shangzhuang, Dongxiang, Hujialiang, Liushu, Jishi, and Wucheng formations. Qiu et al. (2007) established a new deinothere, *Prodeinother-*

ium sinense, on the basis of a mandible collected from Bantu, in a horizon that is identical to the fossiliferous bed of the Liushu Formation at Guonigou. The deinothere site is one of a few grayish-yellow coarse sandstone lenses, with particularly rich turtle shells, about 100 m up a south-facing slope. These lenses are located in the very basal part of the Liushu Formation. Unconformably underlying the lenses is a layer of conglomerates, a widespread unit highly characteristic of the top part of the middle Miocene Hujialiang Formation. The unusual combination of both primitive and advanced characters in *P. sinense* may suggest a separate central Asian lineage split from the main deinothere stem in the early Miocene. If so, *P. sinense* may represent a late segment of that lineage.

Zhongmajia (Dalang, Hezheng)

Zhongmajia (LX 0049; 35°24'47.0"N, 103°24'30.6"E; 2218 m) is a gully 6 km southeast of the Hezheng County seat, where an almost complete skull with its articulated broken mandible of a small and primitive species of *Chilotherium* was found. Deng (2006a) described it as a new species, *Chilotherium primigenius*, the earliest known member of this genus. The fossiliferous bed at Zhongmajia is also within a grayish-green sandstone and conglomerate lens, very similar to the Bantu *Prodeinotherium sinense* site, and near the base of the Liushu Formation at both localities. At the moment, Zhongmajia is the locality with the oldest unambiguous remains of the genus *Chilotherium*. At Zhongmajia, a bed bearing *C. wimani* remains is above the *C. primigenius*–containing bed. Due to the presence in China of these most primitive species of *Chilotherium*, *C. primigenius*, and *C. wimani*, an East Asian origin for this genus is suggested.

Dashengou (Xinzhuang, Hezheng)

Dashengou (LX 0011; 35°21'23.9"N, 103°21'22.4"E; 2276 m) is situated near the southern border of the Linxia Basin. Mostly covered by vegetation, only patches of exposures can be observed. The basal conglomerate of the Hewangjia Formation is exposed close to the hilltop, and the fossiliferous bed is within the red clays under a calcareous concretion of hard cement of the Liushu Formation. Dashengou is the type locality of the Dashengou Fauna, and its mammalian fossils are very rich, including a new genus of pararhizomyine, *Promephitis parvus*, *Dinocrocuta gigantea*, *Tetralophodon exoletus*, *Hipparion chiai, Acerorhinus hezhengensis, Chilotherium wimani, Samotherium* sp., *Honanotherium schlosseri, Gazella* sp., *Hezhengia bohlini,* and *Miotragocerus* sp. (see table 9.5 and figure 9.3).

Qiu, Xie, and Yan (1987) described a new acerathere, *Acerorhinus hezhengensis,* based on an associated skull and mandible from Dashengou. In comparison with *A. fuguensis* from Fugu, Shaanxi (Deng 2000), *A. hezhengensis* seems to be advanced. In 1983, J. Y. Xie discovered a well-preserved hyaenid skull from Dashengou, and Qiu, Xie, and Yan (1988) described it as *Dinocrocuta gigantea. D. gigantea* is a representative of various *Hipparion* faunas in Asia and Europe, and it is known from the Bahean and Baodean ages in Asia and the Vallesian and Turolian of Europe (Howell and Petter 1985; Werdelin 1996). Fossils of *Dinocrocuta* were found from the Bahe Formation in Lantian, Shaanxi (Liu, Li, and Zhai 1978; Zhang et al. 2002; Zhang 2005), and Bulong in Biru, Tibet (Zheng 1980), and these also came from strata corresponding to Vallesian age (Li, Wu, and Qiu 1984; Qiu and Qiu 1995). However, the Bahe *Dinocrocuta,* the best magnetically calibrated record in China so far known, is restricted to a short time span of 8.07–7.84 Ma in the top part of the Bahe Formation (Zhang 2005; Zhang et al., chapter 6, this volume), consistent with the Laogaochuan record in Fugu, Shaanxi (Xue, Zhang, and Yue 1995; Zhang and Xue 1996).

Shanzhuang (Maijiaxiang, Guanghe)

Shanzhuang (LX 0027; 35°25'57.6"N, 103°25'20.4"E; 2245 m) is a small village in the middle of a mountain south of the Guangtong River. Fang et al.'s (1997:table 1, 2003:table 1) "Shanzhuang Fauna" was supposed to have come from the sandstones of the lower part of the Shangzhuang Formation and to contain *Hipparion, Platybelodon,* and other mammalian fossils, but its composition and horizon are incorrect. Fang and colleague's "Shanzhuang Fauna" (Fang et al. 1997; Fang et al. 2003) is apparently a mixture of middle and late Miocene elements. Our field observations show that the majority of the late Miocene components were actually collected from the red clays in the middle part of the Liushu Formation, instead of the underlying sandstones of the Hujialiang Formation or mudstones of the Shangzhuang Formation.

In fact, the fossiliferous bed at Shanzhuang is within the red clays, 80 m above the Hujialiang Formation. We also found many fossil mammals from Shanzhuang, including *Promephitis parvus, Melodon majori, Hyaenictitherium wongii, H. hyaenoides, Dinocrocuta gigantea, Machairodus palanderi, Felis* sp., *Tetralophodon* sp., *Acerorhinus hezhengensis, Chilotherium wimani, Iranotherium morgani, Honanotherium schlosseri, Samotherium* sp., *Gazella* sp., and *Hezhengia bohlini* (see table 9.5 and figure 9.3).

Table 9.5
Mammals from the Middle Part of the Liushu Formation (Middle Late Miocene) of the Linxia Basin

Rodentia	*Hyaenictitherium wongii*	*Chilotherium wimani*
Spalacidae	*Hyaenictitherium hyaenoides*	*Iranotherium morgani*
Prosiphneus sp.	*Dinocrocuta gigantea*	*Diceros gansuensis*
Tachyoryctoididae	Felidae	Artiodactyla
Pararhizomys hipparionum	*Machairodus palanderi*	Suidae
Pararhizomyinae gen. et sp. nov.	*Felis* sp.	*Chleuastochoerus stehlini*
Carnivora	Proboscidea	Cervidae
Ursidae	Gomphotheriidae	*Metacervulus* sp.
Agriotherium sp.	*Tetralophodon exoletus*	Giraffidae
Indarctos sp.	Perissodactyla	*Samotherium* sp.
Mustelidae	Equidae	*Honanotherium schlosseri*
Sinictis sp.	*Hipparion chiai*	Bovidae
Parataxidea sinensis	*Hipparion weihoense*	*Shaanxispira* sp.
Melodon majori	*Hipparion dermatorhinum*	*Gazella* sp.
Promephitis parvus	Chalicotheriidae	*Miotragocerus* sp.
Promephitis hootoni	*Chalicotherium* sp.	*Hezhengia bohlini*
Hyaenidae	Rhinocerotidae	Bovinae gen. et sp. nov.
Ictitherium sp.	*Acerorhinus hezhengensis*	

Previously, *Iranotherium morgani* was discovered only from Maragheh and Kerjavol in Iran (Antoine 2002). Bernor et al. (1996) suggested that *I. morgani* might have first appeared in MN 11. Because the fauna that bears *I. morgani* from Linxia can be correlated to the late Vallesian (late MN 9 and MN 10), *I. morgani* appeared in Linxia earlier than in Maragheh. *I. morgani* is likely to have first appeared in northwestern China and then dispersed westward to western Asia (Deng 2005b).

Sigou (Alimatu, Guanghe)

Sigou (or Gucheng [Wang and Qiu 2004]; LX 0007; 35°29'41.8"N, 103°30'06.1"E; 2140 m) is a gully on the north bank of the Guangtong River. Fang et al.'s (1997:table 1, 2003:table 1) "Sigou Fauna" was supposed to have come from the sandstones of the Dongxiang Formation containing fossils of the *Hipparion* fauna, but its horizon was incorrect, as in the case of the "Shanzhuang Fauna." Actually, the Sigou fossils were collected from the red clays of the lower part of the Liushu Formation. At Sigou, the sandstones and conglomerates of the Hujialiang Formation underlie the Liushu Formation, and below them are the brownish-red mudstones of the Dongxiang and Shangzhuang formations. HPM collections from the Liushu Formation at Sigou include *Pararhizomys hipparionum*, *Promephitis parvus*, *Ictitherium* sp., *Hyaenictitherium wongii*, *Dinocrocuta gigantea*, *Tetralophodon* sp., *Hipparion* sp., *Acerorhinus hezhengensis*, *Chilotherium wimani*, *Chleuastochoerus stehlini*, and *Miotragocerus* sp. (see table 9.5 and figure 9.3). Micromammals collected previously by the IVPP from the Shangzhuang Formation at Sigou (LX 0051) include *Mioechinus* sp., *Amphechinus* sp., *Sinolagomys ulunguensis*, *Atlantoxerus* sp., *Paracricetulus?* sp., *Heterosminthus* sp., *Protalactaga* sp., *Litodonomys* sp., *Sayimys* sp., and *Tachyoryctoides* sp., and they belong to the early Miocene Shanwangian age (see table 9.2).

Promephitis parvus is a skunk species described by Wang and Qiu (2004), and it is very common from many localities in the Linxia Basin, including Sigou. This species occurs in the red clays of the Liushu Formation, not the Dongxiang Formation as reported by Wang and Qiu (2004). In dental measurements, *P. parvus* is rather close to the European *P. pristinidens*, a late Vallesian species. In overall dental proportions and size, *P. parvus* is also like *P. majori*, the smallest *Promephitis* in the Turolian of Europe. *P. parvus* is the most primitive *Promephitis* in China, only slightly more derived than *P. pristinidens*.

Houshan (Alimatu, Guanghe)

Houshan (LX 0008; 35°29'07.6"N, 103°28'28.0"E; 2227 m) is located northwest of Guanghe County seat, and its strata include the Dongxiang, Hujialiang, and Liushu formations. In the middle part of the Liushu Formation, many mammalian fossils were found, such as *Promephitis parvus*, *Parataxidea sinensis*, *Hyaenictitherium wongii*, *Dinocrocuta gigantea*, *Tetralophodon exoletus*, *Hipparion chiai*, *H. weihoense*, *Acerorhinus hezhengensis*, *Chilotherium wimani*, *Iranotherium morgani*, *Diceros gansuensis*, *Cervavitus novorossiae*, *Metacervulus* sp., *Samotherium* sp., *Honanotherium schlosseri*, *Hezhengia bohlini*, *Gazella* sp., and *Miotragocerus* sp. (see table 9.5 and figure 9.3).

Among the abundant late Miocene ungulate fossils in the Linxia Basin, many skulls are referable to *Hezhengia bohlini*, the second most common taxon in terms of numbers of specimens, next only to *Chilotherium wimani*. *H. bohlini* is a medium-size ovibovine, one of the most characteristic forms of the *Hipparion* fauna in the Linxia Basin (Qiu, Wang, and Xie 2000), and it is the most abundant at Houshan. Deng and Qiu (2007) established a new dicerotine species, *Diceros gansuensis*, based on several skulls and mandibles from Houshan, and it is the first fossil species of the *Diceros* lineage ever discovered in East Asia.

Yangwapuzi (Zhuangheji, Guanghe)

Yangwapuzi (LX 0018; 35°24'49.5"N, 103°30'28.6"E; 2160 m) is located southwest of the Guanghe County seat, where only the Liushu Formation is partially exposed. Liu, Kostopoulos, and Fortelius (2004) described an almost complete skull of *Microstonyx major* from the Liushu Formation at Yangwapuzi. The new material provides the first relatively complete suid specimen other than *Chleuastochoerus* from the Chinese late Miocene. The skulls and mandible from Linxia are very similar to those of *Microstonyx* from Pikermi and Kalimanci (Bulgaria), except for the slightly smaller size, approaching even more closely the material from Maragheh in Iran. Other mammal fossils from Yangwapuzi include *Promephitis hootoni*, *Hyaenictitherium wongii*, *Adcrocuta eximia*, *Tetralophodon* sp., *Hipparion coelophyes*, *Chilotherium wimani*, *Chleuastochoerus stehlini*, *Palaeotragus microdon*, and *Gazella* sp. (table 9.6; see figure 9.3).

A pelvic skeleton, recognized as a new species and an early representative of the ostrich (Hou et al. 2005),

Table 9.6

Mammals from the Upper Part of the Liushu Formation (Late Late Miocene) of the Linxia Basin

Rodentia	*Hyaenictitherium hyaenoides*	*Shansirhinus ringstroemi*
Tachyoryctoididae	*Adcrocuta eximia*	*Chilotherium wimani*
Pararhizomys hipparionum	Felidae	*Chilotherium anderssoni*
Pararhizomyinae gen. et sp. nov.	*Metailurus minor*	*Dicerorhinus ringstroemi*
Muridae	*Metailurus* sp.	Artiodactyla
Micromys sp.	*Machairodus palanderi*	Suidae
Hystricidae	*Felis* sp.	*Chleuastochoerus stehlini*
Hystrix gansuensis	Hyracoidea	*Microstonyx major*
Carnivora	Procaviidae	Cervidae
Procyonidae	*Pliohyrax* sp.	*Metacervulus* sp.
Simocyon primigenius	Perissodactyla	*Dicrocerus* sp.
Ursidae	Equidae	*Cervavitus novorossiae*
Agriotherium inexpectans	*Sinohippus robustus*	Giraffidae
Indarctos zdanskyi	*Hipparion coelophyes*	*Palaeotragus microdon*
Mustelidae	*Hipparion dermatorhinum*	*Honanotherium schlosseri*
Plesiogulo sp.	*Hipparion platyodus*	Bovidae
Parataxidea sinensis	Chalicotheriidae	*Protoryx* sp.
Promephitis parvus	*Ancylotherium* sp.	*Gazella* sp.
Promephitis hootoni	Tapiridae	*Miotragocerus* sp.
Hyaenidae	*Tapirus hezhengensis*	*Sinotragus* sp.
Ictitherium sp.	Rhinocerotidae	
Hyaenictitherium wongii	*Acerorhinus hezhengensis*	

was collected from the red clays of the Liushu Formation at Yangwapuzi. *Struthio linxiaensis* represents one of the few relatively well-preserved ostrich skeletal fossils, pushing back the history of the ostrich in China by several million years. The discovery of this ostrich from the Linxia Basin indicates that by the late Miocene ostriches were widely distributed in Eurasia.

Wangjiashan (Maijiaji, Hezheng)

Wangjiashan (LX 0501; 35°22'34.3"N, 103°12'48.9"E; 2575 m) is situated near the south margin of the Linxia Basin. Here, the vegetation is very dense, with rare exposures of the deeply brownish red clays of the Liushu Formation. A tapir and an anchithere were found from Wangjiashan. Hou et al. (2007) established a new species for the large anchithere, *Sinohippus robustus*. *Sinohippus* is observably larger than *Anchitherium aurelianense*. The tooth row reduction is also greater in *Sinohippus* than in *A. aurelianense*. *Sinohippus robustus* is larger than *Sinohippus zitteli*, and its teeth appear more robust. The fossil tapirs from the

Neogene deposits are relatively scarce, mainly consisting of isolated teeth. The discovery of *Tapirus hezhengensis* in the Linxia Basin fills a gap in the evolution of Tapiridae in East Asia, preceeding the uncertain origin of the Quaternary tapirs in China (Deng 2008). *T. hezhengensis* is the earliest and most primitive species of *Tapirus* in China, and it is the closest to *T. telleri* from Göriach, Austria (Hofmann 1893), in shape and size.

Yangjiashan (Sanhe, Hezheng)

Yangjiashan (LX 0004; 35°28'37.9"N, 103°21'25.9"E; 2340 m) is 6 km north of the Hezheng County seat, where the Hewangjia Formation is completely absent and the fossiliferous bed is situated in the upper part of the Liushu Formation. The known taxa from Yangjiashan include a new genus of pararhizomyine, *Simocyon primigenius*, *Promephitis hootoni*, *Ictitherium* sp., *Hyaenictitherium wongii*, *H. hyaenoides*, *Machairodus palanderi*, *Metailurus minor*, *Hipparion* sp., *Chilotherium wimani*, and *Cervavitus novorossiae* (see table 9.6 and figure 9.3).

Promephitis hootoni is found from many localities in the upper part of the Liushu Formation, including Yangjiashan (Wang and Qiu 2004). Most measurements of the type of the early Turolian *P. hootoni* from Turkey fall in the ranges of morphological variations of the large Chinese samples. Referral of the Chinese materials to *P. hootoni* implies a great geographic range for this species across much of Asia, as for modern skunks with wide distributions spanning much of the North American continent.

Yinchuan (Yinchuan, Jishishan)

Yinchuan (LX 9805; 35°39'23.0"N, 103°04'57.8"E; 2063 m) is located in a remote village near the western border of the Linxia Basin. Gu et al. (1995a) reported the "Longguang Fauna" from Yinchuan, but the stratigraphy of their fossils is very confused. In table 1 of Fang et al. (1997), this fauna comes from the brownish red mudstones of the upper part of the Dongxiang Formation, but in figure 6 of the same paper, it comes from the Liushu Formation, and probably neither is correct. By our own field observations, the "Longguang Fauna" was probably collected from the red clays of the early Pliocene Hewangjia Formation, instead of the Dongxiang or Liushu formations.

The brown-red clay of the Liushu Formation underlies the Yinchuan section and has a thickness of 60 m. The Hewangjia Formation consists primarily of 30 m of light-red clay above a 5 m basal conglomerate of poorly

rounded and sorted gravel. The Jishi Conglomerate, which overlies the Hewangjia Formation, is 10 m in thickness, well rounded, and poorly sorted, with largest gravel diameter of 30 cm. The Wucheng Loess overlies the Jishi Formation, and this brownish-yellow loess is banded by brownish-red paleosol. In 1998, many new well-preserved mammal fossils were collected from Yinchuan. A complete rhinocerotid skull and articulated mandible came from this locality, along with skulls of *Hipparion licenti* and *Gazella blacki*. The rhinocerotid material is identified as *Shansirhinus ringstroemi* (Deng 2005a).

Shilidun (Chengguan, Guanghe)

The strata at Shilidun (LX 0014; 35°27'38.8"N, 103°34'40.0"E; 2130 m) are well exposed, including the Hujialiang, Liushu, Hewangjia, and Jishi formations. The fossiliferous layer is situated within the red clays 3 m above the basal conglomerate of the Hewangjia Formation, containing *Hystrix gansuensis*, *Alilepus* sp., *Promephitis* sp., *Chasmaporthetes* sp., *Hyaenictitherium wongii*, *Cervavitus novorossiae*, *Palaeotragus* sp., *Samotherium* sp., *Protoryx* sp., *Capricornis* sp., and *Sinotragus* sp. (table 9.7; see figure 9.3). *Alilepus* is known to range from late Miocene to early Pliocene in China. Wang and Qiu (2002) established a new porcupine, *Hystrix gansuensis*, based on materials from Shilidun and other late Miocene localities in the Linxia Basin. This new porcupine not only

Table 9.7

Mammals from the Hewangjia Formation (Early Pliocene) of the Linxia Basin

Lagomorpha	Felidae	Rhinocerotidae
Leporidae	*Felis* sp.	*Shansirhinus ringstroemi*
Alilepus sp.	Proboscidea	Artiodactyla
Rodentia	Gomphotheriidae	Cervidae
Hystricidae	Gomphotheriidae gen. et sp. indet.	*Cervavitus novorossiae*
Hystrix gansuensis	Perissodactyla	Giraffidae
Carnivora	Equidae	*Palaeotragus microdon*
Mustelidae	*Hipparion licenti*	*Samotherium?* sp.
Sinictis sp.	*Hipparion hippidiodus*	Bovidae
Parataxidea sinensis	*Hipparion platyodus*	*Gazella blacki*
Promephitis sp.	*Hipparion pater*	*Sinotragus* sp.
Hyaenidae	Chalicotheriidae	*Capricornis* sp.
Hyaenictitherium wongii	*Hesperotherium* sp.	
Adcrocuta eximia	*Ancylotherium* sp.	
Chasmaporthetes kani		

widened the distribution of Neogene *Hystrix* in China but also shed new light on the evolutionary history of the fossil porcupines.

Longdan (Nalesi, Dongxiang)

Longdan (LX 0010; 35°31'32.2"N, 103°28'58.3"E; 2248 m) is a small village in Nalesi Township in the south of Dongxiang County. The major fossiliferous levels are those of layers 5 and 9 in the Longdan loess section (Qiu, Deng, and Wang 2004:fig. 2). The lithology of both layers is about the same: yellowish hard loess with layers of calcite concretions. Both are ~5 m in thickness. A preliminary study revealed that the Longdan Fauna was about 2 Ma in age and embedded in loess deposits. In view of the importance of this discovery, a preliminary report was soon published (Qiu et al. 2002). Deng (2002) described a primitive woolly rhino, *Coelodonta nihowanensis* from Longdan. Wang and Qiu (2003) established the monotypic Aepyosciurinae based on *Aepyosciurus orientalis* from Longdan. Qiu, Deng, and Wang (2004) monographed 300 specimens representing 31 mammalian taxa collected mainly

from Longdan, but also from other localities, such as Shitougou, Keshijian, and Xijia. Most of the collection has already been fully prepared: it includes 165 specimens, among which more than 100 are well-preserved skulls, maxillaries, and mandibles. Wang (2005) reported the beaver *Castor anderssoni* from Longdan, and Qiu, Deng, and Wang (2009) reported the bear *Protarctos yinanensis* from Longdan as well. As the research continues, additional new taxa of the Longdan Fauna will be described, such as the hyrax, pig, and takin.

Of the 33 known species of the Longdan fauna, there are 6 small mammals, 2 primates, 17 carnivores, 4 perissodactyls, and 4 artiodactyls, including two new genera and 12 new species (table 9.8; see figure 9.3). The composition of the Longdan Fauna seems rather unique in several respects:

1. The number of the small mammals is unexpectedly low.
2. The number of carnivore species is high, and specimens of carnivores are particularly numerous, the carnivores outnumbering even the total number of the herbivores.
3. The artiodactyls, especially cervids, usually abundant in early Pleistocene faunas in Eurasia, are poorly repre-

Table 9.8

Mammals from the Wucheng Formation (Early Pleistocene) of the Linxia Basin

Primates	*Canis teilhardi*	Elephantidae gen. et sp. indet.
Cercopithecidae	*Canis longdanensis*	Hyracoidea
Macaca cf. *anderssoni*	*Canis brevicephalus*	Procaviidae
Paradolichopithecus gansuensis	*Sinicuon* cf. *dubius*	*Postschizotherium* sp.
Lagomorpha	Ursidae	Perissodactyla
Leporidae	*Protarctos yinanensis*	Equidae
Serricolagus brachypus	Mustelidae	*Hipparion sinense*
Rodentia	*Eirictis robusta*	*Equus eisenmannae*
Sciuridae	*Meles teilhardi*	Chalicotheriidae
Aepyosciurus orientalis	Hyaenidae	*Hesperotherium* sp.
Marmota parva	*Chasmaporthetes progressus*	Rhinocerotidae
Castoridae	*Pachycrocuta licenti*	*Coelodonta nihowanensis*
Castor anderssoni	*Crocuta honanensis*	Artiodactyla
Spalacidae	Felidae	Suidae
Myospalax sp.	*Homotherium crenatidens*	*Sus* sp.
Arvicolidae	*Megantereon nihowanensis*	Cervidae
Mimomys cf. *gansunicus*	*Sivapanthera linxiaensis*	*Nipponicervus longdanensis*
Cricetidae	*Panthera palaeosinensis*	Bovidae
Bahomys sp.	*Felis teilhardi*	*Gazella* cf. *blacki*
Carnivora	*Lynx shansius*	*Budorcas* sp.
Canidae	Proboscidea	*Leptobos brevicornis*
Vulpes chikushanensis	Elephantidae	*Hemibos gracilis*

sented in the Longdan Fauna in both numbers of species and specimens.

4. Raccoon-dogs and camels, common elements in early Pleistocene faunas in Eurasia, are totally absent from the Longdan Fauna.

Except for possible collecting bias, the differences in composition between the Longdan and the Nihewan faunas may reflect differences in geologic age and paleoenvironment.

Shitougou (Nalesi, Dongxiang)

Shitougou (LX 0201; 35°31'11.3"N, 103°29'19.7"E; 2228 m) is situated 600 m south of Longdan. The Wucheng Loess is covered by the late Pleistocene Malan Loess, and the underlying Jishi Conglomerate forms a vertical cliff. The lithostratigraphy and bedding sequence are nearly identical to those of Longdan, with only minor differences, especially in more layers of calcite concretions. The fossiliferous levels in the Shitougou section are stratigraphically comparable with those of the Longdan section. Qiu, Deng, and Wang (2004) described a new cheetah, *Sivapanthera linxiaensis*, based on specimens from Shitougou and Longdan. *S. linxiaensis* is a large cheetah, and it is clearly different from the European *S. pardinensis*.

LITHOSTRATIGRAPHIC CORRELATIONS

As outlined, there are many mammalian fossil localities in the Linxia Basin distributed over a large area. Stratigraphic relationships for some sections and localities have been studied (Li et al. 1995; Fang et al. 1997, 2003; Deng et al. 2004a, 2004b), but most need more detailed correlations. Li et al. (1995:fig. 2.3) lithologically correlated six sections in the Linxia Basin, including the Maogou section, where the Tala, Shangzhuang, and Liushu formations were named, the Wangjiashan section, where the Hewangjia Formation was named, and the Dongshanding, Dayuanding, Beiyuan, and Saleshan sections. Fang et al. (1997, 2003) further correlated these sections based on their paleomagnetic measurements and quoting mammalian fossils from other localities not closely tied to their sections. The Maogou and Wangjiashan sections include major Late Cenozoic deposits, and the other four sections include only Quaternary loess and conglomerates. The Maogou section, 443 m thick, is located near the center of the Linxia Basin, and the strata are mainly flat-lying lacus-

trine sediments. The Wangjiashan section, over 1600 m thick, lies in the western part of the Linxia Basin, and its strata are mainly strongly folded lacustrine sediments (Li et al. 1995). However, these two main sections have relatively few strata with mammalian fossils. In order to include mammalian fossil markers for paleomagnetic interpretations, some faunas have been projected into these sections based on estimated horizons, which has caused errors in paleomagnetic dating.

Linxia sediments are fluvial, lacustrine, and eolian, and they include deposits typical of such environments. The mixture of channel sands and finer overbank deposits fluctuated through time in response to regional tectonic and perhaps environmental factors. These fluctuations are used to define the formational boundaries. At least 50 stratigraphic sections, with a cumulative thickness of over 10,000 m, have been measured by us, and more than 30 additional shorter sections have contributed to the development of a comprehensive lithostratigraphic framework. Some of these sections have been tied to samples for paleomagnetic analysis and used to build a comprehensive chronostratigraphy for the Linxia Basin. Most fossils occur in concentrations that are in or adjacent to small-scale channels and crevasse splays on the floodplains, while occurrences in paleosols are rare.

Deng et al. (2004b) correlated 27 mammalian fossil localities of the most fossiliferous areas in the Linxia Basin and partitioned them into seven faunas—the late Oligocene Jiaozigou Fauna; the middle Miocene Laogou Fauna; the late Miocene Guonigou, Dashengou, and Yangjiashan faunas; the early Pliocene Shilidun Fauna; and the early Pleistocene Longdan Fauna. Almost every locality is situated in a different section, so their precise correlations are very difficult. The mammalian fossils of the Jiaozigou, Laogou, Shilidun, and Longdan faunas derive from relatively thin beds, and they must represent faunas of short duration. The late Miocene faunas of the Linxia Basin are collected from the red clays of the Liushu Formation. The thickness of the Liushu Formation is variable in different sections, reflecting different depositional settings. In the center of the Linxia Basin, the red clay deposits of the Liushu Formation are generally about 90 m. Based on position in the sections, the various localities of the Liushu Formation can be divided into three distinct horizons—lower, middle, and upper—which yielded the Guonigou, Dashengou, and Yangjiashan faunas, respectively (Deng et al. 2004b).

Guan (1988:fig. 7) measured eight sections from Guanghe and Hezheng counties, including Maijiaxiang, Shinanu, and Bajia. He conceptualized a Linxia Formation for the late Miocene and Pliocene strata bearing the

Hipparion fauna and applied the middle Miocene Xianshuihe Formation of the Lanzhou Basin for the strata bearing the *Platybelodon* fauna. Cao et al. (1990) reported two layers of middle Miocene mammalian fossils from Shinanu, calling the upper layers Xianshuihe Formation and the lower layers Chetougou Formation, which is characteristic of the Xining Basin in Qinghai Province.

Qiu, Xie, and Yan (1990) described the Jiaozigou section (4 km southeast of the Maogou section) and regarded the section as typical for the Jiaozigou Formation. They recognized 10 layers in this section, making the lower layers 1–6 as the Jiaozigou Formation. Mainly based on the fossils described by Qiu, Xie, and Yan (1990) and the sedimentary cycles, Li et al. (1995) subdivided the lower part of the Maogou section into the Tala and Zhongzhuang formations, which correspond to layers 1–6 of the Jiaozigou section in Qiu, Xie, and Yan (1990). Thus, for the same fossiliferous beds, different formation names have been proposed. It is evident that Jiaozigou of Qiu, Xie, and Yan (1990) has priority over Zhongzhuang of Li et al. (1995) and should be adopted in the future. However, the Jiaozigou Formation should be restricted to the upper part of its original meaning (layer 5–6 in Qiu et al's section). The lower part (layers 1–4 in Qiu et al.'s section) is distinct and recognized as the Tala Formation (Qiu, Wang, and Deng 2004b); its age is late Eocene to early Oligocene.

Xie (1991) reported the Galijia section in Dongxiang and found middle Miocene mammalian fossils in the middle part of this section, which corresponds to the layers 7–9 of the Jiaozigou section. Xie named the fossiliferous layers the Dongxiang Bed. Li et al. (1995) promoted the Dongxiang Bed to the Dongxiang Formation. Qiu, Wang, and Deng (2004b) described the Yagou section and subdivided the section as the Tala to the Dongxiang formations.

Qiu, Xie, and Yan (1991) described the Jiegou section in Pingzhuang of Dongxiang and established a new bear species, *Agriotherium inexpectans*, from this section. The fossil-bearing deposits, about 50 m of red clay with calcareous nodules, have been correlated with the fourth member of the Linxia Formation by local geologists. Based on lithology and the *Hipparion* fauna fossils they found during the 1970s, the local geologists always assigned this unit to the Baodean Stage. Li et al. (1995) named the Liushu Formation at the Maogou section, which corresponds to fourth member of the Linxia Formation.

Deng et al. (2004b) described and correlated more sections in the Linxia Basin, including Laogou, Guonigou, Wangji, Dashengou, Yangjiashan, Heilinding, Shil-

idun, Yinchuan, Shanzhuang, and Sigou. Except Laogou, these sections produce fossils of the *Hipparion* fauna. At Laogou and several other localities, we found the lithology of the beds bearing middle Miocene *Platybelodon* fauna to be completely different from the underlying Dongxiang and overlying Liushu formations. As a result, a new lithological unit, the Laogou Formation, was established for these beds. Later, the Laogou Formation was renamed as the Hujialiang Formation because of a conflict with a previously established Jurassic formation (Deng 2004b). The Hujialiang Formation is a complex of fluviolacustrine deposits dominated by sandstone and conglomerates.

The *Hipparion* fauna occurs in red clay deposits. Elsewhere on the Chinese Loess Plateau, sedimentological and geochemical studies suggest that the Neogene red clays are eolian in origin, as is the overlying Pleistocene loess (Ding et al. 1998; Guo et al. 2001). One of the most striking features of the red clay sequence is the existence of many carbonate nodule horizons. The thickness of these horizons ranges from 10 cm to >100 cm, with most of the nodules being <10 cm in diameter. The cement matrix within most of the nodule horizons is a reddish, weathered soil material.

Qiu et al. (2002) published a preliminary report about the early Pleistocene Longdan Fauna in Nalesi, Dongxiang, and described the lithological sequence at this locality. The section can be divided into upper and lower parts. The lower part consists of Neogene strata producing *Platybelodon* and *Hipparion* faunas; the upper part is loess deposits. Below the loess is a thick layer of conglomerate (Jishi Formation), which is likely in unconformable contact with beds both above and below. Fang et al. (2003) considered that the Longdan section cannot be easily correlated with the Wangjiashan and Maogou sections based on lithology, and they therefore only provided information on the variation in stratigraphic thickness and grain size across the Linxia Basin. However, we recognized the loess beds in the Longdan section as the Wucheng Loess, and the Longdan section corresponds to only the upper part of the Wangjiashan and Maogou sections. Qiu, Deng, and Wang (2004) described the Wucheng Loess at Longdan in detail and indicated that it is an eolian sequence, different from the contemporaneous lacustrine Dongshan Formation at the Wangjiashan section.

The particularly bluish-gray or grayish-white marly beds of the Dongxiang Formation are very distinct markers for stratigraphical correlation in the central part of the Linxia Basin, but their number decreases and their color becomes grayish green in the eastern part where the fossil localities are very numerous. On the other hand, the conglomerates of the Hujialiang Formation are very stable in the eastern part, so they are also good markers for correlations.

MAGNETOSTRATIGRAPHY

An extensive set of paleomagnetic analyses has provided good correlation within and between the various Linxia sections and with the geomagnetic polarity time scale. We currently have a reasonable understanding of local- and regional-scale depositional environments and the ways in which they change through time.

Li et al. (1995) and Fang et al. (1997, 2003) made magnetostratigraphic studies for the Linxia sequence, including the Maogou, Wangjiashan, and Dongshanding sections. However, we disagree with their age assignments of mammalian faunas in Fang et al. (2003:table 1). For example, mammal layer 7 should be late Oligocene, and mammal layer 6 (Shanzhuang Fauna) should be early late Miocene. There are also some mistakes in the positions of mammalian faunas in Fang et al. (2003:fig. 3). For example, their Sigou and Shanzhuang faunas most likely were collected from the lower part of the Liushu Formation. We reinterpreted the magnetostratigraphy of Fang et al. (2003). According to our correlation (see figure 9.2), N14 in their Wangjiashan section and N8 in their Maogou section should correspond to C5n.2n. Recent new paleomagnetic analysis supports these revisions (X.-m. Fang, pers. comm.).

All mammal fossils of the *Dzungariotherium* fauna from the Linxia Basin show clearly that they belong to the late Oligocene (Deng et al. 2004a, 2004b; Qiu, Wang, and Deng 2004a, 2004b; Qiu and Wang 2007). Fang et al. (2003) gave a magnetic age of ~21 Ma for this fauna. Because this fauna can be correlated to the Chinese Tabenbulukian mammal age, its reversed paleomagnetic polarity best matches chron C8r, with an age near 27 Ma (see figure 9.2).

The Tunggurian fossils of the Linxia Basin are typical middle Miocene forms, and they are represented by three faunas. According to our reinterpretation for the magnetostratigraphy of Fang et al. (2003:fig. 3), the Shinanu Fauna is within C5Br, with an age of 15.5 Ma; the Zengjia Fauna is within C5Ar.1r, with an age of 12.6 Ma; and the Laogou Fauna is within C5An.2n, with an age of 12.3 Ma. On the other hand, sandstones and conglomerates bearing the Tunggurian fossils have rapid or varied sedimentary rates, and their paleomagnetic ages are difficult to determine.

The fossils of the *Hipparion* faunas of the Linxia Basin can be divided into five distinct horizons according to their positions within the red clays of the Liushu and Hewangjia formations, typified by faunas from the Guonigou, Dashengou, Yangjiashan, Qingbushan, and Shilidun localities and their lateral equivalents. The first three horizons belong to the Bahean Age, the Qingbushan horizon belongs to the Baodean Age, and Shilidun is Gaozhuangian Age. According to our paleomagnetic measurements and correlations, the Guonigou Fauna is at the base of chron C5n.2n, with an age of 11 Ma; the Dashengou Fauna is within C4A, with an age of about 9 Ma; the Yangjiashan Fauna is within C4n.2n, with an age of about 8 Ma; the Qingbushan Fauna is within C3An.2n, with an age of 6.5 Ma; and the Shilidun Fauna is within C3n.4n, with an age of 5.0 Ma (see figure 9.2).

The presence of true horse and the absence of typical middle Pleistocene forms place the *Equus* fauna of the Linxia Basin in the early Pleistocene. The paleomagnetic section also shows a long duration of time. The Matuyama/Gauss boundary is determined to be 0.5 m below the lower fossiliferous bed of the *Equus* fauna in the Linxia Basin, whereas the lower boundary of the Reunion subchron is just above the upper fossiliferous bed. This would place the fossil unit in the 2.55–2.16 Ma interval (Qiu, Deng, and Wang 2004).

BIOCHRONOLOGY

Due to their relatively high densities in the fossil record, the mammal fossils of the Linxia Basin are well suited for fine-scale regional stratigraphic correlations, despite the fact that most of the fossils were collected by private collectors. When considered in aggregate, based on a combination of lithologic correlations and fossil characterizations, these fossils provide reasonable evidence to divide and correlate the Late Cenozoic strata of the Linxia Basin. However, understanding of the stratigraphical sequence and age has been controversial, with conflicting lithologic names and confused fossil evidence. In our fieldwork in recent years, we reevaluated the sedimentary sequence and redetermined the corresponding geologic age of each lithologic unit on the basis of expanded evidence from mammalian fossils.

Tabenbulukian Fauna

The Jiaozigou Fauna comes from the sandstones of the Jiaozigou Formation at Jiaozigou, Yagou, and Tala. The giant rhino is a representative mammal in Asia and infrequently discovered in Eastern Europe (see table 9.1 and figure 9.3). Giant rhinos diversified in the middle Oligocene and became very advanced in the late Oligocene. *Dzungariotherium orgosense* was first found from the Junggar Basin in Xinjiang, with a large size, rudimentary lower incisors, well-developed antecrochets, and wide foot bones. In Xinjiang, *D. orgosense* coexists with *Lophiomeryx*, and the last record of the latter in Europe is from the middle late Stampian Age (Qiu, Xie, and Yan 1990). *Allacerops*, another rhinocerotid in the Oligocene of Asia, was found from the Oligocene of Lanzhou Basin adjacent to the Linxia Basin (Qiu and Wang 1999). *Schizotherium* was also a characteristic Oligocene form in China, and it was found from the Oligocene Nanpoping Fauna of Lanzhou Basin (Qiu, Wang, and Xie 1998). *Aprotodon* was previously found only from Pakistan, Kazakhstan, and the early Miocene Zhangjiaping local fauna of Lanzhou Basin, and it coexisted with the giant rhino in these three regions (Qiu and Xie 1997) as in the Jiaozigou Formation of the Linxia Basin. *Ronzotherium* was found only in the Oligocene of Eurasia (Heissig 1969). Entelodonts were most diversified during the Sannoisian and the early–middle Stampian ages, and *Paraentelodon macrognathus* was very abundant in the Jiaozigou Fauna. *Tsaganomys* appeared first in the late Ulantatalian Age, and definitive records of *Tsaganomys* ended in the early Tabenbulukian. As a result, *Tsaganomys* is one of the index fossils for the Asian Oligocene (Wang 2001). *Tsaganomys* was found from the Oligocene Nanpoping Fauna in the Lanzhou Basin. Apparently, the age of the Jiaozigou Fauna is Tabenbulukian, late Oligocene.

Tunggurian Faunas

The middle Miocene mammals of the Linxia Basin include the Shinanu Fauna from the sandstones of the Dongxiang Formation at Dalanggou and Shinanu (Guan 1988; Cao et al. 1990), and the Zengjia and Laogou faunas from the sandstones and conglomerates of the Hujialiang Formation at Laogou, Hujialiang, Shinanu, and Zhujiachuan (Deng 2003) (see table 9.3 and figure 9.3). The fossils are typical middle Miocene forms, represented by Tunggurian *Platybelodon* and *Anchitherium* (Qiu and Qiu 1995). The shared genera in the Tunggurian faunas between the Linxia Basin and the Tunggur area in Inner Mongolia include *Alloptox*, *Hemicyon*, *Amphicyon*, *Platybelodon*, *Zygolophodon*, *Anchitherium*, *Hispanotherium*, *Kubanochoerus*, *Listriodon*, *Palaeotragus*, and *Turcocerus*. *Alloptox* was widespread in the middle Miocene faunas in China, such as Tunggur in Inner

Mongolia (Young 1932), Lengshuigou in Lantian, Shaanxi (Li 1978), Qijia in Minhe, Qinghai (Qiu, Li, and Wang 1981), and Dingjiaergou in Tongxin, Ningxia (Wu, Ye, and Zhu 1991). *Pliopithecus* existed during MN 5–9 in Europe, and it was found from the Dingjiaergou, Halamagai (Junggar, Xinjiang), and Damiao (Siziwangqi, Inner Mongolia) faunas of the middle Miocene in China (Qiu and Guan 1986; Wu, Meng, and Ye 2003; Zhang and Harrison 2008). In Eurasia, *Hispanotherium matritense* was found in Spain, Portugal, and France in Europe, and Turkey, Pakistan, Mongolia, and China in Asia. In China, *H. matritense* is distributed widely among the middle Miocene Dingjiaergou, Olonbuluk (Qaidam Basin, Qinghai), Lengshuigou, and Erlanggang (Fangxian, Hubei) faunas (Zhai 1978; Yan 1979; Guan 1988; Deng and Wang 2004). *H. matritense* is smaller than *H. tungurense* of the Tunggur Fauna (Cerdeño 1996; Deng 2003). *Alicornops* is widely distributed in Europe during MN 6–10 (Cerdeño and Sánchez 2000), and it was found from middle Miocene strata in Turkey (Heissig 1976). *Kubanochoerus* appeared in the strata of MN 6 in Caucasia, and it was found at Dingjiaergou, Tunggur, and Koujiacun (Lantian, Shaanxi). Therefore, *Kubanochoerus* also is a typical middle Miocene form.

Early Bahean Fauna

The Guonigou Fauna comes from the lower part of the Liushu Formation at Guonigou, Wangji, Bantu, and Zhongmajia (see table 9.4 and figure 9.3). Its *Dinocrocuta gigantea* probably represents an early occurrence relative to that in Bahe in Lantian, Shaanxi (Liu, Li, and Zhai 1978; Zhang et al. 2002; Zhang 2005), and Laogaochuan in Fugu, Shaanxi (Xue, Zhang, and Yue 1995; Zhang and Xue 1996). The fossiliferous deposits at Guonigou are correlated to the base of C5n.2n (11.04 Ma) or within C5r.1r (11.04–11.118 Ma) (Qiu et al., chapter 1, this volume). Equids also indicate early Bahean age. The size of *Hipparion dongxiangense* is smaller than that of *H. parvum*, otherwise the smallest known species of *Hipparion* in China, and its characteristic structures of the hypocone and the hypocone groove frequently occur among middle Miocene hipparionines from North America but are infrequent among later hipparionines. As a result, *H. dongxiangense* indicates an early late Miocene age (Qiu and Xie 1998). Guonigou *Parelasmotherium* is more primitive than *Sinotherium*, a representative taxon in the Baode Fauna, and the former is the earliest member of giant elasmotheres beginning to develop hypsodont teeth (Qiu and Xie 1998; Deng 2001c, 2007). *Shaanxispira* also appeared in the Bahe Fauna (Liu, Li, and Zhai 1978; Zhang et al. 2002). Relative to other Bahean faunas, the age of the Guonigou Fauna should be early Bahean. According to this correlation, the Toson Nor Fauna of the Qaidam Basin (Bohlin 1937; Deng 2006a; Wang et al. 2007) may have the same evolutionary level as the Guonigou Fauna (Qiu and Qiu 1995; Qiu, Wu, and Qiu 1999). However, neither fauna has many taxa, and the shared genera include only *Tetralophodon* and *Hipparion*, with materials too rare to be certain about specific status.

Middle Bahean Fauna

The Dashengou Fauna comes from the middle part of the Liushu Formation at Dashengou, Shanzhuang, Houshan, Sigou, Shuanggongbei, and elsewhere (see table 9.5 and figure 9.3). It retains early Bahean taxa, such as *Dinocrocuta gigantea*, which first appeared in the Guonigou Fauna. Other important components of the Bahe Fauna (Lantian, Shaanxi) are relatively richly represented in the Dashengou Fauna, such as *Hipparion weihoense* and *H. chiai*. These two species of *Hipparion* are large, with deep preorbital fossae far from the orbit and narrow and long protocones. These characters show that both of them apparently belong to the *H. primigenium* group, and the hipparionines of this group in Europe and Africa are predominantly Vallesian in age (Qiu, Huang, and Guo 1987). The shared species in the Bahe and Dashengou faunas also include *Tetralophodon exoletus* and *Chleuastochoerus stehlini* (Liu, Li, and Zhai 1978). *Hezhengia bohlini* is one of the most typical taxa in the Dashengou Fauna. The horncores of *Hezhengia* are obviously less specialized than those of the late Bahean and Baodean ovibovines, such as *Plesiaddax*, and its premolars are relatively long, with strong ribs and styles. These primitive characters of *H. bohlini* imply that its age should be earlier than that of the late Bahean and Baodean ovibovines (Qiu, Wang, and Xie 2000). *Acerorhinus hezhengensis* has a very narrow mandibular symphysis and close parietal crests forming a high sagittal crest, and thus it is close to *A. tsaidamensis* in the Toson Nor Fauna but different from *A. palaeosinensis* in the Baode Fauna (Qiu, Xie, and Yan 1987). The whole of the Dashengou Fauna indicates it should be correlated to middle Bahean age. The Lamagou Fauna in Fugu, Shaanxi (Xue, Zhang, and Yue 1995), also may be contemporaneous with the Dashengou Fauna in that they share *Dinocrocuta gigantea*, *Hyaenictitherium wongii*, *Hipparion chiai*, *Chilotherium wimani*, *Samotherium* sp., and *Miotragocerus* sp. Moreover, *Acerorhinus fuguensis* from Lamagou is very close to *A. hezheng-*

ensis from the Dashengou Fauna (Deng 2000). However, paleomagnetic study by Xue, Zhang, and Yue (1995) placed the Lamagou Fauna later, about 8–7 Ma.

Late Bahean Fauna

The Yangjiashan Fauna comes from the upper part of the Liushu Formation at Yangjiashan, Yangwapuzi, Shanjiawan, Heilinding, Songshugou, Shilei, Huaigou, Nanmiangou, Longjiawan, and so on (see table 9.6 and figure 9.3). In this fauna, typical elements in the Dashengou Fauna such as *Dinocrocuta gigantea* and *Hezhengia bohlini* have disappeared. The Yangjiashan Fauna is similar to the Baode Fauna, sharing *Simocyon primigenius*, *Plesiogulo* sp., *Parataxidea sinensis*, *Promephitis hootoni*, *Hyaenictitherium wongii*, *H. hyaenoides*, *Adcrocuta eximia*, *Machairodus palanderi*, *Metailurus minor*, *Chleuastochoerus stehlini*, *Microstonyx major*, *Cervavitus novorossiae*, *Palaeotragus microdon*, and *Honanotherium schlosseri*. Like the Baode Fauna, rhinocerotids are absolutely dominant in the Yangjiashan Fauna, although both faunas are often taken as typical of *Hipparion* faunas. On the other hand, the more primitive *Chilotherium wimani* of the Yangjiashan Fauna is replaced by the more derived *C. anderssoni* in the Baode Fauna. The primitive characters of *C. wimani* include the low position of orbit, well-developed supraorbital tubercle, weak postorbital process, concave dorsal skull profile, closely placed parietal crests, narrow braincase, and strong paracone rib on premolars (Deng 2001a, 2001b), in contrast to the derived characters of *C. anderssoni* including the high position of orbit, absence of supraorbital tubercle, well-developed postorbital process, flat dorsal skull profile, broadly separate parietal crests, rounded braincase, and weak or absent paracone rib on premolars (Ringström 1924). Therefore, the Yangjiashan Fauna would be late Bahean, slightly earlier than Baodean faunas.

On the other hand, some localities near the southern border of the Linxia Basin and at the top of the Liushu Formation, such as Qingbushan, Wangjiashan, Jinchanggou, and Bancaoling, produce *Chilotherium anderssoni*, *Dicerorhinus ringstroemi*, *Sinohippus robustus*, and *Indactos zdanskyi*, indicating a Baodean Age. If so, the Liushu Formation may be diachronous within the Linxia Basin.

Gaozhuangian Fauna

The Shilidun Fauna comes from the Hewangjia Formation at Shilidun, Yinchuan, and Duikang (Deng et al.

2011; see table 9.7 and figure 9.3). *Chasmaporthetes kani*, *Hipparion pater*, and *Gazella blacki* newly appeared in the Shilidun Fauna, whereas the extremely dominant *Chilotherium* of the Dashengou and Yangjiashan faunas disappeared from this fauna. Other members of the late Miocene *Hipparion* fauna, such as *Ictitherium* and *Chleuastochoerus*, are absent in the Shilidun Fauna. *Chasmaporthetes* has a widespread distribution in the latest Miocene and early Pliocene across the world, from Europe, Asia, and Africa to North America. In China, *Chasmaporthetes* was discovered from Yushe and Shouyang in Shanxi, Nihewan in Hebei, and Mianchi in Henan (Qiu 1987). In the Yushe Basin, *Chasmaporthetes kani* appeared first in the late Miocene Mahui Fauna with a paleomagnetic age of about 6 Ma, when *Chilotherium* disappeared (Qiu 1987). *Chasmaporthetes kani* is typical of the Gaozhuang Fauna correlated to MN14–15 European mammal units (Qiu and Qiu 1995). *Hipparion pater* was collected from the Gaozhuang Formation at many localities in the Yushe Basin, Shanxi Province. *H. pater* is smaller than the Nihewanian *Hipparion sinense*, and the former has shorter skull length, longer nasal bones, larger orbits, lower tooth crowns, less plications, narrower parastyles and mesostyles, and more convex labial walls of lower teeth than the latter. *H. pater* was also found from Loc. 5 in the northern Baode area and Youhe in Weinan, Shaanxi, Gaozhuangian age for the former and Mazegouan age for the latter (Qiu, Huang, and Guo 1987). Some common species of the late Miocene *Hipparion* fauna, such as *Hystrix gansuensis*, *Parataxidea sinensis*, *Hyaenictitherium wongii*, *Adcrocuta eximia*, *Hipparion platyodus*, *Palaeotragus microdon*, *Cervavitus novorossiae*, and *Sinotragus* sp., persisted in the Shilidun Fauna, whereas derived taxa from high in the Gaozhuang Formation of Yushe, such as *Ursus*, *Nyctereutes*, *Canis*, *Hipparion houfenense*, and *Sus*, have not been found in the Shilidun Fauna. As a result, the age of the Shilidun Fauna would be Gaozhuangian, early Pliocene.

Nihewanian Fauna

The Longdan Fauna comes from the Wucheng Loess at Longdan, Shitougou, Keshijian, and Xijia (see table 9.8 and figure 9.3). The existence of true horse and the absence of typical middle Pleistocene forms, like *Pachycrocuta sinensis*, *Sinomegaceros*, *Bos*, and *Bubalus*, put the Longdan Fauna in the early Pleistocene Nihewanian Age. The Longdan Fauna is slightly older than that of the Nihewan Fauna from Yangyuan, Hebei (Teilhard de Chardin and Piveteau 1930), based on the following ob-

servations. Although congeneric, the evolutionary levels differ in *Meles teilhardi*, *Sivapanthera linxiaensis*, and *Gazella* cf. *blacki* of the Longdan Fauna and are certainly more primitive than their counterparts in the Nihewan Fauna. Some primitive forms occur in the Longdan Fauna but are absent in the Nihewan Fauna, such as *Panthera palaeosinensis*, *Leptobos brevicornis*, and *Nipponicervus longdanensis*. At Nihewan, no *Panthera* or *Leptobos* occur, and *Cervus elegans* may represent a more advanced species of *Nipponicervus*. On the other hand, a large number of more advanced forms, especially cervids, appear in the Nihewan Fauna, such as *Axis shansius*, *Elaphurus bifurcatus*, *Eucladoceros boulei*, *Bison palaeosinensis*, and *Paleoloxodon namadicus*. The extremely large horse, *Equus eisenmannae* from Longdan, may be a species that appeared earlier than *E. sanmeniensis* from Nihewan. Similar cases can be seen in Europe, where the earlier horse, *E. livenzovensis*, is larger than the typical Villafranchian *E. stenonis*.

ACKNOWLEDGMENTS

We are grateful to Xijun Ni, Guangpu Xie, Zhong Liang, Shaokun Chen, Qinqin Shi, Shiqi Wang, Libo Pang, Zhijie Jack Tseng, and Shanqin Chen for their participation in fieldwork. This work is supported by the Ministry of Science and Technology of China (2012CB821906, 2006FY120300, 2006CB806400), the Knowledge Innovation Program of the Chinese Academy of Sciences (KZCX-YW-Q09, KZCX2-YW-120), the National Natural Science Foundation of China (40730210, 40232023), and the China National Commission on Stratigraphy.

REFERENCES

Antoine, P.-O. 2002. Phylogénie et évolution des Elasmotheriina (Mammalia, Rhinocerotidae). *Mémoires du Muséum National d'Histoire Naturelle* 188:1–359.

Bernor, R. L., N. Solounias, C. C. Swisher III, and J. A. van Couvering. 1996. The correlation of three classical "Pikermian" mammal faunas—Maragha, Samos and Pikermi—with the European MN unit system. In *The Evolution of Western Eurasian Neogene Mammal Faunas*, ed. R. L. Bernor, V. Fahlbusch, and H.–W. Mittmann, pp. 137–156. New York: Columbia University Press.

Bohlin, B. 1937. Eine Tertiäre Säugetier-Fauna aus Tsaidam. *Palaeontologia Sinica*, n.s. C 14(1):1–111.

Cao, Z.-x., H.-j. Du, Q.-q. Zhao, and J. Cheng. 1990. Discovery of the middle Miocene fossil mammals in Guanghe district, Gansu and their stratigraphic significance. *Geoscience* 4(2):16–29 (in Chinese with English abstract).

Cerdeño, E. 1996. Rhinocerotidae from the middle Miocene of the Tung-gur Formation, Inner Mongolia (China). *American Museum Novitates* 3184:1–43.

Cerdeño, E. and B. Sánchez. 2000. Intraspecific variation and evolutionary trends of *Alicornops simorrense* (Rhinocerotidae) in Spain. *Zoologica Scripta* 29:275–305.

Chen, H.-l., X.-m. Fang, J.-j. Li, and S.-c. Kang. 1996. Fission track dating of Cenozoic strata in Linxia Basin. *Nuclear Techniques* 19:632–634.

Deng, T. 2000. A new species of *Acerorhinus* (Perissodactyla, Rhinocerotidae) from the late Miocene in Fugu, Shaanxi, China. *Vertebrata PalAsiatica* 38:203–217.

Deng, T. 2001a. Cranial ontogenesis of *Chilotherium wimani* (Perissodactyla, Rhinocerotidae). In *Proceedings of the Eighth Annual Meeting of the Chinese Society of Vertebrate Paleontology*, ed. T. Deng and Y. Wang, pp. 101–112. Beijing: China Ocean Press.

Deng, T. 2001b. New materials of *Chilotherium wimani* (Perissodactyla, Rhinocerotidae) from the late Miocene of Fugu, Shaanxi. *Vertebrata PalAsiatica* 39:129–138.

Deng, T. 2001c. New remains of *Parelasmotherium* (Perissodactyla, Rhinocerotidae) from the late Miocene in Dongxiang, Gansu, China. *Vertebrata PalAsiatica* 39:306–311.

Deng, T. 2002. The earliest known wooly rhino discovered in the Linxia Basin, Gansu Province, China. *Geological Bulletin of China* 21:604–608.

Deng, T. 2003. New material of *Hispanotherium matritense* (Rhinocerotidae, Perissodactyla) from Laogou of Hezheng County (Gansu, China), with special reference to the Chinese middle Miocene elasmotheres. *Geobios* 36:141–150.

Deng, T. 2004a. A new species of the rhinoceros *Alicornops* from the middle Miocene of the Linxia Basin, Gansu, China. *Palaeontology* 47:1427–1439.

Deng, T. 2004b. Establishment of the middle Miocene Hujialiang Formation in the Linxia Basin of Gansu and its features. *Journal of Stratigraphy* 28:307–312.

Deng, T. 2005a. New cranial material of *Shansirhinus* (Rhinocerotidae, Perissodactyla) from the lower Pliocene of the Linxia Basin in Gansu, China. *Geobios* 38:301–313.

Deng, T. 2005b. New discovery of *Iranotherium morgani* (Perissodactyla, Rhinocerotidae) from the late Miocene of the Linxia Basin in Gansu, China and its sexual dimorphism. *Journal of Vertebrate Paleontology* 25:442–450.

Deng, T. 2006a. A primitive species of *Chilotherium* (Perissodactyla, Rhinocerotidae) from the late Miocene in the Linxia Basin (Gansu, China). *Cainozoic Research* 5:93–102.

Deng, T. 2006b. Neogene rhinoceroses of the Linxia Basin (Gansu, China). *Courier Forschungsinstitut Senckenberg* 256:43–56.

Deng, T. 2007. Skull of *Parelasmotherium* (Perissodactyla, Rhinocerotidae) from the upper Miocene in the Linxia Basin (Gansu, China). *Journal of Vertebrate Paleontology* 27:467–475.

Deng, T. 2008. A new elasmothere (Perissodactyla, Rhinocerotidae) from the late Miocene of the Linxia Basin in Gansu, China. *Geobios* 41:719–728.

Deng, T., S.-k. Hou, Q.-q. Shi, S.-k. Chen, W. He, and S.-q. Chen. 2011. Terrestrial Mio–Pliocene boundary in the Linxia Basin, Gansu, China. *Acta Geologica Sinica* 85:452–464.

Deng, T. and Z.-x. Qiu. 2007. First discovery of *Diceros* (Perissodactyla, Rhinocerotidae) in China. *Vertebrata PalAsiatica* 45:287–306.

Deng, T. and X.-m. Wang. 2004. New material of the Neogene rhinocerotids from the Qaidam Basin in Qinghai, China. *Vertebrata PalAsiatica* 42:216–229.

Deng, T., X.-m. Wang, X.-j. Ni, and L.-p. Liu. 2004a. Sequence of the Cenozoic mammalian faunas of the Linxia Basin in Gansu, China. *Acta Geologica Sinica* 78:8–14.

Deng, T.,-x. M. Wang, X.-j. Ni, L.-p. Liu, and Z. Liang. 2004b. Cenozoic stratigraphic sequence of the Linxia Basin in Gansu, China and its evidence from mammal fossils. *Vertebrata PalAsiatica* 42:45–66.

Ding, Z.-l., J.-m. Sun, T.-s. Liu, R.-x. Zhu, S.-l. Yang, and B. Guo. 1998. Wind-blown origin of the Pliocene red clay formation in the central Loess Plateau, China. *Earth and Planetary Science Letters* 161:135–143.

Fang, X.-m., C. Garzione, R. Van der Voo, J.-j. Li, and M.-j. Fan. 2003. Flexural subsidence by 29 Ma on the NE edge of Tibet from the magnetostratigraphy of Linxia Basin, China. *Earth and Planetary Science Letters* 210:545–560.

Fang, X.-m., J.-j. Li, J.-j. Zhu, H.-l. Chen, and J.-x. Cao. 1997. Division and age dating of the Cenozoic strata of the Linxia Basin in Gansu, China. *Chinese Science Bulletin* 42:1457–1471 (in Chinese).

Fang, X.-m., X.-h. Xu, Q.-q. Meng, C.-h. Song, W.-x. Han, and M. Torii. 2007. High-resolution rock magnetic records of Cenozoic sediments in the Linxia Basin and their implications on drying of Asian inland. *Quaternary Sciences* 27:989–1000.

Gabunia, L. K. 1964. *Benara Fauna of Oligocene Vertebrates*. Tbilissi: Metsniereba Press (in Russian).

Gong, H.-j., Y.-x. Zhang, and L. Huang. 2005. Paleoenvironment significance of grain-size composition of Neogene red clay in Linxia Basin, Gansu Province. *Acta Sedimentologica Sinica* 23:260–267.

Gradstein, F. M., J. G. Ogg, and A. G. Smith. 2004. *A Geological Time Scale 2004*. Cambridge: Cambridge University Press.

Gu, Z.-g., S.-h. Wang, X.-y. Hu, and D.-t. Wei. 1995a. Research progress on biostratigraphy of Tertiary red beds in Linxia Basin, Gansu Province. In *Study on the Formation and Evolution of the Qinghai-Xizang Plateau, Environmental Change and Ecological System*, ed. The Expert Committee on Qingzang Program, pp. 91–95. Beijing: Science Press (in Chinese with English abstract).

Gu, Z.-g., S.-h. Wang, X.-y. Hu, and D.-t. Wei. 1995b. Discovery of *Giraffokeryx* in China and the Tertiary chronostratigraphy of Linxia, Gansu Province. *Chinese Science Bulletin* 40:758–760.

Guan, J. 1988. The Miocene strata and mammals from Tongxin, Ningxia and Guanghe, Gansu. *Memoirs of Beijing Natural History Museum* 42:1–21 (in Chinese with English summary).

Guan, J. and X. Zhang. 1993. The middle Miocene mammals from Guanghe and Hezheng in northwestern China. *Memoirs of Beijing Natural History Museum* 53:237–251.

Guo, Z.-t., S.-z. Peng, Q.-z. Hao, P. E. Biscaye, and T.-s. Liu. 2001. Origin of the Miocene-Pliocene red-earth formation at Xifeng in northern China and implications for paleoenvironments. *Palaeogeography, Palaeoclimatology, Palaeoecology* 170:11–26.

Heissig, K. 1969. Die Rhinocerotidae (Mammalia) aus der oberoligozänen Spaltenfüllung von Gaimersheim bei Ingolstadt in Bayern und ihre phylogenetische Stellung. *Bayerische Akademie der Wissenschaften, Mathematisch-Naturwissenschaftliche Klasse, Abhandlungen*, n.s. 138:1–133.

Heissig, K. 1976. Rhinocerotidae (Mammalia) aus der *Anchitherium*-Fauna Anatoliens. *Geologisches Jahrbuch* 19:1–121.

Hofmann, A. 1893. Fauna von Göriach. *Abhandlungen Der K. K. Geologischen Reichsanstalt* 15(6):1–87.

Hou, L.-h., Z.-h. Zhou, F.-c. Zhang, and Y. Wang. 2005. A Miocene ostrich fossil from Gansu Province, northwest China. *Chinese Science Bulletin* 50:1286–1288.

Hou, S.-k., T. Deng, W. He, and S.-q. Chen. 2007. New materials of *Sinohippus* from Gansu and Nei Mongol, China. *Vertebrata PalAsiatica* 45:213–231.

Howell, F. C. and G. Petter. 1985. Comparative observations on some middle and upper Miocene hyaenids. *Geobios* 18:419–476.

Li, C.-k. 1978. Two new lagomorphs from the Miocene of Lantian, Shensi. *Professional Papers of Stratigraphy and Palaeontology* 7:143–148 (in Chinese).

Li, C.-k., W.-y. Wu, and Z.-d. Qiu. 1984. Chinese Neogene: Subdivision and correlation. *Vertebrata PalAsiatica* 22:163–178 (in Chinese with English summary).

Li, J.-j. et al. 1995. *Uplift of Qinghai-Xizang (Tibet) Plateau and Global Change, A Contribution to XIV INQUA Congress, 1995, Berlin.* Lanzhou: Lanzhou University Press.

Liu, L.-p., D. S. Kostopoulos, and M. Fortelius. 2004. Late Miocene *Microstonyx* remains (Suidae, Mammalia) from northern China. *Geobios* 37:49–64.

Liu, T.-s., C.-k. Li, and R.-j. Zhai. 1978. Pliocene vertebrates of Lantian, Shensi. *Professional Papers of Stratigraphy and Palaeontology* 7:149–200 (in Chinese).

Ma, Y.-z., J.-j. Li, and X.-m. Fang. 1998. Records of the climatic variation and pollen flora from the red beds at 30.6–5.0 Ma in Linxia district. *Chinese Science Bulletin* 43:301–304 (in Chinese).

Qiu, Z.-d., C.-k. Li, and S.-j. Wang. 1981. Miocene mammalian fossils from Xining Basin, Qinghai. *Vertebrata PalAsiatica* 19:156–173 (in Chinese with English summary).

Qiu, Z.-x. 1987. Die Hyaeniden aus dem Ruscinium und Villafranchium Chinas. *Münchner Geowissenschaftliche Abhandlungen*, Reihe A 9:1–110.

Qiu, Z.-x., T. Deng, and B.-y. Wang. 2004. Early Pleistocene mammalian fauna from Longdan, Dongxiang, Gansu, China. *Palaeontologia Sinica*, n.s. C 27:1–198.

Qiu, Z.-x., T. Deng, and B.-y. Wang. 2009. Discovery of *Protarctos yinanensis* from Longdan, Dongxiang, Gansu: Addition to the Longdan mammalian fauna (2). *Vertebrata PalAsiatica* 47:245–264.

Qiu, Z.-x. and J. Guan. 1986. A lower molar of *Pliopithecus* from Tongxin, Ningxia Hui Autonomous Region. *Acta Anthropologica Sinica* 5:201–207 (in Chinese with English abstract).

Qiu, Z.-x., W.-l. Huang, and Z.-h. Guo. 1987. The Chinese hipparionine fossils. *Palaeontologia Sinica*, n.s. C 25:1–250 (in Chinese with English summary).

Qiu, Z.-x. and Z.-d. Qiu. 1995. Chronological sequence and subdivision of Chinese Neogene mammalian faunas. *Palaeogeography, Palaeoclimatology, Palaeoecology* 116:41–70.

Qiu, Z.-x. and B.-y. Wang. 1999. *Allacerops* (Rhinocerotoidea, Perissodactyla), its discovery in China and its systematic position. *Vertebrata PalAsiatica* 37:48–61 (in Chinese with English summary).

Qiu, Z.-x. and B.-y. Wang. 2007. Paracerathere fossils of China. *Palaeontologia Sinica*, n.s. C 29:1–396 (in Chinese with English summary).

Qiu, Z.-x., B.-y. Wang, and T. Deng. 2004a. Indricotheres (Perissodactyla, Mammalia) from Oligocene in Linxia Basin, Gansu, China. *Vertebrata PalAsiatica* 42:177–192.

Qiu, Z.-x., B.-y. Wang, and T. Deng. 2004b. Mammal fossils from Yagou, Linxia Basin, Gansu, and related stratigraphic problems. *Vertebrata PalAsiatica* 42:276–296.

Qiu, Z.-x., B.-y. Wang, T. Deng, X.-j. Ni, and X.-m. Wang. 2002. Notes on the mammal fauna from the bottom of loess deposits at Longdan, Dongxiang County, Gansu Province. *Quaternary Sciences* 22:33–38.

Qiu, Z.-x., B.-y. Wang, H. Li, T. Deng, and Y. Sun. 2007. First discovery of deinothere in China. *Vertebrata PalAsiatica* 45:261–277.

Qiu, Z.-x., B.-y. Wang, and G.-p. Xie. 2000. Preliminary report on a new genus of Ovibovinae from Hezheng district, Gansu, China. *Vertebrata PalAsiatica* 38:128–134 (in Chinese with English summary).

Qiu, Z.-x., B.-y. Wang, and J.-y. Xie. 1998. Mid-Tertiary chalicothere (Perissodactyla) fossils from Lanzhou, Gansu, China. *Vertebrata PalAsiatica* 36:297–318 (in Chinese with English summary).

Qiu, Z.-x., W.-y. Wu, and Z.-d. Qiu. 1999. Miocene mammal faunal sequence of China: Palaeozoogeography and Eurasian relationships. In *The Miocene Land Mammals of Europe*, ed. G. E. Rössner and K. Heissig, pp. 443–455. Munich: Dr. Friedrich Pfeil.

Qiu, Z.-x. and J.-y. Xie. 1997. A new species of *Aprotodon* (Perissodactyla, Rhinocerotidae) from Lanzhou Basin, Gansu, China. *Vertebrata PalAsiatica* 35:250–267 (in Chinese with English summary).

Qiu, Z.-x. and J.-y. Xie. 1998. Notes on *Parelasmotherium* and *Hipparion* fossils from Wangji, Dongxiang, Gansu. *Vertebrata PalAsiatica* 36:13–23 (in Chinese with English summary).

Qiu, Z.-x., J.-y. Xie, and D.-f. Yan. 1987. A new chilothere skull from Hezheng, Gansu, China, with special reference to the Chinese "Diceratherium." *Scientia Sinica* (5):545–552 (in Chinese).

Qiu, Z.-x., J.-y. Xie, and D.-f. Yan. 1988. Discovery of the skull of *Dinocrocuta gigantea*. *Vertebrata PalAsiatica* 26:128–138 (in Chinese with English summary).

Qiu, Z.-x., J.-y. Xie, and D.-f. Yan. 1990. Discovery of some early Miocene mammalian fossils from Dongxiang, Gansu. *Vertebrata PalAsiatica* 28:9–24 (in Chinese with English summary).

Qiu, Z.-x., J.-y. Xie, and D.-f. Yan. 1991. Discovery of late Miocene *Agriotherium* from Jiegou, Gansu, and its taxonomic implications. *Vertebrata PalAsiatica* 29:286–295 (in Chinese with English summary).

Ringström, T. 1924. Nashörner der *Hipparion*-Fauna Nord-Chinas. *Palaeontologia Sinica*, ser. C 1(4):1–159.

Song, C.-h., J.-f. Bai, Y.-d. Zhao, H.-b. Jing, and Q.-q. Meng. 2005. The color of lacustrine sediments recorded climatic changes from 13 to 4.5 Myr in Linxia Basin. *Acta Sedimentologica Sinica* 23:507–513.

Song, C.-h., X.-c. Lu, Q. Xing, Q.-q. Meng, W.-m. Xia, P. Liu, and P. Zhang. 2007. Late Cenozoic element characters and palaeoclimatic change of the lacustrine sediments in Linxia Basin, China. *Acta Sedimentologica Sinica* 25:409–416.

Teilhard de Chardin, P. and J. Piveteau. 1930. Les mammifères fossils de Nihowan (Chine). *Annales de Paléontologie* 19:1–134.

Wang, B.-y. 2001. On Tsaganomyidae (Rodentia, Mammalia) of Asia. *American Museum Novitates* 3317:1–50.

Wang, B.-y. 2005. Beaver (Rodentia, Mammalia) fossils from Longdan, Gansu, China: Addition to the early Pleistocene Longdan mammalian fauna (1). *Vertebrata PalAsiatica* 43:237–242.

Wang, B.-y. and Z.-x. Qiu. 2002. A porcupine from late Miocene of Linxia Basin, Gansu, China. *Vertebrata PalAsiatica* 40:23–33 (in Chinese with English summary).

Wang, B.-y. and Z.-x. Qiu. 2003. Aepyosciurinae: A new subfamily of Sciuridae (Rodentia, Mammalia) from basal loess deposits at the northeastern border of Tibetan Plateau. *Chinese Science Bulletin* 48:691–695.

Wang, J.-l. and X.-m. Fang. 2000. Eolian sand deposition and its environmental significance in Linxia Basin since middle Miocene. *Scientia Geographica Sinica* 20:259–263.

Wang, J.-l., J.-j. Li, X.-m. Fang, and J.-j. Zhu. 1998. Tectonic significance deduced from grain size characteristics in Linxia Basin in 30 million years. *Geographical Research* 17:39–47.

Wang, S. 2007. Pathological microstructure of a Miocene ostrich eggshell from Asia. *Acta Geologica Sinica* 81:697–702.

Wang, S. 2008. Reexamination of taxonomic assignment of "*Struthio linxiaensis* Hou et al., 2005." *Acta Palaeontologica Sinica* 47:362–368.

Wang, S.-q. and T. Deng. 2011. The first *Choerolophodon* (Proboscidea, Gomphotheriidae) skull from China. *Science China, Earth Science* 54:1326–1337.

Wang, X.-m. and Z.-x. Qiu. 2004. Late Miocene *Promephitis* (Carnivora, Mephitidae) from China. *Journal of Vertebrate Paleontology* 24:721–731.

Wang, X.-m., Z.-d. Qiu, Q. Li, B.-y. Wang, Z.-x. Qiu, W. Downs, G.-p. Xie, J.-y. Xie, T. Deng, G. Takeuchi, Z. J. Tseng, M.-m. Chang, J. Liu, Y. Wang, D. Biasatti, Z.-c. Sun, X.-m. Fang, and Q.-q. Meng. 2007. Vertebrate paleontology, biostratigraphy, geochronology, and paleoenvironment of Qaidam Basin in northern Tibetan Plateau. *Palaeogeography, Palaeoclimatology, Palaeoecology* 254: 363–385.

Werdelin, L. 1996. Carnivores, exclusive of Hyaenidae, from the later Miocene of Europe and Western Asia. In *The Evolution of Western Eurasian Neogene Mammal Faunas*, ed. R. Bernor, V. Fahlbusch, and H.-W. Mittmann, pp. 271–289. New York: Columbia University Press.

Wu, W.-y., J. Meng, and J. Ye. 2003. The discovery of *Pliopithecus* from northern Junggar Basin, Xinjiang. *Vertebrata PalAsiatica* 41:76–86.

Wu, W.-y., J. Ye, and B.-c. Zhu. 1991. On *Alloptox* (Lagomorpha, Ochotonidae) from the middle Miocene of Tongxin, Ningxia Hui Autonomous Region, China. *Vertebrata PalAsiatica* 29:204–229 (in Chinese with English summary).

Xie, J.-y. 1991. The Late Tertiary strata and mammalian fossils of Gansu, China. *Journal of Stratigraphy* 15:36–41 (in Chinese).

Xu, X.-h., X.-m. Fang, C.-h. Song, M.-j. Fan, and J. Shen. 2008. Grain-size records of Cenozoic lacustrine sediments from Linxia Basin and the aridification of Asian inland. *Journal of Lake Science* 20:65–75.

Xue, X.-x., Y.-x. Zhang, and L.-p. Yue. 1995. Discovery and chronological division of the *Hipparion* fauna in Laogaochuan Village, Fugu County, Shaanxi. *Chinese Science Bulletin* 40:926–929.

Yan, D.-f. 1979. Einige der Fossilen Miozänen Säugetiere der Kreis von Fangxian in Der Provinz Hupei. *Vertebrata PalAsiatica* 17:189–199 (in Chinese with German summary).

Young, C. C. 1932. On a new ochotonid from north Suiyuan. *Bulletin of the Geological Society of China* 11:255–258.

Zhai, R.-j. 1978. A primitive elasmothere from the Miocene of Lintung, Shensi. *Professional Papers of Stratigraphy and Palaeontology* 7:122–126 (in Chinese with English summary).

Zhang, Y.-x. and X.-x. Xue. 1996. New materials of *Dinocrocuta gigantea* found in Fugu County, Shaanxi Province. *Vertebrata PalAsiatica* 34:18–26 (in Chinese with English abstract).

Zhang, Z.-h., X.-t. Zheng, G.-m. Zheng, and L.-h. Hou. 2010. A new Old World vulture (Falconiformes: Accipitridae) from the Miocene of Gansu Province, northwest China. *Journal of Ornithology* 151:401–408.

Zhang, Z.-q. 2005. New materials of *Dinocrocuta* (Percrocutidae, Carnivora) from Lantian, Shaanxi Province, China, and remarks on Chinese late Miocene biochronology. *Geobios* 38:685–689.

Zhang, Z.-q., A. W. Gentry, A. Kaakinen, L.-p. Liu, J. P. Lunkka, Z.-d. Qiu, S. Sen, R. Scott, L. Werdelin, S.-h. Zheng, and M. Fortelius. 2002. Land mammal faunal sequence of the late Miocene of China: New evidence from Lantian, Shaanxi Province. *Vertebrata PalAsiatica* 40:165–176.

Zhang, Z.-q. and T. Harrison. 2008. A new middle Miocene pliopithecid from Inner Mongolia, China. *Journal of Human Evolution* 54:444–447.

Zheng, D.-w., P.-z. Zhang, J.-l. Wan, D.-y. Yuan, D.-m. Li, J.-w. Yin, C. Y. Li, and Z. C. Wang. 2006. Tectonic events, climate and conglomerate: Example from Jishishan Mountain and Linxia Basin. *Quaternary Sciences* 26:64–69.

Zheng, S.-h. 1980. The *Hipparion* fauna of Bulong Basin, Biru, Xizang. In *Palaeontology of Xizang, Book 1*, ed. The Comprehensive Scientific Expedition to the Qinghai-Xizang Plateau, the Chinese Academy of Sciences, pp. 33–47. Beijing: Science Press (in Chinese with English abstract).

Zhong, W., J.-j. Li, J.-j. Zhu, and X.-m. Fang. 1998. The geochemical record of paleoclimate during about 7.0 Ma–0.73 Ma in Linxia Basin, Gansu Province. *Journal of Arid Land Resources and Environment* 12:36–43.

APPENDIX

Mammalian Fossil Localities in the Linxia Basin, Gansu, China

IVPP	Zhao	Locality (Township, County)	GPS	Formation
LX 9801		Galijia (Pingzhuang, Dongxiang)	N35°35'00.2" E103°23'04.6", H2284 m	Hujialiang
LX 9803		Tala (Dongyuan, Dongxiang)	N35°38'26.1" E103°16'54.9", H2253 m	Jiaozigou
LX 9804		Pitiaogou (Dongyuan, Dongxiang)	N35°38'50.6" E103°16'51.5", H2039m	Jiaozigou
LX 9805		Longguang (Yinchuan, Jishishan)	N35°39'23.0" E103°04'57.8", H2063 m	Hewangjia
LX 9806		Shinanu (Shinanu, Guanghe)	N35°29'51.1" E103°32'59.8", H2263 m	Dongxiang
LX 0001	23	Dalanggou (Maijiaxiang, Guanghe)	N35°26'50.7" E103°27'28.5", H2060 m	Shangzhuang
LX 0002	8	Hujiangliang (Maijiaxiang, Guanghe)	N35°26'23.1" E103°26'37.6", H2150 m	Hujialiang
LX 0003	7	Laogou (Sanhe, Hezheng)	N35°28'05.3" E103°24'50.5", H2200 m	Hujialiang
LX 0004	6	Yangjiashan (Sanhe, Hezheng)	N35°28'37.9" E103°21'25.9", H2340 m	Liushu
LX 0005	48	Youhao (Sanhe, Hezheng)	N35°28'49.3" E103°22'20.8", H2340 m	Liushu
LX 0007	11	Sigou (Alimatu, Guanghe)	N35°29'41.8" E103°30'06.1", H2140 m	Liushu
LX 0008	9	Houshan (Alimatu, Guanghe)	N35°29'07.6" E103°28'28.0", H2227 m	Liushu
LX 0009	13	Shuanggongbei (Nalesi, Dongxiang)	N35°31'05.6" E103°28'20.4", H2227 m	Liushu
LX 0010	38	Longdan (Nalesi, Dongxiang)	N35°31'32.2" E103°28'58.3", H2248 m	Wucheng
LX 0011	1	Dashengou (Xinzhuang, Hezheng)	N35°21'23.9" E103°21'22.4", H2276 m	Liushu
LX 0012	2	Zhonggou (Xinzhuang, Hezheng)	N35°18'46.6" E103°20'07.4", H2410 m	Liushu
LX 0013	5	Moshigou (Xinzhuang, Hezheng)	N35°18'22.5" E103°20'49.4", H2426 m	Liushu
LX 0014	40	Shilidun (Chengguan, Guanghe)	N35°27'38.8" E103°34'40.0", H2130 m	Hewangjia
LX 0016	35	Sijiaping (Baiwang, Guanghe)	N35°26'13.8" E103°34'09.8", H2131m	Liushu
LX 0017	22	Nangou (Chengguan, Guanghe)	N35°27'20.6" E103°33'25.0", H2015 m	Hujialiang
LX 0018	34	Yangwapuzi (Zhuangheji, Guanghe)	N35°24'49.5" E103°30'28.6", H2160 m	Liushu
LX 0019	26	Xiaozhai (Zhuangheji, Guanghe)	N35°23'56.6" E103°30'20.8", H2206 m	Liushu
LX 0020	25	Nanmiangou (Zhuangheji, Guanghe)	N35°24'23.1" E103°29'33.8", H2205 m	Liushu
LX 0021	20	Shadigou (Zhuangheji, Guanghe)	N35°25'34.3" E103°29'39.2", H2223 m	Liushu
LX 0022	12	Longjiawan (Maijiaxiang, Guanghe)	N35°24'41.9" E103°25'19.2", H2303 m	Liushu

(continued)

IVPP	Zhao	Locality (Township, County)	GPS	Formation
LX 0023	16	Hetuo (Diaotan, Hezheng)	N35°21'14.8" E103°23'19.7", H2308 m	Liushu
LX 0024	33	Dashanzhuang (Diaotan, Hezheng)	N35°19'30.0" E103°23'22.6", H2392 m	Liushu
LX 0025	18	Maling (Sanhe, Hezheng)	N35°26'58.3" E103°20'14.4", H2340 m	Liushu
LX 0026	44	Nanyangshan (Sanshilipu, Hezheng)	N35°26'14.2" E103°17'34.7", H2401 m	Liushu
LX 0027	24	Shanzhuang (Maijiaxiang, Guanghe)	N35°25'57.6" E103°25'20.4", H2245 m	Liushu
LX 0029	43	Huaigou (Guanfang, Guanghe)	N35°22'51.4" E103°26'54.4", H2240 m	Liushu
LX 0030	10	Songshugou (Guanfang, Guanghe)	N35°25'05.5" E103°28'05.1", H2327 m	Liushu
LX 0031	50	Shilei (Guanfang, Guanghe)	N35°25'19.7" E103°27'30.3", H2263 m	Liushu
LX 0032	45	Zhujiachuan (Maijiaxiang, Guanghe)	N35°26'33.4" E103°26'03.0", H2159 m	Hujialiang
LX 0033	27	Jinchanggou (Guantangou, Hezheng)	N35°22'06.3" E103°19'50.2", H2420 m	Liushu
LX 0034	19	Gaojiashan (Guantangou, Hezheng)	N35°21'21.3" E103°19'00.7", H2459 m	Liushu
LX 0035	30	Heilinding (Guantangou, Hezheng)	N35°22'35.4" E103°19'10.6", H2382 m	Liushu
LX 0036	28	Baituyao (Bujiazhuang, Hezheng)	N35°26'07.0" E103°18'53.2", H2389 m	Liushu
LX 0037	4	Panyang (Guantangou, Hezheng)	N35°23'35.6" E103°20'38.6", H2270 m	Liushu
LX 0041	3	Shancheng (Maijiaji, Hezheng)	N35°22'00.8" E103°15'58.0", H2352 m	Liushu
LX 0042	32	Guonigou (Nalesi, Dongxiang)	N35°33'01.1" E103°26'15.6", H2234 m	Liushu
LX 0043	31	Bantu (Nalesi, Dongxiang)	N35°33'08.4" E103°27'40.7", H2228 m	Liushu
LX 0044	39	Bantuyinshan (Nalesi, Dongxiang)	N35°33'16.5" E103°27'38.9", H2216 m	Liushu
LX 0045	37	Hualinsanshe (Diaotan, Hezheng)	N35°23'36.7" E103°25'46.9", H2318 m	Liushu
LX 0046	17	Hejiazhuang (Maijiaxiang, Guanghe)	N35°23'57.5" E103°25'25.5", H2207 m	Liushu
LX 0047	15	Qiaojia (Maijiaxiang, Guanghe)	N35°24'12.5" E103°25'28.4", H2218 m	Liushu
LX 0048	36	Jizhuwan (Maijiaxiang, Guanghe)	N35°24'41.3" E103°26'09.9", H2281 m	Liushu
LX 0049	14	Zhongmajia (Dalang, Hezheng)	N35°24'47.0" E103°24'30.6", H2218 m	Liushu
LX 0050	29	Bancaoling (Maijiaji, Hezheng)	N35°23'34.9" E103°13'48.5", H2480 m	Liushu
LX 0051		Sigoucun (Alimatu, Guanghe)	N35°29'44.8" E103°30'14.4", H2074 m	Shangzhuang
LX 0201	47	Shitougou (Nalesi, Dongxiang)	N35°31'11.3" E103°29'19.7", H2228 m	Wucheng
LX 0202		Xijia (Zhaojia, Guanghe)	N35°30'48.9" E103°30'20.1", H2199 m	Wucheng
LX 0203	55	Qingbushan (Xinzhuang, Hezheng)	N35°18'59.4" E103°21'41.0", H2452 m	Liushu
LX 0204	53	Niugouwan (Diaotan, Hezheng)	N35°18'33.4" E103°24'53.2", H2426 m	Liushu
LX 0205	54	Baihuancun (Zhuangheji, Guanghe)	N35°23'43.2" E103°29'45.2", H2233 m	Liushu
LX 0206	51	Hujia (Maijiaxiang, Guanghe)	N35°26'40.5" E103°27'01.7", H2099 m	Dongxiang
LX 0207	52	Lima (Sanhe, Hezheng)	N35°26'43.5" E103°19'07.8", H2365 m	Liushu
LX 0209	59	Daoheigou (Dalang, Hezheng)	N35°27'06.1" E103°23'08.1", H2110 m	Dongxiang
LX 0210	21	Citan (Guangfang, Guanghe)	N35°26'43.4" E103°29'02.8", H2075 m	Dongxiang
LX 0213		Yagou (Dongyuan, Dongxiang)	N35°35'47.9" E103°16'39.2", H1973 m	Jiaozigou
LX 0214		Lijiawan (Dalang, Hezheng)	N35°24'39.4" E103°24'41.0", H2275 m	Liushu
LX 0215		Wangji (Wangji, Dongxiang)	N35°37'24.5" E103°30'30.1", H2350 m	Liushu
LX 0216		Qianzhuang (Diaotan, Hezheng)	N35°21'15.0" E103°23'15.3", H2328 m	Hewangjia
LX 0217	57	Keshijian (Nalesi, Dongxiang)	N35°31'58.6" E103°29'55.2", H2196 m	Wucheng
LX 0401		Caojiashan (Sanhe, Hezheng)	N35°28'29.9" E103°21'13.9", H2384 m	Wucheng
LX 0501	69	Wangjiashan (Maijiaji, Hezheng)	N35°22'34.3" E103°12'48.9", H2575 m	Liushu (u.)
LX 0502	69	Songjianao (Maijiaji, Hezheng)	N35°22'28.6" E103°12'51.4", H2496 m	Liushu (l.)
LX 0503	65	Niuzhawan (Sanhe, Hezheng)	N35°29'08.7" E103°21'20.3", H2343 m	Liushu
LX 0601		Mida (Maijiaji, Hezheng)	N35°21'19.7" E103°11'58.0", H2477 m	Liushu
LX 0701	60	Duikang (Zhuangheji, Guanghe)	N35°25'22.5" E103°32'58.6", H2162 m	Hewangjia
LX 0702		Bajia (Alimatu, Guanghe)	N35°29'07.6" E103°28'21.6", H2185 m	Liushu
LX 0703	87	Hualinqishe (Diaotan, Hezheng)	N35°21'43.0" E103°25'03.1", H2324 m	Liushu
LX 0801		Dazhuang (Maijiaji, Hezheng)	N35°19'06.9" E103°17'06.8", H2448 m	Shangzhuang

IVPP	Zhao	Locality (Township, County)	GPS	Formation
LX 0802		Ganchiliang (Maijiaji, Hezheng)	N35°17'59.4" E103°17'30.1", H2530 m	Dongxiang
LX 0803		Shanchengyishe (Maijiaji, Hezheng)	N35°22'13.1" E103°16'32.8", H2362 m	Liushu
LX 0804		Yancaiping (Maijiaji, Hezheng)	N35°22'21.3" E103°17'03.6", H2256 m	Liushu
LX 1002		Shanggou (Nalesi, Dongxiang)	N35°34'30.9" E103°27'20.3", H2287 m	Hujialiang

NOTE: A large portion of these localities were cataloged by a fossil dealer, Yongchang Zhao, and his three sons, who collaborated with us to devise a number system (under the column "Zhao"), which was then incorporated into our IVPP locality number system (under "IVPP").

Chapter 10

Neogene Mammalian Biostratigraphy and Geochronology of the Tibetan Plateau

XIAOMING WANG, QIANG LI, ZHU-DING QIU, GUANG-PU XIE, BAN-YUE WANG, ZHAN-XIANG QIU, ZHIJIE J. TSENG, GARY T. TAKEUCHI, AND TAO DENG

In recognition of its unique importance as a prime example of continent–continent collisions and the resulting mountain building with profound impact on climates and environments, the Tibetan Plateau has become a focus of intense interest for its geologic history and paleoclimatic/paleoenvironmental evolution (e.g., Molnar 2005). Although paleontologic explorations represent some of the earliest studies of the geoscience of the plateau (e.g., Bohlin 1937, 1942, 1946), vertebrate paleontology has generally not been featured prominently in this debate, with the exception of the discoveries of the Gyirong *Hipparion* fauna at the foothills of the Himalayas (Huang et al. 1980; Ji, Hsu, and Huang 1980) that provoked discussions about its bearing on the elevation history of Tibet (e.g., Li 1995). Discovery of the Gyirong Fauna was hailed as a major achievement of the early expeditions by the Chinese Academy of Sciences during the 1960s and 1970s. Unfortunately, these systematic efforts in field vertebrate paleontology were interrupted during the 1980s and much of the 1990s.

Small-scale expeditions, including those by us, resumed in the late 1990s in an attempt to move toward a preliminary assessment of mammalian evolution in the Tibetan Plateau as a reflection of paleoenvironmental changes and as a response to the growth and uplift of the plateau (figure 10.1). Compared to classic localities elsewhere in northern China, vertebrate paleontology in the Tibetan Plateau is still in its infancy, with a single exception of the recent explosive developments in the Linxia Basin (Deng et al., chapter 9, this volume)—glimpses of

certain taxa are beginning to emerge, but a comprehensive picture is still a long way from being achieved.

As a lofty landmass with sharp boundaries and a physical environment often resembling regions in much higher latitudes, the interior Tibetan Plateau features a unique assemblage of living mammals, with close to 50% of them being endemic forms (Hoffmann 1989, 1991). The Himalaya Range and its lateral extensions form the most impenetrable zoogeographic barrier within a continent, as well as mark sharp climatic, environmental, and vegetational boundaries. As a result, the Palearctic and Oriental zoogeographic provinces are divided by the southern rim of the Tibetan Plateau, which, along with the sub-Saharan desert, forms the only intracontinental zoogeographic provincial boundaries in the world.

Important questions naturally arise concerning the timing and extent of mammalian evolution in response to the formation of the proto–Tibetan Plateau. In addition, there is a keen sense of urgency for a better understanding of faunal divergences between successive mammal faunas from the Siwaliks of Pakistan and north of the Himalaya. With sufficient resolution, one hopes to be able to address the question of when the plateau became a formidable zoogeographic barrier, which in turn permits a sense of the rise and growth of the plateau itself. Such can only be a long-term goal, although tantalizing fossils have already emerged that have interesting implications, helping to orient future research.

We briefly summarize Neogene vertebrate faunas and their chronology within and around the margins of the

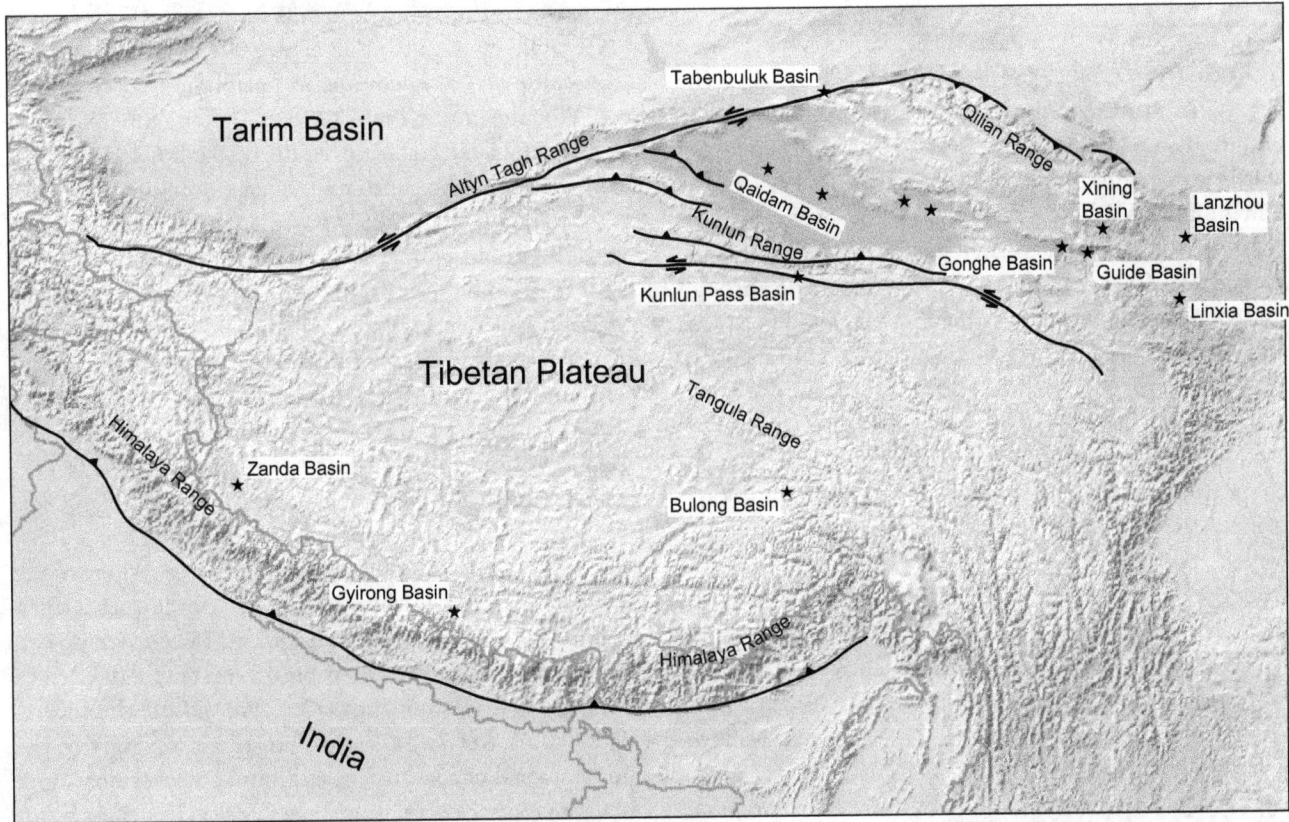

Figure 10.1 Map of known Neogene vertebrate fossil localities. Except for the Qaidam Basin, where multiple fossil-producing regions are noted, all fossiliferous basins are represented by a single black star.

Tibetan Plateau (see figure 10.1). Several localities that produce only fossil fishes are not discussed here. This chapter will not deal with the Linxia Basin, which is the subject of a separate treatment by Deng et al. (chapter 9, this volume), nor are the hominoid-producing sites in Yunnan included, which are summarized by Dong and Qi (chapter 11, this volume). We also did not include a few "Pliocene" fossil localities in the Hengduan Mountain ranges along the eastern edge of the Tibetan Plateau (Zong et al. 1996), none of which have an independent age control (such as magnetostratigraphy).

We use the orbitally tuned Geomagnetic Polarity Time Scale (GPTS) of ATNTS2004 (Lourens et al. 2004). We adopt the Neogene/Quaternary (Plio-Pleistocene) boundary at 2.6 Ma recently passed by the International Commission on Stratigraphy (Mascarelli 2009). Chinese geographic names follow the official Pinyin system, and when more than one is available, we use as a standard the English version of the Atlas of the People's Republic of

China published by the China Cartographic Publishing House.

FAUNAL SUCCESSION AND MAGNETOSTRATIGRAPHY

Lanzhou Basin

Although the Lanzhou Basin had been known to produce vertebrate fossils as early as the 1920s, systematic investigations began only in the late 1980s and tentatively concluded by 1997 (see summary in Qiu et al. 1997; Qiu et al. 2001). A thick sequence of more than 4000 m of terrestrial sediments ranging in age from early Cretaceous to middle Miocene is well exposed along the Limashagou River, north of the city of Lanzhou. Zoogeographically, mammals from the Lanzhou Basin probably are not part of the high plateau faunal province. Nonetheless, because of its current location, it is appropriate to summarize the Lanzhou Basin faunal sequence in the present context.

In an effort to fill in gaps in the early Miocene fossil record in China and to gauge the effect of the rising of the Tibetan Plateau, a multinational team of vertebrate paleontologists and paleomagnetists systematically collected fossils and magnetic samples from 1988 to 1997 in the Lanzhou Basin. These efforts resulted in several publications on fossil mammals ranging in age from late Eocene to middle Miocene, although early Oligocene through middle Miocene records are the most detailed (e.g., Qiu and Xie 1997; Flynn et al. 1999; Qiu 2000, 2001a, 2001b; Wang, Qiu, and Wang 2005; Qiu and Wang 2007). Three Miocene faunas are recognized in the Xianshuihe Formation of the Lanzhou Basin: the early Miocene Zhangjiaping Fauna and the Duitinggou Fauna, and the middle Miocene Quantougou Fauna. The Zhangjiaping Fauna (or local fauna, as in Qiu et al. 2001), as defined by fossils from the basal white sandstones ("white sands" in informal terminology), consists of the following components of Xiejian age with some Oligocene leftovers: *Desmatolagus pusillus, Sinolagomys kansuensis, S. ulunguensis, S. pachygnathus, Tataromys plicidens, Tataromys* sp., *Yindirtemys grangeri, Y. deflexus, Y. ambiguus, Bounomys* sp., *Prodistylomys* sp., *Tachyoryctoides* cf. *T. kokonorensis, Sayimys* sp., *Ansomys* sp., *Anomoemys* sp., *Parasminthus asiae-orientalis, P. tangingoli, Heterosminthus orientalis, Protalactaga* sp., *Cricetodon* sp., *Democricetodon* sp., *Atlantoxerus* sp., *Sinotamias* sp., *Hyaenodon weilini, ?Ictiocyon* cf. *I. socialis, Turpanotherium elegans, Aprotodon lanzhouensis, Phyllotillon huangheensis*, and others. For additional discussions about the Zhangjiaping fauna, see Qiu et al. (chapter 1, this volume). Stratigraphically above the Zhangjiaping Fauna and within alternating sandstones and reddish mudstones, the Duitinggou Fauna contains a Shanwangian assemblage: *?Metexallerix* sp., *Sinolagomys* sp., *Alloptox minor, Bellatona forsythmajori, Megacricetodon* sp., *Democricetodon* sp., *Heterosminthus* sp., *Protalactaga grabaui, Prodistylomys* sp., and *Stephanocemas* sp. At the top of the Xianshuihe Formation, the Quantougou Fauna yields a modestly rich small and large mammal assemblage: *Mioechinus* (?) *gobiensis, Microdyromys wuae, Heterosminthus orientalis, Protalactaga grabaui, Protalactaga major, Mellalomys gansus, Myocricetodon plebius, Plesiodipus leei, Megacricetodon sinensis, Ganocricetodon cheni, Paracricetulus schaubi*, Ochotonidae indet., *Kubanochoerus gigas*, and *Gomphotherium wimani* (Qiu 2000, 2001a, 2001b). For additional discussions about the Zhangjiaping and Duitinggou local faunas, see Qiu et al. (chapter 1, this volume).

Neogene strata were sampled for paleomagnetic studies in the Duitinggou section on the eastern limb of the Limashagou River syncline and in the Dahonggou section where the syncline bends westward (Flynn et al. 1999; Qiu et al. 2001; Yue et al. 2001), but considerable difficulties were encountered correlating these to the GPTS. The more than 1200 m Dahonggou magnetic section was correlated to C5Bn through C25n (~14.8–57.2 Ma). For the shorter Duitinggou section (~420 m), however, Qiu et al. (2001) postulated a long hiatus of almost 8 million yr to account for missing magnetochrons that span from the late Oligocene through early Miocene, in contrast to a correlation scheme by Flynn et al. (1999) that postulates far less of a hiatus. The missing magnetochrons at the Duitinggou section (20 km southeast of the Dahonggou section) correspond to nearly 400 m of lost strata within the lower member of the Xianshuihe Formation. Such a difficulty in correlation is the result of a recognition that the Nanpoping Fauna from the "yellow sands" horizon is of early Oligocene age, whereas the overlying lower "white sands" horizon produced a Xiejian fauna that cannot be older than early Miocene (these two sandstone units were widely used as marker beds to trace stratigraphic relationships). Qiu et al. (2001) thus correlated the lower "white sands" to the lower part of the long normal chron 6n. For additional discussions about paleomagnetic correlations, see Qiu et al. (chapter 1, this volume).

Tabenbuluk Basin

Through a series of fortuitous events, vertebrate paleontologist Birger Bohlin of the Sino-Swedish Expeditions (also known as the Sven Hedin Expeditions) discovered the fossiliferous localities in the Tabenbuluk area, ~15 km southwest of the Subei county seat, in a series of north–south trending major canyons that cut through the basin (Wang, Wang, and Qiu 2008). Although fossils described from this area are largely Oligocene in age, most of which are from the fossiliferous Yandantu (Yindirte) locality in the Yandantu Canyon, Bohlin did suspect the presence of Miocene faunas as well (Bohlin, 1942, 1946). He believed that most of the artiodactyls, rhinos, and primates collected from the Xishuigou Canyon and *Sayimys* from the Tiejianggou Canyon were much younger than the late Oligocene Yindirte Fauna. However, Bohlin was reluctant to formally recognized these faunas of younger ages, and as in the case in the Qaidam Basin (see "Qaidam Basin"), he did not seem to adequately appreciate the long time span represented by the >3,000-m-thick strata. As a result, Neogene faunas from Tabenbuluk largely went unnoticed since Bohlin's initial descriptions.

In 1999 and 2001, we conducted a systematic reinvestigation of the Tabenbuluk area with the aim of reconciling Bohlin's classic fossil localities within a modern biostratigraphy framework (Wang, Wang, and Qiu 2008). Due to the steeply incised exposures that are not conducive to fossil prospecting and general scarcity of vertebrate fossils, we were not able to replicate Bohlin's collections with the exception of the Yandantu locality, which is still highly fossiliferous. However, by combining our own fossil localities with those recorded by Bohlin in archival records, we were able to reconstruct the majority of his fossil localities within a range of historic uncertainties (B.-y. Wang et al. 2003; X. Wang et al. 2003; Wang, Wang, and Qiu 2008), as well as revising previous records and adding new taxa (Wang 2002; Wang and Qiu 2002; Wang 2003; Wang and Qiu 2004; Wang, Qiu, and Wang 2004).

Neogene faunas from the Tabenbuluk strata are still represented by just a handful of taxa, and much additional collecting remains to be done before a comprehensive picture can emerge. Two Miocene faunas can now be recognized: an early Miocene Xishuigou Fauna and a middle Miocene unnamed fauna (poorly represented by fossils and lacking an independent means of chronology, another unnamed fauna from a top sequence of red beds sitting unconformably above the Tiejianggou Formation is not treated in this chapter). The Xishuigou Fauna consists of only a few mammals: *Platybelodon dangheensis*, *Turcocerus* sp., *Amphimoschus* cf. *A. artenensis*, *Kinometaxia guangpui*, "*Kansupithecus*," and *Heterosminthus intermedius*, whereas the overlying unnamed fauna has *Sayimys obliquidens*, *Litodonomys xishuiensis*, *Phyllotillon* sp., Cervidae indet., and Proboscidea indet.

Because of its critical importance in constraining the Altyn Tagh fault as a major strike-slip fault to accommodate the shortening and lateral extrusion of the northern Tibetan Plateau (e.g., Yin et al. 2002; Ritts, Yue, and Graham 2004), the Tabenbuluk strata have been the subject of numerous studies. At least three studies on the magnetostratigraphy have been published (Gilder, Chen, and Sen 2001; Yin et al. 2002; Sun, Zhu, and An 2005). However, two of these studies (Gilder, Chen, and Sen 2001; Yin et al. 2002) suffer from a misinterpretation of the paleontological constraints, resulting in an erroneous correlation that yields a much older age estimate than is indicated by the fossil mammals. Based on our revised biostratigraphic framework, we have reinterpreted Gilder, Chen, and Sen's (2001) Xishuigou magnetic section (X. Wang et al. 2003) (figure 10.2). We now correlate the Xishuigou Fauna (bracketed by localities Dh199910 to Dh199914) to the lower part of chron 5Cr to the middle

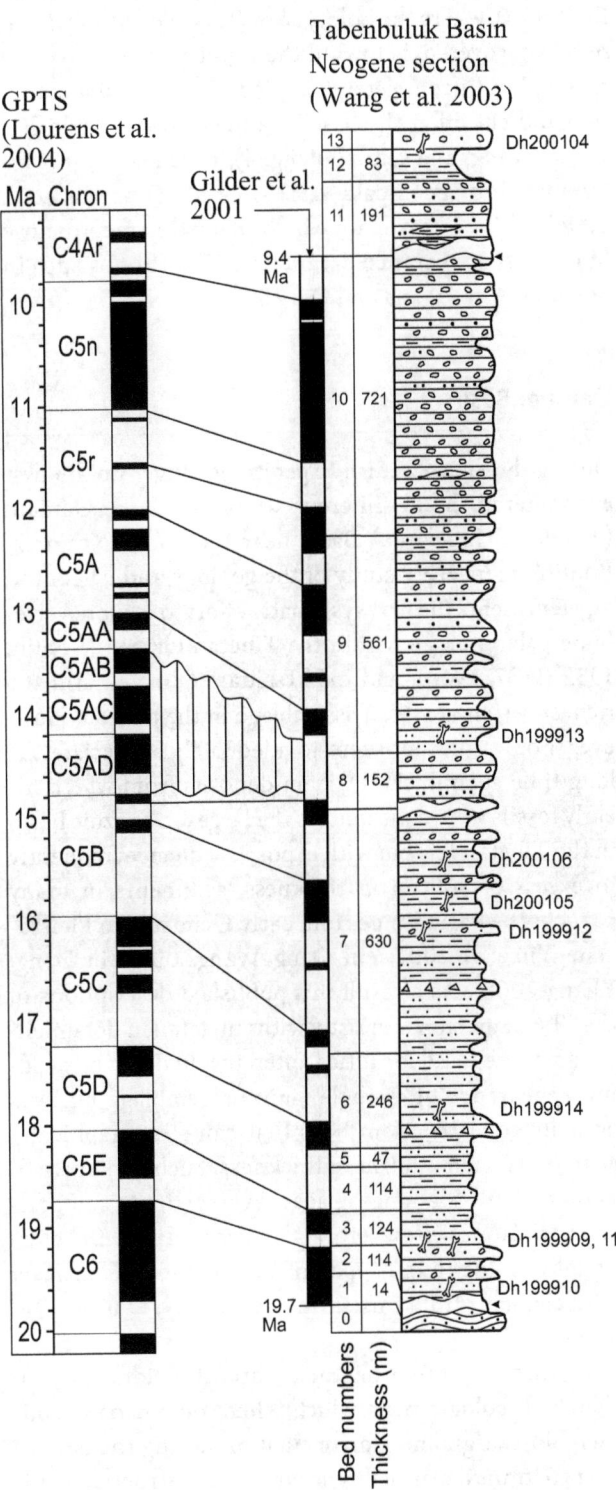

Figure 10.2 Magnetostratigraphy and chronology of Neogene strata in Tabenbuluk Basin. Paleomagnetic section modified from Wang et al. (2003).

part of chron C6n (~17–19.7 Ma). In particular, locality Dh199910, where the holotype of *Platybelodon dangheensis* was recovered, is toward the middle or lower part of chron C6n (incompletely sampled in the Xishuigou section and cut off at the base by a major thrust fault, F0; Wang, Wang, and Qiu 2008:fig. 4), which is roughly interpolated to encompass ~19.2–19.4 Ma (Lourens et al. 2004). This age estimate would place this record close to the first appearance datum (FAD) of Proboscidea in northern Asia (Wang and Qiu 2002).

Qaidam Basin

During the Sino-Swedish Expeditions, Birger Bohlin first encountered the fossiliferous strata in eastern Qaidam (Tsaidam, Chaidamu) Basin near the Tuosu Nor area. Bohlin's pioneering study of the geology and paleontology represents the first systematic effort to explore vertebrate paleontology within the Tibetan Plateau (Bohlin, 1935, 1937, 1960), and his "Tsaidam fauna" is still the richest vertebrate fossil assemblage in the plateau. However, Bohlin did not seem to adequately appreciate the long time span in the Qaidam deposits from which his early fossils were collected (as the largest Cenozoic basin in the Tibetan Plateau with exposed sequence frequently in excess of 5000 m in thickness, sediments in many parts of the basin range from early Cenozoic to Pleistocene; Yin et al. 2007; Yin, Dang, Wang 2008; Yin, Dang, Zhang 2008). As a result, his published descriptions of the "Tsaidam fauna," mostly without detailed documentation of the fossil localities, often resulted in a misleading appearance of a single faunal assemblage but was actually collected from beds that can stratigraphically span more than 2000 m in thickness. Such a mixture of faunas from wide-ranging ages severely limits comparison with faunas elsewhere, and at times causes misinterpretations due to the apparent association of *Hipparion* with elements from earlier time (Qiu 1989; Qiu and Qiu 1995).

Fortunately, Bohlin took careful fieldnotes and sketched geologic maps, which, when combined with our own biostratigraphic documentation during the past 10 years permits a reasonably accurate reconstruction of his original fossil localities (Wang et al. 2011). This new biostratigraphic framework thus permits us to tease out four different faunas ranging in age from middle Miocene to Pliocene (Wang et al. 2007). As a result, the Qaidam vertebrate faunal sequence now provides the best-documented vertebrate record in the Tibetan Plateau (Bohlin 1935, 1937; Wang and Wang 2001; Deng and

Wang 2004a, 2004b; Chen and Liu 2007; Dong 2007; Chang et al. 2008; Qiu and Li 2008; Wang, Xie, and Dong 2009). Despite a relatively long history of exploration and superb exposures, with the exception of three areas (Quanshuiliang, Naoge, and Shengou), vertebrate fossils are generally not very rich and faunas are relatively low in diversity. We can distinguish four major mammalian faunas (an additional four fish faunas are recognizable, which are not discussed here; see Wang et al. 2007); these are the Olongbuluk Fauna, Tuosu Fauna, Shengou Fauna, and Huaitoutala Fauna (Wang et al. 2007; Qiu and Li 2008).

The Olongbuluk Fauna consists of mammals from the foothills of Olongbuluk Mountain (Institute of Vertebrate Paleontology and Paleoanthropology [IVPP] localities CD9809, 9811, 9818, 9826), mainly from the north limb of the Keluke Anticline, and also scattered localities west of Bayin Mountain (IVPP localities CD0231, 0405, 0406) and western Barunya Ula anticline (IVPP localities CD0626, 0630, and others). A small fauna of middle Miocene in character emerges from these localities: *Hispanotherium matritense* (Deng and Wang 2004b), *Acerorhinus tsaidamensis* (Deng and Wang 2004b), *Lagomeryx tsaidamensis*, *Stephanocemas palmatus* (Wang, Xie, and Dong 2009), and possibly *Dicroceros*.

The Tuosu Fauna was named from localities in a narrow band of exposures along the northwestern shore of Tuosu Nor (IVPP localities CD9804–9808, 0238, 0761, 0764). By lateral extension toward the Huaitoutala section, several localities (IVPP localities CD9812, 9815–9817, 9819–9821, 9823–9824), spanning 600 m in strata, are correlative to the Tuosu Nor strata. Furthermore, a highly fossiliferous band south of the Quanshuiliang railroad station (IVPP localities CD0769–07114, 0801–0894, 0896–08102, 08116–08125), informally named the "general strips" by Birger Bohlin, also appears to correlate to this interval (Wang et al. 2011). A composite fauna, combined from localities in these three areas, comprises the majority of what Bohlin (1937) described from the Qaidam Basin: *Ictitherium*, *Eomellivora*, *Chalicotherium brevirostris* (Wang and Wang 2001), *Hipparion teilhardi* (Deng and Wang 2004a), Sivatherinae indet., *Dicroceros* (IVPP V13086), *Euprox* sp. (small), *Olonbulukia tsaidamensis*, *Qurliqnoria cheni*, *Tossunnoria pseudibex*, *Tsaidamotherium hedini*, *Protoryx* sp., *Tetralophodon*, and *Struthio*.

The Shengou Fauna was founded on two major fossiliferous areas from the Naoge localities and Shengou localities. Large mammals consist of *Ictitherium*, *Adcrocuta eximia*, *Plesiogulo*, *Promephitis parvus*, *Acerorhinus tsaidamensis* (Deng and Wang 2004b), *Dicerorhinus ringstromi*

(Deng and Wang 2004b), *Hipparion* cf. *H. chiai* (Deng and Wang 2004a), *Hipparion weihoense* (Deng and Wang 2004a), *Hipparion teilhardi* (Deng and Wang 2004a), *Euprox* sp. (large), *Gazella* sp., and *Amebelodon*. Cervids are particularly rich, possibly related to riparian habitats associated with abundant channels. A modest small mammal fauna was described from a Shengou wash site (IVPP locality CD0227): Talpinae indet., Soricidae indet., *Sinotamias* sp., *Sciurotamias* cf. *S. pusillus*, *Pliopetaurista* sp., Eomyidae indet., *Lophocricetus* cf. *L. xianensis*, *Protalactaga* sp., *Myocricetodon lantianensis*, *Nannocricetus primitivus*, *Sinocricetus* sp., *Huerzelerimys exiguus*, *Pararhizomys* sp., *Ochotonoma primitiva*, and *Ochotona* sp. (Qiu and Li 2008).

The topmost fauna, the Huaitoutala Fauna, is still poorly represented by just six taxa of small mammals plus some ostrich egg shells: *Ochotona*, Leporidae indet., *Orientalomys/Chardinomys*, *Micromys*, *Pseudomeriones*, and an unidentified shrew (Soricidae indet.). Unfortunately, the Pliocene part of the eastern Qaidam section is not very fossiliferous, a period of extreme aridity and drying of local lakes (Chang et al. 2008), and repeated attempts at washing this site for additional fossils have failed.

Of the above four mammalian faunas, three (Olongbuluk, Tuosu, and Huaitoutala faunas) can be directly tied to a ~4600 m paleomagnetic section taken from south of Huaitoutala, which spans from 15.7 Ma to 1.8 Ma (Fang et al. 2007). After recalibrating Fang et al.'s (2007) magnetic correlations based on the more recent GPTS (Lourens et al. 2004), as well as recalculating individual fossil localities relative to Fang et al.'s (2007) section, we provide the following revised chronologic limits to the above named faunas (figure 10.3): C5Ar.1n to C5AB (12.7–13.7 Ma) for the Olongbuluk Fauna (the lower limit of the Huaitoutala section is restricted to the axis of the Keluke Anticline and localities from western Barunya Ula and Bayin Mountain likely will extend the range of this fauna further down), C5n to C5r.1n (9.98–11.15 Ma) for the Tuosu Fauna, and C3n.4n (4.9–5.2 Ma) for the Huaitoutala Fauna (Wang et al. 2011). In particular, Bohlin's Tuosu Nor localities (IVPP localities CD9804–9808) and Quanshuiliang localities (his "general strips") hold the promise of yielding the FAD of *Hipparion* in east Asia (see "Biochronology").

Xining Basin

Fossil mammals from this basin are summarized by Meng et al. (chapter 3, this volume) and Qiu et al. (chapter 1, this volume).

Guide Basin

Isolated reports of fossil mammals from the Guide area date as far back as the late nineteenth century (Nehring 1883), but systematic exploration of the vertebrate fossils was first started in 1977 by researchers from the IVPP (Zheng, Wu, and Li 1985). The Yellow River cuts through the Guide Basin sediments, and fossil mammals were recovered from at least three horizons in a ~200 m section at the Erdaogou area. Zheng, Wu, and Li (1985) thought the presence of *Anancus* and *Myospalax* as significant indicators of a Pliocene age for the Guide strata. Gu et al. (1992), however, led a team from Lanzhou University and recovered additional fossils, including *Kubanochoerus* cf. *K. lantiensis*, a common middle Miocene taxon in the Heixiagou section, which indicated the presence of older strata. More recently, Song et al. (2003) sampled a magnetostratigraphic section at the Erdaogou (He'erjia) section as well as the overlying Lajigai section and interpreted their ages in the range of 3.1–6.5 Ma. Fang et al. (2005) sampled more deeply incised sections at Ganjia and Ashigong farther north from Erdaogou, including Gu et al.'s (1992) *Kubanochoerus* locality. With the expanded coverage, Fang et al. (2005) estimated the Guide Basin to have a depositional history from 1.8 Ma to ~20 Ma.

The potential of Guide vertebrate fossils is still far from being fully realized. Paleontologic work is still very much in its preliminary stage: additional fossils need to be collected, particularly small mammals, and almost no taxon has been fully described.

Bulong Basin

An interesting discovery of a *Hipparion* fauna was made by the Qinghai-Tibetan Comprehensive Scientific Expeditions near Bulong in Biru County, south of Tanggula Mountain in east-central Tibet (Huang et al. 1980; Zheng 1980). In a poorly exposed fluvial bed along the Cuosangqu River valley, mostly covered by vegetation, the team was able to recover a small late Miocene assemblage, including *Brachyrhizomys naquensis*, *Dinocrocuta gigantea xizangensis*, *Metailurus* sp., *Felis* sp., *Chilotherium tanggulaense*, *Hipparion xizangense*, *Samotherium* sp., and *Gazella* sp. Repeated attempts to relocate the site in the early 2000s have failed due to the lack of distinct landmarks in a flat, vegetation-covered river valley.

With such poor exposures, paleomagnetic methods cannot be easily applied, and the age for this fauna is estimated on the basis of faunal considerations only. Zheng (1980) thought that *Brachyrhizomys* from Bulong is most

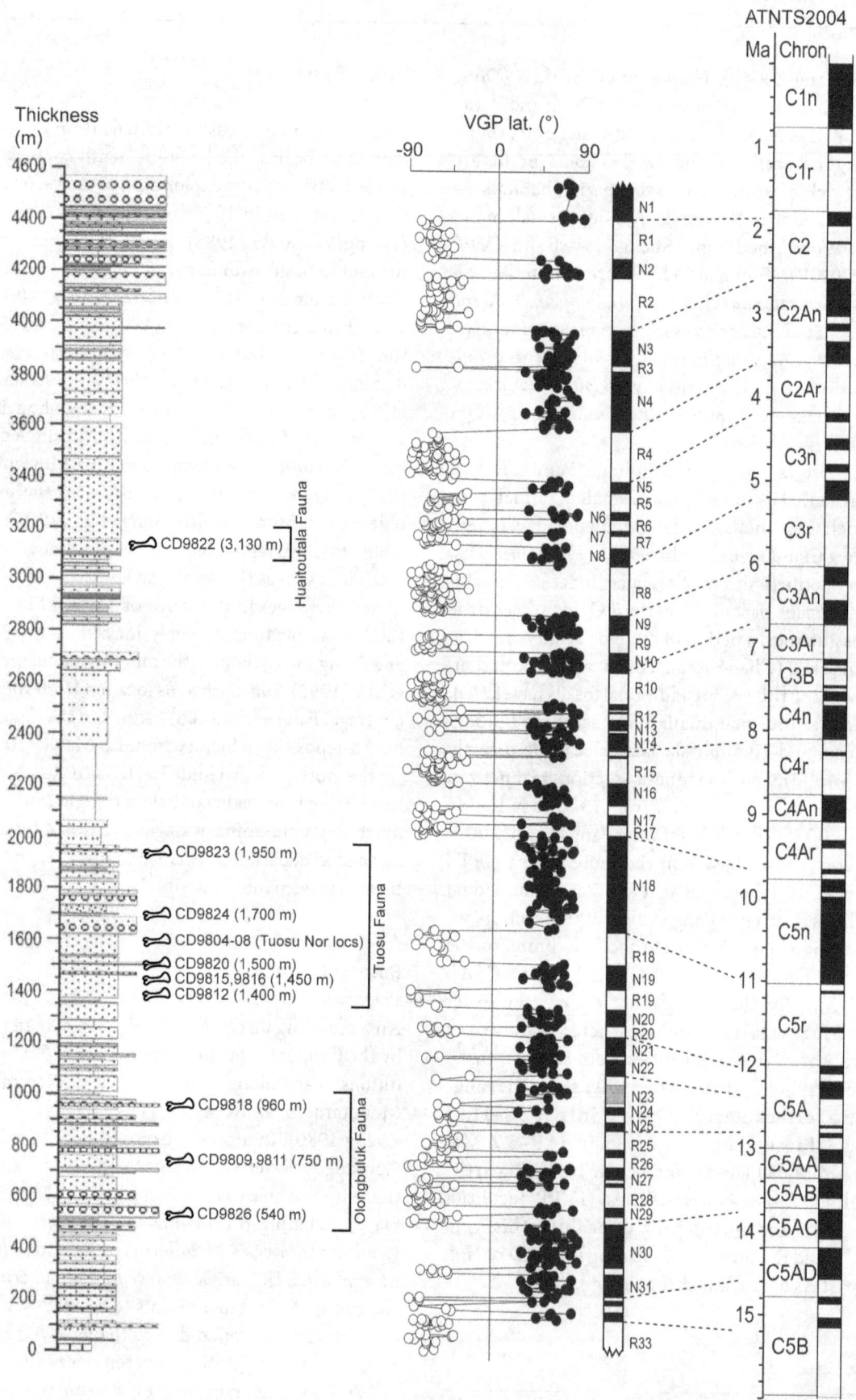

Figure 10.3 Magnetostratigraphy and chronology of Neogene strata in the Qaidam Basin. Modified from Fang et al. (2007:fig. 8). We have re-calibrated the individual fossil localities (e.g., IVPP locality CD9824) relative to GPS locations of paleomagnetic sample sites (unpublished data from Qingquan Meng) and listed estimated numbers of thickness (numbers in meters in parentheses after fossil locality numbers) in Fang et al.'s measured section. Using these new estimates, we arrive at an approximate correlation to individual magnetochrons.

comparable to *Rhizomyoides punjabiensis* from the Chinji Formation of the Siwalik, Pakistan. Qiu, Huang, and Guo (1987:182) remarked on the "obviously primitive" nature of *Hipparion* (*Hippotherium*) *xizangense* from Bulong and considered it to be possibly earlier than those from the Bahe Formation in Lantian. Zhang (2005), on the other hand, thought the Bulong *Dinocrocuta* to be possibly conspecific to *D. senyureki* from the Middle Sinap of Turkey and possibly ancestral to *D. gigantea*. Huang et al. (1980) did not formally give a name for the fauna from the Bulong Basin. The name Bulong Fauna first appears in Li, Wu, and Qiu (1984:22), whereas Zhang (2005:688) used "Bulong fauna" and "Biru fauna" interchangeably. Current consensus seems to be that the Bulong Fauna is early late Miocene in age—that is, Bahean (Asian Land Mammal Age [ALMA]).

Gyirong Basin

One of the great achievements of the Qinghai-Tibetan Comprehensive Scientific Expedition in 1975 was the discovery of a *Hipparion*-producing fauna at the northern foothills of Mount Xixiabangma (Shishapangma) along the Himalayas, about 170 km northwest of Chomolungma (Mt. Everest) (Ji, Hsu, and Huang 1980). Vertebrate fossils were mainly recovered from the lower part of the Oma (Woma) Formation at the Heigou section about 1 km north of Woma (Huang et al. 1980), Gyirong County. A small but distinctly late Miocene fauna includes *Hipparion guizhongensis, Chilotherium xizangensis, Palaeotragus microdon, Metacervulus capreolinus, Gazella gaudryi, Hyaena* sp., *Heterosminthus* sp., and *Ochotona guizhongensis*. Li and Chi (=Ji) (1981) also described two new rodents, *Plesiodipus thibetensis* and *Himalayactaga liui*, which, as pointed out by Qiu (1996), were based on teeth of different locations in the same species and belonged to an endemic Tibetan form. Additional screening of small mammals in recent years (under study by Zhaoqun Zhang at IVPP) confirms Qiu's observation.

The Gyirong *Hipparion* fauna played a significant role among the Chinese earth science community in the debate on the paleoaltitude of the Tibetan Plateau. Using the analogy of maximum elevations of known *Hipparion* faunas elsewhere in the world, Huang et al. (1980) argued that the elevation of the Gyirong Basin must have been substantially lower than it is at the present time, a case that has been further reinforced by additional evidences from Tibetan geology and paleontology (e.g., Li et al. 1981; Xu 1981; Wang et al. 2012). However, the *Hipparion* analogy failed to account for the fact that the living Tibetan wild ass (*Equus kiang*) is quite capable of surviving in very high elevations (up to 5300 m; St-Louis and Côté 2009). More recently, recognition of a series of east–west extensional basins in southern Tibet, including the Gyirong Basin, using the gravitational collapse model, suggests the attainment of substantial elevation before ~8 Ma (Pan and Kidd 1992; Harrison et al. 1995) or even ~14 Ma (Blisniuk et al. 2001).

Shen et al. (1995) first attempted a composite paleomagnetic section in the Gyirong Basin by combining parts of three sections in three different valleys and arrived at a correlation encompassing C2n–C3Bn (1.66–6.54 Ma). They seemed unaware of Gyirong's *Hipparion* fauna, as they neither cited the published work nor used the fossils to constrain their correlation. More recently, Yue et al. (2004) sampled at a much higher density, at intervals of 0.5 m and concentrating on a single section at Longgugou, and were able to detect some of the short magnetochrons missed by Shen et al. (1995). This later study arrived at a correlation similar to that by Shen et al. (1995) toward the lower end of the section, mainly by rooting the fossiliferous horizon in the late Miocene part of the magnetochron, but pulled the top section down so that the total section spans from 2An.2r to 3Br.1n (3.2–7.3 Ma). Within Yue et al.'s (2004) magnetic section, the main Gyirong fossiliferous horizon lies within the lower 20 m of the 160 m section and falls within chron 3Bn (7.14–7.21 Ma in Lourens et al. 2004)—that is, close to the Bahean/Baodean boundary.

Zanda Basin

The Zanda (=Zhada) Basin is located at the northern foothills of the western Himalayas. A spectacularly exposed sequence of fluviolacustrine sediments, more than 800 m in thickness, is preserved in a fault-bounded basin between the South Tibetan Detachment System to the southwest and the Great Counter Thrust to the northeast (Wang, Wang, and Qiu 2008; Kempf et al. 2009; Saylor et al. 2009). A branch of the modern Sutlej River, the Langqên Zangbo, deeply dissects the Zanda sediments down to the Tethyan Mesozoic basement. Vertebrate fossils in the form of a partial maxilla of a fossil giraffe *Palaeotragus* were first discovered during the 1976 Chinese Academy of Sciences expedition (Zhang et al. 1981). Additional isolated discoveries were subsequently reported: a new species of three-toed horse (*Hipparion* [*Plesiohipparion*] *zandaense*; Li and Li 1990), a metatarsal of a rhino (Meng et al. 2004), and an isolated cheek tooth of an ochotonid (Meng et al. 2005).

In 2006–2010, we conducted four short field seasons in an attempt to systematically explore the fossil potential of Zanda Basin. Fossil mammals and fishes were found throughout the section, except in the basal and top conglomerates, and in most parts of the basin. In particular, we have succeeded in finding at least two localities rich in small mammals in the lower part of the basin sequence, further helping to constrain the age relationship. The Zanda fauna documents, for the first time, ancestral forms that later gave rise to elements of the Pleistocene megafauna, such as the case of the primitive woolly rhino, *Coelodonta thibetana* (Deng et al. 2011). Field explorations are still ongoing, and this report is interim in nature.

The Zanda faunal list is still very short, but our continued fieldwork is sure to increase the list. Largely Yushean in nature, the Zanda Fauna (for the moment we choose to keep it undivided although future subdivision is possible) is mostly restricted to 130–625 m of the 800 m section. The composite fauna consists of Soricidae indet., *Nyctereutes* cf. *N. tingi*, *Vulpes* sp., *Panthera* (*Uncia*) sp., *Meles* sp., *Chasmaporthetes* sp., *Hipparion zandaense*, *Coelodonta thibetana*, *Cervavitus* sp., ?*Pseudois* sp., *Antilospira/Spirocerus* sp., *Qurliqnoria* sp., Gomphotheriidae indet., *Aepyosciurus* sp., *Nannocricetus* sp., Cricetidae gen. et sp. nov., *Prosiphneus* cf. *P. eriksoni*, *Mimomys* (*Aratomys*) *bilikeensis*, *Apodemus* sp., *Trischizolagus* cf. *T. mirificus*, *Trischizolagus* cf. *T. dumitrescuae*, and as many as four species of *Ochotona* (Deng et al. 2011:table S1). With the exception of *Coelodonta thibetana*, more precise identifications for most of these taxa will have to wait for detailed systematic studies. Nonetheless, this list gives an impression of mostly Pliocene age, such as *Nyctereutes*, *Antilospira*, *Prosiphneus*, leporid, and *Mimomys*. Furthermore, a certain degree of Tibetan endemism is detectable, such as the blue sheep *Pseudois*, ancestral Tibetan antelope *Qurliqnoria*, as well as the morphologically peculiar *Mimomys* and *Ochotona*.

There have been at least four independent attempts at paleomagnetic age determination of the Zanda strata during the last 10 years, although only three of these provided enough documentation to be evaluated here (Qian 1999; Wang, Wang, and Qiu 2008; Saylor et al. 2009), and Zhu et al. (2005) and Meng et al. (2004) mentioned a paleomagnetic section of their own, but they provided no detailed documentation. All three of these studies measured an 800+ m section for the total thickness of Zanda sediments and arrived at roughly similar magnetic reversal patterns. However, due to a lack of reliable paleontologic constraints, the previous paleomagnetic studies all arrived at different correlations to the GPTS. Our own reinterpretation of the magnetochrons (Deng et al. 2011:fig. S3) to part of C1n to C3An suggests a time span for Zanda sedimentation to range from ~400 Ka to 6.1 Ma in GPTS of ATNTS2004. Since fossils fall within the 130–625 m of the 800 m section, this translates to about 5.5–2.5 Ma for the Zanda Fauna (figure 10.4).

Kunlun Pass Basin

Conveniently located along the Qinghai-Tibetan highway, the Kunlun Pass Basin has been the subject of geologic investigations for its tectonic, seismic, and paleoenvironmental implications (e.g., Kidd and Molnar 1988; Lin et al. 2002; Wang, Wang, and Qiu 2008). Qian and Zhang (1997) first mentioned a partial metapodial of a *Hipparion* horse. Following this lead, we began to recover identifiable vertebrate fossils in 2003, and since then a modest assemblage of fossil mammals and fishes have been collected.

At an elevation of 4786–4923 m, this is one of the highest late Cenozoic vertebrate assemblages so far known. Preliminary explorations during the past eight field seasons have yielded a small collection of fossil mammals in a fluviolacustrine deposit in the lower member of the Qiangtang Formation west of the Kunlun Pass Monument. We name the new mammal assemblage the Yuzhu Fauna, after Mount Yuzhu (also referred to as Burhan Budai), the highest peak in the Kunlun Range east of the Kunlun Pass Basin. Small mammals consist of Soricidae gen. et sp. indet., *Aepyosciurus* sp., *Nannocricetus* sp., *Mimomys* sp., *Prosiphneus* sp., *Chardinomys* sp., and *Ochotona* (two species), whereas large mammals are represented by a rhino, a hipparionine horse, *Qurliqnoria* sp., a felid, *Chasmaporthetes* sp., and *Plesiogulo* sp. Overall, the Yuzhu Fauna is northern China or central Asia in character.

Of these taxa, *Chardinomys* and *Mimomys* are known from the Pliocene of northern China. Morphologically cheek teeth of the Kunlun *Prosiphneus* and *Mimomys* still have roots and have relatively low dentine tracts, comparable to those of the Gaotege Fauna in Inner Mongolia. Kunlun Pass *Aepyosciurus* is also comparable to those from locality ZD1001 of the Zanda Basin. There have been at least three attempts at a paleomagnetic chronology in the Kunlun Pass Basin (Qian and Zhang 1997; Cui et al. 1998; Song et al. 2005), although earlier ones by Qian and Zhang (1997) and by Cui et al. (1998) did not adequately document the magnetic properties of the sediments. All three share broad similarities, but that by Song et al. (2005) attempted a dense sampling that seems

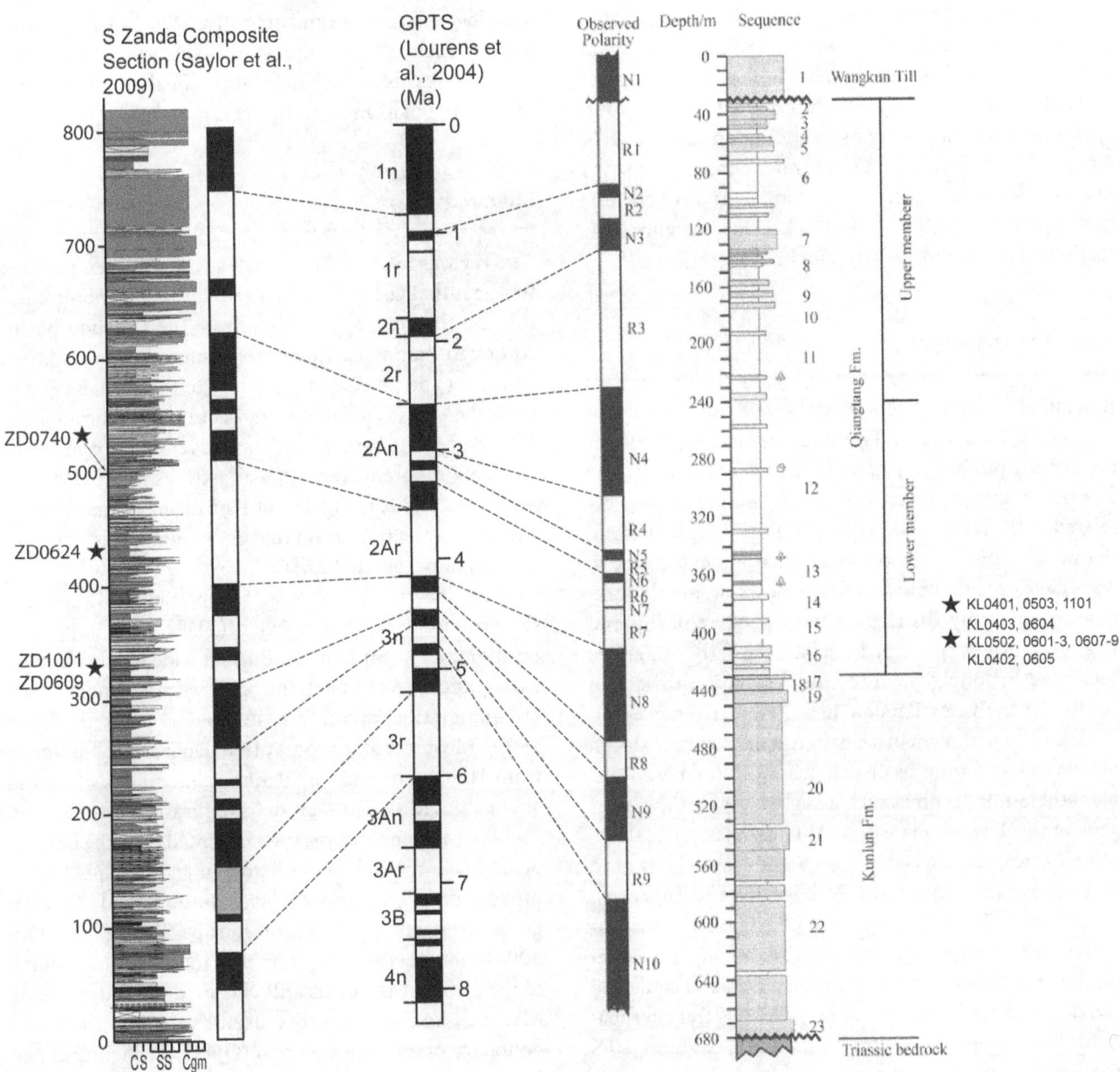

Figure 10.4 Magnetostratigraphy and geochronology of Zanda and Yuzhu faunas: (*left*) three major vertebrate fossil horizons (indicated by stars) are placed in nearest measured sections by Saylor (2008), which are in turn correlated on the basis of intrabasin sequence stratigraphic criteria, and our alternative magnetic correlation (Deng et al. 2011:fig. S3) from that by Saylor et al. (2009); (*right*) main fossil mammal horizon (indicated by star) of the Kunlun Pass Basin and reinterpreted magnetostratigraphy by Song et al. (2005).

to reveal a few short magnetochrons missed by earlier workers. Our small and large mammal localities are produced from dark gray, organic-rich siltstone layers in the lower section of the basin. A precise correlation of the fossil localities to the magnetic section by Song et al. (2005) is difficult due to the general lack of clear bedding planes in many exposures because of severe weathering,

shortage of consistent marker beds, extensive vegetation cover, and the scattered nature of fossil localities. However, by controlling a few key GPS locations (unpublished data from Chunhui Song, January 2012), all of the vertebrate fossil localities can be correlated to beds 14 and 15 in the lower part of the Qiangtang Formation, below a locally prominent lacustrine bed (bed 12 of Song et al.

2005), although a certain degree of uncertainty is inherent in such an estimate. Beds 14 and 15 fall in a long, predominantly reversed chron (R6–R7 of Song et al. 2005:fig. 7), which were correlated to chron 2r by Song et al. (2005). Song et al. (2005) interpreted their Kunlun Pass magnetostratigraphy as including an age range of 3.6–0.5 Ma (Gauss to part of Brunhes chrons). Our own fossil evidence, however, suggests an alternative correlation to part of C1n–C3n.3n (~0.7–4.8 Ma; see figure 10.4; further discussion follows in "Biochronology").

BIOCHRONOLOGY

A detailed biostratigraphic record is still not possible for most of the basins in the Tibetan Plateau due to the short history and preliminary nature of explorations and to the fact that many basins are not very fossiliferous, possibly related to the low productivity of high-plateau biotas and the peculiarities of depositional environments (lack or poor development of paleosols). As a result, fossil localities are unevenly distributed both geographically and stratigraphically, resulting in short fossiliferous intervals separated by long gaps in the fossil record. This lack of continuity in the fossil record in the Tibetan Plateau hinders the establishment of a biostratigraphy comparable in details to those from basins elsewhere in northern Asia. Nonetheless, it becomes increasingly clear that the Neogene of the Tibetan Plateau is of importance, as its strata offers critical sections where two of the most widely used first appearance datum, the Proboscidea and *Hipparion* FADs, occur. The following discussion about the biochronology is restricted to Asian Land Mammal ages represented from basins in the plateau. We also exclude the Linxia Basin from this discussion, which is treated separately in Deng et al. (chapter 9, this volume). Figure 10.5 provides a summary of the chronologic relationships.

Xiejian

This is the only Neogene ALMA that is based on fossil mammals in the Tibetan Plateau, and the Xiejian is the most challenging to characterize due to its relatively poorly defined fauna. Xiejian, as well as the stratotype Xiejia section, is treated in Qiu et al. (chapter 1, this volume), Qiu, Wang, and Li (chapter 5, this volume) and Meng et al. (chapter 3, this volume). Another important fauna, the Zhangjiaping fauna from the Lanzhou Basin, seems to belong to the Xiejian as well. However, faunas from this age are mostly characterized by Oligocene left-

overs, such as *Tataromys, Yindirtemys, Parasminthus, Hyaenodon*, and the like, and definitive early Miocene forms are few, such as *Democricetodon*. A lack of adequate large mammal fossils in Xiejian localities further obscures the true nature of its mammalian composition.

Shanwangian

Two regions in the Tibetan Plateau produce mammals of this age: the Xishuigou Fauna from the Tabenbuluk Basin, and the Duitinggou fauna from the Lanzhou Basin. Although mammals from these faunas are far from as abundant and diverse as compared to the Xiejiahe Fauna from the stratotype Shanwang section in eastern China (Qiu and Qiu, chapter 4, this volume) due to their considerably different depositional settings and paleoenvironments, the Xishuigou and Duitinggou faunas, nonetheless, play an important role in defining the lower limit of the Shanwangian ALMA.

Here, the first appearance of proboscideans, known as the Proboscidean Datum Event (PDE), is an important immigration event that has Eurasia-wide biochronological significance, although the precise timing and number of immigration events are still being debated (Tassy 1990). Most recently, report of a Paleogene proboscidean from the Bugti Hills of Pakistan further casts doubt about a single dispersal event into Eurasia (Antoine et al. 2009). This complexity is further hinted at by the first appearance in the Tabenbuluk area of a rather specialized proboscidean, *Platybelodon dangheensis*, instead of a more generalized gomphothere in Europe (Wang and Qiu 2002). Nonetheless, the proboscidean FAD, at least north of the Himalayas, seems still clustered around the early Miocene, and it may prove useful to define the lower boundary of the Shanwangian (Qiu et al., chapter 1, and Qui, Wang, and Li, chapter 5, both this volume).

As briefly mentioned earlier, *Platybelodon dangheensis* from Tabenbuluk is associated with a reinterpreted paleomagnetic age of ~19.2–19.4 Ma (B.-y. Wang et al. 2003; X. Wang et al. 2003). Another possible proboscidean FAD is in the Upper Red Mudstone Member of the Aoerban Formation in Inner Mongolia (Wang et al. 2009). Proboscidean material from the Aoerban site is so far known from only a few enamel fragments of cheek teeth, but the Inner Mongolian site has the distinct advantage of an associated fauna well represented by diverse small mammals.

Small mammals, increasingly important in determining biochronologies, are generally poorly represented in faunas from the Tibetan Plateau. Appearances of "mod-

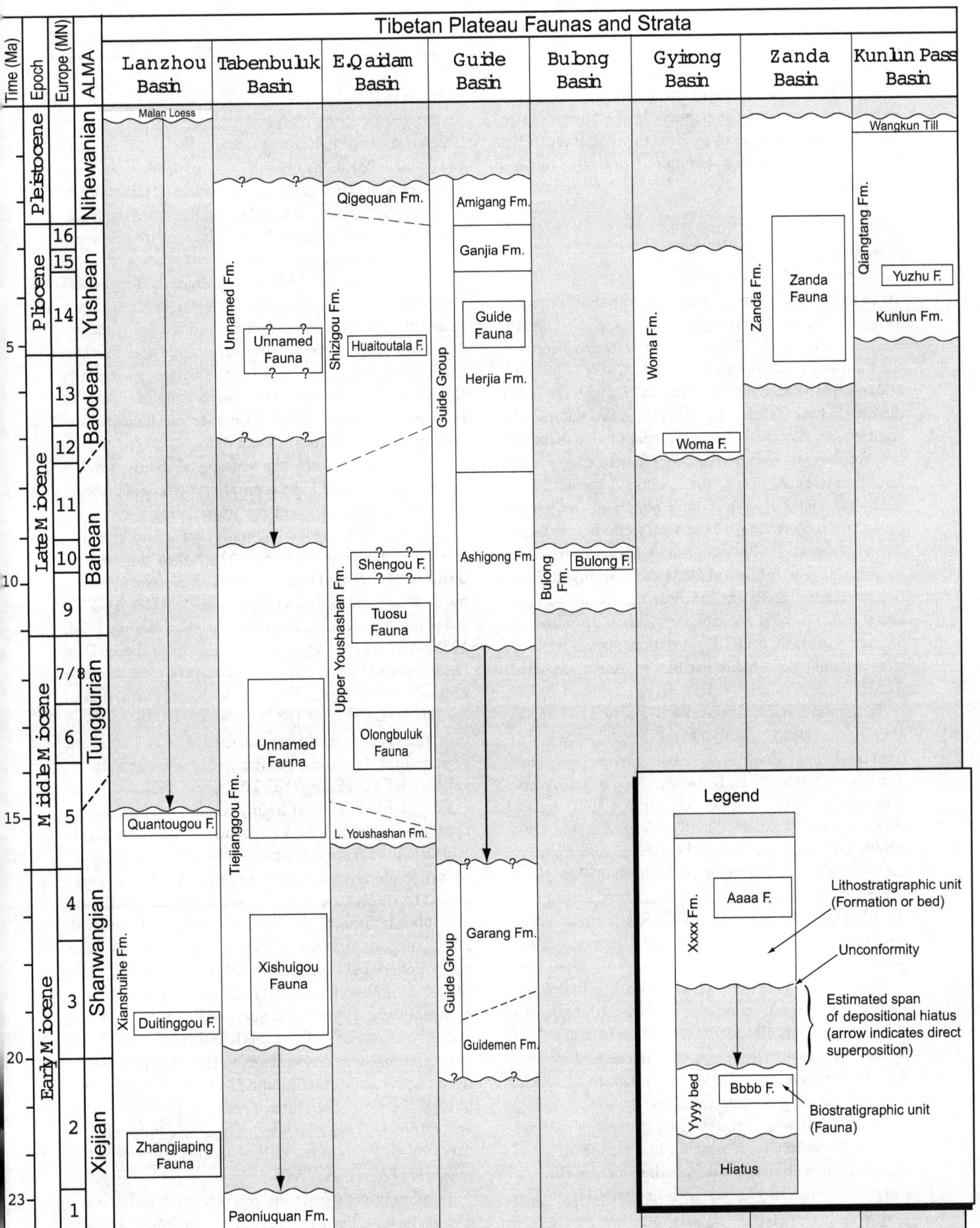

Figure 10.5 Summary of fauna and strata in the Tibetan Plateau and their correlations.

ern cricetids," such as *Democricetodon* and *Megacricetodon*, are important components in the Shanwangian ALMA (Qiu and Qiu, chapter 4, this volume). In the Duitinggou Fauna, for example, cricetids are well established.

Tunggurian

Three basins in and around the plateau produce faunas of Tunggurian age: the Olongbuluk Fauna of the Qaidam Basin (Wang et al. 2007), the Quantougou Fauna of the Lanzhou Basin (Qiu 2000, 2001a, 2001b; Qiu et al. 2001), and the Shinanu, Zengjia, and Laogou faunas of the Linxia Basin (Deng et al., chapter 9, this volume). In addition, an isolated large metacarpal of a chalicothere (*Phyllotillon*) was found from a conglomerate layer in the "middle Miocene" part of the Xishuigou section in Tabenbuluk, and *Kubanochoerus*, a common Tunggurian genus, was reported in the Heixiagou section of the Guide Basin (Gu et al. 1992). With the exception of the spectacular ontogenetic series of *Platybelodon* from the Linxia Basin, none of the basins can rival the classic Tunggur localities from Inner Mongolia in richness and diversity of fossil mammals, and Tunggurian mammals from the plateau do not play a major role in the formulation of this ALMA.

Tunggurian sediments are well exposed in the Qaidam Basin, but so far only a handful of taxa are known from the Olongbuluk Fauna: *Hispanotherium matritense*, *Acerorhinus tsaidamensis*, *Lagomeryx tsaidamensis*, *Stephanocemas palmatus*, and possibly *Dicroceros* (Deng and Wang 2004b; Wang et al. 2007; Wang, Xie, and Dong 2009). Of these, the elasmothere *Hispanotherium* is a Eurasia-wide genus common in Astaracian/Tunggurian strata (Deng 2003). *Lagomeryx* and *Stephanocemas*, on the other hand, first appear earlier in Shanwangian strata, despite their relative abundance and diversity in Tunggurian localities.

The Quantougou Fauna from the Lanzhou Basin is entirely based on small mammals (Qiu 2000, 2001a, 2001b) and is stratigraphically near the top of the Lanzhou Basin Neogene sequence. This fauna is characterized by zapodids (*Heterosminthus*), dipodids (*Protalactaga*), glirids (*Microdyromys*), gerbillids (*Mellalomys*, *Myocricetodon*), and relatively diverse cricetids (*Megacricetodon*, *Ganocricetodon*, *Paracricetulus*, *Plesiodipus*) that are similar in taxonomic composition to those from the Moergen Fauna of Inner Mongolia. There seems little doubt that the Quantougou Fauna is of late Tunggurian age.

Bahean

Tibetan strata of Bahean age include the Tuosu fauna of the Qaidam Basin, the Bulong Fauna of the Bulong Basin, the Guonigou Fauna of the Linxia Basin, and possibly part of the Shengou Fauna of the Qaidam Basin and the Dashengou Fauna of the Linxia Basin. Although most of these faunas are not particularly diverse, fossil horses from the Qaidam, Bulong, and Linxia basins have the best potential for determining the FAD for *Hippotherium/Hipparion*, in addition to other important Bahean elements, such as *Dinocrocuta*, that offer constraints for this age. Furthermore, faunas from the interior of the plateau, particularly those from Qaidam, show a certain measure of endemism in bovids, possibly related to isolation mechanisms created by the surrounding mountains.

In continental basins, the widespread, nearly simultaneous appearance of *Hipparion/Hippotherium* represents an ideal immigrant event for formulating a biochron in early late Miocene land mammal ages of Eurasia. The three-toed horse is relatively well worked out both in terms of its ancestral lineages in North America (Woodburne 2007) and in its FAD approximately 11.1–11.2 Ma (near the top of C5r.2r) in Europe (e.g., Bernor et al. 1988; Woodburne 2009). In southern Asia, the FAD of "*Hipparion* s.l." occurs at 10.7–10.9 Ma (within the long normal magnetochron C5n) in the Nagri Formation of the Siwaliks, Pakistan (Barry et al. 2002). In northern Asia, the Qaidam Basin has emerged to be an important region for a precisely constrained biostratigraphy of *Hipparion* FAD (Wang et al. 2011). In a <300 m and laterally extensive band of highly fossiliferous section in the Quanshuiliang area, we recently recovered multiple *Hipparion*-producing localities. Although a direct magnetostratigraphy is not yet available, lithostratigraphic correlations to an existing magnetic section (Huaitoutala) on the opposite limb of the Keluke Anticline (Fang et al. 2007) suggests that the local *Hipparion* FAD falls near the boundary between chrons C5r.1r and C5r.1n (11.12 Ma in Lourens et al. 2004). The systematics of the Quanshuiliang *Hipparion* is also unclear, as the material mainly consists of isolated teeth and foot bones; however, preliminary comparisons in the enamel patterns in the upper cheek teeth indicate an intermediate morphology between the North American *Cormohipparion* and European *Hippotherium primigenium* (unpublished data), making this site a critical locality to explore the *Hipparion* FAD in East Asia.

Small mammals from the recently established Shengou Fauna have been compared to those from the lower part of the Bahe Formation (Qiu and Li 2008). Of the 12 identifiable genera from Shengou, 7 are shared with those from the Bahe Formation: *Sciurotamias, Lophocricetus, Protalactaga, Myocricetodon, Nannocricetus, Pararhizomys,* and *Ochotona*; however, more progressive status for its *Lophocricetus* and the murids suggests a somewhat younger age than the Bahe small mammals. Increased diversity for hipparionine horses—that is, *Hipparion* cf. *H. chiai, Hipparion weihoense,* and *Hipparion teilhardi* (Deng and Wang 2004a)—also suggests a later age than the Tuosu Fauna. Qiu and Li (2008) considered the Shengou Fauna, which is currently not constrained by magnetic data, to be either late Bahean or early Baodean in age.

The Bulong Fauna is also of potential significance in the characterization of the Bahean ALMA. In particular, *Dinocrocuta gigantea* has been regarded as one of the key taxa for the Bahean age (Qiu, Xie, and Yan 1988), although Zhang (2005) argued for a later occurrence of this genus in China. The Bulong Fauna is also considered to be a Bahean equivalent because of the occurrence of the primitive *Hipparion, H. xizangense* (Qiu, Huang, and Guo 1987).

Baodean

Besides the Yangjiashan Fauna from the Linxia Basin (Deng et al., chapter 9, this volume), only the Gyirong Basin produces a Baodean assemblage, the Gyirong Fauna, in the Tibetan Plateau with common "*Hipparion*" faunal elements such as *Hipparion guizhongensis, Chilotherium xizangensis, Palaeotragus microdon, Metacervulus capreolinus, Gazella gaudryi,* "*Hyaena*" sp., *Heterosminthus* sp., and *Ochotona guizhongensis.* As is the case elsewhere in northern China, the Bahean/Baodean boundary is not easily defined among late Miocene faunas in the Tibetan Plateau, and the Gyirong Fauna, magnetically calibrated to be 7.0–6.7 Ma (Yue et al. 2004), is generally comparable to those from the classic Baode localities. A new small mammal assemblage is currently under study by Zhaoqun Zhang at IVPP.

Yushean

Four basins within the Plateau produced Yushean faunas: the Huaitoutala Fauna of the Qaidam Basin, the Guide Fauna of the Guide Basin, the Zanda Fauna of the Zanda Basin, and the Yuzhu Fauna of the Kunlun Pass Basin. All of these are low-richness and low-diversity fau-

nas, presumably related to environmental deterioration in the Pliocene, and as a result they do not yet have a significant impact on the ALMA system.

Despite our extensive efforts at screen washing, the Huaitoutala Fauna in the Qaidam Basin is still represented by only a handful of taxa: *Ochotona*, Leporidae indet., *Orientalomys/Chardinomys, Micromys, Pseudomeriones,* and Soricidae indet. (Wang et al. 2007). *Chardinomys* is a typical early Pliocene form in northern Asia, suggesting an early Yushean age. The Huaitoutala Fauna is represented by a single locality, IVPP locality CD9822, that falls in a short normal magnetic interval correlated to chron C3n.3n (4.80–4.90 Ma; Fang et al., 2007).

The Guide Fauna has not been worked on in recent years, and Zheng, Wu, and Li's (1985) preliminary composite faunal list still stands as the most current: *Myospalax arvicolinus, Mimomys* sp., Leporidae indet., *Ochotona* sp., *Anancus sinensis, Anancus cuneatus, Felis* sp., *Hipparion* sp. *Chilotherium* sp., *Axis shansius,* and *Gazella kueitensis.* Zheng, Wu, and Li (1985) pointed to the co-occurrence of *Anancus* and *Myospalax*, an association that was also seen in Yushe (Shanxi) and Lingtai (Gansu), as the most indicative of its age relationship. However, recent biostratigraphic work suggests that *Anancus* is present in the Taoyang Member of the Gaozhuang Formation and in the Mazegou Formation (Tedford et al. 1991), whereas *Myospalax* is restricted to the Haiyan Formation (Flynn, Tedford, and Qiu 1991). Guide fossils were recovered from three layers that span more than half of the ~200 m section at Erdaogou, and the fossiliferous horizons fall in three short normal and two reversed magnetic intervals that were correlated to chrons C3n.4n (Thvera) to C3n.1n (Cochiti) (4.187–5.235 Ma) by Song et al. (2003).

The Zanda Basin faunal list is still very much a work in progress. Because of the well-exposed, flat-lying, and continuous strata, a detailed biostratigraphy can be constructed with ranges of individual taxa based on their elevations (to be published elsewhere). So far, the composite Zanda Fauna is represented by at least 24 taxa: Soricidae indet., *Nyctereutes* cf. *N. tingi, Vulpes* sp., *Panthera (Uncia)* sp., *Meles* sp., *Chasmaporthetes* sp., *Hipparion zandaense, Coelodonta thibetana, Cervavitus* sp., ?*Pseudois* sp., *Antilospira/Spirocerus* sp., *Qurliqnoria* sp., Gomphotheriidae indet., *Aepyosciurus* sp., *Nannocricetus* sp., Cricetidae gen. et sp. nov., *Prosiphneus* cf. *P. eriksoni, Mimomys (Aratomys) bilikeensis, Apodemus* sp., *Trischizolagus* cf. *T. mirificus, Trischizolagus* cf. *T. dumitrescuae,* and as many as four species of *Ochotona* (Deng et al. 2011:table S1). Of these, *Mimomys* occurs fairly low in the section (IVPP localities ZD0609 and 0904), represented by three upper first molars and one lower molar. In crown heights, the Zanda *Mimomys* is most comparable to *Mimomys (Aratomys) bilikeensis* from the early Pliocene Bilike locality of Inner Mongolia, which is the earliest representative of arvicoline rodents in China (Qiu and Storch 2000). Arvicoline rodents first appear in the early Pliocene of western Siberia (Repenning 2003) and shortly afterward dispersed to northern Asia (Qiu and Storch 2000), Europe (Fejfar et al. 1997; Chaline et al. 1999), and North America (Repenning 1987; Lindsay et al. 2002; Bell et al. 2004). The appearance of these rodents thus is a highly age-diagnostic event throughout northern continents. IVPP localities ZD0609 and 0904 fall in a relatively long normal chron that we interpret to be C3n.4n—that is, 4.997–5.235 Ma in the early Yushean. Farther up in the section, *Nyctereutes tingi* is also diagnostic as an early Yushean form (Tedford and Qiu 1991), first appearing in the Gaozhuang Formation in the Yushe Basin (Tedford et al. 1991).

At an altitude of 4786–4923 m, the Yuzhu Fauna in the Kunlun Pass Basin is the highest elevation mammal assemblage so far known. Sandwiched between two glacial tills, the Lower Member of the Qiangtang Formation produced a low-diversity assemblage based on very fragmentary material. Of these, the most age diagnostic are *Chasmaporthetes*, *Prosiphneus*, *Mimomys*, and *Chardinomys*, all occurring mostly in the late Miocene to Pliocene of eastern Asia. Kunlun Pass *Prosiphneus* is similar to *Prosiphneus* cf. *P. eriksoni* from Bilike, Inner Mongolia, but more primitive than those from Gaotege, in size and cheek teeth (condition of the root, dentine tract height, m1 crown morphology). More locally (within Tibetan Plateau), the Kunlun Pass *Prosiphneus* seems slightly more derived than that from the Zanda Basin. *Mimomys* is also close to *Aratomys bilikeensis* from Bilike, but different from those from Gaotege, in its small size, rooted cheek teeth, lack of cementum, as well as similar dentine tract height. Kunlun Pass vertebrate fossil localities are mostly restricted to a dark-gray to black, organic-rich siltstone in a reversed magnetic zone correlated to the chron C2r.2r (2.1–2.6 Ma) in the Pleistocene (Song et al. 2005). As discussed previously, however, small mammals from the Kunlun Pass Basin seem to suggest an age more comparable to early to middle Pliocene in Inner Mongolia and the Zanda Basin. This leads us to an alternative correlation of the magnetostratigraphy in placing our Yuzhu Fauna in the lower part of C2Ar (Gilbert Chron 3.6–4.2 Ma). In addition to a better fit for our biochronological assessment, our reinterpretation of Song et al. (2005) also resulted in a faster depositional rate for the glacial moraines (Kunlun Formation) below the fine-grained

Qiangtang Formation (see figure 10.4), in contrast to Song et al.'s correlation that assumed a similar depositional rate for both the Kunlun Formation and the Qiangtang Formation (Song et al. 2005:fig. 7). Kidd and Molnar (1988:fig. 6) identified pyroxenite clasts within the lower moraine (Kunlun Formation of Song et al. 2005) that are sourced 30 km to the west on the north side of the Xidatan-Tuosuohu-Maqu fault. Using this 30 km offset, Kidd and Molnar estimated a slip rate for the Kunlun Pass fault of 13 mm/yr assuming an age of 2.4 Ma for the lower moraine. Our own revised age estimates would translate to a rate of ~ 8 mm/yr, which is close to the lower limit of current estimates (Peter Molnar, pers. comm.).

ACKNOWLEDGMENTS

We greatly appreciate the contributions from numerous field participants, whose hard work and diligent collecting have made possible the current understanding of the biostratigraphy of the Tibetan Plateau: William Downs, Meeman Chang, Yang Wang, Chunfu Zhang, Shanqin Chen, Junyi Xie, Xijun Ni, Juan Liu, Ning Wang, Min Zhao, Sukuan Hou, Gengjiao Chen, Alex Wang, as well as our Chinese and Tibetan drivers Fuqiao Shi, Wenqing Feng, Qiang Li, Qiuyuan Wang, Laba, and Basang. Shifeng Wang and Joel Saylor provided important information about Zanda stratigraphy and tips on fossil localities. We thank Mikael Fortelius, Asta Rosenström-Fortelius, and Håkan Wahlquist for their assistance in translating the Swedish literature and access to the Bohlin archives. Funding for fieldwork and travel are provided by the Chinese National Natural Science Foundation (Nos. 40702004, 49872011, and 40128004), Knowledge Innovation Program of the Chinese Academy of Sciences (KZCX2-YW-Q09), CAS/SAFEA International Partnership Program for Creative Research Teams, Major Basic Research Projects (2006CB806400) of Ministry of Science and Technology of China, Chinese Academy of Science Outstanding Overseas Scholar Fund (KL205208), National Science Foundation (U.S.) (EAR-0446699, EAR-0444073, EAR-0958704), and National Geographic Society (Nos. 6004–97 and 6771–00).

REFERENCES

Antoine, P.-O., J.-L. Welcomme, L. Marivaux, I. Baloch, M. Benammi, and P. Tassy. 2009. First record of Paleogene Elephantoidea (Mammalia, Proboscidea) from the Bugti Hills of Pakistan. *Journal of Vertebrate Paleontology* 23(4):977–980.

Barry, J. C., M. E. Morgan, L. J. Flynn, D. Pilbeam, A. K. Behrens-meyer, S. M. Raza, I. A. Khan, C. Badgley, J. Hicks, and J. Kelley. 2002. Faunal and environmental change in the late Miocene Siwaliks of Northern Pakistan. *Paleobiology* 28(2):1–71.

Bell, C. J., E. L. Lundelius, Jr., A. D. Barnosky, R. W. Graham, E. H. Lindsay, D. R. Ruez, Jr., H. A. Semken, Jr., S. D. Webb, R. J. Za-krzewski, and M. O. Woodburne. 2004. The Blancan, Irvingto-nian, and Rancholabrean mammal ages. In *Late Cretaceous and Cenozoic Mammals of North America: Biostratigraphy and Geo-chronology*, ed. M. O. Woodburne, pp. 232–314. New York: Co-lumbia University Press.

Bernor, R. L., J. Kovar-Eder, D. Lipscomb, F. Rögl, S. Sen, and H. Tobien. 1988. Systematic, stratigraphic, and paleoenvironmental contexts of first-appearing *Hipparion* in the Vienna Basin, Aus-tria. *Journal of Vertebrate Paleontology* 8(4):427–452.

Blisniuk, P. M., B. R. Hacker, J. Glodny, L. Ratschbacher, S. Bi, Z. Wu, M. O. McWilliams, and A. Calvert. 2001. Normal faulting in central Tibet since at least 13.5 Myr ago. *Nature* 412(6847): 628–632.

Bohlin, B. 1935. *Tsaidamotherium hedini*, n. g., n. sp. *Geografiska An-naler* 17:66–74.

Bohlin, B. 1937. Eine Tertiäre säugetier-fauna aus Tsaidam. *Sino-Swedish Expedition Publication (Palaeontologia Sinica, ser. C, 14)* 1:3–111.

Bohlin, B. 1942. The fossil mammals from the Tertiary deposit of Taben-buluk, Western Kansu, Part I: Insectivora and Lagomor-pha. *Sino-Swedish Expedition Publication (Palaeontologia Sinica, n.s. C, No. 8A)* 20:1–113.

Bohlin, B. 1946. The fossil mammals from the Tertiary deposit of Taben-buluk, Western Kansu, Part II: Simplicidentata, Carniv-ora, Artiodactyla, Perissodactyla, and Primates. *Sino-Swedish Expedition Publication (Palaeontologia Sinica, n.s. C, No. 8B)* 28:1–259.

Bohlin, B. 1960. Geological reconnaissances in western Kansu and Kokonor. *Sino-Swedish Expedition Publication* 44:1–79.

Chaline, J., P. Brunet-Lecomte, S. Montuire, L. Viriot, and F. Cou-rant. 1999. Anatomys of the arvicoline radiation (Rodentia): Paleogeographical, palaeoecological history and evolutionary data. *Annales Zoologici Fennici* 36:239–267.

Chang, M., X. Wang, H. Liu, D. Miao, Q. Zhao, G. Wu, J. Liu, Q. Li, Z. Sun, and N. Wang. 2008. Extraordinarily thick-boned fish linked to the aridification of the Qaidam Basin (northern Ti-betan Plateau). *Proceedings of the National Academy of Sciences* 105(36):13246–13251.

Chen, G. and J. Liu. 2007. First fossil barbin (Cyprinidae, Teleostei) from Oligocene of Qaidam Basin in northern Tibetan Plateau. *Vertebrata PalAsiatica* 45(4):330–341.

Cui, Z.-j., Y.-q. Wu, G.-n. Liu, D.-k. Ge, Q.-h. Xu, Q.-q. Pang, and J.-r. Yin. 1998. Records of natural exposures on the Kunlun Shan Pass of Qinghai-Xizang Highroad. In *Uplift and Environmental Changes of Qinghai-Xizang (Tibetan) Plateau in the Late Cenozoic*, ed. Y.-f. Shi, J.-j. Li, and B.-y. Li, pp. 81–114. The Series of Studies on Qinghai-Xizang (Tibetan) Plateau. Guangzhou: Guangdong Sci-ence & Technology Press.

Deng, T. 2003. New material of *Hispanotherium matritense* (Rhinoc-erotidae, Perissodatyla) from Laogou of Hezheng County (Gansu, China), with special reference to the Chinese Middle Miocene elasmotheres. *Geobios* 36(2):141–150.

Deng, T. and X. Wang. 2004a. Late Miocene *Hipparion* (Equidae, Mammalia) of eastern Qaidam Basin in Qinghai, China. *Verte-brata PalAsiatica* 42(4):316–333.

Deng, T. and X. Wang. 2004b. New material of the Neogene rhinoc-erotids from the Qaidam Basin in Qinghai, China. *Vertebrata PalAsiatica* 42(3):216–229.

Deng, T., X. Wang, M. Fortelius, Q. Li, Y. Wang, Z. J. Tseng, G. T. Takeuchi, J. E. Saylor, L. K. Säilä, and G. Xie. 2011. Out of Tibet: Pliocene woolly rhino suggests high-plateau origin of Ice Age megaherbivores. *Science* 333:1285–1288.

Dong, W. 2007. New material of Muntiacinae (Artiodactyla, Mam-malia) from the late Miocene of northeastern Qinghai-Tibetan Plateau, China. *Comptes Rendus Palevol* 6(5):335–343.

Fang, X., M. Yan, R. Van der Voo, D. K. Rea, C. Song, J. M. Parés, J. Gao, J. Nie, and S. Dai. 2005. Late Cenozoic deformation and uplift of the NE Tibetan Plateau: Evidence from high-resolution magnetostratigraphy of the Guide Basin, Qinghai Province, China. *Geological Society of America Bulletin* 117(9):1208–1225.

Fang, X., W. Zhang, Q. Meng, J. Gao, X. Wang, J. King, C. Song, S. Dai, and Y. Miao. 2007. High-resolution magnetostratigraphy of the Neogene Huaitoutala section in the eastern Qaidam Basin on the NE Tibetan Plateau, Qinghai Province, China and its impli-cation on tectonic uplift of the NE Tibetan Plateau. *Earth and Planetary Science Letters* 258:293–306.

Fejfar, O., W.-D. Heinrich, M. A. Pevzner, and E. A. Vangengeim. 1997. Late Cenozoic sequences of mammalian sites in Eurasia: An updated correlation. *Palaeogeography, Palaeoclimatology, Palaeoecology* 133(3–4):259–288.

Flynn, L. J., W. R. Downs, N. O. Opdyke, K. Huang, E. H. Lindsay, J. Ye, G. Xie, and X. Wang. 1999. Recent advances in the small mammal biostratigraphy and magnetostratigraphy of Lanzhou Basin. International Symposium and Field Workshop on Paleo-sols and Climatic Change, Lanzhou, China, 1999.

Flynn, L. J., R. H. Tedford, and Z.-x. Qiu. 1991. Enrichment and sta-bility in the Pliocene mammalian fauna of North China. *Paleobi-ology* 17:246–265.

Gilder, S., Y. Chen, and S. Sen. 2001. Oligo-Miocene magnetostratig-raphy and rock magnetism of the Xishuigou section, Subei (Gansu Province, western China), and implications for shallow inclinations in central Asia. *Journal of Geophysical Research* 106(B12):30505–30521.

Gu, Z.-g., S.-h. Bai, X.-t. Zhang, Y.-z. Ma, S.-h. Wang, and B.-y. Li. 1992. Tertiary stratigraphy of the Guide and Hualong basins, Qinghai Province. *Journal of Stratigraphy* 16(2):96–104.

Harrison, T. M., P. Copeland, W. S. F. Kidd, and O. M. Lovera. 1995. Activation of the Nyainqentanghla Shear Zone: Implications for uplift of the southern Tibetan Plateau. *Tectonics* 14(3):658–676.

Hoffmann, R. S. 1989. The Tibetan Plateau fauna, a high altitude desert associated with the Sahara-Gobi. Fifth International The-riological Congress, Rome, 1989.

Hoffmann, R. S. 1991. The Tibetan Plateau fauna, a high altitude desert associated with the Sahara-Gobi. In *Mammals of the Pa-laearctic Desert: Status and Trends in the Sahara-Gobi Region*, ed. J. A. McNeely and V. Neronov, pp. 285–297. Moscow: Russian Academy of Sciences.

Huang, W.-b., H.-x. Ji, W.-y. Chen, C.-q. Hsu, and S.-h. Zheng. 1980. Pliocene stratum of Guizhong and Bulong Basin, Xizang. In *Paleontology of Tibet, Part 1*, ed. Qinghai-Tibetan Plateau

Comprehensive Scientific Investigation Team of Chinese Academy of Sciences, pp. 4–17. Qinghai-Tibetan Plateau Scientific Investigation Series. Beijing: Science Press.

Ji, H.-x., C.-q. Hsu, and W.-b. Huang. 1980. The *Hipparion* fauna from Guizhong Basin, Xizang. In *Paleontology of Tibet, Part 1*, ed. Qinghai-Tibetan Plateau Comprehensive Scientific Investigation Team of Chinese Academy of Sciences, pp. 18–32. Qinghai-Tibetan Plateau Scientific Investigation Series. Beijing: Science Press.

Kempf, O., P. M. Blisniuk, S. Wang, X. Fang, C. Wrozyna, and A. Schwalb. 2009. Sedimentology, sedimentary petrology, and paleoecology of the monsoon-driven, fluvio-lacustrine Zhada Basin, SW Tibet. *Sedimentary Geology* 222(1–2):27–41.

Kidd, W. S. F., and P. Molnar. 1988. Quaternary and active faulting observed on the 1985 Academia Sinica—Royal Society Geotraverse of Tibet. *Philosophical Transactions of the Royal Society of London*, ser. A, *Mathematical and Physical Sciences* 327(1594): 337–363.

Li, C.-k. and H. Chi. 1981. Two new rodents from Neogene of Chilong Basin, Tibet. *Vertebrata PalAsiatica* 19(3):246–255.

Li, C.-k., W.-y. Wu, and Z.-d. Qiu. 1984. Chinese Neogene: Subdivision and correlation [in Chinese with English abstract]. *Vertebrata PalAsiatica* 22(3):163–178.

Li, F.-l. and D.-l. Li. 1990. Latest Miocene *Hipparion* (*Plesiohipparion*) of Zanda Basin. In *Paleontology of the Ngari Area, Tibet (Xi Zang)*, ed. Z. Yang and Z. Nie, pp. 186–193. Wuhan, China: University of Geological Science Press.

Li, J.-j. 1995. *Uplift of Qinghai-Xizang (Tibet) Plateau and Global Change*. Lanzhou: Lanzhou University Press.

Li, J.-j., B.-y. Li, F.-b. Wang, Q.-s. Zhang, S.-x. Wen, and B.-x. Zheng. 1981. The process of the uplift of the Qinghai-Xizang Plateau. In *Geological and Ecological Studies of Qinghai-Xizang Plateau. Volume 1, Geology, Geological History and Origin of Qinghai-Xizang Plateau*, ed. D.-s. Liu, pp. 111–118. Beijing: Science Press.

Lin, A.-m., B. Fu, J. Guo, Q. Zeng, G. Dang, W. He, and Y. Zhao. 2002. Co-seismic strike-slip and rupture length produced by the 2001 Ms 8.1 Central Kunlun earthquake. *Science* 296:2015–2017.

Lindsay, E. H., Y. U. N. Mou, W. Downs, J. Pederson, T. S. Kelly, C. Henry, and J. I. M. Trexler. 2002. Recognition of the Hemphillian/Blancan boundary in Nevada. *Journal of Vertebrate Paleontology* 22(2):429–442.

Lourens, L., F. Hilgren, N. J. Shackleton, J. Laskar, and J. Wilson. 2004. The Neogene Period. In *A Geologic Time Scale 2004*, ed. F. M. Gradstein, J. G. Ogg, and A. G. Smith, pp. 409–440. Cambridge: Cambridge University Press.

Mascarelli, A. L. 2009. Quaternary geologists win timescale vote. *Nature* 459:624.

Meng, X., D. Zhu, Z. Shao, C. Yang, J.-e. Han, J. Yu, and Q. Meng. 2005. Discovery of fossil teeth of Pliocene *Ochotona* in the Zanda basin, Ngari, Tibet, China. *Geological Bulletin of China* 24(12): 1175–1178.

Meng, X., D. Zhu, Z. Shao, C. Yang, L. Sun, J. Wang, T. Han, J.-j. Du, J.-e. Han, and J. Yu. 2004. Discovery of rhinoceros fossils in the Pliocene in the Zanda basin, Ngari, Tibet. *Geological Bulletin of China* 23(5–6):609–612.

Molnar, P. 2005. Mio-Pliocene growth of the Tibetan Plateau and evolution of East Asian climate. *Palaeontologia Electronica* 8(1):1–23.

Nehring, C. W. A. 1883. Eine fossile *Siphneus*-Art (*Siphneus arvicolinus* n. sp.) aus lacustrinen Ablagerungen am oberen Hoangho. *Sitzungsberichte der Gesellschaft Naturforschender Freunde zu Berlin* 1883:19–24.

Pan, Y. and W. S. F. Kidd. 1992. Nyainqentanglha shear zone: A late Miocene extensional detachment in the southern Tibetan Plateau. *Geology* 20(9):775–778.

Qian, F. 1999. Study on magnetostratigraphy in Qinghai-Tibetan plateau in late Cenozoic. *Journal of Geomechanics* 5(4):22–34.

Qian, F. and J. Zhang. 1997. Study on the magnetic stratigraphy of the Qiangtang Formation and the neotectonism. *Journal of Geomechanics* 3(1):50–56.

Qiu, Z.-d. 1996. *Middle Miocene Micromammalian Fauna from Tunggur, Nei Mongol*. Beijing: Academic Press.

Qiu, Z.-d. 2000. Insectivore, dipodoidean and lagomorph from the middle Miocene Quantougou Fauna of Lanzhou, Gansu. *Vertebrata PalAsiatica* 38(4):287–302.

Qiu, Z.-d. 2001a. Cricetid rodents from the middle Miocene Quantougou Fauna of Lanzhou, Gansu. *Vertebrata PalAsiatica* 39(3): 204–214.

Qiu, Z.-d. 2001b. Glirid and Gerbillid rodents from the middle Miocene Quantougou Fauna of Lanzhou, Gansu. *Vertebrata PalAsiatica* 39(4):297–305.

Qiu, Z.-d. and Q. Li. 2008. Late Miocene Micromammals from the Qaidam Basin in the Qinghai-Xizang Plateau. *Vertebrata PalAsiatica* 46(4):284–306.

Qiu, Z.-d. and G. Storch. 2000. The early Pliocene micromammalian fauna of Bilike, Inner Mongolia, China (Mammalia: Lipotyphla, Chiroptera, Rodentia, Lagomorpha). *Senckenbergiana Lethaea* 80(1):173–229.

Qiu, Z.-x. 1989. The Chinese Neogene mammalian biochronology—its correlation with the European Neogene mammalian zonation. In *European Neogene Mammal Chronology*, ed. E. H. Lindsay, V. Fahlbusch, and P. Mein, pp. 527–556. New York: Plenum Press.

Qiu, Z.-x., W.-l. Huang, and Z.-h. Guo. 1987. The Chinese hipparionine fossils. *Palaeontologia Sinica*, n.s. C, 175:1–250.

Qiu, Z.-x. and Z.-d. Qiu. 1995. Chronological sequence and subdivision of Chinese Neogene mammalian faunas. *Palaeogeography, Palaeoclimatology, Palaeoecology* 116:41–70.

Qiu, Z.-x. and B.-y. Wang. 2007. Paracerathere fossils of China. *Palaeontologia Sinica*, n.s. C, 29:1–396.

Qiu, Z.-x., B.-y. Wang, Z.-d. Qiu, F. Heller, L.-p. Yue, G.-p. Xie, X. Wang, and B. Engesser. 2001. Land mammal geochronology and magnetostratigraphy of mid-Tertiary deposits in the Lanzhou Basin, Gansu Province, China. *Eclogae Geologicae Helvetiae* 94: 373–385.

Qiu, Z.-x., B.-y. Wang, Z.-d. Qiu, G.-p. Xie, J.-y. Xie, and X. Wang. 1997. Recent advances in study of the Xianshuihe Formation in Lanzhou Basin. In *Evidence for Evolution, Essays in Honor of Prof. Chungchien Young on the Hundredth Anniversary of His Birth*, ed. Y.-s. Tong, Y.-y. Zhang, W.-y. Wu, J.-l. Li, and L.-q. Shi, pp. 177–192. Beijing: Ocean Press.

Qiu, Z.-x. and J.-y. Xie. 1997. A new species of *Aprotodon* (Perissodactyla, Rhinocerotidae) from Lanzhou Basin, Gansu, China. *Vertebrata PalAsiatica* 35(4):250–267.

Qiu, Z.-x., J.-y. Xie, and D.-f. Yan. 1988. Discovery of the skull of *Dinocrocuta gigantea*. *Vertebrata PalAsiatica* 26(2):128–138.

Repenning, C. A. 1987. Biochronology of the microtine rodents of the United States. In *Cenozoic Mammals of North America: Geochronology and Biostratigraphy*, ed. M. O. Woodburne, pp. 236–268. Berkeley: University California Press.

Repenning, C. A. 2003. *Mimomys* in North America. In *Vertebrate Fossils and Their Context, Contributions in Honor of Richard H. Tedford*, ed. L. J. Flynn, *Bulletin of the American Museum of Natural History* 279: 469–512.

Ritts, B. D., Y.-j. Yue, and S. A. Graham. 2004. Oligocene-Miocene tectonics and sedimentation along the Altyn Tagh Fault, northern Tibetan Plateau: Analysis of the Xorkol, Subei, and Aksay basins. *Journal of Geology* 112(2):207–230.

Saylor, J. E. 2008. The Late Miocene through Modern evolution of the Zhada Basin, South-Western Tibet. Tucson: Department of Geosciences, University of Arizona.

Saylor, J. E., J. Quade, D. L. Dettman, P. G. DeCelles, P. A. Kapp, and L. Ding. 2009. The late Miocene through present paleoelevation history of southwestern Tibet. *American Journal of Science* 309: 1–42.

Shen, X.-h., F.-b. Wang, J. Zhang, and S.-f. Li. 1995. Late Cenozoic magnetostratigraphy and concerned discussion of Gyirong Basin, South of Tibet. *Annual of Formation and Environmental Evolution of Tibetan Plateau and Research on its Ecosystem* 1995: 103–110.

Song, C., X. Fang, J. Li, J. Gao, D. Sun, J.-s. Nie, and M.-d. Yan. 2003. Analysis of sedimentary evolution of the late Cenozoic Guide Basin and the uplift of Tibetan Plateau. *Geological Review* 40(4):337–345.

Song, C.-h., X.-m. Fang, J.-p. Gao, J.-s. Nie, M.-d. Yan, X.-h. Xu, and D. Sun. 2003. Magnetostratigraphy of Late Cenozoic fossil mammals in the northeastern margin of the Tibetan Plateau. *Chinese Science Bulletin* 48(2):188–193.

Song, C.-h., D.-l. Gao, X.-m. Fang, Z.-j. Cui, J.-j. Li, S.-l. Yang, H.-b. Jin, D. W. Burbank, and J. L. Kirschvink. 2005. Late Cenozoic high-resolution magnetostratigraphy in the Kunlun Pass Basin and its implications for the uplift of the northern Tibetan Plateau. *Chinese Science Bulletin* 50(17):1912–1922.

St-Louis, A. and S. D. Côté. 2009. *Equus kiang. Mammalian Species* 835:1–11.

Sun, J.-m., R.-x. Zhu, and Z.-s. An. 2005. Tectonic uplift in the northern Tibetan Plateau since 13.7 Ma ago inferred from molasse deposits along the Altyn Tagh Fault. *Earth and Planetary Science Letters* 235:641–653.

Tassy, P. 1990. The "Proboscidean Datum Event": How many proboscideans and how many events? In *European Neogene Mammal Chronology*, ed. E. H. Lindsay, V. Fahlbusch, and P. Mein, pp. 159–224. New York: Plenum Press.

Tedford, R. H., L. J. Flynn, Z.-x. Qiu, N. O. Opdyke, and W. R. Downs. 1991. Yushe Basin, China: Paleomagnetically calibrated mammalian biostratigraphic standard for the late Neogene of eastern Asia. *Journal of Vertebrate Paleontology* 11(4):519–526.

Tedford, R. H. and Z.-x. Qiu. 1991. Pliocene *Nyctereutes* (Carnivora: Canidae) from Yushe, Shanxi, with comments on Chinese fossil raccoon-dogs. *Vertebrata PalAsiatica* 29(3):176–189.

Wang, B.-y. 2002. Discovery of late Oligocene *Eomyodon* (Rodentia, Mammalia) from the Danghe area, Gansu, China. *Vertebrata PalAsiatica* 40(2):139–145.

Wang, B.-y. 2003. Dipodidae (Rodentia, Mammalia) from the mid-Tertiary deposits in Danghe area, Gansu, China. *Vertebrata PalAsiatica* 41(2):89–103.

Wang, B.-y. and Z.-x. Qiu. 2002. A new species of *Platybelodon* (Gomphotheriidae, Proboscidea, Mammalia) from early Miocene of the Danghe area, Gansu, China. *Vertebrata PalAsiatica* 40(4):291–299.

Wang, B.-y. and Z.-x. Qiu. 2004. Discovery of early Oligocene mammalian fossils from Danghe area, Gansu, China. *Vertebrata PalAsiatica* 42(2):130–143.

Wang, B.-y., Z.-x. Qiu, X. Wang, G.-p. Xie, J.-y. Xie, W. R. Downs, Z.-d. Qiu, and T. Deng. 2003. Cenozoic stratigraphy in Danghe area, Gansu Province, and uplift of Tibetan Plateau. *Vertebrata PalAsiatica* 41(1):66–75.

Wang, S., W. Zhang, X. Fang, S. Dai, and O. Kempf. 2008. Magnetostratigraphy of the Zanda basin in southwest Tibet Plateau and its tectonic implications. *Chinese Science Bulletin* 53(9):1393–1400.

Wang, X., Z.-d. Qiu, Q. Li, Y. Tomida, Y. Kimura, Z. J. Tseng, and H.-j. Wang. 2009. A new early to late Miocene fossiliferous region in central Nei Mongol: Lithostratigraphy and biostratigraphy in Aoerban strata. *Vertebrata PalAsiatica* 47(2):111–134.

Wang, X., Z.-d. Qiu, Q. Li, B.-y. Wang, Z.-x. Qiu, W. R. Downs, G.-p. Xie, J.-y. Xie, T. Deng, G. T. Takeuchi, Z. J. Tseng, M.-m. Chang, J. Liu, Y. Wang, D. Biasatti, Z. Sun, X. Fang, and Q. Meng. 2007. Vertebrate paleontology, biostratigraphy, geochronology, and paleoenvironment of Qaidam Basin in northern Tibetan Plateau. *Palaeogeography, Palaeoclimatology, Palaeoecology* 254:363–385.

Wang, X., Z.-x. Qiu, and B.-y. Wang. 2004. A new leptarctine (Carnivora: Mustelidae) from the early Miocene of the northern Tibetan Plateau and implications of the phylogeny and zoogeography of basal mustelids. *Zoological Journal of the Linnean Society* 142:405–421.

Wang, X., Z.-x. Qiu, and B.-y. Wang. 2005. Hyaenodonts and Carnivorans from the Early Oligocene to Early Miocene of Xianshuihe Formation, Lanzhou Basin, Gansu Province, China. *Palaeontologia Electronica* 8(1.6A):1–14.

Wang, X. and B.-y. Wang. 2001. New material of a *Chalicotherium* from Tsaidam Basin in northern Qinghai-Tibetan Plateau, China. *Paläontologische Zeitschrift* 75(2):219–226.

Wang, X., B.-y. Wang, and Z.-x. Qiu. 2008. Early explorations of Tabenbuluk region (western Gansu Province) by Birger Bohlin—reconciling classic vertebrate fossil localities with modern stratigraphy. *Vertebrata PalAsiatica* 46(1):1–19.

Wang, X., B.-y. Wang, Z.-x. Qiu, G.-p. Xie, J.-y. Xie, W. R. Downs, Z.-d. Qiu, and T. Deng. 2003. Danghe area (western Gansu, China) biostratigraphy and implications for depositional history and tectonics of northern Tibetan Plateau. *Earth and Planetary Science Letters* 208(3–4):253–269.

Wang, X., G.-p. Xie, and W. Dong. 2009. A new species of crown-antlered deer *Stephanocemas* (Artiodactyla, Cervidae) from middle Miocene of Qaidam Basin, northern Tibetan Plateau, China, and a preliminary evaluation of its phylogeny. *Zoological Journal of the Linnean Society* 156:680–695.

Wang, X., G.-p. Xie, Q. Li, Z.-d. Qiu, Z. J. Tseng, G. T. Takeuchi, B.-y. Wang, M. Fortelius, A. Rosenström-Fortelius, H. Wahlquist, W. R. Downs, C.-f. Zhang, and Y. Wang. 2011. Early explorations of Qaidam Basin (Tibetan Plateau) by Birger Bohlin—Reconciling classic vertebrate fossil localities with modern biostratigraphy. *Vertebrata PalAsiatica* 49(3):285–310.

Wang, Y., T. Deng, L. Flynn, X. Wang, A. Yin, Y. Xu, W. Parker, E. Lochner, C. Zhang, and D. Biasatti. 2012. Late Neogene environmental changes in the central Himalaya related to tectonic uplift and orbital forcing. *Journal of Asian Earth Sciences* 44:62–67.

Wang, Y., X. Wang, Y. Xu, C. Zhang, Q. Li, Z.-j. Tseng, G. T. Takeuchi, and T. Deng. 2008. Stable isotopes in fossil mammals, fish and shells from Kunlun Pass Basin, Tibetan Plateau: Paleoclimatic and paleo-elevation implications. *Earth and Planetary Science Letters* 270:73–85.

Woodburne, M. O. 2007. Phyletic diversification of the *Cormohipparion occidentale* complex (Mammalia; Perissodactyla, Equidae), late Miocene, North America, and the origin of the Old World *Hippotherium* datum. *Bulletin of the American Museum of Natural History* 306:1–138.

Woodburne, M. O. 2009. The early Vallesian vertebrates of Atzelsdorf (Late Miocene, Austria), 9. *Hippotherium* (Mammalia, Equidae). *Annalen des Naturhistorischen Museums in Wien* 111A:585–604.

Xu, R. 1981. Vegetational changes in the past and the uplift of Qinghai-Xizang Plateau. In *Geological and Ecological Studies of Qinghai-Xizang Plateau*. Vol. 1, *Geology, Geological History and Origin of Qinghai-Xizang Plateau*, ed. D.-s. Liu, pp. 139–144. Beijing: Science Press.

Yin, A., Y. Dang, L.-c. Wang, W.-m. Jiang, S.-p. Zhou, X. Chen, G. E. Gehrels, and M. W. McRivette. 2008. Cenozoic tectonic evolution of Qaidam basin and its surrounding regions (Part 1): The southern Qilian Shan-Nan Shan thrust belt and northern Qaidam basin. *Geological Society of America Bulletin* 120(7/8): 813–846.

Yin, A., Y. Dang, M. Zhang, X. Chen, and M. W. McRivette. 2008. Cenozoic tectonic evolution of the Qaidam basin and its surrounding regions (Part 3): Structural geology, sedimentation, and regional tectonic reconstruction. *Geological Society of America Bulletin* 120(7/8):847–876.

Yin, A., Y. Dang, M. Zhang, M. W. McRivette, W. P. Burgess, and X. Chen. 2007. Cenozoic tectonic evolution of Qaidam basin and its surrounding regions (Part 2): Wedge tectonics in southern Qaidam basin and the Eastern Kunlun Range. *Geological Society of America Special Paper* 433:369–390.

Yin, A., P. E. Rumelhart, R. E. Butler, E. Cowgill, T. M. Harrison, D. A. Foster, R. V. Ingersoll, Q. Zhang, X.-q. Zhou, X.-f. Wang, A. Hanson, and A. Raza. 2002. Tectonic history of the Altyn Tagh fault system in northern Tibet inferred from Cenozoic sedimentation. *Geological Society of America Bulletin* 114(10): 1257–1295.

Yue, L.-p., T. Deng, R. Zhang, Z.-q. Zhang, F. Heller, J.-q. Wang, and L.-r. Yang. 2004. Paleomagnetic chronology and record of Himalayan movements in the Longgugou section of Gyirong-Oma Basin in Xizang (Tibet). *Chinese Journal of Geophysics* 47(6): 1135–1142.

Yue, L.-p., F. Heller, Z.-x. Qiu, L. Zhang, G. Xie, Z. Qiu, and Y. Zhang. 2001. Magnetostratigraphy and paleoenvironmental record of Tertiary deposits of Lanzhou Basin. *Chinese Science Bulletin* 46(9):1998–2003.

Zhang, Q.-s., F.-b. Wang, H.-x. Ji, and W.-b. Huang. 1981. The Pliocene stratigraphy of Zhada, Xizang. *Journal of Stratigraphy* 5:216–220.

Zhang, Z. 2005. New materials of *Dinocrocuta* (Percrocutidae, Carnivora) from Lantian, Shaanxi Province, China, and remarks on the late Miocene biochronologic correlation. *Geobios* 38:685–689.

Zheng, S.-h. 1980. The *Hipparion* fauna of Bulong Basin, Biru, Xizang. In *Paleontology of Tibet, Part 1*, ed. Qinghai-Tibetan Plateau Comprehensive Scientific Investigation Team of Chinese Academy of Sciences, pp. 33–47. Qinghai-Tibetan Plateau Scientific Investigation Series. Beijing: Science Press.

Zheng, S.-h., W.-y. Wu, and Y. Li. 1985. Late Cenozoic mammalian faunas of Guide and Gonghe basins, Qinghai Province. *Vertebrata PalAsiatica* 23(2):89–134.

Zhu, D., X. Meng, Z. Shao, C. Yang, J.-e. Han, J. Yu, Q. Meng, and R. Lu. 2005. Redefinition and redivision of the Pliocene-early Pleistocene lacustrine strata in Zanda basin, Ngari, Tibet, China. *Geological Bulletin of China* 24(12):1111–1120.

Zong, G.-f., W.-y. Chen, X.-s. Huang, and Q.-q. Xu. 1996. *Cenozoic Mammals and Environment of Hengduan Mountains Region*. Beijing: Ocean Press.

Chapter 11

Hominoid-Producing Localities and Biostratigraphy in Yunnan

WEI DONG AND GUO-QIN QI

Yunnan Province is located in the southwestern part of China, north of Vietnam and Laos, and northeast of Burma. It is on the Yunnan-Guizhou Plateau, with altitudes of 4000 m in the northern mountain regions and 2000 m on the southern highland. The most conspicuous mountains are the Hengduan Shan in the northwest and the Gaoligong Shan in the west. Over 90% of the province is mountainous. The relative height of mountain peaks above river valleys can reach as much as 3000 m. The climatic features of today span three zones from north to south in the province, temperate, subtropical, and tropical. Annual climate change is very clear, with two distinct seasons: dry from November to April and rainy from May to October. A large number of fault depression basins are sprinkled among various mountain ridges of the plateau. The Neogene hominoid-bearing strata are exposed in such basins as at Kaiyuan, Lufeng, Yuanmou, and Baoshan Counties (figure 11.1). Since the Baoshan hominoid locality has not yet been studied, only those at Kaiyuan, Lufeng, and Yuanmou are discussed here.

KAIYUAN HOMINOID LOCALITY

Kaiyuan hominoid locality is situated in the Xiaolongtan Basin, about 12 km northwest of downtown Kaiyuan and 144 km south of Kunming City. The basin is one of more than 200 lignite-bearing basins of the province and it is the southernmost one. The shape of the basin is oval, with its longitudinal axis lying northeast to southwest. The area of the basin is 21 km². It is narrower in the north and wider in the south. The altitude of the basin varies from 1030 m to 1110 m. The Nanpan River and Yunnan–Vietnam railway cut through the basin from northwest to southeast. The basement and surrounding strata of the basin are composed of Middle and Upper Triassic limestone, dolomitic limestone, and a small amount of sandy shale. Tertiary lacustrine and Quaternary fluvial sediments accumulated in the basin (figure 11.2).

There are two mines with naturally exposed rich lignite beds on each side of the Nanpan River. The one in the north is called the Xiaolongtan Mine (pit certer 23°40'50"N, 103°11'40"E) and another in the south is the Buzhaoba Mine (pit center 23°40'11"N, 103°10'37"E). The lignite resource attracted many geologists to prospect the basin. Lantenois (1907), Deprat (1912), and Mansuy (1912) showed that the lignite beds could be dated as Tertiary. X.-m. Meng, Z.-q. Wang, and Z.-q. Lou demonstrated further from 1936 to 1938 in some unpublished reports that the age of the lignite might be Oligocene after correlation of the Xiaolongtan lignite to the coal layers of Fushun in Liaoning Province and those of Jukousi in Shansi (=Shanxi) Province. Young and Bien (1939) regarded the lignite beds as "Pontian," then considered early Pliocene. Bien (1939) indicated further that the horizon of Xiaolongtan lignite beds was equivalent to the Yunning Measures of Guangxi Autonomous Region. Team 537 of the Geological Bureau of Southwest

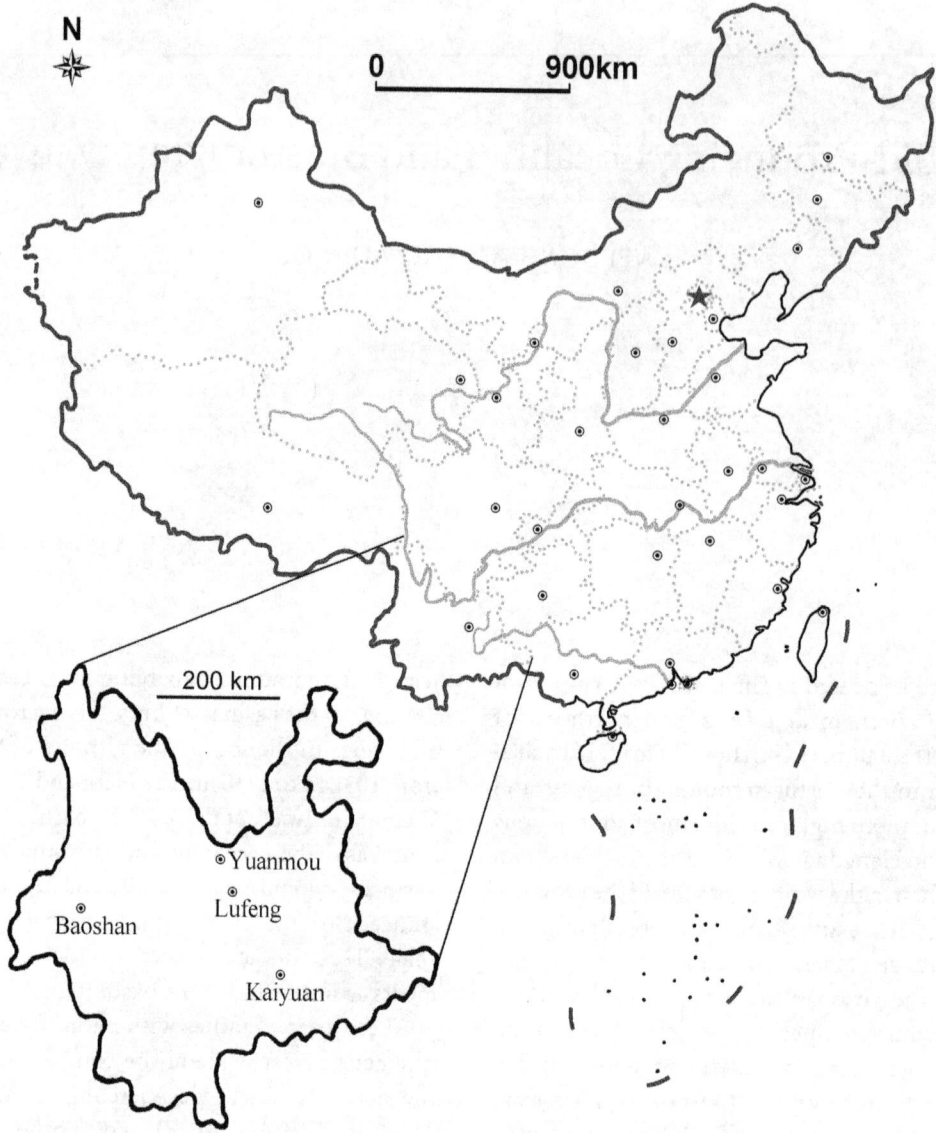

Figure 11.1 Geographic location of Yunnan Hominoid localities.

China carried out a comprehensive geological investigation in the Xiaolongtan Basin from 1955 to 1956, and Xiaolongtan Coal Measures was named for the whole set of lignite beds in the basin, with its geological age estimated as Miocene to Pliocene. The overlying Hetou Coal Measures was named for coal and a marl sandwiched within it and dated as Pliocene (Xiong 1957). It was during the latter investigation that five hominoid cheek teeth and some mammalian fossils were uncov-

ered from the Xiaolongtan Coal Measures (Woo 1957). Further hominoid materials were uncovered and studied by Woo (1958) and Zhang (1987).

The lignite beds of the Xiaolongtan Basin were reinvestigated in 1984 and 1985 for stratigraphic division and correlation (Dong 1987, 2001). The so-called Hetou Coal Measures had been based on some test drillings for the deposits overlying the argillaceous limestone above the lignite (table 11.1). There were six layers of coal, each about

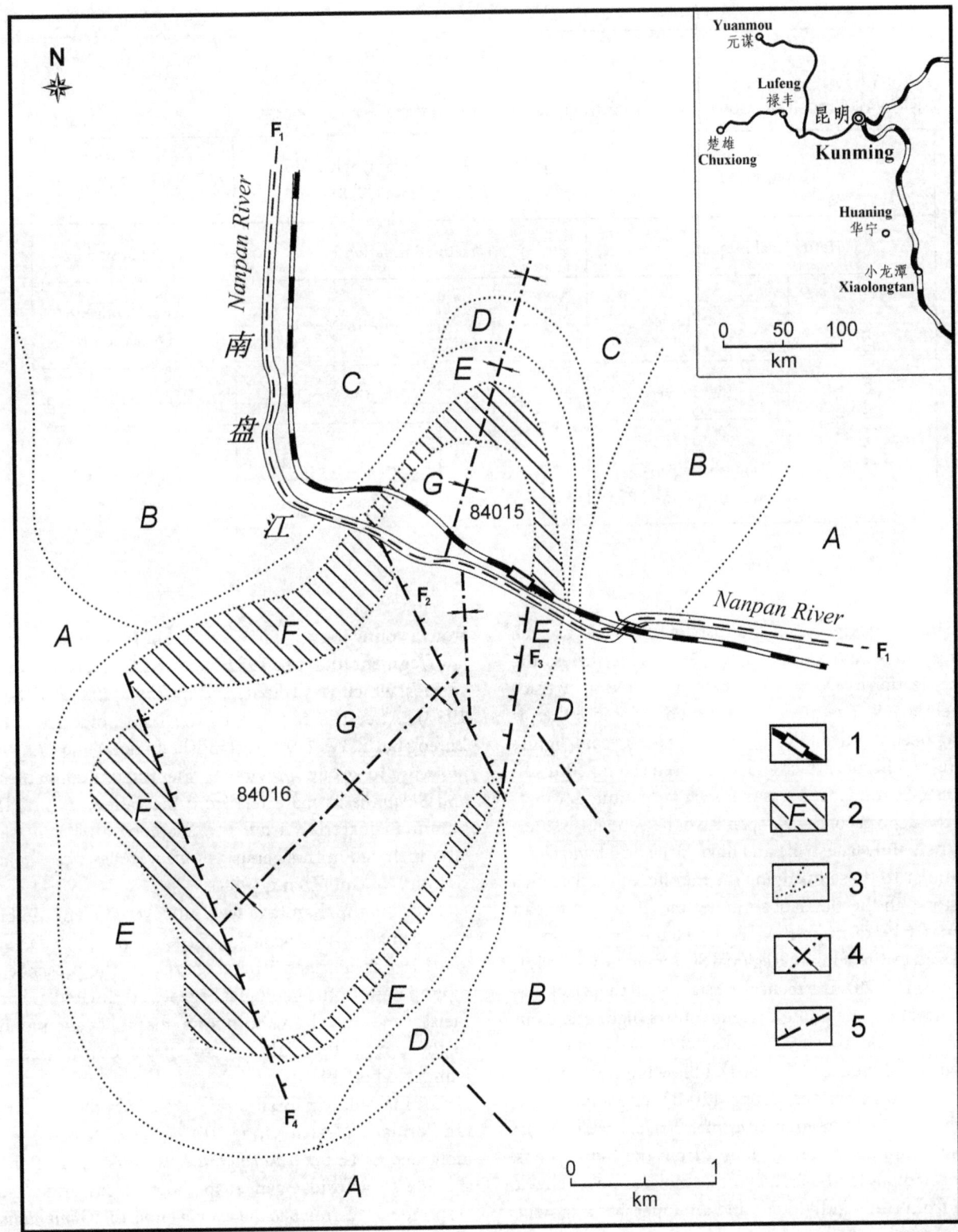

Figure 11.2 Geological map of Xiaolongtan Basin: (*1*) Xiaolongtan railway station; (*2*) hominoid horizon; (*3*) stratigraphic boundary; (*4*) axis of syncline; (*5*) fault; (*A*) Gejiu Formation; (*B*) Falang Formation; (*C*) Huobachong Formation; (*D*) Dongshenqiao Formation; (*E*) Xiaolongtan Formation; (*F*) hominoid horizon; (*G*) Buzhaoba Marlite Formation; 84015 = Xiaolongtan Pit; 84016 = Buzhaoba Pit.

Table 11.1

History of Lithostratigraphic Division for the Late Cenozoic in the Xiaolongtan Basin

Xiong (1957)		Yunnan Regional Stratigraphic Chart Editing Team (1978)		Dong (2001)
Hetou Coal Measures		Hetou Formation		Hetou Formation
Xiaolongtan Coal Measures	Marlite Bed	Xiaolongtan Formation	Marlite Member	Buzhaoba Marlite Formation
	Lignite Bed		Lignite Member	Xiaolongtan Formation
	Dongshengqiao Clay and Sandy Gravel Bed		Clay and Sandy Gravel Member	Dongshengqiao Formation

0.5–5 m in thickness, but the distribution scale of such coal measures was limited (Xiong 1957). Nevertheless, the reinvestigation by Dong (1987) in the basin found no trace of such layers. The measures were considered either as having been removed during more than 26 years' lignite mining or still subsurface and unexposed (Dong 2001). In any case, Hetou Coal Measures were very limited. Moreover, the deposits of the Nanpan River bed and its first terrace are mainly lacustrine and fluvial; the lithologic facies are similar to those overlying the marlite, or argillaceous limestone, in the Buzhaoba mining pit. These strata can reasonably be distinguished as the Hetou Formation, as suggested by the Yunnan Regional Stratigraphic Chart Editing Team (1978). The team also established the Xiaolongtan Formation, including three members: lignite beds in the middle, underlying clay and sandy gravels, as well as overlying marlite (see table 11.1). Following stratigraphic code recommendations, Dong (2001) upgraded these members to lithostratigraphic formations (see table 11.1).

The hominoid materials found from the lignite beds of the Xiaolongtan Basin include 13 isolated lower cheek teeth from three individuals and an upper jaw fragment with 12 teeth. The five isolated cheek teeth from a single individual uncovered in 1956 were attributed to "*Dryopithecus*," with a new species established as *D. keiyuanensis* (Woo 1957). Five new isolated cheek teeth from another individual collected later were also included in the newly named species (Woo 1958), and their different size from the type specimens was regarded as showing sexual dimorphism (Woo 1958). Three isolated molars

from a young individual recovered in 1980 and an upper jaw fragment found in 1982 were attributed to the same species, which was transferred to *Ramapithecus* (Zhang 1987). After the revision of the hominoids from the Lufeng Basin, Wu (1987) established a new genus, *Lufengpithecus*, to group previously identified *Ramapithecus* and *Sivapithecus* materials from the Lufeng Basin. The hominoid materials from the Xiaolongtan Basin were later included in this genus as *Lufengpithecus keiyuanensis* (Wu, Xu, and Zhang 1989).

Four mammalian taxa were uncovered from the Hetou Formation: *Stegodon* sp., *Sus* sp., *Hexaprotodon* sp., and *Rusa* cf. *unicolor*. Characterized by the presence of hippopotamus, the geological age was estimated as early Pleistocene. The Hetou Formation fossils are comparable to those from the early Pleistocene of Burma (Iravaty fauna; Colbert 1938).

The mammalian fossils uncovered from the Xiaolongtan Formation (sensu Dong 2001, i.e., lignite beds) were identified as 12 taxa: *Lufengpithecus*, Mustelidae gen. et sp. indet., Castoridae gen. et sp. indet. 1, Castoridae gen. et sp. indet. 2, *Tetralophodon xiaolongtanensis*, *Gomphotherium* cf. *macrognathus*, *Zygolophodon chinjiensis*, *Tapirus* cf. *yunnanensis*, *Parachleuastochoerus sinensis*, *Propotamochoerus parvulus*, *Hippopotamodon hyotherioides* (= *Dicoryphochoerus* sp.), and *Euprox* (previously identified as *Paracervulus*) sp. The Xiaolongtan Formation is characterized by *Lufengpithecus keiyuanensis*, *Tetralophodon*, *Zygolophodon*, *Tapirus*, *Parachleuastochoerus*, *Hippopotamodon*, and *Euprox*.

LUFENG HOMINOID LOCALITY

The Lufeng Basin is located about 65 km west of Kunming City. It is a small fault depression basin in the east of the Dianzhong (center of Yunnan Province) Plateau. The average altitude of the basin is about 1560 m. The basin extends from south to north, with north–south length of 12 km and east–west width of 3 km. Lufeng hominoid locality is situated on the northern margin of the basin, about 9 km north of downtown Lufeng, and on the southern slope of Miaoshan Hill, which is on the northeast outskirts of Shihuiba Village. Three tributary streams of the Xingxiu River pass through the basin from north, east, and south, respectively, and join in the west to form the Xingxiu River (figure 11.3). The base and surrounding strata of the basin are composed of the Precambrian Kunyang Group of metamorphic limestone and argillaceous shales, and Jurassic and Cretaceous continental redbeds. The Cenozoic strata are distributed only in the center and on some margins of the basin. The exposed Neogene deposits are a series of Late Miocene residual beds, lacustrine-fluvial and limnitic facies, with a thickness of 20–30 m, and Quaternary fluvial gravel and terrace deposits.

During the construction of the Kunming–Chengdu railway in the early 1970s, the upper deposits of the southern slope of Miaoshan Hill, east of Shihuiba Village, were removed. The underlying lignite and the carbonaceous silt were consequently exposed broadly. The exposed lignite offered local villagers a free source of fuel, and they quarried the lignite frequently, sometimes uncovering mammalian fossils. The news attracted paleontologists to investigate the site. A test excavation was carried out in 1975 by the Institute of Vertebrate Paleontology and Paleoanthropology, Chinese Academy of Sciences with close collaboration of the Yunnan Provincial Museum and the Lufeng Cultural Heritage Bureau. Some mammalian fossils—the most significant a hominoid mandible—were unearthed (Xu and Lu 1979). The hominoid horizon was dated as 8 Ma based on small mammals (Flynn and Qi 1982). The geographical coordinates of the site center (labeled 75033) are 25°13'15"N, 102°3'9"E (see figure 11.3). A series of eight excavations followed from 1976 to 1983, with fruitful and significant production of hominoids: 5 incomplete skulls, 6 cranial fragments, 10 mandibles, 41 maxillary and mandibular fragments, 29 upper and lower dentitions, 650 isolated teeth, and some postcranial bones (Xu and Lu 2008). There are two sizes of hominoid mandibles from the Lufeng Basin. The small one was attributed to *Ramapithecus* as a new species, *Ramapithecus lufengensis* (Xu

et al. 1978), and the large one to *Sivapithecus* as another new species, *Sivapithecus yunnanensis* (Xu and Lu 1979; Lu, Xu, and Zheng 1981). The two forms were considered as either two taxa or, later, as two sexes of one species (Xu et al. 1978; Xu and Lu 1979; Wu, Xu, and Lu 1983, 1985, 1986). The two forms were finally regarded as showing sexual dimorphism of a single species (Kelley and Etler, 1989; Kelley and Xu, 1991; Kelley, 1993; Kelley and Plavcan, 1998), and a new genus was named, *Lufengpithecus* (Wu 1987). All hominoid materials from the Lufeng

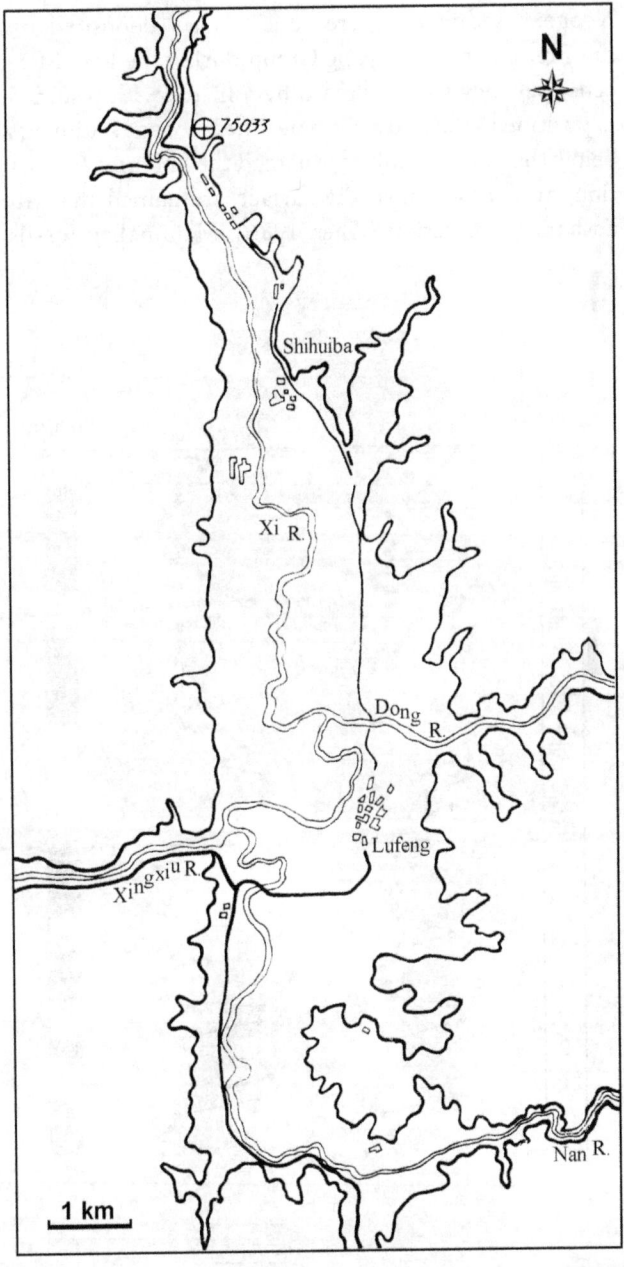

Figure 11.3 Location of hominoid locality 75033 in the Lufeng Basin.

Basin have since been assigned as *Lufengpithecus lufengensis*.

Alongside the track to the quarry on Miaoshan Hill, three stratigraphic sections of the Late Miocene are well exposed and called sections A, B, and C. Hominoid materials have not yet been found in these sections. To the west of section C, fossiliferous layers are exposed on the west side of an excavation area and are labeled section D. All these sections were studied during the excavations from 1976 to 1983 (Qi 1985; Chen 1986; Badgley et al. 1988). Section A is the most comprehensive, and it includes all strata exposed in sections B, C, and D. The Neogene sediments were continuously deposited on the Precambrian Kunyang Group during the Late Miocene, but they were followed by a hiatus with resumed deposition in the Early Pliocene. The Neogene sediments below the unconformity are named the Shihuiba Formation, and those above the surface are named the Miaoshanpo Formation (Chen 1986). Mammalian fossils

such as hipparion, as well as fishes, have been discovered in the Miaoshanpo Formation. All hominoid materials and associated mammal fossils were recovered from the Shihuiba Formation.

The mammalian fossils uncovered from the Shihuiba Formation have been identified and classified into 10 orders, 37 families, and 98 species (see appendix). It is biochronologically characterized by the presence of *Lufengpithecus, Laccopithecus, Prodendrogale yunnanica, Yunoscaptor scalprum, Heterosorex wangi, Sciurotamias wangi, Platacanthomys dianensis, Miorhizomys tetracharax, Kowalskia hanae, Alilepus longisinuosus, Indarctos yangi, Ailurarctos, Sivaonyx bathygnathus, Gomphotherium, Zygolophodon, Tapirus, Yunnanochoerus, Euprox, Muntiacus, Paracervulus,* and *Selenoportax.*

High-resolution paleomagnetic sampling was carried out at sections A and D at Shihuiba (figure 11.4). The results are as follows: section A includes six normal polarity zones and six reversed zones, respectively considered

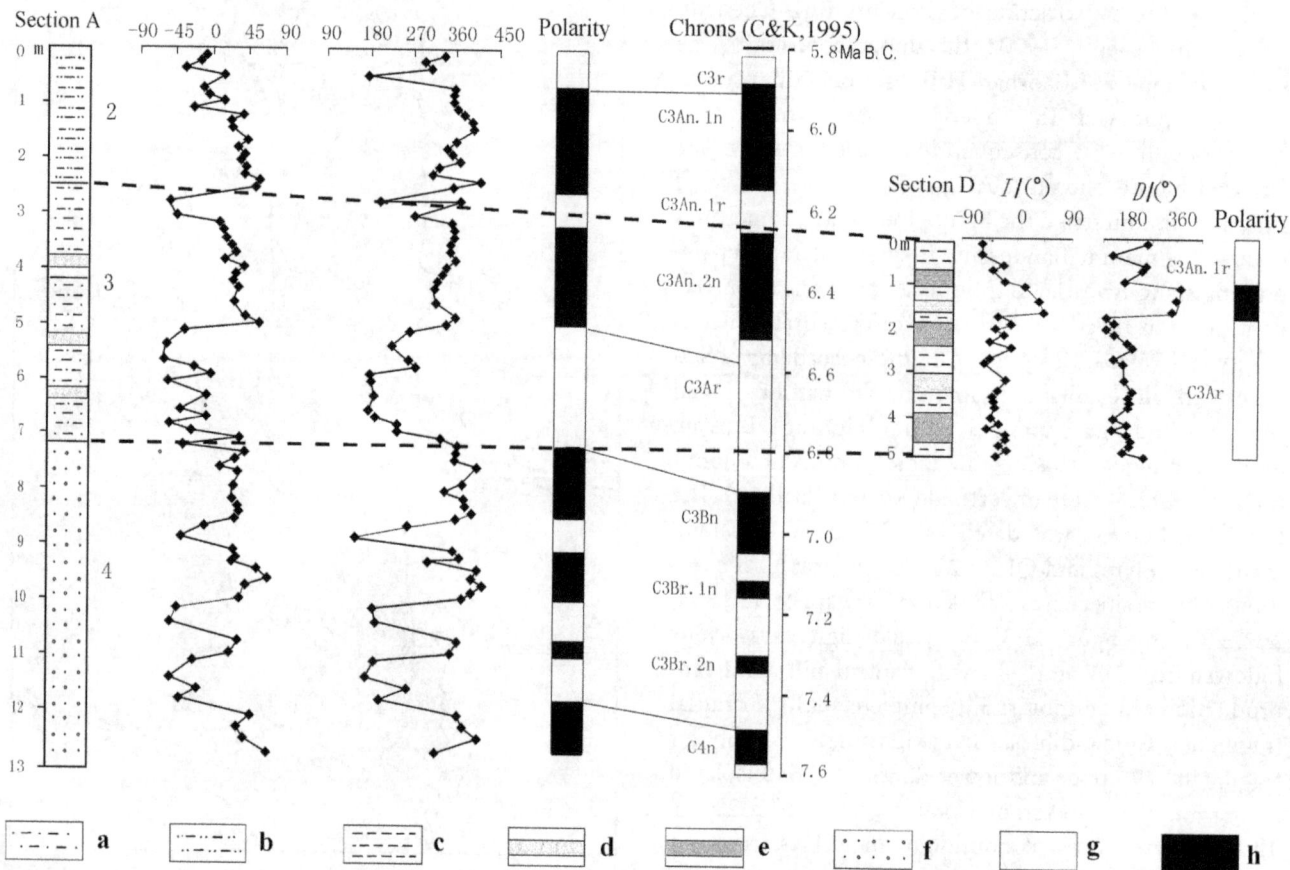

Figure 11.4 Paleomagnetic results of the sections A and D at Lufeng hominoid locality. *a*, sandy clay; *b*, sand with clay; *c*, clay; *d*, carbonaceous clay; *e*, lignite; *f*, sand with gravel; *g*, reversed polarity; *h*, normal polarity.

C3r, C3An.1n, C3An.1r, C3An.2n, C3Ar, C3Bn, C3Br, C3Br.1n, C3Br.1r, C3Br.2n, C3Br.2r, and C4Bn.1n. The uppermost and the lowermost parts of this section are respectively congruent with C3r and C4n.1n, 5.8 to 7.6 Ma on the Cande and Kent (1995) Time Scale. Section D contains C3An.1r, C3An.2n, and C3Ar based on correlation to carbonaceous mudstone strata in the middle of section A. The paleomagnetic age of the *Lufengpithecus lufengensis* layers is therefore ~6.9–6.2 Ma, or late Late Miocene, according to the above correlations (Yue and Zhang 2006).

YUANMOU HOMINOID LOCALITY

The Yuanmou Basin (sensu lato) is also a fault depression basin located in the northern part of Dianzhong (center of Yunnan Province) Plateau, to the south of the Jinsha River. It belongs administratively to Yuanmou County, which is about 112 km northwest of Kunming City (figure 11.5). The geographic coordinates of downtown Yuanmou are about 25°42'16"N, 101°52'13"E. The south–north length of the basin is about 45 km and its largest east–west width about 18 km. The elevation of the basin is relatively high in the south and relatively low in the north, with an average elevation of 1100 m above sea level. The Longchuan River passes through the whole basin from south to north and joins the Jinsha River at Longjie village. The Yuanmou Basin is subdivided by several hills into some smaller basins or subbasins. The southeastern subbasin is the largest, also called the Yuanmou Basin (*sensu stricto*) for Yuanmou City. The other two important subbasins are the Banguo Basin west of Yuanmou Basin (s. s.) and the Wumao Basin northwest of Yuanmou Basin (s. l.). The strata exposed along the eastern edge of the Yuanmou Basin (s. l.) consist of Jurassic and Cretaceous purple-red, yellow-green, and light-gray feldspathic and quartzose sandstone, conglomerate, and mudstone. These strata form the Yuanmou East Mountains. Those exposed in the southern Yuanmou Basin (s. l.) are composed of Cretaceous purple-red and light-gray feldspathic sandstone, mudstone, and conglomerate. And those exposed in the western Yuanmou Basin (s. s.) are mainly Precambrian Kunyang Group, gneiss, quartzite, schist, phyllite, marble, limestone, and Jinnian granite and diorite. They form the western ridges of the basin. The strata in northwestern and northern parts of the Yuanmou Basin (s. l.) are Precambrian Kunyang Group and Cretaceous red beds. The basin basement is mainly Kunyang Group and granite, mostly exposed in the Wumao and Banguo subbasins. The Late Miocene sediments yielding

hominoids are distributed mainly on the slopes of hills in the Banguo and Wumao subbasins. Pliocene sediments without hominoids are distributed both in the northwestern and southern parts of the Yuanmou Basin. The Early Pleistocene hominin sediment is distributed mainly in the Yuanmou subbasin.

To cooperate in geological reconnaissance for Chengdu–Kunming Railway design and construction, the Chinese Academy of Geology launched a series of geological investigations in Yunnan Province. Within the framework of the project, Qian and colleagues investigated the Yuanmou Basin in 1965, and consequently two hominin incisors and some mammalian fossils were uncovered between Danawu and Shangnabang villages (Hu 1973; Qian 1985). The hominin incisors were identified as *Homo erectus* (Hu 1973). More and more mammal fossils were unearthed in the follow-up investigations and excavations. The Quaternary strata yielding the hominin teeth were named the Yuanmou Formation (Huang et al. 1978). In 1986, as a result of his training course for local villagers to recognize and protect the fossils, N.-r. Jiang of the Yunnan Institute of Geology received some fossils from the villagers of Zhupeng. A hominoid maxillary molar was identified from the materials, and a series of investigations and excavations was carried out from 1986 to 1990 in the Xiaohe and Zhupeng areas (Jiang and Zhang 1997). The publication of some research results aroused extensive attention in the scientific community, but many problems required further study. Within the framework of the State Key Project of the ninth five-year plan—Origin of Early Humans and Environmental Background—three main stages of investigations and excavations were carried out from 1998 to 2000 in the Xiaohe, Leilao, and Danawu areas of the Yuanmou Basin (Qi and Dong 2006).

Unlike Kaiyuan hominoids, which are found only in the Xiaolongtan area, and Lufeng hominoids, found only in Shihuiba section D, the Yuanmou hominoids are found in three areas, near the villages of Xiaohe, Zhupeng, and Leilao. The Late Miocene sediments yielding hominoids are classified as the Xiaohe Formation (table 11.2), the Pliocene sediments without hominoids are classified as the Shagou Formation, and the Early Pleistocene sediments are classified as the Yuanmou Formation (with two members).

The hominoid fossils from the Yuanmou Basin were recovered principally from four places: Hudieliangzi, 100–400 m north of the Xiaohe Village; Fangbeiliangzi, 100–450 m south of the Xiaohe Village; Baozidongqing, 500–800 m southwest of the Zhupeng Village; and Dashuqingliangzi, near Leilao Village. A juvenile partial

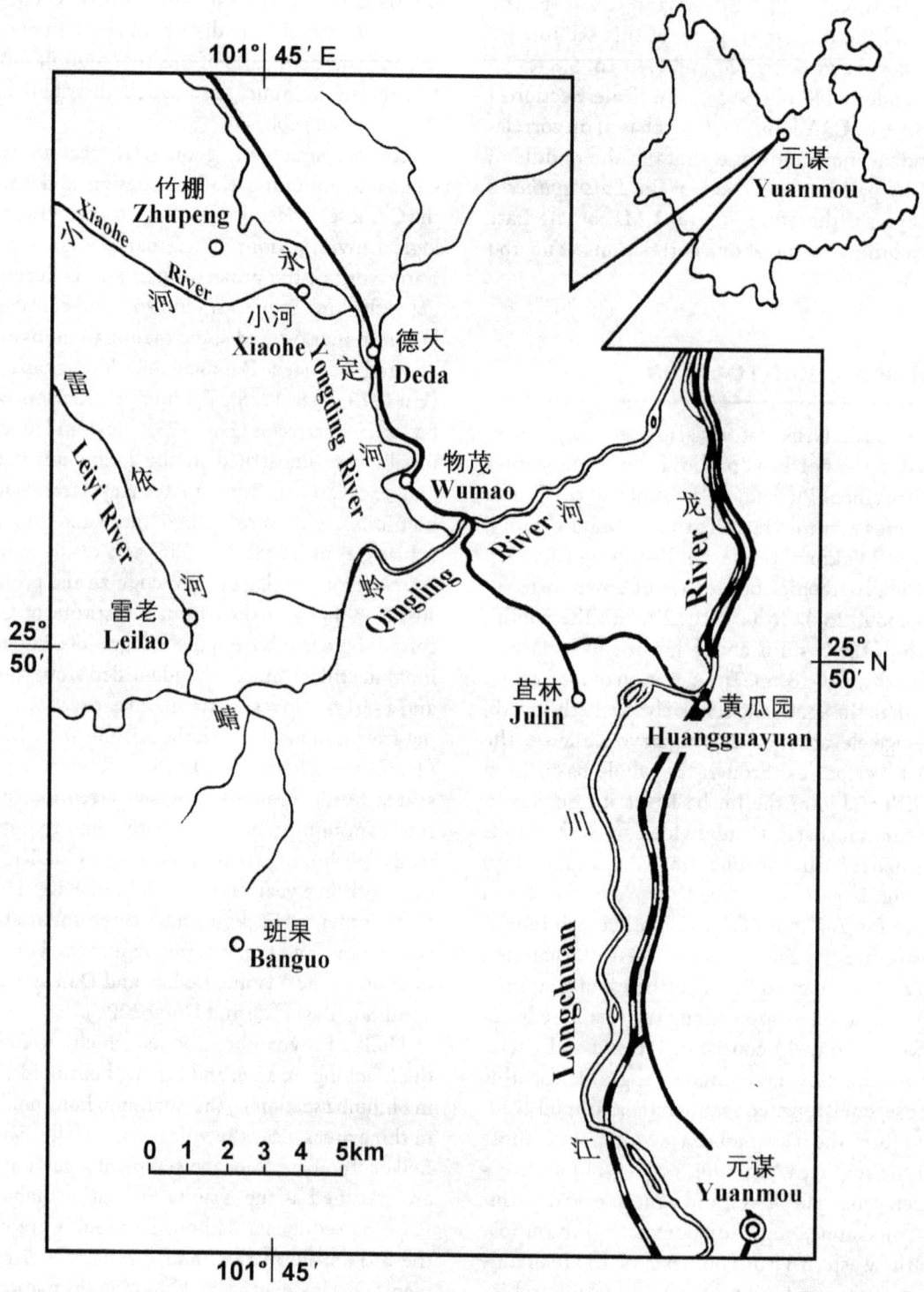

Figure 11.5 Geographic location of Yuanmou hominoid localities.

Table 11.2
History of Lithostratigraphic Division for the Late Cenozoic in the Yuanmou Basin

Chow (1961)	Huang et al. (1978)	Qian and Jiang (1991)	Jiang, Sun, and Liang (1989)	Zong (1996)
Yuanmou Formation	Yuanmou Formation	Yuanmou Formation (4 members)	Yuanmou Formation	Yuanmou Formation (2 members)
			Gantang Formation	
Shagou Formation	Shagou Formation	Longchuan Formation	Shagou Formation	Shagou Formation
			Longchuan Formation	Xiaohe Formation

skull, 8 maxillary and 11 mandibular fragments, and more than 1200 isolated teeth of hominoids were unearthed during the excavations from 1986 to 1990 (Zheng and Zhang 1997); 9 upper and lower jaws, 298 isolated teeth and a phalanx of hominoids were unearthed during the excavations from 1998 to 2000 (Qi and Dong 2006). The first uncovered hominoid maxillary molar from Zhupeng was identified as a new human subspecies, *Homo habilis zhupengensis*, by Jiang, Sun, and Liang (1987). A left maxilla with P3–M2 from Hudieliangzi (in Chinese "Hudie" means butterfly and "Liangzi" means hill) was identified as a new hominoid species, *Ramapithecus hudienensis*, by Zhang et al. (1987a), and six isolated teeth from Zhupeng were assigned to a new human species, *Homo orientalis*, by Zhang et al. (1987b). Some "bone artifacts" from Zhupeng were also reported. The age of *Homo orientalis* and "bone artifacts" from Zhupeng was reported as 2 Ma. No artifacts were found in Hudieliangzi, so the age of *Ramapithecus hudienensis* was estimated as 4–3 Ma. All of these specimens were later considered to belong to a single species of hominoid (Zhang, Zheng, and Gao 1990).

The materials from Leilao show two different sizes regarded as sexual dimorphism (Jiang, Xiao, and Li 1993). In contrast, Zheng and Zhang (1997) regarded such differences as more than expected for sexual dimorphism and suggested that the specimens represent two different species. With the increase of the hominoid collection, it became more and more clear that the hominoid materials represented a member of *Lufengpithecus* instead of *Homo*. However, many questions remained unsettled: the number of species of hominoids of the Yuanmou Basin, the taxonomic positions and phylogenetic status of these hominoids, and their geologic age, among others. It was in an attempt to answer these questions that the 3 years

of excavations were carried out in the Yuanmou Basin (mainly in the Xiaohe and Leilao areas) from 1998 to 2000. After the study of all hominoid materials from the basin and comparison with other hominoids, especially *Dryopithecus* and *Sivapithecus* (=*Ramapithecus*), the conclusion was that the Yuanmou hominoids are closest to the Lufeng hominoids (Liu, Zheng, and Jiang 1999). In addition, the specific name should be *L. hudienensis* (from *Ramapithecus hudienensis*; Zhang et al. 1987a) according to the Law of Priority. The type specimen is a left maxilla fragment with P3–M1 (YV916) housed in the Institute of Cultural Relics and Archaeology of Yunnan Province (Zheng 2006).

The mammalian fossils uncovered from the Xiaohe Formation have been identified and classified into 9 orders, 41 families, and 110 species (see appendix). It is biochronologically characterized by the presence of *Lufengpithecus, Indraloris, Sinoadapis, Prodendrogale yunnanica, Yunoscaptor scalprum, Heterosorex wangi, Sciurotamias wangi, Platacanthomys dianensis, Miorhizomys tetracharax, Miorhizomys blacki, Kowalskia hanae, Amphicyon palaeindicus, Indarctos yangi, Ailurarctos, Sivaonyx bathygnathus, Tetralophodon, Mammut, Tapirus, Yunnanochoerus, Molarochoerus yuanmouensis, Euprox, Muntiacus,* and *Paracervulus*.

The study of small mammals of the *Lufengpithecus hudienensis* fauna was thought to indicate late Miocene (Baodean Chinese Land Mammal Age), corresponding to an age of about 9 Ma, or late Vallesian or early Turolian of European Mammal Age (=upper MN 10 or lower MN 11; Ni and Qiu 2002). Subsequent analysis of sciurids from Leilao shows that the sciurid composition is very similar to that from Shihuiba, Lufeng, with eight taxa in common and all the taxa known from Lufeng occurring in the Leilao Fauna. Differences between the two faunas might indicate a temporal or ecological contrast, but they might also be due to inadequate sampling or different sedimentary environments at the two sites (Qiu and Ni 2006). High endemicity of the fauna, relatively slow evolution of sciurids, and poor knowledge of phylogenetic relations for these animals make it difficult to use Sciuridae to determine precisely the age of the fauna. It is noteworthy in this regard that less change in size and dental morphology in members of the subfamily Sciurinae can be observed than in Pteromyinae during this time. All identifiable genera of Sciurinae from Leilao are extant ones, but no extant genus of Pteromyinae is known from the locality, which may imply that the Pteromyinae evolved more rapidly than the Sciurinae (Qiu and Ni 2006). The less pronounced protoconule and metaconule in *Miopetaurista asiatica*, the less distinct double metaconule on M1/2, and the less striking hypoconulid on m1/2 in *Hylopetodon dianense* from Leilao than from Lufeng might be interpreted as more primitive characters for the flying squirrels from Leilao (Qiu and Ni 2006). That the fauna with *Lufengpithecus hudienensis* from Yuanmou would be older than the fauna with *Lufengpithecus* from Lufeng is supported by the analysis of other small mammals. For example, note that *Miorhizomys tetracharax* is found at both Lufeng and Yuanmou, while *Miorhizomys blacki* is found only at Yuanmou (Ni and Qiu 2002). This work led to interpretation that the *Lufengpithecus hudienensis* assemblage would be only somewhat older than Lufeng.

The small primates found in the *Lufengpithecus hudienensis* localities include *Indraloris progressus, Sinoadapis parvulus* (Sivaladapidae), and *Yuanmoupithecus xiaoyuan* (Pan 2006). *Indraloris* is reported in China for the first time and can be compared with the records of the genus from Pakistan (Flynn and Morgan 2005) and Halitalyangar, India, and may aid in biochronological correlation.

More than 30 taxa of fossil Carnivora from the Xiaohe and Leilao areas (collected in 1998 and 1999, respectively) were identified (Qi 2006). Among them, *Vishnucyon, Pseudarctos,* and *Vishinuictis* are discovered in China for the first time. Some, such as amphicyonids, did not exist beyond MN 9 in most of Europe, surviving to MN 10 or MN 11 only in the areas of Hungary and the Black Sea (Mein 1990; Bernor et al. 1996) and into the "*Selenoportax lydekkeri*" Interval Zone in the Siwaliks of the India–Pakistan Subcontinent (Barry and Flynn 1989; Barry, Lindsay, and Jacobs 1982). Using the revised time scale in Barry et al. (2002), this constrains the age to about 10 Ma to 7 Ma. Others (such as *Adcrocuta eximia*

and *Metailurus parvulus*) continued from MN 10 to MN 13 in Europe. This implies that the time span of the deposits with *Lufengpithecus hudienensis* fossils in the Xiaohe and Leilao areas is earlier than the hominoid-bearing layers at Lufeng, possibly between 10 Ma and 7 Ma. Seeing the Yunnan area as a refuge at that time (Jablonski 2005) explains why some animals survived later there than elsewhere. Analysis of carnivores from Lufeng shows that there are no older elements such as *Amphicyon, Vishinucyon,* and *Pseudarctos,* nor *Adcrocuta*. So the Lufeng deposits only correspond to the younger horizons at Yuanmou. Hipparionines are present in both Yuanmou and Lufeng. The artiodactyls from the *Lufengpithecus hudienensis* localities are suids and ruminants (Dong, Liu, and Pan 2003; Dong, Pan, and Liu 2004). The well-known *Selenoportax* sp. among Lufeng ruminants is absent from Yuanmou, but the primitive *Dorcatherium* is present among Yuanmou ruminants (Pan, Liu, and Dong 2006). Thus, the age of deposits with *L. hudienensis* fossils would be between 9 Ma and 6 Ma.

In high-resolution paleomagnetic sampling carried out at the Xiaohe and Leilao sections (Yue et al. 2004), the samples were taken at 10 cm intervals to reduce the possibility of omitting any short polarity events. The Xiaohe section includes four normal and four reversed polarity zones (figure 11.6) that correspond to the eight polarity zones of C3Br.1n, C3Br.1r, C3Br.2n, C3Br.2r, C4n.1n, C4n.1r, C4n.2n, and C4n.2r in Cande-Kent95 time scale (see figure 11.6). The top normal polarity zone corresponds to C3Br.1n, so the age of the top boundary of this section is 7.20 Ma. The bottom reversed polarity zone corresponds to C4n.2r (again the Cande-Kent95 time scale), so the age of bottom boundary of this section is 8.20 Ma. Thus, the Xiaohe section covers a time span of 1.0 m.y., from 8.2 Ma to 7.2 Ma. The Xiaohe section consists of two main hominoid levels: the lower layer with an age of ~8.20 Ma and the upper layer (see figure 11.4 [16th layer]) with an age of ~7.20 Ma (Yue and Zhang 2006; Gradstein, Ogg, and Smith 2004 time scale).

The Leilao section records 11 polarity zones (figure 11.7), including C3Br, C3Br.1n, C3Br.1r, C3Br.2n, C3Br.2r, C4n.1n, C4n.1r, C4n.2n, C4n.2r, C4n.3n, and C4n.3r. Reversed zones recorded at the top and at the bottom boundaries correspond to C3Br and C4n.3r. Their ages are 7.10 Ma and 8.50 Ma, respectively, on the Cande and Kent (1995) time scale. The section covers a time span of 1.4 m.y., and has two layers with hominoids: the lower layer at 9905T$_0$ includes C4n.2r with an age of 8.10–8.20 Ma and the upper layer at 9906 includes middle C3Br with an age of 7.10–7.15 Ma (Yue and Zhang

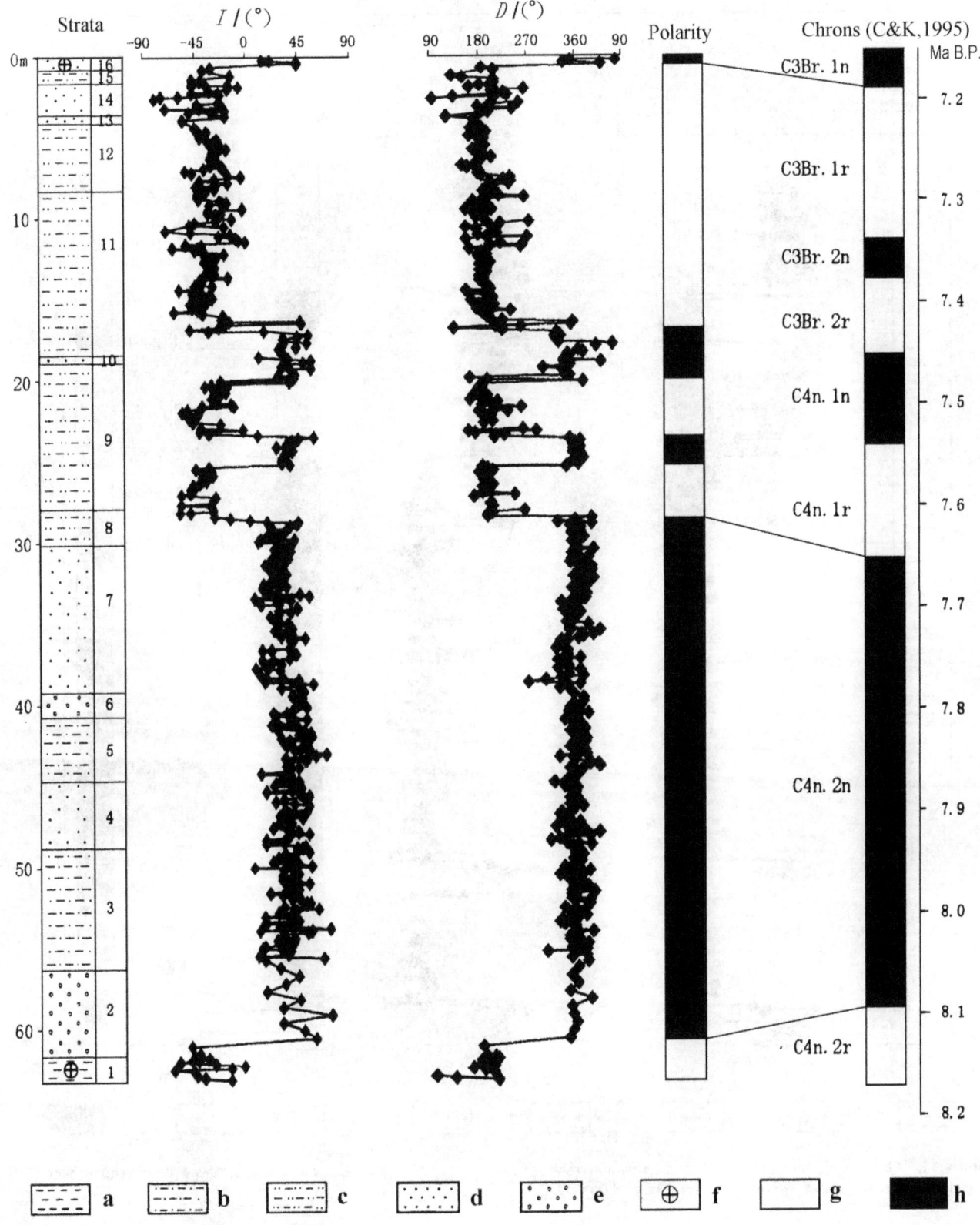

Figure 11.6 Paleomagnetic measurements and correlation results of the Xiaohe section. *a*, clay; *b*, sandy clay; *c*, clayey silt; *d*, sand; *e*, gravel; *f*, hominoid fossil; *g*, reversed polarity; *h*, normal polarity.

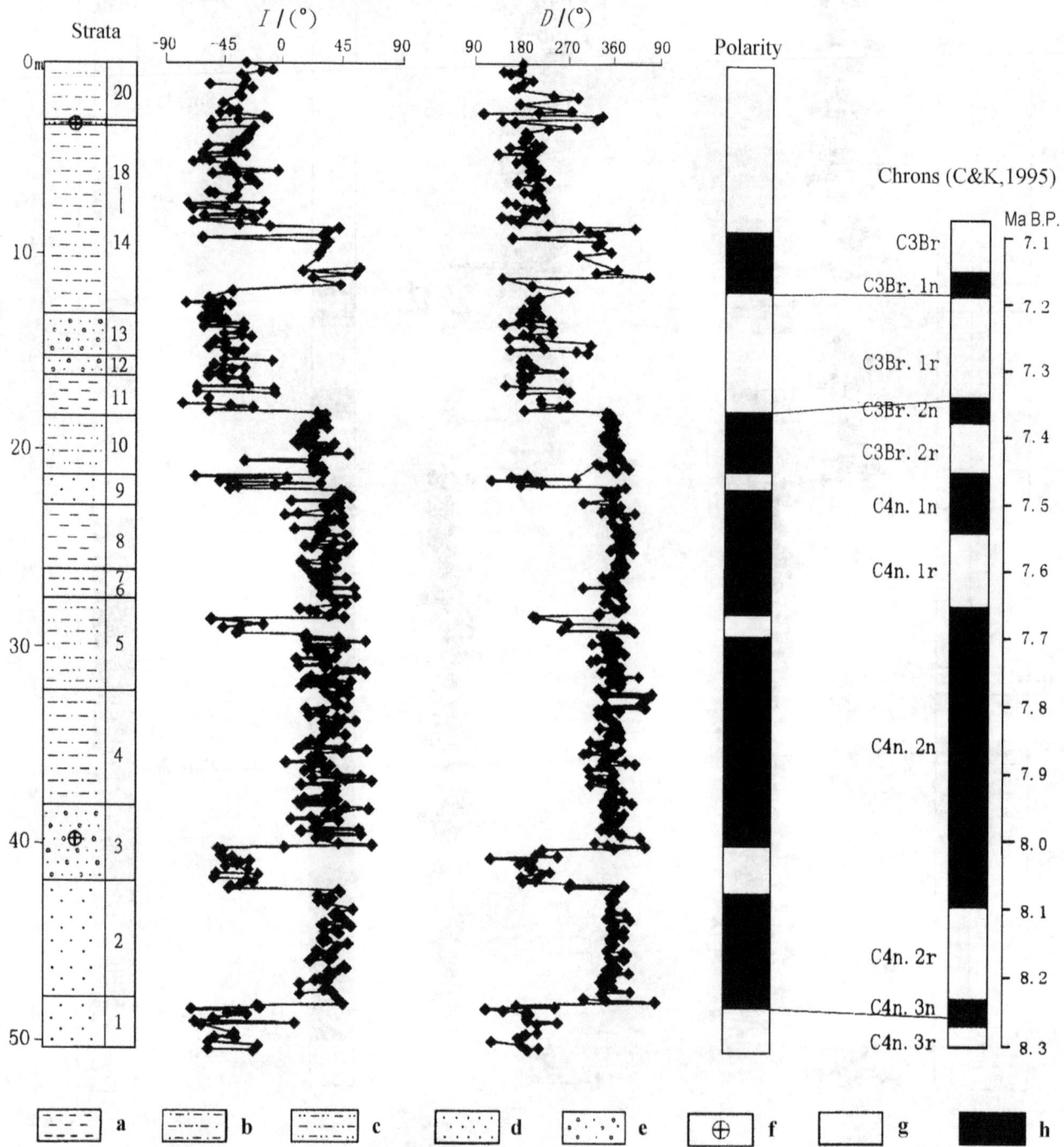

Figure 11.7 Paleomagnetic measurements and correlation results of the Leilao section. *a*, clay; *b*, sandy clay; *c*, clayey silt; *d*, sand; *e*, sand with gravel; *f*, hominoid fossil; *g*, reversed polarity; *h*, normal polarity.

2006). Thus, hominoids from the Xiaohe and Leilao sections occur within the same time interval.

The Zhupeng hominoid-bearing layer is dated to the late Miocene (Zhu et al. 2005), within either polarity chron 3Br.1r (7.34–7.17Ma) or 3Br.2r (7.43–7.38 Ma).

These paleomagnetic correlations from three different areas show that the chronological range of Yuanmou hominoids spans approximately 8–7 Ma, making the Yuanmou hominoids older than those of Lufeng, although the hominoid layers and adjacent strata at Yuanmou have a larger chronological range. Paleomagnetic dating at Zhupeng shows that the Xiaohe Formation as a whole spans 7–11 Ma (Zhu et al. 2005).

BIOSTRATIGRAPHIC CORRELATION

Three hominoid localities in Yunnan are characterized by the presence of the hominoid genus *Lufengpithecus*, represented by three different species, *L. keiyuanensis*, *L. lufengensis*, and *L. hudienensis*. Their associated faunas also show some similarities as well as differences. Due to the uplift of the Himalayas and Yunnan-Guizhou Pla-

teau, the Neogene faunas in southern China differentiate from those of northern China. The Neogene faunas in Yunnan are closer to those of South Asia (Pilbeam et al. 1996) than to those of northern China (Zhang et al. 2002; Deng 2006; Qiu, Wang, and Li 2006; Wang et al. 2009), although there are still major differences with the Siwalik faunas at high taxonomic levels.

The small mammal fauna from Yuanmou hominoid localities shows that their age is older than that of the Lufeng hominoid locality (Ni and Qiu 2002). The large mammal fauna from Yuanmou hominoid localities also shows their age is older than that of Lufeng (Qi and Ni 2006). That Lufeng hominoids are younger than those

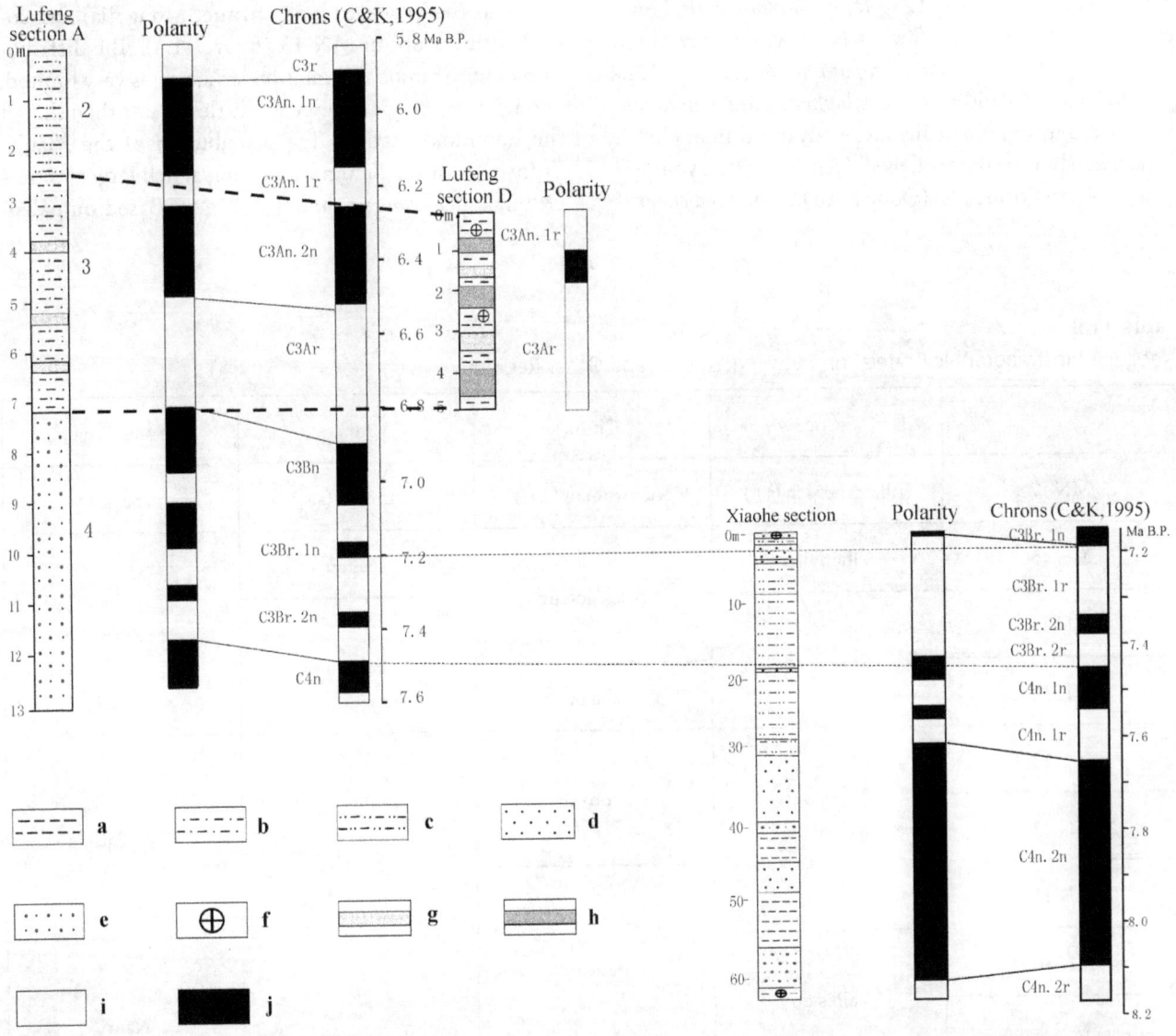

Figure 11.8 Paleomagnetic correlation of the sections A and D in Lufeng with the Xiaohe section. *a*, clay; *b*, sandy clay; *c*, clayey sand; *d*, sand; *e*, sand with gravel; *f*, hominoid layer; *g*, carbonaceous clay; *h*, lignite; *i*, reversed polarity; *j*, normal polarity. Note that the thickness scales for section A and section D are not the same.

from Yuanmou is also supported by paleomagnetic dating (Qi et al. 2006).

The Xiaolongtan Formation is composed of lignite, and it is not suitable for paleomagnetic dating. Its age is therefore mostly based on fauna analysis. *Tetralophodon xiaohensis* of Yuanmou is more advanced than *Tetralophodon xiaolongtanensis* of Kaiyuan in the Xiaolongtan Formation by two extra lophids in the last lower molar; the advanced proboscideans *Stegotetrabelodon primitivum* and *Stegolophodon* and new cervid forms like *Muntiacus* are present at Yuanmou hominoid localities but absent from Kaiyuan, indicating that the Yuanmou hominoid layers are younger than those of Kaiyuan. The Kaiyuan hominoid fauna is similar to that of Chinji horizons of the Siwaliks, particularly by the presence of *Gomphotherium* cf. *macrognathus* and *Zygolophodon chinjiensis*. However, it is also similar to a Siwalik Nagri assemblage in the presence of *Tetralophodon*, *Propotamochoerus*, and *Hippopotamodon*. In addition, *Tetralophodon xiaolongtanensis* appears morphologically more advanced than Chinji *T. falconeri* but similar to *T. punjabiensis* from the younger Dhok Pathan Formation (Dong 2001). *Propotamochoerus*

and *Hippopotamodon* extend into the Dhok Pathan Formation. The age of Xiaolongtan hominoid fauna is therefore likely to be between those of upper Chinji and lower Dhok Pathan assemblages. It is younger than the Middle Miocene Tunggur fauna, but possibly like the early Late Miocene Guonigou fauna of North China, or about European MN 8 to MN 9.

Both Lufeng and Yuanmou hominoid faunas are within the chronological range of the Dhok Pathan Formation (10.1 Ma to ca. 3.5 Ma; see Barry et al. 2002). Furthermore, based on paleomagnetic dating, the Yuanmou hominoid horizons precede early Baode faunas (Kaakkinen et al., chapter 7, this volume) of North China and are roughly equivalent to European MN 12, while the Lufeng hominoid horizon is later Baodean (but not as young as the Ertemte fauna of Inner Mongolia) and coeval with European MN 13 (figure 11.8). The distribution of the Shihuiba Formation at Lufeng is very limited, and its chronological range is a little greater than that at the hominoid section. The distribution of the Xiaohe Formation in the Yuanmou Basin is much larger, and its chronological range is about 7–11 Ma. Based on paleo-

Table 11.3

Neogene Biostratigraphic Correlation Between Yunnan and Other Regions

	Europe	China	Yunnan	South Asia
MN 17	Villafranchian (s.l.)	Nihewanian (s. s.)	Yuanmou (s.s.)	Iravaty
MN 16	Villanyian	Mazegouan	Shagou	Soan
MN 15	Ruscinian	Mazegouan	Zhaotong	Soan
MN 14	Ruscinian	Gaozhuangian	Zhaotong	Soan
MN 13	Turolian	Baodean	Shihuiba	Dhok Pathan
MN 12	Turolian	Baodean	Shihuiba	Dhok Pathan
MN 11	Turolian	Baodean	Xiaohe	Dhok Pathan
MN 10	Vallesian	Bahean	Xiaolongtan	Nagri Chinji
MN 9	Vallesian	Bahean	Xiaolongtan	Nagri Chinji
MN 7+8	Astaracian	Tunggurian		

magnetic results, the time span of the upper Xiaohe Formation overlaps the lower Shihuiba Formation. Although there are no paleomagnetic data from Kaiyuan, the lower Xiaohe Formation likely overlaps the upper Xaiolongtan Formation. The lower Xiaohe Formation is equivalent to Nagri of Siwaliks or Bahe of North China, European MN 9 and MN 10. The biostratigraphic correlation between the hominoid horizons of Yunnan and those of other regions is illustrated in table 11.3.

Kaiyuan hominoid fauna is mostly uncovered from the middle and upper Xiaolongtan Formation, which is mostly lignite. The distribution of the fossils is very scattered, and the fossils are difficult to find. The faunal age is therefore a chronological range estimated from the lowest fossil layer to the highest one, and the time span might be rather large. Lufeng hominoid fauna is on the contrary quite concentrated, uniquely in section D. The fossils at Shihuiba are mostly excavated and the potential to find new taxa and new horizons is not great. The chronological range of Lufeng hominoids is not large. Yuanmou hominoid fauna is in between in terms of its chronological range. The fossils are found in several excavation localities in the Xiaohe, Zhupeng, and Leilao areas. The fossils from the Xiaohe Formation have great potential for further study and biostratigraphic subdivision, and the Yuanmou Basin is still a very interesting place for Neogene research.

ACKNOWLEDGMENTS

The authors would like to thank all members of the Yunnan team for State Key Project of the Ninth Five Year Plan—Origin of Early Human and Environmental Background. They would also like to acknowledge Profs. L. Flynn and J. Kelley for their helpful comments and suggestions to improve the manuscript. The present work was supported by Chinese National Natural Science Foundation (project No. 40772014).

REFERENCES

Badgley, C., G. Qi, W. Chen, and D. Han. 1988. Paleoecology of a Miocene, tropical, upland fauna: Lufeng, China. *National Geographic Research* 4:178–196.

Barry, J. C., L. Flynn. 1989. Key stratigraphical events in the Siwalik sequence. In *European Neogene Mammal Chronology*, ed. E. H. Lindsay and V. Fahlbusch, pp. 557–571. New York: Plenum Press.

Barry, J. C., E. H. Lindsay, and L. L. Jacobs. 1982. A biostratigraphic zonation of the middle and upper Siwaliks of the Potwar plateau of northern Pakistan. *Palaeogeography Palaeoclimatology Palaeobiology* 37:95–139.

Barry, J. C., M. E. Morgan, L. J. Flynn, D. Pilbeam, A. K. Behrensmeyer, S. M. Raza, I. A. Khan, C. Badgley, J. Hicks, and J. Kelley. 2002. Faunal and environmental change in the Late Miocene Siwaliks of Northern Pakistan. *Palaeobiology* 28:1–71.

Bernor, R. L., V. Falbush, P. Andrews, P. De Bruijin, H. Fortelius, M. Rogl, F. Steininger, and F. F. Werdelin. 1996. The evolution of western Eurasian Neogene mammal faunas: A chronologic systematic, biogeographic and paleographic and paleoenvironmental synthesis. In *The Evolution of Western Eurasian Neogene Mammal Faunas*, ed. R. L. Bernor, V. Fahlbush, and M. Hans-Watter, pp. 449–470. New York: Columbia University Press.

Bien, M. N. 1939. Preliminary observation on the Cenozoic geology of Yunnan. *Bulletin of the Geological Society of China* 20:188–189.

Cande, S. C. and D. V. Kent. 1995. Revised calibration of the geomagnetic polarity timescale for the Late Cretaceous and Cenozoic. *Journal of Geophysical Research* 100:6093–6095

Chen, W.-y. 1986. Preliminary studies of sedimental environment and taphonomy in the hominoid fossil site of Lufeng. *Acta Anthropologica Sinica* 5:89–100.

Chow, M.-c. (Zhou, M.-z.) 1961. Occurrence of *Enhydriodon* at Yuanmo, Yunnan. *Vertebrata PalAsiatica* 2:164–167.

Colbert, E. H. 1938. Fossil mammals from Burma. *Bulletin of the American Museum of Natural History* 124:419–424.

Deprat, J. 1912. Étude géologique du Yunnan oriental. *Mémoires du Service Géologique d'Indochine* 1(1):225.

Deng, T. 2006. Chinese Neogene mammal biochronology. *Vertebrata PalAsiatica* 44:143–163.

Dong, W. 1987. Miocene mammalian fauna of Xiaolongtan, Kaiyuan, Yunnan Province. *Vertebrata PalAsiatica* 25:116–123.

Dong, W. 2001. Upper Cenozoic stratigraphy and paleoenvironment of Xiaolongtan Basin, Kaiyuan, Yunnan Province. In *Proceedings of the Eighth Annual Meetings of Chinese Society of Vertebrate Paleontology*, ed. T. Deng and Y. Wang, pp. 91–100. Beijing: China Ocean Press.

Dong, W., J. Liu, and Y. Pan. 2003. A new *Euprox* from the Late Miocene of Yuanmou, Yunnan Province, China, with interpretation of its paleoenvironment. *Chinese Science Bulletin* 48:485–491.

Dong, W., Y. Pan, and J. Liu. 2004. The earliest *Muntiacus* (Artiodactyla, Mammalia) from the late Miocene of Yuanmou, southwestern China. *Comptes Rendus Palevol* 3:379–386.

Flynn, L. J. and M. E. Morgan. 2005. New lower primates from the Miocene Siwaliks of Pakistan. In *Interpreting the Past: Essays on Human, Primate, and Mammal Evolution*, ed. D. E. Lieberman, R. J. Smith, and J. Kelley, pp. 81–101. Boston: Brill.

Flynn, L. J. and G. Qi. 1982. Age of the Lufeng, China, hominoid locality. *Nature* 298:746–747.

Gradstein, F. M., J. G. Ogg, A. G. Smith. 2004. A new Geologic Time Scale, with special reference to Precambrian and Neogene. *Episodes* 27:83–100.

Hu, C., 1973. Ape-man teeth from Yuanmou, Yunnan. *Acta Geologica Sinica* 47:65–71.

Huang, W.-p., J.-w. Wang, Z.-d. Qiu, and Z.-h. Zheng. 1978. Age and correlation of the Yuanmou, Longjie, and Xigeda Formations. *Proceedings of Stratigraphic Paleontology* 7:30–39.

Jablonski, N. G. 2005. Primate homeland: Forests and the evolution of primates during the Tertiary and Quaternary in Asia. *Anthropological Sciences* 113:117–122.

Jiang, C., L. Xiao, and J.-m. Li. 1993. Hominoid teeth fossil from Leilao, Yuanmou, Yunnan. *Acta Anthropologica Sinica* 12: 97–102.

Jiang, N.-r., R. Sun, and Q.-z Liang. 1987. The discovery of Yuanmou early ape man teeth fossils and their significance. *Yunnan Geology* 6:157–162.

Jiang, N.-r., R. Sun, and Q.-z. Liang. 1989. The late Cenozoic stratigraphy and palaeontology in Yuanmou Basin, Yunnan, China. *Yunnan Geology* Supplement 1–107.

Jiang, Z.-w., and X.-y. Zhang. 1997. Discovery and excavations of Yuanmou hominoid fauna. In *Yuanmou Hominoid Fauna*, ed. Z.-q. He, pp. 9–12. Kunming: Yunnan Science and Technology Press.

Kelley, J. 1993. Taxonomic implications of sexual dimorphism in *Lufengpithecus*. In *Species, Species Concepts, and Primate Evolution*, ed. W. Kimbel and L. Martin, pp. 429–457. New York: Plenum Press.

Kelley, J. and D. Etler. 1989. Hominoid dental variability and species number at the late Miocene site of Lufeng, China. *American Journal of Primatology* 18:15–34.

Kelley, J. and J. M. Plavcan. 1998. A simulation test of hominoid species number at Lufeng, China: Implications for the use of the coefficient of variation in paleotaxonomy. *Journal of Human Evolution* 35:577–596

Kelley, J. and Q.-h. Xu. 1991. Extreme sexual dimorphism in a Miocene hominoid. *Nature* 352:151–153.

Lantenois, M. H. 1907. Résultats de la mission géologique et minière du Yunnan méridional. Paris: *Quai des Grand-Augustins*, 49:76–93.

Liu, W., L. Zheng, and C. Jiang. 1999. Statistical analyses of metric data of hominoid teeth found in Yuanmou of China. *Chinese Science Bulletin* 45:936–942.

Lu, Q.-w., Q.-h., Xu, and L. Zheng. 1981. Preliminary research on the cranium of *Sivapithecus yunnanensi*. *Vertebrata PalAsiatica* 19:101–106.

Mansuy, H. 1912. Étude géologique du Yunnan oriental. *Mémoires du Service Géologique de l'Indochine* 1(2):15–17.

Mein, P. 1990. Updating of MN Zones. In *European Neogene Mammal Chronology*, ed. E. H. Lindsay, V. Fahlbusch, and P. Mein. pp. 73–90. New York: Plenum Press.

Ni, X.-j. and Z.-d. Qiu. 2002. The micromammalian fauna from the Leilao, Yuanmou hominoid locality: Implication for Biochronology and Paleoecology. *Journal of Human Evolution* 42:535–546.

Pan, Y.-r. 2006. Primates. In *Lufengpithecus hudienensis Site*, ed. G.- q. Qi and W. Dong, pp. 131–148. Beijing: Science Press.

Pan, Y.-r., J.-h. Liu, and W. Dong. 2006. Artiodactyla. In *Lufengpithecus hudienensis Site*, ed. G.-q. Qi and W. Dong, pp. 195–228. Beijing: Science Press.

Pilbeam, D., M. Morgan, J. C. Barry, and L. Flynn. 1996. European MN Units and the Siwalik faunal sequence of Pakistan. In *The Evolution of Western Eurasian Neogene Mammal Faunas*, ed. R. L. Bernor, V. Fahlbush, and M. Hans-Watter, pp. 96–105. New York: Columbia University Press.

Qi, G.-q. 1985. Stratigraphic summarization of *Ramapithecus* fossil locality, Lufeng, Yunnan. *Acta Anthropologica Sinica* 4:55–69.

Qi, G.-q. 2006. Canivora. In *Lufengpithecus hudienensis Site*, ed. G.-q. Qi and W. Dong, pp. 148–177. Beijing: Science Press.

Qi, G.-q. and W. Dong, eds. 2006. *Lufengpithecus hudienensis Site*. Beijing: Science Press.

Qi, G.-q., W. Dong, L. Zheng, L.-x. Zhao, F. Gao, L.-p. Yue, and Y.-x. Zhang. 2006. Taxonomy, age and environment status of the Yuanmou hominoids. *Chinese Science Bulletin* 51:704–712.

Qi, G.-q. and X.-j. Ni, 2006. Geological age of *Lufengpithecus hudienensis*. In *Lufengpithecus hudienensis Site*, ed. G.-q. Qi and W. Dong, pp. 229–239. Beijing: Science Press.

Qian, F. 1985. On the age of "Yuanmou Man"—A discussion with Liu Tungsheng et al. *Acta Anthropologia Sinica* 4:324–332.

Qian, F. and F.-c. Jiang. 1991. Stratigraphy and correlation. In *Quaternary Geology and Paleoanthropology of Yuanmou, Yunnan, China*, ed. F. Qian and G.-x. Zhou, pp. 65–72. Beijing: Science Press.

Qiu, Z.-d. and X.-j. Ni. 2006. Small mammals. In *Lufengpithecus hudienensis Site*, ed. G.-q. Qi and W. Dong, 113–131. Beijing: Science Press.

Qiu, Z.-d., X.-m. Wang, and Q. Li. 2006. Faunal succession and biochronology of the Miocene through Pliocene in Nei Mongol (Inner Mongolia). *Vertebrata PalAsiatica* 44:165–181.

Wang X.-m., Z.-d. Qiu, Q. Li, Q. Y. Tomida, Y. Kimura, Z. J. Tseng, and H.-j. Wang. 2009. A new early to late Miocene fossiliferous region in central Nei Mongol: Lithostratigraphy and biostratigraphy in Aoerban strata. *Vertebrata PalAsiatica* 47:111–134.

Woo, J. K. (=Wu R.-k.) 1957. *Dryopithecus* teeth from Kaiyuan, Yunnan Province. *Vertebrata PalAsiatica* 1:25–32.

Woo, J. K. (=Wu R.-k.) 1958. New material of *Dryopithecus* from Kaiyuan, Yunnan Province. *Vertebrata PalAsiatica* 2:38–42.

Wu, R.-k. 1987. A revision of the classification of the Lufeng great apes. *Acta Anthropologica Sinica* 6:265–271.

Wu, R.-k., Q.-h. Xu, and Q.-w. Lu. 1983. Morphological features of *Ramapithecus* and *Sivapithecus* and their relationships—Morphology and comparison of the crania. *Acta Anthropologica Sinica* 2:1–10.

Wu, R.-k., Q.-h. Xu, and Q.-w. Lu, 1985. Morphological features of *Ramapithecus* and *Sivapithecus* and their relationships—Morphology and comparison of the teeth. *Acta Anthropologica Sinica* 4:197–204.

Wu, R.-k., Q.-h. Xu, and Q.-w. Lu. 1986. Relationship between Lufeng *Sivapithecus* and *Ramapithecus* and their phylogenetic position. *Acta Anthropologica Sinica* 5:1–30.

Wu, R.-k., X.-z. Wu, and S.-s. Zhang. 1989. *Early Humankind in China*. Beijing: Science Press (in Chinese with English summary).

Xiong, Y.-x. 1957. Geological report on Xiaolongtan Coal Field, Kaiyuan, Yunnan. Kunming: Team 537 of Geological Bureau of Southwest China (unpublished).

Xu, Q.-h. and Q.-w. Lu. 1979. The mandible of *Ramapithecus* and *Sivapithecus*. *Vertebrata PalAsiatica* 17:1–13.

Xu, Q.-h. and Q.-w. Lu. 2008. *Lufengpithecus lufengensis—An Early Member of Hominidae*. Beijing: Science Press.

Xu, Q.-h., Q.-w. Lu, Y.-r. Pan, G.-q. Qi, X.-y. Zhang, and L. Zheng. 1978. The fossil mandible of *Ramapithecus lufengensis*. *Kexue Tongbao* 23:554–556.

Young, C. C. and M. N. Bien. 1939. New horizons of Tertiary mammals in southern China. *Proceedings of the 6th Pacific Congress*, 531–534.

Yue, L.-p. and Y.-x. Zhang. 2006. Results and discussion of paleomagnetic dating of *Lufengpithecus hudienensis*. In *Lufengpithecus hudienensis Site*, ed. G.-q. Qi and W. Dong, pp. 245–255. Beijing: Science Press.

Yue, L.-p., Y.-x. Zhang, G.-q. Qi, A.H. Friedrich, J.-q. Wang, L.-r. Yang, and R. Zhang. 2004. Paleomagnetic age and palaeobiological significance of hominoid fossil strata of Yuanmou Basin in Yunnan. *Science in China*, ser. D, 47:405–411.

Yunnan Regional Stratigraphic Chart Composition Team. 1978. Regional Stratigraphic Chart of Southwest China. Beijing: Geological Publishing House.

Zhang, X.-y. 1987. New materials of *Ramapithecus* from Kaiyuan, Yunnan. *Acta Anthropologica Sinica* 6:81–86.

Zhang, X.-y., Y.-p. Lin, C. Jiang, and L. Xiao. 1987a. A new species of *Ramapithecus* from Yuanmou, Yunnan Province. *Sixiangzhanxian* 3:54–56.

Zhang, X.-y., Y.-p Lin, C. Jiang, and L. Xiao. 1987b. A new species of *Homo* from Yuanmou, Yunnan. *Sixiangzhanxian* 3:57–60.

Zhang, X.-y., L. Zheng, and F. Gao. 1990. New genus *Sinopithecus* and its anthropological significance. *Sixiangzhanxian* 2:53–58.

Zhang, Z.-q., A. W. Gentry, A. Kaakinen, L.-p. Liu, J. P. Lunkka, Z.-d. Qiu, S. Sen, R. S. Scott, and L. Werdelin. 2002. Land mammal faunal sequence of the Late Miocene of China: New evidence from Lantian, Shanxi Province. *Vertebrata PalAsiatica* 40:165–176.

Zheng, L. 2006. Classification and systematic status of *Lufengpithecus hudienensis*. In *Lufengpithecus hudienensis Site*, ed. G.-q. Qi and W. Dong, pp. 101–112. Beijing: Science Press.

Zheng, L. and X.-y. Zhang, 1997. Hominoid fossils. In *Yuanmou Hominoid Fauna*, ed. Z.-q. He, pp. 21–60. Kunming: Yunnan Science & Technology Press.

Zhu, R.-x., Q.-s. Liu, H.-t. Yao, Z.-t. Guo, C.-l. Deng, Y.-x. Pan, L.-q. Lü, Z.-g. Chang, and F. Gao. 2005. Magnetostratigraphic dating of hominoid-bearing sediments at Zhupeng, Yuanmou Basin, southwestern China. *Earth and Planetary Science Letters* 236:559–568.

Zong, G.-f. 1996. The Neogene strata of the Yuanmou Basin, Yunnan. *Journal of Stratigraphy* 20:138–145.

APPENDIX

Faunal Composition and Comparison of Yunnan Hominoid Localities Kaiyuan (Xiaolongtan Formation), Lufeng (Shihuiba Formation), and Yuanmou (Xiaohe Formation)

		Taxon	Kaiyuan	Lufeng	Yuanmou
PRIMATES	Hominoidea	*Lufengpithecus keiyuanensis*	+		
	Hominoidea	*Lufengpithecus lufengensis*		+	
	Hominoidea	*Lufengpithecus hudienensis*			+
	Hylobatidae	*Laccopithecus robustus*		+	
	Sivaladapidae	*Indraloris progressus*			+
	Sivaladapidae	*Sinoadapis parvus*			+
	Sivaladapidae	*Sinoadapis carnosus*		+	
	Sivaladapidae	*Sinoadapis shihuibaensis*		+	
	Incertae familiae	*Yuanmoupithecus xiaoyuan*			+
SCANDENTIA	Tupaiidae	*Prodendrogale yunnanica*		+	+
	Tupaiidae	Ptilocercinae gen. et sp. indet.			+
INSECTIVORA	Erinaceidae	*Hylomys suillus*		cf.	cf.
	Erinaceidae	*Lanthanotherium sanmigueli*		+	
	Talpidae	*Yunoscaptor scalprum*		+	+
	Soricidae	*Heterosorex wangi*		+	+
	Soricidae	*Anourosorex oblongus*		+	+
	Soricidae	*Blarinella*		sp.	
	Soricidae	*Crocidura*		cf.	
	Soricidae	*Sorex*		sp.1	sp.1
	Soricidae	*Sorex*			sp.2
	Soricidae	Soricinae gen. et sp. indet.			+
	Incertae familiae	Gen. et sp. indet.			+
CHIROPTERA	Pteropidae	Gen. et sp. indet.		+	
	Vespertilionidae	*Myotis*		sp.	
	Vespertilionidae	*Eptesicus*		sp.	
	Vespertilionidae	*Pipistrellus*		sp.	

		Taxon	Kaiyuan	Lufeng	Yuanmou
	Vespertilionidae	*Plecotus*		sp.	
	Vespertilionidae	Gen. et sp. indet.			+
	Hipposideridae	Gen. et sp. indet.		+	+
	Incertae familiae	Gen. et sp. indet.			+
RODENTIA	Sciuridae	*Sciurotamias wangi*		+	+
	Sciuridae	*Sciurotamias leilaoensis*			+
	Sciuridae	*Tamiops atavus*		sp.	+
	Sciuridae	*Callosciurus erythraeus*		sp.	sp.
	Sciuridae	*Dremomys primitivus*		+	+
	Sciuridae	*Dremomys pernyi*		sp.	sp.
	Sciuridae	*Ratufa yuanmouensis*			+
	Sciuridae	*Miopetaurista asiatica*		+	+
	Sciuridae	*Hylopetodon dianense*		+	+
	Sciuridae	*Pliopetaurista speciosa*			+
	Sciuridae	*Pliopetaurista rugosa*			sp.
	Sciuridae	*Yunopterus jiangi*		+	+
	Sciuridae	Sciurinae gen. et sp. indet.			+
	Castoridae	Gen. et sp. indet.	+		
	Castoridae	*Steneofiber* sp.		sp.	sp.
	Platacanthomyidae	*Platacanthomys dianensis*		+	+
	Platacanthomyidae	*Typhlomys primitivus*		+	+
	Platacanthomyidae	*Typhlomys hipparionum*		+	+
	Eomyidae	*Leptodontomys*		sp.	sp.
	Eomyidae	Gen. et sp. indet.		+	+
	Rhizomyidae	*Miorhizomys blacki*			+
	Rhizomyidae	*Miorhizomys tetracharax*		+	+
	Rhizomyidae	*Miorhizomys pilgrimi*		cf.	
	Rhizomyidae	*Miorhizomys nagrii*		+	
	Rhizomyidae	*Miorhizomys* sp. nov.		+	+
	Rhizomyidae	Tachyoryctinae gen. et sp. indet.			+
	Rhizomyidae	Rhizomyidae gen. et sp. indet.		+	+
	Cricetidae	*Kowalskia hanae*		+	+
	Cricetidae	Cricetidae gen. et sp. indet.			+
	Muridae	*Linomys yunnanensis*		+	+
	Muridae	*Leilaomys zhudingi*			+
	Muridae	Gen. et sp. indet.			+
	Hystricidae	*Atherurus*			sp.
	Hystricidae	*Hystrix*		sp.	sp.
	Pedetidae	Gen. et sp. indet.			+
LAGOMORPHA	Leporidae	*Alilepus longisinuosus*		+	
	Leporidae	Gen. et sp. indet.			+
CARNIVORA	Amphicyonidae	*Amphicyon palaeoindicus*			+
	Amphicyonidae	*Vishnucyon chinjiensis*			cf.
	Ursidae	*Ursavus depereti*		+	sp.
	Ursidae	*Indarctos yangi*		+	+
	Ursidae	*Indarctos yangi*			cf.

(continued)

	Taxon	Kaiyuan	Lufeng	Yuanmou
Ursidae	Ursidae gen. et sp. indet.		+	+
Incertae familiae	*Pseudarctos bavaricus bavaricus*			+
Ailuropoidae	*Ailurarctos yuanmouensis*			+
Mustelidae	*Martes palaeosinensis*		cf.	
Mustelidae	*Martes zdanskyi*		sp.	cf.
Mustelidae	Mustelinae gen. et sp. indet		+	
Mustelidae	*Eomellivora wimani*		cf.	+
Mustelidae	*Trochotherium yuanmouensis*			+
Mustelidae	Melinae gen. et sp. indet.		+	+
Mustelidae	*Proputorius lufengensis*		+	
Mustelidae	*Proputorius*		sp.	sp.
Mustelidae	*Sivaonyx bathygnathus*		+	+
Mustelidae	*Lutra*		sp.	sp.
Mustelidae	*Parataxidea sinensis*			cf.
Mustelidae	Lutrinae gen. et sp. indet.		+	
Mustelidae	Gen. indet. et sp. 1		+	+
Mustelidae	Gen. indet. et sp. 2		+	+
Mustelidae	Mustelidae gen. et sp. indet.	+		
Viverridae	*Vishnuictis salmontanus*			+
Viverridae	*Viverra*		sp.	
Viverridae	Viverrinae gen. et sp. indet. 1		+	
Viverridae	Viverrinae gen. et sp. indet. 2		+	
Viverridae	*Lufengictis peii*		+	
Viverridae	Gen. et sp. indet. 1		+	+
Viverridae	Gen. et sp. indet. 2		+	+
Hyaenidae	*Ictitherium viverrinum*		+	+
Hyaenidae	*Ictitherium hyaenoides*		sp.	+
Hyaenidae	*Thalassictis wongi*			+
Hyaenidae	*Adcrocuta eximia*			+
Hyaenidae	Gen. et sp. indet. 1			+
Hyaenidae	Gen. et sp. indet. 2			+
Felidae	*Machairodus maximiliani*			cf.
Felidae	*Machairodus*			sp.
Felidae	*Epimachairodus fires*		+	
Felidae	*Metailurus parvulus*			+
Felidae	*Pseudaelurus*		sp.	
Felidae	*Felis*		sp.	sp.1
Felidae	*Felis*			sp.2
Felidae	*Felis*			sp.3
Felidae	*Felis*			sp.4
Felidae	Gen. et sp. indet.			+
Incertae familiae	Gen. et sp. indet.			+
PROBOSCIDEA Gomphotheriidae	*Tetralophodon xiaolongtanensis*	+		
Gomphotheriidae	*Tetralophodon xiaohensis*			+
Gomphotheriidae	*Gomphotherium macrognathus*	cf.	sp.	
Gomphotheriidae	*Serridentinus* sp.		?	
Mummutidae	*Zygolophodon chinjiensis*	+		

		Taxon	Kaiyuan	Lufeng	Yuanmou
	Mummutidae	*Zygolophodon lufengensis*		+	
	Mummutidae	*Mammut zhupengensis*			+
	Elephantidae	*Stegotetrabelodon primitium*			+
	Elephantidae	*Stegolophodon*			sp.
PERISSODACTYLA	Tapiridae	*Tapirus yunnanensis*	cf.	sp.	sp.
	Chalicotheriidae	*Macrotherium yuanmouensis*			+
	Chalicotheriidae	*Macrotherium salinum*		+	
	Chalicotheriidae	*Macrotherium*		sp.	
	Rhinocerotidae	*Subchilotherium intermedium*			+
	Rhinocerotidae	*Chilotherium* sp. nov.		+	
	Rhinocerotidae	*Aceratherium* sp. nov.		+	
	Rhinocerotidae	*Rhinoceros vidali*			cf.
	Rhinocerotidae	Gen. et sp. indet.			+
	Equidae	*Cormohipparion chiai*			cf.
	Equidae	*Hipparion*		sp.1	
	Equidae	*Hipparion*		sp.2	
ARTIODACTYLA	Suidae	*Hyotherium palaeochoerus*		cf.	
	Suidae	*Hyotherium*		sp.	
	Suidae	*Parachleuastochoerus sinensis*	+		
	Suidae	*Propotamochoerus parvulus*	+		
	Suidae	*Propotamochoerus wui*		+	
	Suidae	*Hippopotamodon hyotherioides*	+	+	+
	Suidae	*Yunnanochoerus lufengensis*		+	sp.
	Suidae	*Chleuastochoerus*		sp.	sp.
	Suidae	*Molarochoerus yuanmouensis*			+
	Suidae	Gen. et sp. indet.		+	
	Tragulidae	*Dorcabune progressus*		+	cf.
	Tragulidae	*Dorcatherium*			sp.
	Tragulidae	*Yunnanotherium simplex*		+	sp.
	Moschidae	*Moschus*		sp.	sp.
	Cervidae	*Euprox robustus*	sp.	sp.	+
	Cervidae	*Paracervulus brevis*			+
	Cervidae	*Paracervulus simplex*		cf.	
	Cervidae	*Paracervulus attenuatus*		cf.	cf.
	Cervidae	*Muntiacus nanus*		cf.	+
	Cervidae	*Muntiacus leilaoensis*		sp.	+
	Cervidae	Muntiacinae gen. et sp. indet.		+	+
	Cervidae	Gen. et sp. indet. 1		+	
	Cervidae	Gen. et sp. indet. 2		+	
	Bovidae	*Selenoportax*		sp.	
	Bovidae	Gen. et sp. indet.		+	

NOTE: += the taxon is present at the locality; sp. = the taxon is only identified at generic level at the locality; cf. = the taxon is identified as conformis species at the locality.

Chapter 12

Miocene Land Mammals and Stratigraphy of Japan

YUKIMITSU TOMIDA, HIDEO NAKAYA, HARUO SAEGUSA, KAZUNORI MIYATA, AND AKIRA FUKUCHI

Miocene terrestrial mammal fossils are never abundant in Japan, but the research on those fossils has a rather long history. The first descriptive paper on a Japanese Miocene mammal was the holotype skull of *Desmostylus japonicus* by Yoshiwara and Iwasaki (1902). Considering desmostylians as marine mammals, the first descriptive paper on a terrestrial mammal from the Japanese Miocene was a cervid jaw ("*Amphitragulus minoensis*") by Matsumoto (1918), although a few discovery reports were announced in Japanese before that (e.g., Yoshiwara 1899; Sato 1914; Matsumoto 1916). For nearly a century since then, many fossils have been discovered and described in research papers, and presently they can be classified in at least 7 orders, 13 families, and 18 genera.

Although many papers have been accumulated so far, this chapter is probably the first attempt to compile all the Miocene terrestrial mammals in Japan biostratigraphically and biochronologically with correlations to European and Chinese land mammal zonations. A workshop and symposium entitled "Neogene Terrestrial Mammalian Biostratigraphy and Chronology in Asia" was organized in Beijing in June 2009, and we had an opportunity to join the workshop and to present a paper on the terrestrial mammals and their biostratigraphy of Japanese Miocene. The present chapter is the compilation of that talk.

TERRESTRIAL MAMMALS FROM AND GEOLOGIC AGE OF MIZUNAMI GROUP, GIFU PREFECTURE

The Mizunami Group is distributed in the Kani and Mizunami basins in southern Gifu Prefecture and consists of the Hachiya, Nakamura, and Hiramaki formations in the Kani Basin and the Toki Lignite-bearing, Hongo, and Akeyo formations in the Mizunami Basin (figures 12.1 and 12.2). Except for the Hongo and Akeyo formations, the first four formations are freshwater sediments and have been known to produce terrestrial mammal fossils since the early 1900s; a revised faunal list is in table 12.1.

Rhinocerotids from the Kani Basin

Early Miocene rhinocerotids have been recorded from the Mizunami Group in central Japan. Unfortunately, most of them are so fragmentary that they cannot be precisely identified.

Plesiaceratherium sp. from the Nakamura Formation is a middle-sized acerathere. The absence of the rugosities on the labial wall of the lower premolars indicates that this species is closer to *P. gracile* known from Shanwang, China (Young 1937), than to other species of the genus. However, the premolars of the Japanese species are smaller than those of the Chinese species (Fukuchi and Kawai 2011).

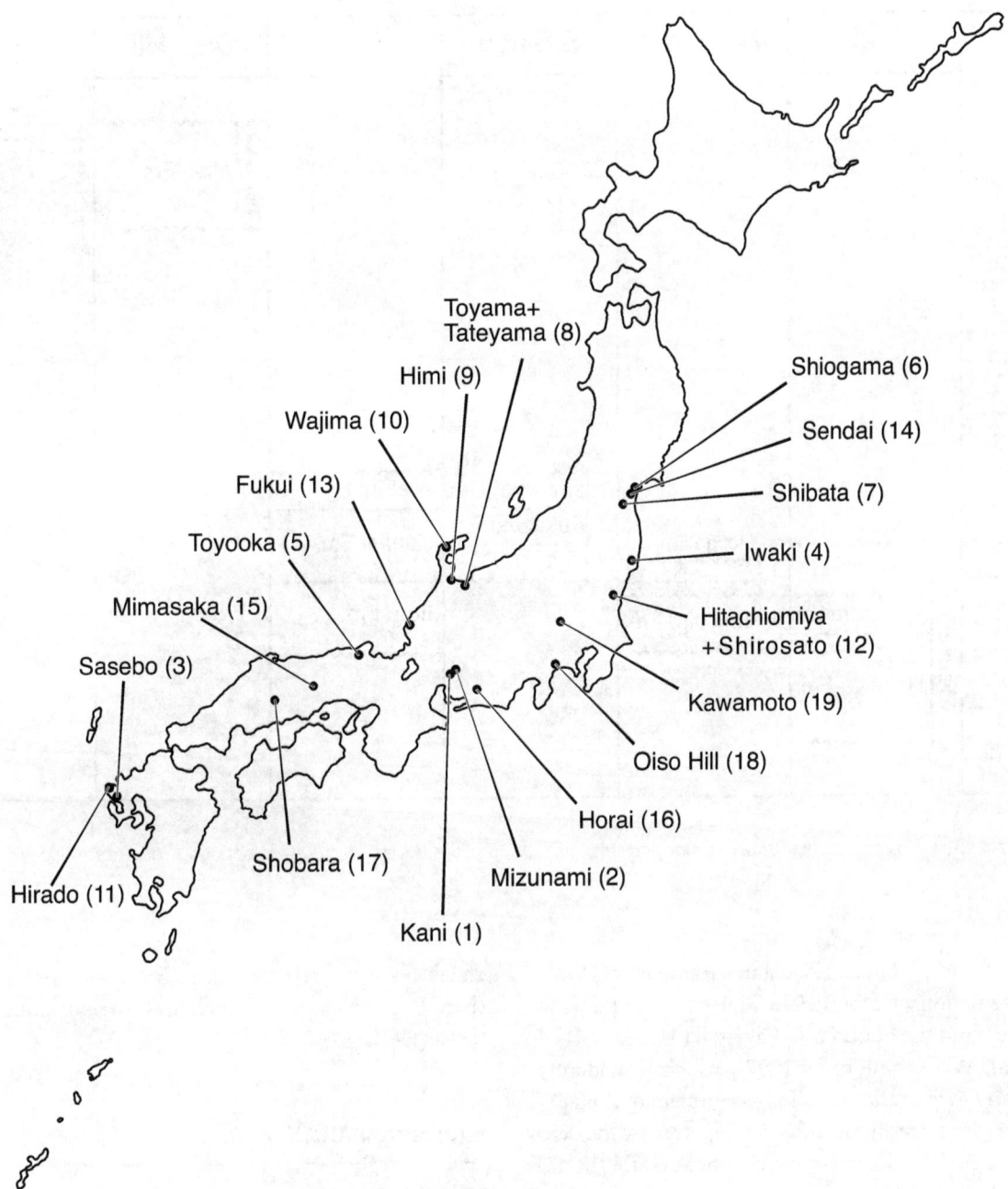

Figure 12.1 Map showing the localities of Miocene terrestrial mammals in Japan.

Brachypotherium pugnator is a large species of the genus and is reported from the Nakamura and Hiramaki formations (e.g., Okumura et al. 1977). Japanese researchers have included the species in the genus *Chilotherium* after Takai (1939), whereas Wang (1965) allocated it to *Plesiaceratherium*. However, this species should not belong to either of these genera for the following reasons.

First, the earliest occurrence of this species from the Nakamura Formation (19.6–18.4 Ma by Shikano 2003; see "(7) Chronology of mammal bearing formations in the Kani and Mizunami basins" section) is earlier than the likely origin of *Chilotherium*, which could date back to the Middle Miocene (ca. 16.0–11.6 Ma; Deng 2006a). Second, this species is similar in size to *B. perimense*

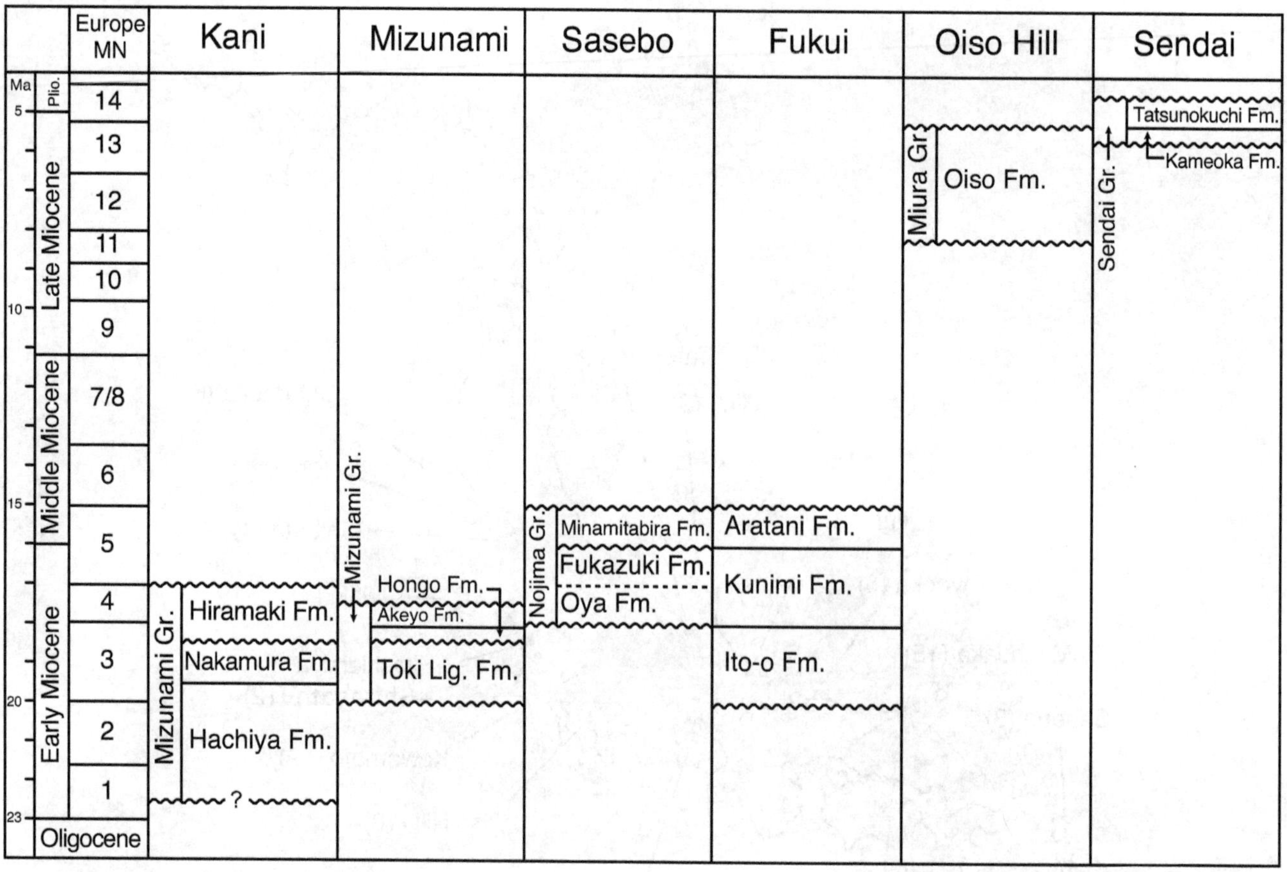

Ma	Europe MN	Kani	Mizunami	Sasebo	Fukui	Oiso Hill	Sendai
Plio. 5	14						Tatsunokuchi Fm.
Late Miocene	13					Oiso Fm.	Kameoka Fm.
	12						
	11						
10	10						
	9						
Middle Miocene	7/8						
	6						
15	5			Minamitabira Fm.	Aratani Fm.		
	4	Hiramaki Fm.	Hongo Fm. / Akeyo Fm.	Fukazuki Fm. / Oya Fm.	Kunimi Fm.		
Early Miocene	3	Nakamura Fm.	Toki Lig. Fm.		Ito-o Fm.		
20	2	Hachiya Fm.					
	1	?					
23	Oligocene						

Figure 12.2 Correlations of Miocene strata of major terrestrial mammal localities (except for the stegolophodont localities) in Japan.

from the Potwar Plateau, Pakistan (Kamlial to Dhok-Pathan formations; Colbert 1935; Heissig 1972; Pilbeam et al. 1996), and the Bugti Hills, Pakistan (MN 3b–MN 4 equivalent; Welcomme et al. 1997), which is evidently larger than *Plesiaceratherium*. The assignment of this species to the genus *Brachypotherium* is supported by the dental features such as the low-crowned cheek teeth, the absence of the coronal cement, a strong metacone bulge in M3, and the constricted protocone in upper cheek teeth. *B. pugnator* has strong protocone constrictions and antecrochets in the upper molars as in *B. fatehjangense*, which has a range in Pakistan similar to that of *B. perimense* (Colbert 1935; Heissig 1972; Pilbeam et al. 1996; Welcomme et al. 1997) and occurs at Chaungtha, Myanmar (Chavasseau et al. 2006). However, *B. perimense* is easily distinguished by its larger size and low-crowned cheek teeth. The increase of hypsodonty can be related to the diet, but it is also known as a general evolutionary trend in the Rhinocerotidae (Heissig 1989). The low-crowned

cheek teeth suggest that *B. pugnator* is more primitive than *B. fatehjangense*, which has subhypsodont cheek teeth (Heissig 1972).

Equids from the Kani Basin

Miocene equid specimens in Japan are known only from the Kani Basin and questionably from the Mizunami Basin, Gifu Prefecture. The three remains described as *Anchitherium "hypohippoides"* were recovered from the upper member of the Hiramaki Formation (ca. 17–18 Ma; Shikano 2003) in the Kani Basin. The first specimen, the holotype described by Matsumoto (1921), is a pair of upper and lower cheek teeth possibly assignable as a right P3 and a left p4. Later, a pair of dentaries with almost complete cheek teeth (p2–m3) and the insufficiently prepared maxillae with cheek teeth interpreted as P3–M3 (misidentified) were described in 1961 and 1977, respec-

Table 12.1
Faunal List of Terrestrial Mammals in Kani and Mizunami Basins

Kani Basin		Mizunami Basin
Hiramaki Formation	Nakamura Formation	Toki Lignite-bearing Formation
Perissodactyla	Perrisodactyla	Proboscidea
Rhinocerotidae	Rhinocerotidae	Gomphotheriidae
Brachypotherium pugnator	*Plesiaceratherium* sp.	*Gomphotherium annectens*
Equidae	*Brachypotherium pugnator*	Rodentia
Anchitherium aff. *A. gobiense*	Tapiridae	Castoridae
Tapiridae	*Plesiotapirus yagii*	*Youngofiber sinensis*
Plesiotapirus yagii	Eulipotyphla	
Proboscidea	Plesiosoricidae	
Gomphotheriidae	*Plesiosorex* sp.	
Gomphotherium annectens	Lagomorpha	
Artiodactyla	Ochotonidae	
Cervoidea fam. indet.	Gen. et sp. indet.	
	Rodentia	
	Castoridae	
	Youngofiber sinensis	
	Gen. et sp. indet.	
	Eucastor ? sp.	
	Eomyidae	
	Megapeomys sp.	
	Gen. et sp. indet. 1	
	Gen. et sp. indet. 2	

NOTE: A few fragmentary specimens identifiable only at the order level are omitted.

tively (Shikama and Yoshida 1961; Okumura et al. 1977). Although the exact locality and horizon of the holotype is unknown, the other two specimens seemed to be collected from the same horizon (Yoshida 1977). However, the name A. "hypohippoides" is problematic, because the holotype was most likely composed of teeth from different species (a lower premolar was from *Anchitherium*, but an upper cheek tooth was from a different perissodactyl), and the type seems to have been lost (Okumura et al. 1977; Miyata and Tomida 2010). Abusch-Siewert (1983) considered A. "hypohippoides" a junior subjective synonym of the most cosmopolitan A. aurelianense, because of the incompleteness of the holotype. Miyata and Tomida (2010) reassigned the maxillary and dentary specimens to *Anchitherium* aff. *A. gobiense* Colbert, 1939 and suggested that early species diversification of *Anchitherium* in East Asia predates a greater diversification

in Europe associated with the Middle Miocene Climatic Optimum.

Kamei and Okazaki (1974) reported a fragment of radius assigned as ?*Anchitherium* sp. from the Togari Sandstone Member (=Togari Member hereafter) of the Akeyo Formation in the adjacent Mizunami Basin. However, the radial fragment lacks definitive equid character, and no justifiable, additional material of *Anchitherium* has been known from the formation.

Tapirids from the Kani and Mizunami Basins

The four Miocene tapirid remains are known from the Kani and Mizunami basins, Gifu Prefecture, although the horizons of the early-discovered specimens are ambiguous due to lack of stratigraphic data. Matsumoto

(1921) established *"Palaeotapirus" yagii* based on an incomplete right dentary with p2–m2 and an isolated m3 from a single individual from the Hiramaki Formation (the holotype seems to be lost; see also Kamei and Okazaki 1974; Okumura et al. 1977). However, Okumura et al. (1977) suspected that the stratigraphic horizon of the type specimen probably belongs to the lower member of the Hiramaki Formation or the lower Nakamura Formation, based on their geological investigation of and around the locality mentioned by Matsumoto (1921). Takai (1949) described a right dentary fragment with m1–2 referred to this species. The referred specimen is also supposed to be from the Hiramaki Formation (Takai 1949), but the horizon of the locality is within the Nakamura Formation in the current local geologic framework (Okumura et al. 1977). Qiu, Yan, and Sun (1991) allocated *"Palaeotapirus" yagii* to a newly erected genus *Plesiotapirus* with three referred specimens including a skull from the Shanwang fauna (NMU 4 and/or 5; Qiu, Wu, and Qiu 1999; Deng 2006b), Shandong Province, China. Except for the two specimens described by Matsumoto (1921) and Takai (1949), only a left calcaneum assigned to the species from the upper part of the Nakamura Formation and two tooth fragments questionably referred to this species (?*P. yagii*) from the Togari Member of the Akeyo Formation, adjacent the Mizunami Basin, were reported (Okumura et al. 1977; Okazaki 1977). Both apparently lack diagnostic tapirid character.

Gomphotherium from the Mizunami Group

Gomphotherium annectens (Matsumoto 1924) from the lower Miocene Hiramaki and Toki Lignite-bearing formations of the Mizunami Group is the earliest proboscidean known from Japan. This species is known from a set of upper and lower jaws, presumably belonging to a single individual, from the Hiramaki Formation within the Kani Basin (Matsusmoto 1926; Makiyama 1938; Kamei et al. 1977) and a fragment of left mandible housing m3 from the Toki Lignite-bearing Formation in the Mizunami Basin (Kamei et al. 1977). A tibia from the lower part of the Hiramaki Formation was assigned to this species (Kamei et al. 1977), but it apparently lacks diagnostic character of the species. The left mandible from the Toki Lignite-bearing Formation is larger than the smaller mandible from the Hiramaki Formation, but both are within the expected range of individual variation for a species of gomphothere exemplified by a sample of *G. angustidens* from En Péjouan, Gers, France (Tassy 1996a). *Gomphotherium annectens* (Matsumoto 1924)

has been considered as the representative of the most primitive stage of the genus. *Gomphotherium "annectens"* group is characterized by the relatively simple crown structure of molars, along with the pyriform cross section of lower tusk (Tassy 1994, 1996b). Species included in this grade have been reported from the early Miocene of East Africa (*G. sp* from Kenya), South Asia (*G. cooperi* from Bugti, Pakistan), Europe (*G. sylvaticum* from MN 4 to MN 5), and Japan. It has thus far not been documented from elsewhere in East Asia other than Japan.

According to recent studies of diatom fossils and magnetostratigraphy of the Mizunami Group (Gladenkov 1998; Hayashida 1986; Hiroki and Matsumoto 1999; Kohno 2000; Ujihara, Irizuki, and Hosoyama 1999), the chronological range of *Gomphotherium annectens* can be roughly assigned to chron C5En (18.056–18.524 MA; Lourens et al. 2004) or slightly older (Saegusa 2008).

Cervoids from the Mizunami Group

Matsumoto (1918) named *Amphitragulus minoensis* based on a fragmentary right mandible with p3–m2, which was found from the Hiramaki Formation. Since then, several additional specimens of similar form have been obtained: left upper molar from the Hiramaki Formation (Nagasawa 1932), fragmentary lower molar and several postcranial bones from the Akeyo Formation (Kamei and Okazaki 1974; Okazaki 1977), and a partial postcranial skeleton from the Hachiya Formation (Shikano and Ando 2000). All of them have been referred, or are questionably referred, to this species or to the genus, but it seems likely that none of the additional specimens has diagnostic characters to characterize the genus. In addition, many cervoid or ruminant taxa have been added in Eurasia even during the Miocene alone, and their classifications have changed drastically since 1918. Therefore, a thorough review of the material is definitely needed for taxonomic identification.

Small Mammals from the Kani and Mizunami Basins

One horizon (Dota locality) of the upper part of the Nakamura Formation has yielded a number of small mammal fossils, and three orders, four families, and eight species have so far been identified (see table 12.1). They are one plesiosoricid insectivore, one ochotonid lagomorph, three castorid rodents, and three eomyid rodents (Tomida 2000).

Plesiosorex sp. is represented by a single jaw with three complete or partial teeth. This genus has not been known from Asia except for the Early Miocene of Kazakhstan (Kordikova 2000). McKenna and Bell (1997:286) listed an Anatolian record, but Ziegler (2009) does not list such a record. Seven species have been known from Europe ranging from the Late Oligocene to Late Miocene (Ziegler 2009), and five species have been known from North America ranging from the latest Arikareean to late Clarendonian (ca. 19.4–9.0 Ma; Gunnell et al. 2008). The ochotonid is known only from an isolated right M2 with typical unilateral hypsodonty. Although an M2 is not diagnostic at the generic level, it is superficially quite similar to the *Amphylagus–Eurolagus* group, which ranges from the Late Oligocene to Late Miocene in Europe (Boon-Kristkoiz and Kristkoiz 1999; McKenna and Bell 1997).

Youngofiber sinensis from the Kani Basin is represented by an isolated P4, which is somewhat smaller and more worn than the holotype and other topotypic specimens, but it can be identified as *Y. sinensis* based mainly on the enamel pattern and tooth height (Tomida et al. 1995). *Y. sinensis* is known from Xiacaowan, Sihong, Jiangsu Province, China (Chow and Li 1978; Li et al. 1983), and its associated fauna is correlated with the early part of MN 4 (Deng 2006b). Castoridae gen. et sp. indet. is the most common element from the Dota locality, represented by nearly complete jaws, fragmentary jaws, and many isolated cheek teeth and incisors. Originally it was identified as a new species of *Anchitheriomys* (Tomida 2000), but it became clear that it does not belong to the genus but rather probably a new genus based on enamel microstructure, surface texture, and morphology of the incisors, although it is still anchitheriomyine (Thomas Mors, pers. comm.). *Eucastor?* sp. is a much smaller species than the other two beavers and is represented by a single lower jaw. Taxonomy and synonymy of *Eucastor* and "*Monosaulax*" are confused, and our identification is tentative.

Megapeomys is a large, peculiar eomyid rodent described first from the Czech Republic in 1998, and an isolated lower molar from Japan was identified as a species of that genus (Fejfar, Rummel, and Tomida 1998). Later, a much larger species was added from North America (Morea and Korth 2002). Direct comparison of the Dota specimen with the European material made it possible to distinguish it from European species (*M. lavocati* and *M. lindsayi*), and it also differs from North American *M. bobwilsoni*. A new species is being described

based on the characters of m1 (Tomida 2011). *Megapeomys* is a rare genus but is restricted to European MN 3–4 zones (Fejfar, Rummel, and Tomida 1998; Engesser 1999; but see Mein 1999) and to the late Hemingfordian (17.3–16 Ma; Flynn 2008). Eomyidae gen. et sp. indet. 1 from the Dota locality is represented by a lower dentition and is diagnosed by bunolophodont cheek teeth with *Pseudotheridomys* occlusal pattern and four complete roots on m1 (and probably also m2). Most eomyid genera have three roots on the lower molars, and the only exceptions have been *Keramidomys* and *Estramomys* that have four complete roots. *Keramidomys* is known from MN 5 to MN 14, and *Estramomys* is known from MN 14 to MN 17 in Europe (Engesser 1999). Both genera differ from the Dota taxon in having more lophodont cheek teeth. *Keramidomys* has recently been known from Gashunyinadege through Shala in Inner Mongolia, China (Qiu, Wang, and Li 2006), which is correlated with MN 4 through 11 (Deng 2006b). Thus, the taxon from Dota locality is very likely ancestral to *Keramidomys* of China and Europe. Another small eomyid is present in Dota, which is represented by an edentulous lower jaw with three root loci on m1–2.

The Toki Lignite-bearing Formation of Mizunami Basin has yielded *Youngofiber sinensis* (Tomida et al. 1995), which is represented by a pair of upper incisors with fragments of premaxilla. The combination of characters (extremely large size, convex anterior surface, and presence of longitudinal ridges and rugose texture on the enamel surface) identifies it as *Y. sinensis*. *Y. sinensis* is known from Xiacaowan, Jiangsu Province, China, which is correlated with MN 4, as mentioned previously (Deng 2006b).

Chronology of Mammal-bearing Formations in the Kani and Mizunami Basins

The terrestrial Miocene sediments in the Kani Basin are associated with volcanic rocks related to the rifting of the Japan Sea basin, which began in the late Oligocene or early Miocene and ceased around 15 Ma (Kano et al. 2002). There is no evidence of marine incursion into the Kani Basin during the Miocene, whereas the adjacent Mizunami Basin has a history of transgression from the Pacific after ca. 18 Ma (Itoigawa 1993). Most of the land mammal remains in the area are recovered from the Nakamura and Hiramaki formations in the Kani Basin and the Hongo and Toki Lignite-bearing formations in the Mizunami Basin. The radiometric dating and investigations of magnetostratigraphy and marine microbiostratigraphy have been carried out for the Miocene sediments, thus the arrangement with available data in the chronological time scale is summarized in figure 12.3.

The oldest and youngest known land mammal fossils in the Kani Basin are collected from the upper part of the Hachiya and the upper member of the Hiramaki formations, respectively (see figure 12.3); the former is the postcranial remains including hindlimb bones of a single cervoid artiodactyl (Shikano and Ando 2000), and the latter is likely remains of *Anchitherium* (Okumura et al. 1977; Miyata and Tomida 2010). The ages of the Hachiya,

Nakamura, and Hiramaki formations are estimated at ca. 24.2–19.6, ca. 19.6–18.4, and ca. 18.4–17.0 Ma, respectively, based on data of the most recent fission-track dating by Shikano (2003). Figure 12.3 also shows the K-Ar dating (Nomura 1986) carried out for the Hachiya Formation. Some discordant fission-track dates from these three formations were previously reported by Kobayashi (1989); we noted but did not use his results primarily because his work predates the recommendations and the zeta calibration method by Hurford (1990). Although the paleomagnetic data in the Kani Basin are

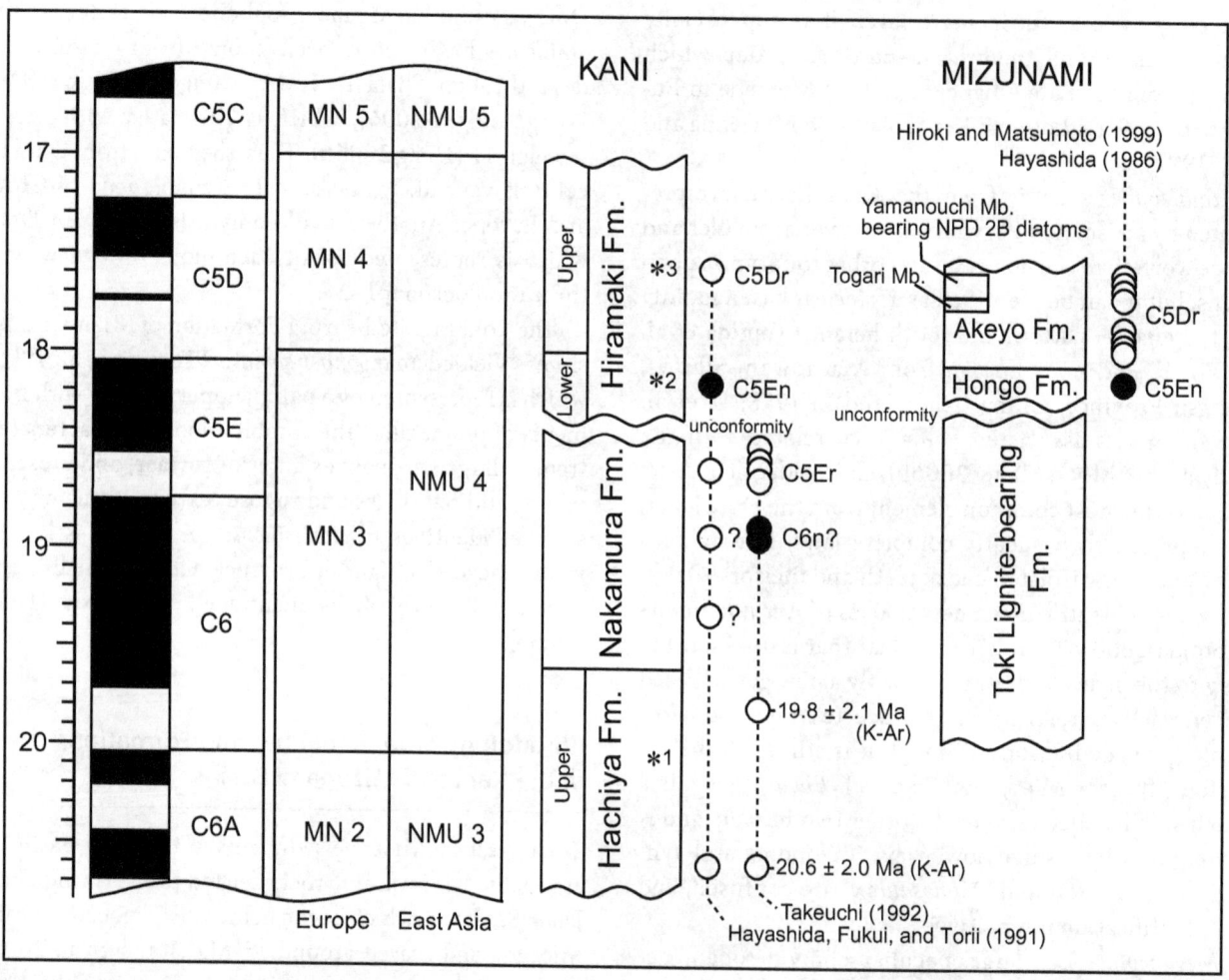

Figure 12.3 Chronological relationships of the Miocene strata bearing mammal fossils in Kani and Mizunami basins, Gifu Prefecture, Japan. Note that each column of the formation does not reflect the thickness, and the framework of each age of the formation is discussed in the text. The black and white circles respectively indicate the normal and reversed polarities of paleomagnetic data from the sediments. Approximate horizon of each paleomagnetic site in Kani Basin is inferred, but the exact horizon and age of each are uncertain. The asterisks with number indicate the approximate positions of the oldest, a cervoid artiodactyl (*1*), and youngest mammal fossils, *Anchitherium* aff. *A. gobiense* (*3*), and *Gomphotherium annectens* (*2*). The relationships of unconformity between Nakamura and Hiramaki formations and between Hongo- and Toki-Lignite-bearing formations follow Itoigawa (1974, 1980), although each hiatus seems to be limited. The geological timescale, the European mammal Neogene zones (MN), and the Chinese Neogene Mammal Faunal Units (NMU) are, respectively, based on Lourens et al. (2004), Steininger (1999), and Deng (2006b). Togari and Yamanouchi members are formally named as the Togari Sandstone Member and Yamanouchi Siltstone Member, respectively.

scattered, the Hiramaki Formation preserves at least one set of reverse (at the upper part) and normal (at the basal part) polarities (Hayashida, Fukui, and Torii 1991), suggesting the correlation to the chron C5Dr/C5En boundary (ca. 18.1 Ma; Lourens et al. 2004) when combining the radiometric data. According to Takeuchi (1992), the upper part of the Nakamura Formation preserves a stable interval of reverse polarity and some normal polarities below the reverse interval, implying the correlation to the chron C5Er/C6n boundary (ca. 18.7 Ma; Lourens et al. 2004). However, Hayashida, Fukui, and Torii (1991) reported three horizons of the Nakamura Formation of reverse polarity. Further study associating radiometric dating with magnetostratigraphy is required to reveal the chronology of the mammal-bearing formations. Nevertheless, we believe that this chronological implication in the Kani Basin suggested from the radiometric and paleomagnetic data is fully worth testing, because a similar chronology is recognized in the sequence of the Mizunami Basin.

The Hongo and Toki Lignite-bearing formations are nonmarine sediments, whereas the Akeyo Formation is marine strata yielding various invertebrates and microfossils with their own biochronologic information (Itoigawa 1993; Irizuki et al. 2004). Especially, the Yamanouchi Siltstone Member (=Yamanouchi Member hereafter) of the Akeyo Formation yields diatoms of the Neogene North Pacific Diatom zone, NPD 2B (Gladenkov 1998; Yanagisawa and Akiba 1998) with an interval estimated to be from 18.3 Ma to 17.0 Ma (Watanabe and Yanagisawa 2005). Several fission-track dates were also obtained from these formations (e.g., Kobayashi 1989; Hayashi and Ohira 2005; and Sasao, Iwano, and Danhara 2006), and a magnetostratigraphic investigation was also carried out. Hiroki and Matsumoto (1999) provided the paleomagnetic polarities from various horizons of the Akeyo Formation and initially assigned a stable reverse interval of the formation and a normal polarity from the Hongo Formation provided by Hayashida (1986) to chron C5Br and C5Cn, respectively; however, the data later were reinterpreted as chron C5Dr and C5En (Irizuki et al. 2004). Therefore, the upper and lower members of the Hiramaki Formation are likely correlated chronologically to the Akeyo Formation and the Hongo Formation, respectively (see figure 12.3). Sasao, Iwano, and Danhara (2006) provided fission-track dates from four different horizons of the middle part of the Toki Lignite-bearing Formation: 20.1 ± 1.0, 19.0 ± 1.2, 20.9 ± 1.3, and 17.2 ± 0.9 Ma, in ascending order. Combining the previous fission-track dates (18.3 ± 1.1 or 18.3 ± 0.6, and 17.1 ± 0.5; Kobayashi 1989; Hayashi and Ohira 2005) from the upper part of the formation, Sasao, Iwano,

and Danhara (2006) considered that the Toki Lignite-bearing Formation most likely extends from ca. 18 Ma to 20 Ma plus undetermined age of the conglomeratic basal part. This inferred age suggests the chronological correlation of the Nakamura Formation and the upper part of the Hachiya Formation (see figure 12.3; Sasao, Iwano, and Danhara 2006).

Miocene Mammal Chronology in the Kani and Mizumani Basins and Correlation to the MN and NMU

As discussed earlier, the chronology of the Early Miocene mammal faunas in the Kani and Mizumani basins suggests the correlation to MN 3 and 4. Following the chronological scheme of Steininger (1999), the boundary between the chron C5Dr and C5En is important in discussing the correlation to the continental faunas, because the boundary is closely related to the boundary of MN 3 and 4. As mentioned before, the upper member of the Hiramaki Formation yielding *Anchitherium* is most likely correlated to chron C5Dr, or MN 4; whereas the lower member of the Hiramaki Formation yielding *Gomphotherium* plus the Nakamura Formation yielding the Dota small mammal fauna (one plesiosoricid insectivore, one ochotonid lagomorph, three castorid rodents, and three eomyid rodents; Tomida 2000) are correlated to MN 3 chronologically (or all are chronologically correlated to MN 3, if using alternative paleomagnetic calibration in western Europe by Agustí et al. 2001). The early forms of *Gomphotherium* (*G. annectens*) and an eomyid (likely ancestral to *Keramidomys*) imply the correlation to MN 3 faunas in Europe, and the evolutionary stages of other small mammals from the Dota locality also might support the correlation (as previously discussed). However, there is no definitive species indicating the faunal association with MN 3.

Chronologically, the Early Miocene faunas in the Kani and Mizumani basins should be correlated to early part of NMU 4, comparable to MN 3. Contrary to the expectation from the chronology, the presence of the two common Chinese species (*Youngofiber sinensis* from the Nakamura Formation and *Plesiotapirus yagii* from the Nakamura Formation and ? lower part of the Hiramaki Formation) rather suggests correlation with the Shanwang (Linqu, Shandong) and Sihong (Xiacaowan, Jiangsu) faunas of the late NMU 4, comparable to MN 4, and/or NMU 5 (Qiu, Wu, and Qiu 1999; Deng 2006b). This chronological discordance between Japan and China leaves room for interpretation, and further materials are required in Japan. Besides, one of the main reasons of the unresolved correlation between Japanese and Chinese faunas is that the early NMU 4 faunas, comparable to MN 3, are very poorly known and not well documented in China compared to the late NMU 4 faunas (Deng 2006b). Further chronological resolution of NMU 4 is also needed.

SMALL MAMMALS FROM THE NOJIMA GROUP, NAGASAKI PREFECTURE

The Nojima Group in the Sasebo area of Nagasaki Prefecture is divided into three formations: the Oya, Fukazuki, and Minamitabira formations, in ascending order (see figures 12.1 [3] and 12.2). The Fukazuki Formation has yielded *Diatomys shantungensis* (Li 1974), which is represented by an isolated left M2 (Kato and Otsuka 1995). *Diatomys* was so unique a rodent that it could be identified without doubt. Although several more genera and species of Diatomyidae have been described recently (Mein and Ginsburg 1985; Flynn, Jacobs, and Cheema 1986; Flynn and Morgan 2005; Flynn 2006, 2007), *Diatomys* can be differentiated from *Fallomus* and *Marymus* by having molars with cusps hardly evident and developing planar wear, and from *Willmus* by having molars much less hypsodont. *D. shantungensis* (Li 1974) can be distinguished from *D. liensis* (Mein and Ginsburg 1985) and *D. chitaparwalensis* (Flynn 2006) by its larger size and other characters. The Oya Formation also has yielded a lower jaw of a beaver, not yet identified to generic level (Kato and Otsuka 1995).

Geologic age of the Nojima Group has been investigated mainly by using fission-track dating and invertebrate fossils (Sakai, Nishi, and Miyachi 1990; Miyachi and Sakai 1991), and the two fission-track ages (18.9 ± 2.9, 18.5 ± 2.3 Ma) obtained from the basal Fukazuki Formation have been used for the approximate age of the rodent fossils mentioned earlier (Kato and Otsuka 1995; Flynn 2006). However, these dates were obtained before the Recommendation by the Fission Track Working Group of the IUGS Subcommission of Geochronology (Hurford 1990). The most recent study on fission-track dating (Komatsubara et al. 2005) suggests that the Nojima Group ranges from 18 Ma to 15 Ma (Oya Formation ranges from 18 Ma to 17 Ma, and Minamitabira Formation from 16 Ma to 15 Ma). This suggests that the age of *Diatomys shantungensis* from the Fukazuki Formation is between 17 Ma and 16 Ma, which is somewhat younger than the type locality.

STEGOLOPHODON FROM VARIOUS LOCALITIES IN JAPAN

Japanese Miocene proboscideans ranging in age from ca. 18 Ma to 16 Ma in the late Early Miocene are represented solely by *Stegolophodon* species of various body sizes. *Stegolophodon* is an extinct elephant-like proboscidean that flourished in Asia from the late Early Miocene to the Pliocene (Saegusa 1996; Saegusa, Thasod, and Ratanasthien 2005). Three stegolophodont species (*S. pseudolatidens*, *S. tsudai*, and *S. miyokoae*; table 12.2) have been described from the Miocene of Japan (Matsumoto 1926; Yabe 1950; Shikama and Kirii 1956; Hatai 1959). Japanese stegolophodont molars are relatively uniform in the structure of loph(id)s, having only two morphological types: a primitive type, which has central conules on two mesial loph(id)s, and a derived type in which the second posterior central conule on the upper molars is much reduced, the main pretrite cusp of the lophid is not displaced distally, the second posterior central conule on the lower molars is absent, and the apex of the cusps is subdivided into fine and pointed mammillae. At the same time, their dimensions vary greatly; those from the Asakawa Formation of Ibaragi Prefecture are the geologically youngest and smallest among the known specimens, being just 60% of the largest and geologically oldest molars reported from the Misawa and Honya formations of Fukushima Prefecture.

Hasegawa, Koda, and Yanagisawa (1984) proposed that the high degree of variability in molar size within Japanese stegolophodonts can be attributed to the sexual dimorphism of a single species. On the other hand, Saegusa (2008) recently argued that this high degree of variability in Japanese stegolophodonts is best explained by insular dwarfism induced by the formation of the proto-Japanese Islands, rather than variation among individuals, sexual dimorphism, or retention of plesiomorphous small dimensions. His argument is based on comparison of the coefficient of variation (CV) of the width of Japanese stegolophodont molars with those of extant and extinct species of Elephantoidea, combined with a review of the stratigraphic distribution of Japanese stegolophodonts. According to his review, most Japanese stegolophodont specimens, including type specimens of the three Japanese species, have been obtained from the formations assignable to North Pacific Diatom (NPD) zone 3 A (17–16.4 Ma), whereas the largest and smallest molars have been found from the formations assignable to NPD 2B (18–17 Ma) and 3B (16.4–16 Ma), respectively (figure 12.4). The CV of the width of molars from NPD

Table 12.2

Stegolophodont Specimens Reported to Date from the Japanese Miocene

Present Classification	Previous Specific Identification	Locality	Location No. in Figure 12.1	Formation	Element (previous identification)	Interval	Reference
S. pseudolatidens stage1	*S.* cf. *tsudai*	Kusebara, Iwaki City, Fukushima Prefecture	4	Misawa Formation	M3	II	Shikama and Yanagisawa (1971)
S. pseudolatidens stage1	*Rhyncotherium* sp.	Taira-Yagawase, Iwaki City, Fukushima Prefecture	4	Honya Formation	Fragment of m3	II	Hasegawa and Koda (1981)
S. pseudolatidens stage1	*Stegolophodon* sp.	Takeno, Toyooka City, Hyogo Prefecture	5	Yoka Formation	Fragment of M3	II	Yasuno (2005)
S. pseudolatidens stage1?	*Stegolophodon* sp.	Taira-Kamitakaku, Iwaki City, Fukushima Prefecture	4	Yoshinoya Formation	Fragmentary mandible, dP4 and two fragmentary molars	III	Hasegawa, Koda, and Yanagisawa (1984)
S. pseudolatidens stage2	*S. pseudolatidens*	Sauramachi, Shiogama City, Miyagi Prefecture	6	Sauramachi Formation	M3	III	Matsumoto (1926)
S. pseudolatidens stage2	*S. pseudolatidens*	Sauramachi, Shiogama City, Miyagi Prefecture	6	Sauramachi Formation	m3	III	Yabe (1950)
S. pseudolatidens stage2	*S. pseudolatidens*	Funaoka, Shibata Town, Miyagi Prefecture	7	Tsukinoki Formation	Fragment of cranium and mandible with M2–3, m2–3 (m1–3)	III	Yabe (1950)
S. pseudolatidens stage2	*S. miyokoae*	Funaoka, Shibata Town, Miyagi Prefecture	7	Tsukinoki Formation	m3	III	Hatai (1959)
S. pseudolatidens stage2	*S. tsudai*	Kasuga, Toyama City, Toyama Prefecture	8	Kurosedani Formation	m1 (M1)	III	Shikama and Kirii (1956)
S. pseudolatidens stage2	*S. tsudai*	Suwara, Toyama City, Toyama Prefecture	8	Kurosedani Formation	M3	III	Shikama and Kirii (1956)
S. pseudolatidens stage2	*S. pseudolatidens*	Tochizu, Tateyama-machi, Toyama Prefecture	8	Kurosedani Formation	M3 (m3)	III	Fujii and Minabe (1964)

(continued)

Table 12.2 (continued)

Present Classification	Previous Specific Identification	Locality	Location No. in Figure 12.1	Formation	Element (previous identification)	Interval	Reference
S. pseudolatidens stage2	*Stegolophodon* sp.	Yatsuomachi-Miyanokoshi, Toyama City, Toyama Prefecture	8	Kurosedani Formation	M1 or M2 (m1)	III	Koda and Hasegawa (2002)
S. pseudolatidens stage2	*S. pseudolatidens*	Harinoki, Himi City, Toyama Prefecture	9	Taniguchi Formatoin	M1	III	Takai and Fujii (1961)
S. pseudolatidens stage2?	*Stegolophodon* sp.	Taira-Kamitakaku, Iwaki City, Fukushima Prefecture	4	Yoshinoya Formation	Complete mandible with m3	III	Koda, Suzuki, and Hasegawa (1986)
S. pseudolatidens stage2?	*Pentalophodon* sp.	Mii, Wajima City, Ishikawa Prefecture	10	Anamizu Formation	Fragment of M3	?	Shikama (1936)
S. pseudolatidens stage2?	*Bunolophodon* sp.	Mii, Wajima City, Ishikawa Prefecture	10	Anamizu Formation	Fragment of M3	?	Kaseno (1955)
S. pseudolatidens stage2?	*S. pseudolatidens*	Himosash, Hirado City, Nagasaki Prefecture	11	Himosashi Andesite	M3	IV?	Kato (1997)
S. pseudolatidens stage3	*S. pseudolatidens*	Kitashioko, Hitachiomiya City, Ibaraki Prefecture	12	Asawaka Formation	M2–3 (M1–dP4)	IV	Kamei and Kamiya (1981)
S. pseudolatidens stage3	*Stegolophodon* sp.	Shimoanosawa, Shirosato Town, Ibaraki Prefecture	12	Asawaka Formation	m2–3 (not identified)	IV	Koda et al. (2003)

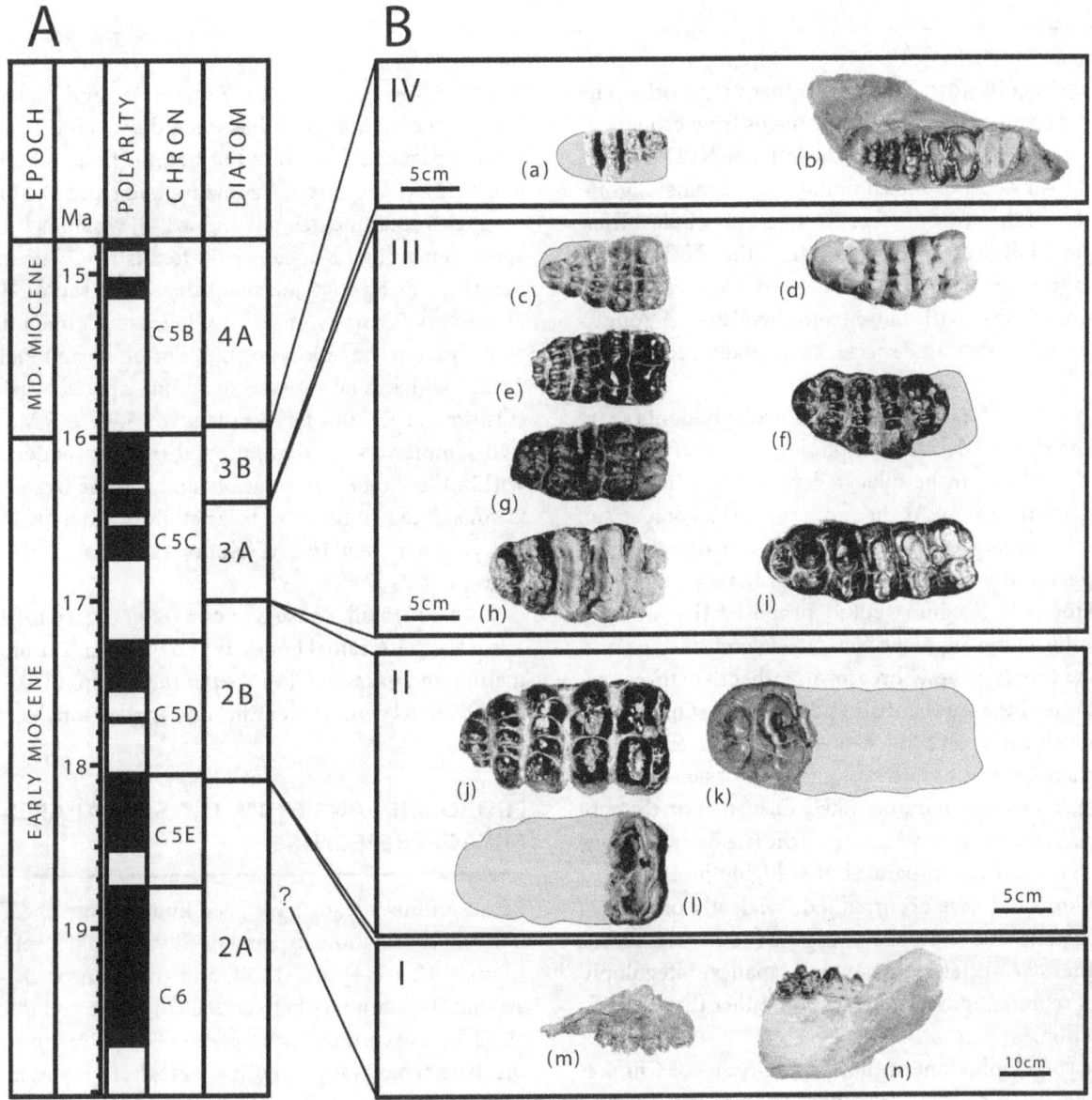

Figure 12.4 Chronological relationship of the Early Miocene proboscidean fossils in various localities in Japan: (*A*) Late Early and early Middle Miocene magnetobiochronologic time scale for Japan. Geological age: Lourens et al. (2004); Magnetic polarity/chron: Ogg and Smith (2004); Diatom zonation: Watanabe and Yanagisawa (2005). (*B*) Upper and lower third molars of proboscideans from the Japanese Early Miocene. Roman numerals indicate the four intervals of Japanese Early Miocene proboscideans. Upper third molars and lower third molars are arranged in the left and right columns, respectively. *Stegolophodon pseudolatidens* stage 3 from interval IV, which is assigned to NPD 3B: (*a*) M3 and (*b*) m2 and 3 on a mandible from Asakawa Formation, Ibaraki Prefecture. *S. pseudolatidens* stage 2 from Interval III, which is assigned to NPD 3A: (*c*) and (*d*) M3 and m3 from the Tsukinoki Formation; (*e*) and (*f*) M3 and m3 from the Sauramachi Formation; (*g*) and (*h*) M3s from the Kurosedani Formation, where (*h*) is the holotype of *S. tsudai*; (*i*) holotype m3 of *S. miyokoae*. *S. pseudolatidens* stage 1 from Interval II, which is assigned to NPD 2B: (*j*) M3 from the Misawa Formation; (*k*) m3 from the Honya Formation; (*l*) M3 from the Yoka Formation. *Gomphotherium annectens* from Interval I, which is assigned to NPD 2A: (*m*) holotype skull fragment from the Hiramaki Formation; (*n*) fragment of mandible from the Toki Lignite-bearing Formation.

3A is not significantly larger than that of the other elephantoid species; however, if specimens from other time intervals are combined with those from NPD 3A, the range of variation of the combined set becomes significantly larger than that of other elephantoid samples. Thus, both the smallest stegolophodont from the NPD 3B zone and largest ones from the NPD 2B zone cannot be grouped together with those from the NPD 3A zone as populations of the same species (for details, see Saegusa 2008).

As mentioned earlier, Japanese stegolophodonts share the same suite of derived morphological traits. This suggests that they can be allocated to a single lineage or monophyletic group. At the same time, the comparison of CV values suggests that they represent three successive species of a single lineage rather than a single species. However, Saegusa (2008) proposed the informal taxonomic name *Stegolophodon pseudolatidens* stage one, two, and three for these three forms rather than three specific names because the distinguishing criterion is body size, which can evolve independently in different species within similar settings; that is, a number of separate populations on the small islands likely formed upon the subsiding crust of the Japan Arc during the late Early Miocene could have evolved in parallel. It is highly probable that dwarfism could have occurred independently on each of these islands. For this reason, Saegusa (2008) considered that the size differences among Japanese stegolophodonts represent grades of evolution rather than specific distinction.

The stegolophodont of the early Miocene of China is represented solely by *Stegolophodon hueiheensis* Chow 1959 from the Sihong Fauna that is roughly correlated with MN 4 of Europe (Qiu and Qiu 1995; Deng 2006b). Thus, this species is contemporaneous with large-sized Japanese stegolophodonts from NPD 2B (18–17 Ma). The only known molar of *S. hueiheensis*, the holotype, is so highly worn that it shows nothing beyond the number of lophs and dimension, making its affinities to Japanese stegolophodonts equivocal, but, at least, it has the same number of lophs and dimension of molars as the large-sized Japanese stegolophodonts from NPD 2B. This may suggest there was an exchange of the population of stegolophodonts between maritime China and the proto-Japanese Islands during 18–17 Ma.

NIU MOUNTAINS AREA, FUKUI PREFECTURE

Miocene sedimentary rocks of the Niu Mountains area are divided into Ito-o, Kunimi, and Aratani formations,

in ascending order (Kano, Yamamoto, and Nakagawa 2007), and a suid fossil has been discovered from the middle part of the Aratani Formation (figures 12.1 [13] and 12.2). It consists of fragmentary left and right lower jaws with a canine, p3, p4, and m1–3, which is the best specimen among Miocene suid fossils from Japan. It is identified as *Hyotherium shanwangense* based on dental characters (Oshima et al. 2008). This specimen differs slightly from the holotype (Liu, Fortelius, and Pickford 2002), which is considered to be intraspecific variation (Oshima et al. 2008). K-Ar dates of 15.7 ± 0.5 Ma from two samples have been obtained from an andesite sill within the Aratani Formation, and the age of the fossil is considered to be close to or slightly older than that, but younger than 16 Ma (Kano, Yamamoto, and Nakagawa 2007).

A "deer" fossil, consisting of a fragmentary mandible and a few postcranial bones, is known from Kunimi Formation and is assigned to *Amphitragulus* sp. (Takeyama 1989), but its generic identification is questionable.

PROBOSCIDEANS FROM THE SENDAI AREA, MIYAGI PREFECTURE

Fossil proboscideans have been known from the Tatsunokuchi Formation in Sendai City, Miyagi Prefecture (figures 12.1 [14] and 12.2), and they can be dated as around the boundary between the Miocene and Pliocene (5.32 Ma, according to Berggren et al. 1995), based on the diatom biostratigraphy, magnetostratigraphy, and the fission-track dates of the underlying tuff layer (Yanagisawa 1990, 1998). Two proboscideans, *Stegodon* and *Sinomastodon*, reported from the Tatsunokuchi Formation are closely similar to those from the Mazegou and Gaozhuang formations of the Yushe Basin in northern China.

Two fragments of upper molars (SSME 13329) from the Tatsunokuchi Formation were described initially as *Stegolophodon* (*Stegolophodon* sp. in Koda et al. 1998), but they actually represent the earliest stegodonts from Japan (Taruno 1999). Although they are so incomplete that the number of lophs cannot be observed, the size of the remaining ridge, the very weak folding, and the weak stufenbildung of the worn enamel surface are comparable to those of *S. zdanskyi* from China (Saegusa, Thasod, and Ratanasthien 2005).

Trilophodon sendaicus (Matsumoto 1924), which is represented by four molars (M1–3 and m3) and a left astragalus from the Tatsunokuchi Formation, was transferred to the genus *Sinomastodon* (Tobien, Chen, and Li

1986), because it shows a combination of the bunodont and zygodont features that are peculiar to the latter genus; bunodont features such as chevron arrangement of the lophid and bulbous cusps are seen along with the typical zygodont feature, the zygodont crest (Kamei 2000). *Sinomastodon sendaicus* has been considered to be specifically distinct from *S. intermedius* (Teilhard de Chardin and Trassaert, 1937) known from the Mazegou and Gaozhuang formations of the Yushe Basin in having crescent central conules on lower molars and a narrower fourth lophid of m3 (Kamei 2000). However, the structure of central conules of m3 of *Sinomastodon sendaicus* is essentially the same as that of *S. intermedius* from the Yushe Basin (compare Matsumoto 1924:pl. III, fig. 3, with Teilhard de Chardin and Trassaert, 1937:pl. II, fig. 2), and the size differences of lophids between Chinese and Japanese specimens are well within the range of the intraspecific variation for a gomphothere species. We therefore propose the synonymy of *Sinomastodon sendaicus* with *S. intermedius*.

OTHER MATERIAL

Early Miocene Suid from Mimasaka City, Okayama Prefecture

Takai (1950, 1954) reported a suid right mandible fragment with m1–2 from Uetsuki Formation of Katsura Group (=Mimasaka coal-bearing bed, late Early Miocene; figure 12.1 [15]). Although it was identified as *Palaeochoerus japonicus* (Takai 1954), the specimen was lost, and recent review suggests that it is best identified as Suidae gen. et sp. indet. (Oshima et al. 2008).

Early Miocene Tapirid from Horai-cho, Aichi Prefecture

A right maxillary fragment with fragmentary P1–4 from Kuroze Formation of Shitara Group in Aichi Prefecture (figure 12.1 [16]) is assigned to *Plesiotapirus* sp. based on comparison with the best specimen of *Pl. yagii* from Shanwang (Kawamura and Fujita 1999). For a review of the genus, see the earlier section regarding the Mizunami Group and Qiu, Yan, and Sun (1991). A fission track date of 18–17 Ma is obtained from the upper part of the Hokusetsu Subgroup (Hoshi, Iwano, and Danhara 2005), and a radiolarian "age" of 20–17 Ma is obtained from the Hokusetsu Subgroup (Hoshi, Ito, and Motoyama 2000). The Shitara Group is subdivided into the Hokusetsu and Nansetsu subgroups, in ascending order, and the Kuroze Formation is included in the top of the Hokusetsu Subgroup or the bottom of the Nansetsu Subgroup depending on stratigraphic studies. Thus, the age of the Kuroze Formation is likely around 17 Ma.

Early Middle Miocene Amphicyonid from Shobara City, Hiroshima Prefecture

An amphicyonid, *Ysengrinia* sp., has been reported from the marine Korematsu Formation of the Bihoku Group in Miyauchi-machi, Shobara City, Hiroshima Prefecture (figure 12.1 [17]; Kohno 1997). It is represented by a single isolated right M1, but it is the first record of the genus in Asia. Geologic age of the formation is somewhat ambiguous, but it is estimated as "early Middle" Miocene, between 16.3 Ma and 15.6 Ma, based on the calcareous nannofossil zonation and marine molluscan biostratigraphy (Kohno 1997). The genus is known from the Late Oligocene to Early Miocene of Europe and from Early Miocene of North America; the Japanese record is the youngest.

Late Miocene Mammals from Oiso Hill, Kanagawa Prefecture

The Oiso Formation of the Miura Group at Oiso Hill, Kanagawa Prefecture, has yielded fragmentary rhinocerotid and suid teeth (figures 12.1 [18] and 12.2). *Brachypotherium* sp. is represented by fragments of M1 or M2 (Zin-Maung-Maung-Thein et al. 2009). This species differs from *B. pugnator* from the Mizunami Group in Gifu Prefecture in having a weakly constricted protocone. A fragment of suid upper molar (M1 or M2) is best identified as Suinae gen. et sp. indet. (Oshima 2007). Geologic age of the Oiso Formation is estimated as N17 planktonic foraminifera zone (ca. 8.2–6.4 Ma; Ibaraki 1978) and CN9 calcareous nannofossil zone (ca. 8.2–5.6 Ma; Kanie, Hirata, and Imanaga 1999).

Late Miocene Rhinocerotid from Kawamoto-machi, Saitama Prefecture

Material of Teleoceratinae gen. et sp. indet. was discovered from the Late Miocene Yagii Formation (Yoshida et al. 1989), in Kawamoto-machi, Saitama Prefecture (figure 12.1 [19]). The specimen is composed of maxillary fragments with left dP1–P4 and right P2–M3 and nearly complete left and right mandibles with i2 and p2–m3.

A fission-track age of 8.13 ± 1.64 Ma for the tuff layer in the Yagii Formation was obtained (Nomura and Kosaka 1987). This result is in harmony with the stratigraphy of the formation that conformably overlies the Tsuchishio Formation, corresponding to the uppermost part of the Neogene North Pacific diatom zone NPD5C (10.1–10.0 Ma; Suto, Takahashi, and Yanagisawa 2003).

Late Miocene Proboscidean from Miyako Island, Okinawa Prefecture

The lower most part of Shimajiri "Formation" of Miyako Island, Okinawa Prefecture (not shown in figure 12.1 but located between the island of Okinawa and Taiwan) has yielded a lingual half of the first lophid of a gomphothere m3, and it was originally identified as *Trilophodon* sp. (Hasegawa, Otsuka, and Nohara 1973). Although rather fragmentary, this specimen can be assigned to the genus *Sinomastodon* on the basis of the combination of bunodont morphology of the cusp and presence of a blunt zygodont crest on the distal wall of the main cusp.

According to Ujiie and Oki (1974), the horizon that yielded the *Sinomastodon* tooth fragment is the lower part of the Nanseien Formation of the Shimajiri Group and is correlated with the N17 planktonic foraminifera zone, which ranges between ca. 8.6 Ma and 5.7 Ma (Lourens et al. 2004). The fact that the Shimajiri Group is marine deposits, and the proboscidean tooth fragment is rather mechanically unworn suggests that the proboscidean was living on nearby land.

DISCUSSION

As described, no terrestrial mammal fossil has been found in Japan between about 15 Ma (early Middle Miocene) and 7–8 Ma (late Late Miocene). Although not mentioned above, a late Late Miocene (ca. 6.2 Ma, based on planktonic foraminifera) terrestrial mammal-bearing deposit, the Aoso Formation, is known from north of Sendai. It has yielded fragmentary molar material of a tetralophodont gomphothere, *Hipparion* [s.l.] sp., and an acerathere rhinocerotid (Kohno et al. 1997).

Disappearance of terrestrial mammal fossil records after 16–15 Ma in Japan coincides with "the climax of the opening of the Japan Sea at ca. 16 Ma with widespread, rapid subsidence of the Japan Arc" (Kano et al. 2002:180–

181). In this context, the early Middle Miocene records of *Hyotherium shanwangense* in Fukui and *Ysengrinia* sp. in Shobara are probably survivors within the subsiding Japanese Arc.

The land connection between the Asian mainland and the Japanese islands may have reappeared during the Late Miocene, as a consequence of a change in the tectonic setting from tension to compression in Northeast Japan (Okumura et al. 1995) and the conspicuous compression on the southern margin of the Sea of Japan (Itoh and Nagasaki 1996). The Late Miocene material described earlier (*Brachypotherium* sp. and Suinae from the Oiso Formation, Teleoceratinae from the Yagii Formation, *Stegodon* cf. *zdanskyi* and *Sinomastodon intermedius* from the Tatsunokuchi Formation, and *Sinomastodon* sp. from the Nanseien Formation), as well as the fragmentary material from Aoso Formation also mentioned earlier, may represent an immigrant wave from the Asian mainland via this corridor. Although their affinities to the continental forms are not clear, the close faunal ties between Japan and North China are demonstrated by the proboscidean fossils from the Tatsunokuchi Formation. Two proboscideans reported from the Tatsunokuchi Formation, *Stegodon* and *Sinomastodon*, are closely related to those from the Mazegou and Gaozhuang formations of the Yushe Basin in north China.

CONCLUSION

As described and discussed, fossil records of terrestrial mammals during the Miocene are never abundant in Japan. However, correlations with marine microfossil biostratigraphy (planktonic foraminifera, radiolarians, calcareous nannoplankton, diatoms, etc.), several fission-track and K-Ar dates as well as paleomagnetic studies support fairly precise correlations with absolute age, and hence European and Chinese mammal ages.

Among those poor records, the fauna of the Mizunami Group is fairly diversified, and the fauna of its lower part is correlated to MN 3 zone, while that of the upper part correlates to MN 4 of the European land mammal zonation. Both the lower and upper parts correlate to NMU4 of the Chinese Neogene mammal faunal units.

The fossil records of the genus *Stegolophodon* in Japan are relatively abundant geographically and stratigraphically during the late Early Miocene (ca. 18–16 Ma). Their restudy in detail suggests that three forms with size reduction but without much morphological change through time (three time intervals) represent a dwarfism of a single lineage and the grade of evolution rather than specific distinction, and that they should be interpreted as informal taxonomic units—*Stegolophodon pseudolatidens* stage one, two, and three—rather than three different species.

The lack of the fossil records of terrestrial mammals between some 15 Ma and 8–7 Ma can be interpreted as follows. Japan was fairly well connected with the Asian mainland until about 17 Ma, but the opening of the Japan Sea climaxed at ca. 16 Ma with rapid subsidence of the Japanese Arc may have led to an extinction of Japanese terrestrial mammals by about 15 Ma. A land connection between the Asian mainland and the Japanese islands likely reappeared during the Late Miocene, and the Late Miocene terrestrial fossil records in Japan may represent an immigrant wave from the Asian mainland via this corridor.

Among the late Miocene terrestrial mammals, *Sinomastodon sendaicus* from the Tatsunokuchi Formation in Japan should be synonymized with *Sinomastodon intermedius* from the Yushe Basin, China.

ACKNOWLEDGMENTS

We are grateful to X. Wang for inviting us to contribute this chapter. We also thank him, N. Kohno, and L. J. Flynn for critiquing a draft of this manuscript. The present compilation is based mainly on our talk at the symposium held in Beijing on June 8–10, 2009, which was supported in part by the National Museum of Nature and Science, project # 20092021 (Studies on the Geography and Evolution of Biodiversity in Japan).

REFERENCES

Abusch-Siewert, S. 1983. Gebißmorphologische Untersuchungen an eurasiatischen Anchitherien (Equidae, Mammalia) unter besonderer Berücksichtigung der Fundstelle Sandelzhausen. *Courier Forschungsinstitut Senckenberg* 62:1–361.

Agustí, J., L. Cabrera, M. Garcés, W. Krijgsman, O. Oms, and J. M. Parés. 2001. A calibrated mammal scale for the Neogene of Western Europe: State of the art. *Earth-Science Reviews* 52:247–260.

Berggren, W. A., D. V. Kent, C. C. Swisher III, and M. P. Aubry. 1995. A revised Cenozoic geochronology and chronostratigraphy. In *Geochronology, Time Scales, and Global Stratigraphic Correlation: A Unified Temporal Framework for a Historical Geology*, ed. W. A. Berggren, D. V. Kent, M. P. Aubry, and J. Hardenbol, pp. 129–212. Society of Economic Paleontologists and Mineralogists Special Publication 54. Tulsa: Society for Sedimentary Geology.

Boon-Kristkoiz, E. and A. R. Kristkoiz, 1999. Order Lagomorpha. In *The Miocene Land Mammals of Europe*, ed. G. E. Rossner and K. Heissig, pp. 259–262. Munich: Dr. Friedrich Pfeil.

Chavasseau, O., Y. Chaimanee, Soe-Thura-Tun, Aung-Naing-Soe, J. C. Barry, B. Marandat, J. Sudre, L. Marivaux, S. Ducrocq, and J. J. Jaeger. 2006. Chaungtha, a new Middle Miocene mammal locality from the Irrawaddy Formation, Myanmar. *Journal of Asian Earth Sceinces* 28:354–362.

Chow, M. 1959. New species of fossil Proboscidea from South China. *Acta Palaeontologica Sinica* 7:251–258, 4 pls. (in Chinese with English abstract).

Chow, M.-c. and C.-k. Li. 1978. "The Xiacaowan System," "Trogontherium," "Huaihe River regional transition phase"—A correction of historical misinterpretation. *Acta Stratigraphica Sinica* 2(2): 122–130, pl. 1 (in Chinese).

Colbert, E. H. 1935. Siwalik mammals in the American Museum of Natural History. *Transactions of the American Philosophical Society* 24:1–401.

Colbert, E. H. 1939. A new anchitheriine horse from the Tung Gur Formation of Mongolia. *American Museum Novitates* 1019:1–9.

Deng, T. 2006a. A primitive species of *Chilotherium* (Perissodactyla, Rhinocerotidae) from the Late Miocene of the Linxia Basin (Gansu, China). *Cainozoic Research* 5(1–2):93–102.

Deng, T. 2006b. Chinese Neogene mammal biochronology. *Vertebrata PalAsiatica* 44(2):143–163.

Engesser, B. 1999. Family Eomyidae. In *The Miocene Land Mammals of Europe*, ed. G. E. Rossner and K. Heissig, pp. 319–335. Munich: Dr. Friedrich Pfeil.

Fejfar, O., M. Rummel, and Y. Tomida 1998. New eomyid genus and species from the Early Miocene (MN Zones 3–4) of Europe and Japan related to *Apeomys* (Eomyidae, Rodentia, Mammalia). *National Science Museum Monographs* 14:123–143.

Flynn, L. J. 2006. Evolution of the Diatomyidae, an endemic family of Asian rodents. *Vertebrata PalAsiatica* 44(2):182–192.

Flynn, L. J. 2007. Origin and evolution of the Diatomyidae, with clues to paleoecology from the fossil record. *Bulletin of the Carnegie Museum of Natural History* 39:173–181.

Flynn, L. J. 2008. Eomyidae. In *Evolution of Tertiary Mammals of North America* Vol. 2, ed. C. M. Janis, G. E. Gunnell, and M. D. Uhen, pp. 415–427. Cambridge: Cambridge University Press.

Flynn, L. J., L. L. Jacobs, and I. U. Cheema. 1986. Baluchimynae, a new ctenodactyloid rodent subfamily from the Miocene of Baluchistan. *American Museum Novitates* 2841:1–58.

Flynn, L. J. and M. E. Morgan. 2005. An unusual diatomyid rodent from an infrequently sampled Late Miocene interval in the Siwaliks of Pakistan. *Palaeontologia Electronica* 8(1): 1:17A (10 pp.); http://palaeo-electronica.org/palaeo/2005_1/flynn17/issue1_05.htm.

Fujii, S. and H. Minabe. 1964. *Stegolophodon pseudolatidens* from the Miocene Kurosedani Formation in Toyama Prefecture, Japan. *Toyama-ken Chigaku Chirigaku Kenkyu Ronshu* 4:98–101, 1 pl. (in Japanese with English abstract).

Fukuchi, A. and K. Kawai. 2011. Revision of fossil rhinoceroses from the Miocene Mizunami Group, Japan. *Palaeontological Research* 15:247–257.

Gladenkov, A. Yu. 1998. Oligocene and lower Miocene diatom zonation in the North Pacific. *Stratigraphy and Geological Correlation* (official English translation of *Stratigrafiya, geologicheskaya korrelyatsiya*) 6:150–163.

Gunnell, G. F., T. M. Bown, J. H. Hutchison, and J. I. Bloch. 2008. Lipotyphla. In *Evolution of Tertiary Mammals of North America*

Vol. 2, ed. C. M. Janis, G. E. Gunnell, and M. D. Uhen, pp. 89–125. Cambridge: Cambridge University Press.

Hasegawa, Y. and Y. Koda. 1981. [Elephant fossil from Taira-Yagawase, Iwaki City (preliminary report)]. [*Annual Bulletin of Association of Taira Geological Club*] *Taira Chigaku Dokokai Kaiho* 14:16–17, 1 pl. (in Japanese).

Hasegawa, Y., Y. Koda, and I. Yanagisawa. 1984. On four alleged occurrences of *Stegolophodon pseudolatidens* (Yabe) from the Miocene bed of the Iwaki New Town area, Fukushima Prefecture. *Science Report of Yokohama National University Sec. II* 31:51–63, 2 pls. (in Japanese with English abstract).

Hasegawa, Y., H. Otsuka, and T. Nohara. 1973. Fossil vertebrates from the Miyako Island (Studies of the palaeovertebrate fauna of Ryukyu Islands, Japan. Part I). *Memoir of the National Science Museum* 6:39–52, pls. 6–7 (in Japanese with English summary).

Hatai, K. 1959. Discovery of a Miocene elephant molar from the Sennan district, Miyagi Prefecture, northeast Japan. *Saito Ho-on Kai Museum Research Bulletin* 28:1–4, 1 pl.

Hayashi, J. and H. Ohira. 2005. Fission track ages from the Miocene Mizunami Group (preliminary result). Abstracts, 112th Annual Meeting of the Geological Society of Japan, p. 220 (in Japanese).

Hayashida, A. 1986. Timing of rotational motion of Southwest Japan inferred from paleomagnetism of the Setouchi Miocene Series. *Journal of Geomagnetism and Geoelectricity* 38:295–310.

Hayashida, A., T. Fukui, and M. Torii. 1991. Paleomagnetism of the Early Miocene Kani Group in southwest Japan and implication for the opening of the Japan Sea. *Geophysical Research Letters* 18:1095–1098.

Heissig, K. 1972. Paläontologische und geologische Untersuchungen im Tertiär von Pakistan 5. Rhinocerotidae (Mamm.) aus den unteren und mittleren Siwalik-Schichten. *Bayerische Akademie der Wissenschaften Mathematiche-Naturwissenschaftliche Klasse Abhandlungen-Neue Folge* 152:7–112.

Heissig, K. 1989. The Rhinocerotidae. In *The Evolution of Perissodactyls*, ed. D. R. Prothero and R. M. Schoch, pp. 399–417. New York: Oxford University Press.

Hiroki, Y. and R. Matsumoto. 1999. Magnetostratigraphic correlation of Miocene regression-and-transgression boundaries in central Honshu, Japan. *Journal of the Geological Society of Japan* 105:87–107.

Hoshi, H, N. Ito, and I. Motoyama. 2000. Geology, radiolarians, and geologic age of the Hokusetsu Subgroup in the Shitara area, Aichi Prefecture, central Japan. *Journal of the Geological Society of Japan* 106(10):713–726 (in Japanese with English abstract).

Hoshi, H., H. Iwano, and T. Danhara. 2005. Miocene tectonics of the Southwestern Japan Arc inferred from geology, paleomagnetism, and FT ages of the Shitara Group. *Fission Track News Letter* 18:47–49 (in Japanese).

Hurford, A. J. 1990. Standardization of fission track dating calibration: Recommendation by the Fission Track Working Group of the I.U.G.S. Subcommission on Geochronology. *Chemical Geology* 80:171–178.

Ibaraki, M. 1978. Notes on planktonic foraminifera from the "Nishikoiso" and "Oiso" formations, Kanagawa Prefecture. *Shizuoka University Geosciences Reports* 3:1–8 (in Japanese with English abstract).

Irizuki, T., K. Yamada, T. Maruyama, and H. Ito. 2004. Paleoecology and taxonomy of Early Miocene Ostracoda and paleoenvironments of the eastern Setouchi Province, central Japan. *Micropaleontology* 50:105–147.

Itoh, Y. and Y. Nagasaki. 1996. Crustal shortening of southwest Japan in the late Miocene. *The Island Arc* 5: 337–353.

Itoigawa, J. 1974. Geology of the Mizunami Group. *Bulletin of the Mizunami Fossil Museum* 1:9–42 (in Japanese with separated English abstract).

Itoigawa, J. 1980. Geology of the Mizunami district, central Japan. *Monograph of the Mizunami Fossil Museum* 1:1–50 (in Japanese).

Itoigawa, J. 1993. Miocene palaeogeography of the Mizunami Group of the Tono region, central Japan. *Palaeogeography, Palaeoclimatology, Palaeoecology* 100:209–215.

Kamei, T. 2000. On Japanese proboscidean fossils and views of the studies after that. *Earth Science* (*Chikyu Kagaku*) 54:221–230 (partly in Japanese).

Kamei, T. and H. Kamiya. 1981. On the fossil teeth of *Stegolophodon pseudolatidens* (Yabe) from the Miocene bed of the Abukuma Mountains. *Memoirs of the Faculty of Science Kyoto University, Series of Geology and Mineralogy* 47(2):165–176, 2 pls.

Kamei, T. and Y. Okazaki. 1974. Mammalian fossils from the Mizunami Group. *Bulletin of the Mizunami Fossil Museum* 1:263–291 (in Japanese).

Kamei, T., Y. Okazaki, I. Nonogaki, and Paleontology Club, Aichi Gakuin Univeristy. 1977. On some new material of *Gomphotherium annectens* (Matsumoto) from the Mizunami Group, Central Japan. *Bulletin of the Mizunami Fossil Museum* 4:1–8, 2 pls. (in Japanese with English abstract).

Kanie, Y., D. Hirata, and I. Imanaga. 1999. Geology and microfossil age of the formations composing the Oiso Hill and Okinoyama Bank Chain, southern-central Japan. *Research Reports of Kanagawa Prefectural Museum of Natural History* 9:95–110 (in Japanese with English abstract).

Kano, K., H. Yamamoto, and T. Nakagawa. 2007. *Geology of the Fukui District. With Geological Sheet Map 1:50,000, Fukui.* Tsukuba: Geological Survey of Japan, AIST, pp. 1–68 (in Japanese with English abstract).

Kano, K., T. Yoshikawa, Y. Yanagisawa, K. Ogasawara, and T. Danhara. 2002. An unconformity in the early Miocene syn-rifting succession, northern Noto Peninsula, Japan: Evidence for short-term uplifting precedent to the rapid opening of the Japan Sea. *The Island Arc* 11:170–184.

Kaseno, Y. 1955. [Geology of the central part of Noto Peninsula]. In [*Geology of Ishikawa Prefecture*], ed. Hokuriku-bukai of the Geological Society of Japan. Kanazawa: Hokuriku-bukai of the Geological Society of Japan, pp. 15–18 (in Japanese).

Kato, T. 1997. Discovery of the Miocene fossil Proboscidea, Stegolophodon, from Hiradojima Island, Nagasaki Prefecture. *Proceedings of the Nishinihon Branch, Geological Society of Japan* 111:9 (in Japanese).

Kato, T. and H. Otsuka. 1995. Discovery of the Oligo-Miocene rodents from West Japan and their geological and paleontological significance. *Vertebrata PalAsiatica* 33:315–329, 1 pl.

Kawamura, Y. and M. Fujita. 1999. Preliminary observation of tapir fossils from Horai, Aichi Prefecture, Japan, and from Shanwang,

Shandong Province, China. *Report of Houraijisan Museum of Natural Science* 28:19–28 (in Japanese).

Kobayashi, T. 1989. Geology and uranium mineralization in the eastern part of the Kani basin, Gifu, central Japan. *Mining Geology* 39:79–94 (in Japanese with English abstract).

Koda, Y. and Y. Hasegawa. 2002. *Stegolophodon* tooth from the Miocene in Yatsuo-machi, Toyama Prefecture, Japan. *Bulletin of Ibaraki Nature Museum* 5:105–108, 1 pl. (in Japanese with English abstract).

Koda, Y., T. Suzuki, and Y. Hasegawa. 1986. [On a mastodont mandible from the Miocene Shirado Group]. *Abstract of the 1986 Annual Meeting of the Palaeontological Society of Japan*, p. 23 (in Japanese).

Koda, Y., T. Suzuki, A. Tomita, and Y. Hasegawa. 1998. *Stegolophodon* teeth from the Miocene of Sendai City, Miyagi Prefecture, Japan. *Bulletin of the Ibaraki Nature Museum* 1:3–8, 3 pls. (in Japanese with English abstract).

Koda, Y., Y. Yanagisawa, Y. Hasegawa, H. Otsuka, and M. Aizawa. 2003. A middle Miocene mandible of *Stegolophodon* (Proboscidea, Mammalia) discovered in Katsura Village, Ibaraki Prefecture, eastern Japan. *Earth Science (Chikyu Kagaku)* 57:49–59 (in Japanese with English abstract).

Kohno, N. 1997. The first record of an amphicyonid (Mammalia: Carnivora) from Japan, and its implication for amphicyonid paleobiogeography. *Paleontological Research* 1(4):311–315.

Kohno, N. 2000. A centenary of studies on the holotype (NSM-PV 5600) of *Desmostylus japonicas* Tokunaga and Iwasaki, 1914. *Bulletin of Ashoro Museum of Paleontology* 1:137–151 (in Japanese with English abstract).

Kohno, N., T. Uyeno, H. Kato, T. Sasaki, and P. G. Davis. 1997. Late Miocene vertebrate fauna from the Sendai region, Miyagi Prefectrue, Japan. *Abstract with Programs of the 1997 Annual Meeting of the Palaeontological Society of Japan*, p. 61 (in Japanese).

Komatsubara, J., H. Ugai, T. Danhara, H. Iwano, T. Yoshioka, T. Nakajima, K. Kano, and K. Ogasawara. 2005. Fission-track age and inferred subsidence rate of the Lower to Middle Miocene Nojima Group in NW Kyushu. *Journal of the Geological Society of Japan* 111(6):350–360 (in Japanese with English abstract).

Kordikova, E. G. 2000. Insectivora (Mammalia) from the Lower Miocene of the Aktau Mountains, south-eastern Kazakhstan. *Senckenbergiana Lethaea* 80(1):67–79.

Li, C.-k. 1974. Probable geomyoid reodent from Middle Miocene of Linchu, Shantung. *Vertebrata PalAsiatica* 12:43–53 (in Chinese with English summary).

Li, C.-k, Y.-p. Lin, Y.-m. Gu, L.-h. Hou, W.-y. Wu, and Z.-d. Qiu. 1983. The Aragonian vertebrate fauna of Xiacaowan, Jiangsu. 1. A brief introduction to the fossil localities and preliminary report on the new material. *Vertebrata PalAsiatica* 21(4):313–327 (in Chinese with English summary).

Liu, L.-p., M. Fortelius, and M. Pickford. 2002. New fossil Suidae from Shanwang, Shandong, China. *Journal of Vertebrate Paleontology* 22(1):152–163.

Lourens, L. J., F. J. Hilgen, J. Laskar, N. J. Shackleton, and D. Wilson. 2004. The Neogene Period. In *A Geological Time Scale 2004*, ed. F. M. Gradstein, J. G. Ogg, and A. G. Smith, pp. 409–440. Cambridge: Cambridge University Press.

Makiyama, J. 1938. Japonic Proboscidea. *Memoir of the College of Science, Kyoto Imperial University*, ser. B, 14:1–59.

Matsumoto, H. 1916. [List of mammalian fossil beds in Japan.] *Journal of Geological Society of Tokyo* 23(275):291–299 (in Japanese).

Matsumoto, H. 1918. On a new archetypal fossil cervid from the Prov. of Mino. *Science Report of Tohoku Imperial University*, 2nd ser. (Geology) 3(2):75–81, pl. 23.

Matsumoto, H. 1921. Descriptions of some new fossil mammals from Kani District, Prov. of Mino, with revisions of some Asiatic fossil rhinocerotids. *Science Reports of the Tohoku Imperial University*, 2nd ser. (Geology) 5:75–91.

Matsumoto, H. 1924. Preliminary notes on two new species of fossil mastodon from Japan. *Journal of the Geological Society of Tokyo* 31:395–414 (in Japanese).

Matsumoto, H. 1926. On two new mastodonts and an archetypal *Stegodon* of Japan. *Science Reports of the Tôhoku Imperial University*, 2nd ser., 10:1–11, pls. 1–5.

McKenna, M. C. and S. K. Bell. 1997. Classification of mammals above the species level. New York: Columbia University Press.

Mein, P. 1999. European Miocene mammal biochronology. In *The Miocene Land Mammals of Europe*, ed. G. E. Rossner and K. Heissig, pp. 25–38. Munich: Dr. Friedrich Pfeil.

Mein, P. and L. Ginsburg. 1985. Les rongeurs miocènes de Li (Thaïlande). *Comptes Rendus De l'Académie des Sciences Paris*, 2nd ser., 301:1369–1374.

Miyachi, M. and H. Sakai. 1991. Zircon fission-track ages of some pyroclastic rocks from the Tertiary formations in Northwest Kyushu, Japan. *Journal of the Geological Society of Japan* 97(8):671–674 (in Japanese).

Miyata, K. and Y. Tomida. 2010. *Anchitherium* (Mammalia, Perissodactyla, Equidae) from the Early Miocene Hiramaki Formation, Gifu Prefecture, Japan, and its implication for the early diversification of Asian *Anchitherium*. *Journal of Paleontology* 84(4):763–773.

Morea, M. F. and W. W. Korth. 2002. A new eomyid rodent (Mammalia) from the Hemingfordian (Early Miocene) of Nevada and its relationship to Eurasian Apeomyinae (Eomyidae). *Paludicola* 4(1):10–14.

Nagasawa, J. 1932. A fossil cervid tooth obtained from Hiramakimura, Kani-gun, Province of Mino. *Journal of Geological Society of Tokyo* 39:219–224 (in Japanese).

Nomura, S. and T. Kosaka. 1987. Geologic development of Neogene system in the southwest part of Gunma Prefecture, central Japan. *Gunma Journal of Liberal Arts and Sciences* 21:51–68 (in Japanese with English abstract).

Nomura, T. 1986. Preliminary report of the Miocene Hachiya Formation and its K-Ar ages. *Journal of the Geological Society of Japan* 92:73–76 (in Japanese).

Ogg, J. G. and A. G. Smith. 2004. The geomagnetic polarity time scale. In *A Geological Time Scale 2004*, ed. F. M. Gradstein, J. G. Ogg, and A. G. Smith, pp. 63–86. Cambridge: Cambridge University Press.

Okazaki, Y. 1977. Mammalian fossils from the Mizunami Group, central Japan (Part 2). *Bulletin of the Mizunami Fossil Museum* 4:9–24 (in Japanese with English abstract).

Okumura, K., Y. Okazaki, S. Yoshida, and Y. Hasegawa. 1977. Mammalian fossils from Kani Town. In *Geology and Palaeontology of Kani Town, Central Japan*, ed. Kani Town Education Board, pp. 21–45. Kani, Gifu: Kani Town Education Board (in Japanese).

Okumura, Y., M. Watanabe, R. Morijiri, and M. Satoh. 1995. Rifting and basin inversion in the eastern margin of the Japan Sea. *Island Arc* 4:166–181.

Oshima, M. 2007. On a molar tooth of upper jaw of suidae from the Miocene Oiso Formation, Kanagawa Prefecture, Japan. *Bulletin of Kanagawa Prefectural Museus (Natural Science)* 36:29–32 (in Japanese with English abstract).

Oshima, M., Y. Tomida, T. Araki, and Y. Azuma. 2008. First record of the genus *Hyotherium* (Mammalia, Suidae) from Japan. *Memoir of the Fukui Prefectural Dinosaur Museum* 7:25–32.

Pilbeam, D., M. Morgan, J. C. Barry, and L. J. Flynn. 1996. European MN units and the Siwalik faunal sequence of Pakistan. In *The Evolution of Western Eurasian Neogene Mammal Faunas*, ed. R. L. Bernor, V. Fahlbusch, and H. W. Mittman, pp. 96–105. Columbia University Press, New York.

Qiu, X.-d, X.-m. Wang, and Q. Li. 2006. Faunal succession and biochronology of the Miocene through Pliocene in Nei Mongol (Inre Mongolia). *Vertebrata PalAsiatica* 44(2):164–181.

Qiu, Z.-x. and Z.-d. Qiu. 1995. Chronological sequence and subdivision of Chinese Neogene mammalian faunas. *Palaeogeography, Palaeoclimatology, and Palaeoecology* 116:41–70.

Qiu, Z.-x., W.-y. Wu, and Z.-d. Qiu. 1999. Miocene mammal faunal sequence of China: Palaeozoogeography and Eurasian relationships. In *The Miocene Land Mammals of Europe*, ed. G. E. Rössner and K. Heissig, pp. 443–455. Munich: Dr. Friendrich Pfeil.

Qiu, Z.-x., D.-f. Yan, and B. Sun. 1991. A new genus of Tapiridae from Shanwang, Shandong. *Vertebrata PalAsiatica* 29:119–135.

Saegusa, H. 1996. Stegodontidae: Evolutionary relationships. In *The Proboscidea: Evolution and Palaeoecology of Elephants and Their Relatives*, ed. J. Shoshani and P. Tassy, pp. 178–190. Oxford: Oxford University Press.

Saegusa. H. 2008. Dwarf *Stegolophodon* from the Miocene of Japan: Passengers on sinking boats. *Quaternary International* 182(1):49–62.

Saegusa, H., Y. Thasod, and B. Ratanasthien 2005. Notes on Asian stegodontids. *Quaternary International* 126–128:31–48.

Sakai, H., H. Nishi, and M. Miyachi. 1990. Geologic age of the unconformity between the Sasebo and Nojima Croups, northwest Kyushu and its tectonic significances. *Journal of the Geological Society of Japan* 96(4):327–330.

Sasao, E., H. Iwano, and T. Danhara. 2006. Fission track ages of tuffaceous sandstone from the Toki Lignite-bearing Formation of the Mizunami Group in the Tono district, Gifu Prefecture, central Japan. *Journal of the Geological Society of Japan* 112:459–468 (in Japanese with English abstract).

Sato, D. 1914. [Preliminary note on the *Tetrabelodon* (?) from the Miocene of the Province of Mino.] *Journal of Geography (Chigaku zasshi)* 26(301):21–28, pl. 2 (in Japanese).

Shikama, T. 1936. The first discovery of *Pentalophodon* from Japan. *Proceedings of the Imperial Academy of Tokyo* 12(9):292–295.

Shikama, T. and Y. Kirii. 1956. A Miocene *Stegolophodon* from the Yatsuo Group in Toyama Prefecture. *Transactions and Proceedings of the Palaeontological Society of Japan*, n.s., 24:285–289, 1 pl.

Shikama, T. and I. Yanagisawa. 1971. Fossil proboscidean tooth from Iwaki City, Hukushima Prefecture. *Science Reports of the Yokohama National University*, Sec. II 18:37–42, 2 pls.

Shikama, T. and S. Yoshida. 1961. On an equid fossil from Hiramaki Formation. *Transaction and Proceedings of the Palaeontological Society of Japan*, n.s., 44:171–174.

Shikano, K. 2003. Fission track ages of the Lower Miocene Mizunami Group in the Minokamo Basin, Gifu Prefecture, central Japan. *Memoirs of the Minokamo City Museum* 2:1–8 (in Japanese).

Shikano, K. and Y. Ando. 2000. Geology and occurrence, mammal from Hachiya Formation in Kawabe-cyo, Gifu Prefecture, central Japan. *Bulletin of the Gifu Prefectural Museum* 21:11–16 (in Japanese).

Steininger, F. F. 1999. Chronostratigraphy, geochronology and biochronology of the Miocene "European Land Mammal Mega-Zones" (ELMMZ) and the Miocene "Mammal-Zones" (MN-Zones). In *The Miocene Land Mammals of Europe*, ed. G. E. Rössner and K. Heissig, pp. 9–24. Munich: Dr. Friedrich Pfeil.

Suto, I., M. Takahashi, and Y. Yanagisawa. 2003. Diatom biostratigraphy of the Miocene Tsuchishio Formation in the Hiki Hills area (Aketo Section), Saitama Prefecture, central Japan. *Journal of the Geological Society of Japan* 109(1):48–62 (in Japanese with English abstract).

Takai, F. 1939. The mammalian faunas of the Hiramakian and Togarian stages in the Japanese Miocene. *Jubilee Publication in the Commemoration of Professor H. Yabe, M.I.A. Sixtieth Birthday* 1:189–203.

Takai, F. 1949. Fossil mammals from Katabira-mura, Kani-gun, Gifu Prefecture, Japan. *Japanese Journal of Geology and Geography* 21:285–290.

Takai, F. 1950. [Fossil wild boar from Tsuyama Basin]. *Journal of the Geological Society of Japan* 56:278–279 (in Japanese).

Takai, F. 1954. An addition to the mammalian fauna of the Japanese Miocene. *Journal of the Faculty of Science, University of Tokyo, Section II* 9(2):331–335.

Takai, F. and S. Fujii. 1961. *Stegolophodon pseudolatidens* from the Miocene Yokawa Group in Toyama Prefecture, Japan. In *Professor Jiro Makiyama Memorial Volume*, ed. S. Matsushita. N. Ikebe, M. Morishima, K. Nakazawa, K. Fujita, and S. Ishida, pp. 225–228, 1 pl. Kyoto: Kyoto University.

Takeuchi, T. 1992. Paleomagnetism of the Miocene Mizunami Group in Kani Basin, Gifu Prefecture, Japan. *Bulletin of the Mizunami Fossil Museum* 19:57–65 (in Japanese with English abstract).

Takeyama, K. 1989. On the fossil deer from the Miocene Kunimi Formation of Fukui Prefecture, Japan. *Bulletin of Fukui Prefectural Museum* 3:9–21 (in Japanese with English abstract).

Taruno, H. 1999. The stratigraphic positions of proboscidean fossils from the Pliocene and lower to middle Pleistocene formations of Japanese Islands. *Earth Science (Chikyu Kagaku)* 53:258–264 (in Japanese with English abstract).

Tassy, P. 1994. Gaps, parsimony, and early Miocene elephantoids (Mammalia), with a revaluation of *Gomphotherium annectens* (Matsumoto, 1925). *Zoological Journal of the Linnean Society* 112:101–117.

Tassy, P. 1996a. Growth and sexual dimorphism among Miocene elephantoids: The example of *Gomphotherium angustidens*. In *The Proboscidea: Evolution and Palaeoecology of Elephants and Their Relatives*, ed. J. Shoshani and P. Tassy, pp. 92–100. Oxford: Oxford University Press.

Tassy, P. 1996b. The earliest gomphotheres. In *The Proboscidea: Evolution and Palaeoecology of Elephants and Their Relatives*, ed. J. Shoshani and P. Tassy, pp. 89–91. Oxford: Oxford University Press.

Teilhard de Chardin, P. and M. Trassaert. 1937. The proboscideans of south-eastern Shansi (Yushe Basin). *Palaeontologia Sinica*, ser. C 13(1):1–58, 13 pls.

Tobien, H., G. Chen, and Y. Li. 1986. Mastodonts (Proboscidea, Mammalia) from the late Neogene and early Pleistocene of the People's Republic of China. Part 1. Historical account: The genera *Gomphotherium, Choerolophodon, Synconolophus, Amebelodon, Platybelodon, Sinomastodon. Mainzer Geowissenschaftliche Mitteilungen* 15:119–181.

Tomida, Y. 2000. New taxa of small mammals from the early Miocene of Japan and the origin of *Keramidomys* (Eomyidae). *Journal of Vertebrate Paleontology* 20(3 sup.):74A.

Tomida, Y. 2011 A new species of the genus *Megapeomys* (Eomyidae, Rodentia, Mammalia) from the Early Miocene of Japan. *Palaeontologia Electronica*, Special Issue Honoring Charles Repenning 14.3.25A:6p.

Tomida, Y., K. Kawai, T. Setoguchi, and T. Ozawa. 1995. A new record of *Youngofiber* (Castoridae: Mammalia) from the Early Miocene of Kani City, central Japan. *Bulletin of the National Science Museum*, ser. C 21(3–4):103–109.

Ujihara, A., T. Irizuki, and M. Hosoyama. 1999. Neogene of Tono district, Gifu Prefecture, central Japan. In *Excursion Guidebook. The 106th Annual Meeting of the Geological Society of Japan*, ed. K. Suzuki and M. Takeuchi, pp. 97–116. Tokyo: Geological Society of Japan (in Japanese).

Ujiie, H. and K. Oki. 1974. Uppermost Miocene–Lower Pleistocene planktonic foraminifera from the Shimajiri Group of Miyakojima, Ryukyu Island. *Memoirs of National Science Museum* 7:31–52, pls. 1–6.

Wang, B.-y. 1965. A new Miocene aceratherine rhinoceros of Shanwang, Shantung. *Vertebrata PalAsiatica* 9(1):109–113 (in Chinese with English summary).

Watanabe, M. and Y. Yanagisawa. 2005. Reined Early to Middle Miocene diatom biochronology for the middle- to high-latitude North Pacific. *Island Arc* 14:91–101.

Welcomme, J. L., P. O. Antoine, F. Duranthon, P. Mein, and L. Ginsburg 1997. Nouvelles découvertes de Vertébrés miocènes dans le synclinal de Dera Bugti (Balochistan, Pakistan). *Comptes Rendus de l'Académie des Sciences*, 325:531–536.

Yabe, H. 1950. Three alleged occurrences of *Stegolophodon latidens* (Clift) in Japan. *Proceedings of the Japan Academy* 26:61–65.

Yanagisawa, Y. 1990. Diatom biostratigraphy of the Neogene Sendai Group, northeast Honshu, Japan. *Bulletin of the Geological Survey of Japan* 41:1–25 (in Japanese with English abstract).

Yanagisawa, Y. 1998. Diatom biostratigraphy of the Neogene Tatsunokuchi Formation in the western Kitakami City, Iwate Prefecture, Japan. In *Geology and Ages of the Plio-Pleistocene Formations in the Kitakami Lowland, Northeast Japan*, ed. M. Oishi, pp. 29–36. Research Report of the Iwate Prefectural Museum 14 (in Japanese with English abstract).

Yanagisawa, Y. and F. Akiba. 1998. Refined Neogene diatom biostratigraphy for the northwest Pacific around Japan, with an introduction of code numbers selected diatom biohorizons. *Journal of the Geological Society of Japan* 104:395–414.

Yasuno, T. 2005. Early Miocene fossil assemblages discovered from Takeno Coast, Toyooka City, Hyogo Prefecture, central Japan (1). *Bulletin of the Fukui City Museum of Natural History* 52:43–65 (in Japanese).

Yoshida, K., S. Miyazaki, H. Mishima, T. Kakinuma, and K. Nakamura. 1989. Discovery of a rhinocerotid fossil from the Miocene Yagii Formation in Kawamoto-machi, Saitama Prefecture, Japan. *Chikyu Kagaku* 43(1):43–48 (in Japanese).

Yoshida, S. 1977. Geology in Kani Town. In *Geology and Palaeontology of Kani Town, Central Japan*, ed. Kani Town Education Board, pp. 3–16. Kani, Gifu: Kani Town Education Board (in Japanese).

Yoshiwara, S. 1899. Discovery of rhinocerotid bone from Togari, Mino Province, central Japan. *Journal of Geological Society of Japan* 6(69):236 (in Japanese).

Yoshiwara, S. and J. Iwasaki. 1902. Notes on a new fossil mammal. *Journal of College of Science, Imperial University of Tokyo* 16(6):1–13, 3 pls.

Young, C. C. 1937. On a Miocene mammalian fauna from Shantung. *Bulletin of the Geological Society of China* 17:209–243.

Ziegler, R. 2009. Plesiosoricids from early Oligocene fissure fillings in South Germany, with remarks on plesiosoricid phylogeny. *Acta Palaeontologica Polonica* 54(3):365–371.

Zin-Maung-Maung-Thein, H. Taru, M. Takai, and A. Fukuchi. 2009. A rhinocerotid (Mammalia, Perissodactyla) from the late Miocene Oiso Formation, Kanagawa Prefecture, Japan. *Palaeontological Research* 13(2):207–210.

Chapter 13

Pliocene Land Mammals of Japan

RYOHEI NAKAGAWA, YOSHINARI KAWAMURA, AND HIROYUKI TARUNO

Pliocene land mammal fossils from Japan are rather poor in number and taxonomic diversity in comparison with those from China, but most of them occur in marine and fluvio-lacustrine sediments that are well investigated stratigraphically and well dated by marine microfossils, magnetostratigraphy, and fission-track and tephrochronological methods. Kamei, Kawamura, and Taruno (1988) attempted to arrange Pliocene and Quaternary fossil mammal records hitherto known from all over Japan in a chronological framework with numerical and geomagnetic polarity time scales and proposed a mammalian biozonation of the Pliocene and Quaternary of Japan. This work treated both terrestrial and marine mammals, but it lacks current data from the Pliocene part of the record. Thus, an up-to-date revision of the work is required for Pliocene mammals in the light of new fossil records as well as subsequent advances in taxonomy of mammal fossils and chronology of fossil-bearing sediments. In this chapter, we present an updated overview of Pliocene land mammals of Japan and discuss their taxonomy and biostratigraphy, as well as their faunal and evolutionary implications.

In June 2009, the Executive Committee of the International Union of Geological Sciences ratified the proposal from the International Commission on Stratigraphy that the base of the Pleistocene be recognized at the base of the Gelasian (ca. 2.6 Ma; figure 13.2 [column B]). This definition implies the "short Pliocene" concept (ca. 5.3–2.6 Ma). In this chapter, however, we adopt the "long Pliocene" concept (ca. 5.3–1.8 Ma; figure 13.2 [column A])

and include the Gelasian in the Late Pliocene, because most literature hitherto published in Japan has accepted the "long Pliocene" concept. A change in concept here may generate confusion when we review previous literature. Here, therefore, we use Early, Middle, and Late Pliocene as the equivalents of the Zanclean, Piacenzian, and Gelasian, respectively (see figure 13.2).

FOSSIL LOCALITIES AND CHRONOLOGY

In Japan, Pliocene land mammal fossils have been found in many localities, and their occurrences have been recorded widely, although each of the localities has generally yielded a small number of mammal fossils. Of these, we have selected the 41 localities shown in figure 13.1 and table 13.1 for this review. Our selection has been conducted under the following criteria:

1. The localities yielding the fossil specimens directly observed by us have been selected.
2. The localities yielding the fossil specimens figured in literature have also been selected, even if we have not observed them directly.
3. The localities without clear stratigraphic or chronological information have been omitted.

Of the 41 localities, 38 occur on Honshu, Shikoku, and Kyushu Islands, and the remaining three localities (33, 34, and 41) are on small islands adjacent to these. None of the localities are distributed on Hokkaido Island or the Ryukyu Islands. About half of the localities are concentrated in the central part of the Kinki District.

Geological ages of all the localities are shown in table 13.1 in association with the dating methods for each. Most of the localities can be placed in the more detailed chronological framework with stratigraphic information, as shown in figure 13.2. This figure also shows land mammal taxa occurring from each locality.

TAXONOMIC NOTES

The land mammal fossils from the 41 localities represent six orders: Primates, Lagomorpha, Carnivora, Proboscidea, Perissodactyla, and Artiodactyla. Among them, the orders other than Carnivora and Proboscidea contain only one or two forms, while Carnivora and Proboscidea contain three and five forms, respectively (see table 13.1). Taxonomic notes for each form are given below.

Primates

A colobine monkey skull from the Middle Pliocene of Kosawa (10) is the only primate fossil in the Pliocene of Japan. Iwamoto, Hasegawa, and Koizumi (2005) allocated this skull to the extinct genus *Dolichopithecus* known from the Pliocene of Europe and proposed a new species *D. leptopostorbitalis* on the basis of its distinctive features (e.g., very thin postorbital rims).

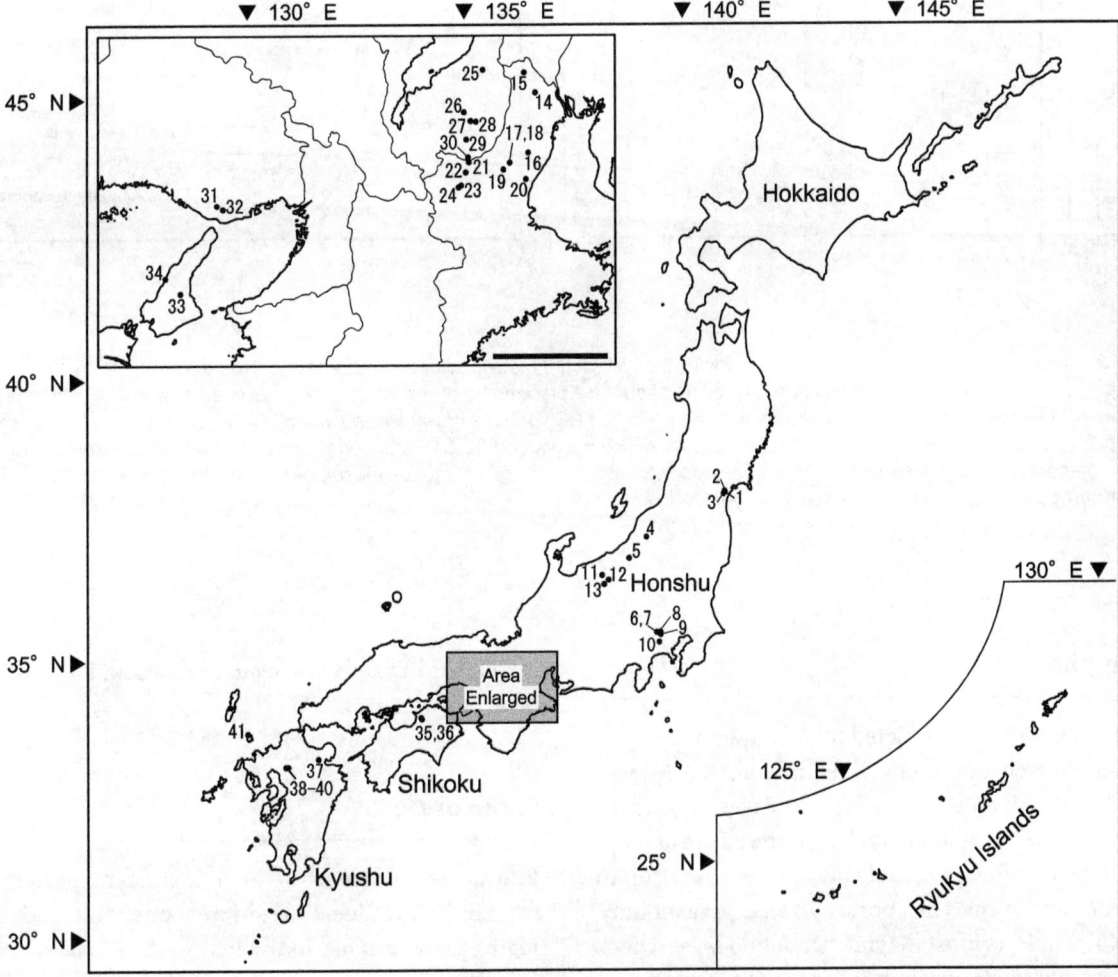

Figure 13.1 Geographic distribution of the representative Pliocene fossil localities of this chapter. The numbers refer to the localities, and detailed information on each locality is given in table 13.1.

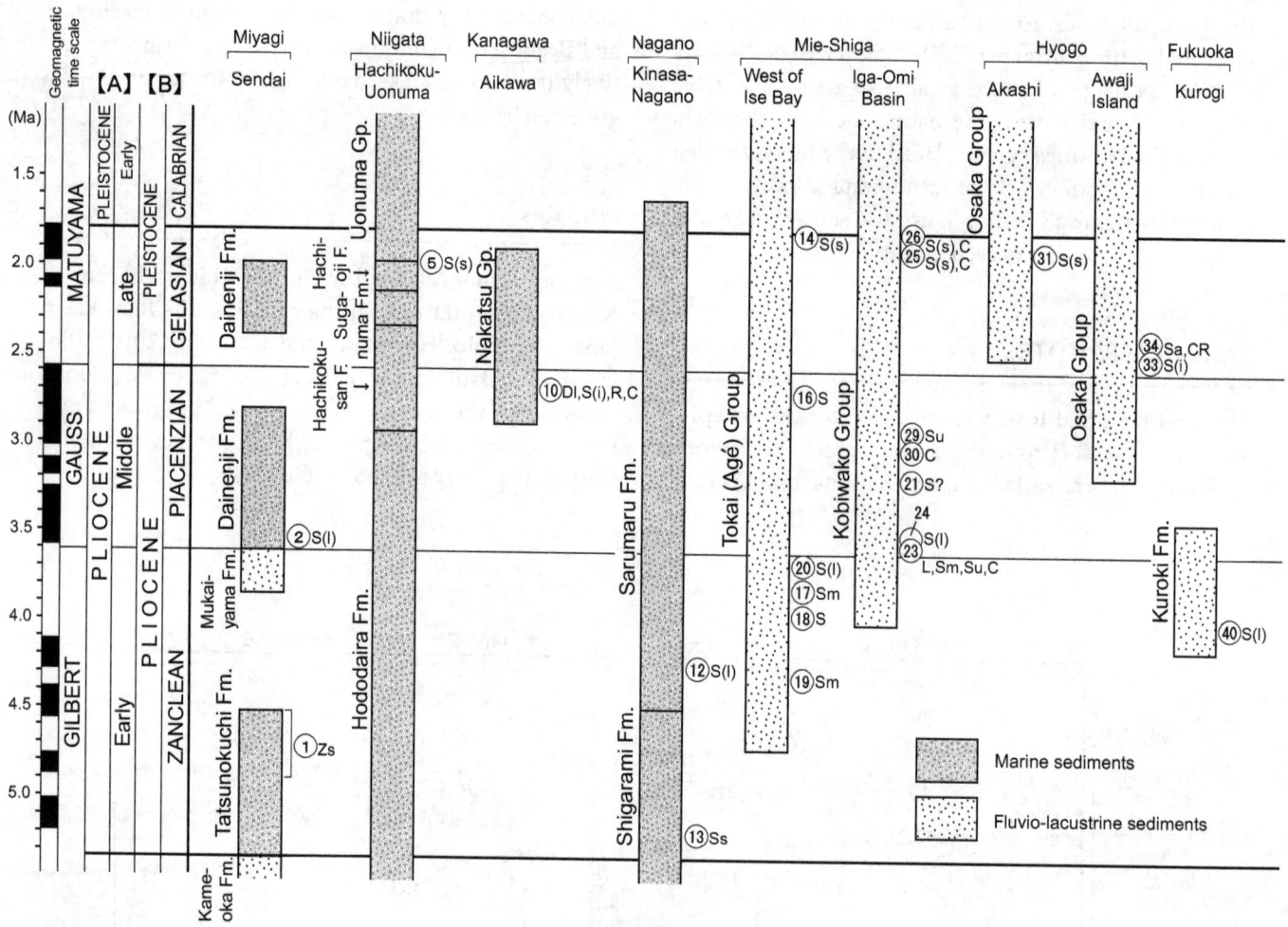

Figure 13.2 Stratigraphic and chronological distribution of the well-dated fossil localities of Pliocene age. Numbers in circles keyed to the numbers in figure 13.1 and table 13.1. Columns A and B reflect alternative placements of the Pliocene–Pleistocene boundary (A is adopted for the text). Abbreviations for mammals at these localities: C = *Cervus* sp.; CR = *Cervus (Rusa) kyushuensis*; Dl = *Dolichopithecus leptopostorbitalis*; L = Leporidae (probably new); R = Rhinocerotidae, gen. et sp. indet.; S = *Stegodon* sp.; Sa = *Stegodon aurorae*; Sca = *S.* cf. *S. aurorae* (probably new form); S(i) = *Stegodon* sp. (intermediate form); S(l) = *Stegodon* sp. (large form); Sm = *Stegodon miensis*; Ss = *Stegodon shinshuensis*; S(s) = *Stegodon* sp. (small form); Su = Suidae, gen. et sp. indet.; Zs = *Zygolophodon sendaicus*.

Lagomorpha

Lagomorph fossils are restricted to skull fragments with teeth and isolated upper cheek teeth from the Middle Pliocene on the riverbed of the Hattori River (23). These specimens show peculiar characters in the palate and upper cheek teeth, which probably permit the allocation to a new genus and species of leporids. In leporid taxonomy, morphology of P_3 is most diagnostic, but no lower cheek tooth is known. Detailed comparisons with leporid genera and species based on the preserved parts are required to determine the precise taxonomic position of the speci-

mens. The limited existing material is under study by one of us (Kawamura).

Carnivora

Pliocene carnivore fossils are few and have been recorded from only three localities. Many bones and teeth belonging to a single canid individual were recovered from the Late Pliocene on the riverbed of the Tama River (9) and were assigned to *Canis falconeri* by Koizumi (2003) on the basis of dental characters. *C. falconeri* is an extinct

Table 13.1

Detailed Data on Pliocene Terrestrial Mammal Localities Selected for This Review

Locality	Horizon	Geological Age (method of dating)	Taxa Revised	Previous Identification and Element (in parentheses)	Literature	Remarks
1. Kitayama and Higashikatsuyama, Aoba-ku, Sendai, Miyagi Pref.	Upper part of the Tatsunokuchi Formation	Early Pliocene (marine microfossils)	*Zygolophodon sendaicus*	*Trilophodon sendaicus* or *Zygolophodon sendaicus* (10 cheek teeth and 1 astragalus)	Matsumoto (1924b, 1926), Hatai and Masuda (1966), Yanagisawa (1990), Kamei (1991)	We follow the allocation by Kamei (1991).
2. Doteuchi, Sunaoshi, Ashinokuchi, Taihakuku, Sendai, Miyagi Pref.	Basal part of the Dainenji Formation	Middle Pliocene (marine microfossils)	*Stegodon* sp. (large form)	*Stegodon insignis* or *S. shin-shuensis* (1 fragmentary M^2)	Shikama (1963), Yanagisawa (1990), Taruno (1991)	Taruno (1991) referred the specimen to *S. shinshuensis*, but it is too fragmentary for specific identification.
3. Tsutsujigaoka, Miyagino-ku, Sendai, Miyagi Pref.	Tatsunokuchi Formation	Early Pliocene (stratigraphic correlation)	*Stegodon* sp. (large form)	*Stegolophodon* sp. (1 fragmentary M^2 and M^3)	Koda et al. (1998)	The allocation and horizon by Koda et al. (1998) were amended by Taruno (1999), which is followed here; specimens represent a single M^3.
4. Kotsunagi, Tochio, Nagaoka, Niigata Pref.	Ushigakubi Formation	Early to Middle Pliocene (marine microfossils and fission-track dating)	*Parailurus* sp.	*Parailurus* sp. (1 P^4)	Kobayashi et al. (1991), Sasagawa et al. (2003)	
5. Kirisawa, Okada, Takayanagi-cho, Kashiwazaki, Niigata Pref.	Lowest part of the Uonuma Group	Late Pliocene (tephrochronology)	*Stegodon* sp. (small form)	*Parastegodon* cf. *akashiensis* or *Stegodon aurorae* (1 mandible with m1 and m2)	Takai (1940), Editorial Committee of "The Uonuma Group" Monograph (1983), Taruno (1991, 1999)	
6. Yamada Dam of the Akigawa River, Akiruno, Tokyo Pref.	Lower part of the Kasumi Gravel Bed or the Yamada Fm.	Middle or Late Pliocene (fission-track dating)	*Stegodon* sp. (large form)	*Stegodon orientalis*, *S. shin-shuensis*, *S. miensis*, or *S.* sp. (1 damaged M$_2$ or M$_3$)	Naora (1954), Taruno (1991), Research Group for Geology of the Western Hills of the Kanto Plain (1995), Taru and Kohno (2002), Taru and Hasegawa (2002), Baba et al. (2005)	

(continued)

Table 13.1 (continued)

Locality	Horizon	Geological Age (method of dating)	Taxa Revised	Previous Identification and Element (in parentheses)	Literature	Remarks
7. Sannotani, Ajiro, Akiruno, Tokyo Pref.	Lower part of the Kasumi Gravel Bed or the Yamada (=Yaoroshi) Formation	Middle or Late Pliocene (fission-track dating)	Stegodon sp. (large form)	Stegodon bombifrons, S. shinshuensis or S. miensis (1 fragmentary M^2 and M^3)	Research Group for Stegodon from Itsukaichi (1980), Taruno (1991), Research Group for Geology of the Western Hills of the Kanto Plain (1995), Taru and Hasegawa (2002), Baba et al. (2005), Aiba, Baba, and Matsukawa (2006)	Research Group for Stegodon from Itsukaichi (1980) reported fragments of a deer from this locality, but they were not figured.
8. Riverbed of the Tama River, Fussa, Tokyo Pref.	Upper part of the Kasumi Gravel Bed	?Late Pliocene	Stegodon sp. (intermediate form)	Stegodon cf. aurorae (2 damaged M_2)	Taru and Hasegawa (2002), Taru (2005)	
9. Riverbed of the Tama River near the Haijima Bridge, Akishima, Tokyo Pref.	Upper part of the Kasumi Gravel Bed	Plio-Pleistocene boundary (tephrochronology and plant fossils)	Canis falconeri	Canis falconeri (nearly complete skeleton of a single individual)	Koizumi (2003)	Koizumi (2003) mentioned the occurrence of Stegodon aurorae and Cervus kazusensis from the same locality, but they have not been figured or described.
10. Kosawa, Aikawa-machi, Kanagawa Pref.	Kanzawa Formation of the Nakatsu Group	Middle Pliocene (marine microfossils and magnetostratigraphy)	Dolichopithecus leptopostorbitalis; Stegodon sp. (intermediate form); Rhinocerotidae gen. et sp. indet.; Cervidae, gen. et sp. indet.	Dolichopithecus leptopostorbitalis (1 skull with canine and cheek teeth); Stegodon sp. (1 fragmentary skull with P^4 and M^1, 2 fragmentary postcrania) Rhinocerotidae, gen. et sp. indet. (1 carpal bone) Cervus sp. (1 fragmentary antler, 4 postcrania)	Hasegawa et al. (1991), Iwamoto et al. (2005)	
11. Hashigozawa, Kinasa, Nagano, Nagano Pref.	Hikage Sandstone and Conglomerate Member of Saramaru Formation	Early Pliocene (tephrochronology)	Stegodon sp. (large form)	Stegodon shinshuensis (1 ill-preserved lower molar)	Taruno and Research Group for Stegodon Fossils from Togakushi (1988), Taruno (1991), Konishi and Takahashi (1999)	Togakushi Museum of Natural History (1993) reported younger fission-track ages for the Sarumaru Formation.

Locality	Formation	Age (dating method)	Original identification	Revised identification (material)	References	Remarks
12. Kawashimo, Togakushi, Nagano, Nagano Pref.	Lower part of the Sarumaru Formation	Early Pliocene (tephrochronology)	*Stegodon* sp. (large form)	*Stegodon shinshuensis* (1 mandible with M_1 and M_2)	Togakushi Museum of Natural History (1993), Konishi and Takahashi (1999), Taruno (1999)	
13. Uranosawa, Kakukura, Nagano, Nagano Pref.	Lower part of the Shigarami Formation (Ichinose Sandstone Mbr)	Early Pliocene (K-Ar dating and tephrochronology)	*Stegodon shinshuensis*	*Stegolophodon shinshuensis* (1 fragmentary skull with M^3)	Fossil Elephant Research Group (1971, 1979) Taruno (1985, 1999), Nagamori (1998), O'shima and Takahashi (2005)	Taruno (1985) amended the generic allocation of the species to *Stegodon*.
14. Kasada, Inabe, Mie Pref.	Lower part of the Oizumi Formation*	Late Pliocene (tephrochronology)	*Stegodon* sp. (small form)	*Stegodon shodoensis akashiensis, Parastegodon akashiensis* or *S. aurorae* (1 M^2 or M^3)	Matsui (1943), Kakuta (1958, 1982), Taruno and Kamei (1993)	Matsui (1943) also showed a photograph of a cheek tooth fragment of *Stegodon*.
15. Tagiri-gawa, Toyashiro, Inabe, Mie Pref.	Oizumi Formation*	Late Pliocene or Early Pleistocene (tephrochronology)	?Cervidae, gen. et sp. indet.	*Cervus* sp. (2 limb bones)	Kakuta (1958) Taruno and Kamei (1993)	
16. Yamabe-cho, Suzuka, Mie Pref.	Tomari Formation*	Late Pliocene (tephrochronology)	*Stegodon* sp.	*Parastegodon* sp., *P. akashiensis* or *Stegodon aurorae* (1 ill-preserved molar)	Kakuta (1958, 1982), Yoshida (1993) Taruno and Kamei (1993)	
17. Kurotani, Nomura, Kameyama, Mie Pref.	Kameyama Formation*	Early Pliocene (tephrochronology)	*Stegodon miensis*	*Stegodon* cf. *elephantoides* or *Stegodon* cf. *shinshuensis* (1 mandible with M_3)	Ikebe (1959), Ikebe, Chiji, and Ishida (1966), Taruno (1991, 1999), Taruno and Kamei (1993)	
18. Mukugawa, Sumiyama-cho, Kameyama, Mie Pref.	Kameyama Formation*	Early Pliocene (tephrochronology)	*Stegodon* sp.	*Stegodon* cf. *elephantoides* (1 fragmentary incisor)	Takemura et al. (1978), Kakuta (1982), Taruno and Kamei (1993)	
19. Hayashi, Geno-cho, Tsu (formerly Akiramura), Mie Pref.	Kameyama Formation*	Early Pliocene (tephrochronology)	*Stegodon miensis*	*Stegodon clifti, Stegodon* cf. *elephantoides* or *Stegodon shinshuensis* (2 mandibles with M_3)	Matsumoto (1924a, 1941), Makiyama (1938), Kakuta (1958, 1982), Ikebe (1959), Ikebe, Chiji, and Ishida (1966), Taruno (1991), Taruno and Kamei (1993), Taru and Kohno (2002)	

(*continued*)

Table 13.1 (continued)

Locality	Horizon	Geological Age (method of dating)	Taxa Revised	Previous Identification and Element (in parentheses)	Literature	Remarks
20. Kitakuroda, Tsu (formerly Kawage-cho), Mie Pref.	Kameyama Formation*	Early Pliocene (tephrochronology)	*Stegodon* sp. (large form)	*Stegodon* cf. *elephantoides* or *Stegodon shinshuensis* (1 fragmentary skull with 2 M^3, 2 mandibles with M_3 and postcranial bones)	Kakuta (1958, 1982), Ikebe (1959), Ikebe, Chiji, and Ishida (1966), Taruno (1991), Taruno and Kamei (1993), Konishi and Takahashi (1999)	In the skull, a molar fragment was subsequently found behind a molar originally considered M^3, indicating that the molars are M^2 and M_2, because all the specimens belong to a single individual.
21. Okutani, Kosugi, Iga, Mie Pref.	Tsuge Member of the Iga Formation[†]	Middle Pliocene (tephrochronology)	?*Stegodon* sp.	?*Stegodon* sp. (1 fragmentary incisor)	Kakuta (1958), Okazaki and Matsuoka (1979), Taruno and Kamei (1993), Konishi and Abe (1997)	
22. Yamanoshita, Midai, Iga, Mie Pref.	Tsuge Member of the Iga Formation[†]	Middle Pliocene (tephrochronology)	?*Stegodon* sp.	?*Stegodon* cf. *elephantoides* (1 incisor fragment)	Kawaguchi (1983), Matsuoka, Okamura, and Tamura (1991), Taruno and Kamei (1993), Konishi and Abe (1997)	
23. Riverbed of the Hattori River, Iga (formerly Oyamada-mura), Mie Pref.	Nakamura Member of the Ueno Formation[†]	Middle Pliocene (tephrochronology)	Leporidae, gen. et sp. indet.; *Stegodon miensis*; Suidae, gen. et sp. indet.; Cervidae, gen. et sp. indet.	Leporidae (1 skull fragment with cheek teeth, 1 skull fragment with an incisor, and 2 isolated cheek teeth); *Stegodon shinshuensis* (1 M_3, 1 premolar fragment and 1 postcranial bone); Suidae (1 mandible with cheek teeth); Cervidae (1 mandibular fragment, isolated cheek teeth and postcranial bones)	Okuyama (1985, 1988, 1989, 1990, 1993), Matsuoka, Okamura, and Tamura (1991), Taruno and Kamei (1993), Kawamura and Okuyama (1995), Konishi and Abe (1997), Kawamura et al. (2009)	The leporid remains are probably a new form.
24. Midoro Dam, Iga (formerly Oyamada-mura), Mie Pref.	Kashikimura Member of the Iga Formation[†]	Middle Pliocene (tephrochronology)	*Stegodon* sp. (large form)	*Stegodon* cf. *elephantoides* or *Stegodon shinshuensis* (1 fragmentary M_3)	Kamei (1984), Matsuoka, Okamura, and Tamura (1991), Taruno (1991), Taruno and Kamei (1993), Konishi and Abe (1997)	

Locality	Formation	Age	Taxa	Material	References	Notes
25. Shide, Taga-cho, Shiga Pref.	Upper part of the Gamo Formation[†]	Late Pliocene (tephrochronology)	Stegodon sp. (small form); Cervidae, gen. et sp. indet.	Stegodon aurorae (most of a single skeleton, including incisors and a mandible with somewhat damaged M_3); Cervidae (1 skull fragment, antlers, 1 mandible, 1 molar and postcranial bones)	Konishi and Otoda (1994), Abe et al. (1994), Amemori et al. (1995), Konishi and Takahashi (1999), Konishi and Abe (1997)	
26. Riverbed of the Sakura River, Hino-cho, Shiga Pref.	Uppermost part of the Gamo Formation[†] or lowermost part of the Kusatsu Formation[†]	Late Pliocene (tephrochronology)	Stegodon sp. (small form) Cervidae, gen. et sp. indet.	Stegodon sp. or Stegodon aurorae (1 ill-preserved upper molar); Equid (1 humerus fragment); Cervidae (postcrania)	Okazaki and Matsuoka (1979), Matsuoka, Okamura, and Tamura (1991), Taruno (1991), Taruno and Kamei (1993), Konishi and Abe (1997), Takahashi and Okamura (1997)	The allocation of the humerus to equids (Takahashi and Okamura, 1997) is problematic, seeming to be a cervid.
27. Kawara, Hino-cho, Shiga Pref.	Upper part of the Gamo Formation[†]	Late Pliocene (tephrochronology)	Cervidae, gen. et sp. indet	Cervidae (antler fragments and postcranial bones)	Matsuoka, Okamura, and Tamura (1991), Taruno and Kamei, (1993), Konishi and Abe (1997)	
28. Nishioji, Hino-cho, Shiga Pref.	Upper part of the Gamo Formation[†]	Late Pliocene (tephrochronology)	Cervidae, gen. et sp. indet.	Cervidae, gen. et sp. indet. (1 femur)	Okazaki and Matsuoka (1979), Matsuoka, Okamura, and Tamura (1991), Konishi and Abe (1997)	
29. Kosaji, Koka-cho, Shiga Pref.	Ayama Formation[†]	Middle Pliocene (tephrochronology)	Suidae, gen. et sp. indet.	Suidae (1 atlas)	Okazaki, Okamura, and Nishide (1983), Matsuoka, Okamura, and Tamura (1991), Konishi and Abe (1997)	
30. Kamimasugi, Konan-cho, Koka, Shiga Pref.	Lower part of the Ayama Formation[†]	Middle Pliocene (tephrochronology)	Cervidae, gen. et sp. indet.	Cervidae, gen. et sp. indet. (1 phalanx)	Matsuoka, Okamura, and Tamura (1991), Konishi and Abe (1997)	
31. Kodera, Igawadani, Nishi-ku, Kobe, Hyogo Pref.	Lower part of the Akashi Formation[‡]	Late Pliocene (tephrochronology)	Stegodon sp. (small form)	Parastegodon akashiensis or Stegodon aurorae (1 fragmentary M_3)	Maeda and Hashimoto (1983),Taruno (1991), Taruno and Kamei (1993), Ikawa and Itihara (1993)	

(continued)

Table 13.1 (continued)

Locality	Horizon	Geological Age (method of dating)	Taxa Revised	Previous Identification and Element (in parentheses)	Literature	Remarks
32. Nagasaka, Igawadani, Nishi-ku, Kobe, Hyogo Pref.	Lower part of the Akashi Formation‡	Late Pliocene or Early Pleistocene (tephrochronology)	*Stegodon* sp. (small form)	*Stegodon akashiensis* or *S. aurorae* (1 dameged M_3)	Komura (1973), Taruno (1991), Taruno and Kamei (1993), Ikawa and Itihara (1993)	
33. Takedani, Ichinomiya-cho, Awaji, Hyogo Pref.	Upper part of the Atago Formation‡	Late Pliocene (tephrochronology)	*Stegodon* sp. (intermediate form)	*Stegodon* sp. (1 M_1)	Taruno (1988, 1991), Taruno and Kamei (1993)	
34. Saizaki or Manzai, Tsushi, Goshiki-cho, Sumoto, Hyogo Pref.	Upper part of the Atago Formation‡	Late Pliocene (tephrochronology and fission-track dating)	*Stegodon* cf. *aurorae* *Cervus (Rusa)* cf. *kyushuensis*	*Stegodon aurorae* or *S.* cf. *aurorae* (1 fragmentary incisor, 1 incisor fragment 1 M^3, 1 fragmentary M^3, 1 fragmentary M^2 or M^3, and 1 M_3); *Elaphurus (Elaphuroides) shikamai* (1 fragmentary antler)	Taruno (1986, 1991), Tsuno and Otsuka (1991), Takahashi et al. (1992), Taruno and Kamei (1993), Katoh, Danhara, and Hyodo (2004), Kawamura and Taruno (2009)	Ikebe (1959) and Ikebe, Chiji, and Ishida (1966) reported tooth and bone remains of *S. shodoensis akashiensis* (=*S. aurorae*) from "Shinzaike", the same place as Saizaki. *S.* cf. *aurorae* is probably a new form.
35. Kameda, Yamamoto-cho, Mitoyo, Kagawa Pref.	Lower part of the Mitoyo Group	Late Pliocene (fission-track dating)	*Stegodon* sp. (small form)	*Stegodon sugiyamai* (1 fragmentary M_2 or M_3)	Furuichi, Bando, and Ishii (1977), Taruno (1991), Akojima et al. (1991)	
36. Iruhi, Saita-cho, Mitoyo, Kagawa Pref.	Lower part of the Mitoyo Group	Late Pliocene (fission-track dating)	*Stegodon* sp. (small form)	*Stegodon sugiyamai* (1 damaged upper molar)	Tokunaga (1935), Taruno (1991), Akojima et al. (1991)	Furuichi, Bando, and Ishii (1977) reported a molar fragment of "*Parastegodon* sp." from this locality.
37. Mori and Tanoku-chi, Ajimu-machi, Usa, Oita Pref.	Lower part of the Tsubusagawa Formation	Early to Middle Pliocene (fission-track dating)	Ursidae, gen. et sp. indet. *Stegodon* sp. indet. Rhinocerotidae, gen. et sp. indet. *Cervus (Rusa)* cf. *kyushuensis*	Ursidae (1 I^3 and 1 phalanx) *Stegodon* cf. *shinshuensis* (1 hyoid and postcranial bones) Rhinocerotidae, gen. et sp. indet. (1 P_4, 1 M_2, and 1 M_3) *Cervus unicolor* (1 skull fragment with 2 antlers, 3 mandibles, postcranial bones)	Takahashi and Kitabayashi (2001), Kato (2001), Hase et al. (2001)	

Locality	Formation	Age		Identification	References
38. Nakabaru, Kurogimachi, Fukuoka Pref.	Lower part of the Kuroki Formation	Early Pliocene (fission-track dating)	Stegodon sp. (large form)	Stegodon cf. bombifrons or S. shinshuensis (1 fragmentary M^3)	Otsuka, Inove, and Takai (1973), Taruno (1991), Urata (1992)
39. Uchikoshi-Kitakawauchi, Joyomachi, Yame, Fukuoka Pref.	Lower part of the Kuroki Formation	Early Pliocene (fission-track dating)	Stegodon sp. (large form)	Stegodon cf. bombifrons or S. shinshuensis (2 mandibles with 1 molar and postcranial bones)	Matsuda, Takemura, Okazaki (1981), Taruno (1991), Urata (1992)
40. Hisaizumi, Hirokawa-machi, Fukuoka Pref.	Lower part of the Kurume Formation	Early Pliocene (fission-track dating)	Stegodon sp. (large form)	Stegodon miensis (1 M_2 fragment)	Sawamura et al. (1983), Okazaki (2004)
41. Tateishinishifure, Katsumoto-cho, Iki, Nagasaki Pref.	Yunomoto Formation of the Ashibe Group	Middle to Late Pliocene (fission-track dating)	Stegodon sp. (large form)	Stegodon sp. (cf. S. yushensis) or S. shinshuensis (2 incisors, 1 molar fragment and postcranial bones)	Geological Research Group of Iki Islands (1973), Taruno (1991), Takeshita (1992)

*Belonging to the Tokai (=Age) Group.

†Belonging to the Kobiwako Group.

‡Belonging to the Osaka Group.

species known from the Late Pliocene and Early Pleistocene of Eurasia and Africa.

On the other hand, isolated P⁴ from the Early to Middle Pliocene of Kotsunagi (4) was referred to *Parailurus* sp. by Sasagawa et al. (2003). *Parailurus* is a rare genus relative of the extant red panda *Ailurus* and was previously known from the Pliocene of Europe and North America. Thus, this tooth confirms the Holarctic distribution of the genus during the Pliocene.

Additionally, an isolated incisor and a phalanx from the Early to Middle Pliocene of Mori (37) were allocated to Ursidae, gen. et sp. indet. by Takahashi and Kitabayashi (2001). Morphological information from these specimens is too limited to evaluate their allocations.

Proboscidea

Proboscidean fossils are recorded from 34 localities out of the 41 and are much more abundant than the fossils assigned to the other five orders. In the taxonomy of these proboscidean fossils, molar morphology plays the most important role, and thus we propose the following key to distinguish genera and species known form the Pliocene and early Early Pleistocene of Japan.

1. Molars with a remarkable central groove distinguishing lingual and buccal cusp rows; each pair of lingual and buccal cusps yoke-shaped in anterior and posterior views: *Zygolophodon*

    Molars with many transverse ridges; no central groove except in the most anterior ridge: *Stegodon* (2)
2. Size large; M³ and M₃ with 8 ridges: *Stegodon shinshuensis*

    Size large; M³ and M₃ with 9 ridges: *S. miensis*

    Size intermediate; M³ and M₃ with 11 ridges: *S.* cf. *aurorae* (probably new form)

    Size small; M³ and M₃ with 12 to 13 ridges: *S. aurorae* (no reliable record in the Pliocene)

Zygolophodon is represented only by the fossils of *Z. sendaicus* that were described by Matsumoto (1924b, 1926) and Hatai and Masuda (1966) from the Early Pliocene of Kitayama and Higashikatsuyama (1). All the other records of proboscidean molars are assigned to *Stegodon*. Among the members of this genus, *S. shinshuensis* was described by Fossil Elephant Reseach Group (1979) as a new species of *Stegolophodon* on the basis of the fossils from the Early Pliocene of Uranosawa (13). *S. miensis* was first proposed as a new subspecies of *S. clifti* by Matsumoto (1941), who designated one of the mandibles from the Early Pliocene of Hayashi (19) as the type specimen. This species also oc-

curred in the Early and Middle Pliocene of Kurotani (17) and the riverbed of the Hattori River (23). Large molars comparable to *S. shinshuensis* or *S. miensis* in size were recovered from the Early and Middle Pliocene of Doteuchi (2), Tsutsujigaoka (3), Yamada Dam (6), Sannotani (7), Hashigozawa (11), Kawashimo (12), Kitakuroda (20), Midoro Dam (24), Nakabaru (38), Uchikoshi-Kitakawauchi (39), Hisaizumi (40), and Tateishinishifure (41). We here refer these molars to *Stegodon* sp. (large form), because in these molars, ridge numbers of M³ and M₃ are unknown or uncountable owing to the state of preservation or the difference in tooth kind.

S. cf. *aurorae* occurred only in the Late Pliocene of Saizaki (34). This form is distinguishable from *S. aurorae* as already pointed out by Taruno and Kamei (1993), and it is probably referred to a new species or subspecies. It is now studied by one of us (Taruno). The molars from the Middle and Late Pliocene of the riverbed of the Tama River (8), Kosawa (10), and Takedani (33) show intermediate size and morphology between *S. miensis* and *S. aurorae* and are referred to *Stegodon* sp. (intermediate form) herein, because ridge number of M³ and M₃ are unknown or uncountable in these molars.

S. aurorae was first described by Matsumoto (1918) on the basis of a molar with uncertain record of its occurrence. Its reliable records are known from the Early Pleistocene of Byobugaura near Kodera (31) and are dated to ca. 1.3 Ma, although the specimens from this locality were originally assigned to *S. shodoensis akashiensis* and *S. sugiyamai* by Makiyama (1938). Small molars comparable to *S. aurorae* in size occurred from the Late Pliocene of Kirisawa (5), Kasada (14), Shide (25), the riverbed of the Sakura River (26), Kodera (31), Nagasaka (32), Kameda (35), and Iruhi (36). These molars are referred to *Stegodon* sp. (small form) herein, again due to lack of well-preserved third molars.

Perissodactyla

Pliocene fossils assigned to perissodactyls are restricted to a few rhinocerotid fossils. Lower cheek teeth from the Early to Middle Pliocene of Tanokuchi (37) described by Kato (2001) undoubtedly belong to Rhinocerotidae. However, their generic and specific allocations seem to be impossible owing to the lack of required diagnostic parts. In addition, a carpal bone from the Late Pliocene of Kosawa (10) was assigned to a rhinocerotid (Hasegawa et al. 1991).

On the other hand, Takahashi and Okamura (1997) reported an "equid" humerus from the Late Pliocene on the riverbed of the Sakura River (26). This humerus is not an equid, because its broken greater tuberosity appears to be originally much higher than its head. Its over-all morphology resembles a large cervid. We believe that no reliable occurrence of Equidae is known from the Pliocene and Pleistocene of Japan, in contrast to their abundant occurrence in sediments of the same ages in China. This suggests that woodlands prevailed in Japan during those times, preventing the distribution of grass-land dwellers such as equids.

Artiodactyla

Artiodactyl fossils from the Pliocene are relatively abun-dant and are assigned to suids and cervids. However, bo-vids have not been recorded, in contrast to their common occurrence in the Pliocene of China. The absence of bo-vids may be explained by the same reason as the absence of equids mentioned earlier.

Among the fossils hitherto recorded, suid fossils are restricted to only two specimens (a mandible and an at-las) from the Middle Pliocene of the riverbed of the Hat-tori River (23) and Kosaji (29), which were described by Kawamura et al. (2009) and Okazaki, Okamura, and Nishide (1983), respectively. More detailed taxonomic positions of these specimens are unknown.

On the other hand, cervid fossils are known from the Early to Late Pliocene of many localities, such as Sanno-tani (7), the riverbed of the Tama River (9), Kosawa (10), Tagiri-gawa (15), Shide (25), the riverbed of the Sakura River (26), Kawara (27), Nishioji (28), Kamimasugi (30), Saizaki (34), and Mori (37). Most of them seem to belong to the medium-size deer or *Cervus*-size deer, in accordance with the allocation of Japanese fossil deer by Kawamura (2009).

Among the cervid fossils, those from Shide, Saizaki, and Mori seem to be identifiable at the species level, be-cause they have antlers that enable their specific alloca-tions. The antlers from Saizaki and Mori are similar to those of the extinct species *Cervus (Rusa) kyushuensis* described by Otsuka (1966) from the Early Pleistocene Kuchinotsu Group of western Kyushu. Thus, the fossils are tentatively referred to *C. (R.)* cf. *kyushuensis* here, al-though Takahashi and Kitabayashi (2001) identified the fossils from Mori as the living species *C. unicolor*.

BIOSTRATIGRAPHY

The well-dated records of the mammalian forms shown in figure 13.2 are rearranged in figure 13.3 to depict the chronological distribution of each form. This figure shows that the primates, lagomorphs, carnivores, and artiodac-

tyls are less valuable in biostratigraphy, because the oc-currence of each form is limited to only one horizon. On the other hand, the proboscideans shown in the figure, comprising several forms of a single genus *Stegodon*, are more useful in biostratigraphic zonation.

Among the proboscidean forms, *S. shinshuensis* appears at 5.2 Ma, while *S. miensis* ranges from 4.3 Ma to 3.5 Ma. *S.* sp. (large form) with morphological similarity to these two species showing a chronological range almost over-lapping that of *S. miensis*. *S.* cf. *aurorae* occurs at 2.4 Ma, and *S.* sp. (intermediate form) with morphological re-semblance to *S.* cf. *aurorae* records a somewhat earlier occurrence (2.7–2.5 Ma). *S. aurorae* appears at 1.3 Ma (after the Pliocene), and *S.* sp. (small form), morphologi-cally similar to *S. aurorae*, ranges from 2.0 Ma to 1.3 Ma or later. *S. shinshuensis*, *S. miensis*, *S.* cf. *aurorae*, and *S. aurorae* are considered to be endemic to Japan, forming a single lineage of *Stegodon*. The size reduction and in-crease in ridge numbers of M^3 and M_3 from *S. shinshuen-sis* to *S. aurorae* seem to represent evolutionary change from primitive to advanced.

In Japan, biozonation of the Pliocene and Pleistocene using proboscidean fossils has been conducted by many authors (Ikebe 1959; Ikebe, Chiji, and Ishida 1966; Kamei 1984; Taruno and Kamei 1993; Taruno 1999; Kawamura and Taruno 2001). Summarizing the chronological dis-tribution discussed previously, we propose the following biozonation to update the previous works (see figure 13.3). In the present biozonation, we tentatively include *S.* sp. (large form), *S.* sp. (intermediate form), and *S.* sp. (small form) in *S. miensis*, *S.* cf. *aurorae*, and *S. aurorae*, respectively.

S. aurorae zone	Later than 2.1 Ma
S. cf. *aurorae* zone	3.1–2.1 Ma
S. miensis zone	4.7–3.1 Ma
S. shinshuensis zone	Earlier than 4.7 Ma

The boundaries between these biozones are of course tentative owing to the lack of fossils recorded around the boundaries. They are placed roughly at midpoints of gaps in the record (see figure 13.3 [dashed lines]). Further dis-cussions on the taxonomy, biostratigraphy, and phylog-eny of the Japanese forms of *Stegodon* will be developed by one of us (Taruno).*

---

* After we submitted this chapter, Aiba, Baba, and Matsukawa (2010) pro-posed a new species name *S. protoaurorae* for the stegodont with teeth inter-mediate between *S. aurorae* and *S. miensis*. However, they used the upper second molar for the holotype. It is not possible to add their report to our discussion, because we use the ridge numbers of the third molar for classifi-cation and *S.* cf. *aurorae* is used as an intermediate.

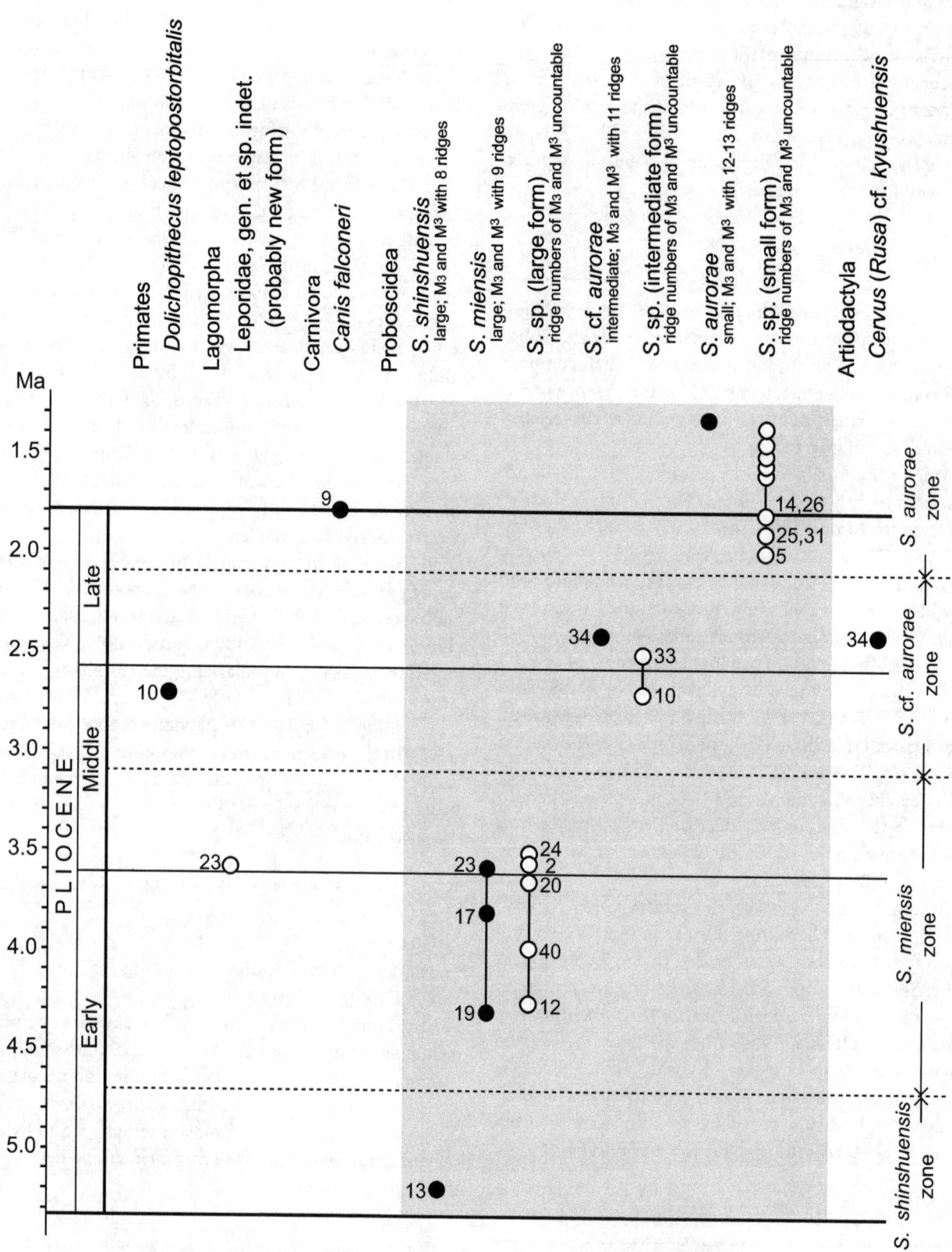

Figure 13.3 Biostratigraphy of the well-dated land mammal forms in the Pliocene of Japan. Numbers at circles correspond to the locality numbers in figure 13.1 and table 13.1. Diagnostic characters are noted for Proboscidea because they play a crucial role in biozonation. Solid circle: form with species-level allocation; open circle: record of a form with genus- or family-level allocation.

Table 13.2

Land Mammals Known from the Pliocene of Japan (indeterminate forms omitted)

Primates
  Dolichopithecus leptopostorbitalis
Lagomorpha
  Leporidae, gen. et sp. indet. (probably new form)
Carnivora
  Canis falconeri
  Ursidae, gen. et sp. indet.
  Parailurus sp.
Proboscidea
  Zygolophodon sendaicus
  Stegodon shinshuensis
  S. miensis
  S. cf. *aurorae* (probably new form)
  S. aurorae
Perissodactyla
  Rhinocerotidae, gen. et sp. indet.
Artiodactyla
  Suidae, gen. et sp. indet.
  Cervus (Rusa) cf. *kyushuensis*

CONCLUSION

Pliocene land mammal fossils from Japan are much fewer and less diversified than those from China. They comprise only 13 forms listed in table 13.2 when we exclude those of taxonomic uncertainty. The absence of equids and bovids is a distinctive feature as contrasted with the Pliocene fossils of China, and it seems to be indicative of forested vegetation during the Pliocene. Another character of these fossils is the abundant occurrence of proboscideans, probably reflecting their large size. Using the proboscidean fossils, we are able to recognize four biozones: (1) *Stegodon shinshuensis* zone (early Early Pliocene, earlier than 4.7 Ma), (2) *S. miensis* zone (late Early to early Middle Pliocene, 4.7–3.1 Ma), (3) *S.* cf. *aurorae* zone (late Middle to early Late Pliocene, 3.1–2.1 Ma), and (4) *S. aurorae* zone (late Late Pliocene, later than 2.1 Ma). On the basis of this chronological division, the faunal content of each time interval is summarized as follows:

4. Late Late Pliocene: *Canis falconeri, Stegodon aurorae,* Cervidae, gen. et sp. indet.
3. Late Middle to early Late Pliocene: *Dolichopithecus leptopostorbitalis, S.* cf. *aurorae,* Rhinocerotidae, gen. et sp. indet., Suidae, gen. et sp. indet., *Cervus (Rusa)* cf. *kyushuensis*
2. Late Early to early Middle Pliocene: Leporidae, gen. et sp. indet., Ursidae, gen. et sp. indet., *Parailurus* sp., *S. miensis,* Rhinocerotidae, gen. et sp. indet., Suidae, gen. et sp. indet., *C. (R.)* cf. *kyushuensis*
1. Early Early Pliocene: *Zygolophodon sendaicus, S. shinshuensis*

REFERENCES

Abe, Y., T. Kobayakawa, K. Amemori, N. Otoda, M. Tamura, A. Kitagawa, T. Arakawa, Y. Taga, T. Tajima, K. Nishikawa, and N. Mitsuya. 1994. Outline and significance of deer fossils from the Kobiwako Group at Shide, Taga-cho. *Research Reports of Cultural Properties and Natural History in Taga Town* 4:33–49 (in Japanese).

Aiba, H., K. Baba, and M. Matsukawa. 2006. *Stegodon miensis* Matsumoto (Proboscidea) from the Pliocene Yaoroshi Formation, Akiruno City, Tokyo, Japan. *Bulletin of Tokyo Gakugei University. Natural Sciences* 58:203–206.

Aiba, H., K. Baba, and M. Matsukawa. 2010. A new species of *Stegodon* (Mammalia, Proboscidea) from the Kazusa Group (Lower Pleistocene), Hachioji City, Tokyo, Japan and its evolutionary morphodynamics. *Palaeontology* 53:471–490.

Akojima, I., M. Furuichi, N. Kashima, and K. Suyari. 1991. Pliocene and Pleistocene Series in foothills of the Asan Mountains. In *Regional Geology of Japan, Part 8 Shikoku,* ed. Editorial Committee of Shikoku, pp. 134–137. Tokyo: Kyoritsu Shuppan (in Japanese).

Amemori, K., T. Kobayakawa, and Taga-cho Elephant Fossils Research Project. 1995. *Stegodon aurorae* (Matsumoto) found from the Kobiwako Group in Taga-cho, Shiga Prefecture, Japan. *Journal of the Geological Society of Japan* 101:743–746 (in Japanese).

Baba, K., H. Ohira, H. Aiba, and M. Matsukawa, 2005. The stratigraphic level of the specimen of *Stegodon miensis* (Proboscidea, Mammalia) in Akiruno city, Tokyo and its fission-track age. *Bulletin of Tokyo Gakugei University, Natural Sciences* 57:185–193 (in Japanese with English abstract).

Editorial Committee of the Monograph "The Uonuma Group." 1983. Development of the sedimentary basin and paleoenvironment of the Uonuma Group in Niigata Prefecture, Central Japan. *Association for the Geological Collaboration in Japan* 28:159–174 (in Japanese with English abstract).

Fossil Elephant Research Group. 1971. A new find of a proboscidean fossil from Nagano Prefecture, central Japan. *Journal of the Faculty of Science, Shinshu University* 6:37–44.

Fossil Elephant Research Group. 1979. New species of *Stegolophodon* found from the Shigarami Formation, northern part of Nagano Prefecture, central Japan. *Earth Science [Chikyu Kagaku]* 33:11–25.

Furuichi, M., Y. Bando, and T. Ishii. 1977. Note on some proboscidean fossils from the Mitoyo Formation in Kagawa Prefecture, southwest Japan. *Memoirs of the Faculty of Education, Kagawa University,* II, 27:29–35 (in Japanese with English abstract).

Geological Research Group of Iki Islands. 1973. Geological and paleontological researches of Iki Islands, Japan. Stratigraphy with notes on fossil elephant and fishes. *Bulletin of the Japan Sea Research Institute, Kanazawa University* 5:89–114, pl. 1–7.

Hase, Y., T. Danhara, M. Shiihara, and E. Kitabayashi. 2001. Stratigraphy and fission-track ages of the Tsubusagawa Formation in the Ajimu area of Northern Kyushu, Japan. *Research Report of the Lake Biwa Museum* 18:5–15 (in Japanese with English abstract).

Hasegawa, Y., A. Koizumi, Y. Matsushima, I. Imanaga, and D. Hirata. 1991. Fossil remains from the Nakatsu Group. *Research Reports of the Kanagawa Prefectural Museum, Natural History* 6:1–98 (in Japanese).

Hatai, K. and K. Masuda. 1966. The stratigraphic position of *Trilophodon sendaicus* Matsumoto in the Mizuho-To of Sendai City, Miyagi Prefecture. *Saito Ho-on Kai Museum Research Bulletin* 35:1–10.

Ikawa, N. and M. Itihara. 1993. Harima Basin: Akashi and adjacent area. In *The Osaka Group*, ed. M. Ichihara, pp. 110–126. Osaka: Sogen-sha.

Ikebe, N. 1959. Stratigraphical and geographical distribution of fossil elephants in Kinki District, central Japan. *Quaternary Research [Daiyonki-Kenkyu]* 1:109–118 (in Japanese with English abstract).

Ikebe, N., M. Chiji, and S. Ishida. 1966. Catalogue of the Late Cenozoic Proboscidea in the Kinki District, Japan. *Journal of Geosciences, Osaka City University* 9:47–56, pls. 1–8.

Iwamoto, M., Y. Hasegawa, and A. Koizumi. 2005. A Pliocene colobine from the Nakatsu Group, Kanagawa, Japan. *Anthropological Science* 113:123–127.

Kakuta, T. 1958. *Paleontology and Geology of the Northern Part of the Ise District*. Sangi Railway Co. and Mie Prefectural Museum (in Japanese).

Kakuta, T. 1982. Stratigraphical and geological distribution of fossil elephants in Ise-Bay. *Kaseikenkyu of Mie Tankidaigaku* 30:105–143 (in Japanese).

Kamei, T. 1984. Fossil mammals: Lake Biwa and fossil mammals: Faunal change since the Pliocene time. In *Lake Biwa*, ed. S. Horie, pp. 475–495. Boston: Dr. W. Junk.

Kamei, T. 1991. Phylogeny and evolution of proboscideans. In *Japanese Proboscidean Fossils*, ed. T. Kamei, pp 22–66. Tokyo: Tsukiji-shokan.

Kamei, T., Y. Kawamura, and H. Taruno. 1988. Mammalian stratigraphy of the late Neogene and Quaternary in the Japanese Islands. *Journal of the Geological Society of Japan* 30:181–204 (in Japanese with English abstract).

Kato, T. 2001. Discovery of the Pliocene Rhinocerotidae (Perissodactyla) from the Tsubusagawa Formation, Oita Prefecture. *Research report of the Lake Biwa Museum* 18:164–168 (in Japanese with English abstract).

Katoh, S., T. Danhara, and M. Hyodo. 2004. Re-examination of the Plio-Pleistocene boundary in the Osaka Group in central Awaji Island, western Japan. *Quaternary Research [Daiyonki-Kenkyu]* 43:213–224 (in Japanese with English abstract).

Kawaguchi, M. 1983. Fossil elephant tusk from the Iga Formation of the Kobiwako Group. *Kansai Shizenkagaku* 34:9–12 (in Japanese).

Kawamura, Y. 2009. Fossil record of sika deer in Japan. In *Sika Deer: Biology and Management of Native and Introduced Populations*, ed.

D. R. McCullough, S. Takatsuki, and K. Kaji, pp. 11–25. New York: Springer.

Kawamura, Y., Y. Matsuhashi, R. Nakagawa, and H. Taruno. 2009. Occurrence of a suid mandible from the Pliocene Ueno Formation, Mie Prefecture, central Japan. *Bulletin of the Osaka Museum of Natural History* 63:15–23.

Kawamura, Y. and S. Okuyama. 1995. Pliocene leporid remains (Lagmorpha, Mammalia) from the Ueno Formation of the Kobiwako Group, at Oyamada, Mie Prefecture, central Japan: The oldest rabbit remains in Japan. *Abstracts of the 1995 Annual Meeting of the Palaeontological Society of Japan* 90 (in Japanese).

Kawamura, Y. and H. Taruno. 2001. Late Neogene and Quaternary terrestrial mammal biostratigraphy in Japan: A revision based on recent data. Program and Late Abstract. *International Symposium on the Assembly and Breakup of Rodinia and Gondwana, and Growth of Asia*: 28–30.

Kawamura, Y. and H. Taruno. 2009. An Early Pleistocene deer antler from the Akashi Formation in Hyogo Prefecture, central Japan. *Bulletin of the Osaka Museum of Natural History* 63:25–34.

Kobayashi, I., M. Tateishi, T. Yoshioka, and M. Shimazu. 1991. *Geology of the Nagaoka District with Geological Map Sheet at 1:50000*. The Geological Survey of Japan, Tsukuba (in Japanese with English abstract).

Koda, Y., T. Suzuki, A. Tomita, and Y. Hasegawa. 1998. *Stegolophodon* teeth from the Miocene in Sendai City, Miyagi Prefecture, Japan. *Bulletin of Ibaraki Nature Museum* 1:3–8 (in Japanese with English abstract).

Koizumi, A. 2003. The first record of the Plio-Pleistocene hypercarnivorous Canid, *Canis (Xenocyon) falconeri* (Mammalia; Carnivora), from the Tama River, Akishima City, Western Tokyo, Japan. *Quaternary Research [Daiyonki-Kenkyu]* 42:105–111 (in Japanese with English abstract).

Komura, R. 1973. Discovery of a fossil elephant tooth-Toban Hills in Hyogo Prefecture. *Chishitsu News* 228:24–29 (in Japanese).

Konishi, S. and Y. Abe. 1997. Vertebrate fossils in the Kobiwako Group. *Journal of Fossil Research* 30:16–23 (in Japanese with English abstract).

Konishi, S. and N. Otoda. 1994. Project for the fossils of *Stegodon aurorae* in the fiscal year 1993. *Research Reports on Cultural Properties and Natural History in Taga Town* 4:21–31 (in Japanese).

Konishi, S. and K. Takahashi. 1999. Mandibular morphology of stegodons from Japan, *Stegodon aurorae* and *Stegodon shinshuensis* (Proboscidea, Mammalia). *Earth Science [Chikyu-Kagaku]* 53:3–18.

Maeda, Y. and I. Hashimoto. 1983. Deciphering the strata in Kobe. *Natural History of Academic City Area of Kobe*. Kobe: Kobe Municipal Institute of Education (in Japanese).

Makiyama, J. 1938. Japonic Proboscidea. *Memoirs of College of Science, Kyoto Imperial University*, ser. B, 14:1–59.

Matsuda, T., K. Takemura, and Y. Okazaki. 1981. Fission trackage of zircon crystals included in the Kurogi formation, Fukuoka Prefecture, Kyushu. *Bulletin of the Kitakyushu Museum of Natural History* 3:85–92, pls. 6–7.

Matsui, H. 1943. Geology of the Yokkaichi-Kuwana area, Mie Prefecture. *Scientific Reports of the Institute of Geology and Mineralogy, Faculty of Science, Kyoto Imperial University* 2:1–11, pls. 1–2 (in Japanese).

Matsumoto, H. 1918. On a new archetypal fossil elephant from Mt. Tomuro, Koga. *Science Reports of the Tohoku Imperial University,* 2nd ser., Geology 3:51–56, pl. 20.

Matsumoto, H. 1924a. Preliminary notes on the species of *Stegodon* in Japan. *Journal of the Geological Society of Japan* 31:323–340 (in Japanese).

Matsumoto, H. 1924b. Preliminary notes on two new species of fossil *Mastodon* from Japan. *Journal of the Geological Society of Japan* 31:395–411 (in Japanese).

Matsumoto, H. 1926. On two new mastodonts and an archetypal stegodont of Japan. *Science Reports of the Tohoku Imperial University,* 2nd ser., Geology 10:1–1, pls.1–5.

Matsumoto, H. 1941. On Japanese Stegodonts and Parastegodonts. *Zoological Magazine* [*Dobutsugaku Zasshi*] 53:385–396 (in Japanese with English abstract).

Matsuoka, C., Y. Okamura, and M. Tamura. 1991. Vertebrate fossils found in Shiga Prefecture. In *Landscape and Environment of Shiga: Scientific Studies of Shiga Prefecture, Japan* (Separate Volume for Geomorphology and Geology), ed. Foundation of Nature Conservation in Shiga Prefecture, Otsu: pp. 543–625 (in Japanese with English abstract).

Nagamori, H. 1998. Molluscan fossil assemblages and paleoenvironment of the Pliocene strata in the Hokushin District, Nagano Prefecture, central Japan. *Earth Science* [*Chikyu-Kagaku*] 52:5–25 (in Japanese with English abstract).

Naora, N. 1954. *Old Stone Age in Japan.* Tokyo: Neiraku-shobo (in Japanese).

Okazaki, Y. 2004. A molar of *Stegodon* from the Kurume Formation (Pliocene), Fukuoka Prefecture. *Bulletin of the Kitakyushu Museum of Natural History and Human History,* Series A 2:65–70.

Okazaki, Y. and C. Matsuoka. 1979. A review of mammalian fossils in Shiga Prefecture, west Japan. In *Land and Life in Shiga: Separate Volume for Geomorphology and Geology and Geologic Map of Shiga Prefecture 1:100,000 in Scale,* ed. Foundation of Nature Conservation in Shiga Prefecture, Otsu: pp. 391–467 (in Japanese).

Okazaki, Y., Y. Okamura, and T. Nishide. 1983. Occurrence of a Pliocene suid (Mammalia, Artiodactyla) from the Sayama Formation, Kobiwako Group, Japan. *Bulletin of the Mizunami Fossil Museum* 10:199–203, pls. 55–56 (in Japanese with English abstract).

Okuyama, S. 1985. *Atlas of Fossils from Iga Basin, Mie Prefecture, Japan 5,* Ueno. (in Japanese).

Okuyama, S. 1988. *Atlas of Fossils from Iga Basin, Mie Prefecture, Japan 8,* Ueno. (in Japanese).

Okuyama, S. 1989. *Atlas of Fossils from Iga Basin, Mie Prefecture, Japan 9,* Ueno. (in Japanese).

Okuyama, S. 1990. *Atlas of Fossils from Iga Basin, Mie Prefecture, Japan 10,* Ueno. (in Japanese).

Okuyama, S. 1993. Molar of *Stegodon shinshuensis* found from the Iga-Aburahi formation in palaeo Lake Biwa group. *Chigaku-Kenkyu* 42:143–148 (in Japanese).

O'shima, H. and K. Takahashi. 2005. Morphological study of a skull of *Stegodon miensis* from Nakajo, Kimiminochi-gun, Nagano Prefecture. *Journal of Fossil Research* 38:90–97 (in Japanese with English abstract).

Otsuka, H. 1966. Pleistocene vertebrate fauna from the Kuchinotsu Group of west Kyushu. Part I. A new species of *Cervus (Rusa).*

Memoirs of the Faculty of Science, Kyushu University, ser. D, Geology 17:251–269, pls. 3–14.

Otsuka, H., M. Inoue and F. Takai. 1973. A molar of *Stegodon* from the Pliocene Yame Group, west Japan. *Reports of the Faculty of Science, Kagoshima University (Earth Science and Biology)* 5–6:1–6, pl. 1.

Research Group for Geology of the Western Hills of the Kanto Plain. 1995. Geology of the western hills of the Kanto Plain, central Japan (1)—Stratigraphy, structure and geological age of Kasumi Hill. *Earth Science* [*Chikyu Kagaku*] 49:391–405 (in Japanese with English abstract).

Research Group for *Stegodon* from Itsukaichi. 1980. Fossils of *Stegodon* from Itsukaichi, Nishitama-gun, Tokyo Prefecture. *Bunkazai-no-hogo* 12:78–91 (in Japanese).

Sasagawa, I., K. Takahashi, T. Sakumoto, H. Nagamori, H. Yabe, and I. Kobayashi. 2003. Discovery of the extinct red panda *Parailurus* (Mammalia, Carnivora) in Japan. *Journal of Vertebrate Paleontology* 23(4):895–900.

Sawamura, M., T. Tanaka, K. Chiashi, and T. Shikada. 1983. Fission track age of zircon in tuff from the lowest member of the Kurume Formation, Fukuoka Prefecture, Kyushu. *Journal of the Geological Society of Japan* 89:129–131 (in Japanese).

Shikama, T. 1963. Note on a *Stegodon* tooth from Sendai. *Science Reports of the Yokohama National University, Section II, Biological and Geological Sciences* 10:67–69.

Takahashi, K. and E. Kitabayashi. 2001. Elephant fossils and other mammalian fossils from Mori, Ajimu-cho, Oita Prefecture, Japan. *Research Report of the Lake Biwa Museum* 18:126–163 (in Japanese with English abstract).

Takahashi, K. and Y. Okamura. 1997. An equid humerus from the Kobiwako Group, central Japan. *Journal of the Geological Society of Japan* 103:391–393 (in Japanese).

Takahashi, Y., A. Sangawa, K. Mizuno, and H. Hattori. 1992. *Geology of the Sumoto District. With Geological Sheet Map at 1:50,000.* Geological Survey of Japan, Tsukuba (in Japanese with English abstract).

Takai, F. 1940. On two teeth of elephant found in Niigata Prefecture. *Journal of the Geological Society of Japan* 47:339–342 (in Japanese with English abstract).

Takemura, K., S. Ishida, T. Kamei, Y. Kawamura, Y. Okazaki, M. Torii, T. Kakuta, H. Nakaya, Y. Tomida, Y. Wada, and J. Yamada. 1978. Fossil elephant tusk from the Muku-gawa River, Kameyama, and its stratigraphic horizon. *Proceedings of the Kansai Branch of the Geological Society of Japan (Proceedings of Nishinihon Branch of the Geological Society of Japan 67)* 84:16 (in Japanese).

Takeshita, H. 1992. Iki. In *Regianal Geology of Japan, Part 9 Kyushu,* ed. Editorial Committee of Kyushu, pp. 123–126. Tokyo: Kyoritsu Shuppan (in Japanese).

Taru, H. 2005. The two teeth of indeterminate form between *Stegodon miensis* and *Stegodon aurorae* from Fussa, Tokyo, Japan. *Journal of Fossil Research* 38:98–107 (in Japanese with English abstract).

Taru, H. and Y. Hasegawa. 2002. The Plio-Pleistocene fossil mammals from the Kasumi and Tama Hills. *Memoirs of the Natural Science Museum* 38:4–56 (in Japanese with English summary).

Taru, H. and N. Kohno. 2002. Redescription and identification of *Stegodon's* tooth from Akiruno-shi, Tokyo, with a reference to the specific name of the large type *Stegodon* from Pliocene of Japan.

Memoirs of the National Science Museum, Tokyo, Japan 38:33–41 (in Japanese with English abstract).

Taruno, H. 1985. Genus *Stegodon* and genus *Stegolophodon*—their criteria and phylogenetic relation—(Mammalia: Proboscidea). *Bulletin of the Osaka Museum of Natural History* 38:23–36, pls. 1–4 (in Japanese with English abstract).

Taruno, H. 1986. Elephant fossils. In *History of Goshiki-cho*, ed. Editorial Committee of "History of Goshiki cho." Goshiki-cho (Sumoto): Town office of Goshiki-cho (in Japanese).

Taruno, H. 1988. Geomorphology and geology. In *History of Tsuna-cho (main volume)*, ed. Editorial Committee of "History of Tsuna-cho," pp. 3–21. Tsuna-cho (Awaji): Town office of Tsuna-cho (in Japanese).

Taruno, H. 1991. Stegodontid fossils from Japan. In *Japanese Proboscidean Fossils*, ed. T. Kamei, pp. 82–99. Tokyo: Tsukiji-shokan, (in Japanese).

Taruno, H. 1999. The stratigraphic positons of proboscidean fossils from the Pliocene and lower to middle Pleistocene formations of Japanese islands. *Earth Science [Chikyu Kagaku]* 53:258–264.

Taruno, H. and T. Kamei. 1993. Vertebrate fossils of the Pliocene and Pleistocene in the Kinki District. In *The Osaka Group*, ed. M. Ichihara, pp. 216–231. Osaka: Sogen-sha.

Taruno, H. and Research Group for Fossils from Togakushi. 1988. Fossils of Stegodont from Togakushi and Kinasa. *Programs and Abstracts of Annual Meeting of Japan Association for Quaternary Research*, pp. 132–133 (in Japanese).

Togakushi Museum of Natural History. 1993. *Record of the Excavation of Stegodon shinshuensis, Tenth Aniversary*. Tokagushi Museum of Natural History (in Japanese).

Tokunaga, S. 1935. A new fossil elephant found in Shikoku, Japan. *Proceeding of the Imperial Academy, Tokyo* 11(10):432–434.

Tsuno, S. and H. Otsuka. 1991. The Plio-Pleistocene Series in Goshiki-cho and its environs on Awaji Island. *Reports of the Faculty of Science, Kagoshima University (Earth Science and Biology)* 24:73–105 (in Japanese with English abstract).

Urata, H. 1992. Kuroki Basin. In *Regional Geology of Japan, Part 9, Kyushu*, ed. Editorial Committee of Kyushu. Tokyo: Kyoritsu Shuppan (in Japanese).

Yanagisawa, Y. 1990. Diatom biostratigraphy of the Neogene Sendai Group, northeast Honshu, Japan. *Bulletin of the Geological Survey of Japan* 41:1–25 (in Japanese, English abstract).

Yoshida, F. 1993. The Tokai Group, the Kentoyama Formation, terrace deposits and alluvium. In *The Osaka Group*, ed. M. Ichihara, pp. 169–189. Osaka: Sogen-sha.

Part II

South and Southeast Asia

Chapter 14

The Siwaliks and Neogene Evolutionary Biology in South Asia

LAWRENCE J. FLYNN, EVERETT H. LINDSAY, DAVID PILBEAM, S. MAHMOOD RAZA,
MICHÈLE E. MORGAN, JOHN C. BARRY, CATHERINE E. BADGLEY,
ANNA K. BEHRENSMEYER, I. U. CHEEMA, ABDUL RAHIM RAJPAR, AND NEIL D. OPDYKE

The extensive, superposed stratigraphic record of the Siwaliks of southern Asia applies broadly to questions in evolutionary biology. In coordination with Barry et al. (chapter 15, this volume), this chapter focuses on the distribution of fossiliferous terrestrial deposits of late Cenozoic age across the Indian Subcontinent, the development of a chronostratigraphic framework in Pakistan and India, and the significance of the Siwalik deposits as a window on a subtropical ecosystem of the Miocene world, a window that was open during the late Paleogene and most of the Neogene. Retrievable information is paleontological, sedimentological, and geochemical, and it is relevant to systematics, biochronology, paleoecology, and paleobiogeography. Tightly constrained spatial and temporal data integrated with paleobiological and geological information provide unifying themes. The wealth of information bears on evolutionary questions from the level of the microevolutionary scale of population changes to the macroevolutionary scale of phenomena such as the dynamics of faunal responses to environmental change.

Vertebrate fossils from southern Asia became known to the Western world as early as the 1830s, primarily from Siwalik deposits of the Indian Subcontinent, and led to the steady accumulation of specimens in natural history collections in Calcutta and London. Virtually all the fossils of the nineteenth century were obtained second- or third-hand by travelers passing through that part of the world. Cautley and Falconer (1835) and Medlicott (1864) applied the term "Sivalik" or "Sewalik" or "Shevallik" to these fossils, and later Pilgrim (1913) defined a succession of strata and fossils (Kamlial, Chinji, Dhok Pathan, Tatrot, and Pinjor) from the Potwar Plateau of Pakistan and the Siwalik Hills in northern India, which produced most of the fossils. This stratigraphic-chronologic framework became used widely by numerous authors (Pilgrim 1926, 1934; Colbert 1935; Lewis 1937; Hussain 1971; Cheema, Raza, and Ahmed 1977). Barry et al. (chapter 15, this volume) interpret the stratigraphy of the Potwar Plateau as it is observed, without prior assumptions of the chronologic significance of Pilgrim's units.

Early in the twentieth century, the ages of Siwalik deposits were estimated from stage of evolution and presence of key mammalian immigrants. The long stratigraphic sequence produced fossils of the horses *Equus* and *Hipparion*, then thought of as indicators of Pleistocene and Pliocene age, respectively. *Equus* occurs in the Pinjor Formation and hipparionine horses appear in the Nagri Formation, but Colbert (1935) believed early collections indicated *Hipparion* in the older Chinji Formation. This led to the idea in the older literature that Chinji strata would be Pliocene in age. We now know that hipparionines do not occur in the Chinji Formation and that these horses entered the Old World in the Late Miocene; the Chinji Formation is mainly Middle Miocene in age, and the underlying Kamlial Formation extends into the Early Miocene. The transformative factor that led to accurate dating of the Siwaliks was the application of magnetostratigraphy to detailed biostratigraphic sections, the development of which we will explore.

The Siwalik sequence of terrestrial deposits of the Indian Subcontinent can be likened to a laboratory in which the evolution of subtropical terrestrial fauna proceeded for over 20 million years. The quality of the Siwalik fossil record resides in its extraordinary time depth, rather than richness of fossil horizons, although specific horizons do contain unusual numbers of well-preserved remains. Fossil assemblages are distributed throughout, and by careful stratigraphic control successive samples can be used to reconstruct a time series that records mammalian community evolution over many millions of years.

TERRESTRIAL DEPOSITS IN THE INDIAN SUBCONTINENT

The term "Siwalik" refers to molasse eroded from the Himalaya-Karakorum ranges and deposited in the foreland basin south of the mountain front. Sediments of the Siwalik Group are kilometers-thick accumulations of terrestrial deposits best exposed along a northwest–southeast belt from northern Pakistan (about 33°N latitude today) through north India and southern Nepal to the Garo Hills of Assam (about 25°N), a distance of over 2000 km (figure 14.1). In total, the Siwaliks record relatively continuous deposition throughout Neogene time. Barry et al. (chapter 15, this volume) present the stratigraphy of the Siwaliks observed on the Potwar Plateau of northern Pakistan and the tectonic setting of accumulation. Collectively, the Siwaliks of northern Pakistan are a clastic wedge that thins southward (Mascle and Hérail 1982). The Irrawaddy Formation of related origin and similar age occurs southeastward in Burma (Myanmar; see Chavasseau et al., chapter 19, this volume). Late Neogene terrestrial deposits in Afghanistan are not part of the Siwaliks, but also result from the tectonics of Indian plate collision.

Siwalik sediments are dominantly sandstones and siltstones, with lesser components of conglomerates and claystones, the latter often red in color. Continuously accumulating floodplain sediments and channel avulsions in the aggrading fluvial systems of the sub-Himalayan foreland basin set the stage for frequent burial and preservation of vertebrate remains. Rapid burial of skeletal remains and favorable post-burial pH conditions resulted in a long sequence of fossil-bearing horizons preserved in partially indurated strata. Most fossils are associated with paleo-channels and abandoned-channel environments (Badgley 1986a; Rajpar 1993; Badgley and Behrensmeyer 1995). Small mammal concentrations occur in abandoned channels and crevasse splay deposits (Badgley, Downs, and Flynn 1998).

Taphonomic analysis of Siwalik fossil assemblages of Pakistan explains the fragmentary nature of the Siwalik vertebrate record. Most vertebrate fossils were preserved in either fluvially transported assemblages of disassociated, fragmentary remains or in attritional, untransported assemblages of biological origin such as sites of repeated predation or other causes of mortality (e.g., waterholes; Badgley 1986a, 1986b; Behrensmeyer 1987). In both taphonomic contexts, vertebrate remains are typically disarticulated and fragmentary, with low levels of association. No instances of mass death followed by rapid burial have been found, except for rare beds of articulated fish remains. The fragmentary nature of the fossil record results from both biological processes—including predation, scavenging, and trampling—and fluvial processes resulting in transport, abrasion, and sorting.

That superposed assemblages of the Potwar Plateau crop out over a limited area presents both limitations and possibilities. Siwalik fossil assemblages there repeatedly sample low-elevation floodplain habitat; they represent communities living near perennial rivers, smaller-scale streams, and ponds. Therefore, the Potwar Siwaliks do not usually preserve faunal elements of the Indian Subcontinent from dry upland or estuarine environments. This is why comparative studies of faunas from distant areas of the Indian Subcontinent representing different environments will be complementary in the future.

The Siwaliks of the Potwar Plateau present a record of how lowland floodplain communities changed through millions of years. Superposed assemblages sample long-term environmental change that affected the broad sub-Himalayan alluvial plain, making the Siwalik record highly applicable to studies of vertebrate evolution. Because temporal sampling is dense, the time scale on which changes can be resolved is short (10^5 yr for the Potwar Plateau; Barry et al. 2002; Badgley et al. 2008). The scale of paleobiological questions that can be addressed through the Siwalik fossil record represents the level of Gould's (1985) "second tier" (see also Jablonski 2008), and helps to "bridge microevolution and macroevolution" (Reznick and Ricklefs 2009).

The Potwar Plateau

Pilgrim's (1913) units were formally designated formations (see Cheema, Raza, and Ahmed 1977) clustered as the Rawalpindi Group and Siwalik Group. The older Rawalpindi Group includes the Kamlial Formation and underlying Murree Formation. The Chinji, Nagri, Dhok Pathan, and (in Pakistan) Soan formations constitute the

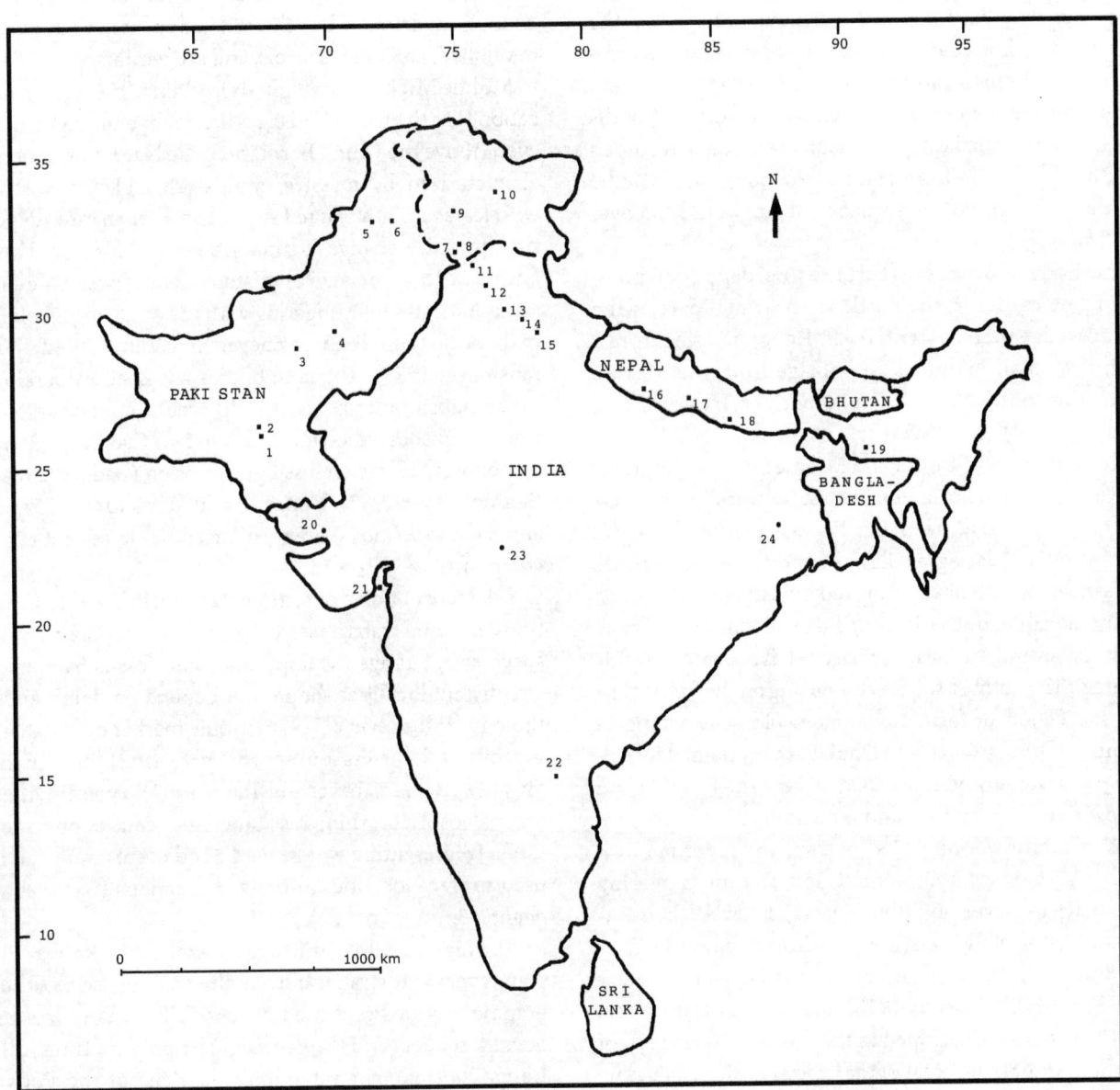

Figure 14.1 Location map for representative fossil localities of the Indian Subcontinent. Sites discussed in the text are the Sehwan (*1*) and Gaj (*2*) sections of the Manchar Formation, Bugti (*3*) and Zinda Pir Dome (*4*) of Balochistan, Kohat (*5*) and the Potwar Plateau (*6*) of the Punjab, Jammu (*7*), Ramnagar (*8*), Karewa (*9*) and Kargil (*10*) in Jammu-Kashmir, Nurpur (*11*), Haritalyangar (*12*), Pinjor (*13*), Saketi (*14*) and Kalagarh (*15*) of northern India, Dang Valley–Surai Khola (*16*), Tinau Khola (*17*) and Rato Khola (*18*) of Nepal, Garo Hills (*19*) in Assam, Kutch (*20*) and Perim Island (*21*) in western India, and the Pleistocene localities Kurnool Caves (*22*), Narmada Valley (*23*), and Bankura, West Bengal (*24*). Some locations, like Bugti (*3*), the Potwar Plateau (*6*), and the Jammu area (*7*), are general positions for multiple fossil sites, clusters of which can be recognized. Latitude and longitude indicated.

Siwalik Group. This sequence is particularly well exposed on the Potwar Plateau, a large area of about 12,000 km² in northern Pakistan, north of the Salt Range (Barry et al., chapter 15, this volume). Unlike the stratigraphy of southern Pakistan, where the Siwaliks grade downward to near-shore marine deposits, the Siwalik clastic wedge of the Potwar Plateau lies unconformably on Eocene ma-

rine limestones and represents alluvial filling of the sub-Himalayan foreland basin formed by the northward movement of the Indian sub-continent and the uplift of the Himalayas along the major boundary faults. Near and west of Islamabad, the thick sands with few vertebrate localities of the Murree Formation characterize the base of the Potwar Siwaliks. This unit is less distinctive nearer the Salt Range and is usually not differentiated there from the overlying Kamlial Formation. In total, the Siwaliks are thousands of meters thick, with some composite sections in the north estimated to be up to 6 or 7 km thick. One detailed study of a continuous section (Willis 1993; not the thickest portion of the wedge) recorded 2100 m of sediment from much of the Lower and Middle Siwaliks; the complete sequence there would be over 3000 m.

The Early Miocene onset of terrestrial deposition probably came earlier in the northern, proximal part of the clastic wedge than nearer the Salt Range, judging from a primitive small mammal assemblage from the Murree Formation of Kohat (de Bruijn, Tasser Hussain, and Leinders 1981). The Kohat area (see figure 14.1 [5]) includes a thick set of Lower and Middle Siwalik sediments above the Murree Formation. In the Potwar Plateau, sediments immediately overlying Eocene marine units are dominated by dark-gray fluvial sandstones, sometimes cross bedded, with intervening red mudstones.

The Siwaliks of the Potwar Plateau contain the type areas for several formations (Barry et al., chapter 15, this volume). The uplifted sequence abutting on the Salt Range is a succession of formations (from oldest to youngest Kamlial, Chinji, Nagri, and Dhok Pathan) named for villages separated by no more than 60 km. The Kamlial and the overlying Chinji formations are lumped in some contexts as the Lower Siwaliks (particularly older publications, e.g., Colbert 1935). The Nagri and interfingering (see Behrensmeyer and Tauxe 1982) Dhok Pathan formations informally constitute the Middle Siwaliks. Upper Siwalik rocks occur on the Potwar Plateau and adjacent Pabbi Hills (Jacobs 1978) and Mangla Dam areas, but they are best developed in the famous Saketi Park of northern India, southeast of the type area of the Pliocene Pinjor Formation. Other bodies of Upper Siwalik rock bear formation names, including the fossiliferous Tatrot Formation of the Potwar Plateau. The Tatrot Formation is local and thin (60 m) and spans a short time (a single magnetozone), and a local unconformity separates it from the subjacent Dhok Pathan Formation (Opdyke et al. 1979). More widespread in northern Pakistan is the Upper Siwalik Soan Formation, named for coarse clastic deposits along the Soan River of the Potwar, which overlies the Dhok Pathan Formation.

The successive Kamlial, Chinji, Nagri, and Dhok Pathan formations span the Miocene without significant hiatus and are distinguished by the proportion and nature of their sand content (Khan 1995). Much of the Kamlial section is sandy, but red siltstones increase toward the top of the unit. The overlying Middle Miocene Chinji For-

mation, dominantly red silts and clays, contains gray cross-bedded sandstones, but the red sediments are very well developed, contain paleosols, and are reminiscent of Early to Middle Miocene age redbeds in China. There is no indication that the Chinji red clays are aeolian in origin, however. Above the Chinji Formation, the Nagri Formation is characterized by massive, multistoried blue-gray sands with lesser amounts of red silt and smaller channel deposits. Upward, a change to more frequent buff-colored sand bodies with associated mudstones and fewer blue-gray sands indicates interfingering with the overlying Dhok Pathan Formation (Behrensmeyer and Tauxe 1982; Behrensmeyer 1987). The sand bodies and their mineralogical composition indicate two different drainage systems (mountain-sourced vs. foothill-sourced) that had differing taphonomic effects on fossil preservation (Badgley 1986b; Behrensmeyer 1987). Upper Siwalik units (Tatrot, Pinjor, and Soan formations) are recognized by significant gravel components.

Like most bodies of sedimentary rock, Siwalik formations are time transgressive. In any one section, they are successive, but age assumptions about fossils from laterally distant localities should not depend on dominant lithology (Pilbeam et al. 1996). Some marker beds extending over kilometers do approximate time lines in the Siwaliks (Behrensmeyer and Tauxe 1982). Typically, these are paleosols or thin, resistant, and continuous sheet sands (representing widespread flood events) and can be used to trace localities into master sections (Barry et al., chapter 15, this volume).

The fundamental key to age control on the Potwar Plateau is magnetostratigraphy applied to precise biostratigraphic sections and used successfully to correlate the Siwalik sequence to the geomagnetic polarity time scale. Long, continuous stratigraphic sections of the Potwar Plateau near the town of Khaur, plus sections in the Hasnot, Jalalpur, and Rohtas areas (Barry et al. 2002:fig. 1), provide the backbone for the biostratigraphy and historical geology of the area (Opdyke et al. 1979; N. M. Johnson et al. 1982; Johnson et al. 1985). The great length of the composite secures correlations to chrons C5Dr to C3An.2n, 17.9 Ma to 6.4 Ma on the Gradstein et al. (2004) time scale. The extensive lateral exposures allow multiple coeval sections, enabling secure correlations across tens of kilometers. Parallel sections recording the same magnetozones and reversal events (time lines) allow correlation of localities to a precision of up to 10^4 yr in some instances (Flynn et al. 1990). The Upper Siwaliks also have magnetostratigraphic control in addition to radiometric dates on tuffs at Kotal Kund, Pabbi Hills, Rohtas Anticline, and elsewhere (Opdyke et al. 1979; G.

D. Johnson et al. 1982), bringing the dated record to less than 1 Ma.

Other Neogene Deposits of the Indian Subcontinent

Other Neogene terrestrial deposits of the Indian Subcontinent may be considered local manifestations of the Siwaliks insofar as their origin is loosely related to the same tectonic history as the formally named units. Usually they bear other formation names (figure 14.2). For example, Pliocene age deposits of the Potwar Plateau are generally undifferentiated as "Upper Siwaliks" or referred to as Soan Formation, but elsewhere as around Mangla Dam (Mirpur and Lehri towns), Upper Siwaliks have been designated as the successive Samwal, Kakra, and Mirpur formations. Important fossils from this area have paleomagnetic control (Johnson et al. 1979) and supplement the Potwar record (e.g., Hussain et al. 1992; Steensma and Hussain 1992; Cheema, Raza, and Flynn 1997; Cheema, Flynn, and Rajpar 2003).

Northern India

The Upper Siwaliks are best exposed and highly fossiliferous along a 400-km, northwest–southeast band of outcrops in northern India and Kashmir. The type area for the Late Pliocene–Early Pleistocene Pinjor Formation is about 10 km northeast of Chandigarh (see figure 14.1 [13]). The Saketi region, 40 km to the southeast and established as a national historical park, has magnetic control (Azzaroli and Napoleone 1981; Tandon et al. 1984). In the Jammu area (see figure 14.1 [7]), the Nagrota Formation is the name usually applied to rocks of the Upper Siwaliks, although other units have been proposed. Basu (2004) related Pliocene faunal assemblages from the Nagrota (=Uttarbani) Formation to a 2.5 Ma tuff. Gupta and Prasad (2001) reported diverse Pliocene microfauna from the Nagrota Formation 10 km north of Jammu. Patnaik (chapter 17, this volume) presents the details of faunal occurrences for this band of Upper Siwaliks.

Gupta and Verma (1988) named the Mansar, Dewal, Mohargarh, Uttarbani, and Dughor formations (oldest to youngest) for an impressive package of molassic sediments 20 km southeast of Jammu, and they noted fossils of late Miocene through Pliocene ages. Recently, Early Miocene deposits north of Jammu have been shown to yield small mammals (Kumar and Kad 2002).

There are also significant high-elevation fossil resources in Kashmir. South of Srinagar (see figure 14.1 [9]), Plio-Pleistocene deposits of the intermontane Karewa Formation, with fission-track dating and a paleomagnetic record for the last 3 myr produce a late Neogene *Equus-Cervus-Canis-Elephas hysudricus* fauna (Kotlia 1990). Micromammals include arvicolines, reflecting high elevation by the late Pliocene (Kotlia and von Koenigswald 1992). Farther east and higher into the hills of Ladakh, the Kargil Formation produces mid-Tertiary fauna including anthracotheres and small mammals (Kumar, Nanda, and Tiwari 1996; see figure 14.1 [10]). A derived diatomyid rodent from the Kargil Formation (Nanda and Sahni 1998) suggests Late Oligocene age, by comparison with Early Oligocene *Fallomus* (see Marivaux and Welcomme 2003).

Middle to Late Miocene age sediments of northern India were at one time considered to represent the formations that are recognized on the Potwar Plateau. The red siltstones of the well-known Ramnagar hominoid locality (40 km east of Jammu; see figure 14.1 [8]) are reminiscent of the Chinji Formation, but they are currently referred loosely to the Lower Siwaliks. The faunal content, including *Sivapithecus*, is very much like that of type-Chinji assemblages, with some notable differences—for example, occurrence at Ramnagar of the rare hystricid *Sivacanthion*, which has not been found anywhere in the Potwar since 1922 (one Chinji area specimen, Colbert 1935). Basu (2004) and Patnaik (chapter 17, this volume) advance the biostratigraphy of Ramnagar significantly and show that it compares well with assemblages of about 13 Ma on the Potwar Plateau.

Exposures near Haritalyangar (100 km north of Chandigarh; see figure 14.1 [12]), yield rich assemblages, including relatively abundant hominoid primates. Despite lithological similarity to the Nagri and Dhok Pathan formations of the Potwar, identical ages should not be assumed, and Indian geologists (Patnaik, chapter 17, this volume) recognize local lithological units for the sequence. The most productive deposits, assigned without differentiation to the Middle Siwaliks, lie above a local sand unit (the Nahan Sandstone; Prasad 1968) that was formerly thought of as equivalent to the Nagri Formation (and probably is of about the same age). The magnetostratigraphy of Haritalyangar (Johnson et al. 1983), supplemented and reinterpreted by Pillans et al. (2005), shows that most of the fossil material derives from an interval of about 9.3 Ma to 8.1 Ma, making this the chronological equivalent of lower part of the Dhok Pathan Formation in the Potwar Plateau.

Patnaik (1994, chapter 17, this volume) reviews the biostratigraphy of Haritalyangar and Dangar and shows the potential for comparing observed temporal ranges of species in North India with those seen 500 km away on

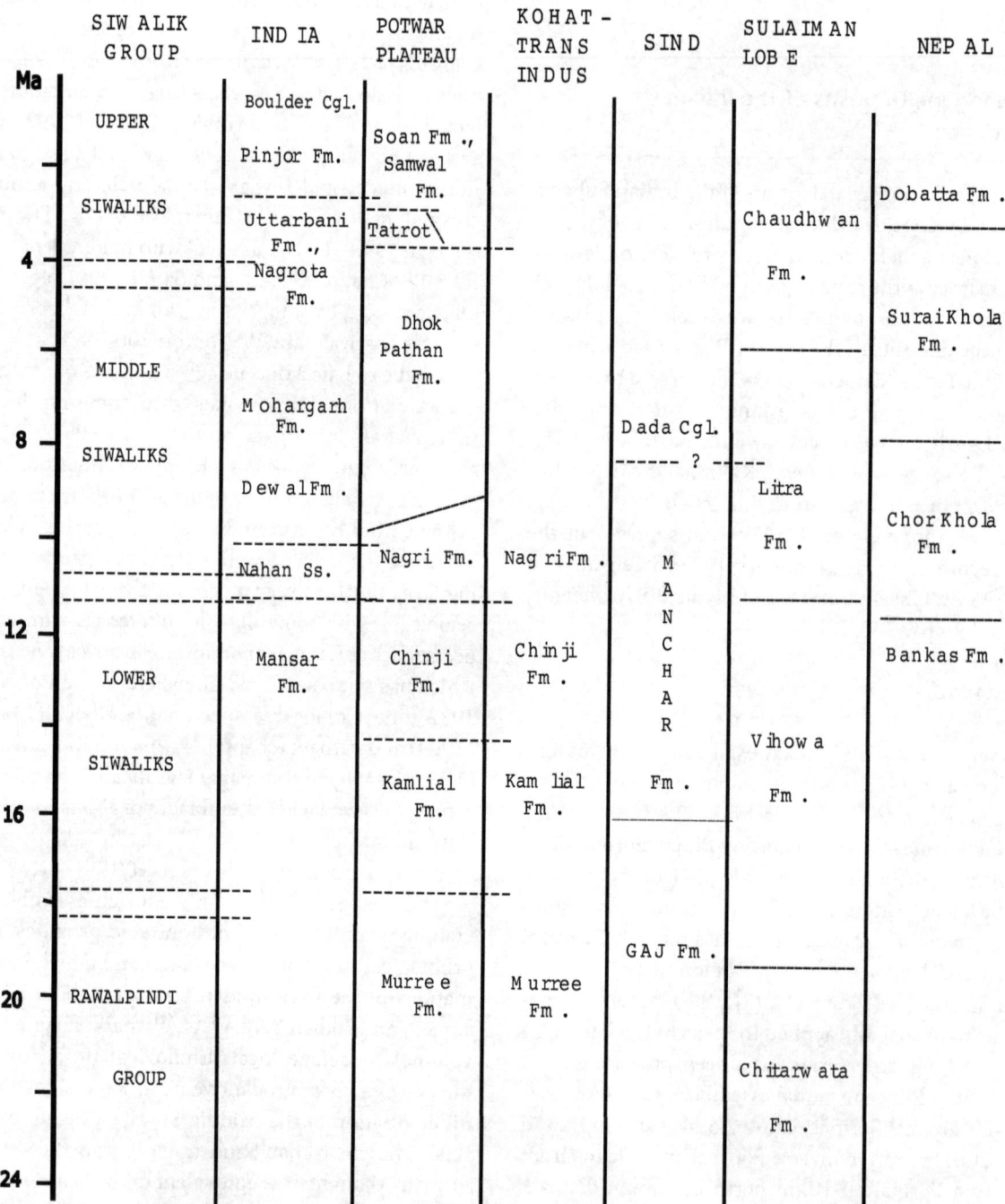

Figure 14.2 Stratigraphic correlation chart for the Indian Subcontinent. Time scale (Ma) adjusted to GTS2004. Geographic areas of the Indian Subcontinent are compared to the generalized succession in the left column. Dashed lines are meant to show that ages of stratigraphic boundaries are approximate and known to transgress time. In northern India, the Boulder Conglomerate is a local unit, and various formation names have been applied in different areas; the rock succession is a composite of units from several areas. On the Potwar Plateau, the Nagri Formation interfingers with the Dhok Pathan Formation, and the Tatrot Formation is local in distribution. In some areas, such as Sind or Kohat, the succession of units is known, but the ages are approximate, according to faunal content. In Nepal, paleomagnetic plus faunal data are the source for the age assessment.

the Potwar Plateau. Any differences will raise interesting questions. Flynn and Morgan (2005) noted that the primate *Sivaladapis* persisted much later at Haritalyangar than in the Potwar Plateau. Does this reflect limitations due to sampling bias, or real differences in paleoecology across the Indian Subcontinent? This may be answerable by ongoing work.

The vertebrate record for Late Miocene deposits of northern India is growing. Recent work near Nurpur and Kalagarh (see figure 14.1 [11 and 15]) complements the biostratigraphic information available from the vicinity of Haritalyangar. Ranga Rao (1993) provides a foundation for beginning direct comparison of the deposits at Nurpur with the Potwar Dhok Pathan Formation. The potential is great to compare local observed biostratigraphies at a high level of resolution when the fossil horizons are tied to a resolved magnetostratigraphy (Patnaik, chapter 17, this volume).

Peninsular India

The later Cenozoic fossiliferous deposits of the Indian Subcontinent are generally northern in distribution, while the fossil record of Peninsular India is poorly represented by Neogene assemblages. The obvious exception for the Miocene is the more southerly assemblage from Perim Island (Falconer 1845; see figure 14.1 [22]), which contains many of the taxa found in northern localities. A recent review of fauna from Kutch (Bhandari et al. 2009:area 21) reveals the same Cetartiodactyla species as encountered in Pakistan. Whether the vertebrate fauna of Peninsular India evolved in an ecological setting similar to that of the north remains to be examined.

Pleistocene localities are distributed broadly in India, including the famous Narmada Valley hominin localities, Bankura District sites, and Pleistocene cave deposits of Kurnool (see figure 14.1; Sahni and Mitra 1980; Patnaik, Badam, and Murty 2008). The Kurnool caves yield a fauna of modern murines, porcupines, and leporids, comparable to that found today in India.

Nepal

Corvinus and Rimal (2001) and West, Hutchinson, and Munthe (1991) developed understanding of the lithostratigraphic sequence and fossil productivity of the Siwaliks in Nepal, focusing on the exposures of Surai Khola, Tinau Khola, and Rato Khola along the southwestern margin of Nepal. The magnetic interpretation of Corvinus and Rimal (2001) places the fossiliferous sediments in the 12 Ma to 2 Ma range, with the most productive portion of the sequence being the Surai Khola and Dobatta formations, about 7.5 Ma to 2 Ma. Fossils are consistent with Potwar assemblages of similar age. Patnaik (chapter 17, this volume) reviews these and other occurrences of vertebrate fossils in Nepal. Corvinus and Rimal (2001) also present important summaries of studies on Nepalese paleofloras, and Hoorn, Ohja, and Quade (2000) use the palynological record to reconstruct a subtropical to tropical ecosystem that gave way to grasslands in the late Miocene. Pliocene conditions in Nepal appear to have involved seasonal flooding and local ponding, which may contrast with a drier, grass-dominated paleohabitat of the Siwalik Hills and northern Pakistan (Quade and Cerling 1995).

Pakistan South of the Potwar

Rocks southwest of the Potwar Plateau are Oligocene and Early Miocene to Late Miocene in age. In Sind Province, vertebrate fossils occur in the Manchar Formation, principally in two areas, Gaj and Sehwan. The long sections there include Miocene faunas comparable to Kamlial, Chinji, and Nagri assemblages (Pilgrim 1912; Raza et al. 1984; van der Made and Hussain 1992), but small mammal fossils show that the early Miocene base of the Gaj section predates the earliest assemblages near the Salt Range (de Bruijn and Hussain 1984; de Bruijn, Boon, and Hussain 1989; Wessels and de Bruijn 2001; Wessels 2009).

In and adjacent to Balochistan in the Sulaiman Lobe of western Pakistan, thick terrestrial accumulations contain Oligocene vertebrate assemblages that pass into early and late Miocene faunas. Welcomme et al. (1999, 2001) established the Oligocene age of the famous Bugti Bone Bed (see figure 14.1 [3]) and demonstrated the Miocene ages of the overlying succession of vertebrate assemblages. Antoine et al. (chapter 16, this volume) resolve the Early Miocene biostratigraphy observed at Bugti and the Dalana area to the east as it pertains to the pattern of emplacement of Siwalik faunas.

The stratigraphy and named stratigraphic units in this part of Pakistan differ from those observed to the north, so distinct stratigraphic units are recognized (Chitarwata and overlying Vihowa formations). The Oligocene to early Miocene Chitarwata Formation represents low-lying terrestrial delta deposits interfingering with lagoonal, brackish water deposits containing sharks and marine shell beds (Downing et al. 1993). The environment of deposition becomes increasingly terrestrial upward. Higher stratigraphically, thick, prominent, gray sandstones with interbedded red mudstones constitute the Vihowa Formation. Lowest Vihowa beds near Dalana are early

Miocene, predating the Kamlial Formation, based on the paleomagnetic and faunal analysis of Lindsay et al. (2005; see following discussion).

Raza et al. (2002) extended knowledge of the Miocene faunal sequence of the Dalana area with explorations through the Vihowa Formation into the overlying Litra and Chaudhwan formations, in all well over 3000 m of sediments recording terrestrial deposition. They showed that the Litra Formation produces fossils consistent with early Late Miocene age, including *Cormohipparion*, *Listriodon*, and a large giraffid perhaps representing *Bramatherium*.

CHRONOLOGY

The wealth of the Siwalik fossil record made it imperative to develop well-resolved temporal control. However, knowledge of the approximate age of the Siwaliks eluded scientists until the later part of the twentieth century. The widespread occurrence of hipparionine horses in upper levels (Nagri Formation and higher) had indicated later Tertiary time, but age control was poor, and the time span of the Siwalik sequence was underestimated. The very few and dispersed tephra of the Siwaliks offered only limited help when radiometric dating became feasible. By the 1970s, the main hope for developing a time frame for the Siwaliks lay in application of magnetostratigraphy in conjunction with biostratigraphy for correlation to the developing geomagnetic polarity time scale.

Magnetochronology

In 1973, Noye Johnson (Dartmouth College) initiated a project funded by the Smithsonian Foreign Currency Program for paleomagnetic correlation of Siwalik deposits in Pakistan with those of North America and the mid-Atlantic ridge system, the standard magnetostratigraphic reference of the time. The magnetic polarity sequence had been established for the Late Neogene (Cox, Doell, and Dalrymple 1963), but at the time it was believed that magnetic reversals older than 5 Ma could not be dated accurately because the limits of radiometric dating would not precisely constrain placement of an unknown magnetozone to a particular magnetic event of that age. Noye Johnson and his colleagues would prove this wrong by determining the magnetic sequence in Siwalik rocks of age greatly exceeding 5 Ma.

Also in 1973, the Yale University (later Harvard University) project in evolutionary biology had begun in the

Potwar Plateau, in part searching for early records of hominoid fossils of southern Asia. In addition, S. Taseer Hussain, Howard University, and colleagues from Universiteit Utrecht, Netherlands, became interested in searching for Cenozoic vertebrate fossils in terrestrial deposits. We soon saw that the geology of Pakistan provided a long terrestrial sequence with multiple fossil horizons that offered documentation and dating of a densely sampled biostratigraphy. Cooperation along lines of mutual interest proved very productive over the next decades.

Key to developing the biochronology of the Potwar Plateau Siwaliks was its dating. The magnetostratigraphic framework used here as the current standard for the GPTS is that of Gradstein, Ogg, and Smith (2004; designated GTS2004), which incorporates numerous refinements of the last 50 years. Earlier time scales were used in many older publications, yielding different ages for localities in some instances. GTS2004 is based on biologic and magnetic data from deep-sea cores, terrestrial magnetostratigraphy, radiometric dating, and astronomical tuning of cyclic phenomena recorded in sediments. The Neogene part of the time scale is thoroughly reviewed by Lourens et al. (2004), who note that one of the least secure parts of GTS2004 is the interval between the Oligocene–Miocene boundary and the Middle Miocene, about 13 Ma.

Potwar Magnetostratigraphy

N. M. Johnson et al. (1982) built their composite magnetostratigraphic framework for the Middle and Upper Siwaliks (figure 14.3) using six long sections correlated with a thick normal magnetozone (N_0) that included a volcanic tuff with fission track age of 9.50 ± 0.63 Ma (G. D. Johnson et al. 1982). Formation boundaries in relation to magnetozones were consistent in each of the sections. The Upper Siwaliks, with late Neogene fauna, were found to include a 2.53 ± 0.35 Ma tuff (G. D. Johnson et al. 1982), and a paleomagnetic reversal pattern characteristic for the Plio-Pleistocene sequence spanning chrons C2n to C3r (current chron designations). The Middle Siwaliks together encompass chrons C3An to C5n.2n. Chron C5n.2n is the long normal magnetozone (N_0). Opdyke et al. (1979) and Tauxe and Opdyke (1982) extended this work to other sections, developing fine scale temporal control over large areas around the towns of Jhelum and Khaur.

The key Siwalik Gabhir Kas/Chita Parwala magnetostratigraphic section (figure 14.4) extended the composite record downward from the Middle Siwaliks through the Chinji Formation to the base of the Kamlial Formation, where Siwalik deposits are underlain by Eocene limestone and marls. The magnetic traverse passes near

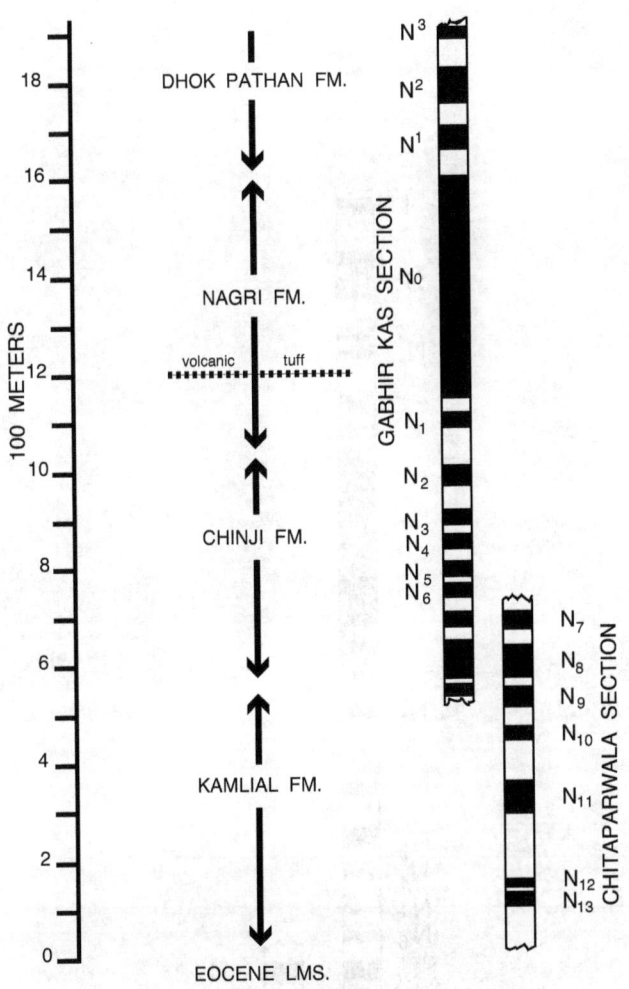

Figure 14.3 Stratigraphic sequence of the Siwaliks, with observed magnetostratigraphy, near the village of Chinji on the Potwar Plateau, Pakistan. Modified from Johnson et al. (1985).

the historical towns of Chinji and Sethi Nagri. Johnson et al. (1985) correlated the base of the Kamlial Formation in their Chita Parwala section with chron 17 of the Mankinen and Dalrymple (1979) time scale (=chron C5Dr), yielding an age for the base of the Siwalik sequence of 18.3 Ma on that time scale, 17.9 Ma on GTS2004.

Chitarwata Magnetostratigraphy

Fieldwork in the Zinda Pir Dome was initiated in 1989 along the Dalana Nala, west of Dera Ghazi Khan in Punjab Province, near Balochistan. This is in the southern part of the area studied in detail by Hemphill and Kidwai (1973) as part of cooperative mapping by the United States Geological Survey and the Geological Survey of Pakistan (GSP). The Zinda Pir Dome is a folded structure

in the eastern foothills of the NE–SW trending Sulaiman Range. The Sulaiman Range bends to the west at its southern extent, flexes to the north, and loops back to the south in a tight fold before it continues southward as the Kirthar Range. Within the loop at the southern end of the Sulaiman Range folded sediments are exposed near the town of Bugti; these are the Bugti beds of early workers beginning with Vickary (1846), and the Chitarwata Formation of Hemphill and Kidwai (1973) and Antoine et al. (chapter 16, this volume). The Chitarwata Formation is overlain by Siwalik-like fluvial sediments that wedge southward from the Himalaya Mountains and eastward from the Sulaiman Range (in ascending order, the Vihowa, Litra, and Chaudhwan formations of Hemphill and Kidwai 1973). Thick exposures of strata assigned to the Chitarwata and Vihowa formations are unconformably superposed on marine Eocene deposits near Dalana in the southern part of the Zinda Pir Dome.

Downing et al. (1993) recognized three units in the Chitarwata Formation, characterizing the lower unit (unconformable on the Eocene marine Kirthar Formation) as representing estuarine habitat, with abundant molluscan borings in highly oxidized sediments located in and above the basal contact (these diminish upward). The middle unit indicated a strandplain habitat, dominated by sands with abundant tabular and trough crossbeds, and the upper unit was characterized as a tidal flat with mostly fine-grained sediments interspersed by shallow channels. Discontinuous but thin shell beds occur throughout the upper unit, which was interpreted to represent storm beds, recording repeated shallow incursions of marine sediments onto the delta plain. The Vihowa Formation disconformably overlies the Chitarwata Formation with a thick, massive sandstone at its base, grading up into silts and thin discontinuous sands.

Friedman et al. (1992) published a paleomagnetic analysis of the Chitarwata Formation and the lower part of the Vihowa Formation near Dalana, noting difficulty in determining reliable results. Lindsay et al. (2005:fig. 4) measured and sampled two more complete sections in the Chitarwata Formation. The three complete Chitarwata sections near Dalana averaged about 420 m thick. Magnetic data from the two new Chitarwata sections were combined with the results of Friedman et al. (1992) to assemble a composite sequence (Lindsay et al. 2005). Again, correlation of the Chitarwata magnetostratigraphy to the Geomagnetic Polarity Time Scale (GPTS) (figure 14.5) is very difficult, due to suspected hiatuses and lack of reliable magnetic samples in some parts of the section (primarily in the middle of the Chitarwata Formation). Lindsay et al. (2005) interpreted a hiatus of unknown

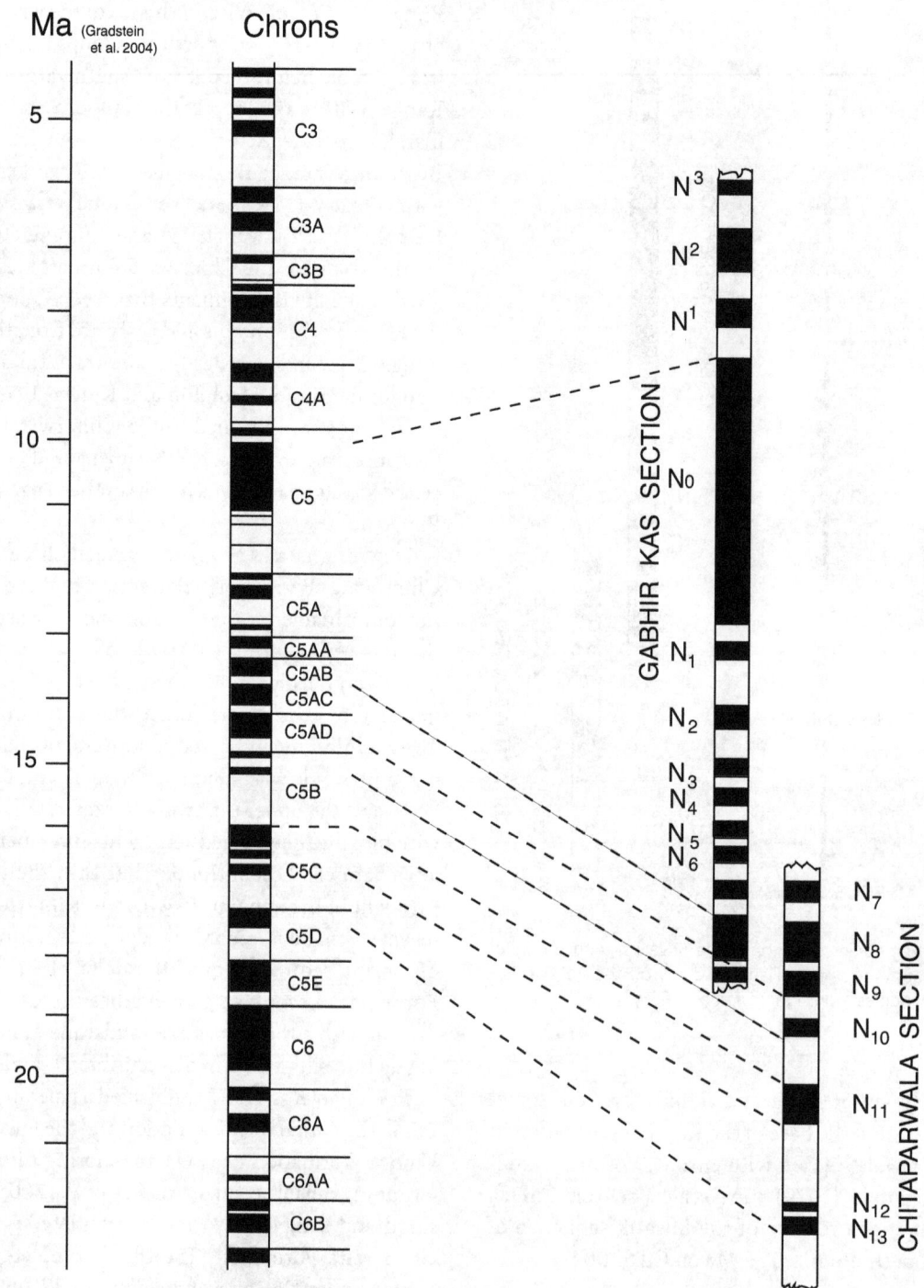

Figure 14.4 Correlation of the Gabhir Kas and Chita Parwala Kas magnetostratigraphic sequence with GTS2004.

duration (and unknown number of reversals) at the Vihowa/Chitarwata unconformity and proposed alternative correlations. Figure 14.5 is a revised correlation that minimizes the hiatus but implies others throughout the section. It is supported by new evidence from associated fossils (Antoine et al., chapter 16, this volume).

The magnetic sequence in the lower part of the Vihowa Formation (magnetozones R_1 through N_4U of Lindsay et al. 2005) in the Zinda Pir Dome matches fairly well with the Early Miocene chrons C5Cr to C6n of GTS2004. Small mammals from low in the Vihowa are consistent with Early Miocene age in that they appear to

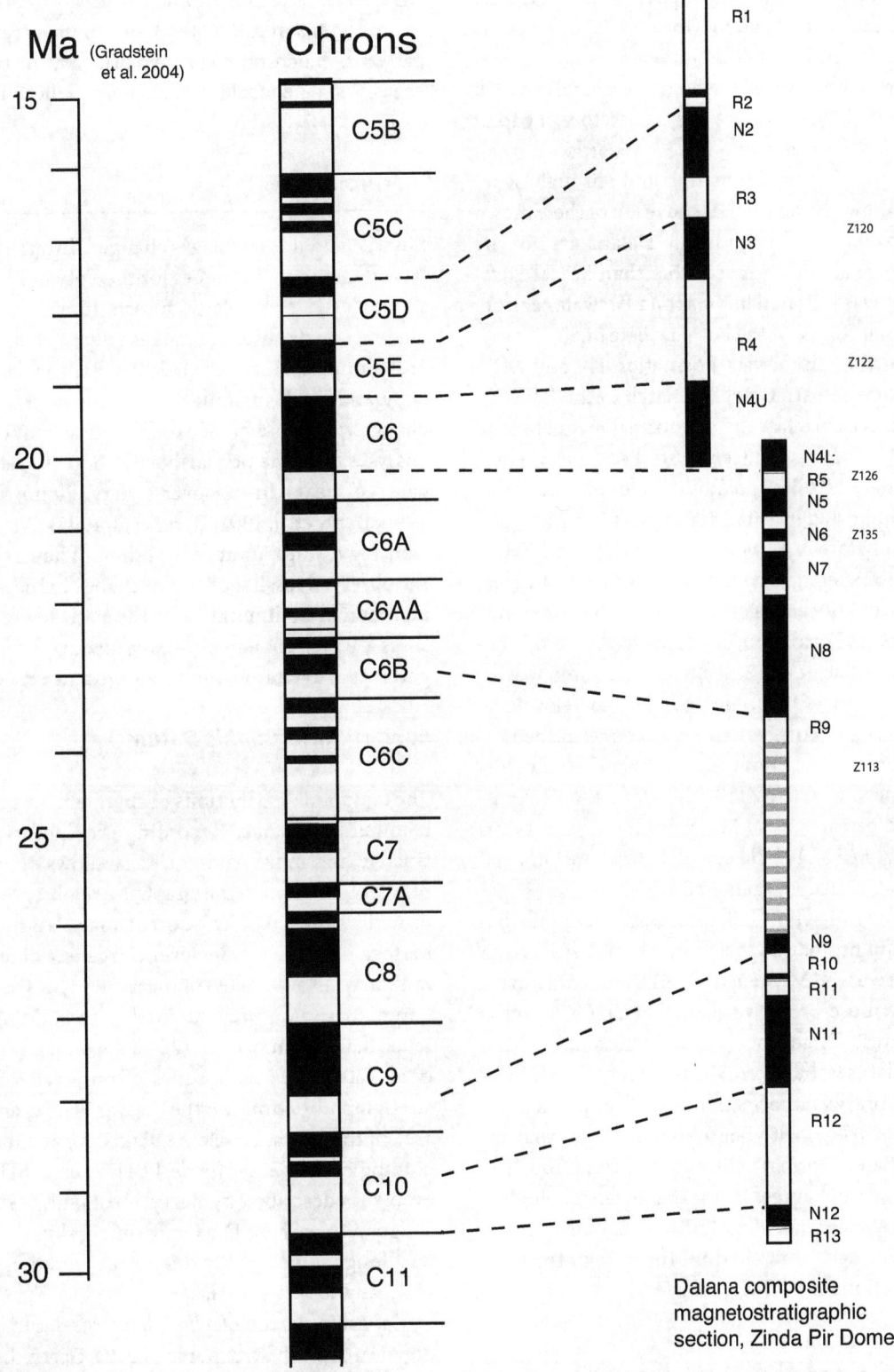

Figure 14.5 Correlation of the lower Vihowa and Chitarwata sections near Dalana, Zinda Pir Dome, with GTS2004. Revised from Lindsay et al. (2005). Key fossil sites, Z numbers, are noted on the right.

be somewhat older than the lowest (chron C5Dr, 17.8 Ma) Kamlial assemblages. Our highest small mammal site in the Vihowa Formation (locality Z120) occurs about 21m above the base of the Vihowa Formation, near the top of magnetozone N_3, which we correlate with chron C5En of GTS2004 (18.4 Ma). Therefore, a gap of 0.6 myr occurs between our lowest records of small mammals in the Kamlial Formation and our highest record in the Vihowa Formation. Deposition of the Vihowa Formation would have begun in the Dalana area of the Sulaiman Range about 1.8 myr earlier than fluvial Siwalik deposition was initiated in the Chita Parwala section, north of the Salt Range of the Potwar Plateau.

The top of the Chitarwata Formation is early Miocene, but its magnetostratigraphic match to the GPTS is controversial. We correlate the top normal magnetozone (N_4L) to part of chron C6n (see figure 14.5), but we previously considered C6Bn a possibility. Our present interpretation is supported by small mammals from Zinda Pir site Z126, just before N_4L, which indicate late Early Miocene age. Antoine et al. (chapter 16, this volume) agree, estimating early Miocene ages for levels of the upper unit of the Chitarwata Formation, based primarily on rhinocerotid fossils. Orliac et al. (2010), studying suoid fossils, project the Miocene–Oligocene boundary to below level M-3bis at Bugti and correlate this horizon with the bottom of the upper member of the Chitarwata, well below Z135 and somewhat above Z113. Therefore, the Z135 microsite should be early Early Miocene, and its short normal magnetozone (N_6) could correlate to a short normal such as chron C6AAr.1n, at about 22.2 Ma.

Z113 (although itself in indeterminate rock) falls below a significant normal magnetozone that likely is chron C6Bn, the base of the Miocene. The Z113 rodent fauna is demonstrably more primitive than that of Z135 and is reasonably latest Oligocene (approximately 23.5 Ma). Significant hiatuses below Z113 are likely. The oldest part of the Chitarwata Formation near Dalana includes site Z108 (base of N_{11}) with Bugti Member fauna that appears to correlate with Early Oligocene chron C10n. The latter interpretation agrees with the biochronological conclusion of Welcomme et al. (2001), who assigned an Early Oligocene age for fauna from the base of the Chitarwata Formation in the Bugti area.

APPLYING THE RECORD

Having developed the Siwalik and pre-Siwalik lithostratigraphy and its magnetochronology in Pakistan, the contained biotic record can be analyzed in detail. Sufficient control is established to develop the biostratigraphy on a regional basis (relating localities separated by up to tens of kilometers) and to interpret occurrence patterns, biochronology, evolutionary histories of lineages, and paleoecological changes in the fauna.

Biochronology

Siwalik faunal assemblages changed through time, and we have documented changes for the Potwar Plateau against its observed chronology. Efforts to analyze change include study of observed ranges (Flynn et al. 1995 for micromammals; Barry et al. 2002; Badgley et al. 2008 for large and small mammals). In these works, we examine change without reference to formation boundaries. Other analyses examine primarily rates of turnover and significance of peaks in turnover (Barry, Flynn, and Pilbeam 1990; Barry et al. 1991; Barry et al. 1995). We are currently working toward an understanding of how representative the observed fossil record is of change in Indian Subcontinent faunal communities of the past, focusing on measures of mammalian diversity, dietary adaptation, and patterns of evolution, immigration, and extinction.

Lowest Stratigraphic Datum

The temporal distributions of abundant taxa reveal paleocommunity change. Recording the appearances of distinctive and more common taxa allows ongoing testing of the observed biostratigraphy: Are observed ranges repeatedly supported? This sort of analysis is most precise if performed at the species level, but genera or taxa of higher rank may also be relevant, particularly if they are immigrants. Based on previous work and building on occurrence data in Barry et al. (2002:appendix 4) and Lindsay et al. (2005), we list a series of observed LSDs (lowest stratigraphic datum) for the Potwar Plateau and Sulaiman area of the Zinda Pir Dome, Pakistan, and date them according to GTS2004 (table 14.1). These LSDs are faunal events, as described by Barry and Flynn (1990); the age assigned to each LSD usually differs slightly from previous designations, partly due to recalibration of the time scale, and to new fossils.

For example, the *Elephas planifrons* datum is extended downward from that perceived by Barry, Lindsay, and Jacobs (1982), following Hussain et al. (1992) to 3.5 Ma (see table 14.1). This is also an example of a datum that coincides approximately with other prominent events—in this case, the entry of Cervidae and *Stegodon* into the Siwalik record. Table 14.1 includes key small

Table 14.1

Lowest Stratigraphic Datums (LSDs) for Selected Mammal Fossils Recorded in Neogene Sediments of the Potwar Plateau and Zinda Pir Dome

Taxon	Age Estimate	Magnetic Chron
Equus sp.	2.6 Ma	chron C2r base
Elephas planifrons	3.5 Ma	chron C2An.3n base
Family Cervidae	3.5 Ma	chron C2An.3n base
Hexaprotodon sivalensis	6.2 Ma	chron C3An.1n bottom
Family Leporidae (*Alilepus*)	7.4 Ma	chron C3Br.3r middle
Hystrix sivalensis	8.1 Ma	chron C4r.1r top
Giraffa punjabiensis	9.1 Ma	chron C4An base
Selenoportax spp.	10.5 Ma	chron C5n.2n middle
Hippotherium s.l.	10.8 Ma	chron C5n.2n lower third
Progonomys hussaini	11.7 Ma	chron C5r.3r top
Sayimys chinjiensis	12.3 Ma	chron C5An.2n middle
Family Gliridae (*Myomimus*)	13.8 Ma	chron C5ACn.1n top
Listriodon pentapotamiae	14.1 Ma	chron C5ACr top
Sayimys sivalensis	15.2 Ma	chron C5Br top
Kochalia geespei	16.1 Ma	chron C5Cn.1n top
Sayimys intermedius (Z122)*	19.0 Ma	chron C5Er lower third
Listriodon guptai (Z124)*	19.1 Ma	chron C6n top
Myocricetodontinae (Z124)*	19.1 Ma	chron C6n top
Prokanisamys arifi (Z126)*	20.1 Ma	chron C6n base
Democricetodon sp. (Z135)*	~22 Ma	?chron C6AAr.1n
Prokanisamys sp. (Z135)*	~22 Ma	?chron C6AAr.1n
Prodeinotherium sp. (Z129)*	~23 Ma	low chron C6Bn
Primus sp. (Z113)*	~23.5 Ma	?below chron C6Br
Prosayimys (Z113)*	~23.5 Ma	?below chron C6Br

*Zinda Pir Dome localities.
SOURCES: Data from Opdyke et al. (1979); Barry, Lindsay, and Jacobs (1982); Hussain et al. (1992); Baskin (1996); Barry et al. (2002); Jacobs and Flynn (2005); Lindsay et al. (2005); Orliac et al. (2010).

mammal appearances, most of which have been presented in previous publications, but ages are refined here. These include Late Miocene appearances of leporids and *Hystrix*, and the Middle Miocene appearance of dormice (Gliridae). Species in a long lineage of ctenodactylid rodents (genus *Sayimys*) are distinctive, as was suggested by de Bruijn, Boon, and Hussain (1989). Key Miocene large-mammal appearances include species of *Listriodon* and *Deinotherium*.

Table 14.1 includes Zinda Pir Dome LSD events that precede the Potwar record. Small mammal sites low in the Vihowa Formation record the appearance of the modern ctenodactyline *Sayimys intermedius* at 19 Ma. Older small mammal LSDs include *Prokanisamys arifi* at about 20 Ma, and the modern grade cricetid *Democricetodon* at Z135, about 22 Ma. We find early *Primus* and *Prosayimys* as old as 23.5 Ma at locality Z113, along with the early spalacid *Eumyarion kowalskii*. If these correlations are correct, then the earlier Oligocene Bugti fauna in the Zinda Pir area, where the diatomyid *Fallomus* is common, would date to about 28.5 Ma (locality Z108 in chron C10n). All first appearances are testable by future field studies in which new localities are placed in dated sections.

Interval Zones and Biochrons

A logical extension of the observed LSD is establishment of biochronological units, after sufficient testing shows the usefulness of key taxa as temporal indicators. Such units are meant to have more widespread time significance, beyond the basin of initial observation, and in the case of the Siwaliks, across a large part of South Asia. Barry, Lindsay, and Jacobs (1982) developed this concept in proposing "interval zones" for upper and middle Siwalik deposits, dating from about 11 Ma to 1 Ma. These were designed as contiguous biostratigraphic interval zones based at that time on mammalian stratigraphic ranges plotted on the N. M. Johnson et al. (1982) paleomagnetic section. The rationale for placing stratigraphic levels into a composite chronologic framework is discussed in Barry et al. (2002:12–14). Siwalik interval zones of Barry, Lindsay, and Jacobs (1982), defined by appearance of the name-sake taxon, include from oldest to youngest (1) the *Hipparion* s.l. interval zone, (2) the *Selenoportax lydekkeri* interval zone, (3) the *Hexaprotodon sivalensis* interval zone, and (4) the *Elephas planifrons* interval zone.

A remaining task is defining the regional extent of the Potwar-based Siwalik biochronology. Biochrons apply to biogeographic provinces and their distinctiveness outside these areas is limited. Siwalik biochrons likely apply throughout the Indian Subcontinent and eastward, with diminishing relevance, toward Myanmar, Thailand, and South China (Yunnan; see Dong and Qi, chapter 11, and Chavasseau et al., chapter 19, both this volume). This region presages the modern Oriental Biogeographic Province. Its boundary is distinct northward where it is defined by mountains. In China, the northern limit of the Oriental Biogeographic Province is less distinct and was inconstant through time, although major rivers played a role. Mountains and aridity are also westward barriers that separate the Oriental province from a distinct Eurasian biogeographic province. During the Late Miocene, this Eurasian province to the west had characteristic vertebrate assemblages, the Pikermian chronofauna (Eronen et al. 2009), which contrasted with Siwalik faunas.

Modes of Deposition and Preservation

Careful attention to microstratigraphy by Badgley (1986a), Behrensmeyer (1987), and Behrensmeyer et al. (2007) showed that vertebrate fossils are frequently associated with a subset of fluvial deposits, especially the fillings of abandoned channels. These low-lying settings adjacent to major streams were characterized by rapid but gentle burial in a protected, well-watered habitat. They preserved time-averaged (on a scale up to thousands of years) vertebrate assemblages sampled from local paleocommunities. One ongoing line of inquiry is the degree of taxonomic variation present in contemporaneous assemblages, both from similar and different sub-habitats. Assuming that the better-sampled assemblages represent local communities, how do taxon associations vary among similar age fossil sites, and is this variation correlated to depositional regime or other ecological factors (e.g., scale of the abandoned channel swale or distance from a major channel)?

Some intermittently flooded settings preserved small mammals abundantly, and these settings are highly comparable throughout the Siwalik sequence, since they repeatedly sample similar environments of deposition (Badgley, Downs, and Flynn 1998). We think they recorded community composition (and possibly relative abundances) with reasonable faithfulness. Small mammal sites are characterized by disassociated, isolated teeth plus occasional jaws and postcrania. The high frequency of isolated teeth suggests a period of subaerial exposure after concentration of small mammal remains by predators and local effects of gentle currents and pooling of water. In some cases, teeth appear to indicate origin from the same individual, but usually this cannot be demonstrated conclusively. Therefore, MNI (minimum number

of individuals) underestimates actual numbers of individuals represented in these paleofaunas.

Through paleomagnetic control, Behrensmeyer (1987) showed that Chinji through Dhok Pathan formation net accumulation rates of sediment increased from 14 to 50 cm/kyr, with correlated change in dominant depositional sources and in fossil productivity. Microstratigraphic control laterally over kilometers allowed Behrensmeyer (1987) to trace individual paleosols with great precision and show that laterally equivalent fossil concentrations had consistent relationships to the soils and were essentially contemporaneous within estimated time spans of 10^3–10^4 yr. Applying sediment accumulation rates calculated from magnetozone thickness, she estimated that durations of individual paleosols averaged no more than 24 kyr. Therefore, reworking of remains accumulated on or in these soils into contemporaneous fluvial channel deposits would represent <24 kyr of time averaging.

Habitat Change and Stable Isotopes

Combining stable isotope geochemistry with precisely controlled stratigraphy and fossil provenance, Quade and Cerling (1995) and Cerling et al. (1997) documented a major shift in carbon isotopes derived from a shift from C3 to C4 plant dominance in the late Miocene Siwaliks, which led to its detection elsewhere on a global scale. The change was initially tracked in soil carbonate nodules, from which both carbon and oxygen isotopes can be extracted and analyzed for paleoclimatic signals. The carbon shift was also recognized in the dental enamel of ungulates, which tracks the dominant vegetation type of the paleohabitat. Vegetation changed in the late Miocene from forest to grassland, and this was reflected in carbon isotopes of both soils and teeth. However, mammalian enamel shows a complex range of diets across the paleohabitat (Morgan, Kingston, and Marino 1994; Barry et al. 2002). Painstaking lateral sampling of sections across the late Miocene carbon shift in the Rohtas Anticline showed that vegetation change was slow and patchy, not sudden and monotonic (Behrensmeyer et al. 2007). Lateral variation in carbon isotopes from large herbivorous mammals sampled across the landscape at about 9.2 Ma (Morgan et al. 2009) provides evidence for an environmental gradient prior to the main shift to C4 plants. This gradient indicates lateral variation in soil moisture and, presumably, dominant plant types of the paleohabitat over tens of kilometers. The consistent association of carbon-isotope values with location on the floodplain implies strong site fidelity during the lifetimes of mammalian ungulates and hominoids (they did not disperse far during their lives).

Isotopic changes in carbonate nodules within the Siwalik paleosols intensified, especially in the interval between 6 Ma and 10 Ma, where $\delta^{13}C$ values record a significant shift in vegetation and habitat. More recent analyses, again made possible only by high-resolution spatial and stratigraphic control of fossil localities against the paleomagnetic time scale, have shown a relationship between regional vegetation change and faunal turnover. Barry et al. (2002) analyzed sustained faunal change over the million years of vegetational shift from dominantly C_3, to mixed C_3/C_4, to dominantly and then exclusively C_4. The change was found to coincide with a peak in faunal turnover. Badgley et al. (2008) took the analysis further to show that dietary preference determined the sequence of lineage extinctions during the vegetational transition. Isotopic evidence from mammalian herbivores showed that in some lineages (e.g., the anthracothere *Merycopotamus* or hipparionine equids) the predominant vegetation in the diet shifted from C_3 to C_4; other taxa such as frugivores feeding exclusively on C_3 plants (e.g., the ape *Sivapithecus* or tragulid artiodactyls) declined in abundance or disappeared.

Change at Faunal and Lineage Levels

Precise biostratigraphy with age control has made study of trends on the scale of 10^5 yr possible for the Siwaliks. Barry et al. (1985) related initial observations on faunal turnover to sea-level curves, proposing that some of the turnover pattern reflected immigration at low sea-level stands. Later studies (Barry, Flynn, and Pilbeam 1990; Barry et al. 1991; Barry et al. 1995; Barry et al. 2002) refined the scale of analysis to distinguish trends in species richness and pulses of appearances (mostly immigration) and extinction on the Potwar Plateau. Peaks in appearances and extinction were separated by longer intervals of low rates of change and did not always coincide. In some cases, elevated rates of change appeared to coincide with global climatic events, but their timing was not always clearly linked to events as then recognized. Given improved understanding of global changes in the marine record (Zachos et al. 2001), we are now reexamining possible correlations, particularly the effects of dramatic cooling on terrestrial faunas at about 14 Ma.

Siwalik data have the potential of resolving macroevolutionary issues in paleobiology. For this subtropical Miocene setting, species properties intrinsic to taxa distinguished by body size may be emerging. Flynn et al. (1995) evaluated lineage duration (residence time) in relation to environmental change and showed that small mammals had shorter species durations than did large mammals in the Siwalik data set. Morgan et al. (1995) examined size increase across taxa, large and small, during the later Miocene, concluding that changes were climatically driven and likely reflected within-group competition.

Flynn et al. (1998) expanded the analysis of trends in species richness in small mammal assemblages throughout the Siwaliks and found evidence for high diversity in the middle Miocene, higher than expected in modern faunas at the same latitude. Faunal change after 11 Ma included decline in species richness over time and suggested that late Miocene species generally had shorter temporal ranges than middle Miocene species (Flynn et al. 1995). The biotic signal in the fossil record, therefore, includes evidence that diversity patterns in deep time differed from those observed today and that rates of species turnover during the Cenozoic (typical longevities observed in lineages) were not constant.

Our agenda of research is targeted at deriving the highest level of precision achievable from the Siwalik biotic record. This level of refinement bridges the gulf between macroevolutionary patterns and microevolutionary processes (cf. Reznick and Ricklefs 2009). We cannot capture actual records of microevolution of species transforming into one or more descendants (except possibly for extraordinary brief stratigraphic intervals), but we can garner data relevant to the microevolutionary process by observing change in intraspecific character variation (e.g., dental variation) in successive populations.

Fine-scale biostratigraphy and dense fossil sampling provide the raw materials for phylogenetic analysis and for interpretation of the tempo and mode of evolutionary processes that are manifested in phenotypic change. Stasis in tooth size and morphology dominated the histories of most lineages of small mammals—for example, species of the primitive bamboo rat genus *Kanisamys*, in which crown height changed in punctuated events after long periods of stability (Flynn 1986; Flynn et al. 1995). More gradual change is evident in other rodents (Flynn 1985).

The Siwaliks fortuitously preserve fossils that show the pattern and timing of the origin and initial diversification of murine and other rodent groups. Multiple superposed small mammal assemblages in the Chinji Formation capture the mode of acquisition of the murine synapomorphy (a third row of cusps on upper molars), and younger sites show diversification with this novelty.

Jacobs and Flynn (2005) extended previous work (Jacobs and Downs 1994) to constrain the timing of murine evolution. This fossil record not only sets minima for dates of genetic splitting but also indicates maxima beyond which certain splits are unlikely. For the *Mus–Rattus* split, for example, an age greater than 12 Ma is highly unlikely because the only ancestor of that age evident in the fossil record from anywhere in the world is *Antemus*, clearly a primitive murine. This age is consistent with the molecular estimate of Steppan, Adkins, and Anderson (2004); a younger age for this split (10–11 Ma) is quite conceivable from the viewpoint of the fossil record.

CONCLUSION

The Siwalik Group is a thick and laterally expansive wedge of sub-Himalayan nonmarine clastic deposits best developed in the northern Indian Subcontinent, extending southeastward through Nepal and into Assam. The Siwaliks span the Neogene, but the most fossiliferous parts are late Early Miocene to Pliocene in age. Temporal equivalents and older deposits occur southward in Pakistan toward distal parts of the clastic wedge. Pliocene deposits are best developed in northern India. Biostratigraphies and age control are refined to the scale of 10^5 yr, thanks to a relatively complete sedimentary record conducive to paleomagnetic analysis.

The Siwaliks are not only a long, continuous sequence, but they are characterized by abundant, superimposed fossiliferous strata that invite analyses of fossil productivity correlated to depositional setting and lithology. Most productive fossil sites of similar age in the Siwaliks yield comparable assemblages because they derive from generally similar depositional regimes: floodplain settings often affiliated with crevasse splay deposits and fine-grained filling of abandoned channels.

Because the Siwaliks record successive faunal assemblages, the sequence invites biochronological analysis. We use interval zones to demarcate times characterized by key taxa. The Siwalik biogeographic province is South Asian in scale, from Pakistan and India on the west, to Yunnan in the east, where the biogeographic identity weakens. This province, a predecessor of the Oriental Biogeographic Province, is sharply bounded by mountains and high elevations to the north, and mountainous, dry habitat to the west.

The rich scale of fossil representation in the Siwaliks, its temporal duration, and its high density of fossil levels in many intervals, make the sequence an ideal laboratory for examining the evolution of terrestrial vertebrates. Data are relevant for both macroevolutionary phenomena and for finer scale issues, those that form a bridge to microevolution. Observations at the macroevolutionary level concern faunal composition, species richness, longevity patterns, turnover, and size trends. Data reflecting microevolutionary processes include changing frequencies of characters within successive samples of single lineages, which approximate transformation of the phenotype. Given tight temporal control, the Siwalik sequence is relevant for constraining various aspects of the evolutionary time scale, specifically estimating times of lineage splitting.

ACKNOWLEDGMENTS

Our long-running field research program in Pakistan would not have been possible without the cooperation and collaboration of our institutions with the Geological Survey of Pakistan and the Pakistan Museum of Natural History. We appreciate the efforts of all support personnel, and of our many valued colleagues, some of whom are departed. Financial support has derived from the U.S. National Science Foundation, from the National Geographic Society, and from the Smithsonian Foreign Currency Program, with special thanks to Francine Berkowitz of the Smithsonian for enduring support of our research in Pakistan.

REFERENCES

Azzarolli, A. and G. Napoleone. 1981. Magnetostratigraphic investigation of the Upper Sivaliks near Pinjor, India. *Rivista Italiana di Paleontoligia e Stratigrafia* 87(4):739–762.

Badgley, C. 1986a. Taphonomy of mammalian fossil remains from Siwalik rocks of Pakistan. *Paleobiology* 12:119–142.

Badgley, C. 1986b. Counting individuals in mammalian fossil assemblages from fluvial environments. *Palaios* 1:328–338.

Badgley, C., J. C. Barry, M. E. Morgan, S. V. Nelson, A. K. Behrensmeyer, T. E. Cerling, and D. Pilbeam. 2008. Ecological changes in Miocene mammalian record show impact of prolonged climatic forcing. *Proceedings of the National Academy of Sciences* 105(34):12145–12149.

Badgley, C. and A. K. Behrensmeyer. 1995. Preservational, paleoecological, and evolutionary patterns in the Paleogene of Wyoming—Montana and the Neogene of Pakistan. *Palaeogeography, Palaeoclimatology, Palaeoecology* 115:319–340.

Badgley, C., W. Downs, and L. J. Flynn. 1998. Taphonomy of small-mammal fossil assemblages from the middle Miocene Chinji Formation, Siwalik Group, Pakistan. In *Advances in Vertebrate Paleontology and Geochronology*, ed. Y. Tomida, L. J. Flynn, L. L. Jacobs, pp. 145–166. National Science Museum Monograph 14. Tokyo: National Science Museum.

Barry, J. C. and L. J. Flynn. 1990. Key biostratigraphic events in the Siwalik Sequence. In *European Neogene Mammal Chronology*, ed. E. H. Lindsay, V. Fahlbusch, and P. Mein, 557–571. New York: Plenum Press.

Barry, J. C., L. J. Flynn, and D. Pilbeam. 1990. Faunal diversity and turnover in a Miocene terrestrial sequence. In *Causes of Evolution: A Paleontological Perspective*, ed. R. M. Ross and W. D. Allmon, pp. 381–421. Chicago: University of Chicago Press.

Barry, J. C., N. M. Johnson, S. M. Raza, and L. L. Jacobs. 1985. Neogene mammalian faunal change in southern Asia: Correlations with climatic, tectonic and eustatic events. *Geology* 13:637–640.

Barry, J. C., E. H. Lindsay, and L. L. Jacobs. 1982. A biostratigraphic zonation of the Middle and Upper Siwaliks of the Potwar Plateau of northern Pakistan. *Palaeogeography, Palaeoclimatology, Palaeoecology* 37:95–130.

Barry, J. C., M. E. Morgan, L. J. Flynn, D. Pilbeam, A. K. Behrensmeyer, S. Mahmood Raza, A. Imran Khan, C. Badgley, J. Hicks, and J. Kelley. 2002. Faunal and environmental change in the late Miocene Siwaliks of northern Pakistan. *Paleobiology* Memoir 3:1–71.

Barry, J. C., M. E. Morgan, L. J. Flynn, D. Pilbeam, L. L. Jacobs, E. H. Lindsay, S. Mahmood Raza, and N. Solounias. 1995. Patterns of faunal turnover and diversity in the Neogene Siwaliks of northern Pakistan. *Palaeogeography, Palaeoclimatology, Palaeoecology* 115:209–226.

Barry, J. C., M. E. Morgan, A. J. Winkler, L. J. Flynn, E. H. Lindsay, L. L. Jacobs, and D. Pilbeam. 1991. Faunal interchange and Miocene terrestrial vertebrates of southern Asia. *Paleobiology* 17(3):231–245.

Baskin, J. A. 1996. Systematic revision of Ctenodactylidae (Mammalia, Rodentia) from the Miocene of Pakistan. *Palaeovertebrata* 25:1–49.

Basu, P. K., 2004. Siwalik mammals of the Jammu Sub-Himalaya, India: An appraisal of their diversity and habitats. *Quaternary International* 117:105–118.

Behrensmeyer, A. K. 1987. Miocene fluvial facies and vertebrate taphonomy in northern Pakistan. *Society of Economic Paleontologists and Mineralogists, Special Publication* 39:169–176.

Behrensmeyer, A. K., J. Quade, T. E. Cerling, J. Kappelman, I. A. Khan, P. Copeland, L. Roe, J. Hicks, P. Stubblefield, B. J. Willis, and C. Latorre. 2007. The structure and rate of late Miocene expansion of C4 plants: Evidence from lateral variation in stable isotopes in paleosols of the Siwalik Group, northern Pakistan. *Geological Society of America Bulletin* 119:1486–1505.

Behrensmeyer, A. K. and L. Tauxe 1982. Isochronous fluvial systems in Miocene deposits of northern Pakistan. *Sedimentology* 29:331–352.

Bhandari, A., D. M. Mohabey, S. Bajpai, B. N. Tiwari, and M. Pickford. 2009. Early Miocene mammals from central Kutch (Gujarat), Western India: Implications for geochronology, biogeography, eustacy and intercontinental dispersals. *Neues Jahrbuch für Geologie und Paläontologie Abhandlungen* 256(1):69–97.

Cautley, P. T. and H. Falconer, 1835. Synopsis of fossil genera and species from the upper deposits of the Tertiary strata of the Sivá-lik Hills, in the collection of the authors. *Journal of the Asiatic Society* 4:706–707.

Cerling, T. E., J. M Harris, B. J. MacFadden, M. G. Leakey, J. Quade, V. Eisenmann, and J. R. Ehleringer. 1997. Global vegetation change through the Miocene/Pliocene boundary. *Nature* 389:153–158.

Cheema, I. U., L. J. Flynn, and A. R. Rajpar. 2003. Late Pliocene murid rodents from Lehri, Jhelum District, northern Pakistan. In *Advances in Vertebrate Paleontology "Hen to Panta": A Tribute to Constantin Rădulescu and Petre Mihai Samson*, ed. A. Petculescu and E. Ştiudcă, pp. 85–92. Bucharest: Roumanan Academy Emil Racovită Institute of Speleology.

Cheema, I. U., S. Mahmood Raza, and L. J. Flynn. 1997. Note on Pliocene small mammals from the Mirpur District, Azad Kashmir, Pakistan. *Geobios* 30:115–119.

Cheema, M. R., S. M. Raza, and H. Ahmed. 1977. Cainozoic. In *Stratigraphy of Pakistan*, ed. S. M. I. Shah. *Memoirs of the Geological Survey of Pakistan* 12:56–98.

Colbert, E. H. 1935. Siwalik mammals in the American Museum of Natural History. *Transactions of the American Philosophical Society* 27:1–401.

Corvinus, G. and L. N. Rimal. 2001. Biostratigraphy and geology of the Neogene Siwalik Group of the Surai Khola and Rato Khola areas in Nepal. *Palaeogeography, Palaeoclimatology, Palaeoecology* 165:251–279.

Cox, A., R. R. Doell, and G. B. Dalrymple. 1963. Geomagnetic polarity epochs and Pleistocene geochronometry. *Nature* 198 (4885):1049–1051.

de Bruijn, H., E. Boon, and S. T. Hussain. 1989. Evolutionary trends in *Sayimys* (Ctenodactylidae, Rodentia) from the Lower Manchar Formation (Sind, Pakistan). *Proceedings of the Koninklijke Nederlandse Akademie van Wetenschappen* B92(3):191–214.

de Bruijn, H. and S. Taseer Hussain. 1984. The succession of rodent faunas from the lower Manchar Formation, southern Pakistan, and its relevance for the biostratigraphy of the Mediterranean Miocene. Montpellier: *Paléobiologie Continentale* 14(2):191–204.

de Bruijn, H., S. Taseer Hussain, and J. J. M. Leinders. 1981. Fossil rodents from the Murree Formation near Banda Daud Shah, Kohat, Pakistan. *Proceedings of the Koninklijke Nederlandse Akademie van Wetenschappen* B84(1):71–99.

Downing, K. F., E. H. Lindsay, W. R. Downs, and S. E. Speyer. 1993. Lithostratigraphy and vertebrate biostratigraphy of the Early Miocene Himalayan Foreland, Zinda Pir Dome, Pakistan. *Sedimentary Geology* 87:25–37.

Eronen, J. T., M. Mirzaie Ataabadi, A. Micheels, A. Karme, R. L. Bernor, and M. Fortelius. 2009. Distribution history and climatic controls of the Late Miocene Pikermian chronofauna. *Proceedings of the National Academy of Sciences* 106(29):11867–11871.

Falconer, H. 1845. Description of some fossil remains of *Dinotherium*, giraffe, and other Mammalia, from the Gulf of Cambay, western coast of India, chiefly from the collection presented by Captain Fulljames of the Bombay Engineers. *Quarterly Journal of the Geological Society of London* 1:356–372

Flynn, L. J. 1985. Evolutionary patterns and rates in Siwalik Rhizomyidae. In *Acta Zoologica Fennica* 170:140–144.

Flynn, L. J. 1986. Species longevity, stasis, and stairsteps in rhizomyid rodents. In *Vertebrates Phylogeny, and Philosophy: Contributions to Geology*, ed. K. M. Flanagan and J. A. Lillegraven, pp. 273–285. Special Paper 3. Laramie: University of Wyoming.

Flynn, L. J., J. C. Barry, M. E. Morgan, D. Pilbeam, L. L. Jacobs, and E. H. Lindsay. 1995. Neogene Siwalik mammalian lineages: Species longevities, rates of change, and modes of speciation. *Palaeogeography, Palaeoclimatology, Palaeoecology* 115:249–264.

Flynn, L. J., W. Downs, M. E. Morgan, J. C. Barry, and D. Pilbeam. 1998. High Miocene species richness in the Siwaliks of Pakistan. In *Advances in Vertebrate Paleontology and Geochronology*, ed. Y. Tomida, L. J. Flynn, and L. L. Jacobs, pp. 167–180. National Science Museum Monograph 14. Tokyo: National Science Museum.

Flynn, L. J., and M. E. Morgan. 2005. New lower primates from the Miocene Siwaliks of Pakistan. In *Interpreting the Past: Essays on Human, Primate, and Mammal Evolution in Honor of David Pilbeam*, ed. D. E. Lieberman, R.J. Smith, and J. Kelley, pp. 81–101. American School of Prehistoric Research Monograph 5. Boston: Brill.

Flynn, L. J., D. Pilbeam, L. L. Jacobs, J. C. Barry, A. K. Behrensmeyer, and J. W. Kappelman. 1990. The Siwaliks of Pakistan: Time and faunas in a Miocene terrestrial setting. *Journal of Geology* 98:589–604.

Friedman, R., J. Gee, L. Tauxe, K. Downing, and E. Lindsay. 1992. The magnetostratigraphy of the Chitarwata and lower Vihowa Formations of the Dera Ghazi Khan area, Pakistan. *Sedimentary Geology* 81:253–268.

Gould, S. J. 1985. The paradox of the first tier: An agenda for paleobiology. *Paleobiology* 11:2–12.

Gradstein, F. M., J. G. Ogg, and A. G. Smith. 2004. *A Geologic Time Scale 2004.* Cambridge: Cambridge University Press.

Gupta, S. S. and G. V. R. Prasad. 2001. Micromammals from the Upper Siwalik subgroup of the Jammu region, Jammu and Kashmir State, India: Some constraints on age. *Neues Jahrbuch für Geologie und Paläontologie Abhandlungen* 220(2):153–187.

Gupta, S. S. and B. C. Verma. 1988. Stratigraphy and vertebrate fauna of the Siwalik Group, Mansar-Uttarbaini section, Jammu District, J&K. *Journal of the Palaeontological Society of India* 33:117–124.

Hemphill, W. R. and A. H. Kidwai. 1973. Stratigraphy of the Bannu and Dera Ismail Khan areas, Pakistan. *U.S. Geological Survey Professional Paper* 716B:1–36.

Hoorn, C., T. Ohja, and J. Quade. 2000. Palynological evidence for vegetation development and climatic change in the Sub-Himalayan Zone (Neogene, Central Nepal). *Palaeogeography, Palaeoclimatology, Palaeoecology* 163:133–161.

Hussain, S. T. 1971. Revision of *Hipparion* (Equidae, Mammalia) from the Siwalik Hills of Pakistan and India. *Abhandlungen Bayerische Akademie der Wissenschaften, Mathematisch-Naturwissenschaftliche Klasse (Neue Folge)* 147:1–68.

Hussain, S. T., G. D. van den Bergh, K. J. Steensma, J. A. de Visser, J. de Vos, M. Arif, J. van Dam, P. Y. Sondaar, and S. B. Malik. 1992. Biostratigraphy of the Plio-Pleistocene continental sediments (Upper Siwaliks) of the Mangla-Samwal Anticline, Asad Kashmir, Pakistan. *Proceedings of the Koninklijke Nederlandse Akademie van Wetenschappen* 95:65–80.

Jablonski, D. 2008. Biotic interactions and macroevolution: Extensions and mismatches across scales and levels. *Evolution* 62:715–739.

Jacobs, L. L. 1978. Fossil rodents (Rhizomyidae and Muridae) from Neogene Siwalik deposits, Pakistan. *Museum of Northern Arizona Press Bulletin Series* 52:1–103.

Jacobs, L. L., and W. Downs. 1994. The Evolution of murine rodents in Asia. In *Rodent and Lagomorph Families of Asian Origins and Diversification*, ed. Y. Tomida, C. Li, and T. Setoguchi, pp. 149–156. National Science Museum Monograph 8. Tokyo: National Science Museum.

Jacobs, L. L., and L. J. Flynn. 2005. Of mice . . . again: The Siwalik rodent record, murine distribution, and molecular clocks. In *Interpreting the Past: Essays on Human, Primate, and Mammal Evolution in Honor of David Pilbeam*, ed. D. E. Lieberman, R. J. Smith, and J. Kelley, pp. 63–80. American School of Prehistoric Research Monograph 5. Boston: Brill.

Johnson, G. D., N. M. Johnson, N. D. Opdyke, and R. A. K. Tahirkheli. 1979. Magnetic reversal stratigraphy and sedimentary tectonic history of the Upper Siwalik group, eastern Salt Range and southwestern Kashmir. In *Geodynamics of Pakistan*, ed. A. Farah and K. A. de Jong, pp. 149–165. Memoirs of the Geological Survey of Pakistan 11. Islamabad: Geological Survey of Pakistan.

Johnson, G. D., N. D. Opdyke, S. K. Tandon, and A. C. Nanda. 1983. The magnetic polarity stratigraphy of the Siwalik Group at Haritalyangar (India) and a new last appearance datum for *Ramapithecus* and *Sivapithecus* in Asia. *Palaeogeography, Palaeoclimatology, Palaeoecology* 44:223–249.

Johnson, G. D., P. Zeitler, C. W. Naeser, N. M. Johnson, D. M. Summers, C. D. Frost, N. D. Opdyke, and R. A. K. Tahirkheli. 1982. The occurrence and fission-track ages of late Neogene and Quaternary volcanic sediments, Siwalik Group, Northern Pakistan. *Palaeogeography, Palaeoclimatology, Palaeoecology* 37:63–93.

Johnson, N. M., N. D. Opdyke, G. D. Johnson, E. H. Lindsay, and R. A. K. Tahirkheli. 1982. Magnetic polarity stratigraphy and ages of Siwalik Group rocks of the Potwar Plateau, Pakistan. *Palaeogeography, Palaeoclimatology, Palaeoecology* 37:17–42.

Johnson, N. M., J. Stix, L. Tauxe, P. F. Cerveny, and R. A. K. Tahirkheli. 1985. Paleomagnetic chronology, fluvial processes, and tectonic implications of the Siwalik deposits near Chinji Village, Pakistan. *Journal of Geology* 93:27–40.

Khan, I. A. 1995. Complexity in stratigraphic division of fluviatile successions. *Pakistan Journal of Mineral Sciences* 7:23–34.

Kotlia, B. S. 1990. Large mammals from the Plio-Pleistocene of Kashmir Intermontane Basin, India, with reference to their status in magnetic polarity time scale. *Eiszeitalter und Gegenwart* 40:38–52.

Kotlia, B. S. and W. von Koenigswald. 1992. Plio-Pleistocene arvicolids (Rodentia, Mammalia) from Kashmir Intermontane Basin, Northwestern India. *Palaeontolgraphica* A223:103–135.

Kumar, K. and S. Kad. 2002. Early Miocene cricetid rodent (Mammalia) from the Murree Group of Kalakot, Rajauri District, Jammu and Kashmir, India. *Current Science* 82(6):736–740.

Kumar, K., A. C. Nanda, and B. N. Tiwari. 1996. Rodents from the Oligo-Miocene Kargil Formation, Ladakh, India: Biochronologic and paleobiogeographic implications. *Neues Jahrbuch für Geologie und Paläontologie Abhandlungen* 202(3):383–407.

Lewis, G. E. 1937. A new Siwalik correlation (India). *American Journal of Science*, 5th ser., 33:191–204.

Lindsay, E. H., L. J. Flynn, I. U. Cheema, J. C. Barry, K. F. Downing, A. R. Rajpar, and S. Mahmood Raza. 2005. Will Downs and the Zinda Pir Dome. *Palaeontologia Electronica* 8(1)19A:1–19, 1MB; http:palaeo-electronica.org/toc.htm.

Lourens, L., F. Hilgen, N. J. Shackleton, J. Laskar, and D. Wilson. 2004. The Neogene Period. In *A Geologic Time Scale*, ed. F. Gradstein, J. Ogg, and A. Smith, pp. 409–440. Cambridge: Cambridge University Press.

Mankinen, E. A. and G. D. Dalrymple. 1979. Revised geomagnetic polarity time scale for the interval 0–5 m.y. B.P. *Journal of Geological Research* 84:615–626.

Marivaux, L. and J.-L. Welcomme. 2003. New diatomyid and bal-uchimyine rodents from the Oligocene of Pakistan (Bugti Hills, Balochistan): Systematic and paleobiogeographic implications. *Journal of Vertebrate Paleontology* 23(2):420–434.

Mascle, G. and G. Hérail. 1982. Les Siwaliks: le prisme d'accrétion tectonique associé à la seduction intracontinentale himalayenne. *Géologie Alpine* 58:95–103.

Medlicott, H. B. 1864. On the geological structure and relations of the southern portions of the Himalayan ranges between the Rivers Ganges and Ravee. *Memoirs of the Geological Survey of India* 3(4):1–206.

Morgan, M. E., C. Badgley, G. F. Gunnell, P. D. Gingerich, J. W. Kappelman, and M. C. Maas. 1995. Comparative paleoecology of Paleogene and Neogene mammalian faunas: Body-size structure. *Palaeogeography, Palaeoclimatology, Palaeoecology* 115:287–317.

Morgan, M. E., A. K. Behrensmeyer, C. Badgley, J. C. Barry, S. Nelson, and D. Pilbeam. 2009. Lateral trends in carbon isotope ratios reveal a Miocene vegetation gradient in the Siwaliks of Pakistan. *Geology* 37(2):103–106.

Morgan, M. E., J. D. Kingston, and B. D. Marino. 1994. Carbon isotopic evidence for the emergence of C4 plants in the Neogene from Pakistan and Kenya. *Nature* 367:162–165.

Nanda, A. C. and A. Sahni. 1998. Ctenodactyloid rodent assemblage from Kargil Formation, Ladakh molasse group: Age and palaeobiogeographic implications for the Indian Subcontinent in the Oligo-Miocene. *Geobios* 31(4):533–544.

Opdyke, N. D., E. Lindsay, G. D. Johnson, N. Johnson, R. A. K. Tahirkheli, and M. A. Mirza. 1979. Magnetic polarity stratigraphy and vertebrate paleontology of the Upper Siwalik Subgroup of Northern Pakisatan. *Palaeogeography, Palaeoclimatology, Palaeoecology* 27:1–34.

Orliac, M. J, P.-O. Antoine, G. Roohi, and J.-L. Welcomme. 2010. Suoidea (Mammalia, Cetartiodactyla) from the Early Oligocene of the Bugti Hills, Balochistan, Pakistan. *Journal of Vertebrate Paleontology* 30(4):1300–1305.

Patnaik, R. 1994. Find of a *Ramapithecus punjabicus* premolar from a palaeosol unit exposed near Village Dangar, Dist. Bilaspur, HP. *Current Science* 67(7):538–539.

Patnaik, R., G. L. Badam, and M. L. K. Murty. 2008. Additional vertebrate remains from one of the Late Pleistocene-Holocene Kurnool Caves (Muchchatla Chintamanu Gavi) of South India. *Quaternary International* 192:43–51.

Pilbeam, D., M. Morgan, J. C. Barry, and L. J. Flynn. 1996. European MN units and the Siwalik faunal sequence in Pakistan. In *The Evolution of Western Eurasian Neogene Mammal Faunas*, ed. R. L. Bernor, V. Fahlbusch, and H.-W. Mittman, pp. 96–105. New York: Columbia University Press.

Pilgrim, G. E. 1912. The vertebrate fauna of the Gaj Series in the Bugti Hills and the Punjab. *Geological Survey of India Memoirs, Palaeontologia Indica*, n. s., 4(2):1–83.

Pilgrim, G.E. 1913. The correlation of the Siwaliks with mammal horizons of Europe. *Records of the Geological Survey of India* 43:264–326.

Pilgrim, G. E. 1926. The fossil Suidae of India. *Memoirs of the Geological Survey of India*, n. s., 8:1–65.

Pilgrim, G. E. 1934. Correlation of the ossiferous sections in the Upper Cenozoic of India. *American Museum of Natural History Novitates* 704:1–5.

Pillans, B., M. Williams, C. Cameron, R. Patnaik, J. Hogarth, A. Sahni, J. C. Sharma, F. Williams, and R. L. Bernor. 2005. Revised correlation of the Haritalyangar magnetostratigraphy, Indian Siwaliks: Implications for the age of the Miocene hominids *Indopithecus* and *Sivapithecus*, with a note on a new hominid tooth. *Journal of Human Evolution* 48:507–515.

Prasad, K. N. 1968. The vertebrate fauna from the Siwalik beds of Haritalyangar, Himachal Pradesh. *Palaeontologia Indica*, n.s., 39:1–55.

Quade, J. and T. E. Cerling. 1995. Expansion of C₄ grasses in the Late Miocene of northern Pakistan: Evidence from stable isotopes in paleosols. In *Long Records of Continental Ecosystems*, ed. C. Badgley and A. K. Behrensmeyer, pp. 91–116. *Palaeogeography, Palaeoclimatology, Palaeoecology* 115.

Rajpar, A. R. 1993. Taphonomy of a Pliocene quarry in the Upper Siwalik, near Mirpur, Azad Jammu Kashmir. *Pakistan Journal of Zoology* 25(3):243–247.

Ranga Rao, A. 1993. Magnetic-polarity stratigraphy of Upper Siwalik of north-western Himalayan foothills. *Current Science* 64(11–12):863–873.

Raza, S. M., J. C. Barrry, G. E. Meyer, and L. Martin. 1984. Preliminary report on the geology and vertebrate fauna of the Miocene Manchar Formation, Sind, Pakistan. *Journal of Vertebrate Paleontology* 4:584–599.

Raza, S. M., I. U. Cheema, W. R. Downs, A. R. Rajpar, and S. C. Ward. 2002. Miocene stratigraphy and mammal fauna from the Sulaiman Range, Southwestern Himalayas, Pakistan. *Palaeogeography, Palaeoclimatology, Palaeoecology* 186:185–197.

Reznick, D. N. and R. E. Ricklefs. 2009. Darwin's bridge between microevolution and macroevolution. *Nature* 457:837–842.

Sahni, A. and H. C. Mitra. 1980. Neogene paleobiogeography of the Indian Subcontinent with special reference to fossil vertebrates. *Journal of Human Evolution* 31:39–62.

Steensma, K. J. and S. Taseer Hussain. 1992. *Merycopotamus dissimilis* (Artiodactyla, Mammalia) from the Upper Siwalik Subgroup and its affinities with Asian and African forms. *Proceedings of the Koninklijke Nederlandse Akademie van Wetenschappen* 95(1):97–108.

Steppan, S. J., R. M. Adkins, and J. Anderson. 2004. Phylogeny and divergence-date estimates of rapid radiations in muroid rodents based on multiple nuclear genes. *Systematic Biology* 53(4):533–553.

Tandon, S. K., R. Kumar, M. Koyama, and N. Niitsuma. 1984. Magnetic polarity stratigraphy of the Upper Siwalik subgroup, east of Chandigarh, Punjab, Sub-Himalaya, India. *Journal of the Geological Society of India* 25(1):45–55.

Tauxe, L. and N. D. Opdyke. 1982. A time framework based on magnetostratigraphy for the Siwalik sediments of the Khaur area, northern Pakistan. *Palaeogeography, Palaeoclimatology, Palaeoecology* 37:43–61.

van der Made, J. and S. Taseer Hussain. 1992. Sanitheres from the Miocene Manchar Formation of Sind, Pakistan, and remarks on sanithere taxonomy and stratigraphy. *Proceedings of the Koninklijke Nederlandse Akademie van Wetenschappen* 95(1):81–95.

Vickary, N. 1846. Geological report on a portion of the Baluchistan Hills. *Quarterly Journal of the Geological Society of London* 2:260–265.

Welcomme, J.-L., M. Benammi, J-Y. Crochet, L. Marivaux, G. Metais, P.-O. Antoine, and Ibrahim Baloch. 2001. Himalayan forelands: Paleontological evidence for Oligocene detrital deposits in the Bugti Hills (Balochistan, Pakistan). *Geological Magazine* 138(4):397–405.

Welcomme, J.-L., L. Marivaux, P.-O. Antoine, and M. Bennami. 1999. Mammifères fossiles des collines Bugti (Balochistan, Pakistan): Nouvelles données. *Bulletin de la Société Histoire Naturelle* (Toulouse) 135:135–139.

Wessels, W. 2009. Miocene rodent evolution and migration: Muroidea from Pakistan, Turkey, and northern Africa. *Geologica Ultraiectina* 307:1–290.

Wessels, W. and H. de Bruijn. 2001. Rhizomyidae from the Lower Manchar Formation (Miocene, Pakistan). *Annals of Carnegie Museum* 70:143–168.

West, R. M., J. H. Hutchinson, and J. Munthe. 1991. Miocene vertebrates from the Siwalik Group, western Nepal. *Journal of Vertebrate Paleontology* 11(1):108–129.

Willis, B. J. 1993. Evolution of Miocene fluvial systems in the Himalayan foredeep through a two kilometer-thick succession in northern Pakistan. *Sedimentary Geology* 88:77–121.

Zachos, J., M. Pagani, L. Sloan, E. Thomas, and K. Billups. 2001. Trends, rhythms, and aberrations in global climate 65 Ma to Present. *Nature* 292:686–693.

Chapter 15

The Neogene Siwaliks of the Potwar Plateau, Pakistan

JOHN C. BARRY, ANNA K. BEHRENSMEYER, CATHERINE E. BADGLEY, LAWRENCE J. FLYNN,
HANNELE PELTONEN, I. U. CHEEMA, DAVID PILBEAM, EVERETT H. LINDSAY,
S. MAHMOOD RAZA, ABDUL RAHIM RAJPAR, AND MICHÈLE E. MORGAN

The Siwalik formations of the Indian Subcontinent comprise fluvial sediments of Miocene through Pleistocene age deposited in a series of basins along the southern margin of the collision zone between peninsular India and Asia. The deposits are thick and fossiliferous, with a diverse fauna of terrestrial and freshwater vertebrates and for well over 170 years they have been the subject of considerable scientific interest.

Research on Siwalik fossils and sediments has touched on diverse subjects, but a recent focus has been on a fundamental problem in paleobiology, the relationship between biotic evolution and environmental change. This is a line of inquiry we have been involved with for over 30 years and one that has had a major influence on our collective research agenda. Our primary focus has been on documenting patterns of faunal turnover and ecological change, as well as on investigating the relationships between observed faunal dynamics and changes in the fluvial system, climate, and local habitats. This, however, can only be achieved by first creating a comprehensive chronostratigraphic framework and understanding the biases inherent in the Siwalik fossil record. Therefore, creation of a chronostratigraphic framework, together with determining the taphonomic characteristics of the fossil assemblages, is a critical component of our research program. To that end, this chapter and ongoing related research are steps toward those goals.

In the following, we discuss aspects of Siwalik stratigraphy, with a focus on the depositional system in which the sediments and fossil sites formed. We begin with an overview of the distribution of Siwalik sediments on the Indian Subcontinent and discussion of recent ideas about the depositional system. We then present an outline of the chronological framework constructed for the Siwalik formations on the Potwar Plateau in northern Pakistan and our methods of estimating the ages of the fossil localities. That is followed by preliminary analyses of the temporal distribution of the fossil sites, their taphonomic characteristics, and the implications for biostratigraphy. We end with a brief summary of some of the important changes in the faunas as documented on the Potwar Plateau. All ages are based on the Ogg and Smith (2004) calibration of the Geomagnetic Polarity Time Scale, which is commonly referred to as the "2004 Gradstein timescale."

OVERVIEW OF THE SIWALIKS

The term "Siwalik" is derived from the name of a range of hills lying south of the main ranges of the Himalayas in northwestern India and composed principally of Neogene fluvial sediments. The name has been loosely applied to all Neogene fluvial sediments along the southern margin of the Himalayas and adjacent ranges in Nepal, India, and Pakistan. While the most intensively studied exposures are those of the Punjab region of northern India and northern Pakistan, Siwalik deposits are widely exposed throughout a broad belt stretching eastward from the trans-Indus region of Pakistan (Hussain et al. 1979; Munthe et al. 1979; Swie-Djin and Hussain 1981) through northern India into central Nepal and the Garo Hills of northwest Assam (Pentland 1828), as well as southward to Piram (Perim) Island in the Gulf of Khambhat (Cambay) (see Flynn et al., chapter 14, this volume:fig. 14.1). While not traditionally considered to be "Siwaliks," other areas with Neogene fluvial sediments include the Gaj and Manchar formations in Sind and the Chitarwata and Vihowa formations in Baluchistan and southwestern Punjab. Finally, although in a different tectonic setting, the molassic Irrawaddy Formation in Myanmar (Burma) is often treated as a Siwalik equivalent, especially in the older literature and in conjunction with discussions of fossil material (e.g., Pilgrim 1926; Colbert 1935).

Although typically mantled by Late Pleistocene deposits, in the more arid regions of the Subcontinent the underlying Siwalik formations are often widely exposed as shallowly dipping and undeformed rocks with relatively simple stratigraphic relationships. Particularly on the Potwar Plateau of northern Pakistan, the sediments are exposed as broad belts of outcrop. As a consequence, numerous sections along small cross-cutting streams can be measured, sampled, and correlated laterally by tracing distinctive lithological units that magnetostratigraphic field studies have shown to approximate isochronous horizons over distances of as much as 30 km (McMurtry 1980; Behrensmeyer and Tauxe 1982; Sheikh 1984; Johnson et al. 1988; Badgley and Tauxe 1990; McRae 1990; Kappelman et al. 1991; Friend et al. 2001). Siwalik lithofacies include sandstones, siltstones, mudstones, and rare conglomerates and marls. Formations have been differentiated on the basis of the relative proportions of sandstones and fine-grained units, as well as the thickness of the large sandstone bodies and in some cases mineralogical characteristics. The Potwar sequence can be divided into seven formations (Cheema, Raza, and Ahmed 1977; Hussain et al. 1992), which are listed in table 15.1 along with estimated ages for their boundaries on the Potwar Plateau. In table 15.1, the overlap in ages between formations represents the approximate degree of known time transgression, but all the formations are likely to be time transgressive to some degree.

The Potwar Siwalik formations, exposed as a tilted and slightly deformed belt bounded by the Margala Hills and Kala Chitta Hills to the north and the Salt Range to the south (figure 15.1), have had a central role in the development of the lithologic and biostratigraphic nomenclature of the Subcontinent. First used by Pilgrim (1913, 1934) and subsequently by others (Cotter 1933; Colbert 1935; Lewis 1937), five stratigraphic units were recognized on the Potwar Plateau (Kamlial, Chinji, Nagri, Dhok Pathan, and Tatrot "zones") and were an important

Table 15.1

Siwalik Formations and Their Ages on the Potwar Plateau

Formation	Age Range (Ma)
Tatrot	3.5 to 3.3
Samwal[a]	ca. 3.6 to ca. 1.5
Dhok Pathan	9.8 to ca. 3.5
Nagri	11.5 to 9.0
Chinji	14.0 to 11.4
Kamlial/Murree[b]	18.0 to 14.0
Murree	? to 18.0

[a]For a discussion of this unit and the problem of Upper Siwalik Formation nomenclature, see Hussain et al. (1992).

[b]In the Southern Potwar, the Kamlial and Murree formations are not differentiated.

advance in devising a nomenclature for what came to be recognized as a series of successive faunas. These, together with additional units established in rocks to the east in India, were at first conceived of as faunal units or "zones" that were conceptually most similar to "stages" as used in modern codes of stratigraphic nomenclature and were intended to be used for long-distance correlations throughout the Subcontinent. Thus, Pilgrim referred to fossil collections from the region around Haritalyangar in northwestern India as being of "Nagri" or of "Dhok Pathan" age, although this area is nearly 600 km to the east of the Potwar. Subsequently, Pilgrim's faunal units came to be used as both lithostratigraphic formations and as chronostratigraphic zones, with the distinction not always being made clear and often leading to considerable confusion. A summary of the stratigraphy of Pakistan by Cheema, Raza, and Ahmed (1977) establishes the Potwar units as "formations" with distinctions among them based on lithological characteristics, but they still tend to be used as units having chronostratigraphic significance. Fortunately the most recent practice has been to recognize and give different names to local lithological units in distant regions that may be more or less time equivalent. (For further discussion and additional references, see Flynn et al., chapter 14, this volume; for a still relevant discussion of the early history of Siwalik stratigraphic nomenclature, see Colbert 1935:7–19.)

Used in a very broad and loose sense as biostratigraphic units, Pilgrim's faunal zones are still useful, although there are several problems in adapting them to the more demanding requirements of current studies. These problems are separate from the confusion over litho- and biostratigraphic units and stem from the imprecision with

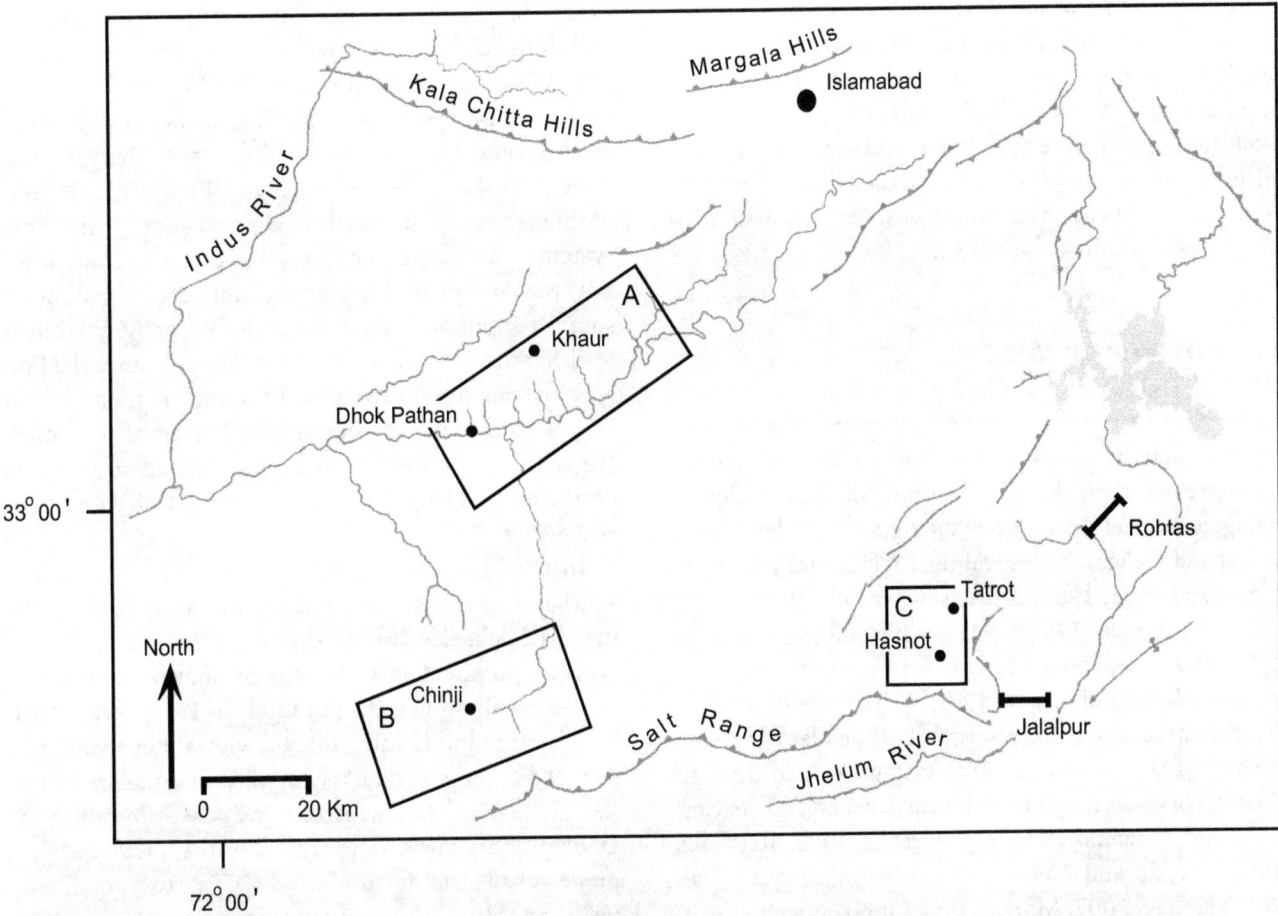

Figure 15.1 Map of the Potwar Plateau in northern Pakistan. Boxes are areas with regional networks of stratigraphic sections: (*A*) northern limb of the Soan Synclinorium along the Soan River; (*B*) southern limb of the Soan Synclinorium near the Salt Range; (*C*) eastern edge of the Potwar Plateau near Hasnot. Isolated sections at Rohtas and Jalalpur are also shown. Barbed lines are the boundaries of thrusts.

which the faunal zones were initially defined. First, only two (the Chinji and Nagri "zones") are in direct superposition in what can be considered to be their type areas. Consequently, it has proved difficult to define the boundaries between the "zones" in precise terms. Second, since some of the areas Pilgrim used to characterize his zones are poorly or only modestly fossiliferous, he used fossils from other regions to supplement his faunas. In doing this, Pilgrim must have assumed that his lithological correlations were also chronological correlations, which we now know to be an error. Chief among the problematic units are the Kamlial and Nagri zones. The former was characterized by species from Dera Bugti and the Manchars some 500 km to the south, while the Nagri "fauna" came to include species from Haritalyangar that are now believed to be significantly younger than those of the Nagri Formation on the Potwar Plateau (Johnson et al.

1983; Pillans et al. 2005). Citing these problems, Barry, Lindsay, and Jacobs (1982) established a series of interval zones based on rocks and faunas from the Potwar Plateau, and while the zones were subsequently refined and extended by Hussain et al. (1992), there has not been much further work in this important area of research.

As might be expected for collections formed over 170 years, fossils from the Siwaliks are scattered among many museums and other institutions and are often without useful stratigraphic data. The initial collections, containing many of the types for Siwalik taxa, are in the British Museum of Natural History and the collections of the Indian Geological Survey in Calcutta. For our work we have primarily relied on material in collections held by the Geological Survey of Pakistan and the Pakistan Museum of Natural History. In addition a number of fossils collected by Barnum Brown and G. Edward Lewis in

1922 and 1932 and now in the American Museum of Natural History and Yale Peabody Museum have been of use, as well as fossils collected in 1935 and divided among the Yale Peabody Museum, the American Museum, and the Geological Survey of India. Other important collections made just before and after World War II are in Munich, Germany, and Utrecht, the Netherlands. (For some discussion of these latter collections, see Dehm, Oettingen-Spielberg, and Vidal 1958 and Hussain 1971.)

THE POTWAR SIWALIK DEPOSITIONAL SYSTEM

Environmental reconstructions of the Siwalik formations are based primarily on sedimentological evidence, using as models the modern analogues provided by the Indus and Ganges Rivers and their tributaries (Behrensmeyer and Tauxe 1982; Behrensmeyer 1987; Badgley and Tauxe 1990; Willis 1993a, 1993b; Willis and Behrensmeyer 1994, 1995; Badgley and Behrensmeyer 1995; Behrensmeyer, Willis, and Quade 1995; Zaleha 1997a, 1997b). Furthermore, while there is essentially no plant or pollen record for the Potwar Siwaliks, insights gained through analysis of stable isotopes and the paleosols (Quade, Cerling, and Bowman 1989; Quade et al. 1995; Retallack 1991; Quade and Cerling 1995; Behrensmeyer et al. 2007; Nelson 2007; Morgan et al. 2009), as well as scattered ecomorphological analyses of selected mammals (e.g., Nelson 2005; Belmaker et al. 2007), have been used to make inferences about the climate and vegetation of the ancient floodplains. Together these studies have been used to develop a conceptual model of local Siwalik deposition and paleoenvironments that potentially can serve as a model for the Neogene fluvial deposits beyond the Potwar Plateau. This model, however, does not apply universally and especially not to the Oligo-Miocene Murree Formation (Najman et al. 2003) or the Oligo-Miocene Chitarwata Formation (Downing et al. 1993; Welcomme et al 2001; Downing and Lindsay 2005; Métais et al. 2009; Flynn, chapter 14, and Antoine et al., chapter 16, both this volume), and most likely not to the Early Miocene Gaj Formation in the Lower Indus Basin of Sind (Raza et al 1984). Each of these three latter formations apparently includes facies transitional between marine, estuarine, and fluvial environments, and thus include additional and more varied depositional environments.

The Miocene Siwaliks were deposited as part of a very large river system, one the size of the modern Indus or Ganges systems. While modern rivers of the Punjab region drain southwest into the Indus and contribute to the Indus fan, paleocurrent directions indicate the proto-Ganges system may have extended farther west and drained the Potwar region in the Miocene (Raynolds 1981; Beck and Burbank 1990). Such an ancient proto-Ganges system would have extended several thousand kilometers eastward to the Bay of Bengal, with floodplain widths on the order of 100–500 km. Even if drainage was to the south as a proto-Indus, the whole system would have been very large as it included not only the floodplains but also the more distant mountain source regions and even the deltaic fans of the axial river in either the Arabian Sea or Bay of Bengal. Since the Potwar Plateau is only about 30,000 km², it encompasses only a comparatively very small fraction of this much larger area and thus can give us only limited information on the entire Neogene Siwalik system (Willis and Behrensmeyer 1995).

In the Miocene, the Potwar was some 100–200 km southeast of the mountain front where its rivers originated (Zaleha 1997b). At that time the Potwar region was at approximately 29°N latitude, about 4° south of its current position (Tauxe and Opdyke 1982). We infer it had a warm and humid climate, with a monsoonal pattern of circulation and likely marked seasonality. Analysis of Late Miocene paleosols, which are primarily in the Dhok Pathan Formation, indicates they are most like modern soils that form under 25°C mean annual temperature with ca. 1400 mm/yr precipitation (Retallack 1991; Behrensmeyer, Willis, and Quade 1995). At all levels, the paleosols indicate intense oxidation and seasonal differences in the height of the water table, with waterlogging followed by leaching and precipitation of carbonates. The stable isotopes of both the paleosols and teeth indicate there were progressive and marked changes in vegetation and climate, with the most rapid in the latest Miocene (Quade and Cerling 1995; Barry et al. 2002, Behrensmeyer et al. 2007). Between about 8 Ma and 4.5 Ma there was a shift from floodplain vegetation dominated by C3 plants, which presumably encompassed largely forested habitats of various types, to vegetation dominated by C4 plants, presumably more open grasslands. The transition, however, was complex, with C3 and C4 habitats coexisting on the floodplain in close proximity for several million years (Behrensmeyer et al. 2007).

The various reconstructions of the ancient Potwar rivers all portray them as part of a single large drainage system with a major trunk river the size of the modern Indus running to either the Arabian Sea or the Bay of Bengal. Deposits of the first-order axial trunk river (figure 15.2 [1]) are not preserved, either because it flowed eastward and did not cross the Potwar Plateau or, if flowing southward, was

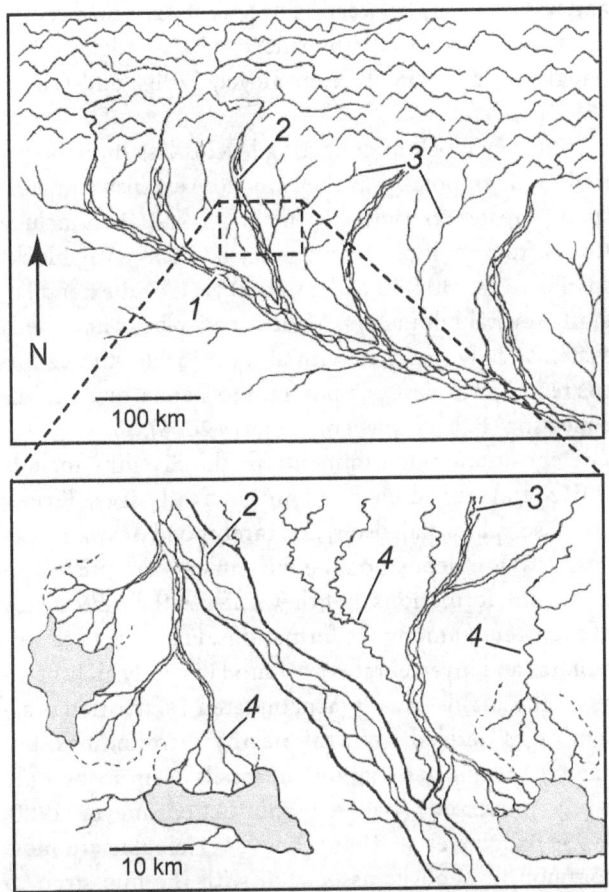

Figure 15.2 Plan view of a reconstructed Siwalik fluvial system. The numbers refer to categories of streams discussed in text: (*1*) main trunk river, shown as flowing eastward and not preserved on the Potwar Plateau; (*2*) emergent river with origin in northern mountains; (*3*) interfan rivers with origins on broad floodplain; (*4*) small floodplain streams, some tributary to larger rivers and some draining into swamps or ponds indicated by gray stipple. The outlined box is approximately the size of the Potwar Plateau, indicating the magnitude of the system.

confined to the distant margin of the depositional basin (Willis 1993a, 1993b; Willis and Behrensmeyer 1995).

The channel and floodplain deposits of tributaries to the trunk river are preserved, however. These tributaries are of two types (Willis 1993b; Zaleha 1997b). One type (see figure 15.2 [2]) includes deposits of large rivers the size of the modern Jhelum. These Miocene rivers apparently emerged from the mountains to the north and west at widely spaced intervals and flowed hundreds of kilometers southeast to join the trunk river (see figure 15.2). These second-order rivers, which are referred to as "emergent" or "upland sourced," carried relatively unweathered sediment from the mountains and deposited it in the basin as large, low-gradient fans. The braided channel belts of the emergent rivers were typically more than 5 km

wide, with individual channels of 200–400 m. The emergent rivers probably had higher discharge during the spring melt and summer monsoon seasons. Because they were prone to frequent avulsions during flood events, the channels did not migrate laterally to form extensive sand sheets (Behrensmeyer, Willis, and Quade 1995). The Jhelum and Chenab are modern analogues of such rivers.

The second type of tributary (see figure 15.2 [3]) includes deposits of smaller rivers that were tributaries of either the trunk river or the emergent rivers (Willis 1993b; Zaleha 1997b). They were braided streams, with channel belts 1–2 km wide and channels of 70–200 m. While a few may have had mountain sources, others had sources in the foothills or the floodplain. Because they were confined by the fan deposits of the larger emergent rivers and carried fine, reworked material eroded from the floodplain, they are referred to as "interfan" or "lowland-source" streams. Since the source of their water was close to or on the floodplains, their flow varied more throughout the year than in the emergent streams, with presumably higher discharge during the summer monsoon (Willis 1993b; Zaleha 1997b). Across the floodplain, floods must have disturbed nearby vegetation, perhaps keeping extensive areas in early stages of ecological succession. The Soan River is a potential modern analogue of an interfan river.

The floodplain was flat, with perhaps at most a few hundred meters difference in elevation across it and local relief of about 10 m (Willis 1993b). Low areas could have been permanent or seasonal ponds and swamps that would have eventually filled with sediment from the second- and third-order channels or from still smaller fourth-order floodplain streams (see figure 15.2 [4]). Preserved floodplain deposits consist chiefly of mudstones, along with minor contributions from crevasse-splays, levees, and smaller floodplain channels. Features of the smallest fourth-order streams indicate that they were 10–100 m wide, smaller than the individual channels of the third-order interfan rivers. Flow in these smallest channels was episodic, with at some times of year standing water in the channels (Willis 1993a). In all the Siwalik formations, the smaller channels were important sites for fossil bone accumulation, as were the fine-grained fills in the upper parts of the larger channels (Badgley and Behrensmeyer 1980, 1995; Behrensmeyer 1987; Badgley et al. 1995; Behrensmeyer, Willis, and Quade 1995; Willis and Behrensmeyer 1995).

Although not studied extensively, paleosols are common throughout the whole Siwalik sequence. Retallack (1991) described and named nine paleosol types ("series") from two short sections in the lower Dhok Pathan Formation. In these two sections, which are within (9.5–9.4 Ma) and just above (8.8–8.7 Ma) the Nagri–Dhok Pathan transition, he documented differences in the frequencies of soil types that he interpreted as resulting primarily from differences in parent material, local topography, and drainage, as well as local vegetation and the amount of precipitation. Retallack (1991) found no evidence for extensive grasslands in either section, although he suggested that two rare types formed under waterlogged, grassy woodlands. Fewer paleosol types were reported in the younger section, which Retallack (1991) interpreted as evidence for a less varied landscape at that time. Nevertheless, considering the smaller number of recognized soil horizons in the upper section (four paleosol types from 31 horizons versus eight types from 80 horizons in the lower section), the proportions are not different. Nevertheless, between and within stratigraphic levels there is always considerable diversity in paleosol types and by inference considerable diversity in vegetation. Low areas contained seasonal swamps or ephemeral ponds, while other areas had extensive wet upland and lowland forests, dry old-growth deciduous woodlands, tall gallery forest with lush undergrowth, and wet grassy wooded meadows. There were also tracts with secondary growth and early succession (Retallack 1991). These vegetation types would have been common, coexisting elements of the landscape.

Temporal changes in the Siwalik fluvial system have been documented, but whether they are related to climate or subsidence, or simply due to the autocyclic dynamics of the fluvial system, is not clear. The transition between the Chinji and Nagri formations has been interpreted as resulting from the progradation of a second-order emergent system over a smaller, third-order interfan system (Willis 1993b; Zaleha 1997b). The Nagri to Dhok Pathan transition, on the other hand, seems a case of prolonged coexistence of two contemporaneous systems ending with displacement of an emergent system by an interfan system (Behrensmeyer and Tauxe 1982). Although we do not yet fully understand the underlying causes, both of these transitions, and that between the Kamlial and Chinji formations, are likely to be local events without chronostratigraphic significance.

In previous publications (e.g., Barry et al. 2002), we used two terms to designate the occurrence of fossils: locality and survey block. In our usage, a locality is limited temporally and spatially. In the Siwaliks, concentrations of fossils typically occur in small areas of outcrop of perhaps a few tens or hundreds of square meters extent, with a single sedimentary body as the source of the fossils. Between patches there is little or no fossil bone, so that it is usually possible to delineate localities spatially as distinct entities. The duration of accumulation of the fossil

material in a given locality is thought to have been brief and was generally between a few tens of years to, rarely, at most 50,000 years (Behrensmeyer 1982; Badgley 1986; Badgley et al. 1995; Behrensmeyer, Willis, and Quade 1995).

A survey block or collecting level is less limited spatially and temporally, as they are more extensive in area, span a greater stratigraphic thickness, and often include fossils from multiple sedimentary layers. Survey blocks span between 30,000 and 350,000 years and extend laterally several kilometers. Most survey blocks have been systematically searched with all identifiable surface fossils recorded as part of a program to standardize surface collecting (Behrensmeyer and Barry 2005).

Depositional environments in the Siwaliks include active and abandoned channels of all sizes, levees, crevasse-splays, paleosols, and rarely pond or swamp deposits. These depositional environments are present in all Siwalik formations, but they differ in their frequency of occurrence among the formations. Fossils of both vertebrates and invertebrates are found in nearly all lithologies. The fossils mostly accumulated as attritional assemblages derived from the nearby surroundings, but habitat specific associations of fossils seem to be only rarely preserved (Badgley and Behrensmeyer 1980, 1995; Badgley et al. 1995). Fossil vertebrates are most common in deposits associated with the mid-sized to small floodplain channels, where they occur as concentrations of disarticulated, often fragmentary bones. A very few such concentrations contain 1,000 or more specimens, but the majority have only 5 to 200 fossils each. Because other facies contain only isolated specimens or small, low-density scatters, collections of fossils are biased toward overrepresentation of small channel floodplain sites. This bias, in addition to those introduced by taphonomic processes common to the different facies, may place limits on reconstructions of the living communities (Badgley 1986). This bias may also severely hamper attempts to determine the true temporal ranges of taxa and therefore development of biostratigraphic zonations.

As an aid in paleoecological and biostratigraphic analysis, a classification system has recently been developed for the depositional environments of Siwalik localities (Behrensmeyer et al. 2005). The classification recognizes four major groups (major channel, floodplain channel, sheet deposits, and floodplain), with all but the sheet deposit category being further subdivided (table 15.2). The criteria for classification are based on physical characteristics, such as lithology or sedimentary structures, and the geometry and relationships of the beds, not proper-

Table 15.2

Depositional Categories

Abbreviation	Primary Context	Secondary Context	Lithologies/Depositional Setting	Number of Localities
MC-L	Major channel	Lower two-thirds of channel	Basal lag and bar	3
MC-U1	Major channel	Upper third of channel	Channel bar	8
MC-U2a	Major channel	Upper third of channel	Large scale, coarse fill: gravel and coarse sand lens, mudclasts, reworked carbonate nodules	11
MC-U2b	Major channel	Upper third of channel	Large scale, fine fill: laminated and cross-stratified sands and silts, fining upward into mudstones	12
MC-U3a	Major channel	Upper third of channel	Small scale, coarse fill: sand, grit	12
MC-U3b	Major channel	Upper third of channel	Small scale, fine fill: mudstones, siltstone	6
FC-C1	Floodplain channel	Complex fill	Basal lag and bar in lower part of channel	19
FC-C2	Floodplain channel	Complex fill	Cross cutting lens of mixed lithologies within channel	43
FC-S1	Floodplain channel	Simple fill	Basal lag and bar in active channels	20
FC-S2	Floodplain channel	Simple fill	Mixed lithologies with inclined bedding, silts and fine sand, mudclast gravels	37
FC-S3	Floodplain channel	Simple fill	Fine grained mudstones and clays, laminated or pedogenically altered	16
FP-P	Floodplain	Patchy	Laterally continuous paleosols with concentrations of bone	37
FP-C	Floodplain	Continuous	Laterally continuous paleosols with extensive bone throughout	10
FP-L	Floodplain	Laminated	Temporary or seasonal water bodies	4
SD	Sheet deposit		Crevasse-splay, levee, or sheet wash deposits	17

NOTE: The total number of localities in the last column is fewer than 321 because the secondary context of all localities is not known.

ties of the fossil assemblages such as taphonomic character or taxonomic composition.

Most major channel (MC) and floodplain channel (FC) localities are part of infillings of abandoned channels. Major channels approximately correspond to the second- and third-order channels of the emergent and interfan system deposits, while the floodplain channels correspond to the small, fourth-order channels of the interfan floodplain. The floodplain (FP) localities mostly occur in paleosols, as either discrete patches or dispersed concentrations, or very rarely as laminated deposits in ponds or swales. The localities of the sheet deposits (SD) formed in crevasse-splays, levees, or in sheet-wash deposits.

Research based on this classification of depositional settings is just beginning. In a later section, we present some new data on the relative frequency of the types and subtypes and discuss some of the implications for biochronology.

THE CHRONOLOGICAL FRAMEWORK AND ESTIMATION OF AGES OF LOCALITIES

The Siwaliks comprise not only a thick sequence but also one that had relatively continuous deposition of sediment with fossils at many horizons and easily determined superpositional relationships between individual localities. The thickness and continuity of deposition has allowed the development of a detailed paleomagnetic reversal stratigraphy and as a consequence a reliable chronostratigraphic framework for much of the region (Barndt 1977; Barndt et al. 1978; Opdyke et al. 1979; Barry, Behrensmeyer, and Monaghan 1980; McMurtry

1980; Behrensmeyer and Tauxe 1982; Johnson et al. 1982; Stix 1982; Tauxe and Opdyke 1982; Johnson et al. 1985; Tauxe and Badgley 1988; Willis 1993a, 1993b). See also Barry et al. (2002:appendix 1).

At present, our chronostratigraphic framework for the Potwar Plateau is based on 29 measured sections between 250 m and 3200 m thick, as well as 34 shorter sections. Twenty-four of the 29 long sections and 23 of the 34 short sections have determined magnetic polarity stratigraphies. Except for outliers at Rohtas and Jalalpur (Opdyke et al. 1979; Johnson et al. 1982; Behrensmeyer et al. 2007), the 63 sections form three regional networks (see figure 15.1) corresponding to the classic Potwar areas of Pilgrim and subsequent collectors. Two networks are on the northern and southern limbs of the Soan Synclinorium, and the third lies at the eastern edge of the Potwar Plateau near Hasnot. In each region, exposures are vertically and laterally continuous, and individual sections can be correlated lithologically as well as by magnetic polarity. However, because of the absence of connecting exposures between regions, correlations among the regions depend on the magnetic polarity stratigraphy. No sections on the Potwar have been correlated on the basis of biostratigraphy.

In the Potwar Siwaliks, individual measured sections over 250 m usually have at least six or seven magnetic polarity transitions and can be independently correlated to the Geomagnetic Polarity Time Scale (GPTS; Johnson and McGee 1983). The shorter sections having fewer polarity zones typically cannot be directly matched to the GPTS, but they can still be reliably placed in the stratigraphic framework by determining their relationships to the long sections. We have correlated between adjacent sections by either tracing magnetic polarity transitions or by tracing sandstone bodies or paleosols; lithological units that have been shown in different parts of the sequence to be approximately isochronous (McMurtry 1980; Behrensmeyer and Tauxe 1982; Johnson et al. 1988; Badgley and Tauxe 1990; McRae 1990; Kappelman et al. 1991, Friend et al. 2001).

When correlating to the GPTS, it is not possible to match every observed magnetic transition to a GPTS magnetic boundary (Johnson and McGee 1983; Kappelman et al. 1991). Nevertheless, the algorithm we use to estimate dates for localities is based on interpolation between two points of known age, and these need not be successive geomagnetic transitions. In principle, radiometric dates, isotopic events, or even biostratigraphic zone boundaries of known age could also be used. Here,

all ages are based on the ages for geomagnetic chrons in the Ogg and Smith (2004) calibration of the GPTS.

Detailed analyses of lithological facies, in several cases accompanied by tracing paleomagnetic transitions laterally between multiple adjacent sections, have been made at five horizons in the Potwar Siwaliks (McMurtry 1980; Behrensmeyer and Tauxe 1982; Sheikh 1984; Kappelman 1986; Johnson et al. 1988; McRae 1990; Kappelman et al. 1991; Friend et al. 2001; Behrensmeyer et al. 2007). These five more detailed studies demonstrate that there was considerable lithological and paleoenvironmental variation across the floodplain at all times. Irregular channel and floodplain deposition combined with contemporaneous periods of nondeposition and erosion to create a complex stratigraphic architecture and on very short timescales produced highly variable rates of sediment accumulation. Consequently, adjacent sections may either record or miss particular magnetic transitions, have differences in apparent thickness of geomagnetic chrons, or even capture otherwise unknown brief geomagnetic events (Johnson et al. 1985; Kappelman 1986; McRae 1990; Kappelman et al. 1991). On longer time scales, however, the sediment accumulation rate was steadier and a more faithful recorder of the history of geomagnetic reversals (Johnson and McGee 1983; Johnson et al. 1988; McRae 1990).

Correlation of the Late Miocene and Pliocene Siwalik sections to the GPTS was discussed in Barry et al. (2002). Other than for publication of a section at Rohtas (Behrensmeyer et al. 2007), we have not altered those stratigraphic correlations and consider them to still be reliable. We have, however, changed the calibration of the time scale. Previously we used the Cande and Kent (1995) calibration of the GPTS (Barry et al. 2002). It is, of course, straightforward to substitute one calibration for another and recalculate ages of localities using the original stratigraphic data.

Our previous work focused on the Late Miocene part of the Siwalik sequence (Barry et al. 2002). The current focus of our efforts is now on the Kamlial and Chinji formations, which comprise the Early and Middle Miocene portions of the sequence. We have extended the correlations made in Barry et al. (2002), developed new ones for the older formations, and added age estimates for 471 localities to our database. We analyze this nearly doubled sample of dated localities in the next section.

Compared to the Nagri and Dhok Pathan formations, correlations of the older Chinji and Kamlial formations to the GPTS are made more difficult by the overall lower rates of sedimentation and possibly less steady rates of accumulation (McRae 1990). The middle of the Chinji Formation has been particularly problematic. In the upper part of the formation, chrons C5r and C5An (11.040–12.415 Ma using the Ogg and Smith (2004) calibration) are consistently identified (Johnson et al. 1985, 1988; Kappelman et al. 1991), as are C5AB and C5ACn (13.369–14.095 Ma) in the lower Chinji (Sheikh 1984; McRae 1990). The intervening intervals C5Ar and C5AA (12.415–13.369 Ma), however, are more irregular in their expression (Johnson et al. 1985, 1988), and it is only at the western edge of the Chinji exposures that an easily interpretable sequence has been documented (Kappelman et al. 1991). Similarly at the base of the Chinji Formation and the top of the Kamlial Formation, chron C5ACr (14.095–14.194 Ma) is only occasionally recorded, and researchers have found an otherwise unknown very short reversed cryptochron within C5ACn (13.734–14.095 Ma) that can be confused with C5ACr. Recognition of this new reversal has altered the correlations of the eastern lower Chinji sections (McRae 1990) slightly, but not that of the Chitaparwala section (Johnson et al. 1985). We still place the local base of the Kamlial Formation in chron C5Dr, but now estimate its age at approximately 17.9 Ma using the Ogg and Smith (2004) calibration of the GPTS.

Fossil sites are typically at most only a few kilometers away from the nearest measured section and can be placed on that reference section by laterally tracing lithological markers. As noted, these marker horizons are types of lithologies that have been demonstrated to be isochronous at timescales of 10^3 years over the distances involved (Behrensmeyer and Tauxe 1982), but the correlations have various degrees of uncertainty that are specific for each locality. In order to preserve the uncertainty, we recorded the stratigraphic position of each locality as bracketed by upper- and lowermost bounds in the section. These are the highest and lowest possible levels where the locality could occur in the reference section (figure 15.3). We have been conservative in these judgments, giving localities a wide range of possible ages. We assume that the locality could occur with equal likelihood at any horizon within the bracketed interval.

To order localities temporally, each is assigned an absolute age determined from its stratigraphic position relative to the Geomagnetic Polarity Time Scale. Ages are estimated with varying degrees of accuracy depending on both the precision of the stratigraphic position of the locality and how precisely the stratigraphic levels of the magnetic polarity transitions have been located in the stratigraphic sections. Most ages are based on interpolations between two levels of known age—that is, identified magnetic polarity transitions of the GPTS. However, some localities are in magnetozones truncated at either the top or bottom. In such cases the ages are extrapolated, using the rate of sediment accumulation of sub- or superjacent magnetic intervals.

Ages are determined for both the upper and lower limits of each locality's stratigraphic position, a practice that incorporates the uncertainty in the correlation of the locality to a stratigraphic section. In addition, although it is conventional to place magnetic polarity transitions midway between sites of differing polarity, in calculating ages for localities we instead use the highest and lowest possible positions of the magnetic transitions in the lithological sections (see figure 15.3). By so doing, we can incorporate uncertainties in the stratigraphic positions of the magnetic transitions as well. When a locality's stratigraphic level is precisely determined and the magnetic transitions are stratigraphically tightly constrained, the oldest and youngest age estimates will converge. Taken together, these two refinements produce conservative estimates for a pair of maximum and minimum ages for each locality and thus the largest reasonable range of possible ages. In the following analyses and discussion, we average the oldest and youngest ages for each locality ("midpoint age") and calculate the difference between them ("delta age").

Both interpolation and extrapolation assume constant rates of sediment accumulation within polarity chrons. This assumption is certainly incorrect, as fluvial sequences are by nature episodic with many short intervals of erosion and nondeposition that leave hiatuses. However, as long as the hiatuses are short relative to the duration of the geomagnetic polarity chron and are evenly distributed within it, the approach will produce reliable results (Johnson and McGee 1983; Johnson et al. 1988).

ANALYSIS OF THE LOCALITY DATA

The variety of depositional environments and the biased preservation of fossils in those different settings (Behrensmeyer et al. 2005) suggest the possibility of imposed biases on the representation of taxa in our collections, with implications for both paleoecology and biostratigraphy. In the following, we first document how the localities and various depositional types are temporally distributed, and then use that information to examine the relationships between the depositional environment and faunal composition of the localities.

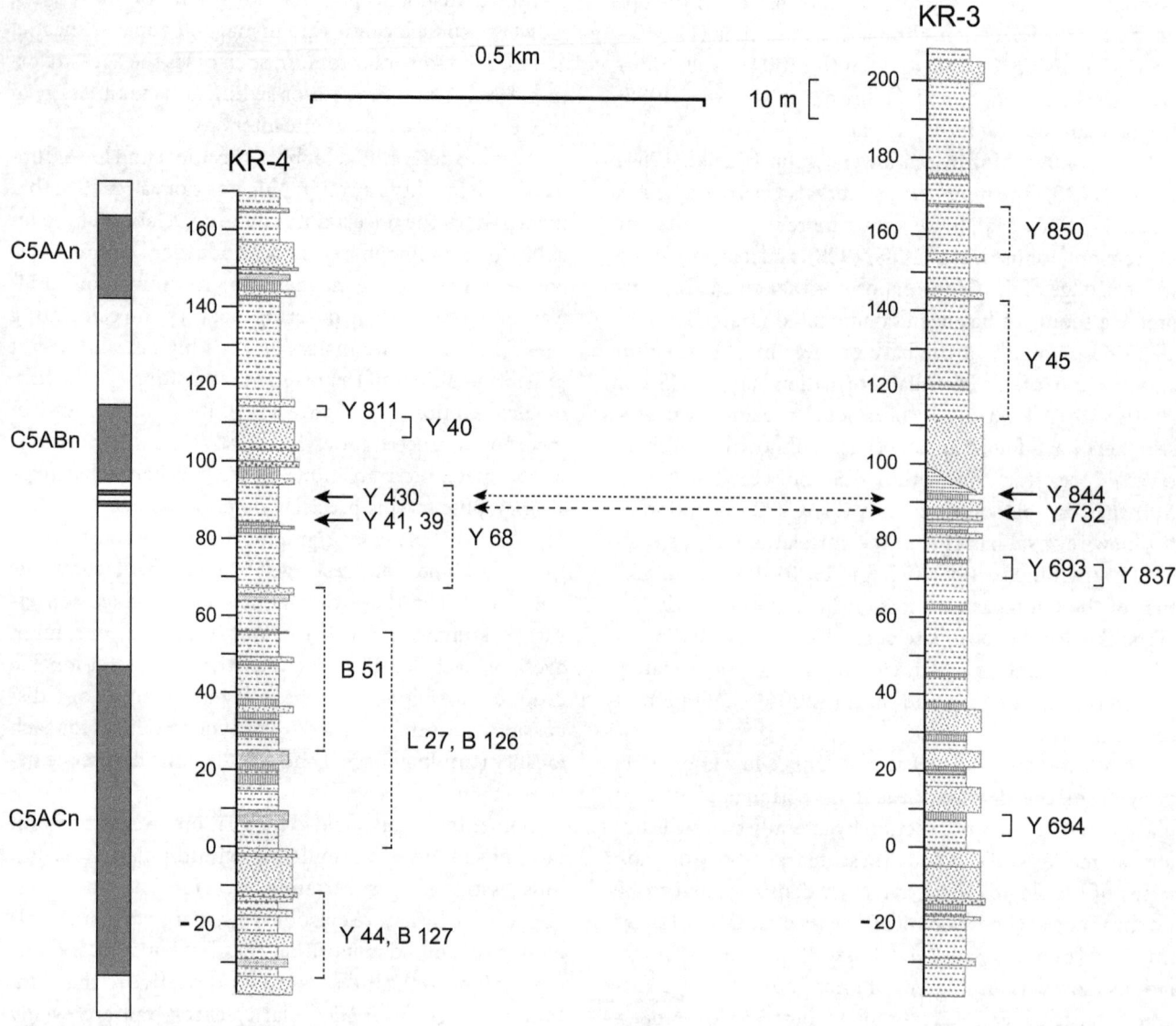

Figure 15.3 Two lower Chinji Formation reference sections (KR-4 and KR-3) with correlated localities shown by letters and numbers. Dashed connecting lines show levels of two paleosols traced between the sections. Correlation of the local magnetostratigraphy of section KR-4 to the Geomagnetic Polarity Time Scale shown on left. Locality stratigraphic positions shown by brackets or arrows.

Our Siwalik database currently has 1375 Miocene localities from northern Pakistan and India, plus another 97 localities from the Chitarwata, Vihowa, Gaj, and Manchar formations in southern Pakistan. The northern group, the majority of which are on the Potwar Plateau, include localities discovered by B. Brown and G. E. Lewis in 1922 and 1932, localities discovered by the Dartmouth–Peshawar University project, and localities discovered by parties of the Harvard–Geological Survey of Pakistan project. Not included are the many localities located in northern India or Jammu and Kashmir, a substantial

number of which are of Late Pliocene and Pleistocene age, time periods not well represented on the Potwar Plateau. (For references and discussion, see Flynn et al., chapter 14, this volume.)

Thirteen hundred of the northern localities have been assigned to one of the Potwar Siwalik formations (see table 15.1), and of these, 1026 (75%) currently have estimated absolute ages. We anticipate that eventually we will be able to incorporate an additional 100 to 200 of the undated residue into our chronostratigraphic framework, at least half of which will be from the Chinji Formation.

Thus, we eventually expect to have 85–90% of the localities in our database with an estimate of absolute age. Nevertheless, there remains a substantial number of localities (and therefore fossils of interest) that are either too poorly located geographically and stratigraphically or else lie outside the area in which we have focused our research. Some of these, especially localities near Harita-lyangar, India, potentially could be incorporated into a chronostratigraphic framework such as that of Pillans et al. (2005). (For discussion and additional references on this topic, see Patnaik, chapter 17, this volume.)

Figure 15.4 shows the total number of localities in each formation, from which it is apparent that while the time encompassed by each formation (see table 15.1) differs by at most only a factor of 3, there is as much as a 9-fold difference in the number of localities among formations. This difference is also reflected in the number of fossils from different horizons (Barry et al. 2002) and is related to the megafacies of each formation. The channel-dominated Kamlial and Nagri formations have few localities, while the floodplain-dominated Chinji and Dhok Pathan formations have many. Figure 15.4 also shows relative proportions of dated and undated sites for each formation. The intensively studied Nagri and Dhok Pathan formations presently have the largest proportion of dated sites, but we expect the other formations to eventually match them.

The youngest dated Potwar site in our database is D 54, the Haro River Quarry of Saunders and Dawson (1998), a site also excavated by de Terra and Teilhard de Chardin in 1935. It has a midpoint age of 1.420 Ma.

(Midpoint ages are expressed to the nearest thousand years. This is only to preserve the fine-scale superposition of localities in the measured sections, not because the age estimates have that degree of precision. The upper and lower estimated ages of D 54, for example, are 1.070 Ma and 1.770 Ma—hardly 10^3 precision!) The oldest dated localities are Y 590 and Y 739, at the base of the Kamlial Formation in the Salt Range. Both have midpoint estimates of 17.899 Ma (17.736–18.061 Ma). There is, however, a Murree Formation locality (Y 405) on the northern flank of the Khairi-Murat Ridge south of Fatehjang that is likely to be still older (Barry and Cheema 1984). Unfortunately, it cannot be inserted into any of our stratigraphic sections in that area.

The precision of our age estimates—that is, the difference between the maximum and minimum estimates for each locality ("delta age")—varies considerably. The longest span is over 1 myr duration, but the ages of most localities are much more tightly constrained (figure 15.5). Fifty-four percent ($n = 554$) of the 1026 dated localities have a temporal resolution of 100,000 yr or less, while 80% ($n = 822$) have a resolution of 200,000 yr or less.

Following Barry et al. (2002) we have assigned each locality from the Potwar Plateau to a 100,000-yr-long interval using the following protocol. Two hundred and forty-eight (24%) of the 1012 dated localities have both minimum and maximum age estimates falling in the same 100,000 yr interval and can be unambiguously assigned to that interval (figure 15.6 [Class I]). Another 306 (30%) localities have a difference between maximum and minimum estimates of less than or equal to 100,000 yr but straddle two intervals (see figure 15.6 [Class II]). This class of localities cannot be assigned unambiguously to a single interval. An additional class containing 268 (26%) localities have maximum–minimum differences between 100,000 and 200,000 yrs and straddle either two or three 100,000 yr intervals (see figure 15.6 [Class III]). These localities are also ambiguous as to interval. Nevertheless, the Class II and III localities do have at least 50% of their range in a single 100,000 yr interval. If the percent of overlap is equivalent to the probability that the true age lies within the interval, then localities of both Classes II and III can be assigned to the interval with the greatest overlap—which will be the interval containing the midpoint age. Two hundred and four (20%) localities fail to meet this criterion (see figure 15.6 [Class IV]) and are left unassigned and are excluded from most of the following synthesis. The unassigned localities may have important fossils and can enter into discussions of specific taxa. However, they are not useful in investigations of turnover dynamics.

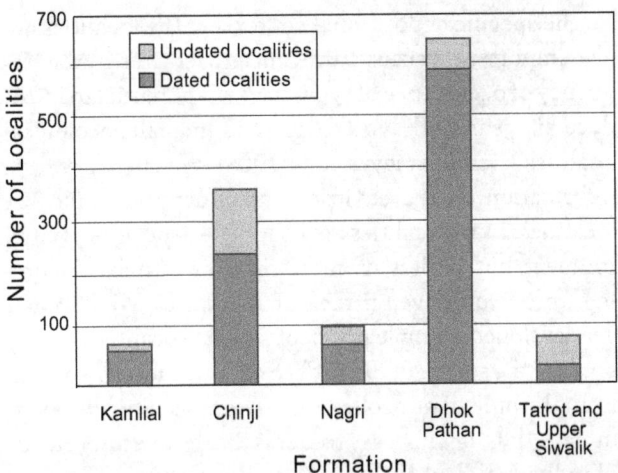

Figure 15.4 Number of localities by formation. Dark gray indicates the number of localities with absolute ages; light gray the number of localities without absolute ages.

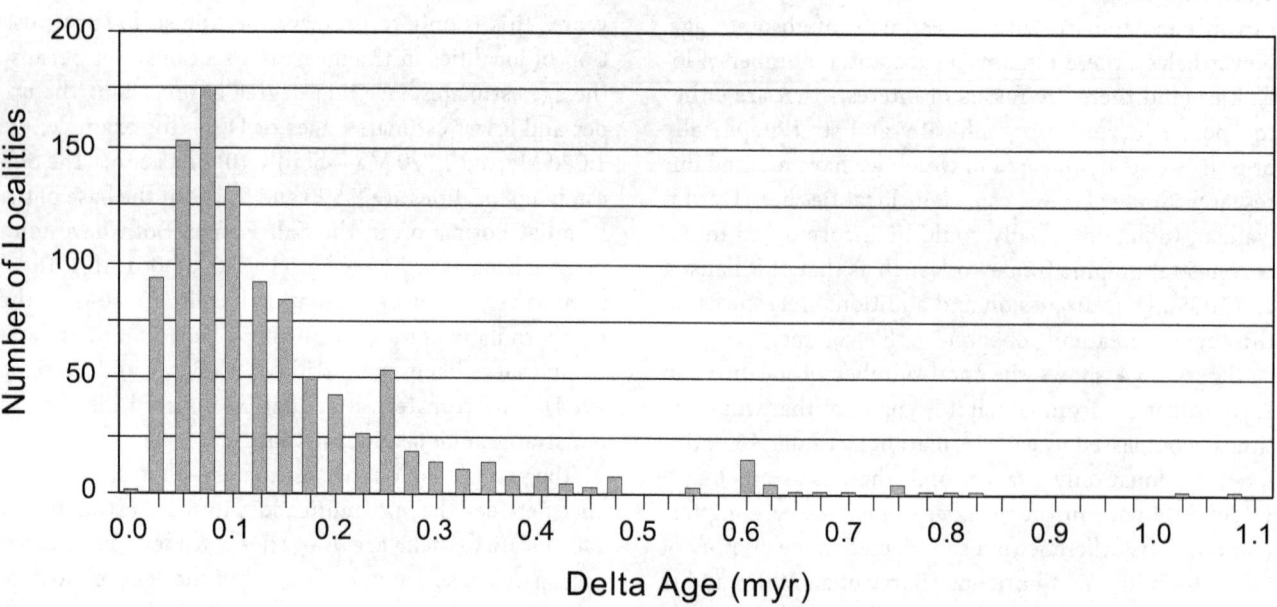

Figure 15.5 Differences between estimated maximum and minimum ages (delta age) of 1026 localities. The ages were calculated using the Ogg and Smith (2004) calibration of the Geomagnetic Polarity Time Scale.

The number of localities meeting the above criterion and assigned to a 100 kyr interval ($n = 822$) is shown in figure 15.7. The figure shows that the number of localities per interval varies from 0 to nearly 70 throughout the sequence, with the richest intervals between the upper Kamlial Formation (ca. 14.3 Ma) to the middle of the Dhok Pathan Formation (ca. 6.0 Ma). In this span of more than 8 myr, all but three (6.2, 6.7, and 11.0) of the 100 kyr intervals have at least one fossil locality. Before 14.3 Ma and after 6.0 Ma, the record is sparse and dominated by clusters of localities near 17 Ma and 3.4 Ma—the latter occurring in the Tatrot Formation. The variation in numbers of localities (and by inference numbers of fossils) is due to numerous factors. Some, such as the size of the animals, are biotic in origin. From a geological and sampling perspective, however, critical factors include the area of outcrop, dip of the beds, amount of vegetation on the exposures, ease of access to outcrops, collecting effort, and the types and frequency of depositional environments. Among these factors, the variety of depositional environments and their relative frequencies have the greatest influence in determining the patterns of locality and fossil occurrences over time. These two factors are examined in more detail in a later section.

The Chinji and Dhok Pathan formations are both characterized by abundant and diverse overbank and floodplain channel deposits. The more modestly fossiliferous Kamlial and Nagri formations, on the other hand, are dominated by the multiple, stacked channels of large emergent and interfan streams. These, in contrast to the overbank deposits, offer few opportunities for preservation of fossils (Badgley and Behrensmeyer 1995; Behrensmeyer et al. 2005). The upper part of the Dhok Pathan Formation (ca. 6.0–3.5 Ma) is a special case. In these rocks, lithologies that are broadly similar to normally productive lithologies are nearly barren.

The specific depositional contexts of the localities are also important, because differences over time in the frequency of occurrence of types have the potential to introduce biases in both biostratigraphic and paleoecological analyses (Behrensmeyer et al. 2005). We currently have information on the environment of deposition for 321 localities. Most localities are in a single depositional environment, but 18 sites are more complex with fossils coming from two or even three different facies. We have not yet developed a simple protocol for such localities, largely because the source of the surface collected fossils—which form the bulk of our collections—cannot easily be determined. Thus, it is not easy to assess the relative importance of individual depositional facies in complex localities—if in fact one facies predominates over the others. However, there are no apparent patterns in the combinations of depositional environments in complex localities. Environments belonging to different major categories co-occur (e.g., Loc Y 310: MC-U1 [major channel bar] and FC-S [floodplain channel, simple fill]) as well as ones belonging to the same category (e.g., Loc. Y 735: MC-U1 [major channel bar] and MC-U3b [major channel with small scale fine fill of a sub-channel]). In the following, we have assigned the 18 complex localities to the type that seems most important.

Grouping the localities by major depositional categories (table 15.3; figure 15.8), floodplain channels (FC) comprise nearly half the localities, followed by localities associated with major channels (MC) and those in floodplain soils (FP). Localities in sheet deposits (SD) are notably rare in all formations, although crevasse-splay deposits in general are common in the floodplain sequences of the Chinji and Dhok Pathan formations. With the exception of sheet deposits in the Kamlial For-

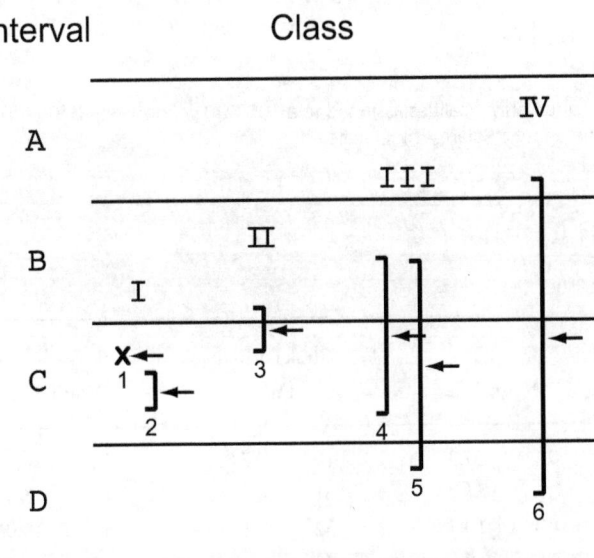

Figure 15.6 Hypothetical examples illustrating different classes of localities. Intervals A–D are of 100,000 yr duration, and midpoint ages of the localities, as indicated by small arrows, are all in interval C. (*1*) Locality with maximum estimated age the same as the minimum estimated age; (*2*) locality with difference between maximum and minimum estimated ages (delta age) less than 100,000 yr and both falling in the same 100,000 yr interval; (*3*) locality with difference between maximum and minimum estimated ages also less than 100,000 yr, but falling in different 100,000 yr intervals; (*4*) locality with difference between maximum and minimum estimated ages greater than 100,000 yr and less than 200,000 yr and falling in adjacent 100,000 yr intervals; (*5*) locality with difference between maximum and minimum estimated ages greater than 100,000 yr and less than 200,000 yr, but not falling in adjacent 100,000 yr intervals; (*6*) locality with difference between maximum and minimum estimated ages greater than 200,000 yr and spanning three or more 100,000 yr intervals. Localities 1, 2, 3, 4, and 5 would be assigned to Interval C. Locality 6 would not be assigned to any interval, although its midpoint lies in Interval C.

mation, this pattern holds both within and between formations (figure 15.8B). As is true for the localities overall, the Chinji and Dhok Pathan formations have many more of the subset of classified localities than either the Kamlial or Nagri formations.

In table 15.4 and figure 15.9, we have rearranged the subtypes of the major depositional environments to reflect cross-category depositional similarities, with the basal lag and channel bar facies (MC-L, FC-C1, FC-S1, MC-U1), upper channel fills (MC-U2a,b, MC-U3a,b, FC-C2, FC-S2, FC-S3), floodplain (FP-P, FP-C, FP-l), and sheet deposits (SD) forming four groups. The temporal distribution of these depositional subtypes is shown in figure 15.9, from which it is apparent that, after allowing for sample size differences among the formations, most depositional environments are present throughout the sequence. For example, while the small floodplain channel deposits (FC-C2 and FC-S2) have many more localities in the floodplain-dominated Chinji and Dhok Pathan formations, these two highly productive facies are also present in the major channel-dominated Kamlial and Nagri formations. There are, however, significant exceptions. Sites in sheet deposits (SD) are absent between 13.8 Ma and 9.8 Ma, a period that encompasses the Chinji Formation and much of the lower Nagri Formation. Other exceptions include the near absence after 9 Ma of basal lag and channel bar sites in floodplain channels (FC-C1 and FC-S1), the sparse record between approximately 12 Ma and 10 Ma of patchy floodplain sites (FP-P), and the relatively few Chinji Formation (ca. 14–11 Ma) sites in the upper parts of the larger major channels (MC-U2a and MC-U2b). Most of these exceptions are only modestly productive facies, but patchy floodplain sites (FP-P) are the third most productive facies overall and are of major importance.

The distribution of localities among depositional subtypes is contrasted in figure 15.10 for the Chinji and Dhok Pathan formations, which have the largest numbers of localities. Overall the figure shows that the profiles of the two floodplain-dominated formations are similar, with by far the largest number of localities being infillings in the top of smaller floodplain channels. These are typically channels that have either been abandoned (FC-S2, FC-S3) or are in late stages of flow activity (FC-C2) (Behrensmeyer et al. 2005). Such localities would be ones most likely to preserve autochthonous (nontransported) bone assemblages with relatively minor allochthonous (transported) constituents and temporal and spatial averaging on the order of 10^2 to 10^4 yr and 10^2 to 10^3 m^2 (Behrensmeyer et al. 2005). Both formations also have large numbers of patchy floodplain localities (FP-P), which

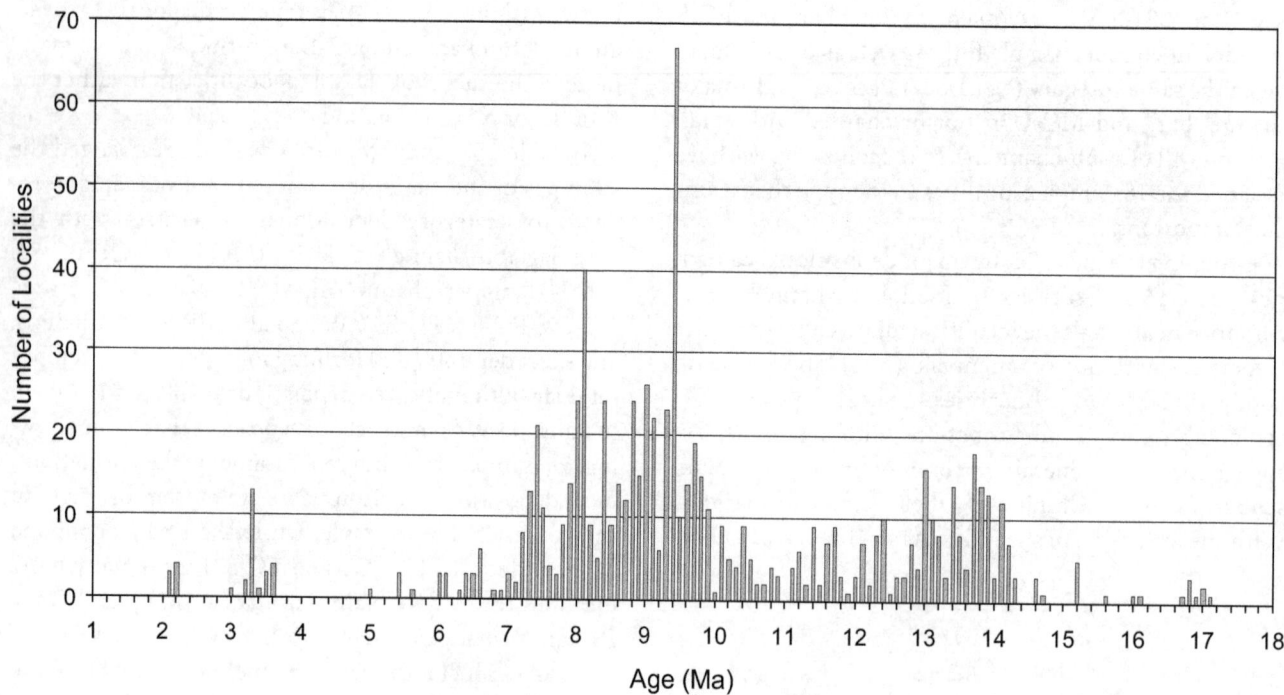

Figure 15.7 Number of localities per 100,000 yr interval. Data selected to include only localities assigned to a 100,000 yr interval (*n* = 822). The ages were calculated using the Ogg and Smith (2004) calibration of the Geomagnetic Polarity Time Scale.

Table 15.3

Number of Localities by Column and Primary Depositional Environment

	Kamlial Fm.		Chinji Fm.		Nagri Fm.		Dhok Pathan Fm.		Row Totals	
Major channel	6	24.0%	35	28.5%	11	37.9%	40	27.8%	92	28.7%
Floodplain channel	10	40.0%	62	50.4%	15	51.7%	61	42.4%	148	46.1%
Floodplain	4	16.0%	25	20.3%	3	10.3%	32	22.2%	64	19.9%
Sheet deposit	5	20.0%	1	0.8%	0	0.0%	11	7.6%	17	5.3%
Totals	25		123		29		144		321	

NOTE: Percentages are for within each formation.

typically have few individuals but high proportions of autochthonous faunal elements. Nevertheless, both formations also have substantial numbers of channel lag and bar localities (FC-C1, FC-S1). These, having formed in high-energy regimes with much reworking, are expected to have fossil assemblages dominated by allochthonous components.

Although similar in many aspects there are also noteworthy differences in the representation of depositional facies in the Chinji and Dhok Pathan formations (see

figure 15.10 and table 15.4). Small, complex floodplain channels with mixed lithologies (FC-C2) are the first-ranked facies in the Chinji Formation, comprising 22.8% of the localities. In the Dhok Pathan Formation, they are the fourth-ranked facies, at 10.4%. Small, simple floodplain channels with fine-grained fill (FC-S3) are the fourth-ranked Chinji Formation facies, comprising 9.9% of localities, while in the Dhok Pathan Formation they are only the tenth ranked. Sheet deposits (SD) are the third-ranked Dhok Pathan Formation facies at 10.4%,

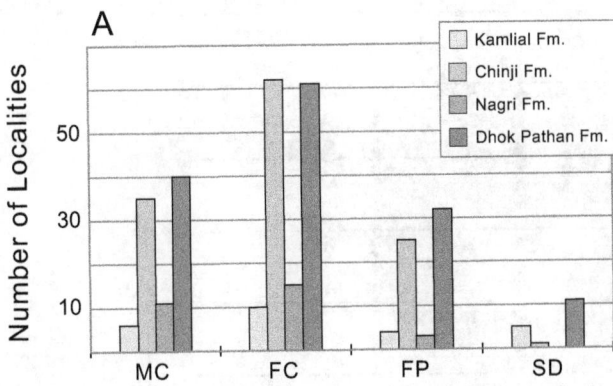

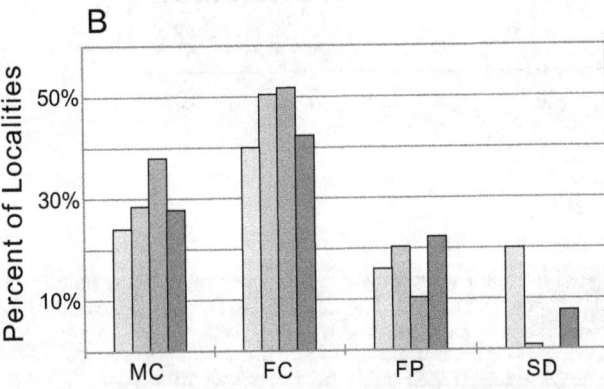

Figure 15.8 Number of localities in each major depositional category. (*A*) Number of localities. (*B*) Percent of localities. MC = major channel; FC = floodplain channel; FP = floodplain; SD = sheet deposit. Data from table 15.3 (*n* = 321).

Table 15.4

Number of Localities by Formation and Secondary Context

	Kamlial Fm.		Chinji Fm.		Nagri Fm.		Dhok Pathan Fm.	
MC-L	0	0.0%	1	1.0%	2	8.0%	0	0.0%
FC-C1	3	13.0%	5	5.0%	2	8.0%	9	8.5%
FC-S1	1	4.3%	9	8.9%	2	8.0%	8	7.5%
MC-U1	0	0.0%	3	3.0%	1	4.0%	4	3.8%
MC-U2a	0	0.0%	1	1.0%	3	12.0%	7	6.6%
MC-U2b	0	0.0%	4	4.0%	2	8.0%	6	5.7%
MC-U3a	2	8.7%	5	5.0%	0	0.0%	5	4.7%
MC-U3b	2	8.7%	2	2.0%	0	0.0%	2	1.9%
FC-C2	1	4.3%	23	22.8%	8	32.0%	11	10.4%
FC-S2	4	17.4%	14	13.9%	1	4.0%	18	17.0%
FC-S3	1	4.3%	10	9.9%	1	4.0%	4	3.8%
FP-P	4	17.4%	16	15.8%	2	8.0%	15	14.2%
FP-C	0	0.0%	5	5.0%	0	0.0%	5	4.7%
FP-L	0	0.0%	2	2.0%	1	4.0%	1	0.9%
SD	5	21.7%	1	1.0%	0	0.0%	11	10.4%
Totals	23		101		25		106	

NOTE: The total of the number of localities in the table is fewer than 321 because the secondary context of all localities is not known. Percentages are for each formation.

but in the Chinji Formation they surprisingly comprise only 1.0% of localities. The coarse fills of large channels (MC-U2a) are a minor component of Dhok Pathan Formation localities at 6.6%, but in the Chinji only an insignificant 1.0%. And, finally, in the Dhok Pathan Formation, basal lag and bar deposits of complex floodplain channels (FC-C1) make up 8.5% of the localities, while in the Chinji Formation they comprise 5.0%, a difference of 3.8%.

Interpretation of these differences is inconclusive, as there is no consistent pattern. However, the differences in the frequency of occurrence of depositional facies between all four formations have the potential to bias both biostratigraphic and paleoecological analyses, including patterns of diversity and faunal turnover. Chief among potential biases are those affecting the size spectrum of species and the degree of allochthonous versus autochthonous mixing of species from different habitats on the floodplain. Examination of such biases is a research priority, and the next two sections are steps in that direction.

The 321 fossil sites with depositional data can be divided into four groups based on the number of large and small mammal specimens they contain: (I) sites with few large or small mammals, (II) sites with many small mammals but few large mammals, (III) sites with few small mammals but many large mammals, and (IV) sites with both many small and many large mammals. (Small mammals include all rodents, bats, insectivores, and strepsirhine primates. Large mammals include all other taxa.) The data are presented in table 15.5. Localities with more than 24 large mammal specimens are classified as having many large mammals, and those with more than 9 small mammal specimens are classified as having many small mammals. Localities with 10 or more small mammal specimens are all productive screen-wash sites. Those with fewer than 5 small mammal specimens are typically sites where the small mammals were surface collected, while those with 5 to 9 specimens are mostly less productive screenwash sites. (Screen-washing is a technique for extracting very small fossils from bulk samples of sediment. Large specimens are rare in such bulk samples compared to the number of small fossils, which is the main criterion for our subdivisions.)

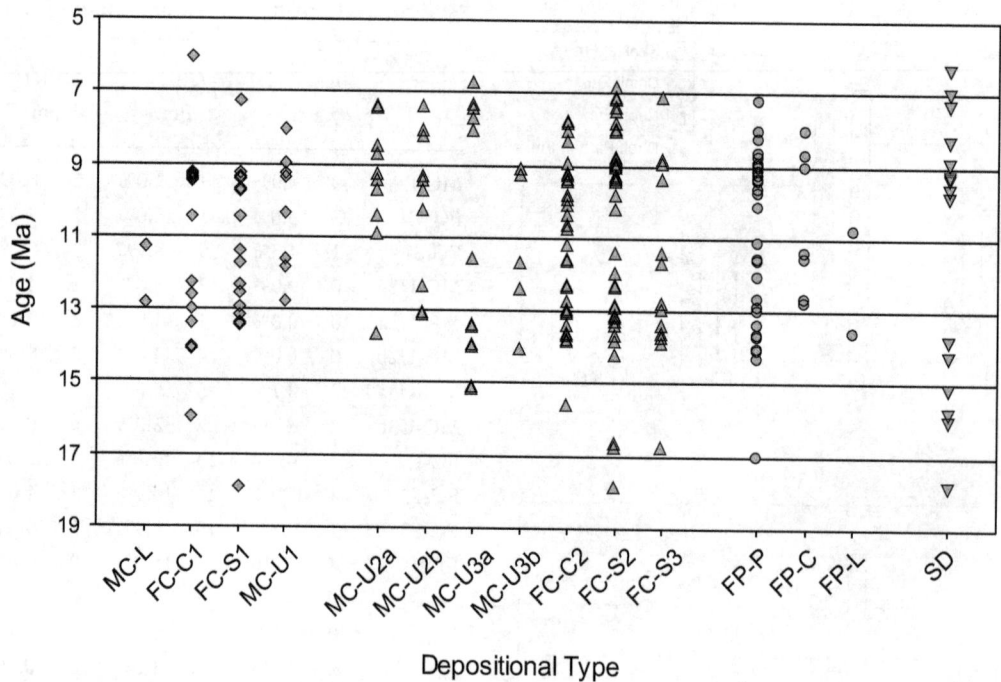

Figure 15.9 Distribution of finer depositional subcategories over time, ordered from left to right by cross-category depositional similarities. Abbreviations as in table 15.2. ⬧ = basal lag and channel bar facies; △ = upper channel fill facies; ○ = floodplain facies; ▽ = sheet deposits. Figure shows only localities with a determined secondary depositional context and known age. Data taken selectively from table 15.4 (*n* = 210).

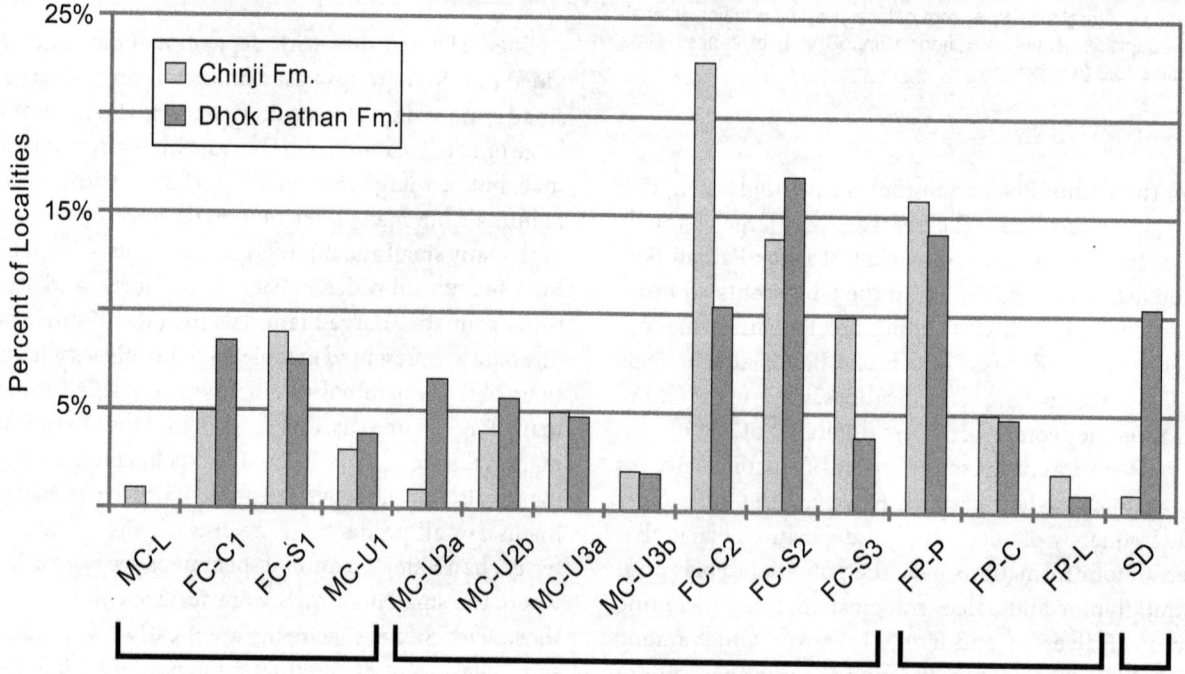

Figure 15.10 Comparison of depositional environments of the Chinji and Dhok Pathan formations. Data from table 15.4. Abbreviations listed in table 15.2; brackets at bottom mark the four cross-category depositional groups, from left to right: basal lag and channel bar facies, upper channel fill facies, floodplain facies, sheet deposits.

Table 15.5
Number of Localities by Major Depositional Environment and Number of Large or Small Mammal Specimens

Type	Description	MC		FC		FP		SD		Row Totals
I	Few small or large mammals	48	25.8%	84	45.2%	46	24.7%	8	4.3%	186
II	Many small, few large mammals	4	26.7%	3	20.0%	6	40.0%	2	13.3%	15
II	Few small, many large mammals	27	32.1%	41	48.8%	11	13.1%	5	6.0%	84
IV	Many small and large mammals	13	36.1%	20	55.6%	1	2.8%	2	5.6%	36
Totals		92		148		64		17		321

NOTE: Percentages are for within each type.

These four groups can be compared in terms of depositional facies (see table 15.5), with the expectation that sites with many small mammals occur in depositional environments different from those containing many large mammals and that sites with many fossils occur in different depositional environments than those with few fossils. Inspection of table 15.5 indicates that the first expectation is borne out, as there are large contrasts between the predominantly small mammal (type II) and predominantly large mammal (type III) sites. Because of the large differences in the number of sites in each category this is best seen in the relative frequencies (%). Sites with many large and few small mammals (type III) are mainly in the small channels of the floodplain (48.8% vs. 20.0%), while those with many small and few large mammals (type II) are largely in the floodplain paleosols (40.0% vs. 13.1%). An even larger difference is seen in the group of sites with both many large and many small mammals (type IV) (see table 15.5), where almost 56% of the sites are in the small floodplain channels and less than 3% are in floodplain paleosols. The differences among all four types and all four depositional categories (see table 15.5) are significantly different ($\chi^2 = 19.397$, $p = 0.022$). Even stronger differences are apparent if the depositional categories are collapsed into channel (MC + FC) and nonchannel (FP + SD) contexts, where again sites with many large mammals (types III + IV) are found mostly in channel settings (101 of 120) rather than nonchannel settings (19 of 120), while the predominantly small mammal sites (type II) are as likely to be in nonchannel (8 of 15) as channel settings (7 of 15). The differences are highly significant ($\chi^2 = 14.854$, $p = 0.002$).

From an analysis of the depositional settings of 13 sites with small mammals in the lower third of the Chinji Formation, Badgley, Downs, and Flynn (1998) concluded that the small mammals were mostly coming from localities in the fill of small-to-large abandoned channels; a conclusion apparently at variance with that reached here. However, with the one exception of locality Y 430, the 13 localities of that study also had large mammals as Badgley, Downs, and Flynn (1998) were contrasting sites with many small mammals to those without, irrespective of the number of large mammals. Most of their sites, therefore, were in the group of sites with both many small and many large mammals, a group that shows a very strong tendency to occur in channel contexts.

The second expectation, that sites with few fossils differ in depositional context from those with many, is only weakly supported. Collapsing the data of table 15.5 into sites with few fossils overall (type I) and sites with many (types II, III, and IV), shows small differences between the depositional settings, with the most marked being the difference in occurrence in the floodplain paleosols (46 of 186 for type I vs. 18 of 135 for types II–IV; $\chi^2 = 7.266$, $p = 0.064$), a difference that while large is not significant at a .05 level.

The preceding two cases focused on size-class distribution of the fossils, an attribute that is strongly influenced by taphonomic factors and collecting technique. It is, however, also of interest to examine potential biases in preservation of individual species, biases that are likely to vary among species and possibly are connected to a species' ecology. In order to explore possible biases, we have selected four suoids of moderate body mass as subjects. The observed range, estimated body mass, and number of sites with the species in each depositional environment ("observed") are listed in table 15.6. Three species (*Listriodon pentapotamiae*, *Conohyus sindiensis*, and *Merycopotamus nanus*) overlap considerably in their observed stratigraphic ranges, while the fourth (*Merycopotamus medioximus*) is closely related to *M. nanus* but is younger. Although not abundant, the four species are

Table 15.6

Distribution of Occurrences by Depositional Environment for Four Suoid Species

	Listriodon pentapotamiae		*Conohyus sindiensis*		*Merycopotamus nanus*		*Merycopotamus medioximus*	
Observed Range	14.1–10.4 Ma		14.6–10.4 Ma		15.1–11.4 Ma		10.2–8.5 Ma	
Estimated Body Mass	80 kg		40 kg		110 kg		150 kg	
	Obs.	Overall	Obs.	Overall	Obs.	Overall	Obs.	Overall
MC-L		2		2		1		
FC-C1	2	7	3	7	2	6	6	9
FC-S1	4	10	2	10	2	9	3	7
MC-U1	2	3	2	4	3	3	1	3
MC-U2	14	18	11	18	16	16	1	10
MC-U3	2	9	3	9	3	10	1	3
FC-C2	13	27	11	27	12	23	4	10
FC-S2	7	12	8	12	6	12	3	11
FC-S3	6	10	2	10		10		3
FP-P	9	19		18	3	17	1	13
FP-C		5		5	1	5	1	3
FP-L		3		3		2		
SD	1	2		2	2	3	2	6
Totals	60	127	42	127	50	117	23	78
Number of categories	10	13	9	13	10	13	10	11

NOTE: For each species, the tabulated numbers are: *Observed*, the number of sites with the species for each depositional category; *Overall*, the number of sites, with or without the species, for each depositional category tallied over the known stratigraphic range of the species. Abbreviations as in table 15.2.

fairly common within their observed ranges. *Conohyus sindiensis* is the rarest, as well as smallest, species.

If the fossil occurrences of a species are unrelated to the depositional context (i.e., its occurrences are random with respect to the depositional facies), then the observed distribution of a species' occurrences should mimic the related "overall" distribution. As a result, for each species in table 15.6 the "overall" distribution can be used to calculate an "expected" frequency distribution, against which the "observed" frequencies can be compared.

Figure 15.11 shows the distribution for each species by depositional environments as well as the "expected" distribution for sites within the observed range of each species. To make comparisons among the species easier, both distributions have been recast as percentages, based on the frequencies in the "observed" and "overall" columns of table 15.6. Comparisons of the panels of figure 15.11 then indicate the following:

1. Each species is found in a variety of depositional environments, rather than being mostly concentrated in one. Nevertheless, three or four categories together typically have nearly 70% of the occurrences.

2. The observed and expected for each category are generally similar, although all species have at least one facies where either the observed greatly exceeds the expected or the expected exceeds the observed.

3. *Listriodon pentapotamiae*, *Conohyus sindiensis*, and *Merycopotamus nanus* have generally similar patterns of occurrence that differ markedly from that of *Merycopotamus medioximus*. Presumably, their profiles are similar because the three species ranges overlap considerably and they come from much the same suite of localities. There are, however, important differences in the pattern of occurrences among the four species.

4. The observed occurrences of *L. pentapotamiae* come the closest to matching the expected occurrences,

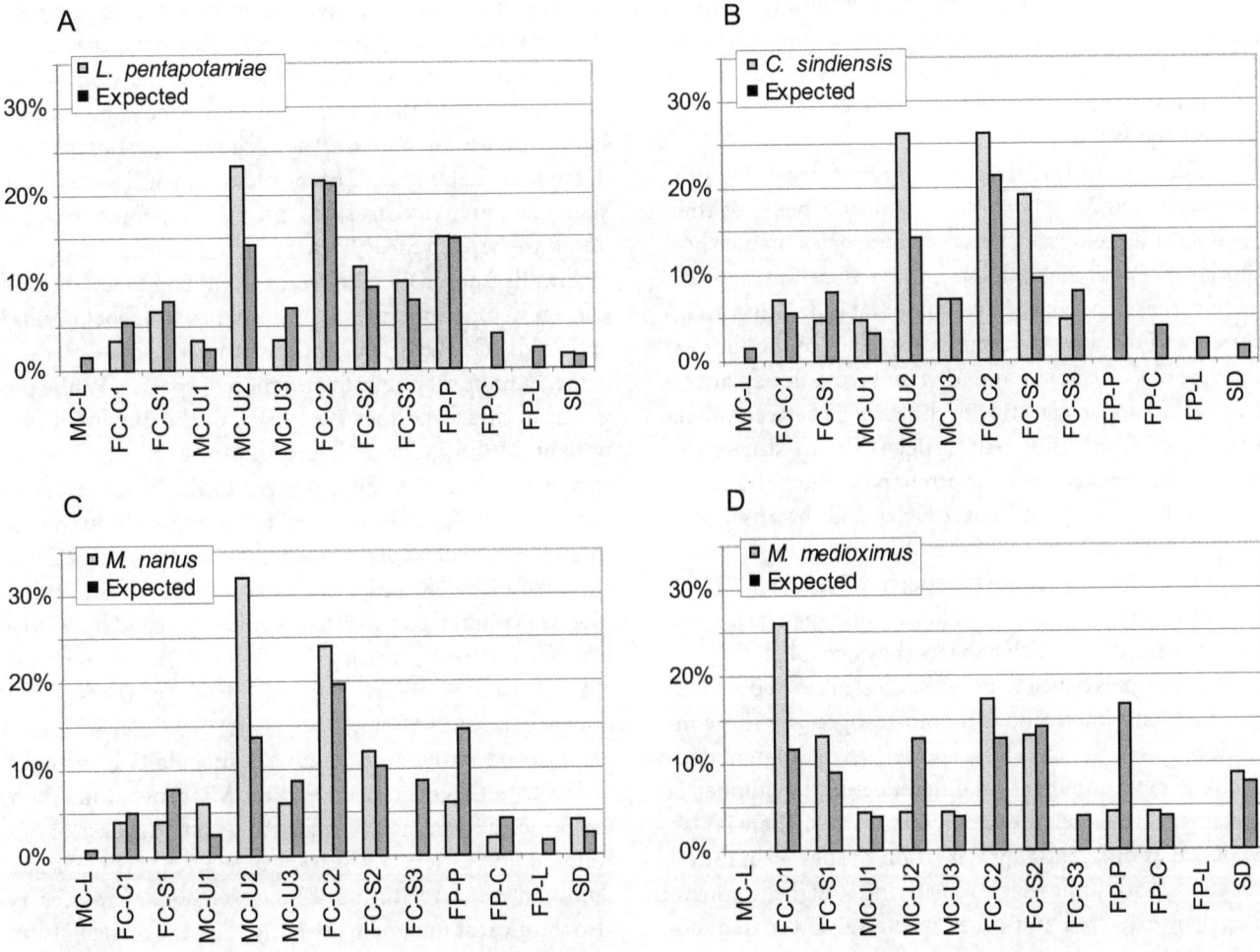

Figure 15.11 Distribution of occurrences by depositional environment for four suoid species. Percentages are based on the frequencies in the "Observed" and "Overall" columns of table 15.6. Abbreviations as in table 15.2.

although there is a major discrepancy in its being over-represented in the MC-U2 facies. The other suid, *C. sindiensis*, is also overrepresented in the MC-U2 facies, but unlike *L. pentapotamiae* it is not present in the FP-P facies at all and is overrepresented in the FC-S2 facies.

5. *M. nanus* is similar to the two suids but is even more strongly overrepresented in the MC-U2 facies and slightly underrepresented in the FP-P facies. *M. medioximus* departs considerably from the suids and the closely related *M. nanus*. It generally matches its expected distribution, occurring more frequently in the more energetic facies and less frequently in the lower-energy facies. Unlike its close relative, *M. medioximus* is overrepresented in the FC-C1 facies and underrepresented in MC-U2. *M. medioximus* is also

underrepresented in the FP-P facies, a similarity shared with *M. nanus*.

There is clearly a strong signal in the species distributions related to the frequency and productivity of different facies in the formations. The overrepresentation of *L. pentapotamiae*, *C. sindiensis*, and *M. nanus* in the MC-U2 facies, which includes localities that formed in the complex fills at the top of large channels, may simply reflect how productive these settings are of large fossil accumulations in the Chinji Formation. Such large accumulations are more likely to have many taxa, including the less common species such as suoids. In the Dhok Pathan Formation, localities in the same setting occur as frequently as in the Chinji Formation, but they are not so productive and rarely if ever host very large assemblages of fossils. Consequently, in the Dhok Pathan Formation relatively uncommon taxa such as *M. medioximus* are not frequently found in this particular facies and the facies does not contribute a significant number of specimens to the *M. medioximus* collection.

Differences in the patterns of occurrence among the two suids and *M. nanus* should reflect aspects of their habitats and ecology. Otherwise, the species should show similar preservational biases, since they are approximately the same size, have comparable relative abundance, and overlap stratigraphically. The absence or near absence of *C. sindiensis* and *M. nanus* from floodplain facies is the most suggestive of differences between them and *L. pentapotamiae*. Also suggestive of differences among the species are the overrepresentation of *C. sindiensis* in the FC-S2 and FC-C2 facies and the absence of *M. nanus* from the FC-S3 facies. In contrast to *L. pentapotamiae*, which is found at expected frequencies in all facies but one, *C. sindiensis* and *M. nanus* may have been more restricted in which habitats they occupied.

Our analyses have identified several areas of potential bias that have implications for biostratigraphy. These include biases in the size classes preserved in different depositional environments and differences in the number of specimens from localities in different facies. There is also evidence of differential preservation of species in different facies, with some species occurring at higher frequencies than expected and other species at lower frequencies—a conclusion we expect will prove more general once a larger number of species has been examined. These differences in the temporal distribution of depositional facies potentially could alter the observed stratigraphic range of the species, although the effects are likely to be subtle.

FAUNAL TURNOVER AMONG THE ARTIODACTYLA

Because they are common with numerous and distinctive species, artiodactyls are potentially very useful for defining biostratigraphic zonations. As a step toward defining biostratigraphic zones, in the following we present a brief summary of what is currently known about Siwalik artiodactyls. However, several groups, including all the ruminant families, are undergoing major revision and we expect there will be significant changes in the species lists. In addition, we do not attempt to infer first and last appearances from the fossil occurrence data, as

was done in Barry et al. (2002), although determining accurate first and last appearances is central to any biostratigraphic study. Even with these caveats, the broad outlines of our taxonomic and stratigraphic studies are now evident and the general patterns should be robust. Unless otherwise noted, all ages cited are first or last occurrences, that is the age of the oldest or youngest known specimen. In many cases, the inferred first or last appearances are beyond those limits.

Siwalik suoids include five subordinate taxa: anthracotheres, hippopotamuses, sanitheres, palaeochoerids, and suids. Of these five, anthracotheres consist of a diverse, but not very abundant, group of species. While gigantic anthracotherines are typical of the upper Chitarwata and Vihowa formations (Antoine et al., chapter 16, this volume), they are not present in the lower levels of the Kamlial Formation. The long-ranging and much smaller *Microbunodon silistrensis* is, however, present from the base of the Kamlial Formation (ca. 17.8 Ma) until 11.4 Ma (Lihoreau et al. 2004a). It is succeeded by *Microbunodon milaensis*, which is best known from sites dated at 9.3 Ma and 9.4 Ma, but extends down to 10.4 Ma (Lihoreau et al. 2004a).

Bothriodontine anthracotheres include two lineages in the Late Oligocene and earliest Miocene Chitarwata and Vihowa formations that continue into the base of the Kamlial Formation. *Sivameryx palaeindicus*, which is the dominant Chitarwata and Vihowa selenodont species, is also the most abundant in the Early Miocene of the Potwar Plateau, with first and last occurrences at 17.8 Ma and 15.6 Ma. The much larger *Hemimeryx blanfordi*, on the other hand, is only known from postcranials at one Potwar Plateau site close to the base of the Kamlial Formation (17.8 Ma). Finally, species of *Merycopotamus* are first known from the upper Kamlial Formation (15.1 Ma) and continue well into the Pliocene Tatrot Formation (3.3 Ma). Three successive species were recognized by Lihoreau et al. (2004b, 2007); *M. nanus* (15.1–11.4 Ma), *M. medioximus* (10.2–8.5 Ma), and *M. dissimilis* (7.8–3.3 Ma). The two gaps between the species' ranges are artifacts, as there are nondiagnostic fossils of *Merycopotamus* from both.

Hippopotamuses are represented in the latest Miocene and Pliocene by *Hexaprotodon sivalensis*. The species' first occurrence is most definitely documented at 6.2 Ma. However, a fragment of what might be a *Hexaprotodon* molar has been collected from a site that is between 6.5 Ma and 6.7 Ma, while a second, more securely identified molar in the Yale collections could be as old as 7.3 Ma. Hippopotamuses are typically very common as well as easily recognized fossils, but other than these two doubtful records, there are no specimens older than 6.2 Ma. *Hexaprotodon sivalensis* and *Merycopotamus dissimilis* overlap stratigraphically.

On the Potwar Plateau, sanitheres are represented by a single species, *Sanitherium schlagintweiti*, first known from the base of the Kamlial Formation (17.9 Ma) and common until it disappears between 14.2 Ma and ca. 14.0 Ma in the uppermost Kamlial Formation or lowermost Chinji Formation. Palaeochoerids are also a family with few species. While *Pecarichoerus* is recorded in the Early Oligocene Chitarwata Formation (Orliac et al. 2010), the only known Potwar Plateau specimen is the holotype of *Pecarichoerus orientalis* described by Colbert (1933) from a ca. 14.0 Ma site near the base of the Chinji Formation. A second palaeochoerid, *Schizochoerus gandakasensis*, has inferred first and last appearances at 11.3 Ma and 8.1 Ma (Barry et al. 2002).

Suids comprise a species-rich family, with at least nine distinctive lineages. They are also relatively common as fossils, giving them considerable biostratigraphic potential. *Listriodon* is first represented by sublophodont species in the earliest Miocene part of the upper Chitarwata and Vihowa formations (Orliac et al. 2010; Antoine et al., chapter 16, this volume), with *Listriodon guptai* subsequently found in the Kamlial Formation. The material is very sparse and fragmentary, but *Listriodon guptai* seems to range from 17.8 Ma to at least 15.8 Ma. The fully lophodont *Listriodon pentapotamiae*, which is very common within its time range, first occurs at 14.1 Ma and perhaps as early as 14.3 Ma. Systematic revision may alter these dates, as the earliest specimens are similar to *Listriodon guptai*. The inferred last appearance of *Listriodon pentapotamiae* is at 10.4 Ma, overlapping with the first appearance of equids (Barry et al. 2002).

Siwalik tetraconodontines include the small *Conohyus sindiensis*, which resembles *Listriodon pentapotamiae* in both being very common and overlapping with equids. It has a first occurrence in the upper Kamlial formation at 14.6 Ma and a last occurrence of at least 10.4 Ma, and possibly younger. The gigantic tetraconodontines *Tetraconodon magnus* and *Sivachoerus prior*, in contrast, have much more restricted ranges. Their known ranges are 10.0 Ma to 9.4 Ma and 3.5 Ma to 2.1 Ma, respectively.

Suines are the most diverse of the Siwalik suids. The oldest species is "*Hyotherium*" *pilgrimi*, which first occurs at 13.7 Ma and is last known at 11.1 Ma. Succeeding "*Hyotherium*" *pilgrimi* in the Late Miocene are species of *Hippopotamodon*, *Propotamochoerus*, *Hippohyus*, *Potamochoerus*, and early *Sus*. Species of *Hippopotamodon* were inferred by Barry et al. (2002) to range from 11.5 Ma to

7.3 Ma, but the genus might still have been present at ca. 6.3 Ma. The earliest *Hippopotamodon* is a relatively small form, likely to be a different species from later *Hippopotamodon sivalense*. The transition between the two forms is at ca. 10.4 Ma. Material presently identified as *Hippopotamodon sivalense* shows a strong trend toward increasing size and may well encompass two distinct species. Similarly, material identified as *Propotamochoerus hysudricus*, which has an inferred range of 10.3 Ma to 6.7 Ma, may belong to more than one species. Species of *Hippohyus*, *Potamochoerus*, and *Sus* all appear much later at the end of the Miocene, but their first occurrences are difficult to determine because of the sparse latest Miocene record and fragmentary material. The oldest record of *Hippohyus* is between 6.6 Ma and 6.4 Ma, that of *Potamochoerus* is at 6.0 Ma, and our oldest record of *Sus* is at 6.5 Ma. These three taxa are still present in the Tatrot Formation at about 3.3 Ma, but they might well have persisted until much later. Pliocene suids include a taxon similar to *Kolpochoerus* and the small *Sivahyus punjabiensis*, both of which are present in the Tatrot Formation (ca. 3.5–3.3 Ma). In addition, a few other, typically very small, taxa are present. In most cases, these are only known from a few specimens of uncertain identity. Most notable among them are a very small suine known from six localities that range in age from 10.1 Ma to 7.5 Ma and a tiny lophodont species from a single locality at 11.4 Ma. Both likely weighed only around 10–15 kg.

Fossil camels are not known in the Miocene of northern Pakistan, but West (1981) reports an Upper Pliocene occurrence in the Marwat Formation. The age is based on faunal correlations and the specimen should be between 3 and 2 million years old.

Common Siwalik ruminants include tragulids, giraffoids, and bovids. Cervids are also known, but not until the Pliocene, with a first occurrence at 3.5 Ma (Flynn et al., chapter 14, this volume). Although common in the earliest Miocene of the upper Chitarwata Formation, species of the enigmatic *Bugtimeryx* are not known from the Potwar Plateau. The genus, however, can only be recognized on the basis of its teeth because the postcranials are virtually identical with those of early bovids.

Tragulids are known from the Early Oligocene of the lower Chitarwata Formation (Antoine et al., chapter 16, this volume) and presumably have a continuing, but poorly documented, presence on the Subcontinent. On the Potwar Plateau, at least one small species of *Dorcabune* is present throughout the Kamlial Formation. The oldest record is at 16.8 Ma, but Antoine et al. (chapter 16, this volume) record *Dorcabune* in the lowermost Vihowa

Formation at Dera Bugti, which is probably slightly older than the base of the Kamlial Formation. Small *Dorcabune* range up into the Chinji Formation, with a last occurrence at 13.6 Ma. The very much larger *Dorcabune anthracotherioides* first occurs at 13.6 Ma and persists until 10.6 Ma, after which the much smaller *Dorcabune nagrii* first occurs. This latter species is the last of the *Dorcabune* lineage, with an inferred last appearance of about 8.4 Ma (Barry et al. 2002).

The systematics of *Dorcatherium* is more complex. There are numerous species that generally have to be distinguished on the basis of size, while the relationships between the smaller species of *Dorcatherium* and *Siamotragulus* have yet to be resolved. *Dorcatherium* (or perhaps *Siamotragulus*) is present in the latest Oligocene and earliest Miocene of the Chitarwata and Vihowa formations, while one medium-size species is present at the base of the Kamlial Formation (17.8 Ma). More than two coexisting species of *Dorcatherium* are not known until about 14 Ma, but throughout the rest of the Miocene there are typically three species present; a very small, a medium, and in the Late Miocene what becomes a very large species. The smallest forms constitute three or four separate species that are not likely ancestor–descendent pairs. By ca. 12 Ma, they have a foot structure similar to that of *Tragulus*. Tragulids are a major component of Middle Miocene Siwalik faunas, at some sites being nearly as abundant as bovids. After about 9.5 Ma, they have become markedly rarer as fossils. The younger species tend toward being high crowned, and the youngest species in the Tatrot Formation (3.3 Ma) could well be placed in a different genus.

Giraffoids have a long Siwalik record and, while having low species richness, are consistently found starting in the earliest Miocene part of the Chitarwata Formation and continuing well into the Pliocene. *Progiraffa exigua*, which we consider a primitive giraffoid, is the common large ruminant of the upper Chitarwata and Vihowa formations. It is also present throughout the Kamlial Formation, although the evidence suggests there may be as many as three time-successive species that differ in size (Barry et al. 2005). The youngest specimens assigned to *Progiraffa* are at least 13.6 Ma and maybe as young as 13.2 Ma. The Middle and Late Miocene giraffids of the Potwar Plateau are currently under study by N. Solounias. While in the past the material has been treated as comprising a small earlier species (*Giraffokeryx punjabiensis*) and a later large species (*Bramatherium megacephalum*; e.g., Barry et al. 2002), Solounias suggests there is evidence for more species, some of which may coexist (N. Solounias, pers. comm., February 2011). Spe-

cifically, the material from the uppermost Kamlial Formation (ca. 14.0 Ma) to the lower Nagri Formation (10.4 Ma), which had been referred to *Giraffokeryx punjabiensis*, must belong to at least two species. Similarly, the large sivatheres of the Nagri through Dhok Pathan Formations (10.4–7.3 Ma) appear to be a pair of time-successive species, with the younger form being considerably larger than the older. Finally, *Giraffa punjabiensis* has been inferred to have a first appearance of 9.1 Ma (Barry et al. 2002). The youngest fossils of it are 7.2 Ma.

Bovids constitute a very abundant and diverse set of species that are currently undergoing major revision. They have a complex and not-yet-well-understood history on the Subcontinent, and here we only make the most general statements about them.

As the name suggests, hypsodontines are characterized by extremely high-crowned teeth. While there is not a consensus as to their phylogenetic affinities, they are generally placed with the Bovidae because they possess unbranched horncores covered by a sheath. *Palaeohypsodontus zinensis* is known from the Early and Late Oligocene of the Chitarwata Formation and earliest Miocene Vihowa Formation (Antoine et al., chapter 16, this volume). Hypsodontines are not, however, known on the Potwar Plateau until 15.2 Ma, undoubtedly because of the dearth of fossils in the Kamlial Formation. Many horncores and teeth of *Hypsodontus sokolovi* have been recovered from the uppermost Kamlial Formation /lowermost Chinji Formation. Most of the material comes from a single site between 14.2 Ma and 14.0 Ma and is comparable in size to the Kamlial material. A larger second species, *Hypsodontus pronaticornis*, occurs between 13.6 Ma and 12.1 Ma.

Bovids other than hypsodontines first occur at early Miocene Vihowa Formation sites on the order of 19 Ma (Barry et al. 2005; Flynn et al., chapter 14, and Antoine et al., chapter 16, both this volume). Antoine et al. (chapter 16, this volume) also reports bovids from the earliest Miocene of the Chitarwata Formation, although the records are based on postcranials that are difficult to distinguish from those of *Bugtimeryx*. The Vihowa bovids include at least two species belonging to divergent taxa. One referred to *Eotragus* has round horncores, while the other has a more compressed horn. On the Potwar Plateau, *Eotragus* occurs at the base of the Kamlial Formation at 17.8 Ma. Teeth and postcranials indicate the presence of several small unidentified species between 17.8 Ma and 15.1 Ma.

The fossil record improves markedly at the base of the Chinji Formation (ca. 14.0 Ma), and at that juncture there is much greater richness of small- to medium-size bovid species that are well represented by horncores, teeth, and postcranials—an increase that is certainly an artifact of the record. Between 14 Ma and 11 Ma, species of four medium-size genera (*Strepsiportax*, *Sivaceros*, *Sivoreas*, and *Helicoportax*) are common. Additionally, there are two or even three antilopines, a small boselaphine, and a few specimens of *Protoryx* and *Tethytragus*. All of these taxa disappear from the Potwar Plateau Siwalik record by 11.3 Ma, after which time *Selenoportax vexillarius*, "*Tragocerides*" *pilgrimi*, *Elachistoceras khauristanensis*, and a small *Gazella* are most common. During the Late Miocene, there is turnover of species in both *Selenoportax* and "*Tragocerides*," and additional species appear, including antilopines other than *Gazella* and reduncines. There is a noticeable trend toward increase in size among the Late Miocene bovids, and several species have very brief durations compared to the more common species.

Further changes in the bovid fauna at the end of the Miocene are apparent, but they are much obscured by the deteriorating quality of the fossil record on the Potwar Plateau. We therefore do not review them here.

CONCLUSION

Neogene age sediments exist over a wide area in Pakistan, India, and Nepal, with contemporaneous and equivalent sediments in Myanmar (Burma). Throughout this area the Siwaliks are the fluvial phase of infilling of foreland basins created by the collision of peninsular India with mainland Asia. In some regions, sediments exist that represent a transitional phase between marine and fluvial deposition. These are best represented by the Chitarwata Formation and perhaps the Gaj and Murree formations, all in Pakistan.

On the Potwar Plateau of northern Pakistan, and probably elsewhere, the fluvial Siwaliks were deposited by a very large river system. This ancient system bears many similarities to the modern Indus and Ganges systems, with their broad floodplains and multitudes of tributaries. The preserved parts of the system—at least on the Potwar Plateau—include large and medium size tributary rivers that originated in the adjacent northern mountains or on the floodplain itself. These tributary rivers left thick sandstones and minor associated overbank deposits. Such deposits are major constituents of all four formations, but they are particularly important in the Nagri and Kamlial formations. There were, in addition, smaller floodplain streams and their associated levees, crevasse-splays, and paleosols. These predominate in the Chinji and Dhok Pathan formations.

Vertebrate and invertebrate fossils typically occur in the Siwaliks as distinct clusters of small areal extent that originate from a very limited stratigraphic horizon. Such clusters or localities may contain from a very few to, in rare cases, over a thousand fossils. The fossils are usually the disarticulated remains of terrestrial and aquatic vertebrates, but bivalves, gastropods, and crabs also occur. Plant remains, whether pollen or body fossils, are very rare. Paleosols and stable isotopes provide indirect evidence of the vegetation. Localities with more than a few fossils ordinarily have several species.

The fossils mostly occur in the infillings at the top of channels and especially the channels of the smaller streams. However, fossils are common in other lithologies, including paleosol, crevasse-splay, and levee deposits and even in the basal lag of the very largest sandstones. A classification of depositional facies recognizes four main types and 15 subcategories. Of the subcategories, the small floodplain channels (FC-C2 and FC-S2) and paleosols with patchy concentrations (FP-P) are consistently the most productive facies. Sheet deposits (SD) are also important in the Kamlial and Dhok Pathan formations. A preliminary analysis indicates that localities with only fossils of small mammal species are most often found in the floodplain paleosols, while large species (with or without many small species) are more often in the various small floodplain channel facies. In addition, there is strong evidence that the occurrences of individual species are not random with respect to the depositional facies. In some cases, this is a result of large differences of the number of fossils in highly productive facies, which entails more frequent recovery of rarer species. In other cases, the patterns of occurrence among species are likely to reflect aspects of their habitats and ecology.

At present our Siwalik database includes over 1400 localities, the majority of which are on the Potwar Plateau. Of these, 810 localities have absolute ages precise enough to be assigned to a 100 kyr interval. Ages of localities range from 1.4 Ma to 17.9 Ma, but there are significantly older localities in the Murree, Chitarwata, and perhaps Gaj formations that have not yet been given absolute ages. On the Potwar Plateau, the segment between 14.3 Ma and 6.0 Ma has the most localities and is the most complete, with only three 100 kyr intervals having no localities.

The long and well-dated sequence of Siwalik sediments and fauna offers an opportunity to document changes in diversity and patterns of faunal turnover in sufficient detail that they can be used to test the relationships between environmental change and the ecological and evolutionary responses of the biota. Long-term research goals therefore include study of those relationships as well as the usual focus on the evolutionary histories of species and clades and reconstruction of the paleoecology of individual species. In all of these endeavors, special concerns are uneven sampling through time, differential preservation of larger-bodied animals and durable parts such as teeth, errors in age dating imposed by uncertainties in correlation and geomagnetic polarity time-scale calibrations, and uneven taxonomic treatment across groups.

ACKNOWLEDGMENTS

We have been fortunate in having a long, productive collaboration with the Geological Survey of Pakistan that has extended from 1973 to 2011. Various Directors General have supported the project over the years, and many more junior members have been active and valued participants. The contributions of Dr. Imran Khan, presently the DG of the Survey, Dr. S. Ibrahim Shah, M. Anwar, and K. A. Sheikh are especially acknowledged, as are those of the late Will Downs and Noye Johnson.

Financial support has come through a number of grants from the National Science Foundation, the Smithsonian Foreign Currency Program, the Wenner-Gren Foundation for Anthropological Research, and the American School for Prehistoric Research. We also thank Ms. Francine Berkowitz of the Smithsonian Office of International Relations.

REFERENCES

Badgley, C. E. 1986. Taphonomy of mammalian fossil remains from Siwalik rocks of Pakistan. *Paleobiology* 12:119142.

Badgley, C. E., W. S. Bartels, M. E. Morgan, A. K. Behrensmeyer, and S. M. Raza. 1995. Taphonomy of vertebrate assemblages from the Paleogene of northwestern Wyoming and the Neogene of northern Pakistan. In *Long Records of Continental Ecosystems*, ed. C. E. Badgley and A. K. Behrensmeyer, pp. 157–180. *Palaeogeography, Palaeoclimatology, Palaeoecology* 115.

Badgley, C. E. and A. K. Behrensmeyer. 1980. Paleoecology of Middle Siwalik sediments and faunas, northern Pakistan. *Palaeogeography, Palaeoclimatology, Palaeoecology* 30:133–155.

Badgley, C. E. and A. K. Behrensmeyer. 1995. Preservational, paleoecological and evolutionary patterns in the Paleogene of Wyoming-Montana and the Neogene of Pakistan. In *Long Records of Continental Ecosystems*, ed. C. Badgley and A. K. Behrensmeyer. *Palaeogeography, Palaeoclimatology, Palaeoecology* 115:319–340.

Badgley, C. E., W. Downs, and L. J. Flynn. 1998. Taphonomy of small-mammal fossil assemblages from the Middle Miocene Chinji Formation, Siwalik Group, Pakistan. In *Advances in Vertebrate Paleontology and Geochronology*, ed. Y. Tomida, L. J. Flynn, and L. L. Jacobs, pp. 145–166. National Science Museum Monographs 14.

Badgley, C. E., and L. Tauxe. 1990. Paleomagnetic stratigraphy and time in sediments: Studies in alluvial Siwalik rocks of Pakistan. *Journal of Geology* 98:457–477.

Barndt, J. 1977. The magnetic polarity stratigraphy of the type locality of the Dhok Pathan Faunal Stage, Potwar Plateau, Pakistan. M.S. thesis, Dartmouth College.

Barndt, J., N. M. Johnson, G. D. Johnson, N. D. Opdyke, E. H. Lindsay, D. Pilbeam, and R. A. H. Tahirkheli. 1978. The magnetic polarity stratigraphy and age of the Siwalik Group near Dhok Pathan Village, Potwar Plateau, Pakistan. *Earth and Planetary Science Letters* 41:355–364.

Barry, J. C., A. K. Behrensmeyer, and M. Monaghan. 1980. A geologic and biostratigraphic framework for Miocene sediments near Khaur Village, northern Pakistan. *Postilla* 183:1–19.

Barry, J. C. and I. U. Chemma. 1984. Notes on a small fossil collection from near Gali Jagir on the Potwar Plateau of Pakistan. *Memoirs of the Geological Survey of Pakistan* 11:15–18.

Barry, J. C., S. Cote, L. MacLatchy, E. H. Lindsay, R. Kityo, and A. R. Rajpar. 2005. Oligocene and Early Miocene ruminants (Mammalia, Artiodactyla) from Pakistan and Uganda. *Palaeontologia Electronica* 8(1); 22A:29 pp.; http://palaeo-electronica.org/paleo/2005_1/barry22/issue1_05.htm.

Barry, J. C., E. H. Lindsay, and L. L. Jacobs. 1982. A biostratigraphic zonation of the middle and upper Siwaliks of the Potwar Plateau of northern Pakistan. *Palaeogeography, Palaeoclimatology, Palaeoecology* 37:95–130.

Barry, J. C., M. E. Morgan, L. J. Flynn, D. Pilbeam, A. K. Behrensmeyer, S. M. Raza, I. A. Khan, C. Badgley, J. Hicks, and J. Kelley. 2002. Faunal and environmental change in the Late Miocene Siwaliks of Northern Pakistan. *Paleobiology Memoirs* 28:1–71.

Beck, R. A. and D. W. Burbank. 1990. Continental-scale diversion of rivers: A control of alluvial stratigraphy. *Geological Society of America, Abstracts and Programs* 22:238.

Behrensmeyer, A. K. 1982. Time resolution in fluvial vertebrate assemblages. *Paleobiology* 8:211–227.

Behrensmeyer, A. K. 1987. Miocene fluvial facies and vertebrate taphonomy in northern Pakistan. In *Recent Developments in Fluvial Sedimentology*, ed. F. G. Ethridge, R. M. Flores, and M. D. Harvey, pp. 169–176. Society for Economic Paleontology and Mineralogy Special Publication 39.

Behrensmeyer, A. K., C. E. Badgley, J. C. Barry, M. E. Morgan, and S. M. Raza. 2005. The paleoenvironmental context of Siwalik Miocene vertebrate localities. In *Interpreting the Past: Essays on Human, Primate, and Mammal Evolution in Honor of David Pilbeam*, ed. D. E. Lieberman, R. J. Smith, and J. Kelley, pp. 47–61. Boston: Brill.

Behrensmeyer, A. K. and J. C. Barry. 2005. Biostratigraphic surveys in the Siwaliks of Pakistan: A method for standardized surface sampling of the vertebrate fossil record. *Palaeontologia Electronica* 8(1); 15A:24 pp.; http://palaeo-electronica.org/2005_1/behrens15/issue1_05.htm.

Behrensmeyer, A. K., J. Quade, T. E. Cerling, J. Kappelman, I. A. Khan, P. Copeland, L. Roe, J. Hicks, P. Stubblefield, B. J. Willis, and C. Latorre. 2007. The structure and rate of late Miocene expansion of C4 plants: Evidence from lateral variation in stable isotopes in paleosols of the Siwalik Group, northern Pakistan. *Geological Society of America Bulletin* 119:1486–1505.

Behrensmeyer, A. K. and L. Tauxe. 1982. Isochronous fluvial systems in Miocene deposits of northern Pakistan. *Sedimentology* 29:331–352.

Behrensmeyer, A. K., B. J. Willis, and J. Quade. 1995. Floodplains and paleosols of Pakistan Neogene and Wyoming Paleogene deposits: Implications for the taphonomy and paleoecology of faunas. In *Long Records of Continental Ecosystems*, ed. C. Badgley and A. K. Behrensmeyer. *Palaeogeography, Palaeoclimatology, Palaeoecology* 115:37–60.

Belmaker, M., S. Nelson, M. Morgan, J. Barry, and C. Badgley. 2007. Mesowear analysis of ungulates in the middle to late Miocene of the Siwaliks, Pakistan: Dietary and paleoenvironmental implications. *Journal of Vertebrate Paleontology* 27 (suppl. 3):46A.

Cande, S. C. and D. V. Kent. 1995. Revised calibration of the geomagnetic polarity timescale for the Late Cretaceous and Cenozoic. *Journal of Geophysical Research* 100B: 6093–6095.

Cheema, M. R., S. M. Raza, and H. Ahmed. 1977. Cainozoic. In *Stratigraphy of Pakistan*, ed. S. M. I. Shah, pp. 56–98. Memoirs of the Geological Survey of Pakistan 12.

Colbert, E. H. 1933. An upper Tertiary peccary from India. *American Museum Novitates* 635:1–9.

Colbert, E. H. 1935. Siwalik mammals in the American Museum of Natural History. *Transactions of the American Philosophical Society*, n.s., 26:1–401.

Cotter, G. de P. 1933. The geology of the part of the Attock District west of longitude 72° 45' E. *Memoirs of the Geological Survey of India* 55:63–161.

Dehm, R., Therese Prinzessin zu Oettingen-Spielberg, and H. Vidal. 1958. Palaontologische und geologische Untersuchungen im Tertiar von Pakistan. I. Die Munchner Forschungsreise nach Pakistan, 1955–1956. *Abhandlungen Bayerische Akademie der Wissenschaften, Mathematisch-Naturwissenschaftliche Klasse*, n.s., 90:1–13.

Downing, K. F. and E. H. Lindsay. 2005. Relationship of Chitarwata Formation paleodrainage and paleoenvironments to Himalayan tectonics and Indus River paleogeography. *Palaeontologia Electronica* 8(1); 20A:12 pp.; http://palaeo-electronica.org/2005_1/downing20/issue1_05.htm

Downing, K. F., E. H. Lindsay, W. R. Downs and S. E. Speyer. 1993. Lithostratigraphy and vertebrate biostratigraphy of the early Miocene Himalayan foreland Zinda Pir Dome, Pakistan. *Sedimentary Geology* 87:25–37.

Friend, P. F., S. M. Raza, G. Geehan, and K. A. Sheikh. 2001. Intermediate-scale architectural features of the fluvial Chinji Formation (Miocene), Siwalik Group, northern Pakistan. *Journal of the Geological Society, London* 158:163–177.

Hussain, S. T. 1971. Revision of *Hipparion* (Equidae, Mammalia) from the Siwalik Hills of Pakistan and India. *Abhandlungen Bayerische Akademie der Wissenschaften, Mathematisch-Naturwissenschaftliche Klasse*, n. s., 147:1–68.

Hussain, S. T., J. Munthe, S. M. I. Shah, R. W. West, and J. R. Lukacs. 1979. Neogene stratigraphy and fossil vertebrates of the Daud Khel area, Mianwali District Pakistan. *Memoirs of the Geological Survey of Pakistan* 13:1–27.

Hussain, S. T., G. D. van den Bergh, K. J. Steensma, J. A. de Visser, J. de Vos, M. Arif, J. van Dam, P. Y. Sondaar, and S. B. Malik.

1992. Biostratigraphy of the Plio-Pleistocene continental sediments (Upper Siwaliks) of the Mangla-Samwal Anticline, Asad Kashmir, Pakistan. *Proceedings of the Koninklijke Nederlandse Akademie van Wetenschappen* 95:65–80.

Johnson, G. D., N. D. Opdyke, S. K. Tandon, and A. C. Nanda. 1983. The magnetic polarity stratigraphy of the Siwalik Group at Haritalyangar (India) and a last appearance datum for *Ramapithecus* and *Sivapithecus* in Asia. *Palaeogeography, Palaeoclimatology, Palaeoecology* 44:223–249.

Johnson, N. M. and V. E. McGee. 1983. Magnetic polarity stratigraphy: Stochastic properties of data, sampling problems, and the evaluation of interpretations. *Journal of Geophysical Research* 88B:1213–1221.

Johnson, N. M., N. D. Opdyke, G. D. Johnson, E. H. Lindsay, and R. A. K. Tahirkheli 1982. Magnetic polarity stratigraphy and ages of Siwalik group rocks of the Potwar Plateau, Pakistan. *Palaeogeography, Palaeoclimatology and Palaeoecology* 37:17–42.

Johnson, N. M., K. A. Sheikh, E. Dawson-Saunders, and L. E. McRae. 1988. The use of magnetic-reversal time lines in stratigraphic analysis: A case study in measuring variability in sedimentation rates. In *New Perspectives in Basin Analysis*, ed. K. L. Kleinspehn and C. Paola, pp. 189–200. New York: Springer.

Johnson, N. M., J. Stix, L. Tauxe, P. F. Cerveny, and R. A. K. Tahirkheli. 1985. Paleomagnetic chronology, fluvial processes and tectonic implications of the Siwalik deposits near Chinji Village, Pakistan. *Journal of Geology* 93:27–40.

Kappelman, J. 1986. The paleoecology and chronology of the Middle Miocene hominoids from the Chinji Formation of Pakistan. Ph.D. diss., Harvard University.

Kappelman, J., J. Kelley, D. Pilbeam, K. A. Sheikh, S. Ward, M. Anwar, J. C. Barry, B. Brown, P. Hake, N. M. Johnson, S. M. Raza, and S. M. I. Shah. 1991. The earliest occurrence of Sivapithecus from the Middle Miocene Chinji Formation of Pakistan. *Journal of Human Evolution* 21:61–73.

Lewis, G. E. 1937. A new Siwalik correlation (India). *American Journal of Science*, 5th ser., 33:191–204.

Lihoreau, F., J. Barry, C. Blondel, and M. Brunet. 2004b. A new species of Anthracotheriidae, *Merycopotamus medioximus* nov. sp. from the Late Miocene of the Potwar Plateau, Pakistan. *Comptes rendus Palevol* 3:653–662.

Lihoreau, F., J. Barry, C. Blondel, Y. Chaimanee, J.-J. Jaeger, and M. Brunet. 2007. Anatomical revision of the genus *Merycopotamus* (Artiodactyla; Anthracotheriidae): Its significance for Late Miocene mammal dispersal in Asia. *Palaeontology* 50:503–524.

Lihoreau, F., C. Blondel, J. Barry, and M. Brunet. 2004a. A new species of the genus *Microbunodon* (Anthracotheriidae, Artiodactyla) from the Miocene of Pakistan: Genus revision, phylogenetic relationships and palaeobiogeography. *Zoologica Scripta* 33:97–115.

McMurtry, M. G. 1980. Facies changes and time relationships along a sandstone stratum, Middle Siwalik Group, Potwar Plateau, Pakistan. B.S. thesis, Dartmouth College.

McRae, L. E. 1990. Paleomagnetic isochrons, unsteadiness, and nonuniformity of sedimentation in Miocene fluvial strata of the Siwalik Group, northern Pakistan. *Journal of Geology* 98:433–456.

Métais, G., P.-O. Antoine, S. R. H. Baqri, J.-Y Crochet, D. De Franceschi, L. Marivaux, and J.-L. Welcomme. 2009. Lithofacies, depositional environments, regional biostratigraphy and age of the Chitarwata Formation in the Bugti Hills, Balochistan, Pakistan. *Journal of Asian Earth Sciences* 34:154–167.

Morgan, M. E., A. K. Behrensmeyer, C. Badgley, J. C. Barry, S. Nelson, and D. Pilbeam. 2009. Lateral trends in carbon isotope ratios reveal a Miocene vegetation gradient in the Siwaliks of Pakistan. *Geology* 37:103–106.

Munthe, J., S. T. Hussain, J. R. Lukacs, R. W. West, and S. M. I. Shah. 1979. Neogene stratigraphy of the Daud Khel area, Mianwali District, Pakistan. *Contributions in Biology and Geology, Milwaukee Public Museum* 23:1–18.

Najman, Y., E. Garzanti, M. Pringle, M. Bickle, J. Stix, and I. Khan. 2003. Early-Middle Miocene paleodrainage and tectonics in the Pakistan Himalaya. *Geological Society of America Bulletin* 115:1265–1277.

Nelson, S. V. 2005. Habitat requirements and the extinction of the Miocene ape, *Sivapithecus*. In *Interpreting the Past: Essays on Human, Primate, and Mammal Evolution in Honor of David Pilbeam*, ed. D. E. Lieberman, R. J. Smith, and J. Kelley, pp. 145–166. Boston: Brill.

Nelson, S. V. 2007. Isotopic reconstructions of habitat change surrounding the extinction of *Sivapithecus*, a Miocene hominoid, in the Siwalik Group of Pakistan. *Palaeogeography, Palaeoclimatology, Palaeoecology* 243:204–222

Ogg, J. G. and A. G. Smith. 2004. The geomagnetic polarity time scale. In *A Geological Time Scale 2004*, ed. F. M. Gradstein, J. G. Ogg, and A. G. Smith, pp. 63–86. Cambridge: Cambridge University Press.

Opdyke, N. D., E. H. Lindsay, G. D. Johnson, N. M. Johnson, R. A. K. Tahirkheli, and M. A. Mirza 1979. Magnetic polarity stratigraphy and vertebrate paleontology of the Upper Siwalik Subgroup of northern Pakistan. *Palaeogeography, Palaeoclimatology, Palaeoecology* 27:1–34.

Orliac, M. J., P.-O. Antoine, G. Roohi, and J.-L. Welcomme. 2010. Suoidea (Mammalia, Cetartiodactyla) from the Early Oligocene of the Bugti Hills, Balochistan, Pakistan. *Journal of Vertebrate Paleontology* 30:1300–1305.

Pentland, J. B. 1828. Description of fossil remains of some animals from the North-East border of Bengal. *Transactions of the Geological Society of London* 2:393–394.

Pilgrim, G. E. 1913. The correlation of the Siwaliks with mammal horizons of Europe. *Records of the Geological Survey of India* 43:264–326.

Pilgrim, G. E. 1926. The fossil Suidae of India. *Memoirs of the Geological Survey of India*, n.s., 8:1–65.

Pilgrim, G. E. 1934. Correlation of the ossiferous sections in the Upper Cenozoic of India. *American Museum of Natural History Novitates* 704:1–5.

Pillans, B., M. Williams, D. Cameron, R. Patnaik, J. Hogarth, A. Sahni, J. C. Sharma, F. Williams, and R. L. Bernor. 2005. Revised correlation of the Haritalyangar magnetostrtigraphy, Indian Siwaliks: Implications for the age of the Miocene hominids *Indopithecus* and *Sivapithecus*, with a note on a new hominid tooth. *Journal of Human Evolution* 48:507–515.

Quade, J., J. M. L. Cater, T. P. Ojha, J. Adam, and T. M. Harrison. 1995. Late Miocene environmental change in Nepal and the northern Indian subcontinent: Stable isotopic evidence from paleosols. *Geological Society of America Bulletin* 107:1381–1397.

Quade, J. and T. E. Cerling. 1995. Expansion of C_4 grasses in the Late Miocene of northern Pakistan: Evidence from stable isotopes in paleosols. In *Long Records of Continental Ecosystems*, ed. C. Badgley and A. K. Behrensmeyer, pp. 91–116. *Palaeogeography, Palaeoclimatology, Palaeoecology* 115.

Quade, J., T. E. Cerling, and J. R. Bowman. 1989. Development of Asian monsoon revealed by marked ecological shift during the latest Miocene in northern Pakistan. *Nature* 342:163–166.

Raynolds, R. G. H. 1981. Did the ancestral Indus flow into the Ganges drainage? *University of Peshawar Geological Bulletin* 14:141–150.

Raza, S. M., J. C. Barry, G. E. Meyer, and L. Martin. 1984. Preliminary report on the geology and vertebrate fauna of the Miocene Manchar Formation, Sind, Pakistan. *Journal of Vertebrate Paleontology* 4:584–599.

Retallack, G. J. 1991. *Miocene Paleosols and Ape Habitats of Pakistan and Kenya*. Oxford: Oxford University Press.

Saunders, J. J. and B. K. Dawson. 1998. Bone damage patterns produced by extinct hyena, *Pachycrocuta brevirostris* (Mammalia:carnivora), the Haro River Quarry, northwestern Pakistan. In *Advances in Vertebrate Paleontology and Geochronology*, ed. Y. Tomida, L. J. Flynn, and L. L. Jacobs, pp. 215–242. *Nation Science Museum Monographs* 14.

Sheikh, K. A. 1984. Use of magnetic reversal time lines to reconstruct the Miocene landscape near Chinji Village, Pakistan. M.S. thesis, Dartmouth College.

Stix, J. 1982. Stratigraphy of the Kamlial Formation near Chinji Village, northern Pakistan. A.B. thesis, Dartmouth College.

Swie-Djin, N. and S. T. Hussain. 1981. Sedimentary studies of Neogene/Quaternary fluvial deposits in the Bhittanni Range, Pakistan. In *Neogene/Quaternary Boundary Field Conference, India, 1979, Proceedings*, ed. M. V. A. Sastry, T. K. Kurien, A. K. Dutta, and S. Biswas, pp. 177–183. Calcutta: Geological Survey of India.

Tauxe, L. and C. Badgley. 1988. Stratigraphy and remanence acquisition of a paleomagnetic reversal in alluvial Siwalik rocks of Pakistan. *Sedimentology*. 35:697–715.

Tauxe, L. and N. D. Opdyke. 1982. A time framework based on magnetostratigraphy for the Siwalik sediments of the Khaur area, northern Pakistan. *Palaeogeography, Palaeoclimatology, Palaeoecology* 37:43–61.

Welcomme, J.-L., M. Benammi, J.-Y. Crochet, L. Marivaux, G. Métais, P.-O. Antoine, and I. Baloch. 2001. Himalayan forelands: Palaeontological evidence for Oligocene detrital deposits in the Bugti Hills (Balochistan, Pakistan). *Geological Magazine* 138:397–405.

West, R. M. 1981. Plio-Pleistocene fossil vertebrates and biostratigraphy, Bhittanni and Marwat Ranges, North-West Pakistan. *Neogene/Quaternary Boundary Field Conference, India, 1979, Proceedings*, ed. M. V. A. Sastry, T. K. Kurien, A. K. Dutta, and S. Biswas, pp. 211–215. Calcutta: Geological Survey of India.

Willis, B. J. 1993a. Ancient river systems in the Himalayan foredeep, Chinji Village area, northern Pakistan. *Sedimentary Geology* 88:1–76.

Willis, B. J. 1993b. Evolution of Miocene fluvial systems in the Himalayan foredeep through a two-kilometer-thick succession in northern Pakistan. *Sedimentary Geology* 88:77–121.

Willis, B. J. and A. K. Behrensmeyer. 1994. Architecture of Miocene overbank deposits in northern Pakistan. *Journal of Sedimentary Research* B64:60–67.

Willis, B. J. and A. K. Behrensmeyer. 1995. Fluvial systems in the Siwalik Miocene and Wyoming Paleogene. In *Long Records of Continental Ecosystems*, ed. C. Badgley and A. K. Behrensmeyer, pp. 13–35. *Palaeogeography, Palaeoclimatology, Palaeoecology* 115.

Zaleha, M. J. 1997a. Fluvial and lacustrine palaeoenvironments of the Miocene Siwalik Group, Khaur area, northern Pakistan. *Sedimentology* 44:349–368.

Zaleha, M. J. 1997b. Intra- and extrabasinal controls on fluvial deposition in the Miocene Indo-Gangetic foreland basin, northern Pakistan. *Sedimentology* 44:369–390.

Chapter 16

Mammalian Neogene Biostratigraphy of the Sulaiman Province, Pakistan

PIERRE-OLIVIER ANTOINE, GREGOIRE MÉTAIS, MAEVA J. ORLIAC, J.-Y. CROCHET,
LAWRENCE J. FLYNN, LAURENT MARIVAUX, ABDUL RAHIM RAJPAR,
G. ROOHI, AND JEAN-LOUP WELCOMME

The Sulaiman Range is a north–south-trending band of rugged mountains rising 1000–3400 m above sea level that defines the modern political boundary between Balochistan and Punjab provinces and extends northward into North-West Frontier Province (figure 16.1). Late Mesozoic and Cenozoic sedimentary rocks here are primarily marine and accumulated in the Tethys Sea in what is now the Indus Basin, bounded to the northwest by the Axial Belt (Shah 1977), or Bela–Waziristan Ophiolite Zone (Bannert et al. 1992), and to the southeast by the Indo–Pakistani Subcontinent. The east side of the Sulaiman Range is of particular interest since a relatively thick, well-exposed mid-Cenozoic sequence registers the transition from marine shelf to terrestrial deposition episodes related to the uplift and erosion of the orogenic Himalayan highlands and the related closure of the Tethys Sea. Deposition of these thick detrital units resulted from the collision between the Indian and Asian plates, which started near the Paleocene/Eocene transition, ~55 Ma (Beck et al. 1995; Clift et al. 2001).

STRATIGRAPHIC AND HISTORICAL CONTEXT

The fossiliferous outcrops were recognized early in the nineteenth century (Vickary 1846) and were initially described in detail by Blanford (1883), who assigned the terrestrial conglomerates and sandstones overlying the Eocene "Nummulitic Limestone" to the Upper Nari Formation. These sediments—especially in the southern Sulaiman Range, roughly corresponding to the Bugti Hills—soon became famous for their fossil vertebrate faunas (Lydekker 1884; Pilgrim 1907, 1908, 1910, 1912; Forster-Cooper 1913, 1915, 1924), rich in large terrestrial mammals of Miocene to Pleistocene age. The faunas, as a whole, have long been considered as terrestrial, as they provide clear evidence that the environment of deposition was primarily fluvial. The stratigraphic classification of the sediments of the Sulaiman foothills has evolved with the growth of knowledge, and many different names have been applied, both generally and locally (Hemphill and Kidwai 1973; Shah 1977; Welcomme et al. 2001). The intensive fieldwork on the Siwalik series of the Potwar Plateau greatly influenced biostratigraphical interpretation of the entire Himalayan foothills of Pakistan (Pilgrim 1912; Raza and Meyer 1984). The exceptionally well-exposed succession in the Potwar Plateau area of northern Pakistan provides the best standard succession for almost the entire Neogene, and its biostratigraphy is successfully dated by paleomagnetic methods (Johnson et al. 1985; Barry et al. 2002).

The period encompassing the Oligocene and earliest Miocene has long been considered as lacking in the entire Sulaiman Geological Province (e.g., Raza and Meyer 1984). Traditionally, the depositional history of pre-Siwalik continental strata is very poorly known, from the point of view of age, environment, or burial structure. The Murrees sensu lato, which unconformably overlies the Eocene marine rocks in some areas of the Potwar region, may represent this missing sedimentary record.

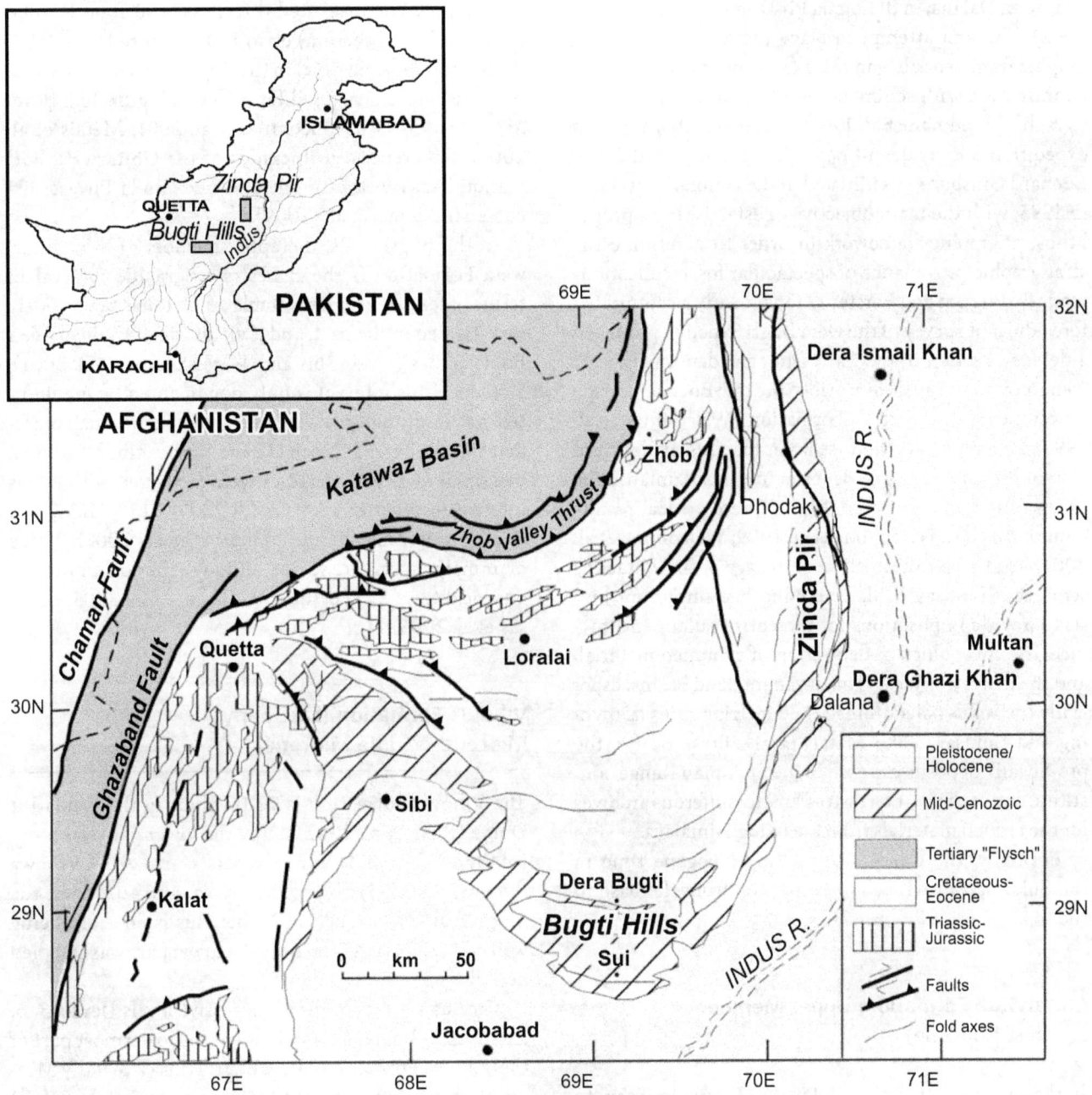

Figure 16.1 Geological map of the Sulaiman Province, Middle Indus Basin, Pakistan. Mammal-yielding localities discussed in the text are located in the Zinda Pir Dome and in the Bugti Hills. Modified after Raza et al. (2002).

However, the Murrees deposits are very poor in fossil remains, and correlations are still delicate to establish with pre-Siwalik deposits of other areas, including those of the Sulaiman Range (Najman and Garzanti 2000; Kumar and Kad 2003).

In the last decades, the continental series of the Sulaiman Range has been investigated as a southern extension of the important framework made in the Potwar Plateau area (Barry, Lindsay, and Jacobs 1982; Barry, Flynn, and Pilbeam 1990; Barry et al. 2002) by a Yale University–Geological Survey of Pakistan (GSP) team since the early 1980s. Important data came with the first collection of small mammals in the Sulaiman foothills (Jacobs, Cheema, and Shah 1981; Flynn, Jacobs, and Cheema

1986). Renewed reconnaissance of the Sulaiman foothills started in Dalana, in the Zinda Pir Dome area (Raza et al. 2002). The first attempt to place the fossils in a stratigraphic framework began there (Downing et al. 1993) in coordination with sediment sampling for magnetic stratigraphy (Friedman et al. 1992). Independently, a French expedition led by Jean-Loup Welcomme and the late Leonard Ginsburg was initiated in the famous Bugti Hills in 1995, with the main objective of establishing a proper lithostratigraphic framework in order to determine the stratigraphic provenance of spectacular fossil collections made by Pilgrim and Forster-Cooper nearly a century before. The first survey of the Dera Bugti Syncline produced a detailed section of the Early Miocene deposits that, at some points, lie unconformably on the Eocene marine limestones of the Kirthar Formation (Welcomme et al. 1997) and provided the first unambiguous evidence of (fossiliferous) Oligocene deposits in the Sulaiman Lobe (Welcomme and Ginsburg 1997). These critical results (summarized in Welcomme et al. 1999; Welcomme et al. 2001), led to a reexamination of the age of the Zinda Pir sequence (Lindsay et al. 2005) and had important biostratigraphic implications for the entire Sulaiman Province, the chronology of deposition of exhumed material, and the history of West Himalayan foreland basins, especially the Indus paleodrainage and its tributaries (Downing and Lindsay 2005; Métais et al. 2009). So far, the pre-Siwalik deposits exposed in the Sulaiman Range constitute one of the best terrestrial and fossiliferous archives for the eroded materials from the rising Himalaya.

Formally, the units documenting Neogene time in the Sulaiman Province are as follows, from the base to the top.

Chitarwata Formation, Upper Member (Earliest Miocene)

In the Zinda Pir area, the Chitarwata Formation is up to 480 m thick (Raza et al. 2002; Lindsay et al. 2005) while it is much more condensed in the Bugti Hills (120–260 m thick; Métais et al. 2009; new data). The lower and middle units recognized in the Zinda Pir area (Lindsay et al. 2005) and the Bugti Member of the Bugti Hills (Métais et al. 2006, 2009) are considered similar faunally and temporally and, being Oligocene in age, are discussed only in reference to younger sites. The upper part of the Chitarwata Formation in both areas is referred to the earliest Miocene (i.e., roughly corresponding to the Aquitanian marine stage or Agenian European Land Mammal Age; Antoine et al. 2010). The thickness of the upper member

varies laterally, but it reaches about 200 m near Dalana (Zinda Pir). The measured thickness is 50 m (at Kumbi, most condensed section) up to 130 m (figure 16.2 [Habib Rahi]; unpublished data) in the Bugti Hills, with two distinct and successive fossiliferous levels (figure 16.3 [level M = 3bis; Q = 4]; Welcomme et al. 2001; Métais et al. 2009). Differences in thickness of the Chitarwata Formation between the Bugti Hills and Zinda Pir are discussed in Métais et al. (2009).

In the Bugti Hills, the upper member of the Chitarwata Formation is the richest stratigraphic interval in terms of fossil vertebrates (table 16.1). It has been widely investigated in the past, and it yields the most diversified Bugti faunas (levels 3bis and 4; Welcomme et al. 2001). However, this interval is only documented by medium-size and large mammals in the Bugti area: no small mammal was recovered. Near Dalana in the Zinda Pir area, five small mammal localities were recovered within this interval (see figure 16.3 [Z113, Z139, Z150, Z135, and Z126, from base to top]; Lindsay et al. 2005). Large mammal remains are scarce, although they occur in various localities (Raza et al. 2002; Barry et al. 2005; Lindsay et al. 2005).

Vihowa Formation (Late Early Miocene–Middle Miocene)

The Vihowa Formation is 720 m thick in the Zinda Pir Dome (Raza et al. 2002). Only the lower part was studied thoroughly (E. H. Lindsay, pers. com., 2010). Vihowa thickness is likely to reach only ca. 100–200 m in the Bugti Hills (see figure 16.3), but this is far from being well constrained. Again, only the lower part was sampled densely.

Various sites from three successive levels (levels 5, 6, and 6sup [=Bugti sup]) document the lowermost part of the Vihowa Formation in the Bugti Hills (late Early Miocene), while a single locality (see figure 16.2 [Lundo W]) is referable to the middle part (Middle Miocene, i.e., considered similar to the Chinji fauna of Potwar Plateau; Welcomme et al. 2001; Métais et al. 2009; Orliac et al. 2009; Antoine et al. 2010). The levels 6 and 6sup yielded both large mammal and micromammal faunas (table 16.2; Welcomme et al. 1997; Welcomme et al. 2001; Métais et al. 2009). For Zinda Pir, three small mammal localities were recovered within this interval (see figure 16.3 [Z124, Z122, and Z120, from base to top]; Lindsay et al. 2005). Large mammal remains are rather scarce, but they occur in various localities (Raza et al. 2002; Barry et al. 2005; Lindsay et al. 2005).

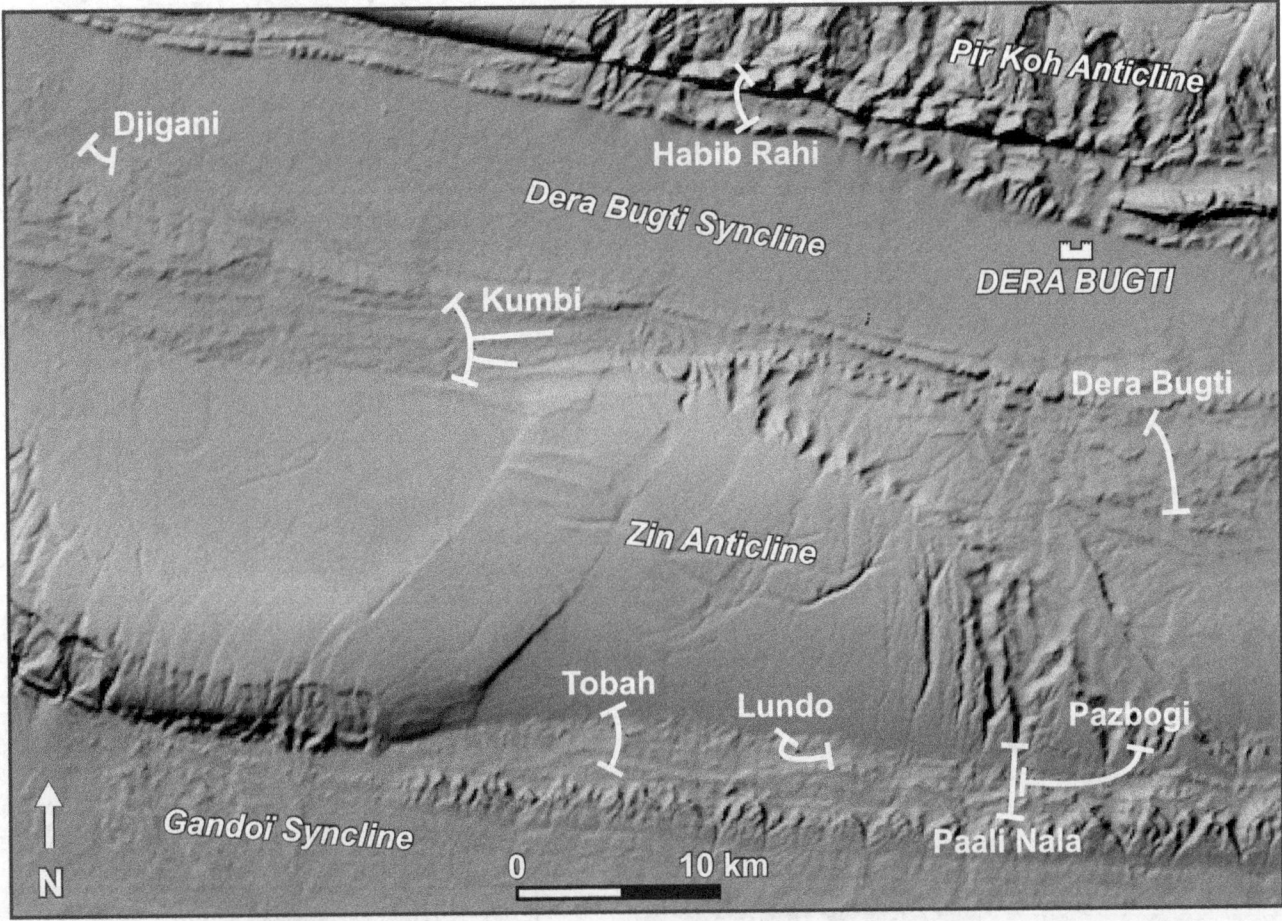

Figure 16.2 Location of main Cenozoic vertebrate-yielding sections from the Bugti Hills, Balochistan, Pakistan.

Litra Formation (Late Miocene)

The Litra Formation, in the Zinda Pir area, is 1700 m thick (Raza et al. 2002); no concentrated survey was done in that interval (E. H. Lindsay, pers. com., 2010). It is most likely much thinner in the Bugti Hills, but this hypothesis is not constrained by formally measured sections, except in the Habib Rahi area (ca. 150 m; see figure 16.2; unpublished data). Two localities referable to this stratigraphic interval are recognized in the Bugti Hills (see figures 16.2 [Djigani] and 16.3 [levels 7 and 7sup]; Welcomme et al. 1997; Welcomme et al. 2001; Antoine, Duranthon, and Welcomme 2003). They have yielded a large mammal fauna similar to what is observed in the lower part of the Middle Siwaliks in the Potwar Plateau, including the First Local Appearance (FLA) of *Hipparion* sensu lato (table 16.3; Antoine, Duranthon, and Welcomme 2003).

Post-Miocene Formations (Chaudhwan Formation, Pliocene–Pleistocene)

The Chaudhwan Formation is ca. 1500 m thick in Zinda Pir (Raza et al. 2002). In the Bugti area, its thickness is not constrained by measured sections, but it appears as highly variable depending on local tectonic context: the corresponding deposits are much deformed and consist essentially of boulder conglomerates and fluvial terraces, mainly observable in the Dera Bugti Syncline and in Habib Rahi (see figures 16.1 and 16.2). Welcomme et al. (1997) mention only egg shells of an unidentified struthioniform (ostrich). Crochet et al. (2009) report prehistoric rock paintings in the vicinity of Lundo (see figures 16.1 and 16.2), with anthropomorphic, geometric, and zoomorphic (e.g., cervid and felid) sketches attesting to favorable climatic conditions in the area around the Last Glacial Maximum and during subsequent periods.

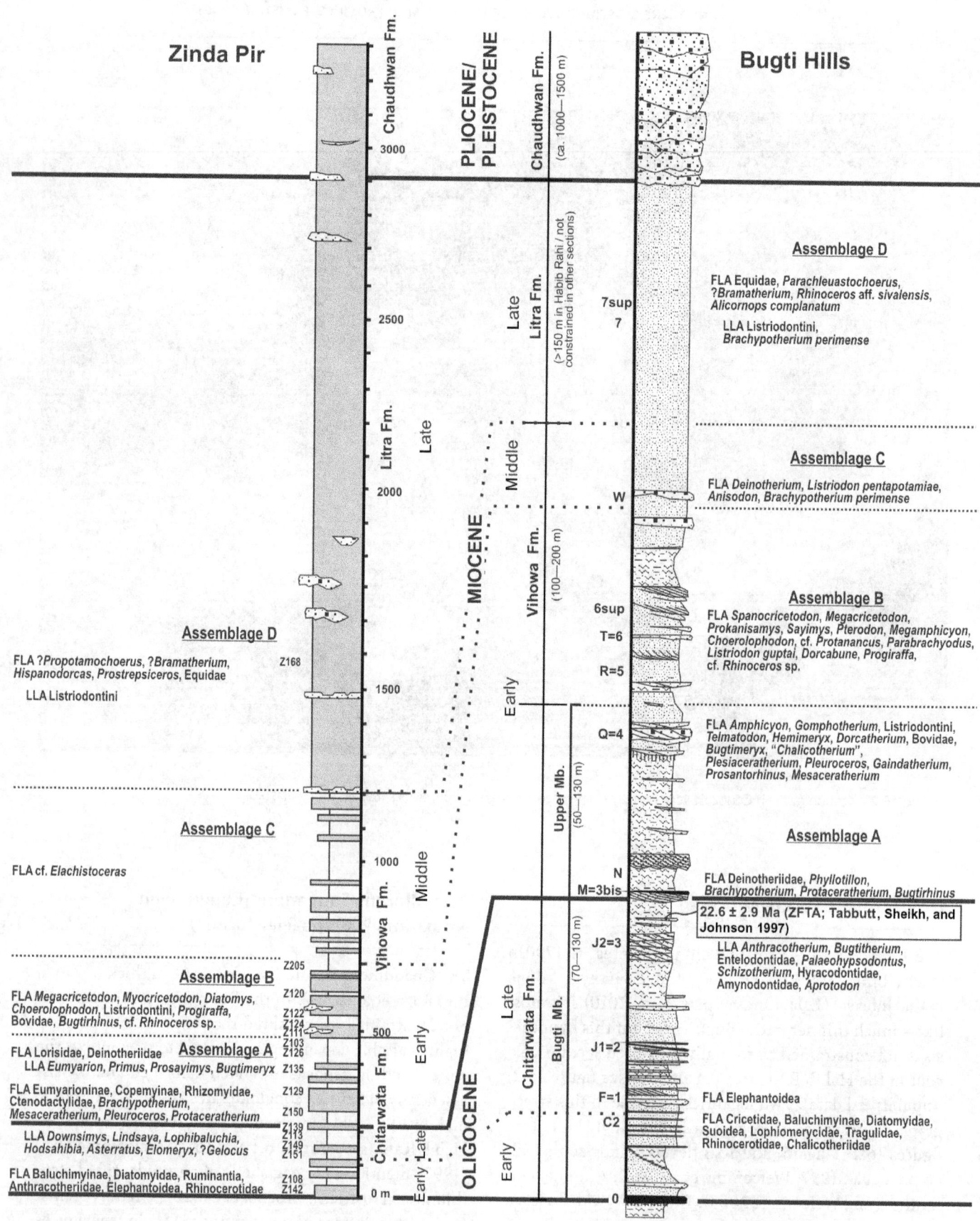

Figure 16.3 Stratigraphic sections of post-Eocene deposits of the Sulaiman Province (Zinda Pir Dome, to the left; Bugti Hills, to the right), with First Local Appearances (FLA) and Last Local Appearances (LLA) of index mammals and tentative correlation to the GPTS. Zinda Pir localities appear as "Z numbers" in the left column. Completed and modified after Marivaux et al. (1999, 2001, 2005, 2006), Welcomme et al. (2001), Raza et al. (2002), Antoine, Duranthon, and Welcomme (2003, 2010), Barry, et al. (2005), Gradstein, Ogg, and Smith (2004), Lindsay et al. (2005), Métais et al. (2009), Métais, Welcomme, and Ducrocq (2009), and Orliac et al. (2009, 2010).

Table 16.1

Mammals from the Upper Member of the Chitarwata Formation (Earliest Miocene) of the Sulaiman Province

Soriidae
  genus and species indet. ZP
Chiroptera
  genus and species indet. ZP
Tupaiidae
  genus and species indet. ZP
Lorisidae
  genus and species indet. ZP
Sciuridae
  genus and species indet. ZP
  Petauristinae gen et sp. indet. ZP
Cricetidae
  Eumyarion kowalskii ZP
  ?*Eumyarion* sp. ZP
  Democricetodon sp. ZP
  Democricetodon sp. A ZP
  Spanocricetodon khani ZP
  Spanocricetodon sulaimani ZP
  Spanocricetodon sp. ZP
  Primus sp. ZP
Rhizomyidae
  Prokanisamys arifi ZP
  Prokanisamys sp. ZP
Ctenodactylidae
  genus and species indet. ZP
  "Z135 *Sayimys* sp." ZP
  Prosayimys flynni ZP
  Prosayimys sp. ZP
  ?baluchimyine indet. ZP
Diatomyidae
  Marymus dalanae ZP
Carnivora
  genus and species indet. ZP
Amphicyonidae
  Amphicyon shahbazi BH
  genus and species indet. ZP
Proboscidea
  genus and species indet. ZP
Deinotheriidae
  Prodeinotherium sp. ZP, BH
  P. pentapotamiae BH
Elephantoidea
  genus and species indet. ZP, BH
  Gomphotherium sp. ZP, BH

Sanitheriidae
  ?*Sanitherium jeffreysi* ZP
  Sanitherium jeffreysi BH
Palaeochoeridae
  Pecarichoerus sp. BH
Suidae
  Listriodon affinis BH
Anthracotheriidae
  Parabrachyodus sp. ZP
  P. hyopotamoides ZP
  Microbunodon silistrense ZP
  Sivameryx palaeindicus ZP, BH
  Hemimeryx cf. *blanfordi* BH
  Telmatodon sp. BH
  ?*Masritherium* sp. ZP
  genus and species indet. ZP
Tragulidae
  genus and species indet. ZP
  Dorcatherium sp. ZP, BH
Giraffidae
  Progiraffa exigua ZP
Bovidae
  Eotragus sp. BH
Ruminantia incertae sedis
  Bugtimeryx pilgrimi ZP, BH
Chalicotheriidae
  Phyllotillon naricus BH
  "*Chalicotherium*" *pilgrimi* BH
  genus and species indet. ZP
Rhinocerotidae
  Protaceratherium sp. ZP, BH
  Plesiaceratherium naricum BH
  Mesaceratherium welcommi ZP, BH
  Pleuroceros blanfordi ZP, BH
  Brachypotherium gajense ZP, BH
  B. fatehjangense ZP, BH
  Prosantorhinus shahbazi BH
  Gaindatherium cf. *browni* BH
  ?cf. *Rhinoceros* sp. BH
  Bugtirhinus praecursor BH
  genus and species indet. ZP

NOTE: ZP, Zinda Pir; BH, Bugti Hills.
SOURCES: Completed and modified after Welcomme et al. (2001), Raza et al. (2002), Ginsburg and Welcomme (2002), Antoine et al. (2004, 2010), Barry et al. (2005), Gradstein, Ogg, and Smith (2004), Lindsay et al. (2005), Métais et al. (2009), Métais, Welcomme, and Ducrocq (2009), Orliac et al. (2009, 2010), references therein, and new data.

Table 16.2

Mammals from the Vihowa Formation (Late Early to Middle Miocene) of the Sulaiman Province

Erinaceidae
  Amphechinus sp. ZP
  Galerix sp. ZP
Soricidae
  genus and species indet. ZP, BH
Chiroptera
  genus and species indet. ZP
Tupaiidae
  genus and species indet. ZP
Sivaladapidae
  Guangxilemur sp. ZP
Sciuridae
  Petauristinae gen. et sp. indet. ZP
  genus and species indet. ZP
Cricetidae
  Democricetodon sp. B ZP
  Democricetodon sp. X ZP
  Spanocricetodon khani BH
  Spanocricetodon sulaimani ZP
  Spanocricetodon sp. BH
  Megacricetodon ?*daamsi* BH
  Megacricetodon sp. ZP
  ?*Megacricetodon* sp. ZP
  Myocricetodon sivalensis ZP
  Myocricetodon sp. 1 ZP
  Myocricetodon sp. 2 ZP
  Myocricetodon sp., large ZP
Rhizomyidae
  Prokanisamys arifi ZP
  Prokanisamys cf. *benjavuni* BH
  Prokanisamys benjavuni BH
  Prokanisamys sp. ZP
Ctenodactylidae
  genus and species indet. ZP
  Sayimys intermedius BH
  Sayimys cf. *intermedius* ZP
  Sayimys sp. ZP
Diatomyidae
  Diatomys sp.
Creodonta
  Hyanailouros sulzeri ZP
  Pterodon bugtiensis BH
  genus and species indet. ZP
Carnivora
  Megamphicyon giganteus BH
  Amphicyon shahbazi BH
  genus and species indet. ZP

Deinotheriidae
  Deinotherium sp. ZP
  Prodeinotherium pentapotamiae BH
Elephantoidea
  Gomphotherium browni BH
  Gomphotherium cooperi BH
  Choerolophodon corrugatus ZP, BH
  cf. *Protanancus chinjiensis* ZP
  genus and species indet. ZP
Sanitheriidae
  genus and species indet. ZP
Suidae
  Listriodon guptai ZP, BH
  ?*Listriodon affinis* ZP
Anthracotheriidae
  Parabrachyodus hyopotamoides BH
  Microbunodon silistrense ZP
  Sivameryx palaeindicus ZP, BH
  ?*Sivameryx palaeindicus* ZP
  genus and species indet.
Tragulidae
  Dorcatherium sp. ZP
  Dorcatherium cf. *parvum* BH
  genus and species indet. ZP
Giraffidae
  Progiraffa exigua ZP, BH
  ?*Giraffokeryx* sp. ZP
Bovidae
  Eotragus noyei ZP
  Eotragus sp. ZP
  cf. *Elachistoceras* sp. ZP
  genus and species indet. ZP
Chalicotheriidae
  Anisodon sp. BH
  genus and species indet. ZP
Rhinocerotidae
  ?*Plesiaceratherium naricum* BH
  Mesaceratherium welcommi BH
  Pleuroceros blanfordi BH
  Brachypotherium fatehjangense BH
  Brachypotherium gajense BH
  Brachypotherium perimense ZP, BH
  cf. *Rhinoceros* sp. ZP, BH
  Bugtirhinus praecursor ZP
  genus and species indet. ZP

NOTE: ZP = Zinda Pir; BH = Bugti Hills.
SOURCES: Completed and modified after Welcomme et al. (2001), Ginsburg and Welcomme (2002), Raza et al. (2002), Barry et al. (2005), Gradstein, Ogg, and Smith (2004), Lindsay et al. (2005), Métais (2009), Métais, Welcomme, and Ducrocq (2009), Orliac et al. (2009), Antoine et al. (2010), references therein, and new data.

Table 16.3

Mammals from the Litra Formation (Late Miocene) of the Sulaiman Province

Deinotheriidae
  Deinotherium sp. BH
Elephantoidea
  Gomphotherium sp. ZP
  Choerolophodon corrugatus ZP, BH
  genus and species indet. BH
Suidae
  Listriodon sp. ZP, BH
  ?*Propotamochoerus* sp. ZP
  Parachleuastochoerus sp. BH
Giraffidae
  ?*Bramatherium perimense* BH
  ?*Bramatherium* sp. ZP
  genus and species indet. ZP
Bovidae
  Prostrepsiceros vinayaki ZP
  Hispanodorcas terrubiae ZP
  Reduncini gen. et sp. indet. ZP
  genus and species indet. BH
Equidae
  Cormohipparion (Sivalhippus) theobaldi ZP
  Hippotherium sp. ZP, BH
Rhinocerotidae
  Alicornops complanatum BH
  Brachypotherium perimense BH
  Rhinoceros aff. *sivalensis* BH
  genus and species indet. ZP

NOTE: ZP = Zinda Pir = BH, Bugti Hills.

SOURCES: Completed and modified after Welcomme et al. (2001), Raza et al. (2002), Antoine, Duranthon, and Welcomme (2003), Antoine et al. (2010), Zouhri and Ginsburg (2003), Gradstein, Ogg, and Smith (2004), references therein, and new data.

MAMMALIAN FAUNAS

Fossil mammals can be split into two categories, roughly corresponding to the collection methods used in the field: prospecting (middle-sized and large mammals) and screening/washing (small mammals).

Small Mammals

Neogene micromammal stes were much more investigated in the Zinda Pir area than in the Bugti Hills (Wel-

comme et al. 2001; Lindsay et al. 2005; figure 16.4A). The Zinda Pir Chitarwata and Vihowa formations notably yielded diversified rodent assemblages (Lindsay et al. 2005). By contrast, no micromammal locality is available in Neogene deposits of the Bugti Hills, with the exception of two successive levels—documenting the same assemblage—in the base of the Vihowa Formation (levels 6 and 6sup, late Early Miocene; Welcomme et al. 1997; Métais et al. 2009). Interestingly, a wide array of micromammals (marsupials, insectivores, bats, dermopterans, primates, and rodents) was recovered in the Bugti Member of the Chitarwata Formation in the Bugti Hills (Paali C2, Early Oligocene; Marivaux et al. 2001 2005). To our knowledge, neither marsupial nor lagomorphs nor dermopteran remains have been found in the Neogene of the Sulaiman Province so far.

Lipotyphla and Chiroptera

Unidentified soricid remains were recovered from the upper member of the Chitarwata Formation in Zinda Pir (Z113 and Z150) and from the lower part of the Vihowa Formation in both the Bugti Hills (level 6; Welcomme et al. 1997) and Zinda Pir (Z122; Lindsay et al. 2005). *Amphechinus* sp., *Galerix* sp., and another unidentified hedgehog occur in the lower part of the Vihowa Formation (Z122; Lindsay et al. 2005). Bats of uncertain affinities occur in Z113 (upper member of the Chitarwata Formation, earliest Miocene) and in Z124–Z122 (lower part of the Vihowa Formation, late Early Miocene; Lindsay et al. 2005).

Euarchontes

By contrast with the rich and diversified primate and dermopteran fauna unearthed in the Bugti member of the Chitarwata Formation in the Bugti area (Paali C2, Early Oligocene; Marivaux et al. 2001; Marivaux et al. 2005, 2006), Neogene deposits of the Sulaiman Province have yielded only scarce euarchontan remains from Zinda Pir, consisting of unidentified Tupaiidae (Z113, upper Chitarwata Formation; Z122, lower Vihowa Formation) and Lorisidae (Z135, upper Chitarwata Formation; Lindsay et al. 2005). The sivaladapiform adapiform *Guangxilemur* sp. occurs in the lower part of the Vihowa Formation (Z122; Lindsay et al. 2005). The corresponding stratigraphic interval is restricted to the Early Miocene period (see tables 16.1 and 16.2 and figure 16.4A). No Neogene euarchontan is identified in the Bugti Hills.

A

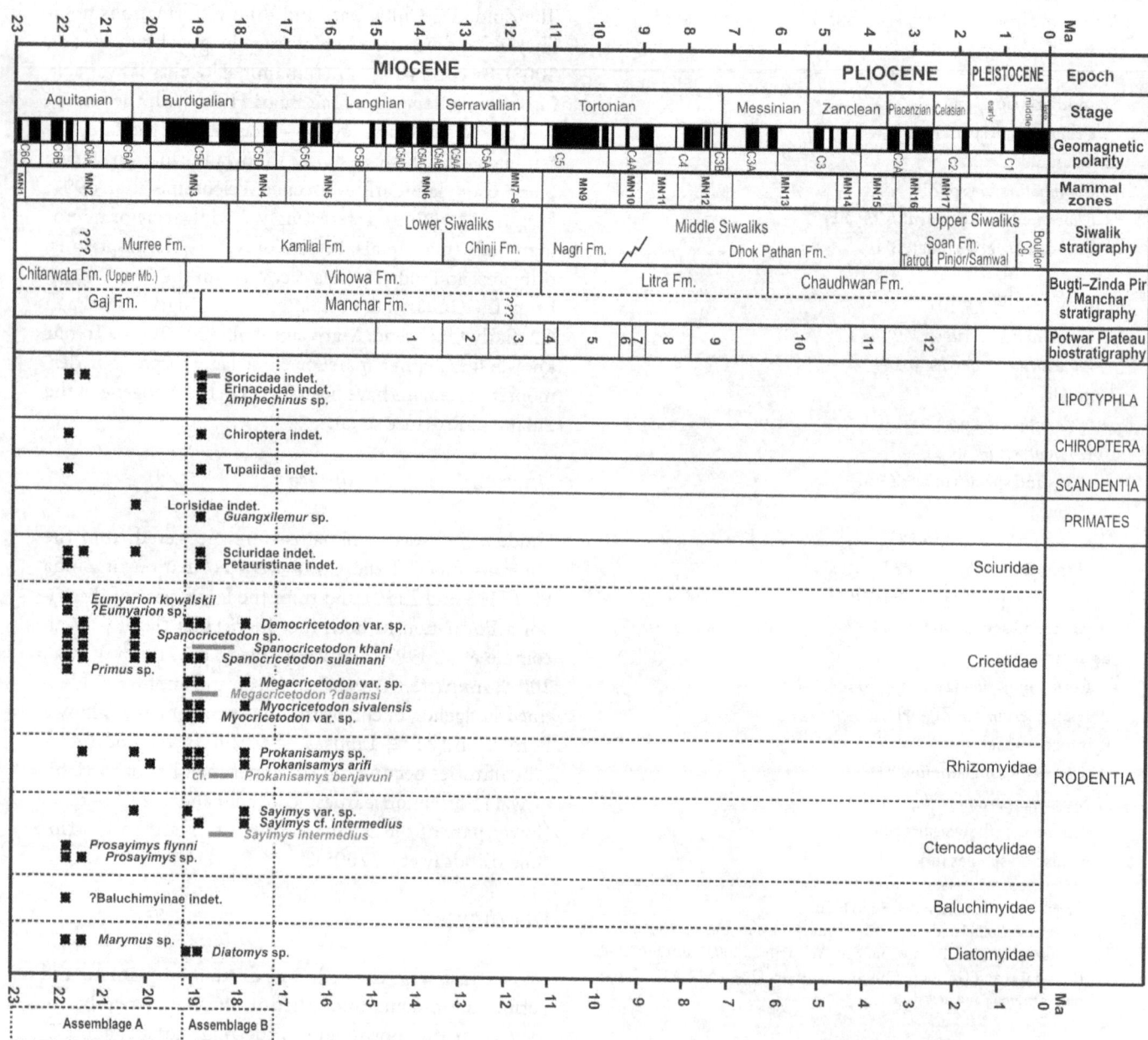

Figure 16.4 (*A*) Biostratigraphical range of small mammals from the Neogene of the Sulaiman Province, Pakistan. Taxa restricted to the Bugti Hills appear in gray. Taxa from Zinda Pir (most of them correlated to the GPTS) and/or from the whole Sulaiman Province (Zinda Pir + Bugti Hills) appear in black. Uncertain ranges are indicated by dashed lines. Taxa are sorted by order and/or family, and then by order of appearance. Completed and modified after Welcomme et al. (2001), Raza et al. (2002), Gradstein, Ogg, and Smith (2004), Lindsay et al. (2005), Métais et al. (2009), and references therein. (*B*) Biostratigraphical range of large mammals from the Neogene of the Sulaiman Province, Pakistan. Taxa restricted to the Bugti Hills appear in gray. Taxa from Zinda Pir (most of them correlated to the GPTS) and/or from the whole Sulaiman Province (Zinda Pir + Bugti Hills) appear in black. Uncertain ranges are indicated by dashed lines. Taxa are sorted by order and/or family, and then by order of appearance. Completed and modified after Welcomme et al. (2001), Ginsburg and Welcomme (2002), Raza et al. (2002), Antoine, Duranthon, and Welcomme (2003), Antoine et al. (2010), Zouhri and Ginsburg (2003), Gradstein, Ogg, and Smith 2004, Lindsay et al. (2005), Métais et al. (2009), Métais, Welcomme, and Ducrocq (2009), Orliac et al. (2009), references therein, and new data.

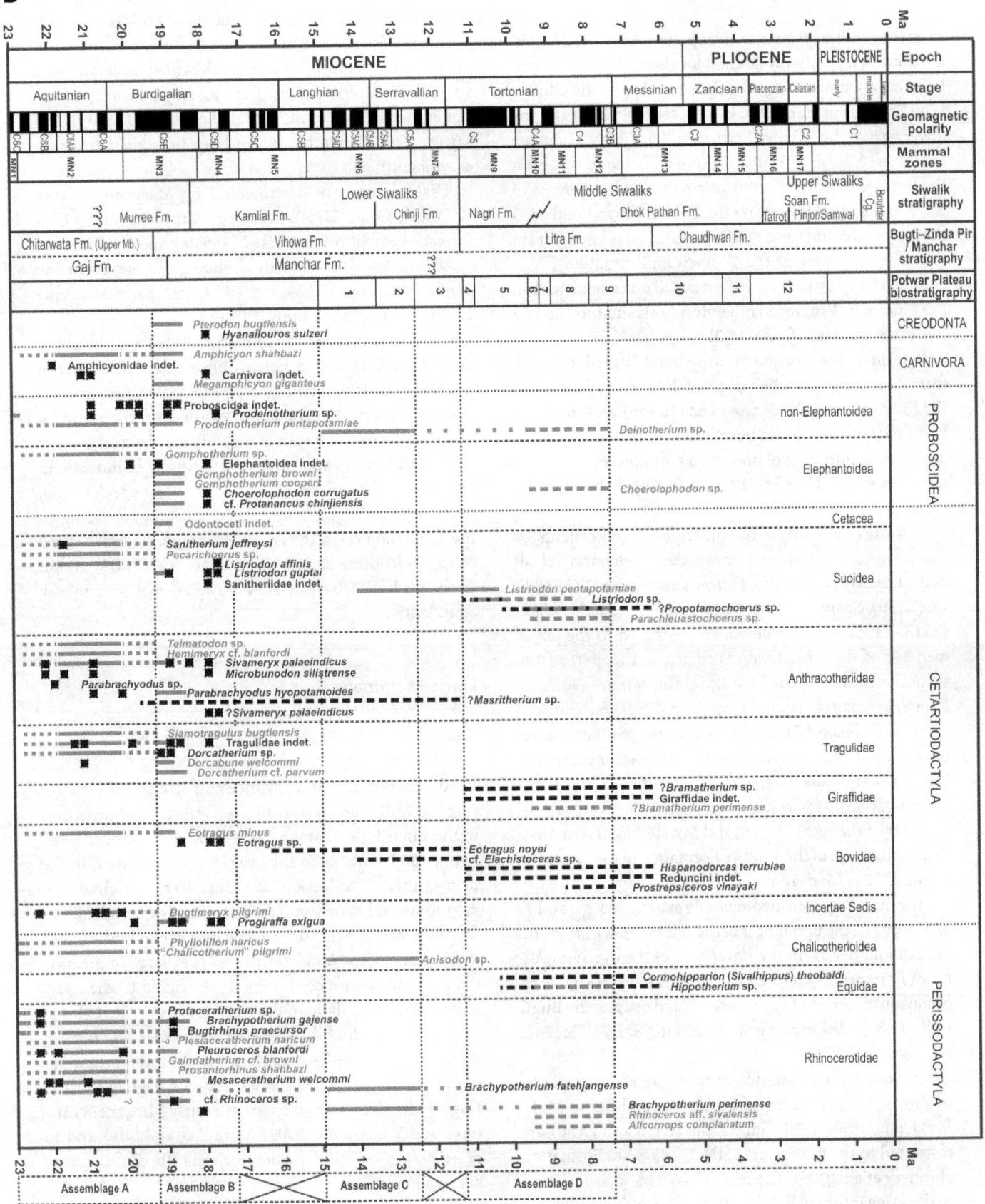

Rodentia

In the Sulaiman Province, Neogene rodents are essentially known by micromammal localities from the Zinda Pir area, encompassing the upper member of the Chitarwata Formation (earliest Miocene; see table 16.1) and the lower part of the Vihowa Formation (late Early Miocene; see table 16.2; Lindsay et al. 2005). Only two levels from the lower part of the Chitarwata have yielded rodents in the Bugti Hills (see figures 16.2 [Dera Bugti] and 16.3 [levels 6 and 6sup]), but no micromammal is known from the upper member of the Chitarwata Formation (Welcomme et al. 2001). Correlation based on rodents within the Sulaiman Province is therefore restricted to the late Early Miocene (see figure 16.3).

Sciuridae are documented by unidentified remains from the upper member of the Chitarwata Formation (Z113, Z139, and Z135, from base to top) and from the lower part of the Vihowa Formation (Z122; Lindsay et al. 2005). A petauristine of uncertain affinities occurs in the same interval (Z113 and Z122; Lindsay et al. 2005), and a ratufine is found at Z135.

Cricetidae are by far the most diversified rodents of this interval, with at least 16 species (Welcomme et al. 2001; Lindsay et al. 2005). *Eumyarion kowalskii* (Z113), *E.* sp. (Z139), *Spanocricetodon* sp. and *Spanocricetodon khani* (Z113–Z135), and *Primus* sp. are restricted to the upper member of the Chitarwata Formation. This part of the section records also the First Local Appearance (FLA) of *Democricetodon* (*D.* sp. and *D.* sp. A, in Z150) and *Spanocricetodon sulaimani* (Z113), while a couple of other species referable to *Democricetodon* appear in the lower part of the Vihowa Formation (*D.* sp. X, Z124; *D.* sp. B, Z122; Lindsay et al. 2005). *Megacricetodon* sp. and *Myocricetodon* (*My.* sp. 1, *My. sivalensis*, *My.* sp. 2, and *My.* sp., large) have their FAD in the base of the Vihowa Formation in the Zinda Pir Dome (Z124; Lindsay et al. 2005).

In Zinda Pir, the rhizomyids *Prokanisamys* sp. and *P. arifi* occur in the upper part of the Chitarwata Formation (Z150) and persist in the lower part of the overlying Vihowa Formation (up to Z122; Lindsay et al. 2005). *P.* cf. *benjavuni* is recorded in the levels 6 and 6sup of the Bugti Hills (lower Vihowa Formation, late Early Miocene; Métais et al. 2009).

Ctenodactylidae are documented by various species in Zinda Pir (see tables 16.1 and 16.2 and figure 16.4). *Prosayimys flynni* (Z113) and *P.* sp. (Z113 and Z150) are restricted to the upper part of the Chitarwata Formation (Lindsay et al. 2005). The FLA of *Sayimys* sp. is recorded in the same interval, but definitive *Sayimys* (*S.* cf. *intermedius*) appears in the lower part of the Vihowa Formation

(Z122; Lindsay et al. 2005), which coincides with the occurrence of *S. intermedius* in the Bugti Hills (level 6sup; Welcomme et al. 2001).

Three enigmatic teeth were identified as *Baluchimys* sp. and *Zindapiria quadricollis* in a single locality of the upper member of the Chitarwata Formation (Z113; Lindsay et al. 2005), but these records are possibly poorly preserved muroids.

Diatomyidae are documented by *Marymus dalanae* (Z113, Z139, and Z150) in the upper member of the Chitarwata Formation (earliest Miocene) and by *Diatomys* sp. in the lower part of the Vihowa Formation in the Zinda Pir Dome (Z124 and Z122, late Early Miocene; Lindsay et al. 2005; Flynn 2007).

The absence of any micromammal site from the upper part of the Chitarwata Formation (earliest Miocene) in the Bugti Hills prevents testing the rodent first appearances as observed in Zinda Pir for the corresponding interval. However, the rodent assemblages from the lower part of the Vihowa Formation are totally homotaxic in both areas (see table 16.2), at the genus or species level, with notably *Megacricetodon* and *Sayimys* appearing in the same interval in the Bugti Hills (level 6 and/or level 6sup; Welcomme et al. 2001; Métais et al. 2009) and in Zinda Pir (Z124, base of the Vihowa Formation; Lindsay et al. 2005).

Large Mammals

Ferae

Pending a revision of a few medium-size specimens unearthed in the upper part of the Chitarwata Formation in the Bugti Hills (Samane, level 4, earliest Miocene; see figure 16.1), and due to the sampling method used in the field—surface collection favoring large specimens—creodonts and carnivore remains are either unidentified to genus and species or referred as hyaenodontid creodonts and amphicyonid carnivores (i.e., large or gigantic taxa). Their fossil record is so far restricted to the Early Miocene in the Sulaiman Province (see figure 16.4B).

The hyaenodontids *Pterodon* and *Hyanailouros* are documented in the lower part of the Vihowa Formation. Corresponding remains are referred to *P. bugtiensis* in the Bugti Hills (levels 6 and 6sup, late Early Miocene; Ginsburg and Welcomme 2002; Métais et al. 2009) and to *H. sulzeri* in coeval deposits of Zinda Pir (Z120; Lindsay et al. 2005).

In the Bugti Hills, the amphicyonid *Amphicyon shahbazi* occurs in several localities encompassing the top of the Chitarwata Formation (level 4, earliest Miocene) and the lower part of the Vihowa Formation (see figure 16.3 [levels 5–6 sup, late Early Miocene]), while its close ally *Megamphicyon giganteus* occurs only in the lower part of the Vihowa Formation (levels 5–6sup, late Early Miocene; Ginsburg and Welcomme 2002; Métais et al. 2009). On the other hand, only unidentified carnivore and amphicyonid remains have been unearthed in coeval Zinda Pir deposits (Lindsay et al. 2005).

Proboscidea

Until recently, the Paleogene history of proboscideans was restricted to Africa, and elephants and their close relatives were supposed to disperse toward Eurasia during the late Early Miocene (Shoshani and Tassy 1996). This dispersal was given strong biostratigraphic significance and named Proboscidean Datum Event (PDE) by Madden and Van Couvering (1976). The recognition of a diagnostic elephantoid incisor from the Bugti Member of the Chitarwata Formation in the Bugti Hills (locality DB-J1, late Oligocene; Antoine, Welcomme, et al. 2003), the discovery of another tusk lower in the same stratigraphic interval in 2004 (level 2 = F; Bugti Member, Chitarwata Formation; unpublished data), and the subsequent mention of an unidentified elephantoid in the lower Chitarwata at Zinda Pir (Oligocene Z108; Lindsay et al. 2005; now rejected as definitive) seriously challenge the concept of PDE as a single event occurring at ca. 17.5 Ma (late Early Miocene; Madden and Van Couvering 1976), and indicate rather that proboscideans dispersed several times from Africa to adjacent plates in Middle Cenozoic times, as suspected by Tassy (Barry, Flynn, and Pilbeam 1990).

Deinotheriidae have their FLA in the upper member of the Chitarwata Formation (earliest Miocene), with *Prodeinotherium* sp. (level 3bis; tusk and dental fragments) and *P. pentapotamiae* (level 4; dental remains) in the Bugti Hills (see figure 16.3 and table 16.1). Scarce remains are reported from the same interval in Zinda Pir (Z129 and upward; Raza et al. 2002; Lindsay et al. 2005). These small deinotheres occur also throughout the Vihowa Formation (late early to middle Miocene) throughout the Sulaiman Province, and in coeval deposits of the Potwar Plateau (Kamlial and Chinji formations; Welcomme et al. 2001; POA, pers. obs.) and Europe (e.g., Antoine, Duranthon, and Tassy 1997). A larger deinotheriid referable to *Deinotherium* sp. occurs in the level 7sup of the Bugti Hills (Djigani, middle late Miocene; new data).

In the Sulaiman Province, the earliest elephantoid remains identified at genus level are referred to *Gomphoth-*

erium sp. They originate from the upper part of the Chitarwata Formation in the Bugti Hills (see figure 16.3 [level 4, earliest Miocene]; Welcomme et al. 2001; Métais et al. 2009). The elephantoid assemblage is much more diversified in the subsequent stratigraphic interval—that is, in the lower part of the Vihowa Formation in the Bugti Hills (levels 6 and 6 sup.; late Early Miocene)—with *Gomphotherium browni*, *G. cooperi*, *Zygolophodon metachinjiensis*, *Choerolophodon* cf. *corrugatus*, and *Protanancus chinjiensis* (Welcomme et al. 2001; Métais et al. 2009). The two latter taxa are also recognized in the middle part of the Vihowa Formation (Z205; Lindsay et al. 2005). *C. corrugatus* persists until the middle Late Miocene in the Bugti Hills (level 7sup at Djigani; new data), as in Potwar Plateau (Dhok Pathan Formation; Pilbeam et al. 1979, 1996).

Cetartiodactyla

Odontoceti

An unidentified odontocete is mentioned in the lower part of the Vihowa Formation (level 6), in the Bugti Hills (Welcomme et al. 1997). The available tooth belongs probably to a river dolphin, given the fluvial depositional environment of the whole formation (Welcomme et al. 2001; Métais et al. 2009).

Ruminantia

Initially, Pilgrim (1908) referred dental remains to a primitive giraffid, *Progiraffa exigua*, with the mention "Dera Bugti Area," which was later refined as "upper Nari" of "Dakko Nala" (Pilgrim 1911). Pilgrim (1912) described two new species of ruminants from the base of the "Gaj" in the Bugti Area, based on sparse dental material; the species were doubtfully assigned as *Gelocus*(?) *gajensis* and *Prodremotherium*(?) *beatrix*. At that time, both genera were common ruminants restricted to the Oligocene of Europe. Despite this weak biochronological evidence of Oligocene age, Pilgrim (1912), probably influenced by the sequence observed in the Sind Province to the south, revised his former opinion concerning the probable occurrence of Oligocene deposits in the Bugti Hills (Pilgrim 1908) and argued for an early Miocene age for the whole Bugti fauna. Later on, there was no further extensive study of the ruminants from the Bugti Hills, principally due to the uncertain geographic and stratigraphic provenance of the described material, and the lack of fieldwork in the Sulaiman Lobe until the early 1980s and late 1990s (Raza and Meyer 1984; Welcomme et al. 1997).

Recent washing and screening in the lower Chitarwata Formation near Dera Bugti led to the discovery of new ruminant material with typically Oligocene taxa (Métais, Welcomme, and Ducrocq 2009). The upper Chitarwata has also yielded ruminant remains whose affinities remain somewhat obscure. Most efforts made by MPFB in the past decade have focused on the lower Chitarwata, and further investigation is necessary to better determine the diversity and stratigraphic ranges of Neogene ruminants of the Bugti Hills. Ginsburg, Morales, and Soria (2001) reassessed the systematics of ruminants collected by Pilgrim (1908, 1911, 1912) and Forster-Cooper (1915), in light of new material collected in 1997–1998 from the early Miocene strata of the Bugti Hills (essentially the levels 4, uppermost Chitarwata, and 6, basal Vihowa), allowing them to describe six taxa, referred to Giraffidae, Tragulidae, or Bovidae. Even if Ginsburg, Morales, and Soria (2001) did not indicate the geographic location of their material, and thus tended to perpetuate errors made by ancient authors, we were able to reassess the stratigraphic allocation of the concerned specimens thanks to their inventory number.

Bugtimeryx pilgrimi was erected by Ginsburg, Morales, and Soria (2001), based on mandibular, dental, and postcranial material collected from level 4 in the Bugti Hills (uppermost Chitarwata; see figures 16.3 and 16.4B and table 16.1). In the same article, *Gelocus* (?) *gajensis* (Pilgrim, 1912) and *Prodremotherium* (?) *beatrix* (Pilgrim 1912) were tentatively referred to *Bugtimeryx*, and this new genus was included within Giraffidae, without further discussion. Barry et al. (2005) and Lindsay et al. (2005) mention *B. pilgrimi* in several localities referred to the upper Chitarwata (Z151–Z133; see figure 16.4B), but rather consider it as a representative of "Pecora, incertae sedis." *Progiraffa exigua* occurs in the uppermost Chitarwata (Z126) and the lower Vihowa in the Zinda Pir (Z124–Z205; see figure 16.4B; Barry et al. 2005; Lindsay et al. 2005) and only in the lower Vihowa deposits in the Bugti Hills (level 6; Ginsburg, Morales, and Soria 2001). The type specimen of *P. exigua* was until recently all that was known of the taxon, but the new collections show that the species is rather common in the Early Miocene deposits of Zinda Pir (see figure 16.4B; Barry et al. 2005). Even if Ginsburg, Morales, and Soria (2001) also consider it as a giraffid closely related to *Canthumeryx* from Gebel Zelten (Libya, late Early Miocene), this taxon is of uncertain affinities within Ruminantia, following Barry et al. (2005).

The earliest Tragulidae from the Bugti Hills occur in the Bugti Member of the Chitarwata Formation at Paali C2 (Early Oligocene; see figure 16.3; Marivaux et al.

2001; Métais, Welcomme, and Ducrocq 2009). Higher in the series, Ginsburg, Morales, and Soria (2001) mention a supposedly primitive species of *Siamotragulus* (*S. bugtiensis*) based on mandibular, dental, and postcranial material from the uppermost Chitarwata (see figure 16.3 [level 4, earliest Miocene]). Small mandibular, dental, and postcranial remains from the base of the Vihowa Formation (see figure 16.3 [level 6, late Early Miocene]) are referred to *Dorcabune welcommi* by Ginsburg, Morales, and Soria (2001). *Dorcabune* was initially described by Pilgrim (1910) in the base of the lower Manchar Formation of Sind—that is, in levels considered as coeval to lower Vihowa deposits from the Sulaiman Province (late Early Miocene; for correlation, see figure 16.4B). In Zinda Pir, Barry et al. (2005) and Lindsay et al. (2005) mention unidentified tragulids in the upper Chitarwata (Z150–Z126, earliest Miocene) and *Dorcatherium* sp. in the lower Vihowa (Z127–Z116, late Early Miocene; see figure 16.4B).

A large taxon documented only by postcranials from the uppermost Chitarwata and the lowermost Vihowa formations in the Bugti area was tentatively referred by Ginsburg, Morales, and Soria (2001) to Tragulidae, under the name "?*Siamotragulus indicus* (Forster-Cooper 1915)." The corresponding remains most probably document an unidentified tragulid, since the type specimen (NHM M15421) of "*Gelocus indicus* Forster-Cooper 1915," for which the stratigraphic provenance is unknown, rather shows close affinities with the lophiomerycid *Nalameryx sulaimani* Métais, Welcomme, and Ducrocq 2009 from the Early Oligocene locality of Paali C2 in the Bugti Hills (Bugti Mb., lower Chitarwata). This family is restricted to Eurasian Paleogene localities (Métais, Welcomme, and Ducrocq 2009).

Interestingly, several Oligocene localities of Zinda Pir have yielded scarce remains of ruminants (Barry et al. 2005). Locality Z108, stratigraphically situated in the lower part of the Chitarwata Formation, has produced an indeterminate lophiomerycid (considered as conspecific to *Nalameryx sulaimani* by Métais, Welcomme, and Ducrocq 2009) and ?*Gelocus gajensis*. The generic status and familial affinities of ?*Gelocus gajensis* are still debated, but either of these taxa can be useful for biochronology. We consider as highly dubious the referral of isolated postcranial remains with a tragulid pattern from the upper Chitarwata Formation and the lower Vihowa Formation in the Bugti Hills (levels 4 to 6, early Miocene; Ginsburg, Morales, and Soria 2001) to the probable Oligocene lophiomerycid "*Gelocus indicus* Forster-Cooper 1915" (see figure 16.3).

At locality Z142, stratigraphically equivalent to Z108 (lower member, Chitarwata Formation, Oligocene; Lindsay et al. 2005), Barry et al. (2005) recognized the bovid-like ruminant *Palaeohypsodontus zinensis*, which is also known in the late Oligocene locality of Lundo Chur J2 in the Bugti area (Métais et al. 2003). There is no mention of this taxon in Miocene deposits of the Bugti area, but it is worth noting that *P. zinensis* is also reported from the locality Z121, situated in the late Early Miocene Vihowa Formation (Lindsay et al. 2005). This occurrence may correspond to the LAD of this puzzling genus.

Bovids are well represented throughout the Early Miocene of the Sulaiman Province. The earliest undisputable remains (horn core, tooth fragments, and postcranials) were unearthed in the uppermost Chitarwata Formation (see figures 16.3 [level 4, earliest Miocene] and 16.4*B*; reinterpreted after Ginsburg, Morales, and Soria 2001) and in the lowermost Vihowa (level 6, late Early Miocene) in the Bugti area. All of them are referred to the minute species *Eotragus minus* (Ginsburg, Morales, and Soria 2001). The occurrence of "Bovidae indet." in the lower Vihowa (see figures 16.3 [Z110–Z120, late Early Miocene] and 16.4*B*) dates back to ca. 19.5 Ma among Zinda Pir occurrences (Lindsay et al. 2005). Higher in the Zinda Pir series, Raza et al. (2002) mention *Eotragus noyei* and cf. *Elachistoceras* sp. (Vihowa Formation, middle Miocene), while the Litra Formation (Late Miocene) records the occurrence of more advanced forms, such as *Prostrepsiceros vinayaki*, *Hispanodorcas terrubiae*, and an unidentified representative of Reduncini. In the Potwar Plateau, *Elachistoceras khauristanensis* occurs in the Late Miocene, with a range of 11.5–7.4 Ma, but its inferred FLA could be ca. 14.0 Ma (Barry et al. 2002). In the same area, cf. *Prostrepsiceros vinayaki* has a short inferred interval (8.6–7.4 Ma) that slightly exceeds its observed time range (8.3–7.9 Ma; Barry et al. 2002). The Reduncini include Recent African reedbucks (*Kobus*) and waterbucks (*Redunca*) and their extinct relatives from the Old World, with a late Miocene to early Pleistocene Asian range (McKenna and Bell 1997).

No fossil cervid is known from the Neogene of the Bugti Hills and Zinda Pir (Welcomme et al. 2001; Raza et al. 2002; Lindsay et al. 2005), even though an undisputable cervid is portrayed with dichotomic branches in the pre-Holocene cave paintings of Lundo (Crochet et al. 2009).

Anthracotheriidae

Historically, the classic "Bugti Bone Beds" yielded abundant remains of anthracotheres, which were described in detail by Pilgrim (1912) and Forster-Cooper (1913, 1924). The generic and specific diversity of anthracothe-riids from Bugti was certainly overestimated by these authors, and upon revision of ancient collections (without any lithostratigraphic control), Pickford (1987) recognized eight species belonging to six genera. Validity of these taxa and implied stratigraphic ranges of the species they would represent in the Bugti Hills remain unclear, making it difficult to use anthracotheriids for biochronologic, paleogeographic, and paleoenvironmental purposes. Anthracothere remains are well represented throughout the Chitarwata Formation in the Bugti Hills, but they tend to be taxonomically more diverse in the upper Chitarwata (Métais et al. 2009). *Anthracotherium* cf. *bugtiense* and another large anthracotheriine, as well as the bothriodontine *Elomeryx* sp., have been found in basal beds of the Chitarwata Formation near Bugti (Early and Late Oligocene; see figure 16.3; Métais et al. 2009). Dental remains of anthracotheres are rare at Paali C2 (basal Chitarwata, Early Oligocene; see figures 16.2 and 16.3), but several postcranial elements suggest the presence of a small species we refer to *Microbunodon* cf. *silistrense*. *Anthracotherium* cf. *bugtiense* ranges up to duricrust J2 (Late Oligocene) but is absent from the overlying levels M and Q (earliest Miocene; see figure 16.3). *Microbunodon* is neither registered in the upper Chitarwata (earliest Miocene) nor in the Vihowa Formation (late Early to Middle Miocene) in the Bugti area, while *Microbunodon silistrense* ranges from the upper Chitarwata (Z151, Z139, Z209) up to the lower Vihowa (Z121, Z120) at Zinda Pir (Lindsay et al. 2005).

In the Bugti area, the upper Chitarwata Formation, referred to the earliest Miocene, has yielded an entirely different assemblage of anthracotheres, mostly documented by fossils from ferruginous duricrust 4 or Q (see figure 16.3), with dental and postcranial remains of *Siv-ameryx palaeindicus*, *Hemimeryx* cf. *blanfordi*, *Parabrachyodus* cf. *hyopotamoides*, and possibly *Telmatodon* sp. (see figure 16.4*b*; Welcomme et al. 2001; Métais et al. 2009). In Zinda Pir, the anthracotheriid assemblage of the upper Chitarwata is somewhat distinct in the absence of *Hemimeryx* and possibly *Telmatodon*, and the occurrence of *Microbunodon*, a genus recorded from the lower Chitarwata in the Bugti area (Lindsay et al. 2005; Métais et al. 2009). *Hemimeryx* cf. *blanfordi* and *Parabrachyodus* cf. *hyopotamoides* are also recorded in the base of the Vihowa Formation (level 6, late Early Miocene) in the Bugti area. Anthracotheres from Zinda Pir have not been described yet, with the exception of *Elomeryx* cf. *borbonicus* from the Oligocene locality Z108 (Ducrocq and Li-horeau 2006). Only a broad revision of the anthracothe-riid material from the Sulaiman Province, encompassing both Zinda Pir and Bugti samples, is likely to provide a satisfactory explanation to such a taxonomic/stratigraphic discrepancy.

Downing et al. (1993) and Raza et al. (2002) reported the presence of an indeterminate anthracotheriid and remains questionably referred to *Masritherium* from the Chitarwata Formation, as well as the occurrence of an unidentified anthracotheriid along with ?*Hyoboops sp.* (=*Sivameryx*) in the Vihowa Formation. To our knowledge, no hippopotamid has been mentioned for Neogene deposits of the Sulaiman Province.

Suoidea

Even though Suoidea are not abundant in terms of number of specimens, all three Eurasian families—that is, Sanitheriidae, Palaeochoeridae ("Old World peccaries"), and Suidae—are represented in mid-Cenozoic deposits of the Sulaiman Province.

Remains referable to Sanitheriidae are scarce, but they occur throughout the Chitarwata and Vihowa formations (Early Oligocene–Middle Miocene; see figure 16.4B; Lindsay et al. 2005; Orliac et al. 2010). A right P3 from Paali C2 (Bugti member, Chitarwata Formation, Early Oligocene; see figure 16.3), strongly reminiscent of the holotype of "*Hyotherium(?) jeffreysi* Forster-Cooper 1913" but with a simpler structure, is identified provisionally as *Sanitherium* sp. (*Hyotherium* is a suid, and in our opinion, Sanitheriidae are monogeneric, *Diamantohyus* Stromer 1926 being a junior synonym of *Sanitherium* Von Meyer 1866; Orliac et al. 2010). A sanitheriid referred to as "?*Diamantohyus jeffreysi*"—here, "?*S. jeffreysi*"—is recorded in both the lower member (Z144, Oligocene) and the upper member of the Chitarwata Formation in Zinda Pir (Z156, earliest Miocene; Lindsay et al. 2005). *S. jeffreysi* occurs in coeval deposits from the Bugti Hills, based on a mandibular fragment with m1–3 found at Samane (see figure 16.3 [level 4, earliest Miocene]); unpublished data). An unidentified sanitheriid is also listed in the late Early Miocene of Zinda Pir (Z120, lower Vihowa; Lindsay et al. 2005).

Palaeochoeridae (sensu van der Made 1997) are documented unquestionably by a single specimen, an M3 referable to *Pecarichoerus* sp., from the upper member of the Chitarwata Formation at Samane in the Bugti Hills (see figure 16.3 [level 4, earliest Miocene]; unpublished data). This specimen strongly recalls the European *Taucanamo grandaevum*, from the late Early Miocene of France (MN 4; Orliac, Antoine, and Duranthon 2006). This family has not been mentioned in Zinda Pir so far (Lindsay et al. 2005). The Late Oligocene locality of Lundo J2, in the Bugti Hills (Bugti Member, Chitarwata

Formation; see figure 16.3), yielded recently a mandibular fragment with m2–3 with morphology and measurements strongly comparable to the holotype of "*Microbunodon sminthos*" figured and described by Forster-Cooper (1913:fig. 5). Pickford (1987) subsequently reassigned this holotype, initially considered as belonging to an anthracotheriid "from the Upper Oligocene deposits of Dera Bugti" (Forster-Cooper 1913:514), to the palaeochoere genus *Pecarichoerus* Colbert 1933, otherwise known from Middle Miocene localities of the Potwar Plateau (Chinji Formation; Colbert 1933), Thailand, and China (Pickford et al. 2004). To date, no specimen documenting this puzzling palaeochoere has been unearthed in Neogene deposits of the Sulaiman Province.

Suidae occur throughout the continental series in the Bugti Hills—that is, from the Bugti member of the Chitarwata Formation at Paali C2 (Early Oligocene) up to the level 7 (Litra Formation, Late Miocene; see figure 16.3 and table 16.3). A few specimens from Paali C2, referrable to the hyotheriine *Hyotherium*, remain unidentified at species level (Orliac et al. 2010). This subfamily occurs only in the base of the Chitarwata Formation so far. Higher in the series, Suidae are only represented by Listriodontinae, Tetraconodontinae, and Suinae.

Listriodontinae have their FLA in the upper member of the Chitarwata Formation in the Bugti Hills, with *Listriodon affinis* (see figure 16.3 [level 4, earliest Miocene]; Orliac et al. 2009); this species is also mentioned from the lower part of the Vihowa Formation in Zinda Pir (Z205, late Early Miocene; Lindsay et al. 2005). *L. guptai* occurs in the base of the Vihowa Formation in both the Bugti Hills (see figures 16.3 [level 6] and 16.4B) and Zinda Pir (Z124, Z210, Z120, and Z205; "*L. gupti*" of Lindsay et al. 2005). The sublophodont listriodont *L. guptai* is very close morphologically to *L. akatikubas* from the late Early Miocene of Maboko, Kenya (ca. 16.5 Ma; Orliac et al. 2009). Fully lophodont listriodonts are found higher in the Vihowa Formation and referred to as either *Listriodon pentapotamiae* in the Bugti Hills (see figure 16.3 [level W, Middle Miocene]; Welcomme et al. 2001) or *Listriodon* sp. in the Vihowa Formation (late Early to Middle Miocene) and the Litra Formation (Late Miocene; Raza et al. 2002). The inferred stratigraphic range of *L. pentapotamiae* in the Potwar Plateau is 14.0–10.3 Ma (Barry et al. 2002:69). As such, the Litra specimens from the Sulaiman Province count among the few attested co-occurrences of listriodontine suids with hipparionine equids and may constitute the Last Local Appearance (LLA) for both *Listriodon* and Listriodontini (Barry et al. 2002; Raza et al. 2002).

Tetraconodontinae are represented by only an m3 from the Litra Formation of the Bugti Hills, identified as *Parachleuastochoerus* sp. (see figures 16.3 [level 7, early to middle Late Miocene] and 16.4*B* and table 16.3; Antoine, Welcomme, et al. 2003). This genus is known from the late Middle Miocene of Europe (MN 7–8; Golpe Posse 1972) and the early Late Miocene of China (correlated with MN10; Pickford and Liu 2001)

Suinae are likely to occur in the Litra Formation (Late Miocene) of Zinda Pir, with ?*Propotamochoerus* sp. (Raza et al. 2002). The only species referred to this suine genus to be documented in the Potwar Plateau is *P. hysudricus*, with a Late Miocene inferred stratigraphic range (10.2–6.5 Ma; Barry et al. 2002:69).

Perissodactyla

Perissodactyls are particularly abundant in the Neogene of the Bugti Hills, and they provide critical information for the age assessment (Welcomme et al. 2001; Antoine, Ducrocq, et al. 2003; Antoine, Duranthon, and Welcomme 2003; Métais et al. 2009).

No equoid is recorded in Chitarwata and Vihowa deposits: Hipparionine equids occur only in the top of the Miocene series—that is, in the Litra Formation (Zinda Pir) and in levels 7 and 7sup (Sartaaf = Djigani; Bugti Hills)—in deposits probably equivalent in age to the Dhok Pathan Formation of the Potwar Plateau (see figure 16.3; DB7; Welcomme et al. 1997; Antoine, Duranthon, and Welcomme 2003; Zouhri and Ginsburg 2003). The hipparionine teeth from Djigani recall those of *Hippotherium nagriense* (Nagri Formation; early Late Miocene) and *Cormohipparion* (*Sivalhippus*) *theobaldi* from the lower part of the Dhok Pathan Formation of Potwar Plateau, which indicates an age earlier than middle Turolian (Zouhri and Ginsburg 2003). *Hipparion* sensu lato in Zinda Pir is recorded at Z168—that is, 1500m above the base of the Chitarwata Formation, at ca. 10.5 Myr (see figure 16.3; Raza et al. 2002). Raza et al. (2002) mention *C.* (*S.*) *theobaldi* in the same formation, without further precision concerning its stratigraphic range.

Ancylopoda are uncommon but present throughout the post-Eocene series in the Bugti Hills (Métais et al. 2009) and in the Zinda Pir Dome (Raza et al. 2002; Lindsay et al. 2005). In the latter area, all the specimens remain unidentified. In the Bugti Hills, the large schizotheriine *Phyllotillon naricus* is documented by dental and postcranial remains from the upper part of the Bugti member of the Chitarwata Formation (Oligocene in age) up to the top of the formation (level 4 (=Q); earliest Miocene), while the smaller chalicotheriine "*Chalicothe-*

rium" *pilgrimi* seems to be restricted to the upper member, in levels 3bis (=M) and 4 (Métais et al. 2009). Higher in the series, a few postcranial and dental specimens unearthed from the middle part of the Vihowa Formation in the Bugti Hills (level W, Middle Miocene) are similar to specimens from Sansan in France and thus referable to *Anisodon* sp. (Anquetin, Antoine, and Tassy 2007).

In the Sulaiman Province, the bulk of Neogene perissodactyls is constituted by rhinocerotids, for which nine or ten species are recognized in the upper part of the Chitarwata Formation, seven or eight species in the Vihowa Formation, and at least three in Litra Formation or coeval deposits (see tables 16.1–16.3 and figure 16.4; Welcomme et al. 1997; Raza et al. 2002; Lindsay et al. 2005; Antoine et al. 2010). Most rhinocerotid suprageneric groups recognized in the Old World occur in the Neogene of the area (see figure 16.4). Rhinocerotinae are much diversified throughout the concerned period, with 12 species referred to Rhinocerotina, Aceratheriina, Teleoceratina, and Rhinocerotinae incertae sedis; on the other hand, Elasmotheriinae are represented by a single species (Antoine, Duranthon, and Welcomme 2003; Antoine et al. 2010).

Four basal offshoots of the Rhinocerotinae were unearthed in the upper Chitarwata Formation of both the Bugti Hills and Zinda Pir (earliest Miocene): *Protaceratherium* sp., *Pleuroceros blanfordi*, *Mesaceratherium welcommi*, and *Plesiaceratherium naricum*. The latter may also occur in the basal Vihowa Formation in the Bugti Hills (see table 16.2). *P. naricum* is the earliest representative of a well-known Eurasian genus, so far restricted to the late Early–early Middle Miocene interval (Yan and Heissig 1986; Antoine, Bulot, and Ginsburg 2000). *Pleuroceros blanfordi* and *Mesaceratherium welcommi* occur in the upper member of the Chitarwata Formation in the Bugti Hills and the Zinda Pir Dome and in the basal Vihowa Formation in the Bugti Hills (Early Miocene; see tables 16.1 and 16.2; Welcomme et al. 2001; Antoine, et al. 2010). Both species are endemic to the Sulaiman Province, but they are sister taxa of the European *P. pleuroceros* and *M. paulhiacense*, respectively, from the earliest Miocene of Western Europe (Antoine et al. 2006, 2010).

Rhinocerotina include all five living rhino species; their fossil record is restricted to the Old World. The earliest representatives of the clade are restricted to Pakistan until the late Early Miocene (*Gaindatherium*; Antoine, Bulot, and Ginsburg 2000; Antoine et al. 2010). *Gaindatherium* cf. *browni*, recognized in the upper member of the Chitarwata Formation in the Bugti Hills (earliest Miocene), widely predates the FLA of other representatives of this one-horned genus elsewhere (Heissig 1972). An early putative representative of the extant genus *Rhinoceros* is recognized in the lower part of the Vihowa Formation (late Early Miocene, the Bugti Hills and Zinda Pir; Antoine et al. 2010). Undescribed specimens recovered from the base of the Kamlial Formation in the Potwar Plateau and referable to the same taxon (POA, pers. obs.) help for correlating both intervals (see figure 16.3). Higher in the series, a partial maxilla with M1–3 unearthed in the level 7 (Djigani, the Bugti Hills; see figure 16.3) is referred to *Rhinoceros* aff. *sivalensis*. Undescribed specimens from the lower Dhok Pathan Formation in the Potwar Plateau document the same taxon. Thus, a middle Late Miocene age can be hypothesized for Djigani locality (Barry et al. 2002).

Teleoceratina are hippo-like extinct rhinos, with shortened limb bones adapted to swamps and riversides. *Prosantorhinus shahbazi* is restricted to the earliest Miocene deposits of the Bugti Hills (see table 16.1; Antoine et al. 2010). Prior to the latter recognition, the genus was only recorded in the late Early and early Middle Miocene of Western Europe (e.g., Antoine, Bulot, and Ginsburg 2000). Three representatives of the large and robust genus *Brachypotherium* are known. *B. fatehjangense* and *B. gajense* are recorded from the upper Chitarwata Formation and the lowest Vihowa Formation in both the Bugti Hills and Zinda Pir (Early Miocene; see table 16.1; Welcomme et al. 2001; Antoine et al. 2010). *B. gajense* is restricted to this stratigraphic interval, while *B. fatehjangense* persists until the Late Miocene in the Potwar Plateau (middle Dhok Pathan Formation; new data). In the Sulaiman Province, *B. perimense* ranges from the base of the Vihowa Formation in Zinda Pir (late Early Miocene; see table 16.1) up to level 7 in the Bugti Hills (middle Late Miocene; Antoine, Duranthon, and Welcomme 2003). New data from the Potwar Plateau provide similar ranges (LAD at ca. 7.1 Ma; Barry et al. 2002; POA, pers. obs.). Bugti and Zinda Pir remains document the earliest occurrences of *Prosantorhinus* and *Brachypotherium* at Eurasian and Old World scales, respectively.

Aceratheriina are extinct hornless rhinos, widespread in the Miocene of North America and Eurasia and in the Miocene of Africa. The only aceratheriine sensu stricto recognized in the Sulaiman Province is *Alicornops complanatum*, which occurs in level 7 of the Bugti Hills (Litra Formation, middle Late Miocene; Antoine, Duranthon, and Welcomme 2003). This taxon is abundant throughout the Late Miocene Dhok Pathan Formation in the Potwar Plateau (Colbert 1935; Heissig 1972; new data).

The Elasmotheriinae are the sister group of Rhinocerotinae. They are well represented in the Neogene of Eurasia and the Miocene of Africa (Antoine 2002). *Bugtirhi-*

nus praecursor is the earliest elasmotheriine known so far. This primitive species is restricted to the upper Chitarwata deposits of the Bugti Hills (levels 3bis and 4, earliest Miocene; see table 16.1) and to the base of the overlying Vihowa Formation of Zinda Pir (Z116, late Early Miocene; see table 16.2 and figure 16.4; POA, pers. obs.).

NEOGENE FAUNAL SUCCESSION IN THE SULAIMAN PROVINCE AND BIOSTRATIGRAPHICAL CORRELATION

In the present work, biostratigraphical correlation between the Bugti and Zinda Pir areas is mostly based on First Local Appearances (FLA) and observed ranges of hoofed mammals (rhinocerotids, proboscideans, artiodactyls, and hipparionine equids) for the whole Oligocene-Miocene series, as well as rodents for the early Miocene period (see figures 16.3 and 16.4). These assemblages are widely homotaxic, at the generic and/or species level (see tables 16.1–16.3).

In the Sulaiman Province, the best-documented stratigraphic interval spans the Chitarwata Formation and overlying Vihowa Formation (Oligocene to Middle Miocene), while the Potwar Plateau in northern Pakistan yields essentially Neogene deposits, among which Middle and Late Miocene faunas (Chinji, Nagri, and Dhok Pathan formations) are far better known than Early Miocene faunas (Murree and Kamlial formations; Pilbeam et al. 1979; Barry, Lindsay, and Jacobs 1982; Barry et al. 2002). Such a situation does not facilitate correlating the concerned areas.

The tentative correlation to GPTS for Neogene deposits of the Sulaiman Province is primarily based on the "Interpretation B" of Lindsay et al. (2005:fig. 6B), with revised ages for chrons 6C to 5C (Gradstein, Ogg, and Smith 2004). This interpretation is by far the most satisfactory for concerned mammal assemblages (i.e., without hiatus between the top of the Chitarwata Formation and the base of the Vihowa Formation) within C5En.2n, at ca. 19.4 Ma ("C6n" in Lindsay et al. 2005).

As a result, four successive faunal assemblages (A to D from old to young) are recognized in the Neogene of the Sulaiman Province, mainly constrained by perissodactyls (rhinocerotids and hipparionine equids) and rodents (cricetids, rhizomyids, and ctenodactylids) and, to a lesser degree, by deinotheriid proboscideans, sanitheriid and listriodontine suoids, anthracotheriids, and most ruminant groups (tragulids, bovids, and Pecora incertae sedis; see figure 16.4).

Assemblage A (Upper Chitarwata Formation Assemblage: Earliest Miocene)

A major turnover is observed within the Chitarwata Formation in the Sulaiman Province. Lower in the series—that is, in the lower Member (Zinda Pir) and Bugti Member (Bugti Hills)—occurs the Last Local Appearance (LLA) of many groups and genera of Oligocene affinities (see figure 16.3): the rodent *Downsimys* and most baluchimyines (*Lindsaya, Lophibaluchia, Hodsahibia,* and *Asterattus*; Lindsay et al. 2005), Entelodontidae, the anthracotheriids *Anthracotherium, Bugtitherium,* and *Elomeryx,* the chalicotheriid *Schizotherium,* Hyracodontidae (*Paraceratherium bugtiense*), Amynodontidae (*Cadurcotherium indicum*), and early rhinocerotids such as *Epiaceratherium* cf. *magnum, Aprotodon smithwoodwardi,* "*Dicerorhinus*" *abeli,* and a close ally of *Diceratherium* (Antoine, Duranthon, and Welcomme 2003; Antoine et al. 2004; Lindsay et al. 2005; Métais et al. 2009). This assemblage is assumed to predate the Oligocene–Miocene transition.

Assemblage A coincides with the appearance of a totally renewed assemblage in the upper Member of the Chitarwata Formation (Z113 and Z139 and localities above in Zinda Pir; levels 3bis and 4 in the Bugti area; Lindsay et al. 2005; Métais et al. 2009; Antoine et al. 2010). The rodent fauna is broadly renewed, with FLAs of *Eumyarion*, Copemyinae (*Democricetodon, Spanocricetodon,* and *Primus*), Rhizomyinae (*Prokanisamys*), and Ctenodactylinae (*Prosayimys*). Large mammals having their FLA in this assemblage are the large carnivore *Amphicyon*, Deinotheriidae (*Prodeinotherium*), *Gomphotherium*, Listriodontini suids (*Listriodon*), the anthracotheriids *Telmatodon* and *Hemimeryx*, the tragulid *Dorcatherium*, Bovidae (*Eotragus*), the ruminant *Bugtimeryx*, the chalicotheriids "*Chalicotherium*" and *Phyllotillon*, and the rhinocerotids *Protaceratherium, Mesaceratherium, Pleuroceros, Plesiaceratherium, Brachypotherium, Prosantorhinus, Gaindatherium,* and *Bugtirhinus* (see figure 16.3).

This stratigraphic interval resembles the Gaj deposits in Sind (Métais et al. 2009) and perhaps the poorly documented base of the Murree Formation in the Potwar Plateau (Barry et al. 2002). Given the faunal content and favored correlation to GPTS (see figure 16.4), this assemblage is tentatively correlated to the earliest Neogene standard age, the Aquitanian, and roughly correlated with the Xiejian Chinese Land Mammal Age (CLMA) and the Agenian European Land Mammal Age (ELMA), MN 1–2.

Assemblage B (Lower Vihowa Formation Assemblage: Late Early Miocene)

The Chitarwata-Vihowa formation transition is marked by the LLA of the cricetids *Eumyarion* and *Primus,* the ctenodactylid *Prosayimys,* the sanitheriid suoid *Sanitherium,* the ruminant *Bugtimeryx,* and several rhinocerotids, such as *Protaceratherium* sp., *Plesiaceratherium naricum, Prosantorhinus shahbazi,* and *Gaindatherium* cf. *browni,* which only occur in the Chitarwata Formation (see figures 16.3 and 16.4). This transition is supposedly coeval to both the Aquitanian-Burdigalian transition (marine standard scale) and the Agenian-Orleanian European Land Mammal Ages transition.

Assemblage B (lower Vihowa Formation) documents the FLA of key taxa such as the derived muroids *Megacricetodon* and *Myocricetodon,* the ctenodactylid *Sayimys intermedius,* the diatomyid *Diatomys,* the creodonts *Pterodon* and *Hyanailouros,* the carnivore *Megamphicyon,* the elephantoids *Protanancus* and *Choerolophodon, Listriodon guptai,* the tragulid *Dorcabune,* the "giraffoid" *Progiraffa,* and the rhinocerotids cf. *Rhinoceros* sp. and *Brachypotherium perimense.* Based on rhinocerotids and early bovids, this stratigraphic interval resembles the lower Manchar Formation in Sind (Raza and Meyer 1984) and the Kamlial Formation in the Potwar Plateau (Barry et al. 2002). Given the faunal content and favored correlation to GPTS (see figure 16.4), this assemblage is considered late Early Neogene standard age, the Burdigalian, and roughly correlated with the Shanwangian ALMA and the Orleanian ELMA (MN 3–5).

Assemblage C (Upper Vihowa Formation Assemblage: Middle Miocene)

Assemblage C is documented by only a few large mammals, and as such, it is not well constrained in terms of biostratigraphy. The concerned interval yields the FLA of *Deinotherium, Listriodon pentapotamiae,* the bovid cf. *Elachistoceras,* and the chalicotheriid *Anisodon* (see figure 16.3). This stratigraphic interval resembles the upper Manchar Formation in Sind and the Chinji Formation in the Potwar Plateau (Barry et al. 2002). Given the faunal content and favored correlation to GPTS (see figure 16.4B), this assemblage is tentatively considered early Middle Miocene, late Langhian–early Serravallian standard ages, and roughly correlated with the Tunggurian ALMA and the Astaracian ELMA (MN 6).

Assemblage D (Litra Formation Assemblage: Late Miocene)

In stratigraphic terms, the Vihowa–Litra formation transition is not well constrained, due to the scarcity of available localities and remains across the whole Sulaiman Province (Welcomme et al. 2001; Raza et al. 2002; Antoine, Welcomme, et al. 2003; Zouhri and Ginsburg 2003).

This assemblage consists only of large mammals (see figures 16.3 and 16.4B), with the FLA of tetraconodontine and suine suids (*Parachleuastochoerus* and ?*Propotamochoerus*, respectively), of giraffids sensu stricto (?*Bramatherium*), of advanced bovids (*Hispanodorcas*, *Prostrepsiceros*, and an unidentified reduncine), of hipparionine equids (*Cormohipparion* (*Sivalhippus*) *theobaldi* and *Hippotherium* sp.), and of the rhinocerotids *Alicornops complanatum* and *Rhinoceros* aff. *sivalensis* and the LLA of listriodontines (*Listriodon* sp.).

The concerned fauna strongly resembles the magnetostratigraphically constrained assemblages recognized in the upper part of the Nagri Formation and the lower part of the Dhok Pathan Formation in the Potwar Plateau (Heissig 1972; Pilbeam et al. 1979; Barry et al. 2002). As such, this assemblage might be early Late Miocene standard age (late Tortonian) and is tentatively correlated with the late Bahean ALMA interval and the late Vallesian–early Turolian ELMA interval (MN 10–12).

FIRST LOCAL APPEARANCES AS "DATUMS"

Several mammalian taxa have their earliest occurrences in mid-Cenozoic deposits of the Sulaiman Range, either at the tribe, family, or even order level. In this section, we have chosen to focus on key taxa used at a large scale for biochronology and dispersal events, and well represented in the Neogene of the Sulaiman Province, such as proboscideans, suoids, bovids, and rhinocerotids.

Proboscidean Datum Event(s) (Bugti Hills/Zinda Pir)

Early specimens referable to elephantoids discovered in the Bugti Hills originate from the Lundo section (Welcomme et al. 2001; Antoine, Welcomme, et al. 2003; Métais et al. 2009). The locality DB-J1 where the first tusk was found (Antoine, Welcomme, et al. 2003) is located ca. 40 m below the classical Lundo locality (level J2 = "Chur Lando" of Pilgrim 1908, 1910; Forster-Cooper 1924, 1934), and about 55 m below the Chitarwata–Vihowa formation

transition (see figure 16.3). Tabbutt, Sheikh, and Johnson (1997) provided a fission track date of 22.6 ± 2.9 Myr for the yellow sands of "Chur Lando" (i.e., level J2). Locality DB-J1 is necessarily older than 19.7 Myr (and may date back to 25.4 Myr). The second tusk was found in a still older locality, referred to the level F, ca. 20 m lower in the same section (see figure 16.3).

In the Zinda Pir area, Lindsay et al. (2005) mention both "Elephantoidea Indet. genus, indet. species" and "Proboscidea Indet. genus, indet. species" in locality Z108, located in the lower member of the Chitarwata Formation, but we now think these specimens are not definitive. A younger specimen at locality Z154 is clearly an elephantoid and found midway in a significant normal magnetochron, currently considered chron C6Br, and therefore likely somewhat younger than 23 Ma (Flynn et al., chapter 14, this volume). These early occurrences alter considerably the concept of Proboscidean Datum Event as documenting a single dispersal of proboscideans out of Africa in the late Early Miocene (ca. 17.5 Ma; Madden and Van Couvering 1976). Proboscideans rather dispersed several times from Africa to Eurasia during the Oligocene and the Early Miocene, as already argued by Tassy (1990).

The earliest "diversified proboscidean fauna" from the Sulaiman Province occurs in the upper Chitarwata Formation, with the FLA of both Deinotheriidae (see figure 16.3 [levels 3bis and 4 in the Bugti Hills, Z129 in Zinda Pir]) and *Gomphotherium* sp. (see figure 16.3 [level 4 in the Bugti Hills]; Welcomme et al. 2001; Métais et al. 2009). Lindsay et al. (2005) correlate Z129 with C6Bn (ca. 22.5 Ma; Gradstein, Ogg, and Smith 2004) and the level 4 may date back to ca. 21 Ma (see figure 16.4B).

Sanitheriidae

The sanitheriid *Sanitherium* sp. was recently recognized in the Bugti member of the Chitarwata Formation at Paali C2 (Early Oligocene, ca. 30 Ma; see figure 16.3; Orliac et al. 2010). *Sanitherium jeffreysi* is documented in the upper Member of the Chitarwata Formation at Samane 4 (Early Miocene, ca. 21 Ma; see figure 16.3; Orliac et al. 2010). "?*Diamantohyus jeffreysi*" is mentioned from coeval deposits of the lower Chitarwata Formation in Zinda Pir (Z144; Lindsay et al. 2005). Z144 is located in the same chron as Z108, but higher (C7n or C10n.2n; Lindsay et al. 2005), dating to 24.6 Ma or even 28.6 Ma (see section "Proboscidean Datum Event[s]").

Both Oligocene occurrences widely predate the previous worldwide FAD of the family, so far considered as occurring during the Early Miocene in Africa (Sperrgebiet, Namibia, ~21–19 Ma; Pickford and Senut 2000, 2002). Such an early settlement of sanitheres in the Indian Subcontinent strongly supports the provocative hypothesis of a late Oligocene–earliest Miocene dispersal event *from* South Asia *toward* Africa, rather than in the opposite direction (Orliac et al. 2010).

Bovidae

The lowest bovid-yielding locality in Zinda Pir is Z120 (lowermost Vihowa), which is correlated with C5En (Lindsay et al. 2005)—that is, estimated at ca. 18.4 Myr (see figure 16.4B; Gradstein, Ogg, and Smith 2004). The horncore referred to *Eotragus minus* in the Bugti area (Ginsburg, Morales, and Soria 2001) was recovered from coeval deposits (level 6, lowermost Vihowa Formation; see figure 16.4B). Both occurrences predate the base of the Kamlial Formation in the Siwalik Group (~18 Ma), where Solounias et al. (1995) have described *Eotragus noyei*, the oldest representative of the family then known.

Furthermore, the uppermost Chitarwata Formation in the Bugti Hills (level 4) yields several unambiguous bovid postcranials (Ginsburg, Morales, and Soria 2001). Given the favored correlation hypothesis between Zinda Pir and the Bugti area, these remains may date back to ca. 21 Ma (see section "Proboscidean Datum Event[s]" and figure 16.4B).

Early Miocene Rhinocerotidae and the "African Rhinocerotid Datum"

The rhinocerotid fauna from assemblage A (upper Chitarwata Formation, earliest Miocene: 23–19.4 Ma; see figure 16.4B) is exceptionally diversified, with nine co-occurring species in the Bugti Hills: the early elasmotheriine *Bugtirhinus praecursor* and the hornless rhinocerotines *Protaceratherium* sp. and *Plesiaceratherium naricum*; the basal rhinocerotines *Pleuroceros blanfordi* and *Mesaceratherium welcommi*; the teleoceratines *Brachypotherium gajense*, *B. fatehjangense*, and *Prosantorhinus shahbazi*; and the rhinocerotine *Gaindatherium* cf. *browni* (Antoine and Welcomme 2000; Antoine et al. 2010). Coeval homotaxic rhinocerotid assemblages (at genus level) are recorded from the Agenian ELMA of France (MN 1 to MN 2: 23–20 Ma interval; Gradstein, Ogg, and Smith 2004), with *Protaceratherium minutum*, *Plesiaceratherium aquitanicum*, *Pleuroceros pleuroceros*, and *Mesaceratherium paulhiacense* (de Bonis 1973; Antoine et al. 2006, 2010).

Interestingly, the earliest representatives of Rhinocerotidae in Africa are strongly comparable to those of assemblage A, which persist into assemblage B (see figure 16.4B): *Brachypotherium heinzelini* and *Aceratherium acutirostratum*, recognized in Napak II and Songhor (Hooijer 1966, 1973; Hooijer and Patterson 1972), as well as *Ougandatherium napakense* from Napak I (Guérin and Pickford 2003), are strongly comparable to the teleoceratine *Brachypotherium fatehjangense*, the acerathere *Mesaceratherium welcommi*, and the earliest elasmotheriine *Bugtirhinus praecursor*, respectively (Antoine et al. 2010). The radiometric age of Songhor is ~19.5 Ma (Pickford 1986; Cote et al. 2007) and Napak might be slightly older (Tassy 1986; Cote et al. 2007), which coincides with the assemblage A–assemblage B transition in terms of age (ca. 19.4 Ma; see figure 16.4B).

The close affinities of South Asian, African, and European rhinocerotid assemblages confirm both the presence of land bridges and the absence of ecological barriers between these continental areas during the Early Miocene.

ACKNOWLEDGMENTS

The authors warmly thank Xiaoming Wang, Deng Tao, Li Xiang, Mikael Fortelius, and all the Organizing Committee of the symposium "Neogene Terrestrial Mammalian Biostratigraphy and Chronology in Asia" held in Beijing in June 2009. Everett H. Lindsay provided highly valuable comments on a previous version of the manuscript. We are grateful to Francis Duranthon, Mouloud Benammi, Jean-Jacques Jaeger, Yaowalak Chaimanee, and Dario De Franceschi for their participation in fieldwork, and to Michèle E. Morgan, John C. Barry, David Pilbeam, Everett H. Lindsay, and Iqbal U. Cheema and Jelle Zijlstra for their valuable help and discussion. This article is dedicated to the memory of Will Downs, Nawab M. A. K. Bugti, and Léonard Ginsburg. This research was supported by the French ANR-PALASIAFRICA Program (ANR-08-JCJC-0017 - ANR-ERC). MPFB Publication no. 39.

REFERENCES

Anquetin, J., P.-O. Antoine, and P. Tassy. 2007. Middle Miocene Chalicotheriinae (Mammalia, Perissodactyla) from France, with a discussion on chalicotheriine phylogeny. *Zoological Journal of the Linnean Society* 151:577–608.

Antoine, P.-O. 2002. Phylogénie et évolution des Elasmotheriina (Mammalia, Rhinocerotidae). *Mémoires du Muséum National d'Histoire Naturelle* (Paris) 188.

Antoine, P.-O., C. Bulot, and L. Ginsburg. 2000. Les rhinocérotidés (Mammalia, Perissodactyla) de l'Orléanien (Miocène inférieur) des bassins de la Garonne et de la Loire: Intérêt biostratigraphique. *Comptes Rendus de l'Académie des Sciences, Sciences de la Terre et des Planètes* (Paris) 330:571–576

Antoine, P.-O., K. F. Downing, J.-Y. Crochet, F. Duranthon, L. J. Flynn, L. Marivaux, G. Métais, A. R. Rajpar, and G. Roohi. 2010. A revision of *Aceratherium blanfordi* Lydekker, 1884 (Mammalia: Rhinocerotidae) from the early Miocene of Pakistan: Postcranials as a key. *Zoological Journal of the Linnean Society* 160:139–194.

Antoine, P.-O., S. Ducrocq, L. Marivaux, Y. Chaimanee, J.-Y. Crochet, J.-J. Jaeger, and J.-L. Welcomme. 2003. Early rhinocerotids (Mammalia, Perissodactyla) from South Asia and a review of the Holarctic Paleogene rhinocerotid record. *Canadian Journal of Earth Sciences* 40:365–374.

Antoine, P.-O., F. Duranthon, S. Hervet, and G. Fleury. 2006. Vertébrés de l'Oligocène terminal (MP30) et du Miocène basal (MN1) du métro de Toulouse (SW de la France). *Comptes Rendus Palevol* 5:875–884.

Antoine, P.-O., F. Duranthon, and P. Tassy. 1997. L'apport des grands mammifères (Rhinocérotidés, Suoidés, Proboscidiens) à la connaissance des gisements du Miocène d'Aquitaine (France). In *Actes du Congrès BiochroM'97*, ed. J.-P. Aguilar, S. Legendre, and J. Michaux, pp. 581–590. Mémoires et Travaux E.P.H.E., Institut de Montpellier, 21.

Antoine, P.-O., F. Duranthon, and J.-L. Welcomme. 2003. *Alicornops* (Mammalia, Rhinocerotidae) dans le Miocène supérieur des Collines Bugti (Balouchistan, Pakistan): Implications phylogénétiques. *Geodiversitas* 25:575–603.

Antoine, P.-O., S. M. I. Shah, I. U. Cheema, J.-Y. Crochet, D. de Franceschi, L. Marivaux, G. Métais, and J.-L. Welcomme. 2004. New remains of the baluchithere *Paraceratherium bugtiense* (Pilgrim, 1910) from the Late/latest Oligocene of the Bugti Hills, Balochistan, Pakistan. *Journal of Asian Earth Sciences* 24:71–77.

Antoine, P.-O. and J.-L. Welcomme. 2000. A new rhinoceros from the lower Miocene of the Bugti Hills, Baluchistan, Pakistan: The oldest elasmotheriine. *Palaeontology* 43:795–816.

Antoine, P.-O., J.-L. Welcomme, L. Marivaux, I. Baloch, M. Benammi, and P. Tassy. 2003. First record of Paleogene Elephantoidea (Mammalia, Proboscidea) from the Bugti Hills of Pakistan. *Journal of Vertebrate Paleontology* 23:978–981.

Bannert, D., A. Cheema, A. Ahmed, and U. Schäffer. 1992. The structural development of the Western Fold Belt, Pakistan. *Geologisches Jahrbuch* B80:3–60.

Barry, J. C., S. Cote, L. MacLatchy, E. H. Lindsay, R. Kityo, and A. R. Rajpar. 2005. Oligocene and early Miocene ruminants (Mammalia, Artiodactyla) from Pakistan and Uganda. *Palaeontologia Electronica* 8(1); 20A: 29 pp; http://palaeo-electronica.org/2005 _1/barry22/issue1_05.htm.

Barry, J. C., L. J. Flynn, and D. Pilbeam. 1990. Faunal diversity and turnover in a Miocene terrestrial sequence. In *Causes of Evolution: A Paleontological Perspective*, ed. R. M. Ross and W. D. Allmon, pp. 381–421. Chicago: University of Chicago Press.

Barry, J. C., E. H. Lindsay, and L. L. Jacobs. 1982. A biostratigraphic zonation of the middle and upper Siwaliks of the Potwar Plateau of Northern Pakistan. *Palaeogeography, Palaeoclimatology, Palaeoecology* 37:95–130.

Barry, J. C., M. E. Morgan, L. J. Flynn, D. Pilbeam, A. Behrensmeyer, S. M. Raza, I. Khan, C. Badgley, J. Hicks, and J. Kelley. 2002. Faunal and environmental change in the late Miocene Siwaliks of northern Pakistan. *Paleobiology Memoir* 3 (supplement to 2):1–71.

Beck, R. A., D. W. Burbank, W. J. Sercombe, G. W. Riley, J. K. Barndt, J. R. Berry, J. Afzal, A. M. Khan, H. Jurgen, J. Metje, A. Cheema, N. A. Shafique, R. D. Lawrence, and M. A. Khan. 1995. Stratigraphic evidence for an early collision between northwest India and Asia. *Nature* 373:55–58.

Blanford, W. T. 1883. Geological notes on the hills in the neighbourhood of the Sind and Punjab Frontier between Quetta and Dera Ghazi Khan. *Memoirs of the Geological Survey of India* 20:1–136.

Clift, P. D., N. Shimizu, G. D. Layne, and J. Blusztajn. 2001. Tracing patterns of erosion and drainage in the Paleogene Himalaya through ion probe Pb isotope analysis of detrital K-feldspars in the Indus Molasse, India. *Earth and Planetary Science Letters* 188:475–491.

Colbert, E. H., 1933. An Upper Tertiary peccary from India. *American Museum Novitates* 635:1–9.

Colbert, E. H., 1935. Siwalik mammals in the American Museum of Natural History. *Transactions of the American Philosophical Society* 26:1–401.

Cote, S., L. Werdelin, E. R. Seiffert, and J. C. Barry. 2007. Additional material of the enigmatic Early Miocene mammal *Kelba* and its relationship to the order Ptolemaiida. *Proceedings of the National Academy of Sciences* 104:5510–5515.

Crochet, J.-Y., P.-O. Antoine, L. Marivaux, G. Métais, and J.-L. Welcomme. 2009. Premières descriptions de peintures et gravures rupestres au Baloutchistan (Pakistan). *International Newsletter On Rock Art* 55.

de Bonis, L. 1973. Contribution à l'étude des mammifères de l'Aquitanien de l'Agenais. Rongeurs-Carnivores-Périssodactyles. *Mémoires du Muséum National d'Histoire Naturelle* (Paris) 28:1–192.

Downing, K. F. and E. H. Lindsay. 2005. Relationship of Chitarwata Formation paleodrainage and paleoenvironments to Himalayan tectonics and Indus River paleogeography. *Palaeontologia Electronica* 8(1):20A; 12 pp; http://palaeo-electronica.org/2005_1/downing20/issue1_05.htm.

Downing, K. F., E. H. Lindsay, W. R. Downs, and S. E. Speyer. 1993. Lithostratigraphy and vertebrate biostratigraphy of the early Miocene Himalayan Foreland, Zinda Pir Dome, Pakistan. *Sedimentary Geology* 87:25–37.

Ducrocq, S. and F. Lihoreau. 2006. The occurrence of bothriodontines (Artiodactyla, Mammalia) in the Paleogene of Asia with special reference to *Elomeryx*: Paleobiogeographical implications. *Journal of Asian Earth Sciences* 27:885–891.

Flynn, L. J. 2007. Origin and evolution of the Diatomyidae, with clues to paleoecology from the fossil record. In *Mammalian Paleontology on a Global Stage: Papers in Honor of Mary R Dawson*, ed. K. C. Beard and Z.-x. Luo, pp. 173–181. *Bulletin of Carnegie Museum of Natural History* 39.

Flynn, L. J., L. L. Jacobs, and I. U. Cheema. 1986. Baluchimyinae, a new ctenodactyloid rodent subfamily from the Miocene of Baluchistan. *American Museum Novitates* 284:1–58.

Forster-Cooper, C., 1913. New anthracotheres and allied forms from Baluchistan. *Annals Magazine of Natural History* 19:514–522.

Forster-Cooper, C. 1915. New genera and species of mammals from the Miocene deposits of Baluchistan. *Annals and Magazine of Natural History* 16:404–410.

Forster-Cooper, C. 1924. The Anthracotheriidae of the Dera Bugti deposits in Baluchistan. *Memoirs of the Geological Survey of India* 4(2):1–59.

Forster-Cooper, C. 1934. XIII. The Extinct Rhinoceroses of Baluchistan. *Philosophical Transactions of the Royal Society of London*, ser. B 223:569–616.

Friedman, R., J. Gee, L. Tauxe, K. Downing, and E. H. Lindsay. 1992. The magnetostratigraphy of the Chitarwata and lower Vihowa formations of the Dera Ghazi Khan area, Pakistan. *Sedimentary Geology* 81:253–268.

Ginsburg, L., J. Morales, and D. Soria. 2001. Les Ruminantia (Artiodactyla, Mammalia) du Miocène des Bugti (Balouchistan, Pakistan). *Estudios Geologicos* 57:155–170.

Ginsburg, L. and J.-L. Welcomme, 2002. Nouveaux restes de créodontes et de carnivores des Bugti (Pakistan). *Symbioses* 7:65–68.

Golpe Posse, J. M. 1972. Suiformes del Terciario español y sus yacimientos. *Paleontologia i Evolucio* 2:1–197.

Gradstein, F. M., J. G. Ogg, and A. G. Smith. 2004. *A Geological Time Scale 2004.* Cambridge: Cambridge University Press.

Guérin, C. and M. Pickford. 2003. *Ougandatherium napakense* nov. gen. nov. sp., le plus ancien Rhinocerotidae Iranotheriinae d'Afrique. *Annales de Paléontologie* 89:1–35.

Heissig, K. 1972. Paläontologische und geologische Untersuchungen im Tertiär von Pakistan. 5. Rhinocerotidae (Mamm.) aus den unteren und mittleren Siwalik-Schichten. *Abhandlungen der Bayerischen Akademie der Wissenschaften Mathematisch-Naturwissenschaftliche Klasse* 152:1–112.

Hemphill, W. R. and A. H. Kidwai. 1973. Statigraphy of the Bannu and Dera Ismail Khan Areas, Pakistan. *United States Geological Survey Professional Paper* 716B:1–36.

Hooijer, D. A. 1966. Miocene rhinoceroses of East Africa. *Bulletin of the British Museum (Natural History) Fossil Mammals of Africa* 21:117–190.

Hooijer, D. A. 1973. Additional Miocene to Pleistocene rhinoceroses of Africa. *Zoological Mededelingen* 46:149–178.

Hooijer, D. A. and B. Patterson. 1972. Rhinoceroses from the Pliocene of northwestern Kenya. *Bulletin of the Museum of Comparative Zoology* (Cambridge, Mass.) 144:1–26.

Jacobs, L. L., I. U. Cheema, and S. M. I. Shah. 1981. Zoogeographic implications of early Miocene rodents from the Bugti Beds, Baluchistan, Pakistan. *Geobios* 15:101–103.

Johnson, N. M., J. Stix, L. Tauxe, P. F. Cerveny, and R. A. K. Tahirkheli. 1985. Paleomagnetic chronology, fluvial processes, and tectonic implications of the Siwalik deposits near Chinji village, Pakistan. *Journal of Geology* 93:27–40.

Kumar, K. and S. Kad. 2003. Early Miocene vertebrates from the Murree Group, northwest Himalaya, India: Affinities and age implications. *Himalayan Geology* 24(2):29–53.

Lindsay, E. H., L. J. Flynn, I. U. Cheema, J. C. Barry, K. F. Downing, A. R. Rajpar, and S. M. Raza. 2005. Will Downs and the Zinda Pir Dome. *Palaeontologia Electronica* 8:1–19.

Lydekker, R. 1884. Indian Tertiary and post-Tertiary Vertebrata. *Memoirs of the Geological Survey of India, Palaeontologica Indica*, 10th ser., 10, 2(1–2):1–66.

Madden, C. T. and J. A. Van Couvering. 1976. The Proboscidean Datum Event: Early Miocene migration from Africa. *Geological Society of America Abstracts with Programs* 8:992–993.

Marivaux, L., P.-O. Antoine, S. R. H. Baqri, M. Benammi, Y. Chaimanee, J.-Y. Crochet, D. de Franceschi, N. Iqbal, J.-J. Jaeger, G. Métais, G. Roohi, and J.-L. Welcomme, 2005. Anthropoid primates from the Oligocene of the Bugti Hills (Pakistan): Data on early anthropoid evolution and biogeography. *Proceedings of the National Academy of Sciences* 102:8436–8441.

Marivaux, L., L. Bocat, Y. Chaimanee, J.-J. Jaeger, B. Marandat, P. Srisuk, P. Tafforeau, C. Yamee, and J.-L. Welcomme. 2006. Cynocephalid dermopterans from the Palaeogene of South Asia (Thailand, Myanmar and Pakistan): Systematic, evolutionary and palaeobiogeographic implications. *Zoologica Scripta* 35:395–420.

Marivaux, L., M. Vianey-Liaud, and J.-L. Welcomme. 1999. Première découverte de Cricetidae (Rodentia, Mammalia) oligocènes dans le synclinal sud de Gandoï (Bugti Hills, Balouchistan, Pakistan). *Comptes Rendus de l'Académie des Sciences de Paris*, 2nd ser., 329:839–844.

Marivaux, L., J.-L. Welcomme, P.-O. Antoine, G. Métais, I. Baloch, M. Benammi, Y. Chaimanee, S. Ducrocq, and J.-J. Jaeger. 2001. A fossil lemur from the Oligocene of Pakistan. *Science* 294:587–591.

McKenna, M. C. and S. K. Bell. 1997. *Classification of Mammals Above the Species Level.* New York: Columbia University Press.

Métais, G., P.-O. Antoine, S. R. H. Baqri, M. Benammi, J.-Y. Crochet, D. de Franceschi, L. Marivaux, S. Ducrocq, and J.-L. Welcomme. 2006. New remains of the enigmatic cetartiodactyl *Bugtitherium grandincisivum* Pilgrim, 1908 from the upper Oligocene of the Bugti Hills (Balochistan, Pakistan). *Naturwissenschaften* 93:348–355.

Métais, G., P.-O. Antoine, S. R. H. Baqri, J.-Y. Crochet, D. de Franceschi, L. Marivaux, and J.-L. Welcomme. 2009. Lithofacies, depositional environments, regional biostratigraphy and age of the Chitarwata Formation in the Bugti Hills, Balochistan, Pakistan. *Journal of Asian Earth Sciences* 34:154–167.

Métais, G., P.-O. Antoine, L. Marivaux, S. Ducrocq, and J.-L. Welcomme. 2003. New artiodactyl ruminant mammal from the Late Oligocene of Pakistan. *Acta Palaeontologica Polonica* 48:365–374.

Métais, G., J.-L. Welcomme, and S. Ducrocq. 2009. Lophiomerycid ruminants from the Oligocene of the Bugti Hills (Balochistan, Pakistan). *Journal of Vertebrate Paleontology* 29:1–12.

Najman, Y. and E. Garzanti. 2000. Reconstructing early Himalayan tectonic evolution and paleogeography from Tertiary foreland basin sedimentary rocks, northern India. *Geological Society of America Bulletin* 112:435–449.

Orliac, M., P.-O. Antoine, and F. Duranthon. 2006. The Suoidea (Artiodactyla, Mammalia), exclusive of Listriodontinae, from the Orleanian of Béon 1 (Montréal-du-Gers, SW France, MN4). *Geodiversitas* 28:685–718.

Orliac, M. J., P.-O. Antoine, G. Métais, J.-Y. Crochet, L. Marivaux, G. Roohi, and J.-L. Welcomme. 2009. *Listriodon guptai* Pilgrim, 1926 (Mammalia, Suidae) from the early Miocene of the Bugti Hills, Balochistan, Pakistan: New insights into early Listriodontinae evolution and biogeography. *Naturwissenschaften* doi: 10.1007/s00114-009-0547-4.

Orliac, M. J., P.-O. Antoine, G. Roohi, and J.-L. Welcomme. 2010. Suoidea (Mammalia, Cetartiodactyla) from the early Oligocene of the Bugti Hills, Balochistan, Pakistan. *Journal of Vertebrate Paleontology* 30:1300–1305.

Pickford, M. 1986. A revision of the Miocene Suidae and Tayassuidae (Artiodactyla, Mammalia) of Africa. *Tertiary Research*, Special Paper 7:1–83.

Pickford, M. 1987. Révision des suiformes (Artiodactyla, Mammalia) de Bugti (Pakistan). *Annales de Paléontologie* 73:289–350.

Pickford, M. and Liu Liping. 2001. Revision of the Miocene Suidae of Xiaolongtan (Kaiyuan), China. *Bollettino della Società Paleontologica Italiana* 40:275–283.

Pickford, M., H. Nakaya, Y. Kunimatsu, H. Saegura, A. Fukuchi, and B. Ratanasthien. 2004. Age and taxonomic status of the Chiang Muan (Thailand) hominoids. *Comptes Rendus Palevol* 3:65–75.

Pickford, M. and B. Senut. 2000. Geology and palaeobiology of the Namib Desert, southwestern Africa. *Memoirs of the Geological Survey of Namibia* 18:1–155.

Pickford, M. and B. Senut. 2002. *The Fossil Record of Namibia.* Windhoek: Ministry of Mines and Energy, Geological Survey of Namibia, Namprint.

Pilbeam, D. R., A. K. Behrensmeyer, J. C. Barry, and S. M. I. Shah. 1979. Miocene sediments and faunas of Pakistan. *Postilla* 179:1–45.

Pilbeam, D., M. E. Morgan, J. C. Barry, and L. J. Flynn. 1996. European MN Units and the Siwalik Faunal Sequence of Pakistan. In *The Evolution of Western Eurasian Neogene Mammal Faunas*, ed. R. L. Bernor, V. Fahlbusch, and H.-W. Mittmann, pp. 290–306. New York: Columbia University Press.

Pilgrim, G. E. 1907. Description of some new Suidae from the Bugti Hills, Baluchistan. *Records of the Geological Survey of India* 36:45–56.

Pilgrim, G. E. 1908. The Tertiary and Post-Tertiary freshwater deposits of Baluchistan and Sind with notices of new vertebrates. *Records of the Geological Survey of India* 37:139–167.

Pilgrim, G. E. 1910. Preliminary note on a revised classification of the Tertiary freshwater deposits of India. *Records of the Geological Survey of India* 40:185–205.

Pilgrim, G. E. 1911. The fossil Giraffidae of India. *Memoirs of the Geological Survey of India*, n. s., 4(1):1–29.

Pilgrim, G. E. 1912. The vertebrate fauna of the Gaj series in the Bugti Hills and the Punjab. *Memoirs of the Geological Survey of India*, n. s., 4:1–83.

Raza, S. M., I. U. Cheema, W. R. Downs, A. R. Rajpar, and S. C. Ward. 2002. Miocene stratigraphy and mammal fauna from the Sulaiman Range, southwestern Himalayas, Pakistan. *Palaeogeography, Palaeoclimatology, Palaeoecology* 186:185–197.

Raza, S. M. and G. E. Meyer. 1984. Early Miocene geology and paleontology of the Bugti Hills. *Geological Survey of Pakistan* 11:43–63.

Shah, S. M. I. 1977. Stratigraphy of Pakistan. *Memoirs of the Geological Survey of Pakistan* 12:1–138.

Shoshani, J. and P. Tassy. 1996. Appendix B. A suggested classification of the Proboscidea. In *The Proboscidea, Evolution and Palaeoecology of Elephants and Their Relatives*, ed. J. Shoshani and P. Tassy, pp. 352–353. Oxford: Oxford University Press.

Solounias, N., J. C. Barry, R. L. Bernor, E. H. Lindsay, and S. M. Raza. 1995. The oldest bovid from the Siwaliks, Pakistan. *Journal of Vertebrate Paleontology* 15:806–814.

Tabbutt, K. D., K. A. Sheikh, and N. M. Johnson. 1997. A fission track age for the Bugti bone beds, Baluchistan, Pakistan. In *Third Workshop on Siwaliks of South Asia. Records of the Geological Survey of Pakistan* 109:53.

Tassy, P. 1986. Nouveaux Elephantoidea (Mammalia) dans le Miocène du Kenya. In *Cahiers de Paléontologie: Essai de Réévaluation Systématique*. Paris: Éditions du Centre National de la Recherche Scientifique.

Tassy, P. 1990. The "Proboscidean Datum Event": How many proboscideans and how many events? In *European Neogene Mammal Chronology*, ed. E. H. Lindsay, V. Fahlbusch, and P. Mein, pp. 237–252. New York: Plenum Press.

van der Made, J. 1997. Systematics and stratigraphy of the genera *Taucanamo* and *Schizochoerus* and a classification of the Palaeochoeridae (Suoidea, Mammalia). *Proceedings of the Koninklijke Nederlandse Akademie van Wetenschappen* 100:127–139.

Vickary, N. 1846. Geological report on a portion of the Baluchistan Hills. *Quaterly Journal of the Geological Society* 2:260–265.

Welcomme, J.-L., P.-O. Antoine, F. Duranthon, P. Mein, and L. Ginsburg. 1997. Nouvelles découvertes de vertébrés miocènes dans le synclinal de Dera Bugti (Balouchistan, Pakistan). *Comptes Rendus de l'Académie des Sciences de Paris*, 2nd ser., 325:531–536.

Welcomme, J.-L., M. Benammi, J. Y. Crochet, L. Marivaux, G. Métais, P.-O. Antoine, and I. Baloch. 2001. Himalayan Forelands: Palaeontological evidence for Oligocene detrital deposits in the Bugti Hills (Balochistan, Pakistan). *Geological Magazine* 138(4):397–405.

Welcomme, J.-L. and L. Ginsburg. 1997. Mise en évidence de l'Oligocène sur le territoire des Bugti (Balouchistan, Pakistan). *Comptes Rendus de l'Académie des Sciences de Paris*, 2nd ser., 325:999–1004.

Welcomme, J.-L, L. Marivaux, P. O. Antoine, and M. Benammi. 1999. Mammifères fossiles des Collines Bugti (Balouchistan, Pakistan). Nouvelles données. *Bulletin de la Société d'Histoire Naturelle de Toulouse* 135:135–139.

Yan Defa and K. Heissig. 1986. Revision and autopodial morphology of the Chinese-European Rhinocerotid genus *Plesiaceratherium* Young 1937. *Zitteliana Abhandlungen der Bayerische Staatssammlung für Paläontologie und historisches Geologie*, München 14:81–110.

Zouhri S. and L. Ginsburg. 2003. Les Hipparions du Miocène supérieur des Bugti-Hills (Balouchistan, Pakistan). *Symbioses*, n. s., 9:33–38.

Chapter 17

Indian Neogene Siwalik Mammalian Biostratigraphy
An Overview

RAJEEV PATNAIK

Almost 6000 m thick, Siwalik freshwater deposits exposed all along the Himalayan foothills are famous for their great wealth of mammalian fossils ranging in age from ~18 Ma (Johnson et al. 1985) to 0.22 Ma (Ranga Rao et al. 1988). These deposits were formed by rivers that were precursors to the present-day mighty Indus, Ganges, and Brahmaputra of the Indian subcontinent. The excellent fossil and sediment record of the Siwaliks has attracted the interest of earth scientists and palaeo-anthropologists for the past 150 years. Medlicott (1879) followed by Pilgrim (1910, 1913) classified the Siwaliks as Lower, Middle, and Upper Siwaliks, further divided as the Kamlial, Chinji, Nagri, Dhok Pathan, Tatrot, Pinjor, and Boulder Conglomerate formations, based mainly on their faunal content. Colbert (1935) preferred calling them zones. Over the years, extensive fieldwork has revealed that the boundaries between most of the Siwalik formations are time transgressive and that temporal ranges of mammals are not usually fixed within the time limits of these formations. Therefore, these formations are now largely referred as lithounits (figure 17.1). Type sections of the Lower and Middle Siwalik formations lie in Pakistan, and those of Upper Siwaliks are mostly in India (figure 17.2). In recent years, extensive multidisciplinary studies, particularly in Pakistan, have resolved several issues concerning biotic changes in the Late Miocene and how they interplay with fluvial dynamics and climatic trends (Keller et al. 1977; Pilbeam et al. 1977; Opdyke et al. 1979; Badgley and Behrensmeyer 1980; Barry, Lindsay, and Jacobs 1982; Flynn and Jacobs 1982; Johnson et al. 1982; Barry et al. 1985; Johnson et al. 1985; Quade, Cerling, and Bowman 1989; Barry and Flynn 1990; Willis 1993; Behrensmeyer, Willis, and Quade 1995; Flynn et al. 1995; Quade and Cerling 1995; Barry et al. 2002; Badgley et al. 2005; Badgley et al. 2008).

India has some of the classic Siwalik fossil mammal localities (Pilgrim 1910, 1913; Lewis 1934; Colbert 1935; Prasad 1970), situated along the Himalayan foothills around Jammu (state of Jammu and Kashmir), Nurpur and Haritalyangar (Himachal Pradesh), Chandigarh (Union Territory), Kala Amb-Saketi (Haryana and Himachal Pradesh), and Kalagarh (Uttar Pradesh). In India, several significant contributions have been made to understand the fauna (Sahni and Khan 1964; Nanda and Sehgal 1993), geology (Karunakaran and Ranga Rao 1979; Parkash, Sharma, and Roy 1980; Kumar and Tandon 1985; Vasishat 1985), magnetostratigraphy (Johnson et al. 1983; Tandon et al. 1984; Ranga Rao et al. 1988; Kotlia et al. 1999; Sangode and Kumar 2003), and tephrochronology (Mehta et al. 1993). Barring a few instances where magnetostratigraphy has been tied to fossil occurrences (Ranga Roa et al. 1988; Agarwal et al. 1993; Basu 2004; Patnaik and Nanda 2010), the mammalian faunal assemblages are largely differentiated only according to formation (Vasishat et al. 1983; Nanda and Sehgal 1993, 2005).

While Neogene mammal fossils of India occurring in the Siwaliks can be assigned to lithostratigraphic units, nomenclature of the units is problematic, and utilization of classic formation names sometimes leads to errors. As most of the formation boundaries have been found to be

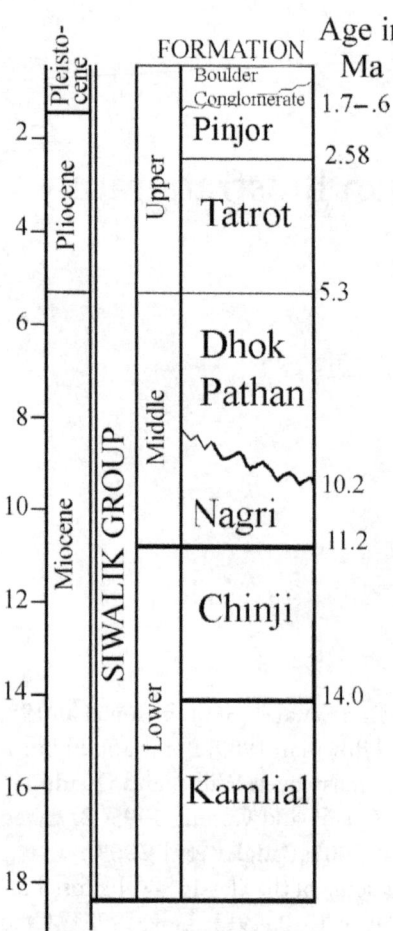

Figure 17.1 A generalized stratigraphic framework of the Siwalik sequence (not applicable everywhere). Modified after Behrensmeyer and Barry (2005); dates from Johnson et al. (1985), Ranga Rao et al. (1988), and Barry et al. (2002).

time transgressive and faunal ranges often extend beyond these boundaries, the common practice of assigning faunal assemblages after various formations, as in "Chinji Fauna" or "Pinjor Fauna," has hindered developing a high-resolution biostratigraphic framework. The present chapter makes an attempt to reevaluate the mammalian biostratigraphy of the Indian Siwaliks by integrating the available mammalian locality data to sections with independent dating using magnetostratigraphy and tephrochronology. A few other important sections that have not been geochronologically dated are placed here in the temporal framework using mammal biochronology already well established in the Pakistan Siwaliks. The present exercise of compiling the data and putting the mammal record in a chronological framework has facilitated a revision of the first and last appearances of several Neogene Siwalik mammals, including those of the Mio-

cene apes, allowing a better intraregional correlation with localities situated in Pakistan and Nepal. Two major faunal turnovers—one at the Late Miocene (8–9 Ma) and the other at the Gauss-Matuyama boundary (2.6 Ma)—coincide with global climate and local tectonic events. Many gaps still exist in the Indian fossil record, and several mammalian lineages are yet to be taxonomically resolved; therefore, interpretations made here are tentative.

The present approach is to reassess the various key sections: (1) Lower Siwalik Ramnagar section in the Jammu sub-Himalaya, (2) Middle Siwalik Haritalyangar and Nurpur sections in the Himachal Himalayas, and (3) Upper Siwalik Uttarbeni–Parmandal section in the Jammu and Haripur Khol, Khetpurali, Markanda, Patiali Rao, and Ghaggar River sections exposed near Chandigarh. Other important assemblages recorded from various Siwalik sections exposed in India and Nepal are also reviewed.

BIOSTRATIGRAPHIC APPRAISAL

Lower Siwalik Ramnagar Locality

The Ramnagar locality is situated in the Jammu region where the Lower, Middle, and Upper Siwaliks are fairly well exposed (figure 17.3). Ramnagar is famous for its Miocene apes (Brown et al. 1924; Lewis 1934; Colbert 1935; Dutta, Basu, and Sastri 1976; Vasishat, Gaur, and Chopra 1978), and in recent times many workers have contributed significantly toward understanding its vertebrate palaeontology and stratigraphy (Thomas and Verma 1979; Nanda and Sehgal 1993; Verma and Gupta 1997; Basu 2004; Parmar and Prasad 2006; Sehgal and Patnaik 2012).

A diverse fossil assemblage is now known from the upper 350 m of the Ramnagar section (figure 17.4; table 17.1). The large mammalian fauna, which includes Miocene apes such as *Sivapithecus sivalensis* and *Sivapithecus* cf. *simonsi*, and the lithology of Ramnagar show striking resemblance to those of the Chinji type area in the Potwar Plateau (Vasishat, Gaur, and Chopra 1978; Nanda and Sehgal 1993; Verma and Gupta 1997; Basu 2004). This strong faunal similarity (see table 17.1) has been used for tentatively placing the Ramnagar assemblage between 11 Ma and 13 Ma (Vashishat, Gaur, and Chopra 1978; Nanda and Sehgal 1993; Basu 2004). Recently, a report of the short-ranging rhizomyid rodent *Kanisamys* cf. *potwarensis* from near Dehari (Parmar and Prasad 2006) suggests that these deposits could be a little older than previously estimated, because the range of this species in Pakistan is about 14.2 Ma to 13.4 Ma on the time scale of Gradstein, Ogg, and Smith (2004) (see figure 17.4).

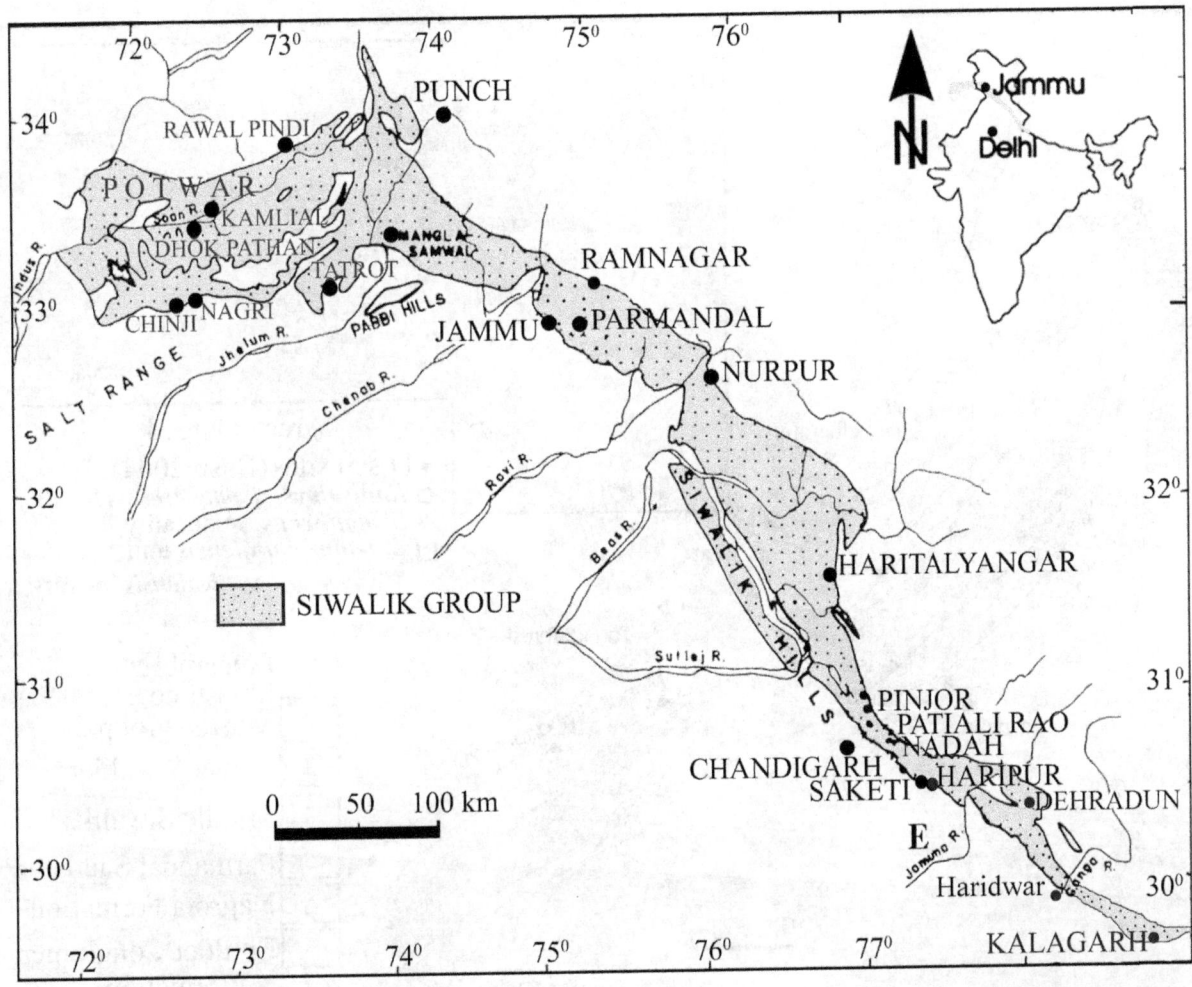

Figure 17.2 Distribution of the Siwaliks showing the localities discussed in this chapter as well as the type localities of various formations in Pakistan.

Sehgal and Patnaik (2012) discuss index fossils that support this older age estimate, the murid rodent *Antemus chinjiensis* and the cricetid rodent *Megacricetodon* cf. *sivalensis* from the stratigraphic level of *Sivapithecus* at Ramnagar. The Ramnagar assemblage can now be constrained between ~13.2 Ma and ~13.8 Ma using the well-established high-resolution Siwalik rodent biochronology. This biochronological framework has the potential of increasing the first appearance datum (FAD) of *Sivapithecus* from the presently estimated 12.8 Ma to at least 13.2 Ma (see figure 17.4).

Kalagarh

The so-called Lower Siwalik fauna reported from Kalagarh, Uttaranchal (Tiwari 1983) includes *Sivapithecus indicus, Viverra chinjiensis, Sivameryx (=Hyoboops) minor, Hipparion (Cormohipparion) antelopinum, Deinotherium*

sp., *Dorcadoxa porrecticornis, Conohyus* sp., *Giraffokeryx* sp., *Dicoryphochoerus* sp., *Propotamochoerus* sp., *Dorcatherium nagrii*, and *Dorcatherium majus*. Nanda and Sehgal (1993) are of the opinion that this assemblage contains both Lower and Middle Siwalik elements. The short-ranging bovid *Dorcadoxa porrecticornis* (in Pakistan) may constrain this locality between 9.3 Ma and 8 Ma. This estimate is well supported by age ranges of other associated species such as *Hipparion* spp. (10.7–5.8 Ma), *Dorcatherium nagrii* (9.3–6.8 Ma), and *Dorcatherium majus* (10.4–7 Ma; dates unmodified from Barry et al. 2002). Therefore, Kalagarh appears to be of late Miocene age.

Dang Valley, Nepal

Farther east, the Dang Valley sections of Nepal have yielded both Lower and Middle Siwalik mammalian

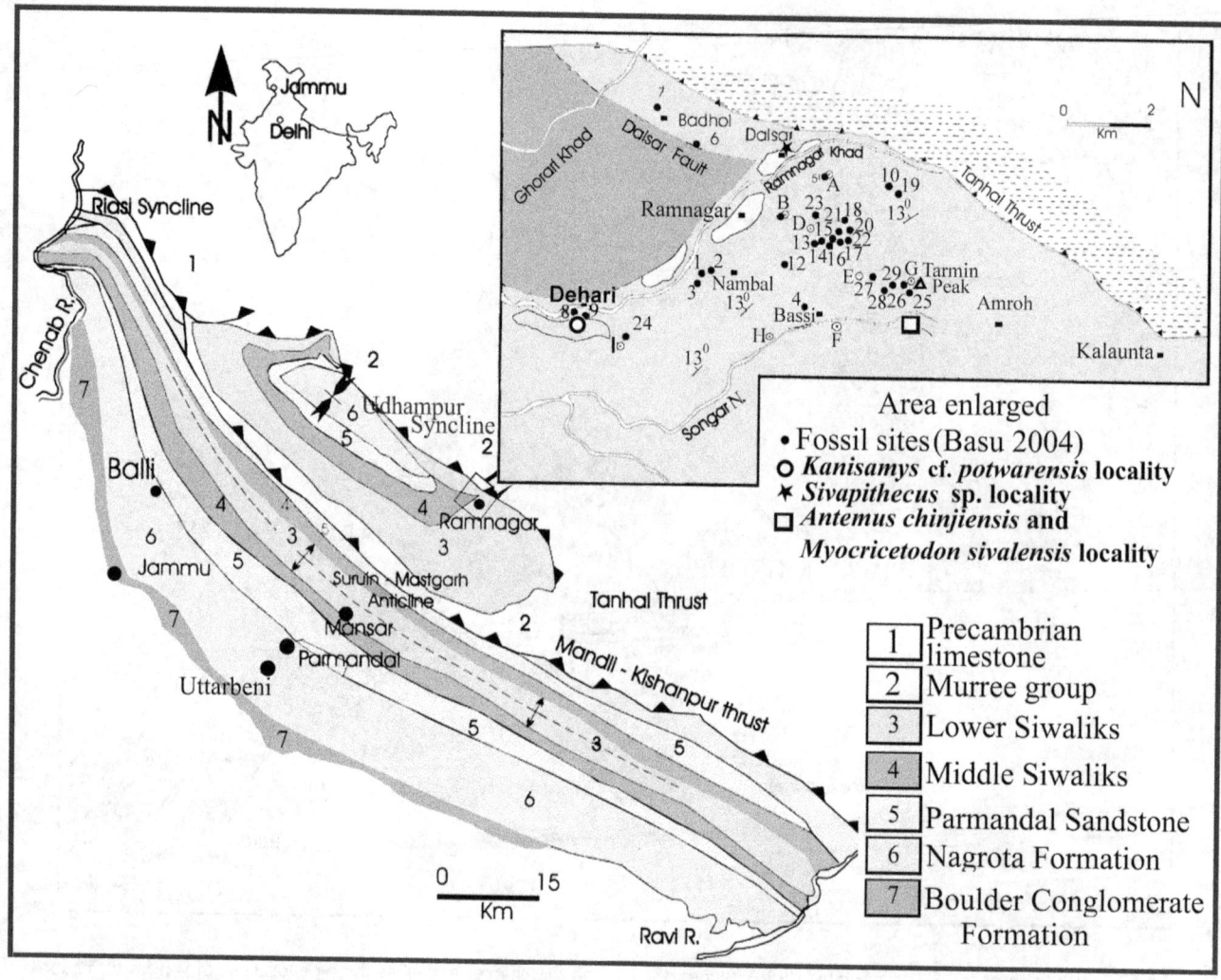

Figure 17.3 Location and geological map of the Siwaliks of Jammu sub-Himalaya. Modified after Gupta and Verma (1988); Basu (2004). The subgroups and the lithostratigraphic units of the Siwaliks and the pre-Siwaliks are indicated as follows: (*1*), Great Limestone (Precambrian); (*2*), Murree Group; (*3*), Lower Siwalik; (*4*), Middle Siwalik; (*5*), Parmandal Sandstone; (*6*), Nagrota Formation; and (*7*), Boulder Conglomerate Formation. Numbers *1–29* on the inset represent the fossil localities tied to the section as shown in figure 17.4. After Basu (2004).

assemblages (West et al. 1978; West, Hutchison, and Munthe 1991). Both magnetostratigraphic (Munthe et al. 1983) and biostratigraphic zonations (pre-*Hipparion* and *Hipparion* Zone of Barry, Lindsay, and Jacobs 1982) have been used to assign these rocks to the Lower and Middle Siwaliks. Taxa that have been found from Dang Valley include *Sivapithecus punjabicus, Amphicyon palaeindicus, Deinotherium pentapotamiae, Brachypotherium perimense, Conohyus sindiensis, Hemimeryx pusillus, Giraffokeryx punjabiensis, Protragocerus gluten, Hipparion* sp., *Dorcabune* sp., *Dorcatherium* sp., and *Pachyportax* sp. (West et al. 1978; West, Hutchison, and Munthe 1991).

Middle Siwalik Haritalyangar Locality

Haritalyangar is one of the best-studied Siwalik localities in India (figure 17.5). The reasons for such attention are obvious. The 1600-m-thick section has yielded some of the best primate specimens anywhere from the Siwaliks. These include apes such as *Sivapithecus sivalensis, Sivapithecus indicus,* and *Indopithecus bilaspurensis,* the small primates *Pliopithecus (Dendropithecus) krishnaii, Indraloris himalayensis,* and *Sivaladapis palaeindicus,* and the tupaiid *Palaeotupia sivalicus* (Simons and Chopra 1969; Simons and Ettel 1970; Simons and Pilbeam 1973; Chopra and Kaul 1979; Chopra and Vasishat 1979; Gingerich

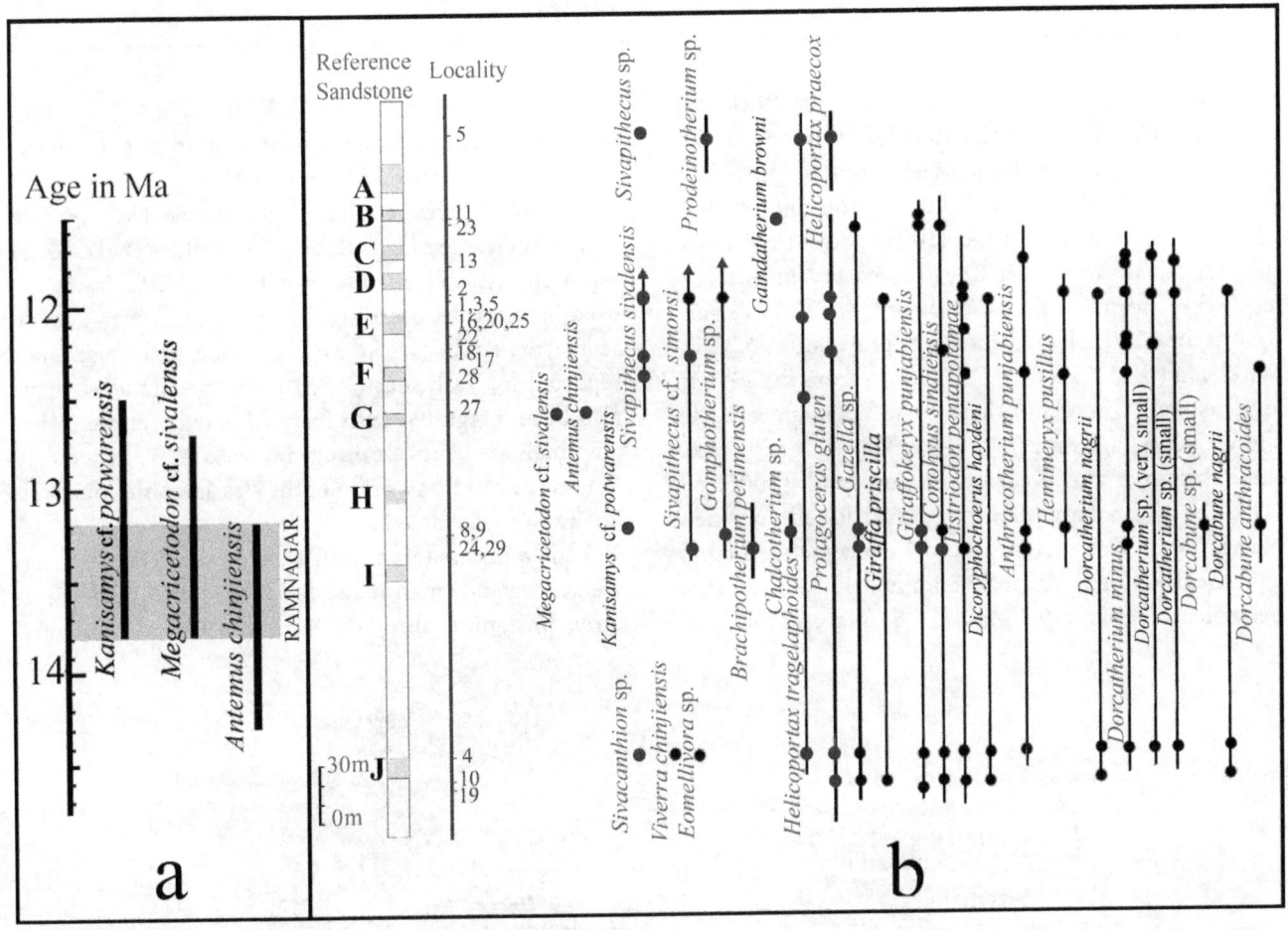

Figure 17.4 Ramnagar biostratigraphy: (*a*) stratigraphic ranges of key rodents observed in the Siwaliks of the Potwar Plateau, Pakistan (Flynn et al. 1995; Jacobs and Flynn 2005); (*b*) Ramnagar section with biostratigraphic occurrences of mammals in the upper 350 m interval of the Lower Siwaliks (modified after Basu 2004). Occurrence levels of species are indicated by black dots (data from Nanda and Sehgal 1993; Basu 2004; Parmar and Prasad 2006). This figure is adopted from Sehgal and Patnaik (2012:fig. 3).

Table 17.1

Mammalian Species Recorded from the Upper Interval of the Lower Siwaliks, Ramnagar

Primates	*Sivaladapis palaeindicus, Sivapithecus indicus, S. sivalensis, S.* cf. *S. simonsi*
Rodentia	*Sivacanthion complicatus, Rhizomyides sivalensis, Rhizomyides punjabiensis, Sayimys sivalensis, Kanisamys* cf. *K. potwarensis, Antemus chinjiensis,* cf. *Myocricetodon* sp.
Carnivora	*Dissopsalis* sp., *Dissopsalis carnifex, Amphicyon* sp., *Viverra chinjiensis,* ?*Eomellivora* sp., *Eomellivora necrophila, Vishnufelis* sp., *Percrocuta carnifex, Vishnuonyx chinjiensis*
Proboscidea	*Prodeinotherium* sp., *Gomphotherium* sp., *Tetralophodon* sp. *Deinotherium pentapotamiae*
Perissodactyla	*Aceratherium perimense, Gaindatherium browni, Brachypotherium* sp. *Brachypotherium perimense, Chilotherium*? *intermedium, Chilotherium* sp., *Chalicotherium* sp.
Artiodactyla	*Listriodon pentapotamiae, Dicoryphochoerus haydeni, Conohyus chinjiensis, Conohyus sindiensis, Propotamochoerus* sp., *Sus* sp., *Hippopotamodon haydeni, Anthracotherium punjabiense, Hemimeryx pusillus, Dorcabune anthracotherioides, Dorcabune nagrii, Dorcatherium majus, D. minus, D. nagrii, D.* sp. (*small*), *D.* sp. (*very small*), *Giraffokeryx punjabiensis, Progiraffa priscilla, Progiraffa* sp., *Protragocerus gluten, Miotragocerus gradiens, Gazella* sp., *Helicoportax tragelaphoides, H. praecox, Miotragocerus* sp.?, *Kubanotragus sokolovi*

SOURCES: After Vasishat, Gaur, and Chopra (1978), Gaur and Chopra (1983), Nanda and Sehgal (1993), Sehgal (1998), Sehgal and Nanda (2002), Basu (2004), and Parmar and Prasad (2006).

and Sahni 1979, 1984; Patnaik and Cameron 1997; Patnaik et al. 2005; Pillans et al. 2005). Johnson et al. (1983) carried out a detailed palaeomagnetic stratigraphy of the Haritalyangar region and found that the *Sivapithecus* fossils occur between 850 m and 1100 m above the base of the Haritalyangar section. They placed *Indopithecus bilaspurensis* (formerly *Gigantopithecus*, includes *G. giganteus*) at the 1200 m level, which they projected as 6.3 Ma. Sankhyan (1985) recorded *Sivapithecus* from the 1500 m level as well. Therefore, these fossil apes appeared to show a young age range (~7.5–5.5 Ma) with respect to the Pakistan record. Recently, Pillans et al. (2005) revised the chronology of Haritalyangar based on a new correlation of this section to the Geomagnetic Polarity Time Scale (GPTS). The main hominoid interval has now been placed between 9.2 Ma and 8.85 Ma, and the *Indo-*

pithecus bilaspurensis level at 8.6 Ma (figure 17.6). With these new dates, the full range of hominoids at Haritalyangar falls between 9.2 Ma and 8.1 Ma.

Beside primates, Haritalyangar has yielded a diverse mammalian assemblage (table 17.2). Pillans et al. (2005) reported some faunal material from specific horizons, used here for determining first and last occurrences of several mammals (see figure 17.6). Some of the mammals reported by Vashishat (1985) from specific levels have also been integrated with the well-dated section. Other mammals are of uncertain age between 9.23 Ma and 8.10 Ma, as their exact position in the stratigraphic column is not known (see table 17.2).

Ladhyani, a site situated east of Haritalyangar, has yielded several small mammals. These include *Sivalikosorex prasadi* (Sahni and Khare 1976), *Progonomys*

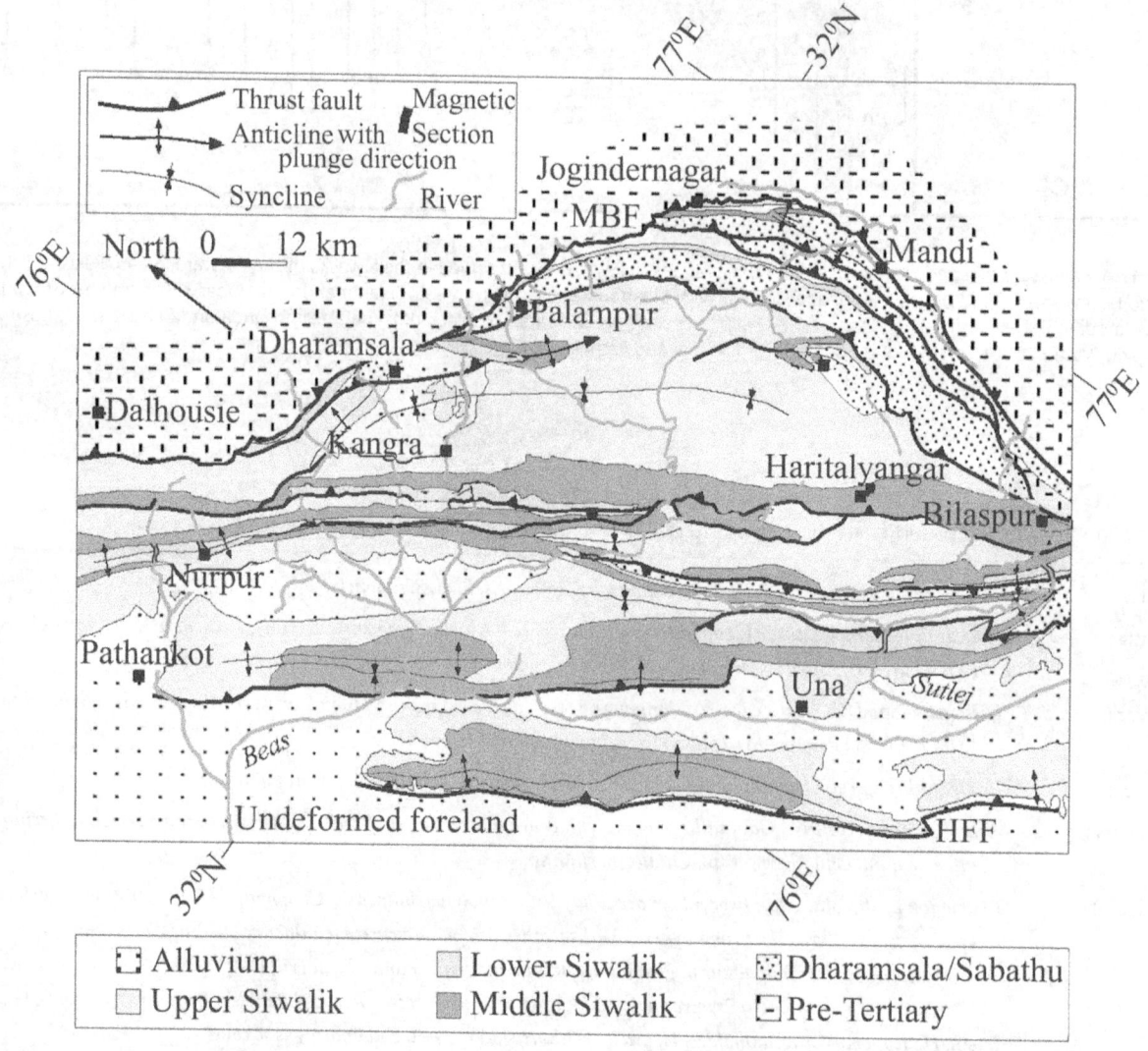

Figure 17.5 General location map of the Himachal Pradesh reentrant. Modified from Brozovic and Burbank (2000). MBF = main boundary fault; HFF = Himalayan frontal fault.

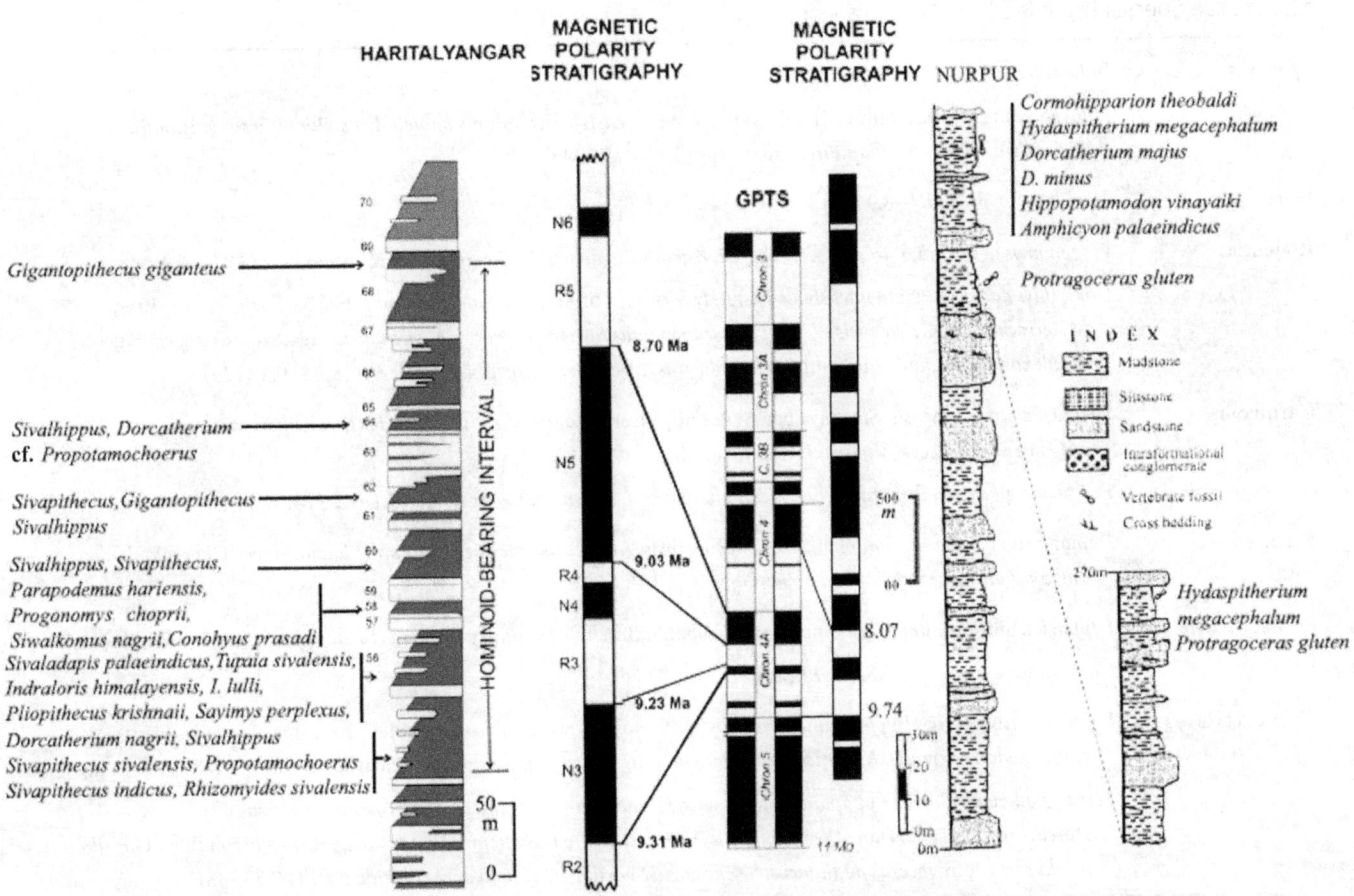

Figure 17.6 Haritalyangar and Nurpur section showing stratigraphic occurrences of mammals. Data from Vashishat (1985), Pillans et al. (2005), Sangode and Kumar (2003), and Nanda and Sehgal (2007).

debruijni, Parapelomys robertsi, Karnimata darwini, Palaeotupia sivalicus (Tiwari 1996), and *Karnimata* cf. *intermedia* (tentative identification of a nondiagnostic M2–3; Flynn et al. 1990). In comparing the murine assemblage in Ladhyani to that of Pakistan (Jacobs and Flynn 2005), the co-occurrence of *Progonomys debruijni* and *Karnimata darwini* suggests an upper age limit of ~8 Ma to 9 Ma for Ladhyani. *Parapelomys robertsi* has a younger age range in the Potwar area (Barry et al. 2002) and should be reevaluated.

Middle Siwalik Nurpur Locality

The lower 270 m of red beds exposed near Nurpur, Himachal Pradesh, have yielded a diverse assemblage (see figure 17.6; table 17.3). Initially, these beds were considered Lower Siwalik (Vasishat et al. 1983; Gaur, Vasishat, and Chopra 1985). Subsequently, based on new findings of

Hipparion, Nanda and Sehgal (1993) placed them in the Middle Siwaliks. The magnetostratigraphy developed for Nurpur section (Ranga Rao 1993), as recently reinterpreted (Sangode and Kumar 2003), indicates that the assemblage should fall in the 9.7–10.1 Ma interval (see figure 17.6). This rough estimate correlates fairly well considering temporal occurrences of *Hipparion, Propotamochoerus hysudricus*, and *Dorcatherium majus* established in Pakistan (Barry et al. 2002; Badgley et al. 2008). As per the sediment accumulation rate at this interval (Sangode and Kumar 2003), the lower 270-m-thick fossil-yielding sequence should represent ~500,000 yr. However, it may be noted that Nanda and Sehgal (2005) placed this assemblage between 8.14 and 7.51 Ma, following Ranga Rao (1993).

A site near Palampur yielding the short-ranging rhizomyid *Miorhizomys micrus* (Tiwari 1990) constrains the site to about 9.2 Ma based on its isolated occurrence in Pakistan (Flynn et al. 1990) (see figure 17.5). Other

Table 17.2

Mammalian Species Recorded from Haritalyangar

Primates	*Indopithecus bilaspurensis* (8.6 Ma)
	Palaeotupaia sivalicus, Indraloris lulli [syn. *I. himalayensis*], *Sivaladapis palaeindicus, Pliopithecus krishnaii, Sivapithecus indicus, Sivapithecus sivalensis* (8.85–9.23 Ma)
Insectivora	*Sivalikosorex prasadi* (~8.9 Ma)
Rodentia	*Progonomys debruijni, Parapelomys robertsi, Karnimata darwini, Karnimata* cf. *K. intermedia* (~8.9 Ma)
	Dakkamys nagrii, Rhizomyides sivalensis, R. lydekkeri, Miorhizomys nagrii, M. pilgrimi, M. harii, M. choristos, M. tetracharax, K. sivalensis, K. nagrii, Parapodemus hariensis, Progonomys choprii, Siwalikomus nagrii, Sayimys perplexus, S. badauni, Sivacanthion complicatus, Mastomys (Karnimata) colberti (8.85–9.23 Ma)
Carnivora	*Sivanasua himalayaensis, Sivaonyx bathygnathus, Viverra nagrii, Vishnuictis hariensis, Ictitherium nagrii, Megantereon praecox, Vinayakia intermedia* (8.85–9.23 Ma)
	Percrocuta gigantea, P. mordax, Enhydriodon falconeri, Lycyaena macrostoma (8.1–9.23 Ma)
Proboscidea	*Deinotherium indicum, Gomphotherium hasnotensis, Anancus sivalensis, Tetralophodon falconeri, Choerolophodon dhokpathanensis, Stegolophodon bombifrons* (8.1–8.6 Ma)
Perissodactyla	*Hipparion antelopinum, Cormohipparion theobaldi, Brachypotherium perimense, Sivalhippus sp.* (8.1–9.23 Ma)
	Gaindatherium browni (8.85–9.23 Ma)
Artiodactyla	*Lophochoerus nagrii, Hippopotamodon robustus, Conohyus prasadi, Sus advena, Anthracotherium punjabiense, Anthracodon hariensis, Anthracothema dangari, Giraffokeryx punjabiensis, Pachyportax nagrii* (8.85–9.23 Ma)
	Tetraconodon mirabilis, Hippopotamodon vagus, Hippopotamodon titanoides, Propotamochoerus uliginosus, Dorcatherium nagrii, Dorcatherium minus, Vishnutherium iravaticum, Hydapsitherium megacephalum, Gazella lydekkeri, Miotragocerus punjabicus, Selenoportax vexillarius, Pachyportax latidens (8.1–9.23 Ma)

SOURCES: After Colbert (1935), Pilgrim (1939), Prasad (1970), Vasishat, Guar, and Chopra (1978), Vasishat (1985), Nanda and Sehgal (2005), and Pillans et al. (2005).

isolated Middle Siwalik mammal occurrences are from the Jammu and Poanta regions (Gupta and Verma 1988; Nanda et al. 1991). Gupta and Verma (1988) reported *Gomphotherium* (=*Trilophodon*) sp., *Gomphotherium* cf. *G. falconeri*, *Tetralophodon* cf. *T. iongirostris*, *Choerolophodon* (=*Synconolophus*) cf. *C. dhokpathanensis*, *Elephas* cf. *E. hysudricus*, and *Hippopotamodon* sp. from the Jammu region. Nanda and Sehgal (2007) opine that of this assemblage only *C. dhokpathanensis* is characteristic of the Middle Siwaliks. On the other hand, Siwaliks exposed west of Paonta Saab have yielded typical Middle Siwalik *Hippopotamodon titan*, as well as *Hippohyus* cf. *H. grandis* and *Propotamochoerus* cf. *P. salinus* (Nanda, Sati, and Mehra 1991).

Upper Siwalik Localities Around Chandigarh

Several localities have yielded diverse fossil mammal assemblages around Chandigarh since the report of Colbert (1935; figure 17.7). Azzaroli and Napoleone (1982) were the first to put them in a chronological framework,

followed by Tandon et al. (1984). A decade later, Ranga Rao et al. (1995) and Sangode, Kumar, and Ghosh (1996) carried out extensive magnetostratigraphic work in the Upper Siwaliks. The Haripur section has been well dated and has yielded large mammals and pollens (Sangode, Kumar, and Ghosh 1996; Phadtare, Kumar, and Ghosh 1994; figure 17.8). Tuffaceous mudstone discovered from the Ghaggar river section (Tandon and Kumar 1984; Mehta et al. 1993) helped in placing the local magnetic reversals in the GPTS. Recently, Kumaravel et al. (2005) have provided the most comprehensive magnetostratigraphic data on sediments exposed around Chandigarh. The Tatrot Formation and Pinjor Formation boundary of the Indian Subcontinent is close, temporally, to the Gauss–Matuyama magnetic reversal dated to 2.58 Ma (Ranga Rao et al. 1995; Cande and Kent 1995). Tatrots are characterized by the presence of thin gray sandstones, variegated mudstones, and siltstones; this lithology gradually changes upward into brownish sandstones and mudstones characterizing the Pinjors. The Pinjor Formation and Boulder Conglomerate boundary has been found

Table 17.3

Mammalian Species Recorded from Nurpur

Carnivorans	*Dissopsalis carnifex, Amphicyon palaeindicus*
Proboscidea	*Tetralophodon* sp., *Deinotherium* sp.
Perissodactyla	*Aceratherium perimense, Gaindatherium* sp., *Hipparion antelopinum, Cormohipparion theobaldi*
Artiodactyla	*Listriodon pentapotamiae, Propotamochoerus hysudricus, Sus* sp., *Hippopotamodon vinayaki, Anthracotherium* sp., *Merycopotamus dissimilis, Dorcabune* sp., *Dorcatherium majus, D. minus, D. nagrii, Giraffokeryx punjabiensis, Hydapsitherium megacephalum, Bramatherium megacephalum minus, Protragocerus gluten*

SOURCES: After Vasishat et al. (1983), Gaur, Vasishat, and Chopra (1985), Nanda and Sehgal (1993), and Sehgal and Nanda (2002).

to be time transgressive, ranging in age from 1.77 Ma to .63 Ma (Ranga Rao et al. 1988; Ranga Rao 1995; Nanda 2002; Kumaravel et al. 2005). An attempt has been made to integrate the mammalian faunal material to well-dated sections (see figure 17.8). The assemblages representing Tatrot and Pinjor formations are given in table 17.4.

Upper Siwalik Localities Around Jammu Region

Upper Siwalik exposures around Parmandal and Nagrota are fairly well studied for magnetostratigraphy, tephrochronology, and vertebrate paleontology (Yokoyama et al. 1987; Gupta and Verma 1988; Ranga Rao et al. 1988; Agarwal et al. 1993; Nanda 1997, 2002; Basu 2004; Nanda and Sehgal 2005; figure 17.9). Nagrota Formation deposits contain two volcanic tuff horizons, dated to 2.8 ± 0.56 Ma and 2.31 ± 0.54 Ma (Agarwal et al. 1993; Ranga Rao et al. 1988). *Anancus* sp., *Stegodon bombifrons, Stegodon insignis, Stegolophodon, Elephas planifrons, Hipparion antelopinum, Cormohipparion theobaldi,*

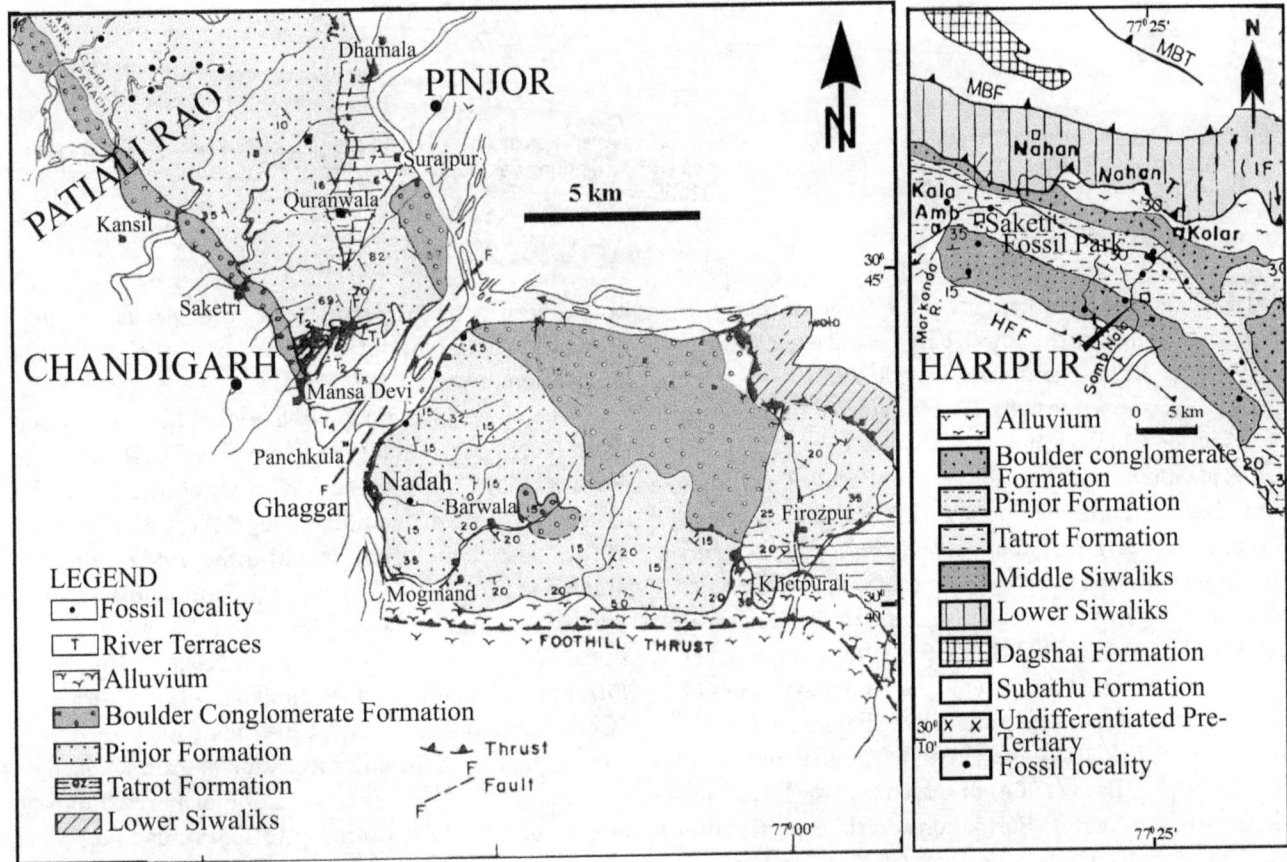

Figure 17.7 Geological map of the Chandigarh region and Haripur Khol area. Modified after Sahni and Khan (1964), Nanda (2002), and Kumar et al. (2002).

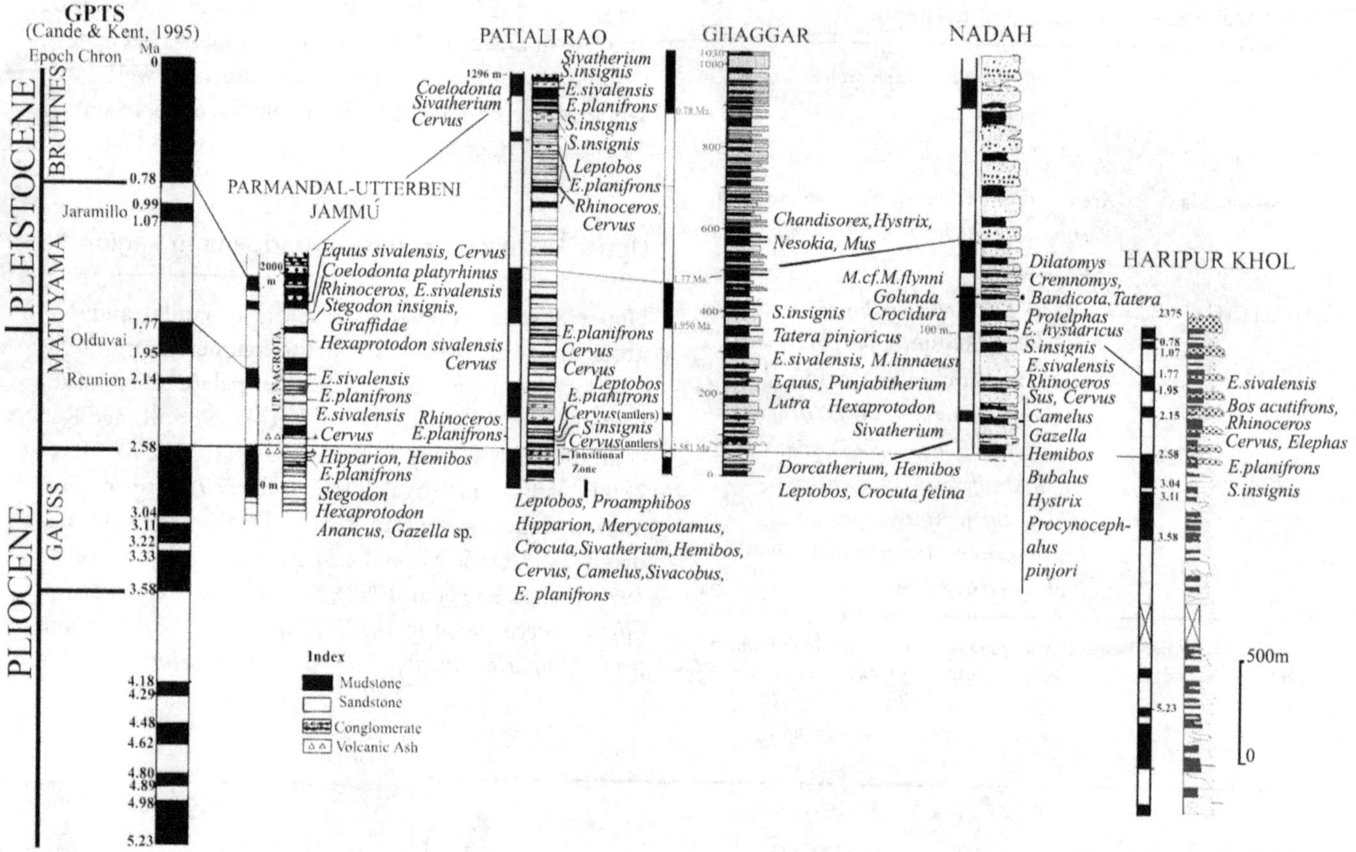

Figure 17.8 Dated Plio-Pleistocene Siwalik sections with mammal-bearing levels exposed near Jammu and Chandigarh. Modified after Patnaik and Nanda (2010), and data compiled from Azzaroli and Napoleone (1982), Tandon et al. (1984), Ranga Rao et al. (1988, 1995), Agarwal et al. (1993), Sangode, Kumar, and Ghosh (1996), and Kumaravel et al. (2005).

and *Propotamochoerus hysudricus* are some of the common taxa found in the lower part (Nanda and Kumar 1999; Basu 2004; Nanda and Sehgal 2005). The upper assemblage, above the tuffs, is a bit depleted in the number of proboscidean remains, with only *Elephas hysudricus*, *E. planifrons*, and *Stegodon insignis* (Basu 2004). Other taxa characterizing the upper zone are *Equus sivalensis*, *Hemibos acuticornis*, *Antilope* sp., *Cervus* sp., *Rhinoceros* sp., *Coelodonta* sp., *Sivatherium giganteum*, *Panthera* sp., *Viverra* sp., *Crocuta feline*, and *Hexaprotodon* (Agarwal et al. 1993; Ranga Rao et al. 1988; Basu 2004). Agarwal et al. (1993) correlated the lower and upper faunal zones of the Nagrota Formation to the Tatrot faunal zone (Potwar Plateau, Pakistan) and the Pinjor faunal zone (Indian Siwaliks). Basu (2004) observed that the lower faunal zone of the Nagrota Formation correlates partly with the *Hexaprotodon sivalensis* Interval Zone and partly with the *Elephas planifrons* Interval Zone proposed by Barry, Lindsay, and Jacobs (1982). Basu (2004) went on

to propose a new interval zone, the "*Antilope-Equus-Bovine*" faunal zone for the upper part of the Nagrota Formation.

Upper Siwalik mammals have also been recovered from Nepal sub-Himalayan exposures. These areas are Rato Khola, Gidhniya-Surai Khola, and the Lokundol Formation of Kathmandu Valley (West and Munthe 1981; West, Hutchison, and Munthe 1991; Corvinus and Nanda 1994; West 1996). The Rato Khola area has yielded pre-Pinjor fauna, including *Stegodon bombifrons*, *Elephas planifrons*, ?*Stegotetrabelodon*, *Hippohyus tatroti*, *H. sivalensis*, and *Proamphibos* cf. *P. lachrymans* (Corvinus and Rimal 2001; Corvinus 2006). The upper part of the Upper Siwalik succession at Rato Khola falls into the *E. planifrons* Interval Zone of Barry, Lindsay, and Jacobs (1982), according to Corvinus and Rimal (2001) and Corvinus (2006). The lower boundary of this interval zone is 3.2 Ma (Flynn et al., chapter 14, this volume).

Table 17.4

Mammalian Species Recovered from Upper Siwaliks of Chandigarh and Jammu Region

Primates	*Procynocephalus pinjori, Theropithecus delsoni*
Insectivora	*Chandisorex punchkulaensis, Crocidura* sp.
	Crocidura sp.*
Rodentia	*Hystrix leucurus, Mus linnaeusi, Mus* cf. *M. flynni, Hadromys* sp., *Cremnomys* cf. *C. blanfordi, Tatera pinjoricus,* "*Rhizomys*" *pinjoricus, Dilatomys* sp., *Golunda* sp., *Bandicota* sp.
	Rhizomyides saketiensis, R. sivalensis*,* cf. *R. sivalensis*, Mus flynni*,* cf. *M. flynni*, M. jacobsi*, Parapelomys robertsi*,* cf. *P. robertsi*, Cremnomys* cf. *C. cutchicus*, Bandicota sivalensis*, Golunda tatroticus*, G. kelleri*, Dilatomys moginandensis*, D. pilgrimi*, Millardia* sp.*, Abudhabia* cf. *A. kabulense*, Hystrix* sp.*
Lagomorpha	*Caprolagus* sp.
	*Pliosiwalagus whitei**
Carnivora	*Canis pinjorensis, Mellivora sivlensis, Crocuta felina, C. colivini, Panthera* cf. *P. cristata, Lutra palaeindica, Amblonyx* sp.
Proboscidea	*Pentalophodon sivalensis, Stegolophodon stegodontoides, Elephas hysudricus, Elephas planifrons, Elephas platycephalus, Stegodon insignis*
	Pentalophodon khetpuraliensis, Stegodon bombifrons*, Stegodon insignis*, Stegodon* sp., **Stegolophodon* sp.*, Elephas planifrons*, Elephas hysudricus*, Anancus* sp.*
Perissodactyla	*Equus sivalensis, Coelodonta platyrhinus, Rhinoceros palaeindicus, Rhinoceros sivalensis*
	Rhinoceros palaeindicus, Rhinoceros* sp.*, Coelodonta* sp.*, Coelodonta platyrhinus*, Chilotherium intermedium*, Cormohipparion theobaldi*, C.* sp.*, Hipparion antelopinum*, Hipparion* sp.*
Artiodactyla	*Potamochoerus theobaldi, Propotamochoerus hysudricus, Sus falconeri, S. hysudricus, S. choprai, S. giganteus, Hippohyus sivalensis, Hexaprotodon sivalensis, Rucervus simiplicidens, Cervus punjabiensis, Sivatherium giganteum, Sivacapra subhimalayaensis, Oryx sivalensis, Damalops palaeindicus, Sivacobus palaeindicus, Hemibos acuticornis, H. triquetricornis, H. antilopinus, Bubalus palaeindicus, B. platyceros, Leptobos falconeri, Bison sivalensis, Bos acutifrons, Camelus sivalensis*
	Merycopotamus dissimilis, Hippohyus tatroti*, Proamphibos kashmiricus*, Probison dehmi*, Cervus* sp.*, Sivatherium giganteum*, Hemibos triquetricornis*, Bos* sp.*, Sivacapra* sp.*, Propotamochoerus* sp.*, P. hysudricus*, Hemibos acuticornis*, Gazella* sp.*, Hippohyus* sp.*, Hexaprotodon sivalensis*, Cervus* cf. *punjabiensis*, Camelus* sp.*, C. sivalensis*, Sivachoerus* sp.*, Hydaspitherium megacephalum**

NOTE: * = Taxa that occur in sediments older than 2.6 Ma.

SOURCES: After Gaur (1987), Yokoyama et al. (1987), Gupta and Verma (1988), Ranga Rao et al. (1988), Nanda (1997, 2002), Nanda and Kumar (1999), Gupta and Prasad (2001), Basu (2004), Nanda and Sehgal (2005, 2007), and Patnaik and Nanda (2010).

Other mammal occurrences including *Elephas hysudricus, Equus sivalensis, Cervus* sp., *Potamochoerus palaeindicus,* cf. *P. theobaldi,* and *Hemibos* cf. *H. acuticornis* have been assigned as "Pinjor Fauna" (Nanda 2002).

DISCUSSION

Gaps in the Fossil Record

The present assessment recognizes several gaps in the Indian Siwalik fossil record. Ramnagar can now be re-garded as the oldest locality yielding a diverse mammalian assemblage. Based on the rodent biochronology established in Pakistan, the Ramnagar locality can now be placed safely in the middle Miocene, in the ~13.2 to ~13.8 Ma range (see figure 17.4). With this age estimation, the first appearance of *Sivapithecus* in the Siwaliks may be older than that perceived in Pakistan, about ~13.2 Ma (Sehgal and Patnaik 2012). However, this is still within the inferred maximum limit (14 Ma) of the *Sivapithecus* FAD (Barry et al. 2002; Badgley et al. 2008). The Nurpur assemblage represents a small window, probably around 9.7 Ma to 10.1 Ma. This estimate accords with age

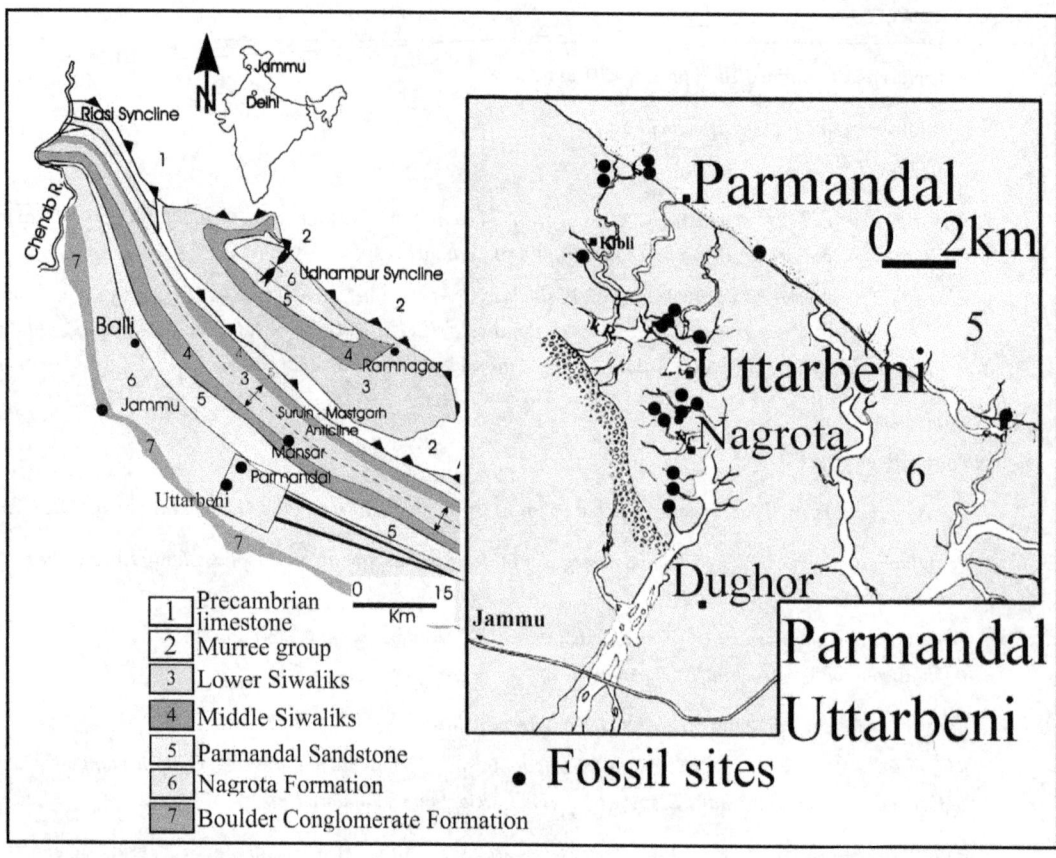

Figure 17.9 Location and geological map of the Upper Siwaliks of Jammu sub-Himalaya. Modified after Gupta and Verma (1988) and Basu (2004).

ranges (established in Pakistan; Barry et al. 2002) of the characteristic fauna it contains: *Hipparion* sp. (10.7–5.8 Ma), *Propotamochoerus hysudricus* (10.2–6.2 Ma), *Bramatherium megacephalum* (10.3–7.1 Ma), *Giraffokeryx punjabiensis* (13.6–10.4 Ma), *Dorcatherium majus* (10.4–7.0 Ma), *Dissopsalis carnifex* (16.3–9.1 Ma), and *Deinotherium* sp. (12.9–8.0 Ma). There exists in India a gap from ~12.5 Ma to ~10.1 Ma, apparently unrepresented by a mammalian record.

Combining the famous Haritalyangar sequence with the nearby Ladhyani site, the late Miocene fossil-producing interval has been constrained to between 10.1 Ma and 8.1 Ma. The age ranges of most mammals in the Haritalyangar assemblage compare very well with their age ranges observed in Pakistan (table 17.5), but there are some differences, mostly explained by incompleteness of the records in the two regions. In India, there is a late Neogene gap in the vertebrate record, without well-dated mammal assemblages between 8 Ma and 4 Ma. In the Potwar, late Neogene deposits are poorly represented, but the Upper Siwalik record of India, combining both the

Jammu and Chandigarh regions, is fairly continuous and well represented from ~4 Ma to .78 Ma. Some remaining differences in taxon ranges are noted here.

Comparison with the Potwar Record

As better chronology of the fauna has been achieved by recent magnetostratigraphic data from India, the mammalian record of India has started to tally fairly well with the observed Pakistan Neogene Mammalian Biochronology (see table 17.5). However, several differences exist between these two regions. For example, the last appearance of *Sivaladapis palaeindicus* in India extends to 9.1 Ma, whereas in Pakistan its range stops at 11.5 Ma (Flynn and Morgan 2005). *Sivapithecus parvada* has not been found in India, and *Sivapithecus simonsi* is yet to be recorded reliably from Pakistan. There is a good match in murine rodent content, but there are several cricetid taxa known from Pakistan yet to be recovered from India.

Table 17.5

First and Last Occurrences Observed in India for Common Mammalian Taxa, Compared with Ranges Observed in Pakistan

	Species	Indian Siwaliks		Pakistan Siwaliks	
		FO	LO	FO	LO
	Primates				
1	*Sivapithecus indicus/sivalensis*	~13	8.85	12.5	8.5
2	*Sivapithecus* cf. *simonsi*	~13.2	8.85		
3	*Sivaladapis palaeindicus*	13	9.1	15.6	11.5
4	*Palaeotupaia sivalicus*	9.1	9.1		
5	Tupaiidae spp.			13	8.1
	Rodentia				
	Ctenodactylidae				
6	*Sayimys perplexus*	9.1	9.1	~18	9.1
7	*Sayimys sivalensis*	13	~9	15.2	12.7
8	*Sayimys chinjiensis*			12.3	10.0
	Rhizomyidae				
9	*Kanisamys indicus*	9.1	9.1	17	10.8
10	*Kanisamys nagrii*	9.1	9.1	11.5	9.6
11	*Kanisamys sivalensis*	9.1	9.1	9.3	7.3
12	*Kanisamys* cf. *potwarensis*	~13.2	~13.2	14.3	13.2
13	*Rhizomyides sivalensis*	9.1	8.9	7.1	6.4
14	*Miorhizomys nagrii*	9.2	9.2	11.5	9.6
15	*Rhizomyides punjabiensis*	~10.0	~10.0	10.4	10.2
16	*Miorhizomys choristos*	9.0	9.0	8.4	8.2
17	*Miorhizomys tetracharax*	9.0	9.0	9.3	7.9
18	*Miorhizomys micrus*	9.2	9.2	9.2	9.2
	Muroidea				
19	*Antemus chinjiensis*	~13	~13	13.8	12.7
20	*Progonomys debruijni*	8.9	8.9	9.3	8.9
21	*Parapelomys robertsi*	8.9	~2.5	7.0	6.4
22	*Karnimata darwini*	8.9	8.9	9.3	8.9
23	*Mus jacobsi*	2.5	2.5		
24	*Mus flynni*	2.5	1.8		
25	*Golunda kelleri*	2.5	2.5	3	2
26	*Golunda* sp.	2.5	2.5		
27	*Hadromys* sp.	1.8	1.8	3	3
28	*Hadromys loujacobsi*			2	1.7
29	*Cremnomys* sp.			2	2
30	*Cremnomys* cf. *C. cutchicus*	4	1.8		
31	*Dakkamys nagrii*	9	9		
32	*Megacricetodon* cf. *sivalensis*	~13	~13	16.1	13
33	*Dakkamys asiaticus*			11.5	10.4
	Leporidae				
34	*Pliosiwalagus whitei*	~4	2.5		
35	*Alilepus* sp.			7.4	6.5
	Carnivora				
36	*Dissopsalis carnifex*	13	9.8	16.3	9.1
37	*Percrocuta carnifex*	13	13	12.7	9.0
38	*Panthera*	2	2		

(continued)

Table 17.5 (continued)

	Species	Indian Siwaliks		Pakistan Siwaliks	
		FO	LO	FO	LO
39	*Panthera uncia*			1	1
40	*Panthera* cf. *cristata*	2	2		
41	*Mellivora sivalensis*	2	2		
42	*Crocuta colivini*	2	2		
43	*Crocuta felina*	2	2		
44	*Crocuta crocuta*			1	1
45	*Canis pinjorensis*	1.5	1.5		
46	*Canis cautleyi*			1.8	1.8
	Proboscidea				
47	*Prodeinotherium* sp.	~12.5	~12.5		
48	*Deinotherium pentapotamiae*	13	13	18.3	6.9
49	*Deinotherium* sp.	13	9.8	12.9	8.0
50	*Choerolophodon* cf. *dhokpathanensis*	8.9	8.6		
51	*Choerolophodon corrugatus*			13.5	6.5
52	*Stegodon* sp.	3.4	1.5	1.8	.78
53	*Stegolophodon stegodontoides*			13.5	4.2
54	*Elephas hysudricus*			1.7	1
	Perissodactyla				
	Rhinocerotidae				
55	*Brachypotherium perimense*	13.2	13.2	16.0	7.1
56	*Brachypotherium* sp.	13	13		
57	*Chilotherium intermedium*	3	3	16.3	7.6
58	*Chilotherium* sp,	13.2	13.2		
59	*Gaindatherium* spp.	13	13	16.0	7.3
60	*Gaindatherium browni*	12.7	8.85		
61	*Gaindatherium vidali*			11.5	8.1
62	*Chalicotherium* sp.	13.2	13.2		
63	*Chalicotherium salinum*			12.9	8.0
64	*Aceratherium perimense*	13	9.8		
65	*Rhinoceros palaeindicus*	3	1.5		
66	*Rhinoceros sivalensis*	2	1	1.7	1
	Equidae				
67	*Hipparion antelopinum*	9.8	3		
68	*Cormohipparion theobaldi*	9.8	3		
69	*Hipparion* spp.			10.7	5.8
70	*Hipparion* sp.	3	2.5	2.5	1.6
71	*Cormohipparion* sp.	3	2.5		
72	*Equus sivalensis*	2.5	.78	2.5	.78
	Artiodactyla				
	Suidae				
73	*Conohyus sindiensis*	13.6	12.7	13.1	10.3
74	*Hippopotamodon sivalense*			10.2	7.2
75	*? Hippopotamodon* Y450 sp. indet			11.2	10.2
76	*Hippopotamodon haydeni*	13	13		
77	*Hippopotamodon vagus*	9.23	8.85		
78	*Hippopotamodon robustus*	9.23	8.85		
79	*Tetraconodon magnus*			10.0	9.3
80	*Tetraconodon mirabilis*	9.23	8.1		

	Species	Indian Siwaliks		Pakistan Siwaliks	
		FO	LO	FO	LO
81	*Listriodon pentapotamiae*	13.6	9.8	13.7	10.3
82	*Propotamochoerus* sp.	13	3	5.8	5.1
83	*Propotamochoerus hysudricus*	9.8	1.5	10.2	6.8
84	*Sus* sp.	13	9.8	6.4	3.3
	Tragulidae				
85	*Dorcabune nagrii*	13.6	12.8	10.4	8.5
86	*Dorcabune anthracotheroides*	13.2	12.9	12.8	10.5
87	*Dorcatherium nagrii*	13.6	12.8	9.3	6.8
88	*Dorcatherium minus*	9.23	8.1	13.9	11.5
89	*Dorcatherium majus*	13	13	10.4	7.0
	Antrhacotheriidae				
90	*Anthracotherium punjabiense*	9.23	8.85	~18	8.3
91	*Hemimeryx pusillus*	13.2	12.8	18	5.5
92	*Hemimeryx* spp.			13.6	6.2
93	? *Merycopotamus dissimilis*			5.8	3.3
94	*Merycopotamus dissimilis*	13	3	13.9	3.3
95	*Hexaprotodon sivalensis*	4	1.5	5.9	3.5
	Giraffidae				
96	*Sivatherium giganteum*	2.5	.75	2.2	1.7
97	*Giraffokeryx punjabiensis*	13.6	9	13.6	10.3
98	*Giraffa punjabiensis*			8.9	7.2
99	*Bramatherium megacephalum*	9.8	9.8	10.3	7.1
	Bovidae				
100	*Gazella lydekkeri*	9.23	8.1	10.2	6.1
101	*Gazella* spp.	13	3	11.3	6.2
102	*Dorcadoxa porrecticornis*	9	9	9.3	8.0
103	*Tragoceridus* spp.			11.2	6.2
104	*Pachyportax nagrii*	9.23	8.85		
105	*Pachyportax* sp.			7.3	7.2
106	*Protragoceras gluten*	13.6	9.8	13.9	10.8
107	*Kubanotragus sokolovi*	13	13	13.8	13.8
108	*Helicoportax tragelaphoides*	13.6	12.6	13.1	10.8
109	*Selenoportax vexillarius*	9.23	8.10	10.2	9.8
110	*Selenoportax falconeri*			9.5	8.9
111	*Selenoportax* spp.			10.2	7.9
112	*Selenoportax giganteus*			8.8	7.1
113	*Hemibos acuticornis*	3	1.5		
114	*Hemibos triquetricornis*	3	3	2.2	.78
115	*Damalops (Proamphibos) palaeindicus*	2	.78	2.2	.78
116	*Proamphibos*	3.5	2.5	2.5	1.6
117	*Proamphibos kashmiricus*	3	3		
118	*Bubalus (Hemibos) palaeindicus*	1.5	1.5		
119	*Gazella* sp.	13	3	1.9	.78
120	*Cervus* sp.	3	1.5	2.5	1.6

SOURCES: First and last occurrence data from Pakistan are from Barry et al. (2002) and Badgley et al. (2008).

A recent revision of the Upper Siwalik biostratigraphy of Pakistan by Dennell, Coard, and Turner (2006) revealed some regional differences with India (figures 17.10 and 17.11). For example, *Theropithecus* and *Camelus* occur in India but appear to be absent from Pakistan. Dennell, Coard, and Turner (2006) perceived the absence of *Megantereon*, *Pachycrocuta*, *Ursus*, and anthracotheres from the Indian part, whereas they have confirmed records in Pakistan. In contrast, it may be noted that *Megantereon falconeri*, *Pachycrocuta*, *Ursus*, and the anthracothere *Merycopotamus dissimilis* have been reported from India (Colbert 1935; Gaur 1987; Lihoreau et al. 2007), but probably some of these records need taxonomic review. The baboon *Procynocephalus subhimalayanus* from the Pinjor deposits of India (Verma 1969; Szalay and Delson 1979) has not been found in Pakistan. Nevertheless, the well-dated mammalian localities in India can now be correlated with those of Pakistan on a nonarbitrary footing (see figure 17.10).

Climate Change vs. Major Faunal Turnover

The great mammalian diversity observed in the Ramnagar and Haritalyangar localities (see tables 17.1 and

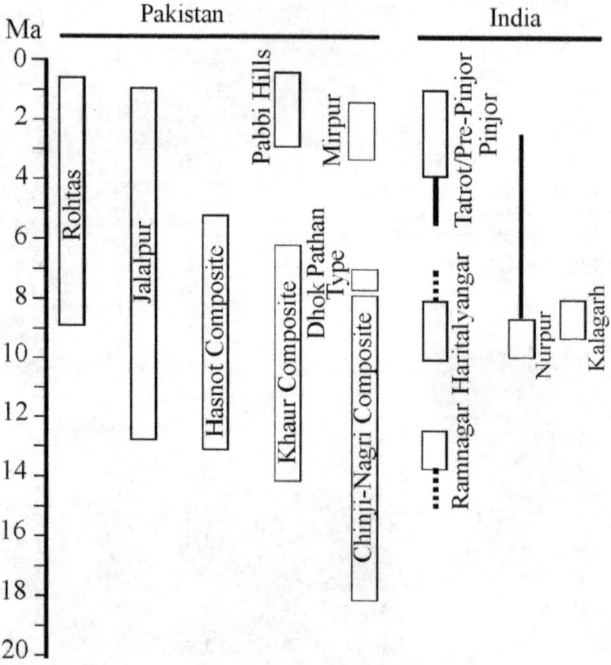

Figure 17.10 Stratigraphic correlation of important mammal-yielding dated sections in Pakistan and India. Modified after Barry (1995). Empty boxes denote known dated mammal occurrences, dark solid bands denote dated sequence but no mammalian assemblages, and dashed bands represent possible age extension.

17.2) does not extend to sediments younger than 7 Ma. Relatively few taxa, such as *Hipparion*, *Propotamochoerus*, *Stegolophodon*, *Anancus*, *Dorcatherium*, *Gazella*, *Merycopotamus*, and *Parapelomys*, characterize the Late Miocene. This cannot be demonstrated in the Indian region, as there is a lack of well-dated localities, but a far better record for this interval available from Pakistan attests to such a faunal change (Barry et al. 2002; Badgley et al. 2008). The kind of resolution Barry et al. (2002) observed in interpreting three very brief periods of high Siwalik faunal turnover at 10.3 Ma, 7.8 Ma, and 7.37–7.04 Ma cannot be achieved in the Indian context at the moment due to gaps among well-dated fossil sites. Paleosol and paleobotanical data suggest warm and humid climatic conditions and evergreen to deciduous tropical forests covering a large part of the northwestern Indian Subcontinent in the Middle Miocene (Ashton and Gunatilleke 1987; Nanda and Sehgal 1993; Prasad 1993; Thomas, Parkash, and Mohindra 2002). The advent of Late Miocene global cooling and the spread of drier conditions caused the forests to shrink (Scott, Kappelman, and Kelley 1999; Kennett and Hodell 1986). An intensification of the Asian monsoon due to late Miocene Tibetan-Hiamlayan uplift would also have contributed to this ecological change (Ruddiman and Kutzbach 1989; Raymo and Ruddiman 1992; Kutzbach, Prell, and Ruddiman 1993; Hay 1996; Ramstein et al. 1997), which eventually led to the replacement of subtropical to temperate broadleafed and tropical forest taxa by grasslands through the late Late Miocene (~8–6.5 Ma; Hoorn, Ojha, and Quade 2000). The Siwalik palaeosols showed marked seasonality in rainfall in the Late Miocene (Retallack 1991, 1995), and their stable isotope content reflected presence of a C3 dominant vegetation prior to 8 Ma and a C4 grass dominated landscape by 7 Ma (Quade, Cerling, and Bowman 1989; Cerling et al. 1997).

A distinct faunal as well as floral turnover is observed at the Gauss-Matuyama boundary, coinciding with the Tatrot–Pinjor boundary in India (Ranga Rao et al. 1988; Verma 1988; Phadatre, Kumar, and Ghosh 1994; Nanda and Sehgal 2005). In the Haripur locality, Phadatre, Kumar, and Ghosh (1994) noticed dominance of a piedmont drainage system and the first appearance of prominent conglomerate facies in the Pinjor lithology. Also, at around 2.4 Ma, volcanic ash beds (tuffaceous mudstones/bentonitic clays) have been found near the Rohtas anticline, Chambal, Mangla Samwal, Jehl Kas, and Bhimbur (Pakistan; Opdyke et al. 1979), Jammu, Chandigarh, Kala Amb, and Haripur (India; Tandon and Kumar 1984; Patnaik 1995; Phadatre, Kumar, and Ghosh 1994). This time coincides with disappearance of *Hipparion* and

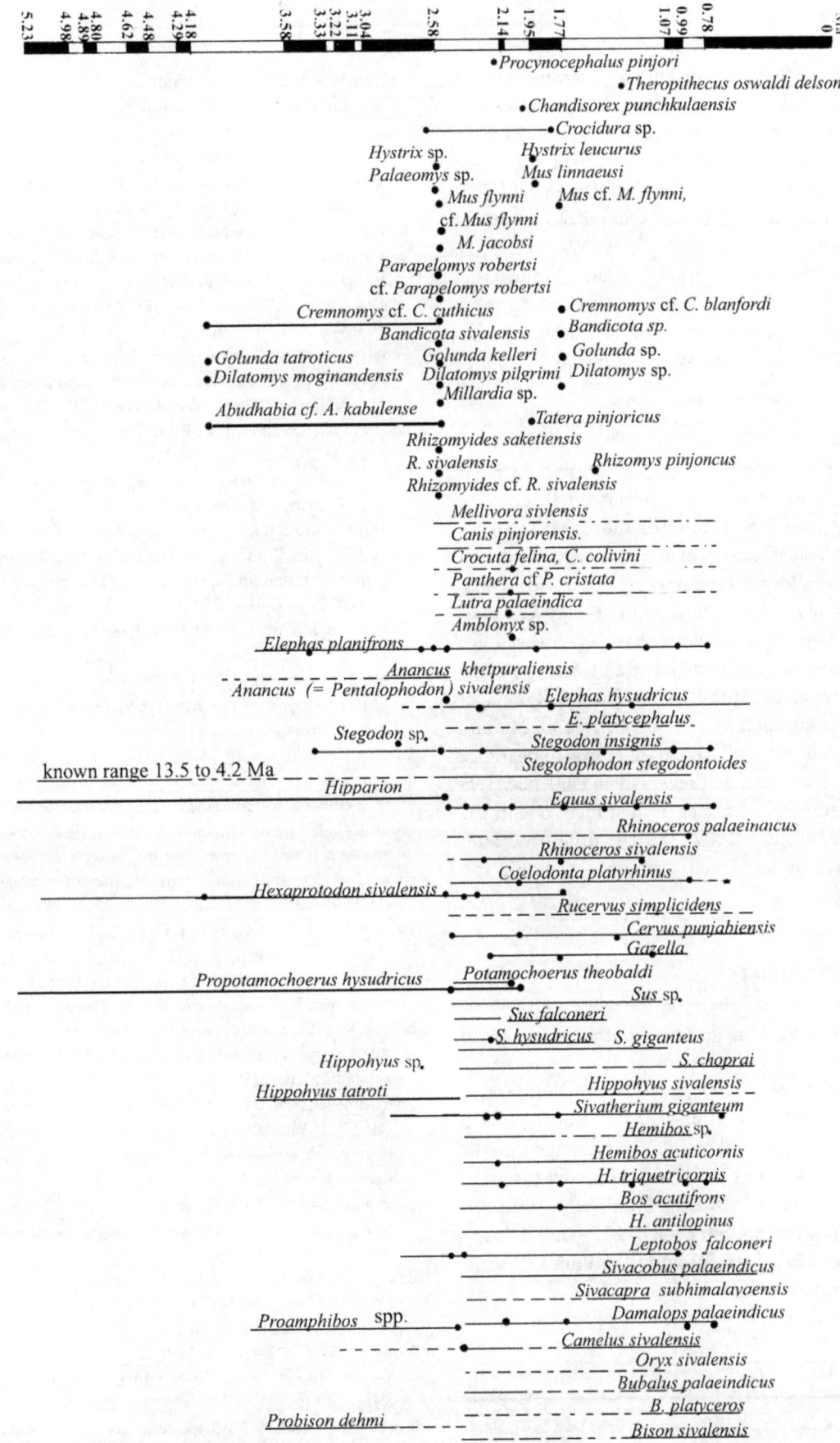

Figure 17.11 Temporal ranges of various Plio-Pleistocene Siwalik mammals of Indo-Pakistan. Data from Nanda (2002), Dennell (2004), and Dennell, Coard, and Turner (2006). Filled dots represent known sites, complete lines represent known ranges, and dotted lines represent uncertain ranges.

the appearance of *Equus*. Several modern carnivore forms appear in Pinjor beds, and murine rodents, especially living genera, diversified around this time (see figure 17.11). At about this time (2.6 Ma), a global change toward a cooler, drier, and more variable climate associated with the Northern Hemisphere glaciations has been observed (Mangerud, Jansen, and Landvik 1996; Williams et al. 1999), and in Africa a major turnover in land mammals occurred after 2.5 Ma (Behrensmeyer et al. 1997). Also, monsoonal circulation was greatly strengthened by the Plio-Pleistocene uplift of the Himalaya (Dewey et al. 1988).

CONCLUSION

Though gaps occur in the Neogene mammal record of India, notably the ~12.5 Ma to ~10.1 and ~8 Ma to ~4 Ma intervals, the present reassessment of Indian Siwalik biochronology has allowed a better correlation with the already well-established Pakistan biochronology. Comparison of these two regions reveals that the observed age ranges of almost 100 taxa overlap (see table 17.5). The rodent biochronology of the Lower Siwalik Ramnagar locality has constrained its age between ~13.2 Ma and ~13.8 Ma, which may indicate an older first appearance datum for *Sivapithecus* than previously in evidence. Two major faunal turnovers have been identified, one in the Late Miocene, around 8 Ma, and the other one after 2.6 Ma, coinciding fairly well with global tectonic and climate changes.

ACKNOWLEDGMENTS

I would like to extend my gratitude to the organizers of the Asian Mammalian Biochronology Workshop, particularly to Drs. Xiaoming Wang, Larry Flynn, and Mikael Fortelius for inviting me to present this work. I would like to thank Dr. Larry Flynn, who encouraged me to write this paper and very kindly helped improve the manuscript. Financial support by DST, New Delhi, to the ongoing Siwalik project (SR/S4/ES-171/2005 and PURSE) and by Wenner-Gren Foundation to the Pinjor Primate Project is thankfully acknowledged.

REFERENCES

Agarwal, R. P., A. C. Nanda, D. N. Prasad, and B. K. Dey. 1993. Geology and biostratigraphy of the Upper Siwalik of Samba area, Jammu Foothills. *Journal of Himalayan Geology* 4:227–236.

Ashton, P. S. and C. V. S. Gunatilleke. 1987. New light on the plant geography of Ceylon: Historical plant geography. *Journal of Biogeography* 14:249–285.

Azzaroli, A. and G. Napoleone. 1982. Magnetostratigraphic investigation of the Upper Siwaliks near Pinjor, India. *Rivista Italiana di Paleontoligia e Stratigrafia* 87:739–762.

Badgley, C., J. C. Barry, M. E. Morgan, S. V. Nelson, A. K. Behrensmeyer, T. E. Cerling, and D. Pilbeam. 2008. Ecological changes in Miocene mammalian record show impact of prolonged climatic forcing. *Proceedings of the National Academy of Sciences* 105(34):12145–12149.

Badgley, C. and A. K. Behrensmeyer. 1980. Palaeoecology of Middle Siwalik sediments and faunas, northern Pakistan. *Palaeogeography, Palaeoclimatology, Palaeoecology* 30:133–155.

Badgley, C., S. Nelson, J. C. Barry, A. K. Behrensmeyer, and T. E. Cerling. 2005. Testing models of faunal turnover with Neogene mammals from Pakistan. In *Interpreting the Past: Essays on Human, Primate, and Mammal Evolution*, ed. D. E. Lieberman, R. H. Smith, and J. Kelley, pp. 29–46. Boston: Brill.

Barry, J. C. 1995. Faunal turnover and diversity in the terrestrial Neogene of Pakistan. In *Paleoclimate and Evolution with Emphasis on Human Origins*, ed. E. S. Vrba, G. H. Denton, T. C. Partridge, and L. H. Burckle, pp. 115–134. New Haven: Yale University Press.

Barry, J. C. and L. J. Flynn. 1990. Key biostratigraphic events in the Siwalik sequence. In *European Neogene Mammal Chronology*, ed. E. H. Lindsay, V. Falbusch, and P. Mein, pp. 557–571. New York: Plenum Press.

Barry, J. C., N. M. Johnson, S. M. Raza, and L. L. Jacobs. 1985. Neogene mammalian faunal change in southern Asia: Correlations with climatic, tectonic, and eustatic events. *Geology* 13:637–640.

Barry, J. C., E. H. Lindsay, and L. L. Jacobs. 1982. A biostratigraphic zonation of the Middle and Upper Siwaliks of the Potwar Plateau of northern Pakistan. *Palaeogeography, Palaeoclimatology, Palaeoecology* 37:95–130.

Barry, J. C., M. E. Morgan, L. J. Flynn, D. Pilbeam, A. K. Behrensmeyer, S. M. Raza, I. Khan, C. Badgley, J. Hicks, J. Kelley. 2002. Faunal and environmental change in the Late Miocene Siwaliks of northern Pakistan. *Paleobiology* 28(*Memoir* 3):1–71.

Basu, P. K. 2004. Siwalik mammals of the Jammu Sub-Himalaya, India: An appraisal of their diversity and habitats. *Quaternary International* 117:105–118

Behrensmeyer, A. and J. Barry. 2005. Biostratigraphic surveys in the Siwaliks of Pakistan: A method for standardized surface sampling of the vertebrate fossil record. *Palaeontologia Electronica* 8(1): 24 pp.

Behrensmeyer, A. K., B. J. Willis, and J. Quade. 1995. Floodplains and paleosols of Pakistan Neogene and Wyoming Paleogene deposits: A comparative study. *Palaeogeography, Palaeoclimatology, Palaeoecology* 115:37–60.

Behrensmeyer A. K., N. E. Todd, R. Potts, and G. E. McBrinn. 1997. Late Pliocene faunal turnover in the Turkana Basin, Kenya and Ethiopia. *Science* 278(5343):1589–1594

Brown, B., W. K. Gregory, M. Hellman. 1924. On three incomplete anthropoid jaws from the Siwaliks, India. *American Museum Novitates* 130:1–8.

Brozovic, N. and D. W. Burbank, 2000. Dynamic fluvial systems and gravel progradation in the Himalayan foreland. *Geological Society of America Bulletin* 112:394–412.

Cande, S. C. and D. V. Kent. 1995. Revised calibration of the geomagnetic polarity timescale for the Late Cretaceous and Cenozoic. *Journal of Geophysical Research* 100:6093–6095.

Cerling, T. E., J. M. Harris, M. G. MacFadden, M. G. Leakey, and J. Quade. 1997. Global vegetation change through the Miocene/Pliocene boundary. *Nature* 389:153–158

Chopra, S. R. K. and S. Kaul. 1979. A new species of *Pliopithecus* from the Indian Siwaliks. *Journal of Human Evolution* 8:475–477.

Chopra, S. R. K. and R. N. Vasishat. 1979. Siwalik fossil tree shrew from Haritalyangar, India. *Nature* 281:214–215.

Colbert, E. H. 1935. Siwalik mammals in the American Museum of Natural History. *Transactions of the American Philosophical Society* 26:1–401.

Corvinus, G. 2006. The Siwalik stratigraphy of Surai Khola, Nepal: A reevaluation of the stratigraphy and original road traverse. *Himalayan Geology* 27:41–71.

Corvinus, G. and A. C. Nanda. 1994. Stratigraphy and palaeontology of the Siwalik Group of Surai Khola and Rato Khola in Nepal. *Neues Jahrbuch für Geologie und Paläontologie Abhandlungen* 191:25–68.

Corvinus, G. and L. N. Rimal. 2001. Biostratigraphy and geology of Neogene Siwalik Group of Surai Khola and Rato Khola areas in Nepal. *Palaeogeography, Palaeoclimatology, Palaeoecology* 165:251–279.

Dennell, R. W. 2004. Early Hominin landscape in northern Pakistan. Investigations in the Pabbi Hills. *BAR International Series* 1265:1–454.

Dennell, R.W., R. Coard, and A. Turner. 2006. The biostratigraphy and magnetic polarity zonation of the Pabbi Hills, northern Pakistan: An Upper Siwalik (Pinjor Stage) Upper Pliocene–Lower Pleistocene fluvial sequence. *Palaeoecology, Palaeoclimatology, Palaeogeography* 234:168–185.

Dewey, J. F., R. M. Shakelton, C. Chengfa, and S. Yiyin. 1988. The tectonic evolution of the Tibetan Plateau. *Philosophical Transactions of the Royal Society London* A327:379–413.

Dutta, A. K., P. K. Basu, and M. V. A. Sastri. 1976. On the new finds of hominoids and additional finds of Pongidae from the Siwaliks of Ramnagar area, Udhampur district, J and K State. *Indian Journal of Earth Science* 3:234–235.

Flynn, L. J., J. C. Barry, M. E. Morgan, D. Pilbeam, L. L. Jacobs, and E. H. Lindsay. 1995. Neogene Siwalik mammalian lineages: Species longevities, rates of change and modes of speciation. *Palaeogeography, Palaeoclimatology, Paleoecology* 115:249–264.

Flynn, L. J. and L. L. Jacobs. 1982. Effects of changing environments on Siwalik rodent faunas of northern Pakistan. *Palaeogeography, Palaeoclimatology, Palaeoecology* 38:129–138.

Flynn, L. J. and M. E. Morgan. 2005. New lower primates from the Miocene Siwaliks of Pakistan. In *Interpreting the Past: Essays on Human, Primate and Mammal Evolution*, ed. D. Lieberman, R. Smith, and J. Kelley, pp. 81–101. Boston: Brill.

Flynn, L. J., A. Sahni, J. J. Jaeger, B. Singh, and S. B. Bhatia. 1990. Additional fossil rodents from the Siwalik Beds of India. *Proceedings of the Koninklijke Nederlandse Akademie van Wetenschappen* 93:7–20.

Gaur, R. 1987. *Environment and Ecology of Early Man in Northwest India*. Delhi: B. R. Publishing.

Gaur, R. and S. R. K. Chopra. 1983. Palaeoecology of the Middle Miocene Siwalik sediments of a part of Jammu and Kashmir State (India). *Paleogeography, Paleoclimatology, Palaeoecology* 43:313–327.

Gaur, R., R. N. Vasishat, and S. R. K. Chopra. 1985. New and additional fossil mammals from the Middle Siwaliks exposed at Nurpur, Kangra (HP) with remarks on Siwalik giraffids. *Journal of the Palaeontolological Society of India* 30:42–48

Gingerich, P. D. and A. Sahni. 1979. *Indraloris* and *Sivaladapis*: Miocene adapid primates from the Siwaliks of India and Pakistan. *Nature* 279:415–416.

Gingerich, P. D. and A. Sahni. 1984. Dentition of *Sivaladapis nagrii* (Adapidae) from the Late Miocene of India. *International Journal of Primatology* 5(1):63–79.

Gradstein, F. M., J. G. Ogg, and A. G. Smith. 2004. *A Geologic Time Scale 2004*. Cambridge: Cambridge University Press.

Gupta, S. S. and G. V. R. Prasad. 2001. Micromammals from the Upper Siwalik subgroup of the Jammu region, Jammu and Kashmir State, India: Some constraints on age. *Neues Jahrbuch für Geologie und Paläontologie* 220:153–187.

Gupta, S. S. and B. C. Verma, 1988. Stratigraphy and vertebrate fauna of the Siwalik Group, Mansar–Uttarbani section, Jammu District, J & K. *Journal of the Palaeontological Society of India* 33:117–124.

Jacobs, L. J. and L. J. Flynn. 2005. Of mice again: The Siwalik rodent record, murine distribution, and molecular clocks. In *Interpreting the Past: Essays on Human, Primate and Mammal Evolution*, ed. D. Lieberman, R. Smith, and J. Kelley, pp. 63–80. Boston: Brill.

Johnson, G. D., N. D. Opdyke, S. K. Tandon, and A. C. Nanda. 1983. The magnetic polarity stratigraphy of the Siwalik Group at Haritalyangar (India) and a new last appearance datum for *Ramapithecus* and *Sivapithecus* in Asia. *Paleogeography, Paleoclimatology, Palaeoecology* 44:223–249.

Johnson, N. M., N. D. Opdyke, G. D. Johnson, E. H. Lindsay, and R. A. K. Tahirkheli. 1982. Magnetic polarity stratigraphy and ages of Siwalik Group rocks of the Potwar Plateau, Pakistan. *Palaeogeography, Palaeoclimatology, Palaeoecology* 37:17–42.

Johnson, N. M., J. Stix, L. Tauxe, P. F. Cerveny, and R. A. K. Tahirkheli.1985. Paleomagnetic chronology, fluvial processes, and tectonic implications of the Siwalik deposits near Chinji village, Pakistan. *Journal of Geology* 93:27–40.

Hay, W. W. 1996. Tectonics and climate. *Geologische Rundschau* 85:409–437.

Hoorn, C., T. Ojha, and J. Quade 2000. Palynological evidence for vegetation development and climatic change in the Sub-Himalayan Zone (Neogene, Central Nepal). *Palaeogeography, Palaeoclimatology, Palaeoecology* 163:133–161.

Karunakaran, C. and A. Ranga Rao. 1979. Status of exploration for hydrocarbon: The Himalayan region. *Geological Survey of India, Miscellaneous Publication* 41:1–66.

Keller, H. M., R. A. K. Tahirkheli, M. A Mirza, G. D. Johnson, N. M. Johnson, and N. D. Opdyke. 1977. Magnetic polarity stratigraphy of the Upper Siwalik deposits, Pabbi Hills, Pakistan. *Earth and Planetary Science Letters* 36:187–201.

Kennett, J. P. and D. A. Hodell. 1986. Major events in Neogene oxygen isotopic records. *South African Journal of Science* 82: 497–498.

Kotlia, B. S., K. Nakayama, B. Phartiyal, S. Tanaka, M. S. Bhalla, T. Tokuoka, and R. N. Pande. 1999. Lithology and magnetostratigraphy of the upper Siwaliks. *Memoir of the Geological Society of India* 44:209–220.

Kumar, R., S. K. Ghosh, S. J. Sangode, and V. C. Thakur. 2002. Manifestation of intra-foreland thrusting in the Neogene Himalayan foreland basin fill. *Journal of the Geological Society of India* 59:547–560.

Kumar, R. and S. K. Tandon. 1985. Sedimentology of Plio-Pleistocene late orogenic deposits associated with interplate subduction. The Upper Siwalik Subgroup of a part of Panjab Sub-Himalaya, India. *Sedimentary Geology* 42:105–158.

Kumaravel, V., S. J. Sangode, R. Kumar, and N. S. Siddaiah. 2005. Magnetic polarity stratigraphy of Plio-Pleistocene Pinjor Formation (type locality), Siwalik Group, NW Himalaya, India. *Current Science* 88(9):1453–1461.

Kutzbach, J. E, W. L. Prell, and W. F. Ruddiman. 1993. Sensitivity of Eurasian climate to surface uplift of the Tibetan Plateau. *Journal of Geology* 101:177–190

Lewis, G. E. 1934. Preliminary notices of new man-like apes from India. *American Journal of Science* 27:161–179.

Lihoreau, F., J. Barry, C. Blondel, Y. Chaimanees, J. J. Jaeger, and M. Brunet. 2007. Anatomical revision of the genus *Merycopotamus* (Artiodactyla; Anthracotheriidae): Its significance for Late Miocene Mammal dispersal in Asia. *Palaeontolgy* 50(20):503–524.

Mangerud, J., E. Jansen, and J. Y. Landvik. 1996. Late Cenozoic history of the Scandinavian and Barents Sea ice sheets. *Global Planetary Change* 12:11–26.

Medlicott, H. B. 1879. Manual of the geology of India. *Geological Survey of India, Publication* 2:1–524.

Mehta, Y. P., A. K. Thakur, Nand Lal, B. Shukla, and S. K. Tandon. 1993. Fission track age of zircon separates of tuffaceous mudstones of the Siwalik Subgroup of Jammu Chandigarh sector of the Punjab Sub-Himalaya. *Current Science* 64:519–521.

Munthe, J., B. Dongol, J. H. Hutchison, W. F. Kean, K. Munthe, and R. M. West. 1983. New fossil discoveries from the Miocene of Nepal include a hominoid. *Nature* 303:331–333.

Nanda, A. C. 1997. Comments on Neogene/Quaternary Boundary and associated faunas in the Upper Siwalik of Chandigarh and Jammu. In *Proceedings of the First Regional GEOSAS Workshop on Quaternary Geology of South Asia*, pp. 120–136. Madras, India: Anna University.

Nanda, A. C. 2002. Upper Siwalik Mammalian faunas of India and associated events. *Journal of Asian Earth Sciences* 21:47–58.

Nanda, A. C. and K. Kumar. 1999. *Excursion Guide. The Himalayan Foreland Basin (Jammu-Kalakot-Udhampur Sector)*. Wadia Institute of Himalayan Geology, Special Publication 2:1–85.

Nanda, A. C., D. C. Sati, and G. S. Mehra. 1991. Preliminary report on the stratigraphy and mammalian faunas of the Middle and Upper Siwalik, west of Yamuna River, Paonta, Himachal Pradesh. *Journal of Himalayan Geology* 2(2):151–158

Nanda, A. C. and R. K. Sehgal. 1993. Siwalik mammalian faunas from Ramnagar (J & K) and Nurpur (H.P.) and lower limit of *Hipparion*. *Journal of the Geological Society of India* 42:115–134.

Nanda, A. C. and R. K. Sehgal. 2005. Recent advances in palaeontology and magnetostratigraphic aspects of the Siwalik group of Northwestern Himalaya. *Himalayan Geology* 26 (1):93–102.

Nanda, A. C. and R. K. Sehgal. 2007. Geology and Mammalian fauna of the Siwalik Group of India. In *Human Origins, Genome and People of India*, ed. A. R. Sankhyan and V. R. Rao, pp. 77–90. New Delhi: Allied Publishers.

Opdyke, N. D., E. H. Lindsay, G. D. Johnson, N. M. Johnson, R. A. K. Tahirkheli, and A. Mirza. 1979. Magnetic polarity stratigraphy and vertebrate paleontology of the Upper Siwalik Subgroup of northern Pakistan. *Palaeogeography, Palaeoclimatology, Palaeoecology* 27:1–34.

Parkash, B., R. P. Sharma, and A. K. Roy. 1980. The Siwalik Group (molasse)-sediments shed by collision of continental plates. *Sedimentary Geology* 25:127–159.

Parmar, V. and G. V. R. Prasad. 2006. Middle Miocene rhizomyid rodent (Mammalia) from the Lower Siwalik Subgroup of Ramnagar, Udhampur District, Jammu and Kashmir, India. *Neues Jahrbuch für Geologie und Paläontologie Mh.* 6:371–384.

Patnaik, R. 1995. Micromammal-based palaeoenvironment of Upper Siwaliks exposed near Village Saketi, H.P. *Journal of the Geological Society of India* 46:429–437

Patnaik, R. and D. W. Cameron. 1997. New Miocene fossil ape locality, Dangar, Hari-Talyangar region, Siwaliks, northern India. *Journal of Human Evolution* 32:93–97.

Patnaik, R., D. Cameron, J. C. Sharma, and J. Hogarth. 2005. Extinction of Siwalik fossil apes: A review based on a new fossil tooth and on palaeoecological and palaeoclimatological evidence. *Anthropological Science* 113:65–72.

Patnaik, R. and A. C. Nanda. 2010. Early Pleistocene Mammalian faunas of India and evidence of connections with other parts of the world. In *Out of Africa I: The First Hominin Colonization of Eurasia*, ed. J. Fleagle, J. Shea, F. E. Grine, A. L. Baden, and R. Leakey, pp. 129–143. Vertebrate Paleobiology and Paleoanthropology Series. New York: Springer.

Phadatre, N. R., R. Kumar, and S. K. Ghosh. 1994. Stratigraphic palynology, floristic succession and the Tatrot/Pinjor boundary in Upper Siwalik sediments of Haripur Khol Area District Sirmuar (H.P.) India. *Himalayan Geology* 15:69–82.

Pilbeam, D., J. C. Barry, G. E. Meyer, S. M. I. Shah, M. H. L. Pickford, W. W. Bishop, H. Thomas, and L. L. Jacobs. 1977. Geology and paleontology of Neogene strata of Pakistan. *Nature* 270:684–689.

Pilgrim, G. E. 1910. Preliminary note on a revised classification of the Tertiary freshwater deposits of India. *Record of the Geological Survey of India* 40:185–205.

Pilgrim, G. E. 1913. The correlation of the Siwaliks with mammal horizons of Europe. *Record of the Geological Survey of India* 43:264–326.

Pilgrim, G. E. 1939. The fossil Bovidae of India. *Memoirs of the Geological Survey of India*, n. s., 26(1):1–356.

Pillans B., M. Williams, D. Cameron, R. Patnaik, J. Hogarth, A. Sahni, J. C. Sharma, F. Williams, and R. Bernor. 2005. Revised correlation of the Haritalyangar magnetostratigraphy, Indian Siwaliks: Implications for the age of the Miocene hominids *Indopithecus* and *Sivapithecus*, with a note on a new hominid tooth. *Journal of Human Evolution* 48:507–515.

Prasad, K. N. 1970. The vertebrate fauna from the Siwalik Beds of Haritalyangar, Himachal Pradesh. *Palaeontologia Indica*, n. s., 39:1–55.

Prasad, M. 1993. Siwalik (Middle Miocene) woods from the Kalagarh area in the Himalayan foot hills and their bearing on paleoclimate and phytogeography. *Review of Paleobotany and Palynology* 76:49–82.

Quade, J. and T. E. Cerling. 1995. Expansion of C4 grasses in the late Miocene of northern Pakistan: Evidence from paleosols. *Palaeogeography, Palaeoclimatology, Palaeoecology* 115:91–116.

Quade, J., T. E. Cerling, and I. R. Bowman. 1989. Development of the Asian monsoon revealed by marked ecologic shift in the latest Miocene of northern Pakistan. *Nature* 342:163–166.

Ramstein, G., F. Fluteau, J. Besse, and S. Joussaume. 1997. Effect of orogeny, plate motion and land-sea distribution on Eurasian climate change over the past 30 million years. *Nature* 386:788–795.

Ranga Rao, A. 1993. Magnetic polarity stratigraphy of the Upper Siwaliks in the NW Himalaya, India. *Current Science* 64:863–873.

Ranga Rao, A., R. P. Agarwal, U. N. Sharma, M. S. Bhalla, and A. C. Nanda. 1988. Magnetic polarity stratigraphy and vertebrate paleontology of the Upper Siwalik Subgroup of Jammu Hills, India. *Journal of the Geological Society of India* 31:361–385.

Ranga Rao, A., A. C. Nanda, U. N. Sharma, and M. S. Bhalla. 1995. Magnetic polarity stratigraphy of the Pinjor Formation (Upper Siwalik) near Pinjor, Haryana. *Current Science* 68:1231–1236.

Raymo, M. E. and W. F Ruddiman. 1992. Tectonic forcing of late Cenozoic climate. *Nature* 359:117–122.

Retallack, G. J. 1991. *Miocene Paleosols and Ape Habitats of Pakistan and Kenya*. New York: Oxford University Press.

Retallack, G. J. 1995. Palaeosols of the Siwalik Group as a 15 My record of South Asian Palaeoclimate. *Memoir Geological Society of India* 32:36–51.

Ruddiman, W. F. and J. E. Kutzbach. 1989. Forcing of late Cenozoic northern hemisphere climate by plateau uplift in Southern Asia and the American West. *Journal of Geophysical Research* 94:18409–18427.

Sahni, A. and S. K. Khare. 1976. Siwalik insectivore. *Geological Society of India* 17:114–116.

Sahni, M. R. and E. Khan. 1964. Stratigraphy, structure and correlation of the Upper Shivaliks east of Chandigarh. *Journal of the Palaeontological Society of India* 4:61–74.

Sangode, S.J. and R. Kumar. 2003. Magnetostratigraphic correlation of the Late Cenozoic fluvial sequences from NW Himalaya. *Current Science* 84(8):1014–1024.

Sangode, S. J., R. Kumar, and S. K. Ghosh. 1996. Magnetic polarity stratigraphy of the Siwalik sequence of Haripur area (H. P.), NW Himalaya. *Journal of the Geological Society of India* 47:683–704.

Sankhyan, A. R. 1985. Late occurrence of *Sivapithecus* in Indian Siwaliks. *Journal of Human Evolution* 14:573–578.

Scott, R., S. J. Kappelman, and J. Kelley. 1999. The paleoenvironment of *Sivapithecus parvada*. *Journal of Human Evolution* 36:245–74.

Sehgal, R. K. 1998. Lower Siwalik carnivores from Ramnagar, Jammu and Kashmir, India. *Himalayan Geology* 19(1):109–118.

Sehgal, R. K. and A. C. Nanda. 2002. Palaeoenvironment and palaeoecology of the Lower and Middle Siwlaik Subgroups of a part of Northwest Himalaya. *Journal of the Geological Society of India* 59:517–529.

Sehgal, R. K. and R. Patnaik. 2012. New muroid rodent and *Sivapithecus* dental remains from the Lower Siwalik deposits of Ramnagar (J&K, India): Age Implication. *Quaternary International* 269:69–73.

Simons, E. L. and S. R. K. Chopra. 1969. *Gigantopithecus* (Pongidae, Hominoidea): A new species from Northern India. *Postilla* 138:1–8.

Simons, E. L. and P. C. Ettel. 1970. *Gigantopithecus*. *Scientific American* 222:77–85.

Simons, E. L. and D. Pilbeam. 1973. A gorilla-sized ape from the Miocene of India. *Science* 173:23–27.

Szalay, F. S. and E. Delson. 1979. *Evolutionary History of the Primates*. New York: Academic Press.

Tandon, S. K. and R. Kumar. 1984. Discovery of tuffaceous mudstones in the Pinjor Formation of Panjab Himalaya, India. *Current Science* 53(18):982–984.

Tandon, S. K., R. Kumar, M. Koyama, and N. Niitsuma. 1984. Magnetic polarity stratigraphy of the Upper Siwalik Subgroup, east of Chandigarh, Punjab Sub-Himalaya, India. *Journal of the Geological Society of India* 25(1):45–55.

Thomas, H. and S. N. Verma. 1979. Découverte d'un primate Adapiforme (Sivaladapinae Subfam. Nov.) dans le Miocène moyen des Siwaliks de la region de Ramnagar (Jammu et Cachemire, Inde). *Comptes Rendus de l'Académie des Sciences Paris* 289:833–836.

Thomas, J. V., B. Parkash and R. Mohindra. 2002. Lithofacies and palaeosol analysis of the Middle and Upper Siwalik Groups (Plio–Pleistocene), Haripur-Kolar section, Himachal Pradesh, India. *Sedimentary Geology* 150(3–4):343–366.

Tiwari, B. N. 1983. Lower Siwalik faunas of the Indian subcontinent with special reference to Kalagarh Fauna. *Publications of the Centre of Advanced Study in Geology, Panjab University, Chandigarh* 13:98–112.

Tiwari, B. N. 1990. Late Miocene rhizomyid rodent-bearing faunule from the Siwalik of Palampur, Himachal Pradesh. *Current Science* 59(9):460–463.

Tiwari, B. N. 1996. Middle Siwalik murid rodents from Ladhyani, Bilaspur, Himachal Pradesh. In *Contributions to XV Indian Colloquium on Micropalaeontology and Stratigraphy*, ed. J. Pandey, R. J. Azmi, A. Bhandari, and A. Dave, pp. 513–518. Dehra Dun: Wadia Institute of Himalayan Geology.

Vasishat, R. N. 1985. *Antecedents of Early Man in Northwestern India: Paleontological and Paleoecological Evidences*. New Delhi: Inter India.

Vasishat, R. N., R. Gaur, and S. R. K. Chopra. 1978. Geology, fauna and palaeoenvironments of Lower Siwalik deposits around Ramnagar, India. *Nature* 275:736–737.

Vasishat, R. N., I. J. Suneja, R. Gaur, and S. R. K. Chopra. 1983. Neogene mammals (equids, rhinocerotids, suids and anthracotheriids). *Publication of the Centre for Advanced Study in Geology* 13:207–215.

Verma, B. C. 1969. *Procynocephalus pinjori* sp. nov., a new fossil primate from the Pinjor beds (Lower Pleistocene) east of Chandigarh. *Journal of the Palaeontological Society of India* 13:53–57.

Verma, B. C. 1988. Search for microvertebrates in the Upper Siwaliks of Markanada Valley, Sirmur District, Himachal Pradesh and development of Upper Siwalik biostratigraphy. *Records of the Geological Survey of India* 122:309–312.

Verma, B. C. and S. S. Gupta. 1997. New light on the antiquity of Siwalik great apes. *Current Science* 72(5):302–303.

West, R. M. 1996. The Cenozoic of Nepal: Mountain evolution and vertebrate evolution. *Journal of the Nepal Geological Society* 14:11–19.

West, R. M., J. H. Hutchison, and J. Munthe. 1991. Miocene vertebrates from the Siwalik Group, Western Nepal. *Journal of Vertebrate Paleontology* 11(1):108–129.

West, R. M., J. R. Luckas, J. Munthe, and S. T. Hussain 1978. Vertebrate fauna from Neogene Siwalik Group, Dang Valley, western
Nepal. *Journal of Paleontology* 52:1015–1022.

West, R. M. and J. Munthe. 1981. Neogene vertebrate paleontology
and stratigraphy of Nepal. *Journal of the Nepal Geological Society*
1:1–14.

Williams, M. D., D. Dunkerley, P. D. Decker, P. Kershaw, and
J. Chappel. 1999. *Quaternary Environments.* New York: Arnold.

Willis, B. J. 1993. Ancient river systems in the Himalayan Foredeep,
Chinji Village area, northern Pakistan. *Sedimentary Geology*
88:1–76.

Yokoyama, T., B. C. Verma, T. Mastuda, S. S. Gupta, and A. P. Tewari.
1987. Fission-track age of a bentonitized ash bed and mammalian
fauna from Nagrota Formation (Upper Siwalik) of Jammu District (J & K). *Indian Minerals* 41:13–18.

Chapter 18

Paleobiogeography and South Asian Small Mammals

Neogene Latitudinal Faunal Variation

LAWRENCE J. FLYNN AND WILMA WESSELS

Our purpose is to present the Neogene small mammal record of South Asia in its biogeographic context. Within the Indian Subcontinent, we observe high faunal similarity among local small mammal assemblages distributed on a scale of 1000 km, and these are distinct from assemblages to the west and northwest and northeast, beyond the subcontinent. Eastward into Thailand, Myanmar (Burma), and Yunnan, China, faunal similarity with the Indian Subcontinent is apparent but weaker than within the subcontinent. This pattern mimics the distribution of the present-day Oriental biogeographic province. Our goal is to study to what extent we can define this paleozoogeographic pattern and to recognize faunal elements that do not follow the rules—specifically those taxa that show wider distributions and point to dispersal events. Ultimately, we hope to be able to specify the theater of evolution of key dispersers and direction of dispersal, but that is a second step. In this chapter, we outline the small mammal fossil record of the Indian Subcontinent in comparison with that of Thailand and Yunnan, and with those from faunas to the west in Anatolia, North Africa, Afghanistan, and northern China.

HISTORY OF RESEARCH

Based on the few remains then known, Colbert (1935) presumed Siwalik fossil rodents to be rare, supposing that small vertebrates would likely be destroyed in rapidly accumulating sediments. Nonetheless, fossil rodents had been found steadily since early records of Siwalik vertebrate remains were published in the 1830s. Jacobs (1978) noted that Cautley (1836) recorded a "rat and a small variety of *Castor*." Already, then, the two dominant groups of rodents had been found. In modern analyses, the "rat" represents Muridae, and the "*Castor*" (living beaver) certainly is what was later recognized as a bamboo rat (Rhizomyinae). Lydekker (1884) studied the bamboo rat and illustrated two kinds of porcupine (Hystricidae), establishing *Hystrix sivalensis*. Another half century passed before significant advances were made.

Hinton (1933) named five rhizomyines, a thryonomyid (cane rat), and a ctenodactylid (gundi). Colbert (1933) described another rhizomyine and the hystricid *Sivacanthion*. Wood (1937) advanced knowledge of Siwalik Rodentia with more rhizomyine and ctenodactylid material. Rodent research then stood dormant into the late 1900s except for Black's (1972) review, in which he figured much of the known material and revised the taxonomy, attributing Siwalik rodents to currently accepted family-level groups. Within a few years of that work, a new approach was undertaken for the Siwaliks. Teams of researchers were attracted to the Siwaliks to develop modern biostratigraphic studies and apply the technique of wet-screening large volumes of fossiliferous rock for small and isolated specimens not normally seen on the surface of the ground. Studies by two teams marked a new era.

Jacobs (1978) broke new ground by systematic screening of multiple horizons in the classic Chinji and Dhok

Pathan formations of the Potwar Plateau and the Upper Siwaliks of the adjacent Pabbi Hills. His work produced shrews and hedgehogs, as well as diverse rodents, including mice and bamboo rats, of course, but also a thryonomyid, Cricetidae (hamster relatives), and Ctenodactylidae (gundis). Jacobs (1978) found good representation of rhizomyines and a wealth of mice representing at least eight genera. Whereas surface finds had produced mostly rhizomyines (larger size bias), screening made clear that murids were truly the dominant rodents. A related study using Jacobs' samples and a growing collection of surface and screened finds from the Potwar Plateau allowed Flynn (1982) to revise the systematics of bamboo rats and add seven new species.

At about this time, associates of S. Taseer Hussain (Howard University) began highly productive screening programs elsewhere in Pakistan. At Daud Khel, near the Indus River, Munthe (1980) found a diverse microfauna of four insectivores, a tree shrew, and six rodent groups (squirrels, gundis, dormice, bamboo rats, hamsters, and mice). In Kohat, west of the Indus River, Wessels et al. (1982) recovered higher species richness (14 rodents) in Chinji Formation sediments. These studies began to reveal the composition of Siwalik microfaunas and a high degree of similarity in northern Pakistan among assemblages through geologic time. They indicated a resilient community of small mammals, one that persisted despite additions and deletions during most of the Miocene.

De Bruijn, Hussain, and Leinders (1981) advanced understanding of the roots of Siwalik small mammal faunas when they discovered an important assemblage in the Early Miocene Murree Formation of Kohat. These deposits antedate the oldest Siwalik rocks near the Salt Range. Not dated directly, the assemblage is comparable to faunas in central Pakistan of about 20 Ma (Lindsay et al. 2005). Higher-taxonomic-level resemblance between this Murree fauna and younger Miocene assemblages lies in Sciuridae, a small ctenodactylid, a primitive rhizomyine, and several species close to modern Muroidea. Elements retained from Paleogene faunas, like Paali Nala C2 (Bugti; see Marivaux and Welcomme 2003), are indicated by a diatomyid (Flynn 2000) and a possible phiomorph (de Bruijn, Hussain, and Leinders 1981:93).

By the 1980s, the Siwaliks were proven to be highly productive of rodent material, with superposed assemblages representing a record approaching 10^5 yr in completeness (5–10 assemblages per million years). Wessels (1996) published diverse Middle Miocene (Chinji Formation) muroids from a range of horizons from Sind Province. Cheema, Sen, and Flynn (1983) and Cheema

et al. (2000) analyzed early Late Miocene rodents from near Jalalpur, Pakistan. These and other studies (Lindsay 1988; Baskin 1996) complemented the record of Miocene Rodentia of the Indian Subcontinent, and key contributions continue today (Patnaik, chapter 17, this volume).

This is the foundation for a preliminary analysis of what is now known to be a rich fossil record of small mammals. The Siwaliks and equivalent strata across the Indian Subcontinent, plus antecedent deposits, contain a densely sampled history of the microfauna, which spans the entire Neogene. It is rich in species diversity and representation through time. It offers sufficient data for comparison with assemblages from elsewhere across southern Asia and beyond. Similarities and differences with faunas from other regions allow recognition of the geographic limits of extinct taxa. The data are maturing to allow evaluation of biogeographic provinciality. What can we learn about dispersal patterns during the Miocene? Do South Asian microfaunas represent the western extent of the modern Oriental Biogeographic Province? Does South Asia record a mix of taxa, of which today's descendants are nonoverlapping? In the following, we summarize the succession of faunas observed in the Indian Subcontinent for comparison with fossiliferous regions across much of Asia and into Africa.

INDIAN SUBCONTINENT

The Early Miocene Siwalik fossil record is known primarily from the Murree assemblage explored by de Bruijn, Hussain, and Leinders (1981), a succession of faunas from the Zinda Pir Dome area, near Dera Ghazi Khan (Lindsay et al. 2005), and low levels in the Gaj and Sehwan sections of Sind (de Bruijn and Hussain 1984; Wessels 2009) (figure 18.1). These show that the precursors to the Middle and Late Miocene Siwalik fauna were already well established at the beginning of the Miocene. Small primates, tupaiids, and erinaceids are in evidence, as well as muroid, ctenodactylid, and sciurid rodents. A few thryonomyid and diatomyid rodent specimens are known also. Muroids are the most diverse rodent group, a pattern that holds true throughout the Miocene. The Murree fauna includes the early cricetid Spanocricetodon, the earliest indication of South Asian Myocricedontinae (Theocharopoulos 2000), and the primitive rhizomyine Prokanisamys.

The following discussion of Miocene microfaunas of the Indian Subcontinent focuses on rodents, but other groups are also informative. The insectivores are never

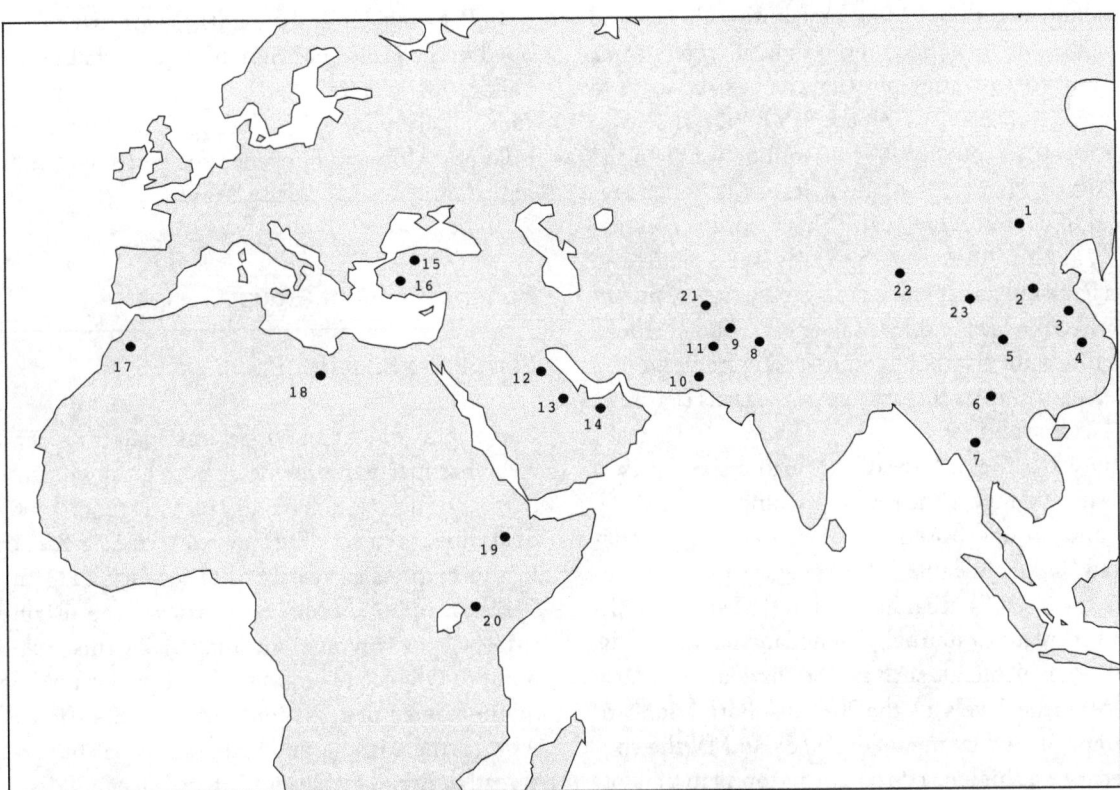

Figure 18.1 Simplified map showing principal Neogene localities used to document occurrences of small mammal taxa. Single points may represent multiple fossil sites. (*1*) Inner Mongolian Tunggur Tableland and younger sites, e.g., Ertemte (Qiu, Wang, and Li, chap. 5, this volume); (*2*) Yushe Basin, Shanxi Province, Late Miocene and Pliocene sites; (*3*) Sihong, Jiangsu Province, late Early Miocene; (*4*) Fangshan, Jiangsu Province (type locality of *Spanocricetodon*); (*5*) Lantian, Bahe Formation, Shaanxi Province, early Late Miocene; (*6*) Yunnan Province Late Miocene–Yuanmou basin complex and Lufeng; (*7*), Thailand–Li Basin, Mae Moh Basin, Pong Basin, Lampang Basin; (*8*) North India Siwaliks Patnaik chap. 17, this volume); (*9*) North Pakistan (Potwar and Trans Indus); (*10*) South Pakistan Sehwan and Gaj, Early to Middle Miocene localities; (*11*) Sulaiman Area, Pakistan–Bugti and Zinda Pir, Oligo-Miocene; (*12*) As Sarrar, Saudi Arabia; (*13*) Saudi Arabia—Al Jadidah and Jabal Miri ash Shamali; (*14*) Abudhabi (UAE), Shuwayhat, Late Miocene; (*15*) Keseköy, Anatolia, Turkey; (*16*) Turkey–Kargı, Kılçak, and Harami; (*17*) Morocco—Beni Mellal; (*18*) Libya–Jebel Zelten (*19*) Ethiopia–Ch'orora; (*20*) Kenya–Fort Ternan, Middle Miocene, and Late Miocene of Baringo area; (*21*) Khurdkabul Basin and other Late Miocene sites of Afghanistan; (*22*) Danghe Area, near Tsaidam, Western Gansu; (*23*) Linxia and Tianzhu of Eastern Gansu.

diverse, but they include shrews of the *Crocidura* group and later *Suncus*, as well as a long lineage of hedgehogs. Tupaiids and several small primates persist until the Late Miocene and show affinity with areas to the east; however, southeast Asian tarsiers remain undocumented in the Indian Subcontinent.

The early Miocene of the Gaj and Zinda Pir Dome areas yields a record of *Democricetodon* nearly as early as that of Harami, Turkey (Theocharopoulos 2000). The older Gaj localities are dominated by rhizomyines, especially *Prokanisamys arifi*, and contain the muroids *Sindemys sehwanensis* and *Democricetodon* sp. and the ctenodactylid *Sayimys intermedius*. These and the Murree assemblages date to around 20 Ma (Wessels 2009). Later assemblages, ca. 18 Ma and younger, show higher diver-

sity and establish as dominant advanced muroids, especially myocricetodontines. There is rare evidence of ochotonid lagomorphs (Flynn et al. 1997; Wessels 2009) at this time. Around 16 Ma, faunal dynamics reassert dominance of more primitive muroids, especially rhizomyines. The low number of Rhizomyinae and high number of Myocricetodontinae around 18 Ma could indicate drier conditions (Clift et al. 2008). A drier phase followed by a Middle Miocene increase in humidity, in combination with cooler global climate (Zachos et al. 2001; Abels 2008:113–132), could have induced observed variations in dominance. Future work can test the reality of these trends throughout southern Asia.

Northern Pakistan preserves a remarkably complete record of sedimentation for the Middle and Late Miocene

Siwaliks. This sequence is best developed in the Potwar Plateau, a large region of 12,000 km², with well-exposed outcrops. Measured sections thousands of meters thick preserve a magnetostratigraphy that can be used to build a time frame. To the south near the Salt Range, the Siwalik sequence spans roughly 18 to 10 million years. In the central Potwar Plateau, the Late Miocene is better preserved, up to ca. 6 million years. This record is complemented by key localities from the Trans-Indus plains and southern Pakistan and is balanced by fossil sites of northern India (Ramnagar and Haritalyangar). The northern India localities are distant by some 600 km from the Potwar and show a similar history in most regards (Patnaik, chapter 17, this volume).

The Potwar series of Siwalik formations comprises, from oldest to youngest, the Kamlial, Chinji, Nagri, Dhok Pathan, and a superposed set of younger sediments designated "Upper Siwaliks." The rodent community recorded in the basal Kamlial Formation (18 Ma) is not diverse, but it includes muroids, ctenodactylids, sciurids, and a few other elements, such as Diatomyidae. The early Middle Miocene levels of the Kamlial Formation are spottily represented by microfaunas. By 16 Ma, the species diversity was higher, a dozen forms found in the Potwar, and up to 18 in some Sehwan assemblages, mostly due to diversification of muroids. Chinji Formation localities (14–12 Ma) are many and record high rodent diversity in the later Middle Miocene, with up to 30 species present (Flynn et al. 1998). There are many muroids, but also sciurids, glirids, ctenodactylids, thryonomyids, and a hystricid. Among other small mammals, Chinji Formation sites contain lorisid and sivaladapid primates, tupaiids, erinaceids, soricids, and bats. By the early Late Miocene (~10 Ma), species richness had declined radically, and decreasing diversity characterized subsequent Siwalik faunal history (Flynn et al. 1998).

There were, however, distinctive changes in the composition of the late Late Miocene rodent fauna (Jacobs, Flynn, and Downs 1989). The Dhok Pathan Formation localities record a flowering of the Murinae (derivatives of *Progonomys* and *Karnimata* and other genera), reduction of Cricetinae and relatives, and an explosion in evolution of subterranean Rhizomyinae. Also distinctive are the appearance after 8 Ma of *Hystrix* and later the leporid *Alilepus*. These changes occur on the backdrop of declining overall species diversity, mainly by disappearance of nonmurine, nonrhizomyine muroids and extinction of ctenodactylids. Perhaps indicating ecological difference, the ctenodactylids appear to have persisted longer at Haritalyangar, northern India, than in the Potwar (Pillans et al. 2005).

Pliocene deposits ("Upper Siwaliks") are known from both Pakistan and India, but these are extensively developed in the Pinjor Formation of Himachal Pradesh. Patnaik (2001; chapter 17, this volume) recovered up to 15 rodents, all muroids (including Rhizomyinae and Gerbillinae). Although Pliocene species diversity is relatively high, family-level diversity is low.

Dominant Rodent Groups

Burrowing Rodents

The origins of Rhizomyinae and Spalacinae remain unclear, but species representing both groups occur in Early Miocene rocks and have distinct biogeographic ranges, Spalacinae in eastern Europe–western Asia, Rhizomyinae in southern Asia. Fossils from locality Z113 in central Pakistan (ca. 23.5 Ma) represent a deep origin for the rhizomyines (Flynn et al., chapter 14, this volume) and point to Oligocene origin of the larger group. The rhizomyine center of radiation and diversification is South Asia in the Middle and Late Miocene, but the group spread to Africa by the late Early Miocene. Tribe Tachyoryctini evolved in South Asia and dispersed by the Early Pliocene to Africa; Rhizomyini (true bamboo rats) spread to southern North China (Yushe Basin) by the end of the Miocene.

Gerbils

The first members of the Myocricetodontinae appear in the Early Miocene of Anatolia and southern Pakistan (20 Ma), and after 18 Ma a rapid diversification occurs through evolutionary change and interregional migrations. The occurrence of different genera of early myocricetodontines at almost the same time in Anatolia and Pakistan suggests immigration into these areas from the East, rather than exchange between Pakistan and Anatolia. In Anatolia (Wessels 2009), they disappear temporarily from the record during the latest part of the Early Miocene, returning in the Middle Miocene. In the Middle Miocene, derived Myocricetodontinae appear in North African assemblages. Myocricetodontines spread to China (10 Ma, Bahe Formation, Lantian Basin; Qiu 2001; Qiu, Zheng, and Zhang 2004b) and to Europe (Late Miocene). These Middle Miocene occurrences point to restocking by immigrants from South or Southeast Asia.

Modern Taterillinae appear in the late Miocene. Named for fossils from the United Arab Emirates, *Abud-*

habia has older records in the Bahe Formation, Lantian Basin (10 Ma; Qiu, Zheng, and Zhang 2004b) and Pakistan (8.7 Ma; Flynn and Jacobs 1999). The genus occurs widely in the Late Miocene and Pliocene (Flynn et al. 2003).

Muridae

It appears that the origin of the Muridae (represented by *Potwarmus*) lies in southern Asia, and by 16 Ma this stem group occurs throughout South Asia. Migration into northern Africa of a murid close to *Potwarmus* occurs shortly after 16 Ma (Wessels et al. 2003). A transition from *Potwarmus* to *Antemus* and subsequently to *Progonomys* (by about 12 Ma) has been well documented in the Potwar Plateau and southern Pakistan (Jacobs and Downs 1994; Wessels 2009). *Progonomys* migrated into Anatolia, North Africa, and Europe between 11 Ma and 10 Ma, and somewhat later (~10 Ma) into China (Bahe Formation, Lantian Basin; Qiu, Zheng, and Zhang 2004a).

Ctenodactylidae

Ctenodactylidae have their origin in the Eocene of Asia, radiate during the Oligocene, and penetrate westward into the Anatolian Oligocene. The modern Subfamily Ctenodactylinae is known from the terminal Oligocene onward in Pakistan (de Bruijn, Hussain, and Leinders 1981; de Bruijn, Boon, and Hussain 1989; Baskin 1996; Flynn et al., chapter 14, this volume). The characteristic latest Oligocene genus *Prosayimys* (Baskin 1996; López-Antoñanzas and Sen 2003) is replaced in the Miocene by the long-lived genus *Sayimys*. Several species are assigned to this genus, including *Sayimys obliquidens* of Gansu, China, which is late Early to early Middle Miocene in age (Qiu et al., chapter 1, and Wang et al., chapter 10, both this volume). *Sayimys* dispersed northward to Mongolia in the late Early Miocene (Daxner-Höck et al., chapter 20, this volume). During the late Early Miocene, *Sayimys* and derivatives migrated to Saudi Arabia, Anatolia, and North Africa (de Bruijn, Boon, and Hussain 1989).

Cricetidae

The first modern cricetids to appear in Asia are species from the latest Oligocene of Anatolia (Theocharopoulos 2000) and Pakistan (Z113; Lindsay et al. 2005), assigned to *Spanocricetodon* and *Primus*. The type species of *Spanocricetodon* from Fangshan, Jiangsu, China (Li 1977), and *Spanocricetodon* from the Murree Formation of Banda Daud Shah (de Bruijn, Hussain, and Leinders

1981) are both late Early Miocene in age. It appears that *Primus* and *Spanocricetodon* migrated from an East Asian origin into Anatolia and Pakistan (Wessels 2009). *Democricetodon* species appear simultaneously at the beginning of the Miocene (ca. 22 Ma) in Anatolia, Pakistan, and China (Maridet et al. 2011; Meng et al., chapter 3, this volume). Representatives of this cricetid genus, beginning about 18 Ma, are known in Europe from the late Early Miocene onward. Somewhat later, *Megacricetodon* joined *Democricetodon* in many assemblages.

Thryonomyidae

De Bruijn (1986) rooted Siwalik *Kochalia* and *Paraulacodus* among primitive ctenodactyloid rodents from Asia and considered the African Miocene thryonomyids as descendants of immigrants from southwestern Asia. In contrast, Flynn and Winkler (1994) placed the origin of thryonomyids in Africa, whence they would have dispersed in two subsequent events to southern Asia, *Kochalia* before 16 Ma and *Paraulacodus* before 13 Ma. That scenario, adopted here, considers the Thryonomyidae as rooted among African Phiomyidae.

Sciuridae

Originating in the Eocene of North America, the Sciuridae migrated to Asia (north of the Himalaya) and Europe during the Oligocene, and at least one lineage penetrated South Asia during that epoch (Marivaux, Vianey-Liaud, and Welcomme 1999). Since the early Miocene, they have been diverse in eastern and southern Asia and appeared in Anatolia and Africa, evidently immigrants from Asia. From the early Middle Miocene through the Late Miocene, Siwalik squirrels became represented by a number of extant terrestrial and arboreal genera. A few of these (*Tamias*, *Atlantoxerus*) indicate incursion to South Asia by northern elements. Most genera are shared with South China and help to characterize the Oriental Biogeographic Province.

Gliridae

The Gliridae are known from Europe since the Eocene, disperse into Anatolia during the Oligocene, and disperse into China during the Early Miocene. The group was apparently blocked from South Asia at this time, but it appeared abruptly in Pakistan after 14 Ma (Middle Miocene). Glirids are known from Northern Africa from the late Middle Miocene onward. A reasonable interpretation sees Gliridae as a group that entered South Asia from an eastern source in China, a pathway followed by many Miocene squirrels.

Diatomyidae

This peculiar group represents a distinct lineage derived from Eocene ctenodactyloid rodents (Flynn 2007). Named for fossils from late Early Miocene diatomites of China (Li 1974), its records are mainly southern Asian, from Pakistan eastward to South China, representing a distributional pattern that reflects the Oriental Biogeographic Province. Recently, López-Antoñanzas (2011) has shown that the group dispersed out of South Asia in the late Early Miocene. *Diatomys* occurs sporadically in the Early Miocene of southern Pakistan and is abundant in the Middle Miocene of Thailand. The extant *Laonastes* occurs in dry, upland terrain of Laos, likely not the preferred habitat of mid-Tertiary diatomyids.

Temporal Succession

To illustrate the succession of Glires (Rodentia plus Lagomorpha) assemblages, we have constructed composite faunal lists for the Indian Subcontinent (table 18.1). Assemblages are censuses of successive 2-million-year (2 m.y.) time horizons. They are not designed to pool taxa of different ages into 2 m.y. bins. They are meant to sample, roughly, the small mammal fauna encountered each 2 m.y. Necessarily, the lists are composite in the sense that not all taxa are found at a single site, although many are, and we merge data from the Manchars, the Chitarwata and Vihowa formations, and the Siwaliks of the Potwar Plateau and Trans-Indus plains. The lists necessarily sample sites of ages that differ by several hundred thousand years. They draw on information from the Indian Siwaliks, but that is best sought in the chapter by Patnaik (chapter 17, this volume). A drawback from this approach is that some assemblages (and taxa) are omitted, because they occur near the midpoints between samples. For example, we miss important records of about 9 Ma (Haritalyangar, and the Potwar U-level), and we omit *Paraulacodus indicus* of about 13 Ma (e.g., at locality HGSP 82-14). Ages are *not* meant to designate appearances (an example is *Alilepus*, appearing at 7.4 Ma but not listed until the 6 Ma census).

We hesitated to construct this table because of the pitfalls in doing so. For the many excellent reasons explored in Barry et al. (chapter 15, this volume), it is very misleading to interpret these lists as a biostratigraphy. As de

Table 18.1

Selected Miocene Glires Assemblages for the Indian Subcontinent

6 Ma (Loc. D013)
- *Alilepus elongatus*
- *Hystrix* sp.
- *Rhizomyides sivalensis*
- *Mus auctor*
- *Karnimata huxleyi*
- *Parapelomys robertsi*

8 Ma (Y24, Y547, Y387)
- *Hystrix sivalensis*
- *Dremomys* large
- *Sciurotamias* large
- *Microdyromys* sp.
- *Dryomys* sp.
- cf. *Sayimys perplexus*
- *Kanisamys sivalensis*
- *Miorhizomys tetracharax*
- *Miorhizomys choristos*
- *Miorhizomys* cf. *M. pilgrimi*
- *Rhizomyides* sp.
- *Democricetodon* sp.
- *Abudhabia pakistanensis*
- *Progonomys debruijni*
- *Karnimata darwini*
- *Parapodemus* sp.

10 Ma (Y450, Y311)
- *Tamias* n. sp.
- *Ratufa sylva*
- *Funambulus* sp.
- *Dremomys* large
- *Sciurotamias* large
- *Tamiops* large
- *Sayimys chinjiensis*
- *Myomimus* n. sp.
- *Peridyromys* sp.
- *Kanisamys nagrii*
- *Rhizomyides punjabiensis*
- *Democricetodon kohatensis*
- *Democricetodon* spp.
- *Dakkamys asiaticus*
- *Paradakkamys chinjiensis*
- *Progonomys* sp.
- *Karnimata* sp.

12 Ma (Y634, Y496)
- *Eutamias urialis*
- *Funambulus* sp.
- *Sciurotamias* small

Tamiops small
Tamiops large
Myomimus sumbalenwalicus
Peridyromys sp.
Sayimys chinjiensis
Sayimys "sp. B"
Kochalia geespei
Prokanisamys "benjavuni"
Kanisamys indicus
Democricetodon kohatensis
Democricetodon spp.
Dakkamys barryi
Dakkamys asiaticus
Paradakkamys chinjiensis
Mellalomys perplexus
Progonomys hussaini

14 Ma (HGSP 82-24, Y709, Y491, lower Ramnagar)
  Pteromyinae indet.
    Tamias sp.
    Atlantoxerus sp.
    Heteroxerus sp.
    Callosciurus sp.
    Dremomys sp.
    Sciurotamias small
    Tamiops large
    Sayimys sivalensis
    Sivacanthion complicatus
    Kochalia geespei
    cf. *Paraulacodus indicus*
    Diatomys sp.
    Myomimus sumbalenwalicus
    Prokanisamys sp.
    Kanisamys indicus
    Kanisamys potwarensis
    Democricetodon kohatensis
    Democricetodon spp.
    Dakkamys sp.
    Punjabemys mikros
    Punjabemys downsi
    Sindemys aguilari
    Myocricetodon sivalensis
    Myocricetodon cf. *M. parvus*
    Mellalomys lavocati
    Potwarmus primitivus
    Antemus chinjiensis

16 Ma (HGSP 81-14, HGSP 84-26, Y591, Y592, Y642, Y682)
  Pteromyinae indet.
    Aliveria sp.
    Ratufa sp.

Tamiops sp.
Sayimys baskini
Sayimys intermedius
Sayimys large sp.
Kochalia sp.
Prokanisamys "benjavuni"
Prokanisamys major
Kanisamys indicus
Democricetodon kohatensis
Democricetodon spp.
Sindemys aguilari
Myocricetodon sivalensis
Mellalomys lavocati
Punjabemys mikros
Potwarmus primitivus
Antemus mancharensis

18 Ma (HGSP 81-14a, HGSP 81-06, Y721, Y747)
  cf. *Alloptox* sp.
  Pteromyinae indet.
    Oriensciurus sp.
    Sayimys intermedius
    Sayimys baskini
    Diatomys chitaparwalensis
    Kochalia sp.
    Prokanisamys arifi
    Prokanisamys major
    Democricetodon sp.
    Punjabemys mikros
    Sindemys sehwanensis
    Sindemys aguilari

20 Ma (Murree at Kohat, Z122, Z124, HGSP 81-07A)
    cf. *Oriensciurus* sp.
    cf. *Tamiops* sp.
    Diatomys shantungensis
    Marymus sp.
    Thryonomyidae (?) indet.
    Sayimys intermedius
    Prokanisamys arifi
    Democricetodon sp.
    Sindemys sehwanensis
    Spanocricetodon khani
    Primus microps

22 Ma (Loc. Z135)
    cf. *Ratufa* sp.
    Prosayimys n. sp.
    Prokanisamys sp.
    Spanocricetodon n. sp.
    Democricetodon sp.

NOTE: Assemblages are drawn from localities that date to near the selected time horizon and in some cases are composites from localities differing as much as 0.5 m.y. in age. Locality numbers with D, Y, and Z prefixes are in the Harvard–Geological Survey of Pakistan database. Many are listed in Barry et al. (2002) and Lindsay et al. (2005). H-GSP localities are in the Howard University–Geological Survey of Pakistan database and are listed in Wessels (2009).

Bruijn, Ünay, and Hordijk (chapter 26, this volume) note, taxa do not make simultaneous appearances in most cases, so associations only mean that species coexist for a time during their ranges. The survey is done to illustrate diversity at specific times (each 2 m.y.) and to provide a time framework for comparisons to other biogeographic regions. After outlining known small mammal records for northern Africa, Anatolia, Afghanistan, and southern Asia, we place dispersal events into this time framework for the Miocene.

REGIONAL COMPARISONS

North and East Africa

Knowledge of Miocene small mammal faunas in North Africa is based on key assemblages of different ages from Jebel Zelten, Libya (see figure 18.1; Wessels et al. 2003). Among African elements, a few specimens attest to late Early Miocene Rhizomyinae and a cricetid. Early Middle Miocene elements include the rhizomyine *Prokanisamys*, myocricetodontines, the murine *Potwarmus*, the ctenodactylid *Sayimys*, and the dipodid *Heterosminthus*. These are all immigrants into Northern Africa during earlier parts of the Miocene (Wessels 2009). Subsequent myocricetodontine and gerbilline evolution in northern Africa is detailed by Jaeger (1977a, 1977b) from a series of late Middle through Late Miocene localities. The late Middle Miocene Pataniak 6 and Beni Mellal sites in Morocco include the glirid *Microdyromys*, the squirrel *Atlantoxerus*, ctenodactylids, and diverse myocricetodontine gerbils.

Key East African sites document the equatorial Neogene faunas of the continent. The many important Kenyan sites demonstrate relative isolation in the Early Miocene, followed by sporadic interchange with southern Asia in the middle and late Miocene. The Middle Miocene hominoid locality Fort Ternan, Kenya, includes a rhizomyine, myocricetodontines, and the hamster *Democricetodon* (Tong and Jaeger 1993). Later Kenyan sites of the Baringo area and Ch'orora in Ethiopia attest to further rhizomyine and gerbil evolution, probably within Africa (Geraads 1998; Winkler, Denys, and Avery 2010).

Arabian Peninsula

Of the few mammal faunas known from the Arabian Peninsula, the most important are Jabal Midra ash Shamali (Early Miocene), As Sarrar (early Middle Miocene),

Al Jadidah (Hofuf Formation), all from Saudi Arabia, and the Shuwayhat sites (Late Miocene) from Abu Dhabi (see figure 18.1; Thomas et al. 1978; Thomas et al. 1982; Ziegler 2001; Saner, Al-Hinai, and Perineck 2005; Whybrow et al. 1982; Whybrow et al. 1990). Faunal lists are given in table 18.2.

The oldest fauna, Jabal Midra ash Shamali (Whybrow et al. 1982), produces the dipodoid *Arabosminthus quadratus* and the myocricetodontine *Shamalina tuberculata*. *Shamalina* seems to be related to *Sindemys* from Pakistan and is comparable in evolutionary development to *S. aguilari* known from Pakistani assemblages younger than 18 Ma (Wessels 2009), thus indicating faunal exchange in the Early Miocene. The interpretation of the paleoenvironment based on the complete fauna is that of a coastal plain with open savannah more landward (Whybrow et al. 1982).

The As Sarrar fauna contains 11 small mammal species (Thomas et al. 1982), an insectivore (Erinaceidae), a lagomorph (Ochotonidae), and nine rodents. The latter show diversity in the ctenodactylid *Sayimys assarrarensis*, the diatomyid *Pierremus*, the dipodoids cf. *Protalactaga* sp. and *Arabosminthus isabellae*, the thryonomyid *Paraphiomys knolli*, pedetids (*Megapedetes* cf. *pentadactylus* and an unidentified form), an unspecified cricetid, and a gerbil. The *Arabosminthus*, *Sayimys*, and *Paraphiomys* species are interpreted to represent local lineages (López-Antoñanzas and Sen 2004, 2005, 2006; López-Antoñanzas 2011). Thomas et al. (1982) consider the paleoenvironment as fluvio-marine, with mangroves and woodlands near the rivers but more open environment inland.

The Middle Miocene occurrences in the Al Jadidah fauna of a sciurid *Atlantoxerus* sp., the ctenodactylid *Sayimys intermedius*, the murid *Potwarmus*, and a myocricetodontine (Sen and Thomas 1979; Thomas et al. 1978; Thomas 1983; López-Antoñanzas and Sen 2004; López-Antoñanzas 2009) point to faunal influx. The late Miocene Shuwayhat sites of Abu Dhabi contain the gerbil *Abudhabia baynunensis*, the murine *Parapelomys* cf. *charkhensis*, and the muroids *Dendromus* and *Myocricetodon* (Whybrow et al. 1990; de Bruijn and Whybrow 1994; de Bruijn 1999), showing additional faunal exchange.

The sequence of these rodent assemblages and the supposed relationships of some species with Pakistani species indicate faunal exchange between these two areas, probably in the late Early Miocene, followed by a long period of separation until renewed interchange late in the Miocene.

Table 18.2

Miocene Glires Assemblages for the Arabian Peninsula and Thailand

---

The Arabian Peninsula

Jabal Midra ash Shamali (Saudi Arabia, Hadrukh Formation, Early Miocene; Whybrow et al. 1982)
- Dipodidae
  - *Arabosminthus quadratus*
- Myocricetodontinae
  - *Shamalina tuberculata*

As Sarrar (Saudi Arabia, continental equivalent of the Dam Formation, early Middle Miocene) (Thomas et al. 1982; López-Antoñanzas and Sen 2004, 2005, 2006; López-Antoñanzas 2011)
- Ochotonidae
- Ctenodactylidae
  - *Sayimys assarrarensis*
- Dipodidae
  - cf. *Protalactaga* sp.
  - *Arabosminthus isabellae*
- Thryonomyidae
  - *Paraphiomys knolli*
- Pedetidae
  - *Megapedetes* cf. *pentadactylus*
  - *Megapedetes* sp.
- Diatomyidae
  - *Pierremus explorator*
- Cricetinae
  - sp. indet.
- Gerbillinae
  - sp. indet.

Al Jadidah (Saudi Arabia, Hofuf Formation, later Middle Miocene) (Thomas et al. 1978; Sen and Thomas 1979; Thomas 1983; López-Antoñanzas and Sen 2004; López-Antoñanzas 2009)
- Sciuridae
  - *Atlantoxerus* sp.
- Ctenodactylidae
  - *Sayimys intermedius*
- Muridae
  - *Potwarmus flynni*
- Myocricetodontinae
  - sp. indet.

Shuwayhat sites (Abu Dhabi, Baynunah Formation, Late Miocene) (Whybrow et al. 1990; de Bruijn and Whybrow 1994; de Bruijn 1999)
- Gerbillinae
  - *Abudhabia baynunensis*
- Murinae
  - *Parapelomys* cf. *charkhensis*

- Dendromurinae
  - *Dendromus* sp.
- Myocricetodontinae
  - *Myocricetodon* sp.
- Dipodidae
  - sp. indet.
- Thryonomyidae
  - sp. indet.

Thailand

Mae Long, Li Basin (Jaeger et al. 1985; Mein and Ginsburg 1985; Ducrocq et al. 1995; Mein, Ginsburg, and Ratanasthien 1990)
- Sciuridae
  - *Ratufa maelongensis*
  - ?*Atlantoxerus* sp.
- Diatomyidae
  - *Diatomys liensis*
- Rhizomyinae
  - *Prokanisamys benjavuni*
- Cricetidae
  - *Democricetodon kaonou*
  - *Spanocricetodon janvieri*
- Muridae
  - *Potwarmus thailandicus*
- Platacanthomyidae
  - *Neocometes orientalis*

Ban San Klang, Pong Basin (Ducrocq et al. 1995)
- Sciuridae
  - sp. indet.
- Diatomyidae
  - *Diatomys* sp.

Had du Dai, Lampang Basin (Ducrocq et al. 1995)
- Sciuridae
  - sp. indet.
- Gliridae
  - sp. indet.
- Diatomyidae
  - *Diatomys* sp.
- Murinae
  - *Potwarmus* sp.

Mae Moh Basin (Chaimanee et al. 2007)
- Platacanthomyidae
  - *Neocometes* cf. *N. orientalis*
- Rhizomyinae
  - *Prokanisamys benjavuni*

Anatolia

The succession of Neogene faunas in Anatolia is becoming well represented by ongoing field efforts. Theocharopoulos (2000), Ünay, de Bruijn, and Sarac (2003), Wessels (2009), and de Bruijn et al. (chapter 26, this volume) document several Early Miocene immigrants, including *Democricetodon, Eumyarion, Heterosminthus, Megacricetodon, Sayimys, Keramidomys,* and *Eomyops. Myocricetodon* arrives in the Middle Miocene, and *Progonomys* distinguishes the early Late Miocene. Late Miocene Turkish faunas become progressively less endemic, with more taxa shared with Europe through time (de Bruijn et al., chapter 26, this volume).

Afghanistan

Small mammal faunas of Late Miocene and Pliocene age are known from several basins of eastern Afghanistan. The oldest assemblage in a biochronological seriation is from Sherullah, thought to approximate the Siwalik "U-level" (Heintz and Brunet 1982), which is now dated at about 9 Ma. Sherullah has a high-crowned *Kanisamys* and a leporid; this combination suggesting a younger date.

Small mammals are well represented from the sequence Sherullah, Ghazgay, Molayan, Pul-e Charkhi, Dawrankhel, Hadji Rona (Brandy 1981; Sen 1983, 2001). The first three complexes (more than one site for some of these) span much of the Late Miocene. Pul-e Charkhi is earliest Pliocene, and Dawrankhel and Hadji Rona are younger. These sites indicate presence of both shrews and modern hedgehogs. Distinctive in their apparent absence are ctenodactylids and sciurids, as well as southern elements (diatomyids, thryonomyids). Absences of uncommon taxa (tree shrews, dormice, flying squirrels) must be evaluated against sample size. Their absences are consistent with drier habitat differences, but any of these could have been missed due to sampling effects. Bamboo rats of the genus *Rhizomyides* in the Afghan record may indicate significant moisture.

Total diversity is not great in the Afghan late Neogene, with only eight rodents seen in the rich Early Pliocene locality (Pul-e Charkhi). In all, the dominant group is the Murinae, including some species apparently related to Siwalik *Karnimata,* and others (*Saidomys*) showing affinity westward to North Africa. Also occurring are two Late Miocene gerbils, *Pseudomeriones* and *Abudhabia,* which show wide distribution. The former is known across Asia from China to the Mediterranean (Sen 1983), and the latter is recorded from North Africa to China, as well as South Asia (Qiu, Zheng, and Zhang 2004b; Flynn and Jacobs 1999). Their dispersal was probably primarily east–west at temperate latitudes. Important records are the porcupine *Hystrix aryanensis* from Molayan (Sen 2001) and Leporidae from several localities. If the age of Sherullah is correctly interpreted, then leporids arrived in Afghanistan before they are known in the Siwalik or Chinese records. That *Trischizolagus* is identified from Pul-e Charkhi may indicate earliest Pliocene affinity westward to southeastern Europe.

Thailand

Of the eight mammal assemblages reviewed from the Miocene of Thailand by Ducrocq et al. (1995), three contain rodents. Ban San Klang (Pong Basin, Middle Miocene) yielded a hedgehog and the diatomyine *Diatomys* sp. Had du Dai (Lampang Basin) produced a hedgehog, bat, squirrel, dormouse, the murine *Potwarmus,* and *Diatomys.* Both sites indicate a moist habitat and Oriental biogeographic affinity.

The Mae Long locality (Li Basin) yielded the richest assemblage by far (see table 18.2). Mein and Ginsburg (1997) documented a didelphid, five insectivores including a mole, nine bats, a tree shrew, a loris, and a tarsier. Note the absence of lagomorphs. The eight rodent species are comparable at the generic level to late Early–Middle Miocene assemblages from Pakistan. These include *Ratufa maelongensis,* ?*Atlantoxerus* sp., *Diatomys liensis, Democricetodon kaonou, Spanocricetodon janvieri, Prokanisamys benjavuni, Potwarmus thailandicus,* and a platacanthomyid assigned as *Neocometes orientalis.* The paleoenvironment was a tropical forest along a very shallow lake (Mein and Ginsburg 1997). Taxa such as the tree shrew, primates, insectivore genera *Hylomys, Neotetracus,* and *Scapanulus,* and *Diatomys* indicate the Oriental Biogeographic Province. *Prokanisamys* and platacanthomyids also appear to prefer tropical habitat.

Chaimanee et al. (2007) added fauna from the Mae Moh Basin (*Neocometes* and *Prokanisamys*) and noted new data that suggest that all of these deposits are Middle Miocene in age, not late Early Miocene. Ducrocq et al. (1995) had suggested an age of 16 Ma, while Mein and Ginsburg (1997) preferred an age of around 18 Ma. Wessels (2009) supported the latter age assignment because of primitive characters in *Potwarmus thailandicus,* which is interpreted as more primitive than *Potwarmus primitivus,* a species characterizing Potwar assemblages at 16 Ma.

Small mammal assemblages of Thailand share elements with both Siwalik and Yunnan faunas. Taxa held in common with the Siwaliks include Tupaiidae, Lorisidae, *Ratufa, Atlantoxerus, Diatomys, Democricetodon, Spanocricetodon, Prokanisamys, Potwarmus,* and possibly a platacanthomyid. Except for the last rodent and the tupaiid and loris, species in the two areas are close, although not identical.

China

Yunnan

Small mammals from the late Miocene localities of Lufeng and Yuanmou basins are well documented by Yunnan and IVPP scholars (Dong and Qi, chapter 11, this volume). These represent South China and the eastern portion of the Oriental Biogeographic Province. Miocene assemblages from the Yuanmou Basin antedate those of Lufeng, and paleomagnetic evidence places them in the range of 8.3 Ma to 7.2 Ma (Qi et al. 2006; Gradstein, Ogg, and Smith 2004 time scale). Small mammals of Yuanmou are diverse (Ni and Qiu 2002; Qiu 2006). There are tupaiids and insectivores including a hedgehog, mole, and shrews, diverse squirrels including volant forms and the tree squirrels *Ratufa, Callosciurus,* and *Tamiops,* plus *Dremomys* and *Sciurotamias.* There are platacanthomyids and bamboo rats, which indicate dense growth, as well as a beaver, hamster, porcupines, and two murids. The Lufeng assemblage includes two hedgehogs, a talpid, several shrews, a tupaiid, diverse squirrels, fossorial bamboo rats (*Miorhizomys*), other elements similar to Yuanmou species (hamster, murids, beaver, platacanthomyids, porcupine), and a new addition, the hare *Alilepus* (Storch and Qiu 1991; Qiu and Han 1986). Taxa in common with Siwalik assemblages include the small catarrhines, tupaiids, tree squirrels noted previously, *Miorhizomys, Alilepus,* and *Hystrix.*

Like the faunal assemblages from Thailand, the Yunnan assemblages indicate a moist forested environment. The assemblages are distinct taxonomically from faunas of North China in the occurrence of tree shrews, some of the insectivores, diverse arboreal squirrels, platacanthomyids, and (until the end of the Miocene) bamboo rats. The distinctions are not absolute: *Rhizomys* entered the Yushe Basin (southern margin of North China) in the Late Neogene, and "northern" beavers penetrated Yunnan. Nonetheless, the Yunnan faunas illustrate a Late Miocene biogeographic pattern of contrasts to faunas north of the Yangtze River, but similarities to Thai faunas and, to a lesser extent, to Siwalik assemblages. Qiu et al. (chapter 1, this volume) discuss faunal correlations and

aspects of biotic provinces in China, noting that the position of the boundary between North and South China has not been static throughout the Neogene. However, we endorse the reality of the Oriental Biogeographic Province through the Neogene and see commonalities among small mammals with the faunas of Pakistan and India.

North China

Various chapters of this volume characterize the small mammal faunas of North China, and the data are summarized by Qiu et al. (chapter 1, this volume). Northern faunas are dominated by hedgehogs and moles, ochotonid lagomorphs, diverse dipodoids, primitive muroids that are replaced progressively through the Miocene, squirrels (but mostly different genera than in Yunnan) and dormice, beavers, and eomyids. Among the more progressive muroids, the zokors (myospalacines) radiate from the Late Miocene onward and set northern assemblages apart distinctly. Most of the taxa above differ at high levels from Oriental Province components. A few genera are shared with West Asian and even South Asian areas, attesting to long-distance dispersal.

During the early Neogene, a relatively open connection to the west is apparent, with a number of elements shared widely. Most prominent is west–east exchange with almost no north–south exchange. Beginning with the late Early Miocene, some connection to southern areas is in evidence, and at the same time, exchange to the west decreased, probably due to tectonic obstructions (higher elevations of the mountain ranges: Pivnik and Wells 1996; Métais et al. 2009). During the Late Miocene, North China saw the distinctive immigration of the murine *Progonomys,* possibly from the west. The gerbils *Abudhabia* and *Pseudomeriones* attest to communication across Asia, possibly via the Siwaliks in the case of *Abudhabia,* but the direction of dispersal is unknown. It is possible that these gerbils evolved in Asia from myocricetodontine stock. Many Pliocene members of the North China assemblages appear to be Holarctic in affinity, especially the arvicolines, beavers, hamsters, and mice.

Appearances Due to Miocene Dispersal Events

Complementing the records of India and Pakistan, small mammal faunas from elsewhere in Asia and Africa have become well-enough known to recognize exotic introductions into the Indian Subcontinent. They are exotic because preceding assemblages in the area of interest lack close outgroups with common ancestry. These exotics can be interpreted as immigrants, but fossil records are generally too weak to specify place of origin and route of immigration. The following discussion focuses on a number of appearance events that represent dispersal, and we explore direction of dispersal where that is feasible.

Neogene faunal exchange from and into the Indian Subcontinent was limited. Rodent species diversity increased gradually during the Early Miocene, reaching a peak around 13–14 Ma (Flynn et al. 1998), after which the diversity decreased to less than 10 rodent species at the end of the Miocene (see table 18.1). This increase is due largely to evolution within the biogeographic region, and secondarily to immigration. In the earliest Miocene, the modern cricetids *Democricetodon*, *Primus*, and *Spanocricetodon* appear in the fossil record, followed by myocricetodontines at around 20 Ma. Their origin is unclear, quite possibly East Asian, but the pathway of arrival is undemonstrated. *Ratufa*, another Early Miocene appearance, is an example of an element shared broadly

throughout the Oriental Biogeographic Province, and it may have spread to the subcontinent by range extension. Figure 18.2 notes some of these key immigrants; its emphasis is dispersal, not distribution.

Early in the Miocene outside South Asia, rhizomyines (*Prokanisamys*) migrated to Saudi Arabia and northern Africa together with a myocricetodontine, an early murid (*Potwarmus*), and the ctenodactyline *Sayimys* (see figure 18.2 [westward dispersal]). Somewhat later, some of these elements entered Anatolia. In the late Early Miocene within the Indian Subcontinent, the thryonomyid *Kochalia* and an ochotonid lagomorph near *Alloptox* appeared in the record. *Kochalia* is thought to be an immigrant from Africa, where its outgroup resided (Flynn and Winkler 1994), but the ochotonid likely indicates invasion from northern Asia. Somewhat later, near the early Miocene–middle Miocene boundary, *Sayimys* dispersed into northern China, but whether from the south or west is undemonstrated.

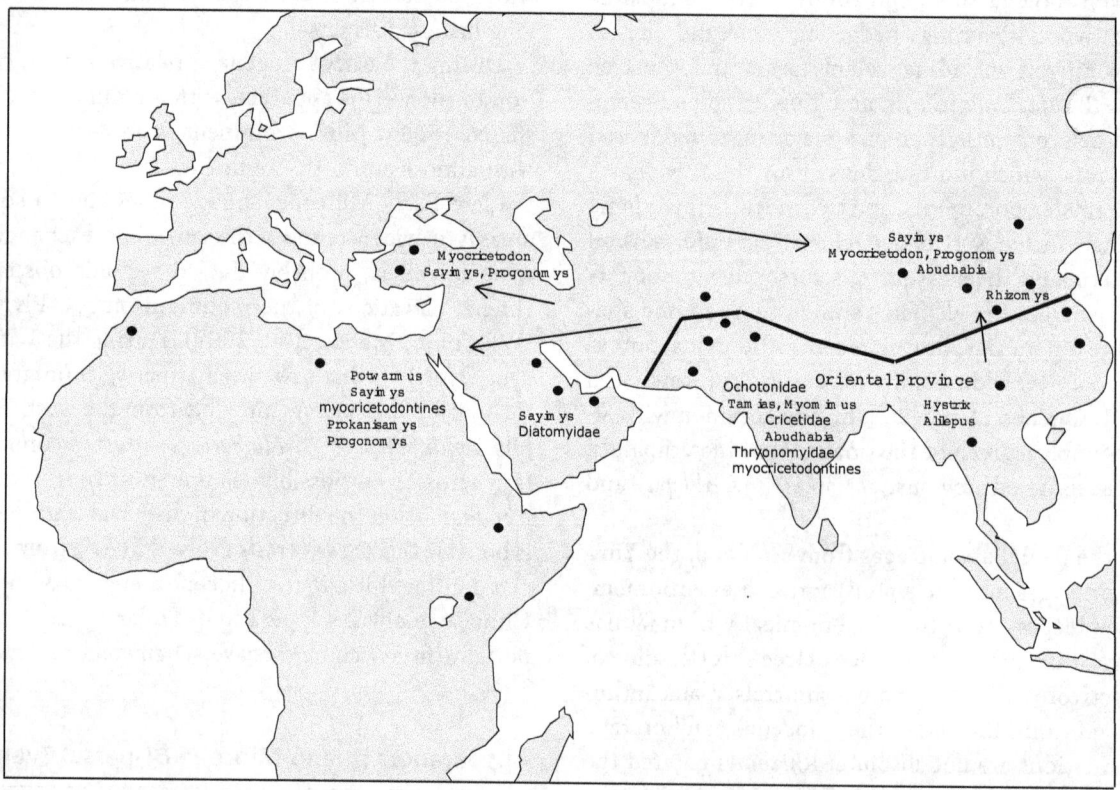

Figure 18.2 The zoogeographic setting of South Asian localities. Miocene faunas of the Indian Subcontinent represent the western extent of the Oriental Biogeographic Province of the time. This map uses figure 18.1 as its base and adds key immigrant taxa; it does not attempt to show distributions of taxa, but rather emphasizes dispersal events. Some immigrants for the Indian Subcontinent are noted, but they do not arrive simultaneously, and point of origin for most is uncertain; see the text for data on likely time and direction of introduction for some of these. For select immigrants to North China, western Asia, the Arabian Peninsula, and Africa, direction of introduction is indicated with some confidence. There is a low level of interchange between Africa and the Indian Subcontinent, and relatively little north–south interchange within Asia. Most immigrants to North China, other than *Rhizomys*, likely arrived by a route from temperate West Asia, rather than from South China.

The high Middle Miocene rodent diversity in Pakistan includes a few immigrants. Examples are glirids (*Myomimus*) and chipmunks (*Tamias*), either from northern or western sources. The successful murids evolving in southern Asia produced *Progonomys*, which became a hallmark migrant of the Late Miocene into western Asia and Europe (see figure 18.2). This genus and *Myocricetodon* migrated to China somewhat later, possibly from the west.

At the end of the Miocene, the Siwalik species richness decreased, but exotic elements continued to appear. The gerbil *Abudhabia* appeared in the Siwaliks shortly after 9 Ma, later than the first record in China, and therefore quite possibly entered the Indian Subcontinent via a northeastern route. *Hystrix* appeared in the Siwalik record about 8 Ma, followed by the leporid *Alilepus*, but their dispersal path is not yet clear. Both elements are shared with Late Miocene faunas of Yunnan and therefore reflect spread throughout the Oriental Province. The center of evolution of modern Rhizomyinae was the Oriental Province. It was not until the latest Miocene that *Rhizomys* expanded northward into North China.

CONCLUSION

The Neogene of the Indian Subcontinent shows a high degree of faunal similarity throughout that region, with faunal similarity also through time, despite significant turnover, and increasing seasonality of climate in the later Miocene. Evidence from Yunnan, Thailand, and Myanmar shows that many small mammal genera are shared throughout southern Asia. This supports the existence of the Oriental Biogeographic Province seen by Qiu et al. (chapter 1, this volume) from the Early Miocene onward. At times, the boundary of the Oriental Province extended northward—for example, in the late Early Miocene to the Korean Peninsula (where *Neocometes* has been found; Lee and Jacobs 2010) and Kyushu (where *Diatomys* occurred; Kato and Otsuka 1995).

Faunal differences to the north of the Oriental Province are significant, both with North China and with Afghan basins, the latter only 200 km away from Siwalik localities. This reflects the importance of mountain ranges in defining biogeographic boundaries (Heintz and Brunet 1982). Our analysis reveals faunal distinctions west of Pakistan, with few elements shared by the Siwalik faunas and Anatolia, for example. Arabia and Africa as the Ethiopian Biogeographic Province also are mostly distinct, but these areas record appearances of South Asian elements throughout the Neogene. Dispersal to Africa is balanced by dispersal from Africa to the Indian Subcontinent by diverse taxa, including catarrhines and thryonomyids. Faunal similarity appears to be marginally greater among low-latitude areas of Africa and southern Asia than with more temperate regions to the north, which illustrates the distinction of the temperate Greco-Irano-Afghan faunas (de Bonis et al. 1992) and (through time) the Pikermian chronofauna (Eronen et al. 2009).

While the origin and direction of dispersal of taxa can be specified in ideal circumstances (Flynn and Winkler 1994), data are too sparse to specify this for others. Where Early Miocene *Democricetodon* and myocricetodontines originated is uncertain. *Democricetodon* appeared almost simultaneously in various parts of Asia without clear indication of origin by older outgroups. Myocricetodontines do not appear to have originated in Africa, but rather in the Oriental Province; the genus *Myocricetodon*, however, is a later migrant to Anatolia and China, possibly spreading there from lower latitudes. With respect to the Indian Subcontinent, a few elements appear to have dispersed there from temperate parts of Asia, either by a western or an eastern route. Early examples are Ochotonidae, the chipmunk *Tamias*, and, in the Middle Miocene, Gliridae. Early ctenodactylines spread out of South Asia, but the pathway of arrival into China for *Sayimys* is unknown, possibly from the south, or quite reasonably from the west, after dispersal to Anatolia (see figure 18.2). Chinese *Progonomys*, too, may have dispersed from the west rather than from the south. Late Miocene *Abudhabia* presents an interesting pattern of appearance in China well before its record in the Siwaliks; it could have entered South Asia from the north or west. Only latest Miocene *Rhizomys* appears to show a clear expansion northward from the Oriental Province.

The composition of Miocene Siwalik fossil assemblages is similar to that of coeval faunas of southeastern Asia, despite differences. While the paleo-Oriental Province was subtropical and moist in general, the western extent in Pakistan may have been somewhat more seasonal in climate, yielding minor faunal differences (absence of tarsiers, fewer and less diverse insectivores, rare platacanthomyids; Flynn 2003). These differences suggest subprovincial faunal variation. The fauna of the Pakistan Siwaliks included thryonomyids and ctenodactylines, which today are African, and tree shrews, varied tree squirrels, and bamboo rats, which today are restricted eastward from peninsular India. Among the bamboo rats, Tachyoryctini (African today) coexisted with the fossorial Rhizomyini. The interplay of evolution and immigration led to Miocene coexistence of groups that today have disjunct distributions. The Neogene history of the Pakistan Siwaliks is that of a regional variant of the

Oriental Biogeographic Province, punctuated by introductions from western (Ethiopian) as well as northern sources, and experiencing more greatly increased seasonality of climate than that affecting southeastern Asia.

ACKNOWLEDGMENTS

We have synthesized this study from a great deal of work published by many colleagues who conduct research in Asia. We appreciate their pioneering efforts. The ideas presented here grew out of discussions with David Pilbeam, John Barry, Michèle Morgan, Zhuding Qiu, Zhanxiang Qiu, and Richard H. Tedford—namely, that the Oriental Biogeographic Province can be traced through time, and that as fossil records mature, it will become increasingly feasible to stipulate when and from what source immigrants appear in the biostratigraphic record. Sevket Sen, Everett Lindsay, and Louis Jacobs kindly supplied information used in this analysis. We thank Xiaoming Wang and Mikael Fortelius for their editorial efforts, and sincerely thank Hans de Bruijn for his numerous helpful comments on the manuscript.

REFERENCES

Abels, H. A. 2008. Long period orbital climate forcing: Cyclostratigraphic studies of Cenozoic continental and marine succession in Europe. *Geologica Ultraiectina* 297:1–178.

Barry, J. C., M. E. Morgan, L. J. Flynn, D. Pilbeam, A. K. Behrensmeyer, S. Mahmood Raza, A. Imran Khan, C. Badgley, J. Hicks, and J. Kelley. 2002. Faunal and environmental change in the late Miocene Siwaliks of northern Pakistan. *Paleobiology Memoir* 3:1–71.

Baskin, J. A. 1996. Systematic revision of Ctenodactylidae (Mammalia, Rodentia) from the Miocene of Pakistan. *Palaeovertebrata* 25:1–49.

Black, C. C. 1972. Review of fossil rodents from the Neogene Siwalik beds of India and Pakistan. *Palaeontology* 15(2):238–266.

Brandy, L. D. 1981. Rongeurs muroïdés du Néogène supérieur d'Afghanistan. Évolution, Biogéographie, Corrélations. *Palaeovertebrata* 11(4):133–179.

Cautley, P. T. 1868. On the structure of the Sewalik Hills, and the organic remains found in them. In *Fauna Antiqua Sivalensis*, ed. H. Falconer, pp. 30–42. Palaeontological Memoirs 1 (series ed. C. Murchison). London: Robert Hardwicke.

Cheema, I. U., S. M. Raza, L. J. Flynn, A. R. Rajpar, and Y. Tomida. 2000. Miocene small mammals from Jalalpur, Pakistan, and their biochronologic implications. *Bulletin of the National Science Museum, Tokyo* C26(1–2):57–77.

Cheema, I. U., S. Sen, and L. J. Flynn. 1983. Early Vallesian small mammals from the Siwaliks of northern Pakistan. *Bulletin du Muséum National d'Histoire Naturelle* 4(5)C3:267–286.

Chaimanee, Y., C. Yamee, B. Marandat, and J.-J. Jaeger. 2007. First Middle Miocene rodents from the Mae Moh Basin (Thailand): Biochronological and paleoenvironmental implications. *Bulletin of Carnegie Museum of Natural History* 39:157–163.

Clift, P. D., K. V. Hodges, D. Heslop, R. Hannigan, H. V. Long, and G. Calves. 2008. Correlation of Himalayan exhumation rates and Asian monsoon intensity. *Nature Geosciences* 1:875–880.

Colbert, E. H. 1933. Two new rodents from the lower Siwalik beds of India. *American Museum Novitates* 633:1–6.

Colbert, E. H. 1935. Siwalik mammals in the American Museum of Natural History. *Transactions American Philosophical Society* 27:1–401.

de Bonis, L., M. Brunet, E. Heintz, and S. Sen. 1992. La province greco-irano-afghane et la répartition des faunes mammaliennes au Miocène supérieur. *Paleontologia y Evolució* 24–25:103–112.

de Bruijn, H. 1986. Is the presence of the African Thryonomyidae in the Miocene deposits of Pakistan evidence for fauna exchange? *Proceedings of the Koninklijke Nederlandse Akademie van Wetenschappen* B89(2):125–134.

de Bruijn, H. 1999. A late Miocene insectivore and rodent fauna from the Baynunah Formation, Emirate of Abu Dhabi, United Arab Emirates. In *Fossil Vertebrates of Arabia*, ed. P. J. Whybrow and A. Hill, pp. 186–197. New Haven: Yale University Press.

de Bruijn, H., E. Boon, and S. T. Hussain. 1989. Evolutionary trends in *Sayimys* (Ctenodactylidae, Rodentia) from the Lower Manchar Formation (Sind, Pakistan). *Proceedings of the Koninklijke Nederlandse Akademie van Wetenschappen* B92(3):191–214.

de Bruijn, H. and S. Hussain. 1984. The succession of rodent faunas from the lower Manchar Formation, southern Pakistan, and its relevance for the biostratigraphy of the Mediterranean Miocene. *Paléobiologie Continentale* 14(2):191–204.

de Bruijn, H., S. Hussain, and J. J. M. Leinders. 1981. Fossil rodents from the Murree Formation near Banda Daud Shah, Kohat, Pakistan. *Proceedings of the Koninklijke Nederlandse Akademie van Wetenschappen* B84(1):71–99.

de Bruijn, H. and P. J. Whybrow. 1994. A late Miocene rodent fauna from the Baynunah Formation, Emirate of Abu Dhabi, United Arab Emirates. *Proceedings of the Koninklijke Nederlandse Akademie van Wetenschappen* B97(3):407–422.

Ducrocq, S., Y. Chaimanee, V. Suteethorn, and J.-J. Jaeger. 1995. Mammalian faunas and the ages of the continental Tertiary fossiliferous localities from Thailand. *Journal of Southeast Asian Earth Sciences* 12:65–78.

Eronen, J. T., M. Mirzaie Ataabadi, A. Micheels, A. Karme, R. L. Bernor, and M. Fortelius. 2009. Distribution history and climatic controls of the Late Miocene Pikermian chronofauna. *Proceedings of the National Academy of Sciences* 106(20):11867–11871.

Flynn, L. J. 1982. Systematic revision of Siwalik Rhizomyidae (Rodentia). *Geobios* 15:327–389.

Flynn, L. J. 2000. The Great Small Mammal Revolution. *Journal of Himalayan Geology* 21:39–42.

Flynn, L. J. 2003. Small mammal indicators of forest paleoenvironment in the Siwalik deposits of the Potwar Plateau, Pakistan. DEINSEA 10:183–195.

Flynn, L. J. 2007. Origin and evolution of the Diatomyidae, with clues to their paleoecology from the fossil record. *Bulletin of Carnegie Museum of Natural History* 39:173–181.

Flynn, L. J., J. C. Barry, W. Downs, J. A. Harrison, E. H. Lindsay, M. E. Morgan, and D. Pilbeam. 1997. Only ochotonid from the Neogene of the Indian Subcontinent. *Journal of Vertebrate Paleontology* 17(3):627–628.

Flynn, L. J., W. Downs, M. E. Morgan, J. C. Barry, and D. Pilbeam. 1998. High Miocene species richness in the Siwaliks of Pakistan. In *Advances in Vertebrate Paleontology and Geochronology*, ed. Y. Omida, L. J. Flynn, and L. L. Jacobs, pp. 167–180. National Science Museum Monograph 14. Tokyo: National Science Museum.

Flynn, L. J. and L. L. Jacobs. 1999. Late Miocene small mammal faunal dynamics: The crossroads of the Arabian Peninsula. In *Vertebrate fossils of the Arabian Peninsula*, ed. P. Whybrow and A. Hill, pp. 410–419. New Haven: Yale University Press.

Flynn, L. J. and A. J. Winkler. 1994. Dispersalist implications of *Paraulacodus indicus*, a south Asian rodent of African affinity. *Historical Biology* 9:223–235.

Flynn, L. J., A. Winkler, L. Jacobs and W. Downs. 2003. Tedford's gerbils from Afghanistan. In *Vertebrate Fossils and Their Context: Contributions in Honor of Richard H. Tedford*, ed. L. J. Flynn, pp. 603–624. *Bulletin of the American Museum of Natural History* 279.

Geraads, D. 1998. Rongeurs du Miocène supérieur de Chorora (Ethiopie): Cricetidae, Rhizomyidae, Phiomyidae, Thryonomyidae, Sciuridae. *Palaeovertebrata* 27(3–4):203–216.

Gradstein, F. M., J. G. Ogg, and A. G. Smith. 2004. *A Geologic Time Scale 2004*. Cambridge: Cambridge University Press.

Heintz, E. and M. Brunet. 1982. Une barrière géographique entre le sous-continent Indien et l'Eurasie occidentale pour les faunes continentales du Miocène supérieur. *Comptes Rendus de l'Académie des Sciences* 2(294):477–480.

Hinton, M.A.C. 1933. Diagnoses of new genera and species of rodents from the Indian Tertiary deposits. *Annals and Magazine of Natural History* 10:620–622.

Jacobs, L. L. 1978. Fossil rodents (Rhizomyidae and Muridae) from Neogene Siwalik deposits, Pakistan. *Museum of Northern Arizona Press Bulletin* 52:1–103.

Jacobs, L. L., L. J. Flynn, and W. R. Downs. 1989. Neogene rodents of southern Asia. In *Papers on Fossil Rodents in Honor of Albert Elmer Wood*, ed. C. C. Black and M. R. Dawson, pp. 157–178. Los Angeles: Natural History Museum of Los Angeles County.

Jacobs, L. L. and W. Downs. 1994. The Evolution of murine rodents in Asia. In *Rodent and Lagomorph Families of Asian Origins and Diversification*, ed. Y. Tomida, C. Li, and T. Setoguchi, pp. 149–156. National Science Museum Monograph 8. Tokyo: National Science Museum.

Jaeger, J.-J. 1977a. Rongeurs (Mammalia, Rodentia) du Miocène de Beni-Mellal. *Palaeovertebrata* 7(4):91–125, 2 pls.

Jaeger, J.-J. 1977b. Les rongeurs du Miocène moyen et supérieur du Maghreb. *Palaeovertebrata* 8(1):1–166, 7 pls.

Jaeger, J.-J., H. Tong, E. Buffetaut, and R. Ingavat. 1985. The first fossil rodents from the Miocene of northern Thailand and their bearing on the problem of the origin of the Muridae. *Revue de Paléobiologie* 4(1):1–7.

Kato, T. and H. Otsuka. 1995. Discovery of the Oligo-Miocene rodents from west Japan and their geological and paleontological significance. *Vertebrata PalAsiatica* 33:315–329.

Lee, J.-N. and L. Jacobs. 2010. The platacanthomyine rodent *Neocometes* from the Miocene of South Korea and its paleobiogeographical implications. *Acta Palaeontologica Polonica* 55(4):581–586.

Li, C.-k. 1974. A probable geomyoid rodent from middle Miocene of Linchu, Shantung. *Vertebrata PalAsiatica* 12:43–53.

Li, C.-k. 1977. A new cricetodont rodent of Fangshan, Nanking. *Vertebrata PalAsiatica* 15:67–75.

Lindsay, E. H. 1988. Cricetid rodents from Siwalik deposits near Chinji Village. I: Megacricetodontinae, Myocricetodontinae, and Dendromurinae. *Palaeovertebrata* 18:95–154.

Lindsay, E. H., L. J. Flynn, I. U. Cheema, J. C. Barry, K. F. Downing, A. R. Rajpar, and S. Mahmood Raza. 2005. Will Downs and the Zinda Pir Dome. *Palaeontologia Electronica* 8(1)19A:1–19, 1MB; http://palaeo-electronica.org/2005_1/lindsay19/lindsay19.pdf.

López-Antoñanzas, R. 2009. First *Potwarmus* from the Miocene of Saudi Arabia and the early phylogeny of murines (Rodentia: Muroidea). *Zoological Journal of the Linnean Society* 156:664–678.

López-Antoñanzas, R. 2011. First diatomyid rodent from the Early Miocene of Arabia. *Naturwissenschaften* 98:117–123.

López-Antoñanzas, R. and S. Sen. 2003. Systematic revision of Mio-Pliocene Ctenodactylidae (Mammalia, Rodentia) from the Indian Subcontinent. *Eclogae Geologicae Helvetiae* 96:521–529.

López-Antoñanzas R. and S. Sen. 2004. Ctenodactylids from the Lower and Middle Miocene of Saudi Arabia. *Palaeontology* 47:1477–1494.

López-Antoñanzas R. and S. Sen. 2005. New species of Paraphiomys (Rodentia, Thryonomyidae) from the Lower Miocene of As-Sarrar, Saudi Arabia. *Palaeontology* 48:223–233.

López-Antoñanzas R. and S. Sen. 2006. New Saudi Arabian Miocene jumping mouse (Zapodidae): Systematics and phylogeny. *Journal of Vertebrate Paleontology* 26:170–181.

Lydekker, R. L. 1884. Rodents and new ruminants from the Siwaliks and synopsis of Mammalia, *Paleontographica Indica* 10:134–185.

Maridet, O., W.-y. Wu, J. Ye, S.-d. Bi, X.-j. Ni, and J. Meng. 2011. Earliest occurrence of *Democricetodon* in China, in the Early Miocene of the Junggar Basin (Xinjiang, China) and comparison with the genus *Spanocricetodon*. *Vertebrata PalAsiatica* 49(4):393–405.

Marivaux, L., M. Vianey-Liaud, and J.-L. Welcomme. 1999. Première découverte de Cricetidae (Rodentia, Mammalia) oligocènes dans le synclinal sud de Gandoï (Bugti Hills, Balochistan, Pakistan). *Comptes Rendus de l'Académie des Sciences de la terre et des planètes* 329:839–844.

Marivaux, L. and J.-L. Welcomme. 2003. New diatomyid and baluchimyine rodents from the Oligocene of Pakistan (Bugti Hills, Balochistan): Systematic and paleobiogeographic implications. *Journal of Vertebrate Paleontology* 23(2):420–434.

Mein, P. and L. Ginsburg. 1985. Les rongeurs Miocènes de Li (Thaïlande). *Comptes Rendus de l'Académie des Sciences* 301(19):1369–1374.

Mein, P. and L. Ginsburg. 1997. Les mammifères du gisement Miocène inférieur de Li Mae Long, Thaïlande: systématique, biostratigraphie, et paléoenvironement. *Géodiversitas* 19(4):783–844.

Mein, P., L. Ginsburg, and B. Ratanasthien. 1990. Nouveaux rongeurs du Miocène de Li (Thaïlande). *Compte Rendus de l'Académie des Sciences*, 2nd ser., 310:861–865.

Métais, G., P.-O. Antoine, S.R.H. Hassan, J.-Y. Crochet, D. De Franceschi, L. Marivaux, and J.-L. Welcomme. 2009. Lithofacies, depositional environments, regional biostratigraphy and age of the Chitarwata Formation in the Bugti Hills, Balochistan, Pakistan. *Journal of Asian Earth Sciences* 34:154–167.

Munthe, J. 1980. Rodents of the Miocene Daud Khel Local Fauna, Mianwali District, Pakistan. Part II. Sciuridae, Gliridae, Ctenodactylidae and Rhizomyidae. *Milwaukee Public Museum, Contributions in Biology and Geology* 34:1–36.

Ni, X.-j. and Qiu Z.-d. 2002. The micromammalian fauna from Leilao, Yuanmou hominoid locality: Implications for biochronology and paleoecology. *Journal of Human Evolution* 42:535–546.

Patnaik, R. 2001. Late Pliocene micromammals from Tatrot Formation (Upper Siwaliks) exposed near village Saketi, Himachal Pradesh, India. *Palaeontographica*, part A, 261:55–81.

Pillans, B., M. Williams, C. Cameron, R. Patnaik, J. Hogarth, A. Sahni, J. C. Sharma, F. Williams, and R. L. Bernor. 2005. Revised correlation of the Haritalyangar magnetostratigraphy, Indian Siwaliks: Implications for the age of the Miocene hominids *Indopithecus* and *Sivapithecus*, with a note on a new hominid tooth. *Journal of Human Evolution* 48:507–515.

Pivnik, D. A. and N. A. Wells 1996. The transition from Tethys to the Himalaya as recorded in northwest Pakistan. *Bulletin of the Geological Society of America* 108(10):1295–1313.

Qi, G.-q., W. Dong, L. Zheng, L. Zhao, F. Gao, L.-p. Yue, and Y.-x. Zhang. 2006. Taxonomy, age and environment status of the Yuanmou hominoids. *Chinese Science Bulletin* 51(6):704–712.

Qiu, Z.-d. 2001. Glirid and gerbillid rodents from the middle Miocene Quantougou Fauna of Lanzhou, Gansu. *Vertebrata PalAsiatica* 39(4):297–305.

Qiu, Z.-d. 2006. Small mammals. In *Lufengpithecus hudiensis Site*, ed. G. q. Qi and W. Dong, pp. 113–131, 308–318. Beijing: Science Press.

Qiu, Z.-d. and D.-f. Han. 1986. Fossil Lagomorpha from the hominoid locality of Lufeng, Yunnan. *Acta Anthropologica Sinica* 5(1):41–53.

Qiu, Z.-d., S.-h. Zheng, and Z.-q. Zhang. 2004a. Gerbillids from the Late Miocene Bahe Formation, Lantian, Shaanxi. *Vertebrata PalAsiatica* 42(7):193–204.

Qiu, Z.-d., S.-h. Zheng, and Z.-q. Zhang. 2004b. Murids from the Late Miocene Bahe Formation, Lantian, Shaanxi. *Vertebrata PalAsiatica* 42(1):67–76.

Saner, S., K. Al-Hinai, and D. Perincek. 2005. Surface expressions of the Ghawar structure, Saudi Arabia. *Marine and Petroleum Geology* 22:657–670.

Sen, S. 1983. Rongeurs et Lagomorphes du gisement Pliocène de Pul-e Charki, bassin de Kabul, Afghanistan. *Bulletin du Museum National d'Histoire Naturelle*, 5th ser., 5C(1):33–74.

Sen, S. 2001. Rodents and Insectivores from the Late Miocene of Malayan in Afghanistan. *Palaeontology* 44:913–932.

Sen, S. and H. Thomas. 1979. Découverte de rongeurs dans le Miocène moyen de la Formation Hofuf (Province du Hasa, Arabie Saoudite). *Comptes Rendus sommaires de la Société Géologique de France* 1:4–37.

Storch, G. and Z.-d. Qiu. 1991. Insectivores (Mammalia: Erinaceidae, Soricidae, Talpidae) from the Lufeng hominoid locality, Late Miocene of China. *Geobios* 24(5):601–621.

Theocharopoulos, K. D. 2000. Late Oligocene–Middle Miocene *Democricetodon*, *Spanocricetodon*, and *Karydomys* n.gen. from the eastern Mediterranean area. *Gaia* 89:1–116

Thomas, H. 1983. Les Bovidae (Artiodactyla, Mammalia) du Miocène Moyen de la Formation Hofuf (Province du Hasa, Arabie Saoudite). *Palaeovertebrata* 13:157–206.

Thomas, H., S. Sen, M. Khan, B. Battail, and C. Ligabue. 1982. The Lower Miocene fauna of Al-Sarrar (Eastern Province, Saudi Arabia). *ATLAL, Journal of Saudi Arabian Archeology* 4:109–136.

Thomas, H., P. Taquet, G. Ligabue, and C. Del'Agnola. 1978. Découverte d'un gisement de vertébrés dans les dépots continentaux du Miocène Moyen du Hasa (Arabie Saoudite). *Compte Rendu sommaire des séances de la Société géologique de France* 2:69–72.

Tong, H. and J.-J. Jaeger. 1993. Muroid rodents from the Middle Miocene Fort Ternan locality (Kenya) and their contribution of the phylogeny of muroids. *Palaeontographica*, part A, 229(1–3):51–73.

Ünay, E., H. de Bruijn, and G. Sarac. 2003. A preliminary zonation of the continental Neogene of Anatolia based on rodents. In *Distribution and Migration of Tertiary Mammals in Eurasia*, ed. J. W. F. Reumer and W. Wessels, pp. 539–548. DEINSEA 10.

Wessels, W. 1996. Myocricetodontinae from the Miocene of Pakistan. *Proceedings of the Koninklijke Nederlandse Akademie van Wetenschappen* 99:253–312.

Wessels, W. 2009. Miocene rodent evolution and migration: Muroidea from Pakistan, Turkey, and northern Africa. *Geologica Ultraiectina* 307:1–290.

Wessels, W., H. de Bruijn, S. T. Hussain, and J. J. M. Leinders. 1982. Fossil rodents from the Chinji formation, Banda Daud Shah, Kohat, Pakistan. *Proceedings of the Koninklijke Nederlandse Akademie van Wetenschappen* B85(3):337–364.

Wessels, W., O. Fejfar, P. Pelaez-Campomanes, A. J. van der Meulen, and H. de Bruijn. 2003. Miocene small mammals from Jebel Zelten, Libya. In *Surrounding Fossil Mammals: Dating, Evolution and Paleoenvironment*, ed. N. López-Martínez, P. Peláez- Campomanes, and M. Hernández Fernández, pp. 699–715. *Coloquios de Paleontologia*, Volumen Extraordinario 1, en Homenaje al Dr. Remmert Daams.

Whybrow, P. J., M. E. Collinson, R. Daams, A. W. Gentry, and H. A. McClure. 1982. Geology, fauna (Bovidae, Rodentia) and flora of the early Miocene of eastern Saudi Arabia. *Tertiary Research* 4(3):105–120.

Whybrow, P. J., A. Hill, W. Yasin al-Tikriti, and E. A. Hailwood. 1990. Late Miocene primate fauna, flora and initial palaeomagnetic data from the Emirate of Abu Dhabi, United Arab Emirates. *Journal of Human Evolution* 19:583–588.

Winkler, A. J., C. Denys, and D. M. Avery. 2010. Rodentia. In *Cenozoic Mammals of Africa*, ed. L. Wedelin and W. J. Sanders, pp. 263–304. Berkeley: University of California Press.

Wood, A. E. 1937. Fossil rodents from the Siwalik beds of India. *American Journal of Science* 36:64–76.

Zachos, J., M. Pagani, L. Sloan, E. Thomas, and K. Billups. 2001. Trends, rhythms, and aberrations in global climate 65 Ma to Present. *Nature* 292:686–693.

Ziegler, M. A. 2001. Late Permian to Holocene Paleofacies evolution of the Arabian Plate and its hydrocarbon occurrences. *GeoArabia* 6(3):445–504.

Chapter 19

Advances in the Biochronology and Biostratigraphy of the Continental Neogene of Myanmar

OLIVIER CHAVASSEAU, AUNG AUNG KHYAW, YAOWALAK CHAIMANEE, PAULINE COSTER, EDOUARD-GEORGES EMONET, AUNG NAING SOE, MANA RUGBUMRUNG, SOE THURA TUN, AND JEAN-JACQUES JAEGER

Chronology has always played a critical role in the perspective of interpreting the fossil record. In the Cenozoic, fossil mammals have long demonstrated their particular usefulness for establishing chronological frameworks for paleontologists and geologists. The Neogene is a particularly rich period that witnessed the appearance of several extant mammalian groups (e.g., giraffids, bovids, suids, murids, cervids) and major evolution in some preexisting groups (e.g., hominoids, rhinocerotids, proboscideans). In southern Asia, numerous Neogene paleontological challenges are in need of answers. For mammals, the evolution of the hominoid primates in this region of the world is perhaps the most important of them. The hominoids seem to have radiated in southern Asia in the Middle Miocene, which led to the appearance of several genera (e.g., *Sivapithecus*, *Indopithecus*, *Khoratpithecus*, *Lufengpithecus*). However, the history of this group is still poorly known. While additional fossils are indispensable to progress, a solid chronological framework is also necessary to allow testing of the evolutionary hypotheses related to this group and, in a more general way, to all mammalian taxa. The Indian Subcontinent possesses a very good Neogene record in the Siwaliks of Pakistan, India, and Nepal. The chronology of this record is robust overall, especially in Pakistan, where it is extremely well constrained. Nevertheless, the advanced paleontological knowledge in the Indian Subcontinent is not enough to resolve the Neogene evolution of mammals at the scale of Southern Asia. Southeast Asia represents a vast area in which major evolutionary events have probably occurred.

The discovery of *Khoratpithecus* in Thailand and Myanmar, with its implications for the evolution of the Asian hominoids (Chaimanee et al. 2003, 2004; Jaeger et al. 2011), is an excellent example. While the study of the Neogene of Thailand suffers from the lack of outcrops (most of the sites are located in mines), Myanmar possesses extensive Neogene exposures (Bender 1983). Moreover, Myanmar is intermediate in longitude between the Indian Subcontinent and Thailand and has a boundary with China. Hence, the fossil record of Myanmar is of crucial importance to the aim of understanding mammalian biochronology and biostratigraphy in Southern Asia as well as improving correlations between southern Asia and China.

MYANMAR GEOLOGY AND PALEONTOLOGY

Greatly simplifying the geology of Myanmar, three structural units can be recognized along a west–east cross section (at intermediate latitudes) of the country (figure 19.1):

- The Indo-Burmese ranges: situated at the western margin of the country, they are an accretion wedge formed by the westerly subduction of the ocean crust of the Bengal Basin (Socquet et al. 2002). The Indo-Burmese ranges are mainly Early Tertiary rocks (Bender 1983).
- The Central Basin: a large syncline containing Tertiary sedimentary rocks.

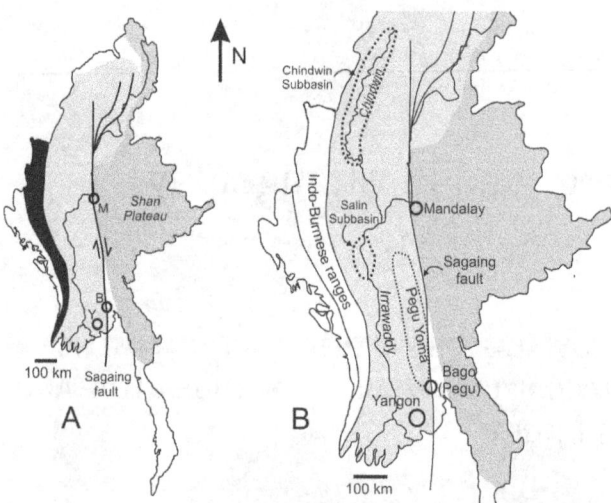

Figure 19.1 (*A*) The three main geological terrains of Myanmar: (*black*) Indo-Burmese ranges; (*light gray*) Central Basin; (*dark gray*) eastern unit (Shan Plateau). Y = Yangon; B = Bago; M = Mandalay. (*B*) Enlarged map of Myanmar displaying details of the Central Basin. The sub-basin of Salin, which contains major Irrawaddy Formation outcrops, is noted.

- The eastern unit: mainly the Shan Plateau, which forms the Shan-Thai tectonic block together with the western part of Thailand. The Shan Plateau is made of sedimentary, metasedimentary, and metaintrusive rocks. The Plateau is separated from the Central Basin by the large dextral, strike-slip Sagaing fault.

The Central Basin of Myanmar was probably opened in the lower Eocene by transtension induced by India-Asia subduction. Its hydrographic network is mainly composed of the Irrawaddy River (Ayeyarwady) and the Chindwin River in the northwestern part of the basin. The Central Basin rocks are divided into different units ranging from the Eocene to the Pleistocene. The Eocene to Middle Miocene sediments were deposited predominantly under marine conditions. A Miocene marine regression led to the development of continental detritus sedimentation in the Central Basin, which is documented by extensive outcrops. Although variable, the transition between the Neogene marine Pegu Group and the continental Irrawaddy Formation is sometimes marked by an angular unconformity accompanied by a ferruginous paleosol at the base of the latter (Chhibber 1934; Pascoe 1964; Bender 1983).

Originally named the "Fossil Wood Group" in the nineteenth century because of their abundant silicified wood remains (Theobald 1873), these continental deposits have been baptised under various appellations including Irrawaddy Group (e.g., Bender 1983) and Irrawaddy

Formation (e.g., Khin and Win 1969; Khin and Myitta 1999). The term "Irrawaddy Formation" is employed here. It is worth noting, however, that our usage has a broader scope than what is classically attributed to a body of rock of formation status. This is particularly true in this chapter, where it represents both the typical Irrawaddy Formation series and sequences that are more probably northern equivalents of the Irrawaddy Formation with different sedimentological, chronological, and faunal characteristics.

Since the nineteenth century, the Irrawaddy Formation has yielded fossil mammals (e.g., Buckland 1829). This fossiliferous content, which comprises notably proboscideans, suids, rhinocerotids, giraffids, and anthracotheres, has rapidly easily been recognized as Neogene. Nevertheless, a major concern regarding the Irrawaddy Formation fauna has been the inaccurate location of fossils (both in stratigraphy and space) and the lack of formal descriptions. Consequently, fossils from various horizons have often been taken into consideration together. Such imprecision is well exemplified by early faunal lists established for the formation that were often given globally, ignoring problems linked to the stratigraphy and the time span of the formation (table 19.1).

Subsequent authors have proposed a more detailed vision of the faunal succession of the formation. Instead of considering the fossils as belonging to a single faunal unit, some authors (e.g., Stamp 1922; Colbert 1938, 1943) have proposed a separation of two faunal units based on the fossil record of Yenangyaung (see the following): the faunas of the "Lower Irrawaddy" and the "Upper Irrawaddy." This treatment, which can be considered as a primary biostratigraphical subdivision of the Irrawaddy Formation, is still adopted in recent literature (e.g. Takai et al. 2006).

Used as a biostratigraphical unit, the Lower Irrawaddy is very imprecise since it encompasses the totality of the Late Miocene for Bender (1983) and most of both the Late Miocene and Pliocene according to Takai et al. (2006). The faunal list of the latter authors (table 19.2) reflects their vision of the Lower Irrawaddian with a very long temporal extension. Species that can be found in their list include *Tetraconodon minor*, a likely early Late Miocene suid (Pickford 1988), hippopotamids, which are classically thought to appear in southern Asia in the second half or the end of the Late Miocene (e.g., Boisserie and Lihoreau 2006), and taxa such as *Sivachoerus prior*, a suid known from the Pliocene of the Siwaliks of Pakistan and India (Pickford 1988).

It is likely that the long temporal extension of the lower Irrawaddy is connected with the problems of dating the Yenangyaung fauna. The age and composition of this fauna, whose taxa represent a significant proportion of the list of the "Lower Irrawaddy" fauna (see table 19.2),

Table 19.1

Four Early Faunal Lists for the Irrawaddy Formation

Buckland (1829)	Theobald (1873)
Mastodon latidens	*Mastodon latidens*
Mastodon elephantoides	*Elephas cliftii*
Hippopotamus sp.	*Mastodon elephantoides*
Rhinoceros sp.	*Rhinoceros* sp.
Sus sp.	*Equus* sp.
Tapirus sp.	*Hippopotamus (Hexapro-*
Bos sp.	*todon) irravadicus*
Cervus sp.	*Merycopotamus dissimilis*
Antelope sp.	*Sus* sp.
Crocodylus sp. aff. *vulgaris*	*Tapirus* sp.
Leptorhynchus sp. (*Garialis* sp.)	*Bos* sp.
Trionyx sp.	*Cervus* sp.
Emys sp.	*Antelope* sp.
	Crocodylus sp.
	Leptorhynchus sp.
	Emys sp.
	Trionyx sp.
	Colossochelys sp.

Noetling (1895)	Chhibber (1934)
Mastodon latidens	*Mastodon latidens*
Stegodon cliftii	*Mastodon elephantoides*
Acerotherium perimense	*Stegodon clifti*
Rhinoceros sivalensis	*Hippopotamus sivalensis*
Hippopotamus irravadicus	*Hippopotamus irravaticus*
Sus titan	*Rhinoceros sivalensis*
Bubalus sp.	*Aceratherium lydekkeri*
Boselaphus sp.	*Tetraconodon minor*
Hippotherium antelopinum	*Merycopotamus dissimilis*
Cervus sp.	*Vishnutherium iravaticum*
Lutra (?) sp.	*Hydaspitherium birmanicum*
Crocodylus cf. *bicorpatus*	*Cervus* sp.
Gavialis sp. cf. *gangeticus*	*Taurotragus latidens*
Emyda palaeindica	*Hipparion punjabiense*
Trionyx sp.	*Colossochelys atlas*
Colossochelys atlas	*Gharialis gangeticus*
Testudo sp.	*Gharialis leptodus*
Emys sp.	*Carcharodon* sp.
Carcharias sp.	

NOTE: Buckland (1829) and Noetling (1895) identified fossils collected around Yenangyaung without further stratigraphic indications. The list of Theobald (1873) was supposed by Noetling (1895) to arise from the identification of the fossils collected during the expedition to Ava (Oldham, in Yule 1858:343, gives no indication of location for these fossils). The faunal list given by Chhibber (1934), probably compiled from the literature, is global and lacks geographic and stratigraphic data. Taxon names preserve original spelling as published.

Table 19.2

Faunal List of the "Lower Irrawaddian" as Given by Takai et al. (2006)

Taxon	Listed in Yenangyaung (source)
Proboscidea	
Stegodontidae	
Stegolophodon latidens	Yes (N, Pa)
Stegolophodon stegodontoides	No
Stegodon sp.	No
Stegodon elephantoides	Yes (C)
Gomphotheriidae	
Sinomastodon sp.	No
Artiodactyla	
Suidae	
Tetraconodon minor	Yes (B, C, Pa, Pi)
Tetraconodon cf. *magnus*	No
Tetraconodon cf. *intermedius*	No
Sivachoerus prior	No
Parachleuastochoerus sp.	No
Propotamochoerus hysudricus	No
Anthracotheriidae	
Merycopotamus dissimilis	Yes (B, Pa)
Hippopotamidae	
Hexaprotodon iravaticus	Yes (N, N2)
Hexaprotodon sivalensis	No
Giraffidae	
Hydaspitherium birmanicum	Yes (B, Pa)
Vishnutherium iravaticum	Yes (B, Pa)
Bovidae	
Pachyportax latidens	Yes (B, Pa)
Proleptobos birmanicum	Yes (B)
Hemibos sp.	No
Perissodactyla	
Rhinocerotidae	
Aceratherium lydekkeri	Yes (B, N, Pa, S)
Brachypotherium sp.	No
Equidae	
Hipparion antelopinum	Yes (B, C, N, Pa, S)

SOURCES: The taxa identified in Yenangyaung are indicated as follows: B = Bender (1983); C = Colbert (1938); N = Noetling (1895); N2 = Noetling (1897); Pa = Pascoe (1964); Pi = Pilgrim (1926); S = Stamp (1922).

has long been debated. The fossils of Yenangyaung have also often been considered as belonging to a unique faunal unit. For instance, Pilgrim (1910) mixed taxa from the Lower and Upper Irrawaddy of Yenangyaung according to Colbert (1938). Table 19.1 also provides evidence of the faunal mixing made by Noetling (1895). Other authors (e.g., Stamp 1922; Colbert 1938) have recognized the existence at Yenangyaung of two faunal sets. The first, the Lower Irrawaddy fauna, comprises for instance "*Hipparion*," the rhinocerotid "*Aceratherium*," the suid *Tetraconodon*, and the giraffids *Vishnutherium* and *Hydaspitherium*. The second fauna, the Upper Irrawaddy assemblage, comprises the proboscideans *Stegolophodon latidens*, *Stegodon elephantoides*, and hippopotamids (*Hippopotamus* or *Hexaprotodon* depending on the authors). Even with the recognition of two assemblages, two ages were proposed for the Yenangyaung faunas:

1. Pilgrim (1927:160) and Colbert (1938:277) defended an "Upper Pontian" or "Post-Pontian" age for the oldest assemblage, which would correspond in modern sense to a latest Miocene/Pliocene age. The youngest assemblage was considered Lower Pleistocene by Colbert (1938).

2. Stamp (1922) proposed alternative ages for both faunas. According to him, the first assemblage would be equivalent in age to the Nagri faunas of the Siwaliks of Pakistan (the Nagri Formation is now approximately 11.2–9.0 Ma, the younger boundary transgressing time greatly; Barry et al. 2002). Stamp dated the second fauna by correlation to the Pliocene Tatrot fauna of the Siwaliks of Pakistan, which is now dated between 3.5 Ma and 3.3 Ma (Barry et al. 2002).

The age of the oldest Yenangyaung assemblage will be discussed in a later section on the basis of newly collected materials. As it is presently defined, the "Lower Irrawaddy" appears to be an imprecise biostratigraphical unit because it contains taxa that potentially document a temporal interval of several million years. This chapter aims, therefore, to first discuss the presently accepted biostratigraphical concept of the Irrawaddy Formation, and second, to improve the biochronology and biostratigraphy of the Irrawaddy Formation with new data acquired in the last years.

THE MIDDLE MIOCENE FAUNA OF THE IRRAWADDY FORMATION

Continental vertebrates dating from the Middle Miocene are scarce in Myanmar. A few taxa were reported from the freshwater deposits of the Pegu Group (e.g., Colbert 1938),

while faunal lists without formal description from supposed Middle Miocene deposits (e.g., south Chindwin Basin) were published by Bender (1983). Thus, additional data on the Middle Miocene faunas of Myanmar were needed to provide a better biochronological framework.

Following prospecting by Burmese academics of the University of Mandalay, a Myanmar-French paleontological team (composed of researchers from Montpellier 2, Yangon, and Pa-an Universities) surveyed Irrawaddy Formation outcrops north of Mandalay in 2002 and subsequently described a Middle Miocene assemblage (Chavasseau et al. 2006). The outcrops are situated in the Sagaing division near the village of Chaungtha, which is located on the west bank of the Irrawaddy River a few kilometers away from the city of Male and the pressure ridge of the Sagaing fault (figure 19.2). The fossils come from badlands southeast of Chaungtha formed by the incision of local tributaries of the Irrawaddy River into the Tertiary series. Two main spots ("Gyet Pyi Gye" and "Kangyi") separated by only 800 m have produced fossils. The Irrawaddian exposures, which are about 100 m thick, are mostly composed of silts and clays interstratified with rarer sandstone beds and were deposited under fluviatile conditions (Chavasseau et al. 2006).

Mammals

The fauna of the Chaungtha area has yielded only large mammals. The faunal assemblage is mainly formed by

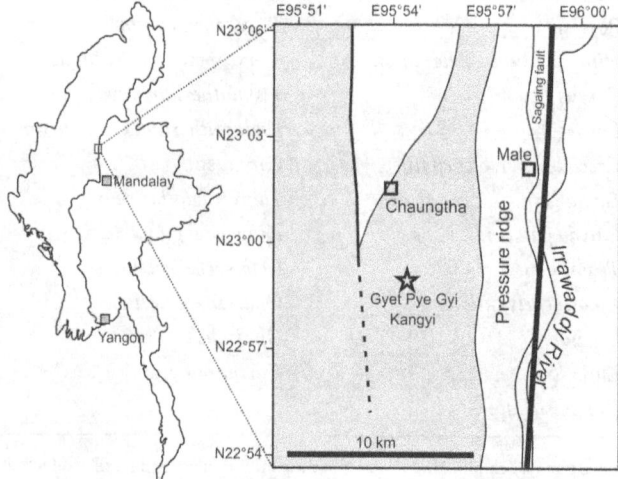

Figure 19.2 Detailed map of Myanmar showing the position of the Middle Miocene fossiliferous spots around the village of Chaungtha in the Sagaing division. Gray: Irrawaddy Formation outcrops. White: Pegu Group rocks (or equivalent). After Bender (1983).

the suid *Conohyus thailandicus*, the rhinocerotid *Brachy-potherium fatehjangense*, two tragulids referable to the genus *Siamotragulus*, and a low-crowned bovid (Chavasseau et al. 2006). No hipparionine specimens have been discovered so far. Other mammals discovered in 2002 include a gomphothere and a second, indeterminate rhinocerotid. Recent fieldwork (figure 19.3) extended the large mammal faunal list to 10 taxa (table 19.3) with the discovery of a chalicothere, a viverrid, and an indeterminate anthracothere.

Chronology

Although scarce, the mammal remains of the Chaungtha area are critical for understanding of the biochronology of the Central Basin of Myanmar. The mammalian association is most likely Middle Miocene or perhaps earliest Late Miocene in age: *Conohyus thailandicus*, known from northern Thailand (e.g., Ducrocq et al. 1997), is very close to *Conohyus sindiensis*, whose range is dated between 14.0 Ma and 10.3 Ma in Pakistan (Barry et al. 2002). In addition, *Brachypotherium fatehjangense* is present in the Miocene of the Indian Subcontinent, where it possesses a long temporal range. The material conforms metrically to the Middle Miocene material of Pakistan. The tragulids are close to *Siamotragulus*, a genus only known from the Middle Miocene of Thailand, one of the two species representing possibly a species described in Thailand (*S. sanyathanai*). Combined with the notable absence of "*Hipparion*," the fauna of the Chaungtha area can be bracketed between 14 Ma and 11 Ma.

The fauna of the Chaungtha area demonstrates that continental sedimentation in the north of the Central Basin of Myanmar was already initiated in the second half of the Middle Miocene and confirms previous assertions (Bender 1983) of a Middle Miocene age for the beginning of the Irrawaddy Formation. The discovery of this Middle Miocene fauna undermines the use of the term "Lower Irrawaddy."

Figure 19.3 Photograph of the Middle Miocene Irrawaddy Formation outcrops of the Chaungtha area taken at Gyet Pye Gyi. Pauline Coster (indicated by an arrow) gives the scale.

Table 19.3

Faunal List (Mammalian Taxa Only) of the Chaungtha Area

Proboscidea
   Gomphotheriidae
      Gen. et sp. indet.
Perissodactyla
   Rhinocerotidae
      Brachypotherium fatehjangense
      Gen. et sp. indet.
   Chalicotheriidae
      Chalicotheriinae indet.
Artiodactyla
   Suidae
      Conohyus thailandicus
   Anthracotheriidae
      Gen et sp. indet.
   Bovidae
      Low-crowned form
   Tragulidae
      cf. *Siamotragulus sanyathanai*
      cf. *Siamotragulus* sp.
Carnivora
   Viverridae
      Gen. et sp. indet.

SOURCE: Updated from Chavasseau et al. (2006) by January 2009 fieldwork.

LATE MIOCENE FAUNAS OF MYANMAR

Since 2006, a Myanmar–French paleontological team composed of academics from Dagon and Yangon Universities, the Department of Archeology of the National Museum of Mandalay (a part of the Ministry of Culture of Myanmar), the University of Poitiers, and the Department of Mineral Resources (Thailand) has surveyed the Irrawaddy Formation deposits of the Salin sub-basin. This sub-basin is situated in the province of Magway (Magway Division) and is bisected by the Irrawaddy River. In this region, the Irrawaddy Formation deposits crop out on both sides of the Irrawaddy, the best exposures being preserved around the cities of Yenangyaung and Chauk. It is to be noted that these areas have been the focus of many of the historical paleontological reports about the Irrawaddy Formation (e.g., Noetling 1895, 1897; Stamp 1922; Pilgrim 1927), and in particular that the Yenangyaung outcrops have been considered as the "type region" of the formation.

New Data on the Fauna of Yenangyaung

The type region of the Irrawaddy Formation lies within a 25 km × 8 km rectangle of NNW orientation on the east bank of the Irrawaddy River. The north end of this rectangle contains the city of Yenangyaung and its oilfield (figure 19.4). Structurally, this area is an anticline of NNW–SSE axis with the oil-bearing Pegu Group rocks of its core exposed at the level of the oilfield and unconformably covered by the Irrawaddy Formation sequence (Bender 1983).

The exposed Pegu Group rocks are attributed to the Miocene-age Obogon Formation. These beds were deposited in a marine environment, except at the top of the formation where deposits yielding the corbiculid bivalves "*Batissa (Cyrena) crawfurdi*" and "*Batissa (Cyrena) petrolei*" are observable (Noetling 1895, 1900, 1901; Pilgrim 1904; Pascoe 1908). The "*Batissa*" layer, a very soft white clay of about 15 m thickness (figure 19.5), characterizes a brackish environment (e.g., Bender 1983). The softness is due to the presence of kaolin according to Stamp (1922) and Pascoe (1964).

The base of the Irrawaddy Formation is marked by a meter-thick ferruginous paleosol resting unconformably on the Obogon Formation (see figure 19.5). This "red bed" has been noticed since the nineteenth century (e.g., Chhibber 1934; Pascoe 1964). Over this bed abruptly begins a classical fluvial sequence for the Irrawaddy For-

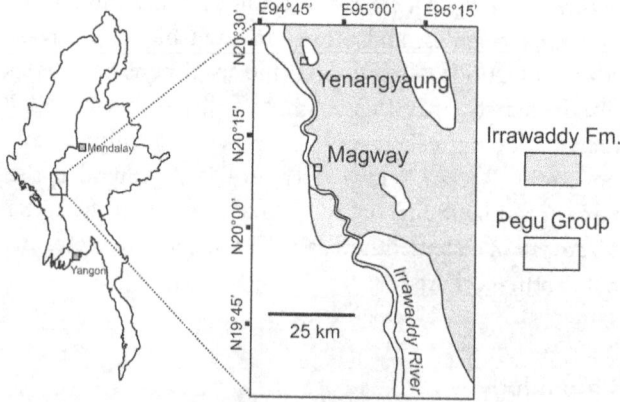

Figure 19.4 Map of the Magway Province showing the Irrawaddy outcrops around the cities of Magway and Yenangyaung. After Bender (1983).

mation in the Salin sub-basin. Soft sandstones or sometimes sands containing gypsum interstratified with harder, decimeter-thick sandstones and ferruginous conglomerates. Most of the fossils found were in a thick ferruginous conglomerate layer (1–2 m thick) situated a few meters over the base of the formation. This level probably yielded a significant part of the nineteenth- and early-twentieth-century "Lower Irrawaddy" fauna because descriptions of Oldham (in Yule 1858:315) or Pascoe (1908:143) clearly indicate that the present area of study has long been prospected by paleontologists.

The section at Yenangyaung demonstrates a gradual transition from marine to freshwater environment. The hiatus between the marine/brackish deposits of the Obogon Formation and the fluvial Irrawaddy Formation was certainly short considering the Middle to Late Miocene age attributed to the Obogon Formation (Bender 1983; Khin and Myitta 1999). Thus, this change in depositional environment reasonably preceded 10 Ma (age of the fauna of the basal beds of the Irrawaddy Formation at Yenangyaung, as demonstrated in the next section) and was latest Middle Miocene or early late Miocene. The transition between marine and fluvial sedimentary environments is diachronous and variably expressed at the scale of the Central Basin. The brackish water facies is often absent, and the red paleosol, which demonstrates an episode of emersion/erosion, overlaps the Pegu Group in angular unconformity (or disconformity) and is also frequently absent (Pascoe 1964; Bender 1983).

The rather gradual transition between marine and fluvial sediments in the Central Basin contrasts noticeably

Figure 19.5 Photograph looking toward the north of the Pegu Group-Irrawaddy Formation transition at the place name "Minlindaung" (southeast of Yenangyaung). (*A*) White clays containing the bivalve "*Batissa*" (Pegu Group); (*B*) ferruginous paleosol marking the beginning of the Irrawaddy Formation; (*C*) sands or weakly coherent sandstones with a few resistant decimeter sandstone layers of the Irrawaddy Formation. The foreground represents the western side of the Yenangyaung anticline, while the series is repeated in the background on the eastern side of the fold.

to the evolution of paleoenvironments in the Tertiary sequences of Bugti and Zinda Pir in Pakistan. Both sequences there show a more abrupt transition with a major unconformity between marine and overlying clastic sediments (e.g., Métais et al. 2009).

The thickness of the Irrawaddy Formation has been variously interpreted in the literature. Chhibber (1934) reports thickness estimations between 4000 feet and 20,000 feet, while Bender (1983) indicates a thickness of more than 2 km. In the type region of Yenangyaung, the thickness of the formation has often been estimated at several thousands of feet from the heart of the anticline to the border of the Irrawaddy River (e.g., 4500 feet for Colbert 1938). Our observation of strongly dipping strata on the west flank of the anticline confirms this important thickness. Nevertheless, the thickness of the Irrawaddy Formation available for study is far less in most other places: Khin and Myitta (1999) documented a thickness of 386 m in the Irrawaddy Formation sequence of the Bago (Pegu) Yoma (see figure 19.1). Our data also testify to a much thinner series in Chaungtha and in the

surroundings of Magway as a result of low strata dip and limited outcrops.

Faunal Assemblage

Recent mammalian findings in Yenangyaung only include large mammals (table 19.4). Hipparionin teeth and postcranial remains are common in the sequence. Two pigs, the tetraconodont *Tetraconodon minor*, for which Yenangyaung is the type locality, and the suine *Propotamochoerus hysudricus*, a common taxon in the Late Miocene of the Indian Subcontinent (Pilgrim 1926; Pickford 1988), were also discovered. In addition, two anthracotheres are part of the fauna. One of them is *Microbunodon milaensis*, an early Late Miocene taxon originating from Pakistan (Lihoreau, Blondel et al. 2004). The other, only known by fragmentary remains, represents an indeterminate species of the genus *Merycopotamus*, known in South Asia from the Middle Miocene to the Pleistocene (Lihoreau et al. 2007). Remains of rhinocerotids, bovids, and a tragulid complete the

Table 19.4

Taxa Recovered by the Myanmar-French Paleontological Team Since 2006 at the Base of the Irrawaddy Formation in Yenangyaung

---

Mammalia

Perissodactyla
  Rhinocerotidae
    Gen. et sp. indet.
  Equidae
    "*Hipparion*" sp.
Artiodactyla
  Suidae
    Propotamochoerus hysudricus
    Tetraconodon minor
  Anthracotheriidae
    Microbunodon milaensis
    Merycopotamus sp.
  Bovidae
    Several indet. species
  Tragulidae
    Gen. et sp. indet.

Reptilia

Testudines
  Trionychidae
    Trionyx sp.
Crocodilia
  Crocodilidae
    Gen. et sp. indet.

---

mammalian faunal list. The rhinocerotid remains are probably in part distinct from "*Aceratherium lydekkeri*" of older faunal lists, a species synonymized with *Brachypotherium perimense* (Zin-Maung-Maung-Thein et al. 2010) since a recently collected p2 has a spur-like paraconid unlike *B. perimense*.

Age of the Yenangyaung Fauna

The presence of hipparionin fossils in the fauna demonstrates that it is obviously Late Miocene in age since the representatives of this group have first appearances of 10.7 Ma both in Pakistan (Barry et al. 2002) and Turkey (Kappelman et al. 2003). The species *Propotamochoerus hysudricus* and *Microbunodon milaensis* have known ranges in the Siwaliks of Pakistan of 10.2–7.1 Ma and 10.3–9.2

Ma, respectively (Barry et al. 2002; Lihoreau, Blondel, et al. 2004). Moreover, *Tetraconodon minor* is a close ally of *Tetraconodon magnus*, reported in a short time interval of the Siwaliks (10.0–9.3 Ma).

These new mammal fossils from Yenangyaung suggest that the base of the Irrawaddy Formation is most probably early Late Miocene in age. An estimate of 10–9 Ma is proposed for this fauna considering the available data. As mentioned earlier, Stamp (1922) had treated the Yenangyaung fossils as belonging to two distinct faunal sets, one of early Late Miocene and the other of Pliocene age. Conversely, Colbert (1938) and Pilgrim (1927) attributed a latest Miocene/Pliocene age to the oldest fauna and an Early Pleistocene age to the second assemblage. Thus, these new data give support to Stamp's age estimate for the fauna from the base of the Irrawaddy Formation.

Concerning the younger fauna of Yenangyaung, no remains have been found *in situ* by the Myanmar-French team. The villagers living at the border of the Irrawaddy River showed us fossils of this fauna (mainly the proboscidean *Stegodon elephantoides*), confirming the stratigraphical indications given by Stamp (1922:497). Taking into account the advanced morphology displayed by *Stegodon elephantoides* (Colbert 1938), there is little doubt the fauna is Plio-Pleistocene in age. With the present data, however, it is impossible to know whether this fauna dates from the Middle Pliocene or the Early Pleistocene.

The Fauna of Magway

In parallel with prospecting in the area of Yenangyaung, every year since 2006 the Myanmar–French team has surveyed the Irrawaddy Formation deposits around the city of Magway (see figure 19.4). Most of the outcrops are situated south of Magway in a rectangle of at least 50 km north–south and 20 km east–west. An area situated approximately 20 km southeast of Magway yielded a new fauna with an interesting potential for biostratigraphy. The deposits generally do not exceed 100 m in thickness, which contrasts with the 4000 to 20,000 feet attributed to the Irrawaddy Formation in the literature. The sediments in this area are most similar to those observed at the base of the Irrawaddy Formation in Yenangyaung: sands or weakly coherent sandstones with common cross-beddings are dominant along the section. These sandstones sometimes have a significant clay content. Clay-dominated or claystone layers occur but are rarer. The sandstones contain several conglomeratic levels that are rich in iron oxides. Fossils occur predominantly within the conglomerates and, to a lesser extent, within sandstones.

Faunal Assemblage

Nonmammalian fauna comprise, as is typical in the Irrawaddy Formation, numerous trionychid turtle remains, as well as crocodile teeth, skeletal fragments, and vertebrae. Teeth of a freshwater shark species are also common along the section.

Mammalian taxa include the equid *"Hipparion,"* whose remains are the most abundant in the deposits, and a proboscidean that is either a primitive member of the genus *Stegolophodon* or a species of *Tetralophodon* (table 19.5). Among artiodactyls, three pigs are present: the tetraconodont *Tetraconodon minor*, a very large suine similar to the Siwalik *Hippopotamodon sivalense*, and an indeterminate species of *Propotamochoerus*. A palaeochoerid that displays morphological affinities with *Schizochoerus gandakasensis* was also discovered. Other artiodactyls include the anthracothere *Merycopotamus medioximus*, a form described from Pakistan (Lihoreau, Barry, et al. 2004; Lihoreau et al. 2007), several species of bovids, a giraffid, and a tragulid. Perissodactyls are represented by a rhinocerotid and a chalicotheriine. Finally, a new species of hominoid, *Khoratpithecus ayeyarwadyensis*, has been discovered in this area (Jaeger et al. 2011).

Age of the Magway Fauna

The ubiquitous presence in the section of the equid *"Hipparion"* denotes, as in Yenangyaung, an undoubted Late Miocene age for the fauna. The association of two suids belonging to the Suinae subfamily also points toward a Late Miocene age, this group becoming common after the early Late Miocene. In Southern Asia, the Suinae representatives are *Hippopotamodon* and *Propotamochoerus*, an association that is extremely common in Pakistan after 10.2 Ma (Pickford 1988; Barry et al. 2002). Both genera seem to occur in the Magway outcrops. The anthracothere *Merycopotamus medioximus* indicates, as does *Microbunodon milaensis*, an early Late Miocene age. This taxon, which is considered as originating on the Indian Subcontinent, ranges from 10.4 Ma to 8.6 Ma in the Siwaliks of Pakistan (Lihoreau, Barry et al. 2004; Lihoreau et al. 2007). The suid *Tetraconodon minor*, a likely early Late Miocene species, is held in common with the Miocene Yenangyaung fauna. The *Schizochoerus* is close to *S. gandakasensis* known in Pakistan in the early

Table 19.5

Faunal List for Material Collected by the Myanmar-French Paleontological Team Since 2006 in the Irrawaddy Formation Outcrops South of the City of Magway

Mammalia

Primates
  Khoratpithecus ayeyarwadyensis
Proboscidea
  Stegolophodon/Tetralophodon sp.
Artiodactyla
  Suidae
    Tetraconodon minor
    cf. *Hippopotamodon sivalense*
    Propotamocheorus cf. *hysudricus*
  Paleochoeridae
    Schizochoerus sp.
  Tragulidae
    Gen. et sp. indet.
  Bovidae
    Antilopini indet.
    Gen. et sp. indet.
    Gen. et sp. indet.
  Anthracotheriidae
    Merycopotamus medioximus
Perissodactyla
  Equidae
    Hipparion (s.l.) sp.
  Rhinocerotidae
    Gen. et sp. indet.
  Chalicotheriidae
    Gen. et sp. indet.
  Carnivora
    Fam. et gen. indet.

Reptilia

Testudines
  Trionychidae
    Trionyx sp.
Crocodilia
  Crocodilidae
    Gen. et sp. indet.

Chondrichthyes

Carcharhiniformes
  Carcharhinidae
    Glyphis pagoda

(continued)

Table 19.5 (*continued*)

Ichthyes

Siluriformes
  Fam. et gen. indet.

Late Miocene (Pickford 1976), with an observed range of 10.1–8.7 Ma (Barry et al. 2002).

All together, these data suggest that the age of the Magway fauna is likely to be between 10.5–8.5 Ma. This age is close to that of the fauna of the base of the Irrawaddy Formation near Yenangyaung described previously. Both faunas show a *Merycopotamus*–"*Hipparion*"–*Tetraconodon minor* association. The degree of faunal similarity between the two areas is probably underestimated by the scarcity of the remains of some taxa. For instance, the *Merycopotamus* remains of Yenangyaung most likely belong to *M. medioximus*. The study of the bovid remains (several dental remains and horncores) might also reveal a higher similarity between the two faunas.

TURNOVER BETWEEN MIDDLE AND LATE MIOCENE FAUNAS OF MYANMAR

Comparisons among the Chaungtha, Yenangyaung, and Magway faunas reveals that marked faunal turnover occurred between the end of the Middle Miocene and the beginning of the Late Miocene. The recognition of a faunal change is obvious considering that there is no species in common between the pre-"*Hipparion*" fauna of Chaungtha and the "*Hipparion*" faunas of Yenangyaung and Magway. The faunal turnover/change is notably characterized by the following:

- The appearance of the equid "*Hipparion*."
- Among the suids, the replacement of the primitive tetraconodont *Conohyus* by the advanced tetraconodont *Tetraconodon* and at least two suine species including the well-known *Propotamochoerus hysudricus*.
- The appearance of the anthracothere *Merycopotamus medioximus*. According to Lihoreau, and colleagues (Lihoreau, Barry, et al. 2004; Lihoreau et al. 2007), *M. medioximus* evolved in the Indian Subcontinent from *M. nanus* during the very beginning of the Late Miocene.
- The appearance of the anthracothere *Microbunodon milaensis*. As with *Merycopotamus*, this species most

likely evolved from a Middle Miocene form known in the Indian Subcontinent, *Microbunodon silistrense* (Lihoreau, Blondel, et al. 2004).

When comparing the evolution of Burmese faunas with other Asian faunas, it appears that the faunal change observed in Myanmar is very similar to what is known in other parts of southern Asia. As indicated previously, the anthracothere turnover in Myanmar follows that of the Siwaliks of Pakistan. In Thailand, however, no Middle Miocene species of *Microbunodon* or *Merycopotamus* have been recorded, while the Late Miocene is marked by the immigration of these genera (Chaimanee et al. 2006; Lihoreau et al. 2007; Hanta et al. 2008). The suids also display similar turnover within southern Asia. The Middle Miocene suid assemblages are dominated by *Conohyus* in Pakistan (Pilgrim 1926; Pickford 1988), Thailand (Ducrocq et al. 1997), and probably in Myanmar. In the Late Miocene, the genera *Propotamochoerus* and *Hippopotamodon* became dominant in Pakistan (Pilgrim 1926; Pickford 1988), Myanmar, and Thailand (Chaimanee et al. 2006). For instance, *Propotamochoerus hysudricus* seems to display a pan–South Asian distribution. In addition, the "*Hipparion*" faunas of both Pakistan and Myanmar witness the appearance of *Tetraconodon*, and the genus may even be present in the Late Miocene of Thailand (Chavasseau 2008).

It seems thus that biogeographical patterns within Southern Asia, which are intimately linked to a Himalayan barrier during the Neogene, influenced the faunal successions of Myanmar. This hypothesis accords with that proposed by Lihoreau et al. (2007) on the basis of the evolution of the anthracothere *Merycopotamus*. According to these authors, the abrupt drop of the ocean level recorded around 11 Ma (Haq, Hardenbol, and Vall 1987) allowed a dispersal of *Merycopotamus* toward Southeast Asia. In a more general way, these authors stressed that the biogeographical events conditioned by the eustatic sea level change may have triggered the 10 Ma faunal turnover observed in Pakistan.

DISCUSSION

Faunal Successions in the Continental Neogene of Myanmar

Figure 19.6 sums up the age of the Irrawaddy Formation faunas discussed in this chapter with their most characteristic taxa. Our data indicate that the Chaungtha fauna is correlative to the Middle Miocene Chinji Formation of Pakistan and the Middle Miocene localities of northern Thailand. The Magway and Yenangyaung faunas are equivalent to the Nagri Formation and lower part of the Dhok Pathan Formation of the Potwar Plateau. For the Late Miocene, at least the *"Hipparion"* fauna of Magway partly overlaps chronologically with the *Khoratpithecus* fauna of northeastern Thailand with an age estimate of 9–7 Ma (Chaimanee et al. 2004) or 9–6 Ma (Chaimanee et al. 2006). In addition to these two faunal complexes, there is the Plio-Pleistocene assemblage described in the Salin sub-basin (e.g., Yenangyaung) by previous authors. The morphological traits of the stegodontid proboscideans (high number of lophs, developed cementodonty, sometimes high number of cusps per loph) as well as their association with high-crowned elephantids (Colbert 1938) are in ac-

cordance with a Plio-Pleistocene age for this younger fauna.

Figure 19.6 clearly illustrates that faunas from the second part of the Late Miocene and Pliocene are still unknown from the Irrawaddy Formation, the base of the formation having produced only an early Late Miocene assemblage in the region of Magway. Considering the importance of the Irrawaddy fluvial system, it is likely that faunal assemblages of this age have been preserved somewhere in the Central Basin. However, the lack of Late Miocene–Pliocene faunas in the Magway region suggests these faunas were not preserved there. Under this hypothesis, the discontinuity of the faunal succession reflects either an important gap driven by long intervals without accumulation of sediment or an erosional gap.

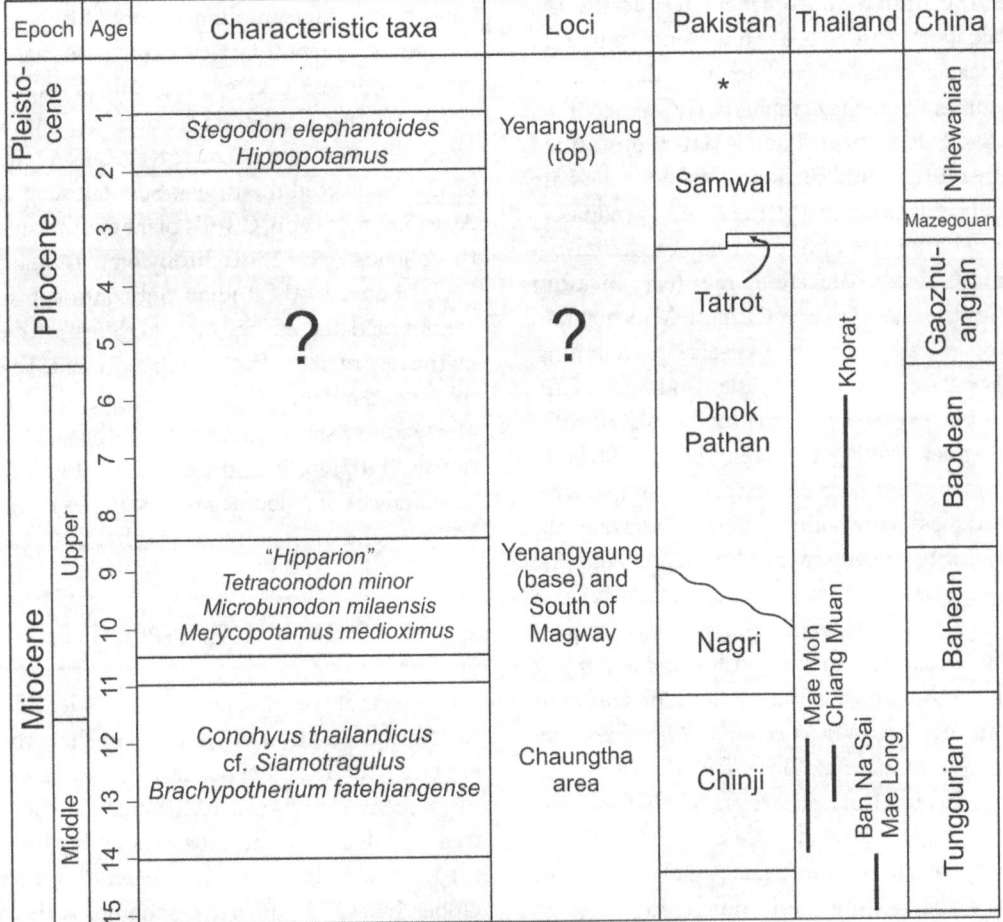

Figure 19.6 Summary of the biochronological assignment of the Irrawaddy Formation faunas and correlation with the stratigraphic record of the Potwar Plateau of the Siwaliks of Pakistan (after Barry et al. 2002) and with the fossil record of Thailand. After Chaimanee et al. (2006), Chavasseau (2008), and Coster et al. (2010). The Chinese mammal ages for the Neogene (after Deng 2006) are also indicated to facilitate comparisons. The Plio-Pleistocene Upper Irrawaddy fauna, which is not revised here, was placed approximately between 2 Ma and 1 Ma, given the Lower Pleistocene estimate of Colbert (1943).

Scaling the Burmese Faunas Within Asia

Importance of Biogeographical Patterns

When comparing the Burmese faunas to other Miocene faunas of Eurasia, strong affinities can be found with the faunas of Pakistan, India, and Thailand. Theses affinities are identified by the co-occurrence in Myanmar and Thailand/Pakistan/India of several genera and species such as *Tetraconodon*, *Brachypotherium fatehjangense*, *Conohyus thailandicus*, *Merycopotamus medioximus*, or *Microbunodon milaensis*. Common taxa date both from the Middle and Late Miocene, suggesting that large-scale species distributions and faunal dispersions were possible within southern Asia. The common "faunal pool" presented by South Asia offers interesting opportunities of biochronological and biostratigraphical correlations. Hence, the Irrawaddy Formation can be correlated to the Indian Subcontinent and Thailand, where numerous faunal and stratigraphic data are available. The Indian Subcontinent (especially the Siwaliks of Pakistan) represents a more robust source for correlations because of its long and precisely dated series, whereas data for Thailand are comparatively less accurate. The Miocene fossil record of Thailand is better constrained chronologically for the Middle Miocene than for the Late Miocene period, as documented by the Khorat sandpits.

Even if faunal affinities have been demonstrated between Thai and Chinese Miocene faunas (e.g., Pickford et al. 2004), Southeast Asian and Chinese Miocene faunas remain too different to allow precise correlations. Low similarity between Southeast Asian and South Chinese faunas can be explained in part by the lack of sufficient chronological overlap between the localities of these areas and by the lack of extensive comparative studies. Nevertheless, true faunal differences have probably existed owing to the presence in Yunnan of endemic and/or highly specialized taxa (e.g., *Lufengpithecus* and *Sinoadapis* for primates, *Mammut zhupengensis* for proboscideans, *Molarochoerus* for suids, *Yunnanotherium* for tragulids...) combined with the absence of common taxa in southern Asia (e.g., *Merycopotamus*, *Microbunodon*, *Tetraconodon*, *Conohyus*...). These faunal differences may have arisen due to dispersal barriers of relief and major rivers. These barriers depend more or less directly on tectonic activity resulting from the Himalayan uplift, which became important by the Middle Miocene (e.g., Molnar 2005). For instance, the southeastern margin of the Tibetan Plateau, which delimits the northern margin of Myanmar and northern Yunnan Province in China, knew significant uplift before 13 Ma, as deduced from rapid river incision between 13 Ma and 9 Ma (Clark et al. 2005).

Thus, the biogeographical specificity of southern Asia, which shows a tendency to provincialism relative to the rest of Asia, limits the possibilities of direct correlations. In a more general way, the case of Myanmar illustrates that biogeographical patterns are crucial when considering continental-scale correlations.

Precision of Dating and Diachrony

The dating of the Neogene faunas of Myanmar is presently based on correlations to other Asian faunas, but with few mammalian occurrences. This lack of data limits the precision of the correlation, as large faunal sets provide more robust and precise correlations (e.g., Alroy 1998). Hence, increasing the number of localities and completing the faunal lists is a priority to improve the precision of age control on the Neogene of Myanmar.

Another potential problem is the diachrony of mammalian events. Although a significant part of diachrony is caused by undersampling effects (Alroy 1998), climatic and environmental factors may cause real diachrony on taxon FADs and LADs at the scale of southern Asia. This hypothesis is considered here because, since the Late Miocene, different climatic and vegetation differences have probably affected the Indian Subcontinent and Southeast Asia (e.g., monsoon, C3/C4 plant equilibrium). If existent, this phenomenon of diachrony on mammalian FADs and LADs between the Indian Subcontinent and Myanmar cannot be detected, because the Burmese dating depends on the record of the Indian Subcontinent. To become more robust and independent, the dating of the Neogene faunas of Myanmar should benefit from the acquisition of additional stratigraphic and paleomagnetostratigraphic data. The analysis of paleomagnetic samples collected south of Magway and in Chaungtha has already begun.

CONCLUSION AND PERSPECTIVES

Three faunal assemblages can be identified in the Irrawaddy Formation of Myanmar. Their respective ages are late Middle Miocene, early Late Miocene, and Plio-Pleistocene (see figure 19.6). Our data suggest that the traditional use of the biostratigraphical subdivision of the Irrawaddy Formation between Lower Irrawaddy and Upper Irrawaddy does not encompass the Middle Miocene fauna of Chaungtha, which predates the Lower Irrawaddy. If conserved, the latter biostratigraphic unit should be extended and subdivided into two parts to take into account the discoveries in the Middle Miocene and lose its present ambiguity. Moreover, the lack of Latest Miocene–Pliocene faunas in Yenangyaung is in conflict with the accepted range of the Lower Irrawaddy. The Lower Irrawaddy may in fact only represent early late Miocene sediments in Yenangyaung, the type region of the Irrawaddy Formation. Hence, these data suggest that a complete redefinition of the biostratigraphy of Myanmar will be necessary. At present, however, because of insufficient stratigraphical data, a biochronological system is employed (see figure 19.6) and will serve as a basis for testing future biostratigraphical work.

ACKNOWLEDGMENTS

Many thanks to the organizers of the "Neogene Terrestrial Mammalian Biostratigraphy and Chronology in Asia" workshop and symposium in Beijing, 2009, for inviting the first author and for financial support to participate in this symposium. The language improvements and editing made by Lawrence Flynn on the manuscript and the useful comments of John Barry were greatly appreciated. This work would not have been possible without the cooperation of the Ministry of Culture of the Union of Myanmar. Camille Grohé and Stéphane Peigné are thanked for their preliminary comments on the Chaungtha carnivore. This study was funded by the ANR-BLANC-0235 program, the "Eclipse 2" program of CNRS and a grant of the Fyssen Foundation.

REFERENCES

Alroy, J. 1998. Diachrony of mammalian appearance events: Implications for biochronology. *Geology* 26:23–26.

Barry, J. C., M. E. Morgan, L. J. Flynn, D. Pilbeam, A. K. Behrensmeyer, S. Mahmood Raza, I. A. Khan, C. Badgley, J. Hicks, and J. Kelley. 2002. Faunal and environmental change in the Late Miocene Siwaliks of northern Pakistan. *Paleobiology Memoirs* 3:1–72.

Bender, F. 1983. *Geology of Burma*. Berlin: Gebrüder Borntraeger.

Boisserie, J.-R. and F. Lihoreau. 2006. Emergence of Hippopotamidae: new scenarios. *Comptes Rendus Palevol* 5:749–756.

Buckland, W. 1829. Geological account of a series of animal and vegetable remains and of rocks collected by T. Crawfurd Esq. on a voyage up the Irrawadi to Ava in 1826/29. *Transactions of the Geological Society of London*, 2nd ser., 2:377–392.

Chaimanee, Y., D. Jolly, M. Benammi, P. Tafforeau, D. Duzer, I. Moussa and J.-J. Jaeger. 2003. A Middle Miocene hominoid from Thailand and orangutan origins. *Nature* 422: 61–65.

Chaimanee, Y., V. Suteethorn, P. Jintasakul, C. Vidthayanon, B. Marandat, and J.-J. Jaeger. 2004. A new orang-utan relative from the Late Miocene of Thailand. *Nature* 427 439–441.

Chaimanee, Y., C. Yamee, P. Tian, K. Khaowiset, B. Marandat, P. Afforeau, C. Nemoz, and J.-J. Jaeger. 2006. *Khoratpithecus piri-*

yai, a Late Miocene hominoid of Thailand. *American Journal of Physical Anthropology* 131:311–323.

Chavasseau, O. 2008. Les faunes miocènes de grands mammifères d'Asie du Sud-Est: biochronologie et biogéographie. Ph.D. diss., Université Montpellier 2.

Chavasseau, O., Y. Chaimanee, S. Thura Tun, A. N. Soe, J. C. Barry, B. Marandat, J. Sudre, L. Marivaux, S. Ducrocq, and J.-J. Jaeger. 2006. Chaungtha, a new Middle Miocene mammal locality from the Irrawaddy Formation, Myanmar. *Journal of Asian Earth Sciences* 28:354–362.

Chhibber, H. L. 1934. *The Geology of Burma*. London: Macmillan.

Clark, M. K., M. A. House, L. H. Royden, B. C. Burchfiel, K. X. Whipple, X. Zhang, and W. Tang. 2005. Late Cenozoic uplift of southeastern Tibet. *Geology* 33:525–528.

Colbert, E. H. 1938. Fossil mammals from Burma in the American Museum of Natural History. *Bulletin of the American Museum of Natural History* 74:255–436.

Colbert, E. H. 1943. Pleistocene vertebrates collected in Burma by the American Southeast Asiatic Expedition. *Transactions of the American Philosophical Society* 32:395–429.

Coster, P., M. Benammi, Y. Chaimanee, O. Chavasseau, E.-G. Emonet, and J. J. Jaeger. 2010. A complete magnetic-polarity stratigraphy of the Miocene continental deposits of Mae Moh Basin, northern Thailand, and a reassessment of the age of hominoid-bearing localities in northern Thailand. *Geological Society of America Bulletin* 122:1180–1191.

Deng, T. 2006. Chinese Neogene mammal biochronology. *Vertebrata PalAsiatica* 44:143–163.

Ducrocq, S., Y. Chaimanee, V. Suteethorn, and J.-J. Jaeger. 1997. A new species of *Conohyus* (Suidae, Mammalia) from the Miocene of northern Thailand. *Neues Jahrbuch für Geologie und Paläontologie, Monatshefte* H.6: 348–360.

Hanta, R., B. Ratanasthien, Y. Kunimatsu, H. Saegusa, S. Nagaoka, and P. Jintasakul. 2008. A new species of Bothriodontinae, *Merycopotamus thachangensis* (Cetartiodactyla, Anthracotheriidae) from the Late Miocene of Nakhon Ratchasima, Northeastern Thailand. *Journal of Vertebrate Paleontology* 28:1182–1188.

Haq, B., J. Hardenbol, and P. Vail. 1987. Chronology of fluctuating sea levels since the Triassic. *Science* 235:1156–1167.

Jaeger, J.-J., A. Naing Soe, O. Chavasseau, P. Coster, E.-G. Emonet, F. Guy, R. Lebrun, A. Maung, A. Aung Khyaw, H. Shwe, S. Thura Tun, K. Linn Oo, M. Rugbumrung, H. Bocherens, M. Benammi, K. Chaivanich, P. Tafforeau, and Y. Chaimanee. 2011. First hominoid from the Late Miocene of the Irrawaddy Formation (Myanmar). *PLoS ONE* 6: e17065.

Kappelman, J., A. Duncan, M. Feseha, J.-P. Lunkka, D. Ekart, F. McDowell, T. M. Ryan, and C. C. Swisher III. 2003. Chronology. In *Geology and Paleontology of the Miocene Sinap Formation, Turkey*, ed. M. Fortelius, J. Kappelman, S. Sen, and R. Bernor, pp. 41–66. New York: Columbia University Press.

Khin, A. and K. Win. 1969. Geology and hydrocarbon prospects of the Burma Tertiary Geosyncline. *Union of Burma Journal of Science and Technology* 2:53–81.

Khin, K. and Myitta. 1999. Marine trangression and regression in Miocene sequences of northern Pegu (Bago) Yoma, Central Myanmar. *Journal of Asian Earth Sciences* 17:369–393.

Lihoreau, F., J. C. Barry, C. Blondel, and M. Brunet. 2004. A new species of Anthracotheriidae, *Merycopotamus medioximus* nov. sp.

from the Late Miocene of the Potwar Plateau, Pakistan. *Comptes Rendus Palevol* 3:653–662.

Lihoreau, F., J. C. Barry, C. Blondel, and Y. Chaimanee, J.-J. Jaeger and M. Brunet. 2007. Anatomical revision of the genus *Merycopotamus* (Artiodactyla; Anthracotheriidae): Its significance for Late Miocene mammal dispersal in Asia. *Palaeontology* 50:503–524.

Lihoreau, F., C. Blondel, J. C. Barry, and M. Brunet. 2004. A new species of the genus *Microbunodon* (Mammalia, Artiodactyla) from the Miocene of Pakistan: Phylogenetic relationships and palaeobiogeography. *Zoologica Scripta* 33:97–115.

Métais, G., P.-O. Antoine, S. R. H. Baqri, J.-Y. Crochet, D. De Franceschi, L. Marivaux, and J.-L. Welcomme. 2009. Lithofacies, depositional environments, regional biostratigraphy and age of the Chitarwata Formation in the Bugti Hills, Balochistan, Pakistan. *Journal of Asian Earth Sciences* 34:154–167.

Molnar, P. 2005. Mio-Pliocene growth of the Tibetan Plateau and evolution of East Asian climate. *Palaeontologia Electronica* 8:2A.

Noetling, F. 1895. The development and sub-division of the Tertiary system in Burma. *Records of the Geological Survey of India* 28:59–86.

Noetling, F. 1897. Note on a worn femur of *Hippopotamus irravadicus*, Caut. and Falc., from the Lower Pliocene of Burma. *Records of the Geological Survey of India* 60:243–249.

Noetling, F. 1900. The Miocene of Burma. *Verhandelingen der Koninklijke Akademie van Wetenschappen te Amsterdam, Afdeeling Natuurkunde*, 2nd ser., 7:1–131.

Noetling, F. 1901. Fauna of the Miocene beds of Burma. *Paleontologia Indica*, n. s., 1:1–378.

Pascoe, E. H. 1908. Marine fossils in the Yenangyaung oil-field, Upper Burma. *Records of the Geological Survey of India* 36:135–148.

Pascoe, E. H. 1964. *A Manual of Geology of India and Burma*. Calcutta: Government of India Press.

Pickford, M. 1976. A new species of *Taucanamo* (Mammalia: Artiodactyla) from the Siwaliks of the Potwar plateau, Pakistan. *Pakistan Journal of Zoology* 8:13–20.

Pickford, M. 1988. Revision of the Miocene Suidae of the Indian Subcontinent. *Münchner Geowissenschaftliche Abhandlungen Reihe A: Geologie und Paläontologie* 12:1–92.

Pickford, M., H. Nakaya, Y. Kunimatsu, H. Saegusa, A. Fukuchi, and B. Ratanasthien. 2004. Age and taxonomic status of the Chiang Muan (Thailand) hominoids. *Comptes Rendus Palevol* 3:65–75.

Pilgrim, G. E. 1904. Fossils from the Yenangyoung Oil-Field, Burma. *Records of the Geological Survey of India* 31:103–104.

Pilgrim, G. E. 1910. Preliminary note on a revised classification of the Tertiary freshwater deposits of India. *Records of the Geological Survey of India* 40:185–205.

Pilgrim, G. E. 1926. The fossil Suidae of India. *Memoirs of the Geological Survey of India* 8:1–68.

Pilgrim, G. E. 1927. The lower canine of *Tetraconodon*. *Records of the Geological Survey of India* 60:160–163.

Socquet, A., B. Goffé, M. Pubellier, and C. Rangin. 2002. Le métamorphisme Tardi-Crétacé à Éocène des zones internes de la chaîne Indo-Birmane (Myanmar occidental): Implications géodynamiques. *Comptes Rendus Geoscience* 334:573–580.

Stamp, L. D. 1922. An outline of the Tertiary geology of Burma. *Geological Magazine* 59:481–501.

Takai, M., H. Saegusa, Thaung-Htike, and Zin-Maung-Maung-Thein. 2006. Neogene mammalian fauna in Myanmar. *Asian Paleoprimatology* 4:143–172.

Theobald, W. 1873. On the geology of Pegu. *Memoirs of the Geological Survey of India* 10:189–359.

Yule, H. 1858. *A Narrative of the Mission Sent by the Governor-General of India to the Court of Ava in 1855, with Notices of the Country, Government, and People*. London: Smith, Elder.

Zin-Maung-Maung-Thein, M. Takai, T. Tsubamoto, N. Egi, Thaung-Htike, T. Nishimura, Maung-Maung, and Zaw-Win. 2010. A review of fossil rhinoceroses from the Neogene of Myanmar with description of new specimens from the Irrawaddy Sediments. *Journal of Asian Earth Sciences* 37:154–165.

Part III

North and Central Asia

Chapter 20

Miocene Mammal Biostratigraphy of Central Mongolia (Valley of Lakes)

New Results

GUDRUN DAXNER-HÖCK, DEMCHIG BADAMGARAV, MARGARITA ERBAJEVA, AND URSULA BETTINA GÖHLICH

The Valley of Lakes is one of the Pre-Altai depressions in Mongolia. It is situated between the Gobi Altai Mountains in the south and the Khangai Mountains in the north. It extends ~500 km in the east–west direction in Central Mongolia and is filled, above a Proterozoic and Paleozoic basement, with continental sediments ranging from the Cretaceous to the Quaternary. A Cenozoic sedimentary sequence interlayered with several basalts is exposed in the Taatsiin Gol and Taatsiin Tsagaan Nuur area. Basaltic volcanism is restricted to a relatively small, approximately south–north corridor in Central Mongolia between 98°E and 104°E longitude (Kepezhinskas 1979). It extends from China to Siberia. The Valley of Lakes is positioned within this corridor and is well known for its basalt layers interfingered with Paleogene and Neogene sediments. It is among the best places in Mongolia where fossiliferous terrestrial sediments are associated with basalts (figure 20.1).

During three field seasons, from 1995 to 1997, a joint Austrian–Mongolian project was carried out in this region, and in 2001, 2004, and 2006 additional field trips were conducted to investigate new fossil localities. Sediment sequences of the Hsanda Gol and the Loh Formations allow a stratigraphic concept based on the evolution of mammals and on the age determination of basalts as elaborated by Daxner-Höck et al. (1997), Höck et al. (1999) and Daxner-Höck and Badamgarav (2007). The historical background of previous stratigraphic investigations was given by Höck et al. (1999).

REVIEW OF THE INTEGRATED OLIGOCENE/ MIOCENE STRATIGRAPHY FROM THE TAATSIIN GOL AND TAATSIIN TSAGAAN NUUR AREA

The basalt occurrences in the study area have been dated by the $^{40}Ar / ^{39}Ar$ method (whole rock samples; Höck et al. 1999), providing at least three groups of ages. The first is the Early Oligocene basalt I group of around 31.5 Ma (individual ages ranging from 30.4 Ma to 32.2 Ma; errors varying from 0.3 Ma to 0.8 Ma). Magnetostratigraphic data of basalt I and the Hsanda Gol sediments below basalt I show reversed polarity (Kraatz and Geisler 2010) and are correlated to the upper part of magnetochron C12r (ca. 31.1–33.3 Ma; Gradstein and Ogg 2004), which is consistent with the radioisotopic dates. The second is the Late Oligocene basalt II group, which yielded age differences of at least 1 m.y. between western and more eastern occurrences. The eastern basalt II flows are around 28 Ma in age (ranging from 27 Ma to 29 Ma), and the western flows, around 26.5 Ma, range from 25 Ma to 27 Ma. The third, the Middle Miocene basalt III group, yields a well-defined age of 13 Ma, ranging from 12.2 Ma to 13.2 Ma (errors varying from 0.2 Ma to 0.7 Ma). Basalt III forms the top of the plateaus in the Taatsiin Gol area and northwest of there. Basalt III is 10–20 m thick in the north, but it thins to less than 1 m in the south.

The paleontological focus was on sediments of the Hsanda Gol Formation and the Loh Formation, which are rich in mammal remains. The stratigraphic concept is

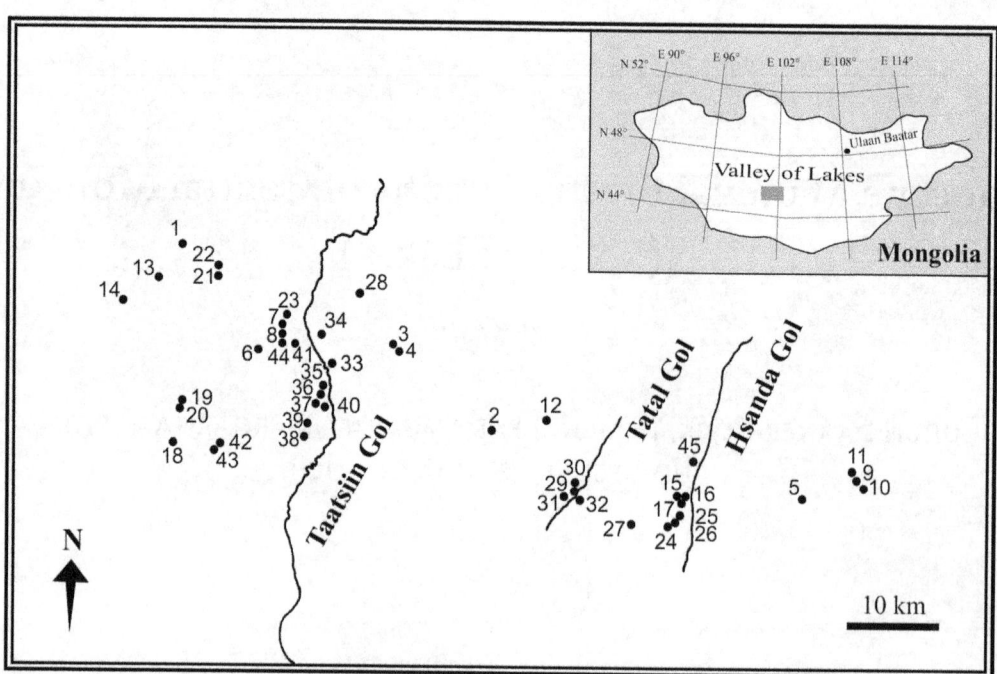

Figure 20.1 Location of Oligocene and Miocene fossil sites from the Valley of Lakes, Uvurkhangai, Central Mongolia: (*1*) Abzag Ovo (ABO-A); (*2*) Builstyn Khudag (BUK-A); (*3–4*) Del (DEL-A, DEL-B); (*5*) unnamed locality west of Ikh Argalatyn Nuruu (GRAB-II); (*6*) Khongil (HL-A); (*7*) Hotuliin Teeg east (HTE); (*8*) Hotuliin Teeg southeast (HTSE); (*9–11*) Ikh Argalatyn Nuruu (IKH-A, IKH-B, IKH-C); (*12*) unnamed locality (K-II); (*13–14*) Luugar Khudag (LOG-A, LOG-B); (*15–17*) Loh (LOH-A, LOH-B, LOH-C); (*18–20*) Menkhen Teeg (MKT-W; MKT-1, MKT-2); (*21–22*) Olon Ovoony Khurem (ODO-A, ODO-B); (*23*) Huch Teeg (RHN-A) = Tavan Ovoony Deng in Höck et al. (1999:table 1); (*24–27*) Hsanda Gol (SHG-A, SHG-B, SHG-AB, SHG-C); (*28*) Unzing Khurem (TAR-A) = Tarimalyn Khurem in Höck et al. (1999:table 1); (*29–32*) Tatal GOL (TAT-B, TAT-C, TAT-D, TAT-SE); (*33–34*) Taatsiin Gol left side (TGL-A, TGL-B); (*35–41*) Taatsiin Gol right side (TGR-A, TGR-B, TGR-AB, TGR-C, TGR-D, TGR-ZO, TGR-1564; (*42–43*) Toglo-rhoi (TGW-A, TGW-D) = Taatsiin Gol west in Höck et al. (1999); (*44*) Unkheltseg (UNCH-A); (*45*) Ulaan Tolgoi (UTO-A). From Daxner-Höck and Badamgarav (2007).

based on eight informal rodent biozones A, B, C, C1, D, D1/1, D1/2, E and on the radiometric ages of the basalts (Daxner-Höck et al. 1997; Höck et al. 1999). The Oligocene part of the stratigraphic sequence is characterized by fossils of the biozones A, B, C, C1 and contains basalts I and II. The Miocene part of the sequence is characterized by fossils of biozones D, D1/1, D1/2, E and contains basalt III.

The Hsanda Gol Formation consists of brick-red clays and silts divided by the 5–20-m-thick basalt I. The red beds of the Hsanda Gol Formation below basalt I yield fossils of biozone A, and above basalt I they contain fossils of biozone B. The age is Early Oligocene as indicated by basalt I (ca. 31.5 Ma). The upper part of the Hsanda Gol sediments extends into the Late Oligocene, as indicated by fossils of biozones C and C1. Simultaneously, sediments of the Loh Formation were deposited laterally—that is, sands, silts, and gravels, along with basalt II. This lower part of the Loh Formation locally displays fossils of biozone C immediately below or above basalt II flows. In Abzag Ovo (ABO-A/3), the fossil bed of biozone C is lo-

cated directly below basalt II. In the section Unzing Khurem (TAR-A/2), fossils of biozone C were recovered directly above basalt II (see figure 20.1).

MIOCENE DEPOSITS

The upper part of the Loh Formation was deposited in the Miocene, as indicated by its position above Hsanda Gol sediments, by its characteristic fossils of biozones D, D1/1, D1/2, and E, and by basalt III (around 13 Ma, Middle Miocene). Note that all Miocene small mammal fossils (biozones D–E) derive from sediments of the Loh Formation.

1. Fossils of biozone D are evidenced from the following localities (sections/layers; table 20.1): Luugar Khudag (LOG-A/1; N 45°32'18" E 101°00'48"), Huch Teeg (RHN-A/12; N45°29'37" E 101°12'17"), Hotuliin Teeg (HTE-3-12, HTE-O; N 45°29'15" E 101°11'85" and HTSE-O;

Table 20.1

Investigated Localities (Sections/Fossil Layers) and the Respective Formations and Biozones from the Valley of Lakes

Locality	Section/Layer	Formation	Biozone
Menkhen Teeg	MKT-W/O	Hsanda Gol	A
Khongil	HL-A/2	Hsanda Gol	A
Khongil	HL-A/1	Hsanda Gol	A
Taatsiin Gol right	TGR-A/14	Hsanda Gol	A
Taatsiin Gol right	TGR-A/13	Hsanda Gol	A
Taatsiin Gol left	TGL-A2	Hsanda Gol	A
Taatsiin Gol left	TGL-A/1	Hsanda Gol	A
Del	DEL-B/2	Hsanda Gol	A
Unnamed	KII-O	Hsanda Gol	A
Tatal Gol	TAT-C/3	Hsanda Gol	A
Tatal Gol	TAT-C/2	Hsanda Gol	A
Tatal Gol	TAT-C/1	Hsanda Gol	A
Tatal Gol	TAT-D/1/Hü6	Hsanda Gol	A
Tatal Gol	TAT-D/1/Hü5	Hsanda Gol	A
Tatal Gol	TAT-D/1/Hü4	Hsanda Gol	A
Tatal Gol	TAT-D/1/Hü3	Hsanda Gol	A
Tatal Gol	TAT-D/1/Hü2	Hsanda Gol	A
Tatal Gol	TAT-D/1/Hü1	Hsanda Gol	A
Hsanda Gol	SHG-C/2	Hsanda Gol	A
Hsanda Gol	SHG-C/1	Hsanda Gol	A
Unnamed	GRAB-II/3	Hsanda Gol	A
Unnamed	GRAB-II/2	Hsanda Gol	A
Unnamed	GRAB-II/1	Hsanda Gol	A
Unkheltseg	UNCH-A/5 (mixed with D)	Hsanda Gol (Loh)	B
Unkheltseg	UNCH-A/4 (mixed with D)	Hsanda Gol (Loh)	B
Unkheltseg	UNCH-A/3 (mixed with D)	Hsanda Gol (Loh)	B
Unkheltseg	UNCH-A/O (mixed with D)	Hsanda Gol (Loh)	B
Taatsiin Gol right	TGR-AB/22	Hsanda Gol	B
Taatsiin Gol right	TGR-AB/21	Hsanda Gol	B
Taatsiin Gol right	TGR-ZO/2	Hsanda Gol	B
Taatsiin Gol right	TGR-ZO/1	Hsanda Gol	B
Taatsiin Gol right	TGR-1564	Hsanda Gol	B
Taatsiin Gol right	TGR-B/1	Hsanda Gol	B
Taatsiin Gol left	TGL-D/O	Hsanda Gol	B
Taatsiin Gol left	TGL-A/11c	Hsanda Gol	B
Taatsiin Gol left	TGL-A/11b	Hsanda Gol	B
Taatsiin Gol left	TGL-A/11a	Hsanda Gol	B
Taatsiin Gol left	TGL-A/11	Hsanda Gol	B
Taatsiin Gol left	TGL-A/O	Hsanda Gol	B
Del	DEL-A/11	Hsanda Gol	B
Del	DEL-A/O	Hsanda Gol	B
Del	DEL-B/7	Hsanda Gol	B
Del	DEL-B/6	Hsanda Gol	B
Del	DEL-B/O	Hsanda Gol	B
Builstyn Khudag	BUK-Shg.	Hsanda Gol	B

(continued)

Table 20.1 *(continued)*

Locality	Section/Layer	Formation	Biozone
Tatal Gol	TAT-D/2	Hsanda Gol	B
Hsanda Gol	SHG-AB/20->	Hsanda Gol	B
Hsanda Gol	SHG-AB/17-20	Hsanda Gol	B
Hsanda Gol	SHG-AB/12	Hsanda Gol	B
Hsanda Gol	SHG-AB/O	Hsanda Gol	B
Hsanda Gol	SHG-A/20	Hsanda Gol	B
Hsanda Gol	SHG-A/18+20	Hsanda Gol	B
Hsanda Gol	SHG-A/15+20	Hsanda Gol	B
Hsanda Gol	SHG-A/15	Hsanda Gol	B
Hsanda Gol	SHG-A/10	Hsanda Gol	B
Hsanda Gol	SHG-A/9	Hsanda Gol	B
Hsanda Gol	SHG-A/6	Hsanda Gol	B
Hsanda Gol	SHG-A/4	Hsanda Gol	B
Hsanda Gol	SHG-A/1	Hsanda Gol	B
Hsanda Gol	SHG-A/O	Hsanda Gol	B
Ikh Argalatyn Nuruu	IKH-C/O	Hsanda Gol	B
Ikh Argalatyn Nuruu	IKH-B/2	Hsanda Gol	B
Ikh Argalatyn Nuruu	IKH-B/1+2 (=IKH-B/a)	Hsanda Gol	B
Ikh Argalatyn Nuruu	IKH-B/O	Hsanda Gol	B
Ikh Argalatyn Nuruu	IKH-A/3-4	Hsanda Gol	B
Ikh Argalatyn Nuruu	IKH-A/2	Hsanda Gol	B
Ikh Argalatyn Nuruu	IKH-A/1	Hsanda Gol	B
Ikh Argalatyn Nuruu	IKH-A/O	Hsanda Gol	B
Abzag Ovoo	ABO-A/3	Loh	C
Toglorhoi	TGW-A/2b	Loh	C
Toglorhoi	TGW-A/2a	Loh	C
Toglorhoi	TGW-A/2	Loh	C
Toglorhoi	TGW-A/1	Loh	C
Toglorhoi	TGW-A/O	Loh	C
Huch Teeg	RHN-A/6	Loh	C
Taatsiin Gol right	TGR-C/1O	Hsanda Gol	C
Taatsiin Gol right	TGR-C/5	Hsanda Gol	C
Taatsiin Gol right	TGR-C/2	Hsanda Gol	C
Taatsiin Gol right	TGR-C/1	Hsanda Gol	C
Taatsiin Gol right	TGR-C/Bad.6	Hsanda Gol	C
Taatsiin Gol right	TGR-C/Bad.5	Hsanda Gol	C
Taatsiin Gol right	TGR-C/O	Hsanda Gol	C
Unzing Khurem	TAR-A/2	Loh	C
Tatal Gol	TAT-SE (mixed with C1?)	Hsanda Gol	C
Toglorhoi	TGW-A/5	Loh	C1
Huch Teeg	RHN-A/10	Loh	C1
Huch Teeg	RHN-A/9	Loh	C1
Huch Teeg	RHN-A/8	Loh	C1
Huch Teeg	RHN-A/7	Loh	C1
Del	DEL-B/12	Hsanda Gol	C1
Tatal Gol	TAT-SE (mixed with C)	Hsanda Gol	C1
Tatal Gol	TAT-B/10	Hsanda Gol	C1

Locality	Section/Layer	Formation	Biozone
Tatal Gol	TAT-D/3	Hsanda Gol	C1
Loh	LOH-C/1	Hsanda Gol	C1
Loh	LOH-B/3	Hsanda Gol	C1
Ikh Argalatyn Nuruu	IKH-B/5	Hsanda Gol	C1
Ikh Argalatyn Nuruu	IKH-A/5	Hsanda Gol	C1
Luugar Khudag	LOG-A/1	Loh	D
Huch Teeg	RHN-A/12	Loh	D
Hotuliin Teeg	HTSE-10	Loh	D
Hotuliin Teeg	HTE-12	Loh	D
Hotuliin Teeg	HTE-9	Loh	D
Hotuliin Teeg	HTE-8	Loh	D
Hotuliin Teeg	HTE-7	Loh	D
Hotuliin Teeg	HTE-5	Loh	D
Hotuliin Teeg	HTE-4	Loh	D
Hotuliin Teeg	HTE-3	Loh	D
Hotuliin Teeg	HTE-O	Loh	D
Unkheltseg	UNCH-A/4 (mixed with B)	Loh (Hsanda Gol)	D
Unkheltseg	UNCH-A/3 (mixed with B)	Loh (Hsanda Gol)	D
Unkheltseg	UNCH-0 (mixed with B)	Loh (Hsanda Gol)	D
Luugar Khudag	LOG-B/1	Loh	D1/1
Uolon Ovoony Khurem	ODO-A/6	Loh	D1/1
Uolon Ovoony Khurem	ODO-A/2	Loh	D1/1
Uolon Ovoony Khurem	ODO-B/1	Loh	D1/1
Ulaan Tolgoi	UTO-A/6 (=UTO-c)	Loh	D1/2
Ulaan Tolgoi	UTO-A/5 (=UTO-b)	Loh	D1/2
Ulaan Tolgoi	UTO-A/3 (=UTO-a)	Loh	D1/2
Loh	LOH-A/2 (=LOH-b)	Loh	D1/2
Loh	LOH-A/O	Loh	D1/2
Builstyn Khudag	BUK-A/12+14	Loh	E
Builstyn Khudag	BUK-O	Loh	E?

NOTE: Biozones C and C1 are correlative with the Late Oligocene, the biozones D to E with the Miocene. O = collected from the surface lateral to the section.

SOURCE: Modified after Daxner-Höck and Badamgarav (2007:table 2).

45°28'74" E 101°12'37"), and Unkheltseg (UNCH-A/3+4 and UNCH-O; N 54°27'41" E 101°12'05").

2. Fossils of biozone D1/1 are evidenced from the following localities (sections/layers): Luugar Khudag (LOG-B/1; N 45°30'48" E 100°58'22") and Olon Ovoony Khurem (ODO-A/2, 6; N 45°32'24" E 101°08'17" and ODO-B/1; N 45°32'58" E 101°08'16").

3. Fossils of biozone D1/2 are evidenced from the following localities (sections/layers): Ulaan Tolgoi (UTO-A/3, O; N 45°20'49" E 101°50'16"; UTO-A/5; N45°20'41" E 101°50'28" and UTO-A/6; N45°20'32" E 101°50'12") and Loh (LOH-A/2, O; N 45°17'22" E 101°47'04").

4. Fossils of biozone E are evidenced from the locality (section/layers) Builstyn Khudag (BUK-A/12+14; N 45°23'03" E 101°30'44").

EMENDED BIOSTRATIGRAPHY AND BIOCHRONOLOGY OF INFORMAL BIOZONES D TO E

The Neogene biozones D, D1/1, D1/2, and E have been updated since new field activities were conducted in the years 2001, 2004, and 2006, and since detailed systematic-taxonomic investigations on different vertebrate groups have become available (Daxner-Höck 2000, 2001; Erbajeva and Daxner-Höck 2001; Morlo and Nagel 2002, 2006, 2007; Vislobokova and Daxner-Höck 2002; Daxner-Höck and Wu 2003; Erbajeva 2003; Nagel and Morlo 2003; Böhme 2007; Daxner-Höck and Badamgarav 2007; Erbajeva 2007; Göhlich 2007; Heissig 2007; Koenigwald and Kalthoff 2007; Schmidt-Kittler, Vianey-Liaud, and Marivaux 2007; Ziegler, Dahlmann, and Storch 2007). Here, we document the assemblages observed in the study area, this being the necessary step that precedes formal description of mammalian biochrons. Each biozone is characterized by its small mammals, their first and/or last records, and the most abundant species. The known time ranges of the biozones are given in figures 20.2 and 20.3 and tables 20.2–20.4.

The first collections from bulk samples of Unkheltseg (UNCH-A/3+4) contained mixed assemblages. There, a red silt layer with reworked carbonate nodules and fossils of biozone D is located directly on top of Hsanda Gol sediments with fossils of biozone B. As a result, some fossils of biozone B (e.g., Tsaganomyidae) were erroneously included in the original description of biozone D (Daxner-Höck et al. 1997; Höck et al. 1999). Meanwhile, homogeneous assemblages from Huch Teeg, Luugar Khudag, and Hotuliin Teeg allow correct characterization of biozone D, excluding species of biozone B (see tables 20.2–20.4).

Biozone D

Characteristic Insectivora, Lagomorpha, and Rodentia

Exallerix sp. (small), *Amphechinus* aff. *taatsiingolensis* (Insectivora), *Amphilagus, Sinolagomys ulungurensis, Bellatona* (3 species), *Alloptox minor* (Lagomorpha), *Tachyoryctoides kokonorensis, Heterosminthus firmus, Plesiosminthus barsboldi, Litodonomys* sp. 1–3, *Yindirtemys suni, Prodistylomys, Distylomys,* and *Kherem* (Rodentia).

Large Mammal Occurrences

cf. *Hoploaceratherium gobiense* (Rhinocerotidae; Heissig 2007), *Nimravus mongoliensis* (Nimravidae; Morlo and Nagel 2007).

Biotic Events

First record: *Amphilagus, Sinolagomys ulungurensis, Bellatona* aff. *kazakhstanica, Bellatona yanghuensis, Alloptox, Litodonomys* sp. 2 and *Litodonomys* sp. 3, *Yindirtemys suni, Tachyoryctoides kokonorensis, Heterosminthus firmus, Plesiosminthus barsboldi, Prodistylomys, Distylomys,* several Sciuridae, and *Democricetodon.* Last record: *Exallerix, Amphechinus, Desmatolagus, Bohlinotona, Sinolagomys, Plesiosminthus, Litodonomys, Yindirtemys suni, Tachyoryctoides kokonorensis, Heterosminthus firmus,* Didymoconidae (*Didymoconus berkey*). Most abundant taxa: *Amphechinus* aff. *taatsiingolensis, Sinolagomys ulungurensis, Tachyoryctoides kokonorensis, Yindirtemys suni, Heterosminthus firmus.*

Lithostratigraphy and Biochronology

Sediments containing fossils of biozone D are red, rose, and white silts and sands of the Loh Formation and calcrete layers (paleosol horizons) enriched with fossils. There is no age control of zone D by basalt ages. However, in the Huch Teeg section (RHN-A/6–12), sediment layers with fossils of biozones C-C1-D are in stratigraphic sequence. There, biozone D is evidenced from the uppermost level (RHN-A/12, Early Miocene)—that is, on top of biozones C-C1 (see figure 20.2 [RHN-A/6-A/10], Late Oligocene).

Although several Oligocene genera persist in biozone D, a Miocene age is suggested by the presence of ?*Hoploaceratherium, Democricetodon, Alloptox, Amphilagus,* and *Bellatona* and by advanced species of *Amphechinus* (*A.* aff. *taatsiingolensis*), *Sinolagomys* (*S. ulungurensis*), *Yindirtemys* (*Y. suni*), *Tachyoryctoides* (*T. kokonorensis*), and *Plesiosminthus* (*P. barsboldi*). The estimated age is early Early Miocene, which is supported by the stratigraphic position above biozone C1 and below D1/1.

Biozone D 1/1

Characteristic Insectivora, Lagomorpha, and Rodentia

Parvericius buk (Insectivora), *Bellatona* cf. *yanghuensis, Alloptox minor* and *Alloptox gobiensis* (Lagomorpha), *Heterosminthus mongoliensis, Democricetodon* sp. 2, and *Megacricetodon* sp.1 (Rodentia).

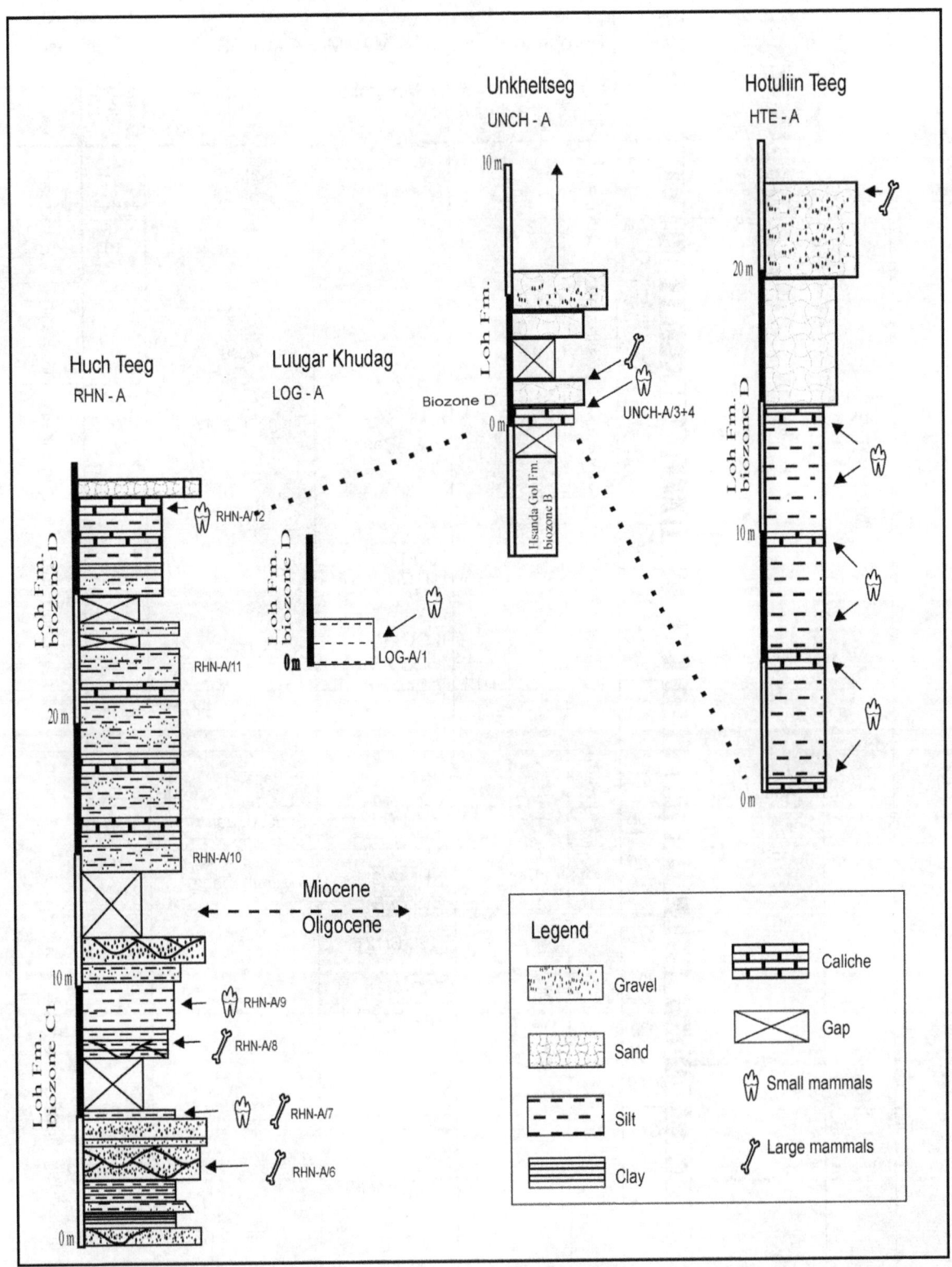

Figure 20.2 Correlation of the sections Huch Teeg (RHN-A), Hotuliin Teeg (HTE-A), Unkheltseg (UNCH-A), and Luugar Khudag (LOG-A) from the Taatsiin Gol area. The sections display the Oligocene/Miocene transition (biozone C1/D) and the Early Miocene (biozone D), respectively.

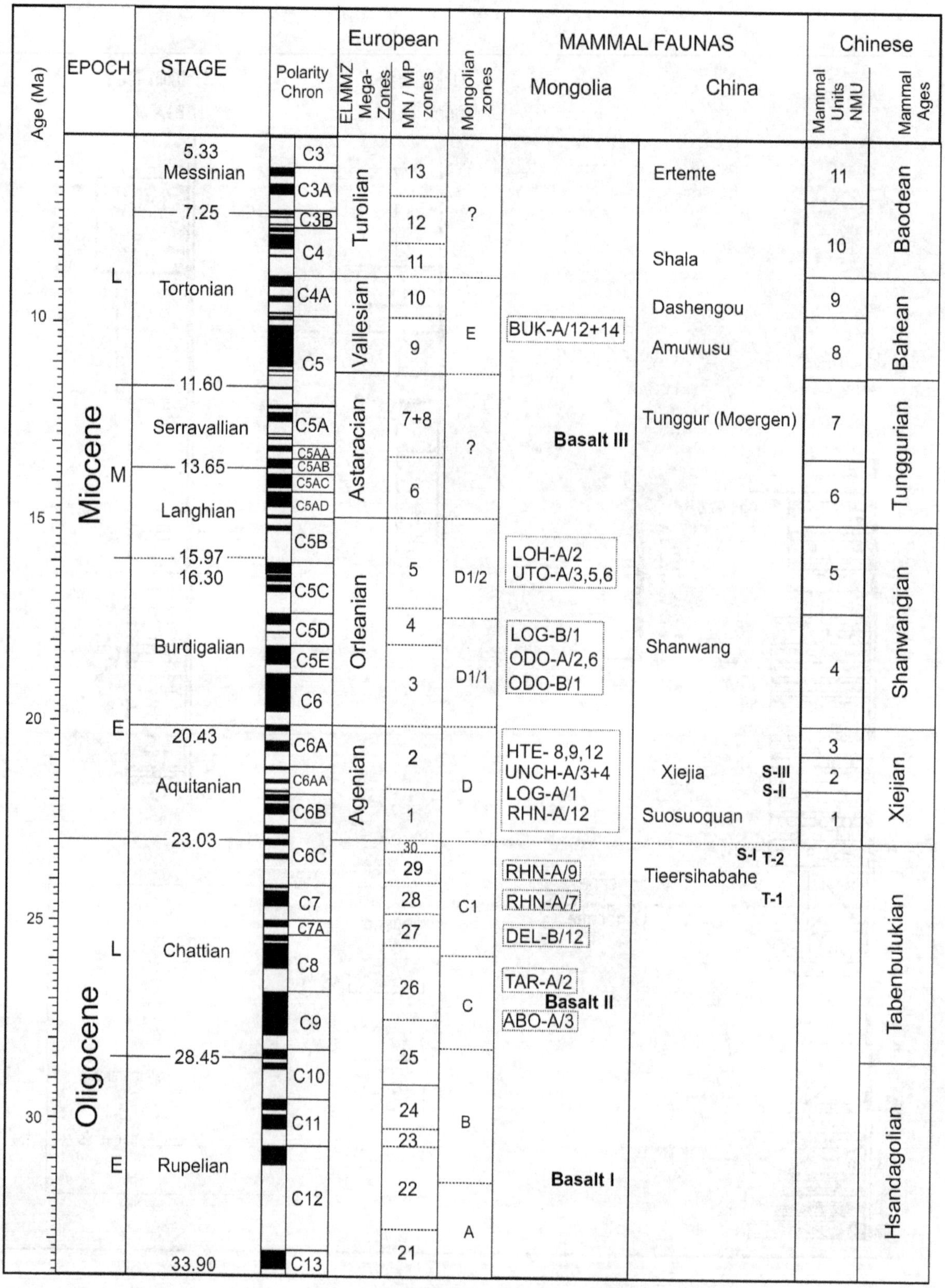

Age (Ma)	EPOCH	STAGE	Polarity Chron	European ELMMZ Mega-Zones	European MN/MP zones	Mongolian zones	MAMMAL FAUNAS Mongolia	MAMMAL FAUNAS China	Chinese Mammal Units NMU	Chinese Mammal Ages
5.33	Miocene L	Messinian	C3	Turolian	13	?		Ertemte	11	Baodean
			C3A	Turolian						
7.25			C3B	Turolian	12			Shala	10	
		Tortonian	C4	Turolian	11					
			C4A	Vallesian	10	E		Dashengou	9	Bahean
10				Vallesian	9		BUK-A/12+14	Amuwusu	8	
			C5	Vallesian						
11.60	Miocene M	Serravallian	C5A	Astaracian	7+8	?	Basalt III	Tunggur (Moergen)	7	Tunggurian
13.65			C5AA / C5AB / C5AC / C5AD	Astaracian	6				6	
		Langhian	C5B	Orleanian	5	D1/2	LOH-A/2 UTO-A/3,5,6		5	Shanwangian
15.97 / 16.30			C5C	Orleanian						
	Miocene E	Burdigalian	C5D	Orleanian	4		LOG-B/1 ODO-A/2,6 ODO-B/1	Shanwang	4	
			C5E	Orleanian	3	D1/1				
			C6	Orleanian						
20.43			C6A	Agenian	2	D	HTE- 8,9,12 UNCH-A/3+4 LOG-A/1 RHN-A/12	Xiejia S-III / S-II	3 / 2	Xiejian
		Aquitanian	C6AA / C6B	Agenian	1			Suosuoquan	1	
23.03			C6C		30 / 29	C1	RHN-A/9	Tieersihabahe S-I T-2		Tabenbulukian
	Oligocene L		C7		28		RHN-A/7	T-1		
			C7A		27		DEL-B/12			
		Chattian	C8		26	C	TAR-A/2 ABO-A/3			
			C9				Basalt II			
28.45			C10		25					
			C11		24	B				Hsandagolian
30					23					
	Oligocene E	Rupelian	C12		22	A	Basalt I			
33.90			C13		21					

Figure 20.3 Geologic time scale after Gradstein and Ogg (2004); European Land Mammal Megazones (ELMMZ) and Neogene NM Zones after Steininger (1999); Paleogene MP Zones after Luterbacher et al. (2004); Chinese Mammal Ages and Neogene Mammal Units (NMU) after Deng (2006); and Chinese Neogene localities after Deng (2006), Meng et al. (2006), and Qiu, Wang, and Li (2006).

Table 20.2

Late Oligocene and Miocene Insectivora from the Valley of Lakes

	Upper Oligocene Biozones		Miocene Biozones			
	C	C1	D	D1/1	D1/2	E
Erinaceidae						
Brachyericinae						
Exallerix pustulatus Ziegler, Dahlmann, and Storch 2007	x					
Exallerix sp.			x			
Erinaceinae						
Palaeoscaptor gigas Lopatin 2002	x					
Palaeoscaptor cf. *rectus* Matthew and Granger 1924	x	x	x			
Palaeoscaptor acridens Matthew and Granger 1924	x	x				
Palaeoscaptor tenuis Ziegler, Dahlmann, and Storch 2007	x					
Amphechinus taatsiingolensis Ziegler, Dahlmann, and Storch 2007	x					
Amphechinus aff. *taatsiingolensis* Ziegler, Dahlmann, and Storch 2007			x			
Amphechinus minutissimus Ziegler, Dahlmann, and Storch 2007		x				
Amphechinus major Ziegler, Dahlmann, and Storch 2007		x				
Parvericius buk Ziegler, Dahlmann, and Storch 2007				x	x	x
Erinaceinae gen. et sp. indet II			x			
Soricidae						
Heterosoricinae						
Gobisorex kingae Sulimsky 1970	x	x				
Heterosoricinae gen. et sp. indet. 3		x				
Crocidosoricinae						
Tavoonyia altaica Ziegler, Dahlmann, and Storch 2007			x			
Crocidosoricinae gen. et sp. indet. 7	x	x				
Crocidosoricinae gen. et sp. indet. 8	x					
Crocidosoricinae gen. et sp. indet. 9			x			
Crocidosoricinae gen. et sp. indet. 10			x			
Crocidosoricinae gen. et sp. indet. 11			x			
Soricinae						
Sorex sp. ?						x
Builstynia fontana Ziegler, Dahlmann, and Storch 2007						x
Talpidae						
Talpidae gen et spec. indet. 7	x					
Talpidae gen et spec. indet. 8		x				
Talpidae gen et spec. indet. 9			x			

SOURCE: Ziegler, Dahlmann, and Storch (2007:table 15).

Table 20.3

Late Oligocene and Miocene Lagomorpha from the Valley of Lakes

	Upper Oligocene Biozones		Miocene Biozones			
	C	C1	D	D1/1	D1/2	E
Palaeolagidae						
Palaeolaginae						
Desmatolagus gobiensis Matthew and Granger 1923	x					
Desmatolagus cf. *gobiensis* Matthew and Granger 1923	x	x				
Desmatolagus cf. *chinensis* (Erbajeva and Sen 1998)		x				
Desmatolagus aff. *chinensis* (Erbajeva and Sen 1998)	x	x				
Desmatolagus aff. *simplex* (Argyropulo 1940)	x		x			
Desmatolagus sp.	x	x	x			
Desmatolagus sp. A	x					
Desmatolagus cf. *orlovi* (Gureev 1960)	x					
Desmatolagus cf. *shargaltensis* Bohlin 1937	x					
Desmatolagus sp. 1	x					
Desmatolagus sp. 2	x					
Amphilaginae						
Amphilagus sp.			x			
Leporidae						
Ordolagus cf. *teilhardi* (Burke 1941)	x					
Ochotonidae						
Sinolagomyinae						
Bohlinotona sp. (small)	x	x	x			
Sinolagomys cf. *kansuensis* Bohlin 1937	x	x				
Sinolagomys sp.	x	x	x			
Sinolagomys cf. *tatalgolicus* Gureev 1960		x				
Sinolagomys aff. *major* Bohlin 1937		x				
Sinolagomys cf. *major* Bohlin 1937		x	x			
Sinolagomys cf. *ulungurensis* Tong 1989		x				
Sinolagomys ulungurensis Tong 1989			x			
Bellatona yanghuensis Zhou 1988			x			
Bellatona sp.			x			
Bellatona aff. *kazakhstanica* Erbajeva 1988			x			
Bellatona cf. *yanghuensis* Zhou 1988				x		
Ochotoninae						
Alloptox minor Li 1978			x	x		
Alloptox gobiensis (Young 1932) (archaic type)				x		
Alloptox gobiensis (Young 1932) (advanced type)					x	
Alloptox cf. *gobiensis* (Young 1932)					x	
Alloptox sp.					x	
Ochotona lagreli Schlosser 1924						x
Ochotona minor Bohlin 1942						x

SOURCE: Updated after Erbajeva (2007:table 1).

Table 20.4

Late Oligocene and Miocene Rodentia from the Valley of Lakes

	Upper Oligocene Biozones		Miocene Biozones			
	C	C1	D	D1/1	D1/2	E
Aplodontidae						
Prosciurus sp. 4	x	x				
Aplodontidae indet.			x			
Ansomys sp.					x	
Sciuridae						
Spermophilinus sp.	x		x		x	
Sciuridae indet. 1			x			
Kherem sp.		x	x			
Sciuridae indet. 2			x			
Sciuridae indet. 3				x		
Sciuridae indet. 4						x
Prospermophilus cf. orientalis (Qiu 1991)						x
Gliridae						
Microdyromys sp.					x	
Eomyidae						
Eomyidae indet. 2	x	x	x			
Keramidomys sp.					x	
Eomyops sp.					x	
Eomyidae indet. 3					x	
Eomyidae indet. 4						x
Ctenodactylidae						
Huangomys frequens Schmidt-Kittler, Vianey-Liaud, and Marivaux 2007	x					
Tataromys sigmodon Matthew and Granger 1923	x					
Tataromys minor longidens Schmidt-Kittler, Vianey-Liaud, and Marivaux 2007	x	x				
Tataromys plicidens Matthew and Granger 1923		x				
Yindirtemys aff. ulantatalensis (Huang 1985)	x					
Yindirtemys deflexus (Teilhard de Chardin 1926)		x				
Yindirtemys suni (Li and Qiu 1980)			x			
Prodistylomys sp.			x			
Distylomys sp.			x			
Sayimys sp. cf. intermedius (Sen and Thomas 1979)					x	
Tsaganomyidae						
Cyclomylus sp.	x					
Tsaganomys altaicus Matthew and Granger, 1923	x	x				
Zapodidae/Dipodidae						
Heosminthus sp.1	x					
Heosminthus cf. minutus Daxner-Höck 2001			x			
Plesiosminthus sp. A	x					

(continued)

Table 20.4 (*continued*)

	Upper Oligocene Biozones		Miocene Biozones			
	C	C1	D	D1/1	D1/2	E
Plesiosminthus cf. *asiaticus* Daxner-Höck and Wu 2003		x				
Plesiosminthus promyarion Schaub 1930		x				
Plesiosminthus sp. B		x				
Plesiosminthus barsboldi Daxner-Höck and Wu 2003			x			
Heterosminthus cf. *firmus* Zazhigin and Lopatin 2000		x	x			
Heterosminthus firmus Zazhigin and Lopatin 2000			x			
Heterosminthus mongoliensis Zazhigin and Lopatin 2000				x	x	
Heterosminthus gansus Zheng 1982						x
Parasminthus sp.1	x					
Parasminthus cf. *tangingoli* Bohlin 1946	x	x				
Parasminthus cf. *asiae-centralis* Bohlin 1946	x	x				
Parasminthus cf. *debruijni* Lopatin 1999	x	x				
Bohlinosminthus parvulus (Bohlin 1946)	x	x				
Bohlinosminthus sp.	x	x				
Litodonomys sp. 1	x	x	x			
Litodonomys sp. 2			x			
Litodonomys sp. 3			x			
Dipodidae indet. 1					x	
Protalactaga sp.						x
Eozapus sp.						x
Dipodidae indet. 2						x
Cricetidae						
Eucricetodon sp. 1	x					
Eucricetodon sp. 2	x					
Eucricetodon sp. 3	x					
Aralocricetodon sp.	x	x				
Tachyoryctoides sp. 1	x					
Tachyoryctoides obrutschewi Bohlin 1937	x	x				
Tachyoryctoides sp. 2			x			
Tachyoryctoides kokonorensis Li and Qiu 1980			x			
Democricetodon sp. 1			x			
Democricetodon sp. 2				x		
Democricetodon cf. *lindsayi* Qiu 1996					x	
Democricetodon cf. *tongi* Qiu 1996					x	
Democricetodon sp. 3						x
Megacricetodon sp. 1				x		
Megacricetodon sp. 2					x	
? *Allocricetus* sp.						x
Prosiphnaeus sp.						x
Myocricetodon sp.						x
Gerbillinae indet.						x

SOURCE: Daxner-Höck and Badamgarav (2007:table 3); Schmidt-Kittler, Vianey-Liaud, and Marivaux (2007).

First Record

Parvericius buk, Alloptox gobiensis, Bellatona cf. *yanghuensis, Heterosminthus mongoliensis,* and *Megacricetodon.*

Lithostratigraphy and Biochronology

The fossil record of this biozone is poor. However, a few small faunules of biozone D1/1 were recovered from red silty sands of the Loh Formation. In one section of Olon Ovoony Khurem (ODO-B/1), the fossil horizon is sandwiched between basalt IIb and basalt III. In the sections (ODO-A/1–6) and Luugar Khudag (LOG-B/1), the fossil layers are situated below basalt III. As controlled by basalt III, the fossils are older than 13 Ma. According to the first occurrences of *Parvericius, Alloptox gobiensis,* and *Megacricetodon* and the advanced species of *Heterosminthus* (*H. mongoliensis*), the estimated age is Early Miocene.

Biozone D1/2

Characteristic Insectivora, Lagomorpha, and Rodentia

Parvericius buk (Insectivora), *Alloptox gobiensis* (Lagomorpha), *Heterosminthus mongoliensis, Democricetodon* cf. *lindsayi, Democricetodon* cf. *tongi, Megacricetodon* sp. 2, *Ansomys* sp., *Microdyromys* sp., *Keramidomys* sp., *Eomyops* sp.1, and *Sayimys* sp. (Rodentia).

Large Mammal Occurrences

Gobitherium sp., cf. *Caementodon* sp. (Rhinocerotidae; Heissig 2007), *Anchitherium* (Equidae), *Amphitragulus, Eotragus* and *Lagomeryx* (Ruminantia; Vislobokova and Daxner-Höck 2002), cf. *Gomphotherium mongoliense,* Gomphotheriidae indet. (Proboscidea; Göhlich 2007).

First Record

Ansomys, Microdyromys, Keramidomys, Eomyops, Sayimys.

Last Record

Alloptox gobiensis (advanced), *Heterosminthus mongoliensis.*

Lithostratigraphy and Biochronology

Sediments containing fossils of biozone D1/2 are white, rose, gray, brown, red silts and sands of the Loh Formation. There is no age control by basalts. The estimated age is late Early Miocene, as indicated by the first occurrences of the rodents *Ansomys, Microdyromys, Keramidomys, Eomyops,* and *Sayimys,* by advanced species of *Democricetodon* (*D.* cf. *lindsayi, D.* cf. *tongi*) and *Megacricetodon* sp. (*M.* cf. *sinensis*), and by the ruminant occurrences *Lagomeryx, Eotragus,* and *Amphitragulus.*

Biozone E

Characteristic Insectivora, Lagomorpha, and Rodentia

Parvericius buk and *Builstynia fontana* (Insectivora), *Ochotona* (Lagomorpha), *Heterosminthus gansus, Prosiphneus* sp., *Protalactaga* sp., *Prospermophilus* cf. *orientalis, Eozapus* sp., *Democricetodon* sp. 3, and ?*Allocricetus* sp. (Rodentia).

Large Mammal Occurrences

Nimravus mongoliensis (Carnivora; Morlo and Nagel 2007), cf. *Iranotherium* sp. (Rhinocerotidae; Heissig 2007), Gomphotheriidae indet. (Proboscidea; Göhlich 2007).

First Record

Builstynia, Ochotona lagreli, Ochotona minor, Heterosminthus gansus, ?*Allocricetus* sp., *Prosiphneus* sp.

Most Abundant Taxa

Builstynia, Parvericius, Ochotona, Heterosminthus gansus, and ?*Allocricetus* sp.

Lithostratigraphy and Biochronology

Fossils of biozone E are known from rose-red silts of the Loh Formation in the section Builstyn Khudag (BUK-A). Only one horizon in a high position of the section (BUK-A/12+14) yielded an incisor (I3) of *Nimravus mongoliensis* (Morlo and Nagel 2007) and the small mammal fossils characterizing biozone E. Isolated large mammal remains (cf. *Iranotherium* sp. and Gomphotheriidae indet.; see Heissig 2007 and Göhlich 2007) were recovered from the basal part of the section. There is no age control by basalt. However, the first occurrences of *Builstynia fontana*, *Ochotona* (*O. lagreli* and *O. minor*), *Eozapus* sp., *Prosiphneus* and ?*Allosminthus*, the advanced species of *Heterosminthus* (*H. gansus*), and the occurrence of *Iranotherium* indicate a Late Miocene age.

CORRELATIONS

The Oligocene/Miocene Transition

The Oligocene/Miocene transition can be recognized in Huch Teeg only (see figure 20.2 [section RHN-A]). In this region, a sequence of Loh Formation sediments is exposed, ranging from the Late Oligocene to the Early Miocene, as indicated by the fossils of the biozones C1 and D. The Early Miocene assemblage (biozone D) is in evidence at three other localities of the Taatsiin Gol area—that is, Hotuliin Teeg, Unkheltseg, and Luugar Khudag (see figures 20.1 and 20.2).

Previously (Daxner-Höck and Badamgarav 2007 and references within), the Mongolian faunas of biozone D were correlated to the small fauna Xiejia, Qinghai Province, China (Li and Qiu 1980), which gave the name to the Chinese Mammal Age Xiejian (Tong, Zheng, and Qiu 1995). Although the Xiejia assemblage (Li and Qiu 1980) is not very diverse, it corresponds remarkably well with the most abundant taxa of biozone D in Mongolia—for example, *Tachyoryctoides kokonorensis*, *Yindirtemys suni*, *Litodonomys* sp. 2 (? =*Plesiosminthus lajeensis* sensu Li and Qiu 1980) (see tables 20.2–20.4 and figure 20.3). Following the correlation chart of Deng (2006), the Xiejian is correlative with the Chinese Neogene Mammal Units NMU1–3, and the Xiejia fauna represents NMU2. Deng, Wang, and Yue (2006) correlate the Xiejian with the European Land Mammal Mega-Zone Agenian (Steininger 1999), which covers the Neogene Mammal Zones MN 1 and MN 2. Consequently, the Xiejia fauna and the Mongolian assemblages of zone D would correlate approximately with MN 2.

In China, the Tieersihabahe section in the Northern Junggar Basin, Xinjiang, is one of the rare sections bridging the Oligocene/Miocene boundary (Ye, Meng, and Wu 2003). Here, the Tieersihabahe and Suosuoquan formations display Late Oligocene and Early Miocene fossils, which enabled the establishment of five mammal assemblage zones (Meng et al. 2006), two Tieersihabahe Mammal Assemblage Zones (= T-1 and T-2 zones) and three Suosuoquan Mammal Assemblage Zones (= S-I zone, S-II zone, S-III zone). Furthermore, magnetostratigraphic investigations allow age calibration (Meng et al. 2006:fig. 4):

T-1: C7n.2n and C7n.1r (24.4–24.15 Ma), T-2: part of C6Cn.2r (23.2–23.1 Ma), S-I: C6Cn.2r (close to 23.03 Ma), S-II: C6Bn.1r and C6AAr.3r (21.9–21.7 Ma), S-III: C6AAr.3r up to C6An.1r (21.7–21.16 Ma). Following Meng et al. (2006), the Tieersihabahe Zones T-1 and T-2 and the Sousuoquan Zone S-I are correlated to the Late Oligocene, and the Sousuoquan Zones S-II and S-III to the Early Miocene. S-II would be correlative with the Xiejia fauna and with biozone D in Mongolia (Meng et al. 2006:fig.3).

Though the small mammal assemblages of Suosuoquan Zone S-II and the Mongolian zone D (see tables 20.2–20.4) share some genera, there are remarkable differences; for example, the "Oligocene" taxa *Parasminthus asiae-centralis*, *Parasminthus tangingoli*, and *Bohlinosminthus parvulus* from S-II do not range beyond the Late Oligocene (biozone C1) in Mongolia, but the advanced species of biozone D are missing in Zone S-II from Xinjiang (*Plesiosminthus barsboldi*, *Tachyoryctiodes kokonorensis*, *Yindirtemys suni*, *Heterosminthus firmus*, and the modern lagomorphs *Amphilagus*, *Alloptox*, and *Bellatona*). Consequently, the Mongolian assemblages of biozone D should be younger (Chinese Mammal Unit NMU2–?3) than Suosuoquan S-II from China (NMU2) (see figure 20.2). Future investigations will have to clarify whether these differences express time differences or ecological factors.

The biostratigraphic agreement of the Tieersihabahe (T-1, T-2, and S-I) and the Mongolian C1 zones is high (*Plesiosminthus asiaticus*, *Parasminthus asiae-centralis*, *Parasminthus tangingoli*, *Bohlinosminthus parvulus*, *Tachyoryctoides obrutschewi*, *Sinolagomys major*, *Desmatolagus gobiensis*, and other taxa shared). The assemblage T-1 seems to be correlative with the Mongolian assemblage RHN-A/7 (biozone C1), as indicated by *Plesiosminthus asiaticus*. However, the Chinese assemblages T-1, T-2, and S-I lack several taxa characteristic for assemblage C1 in Mongolia (i.e., *Plesiosminthus promyarion*, *Bohlinotona*, *Aralocricetodon*, *Yindirtemys deflexus*, *Tataromys minor longidens* and *Heterosminthus* cf. *firmus*), whereas in Mongolia *Eucricetodon* does not extend beyond assemblage C.

The Late Oligocene Mongolian zones C1 and C are correlative with the Chinese Mammal Age Tabenbulukian (see figure 20.3 and tables 20.2–20.4). A number of rodents, (e.g., Eomyidae, *Eucricetodon* and *Plesiosminthus*) allow correlation between Mongolia and Europe. In Europe, *Plesiosminthus* is known from MP 26–30 and MN 1–2, in Mongolia from the higher part of zone C, in C1 and D, which turns out to be the identical time range (see figure 20.3). Moreover, Daxner-Höck and Wu (2003) found the species *Plesiosminthus promyarion* in Mongolia from the higher part of zone C1 in Huch Teeg (see figure

20.2 [RHN-A/9]). This occurrence is within the European range of *P. promyarion* (MP 26–29).

The vertebrate faunas from the Aral Formation in Kazakhstan (i.e., Altynshokysu bone beds 1–4, Akespe, Aktau, and others) are described as Early Miocene (NMU1) in age (Lopatin 2004; Akhmetiev et al. 2005). Previously they were considered Late Oligocene (Russell and Zhai 1987; Lucas, Kordikova, and Emry 1998), and recently the Oligocene age of the Aral faunas was confirmed (Bendukidze, de Bruijn, and van den Hoek Ostende 2009). As indicated by the presence of several *Desmatolagus* species, the diversity of Ctenodactylidae, Cricetidae (*Eucricetodon* spp. and *Aralocricetodon*), and Zapodidae (*Bohlinosminthus* and *Parasminthus debruijni*), and the absence of *Bellatona* and *Alloptox*, the Aral faunas seem very close to the Mongolian faunas of biozones C and C1 (see tables 20.2–20.4), which are dated as Late Oligocene based on basalt II. There, the assemblage of ABO-A/3 (biozone C) is located immediately below basalt II (see figure 20.3). Between the Aral faunas and the Mongolian faunas of biozone D, almost no affinities exist. To our present knowledge, the Mongolian faunas of biozone D are Early Miocene in age (NMU2–?3) and are definitely younger than the Aral faunas.

The Zones D1/1 and D1/2 of the Early Miocene and Zone E of the Late Miocene

The Mongolian mammal assemblages indicating the zones D1/1 and D1/2 are rather poor and display only limited species variability. However, the co-occurrences of *Democricetodon*, *Megacricetodon*, and *Alloptox* allow correlation with the Chinese Shanwangian (see figure 20.3). The D1/1 assemblages are characterized by the first occurrences of *Megacricetodon*, *Heterosminthus mongoliensis*, and *Parvericius buk*. A correlation of D1/1 with the European Land Mammal Mega-Zone Orleanian (MN 3–4) and the Shanwangian (NMU4) is probable.

The D1/2 assemblages are characterized by more advanced species of *Democricetodon* and *Megacricetodon* and by the first occurrences of *Ansomys*, *Keramidomys*, *Eomyops*, *Microdyromys*, *Sayimys*, and *Gomphotherium*. These taxa and advanced species of Lagomorpha would not exclude the Middle Miocene age, but the co-occurring ruminants *Lagomeryx*, *Eotragus*, and *Amphitragulus* hint at an Early Miocene age (Vislobokova and Daxner-Höck 2002). D1/2, therefore, correlates with the Chinese Mammal Age Shanwangian (NMU4–5) and the European Land Mammal Mega-Zone Orleanian (MN 4–5).

Zone E is thought to be of Late Miocene age, as indicated by *Prosiphneus* in an early stage of evolution, by the occurrences of *Eozapus* sp., *Heterosminthus gansus* (=*Lophocricetus gansus* sensu Qiu 1985), the soricid *Builstynia fontana*, two species of *Ochotona* representing Lagomorpha, and the modern rhino *Iranotherium*. Most of the Middle Miocene taxa are missing, as compared with the late Middle Miocene fauna from Tunggur at Moergen (Qiu 1996). The fauna seems to be intermediate between Amuwusu in Nei Mongol (Qiu, Wang, and Li 2006) and Dashengou in Gansu (Deng 2005; Deng et al. 2006) because *Prosiphneus* and *Iranotherium* are already present, but *Progonomys*, *Kowalskia*, *Sinocricetus*, and other more advanced taxa are absent. The Mongolian biozone E correlates with the Chinese Mammal Age Bahean and with the Vallesian of Europe.

CONCLUSION

The Taatsiin Gol and Taatsiin Tsagaan Nur area comprises a sequence of terrestrial sediments that can be divided into two lithologically well-defined formations, the Hsanda Gol and the Loh formations. These formations are interbedded by three different basalt layers (basalt I–III) and have yielded many superposed fossil assemblages. These data establish a robust stratigraphic concept that yielded a reliable calibration of the Oligocene informal biozones A → C1 (indicated by ^{40}Ar/^{39}Ar data of basalt I and II), which potentially could be used to recognize biochrons. Compared with the Oligocene, the Miocene fossiliferous beds are rather rare and do not provide a continuous evolutionary record. Moreover, not one of the Miocene fossil beds is in direct contact with the Middle Miocene basalt that forms the top of the plateaus in the study area (Höck et al. 1999). The Miocene biozones D → E are therefore based solely on faunal data.

The age of 24 ± 1.0 Ma (Devjatkin 1993) of the uppermost basalt layer of Taatsiin Gol (right side) could not be verified. No reliable basalt data around the Oligocene/Miocene transition are available from the area. The Huch Teeg section, however, displays a number of superposed fossil layers. It is the sole section that bridges the Oligocene/Miocene transition (see figure 20.2). For the time being, the correlation of the Mongolian biozone D with the Xiejian Chinese Mammal Age and with the European Mega-Zone Agenian is confirmed based on biostratigraphic data.

ACKNOWLEDGMENTS

We thank the Natural History Museum of Vienna, our Mongolian partners R. Barsbold and Y. Khand from the Mongolian Academy of Sciences, the Mongolian driver A. Radnaa, and the family of Ulzibaataar from the Taatsiin Gol for scientific, logistic, and technical support during the fieldwork in Mongolia. We thank M. Stachowitsch, who helped improve the English, and the reviewer H. de Bruijn for constructive comments.

Our investigations were financially supported by the FWF-projects: P-10505-GEO and P-15724-N06 and by an exchange programme of the Austrian and Chinese Academy of Sciences.

REFERENCES

Akhmetiev, M. A., A. V. Lopatin, E. K. Sytchevskaya, and S. V. Popov. 2005. Biogeography of the Northern Peri-Tethys from the Late Eocene to the Early Miocene: Part. 4. Late Oligocene–Early Miocene: Terrestrial biogeography, conclusions. *Paleontological Journal* 39:1–554.

Argyropulo, A. I. 1940. A review of finds of Tertiary rodents on the territory of USSR and adjacent regions of Asia. *Priroda* 12:74–82 (in Russian).

Bendukidze, O. G., H. de Bruijn, and L. W. van den Hoek Ostende. 2009. A revision of Late Oligocene associations of small mammals from the Aral Formation (Kazakhstan) in the National Museum of Georgia, Tiblisi. *Palaeodiversity* 2:343–377.

Bohlin, B. 1937. Oberoligozäne Säugetiere aus dem Shargaltain-Tal (Western Kansu). *Paleontologica Sinica*, n. s. C 3:1–66.

Bohlin, B. 1942. A revision of the fossil Lagomorpha in the Palaeontological Museum Uppsala. *Geological Institute Uppsala Bulletin* 30(6):117–154.

Bohlin, B. 1946. The Fossil Mammals from the Tertiary Deposit of Tabenbuluk, Western Kansu. Part II: Simplicidentata, Carnivora, Artiodactyla, Perissodactyla, and Primates. *Vertebrate Palaeontology* 4:1–259.

Böhme, M. 2007. 3. Herpetofauna (Anura, Squamata) and paleoclimatic implications: Preliminary results. In *Oligocene-Miocene Vertebrates from the Valley of Lakes (Central Mongolia): Morphology, Phylogenetic and Stratigraphic Implications*, ed. G. Daxner-Höck, pp. 43–52. *Annalen des Naturhistorischen Museums Wien* 108A.

Burke, J. J. 1941. New fossil Leporidae from Mongolia. *American Museum Novitates* 1117:1–23.

Daxner-Höck, G. 2000. *Ulaancricetodon badamae* n. gen. n. sp. (Mammalia, Rodentia, Cricetidae) from the Valley of Lakes in Central Mongolia. *Paläontologische Zeitschrift* 74(1/2):215–225.

Daxner-Höck, G. 2001. New zapodids (Rodentia) from Oligocene-Miocene deposits in Mongolia. Part 1. *Senckenbergiana Lethaea* 81(2):359–389.

Daxner-Höck, G. and D. Badamgarav. 2007. 1. Geological and stratigraphical setting. In *Oligocene-Miocene Vertebrates from the Valley of Lakes (Central Mongolia): Morphology, Phylogenetic and Stratigraphic Implications*, ed. G. Daxner-Höck, pp. 1–24. *Annalen des Naturhistorischen Museums Wien* 108A.

Daxner-Höck, G., V. Höck, D. Badamgarav, G. Furtmüller, W. Frank, O. Montag, and H. P. Schmid. 1997. Cenozoic stratigraphy based on a sediment-basalt association in Central Mongolia as requirement for correlation across Central Asia. In *Actes du Congrès BiochroM'97*, ed. J.-P. Aguilar, S. Legendre, and J. Michaux, pp. 163–176. Mémoires et Travaux E.P.H.E., Institut de Montpellier, 21.

Daxner-Höck, G. and W.-y. Wu. 2003. *Plesiosminthus* (Zapodidae, Mammalia) from China and Mongolia: Migrations to Europe. In *Distribution and Migration of Tertiary Mammals in Europe*, ed. J. W. F. Reumer and W. Wessels, pp. 127–151. DEINSEA 10.

Deng, T. 2005. New discovery of *Iranotherium morgani* (Perissodactyla, Rhinocerotidae) from the Late Miocene of the Linxia Basin in Gansu, China and its sexual dimorphism. *Journal of Vertebrate Paleontology* 25:442–450.

Deng, T. 2006. Chinese Neogene mammal biochronology. *Vertebrata PalAsiatica* 44(2):143–163.

Deng, T., W.-m. Wang, and L.-p. Yue. 2006. The Xiejian stage of the continental Miocene series in China. *Journal of Stratigraphy* 30:315–322.

Devjatkin, E. V. 1993. Paleomagnetic and geochronological studies of Paleogene and Miocene deposits of Mongolia. In *Oligocene-Miocene Boundary in Mongolia: Excursion Guidebook*, ed. R. Barsbold, M. A. Akhmetiev, and V. Y. Reshetov, pp. 28–35. Ulaan-Baator.

Erbajeva, M. A. 1988. Cenozoic pikas (taxonomy, systematics, phylogeny). *Nauka*, 1–224 (in Russian).

Erbajeva, M. A. 2003. Late Miocene ochotonids (Mammalia, Lagomorpha) from Central Mongolia. *Neues Jahrbuch für Geologie und Paläontologie, Monatshefte* 2003(4):212–222.

Erbajeva, M. A. 2007. 5. Lagomorpha (Mammalia): Preliminary results. In *Oligocene-Miocene Vertebrates from the Valley of Lakes (Central Mongolia): Morphology, Phylogenetic and Stratigraphic Implications*, ed. G. Daxner-Höck, pp. 165–171. *Annalen des Naturhistorischen Museums Wien* 108A.

Erbajeva, M. A. and G. Daxner-Höck. 2001. Paleogene and Neogene lagomorphs from the Valley of Lakes, Central Mongolia. *Lynx*, ISSN 0024-7774, n. s. 32:55–65.

Erbajeva, M. A. and S. Sen. 1998. Systematic of some Oligocene Lagomorpha (Mammalia) from China. *Neues Jahrbuch für Geologie und Paläontologie Monatshefte* 2:95–105.

Göhlich, U. B. 2007. 9. Gomphotheriidae (Proboscidea, mammalian). In *Oligocene-Miocene Vertebrates from the Valley of Lakes (Central Mongolia): Morphology, Phylogenetic and Stratigraphic Implications*, ed. G. Daxner-Höck, pp. 273–289. *Annalen des Naturhistorischen Museums Wien* 108A.

Gradstein, F. M. and J. G. Ogg. 2004. Geologic Time Scale 2004: Why, how, and where next! *Lethaea* 37:175–181.

Gureev, A. A. 1960. Lagomorphs from the Oligocene of Mongolia and China. *Trudy Paleotologichieskogo Institute Akademia Nauk SSSR* 77(4):5–34 (in Russian).

Heissig, K. 2007. 8. Rhinocerotidae (Perissodactyla, Mammalia). In *Oligocene-Miocene Vertebrates from the Valley of Lakes (Central Mongolia): Morphology, Phylogenetic and Stratigraphic Implications*, ed. G. Daxner-Höck, pp. 233–269. *Annalen des Naturhistorischen Museums Wien* 108A.

Höck, V., G. Daxner-Höck, H. P. Schmid, D. Badamgarav, W. Frank, G. Furtmüller, O. Montag, R. Barsbold, Y. Khand, and J. Sodov. 1999. Oligocene-Miocene sediments, fossils and basalts from the

Valley of Lakes (Central Mongolia)—an integrated study. *Mitteilungen der Geologischen Gesellschaft Wien* 90:83–125.

Huang, X.-s. 1985. Middle Oligocene Ctenodactylids (Rodentia, Mammalia) of Ulantatal, Nei Mongol. *Vertebrata PalAsiatica* 23(1):27–38.

Kepezhinskas, V. V. 1979. Cenozoic Alkaline Basaltoids of Mongolia and related deep inclusions. *Joint Soviet-Mongolian Scientific-Research Geological Expedition, Transactions* 25:1–311 (in Russian).

Koenigswald W. V. and D. Kalthoff. 2007. The enamel microstructure of molars and incisors of Paleogene and Neogene rodents from Mongolia. In *Oligocene-Miocene Vertebrates from the Valley of Lakes (Central Mongolia): Morphology, Phylogenetic and Stratigraphic Implications*, ed. G. Daxner-Höck, pp. 291–312. *Annalen des Naturhistorischen Museums Wien* 108A.

Kraatz, B. P. and J. H. Geisler. 2010. Eocene-Oligocene transition in Central Asia and its effects on mammalian evolution. *Geology* 38(2):111–114.

Li, C.-k. 1978. Two new Lagomorpha from the Miocene of Lantian, Shensi. *Professional Papers on Stratigraphy and Palaeontology* 7:143–146 (in Chinese).

Li, C. and Z.-d. Qiu. 1980. Early Miocene mammalian fossils of Xining Basin, Quinghai. *Vertebrata PalAsiatica* 18(3):198–214.

Lopatin, A. V. 1999. New Early Miocene Zapodidae (Rodentia, Mammalia) from the Aral Formation of the Altynshokysu Locality (North Aral Region). *Paleontological Journal* 33(4):429–438.

Lopatin, A. V. 2002. The largest Asian *Amphechinus* (Erinaceidae, Insectivora, Mammalia) from the Oligocene of Mongolia. *Paleontological Journal* 36(3):302–306.

Lopatin, A. V. 2004. Early Miocene small mammals from the North Aral Region (Kazakhstan) with special reference to their biostratigraphic significance. *Paleontological Journal* 38:217–323.

Lucas, S. G., E. G. Kordikova, and R. J. Emry. 1998. Oligocene stratigraphy, sequence stratigraphy, and mammalian biochronology north of the Aral Sea, western Kazakhstan. *Bulletin of Carnegie Museum of Natural History* 34:313–348.

Luterbacher, H. P., J. R. Ali, H. Brinkhuis, F. M. Gradstein, J. J. Hooker, S. Monechi, J. G. Ogg, J. Powell, U. Röhl, A. Sanfilippo, and B. Schmitz. 2004. The Paleogene Period. In *A Geologic Time Scale 2004*, ed. F. Gradstein, J. Ogg, and A. Smith, pp. 384–408. Cambridge: Cambridge University Press.

Matthew, W. D. and W. Granger. 1923. Nine new rodents from the Oligocene of Mongolia. *American Museum Novitates* 102:1–10.

Matthew, W. D. and W. Granger. 1924. New insectivores and ruminants from the Tertiary of Mongolia, with remarks on the correlation. *American Museum Novitates* 2311:1–11.

Meng, J., J. Ye, W.-y. Wu, L.-p. Yue, and X.-J. Ni. 2006. A recommended Boundary Stratotype section for Xiejian stage from northern Junggar Basin: Implications to related bio-chronostratigraphy and environmental changes. *Vertebrata PalAsiatica* 44(3):205–236 (in Chinese with English summary).

Morlo, M. and D. Nagel. 2002. New Didymoconidae (Mammalia) from the Oligocene of Central Mongolia and first information on toot eruption sequence of the family. *Neues Jahrbuch für Geologie und Paläontologie, Abhandlungen* 223(1):123–144.

Morlo, M. and D. Nagel. 2006. New remains of Hyaenodontidae (Creodonta, Mammalia) from the Oligocene of Central Mongolia. *Annales de Paléontologie* 92(3):305–321.

Morlo, M. and D. Nagel. 2007. 7. The carnivore guild of the Taatsiin Gol area: Hyaenodontidae (Creodonta, Carnivora and Didymoconidae). In *Oligocene-Miocene Vertebrates from the Valley of Lakes (Central Mongolia): Morphology, Phylogenetic and Stratigraphic Implications*, ed. G. Daxner-Höck, pp. 217–231. *Annalen des Naturhistorischen Museums Wien* 108A.

Nagel, D. and M. Morlo. 2003. Guild structure of the carnivorous mammals (Creodonta, Carnivora) from the Taatsiin Gol area, Lower Oligocene of Central Mongolia. In *Distribution and Migration of Tertiary Mammals in Europe*, ed. J. W. F. Reumer and W. Wessels, pp. 419–429. DEINSEA 10.

Qiu, Z.-d. 1985. The Neogene Mammalian faunas of Ertemte and Harr Obo in Inner Mongolia (Nei Mongol), China. 3. Jumping mice. *Senckenbergiana Lethaea* 66:39–67.

Qiu, Z.-d. 1991. The Neogene mammalian faunas of Ertemte and Harr Obo in Inner Mongolia (Nei Mongol), China. 8. Sciuridae (Rodentia). *Senckenbergiana Lethaea* 71:223–255.

Qiu, Z.-d. 1996. *Middle Miocene Micromammalian Fauna from Tunggur, Nei Mongol*. Beijing: Science Press (in Chinese with English summary).

Qiu, Z.-d., X. Wang, and Q. Li. 2006. Faunal succession and biochronology of the Miocene through Pliocene in Nei Mongol (Inner Mongolia). *Vertebrata PalAsiatica* 44(2):164–181 (in Chinese with English summary).

Russell, D. and R. Zhai. 1987. The Paleogene of Asia: Mammals and stratotype. *Mémoires du Muséum National d'Histoire Naturelle* 52:1–488.

Schaub, S. 1930. Fossile Sicistinae. *Eclogae Geologicae Helvetiae* 23 (2): 616–636.

Schlosser, M. 1924. Tertiary vertebrates from Mongolia. *Paleontologia Sinica*, C 1(1):45–53.

Schmidt-Kittler, N., M. Vianey-Liaud, and L. Marivaux. 2007. 6. Ctenodactylidae (Rodentia, Mammalia). In *Oligocene-Miocene Vertebrates from the Valley of Lakes (Central Mongolia): Morphology, Phylogenetic and Stratigraphic Implications*, ed. G. Daxner-Höck, pp. 173–215. *Annalen des Naturhistorischen Museums Wien* 108A.

Sen, S. and H. Thomas. 1979. Découverte de Rongeurs dans le Miocène moyen de la Formation Hofuf (Province du Hasa, Arabie Saoudite). *C. R. Somm. Séances Société Géologique de France* 1:34–37.

Steininger, F. F. 1999. The continental European Miocene. Chronostratigraphy, geochronology and biochronology of the Miocene "European Land Mammal Mega-Zones"(ELMMZ) and the Miocene "Mammal-Zones" (MN-Zones). In *The Miocene Land Mammals of Europe*, ed. G. R. Rössner and K. Heissig, pp. 9–38. Munich: Dr. Friedrich Pfeil.

Sulimsky, A. 1970. On some Oligocene insectivore remains from Mongolia. *Palaeontologica Polonica* 21:53–70.

Teilhard de Chardin, P. 1926. Mammifères tertiaires de Chine et de Mongolia. *Annales de Paléontologie* 15:1–51.

Tong, Y. 1989. A new species of Sinolagomys (Lagomorpha, Ochotonidae) from Xinjiang. *Vertebrata PalAsiatica* 27(4):103–116 (in Chinese, with English summary).

Tong, Y., S. Zheng, and Z.-d. Qiu.1995. Cenozoic mammal ages of China. *Vertebrata PalAsiatica* 33:290–314 (in Chinese with English summary).

Vislobokova, I. and G. Daxner-Höck. 2002. Oligocene–Early Miocene Ruminants from the Valley of Lakes (Central

Mongolia). *Annalen des Naturhistorischen Museums Wien* 103A:213–235.

Ye, J., J. Meng, and W.-y. Wu. 2003. Oligocene/Miocene beds and faunas from Tieersihabahe in the northern Junggar Basin of Xinjiang. *Bulletin of the American Museum of Natural History* 13(279):568–585.

Young, C. C. 1932. On a new ochotonid from north Suiyan. *Bull. Geological Society of China* 11:255–258.

Zazhigin, V. S. and A. V. Lopatin. 2000. The History of the Dipodoidea (Rodentia, Mammalia) in the Miocene of Asia: 1. *Heterosminthus* (Lophocricetinae). *Palaeontological Journal* 34:319–332.

Zheng, S. 1982. Middle Pliocene micromammals from the Tianzhu Loc. 80007 (Gansu Province). *Vertebrata PalAsiatica* 20:138–147.

Zhou, X.-y. 1988. Miocene ochotonid (Mammalia, Lagomorpha) from Xinzhou, Shanxi. *Vertebrate PalAsiatica* 26(2):139–148 (in Chinese, with English summary).

Ziegler, R., T. Dahlmann, and G. Storch. 2007. Marsupialia, Erinaceomorpha and Soricomorpha. In *Oligocene-Miocene Vertebrates from the Valley of Lakes (Central Mongolia): Morphology, Phylogenetic and Stratigraphic Implications*, ed. G. Daxner-Höck, pp. 53–164. *Annalen des Naturhistorischen Museums Wien* 108A.

Chapter 21

Late Cenozoic Mammal Faunas of the Baikalian Region

Composition, Biochronology, Dispersal, and Correlation with Central Asia

MARGARITA ERBAJEVA AND NADEZHDA ALEXEEVA

The Baikalian region is located in the middle of the continental interior of Asia and includes two areas—Prebaikalia to the west and Transbaikalia to the east of Lake Baikal. At present, the modern mammal faunas of these territories differ significantly because they belong to different zoogeographical subregions, a modern Europe-Siberian subregion (Prebaikalia) and a central Asian subregion (Transbaikalia).

Comparative analysis of the assemblages from Transbaikalia and Prebaikalia (including Olkhon Island) demonstrates that the faunas included a number of taxa in common with central Asia and Europe. This indicates significant interchange of European and Asian elements during the Miocene and continuing through the Early Pliocene, when the paleoenvironmental conditions of northern Eurasia were favorable to the wide distribution of mammals. It should be stressed that within the region, the Pliocene through Early Pleistocene mammal faunas of Prebaikalia and Transbaikalia were similar.

The study area experienced a complicated geological history during the Neogene. Uplift of mountain ranges of central Asia led to the development of the Lake Baikal Depression, which separated Prebaikalia and Transbaikalia and ultimately became a barrier between them. Available paleontological data based mostly on small mammals refines the biochronology and elucidates patterns of faunal turnover in relation to paleoenvironmental and climatic changes from the Miocene through the Pleistocene.

The faunas and floras of these areas are reviewed and reexamined based on the study of new fossil loclities and on the analysis of all known fossil mammal data. This is the basis for discussing the mammal successions and the key features of faunal turnover, establishing the biochronology in relation to paleoenvironmental and climatic changes, and tracing a number of dispersal events during the Neogene.

SUCCESSION OF FAUNAS

Limited data on Paleogene small mammals is reported by Pokatilov (1994) in his study of the locality Ulariya in Olkhon Island (figure 21.1). The scarce specimens represent *Desmatolagus* cf. *gobiensis*, *Cricetops* cf. *dormitor*, and Leporidae indet. Equivalent taxa are known from Oligocene localities in Mongolia (Mellett 1968). The Miocene mammals of the region are known mainly from the sites Tagay and Sarayskoe (section 1), located on Olkhon Island and on the western shore of Lake Baikal (Aya Cave). The early Miocene fauna from the site in Tagay Bay of Olkhon Island includes remains of both small and large mammals (table 21.1; Logatchev, Lomonosava, and Klimanova 1964; Vislobokova 1994).

According to Vislobokova (1994), this fauna contains some species in common with European Early Orleanian (MN 3–4) localities (such as *Amphitragulus*, *Lagomeryx*, *Stephanocemas*), suggesting a united Europe-Siberian paleozoogeographical subregion at that time. The Tagai fauna likely correlates with the Shanwangian of China because these faunas contain genera in

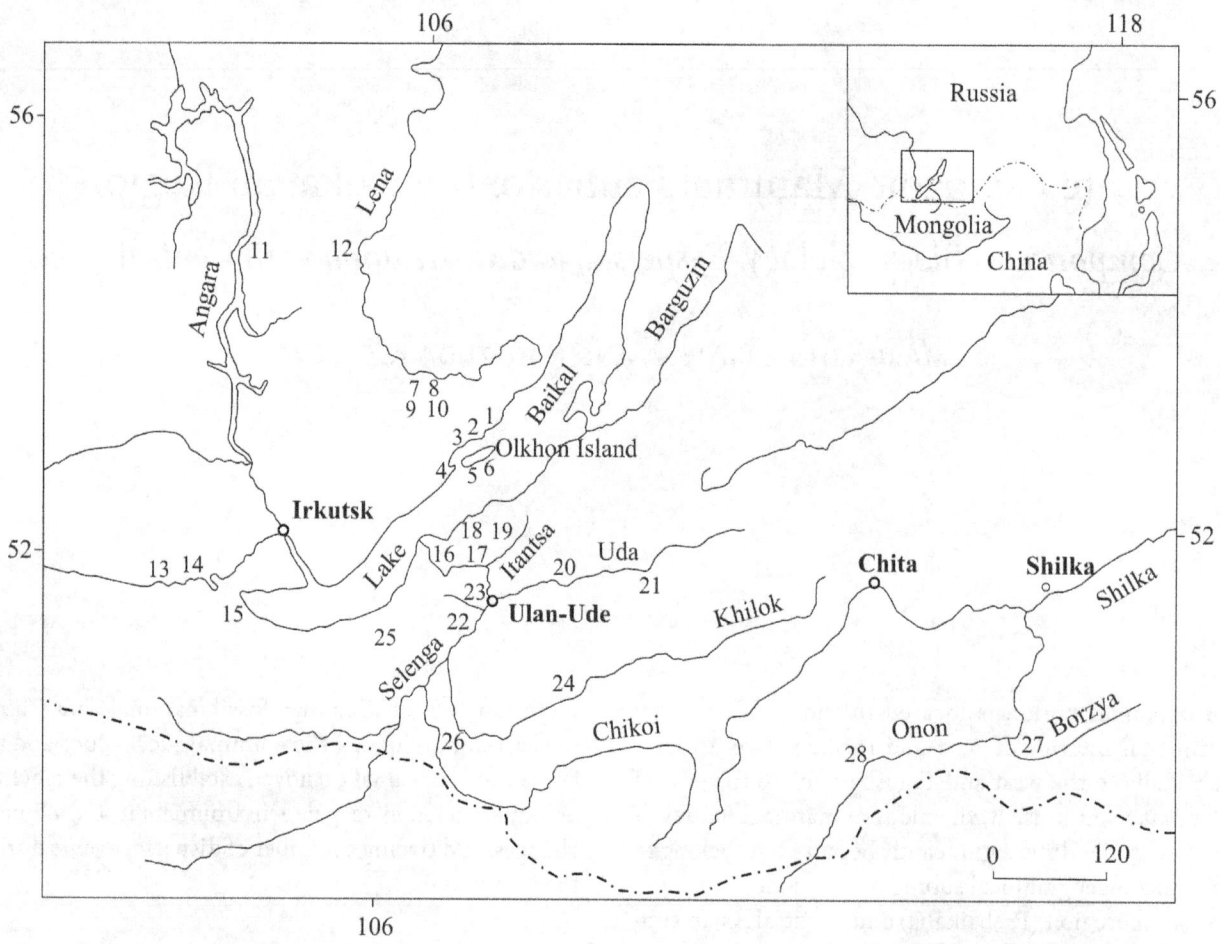

Figure 21.1 Sketch map of the main fossil mammal localities of the Baikalian region. *Oligocene*: (*1*) Ulariya. *Miocene*: (*2*) Sarayskoe (section 1, horizon 5); (*3*) Tagai; (*4*) Aya. *Early Pliocene*: (*2*) Sarayskoe (section 1, horizon 6). *Middle Pliocene*: (*22*) Tologoi 1; (*25*) Udunga; (*26*) Beregovaya; (*27*) Gryazi. *Late Pliocene*: (*5*) Kharaldai; (*6*) Sasa; (*7*) Podtok; (*8*) Cherem Khaem; (*9*) Malye Goly1; (*10*) Malye Goly 2; (*13*) Ilcha; (*14*) Anchuk; (*15*) Shankhaikha; (*16*) Klochnevo 1; (*17*) Klochnevo 2; (*18*) Zasukhino 1. *Pleistocene*: (*11*) Zayarsk; (*12*) Podymakhino; (*19*) Zasukhino 2, 3; (*20*) Dodogol 1, 2; (*21*) Kudun; (*23*) Tologoi 2; (*24*) Ust-Obor; (*28*) Nyzhnii Tsasuchei.

common, taxa such as *Lagomeryx* and *Stephanocemas* (table 21.2).

Most of the mammal taxa from the Tagai fauna were the inhabitants of woodland biotopes. Pollen data, obtained by Logatchev, Lomonosava, and Klimanova (1964) in Tagay Bay, show that in the surrounding areas broadleaved/coniferous forests were widely distributed. *Ulmus* sp., *Pinus* s/g. *Haploxylon, Dyploxylon, Picea* sp., *Abies* sp., *Tsuga* sp., *Juglans* sp., *Tilia* sp., and *Corylus* sp. were characteristic taxa, and *Pterocarya* sp., *Alnus* sp., *Celtis* sp., *Carpinus* sp., *Liquidambar* sp., and *Carya* sp. also existed there but were scarce. Among the herbaceous plants, Gramineae, Cyperaceae, Chenopodiaceae, and composites were recognized. The analysis of the species composition of both flora and fauna indicates that

forest-steppe and restricted steppe landscapes existed during the early Miocene as well. Based on the mammalian fauna, malacofauna, pollen flora, and geology, the Early Miocene climate of the region was warm and moist (Logatchev, Lomonosava, and Klimanova 1964; Popova et al. 1989; Belova 1975, 1985).

The middle Miocene is represented by the fauna from the Aya Cave locality, located on the western shore of Lake Baikal. This fauna includes *Succinea* ex gr. *oblonga* Draparnaud, *Leuciscus* sp., *Rutilus* sp., *Carassius (Paleocarassius)* sp., *Esox* sp., *Channa* sp., *Leobergia* sp., *Rana* sp., *Trionyx* sp., Aves gen indet., Chiroptera gen indet., Insectivora gen. indet., *Amphilagus* cf. *fontannesi* (Deperet), *Heterosminthus erbajevae* Lopatin, *Gobicricetodon filippovi*, and Aplodontidae gen. indet. (Filippov, Erbajeva,

Table 21.1

Correlation of the Stratigraphic Position of Faunas (F) and Faunal Complexes (FC) of the Baikalian Region and Adjacent Asian Territories

		Prebaikalia and Olkhon Island[a]	Transbaikalia[b]	Northern Mongolia[c]	Northern China Mammal Age
Pleistocene	Middle		Ust-Kiranian F. Ivolginian F.		Zhoukoudianian
		Begulian F. Nyurganian F.	Tologoi FC.	Nalaikhan F.	
	Early	Zaglinian F. Khogotian F.	Zasukhinian F. Kudunian F.		Nihewanian
		Elgan F. Nikiley F.	Ust-Oborian F. Dodogolian F.	Bulganian FC	
Pliocene	Late	Kharantsinian FC	Itantsinian FC	Bural-Obonian FC	Yushean
	Middle		Chikoian FC	Chikoian FC	
			Udunginian FC	Orkhon 1B F	
	Early	Khuzirian F. Olkhonian F.		Orkhon 1A and Chono-Khariakh 2 F.	
Miocene	Late	Odonimskii FC Sarayskii FC		Khirgis Nur II	Baodean
				Biozone E	Bahean
	Middle	Aya cave F.		Ulan Tologoi F.	Tunggurian
	Early	Tagai Loc. F.		Biozone D1/1	Shanwangian
					Xiejian
Oligocene		Fauna of Loc. Ulariya		Biozone C1	Tabenbulukian

[a]Adamenko (1977); Pokatilov (1994); Logatchev, Lomonosava, and Klimanova (1964).
[b]Vangengeim (1977); Erbajeva and Alexeeva (2000); Alexeeva (2005).
[c]Zazhigin (1989); Höck et al.(1999); Erbajeva (2007).

and Khenzykhenova 1995; Erbajeva and Alexeeva 1997; Erbajeva and Filippov 1997; Filippov, Erbajeva, and Khenzykhenova 2000; Lopatin 2001; Sen and Erbajeva 2011). Previously, this fauna was referred either to the early Middle Miocene (Erbajeva and Alexeeva 1997; Erbajeva and Filippov 1997) or to the middle Late Miocene (Filippov, Erbajeva, and Khenzykhenova 1995). However, tooth morphology and the evolutionary stage of the genera *Heterosminthus* and *Gobicricetodon* suggest that the fauna is Middle Miocene in age (Lopatin 2001; Sen and Erbajeva 2011). The species composition of this fauna includes both European and Asian elements, which were distributed widely in Eurasia at that time. The genus *Am-*

philagus occupied vast areas of Eurasia from Europe to Japan (Tomida and Goda 1993), and *Heterosminthus* and *Gobicricetodon* were widely distributed in Asia (Qiu 1996; Zazhigin and Lopatin 2000). Another ochotonid group, although absent from Aya Cave, indicates an equally wide Asian distribution: far to the west and spanning the Asian continent, *Alloptox anatoliensis* from Turkey is found to be very close in tooth morphology to *A. cinghaiensis* from China. Thus, during the end of the middle Miocene, the lagomorphs *Alloptox* and *Bellatonoides* (Qiu 1996; Sen 2003) had wide distributions from Asia Minor to China. At that time, unobstructed terrestrial connection between Europe and Asia allowed migration for mammalian

Table 21.2

Miocene–Pleistocene Faunal Succession of the Baikalian Region and Adjacent Territory of Mongolia

		Olkhon Island[a]	Prebaikalia[b]	Transbaikalia[c]	Northern Mongolia[d]
Pleistocene	Middle		Ochotona sp., Spermophilus undulatus, Cricetulus barabensis, Clethrionomys sp., Lagurus lagurus, Microtus gregalis, M. oeconomus, Dicrostonyx cf. simplicior	Sorex sp., Lepus cf. tolai, Ochotona cf. daurica, Marmota sibirica, Spermophilus undulatus, Allactaga sp., Eolagurus cf. luteus, Lagurus lagurus, Lasiopodomys brandti, Microtus fortis, M. gregalis	
		Begulian and Nyurganian Faunas Ochotona sp., Spermophilus undulates, Ellobius talpinus, Eolagurus simplicidens, Myospalax wongi, Pitymys ex gr. arvaloides, Microtus oeconomus, Microtus sp.		Tologoi Complex Ochotona gureevi, Spermophilus gromovi, Marmota nekipelovi, Cricetulus barabensis, Allactaga sibirica transbaikalica, Ellobius tancrei, Meriones unguiculatus, Eolagurus simplicidens, Microtus gregalis, M. mongolicus, Lasiopodomys brandti, Myospalax wongi, Coelodonta tologoijensis, Bison sp., Archidiskodon sp., Equus sanmeniensis, Spirocerus cf. peii, Cervus ex gr. elaphus	Nalaikhan Complex Ochotona sp., Spermophilus sp., Marmota sp., Allactaga sp., Mammuthus sp., Coelodonta tologojensis, Equus ex. gr. sanmeniensis, Equus (Hemionus) sp., Spirocerus kiakhtensis, Canis variabilis, Xenocuon cf. lycaonoides, Ursus ex.gr. deningeri, Hyaena brevirostris sinensis, Panthera cf. tigris paleosinensis, Felis sp.
	Early	Zaglinian Fauna Spermophilus ex gr. undulatus, Prolagurus cf. ternopolitanus, Allophaiomys cf. deucalion, Prosiphneus youngi	Khogotian Fauna Ochotona sp., Lepus sp., Allophaiomys pliocaenicus, Spermophilus sp., Microtinae gen.	Zasukhinian Fauna Crocidura sp., Ochotona tologoica, O.zasichini, O. cf. sibirica, Ochotonoides complicidens, Marmota sp., Spermophilus cf. undulatus, S. cf. tologoicus, Cricetulus barabensis, Cricetinus sp., Terricola ex gr. hintoni-gregaloides, Myopus sp., Clethrionomys sp., Allactaga sp., Allophaiomys pliocaenicus, Lasiopodomys brandti, Microtus cf. oeconomus, Alticola sp., Prosiphneus ex gr. youngi, Equus sanmeniensis, Coelodonta. tologojensis, Alces latifrons, Spirocerus wongi, Pachycrocuta brevirostris, Canis variabilis	
		Elgan Fauna Spermophilus cf. tologoicus, Mimomys pusillus, Borsodia hungarica, Borsodia laguriformes, Prosiphneus sp.	Fauna of Nikilei locality Ochotona sp., Spermophilus sp., Cricetulus sp., Mimomys aff. coelodus, Cromeromys intermedius, Villanyia sp., Lagurodon arankae, Prolagurus ternopolitanus, Allophaiomys pliocaenicus	Ust-Oborian Fauna Ochotona ustoborica, Spermophilus sp., Lagurodon arankae, Prolagurus cf. ternopolitanus, Allophaiomys cf. deucalion, Clethrionomys sp., Prosiphneus cf. youngi, Equus ex gr. sanmeniensis, Coelodonta cf. tologojensis, Spirocerus wongi	Bulganian Complex Beremendia sp., Shikamainosorex sp., Sorex sp., Ochotona cf. zazhigini, Cricetulus cf. barabensis, Allactaga sp., Scirtopoda sp., Alactagulus sp., Villanyia chinensis, Mesosiphneus ex. gr. praetingi

Epoch	Age	Fauna / Complex	Taxa
Pliocene	Late	Zayarskian Fauna	Ochotona cf. whartoni, O. filippovi, Spermophilus sp., Mimomys cf. pusillus, Lemmus cf. kowalskii, Clethrionomys sp., Allophaiomys cf. deucalion, Borsodia sp.
	Late	Dodogolian Fauna	Ochotona cf. nihewanica, O. bazarovi, Spermophilus sp., Borsodia chinensis laguriformes, Mimomys cf. pusillus, Allophaiomys deucalion, Prosiphneus youngi, Equus sp., Coelodonta sp., Rhinocerotidae indet., Spirocerus wongi, Homotherium sp.
	Late	Bural-Obonian Complex	Scaptochirus cf. primaevus, Sorex sp., Hypolagus sp., Ochotona zazhigini, Cricetulus cf. barabensis, Allactaga sp., Sicista cf. pliocaenica, Orientalomys cf. sibiricus, Micromys sp., Villanyia orchonensis, Mimomys ex gr. hintoni-coelodus, Synaptomys cf. mongolicus, Mesosiphneus ex. gr. praetingi
	Late	Kharantsinian Complex	Sorex sp., S. cf. palaeosibiriensis, Ochotona sp., Ochotonoides complicidens, Cricetulus sp., Villanyia cf. petenyi, V. ex gr. chinensis, Mimomys cf. coelodus, M. reidi, M. sp., Mesosiphneus praetingi
	Late	Fauna of Podtok Site	Sorex sp., Desmana sp., Hypolagus sp., Pliolagomys sp., Ochotona sp., Sciuridae indet., Spermophilus sp., Allactaga sp., Clethrionomys cf. rutilus., Villanyia petenyi, V. lenensis, V. angensis, V. cf. chinensis, Mimomys aff. polonicus, M. pliocaenicus, M. reidi, M. aff. newtoni, M. parapolonicus, Cromeromys intermedius
	Late	Itantsinian Complex	Ochotona cf. intermedia, Ochotona cf. nihewanica, Spermophilus itancinicus, S. tologoicus, Marmota sp., Castor sp., Allactaga sp., Cricetulus cf. barabensis, Cricetinus cf. varians, Clethrionomys sp., Villanyia klochnevi, Mimomys cf. reidi, M. burgondiae, M. cf. pusillus, M. cf. pseudintermedius, Mesosiphneus youngi, Equus sanmeniensis, Itantzatherium angustirostre, Gazella cf. sinensis,
	Late	Chikoian Complex	Beremendia sp., Sorex sp., Neomys sp., Hypolagus multiplicatus, H. transbaicalicus, Ochotonoides complicidens, Ochotona cf. gromovi, O. cf. intermedia, O. sibirica, Castor cf. zdanskyi, Allactaga cf. saltator, Sicista pliocaenica, Orientalomys sibiricus, Cricetinus cf. varians, Cricetulus sp., Promimomys sp., Villanyia eleonorae, Mimomys ex gr. hintoni-coelodus, Synaptomys cf. mongolicus, Mesosiphneus praetingi, Hipparion tchikiocum, H. houfenense, Gazella cf. sinensis, Orchonoceros gromovi, Canis cf. chihliensis, Hyaena cf. licenti, Acinonyx sp., Nyctereutes sinensis, Euryboas cf. lunensis, Lynx shansius, Presbytis eohanuman
	Middle	Chikoian Complex	Beremendia fissidens, Petenyia hungarica, Sorex sp., Hypolagus multiplicatus, H. transbaicalicus, Ochotonoides complicidens, Ochotona gromovi, O. intermedia, O. sibirica, Castor cf. zdanskyi, Sicista pliocaenica, Orientalomys sibiricus, Cricetinus cf. varians, Cricetulus sp., Villanyia eleonorae, Mimomys minor, M. cf. reidi, M. pseudintermedius, Mesosiphneus praetingi, Hipparion chikoicum, H. houfenense, Gazella sinensis, Canis cf. chihliensis, Hyaena cf. licenti, Nyctereutes sinensis, Euryboas cf. lunensis, Lynx issiodorensis, Acinonyx cf. pardinensis
	Middle	Fauna of Tologoi 1	Petenyia hungarica, Sorex sp., Ochotonoides complicidens, Ochotona intermedia, Marmota tologoica, Orientalomys sibiricus, Cricetulus sp., Cricetinus cf. varians, Promimomys gracilis, Pitymimomys koenigswaldi, Villanyia eleonorae, Mimomys minor, Mesosiphneus praetingi, Hipparion chikoicum

(continued)

Table 21.2 (continued)

	Olkhon Island[a]	Prebaikalia[b]	Transbaikalia[c]	Northern Mongolia[d]
	Khuzirian Fauna Ochotona sp., Sicista cf. pliocaenica, Apodemus sp., Promimomys cf. gracilis, Promimomys sp., Prosiphneus chuzhirica		Udunginian Complex Hypolagus multiplicatus, H. transbaicalicus, Ochotonoides complicidens, Ochotona aff. sibirica, Ochotona sp., Kowalskia sp., Gromovia daamsi, Cricetinus cf. varians, Orientalomys cf. sibiricus, Castor sp., Promimomys cf. gracilis, P. cf. stehlini, Mesosiphneus praetingi, Villanyia ex gr. eleonorae, Mimomys cf. minor, Pliocrocuta pyrenaica, Lynx issiodorensis, Homotherium crenatidens, Parameles suillus, Ursus minimus, Parailurus baikalicus, Arctomeles sp., Nyctereutes sp., Gulo minor, Hipparion houfenense, H. chikoicum, Archidiskodon sp., Orchonoceros gromovi, Axis shansius, Capreolus constantini, Gazella sinensis, Ovis sp. Antilospira zdanskyi, Presbytis eohanuman	Fauna of Orkhon-1A and Chono-Khariakh-2 Erinaceus sp., Sorex sp., Anourosorex sp., Hypolagus sp., Ochotona sp., Apodemus sp., Micromys sp., Kowalskia cf. magna, Mesocricetus cf. primaevus, Cricetulus sp., Microtodon cf. progressus, Aratomys multifidus, Prosiphneus ex gr. eriksoni
Early	Olkhonian Fauna Hypolagus sp., Ochotona sp., Ochotonoides complicidens, Eozapus sp., Kowalskia sp., Micromys sp., Stachomys ex gr. trilobodon, Promimomys insuliferus, Microtodon sp., Prosiphneus olchonicus			

			Khirgis Nur II
Miocene	Late	**Odonimskii Complex**[a] *Hypolagus* sp., *Alilepus?* sp., *Proochotona* sp., *Ochotona* sp., *Paralophocricetus* progressus, *Lophocricetus orientalis*, *Microtoscoptes praetermissus*, *M.* cf. *tjuvanensis*, *Microtodon* cf. *atavus*, *Prosiphneus eriksoni*, *Hipparion* sp. **Sarayskii Complex** Soricidae gen., *Proochotona* sp., *Ochotona* sp., *Paralophocricetus saraicus*, *Monosaulax* sp., *Prosiphneus licenti*, *Moschus* cf. *grandaevus*, *Pavlodaria* sp.		*Ochotona* sp.
	Middle	**Fauna of Aya Cave**[b] *Amphilagus* cf. *fontannesi*, *Heterosminthus erbajevae*, *Gobicricetodon* n. sp.	**Fauna of Ulan Tologoi** *Alloptox gobiensis*, *Bellatona forsythmajori*, *Protalactaga* cf. *tungurensis*, *Cricetodon* sp., *Lagomeryx* sp., *Zyglophodon gromovae*, *Serridentinus tologojensis*, *Gobiceros mongolicus*, *Stephanocemas* sp.
	Early	**Fauna of Tagai Loc.**[c] *Proscalops* sp., *Talpa* sp., *Monosaulax* sp., *Cricetodon* cf. *sansaniensis*, Mustelidae gen., Felidae gen., *Anchitherium* sp., *Dicerorhinus* sp., *Amphitragulus boulangeri*, *Lagomeryx parvulus*, *Stephanocemas* sp., *Palaeomeryx* cf. *kaupi*, *Orygotherium* aff. *escheri*, *Brachyodus intermedius*	**Fauna of Biozone D1/1** *Heterosminthus mongoliensis*, *Democricetodon* sp., *Megacricetodon* sp., *Bellatona* cf. *yanghuensis*, *Alloptox minor*, *Alloptox gobiensis*
Oligocene		**Fauna of Ulariya Loc.**[d] *Desmatolagus* cf. *gobiensis*, *Cricetops* cf. *dormitor*, Leporidae indet.	**Fauna of Biozone C, C1** *Desmatolagus* cf. *gobiensis*, *Bohlinotona* sp., *Sinolagomys* cf. *major*, *Sinolagomys* sp., *Tsaganomys altaicus*, *Tataromys parvus*, *Parasminthus* sp., *Eucricetodon* sp.

[a] Pokatilov (1994); Logatchev, Lomonosava, and Klimanova (1964).

[b] Adamenko (1977); Filippov, Erbajeva, and Khenzykhenova (1995); Erbajeva (1998).

[c] Vangengeim (1977); Erbajeva and Alexeeva (2000); Alexeeva (2005).

[d] Zhegallo et al. (1982); Zazhigin (1989); Höck et al. (1999); Erbajeva (2007).

faunas. Such connections existed because the Paratethys ceased to exist and there were no high mountains to serve as barriers to dispersal events (Ilyina et al. 2004). The Aya Cave fauna may be correlated with Tunggur (Moergen) in China (Qiu 1996), and Ulan-Tologoi in Mongolia, based on common occurrence of *Gobicricetodon* (Tunggur) and *Amphilagus* (Ulan Tologoi).

The late Miocene fauna is represented at the Sarayskoe locality on Olkhon Island in Saray Bay of Lake Baikal. This site is multilayered, and mammalian remains were recovered from several successive fossiliferous horizons of different geological ages. They contain small mammal assemblages, which differ from each other in species composition. These different assemblages were proposed as independent faunal complexes (Mats et al. 1982).

The early stage of the late Miocene fauna is the Sarayskii Faunal Complex (type locality Sarayskoe, section 1, horizon 3; see table 21.1), including insectivores, ochotonids, and rodents as well as *Moschus* cf. *grandaevus* Schlosser and *Pavlodaria* sp. (Pokatilov 1994). This fauna is characterized by the first appearance of a number of typical Central Asian taxa (*Ochotona*, cricetines, siphneines, *Moschus*).

The succeeding fauna is the Odonimskii Faunal Complex, representing the latest stage of the Late Miocene fauna. The type locality of this fauna is at Sarayskoe as well (section 1, horizon 5). This fauna contains mainly small mammals represented by lagomorphs and rodents (see table 21.1; Pokatilov 1994). The fauna is characterized by the predominance of Central Asian elements such as *Proochotona*, *Ochotona*, and *Prosiphneus*, as well as by the first appearances of the Asian genera *Microtodon* and *Microtoscoptes*. Also useful are the New World leporid genera *Hypolagus* and *Alilepus*, which indicate further Holarctic distribution. These genera in Siberia indicate that faunal interchange continued between Europe and Asia, as well as between Asia and North America. The Miocene faunas of the Baikalian region should be correlated generally with the Tunggurian and Baodean faunas of Northern China due to the presence of common genera such as *Gobicricetodon* and *Heterosminthus* (Tunggurian), as well as *Lophocricetus*, siphneines, and ochotonids (Baodean). Elements in the Odonimskii Faunal Complex such as *Microtoscoptes*, *Microtodon*, and *Myotalpavus* indicate correlation with the latest Miocene Ertemte fauna of Inner Mongolia. In the Late Miocene, the climate continued to be warm and rather humid, but a trend toward cooling and aridification is evident. Broadleaved and broadleaved/coniferous forests were gradually replaced by open landscapes distributed widely, as demonstrated by the abundance and relative diversity of

dipodids among rodents (Logatchev, Lomonosava, and Klimanova 1964; Belova 1985; Pokatilov 1994).

Gradual cooling and intensified orogenic processes in the northern latitudes characterized the Pliocene. As a result, considerable changes in the paleoenvironment and in biogeocenoses are reflected in particular among small mammals and floral composition.

Early Pliocene faunas are known from Olkhon Island. The Olkhonian fauna (MN 14, early Ruscinian) is known from locality Sarayskoe (section 1, horizon 6). The Olkhonian faunal stage includes a shrew, the lagomorphs *Hypolagus* sp., *Ochotonoides complicidens* (Boule and Teilhard de Chardin) and *Ochotona* sp., and the rodents *Eozapus* sp., *Kowalskia* sp., *Micromys* sp., *Stachomys* ex gr. *trilobodon* Kowalski, *Microtodon* sp., and *Prosiphneus olchonicus* Pokatilov (1994). This fauna differs from the preceding one by the loss of the genus *Microtoscoptes*, by the reduction in abundance of *Microtodon* and dipodids, and by more advanced siphneines, which became abundant among small mammals and composed up to 40% of the total number of species. New lagomorphs and rodents (*Ochotonoides*, *Kowalskia*, *Stachomys*, and *Promimomys*) appeared for the first time. The later Ruscinian (MN 15) is represented by the Khuzhirian fauna, discovered also from Sarayskoe (section 5, horizon 8). This fauna includes *Ochotona* sp., *Sicista* sp., *Apodemus* sp., *Promimomys* cf. *gracilis* (Kretzoi), *Promimomys* sp., and *Prosiphneus chuzhirica* Pokatilov (1994). In these faunas, the presence of both European (*Kowalskia*, *Stachomys*, and *Promimomys*) and Asian (*Ochotonoides*, *Ochotona*, *Microtodon*, and siphneine) taxa indicate continued intercontinental faunal exchange across the region. The stratigraphic equivalents of these faunal assemblages in China are early Yushean faunas from Gaotege and Bilike, Inner Mongolia, and Gaozhuang, Yushe Basin, Shanxi, which contain a number of taxa in common (Deng 2006; Qiu, Wang, and Li 2006). In Mongolia, faunas of this stage are known from the sites Chono-Khariakh-2 and Orkhon 1A (Zazhigin 1989).

The small mammals, in conjunction with the flora including *Alnus* sp., *Picea* sp., *Betula* sec., *Albae*, *Tilia* sp., *Corylus* sp,. and *Juglans* sp., suggest that during the early Pliocene the climate became slightly cooler and drier than in the latest Late Miocene. The landscapes of the region were of mosaic type (Belova 1985; Erbajeva and Alexeeva 1997).

At the beginning of the middle Pliocene, the climate in the region became slightly cooler than at the preceding epoch and was still relatively warm and humid, and woodland predominated. The most representative mammal faunas of that time are the Udunginian and Chikoian

Faunal Complexes known from the Transbaikal area and referred, respectively, to mammal zones MN 16a and MN 16b. These faunal complexes reflect two stages in the evolutionary development of the middle Pliocene faunas. The earlier stage, known in the Udunga site (MN 16a), includes lagomorphs and rodents of small mammals and different large mammal taxa (see table 21.1) (Sotnikova and Kalmykov 1991; Vislobokova, Erbajeva, and Sotnikova 1993, 1995; Erbajeva, Alexeeva, and Khenzykhenova 2003; Alexeeva 2005; Kalmykov and Maschenko 2005; Sotnikova 2006). The small mammal association at Udunga shows a predominance of leporids (*Hypolagus*, up to 45% of the fauna) and zokors (*Mesosiphneus*, ~ 20% of the fauna). Pikas of the genus *Ochotonoides* also were rather abundant. Siphneines were numerous, as in the preceding early Pliocene faunas known from Olkhon Island. The Ruscinian rooted voles of the genus *Promimomys* and the hamster *Kowalskia* were still present. *Gromovia daamsi*, a hamster with peculiar, high-crowned teeth, is a characteristic form of this fauna. Rooted voles of the genera *Villanyia* and *Mimomys* appeared for the first time in the region, but they were very scarce (less than 5% of the total number of specimens). Another peculiarity of the Udunginian fauna is the composition of carnivore assemblages due to joint occurrence of the boreal inhabitant *Gulo* and warm-loving, forest elements like *Parailurus*, *Parameles*, and *Ferinestrix*. No analogue of this assemblage has ever been noted among the coeval faunal assemblages of Eurasia and North America (Sotnikova 2006). Moreover, for the first time in Transbaikalia, this locality records the lesser panda *Parailurus* (Sotnikova 2008); it coexisted with common Eurasian Late Ruscinian–Villafranchian ursid, badger (*Parameles*), hyaenid, and felid taxa (Sotnikova 2006). At the Udunga locality *Archidiskodon* sp. is encountered for the first time in Transbaikalia (Kalmykov and Maschenko 2005).

The species composition of both the large and small mammal faunas indicates that the inhabitants of the forest biotopes were abundant at that time, whereas the proportion of steppe and meadow dwellers was relatively less. Among the large mammals in the Udunga fauna, artiodactyls and the genera *Parapresbytis* and *Parailurus* (extant representatives of some of these are forest inhabitants) were predominant, as were leporids, murids, and castorids among the small mammals (Sotnikova and Kalmykov 1991; Vislobokova, Erbajeva, and Sotnikova 1995; Erbajeva, Alexeeva, and Khenzykhenova 2003). The plant communities of the region of that time show the predominance of forest taxa of the genera *Tsuga*, *Juglans*, *Carpinus*, *Corylus*, *Ulmus*, *Tilia*, *Quercus*, *Acer*, *Larix*,

and *Pinus* (Kalmykov and Malaeva 1994). Thus, both paleofaunal and paleofloral data obtained at the Udunga site demonstrate that forests were widely distributed at the beginning of the middle Pliocene. The meadow and steppe landscapes were subordinate. Analysis of the mammal fauna composition and paleovegetation indicates a warm and relatively humid climate in the area, similar probably to the modern environment of South China (Erbajeva, Alexeeva, and Khenzykhenova 2003; Alexeeva 2005). This is confirmed by mineralogical analysis (Dergausova, Rezanova, and Baldaeva 1991). Although the climate continued to be relatively warm and humid, there still appears to be a trend toward increasing aridity and cooling compared to the early Pliocene.

The mammalian fauna of the Chikoian Faunal Complex (MN 16b) suggests a predominance of open landscapes, reflecting a cooler, drier climate compared to that during the existence of the Udunginian fauna. The Chikoian complex is known from the Tologoi 1 and Beregovaya localities. The detailed analysis of the faunas of these two sites, especially new paleontological data from Tologoi 1, reveals that the faunas of Tologoi 1 and Beregovaya differ by the species composition and by the evolutionary stage of some arvicolid taxa. Thus, in the Chikoian fauna two successive substages might be recognized. The first one is represented by the fauna of Tologoi 1 site. In this fauna, the abundance of leporids, siphneines, and *Ochotonoides* decreases significantly compared with the Udunga locality. However, this substage is characterized by the last occurrence of the archaic arvicolid *Promimomys* and the first appearance of the European genera *Petenyia* and *Pitymimomys*. The genus *Marmota* is recorded also for the first time in the Baikalian region, demonstrating the immigration of this rodent from North America to Asia in the Pliocene (Erbajeva and Alexeeva 2009). Arvicolids of the genera *Villanyia* and *Mimomys* are represented but still are not abundant.

The next substage of the Chikoian Faunal Complex (see table 21.1) is known from the Beregovaya site (Erbajeva 1974; Vangengeim 1977; Erbajeva and Alexeeva 2000; Sotnikova 2006). The fauna from Beregovaya is characterized by the abundance of arvicolids of the genera *Mimomys* and *Villanyia*, especially *Villanyia eleonorae* (up to 70% of the total number of small mammals). Moreover, lagomorphs are highly diversity and abundant (seven taxa), and ochotonids predominate. The Chikoian Faunal Complex as a whole differs significantly from the Udunginian one by the predominance of open landscape inhabitants, by the reduction in number of leporids and siphneines, and by the increase in diversity of ochotonids and rooted voles. Ochotonids and rooted voles *Villanyia* and *Mimomys* are represented by more evolved forms than those of the Udunginian fauna.

The species composition of the Chikoian fauna shows that open landscapes expanded and savanna-like forest-steppes and steppes became widespread due to increasing aridity and coolness, but the climate of the region was still mild. Thus, during the middle Pliocene, climatic and paleoenvironmental changes occurred: at the beginning of this time, the landscapes were dominated by broad-leaved forest, which were replaced by forest-steppe areas, becoming savanna-like at the end. Along river valleys woodland persisted, and some parts of the territory were occupied by meadows, bushes, and thickets.

The middle Pliocene faunas of the Baikalian region contain a number of taxa in common with central Asia (*Ochotonoides, Ochotona, Mesosiphneus, Orientalomys, Cricetinus, Mimomys, Villanyia*), which allows correlation with coeval faunas, particularly those of northern Mongolia and China. In Mongolia, successive mammal associations like the Udungan and Chikoian assemblages are recognized. The fauna of the Orkhon 1B site might be considered as the equivalent of the Udungan fauna; the mammal assemblages from the lower level of the Shamar site correspond to the fauna of Tologoi 1, and those from the upper level of Shamar are the equivalent of assemblages from Beregovya (Zazhigin 1989; Sotnikova 2006). In China, the coeval faunas are late Yushean in age, known from a number of localities in Shansi and Gansu (Lingtai; Flynn et al. 1995; Zheng and Zhang 2001; Deng 2006).

At the beginning of the Late Pliocene, the trends of climatic cooling and aridification continued in the Baikalian region. The most thermophilic elements of the East Siberian flora had already disappeared, and there was significant reorganization in mammal faunas.

Late Pliocene successive faunal assemblages in the Baikalian region are known from western Transbaikalia (Klochnevo 1, Klochnevo 2, and Zasukhino 1 localities), Prebaikalia (Podtok, Cherem Khaem, Malye Goly 1, Malye Goly 2, Shankhaikha, Aerkhan, Ilcha, and Anchuk), and from the Olkhon Island (sections along Kharaldai, Bayan Shungen, and Saray bays). The faunas are referred to MN 17 Mammal Zone (Adamenko 1977; Pokatilov 1994; Vislobokova, Erbajeva, and Sotnikova 1995; Erbajeva and Alexeeva 2000; Alexeeva 2005).

The most representative fauna of the Late Pliocene in the Transbaikal area is represented by the Itantsinian Complex (see table 21.1; Vangengeim 1977; Erbajeva and Alexeeva 2000). Characteristic are species of *Ochotona*, ground squirrels, hamsters, moderately advanced voles, derived zokors, and an *Equus* large mammal fauna.

In Prebaikalia, the mammalian faunas are represented mostly by small mammal associations, which are known from several localities, in particular from Podtok site (Adamenko 1977; Adamenko and Adamenko 1986). The Late Pliocene fauna on Olkhon Island is represented by the mammals of the Kharantsinian Complex (Pokatilov 1994; see table 21.1). The most remarkable feature of the Baikalian faunas of this age is the first appearance of the genera *Clethrionomys, Spermophilus, Allactaga,* and *Equus,* both in the Transbaikal and Prebaikal regions. In the Transbaikal area, the rooted voles of the genera *Villanyia* and *Mimomys* are less abundant than in the preceding Chikoian fauna. However, they are highly diversified in the faunas of Olkhon Island and Prebaikalia. It should be noted that the faunas of these three areas (Transbaikalia, Prebaikalia, and Olkhon Island) share several typical Asian elements such as *Ochotona* and *Ochotonoides,* but they differ from each other mainly in the species of the rooted voles *Villanyia* and *Mimomys.* The Prebaikalian arvicolids are represented by European species such as *Villanyia* cf. *petenyi, Mimomys* cf. *coelodus,* and *M. reidi.* The Transbaikalian faunas includes endemic forms such as *Mimomys* cf. *pseudintermedius* and *Villanyia klochnevi.*

A significant reorganization in small mammal faunas occurred at the beginning of the early Pleistocene. The Pliocene genera *Pitymimomys, Promimomys,* and *Villanyia* disappeared completely; *Villanyia* was replaced by *Borsodia.* The genus *Mimomys* is scarce in early Pleistocene faunas, and it appears to be extinct shortly after the beginning of this epoch. *Spermophilus, Cricetulus,* and derived siphneines are typical; ochotonids are numerous and diverse. Some forest inhabitants, such as *Clethrionomys, Alticola,* and *Myopus* are also present.

The main characteristic of the Early Pleistocene small mammal faunas of the Baikalian region is the first appearance of the genera *Borsodia, Allophaiomys, Lagurodon, Prolagurus, Lemmus, Eolagurus, Terricola, Lasiopodomys,* and *Microtus.* Most of these taxa were widely distributed in Eurasia (Alexeeva and Erbajeva 2005). The analysis of faunal succession shows that the rates of evolution in microtines at that time were relatively high (Erbajeva and Alexeeva 2000; Alexeeva and Erbajeva 2005). During the Early Pleistocene, the genus *Borsodia* was replaced by the rootless, cementless genera *Lagurodon* and *Prolagurus,* which were known from western Europe to eastern Asia. It is assumed that in Asia *Lagurodon* evolved into *Eolagurus* and that in Eurasia *Prolagurus* evolved into *Lagurus.* The latest *Mimomys* coexisted with *Allophaiomys,* but both were extinct by the end of the Early Pleistocene. However, at this time the genera *Microtus* and *Lasiopodomys* diversified. Siberian faunas of the Early

Pleistocene may be correlated with Chinese Nihewanian assemblages from the Nihewan Basin and the Haiyan Formation of the Yushe Basin on the basis of species composition. During the Early Pleistocene, Baikalian landscapes were of the mosaic type; relatively humid meadows, meadows with steppe elements, and dry steppes likely coexisted. The mountain slopes were covered by mixed forests and open grasslands. The climate had dried overall but still was not cold (Alexeeva and Erbajeva 2004).

Due to tectonic movements and the continuation of orogenic processes, the mountains surrounding Lake Baikal continued to uplift since the Pliocene and became a major orographic barrier in the Middle Pleistocene. As a result, the Transbaikal area was isolated from the influence of humid Atlantic cyclones from the west. This is considered to be the main reason for intensified aridification in the Transbaikal region (Bazarov 1986; Alexeeva 2005). The paleoenvironment and faunas of Prebaikalia and Transbaikalia became significantly different.

The most typical Transbaikalian fauna of the Middle Pleistocene is the Tologoi Faunal Complex known from Tologoi 2 site (see table 21.1). The species composition is similar to that of Zhoukoudian 1, China (Young 1934; Erbajeva 1970), but Mongolian faunas of this stage are not known yet. As evident from table 21.1, the mammalian fauna from the Tologoi complex includes inhabitants of dry steppes, subdesert, and desert landscapes. The climate of the region was moderately warm and more arid than in the Early Pleistocene (Alexeeva 2005). The fauna based on the Tologoi 2 assemblage of the late Middle Pleistocene, includes mostly the Central Asian elements such as *Ochotona daurica*, *Marmota sibirica*, *Allactaga* sp., *Eolagurus* cf. *luteus*, *Lasiopodomys brandti*, *Microtus fortis*, and *M.* cf. *mongolicus*. In contrast to the fauna from Transbaikalia, the Prebaikalian fauna included nonanalogous representatives such as *Ochotona* sp., *Spermophilus undulatus*, *Cricetulus barabensis*, *Clethrionomys* sp., *Lagurus lagurus*, *Microtus gregalis*, *M. oeconomus*, and *Dicrostonyx* cf. *simplicior* that are the inhabitants of tundra/forest/steppe, the so-called mammoth fauna (Khenzykhenova 2003). By this time, the climate was relatively cold and continental (Khenzykhenova 2003, 2008). Differences between small mammal faunas of these two regions strengthened during the Late Pleistocene and continued to the present time (Alexeeva and Erbajeva 2008).

CONCLUSION

The study of new paleontological data and the analysis of known published materials show that the Late Neogene–

Quaternary mammal associations of the Baikalian region include a number of taxa in common with Europe and Central Asia due to the intensive interchange of European and Asian faunas through the Miocene and Pliocene. The paleoenvironment of northern Eurasia at that time was favorable to wide distribution of mammals. Floral data indicate warm and moist mesophytic vegetation through the middle Miocene (Akhmetiev 1993). Faunal resemblance with Mongolian and Chinese faunas to the south appears to be quite strong through the Miocene.

The species composition of small mammals of the studied areas of western Prebaikalia and eastern Transbaikalia remained mostly similar during the Pliocene and Early Pleistocene. Tectonic movements and intensive orogenic processes led to the development of the Lake Baikal Depression, which separated Prebaikalia and Transbaikalia. Further uplift of the mountains surrounding Lake Baikal established a major orographic barrier, isolating the Transbaikal area from the influence of humid Atlantic cyclones from the west. As a result, high aridification in the Transbaikal region occurred in the Middle Pleistocene, while in Prebaikalia the contemporaneous paleoenvironmental condition was more humid and cool. From that period onward, the small mammal faunas of these two regions differed significantly.

ACKNOWLEDGMENTS

We are grateful to Xiaoming Wang, Deng Tao, Li Qiang, and all of the colleagues who organized the Neogene Conference in Beijing. We thank them for the opportunity to participate at the meeting and to contribute to this volume. We thank Alisa Winkler for her valuable comments and Lawrence J. Flynn for the kindness to improve our English. This study has been supported financially by the project "RFBR-Siberia," current grant N 08-05-98033, and in part by the project "RFBR-NSF, China," current grant N 08-05-97212.

REFERENCES

Adamenko, R. S. 1977. Late Pliocene small mammals of Pribaikalie. Ph.D. diss., Institute of Zoology, Kiev (in Russian).

Adamenko, O. M. and R. S. Adamenko. 1986. Pribaikalie. In *The Stratigraphy of the USSR: Neogene System*, vol. 2, ed. M. V. Muratov and L. A. Nevesskaya, pp. 88–97. Moscow: Nedra Press (in Russian).

Akhmetiev, M. A. 1993. *Phytostratigraphy of the Paleogene and Miocene Continental Deposits of the Out-Tropical Asia.* Geological Institute Russian Academy of Sciences Trudy 475. Moscow: Nauka Press (in Russian).

Alexeeva, N. V. 2005. *Environmental Evolution of Late Cenozoic of West Transbaikalia (Based on Small Mammal Fauna)*. Moscow: GEOS (in Russian).

Alexeeva, N. V. and M. A. Erbajeva. 2004. Quaternary environment changes in the Western Transbaikal Area. In *Miscelanea en homenaje a Emiliano Aguirre*. Zona Arqueologica 4, Volume Geologia I:14–19. Madrid: Museo Arqueologico Regional de la Comunidad de Madrid.

Alexeeva, N. V. and M. A. Erbajeva. 2005. Changes in the fossil mammal faunas of Western Transbaikalia during the Pliocene-Pleistocene boundary and the Early-Middle Pleistocene transition. *Quaternary International* 131:109–115.

Alexeeva, N. V. and M. A. Erbajeva. 2008. Diversity of Late Neogene-Pleistocene small mammals of the Baikalian region and implications for paleoenvironment and biostratigraphy: An overview. *Quaternary International* 179:190–195.

Bazarov, D. B. 1986. *The Cenozoic of the Prebaikalia and Western Transbaikalia*. Novosibirsk: Nauka Press (in Russian).

Belova, V. A. 1975. *The History of Vegetation Development of the Basins of Baikal Rift Zone (on the Example of Baikal and Verkhnecharsky Basins)*. Novosibirsk: Nauka Press (in Russian).

Belova, V. A. 1985. *Plants and Climate of the Late Cenozoic in the South of the Eastern Siberia*. Novosibirsk: Nauka Press (in Russian).

Deng Tao. 2006. Chinese Neogene mammal biochronology. *Vertebrata PalAsiatica* 44(2):143–63.

Dergausova, M. I., V. P. Rezanova, and G. P. Baldaeva. 1991. Geology and palynology of the Udunginian sections (Western Transbaikalia). In *The Problems of the Cenozoic Geology of Prebaikalia and Transbaikalia*, pp. 83–96. Ulan-Ude (in Russian).

Erbajeva, M. A. 1970. *The History of the Anthropogene Lagomorph and Rodent Faunas of the Selenginian Midland*. Moscow: Nauka Press (in Russian).

Erbajeva, M. A. 1974. Villafranchian fauna of small mammals of Western Transbaikal. In *V Congrès du Néogene Mediterraneen, Lyon, 1971. Mémoires du Bureau des Recherches Géologiques et Minieres* 78(1):137–139.

Erbajeva, M. A. 1998. *Allophaiomys* in the Baikalian region. *Paludicola* 2(1):20–27.

Erbajeva, M. A. 2007. 5. Lagomorpha (Mammalia): Preliminary results. In *Oligocene-Miocene Vertebrates from the Valley of Lakes (Central Mongolia): Morphology, Phylogenetic and Stratigraphic Implications*, ed. G. Daxner-Höck, pp. 165–171. *Annalen des Naturhistorischen Museums Wien* 108A.

Erbajeva, M. A. and N. V. Alexeeva. 1997. Neogene mammalian sequence of the Eastern Siberia. In *Actes du Congrès BiochroM'97*, ed. J.-P. Aguilar, S. Legendre, and J. Michaux, pp. 241–248. Mémoires et Travaux E.P.H.E., Institut de Montpellier, 21.

Erbajeva, M. A. and N. V. Alexeeva. 2000. Pliocene and Pleistocene biostratigraphic succession of Transbaikalia with emphasis on small mammals. In *Honorarium: Nat Rutter*, ed. N. R. Catto, P. T. Bobrovsky, and D. G. E. Liverman, pp. 67–75. *Quaternary International* 68/71.

Erbajeva, M. A. and N. V. Alexeeva. 2009. Pliocene—recent Holarctic marmots: Overview. *Ethology, Ecology & Evolution* 21(3–4):339–348.

Erbajeva, M. A., N. V. Alexeeva, and F. I. Khenzykhenova. 2003. Pliocene small mammals from the Udunga site of the Ttransbaikal area. *Colloquios de Paleontologia* 1:133–145.

Erbajeva, M. A. and A. G. Filippov. 1997. Miocene small mammalian faunas of the Baikalian region. In *Actes du Congrès BiochroM'97*, ed. J.-P. Aguilar, S. Legendre, and J. Michaux, pp. 249–259. Mémoires et Travaux E.P.H.E., Institut de Montpellier, 21.

Filippov, A. G., M. A. Erbajeva, and F. I. Khenzykhenova. 1995. *The Application of the Late Cenozoic Small Mammals of South East Siberia to Stratigraphy*. Irkutsk: VostSibNIIGIMS (in Russian).

Filippov, A. G., M. A. Erbajeva, and E. K. Sychevskaya. 2000. Miocene deposits in Aya Cave near Baikal. *Russian Geology and Geophysics* 41(5):755–764.

Flynn, L. J., Z.-x. Qiu, N. O. Opdyke, and R. H. Tedford. 1995. Ages of key fossil assemblages in the late Neogene terrestrial record of northern China. In *Geochronology, Time Scales, and Global Stratigraphic Correlation: A Unified Framework for a Historical Geology*, ed. W. A. Berggren, D. V. Kent, M.-P. Aubrey, and J. Hardenbol, pp. 317–333. Society for Economic Paleontologists and Mineralogists Special Publication 54. Tulsa: Society for Sedimentary Geology.

Höck, V., G. Daxner-Höck, H. P. Schmid, D. Badamgarav, W. Frank, G. Furtmüller, O. Montag, R. Barsbold, Y. Khand, and J. Sodov. 1999. Oligocene-Miocene sediments, fossils and basalts from the Valley of Lakes (Central Mongolia)—An integrated study. *Mitteilungen der Geologischen Gesellschaft Wien* 90:83–125.

Ilyina, L. B., I. G. Shcherba, S. O. Khondkarian, and I. A. Goncharova. 2004. Map 6: Mid-Middle Miocene (Middle Serravalian, Late Badenian, Konkian). In *Lithological-Paleogeographic Maps of Paratethys: Ten Maps, Late Eocene to Pliocene*, ed. S. V. Popov, F. Roegl, A. Y. Rozanov, F. F. Steininger, I. G. Shcherba, and M. Kovac, pp. 1–46. Frankfurt am Main: *Courier Forsch.-Inst. Senckenberg* 250.

Kalmykov, N. P. and E. M. Malaeva. 1994. Lower Pliocene continental biota of the Western Transbaikalia. *Doklady Academii Nauk* 339(6):785–788 (in Russian).

Kalmykov, N. P. and E. N. Maschenko. 2005. The oldest representative of Elephantidae (Mammalia, Proboscidea) in Asia. *Paleontological Journal* 6:1–7.

Khenzykhenova, F. I. 2003. Middle Pleistocene through Early Holocene small mammals of the Baikalian region. Ph.D. diss., Novosibirsk State University (in Russian).

Khenzykhenova, F. I. 2008. Paleoenvironments of Paleolithic humans in the Baikal region. *Quaternary International* 179:53–58.

Logatchev, N. A., T. K. Lomonosova, and V. M. Klimanova. 1964. *Cenozoic Deposits of the Irkutsk Amphitheater*. Moscow: Nauka Press (in Russian).

Lopatin, A. V. 2001. A new species of *Heterosminthus* (Dipodidae, Rodentia, Mammalia) from the Miocene of the Baikal region. *Paleontological Journal* 2:93–96 (in Russian).

Mats, V. D., A. G. Pokatilov, S. M. Popova, A. Y. Kravchinskii, N. V. Kulagina, and M. K. Shimaraeva. 1982. *Pliocene and Pleistocene of Baikal*. Novosibirsk: Nauka Press (in Russian).

Mellet, J. 1968. The Oligocene Hsanda Gol Formation of Mongolia. *American Museum Novitates* 2318:1–16.

Pokatilov, A. G. 1994. Biostratigraphy of the Neogene-Quaternary deposits of the southern east Siberia (on small mammals). Ph.D. diss., Novosibirsk State University (in Russian).

Popova, S. M., V. D. Mats, G. P. Chernyaeva, and M. K. Shimaraeva. 1989. *Paleolimnological Reconstructions (Baikal Rift Zone)*. Novosibirsk: Nauka Press (in Russian).

Qiu, Z.-d. 1996. *Middle Miocene Micromammalian Fauna from Tunggur, Nei Mongol*. Beijing: Science Press.

Qiu, Z.-d. X.-m. Wang, and Q. Li. 2006. Faunal succession and biochronology of the Miocene through Pliocene in Nei Mongol (Inner Mongolia). *Vertebrata PalAsiatica* 44(2):164–81.

Sen, S. 2003. Lagomorpha. In *Geology and Paleontology of the Miocene Sinap Formation, Turkey*, ed. M. Fortelius, J. Kappelman, S. Sen, and R. Bernor, pp. 163–77. New York: Columbia University Press.

Sen, S. and M. A. Erbajeva. 2011. New species of hamster from Aya Cave, Baikalian region. *Vertebrata PalAsiatica* 49(3):257–274.

Sotnikova, M. V. 2006. Pliocene–Early Pleistocene Carnivora assemblages of Transbaikalian area, Russia. In *Abstracts of International Conference "Stratigraphy, Paleontology and Paleoenvironment of Pliocene-Pleistocene of Transbaikalia and Interregional Correlations,"* ed. Erbajeva Ulan-Ude, pp. 84–85. Geological Institute, Siberian Branch.

Sotnikova, M. V. 2008. A new species of Lesser Panda *Parailurus* (Mammalia, Carnivora) from the Pliocene of Transbaikalia (Russia) and some aspects of ailurine phylogeny. *Paleontological Journal* 1:92–102.

Sotnikova, M. V. and N. P. Kalmykov. 1991. Pliocene Carnivora from Udunga locality (Transbaikalia, USSR). In *Pliocene and Anthropogene Paleogeography and Biostratigraphy*, ed. E.A. Vangengeim, pp. 146–60. Moscow: GIN AN SSSR.

Tomida, Y. and T. Goda. 1993. First discovery of *Amphilagus*-like ochotonid from the Early Miocene of Japan. In *Annual Meeting Paleontological Society of Japan, Abstract*: 76.

Vangengeim, E. A. 1977. *Paleontological Foundation of the Anthropogene Stratigraphy of Northern Asia (Mammals)*. Moscow: Nauka Press (in Russian).

Vislobokova, I. A. 1994. The Lower Miocene artiodactyls of Tagai bay, Olkhon Island, Lake Baikal (Russia). *Palaeovertebrata* 23(1–4):177–197.

Vislobokova, I. A., M. A. Erbajeva, and M. V. Sotnikova. 1993. The Early Villafranchian stage in the development of the mammalian fauna of Northern Eurasia. *Stratigraphy and Geological Correlation* 1(5):87–96.

Vislobokova, I. A., M. A. Erbajeva, and M. V. Sotnikova. 1995. The Villafranchian mammalian fauna of the Asiatic part of the former USSR. *Il Quaternario, Italian Journal of Quaternary Sciences* 8(2):367–76.

Young, C. C. 1934. On the Insectivora, Chiroptera, Rodentia and Primates other than Sinanthropus from Locality I in Choukoutien. *Palaeontologia Sinica*, ser. C8(3):1–142.

Zazhigin, V. S. 1989. Late Pliocene type sections and their biostratigraphy (based on mammals). Late Cenozoic of Mongolia (stratigraphy and paleogeography). In *The Joint Soviet-Mongolian Scientific-Research Geological Expedition*, ed. N. A. Logachev, pp. 10–24. Transaction 47. Moscow (in Russian).

Zazhigin, V. S. and A. V. Lopatin. 2000. The history of Dipodoidea (Rodentia, Mammalia) in the Miocene Asia. *Paleontological Journal* 3:90–102. Moscow (in Russian).

Zhegallo, V. I., V. S. Zazhigin, G. N. Kolosova, E. M. Malaeva, V. E. Murzaeva, M. V. Sotnikova, I. A. Vislobokova, E. L. Dmitrieva, and I. A. Dubrovo. 1982. Nalaikha as a Lower Pleistocene reference section of Mongolia. In *Stratigraphy and Paleogeography of the Anthropogene*, ed. K. V. Nikiforova, pp. 124–142. Moscow: Nauka (in Russian).

Zheng, S.-h. and Z.-q. Zhang. 2001. Late Miocene–Early Pleistocene biostratigraphy of the Leijiahe area, Lingtai, Gansu. *Vertebrata PalAsiatica* 39(3):215–228 (in Chinese with English summary).

Chapter 22

New Data on Miocene Biostratigraphy and Paleoclimatology of Olkhon Island (Lake Baikal, Siberia)

GUDRUN DAXNER-HÖCK, MADELAINE BÖHME, AND ANNETTE KOSSLER

Lake Baikal, located in the East Siberian Baikal Rift System, is the deepest, most voluminous, and oldest freshwater body on Earth. Its morphology is characterized by three basins, the older Southern and Central basins and the younger Northern Basin. The Southern and Central basins are thought to have existed permanently since the Paleogene, whereas the Northern Basin did not develop before the Miocene. Olkhon Island (Irkutsk region, Russia) is located in the transitional zone between the Central and the Northern basins of Lake Baikal. It is separated from the mainland in the west by a shallow bay of the Northern Basin that extends far to the south (for localization and references, see Kossler 2003:fig. 1–2). From the northwestern part of Olkhon Island, two localities are known to have yielded terrestrial fossils of the Neogene, specifically the Tagay section (Logachev, Lomonosova, and Klimanova 1964; Vislobokova 1990, 1994, 2004) and the Saray section (Mats et al. 1982). Both belong to the Khalagay Formation (Khalagay Suite of Logachev, Lomonosova, and Klimanova 1964:fig. 15).

Vertebrate Assemblages of the Khalagay Formation from former investigations according to Logachev, Lomonosova, and Klimanova (1964:fig. 12), the Tagay section displays a >12-m-thick sequence of sand, clay, silty clay, and calcretes, corresponding to the carbonate-rich upper part of the lower Khalagay Formation. The vertebrate fossils were primarily recovered from the clay layers 3, 5, and 7 of the section, and are listed in Logachev, Lomonosova, and Klimanova (1964:41). Later, the ruminant fossils were reexamined by Vislobokova (1990,

1994, 2004), the fishes by Sytchevskaya (in Filippov, Erbajeva, and Sytchevskaya 2000), the turtles by Khosatzky and Chkikhvadse (1993), and the snakes by Rage and Danilov (2008). An update of the mammal fauna from the Tagay section is given in Erbajeva and Filippov (1997): *Proscalops* sp., *Talpa* sp., *Procaprolagus* sp., *Monosaulax* sp. nov., *Cricetodon* cf. *sansaniensis* Lartet (=referred to *Gobicricetodon* by Filippov, Erbajeva, and Sytchevskaya 2000), Mustelidae, Felidae, *Anchitherium* sp., *Dicerorhinus* sp., Bovidae, *Amphitragulus boulangeri* Pomel, *Lagomeryx parvulus* (Roger), *Stefanocemas* sp., *Palaeomeryx* cf. *kaupi* (Meyer), *Orygotherium tagaiensis* Vislobokova 2004, and *Brachyodus intermedius* (Mayer). The fish fauna comprises the taxa *Rutilus* sp., *Esox* sp., and *Leobergia* sp. (Logachev, Lomonosova, and Klimanova 1964; Filippov, Erbajeva, and Sytchevskaya 2000). Amphibians and reptiles are documented by *Rana* sp., *Bufo* sp., *Baicalemys gracilis*, Boinae indet., *Coluber* s.l. sp. A, *Coluber* s.l. sp. B, Colubrinae indet., and ?*Vipera* sp. (Logachev, Lomonosova, and Klimanova 1964; Khosatzky and Chkikhvadse 1993; Rage and Danilov 2008). The pond turtle *Baicalemys*, represented with more than 10,000 fragments, is by far the most common vertebrate of this locality. Stratigraphic determinations of these old collections range from Middle to Late Miocene (Logachev, Lomonosova, and Klimanova 1964; Vislobokova 1990, 1994; Filippov, Erbajeva, and Sytchevskaya 2000).

From the Saray section in the upper part of the Khalagay Formation, several fossil horizons were described and the fossils listed and dated by Mats et al. (1982). The following mammal assemblages from different horizons of the Saray section were distinguished (Mats et al. 1982; Pokatilov 1994; Erbajeva and Alexeeva 1997).

• Horizon 3 of the Saray section 1 (Saray faunistic complex) contains Soricidae gen., *Proochotona* sp., *Ochotona* sp., Leporidae gen., *Heterosminthus saraicus* Zazhigin, Lopatin et Pokatilov, *Monosaulax* sp., *Prosiphneus licenti* Teilhard, *Plesiogulo* cf. *brachygnathus* (Schlosser), *Moschus grandaevus* Schlosser, and *Pavlodaria* sp. The age is given as Late Miocene.

• Horizon 5 of the Saray section 1 (Odonim faunistic complex) contains *Hypolagus* sp., *Alilepus*? sp., *Proochotona* sp., *Ochotona* sp., *Lophocricetus (Paralophocricetus) progressus* Zazhigin, Lopatin et Pokatilov, *Microtoscoptes praetermissus* Schlosser, *M.* cf. *tjuvanensis* Zazhigin (nom. nud.), *Microtodon* cf. *atavus* (Schlosser), *Prosiphneus eriksoni* (Schlosser), and *Hipparion* sp. According to Pokatilov (1994:table 1), the assemblage is Early Pliocene or Latest Miocene (Pokatilov 1994:table 2). Erbajeva and Alexeeva (1997:243) give the age as Late Miocene.

• Horizon 6 of the Saray section 1 (Olkhonian fauna) contains Soricidae gen., *Hypolagus* sp., *Ochotonoides complicidens* (Boule and Teilhard), *Ochotona* sp., *Kowalskia* sp., *Micromys* sp., *Stachomys* ex gr. *trilobodon* Kowalski, *Promimomys insuliferus* Kowalski, *Microtodon* sp., *Eozapus* sp., and *Prosiphneus olchonicus* Pokatilov. The age is given as Early Pliocene.

• Horizon 8 of Saray section 5 (Khuzhirian fauna) and horizon 8 of section 1 are considered to be identical. There dark-brown clays contain the following fossils: *Ochotona* sp., *Sicista* sp., *Apodemus* sp., *Promimomys* cf. *gracilis* (Kretzoi), *Promimomys* sp., and *Prosiphneus chuzhirica* Pokatilov. The estimated age is Pliocene (Ruscinian, MN 15).

RECENT INVESTIGATIONS

Recently, Olkhon Island has been examined based on sedimentological, stratigraphic, and palaeontological aspects (Kossler 2003). Though several sections were carefully studied, only the Tagay and the Saray I sections (figure 22.1) yielded vertebrate fossils and molluscs. However, new collections turned out to differ considerably from the previous fossil record, making them a valuable contribution to the paleontological puzzle. Lithological descriptions of the sections and the exact position of the fossil layers are available in Kossler (2003:figs. 3–4).

Material and Methods

During the fieldwork, sediment samples from the Tagay and Saray I sections were examined for fossils and clay minerals. Numerous samples were checked for fossils by dry sieving on site. From fossil-bearing layers, bulk samples were taken for screen washing (mesh size 250 µm) and for final treatment in the laboratory. The fossils (i.e., gastropods, remains of fish, amphibians, reptiles and small mammals) were picked out from the residue under a stereomicroscope. This yielded 22 cheek teeth (fragments included) and several incisors of small mammals from the Tagay section and one lower jaw with m1–3 and an isolated incisor from the Saray I section. Almost all available cheek teeth are figured (plate 22.1). In addition to small mammals, the Tagay sample yielded several thousand isolated pharyngeal teeth and vertebrae of fishes, whereas amphibians and reptiles and gastropods are comparatively rare. About 20 plate fragments of turtles make up the most numerous herpetological material. In contrast, the Saray I section lacks fishes and reptiles and contains few bones of frogs. Gastropods from Saray section are abundant and predominantly belong to terrestrial species.

SEM images of fossils were taken by a digital SEM (type Zeiss Supra 40 VP) from the Institute of Geosciences (Paleontology), Freie Universität Berlin, Section of Paleontology, and at the University of Munich, and by a Philips XL 20 scanning microscope from the Biocenter, University of Vienna.

The small mammal teeth are integrated in the collections of the Natural History Museum Vienna, Geological-Palaeontological Department (NHMW), whereas the ectothermic vertebrates are deposited in the Bavarian State Collection for Paleontology and Geology in Munich (BSPG). Gastropod remains are stored in the collections of the Freie Universität Berlin, Section of Paleontology (collection Kossler).

New Investigations of the Tagay and Saray Sections (Olkhon Island)

Lithology and Fossil Content

The Tagay section is located at the northern end of Tagay Bay, approximately 10 km southwest of the village of Khuzhir (see figure 22.1). The investigated profile corresponds approximately to the section excavated by N. A. Logachev and colleagues in 1958 (Logachev, Lomonosova, and Klimanova 1964:fig. 13) and belongs to the

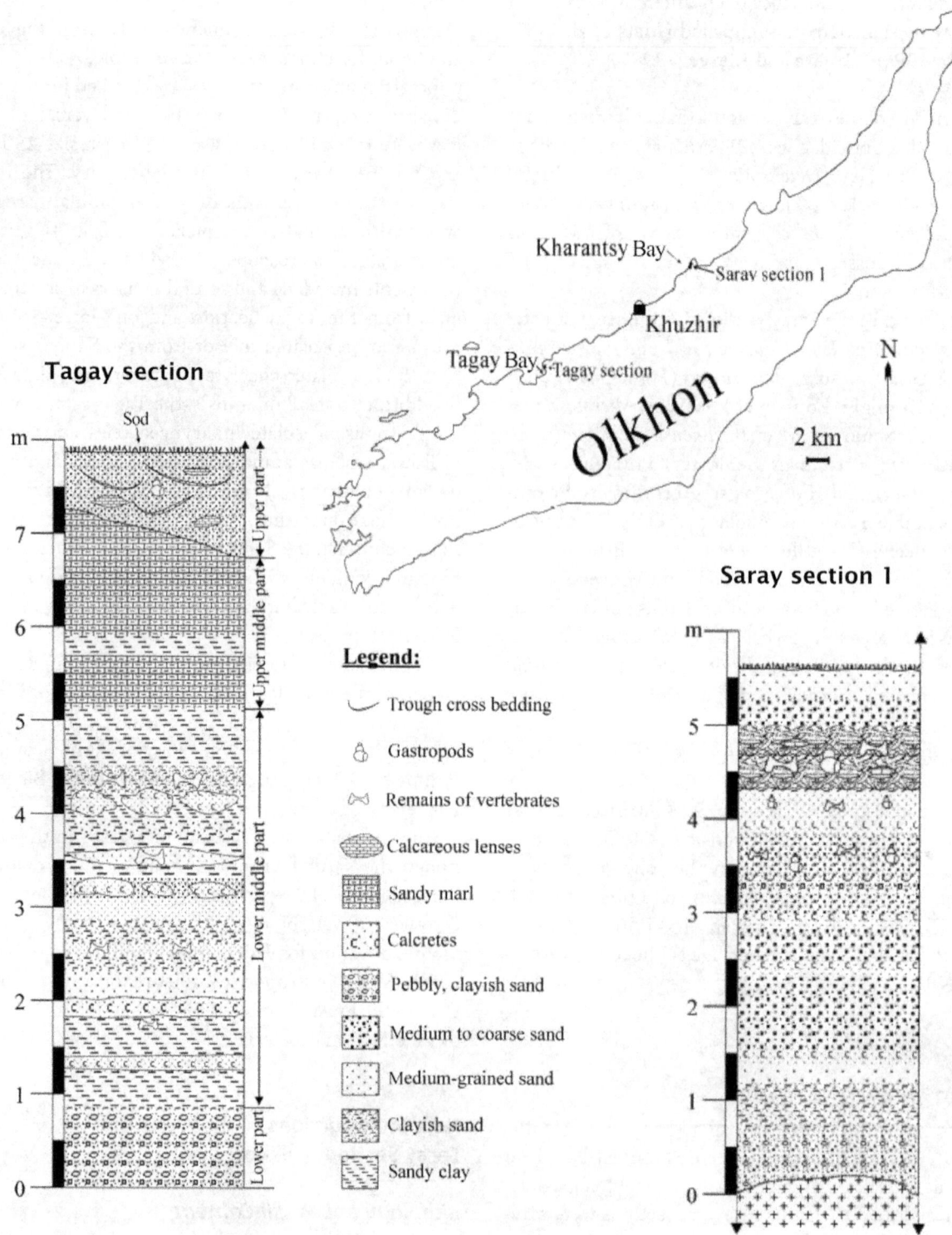

Figure 22.1 Location and description of the Tagay and Saray I sections from Olkhon Island. Modified after Kossler (2003).

carbonate-rich upper part of the lower Khalagay Formation (Tagay sequence of Mats et al. 2000). The exposure displays an 8-m-thick sediment succession. It consists of 1 m clay-sand facies in the lower part, which is followed by 6 m of alternating clay, marl, and sand layers with thick calcareous paleosols in the middle part. On top, 1 m of medium- to coarse-grained sand with calcrete lenses is exposed. The fossils were recovered from three layers in the middle part.

The Saray I section is located along the steep coast at the northern end of Kharantsy Bay. The outcrop is not well exposed. Here, the 5.6 m sediment succession overlies the metamorphic basement. Similar to the Tagay section, lacustrine sediments and paleosol horizons change, although the sequence contains more sand and corresponds to the middle part of the Khalagay Formation (Logachev, Lomonosova, and Klimanova 1964:fig. 15; Sasa sequence of Mats et al. 2000). Fossils were recovered from the higher part of the section (see figure 22.1).

Paleontological Data

TAGAY SECTION

Gastropoda, Pulmonata
  Gastrocopta sp.
  Vallonia sp.
Cypriniformes
  Palaeocarassius sp.
  Palaeotinca sp.
  Leuciscinae indet.
Esociformes
  Esox sp.
Anura
  Rana sp. (*R. temporaria* group)
  Pelophylax sp.
Chelonia
  ? *Baicalemys gracilis* Khosatzky and Chkikhvadse 1993
Squamata
  ? *Chalcides* nov. spec.
  Texasophis sp.
Insectivora
  Desmaninae gen. et spec. indet.
  Erinaceinae gen. et spec. indet.
Rodentia
  Sciurinae gen. et spec. indet.
  Miodyromys sp.
  Keramidomys aff. *mohleri* Engesser 1972 vel *K.* aff. *fahlbuschi* Qiu 1996
  Eomyops oppligeri Engesser 1990
  Democricetodon sp.

SARAY I SECTION

Gastropoda, Pulmonata
  Succineidae gen. et sp. indet.
  Radix sp.
  Carychium sp.
  Vallonia subcyclophorella (Gottschick 1911)
  Vallonia tokunagai Suzuki 1944
  Vertigo (*Ungulidenta*) *uncata* Steklov 1967
  Gastrocopta (*Kazachalbinula*) cf. *ucrainica* Steklov 1966
  Gastrocopta (*Sinalbulina*) *intorta* Steklov 1967
Anura
  Bufo aff. *calamita* (Laurenti 1768)
Rodentia
  Eozapus intermedius (Bachmayer and Wilson 1970)

Remarks

Tagay

Both gastropod genera from the upper part of the Tagay section are common in the Miocene of Europe and Asia (e.g., Steklov 1966; Gerber 1996). The small mammal assemblage from the Tagay section comprises two insectivore taxa and five rodent taxa. The scarce insectivore remains represent a talpid (see plate 22.1 [1–3]) and an erinaceid (see plate 22.1 [4]). They do not allow identification below the subfamily level. Dormice are represented by two teeth of a small, so far unknown species of *Miodyromys* Kretzoi 1943, which atypically displays a high number of very thin lophs (see plate 22.1 [6–7]). The eomyid rodent *Keramidomys* Hartenberger 1966 is well represented by maxillary and mandibular molars (see plate 22.1 [8–15]). Its pronounced pentalophodont dental pattern and the large size indicate close affinities with *K. mohleri* Engesser 1972 (from Anwil, Switzerland) as well as with *K. fahlbuschi* Qiu 1996 (from Moergen II, China). The Siberian material, however, seems to be transitional between these two species in dental pattern and possibly represents a new species. The second eomyid species resembles *Eomyops* Engesser 1979 and *Leptodontomys* Shotwell 1956 in molar pattern. The Siberian specimen (see plate 22.1 [16]) is within the size range and the morphological variability of *Eomyops oppligeri* Engesser 1990 (from Anwil, Switzerland); it is clearly smaller than *Leptodontomys lii* Qiu 1996 (from Moergen II, China). The cricetid *Democricetodon* Fahlbusch 1964 is represented by a very small species (see plate 22.1 [17–19]). Though being smaller, it shares dental characters with two species that are well known from the Middle Miocene of Europe—that is, *D. crassus* Baudelot 1972 and *D. brevis* (Schaub 1925). It is even smaller than *D. tongi* Qiu 1996, the smaller of two *Democricetodon* species from Moergen II in China. Because of the scarce fossil record, the species determination of *Democricetodon* from Tagay remains open.

The ectothermic vertebrate fauna from Tagay contains nine taxa. Most frequent is the cyprinid *Palaeocarassius* sp., which could be conspecific with a species described from the North Alpine Foreland Basin (Böhme 1999, 2010) and which may also occur in different Early to Middle Miocene (MN 4–6) localities in Russia, Kazakhstan (Zaissan Basin, Lake Aral area, Altai, and Lake Baikal) and Mongolia (Sytchevskaya 1989; Filippov, Erbajeva, and Sytchevskaya 2000; Böhme, unpubl.). Both other cyprinids, Leuciscinae indet. and *Palaeotinca*, are comparatively rare. The latter taxon is reported here for the first time from Asia outside Kazakhstan. *Palaeotinca* is a typical minnow of the Late Oligocene and the Early Miocene of central and western Europe (MP 25 to MN 4b; Kvaček et al. 2004; Böhme 2008). It is described by Sytchevskaya (1989) from the late Early Miocene (*Palaeotinca* cf. *egeriana*, Shamangorin Formation) and the late Middle to Late Miocene (*Palaeotinca* sp., Sarybulak Formation) of the Zaissan Basin (Kazakhstan). This genus was also found recently in Kazakh localities from Altyn Chokysu (Aral Formation) and Mynsualmas (Shomyshtin Formation), indicating that *Palaeotinca* was probably a common Central Asian taxon. The predatory fish *Esox* is documented by isolated teeth, cranial bones, and vertebrae. Amphibians are represented by two ranid species, the water frog *Pelophylax* and the brown frog *Rana* (*R. temporaria* group). The oldest known brown frog, *Rana* cf. *temporaria*, is described from the Early Miocene (MN 3) of Germany (Böhme 2001). Members of the *Rana temporaria* group are known in Asia from the early Middle Miocene of Mynsualmas (Lake Aral area; Böhme unpubl.) and from the Late Miocene of Kabutoiwa (Japan; Sanchiz 1998).

The rare finds of squamate reptiles belong to a new species of skink (? *Chalcides* nov. sp.) and to the colubrid snake *Texasophis*. Both taxa are recorded for the first time from Asia. *Texasophis* is known from the latest Oligocene to the late Middle Miocene of Europe and North America (Szyndlar 1991; Holman 2000; Böhme 2008). The Tagay material therefore represents an important biogeographic link and reveals a Holarctic distribution for *Texasophis*.

Saray I

All identified terrestrial gastropod species from the Saray I section are also described from other Miocene localities of Asia (e.g., Steklov 1966, 1967; Gerber 1996). Except for *Vallonia subcyclophorella*, which is distributed over Eurasia, the listed gastropod species seem to be restricted to the Asian faunal realm. The few fragments of *Radix* sp. and of the Succineidae do not allow identification at species level.

From the Saray I section, a mandible with m1–3 and an isolated lower incisor were recovered, which can be identified as *Eozapus intermedius* (Bachmayer and Wilson 1970) (see plate 22.1 [20a and 20b]). The first record of this species is the early Late Miocene of Europe (Austria: Kohfidisch and Eichkogel, MN 11) with first scarce occurrences in the Vallesian of Austria (MN9–10; Daxner-Höck 1996). In the lower Turolian (MN 11), however, it rapidly dispersed all over Europe (Daxner-Höck 1999). *E. intermedius* is the smallest among the three known *Eozapus* species, which are extremely conservative in dental morphology. Fahlbusch (1992) observed a size increase from the oldest toward the youngest species—that is, from *E. intermedius* (Late Miocene, MN 9–11; Europe) to *E. similis* 1992 (Late Miocene, MN 13; China) to *E. setchuanus* (Pousargues 1896; today living in China), along with minor morphological differences.

The only ectothermic vertebrate from the Saray I section is a toad, *Bufo* aff. *calamita* (plate 22.2 [9]), which is osteologically more similar to the extant *Bufo calamita* than to any other compared recent or fossil Eurasian species (*B. bufo*, *B. viridis*; Pliocene Mongolian *B. raddei*; Early and Middle Miocene European *B. viridis* sp. and *B. priscus*). So far, the oldest fossil record of *Bufo calamita* is the early Messinian of Spain (MN 12; Sanchiz 1998).

Biostratigraphy

Tagay

All identifiable rodent genera are well represented in the Miocene of Europe and Asia.

The eomyids *Eomyops oppligeri* and *Keramidomys* aff. *mohleri-fahlbuschi* and the cricetid *Democricetodon* sp. can best be compared with their relatives from the Middle Miocene of Europe and China, respectively. According to the rodent data, the estimated age of the new Tagay fauna is Middle Miocene (~13 Ma). The fauna approximately

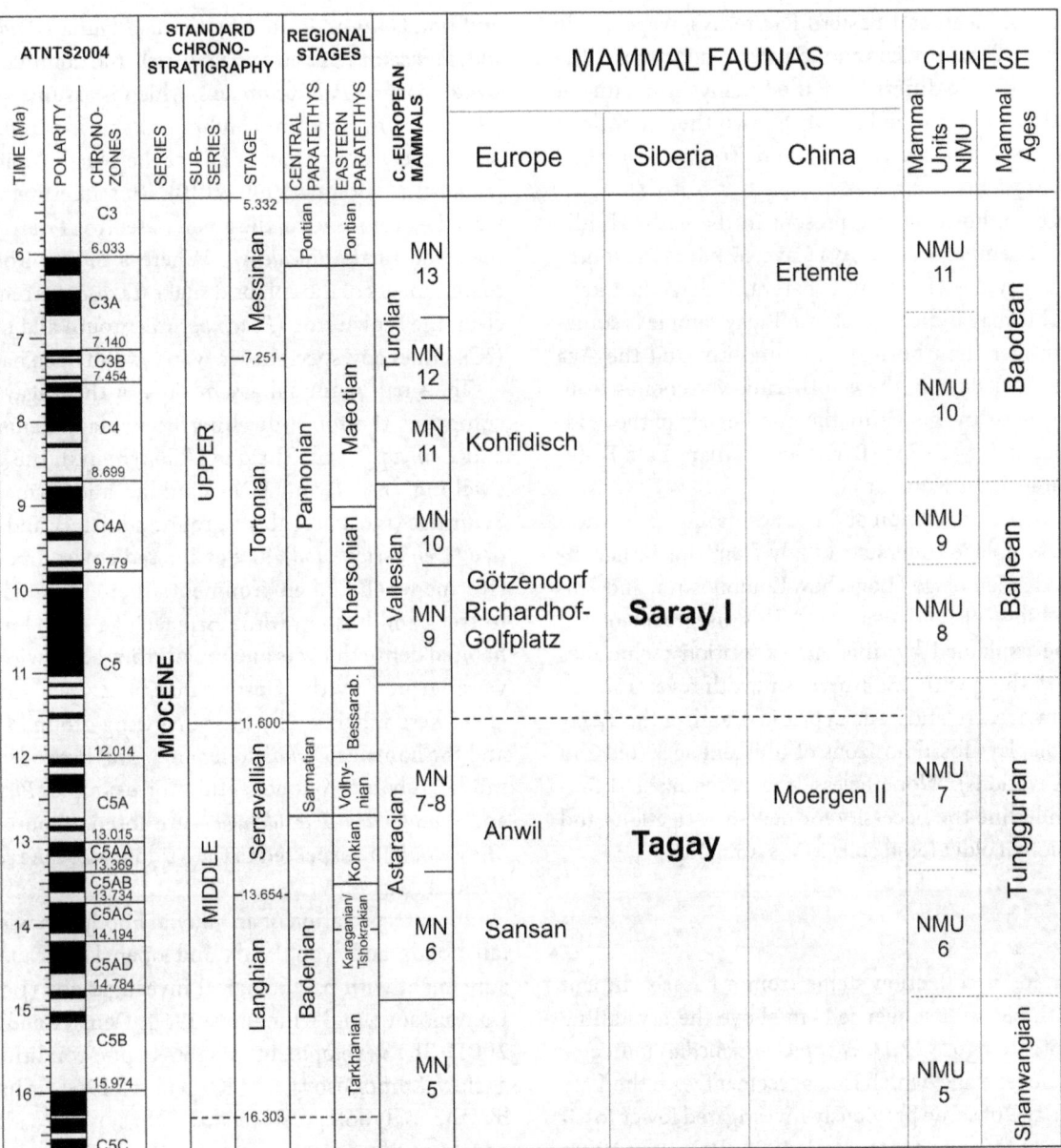

Figure 22.2 Chronostratigraphy and biostratigraphy of the Middle and Late Miocene. Modified after Harzhauser and Piller (2007). Stratigraphic position and correlation of the Tagay and Saray assemblages from Olkhon Island in Lake Baikal (Siberia).

correlates with the European Mammal Zone MN 7+8 and the Chinese Mammal Unit NMU7 (figure 22.2).

The ectothermic vertebrate fauna yield similar results. It shows closer affinities to late Early and Middle Miocene faunas of western Eurasia (Europe, Kazakhstan) than to contemporaneous faunas from eastern Eurasia (Mongolia, China). The comparatively rich Tagay association exhibits the most similarities to (yet undescribed) material from the Lake Aral region (Mynsualmas localities), where *Palaeocarassius*, *Palaeotinca*, Leu-

ciscinae, *Esox*, *Pelophylax*, and *Rana* are present. The Mynsualmas sites (Shomyshtin fauna) are located in the northwestern part of the Ustyurt Plateau, between the Caspian and Aral Sea (SW Kazakhstan; Kordikova, Heizmann, and Pronin 2003:fig. 1). The fossil vertebrate-bearing samples are from the lower part of the Shomyshtin formation. This formation belongs to the lower part of the Tarkhanian Eastern Paratethys regional stage (Kordikova, Heizmann, and Pronin 2003), which correlates to the Langhian and Early Badenian stages of the Mediterranean and Eastern Paratethys, respectively (Andreyeva-Grigorovich and Savytskaya 2000). The Tagay assemblage differs from the Shomyshtin fauna at Mynsualmas due to the lack of the two thermophilous taxa *Channa* (snakehead fish) and *Trionyx* (softshell turtle).

Both taxa, however, are present in the early Middle Miocene assemblage of the Aya Cave, 57 km to the southwest of Tagay Bay (Filippov, Erbajeva, and Sytchevskaya 2000). This may indicate that our Tagay sample is somewhat younger than both the Shomyshtin and the Aya faunas. In conclusion, the ectothermic vertebrates from Tagay point to an age within the middle part of the Middle Miocene (14–13 Ma, Early Serravallian, Late Badenian, Karaganian-Konkian).

The fossil composition and the age dating of the new Tagay assemblage differs strikingly from the earlier investigated assemblage (Logachev, Lomonosova, and Klimanova 1964; Vislobokova 1994). This disagreement can partly be explained by different excavation techniques (screen washing with 250 μm mesh width reveals small-sized eomyids, cricetids, and cyprinids), and/or the Tagay section displays fossil horizons of different ages (old and new excavations). Nonetheless, the recognized differences underline the necessity for new investigations and for revision of older fossil collections from Tagay.

Saray I

The new fossil collection stems from a basal sediment layer of the Saray I sequence (4 m above the crystalline basement; see figure 22.1). *Eozapus intermedius* indicates a Late Miocene age, which is in agreement with the Late Miocene age of some previously investigated lower fossil horizons of this site (Mats et al. 1982; Pokatilov 1994; Erbajeva and Alexeeva 1997).

Paleoenviroment

The sediment sequences of both the Tagay and the Saray I sections are characterized by debris flows of alluvial fans and by floodplain accumulations with calcrete paleosol horizons. These horizons yielded a fossil record from three sediment layers of the Tagay section and from two layers of the Saray I section.

Tagay

The sedimentary succession is interpreted as floodplain deposits (Kossler 2003:57). Shallow-water lakes and ponds are indicated by the fish fauna (*Palaeocarassius, Palaeotinca,* Leuciscinae, *Esox*). The dominant small-sized cyprinid *Palaeocarassius,* which is assumed to tolerate very low oxygen conditions, shows similar mass occurrences in riparian ponds of the North Alpine Foreland Basin (Böhme 1999, 2010). This interpretation fits with the occurrence of the green water frog *Pelophylax* and the pond turtle *Baicalemys.* Whereas the environmental requirements of the colubrid snake *Texasophis* remain unclear, the brown frog (*R. temporaria* group) and the skink (?*Chalcides* nov. spec.) live in wooded and open habitats.

The small mammal assemblage of the Tagay section comprises the forest-dwelling Eomyidae (*Keramidomys* and *Eomyops*) and Gliridae (*Miodyromys*), the ground-dwelling Insectivora (Desmaninae and Erinaceinae), Sciuridae (most probably a ground squirrel), and Cricetidae (*Democricetodon*). The estimated environment was a riverine woodland environment. There, Eomyidae lived in trees and bushes, while Erinaceidae and Gliridae inhabited dense thickets and undergrowth. Flowing waters were favored by the Desmaninae, and open sunny dry areas were inhabited by the burrowing ground squirrels and the hamster *Democricetodon.* Note that rodents with moderately hypsodont teeth—for example, *Plesiodipus* and *Protalactaga-Paralactaga*—are absent in our samples. They would be expected in the Middle Miocene record of Central Asia.

The interpretation of an alluvial floodplain with riparian ponds and woodlands and open landscapes, is in agreement with palynological investigations (Logachev, Lomonosova, and Klimanova 1964; Demske and Kossler 2001). The development of open-steppe conditions (Logachev, Lomonosova, and Klimanova 1964) is indicated by the palynological record—that is, broadleaved/coniferous forests (*Picea, Abies, Pinus, Tsuga, Larix, Alnus, Betula, Quercus, Fagus, Carpinus, Castanea, Tilia, Celtis, Acer, Ulmus, Corylus, Carya, Juglans,* and *Liquidambar*) and a significant part of herbaceous and gramineous vegetation (*Artemisia,* Chenopodiaceae, Poaceae, Cyperaceae, Nympheaceae, Ranunculaceae, Lygopodiaceae, and ferns).

Saray I

Based on sedimentology, this outcrop is also interpreted as a floodplain deposit of an ephemeral river system (Kossler 2003:59), which offered various habitats for the ground-dwelling rodent *Eozapus* and the toad *Bufo* cf. *calamita.* With the exception of *Radix* sp., an aquatic gastropod occurring in standing and flowing waters, the majority of gastropods belong to terrestrial species. Today,

Carychium, Succineidae, and *Vallonia* frequently live in wet to humid habitats. *Gastrocopta* und *Vertigo* are known to live in various habitats, in plant litter, and under stones. According to Stworzewicz (2007), they are often washed into the depositional environment and are very common in flood debris.

Palaeoclimate

Tagay

Among the vertebrates, the skink and the pond turtle are most sensitive for paleotemperature estimation. The Central Asian emydid *Baicalemys* is morphologically related to the extant North American genera *Chrysemys*, *Pseudemys*, and *Trachyemys* (Khosatzky and Chkikvadse 1993). *Chrysemys picta belii* reaches the most northern distribution. It lives in aquatic habitats north of Lake Superior in boreal forest and cold prairie areas across Canada (Holman and Andrews 1993). The mean cold month temperature (mCMT) reaches −17°C at the northern distribution limit, with +2.5°C mean annual temperature (MAT) and +20°C warmest month temperature (WMT).

The extant Scincidae have a worldwide distribution on all continents, except Antarctica. Analyzing the climatic conditions at the margins of their distribution (Hokaido Island, Lake Aral, and Karpathian Mountains in Eurasia; 49°N and 39°S in the Americas, 46°S in New Zealand), their poleward expansion seems limited by temperature, mainly by mean cold month temperature. In Eurasia, the most extreme occurrence of a skink is reached north of Lake Aral, with a mCMT around −12°C and a mean annual temperature around +7°C. Slightly cooler conditions are found in North America at their northernmost distribution around the Great Lakes, where the mCMT reaches up to −15°C and the MAT around +3°C. The warmest month temperature is always well above 20°C. Using the more conservative values of their Eurasian distribution, mainly based on biogeographical considerations, we can estimate −12°C in mCMT, > +20°C in WMT and +7°C in MAT as minimal thermal requirements for the Tagay skink.

Considering the minimal values of both reptile groups, the Middle Miocene Tagay record would indicate warmer conditions than today (Irkutsk MAT −1.2°C, mCMT −20.9°C, WMT +17.5°C; Ulan-Ude MAT −1.7°C, mCMT −25.4°C, WMT 19.4°C; recent climate data from M. J. Müller and D. Hennings, The Global Climate Data Atlas, Climate 1, www.climate-one.de).

The estimated paleotemperatures are at least 4– 8°C warmer in the winter and in the annual mean. These values are certainly underestimated because the high amount of kaolinite in the upper part of the Tagay section, and the calcretes and pedogenic gypsum (Logachev, Lomonosova, and Klimanova 1964:38; Mats et al. 2000:236) indicate alternation of warm-humid and semi-arid climatic conditions (Kossler 2003). This is in agreement with the palynological record.

Saray I

The few fossils recovered from the Saray section contribute little to our understanding of the paleoclimatology. The northern and northeastern limits of the extant distribution of the natterjack *Bufo calamita* is southern Sweden and Estonia (Gasc et al. 2004), where mean annual temperatures reach +5°C, winter temperatures −5°C, and summer temperatures at least +17°C. Again, warmer temperatures are indicated by the clay mineralogy, where the kaolinite content is low but increases toward the top (above the level with fossil vertebrates). This indicates enhanced chemical weathering under warm, humid conditions (Kossler 2003:59).

CONCLUSION

The vertebrate assemblages from the Tagay and Saray I sections show remarkable affinities to assemblages from western Eurasia, especially from Central Europe. Ectothermic vertebrates from these two sections also resemble assemblages from western Kazakhstan. Similar conclusions were drawn by Vislobokova (1994) and Rage and Danilov (2008), who studied the Tagay artiodactyls and snakes. These results support the interpretation that a largely homogeneous vertebrate fauna (at least at the genus level) may have existed in the middle latitudes of Eurasia during the Middle Miocene and perhaps during parts of the Late Miocene. Although no long and continuous sections have been studied, our paleoclimatic analysis indicates that repeated strong fluctuations in humidity affected the Baikal Lake area during both the Middle and Late Miocene.

ACKNOWLEDGMENTS

The authors thank M. Erbajeva and P. Tarasov for some translations of the Russian literature. Fieldwork, sampling, and studies on the gastropods were supported by the German Science Foundation (grant number DFG Ko 2004/1). The studies on the small mammals were supported by the Austrian Science Fund, FWF-project P-15724-N06, and studies on ectotherms by the German Science Foundation (grant number BO 1550/8). We also thank R. Ziegler for identification of the insectivore teeth, E. Höck and B. Schenk for digital image processing of the plates, and M. Stachowitsch for improving the English.

REFERENCES

Andreyeva-Grigorovich, A. S. and N. A. Savytskaya. 2000. Nannoplankton of the Tarkhanian deposits of the Kerch peninsula (Crimea). *Geologica Carpathica* 51(6):399–406.

Bachmayer, F. and R. W. Wilson. 1970. Die Fauna der altpliozänen Höhlen- und Spaltenfüllungen bei Kohfidisch, Burgenland (Österreich). *Annalen des Naturhistorischen Museums Wien* 74:533–587.

Baudelot, S. 1972. Étude des Chiroptères, Insectivores et Rongeurs du Miocène de Sansan (Gers). *Thèse Université Paul Sabatier Toulouse* 496:1–364.

Böhme, M. 1999. Die miozäne Fossil-Lagerstätte Sandelzhausen. 16. Fisch- und Herpetofauna—Erste Ergebnisse. *Neues Jahrbuch für Geologie und Paläontologie Abhandlungen* 214(3):487–495.

Böhme, M. 2001. The oldest representative of a brown frog (Ranidae) from the Early Miocene of Germany. *Acta Palaeontologica Polonica* 46(1):119–124.

Böhme, M. 2008. Ectothermic vertebrates (Teleostei, Allocaudata, Urodela, Anura, Testudines, Choristodera, Crocodylia, Squamata) from the Upper Oligocene of Oberleichtersbach (northern Bavaria, Germany). *Courier Forschungs-Institut Senckenberg* 260:161–183.

Böhme, M. 2010. Ectothermic vertebrates (Osteichthyes, Allocaudata, Urodela, Anura, Crocodylia, Squamata) from the Miocene of Sandelzhausen (Germany, Bavaria): Their implication for environmental reconstruction and palaeoclimate. *Paläontologische Zeitschrift* 84(1):3–41.

Daxner-Höck, G. 1996. Faunenwandel im Obermiozän und Korrelation der MN-"Zonen" mit den Biozonen des Pannons der Zentralen Paratethys. *Beiträge zur Paläontologie Wien* 21:1–9.

Daxner-Höck, G. 1999. Family Zapodidae. In *The Miocene Land Mammals of Europe*, ed. G. E. Roessner and K. Heissig, pp. 337–342. Munich: Dr. Friedrich Pfeil.

Demske, D. and A. Kossler. 2001. Lokale Vegetationsentwicklung am Bajkalsee im späten Neogen als Abbild von Paläoklima-Wechseln in Zentral-Asien. *Terra Nostra 2001* (6):28.

Engesser, B. 1972. Die obermiozäne Säugetierfauna von Anwil (Baselland). *Inauguraldissertation. Tätigkeitsberichte der Naturforschenden Gesellschaft Baselland* 28:37–363.

Engesser, B. 1979. Relationships of some insectivores and rodents from the Miocene of North America and Europe. *Bulletin of Carnegie Museum of Natural History* 14:1–68.

Engesser, B. 1990. Die Eomyidae (Rodentia, Mammalia) der Molasse der Schweiz und Savoyens. *Schweizerische Paläontologische Abhandlungen* 112:1–144.

Erbajeva, M. A. and N. V. Alexeeva. 1997. Neogene Mammalian Sequence of the Eastern Siberia. In *Actes du Congrès BiochroM'97*, ed. J.-P. Aguilar, S. Legendre, and J. Michaux, pp. 241–248. Mémoires et Travaux E.P.H.E., Institut de Montpellier, 21.

Erbajeva, M. A. and A. G. Filippov. 1997. Miocene small mammals of the Baikal region. In *Actes du Congrès BiochroM'97*, ed. J.-P. Aguilar, S. Legendre, and J. Michaux, pp. 249–259. Mémoires et Travaux E.P.H.E., Institut de Montpellier, 21.

Fahlbusch, V. 1964. Die Cricetiden (Mamm.) der Oberen Süßwasser-Molasse Bayerns. *Bayerische Akademie der Wissenschaften, Mathematisch-Naturwissenschaftliche Klasse, Abhandlungen, N.F.* 118:1–136.

Fahlbusch, V. 1992. The Neogene mammalian faunas of Ertemte and Harr Obo in Inner Mongolia (Nei Mongol), China. 10. *Eozapus* (Rodentia). *Senckenbergiana Lethaea* 72:199–217.

Filippov, A. G., M. A. Erbajeva, and E. K. Sytchevskaya. 2000. Miocene deposits in Aya cave near Baikal. *Geologia i Geofisika* 41(5):755–764 (in Russian).

Gasc, J. P., A. Cabela J. Crnobrnja-Isailovic, D. Dolmen, K. Grossenbacher, P. Haffner, J. Lescure, H. Martens, J. P. Martínez Rica, H. Maurin, M. E. Oliveira, T. S. Sofianidou, M. Veith, and A. Zuiderwijk. 2004. *Atlas of Amphibians and Reptiles in Europe*. Collection Patrimoines Naturels 29. Paris: Societas Europaea Herpetologica, Muséum National d'Histoire Naturelle & Service du Patrimoine Naturel.

Gerber, J. 1996. Revision der Gattung *Vallonia* RISSO 1826 (Mollusca: Gastropoda: Vallonidae). *Schriften zur Malakozoologie* 8:1–227.

Gottschick, F. 1911. Aus dem Tertiärbecken von Steinheim a. A. *Jahreshefte des Vereins für Vaterländische Naturkunde in Württemberg* 67:496–534.

Hartenberger, J. L. 1966. Les rongeurs de Vallésien (Miocène supérieur) de Can Llobateres (Sabadell, Espagne): Gliridae et Eomyidae. *Bulletin de la Société géologique de France* 8/7:596–604.

Harzhauser, M. and P. W. E. Piller. 2007. Benchmark data of a changing sea—palaeogeography, palaeobiogeography and events in the Central Paratethys during the Miocene. *Palaeogeography, Palaeoclimatology, Palaeoecology* 253:8–31.

Holman, J. A. 2000. *Fossil Snakes of North America: Origin, Evolution, Distribution, Paleoecology*. Bloomington: Indiana University Press.

Holman, J. A. and K. D. Andrews. 1993. North American Quaternary cold-tolerant turtles: Distributional adaptations and constraints. *Boreas* 23(1):44–52.

Khosatzky, L. I. and V. M. Chkikhvadse. 1993. New data about Miocene turtles of the genus Baicalemys. *Bulletin of the Academy of Science of Georgia* 148(3):155–160 (in Russian).

Kordikova, E. G., E. P. J. Heizmann, and V. G. Pronin. 2003. Tertiary litho- and biostratigraphic sequence of the Ustyurt Plateau area, SW Kazakhstan, with the main focus on vertebrate faunas from the Early to Middle Miocene. *Neues Jahrbuch für Geologie und Paläontologie Abhandlungen* 227(3):381–447.

Kossler, A. 2003. Neogene sediments of Olkhon and Svyatoy Nos (Baikal Rift System, East Siberia): Suggestions about the development of Lake Baikal. *Berliner Paläobiologische Abhandlungen* 4:55–63.

Kretzoi, M. 1943. Ein neuer Muscardinide aus dem Ungarischen Miozän. *Különlenyomat a Földtani Közlöny* 73(1–2):271–273.

Kvaček, Z., M. Böhme, Z. Dvořák, M. Konzalova, K. Mach, J. Prokop, and M. Rajchl. 2004. Early Miocene freshwater and swamp ecosystems of the Most Basin (northern Bohemia) with particular reference to the Bílina Mine section. *Journal of the Czech Geological Society* 49(1–2):1–40.

Laurenti, J. N. 1768. *Specimen Medicum, Exhibens Synopsin Reptilium Emendatam cum Experimentis circa Venena et Antidota Reptilium Austriacorum.* Vienna: J. T. N. de Trattnern.

Logachev, N. A., T. K. Lomonosova, and V. M. Klimanova. 1964. *Cenozoic Deposits of the Irkutsk Amphitheatre.* Moscow: Nauka Press (in Russian).

Mats, V. D., O. M. Khlystov, M. Batist, S. de, Ceramicola, T. K. Lomonosova, and A. Klimansky. 2000. Evolution of the Academician Ridge accommodation zone in the central part of the Baikal rift, from high-resolution reflection seismic profiling and geological field investigation. *International Journal of Earth Science* 89:229–250.

Mats, V. D., A. G. Pokatilov, S. M. Popova, A. Ya. Kravchinsky, N. V. Kulagina, and M. K. Shimaraeva. 1982. *Pliocene and Pleistocene of the Middle Baikal.* Novosibirsk: Nauka Press (in Russian).

Pokalitov, A. G. 1994. Neogene-Quaternary biostratigraphy of the south east Siberia. Ph.D. diss., Novosibirsk State University (in Russian).

Qiu, Z.-d. 1996. *Middle Miocene Micromammal Fauna from Tunggur, Nei Mongol.* Beijing: Science Press (in Chinese with extended English Summary).

Rage, J.-C. and I. G. Danilov. 2008. A new Miocene fauna of snakes from eastern Siberia, Russia. Was the snake fauna largely homogenous in Eurasia during the Miocene. *Comptes Rendus Palevol* 7:383–390.

Sanchiz, B. 1998. Salientia. *Encyclopedia of Paleoherpetology* 4/4, ed. P. Wellnhofer, 1–275. Munich: Dr. Friedrich Pfeil.

Schaub, S. 1925. Die hamsterartigen Nagetiere des Tertiärs und ihre lebenden Verwandten. *Abhandlungen der Schweizerischen Paläontologischen Gesellschaft* 14:1–111.

Shotwell, J. A. 1956. Hemphillian mammalian assemblages from north-eastern Oregon. *Bulletin of the Geological Society of America* 67:717–736.

Steklov, A. A. 1966. *Terrestrial Neogene Molluscs of Ciscaucasia and Their Stratigraphic Importance.* Academy of Sciences of the USSR Transactions 163. Moscow: Nauka Press (in Russian).

Steklov, A. A. 1967. Land molluscs from the Neogene of Touva. *Trudy Zoologicheskogo Instituta Leningrad* 42:269–279 (in Russian).

Stworzewicz, E. 2007. Molluscan fauna (Gastropoda: Pulmonata: Pupilloidea): A systematic review. In *Oligocene-Miocene Vertebrates from the Valley of Lakes (Central Mongolia)*, ed. G. Daxner-Höck, pp. 25–41. *Annalen des Naturhistorischen Museums Wien* 108A.

Suzuki, K. 1944. Quaternary land and fresh-water shells from Kuhsiangtung, Habin, Manchoukou. *Miscellaneous Reports of the Research Institute for Natural Resources* 1(3):319–390.

Sytchevskaya, E. K. 1989. *Neogene Freshwater Fish Fauna of Mongolia.* Trudy Sovmestnaya Sovetsko-Mongolkaya Paleontologitcheskaya Ekspeditija, vol. 39. Moscow: Nauka Press (in Russian).

Szyndlar, Z. 1991. A review of Neogene and Quaternary snakes of Central and Eastern Europe. Part I: Scolecophidia, Boidae, Colubridae. *Estudios Geologicos* 47:103–126.

Vislobokova, I. 1990. On the artiodactyls from the Lower Miocene of Tagai bay of Olhon Island (Baikal). *Paleontological Journal* 2:134–138 (in Russian).

Vislobokova, I. 1994. The Lower Miocene artiodactyls of Tagay Bay, Olhon Island, Lake Baikal (Russia). *Palaeovertebrata* 23(1–4):177–197.

Vislobokova, I. 2004. New species of Orygotherium (Palaeomerycidae, Ruminantia) from the Early and Late Miocene of Eurasia. *Annalen des Naturhistorischen Museums Wien* 106A:371–385.

Part IV

West Asia and Adjacent Regions

Chapter 23

Late Miocene Mammal Localities of Eastern Europe and Western Asia

Toward Biostratigraphic Synthesis

ELEONORA VANGENGEIM AND ALEXEY S. TESAKOV

Numerous localities of the "fauna of Hipparion" are known in the Northern Black Sea region (Ukraine, Moldova, Russia) and in the Transcaucasus (Georgia). They are associated with shallow marine deposits of the Eastern Paratethys spanning the middle Sarmatian through Pontian regional stages, and with synchronous terrestrial formations. Many of these sites have a paleomagnetic record and can be correlated with the magnetochronological time scale based on relatively complete marine sections of the Eastern Paratethys.

The mammal-based biochronological zonation of Pierre Mein (1975, 1989) originally established for the continental Neogene of western and central Europe (MN zones) is widely used in different regions of the Palearctic—in particular, in the south of east Europe and in Transcaucasus.

In the last decade, magnetostratigraphic data have been obtained for many mammal localities in central and southwestern Europe (Bernor et al. 1996b; Krijgsman et al. 1996; Agustí et al. 1997; Garcés, Krijgsman, and Agustí 1998; Agustí, Cabera, and Garcés 2001; Daxner-Höck 2001; and other works). These studies resulted in age estimates for the boundaries of MN zones (figure 23.1).

In this chapter, we give a synthesis of magnetostratigraphy-based correlations of large mammal localities from southern east Europe and the Transcaucasus with MN zones of central and western Europe (Vangengeim, Lungu, and Tesakov 2006; Vangengeim and Tesakov

2008a, 2008b) (figures 23.2 and 23.3). In addition, data on most important small mammal sites and biochronological markers were used, as well. In this study, we attempted to reconsider traditional concepts on faunal correlations between west and east Europe and define biochronological boundaries in southern east Europe in accordance with the known ages of MN zones' boundaries.

The study of the paleogeographic setting for middle to upper Sarmatian and paleomagnetic data has shown a possibility of a somewhat different chronological sequence of localities in this geological interval as compared to views of other authors (figure 23.4). The sequence of localities in the Maeotian-Pontian interval is agreed upon by the majority of Ukrainian and Russian authors (Korotkevich 1988; Pevzner and Vangengeim 1993; Pevzner et al. 2003; Vangengeim, Lungu, and Tesakov 2006; Vangengeim and Tesakov 2008a, 2008b).

In this chapter, the taxonomic nomenclature for hipparions is after Bernor, Koufos, et al. (1996b), and for carnivorans, ungulates, and elephants it is after Gabunia (1986), Godina (1979), Korotkevich (1988), Krakhmalnaya (1996a, 2008), Krokos (1939), Lungu (1990), Lungu and Delinschi (2008), Markov (2008), Pevzner et al. (1987), Semenov (1989, 2001a, 2001b, 2008), Sotnikova (2005), Vangengeim (1993), Werdelin and Solounias (1991), Wolsan and Semenov (1996). Many mammalian groups and faunal lists need revision.

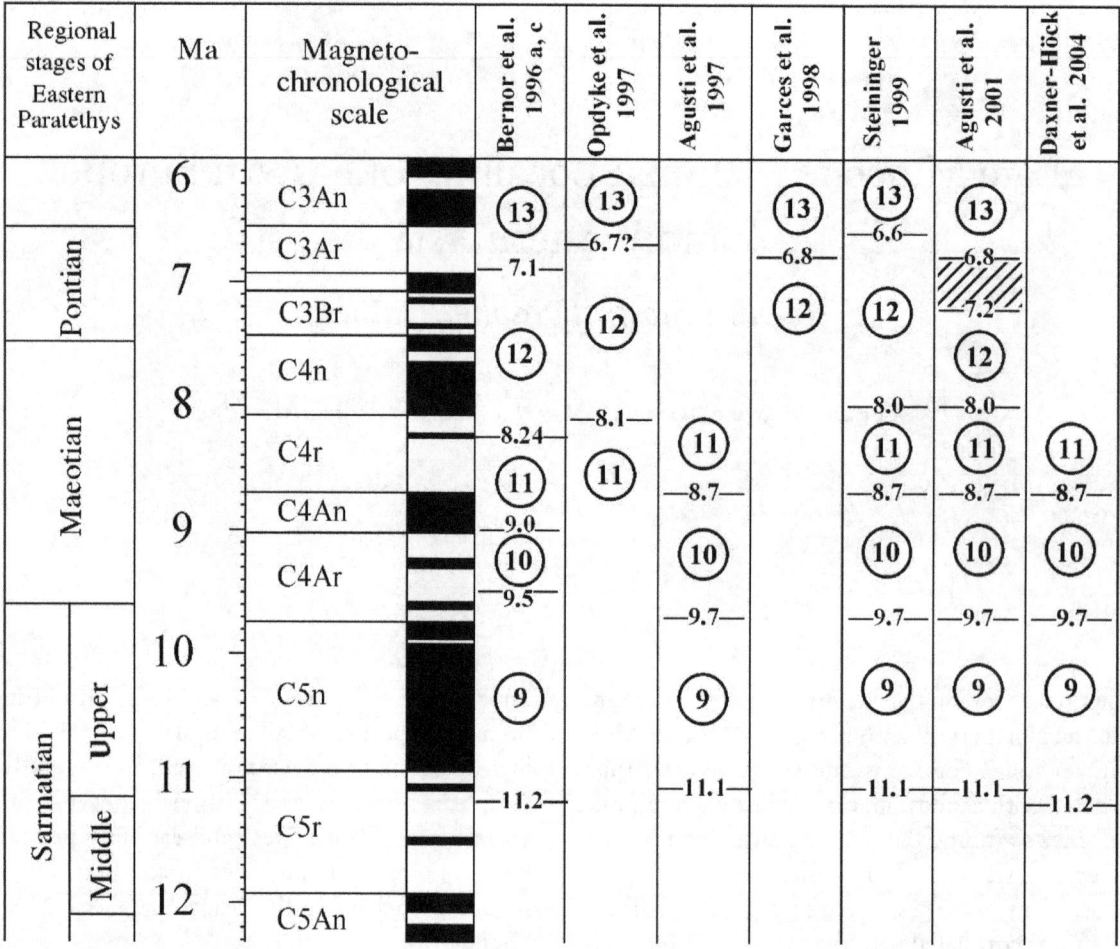

Figure 23.1 Ages of MN zone boundaries in western Europe according by different authors. Zone numbers are in circles. After Vangengeim and Tesakov (2008b).

Ages of magnetochronological boundaries are according to the geomagnetic polarity timescale (GPTS) CK95 following Berggren et al. (1995). The updated scale ATNTS2004 (Lourens et al. 2004) was used for comparisons with most recent data. Deposits of Eastern Paratethys are correlated to the magnetochronological scale based on data of Pevzner (1986; Pevzner, Semenenko, and Vangengeim 2003).

Locations of most important localities are given in figure 23.5 and table 23.1.

PALEOMAGNETIC CHARACTERISTICS FOR MIDDLE SARMATIAN-KIMMERIAN DEPOSITS OF EASTERN PARATETHYS

The representative sections of the marine middle and upper Sarmatian, characterized by malacofauna and stud-

ied paleomagnetically, are known on the Taman Peninsula, and at the Eldari section in eastern Georgia (Pevzner 1986; Vangengeim et al. 1989; Pevzner and Vangengeim, 1993; Pevzner, Semenenko, and Vangengeim 2003; Vangengeim, Lungu, and Tesakov 2006; see figure 23.2).

The lower normally magnetized part of the middle Sarmatian is correlated with the upper part of chron C5An. The middle Sarmatian is 135 m thick in the section east of the cape Panagia on the Taman Peninsula, and about 360 m thick in the Eldari section. This interval is capped by reversely magnetized deposits correlative with larger part of chron C5r. On the Taman Peninsula, the upper Sarmatian deposits have a thickness about 150 m; in Eldari, they are 980 m thick. The boundary between the upper and middle Sarmatian is situated slightly below a normal polarity zone that we correlate with subchron C5r.1n. Its age, estimated near 11.2 Ma, is controlled by the fission-track date 11.19 ± 0.74 Ma

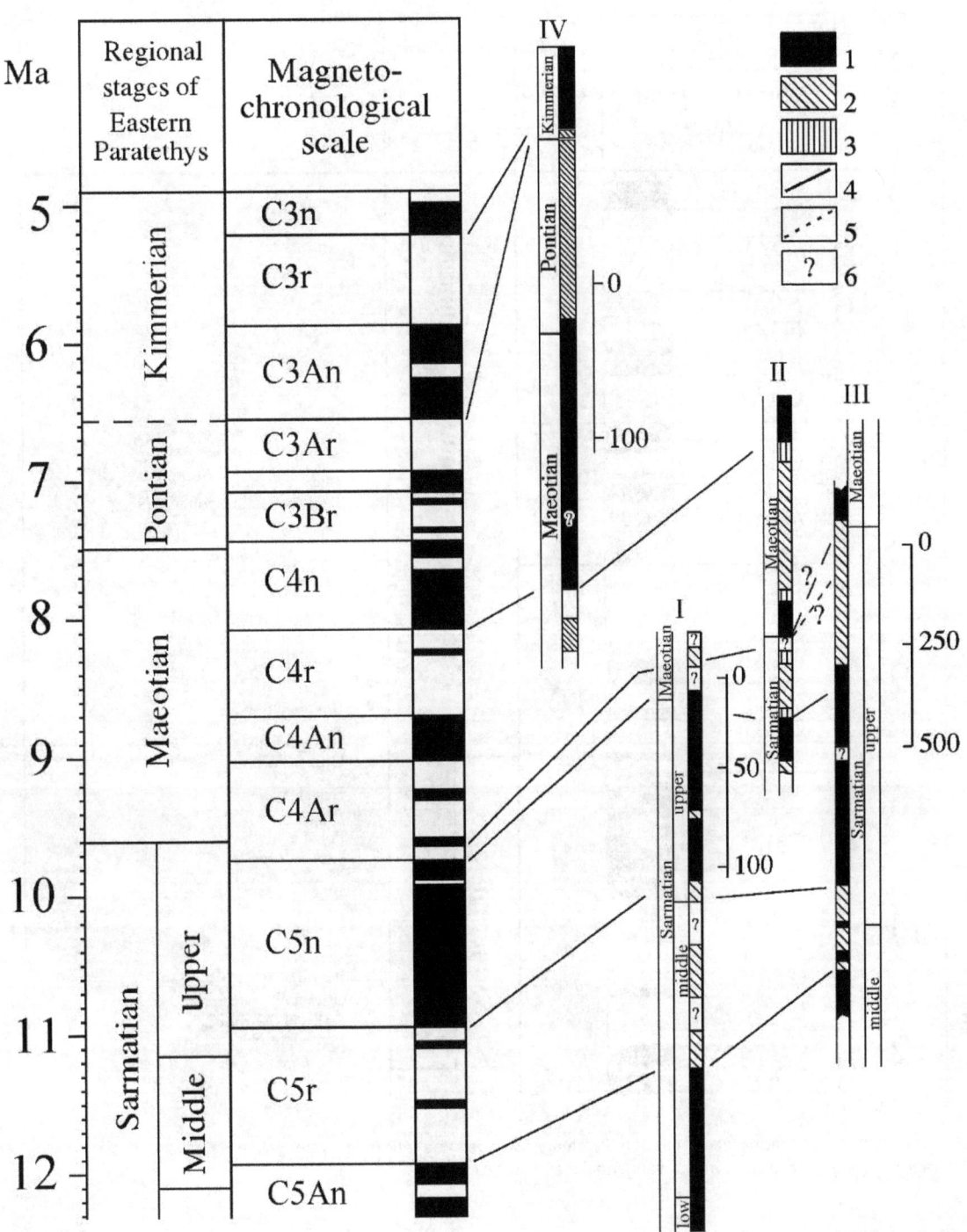

Figure 23.2 Magnetostratigraphic sections of the Sarmatian, Maeotian, and Pontian deposits of eastern Paratethys. Taman Peninsula east (*I*) and west (*II*) of the Panagia Cape; Eldari, eastern Georgia (*III*); Taman Peninsula, Zheleznyi Rog (*IV*). (*1*) normal polarity; (*2*) reversed polarity; (*3*) unstable polarity; (*4*) magnetostratigraphic correlation; (*5*) assumed magnetostratigraphic correlation; (*6*) unsampled interval. Modified after Vangengeim and Tesakov (2008a, 2008b).

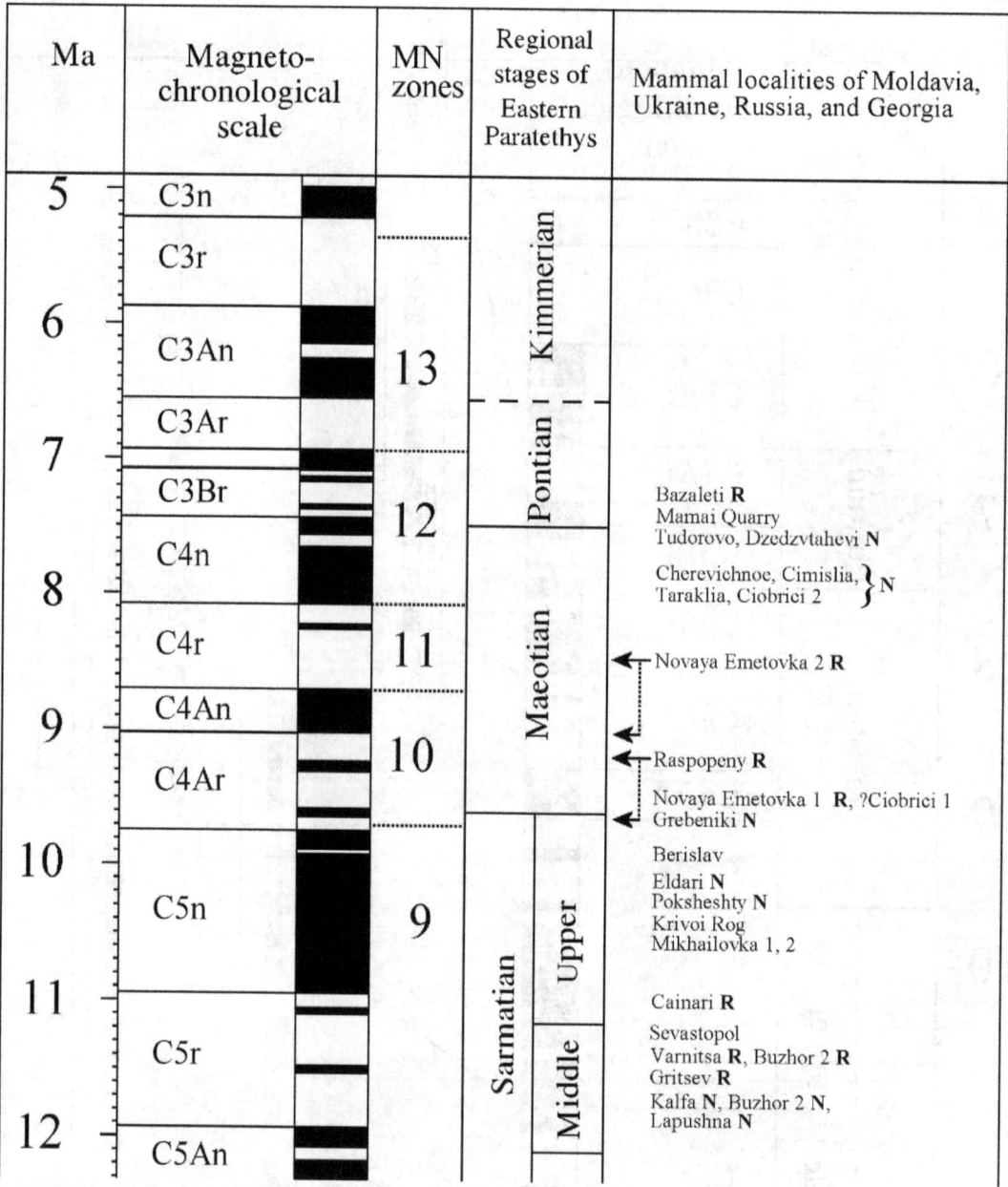

Figure 23.3 Correlation of mammal localities of eastern Paratethys with geomagnetic polarity time scale: R = reversed polarity; N = normal polarity. Modified after Vangengeim et al. (2006) and Vangengeim and Tesakov (2008a, 2008b).

several meters above the base of the Khersonian on the Taman Peninsula (Chumakov, Byzova, and Ganzei 1992; Steininger et al. 1996). The larger part of the upper Sarmatian deposits is normally magnetized (chron C5n), and only the lowermost and uppermost beds provided reversed polarities correlated with C5r.2r and C4Ar.3r (Vangengeim et al. 1989; Pevzner and Vangengeim 1993; Vangengeim and Tesakov 2008a).

A normal polarity zone correlative to subchron C4Ar.2n with the older boundary, dated at 9.6 Ma by Berggren et al.

(1995) or at 9.7 Ma by Lourens et al. (2004), occurs at the base of Maeotian in the Taman section. This defines the age of the Sarmatian–Maeotian boundary. This boundary was estimated at 9.3 Ma based on the fission-track date 9.26 ± 0.74 Ma from the base of the Maeotian in east Azerbaijan (Chumakov, Byzova, and Ganzei 1992). Paleomagnetic data for Maeotian deposits are known from two sections on the Taman Peninsula. The lower part of the Maeotian was studied 2 km west of the Cape Panagia, and the upper part was studied in the Zheleznyi Rog (Iron Horn) section. The

Regional Stages of Eastern Paratethys			Korotkevich (1988), Nesin, Nadachowski (2001) Semenov (2002)		Gabunia (1986)		Lungu, Chemyrtan (1989) Lungu (1990)		Pevzner, Vangengeim (1993)		This contribution	
Pontian	Lower		13	Cherevichnoe Tudorovo	13	Mamai Bazaleti						Mamai Bazaleti
Maeotian	"Middle"	Upper			12	Tudorovo Cherevichnoe					12	Tudorovo, Dzedzvtahevi, Cherevichnoe Belka, Cimislia, Taraklia, Ciobruci u.b.
			12b	Belka, Taraklia Cimislia								
											11	Novaya Emetovka 2 Novoelizavetovka
		Lower	12a	Novaya Emetovka 2 Novoelizavetovka Novaya Emetovka 1	11	Taraklia Cimislia Novoelizavetovka Grebeniki			11		10	Novoukrainka ?Ciobruci l.b., Novaya Emetovka 1 Grebeniki ?Raspopeny
Sarmatian	Upper		11b	Staraya Kubanka Novoukrainka Grebeniki, Ciobruci	10	Cainary	10	Grebeniki Raspopeny Staraya Kubanka Krivoi Rog Poksheshty	10	Cainari, Raspopeny Eldari, Berislav, Staraya Kubanka, Poksheshty Krivoi Rog		?Staraya Kubanka Berislav, Tyaginka Eldari Poksheshty Krivoi Rog, Mikhailovka 1,2 Cainari
			11a	Poksheshty, Raspopeny, Berislav, Tyaginka, Krivoi Rog, Mikhailovka 2		Berislav Eldari		Eldari Tyaginka Berislav Cainari			9	
	Middle		10	Mikhailovka 1 Sevastopol Varnitsa Zheltokamenka Atavaska Kalfa, Buzhor	9	Sevastopol Varnitsa Zheltokamenka Buzhor, Kalfa	9	Sevastopol Buzhor 2, Varnitsa, Gritsev Buzhor 1 Atavaska, Kalfa, Zheltokamenka	9	Sevastopol Gritsev, Varnitsa Buzhor 2 Lapushna Buzhor 1, Kalfa, Kishinev, Atavaska, Zheltokamenka		Sevastopol Varnitsa, Gritsev, Buzhor 2 Lapushna, Buzhor 1 Kalfa, Kishinev, Atavaska, Zheltokamenka
			9	Gritsev, Klimentovichi								

Figure 23.4 Correlation of Late Miocene mammal localities of eastern Paratethys with MN zones according to different authors. Modified after Vangengeim and Tesakov (2008b).

zone of normal polarity in the lower part of Maeotian (17 m, the Panagia Cape section) is followed by a thick zone (60 m) of reversely magnetized sediments correlated to chron C4Ar. On the Kerch Peninsula, deposits of the lower Maeotian contain the nannoplankton boundary of zones

NN9/10 (Bogdanovich and Ivanova 1997) dated at 9.4 Ma (Berggren et al. 1995). In the Panagia record, the higher interval of paleomagnetic section is represented by normally magnetized deposits (20 m) correlative with chron C4An. In Zheleznyi Rog, the reversely magnetized zone (18 m)

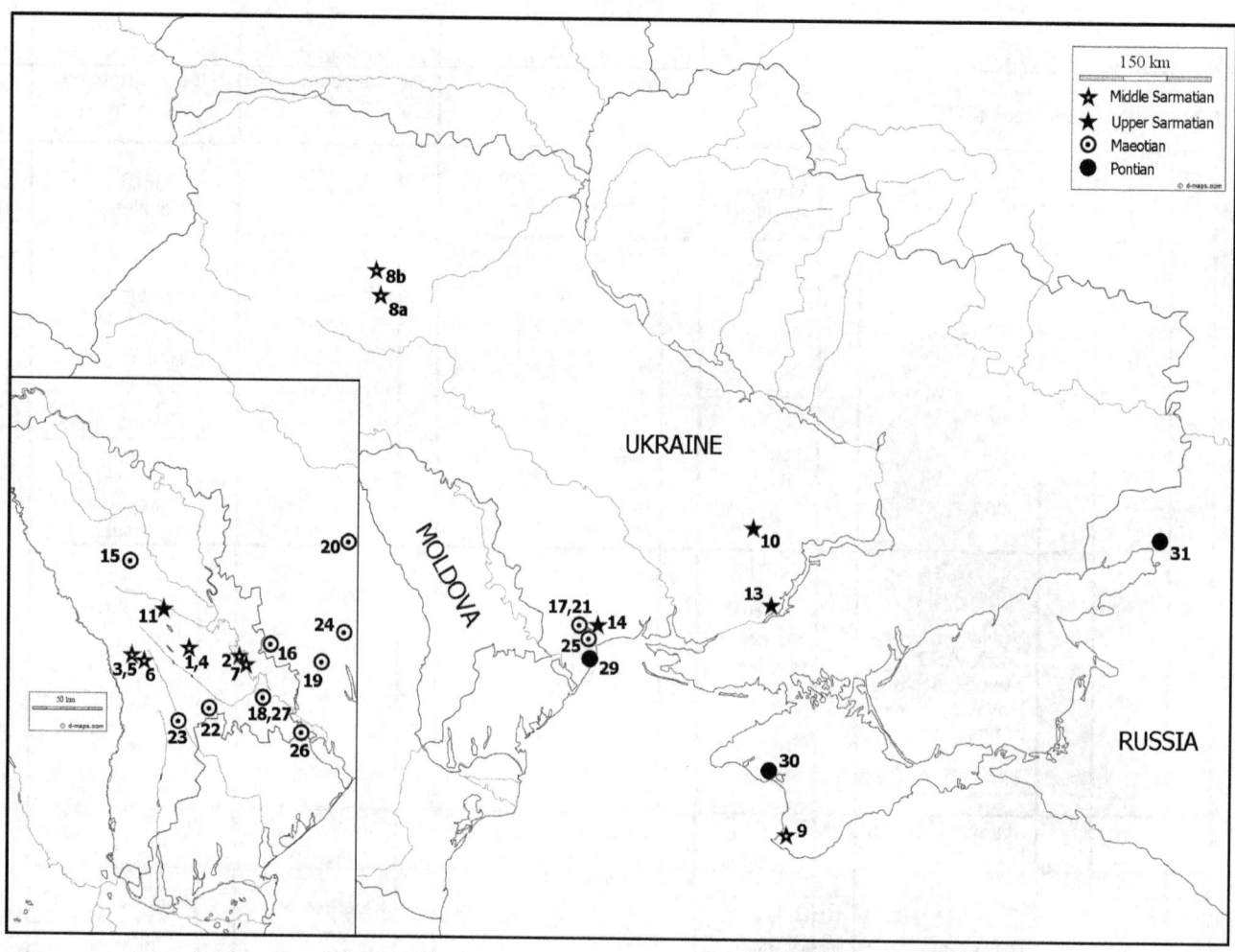

Figure 23.5 Geographic positions of reference Late Miocene mammalian localities in the North Black Sea area. Numbers correspond to locality numbers in tables 23.1–23.3.

and 20-m-thick unsampled interval is capped by a 160-m-thick zone of normally magnetized deposits. This interval corresponds to the larger part of chron C4n. The Maeotian-Pontian boundary in this section lies in the upper part of normal polarity zone (C4n) and is dated at ca. 7.5 Ma. The fission-track date 7.14 ± 0.58 Ma was obtained for presumably lower Pontian continental Shiraki Formation in eastern Georgia (Chumakov, Byzova, and Ganzei 1992). The Shiraki Formation is correlated with undivided Maeotian and Pontian stages. The remaining part of Pontian deposits is mainly reversely magnetized and is correlated with chrons CBr–C3Ar (Pevzner 1986; Pevzner, Semenenko, and Vangengeim 2003). The observed simplified structure of paleomagnetic pattern compared to the complete set of reversed and normal chrons in the correlated interval of GPTS may be accounted for by the known erosional gap between the lower and upper Pontian (Filippova

and Trubikhin 2009). This hiatus can correspond to chron C3Bn (Pevzner et al. 2004). In our view, the upper boundary of the lower Pontian can be dated at 7.1–7.2 Ma, which is, apparently, synchronous with the Messinian lower boundary (chron C3Br.1r).

The upper boundary of the Pontian in the Black Sea basin is not defined, but it is probably not younger than 6.5 Ma (the older boundary of chron C3An). A stratigraphic hiatus occurs between the Pontian and Kimmerian in the Taman Peninsula. Another hiatus is also recognized inside the Azov beds of the Kimmerian. On the Kerch Peninsula, the borehole 15 exposed Pontian deposits with reversed magnetization. The lower part of the Azov beds (about 80 m) have no paleomagnetic characteristics. The higher Azov beds show three alternating zones of reversed and normal polarity interpreted as Thvera, Sidufjall, and Nunivak (Pevzner, Semenenko, and Vangengeim 2003).

Table 23.1

Locations of Late Miocene Reference Localities

No.	Locality	Country	Geographic Position
	Middle Sarmatian (Bessarabian)		
1	Kishinev	Moldova	47°0′0″N 28°51′0″E
2	Kalfa	Moldova	46°54′20″N 29°21′30″E
3	Buzhor 1 (Bujor 1)	Moldova	46°55′41″N 28°16′4″E
4	Atavaska	Moldova	47°0′0″N 28°48′0″E
5	Buzhor 2 (Bujor 2)	Moldova	46°55′41″N 28°16′4″E
6	Lapushna	Moldova	46°53′32″N 28°24′38″E
7	Varnitsa	Moldova	47°12′0″N 28°54′0″E
8a	Gritsev	Ukraine	49°57′57″N 27°12′57″E
8b	Klimentovichi	Ukraine	50°14′20″N 27°7′0″E
9	Sevastopol	Ukraine	44°30′0″N 33°36′0″E
	Upper Sarmatian (Khersonian)		
10	Krivoi Rog	Ukraine	47°54′0″N 28°30′0″E
11	Poksheshty	Moldova	47°14′50N 28°40′55″E
12	Eldari	Georgia	41°18′0″N 45°42′0″E
13	Berislav	Ukraine	46°50′12″N 33°25′25″E
14	Staraya Kubanka	Ukraine	46°42′0″N 30°43′0″E
	Maeotian		
15	Raspopeny	Moldova	47°27′0″N 28°21′36″E
16	Grebeniki	Ukraine	46°53′0″N 29°49′0″E
17	Novaya Emetovka 1	Ukraine	46°39′0″N 30°36′0″E
18	Ciobruci (lower bed)	Moldova	46°36′0″N 29°42′0″E
19	Novoukrainka	Ukraine	48°48′54″N 30°17′12″E
20	Novoelizavetovka	Ukraine	47°09′10″N 30°24′6″E
21	Novaya Emetovka 2	Ukraine	46°39′0″N 30°36′0″E
22	Taraklia	Moldova	46°34′0″N 29°08′0″E
23	Cimislia	Moldova	46°30′10″N 28°48′30″E
24	Belka	Ukraine	46°54′0″N 30°25′55″E
25	Cherevichnoe	Ukraine	46°39′0″N 30°36′0″E
26	Tudorovo	Moldova	46°26′20″N 30°1′30″E
27	Ciobruci (upper bed)	Moldova	46°36′0″N 29°42′0″E
	Lower Pontian		
28	Bazaleti	Georgia	41°54′0″N 45°00′0″E
29	Odessa	Ukraine	46°22′30″N 30°44′35″E
30	Eupatoria	Ukraine	45°12′0″N 33°24′0″E
31	Rostov-on-Don	Russia	47°13′0″N 39°40′0″E

Close age estimates and correlation of Central and Eastern Paratethys strata are used by Austrian geologists (Rögl and Daxner-Höck 1996; Steininger et al. 1996; Steininger 1999; Rögl 2001). The base of the Sarmatian

s.l. (Eastern Paratethys) and the Sarmatian s. Suess (Central Paratethys) is dated at 13–13.6 Ma. The lower boundary of the Pannonian lies in the upper part of the middle Sarmatian in chron C5r. It is estimated close to 11.5 Ma (Vasiliev 2006), and recently it was refined at 11.42 Ma (Lirer et al. 2009). This boundary is close to the lower boundary of the Tortonian in the Mediterranean, occurring inside chron C5r at 11.1 Ma according to Berggren et al. (1995) or at 11.6 Ma according to the astronomical age by Gradstein et al. (2004) and Hilgen, Brinkhuis, and Zachariasse (2006). The Pannonian is correlative to the upper part of the middle Sarmatian, upper Sarmatian s.l., and Maeotian. The Pannonian-Pontian boundary is placed in the interval between 8 Ma and 7 Ma (Rögl and Daxner-Höck 1996:50), or recently at 6.1 Ma (Krijgsman et al. 2010).

The interpretation presented here of paleomagnetic data contrasts with the views of Trubikhin (1989, 1998; Nevesskaya et al. 2004; Popov, Nevesskaya, and Pinchuk 2007): "Pontian and the Azov beds [of the Kimmerian] . . . correspond to the lower part of the Gilbert Chron, and the Maeotian, to Chrons 5 [= chron C3An] and 6 of normal and reversed polarity, respectively. This pair (Pontian and Maeotian) . . . corresponds to the Messinian of the Mediterranean and, probably, to the Pannonian of the Pannon and Vienna Basin" (Trubikhin 1998:13). According to this interpretation, the Maeotian lower boundary has the age of 7.4 Ma (chron C3Br), and that of Pontian, 6.15 Ma (in chron C3n.1n). "The underlying Sarmatian sequence belongs to chrons 7 and 8 (mainly upper Sarmatian) [C4Ar–C4n, our note], 9 (mainly middle Sarmatian) [chron C5n, our note] and, finally, 10 (mainly lower Sarmatian) of the magnetochronological scale" (Trubikhin 1998:16; see also Popov et al. 1996). Correlation of the upper Sarmatian (Khersonian) with chrons C4Ar–C4n with predominantly reversed polarity contradicts the observed mostly normal magnetization of the upper Sarmatian in sections of the Taman Peninsula and east Georgia.

The group of Dutch and Romanian researchers (Snel et al. 2006) used the Turbikhin correlation scheme for the Dacic Basin, at least, for the Maeotian and Pontian, with the lower boundary of Pontian at 6.15 Ma, and the Pontian–Kimmerian boundary at 5.3 Ma. The recent restudy of the Maeotian-Kimmerian interval of the Zheleznyi Rog section (Krijgsman et al. 2010) produced similar results and correlations. Worth mentioning is the recent micropaleontological study of the same section that bracketed the Maeotian-Pontian boundary between 6.1 Ma and 5.9 Ma based on the short-lived presence of index species of oceanic diatoms (Radionova and Golovina 2009).

Zone MN9: Early Vallesian ([?]12–9.7 Ma)

Localities of the Middle Sarmatian

The lower boundary of the middle Sarmatian (Bessarabian), dated at ca. 12 Ma, lies in the upper part of chron C5An (Pevzner 1986; Steininger et al. 1996).

The oldest localities with hipparion remains are known in bioherm limestones of the lower (?) bed of middle Sarmatian near Chisinau (=Kishinev, Moldova; Lungu and Chemyrtan 1989; Lungu 1990). They include *Thalassictis robusta* Nordmann, *Deinotherium* sp., "*Hipparion*" sp., *Alicornops simorrense* (Lartet), and *Microstonyx antiquus* Kaup. The locality Zheltokamenka (Ukraine) of similar geological age yielded *Machairodus* cf. *aphanistus* Kaup, *Gomphotherium* cf. *angustidens* (Cuvier), and *Anchitherium* sp. (a single record). No hipparion remains have been found (Pidoplichko 1956). These sites do not have paleomagnetic records.

The localities Kalfa, Buzhor 1, and Lapushna (Moldova) are associated with marine deposits of the middle Sarmatian. Normal magnetization of bone-bearing beds suggests a correlation with the subchron C5r.2n (dated in ATNTS2004 at 11.554–11.614 Ma), or alternatively with the upper part of chron C5An (Vangengeim, Lungu, and Tesakov 2006). The appearance of hipparions (i.e., that mark the beginning of the Vallesian) in the middle Sarmatian in these sites thus can be dated at 11.5 Ma (or at 12 Ma as an older alternative). This is at least 0.3 myr older than the estimate of this datum (11.2 Ma) in western and central Europe (Steininger et al. 1996; Actes du Congrés BiochroM'97 1997; Vangengeim, Lungu, and Tesakov 2006; Lirer et al. 2009; and others).

Progonomys is known from 12 Ma onward in Pakistan (Flynn et al. 1995). The oldest lower Vallesian records of *Progonomys* in North Africa, Asia Minor, and Europe may be as old as 10.3 Ma (van Dam 1997; Sen 2003; Wessels 2009). Murid material described as *Progonomys cathalai* Schaub from Buzhor 1 (Lungu 1981) probably contains two different forms, with one of them showing a very primitive "*Mus*"-like morphology (Mein, Martín-Suárez, and Agustí 1993) and another representing *Parapodemus* morphology. The similar mixture of *Progonomys* and *Parapodemus* morphologies (lineages?) was described in lower Vallesian of Turkey (Sen 2003).

The observed diachrony of the Vallesian lower boundary could be minimized if the alternative correlation with the GPTS of Can Llobateres (Spain) and Kastellios (Greece), the sites with hipparionine horses and *Progonomys*, is considered. Until recently, these sites were placed in chron C4Ar (Sen 1996; Agustí, Cabera, and Garcés

2001). Aguilar with coauthors (Aguilar et al. 2004; Costeur et al. 2007) provided evidence for their placement in chron C5r. According to this viewpoint, Can Llobateres 1 is correlated to subchron C5r.3r, Can Llobateres 2, to C5r.2n, and Kastellios, to chrons C5r.3r–C5r.1r. See discussion in Wessels (2009) for rejection of this correlation in favor of younger (C4Ar) age of these records.

The upper (and larger) part of the middle Sarmatian belongs to a zone of reversed polarity correlative to chron C5r. This interval includes localities Buzhor 2, Varnitsa, Viveritsa (Moldova), Gritsev, and Klimentovichi (northwestern Ukraine). The Ukrainian paleontologists (e.g., Topachevsky et al. 1996) consider Gritsev and Klimentovichi as the oldest Vallesian sites of eastern Europe. They correlate these localities to the lower middle Sarmatian. The bone-bearing deposits of these sites, however, represent fissure fillings associated with the erosional surface (and, accordingly, a stratigraphic hiatus) on the lower middle Sarmatian deposits.

The Gritsev fauna contains two forms of hipparions: *Hippotherium primigenium* and "*Hipparion*" sp. (Krakhmalnaya 1996b), which is not common for the earliest Vallesian. Small mammals are represented by abundant and diverse soricids and glirids, and carnivorans, by abundant mustelids, indicating widespread forested environment. The fauna contains a significant number of Astaracian forms (Nesin and Topachevsky 1999). These features markedly distinguish the fauna of Gritsev from faunas of more southern Moldavian sites that show drier and more open conditions. This evidently places the two areas into different paleozoogeographic provinces, the central European (Gritsev) and southeastern (Moldavian sites) (Bernor 1996b; Fortelius, Van der Made, and Bernor 1996) or central European and east European (Lungu 1990). The Gritsev fauna shares numerous genera with a slightly younger Rudabánya fauna in Hungary, which also belongs to the central European province (Bernor et al. 2004).

The youngest middle Sarmatian locality, lying close to the boundary with the upper Sarmatian, is Sevastopol in the Crimea (Lungu 1990; Pevzner and Vangengeim 1993; Vangengeim, Lungu, and Tesakov 2006).

The early Vallesian faunas of Eastern Paratethys (tables 23.2 and 23.3) show the appearance of *Machairodus*, *Hippotherium primigenium* (Meyer), *Microstonyx antiquus* (Kaup), a chilothere, and *Progonomys cathalai* Schaub, and many Astaracian elements are retained. Only the middle Sarmatian faunas document the presence of *Eomellivora wimani piveteaui* Ozansoy, *Sansanosmilus*, *Thalassictis*, *Protictitherium crassum* (Deperet), *Metailurus pamiri* (Ozansoy), *Pseudaelurus turnauensis* (Hoernes), *Alicor-*

Table 23.2

Stratigraphic Distribution of Large Mammals (Carnivora) in Eastern Paratethys

	MN 9		MN 10	MN 11	MN 12/?MN 13
	Sarmatian		Maeotian–Lower Pontian		
	Middle	Upper	Lower Maeotian	Middle Maeotian	Upper Maeotian–Lower Pontian
Simocyon primigenius	8 sp.	10	19		
Indarctos sp.	8				
Plesiogulo crassa	8 cf.				25
Plesiogulo brachygnathus	3	11 aff.			
Eomellivora wimani piveteaui	2, 3, 8				
Eomellivora wimani wimani			16	21	
Palaeomeles sp.	8				
Promeles sp.	2				
Promeles palaeattica					23
Proputorius aff. *medius*	3				
Parataxidea aff. *polaki*		11 sp., 14			
Promephitis maeotica			16	20, 21 sp.	28 sp.
Sansanosmilus piveteaui	2, 7, 8				
Dinocrocuta gigantea	2, 3, 8 sp.	11, 12			
Percrocuta robusta	1, 2, 3, 7				
Allohyaena sarmatica	8				
Adcrocuta eximia		10 cf.	16, 17, 19	20, 21	23–26
Miochyaenotherium bessarabicum					23, 24
Ictitherium viverrinum		10	16	20	24, 28
Ictitherium pannonicum				21	25 cf.
Sansanosmilus piveteaui	8				
Dinocrocuta gigantea				20	22–24, 26
Protictitherium crassum	2, 9	11 aff.		20, 21	
Thalassictis robusta	1				
Thalassictis montadai	2, 3				
Hyaenotherium wongii			16 cf., 17 cf., 19 cf.	21 cf.	25
Semigenetta sp.	8				
Machairodus aphanistus	8 cf.	10 cf., 11 sp.	19 cf.	?21	
Machairodus giganteus				20	22, 23, 24
Machairodus laskarevi	2, 3				
Machairodus copei			16		
Pseudaelurus turnauensis	2, 3				
Pseudaelurus sp.	2, 3, 8		19		
Metailurus pamiri	2				
Metailurus sp.		11			25
Metailurus parvulus			16	20	22, 23

Mammal localities. Middle Sarmatian: 1. Kishinev; 2. Kalfa; 3. Buzhor 1 (Bujor 1); 4. Atavaska; 5. Buzhor 2 (Bujor 2); 6. Lapushna; 7. Varnitsa; 8. Gritsev, Klimentovichi; 9. Sevastopol.

Upper Sarmatian: 10. Krivoi Rog; 11. Poksheshty; 12. Eladari; 13. Berislav; 14. Staraya Kubanka.

Maeotian: 15. Raspopeny; 16. Grebeniki; 17. Novaya Emetovka 1; 18. Ciobruci (lower bed); 19. Novoukrainka 1; 20. Novoelizavetovka; 21. Novaya Emetovka 2; 22. Taraklia; 23. Cimislia; 24. Belka; 25. Cherevichnoe; 26. Tudorovo; 27. Ciobruci (upper bed). Lower Pontian: 28. Bazaleti, 29. Sporadic occurrences in lower Pontian limestone in Odessa, Eupatoria, and Rostov-on-Don.

Table 23.3

Stratigraphic Distribution of Large Mammals (Proboscidea, Perissodactyla, Artiodactyla) in Eastern Paratethys

	MN 9		MN 10	MN 11	MN 12/?MN 13
	Sarmatian		Maeotian–Lower Pontian		
	Middle	Upper	Lower Maeotian	Middle Maeotian	Upper Maeotian–Lower Pontian
Gomphotherium angustidens	8				
Choerolophodon spp.	2, 7, 9	12, 13, 14	17, 19	20, 21	
Tetralophodon gr. *longirostris-atticus*	4	11	16	20, 21,	22, 23, 29
"Mammut" gr. *obliquelophus-borsoni*				20	22, 23, 24
Deinotherium giganteum	1 sp., 2, 8, 9 sp.	10, 12, 13 sp.	15, 16, 17 sp.,19 sp.	20	22, 23, 24 sp., 25, 28 sp.
Hippotherium primigenium	1–6, 8, 9 ex gr.	12 ex gr.			
Hippotherium giganteum		11 cf., 14	15 aff., 16, 19		
"Hipparion" verae	7	11 cf., 13 cf. 14	16, 17, 18 cf., 19 cf.		
"Hipparion" sp.	8		17	21	25 (cf. probos-cideum), 28, 29
Cremohipparion moldavicum				20, 21, ?27 cf.	22–26
Chalicotherium goldfussi	8 cf.				
Acerorhinus sp. (= *Aceratherium incisivum* auct)			16	20, 21	22–24, 25 sp. 26, 28 cf.
Aceratherium simplex					26
Chilotherium schlosseri		14 cf.	16, 17, 19 sp.		
Acerorhinus zernowi	7, 9	10 sp., 11 cf. 12			
Chilotherium kowalevskii			15, 16		
Chilotherium sarmaticus		11 aff., 13			
Alicornops simorrensis	1, 2, 3, 4, 5, 8				
"Dicerorhinus" sp.	2, 7, 9				25
Diceros gabuniai		12			
Ceratotherium neumayri					29 cf.
Ancylotherium pentelicum			19 cf.		
Schizochoerus vallesiensis	2				
Microstonyx antiquus	1, 2	12, 13 cf.			
Microstonyx major			15, 16, 17, 19	20, 21	22, 23, 24, 25 sp., 28 sp.
Propotamochoerus palaeochoerus	8				
Lagomeryx flerovi	2–7, 9				
Dorcatherium sp.	8				
Hispanomeryx duriensis	8 aff.				
Amphiprox sp.	8				
Euprox furcatus	2, 3, 4, 8 sp.				
Euprox sarmaticus		10			
Procervulus sp.	8				
Cervavitus sp.		11	15, 18		
Cervavitus variabilis			16, 17, 19	20	22, 23
Cevavitus novorossiae					22, 23
Procapreolus frolovi					25

	MN 9		MN 10	MN 11	MN 12/?MN 13
	Sarmatian		Maeotian–Lower Pontian		
	Middle	Upper	Lower Maeotian	Middle Maeotian	Upper Maeotian–Lower Pontian
Procapreolus ukrainicus		12	16 cf., 19	20 cf.,	24 cf.
Palaeotragus (Achtiaria) expectans	1, 6, 7, 9				
Palaeotragus (Achtiaria) bereslavicus		13			
Palaeotragus (Achtiaria) moldavicus		11 aff.	15, 16 sp., 17 sp.		
Palaeotragus (Achtiaria) borissiaki		12, 14 sp.			
Palaeotragus pavlowae			16		
Palaeotragus rouenii				20, 21	22, 23, 24
Palaeotragus sp.			17., ?18, 19		28
Chersonotherium eminens			19 sp.	20	
Samotherium maeoticum		13 sp., 14 sp.	19 sp.	21	24 sp.
Samotherium boisseri					22, 24 sp.
Helladotherium duvernoyi					22, 23
Procapra capricornis		13 cf., 14	16, 17 cf., 19		23
Procapra rodleri				21 aff.	22, 24
Procapra longicornis					25
Gazella gracile		11, 12, 13	15		
Gazella pilgrimi				20, 21	23
Gazella schlosseri			16, 17, 19		
"*Protragocerus*" *leskewitschi*	9	11, 12 aff.	16, 19		
Miotragocerus pannoniae	2, 3, 8 cf.				
Miotragocerus borissiaki	9				
Tragocerus frolovi			18, 19	20	23, 26
Mesotragocerus citus				21	
Graecoryx valenciennesi		12, 13 cf.			
Graecoryx bonus					22, 24
Protragelaphus skouzesi		14 sp.	16		22
Palaeoreas lindermayeri					23
Procobus melania				21	22
Procobus brauneri					22
Moldoredunca amalthea	7, 9				
Palaeoryx pallasi				21	23, 29
Palaeoryx majori			16, 17 sp.		22, 26
Tragoreas oryxoides					22

NOTE: Mammal localities. Middle Sarmatian: 1. Kishinev; 2. Kalfa; 3. Buzhor 1 (Bujor 1); 4. Atavaska; 5. Buzhor 2 (Bujor 2); 6. Lapushna; 7. Varnitsa; 8. Gritsev, Klimentovichi; 9. Sevastopol.

Upper Sarmatian: 10. Krivoi Rog; 11. Poksheshty; 12. Eldari; 13. Berislav; 14. Staraya Kubanka.

Maeotian: 15. Raspopeny; 16. Grebeniki; 17. Novaya Emetovka 1; 18. Ciobruci (lower bed); 19. Novoukrainka 1; 20. Novoelizavetovka; 21. Novaya Emetovka 2; 22. Taraklia; 23. Cimislia; 24. Belka; 25. Cherevichnoe; 26. Tudorovo; 27. Ciobruci (upper bed). Lower Pontian: 28. Bazaleti; 29. Sporadic occurrences in lower Pontian limestone in Odessa, Eupatoria, and Rostov-on-Don.

nops simorrensis (Lartet), *Lagomeryx, Euprox, Amphiprox, Palaeotragus* (*Achtiaria*) *expectans* (Borissiak), and *Gomphotherium angustidens* (Cuvier). *Propotamochoerus palaeochoerus* (Kaup) is found only in northwestern Ukraine, in the central European province (Lungu 1990; Wolsan and Semenov 1996; Krakhmalnaya 2008).

From the mid-middle Sarmatian, hipparionine horses show a marked differentiation. In addition to the typical *Hippotherium primigenius*, other forms as "*H.*" sp. (Gritsev), "*Hipparion*" cf. *verae* Gabunia (Varnitsa), and *Hippotherium sebastopolitanum* (Borissiak; Sevastopol) appeared. Forsten (1978) considered the latter species as an independent local form, parallel to *H. primigenium*.

Localities of the Late Sarmatian

The middle–late Sarmatian boundary is dated at about 11.2 Ma (below subchron C5r.1n). This transition is marked by a considerable marine regression and possibly related changes in the composition of mammalian fauna (Muratov and Nevesskaya 1986; Pevzner and Vangengeim 1993). The lowermost beds of marine upper Sarmatian contain the small mammal locality Cainari (Moldova) with *Progonomys* cf. *cathalai* Schaub and *P. woelferi* Bachmayer et Wilson (Lungu and Chemyrtan 1989; Lungu 1990). The reversely magnetized deposits in Cainari are correlated with subchron C5r.1r (Vangengeim and Tesakov 2008a).

No reliably dated large mammal sites corresponding to the early late Sarmatian are known yet. The site Krivoi Rog (Ukraine) is located close to northern coastal line of the maximum late Sarmatian marine transgression. The small mammal locality Mikhailovka (Ukraine) has a similar geological age. The lower bone-bearing bed (Mikhailovka 1) occurs in lacustrine-fluviatile sequence upon the erosional surface truncating upper middle Sarmatian deposits. The higher bone-bearing bed of Mikhailovka 2 occurs in limestones of the maximum late Sarmatian transgressions (Nesin and Topachevsky 1999; Vangengeim and Tesakov 2008a). Ukrainian paleontologists refer the lower bed to the upper middle Sarmatian and refer the upper bed to the beginning of the Upper Sarmatian. There are no paleomagnetic data for this site. Because the precise age of the maximum late Sarmatian transgressions is currently unknown, we by convention correlate sites Krivoi Rog and Mikhailovka 1 and 2 to the middle of chron C5n. Both bone beds of Mikhailovka yielded remains of microtoid cricetid *Ischymomys quadriradicatus* Zazhigin (Nesin and Topachevsky 1999).

The later part of marine late Sarmatian is represented by the Eldari (Georgia), Berislav, Staraya Kubanka, Yurievka, and Tyaginka (Ukraine) localities; the Poksheshty (Moldova) locality occurs in continental deposits of the Balta Formation (Gabunia 1986; Korotkevich 1988; Lungu and Chemyrtan 1989; Lungu 1990; Vangengeim and Tesakov 2008a). All these sites belong to the younger part of chron C5n (Vangengeim and Tesakov 2008a). A remarkable occurrence of murids (cf. *Progonomys*) was reported by Steklov (1966) from the Upper Sarmatian deposits of Maikop in the Northern Caucasus, Russia.

Hipparions of the later part of late Sarmatian are represented by the very large form *Hippotherium giganteum* (Gromova), accompanied by smaller forms of the *H. primigenium* group, and "*Hipparion*" *verae* (Gabunia). According to Forsten (1980), the latter species is at the base of the lineage leading to *Cremohipparion mediterraneum*. The fauna documents the last occurrences of *Dinocrocuta* and *Percrocuta*. Among newcomers are *Adcrocuta eximia* (Roth et Wagner), *Ictitherium viverrinum* Roth and Wagner, *Gazella, Procapra, Procapreolus, Graecoryx*, new forms of *Palaeotragus* (*Achtiaria*) (*P. (A.) borissiaki* (Alexejev), *P. (A.) berislavicus* Korotkevich), *Samotherium, Cervavitus* (Korotkevich 1988; Lungu 1990; Werdelin and Solounias 1991). The fauna is dominated by chilotheres and hipparions.

In Western Europe (Spain), the zone MN 9 was subdivided into two subzones: MN9 a (*Megacricedon + Hipparion*) and MN 9b (*Cricetulodon*), with a boundary at 10.4 Ma (Agustí et al. 1997; Agustí 1999). The younger late Sarmatian faunas can be correlated with the subzone MN 9b (Vangengeim and Tesakov 2008a).

Zone MN10: Late Vallesian (9.7–8.7 Ma)

The zone MN 10 is correlated with upper levels of the upper Sarmatian and the lower Maeotian (chron C4Ar; Vangengeim and Tesakov 2008b). The Sarmatian/Maeotian boundary lies at the base of subchron C4Ar.2n, at 9.6 Ma.

Localities of the Early Maeotian

The zone MN 10 includes the Grebeniki, Novaya Emetovka 1, Novoukrainka (Ukraine), and Raspopeny sites and the lower(?) bed of Ciobruci (Moldova). In Grebeniki, the bone-bearing bed is normally magnetized and compared to subchron C4Ar.2n; in Novaya

Emetovka 1 and Raspopeny, the bone beds show reversed polarity correlative to chron C4Ar (Vangengeim and Tesakov 2008b). The Raspopeny locality can belong to the uppermost levels of the upper Sarmatian (subchron C4Ar.3r; Vangengeim and Tesakov 2008a).

The zone MN 10 documents last occurrences of *Hippotherium giganteum* (Gromova), "*Hipparion*" *verae* Gabunia, *Machairodus* ex gr. *aphanistus, Chilotherium,* and *Protragocerus leskewitschi* (Borissiak). The new appearances include that of *Hyaenotherium wongi* (Zdansky), *Aceratherium incisivum* Kaup, and *Cervavitus.* According to Godina (1979), giraffes are represented by transitional forms from *Palaeotragus (Achtiaria)* to *Palaeotragus (P.(A.) moldavicus* Godina, *P. pavlowae* Godina, and *P.* sp.). Sarmatian *Microstonyx antiquus* (Kaup) and *Eomellivora wimani piveteaui* Ozansoy are replaced with *Microstonyx major* Gervais and *Eomellivora wimani wimani* Zdansky (Gabunia 1986; Korotkevich 1988; Krakhmalnaya 1996a; Lungu and Chemyrtan 1989; Wolsan and Semenov 1996). Hipparions and chilotheres are most abundant, followed by gazelles and procapras.

Zone MN 11: Early Turolian (9.7–8.7 Ma to 8.24–8.0 Ma)

Localities of the Middle Maeotian

Localities Novoelizovetovka and Novaya Emetovka 2 (Ukraine) belong to the zone MN 11. Bone-bearing deposits in Novaya Emetovka 2 are reversely magnetized and correlated with the upper part of chron C4Ar or chron C4r (Krakhmalnaya 1996a; Vangengeim and Tesakov 2008b).

Remarkable events in this zone include the appearance of *Cremohipparion moldavicum* (Gromova) and *Palaeotragus rouenii* Gaudry, the replacement of *Machairodus* ex gr. *aphanistus* with *M.* ex gr. *giganteus* Kaup, and the replacement of *Gazella schlosseri* M.Pavlow with *G. pilgrimi* Bohlin. Newly occurring forms include *Ictitherium pannonicum* Kretzoi, *Zygolophodon turicensis* (Schinz), *Chersonotherium, Tragocerus frolovi* M. Pavlow., *Mesotragocerus citus* Korotkevich, *Procobus melania* Khomenko, and *Palaeoryx pallasii* (Wagner). No chilothere remains have been found in this zone. The assemblages are dominated by hipparions; important shares belong to *Aceratherium,* procapras, gazelles, and giraffes (Korotkevich 1988; Krakhmalnaya 1996a; Krakhmalnaya and Forsten 1998).

Zone MN 12: Middle Turolian (8.24–8.0 Ma to 7.1 Ma)

Localities of the Late Maeotian

The zone MN 12 includes localities of the later part of the Maeotian, such as Taraklia, Cimislia, Tudorovo (Moldova), Belka, and Cherevichnoe (Ukraine) (Gabunia 1986; Korotkevich 1988; Semenov 1989; Krakhmalnaya 1996a; Lungu and Delinschi 2008). Fossiliferous deposits in Taraklia, Cimislia, and Cherevichnoe have normal magnetization correlated with chron C4n. Faunas of this zone retain many forms characteristic for MN 11. In addition, the fauna includes *Miochyaenotherium bessarabicum* Semenov, *Aceratherium simplex* Korotkevich (highly specialized form), *Pliocervus, Procapreolus frolovi* Korotkevich, *Helladotherium duvernoyi* Gaudry, *Palaeoreas lindermayeri* Wagner, *Tragoreas oryxoides* Schlosser, *Procapra rodleri* Pilgrim et Hopwood, and *Procobus brauneri* Khomenko. The assemblages are dominated by *Cremohipparion* ex gr. *moldavicum* and procapras.

Localities of the Early Pontian

Deposits of the early Pontian are bracketed in the range of 7.5 Ma to 7.1 Ma, from the end of chron C4n to the beginning of chron C3Bn (Pevzner, Semenenko, and Vangengeim 2003). The upper boundary of the zone MN 12 is estimated at 7.1 Ma. It coincides with the onset of the Messinian and the period of the global carbon shift from C3 to C4 plants (Bernor, Solounais, et al. 1996c; Swisher 1996). Lower Pontian limestone near Odessa and Eupatoria, and Rostov-on-Don yielded sparse remains of large mammals. This fauna includes *Tetralophodon longirostris* Kaup (according to Markov [2008], Turolian tetralophodons possibly belong to the Pikermian *T. atticus* [Wagner]), "*Mammut*" *borsoni* (Hays), *Diceros* cf. *pachygnathus* (Wagner) (=*Ceratotherium neumayri*), *Palaeoryx pallasi* (Wagner), and *P. longicephalus* Sokolov (Gabunia 1986). The hipparion is represented by *Cremohipparion* cf. *mediterraneum* (Gervais) (Gabunia, pers. comm., 1985). The most remarkable event in the early Pontian is the first occurrence of the genus *Paracamelus* (Gabunia 1986; Pevzner, Semenenko, and Vangengeim 2003; Titov and Logvinenko 2006).

The early Pontian age has been presumably assigned to the locality Bazaleti occurring in upper parts of the Dusheti Formation in Georgia. The bone-bearing deposits have reversed polarity correlative with chron C3Br. The fauna includes *Promephitis* sp., *Ictitherium veverrinum* Roth et Wagner (=*Ictitherium ibericum* Meladze), *Deinotherium* sp., *"Hipparion"* sp. (slender-limbed form), *Aceratherium* cf. *incisivum* Kaup, large sivatherine *Karsimatherium bazaleticum* Meladze, *Palaeotragus* sp., *Gazella*, peculiar specialized *Phronetragus*, and other forms (Meladze 1967; Gabunia 1986; Vangengeim et al. 1989; Semenov 2008; Vekua and Lordkipanidze 2008).

Zone MN 13: Late Turolian (7.1–[?]5.3 Ma)

No reliable record of late Pontian large mammals (correlative with zone MN 13, the older boundary 7.1 Ma) is presently known (Gabunia 1986). In Eastern Paratethys, the late Pontian is characterized by a drastic marine regression synchronous with the onset of the Messinian in the Mediterranean (Pevzner, Semenenko, and Vangengeim 2003). In western and central Europe, the zone MN 13 is characterized by a strong reduction in diversity of large mammals (Bernor et al. 1996a).

If accepted, the alternative paleomagnetic correlation of the Maeotian and Pontian (+Azov beds of the Kimmerian) to chrons C3Br–C3r (Trubikhin 1989, 1998; Krijgsman et al. 2010) would suggest a correlation of these Eastern Paratethys units with the mammal zone MN 13 (7.1–5.3 Ma). We cannot currently reconcile this correlation with the faunal evidence. Maeotian mammal fauna of the Eastern Paratethys shares numerous common taxa, faunal abundance, and diversity with middle Turolian (zones MN 12) faunas of central and western Europe. Likewise, the presence of middle Turolian (Pikermian) faunal elements in the lower Pontian deposits seemingly excludes their correlation with C3r.

CONCLUSION

Our study focused on mammalian biochronology controlled by paleomagnetic data. It resulted in a correlation model of the studied mammal faunas of eastern Paratethys to MN zones of western Europe. Vallesian and Turolian faunas of Eastern Europe were characterized. This study revealed a number of common features and distinctions of different paleozoogeographic provinces. The distinctions are caused by dissimilar physiographic conditions in these provinces. The westward increasing aridification accounts for earlier occurrences of some forms in eastern Europe in comparison with the record of cen-

tral and western Europe. The central European region preserved a stable wooded environment longer than in the east and west of the continent.

The southeastern province is importantly distinct in earlier (middle part of MN 9, late middle Sarmatian) differentiation of hipparions compared to the beginning of MN 10 in the west of the continent (excluding the central European province). The appearance of genera *Gazella, Procapra, Cervavitus,* and *Adcrocuta* in the upper Sarmatian (middle part of MN 9) of the southeastern province predates the occurrence of these genera in the west (MN 10). The fossil record of eastern and central Europe documents the extinction of *Microstonyx antiquus* in the middle part of MN 9. In MN 10, this form was replaced by *M. major,* which survived untill the end of the Turolian and spread to Spain. Chilotheres in the studied part of the southeastern province are not known in faunas younger than MN 10. The western limit of their range at that time reached Bulgaria and Greece (Spassov, Tsankov, and Geraads 2006). In contrast, this genus is absent in central Europe and extremely rare in Spain. The fauna of western and central Europe includes *Aceratherium* since the beginnings of MN 9. In contrast, this genus seemingly appeared in the fauna of the southeastern province in MN 10 (but see Codrea and Ursachi 2007).

The beginning of the Turolian is marked throughout Europe by the appearance of the genus *Cremohipparion,* the replacement of *Machairodus* gr. *aphanistus* with *M.* gr. *giganteus,* and the replacement of giraffes *Palaeotragus* (*Achtiaria*) with *Palaeotragus rouenii.* In the middle Turolian (MN 12), *Paracamelus* migrated to southern East Europe. This genus had a short-term westward dispersal only in the Messinian (MN1 3).

ACKNOWLEDGMENTS

The work was supported by the Russian Foundation for Basic Research, project nos. 09-05-00307a and 12-05-00904a. We thank Marina Sotnikova (Moscow) and Alexander Lungu (Chisinau) for valuable advice and discussions. We thank the organizers of the International Asian Mammal Biostratigraphy Conference in Beijing (2009), and particularly Xiaoming Wang (Los Angeles) and Mikael Fortelius (Helsinki) for their invitation to report results of our synthesis and for bridging gaps between scientists. Raymond Bernor (Washington), Nikolay Spassov (Sofia), and two anonymous reviewers are acknowledged for critical comments and suggestions that helped to improve this contribution.

REFERENCES

Aguilar, J.-P., W. A. Berggren., M.-P. Aubry, D. V. Kent, G. Clauzon, M. Benammi, and J. Michaux. 2004. Mid-Neogene Mediterranean marine-continental correlations: An alternative interpretation. *Palaeogeography, Palaeoclimatology, Palaeoecology* 204:165–186.

Aguilar, J.-P., S. Legendre, and J. Michaux, eds. *Actes du Congrès BiochroM'97.* Mémoires et Travaux E.P.H.E., Institut de Montpellier, 21.

Agustí, J. 1999. A critical re-evaluation of Miocene mammal units in Western Europe: Dispersal events and problems of correlation. In *The Evolution of Neogene Terrestrial Ecosystems in Europe,* ed. J. Agustí, L. Rook, and P. Andrews, pp. 84–112. Cambridge: Cambridge University Press.

Agustí, J., L. Cabrera, and M. Garcés. 2001. A calibrated mammal scale for the Neogene of Western Europe: State of the art. *Earth-Science Reviews* 52:247–260.

Agustí J., L. Cabrera, M. Garcés, and J. M. Parés. 1997. The Vallesian mammals succession in the Vallés-Penedés basin (northeast Spain): Paleomagnetic calibration and correlation with global events. *Palaeogeogaphyr, Palaeoclimatology, Palaeoecology* 133:149–180.

Berggren, W. A., D. V. Kent, C. C. Swisher III, and M. P. Aubry. 1995. A revised Cenozoic geochronology and chronostratigraphy. In *Geochronology, Time Scale, and Global Stratigraphic Correlation: A Unified Framework for a Historical Geology,* ed. W. A. Berggren, D. V. Kent, M. P. Aubry, and J. Hardenbol, pp. 129–212. Society of Economic Paleontologists and Mineralogists Special Publication 54. Tulsa: Society for Sedimentary Geology.

Bernor, R. L., V. Fahlbusch, P. Andrews, H. de Brujn, M. Fortelius, F. Rögl, F. F. Steininger, and L. Werdelin. 1996a. The evolution of Western Eurasian Neogene mammal faunas: A chronologic, systematic, biogeographic, and paleoenvironmental synthesis. In *The Evolution of Western Eurasian Neogene Mammal Faunas,* ed. R. L. Bernor, V. Fahlbusch, and H.-W. Mittmann, pp. 449–469. New York: Columbia University Press.

Bernor, R. L., L. Kordos, L. Rook, J. Agusti, P. Andrews, M. Armour-Chelu, D. R. Begun et al. 2004. Recent advances on multidisciplinary research at Rudabánya, Late Miocene (MN9), Hungary: A compendium. *Palaeontographia Italica* 89:3–36.

Bernor, R. L., G. D. Koufos, M. O. Woodburne, and M. Fortelius. 1996b. The evolutionary history and biochronology of European and Southwest Asian Late Miocene and Pliocene hipparionine horses. In *The Evolution of Western Eurasian Neogene Mammal Faunas,* ed. R. L. Bernor, V. Fahlbusch, and H.-W. Mittmann, pp. 307–338. New York: Columbia University Press.

Bernor R. L., N. Solounias, C. C. Swisher III, and J. A. Van Couvering. 1996c. The correlation of three classical "Pikermian" mammal faunas—Maragheh, Samos, and Pikermi—with European MN Unit System. In *The Evolution of Western Eurasian Neogene Mammal Faunas,* ed. R. L. Bernor, V. Fahlbusch, and H.-W. Mittmann, pp. 137–154. New York: Columbia University Press.

Bogdanovich, E. M. and T. A. Ivanova.1997. On newly found planktonic organisms from the Maeotian of Crimea. *Doklady Natsional'na Akademiya Nauk Ukrainy* 6:127–129 (in Russian).

Chumakov, I. S., S. L. Byzova, and S. S. Ganzei. 1992. *Late Cenozoic Geochronology and Correlation of the Paratethys.* Moscow: Nauka (in Russian).

Codrea, V. and L. Ursachi. 2007. The Sarmatian vertebrates from Draxeni (Moldavian platform). *Studia Universitatis Babeş-Bolyai, Geologia* 52(2):19–28.

Costeur, L., S. Legendre, J.-P. Aguilar, and C. Lécuyer. 2007. Marine and continental synchronous climatic records: Towards a revision of the European Mid-Miocene mammalian biochronologic framework. *Geobios* 40:775–784.

van Dam, J. A. 1997. The small mammals from the Upper Miocene of the Teruel-Alfambra region (Spain): Paleobiology and paleoclimatic reconstructions. *Geologica Ultraiectina* 156:1–204.

Daxner-Höck, G. 2001. Early and Late Miocene correlation (Central Paratethys). In *Environmental and Ecosystem Dynamics of the Eurasian Neogene. Berichte des Institutes für Geologie and Paläontologie der Karl-Franzens-Universität Graz* 4:28–33.

Daxner-Höck, G., P. M. Miklas-Tempfer, U. B. Göhlich, K. Huttunen, E. Kazár, D. Nagel, G. E. Roessner, O. Schultz, and R. Ziegler. 2004. Marine and terrestrial vertebrates from the Middle Miocene of Grund (Lower Austria). *Geologica Carpathica* 55(2):191–197.

Filippova N. Y. and V. M. Trubikhin 2009. The question of correlation of upper Miocene deposits in Black Sea and Mediterranean basins. In *Current Issues of Neogene and Quaternary Stratigraphy and Their Discussion at the 33rd Session of the International Geological Congress* (Norway, 2008), ed. A. B. Gladenkov, pp. 142–152. Moscow: Geos (in Russian).

Flynn L. J., J. C. Barry, M. E. Morgan, D. Pilbeam, L. L. Jacobs, and E. H. Lindsay. 1995. Neogene Siwalik mammalian lineages: Species longevities, rates of change and modes of speciation. *Palaeogeography, Palaeoclimatology, Palaeoecology* 115:249–264.

Forsten, A.-M. 1978. *Hipparion primigenium* (v. Meyer, 1829), an early three-toed horse. *Annales Zoologica Fennici* 15:298–313.

Forsten, A.-M. 1980. Hipparions of the *Hipparion mediterraneum* group from southwestern USSR. *Annales Zoologica Fennici* 17:27–38.

Fortelius, M., J. van der Made, and R. L. Bernor. 1996. Middle and Late Miocene *Suoidea* of Central Europe and Eastern Mediterranean: Evolution, biogeography, and paleoecology. In *The Evolution of Western Eurasian Neogene Mammal Faunas,* ed. R. L. Bernor, V. Fahlbusch, and H.-W. Mittmann, pp. 348–375. New York: Columbia University Press.

Gabunia, L. K. 1986. Terrestrial mammals. In *Stratigraphy of the Neogene System,* vol. 2, ed. M. V. Muratov and L. A. Nevesskaya, pp. 310–327. Moscow: Nedra (in Russian).

Garcés, M., W. Krijgsman, and J. Agustí. 1998. Chronology of the late Turolian deposits of the Fortuna Basin (SE Spain): Implications for the Messinian evolution of the eastern Betics. *Earth and Planetary Science Letters* 163:69–81.

Godina, A. Ya. 1979. *Historic Development of Giraffe Genus Palaeotragus.* Moscow: Nauka (in Russian).

Gradstein, F. M., J. G. Ogg, A. G. Smith, W. Bleeker, and L. J. Lourens. 2004. A new Geologic Time Scale, with special reference to Precambrian and Neogene. *Episodes* 27(2):93–100.

Hilgen, F., H. Brinkhuis, and W.-J. Zachariasse. 2006. Unit stratotypes for global stages: The Neogene perspective. *Earth Science Reviews* 74:113–125.

Korotkevich, E. L. 1988. *The History of Hipparion Fauna of Eastern Europe.* Kiev: Naukova Dumka (in Russian).

Krakhmalnaya, T. 1996a. *Hipparion Fauna of the Early Meotian of the Northern Black Sea Coast Area.* Kiev: Naukova Dumka (in Russian).

Krakhmalnaya, T. 1996b. Hipparions of the Northern Black Sea coast area (Ukraine and Moldova): Species composition and stratigraphic distribution. *Acta zoologica cracoviensia* 39(1):261–267.

Krakhmalnaya, T. 2008. Proboscideans and Ungulates of the Late Miocene fauna of Ukraine. In *6th Meeting of the European Association of Vertebrate Palaeontologists*, pp. 51–55. *Volume of Abstracts*. Spisska Nova Ves: Museum Spisa.

Krakhmalnaya, T. and A.-M. Forsten. 1998. The hipparions (Mammalia, Equidae) from the late Miocene of Novaya Emetovka-2, Odessa region, Ukraine. *Neues Jharbuch für Geologie und Paläontologie* 8:449–462.

Krijgsman, W., M. Garcés, C. G. Langereis, R. Daams, A. J. van der Meulen, J. Agustí, and L. Cabrera. 1996. A new chronology for the Middle to Late Miocene continental record in Spain. *Earth and Planetary Science Letters* 142:367–380.

Krijgsman, W., M. Stoica, I. Vasiliev, and V. V. Popov. 2010. Rise and fall of the Paratethys Sea during the Messinian Salinity Crisis. *Earth and Planetary Science Letters* 290:183–191

Krokos, V. I. 1939. Carnivorans from Maeotian deposits near Grebenniki village in Moldavian ASSR. *Geologichnyi Zhurnal* 6(1–2):129–183 (in Russian).

Lirer, F., M. Harzhauser, N. Pelosi, W. E. Piller, H. P. Schmid, and M. Sprovieri. 2009. Astronomically forced teleconnection between Paratethyan and Mediterranean sediments during the Middle and Late Miocene. *Palaeogeography, Palaeoclimatology, Palaeoecology* 275:1–13.

Lourens, L. J., F. J. Hilgen, J. Laskar, N. J. Shackleton and D. Wilson. 2004. The Neogene Period. In *A Geological Time Scale 2004*, ed. F. M. Gradstein, J. G. Ogg, and A. G. Smith, pp. 409–440. Cambridge: Cambridge University Press.

Lungu, A. N. 1981. *The Middle Sarmatian Hipparion Fauna of Moldavia (Insectivores, Lagomorphs and Rodents)*. Kishenev: Shtiintsa (in Russian).

Lungu, A. N. 1990. Early stages of Hipparion fauna development of the Paratethys continental surroundings. Abridged thesis of Ph.D. diss., Geological Institute of the Academy of Sciences of Georgian Soviet Socialist Republic (in Russian).

Lungu, A. N. and G. K. Chemyrtan. 1989. On evolution of the late Sarmatian Hipparion fauna in northern continental surroundings of the eastern Paratethys. *Trudy Gosudarstvennogo Kraevedcheskogo Muzeya MSSR*. 3:48–66 (in Russian).

Lungu, A. N. and A. Delinschi. 2008. Les particularites des orictocénoses de la faune de Hipparion du site du Cimişlia. *Acta Palaeotologica Romaniae* 6:187–193.

Markov, G. N. 2008. The Turolian proboscideans (Mammalia) of Europe: Preliminary observations. *Historia naturalis bulgarica* 19:153–178.

Mein, P. 1975. Résultats de groupe de travail des Vertébrés. In *Report on Activity of the RCMNS Working Group (1971–1975)*. Bratislava: Regional Committee on Mediterranean Neogene Stratigraphy, pp. 78–81.

Mein, P. 1989. Updating of MN zones. In *European Neogene Mammal Chronology*, ed. E. H. Lindsay, V. Fahlbusch, and P. Mein, pp. 73–90. New York: Plenum Press.

Mein P., E. Martín-Suárez, and J. Agustí.1993. *Progonomys* Schaub, 1938 and *Huerzelerimys* gen. nov. (Rodentia): Their evolution in Western Europe. *Scripta Geologica* 103:41–64.

Meladze, G. K. 1967. *Hipparion Fauna of Arkneti and Bazaleti*. Tbilisi: Metsniereba (in Russian).

Muratov, M. V. and L. A. Nevesskaya, eds. 1986. *Stratigraphy of the USSR. Neogene System*. Vol. 1. Moscow: Nedra (in Russian).

Nesin, V. A. and A. Nadachowski. 2001. Late Miocene and Pliocene small mammal faunas (Insetivora, Lagomorpha, Rodentia) of Southeastern Europe. *Acta Zoologica Cracoviensia* 44(2):107–135.

Nesin, V. A. and V. A. Topachevsky. 1999. The Late Miocene small mammals in Ukraine. In *The Evolution of Neogene Terrestrial Ecosystems in Europe*, ed. J Agustí, L. Rook, and P. Andrews, pp. 265–272. Cambridge: Cambridge University Press.

Nevesskaya, L. A., E. I. Kovalenko, E. V. Beluzhenko, S. V. Popov, I. A. Goncharova, G. A. Danukalova, I. Ya. Zhindovinov, et al. 2004. *Explanatory Note to the Standard Regional Stratigraphic Scheme of Neogene Deposits of Southern Regions of the European Part of Russia*. Moscow: Paleontological Institute of the Russian Academy of Sciences (in Russian).

Opdyke, N. P., E. Mein, A. Lindsay, A. Perez-Gonzales, E. Moissenet, and V. L. Norton. 1997. Continental deposits, magnetostratigraphy and vertebrate palaeontology, late Neogene of Eastern Spain. *Palaeogeography, Palaeoclimatology, Palaeoecology* 133:129–148.

Pevzner, M. A. 1986. Stratigraphy of Middle Miocene-Pliocene of south Europe. Ph.D. diss., Geological Institute of the USSR Academy of Sciences, Moscow (in Russian).

Pevzner, M. A., A. N. Lungu, E. A. Vangengeim, and A. E. Basilyan. 1987. Position of the Vallesian localities of hipparion faunas of Moldavia in the magnetochronological scale. *Izvestiya. Akademiya Nauk SSSR, series geol.* 4:50–59 (in Russian).

Pevzner, M. A., V. N. Semenenko, and E. A. Vangengeim. 2003. Position of the Pontian of the Eastern Paratethys in the magnetochronological scale. *Stratigraphy. Geological correlation* 11:482–491.

Pevzner M. A., V. N. Semenenko, E. A. Vangengeim, T. A. Sadchikova, V. A. Kovalenko, and S. A. Lyul'eva. 2004. On the marine genesis and Pontian age of deposits from the Lyubimvka reference section in the Crimea. *Stratigraphy. Geological Correlation* 12, 5:96–106.

Pevzner, M. A. and E. A. Vangengeim. 1993. Magnitochronological age assignments of Middle and Late Sarmatian mammalian localities of the Eastern Paratethys. *Newsletters on Stratigraphy* 29:63–75.

Pidoplichko, I. G. 1956. *Materials to the Study of the Ancient Faunas of Ukraine*. Kiev: Academy of Sciences of Ukrainian Soviet Socialist Republic (in Russian).

Popov, S. V., I. A. Goncharova, T. F. Kozyrenko, E. P. Radionova, M. A. Pevzner, E. K. Sychevskaya, V. M. Trubikhin, and V. I. Zhegallo. 1996. *Neogene Stratigraphy and Palaeonotolgy of Taman and Kerch Peninsulas. Excursion Guidebook*. Moscow: Paleontological Institute of the Russian Academy of Sciences.

Popov, S. V., L. A. Nevesskaya, and T. N. Pinchuk. 2007. Messinian events in the Mediterranean and Eastern Paratethys. In *Paleontological Research in Ukraine: History, Modern Status and Perspectives*, ed. P. F. Gozhik, pp. 36–41. Kiev: Institute of Geological Sciences of the National Academy of Sciences of Ukraine (in Russian).

Radionova, E. P. and Golovina L. A. 2009. Marine transitional sequence between Maeotian and Pontian deposits of the Taman Peninsula: Stratigraphic position and paleogeoraphic interpretation. In *Current Issues of Neogene and Quaternary Stratigraphy and*

Their Discussion at the 33rd Session of the International Gelogical Congress (Norway, 2008), ed. A. B. Gladenkov, pp. 100–109. Moscow: Geos (in Russian).

Rögl, F. 2001. Mid-Miocene Circum-Mediterranean paleogeography. *Berichrte des Institutes für Geologie and Paläontologie der Karl-Franzens-Universität Graz, Österreich* 4:49–59.

Rögl, F. and G. Daxner-Höck. 1996. Late Miocene Paratethys correlations. In *The Evolution of Western Eurasian Neogene Mammal Faunas*, ed. R. L. Bernor, V. Fahlbusch, and H.-W. Mittmann, pp. 47–55. New York: Columbia University Press.

Semenov, Y. 1989. *Ictitheriums and Morphologically Similar Hyaenas of the USSR Neogene*. Kiev: Naukova Dumka (in Russian).

Semenov, Y. 2001a. Stratigraphic distribution of the terrestrial carnivores in the Vallesian and Turolian of Ukraine. *Beiträge zür Paläontologie* 26:139–144.

Semenov, Y. 2001b. Stratigraphic distribution of the terrestrial carnivores in the Late Miocene of Ukraine. *Proceedings of National Natural History Museum of the National Academy of Sciences of Ukraine*, Kiev: pp. 52–67 (in Ukrainian).

Semenov, Y. 2002. Stratigraphic distribution of terrestrial carnivorous mammals in the Late Miocene of Ukraine. *Transactions of the National Museum of Natural History of the National Academy of Sciences of Ukraine* 1:52–67 (in Ukrainian).

Semenov, Y. 2008. Taxonomical reappraisal of "ictitheres" (Mammalia, Carnivora) from the Late Miocene of Kenya. *Comptus Rendus Palevol* 7:520–539.

Sen, S. 1996. Present state of magnetostratigraphic studies in the continental Neogene of Europe and Anatolia. In *The Evolution of Western Eurasian Neogene Mammal Faunas*, ed. R. L. Bernor, V. Fahlbusch, and H.-W. Mittmann, pp. 56–63. New York: Columbia University Press.

Sen, S. 2003. Muridae and Gerbillidae (Rodentia). In *Geology and Paleontology of the Miocene Sinap Formation*, ed. M. Fortelius, J. Kappelman, S. Sen, and R. L. Bernor, pp. 125–140. New York: Columbia University Press.

Snel, E., M. Mărunteanu, R. Macalet, J. E. Meulenkamp, and N. van Vugt. 2006. Late Miocene to Early Pliocene chronostratigraphic framework for the Dacic Basin, Romania. *Palaeogeography, Palaeoclimatology, Palaeoecology* 238:107–124.

Sotnikova, M. V. 2005. Development of Felidae fauna in Late Miocene-Early Quaternary of southern Russia and neighboring regions. In *Problems of Paleontology and Archaeology of Southern Russia and Adjacent Territories*, ed. G. G. Matishov, N. P. Kalmykov, and V. V. Titov, pp. 90–92. Rostov-on-Don: Southern Scientific Centre of the Russian Academy of Sciences (in Russian).

Spassov, N., T. Tsankov, and D. Geraads. 2006. Late Neogene stratigraphy, biochronology, faunal diversity and environments of South-West Bulgaria (Struma River Valley). *Geodiversitas* 28(3):477–498.

Steininger, F. F. 1999. The continental European Miocene: Chronostratigraphy, geochronology and biochronology of the Miocene "European Land Mammal Mega-Zones" (ELMMZ) and the Miocene "Mammal-Zones" (MN-Zones). In *Land Mammals of Europe*, ed. G. E. Rössner and K. Hessig, pp. 9–24. Munich: Dr. Friedrich Pfeil.

Steininger, F. F., W. A. Berggren, D. V. Kent, R. L. Bernor, S. Sen, and J. Agustí. 1996. Circum-Mediterranean Neogene (Miocene and Pliocene) marine-continental chronologic correlations of Euro-pean Mammal Units. In *The Evolution of Western Eurasian Neogene Mammal Faunas*, ed. R. L. Bernor, V. Fahlbusch, and H.-W. Mittmann, pp. 7–46. New York: Columbia University Press.

Steklov, A. A. 1966. *Terrestrial Neogene Mollusks of Ciscaucasia and Their Stratigraphic Importance*. Moscow: Nauka (in Russian).

Swisher III, C. C. 1996. New 40Ar/39Ar dates and their contribution toward a revised chronology for the Late Miocene of Europe and West Asia. In *The Evolution of Western Eurasian Neogene Mammal Faunas*, ed. R. L. Bernor, V. Fahlbusch, and H.-W. Mittmann, pp. 64–77. New York: Columbia University Press.

Titov, V. V. and V. N. Logvinenko. 2006. Early *Paracamelus* (Mammalia, Tylopoda) in Eastern Europe. *Acta zoologica cracoviensia* 49A(1–2):163–178.

Topachevsky, V. A., V. A. Nesin, I. V. Topachevsky, and Y. A. Semenov. 1996. The oldest locality of Middle Sarmatian small mammals in Eastern Europe. *Dopovidi Nationalnoi Akademii Nauk Ukraini* 2:107–109.

Trubikhin, V. M. 1989. Paleomagnetic data for the Pontian. In *Pontien-Pl1 (sensu F. Le Play, N.P. Barbot de Marny, N.I. Andrusov)*. Serie Chronostratigraphie und Neostratotypen, Neogen der Westlichen ("Zentrale") Paratethys 8., ed. P. M. Stevanović, L. A. Nevesskaja, F. Marinescu, A. Sokac, and A. Jambor, pp. 76–79. Zagreb-Belgrade: JAZU and SANU.

Trubikhin, V. M. Paleomagnetic scale and stratigraphy of Neogene and Quaternary deposits of Paratethys. In *Reference Sections of Neogene of Eastern Paratethys (Taman Peninsula)*, pp. 13–17. Abstracts. Volgograd-Taman: Ministry of Natural Resources of the Russian Federation, Volgograd Prospecting Expedition, Paleontological Institute of the Russian Academy of Sciences, Geological Institute of the Russian Academy of Sciences (in Russian).

Vangengeim, E. A., L. K. Gabunia, M. A. Pevzner, and C. V. Ziskarishvili. 1989. The stratigraphic position of the Hipparion fauna sites of Transcaucasia in the light of magnetostratigraphy. *Izvestiya Akademii Nauk SSSR*, ser. geol. 8:70–77 (in Russian).

Vangengeim, E. A., A. N. Lungu, and A. S. Tesakov. 2006. Age of the Vallesian lower boundary (Continental Miocene of Europe). *Stratigraphy and Geological Correlation* 14(6):655–667.

Vangengeim, E. A. and A. S. Tesakov. 2008a. Late Sarmatian mammal localities of the Eastern Paratethys: Stratigraphic position, magnetochronology, and correlation with the European continental scale. *Stratigraphy. Geological Correlation* 16(1):92–103.

Vangengeim, E. A. and A. S. Tesakov. 2008b. Maeotian mammalian localities of Eastern Paratethys: Magnetochronology and position in European Continental Scales. *Stratigraphy. Geological Correlation* 16(4):437–450.

Vasiliev, I. 2006. A new chronology for the Dacian Basin (Romania). *Geologica Ultraiectina* 267:1–191.

Vekua, A. and D. Lordkipanidze. 2008. The history of veretebrate fauna in Eastern Georgia. *Bulletin of the Georgian National Academy of Sciences* 2(3):149–155.

Werdelin, L. and N. Solounias. 1991. The Hyaenidae: Taxonomy, systematics and evolution. *Fossils and Strata* 30:1–104.

Wessels, W. 2009. Miocene rodent evolution and migration: Muroidea from Pakistan, Turkey and Northern Africa. *Geologica Ultraiectina* 307:1–290.

Wolsan, M. and Y. A. Semenov. 1996. A revision of the late Miocene mustelid carnivoran *Eomellivora*. *Acta Zoologica Cracoviensia* 39(1):593–604.

Chapter 24

Late Miocene (Turolian) Vertebrate Faunas from Southern European Russia

VADIM V. TITOV AND ALEXEY S. TESAKOV

Compared with a relatively rich and continuous record of Late Miocene mammals in the southwestern parts of the North Black Sea region, mainly Ukraine and Moldova (Korotkevich 1988; Topachevsky, Nesin, and Topachevsky 1998; Nesin and Nadachowski 2001; Vangengeim and Tesakov, chapter 23, this volume), the fossil record of more eastern areas, including the eastern coast of the Sea of Azov, the lower Don River area, and Northern Caucasus, is very patchy and relatively unstudied. The rare occurrence of fossil vertebrates is due in part to widespread marine deposits of the Eastern Paratethys and a limited distribution of synchronous continental sediments. Fortunately, during the last decade, the authors have managed to organize systematic excavations at a number of previously known and newly discovered localities (figure 24.1). Most localities yield sporadic remains of predominantly large mammals. Below, we briefly review the local faunas that, based on their composition, are attributable to the Turolian European Land Mammal Age. All sites are referred to or correlated to regional Stages of the Eastern Paratethys also, as the most important independent geological framework for the region (figure 24.2). We also indicate correlations with the Late Miocene mammalian complexes developed by Ukrainian paleontologists (Topachevsky, Nesin, and Topachevsky 1998).

EARLY MIDDLE TUROLIAN LOCALITIES (MAEOTIAN)

The oldest Turolian faunas in southern European Russia were found in the North Caucasus. They occur here in continental formations overlying marine Upper Sarmatian deposits of the Eastern Paratethys in the valleys of the Kuban and Belaya rivers. Paleogeographically, both fossiliferous areas occur at the southern shore of the Maeotian marine basin of the Eastern Paratethys. The Armavir Formation exposed on the right bank of the Kuban River near the town of Armavir has been known to yield a *Hipparion* fauna since the 1950s (Alexeeva 1959). The locality Forstadt (45°01'N, 41°10'E; see figure 24.1) at the upper part of the formation is associated with gray micaceous laminated sands and overlying brown clays. The fossiliferous deposits are reversely magnetized. Recent excavations provided new data on large and small mammals: *Amblycoptus* cf. *oligodon*, *Prolagus* sp., *Pseudocricetus* sp., Felidae gen., *Chilotherium* cf. *schlosseri*, Rhinocerotidae gen., cf. *Cremohipparion* sp., *Hippotherium* sp., *Gazella* cf. *pilgrimi*, *Procapreolus* sp., and *Miotragocerus* sp. [determination of artiodactyls by I. Vislobokova (2010)]. The evolutionary level of *Amblycoptus* is earlier or equal to that of *A. oligodon* from middle Turolian faunas of Hungary (Mészáros 1997). Hamsters of the *Pseudocricetus* group are known in Ukraine Maeotian and Pontian faunas correlated to middle and late Turolian. The Forstadt fauna is thus tentatively correlated to middle Turolian, mid-Maeotian.

There is a very similar geological situation for the Gaverdovskii Formation in the Belaya River valley near the city of Maikop. Here, continental variegated "ocherous" beds of sands and gravels overlie the Late

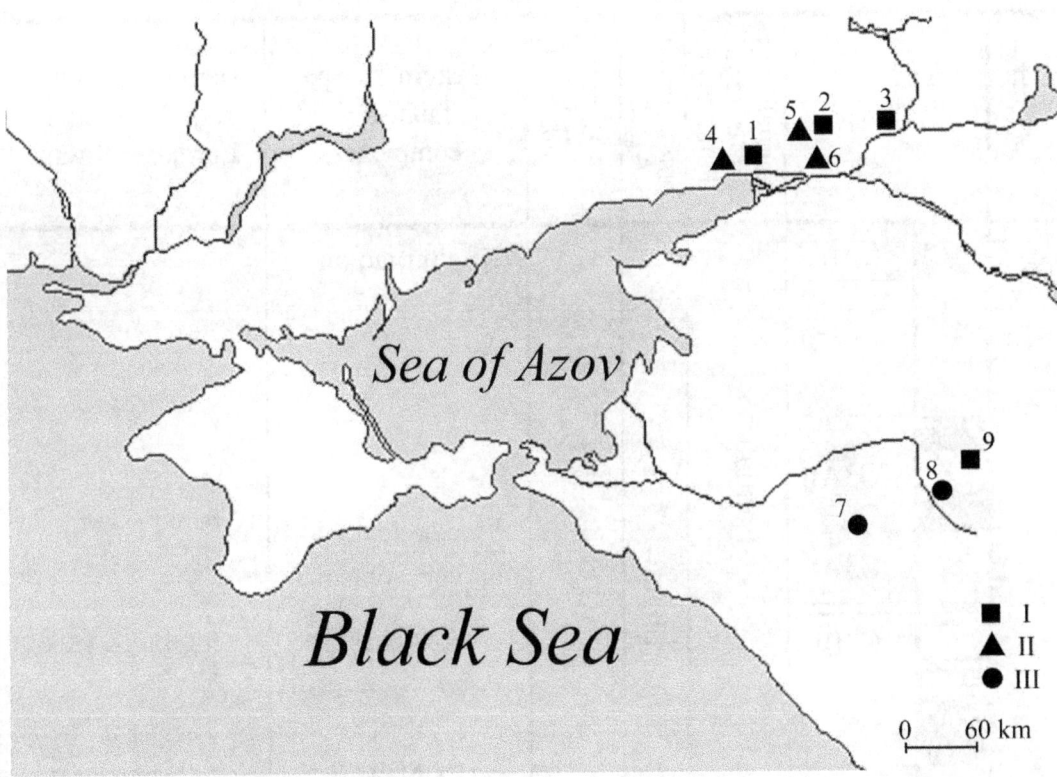

Figure 24.1 Geographic location of some Late Miocene localities in southern European Russia. (*I*) Lower Pontian; (*II*) Upper Maeotian; (*III*) Maeotian s.l. (*1*) Sinyavskaya; (*2*) Novocherkassk; (*3*) Razdorskaya; (*4*) Morskaya 2; (*5*) Yanovka-Obukhovka; (*6*) Kamenka; (*7*) Khanskaya; (*8*) Forstadt; (*9*) Solnechnodolsk.

Sarmatian marine beds. These deposits were dated to the Early–Middle Maeotian (Kolesnikov 1940; Alexeeva 1955; Beluzhenko 2000). Scanty remains of *Deinotherium* sp. and "*Mammut*" cf. *obliquelophus* were found here in the Khanskaya locality (44°38'N, 39°58'E; see figure 24.1). We tentatively date this locality to middle Turolian.

In the northern part of the studied area, north of the maximum Maeotian transgression boundary, a thick sequence of continental deposits with numerous mammal localities occurs between upper Sarmatian and lower Pontian marine deposits (Vangengeim and Tesakov 2008). In the eastern Sea of Azov area, near the city of Rostov-on-Don, these fluviatile or deltaic deposits are included to the Yanov Formation. These deposits yielded an almost complete skeleton of *Deinotherium* cf. *giganteum* found in the Obukhovka sand pit (47°29'N 40°01'E; see figure 24.1) near the city of Novocherkassk. In the neighboring Yanovka sandpit, these deposits also produced large mammal remains (figure 24.3), including Rhinocerotidae gen. indet., "*Palaeoryx*" *longicephalus* (see figure 24.3*a*), and the mastodonts "*Mammut*" *obliquelo-*

phus and *Mammut* cf. *borsoni* (see figure 24.3*e, f*) (Bajgusheva, Titov, and Tesakov 2001; Bajgusheva and Titov 2006).

Another rich site in the Sea of Azov Region is Morskaya 2 near Taganrog city (47°17N 39°06'E; see figure 24.1). This locality occurs in lacustrine greenish clay and gray sand overlying Late Miocene marine dark clays and limestones of the Middle Sarmatian. Fossiliferous deposits are capped by Upper Pliocene–Lower Pleistocene white quartz sands of the Khapry alluvial formation, red clayey beds tentatively referred to the Lower Pleistocene, a thin layer of Upper Pleistocene loess-like loams, and modern soil. This locality yielded shells of freshwater mollusks and bones of fishes, amphibians, turtles, other reptiles, and birds (Titov et al. 2006). Special publications deal with porcupines (Lopatin, Tesakov, and Titov 2003) and bats (Rossina et al. 2006). The mammal association includes *Blarinella* cf. *dubia, Asoriculus gibberodon*, Erinaceidae gen., *Vespertilio* cf. *villanyiensis, Hypolagus igromovi, Hystrix primigenia, Castor* sp., *Trogontherium* cf. *minutum, Tamias* sp., *Nannospalax compositodontus, Prospalax* sp., *Sibirosminthus* cf. *latidens,*

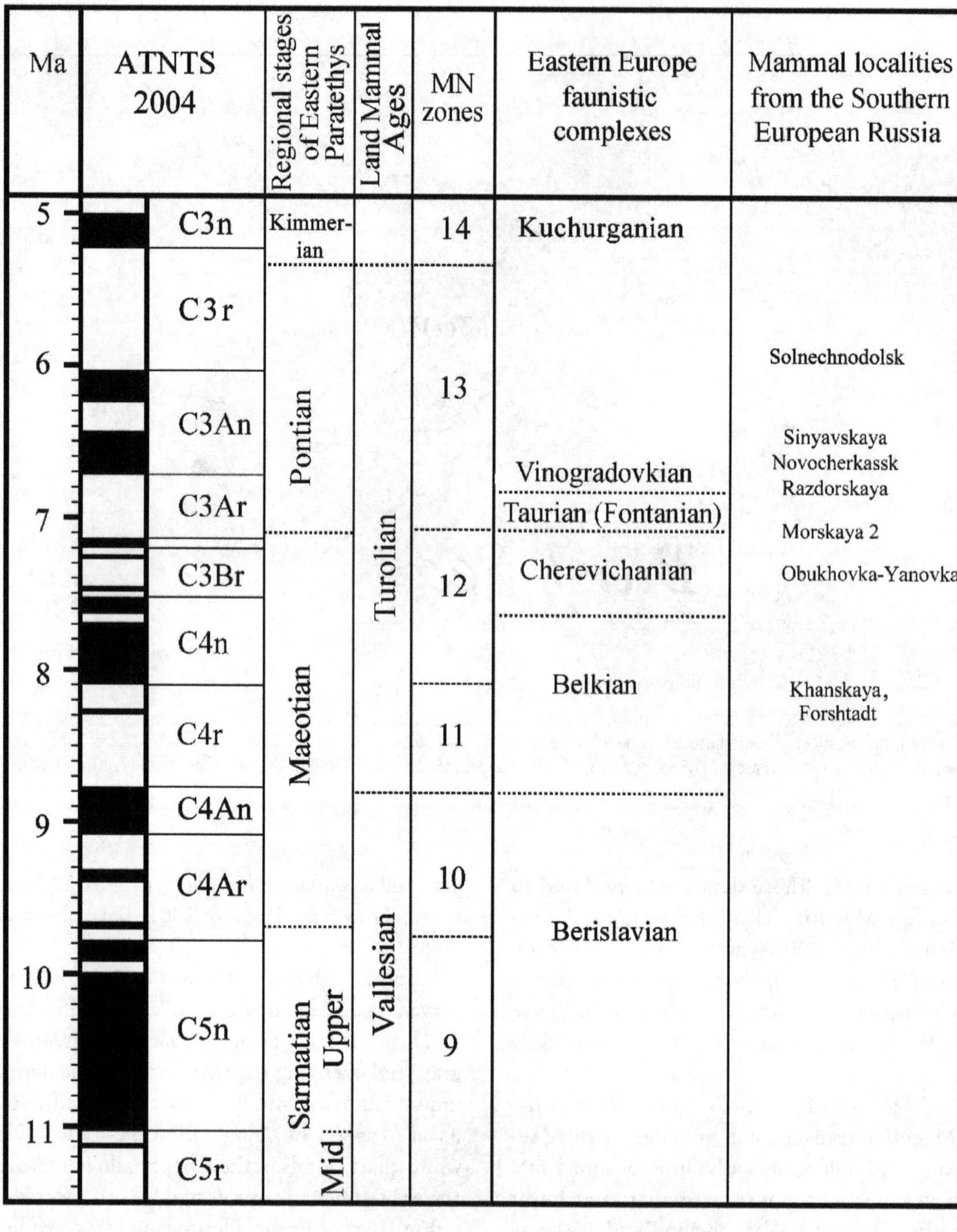

Ma	ATNTS 2004	Regional stages of Eastern Paratethys	Land Mammal Ages	MN zones	Eastern Europe faunistic complexes	Mammal localities from the Southern European Russia
5	C3n	Kimmerian		14	Kuchurganian	
6	C3r					Solnechnodolsk
	C3An	Pontian		13		Sinyavskaya Novocherkassk Razdorskaya
7	C3Ar		Turolian		Vinogradovkian — Taurian (Fontanian)	Morskaya 2
	C3Br			12	Cherevichanian	Obukhovka-Yanovka
8	C4n	Maeotian			Belkian	
	C4r			11		Khanskaya, Forshtadt
9	C4An					
	C4Ar			10		
10	C5n	Sarmatian Upper	Vallesian		Berislavian	
				9		
11	C5r	Mid				

Figure 24.2 Scheme of stratigraphic position of Late Miocene localities in southern European Russia.

Pseudocricetus cf. *kormosi, Kowalskia* sp., *Allocricetus* sp., *Pseudomeriones* cf. *latidens, Apodemus* gr. *gudrunae-gorafensis, Apodemus* gr. *dominans-atavus, Micromys* sp., *Hansdebruijnia* aff. *neutrum,* "*Mammut*" cf. *obliquelophus, Cremohipparion* cf. *moldavicum,* and Cervidae gen. Carnivora (determinations of Marina Sotnikova) include

Martes lefkonensis, Promeles sp., ?*Enhydriodon* sp., *Promephitis* cf. *maeotica, Hyaenotherium wongii, Metailurus parvulus,* and *Felis attica.* The rabbit *Hypolagus igromovi* is the most abundant species in the assemblage. The mastodon is represented by an incomplete skeleton of an adult male (see figure 24.3*b–d*). This is the first record in Rus-

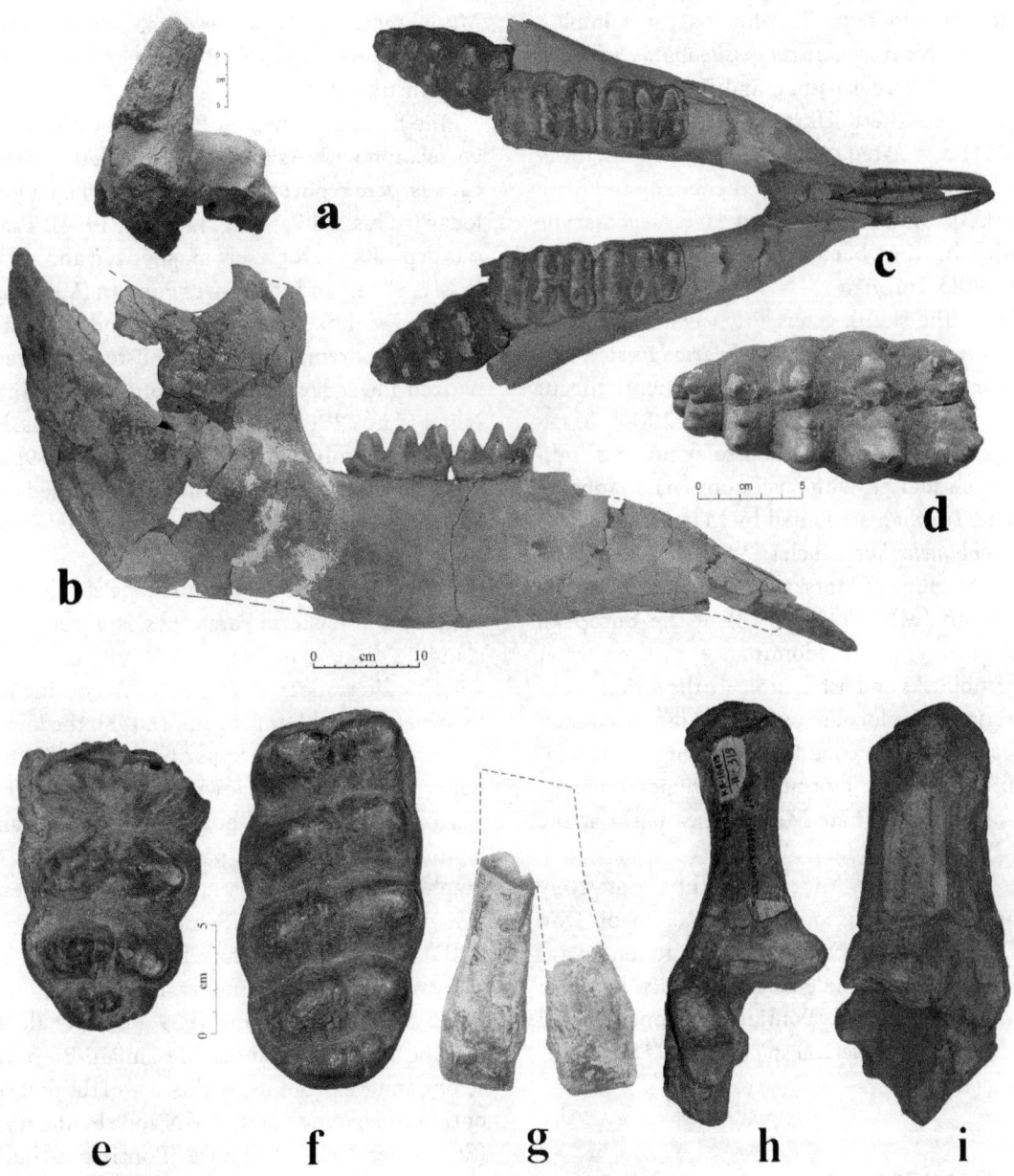

Figure 24.3 Large mammal remains from Turolian localities of the Sea of Azov region (Russia): (*a*) *"Palaeoryx" longicephalus*: specimen ZIN–24638 (coll. Zoological institute RAS), incomplete skull, lateral view, Yanovka-Obukhovka sand pits, Upper Maeotian; (*b*) *"Mammut" obliquelophus*: specimen SSC RAS M-2/63 (coll. Southern Scientific Centre RAS), incomplete lower jaw, lateral view, (*c*) dorsal view, (*d*) right unerupted m3, dorsal view, Morskaya 2, Upper Maeotian; (*e*) *"Mammut" obliquelophus*: specimen NMIDK KP-10589/P-89 (coll. Novocherkassk Museum of Don Cossacks History), fragmental left M3, dorsal view, Yanovka-Obukhovka sand pits, Upper Maeotian; (*f*) *Mammut* cf. *borsoni*: specimen NMIDK KP-10528/P-25 (coll. Novocherkassk Museum of Don Cossacks History), right M3, dorsal view, Yanovka-Obukhovka sand pits, Upper Maeotian; (*g*) *Paracamelus* cf. *aguirrei*: specimen AKM OP-27213/53 (coll. Azov Museum-reserve), distal part of metapodials, cranial view, Sinyavskaya, Lower Pontian; (*h*) *Paracamelus* cf. *aguirrei*: specimen NMHDC–5747, right calcaneus, medial view, (*i*) anterior view, Novocherkassk, Lower Pontian.

sia of an almost complete mandible with p4–m3 tooth rows and small, relatively straight tusks. In this specimen, p4 is nearly shed, and m3 is not erupted. A large share of small mammal material belongs to remains of Muridae typical for the European Turolian. The lower stratigraphic limit of this fauna is controlled by the presence of the genus *Apodemus* with incompletely developed t7, which appeared in southern Europe by the middle Late Miocene (Storch and Dahlmann 1995). The murid *Hansdebruijnia* from this site shows well-developed stephanodonty. The upper stratigraphic limit is most precisely indicated by the presence of two taxa. The first is the complex-toothed mole rat *Nannospalax compositodontus*, which is known from the Cherevichanian and Fontanian faunal assemblages of southern Ukraine (Nesin and Nadachowski 2001) correlated to the upper Maeotian, lower Pontian, and MN 12 (Pevzner, Semenenko, and Vangengeim 2003). The second taxon is *Hyaenotherium wongii*, which has not been recorded in Europe later than the Middle Turolian (MN 12; Werdelin and So-lounias 1991). The skunk genus *Promephitis*, the large *Aonyx*-like otter Lutrinae gen. (*?Enhydriodon* sp.), and the small lynx-like felid (*Felis attica*) frequently occur in Turolian faunas of Eurasia (Semenov 2001). Mastodon molars, tusks, and medium-size symphysis indicate a close similarity with zygodont mastodons of the European Turolian attributed by Markov (2008) to "*Mammut*" *obliquelophus* Mucha 1980. Available data suggest that the fauna of Morskaya 2 can be dated to the middle Turolian (MN 12–?MN 13) of the European biochronological scale. Predominance of freshwater taxa among mollusks and fishes indicate the strictly continental origin of the locality, without a direct contact with the sea. Terrestrial conditions favoring the formation of shallow lakes could occur in the region during a regressive phase of the Late Miocene sea basin at the Maeotian–Pontian boundary.

Another isolated Late Miocene locality, tentatively dated to the Maeotian, was found in the Rostov-on-Don city in the Kamenka district during construction operations. Greenish clays at the erosional surface between Middle Sarmatian and Lower Pontian limestones yielded scanty remains of *Deinotherium* sp., juvenile "*Hipparion*" sp., and *Gazella* sp.

LATE TUROLIAN (PONTIAN)

The late Turolian mammal faunas originate in the region from the shallow-water limestone deposits of the lower Pontian (Novorossian Substage) of the Black Sea marine stratigraphic scheme. They overlie upper Miocene terrestrial and marine sediments (Sarmatian and Maeotian) and are usually overlain by Pliocene continental deposits. The mammal remains in these marine deposits are scanty. Several localities are known from the Azov Sea region and from the lower course of the Don River at the northern shore of the Pontian marine basin: Sinyavskaya (47°17'N 39°18'E), Novocherkassk (47°30'N 40°03'E), and Razdorskaya (47°32'N 40°33'E) (see figure 24.1).

The combined faunal list includes *Hypolagus igromovi*, *Machairodus sp.*, *Hippotherium* sp., *Paracamelus* cf. *aguirrei* (see figure 24.3g, h), Bovinae gen. indet. (Titov and Logvinenko 2006).

The finds of camels in Pontian localities enable their correlation with Asian sites. The oldest remains of large camels were reported from the Kazakhstan locality Pavlodar (=Gusinyi Perelet; Havesson 1954). The fossiliferous deposits are reversely magnetized and are correlated to MN 12–13 and the Lower Pontian (Vangengeim et al. 1993; Vislobokova, Sotnikova, and Dodonov 2001). *Paracamelus* remains were also listed for reversely magnetized lower levels of the Mongolian locality Khirgis Nur-2 (Titov 2008); its age is estimated as the latest Miocene, late Turolian, MN 13 (Pevzner et al. 1982; Vislobokova, Sotnikova, and Dodonov 2001). Based on the similarities of the camelids, the three studied localities could be dated to late Turolian, MN 13.

The position of the Pontian in the chronostratigraphic scale of the Eastern Paratethys is a matter of a longstanding discussion due to ambiguity of its correlations with Mediterranean stratigraphy. According to Pevzner, Semenenko, and Vangengeim (2003), the lower Pontian is correlated with the upper part of the Tortonian; the upper boundary of the lower Pontian coincides with the Tortonian–Messinian boundary. Using nannofossils, fission-track dates, and paleomagnetic data, the stratigraphic range of lower Pontian deposits is determined as 7.5–7.1 Ma (Pevzner, Semenenko, and Vangengeim 2003). The alternative correlation places the Pontian in the lower Gilbert (C3r) interval with the estimated range of 6.1–5.3 Ma (Trubikhin 1989; Popov et al. 2006), correlating it with the upper Messinian. Both views have strong and weak points. In the case of the first correlation option and having in mind MN zone boundary estimates (Steininger 1999), the lower Pontian of the Black Sea region is correlated to MN 12. However, the small mammal assemblages from the lower Pontian deposits at its lectostratotype in Odessa (16th Station of Bolshoy Fontan) and from the Vinogradovka locality (Odessa region, Ukraine) have a substantially late Turolian ("MN 13") appearance with common advanced *Apodemus*, primitive small *Micromys* (Topachevsky et al. 1994; Nesin and Storch 2004), and the characteristic microtoid cricetid *Baranarviomys admirabilis* (Nesin 1996), a form very similar to *Microtodon atavus* from Ertemte.

Likewise, the placement of the whole Pontian close to the Miocene–Pliocene boundary in C3r would imply an unusually fast transformation in many mammalian lineages at the Turolian–Ruscinian transition. The recent dating of this transition in the Ptolemais Basin in Greece (Steenbrink et al. 2006; de Bruijn and Hordijk 2009) at ca. 5.3 Ma gives a very important benchmark both chronologically and regarding evolutionary levels of reference mammalian lineages.

According to the biozonal analysis of the Ukrainian fossil record (Topachevsky, Nesin, and Topachevsky 1998; Nesin and Nadachowski 2001), the Early Pontian small mammals indicate zone MN 13 and can be correlated with the base of Messinian. The "long" Pontian of this scheme ranging from 7.1 to 5.5 Ma seemingly leaves enough space for gradual evolution of late Turolian mammalian assemblages, including the transformation of the Lower Pontian microtoid hamsters into the primitive *Promimomys* of the Lower Ruscinian.

More detailed studies on Pontian small mammal faunas are necessary to provide a stronger evidence for biozonal assignment and chronostratigraphic position of these faunas.

The final site of this review is the recently discovered Solnechnodolsk locality (45'18°N, 41°33'E) situated in the Northern Caucasus, 40 km NW of the city of Stavropol (see figure 24.1). The site is situated at the southern shore of the Pontian marine basin. The mammalian fauna was collected here from fluviatile and lacustrine beds incised in the Middle Sarmatian limestones.

The fauna includes a primate, a talpid, *Blarinella* sp., *Asoriculus gibberodon, Amblycoptus* cf. *jessiae, Prolagus* gr. *michauxi-sorbini, Hypolagus* cf. *igromovi, Hystrix primigenia, Trogontherium* sp., *Nannospalax* cf. *macoveii, Pliopetaurista* sp., *Spermophilinus turolensis, Tamias* sp., *Dryomys* sp., *Micromys* sp., *Apodemus* cf. *gorafensis, A.* gr. *dominans, Hansdebruijnia* sp., *Pseudocricetus* cf. *kormosi, Allocricetus* sp., *Kowalskia* sp., cf. *Microtodon* sp., *Pseudomeriones* cf. *latidens, Parameles* sp., *Baranogale adroveri*, a hyaenid, *Felis* cf. *attica*, a proboscidean, *Hipparion* cf. *Cremohipparion* sp., a rhinocerotid, *Muntiacus* sp., *Cervavitus* sp., *Gazella* sp., and a tragoportacine.

This farthest eastern Turolian fauna in Europe has strong European affinities. The evolutionary level of *Amblycoptus* is very close to that known in Maramena. Spalacids in Solnechnodolsk are more evolved than in faunas correlated with the Maeotian and lowermost Pontian of the Ukraine (Nesin and Nadachowski 2001). *Microtodon*-like hamsters have not been known in the region in deposits older than the Early Pontian of Odessa, Ukraine. We correlate this fauna with the Late Turolian, MN 13, and with the Pontian stage of the Eastern Paratethys.

CONCLUSION

The easternmost European region, where Late Miocene mammals are known in a number of localities, is obviously important for Euro-Asian correlations. The current regional knowledge of the Latest Miocene terrestrial record represented by Turolian mammal faunas indicates the presence of still poorly defined early-middle Turolian faunas and well-expressed faunas of the Late Turolian. In terms of marine regional stages of the Eastern Paratethys, these faunas correspond to the Maeotian and Pontian, ca. 9–5.5 Ma. The majority of the reviewed Late Miocene (Turolian) mammal faunas indicates a savanna-like landscape combining open and wooded areas. Most faunas have strong European affinities, indicated, for example, by mastodons and the lagomorph *Prolagus*. Apart from species and genera that have large trans-Palaearctic ranges, the presence of lophodont lophocricetines (*Sibirosminthus*), microtoid hamsters, and camels (*Paracamelus*) mark immigrations from Asia.

ACKNOWLEDGMENTS

We thank Ray Bernor, George Koufos, and Nikolai Spassov for the useful and professional reviews of the manuscript. We are grateful to the organizers of the International Asian Mammal Biostratigraphy Conference, June 8–14, 2009, in Beijing, Xiaoming Wang, Deng Tao, Li Qiang, Mikael Fortelius, and employees of the Institute of Vertebrate Paleontology and Paleoanthropology for the opportunity to discuss many issues of the mammalian biostratigraphy and biochronology in Eurasia. We are indebted to Larry Flynn for the final polishing of the text. We thank Jan van Dam for important comments on soricids from Armavir, and Marina Sotnikova for the work with carnivoran material. This study was supported in part by the Russian Foundation for Basic Research (projects 07-05-00400-a, 09-05-10024-k, 09-05-00307-a, 12-04-01691-a).

REFERENCES

Alexeeva, L. I. 1955. Late Neogene mastodons from the territory of the USSR. Ph.D. diss., Paleontological Institute, Moscow (in Russian).

Alexeeva, L. I. 1959. The significance of mammal fauna of Armavir suite for stratigraphy of terrestrial strata of Northern Caucasus. *Proceedings of Geological Institute* 32:185–191 (in Russian).

Bajgusheva, V. S. and V. V. Titov. 2006. About teeth of *Deinotherium giganteum* Kaup from Eastern Paratethys. *Hellenic Journal of Geosciences* 41:177–182.

Bajgusheva, V. S., V. V. Titov, and A. S. Tesakov. 2001. The sequence of Plio-Pleistocene mammal faunas from the south Russian Plain (the Azov Region). *Bollettino Società Paleontologica Italiana* 40(2):133–138.

Beluzhenko, E. V. 2000. Subcontinental Upper Miocene-Pliocene deposits of Belaya River (Gaverdovskii Formation). In *Geology and Mineral Resources of North Caucasus: Materials of 9th International Scientific and Practical Geological Conference*, ed. G. I. Arutyunov, pp. 85–94. Essentuki: Ofset (in Russian).

de Bruijn, H. and K. Hordijk. 2009. The succession of rodent faunas from the Mio/Pliocene lacustrine deposits of the Florina-Ptolemais-Servia Basin (Greece). *Hellenic Journal of Geosciences* 44:21–103.

Havesson, J. I. 1954. Tertiary camels of the Eastern Hemisphere (genus *Paracamelus*). *Proceedings of Paleontological Institute* 67:100–161 (in Russian).

Kolesnikov, V. P. 1940. Upper Miocene. In *Stratigraphy of the USSR*, ed. A. D. Arhangelskiy, pp. 229–373. Moscow: AS USSR press (in Russian).

Korotkevich, E. L. 1988. *The History of the Forming of Hipparion Faunas in Eastern Europe*. Kiev: Naukova dumka (in Russian).

Lopatin, A. V., A. S. Tesakov, and V. V. Titov. 2003. Late Miocene–early Pliocene porcupines (Rodentia, Hystricidae) from south European Russia. *Russian Journal of Theriology* 2(1):26–32.

Markov, G. N. 2008. The Turolian proboscideans (Mammalia) of Europe: Preliminary observations. *Historia Naturalis Bulgarica* 19:153–178.

Mészáros, L. G. 1997. *Kordosia*, a new genus for some late Miocene Amblycoptini shrews (Mammalia, Insectivora). *Neues Jahrbuch für Geologie und Paläontologie, Monatshefte* 2:65–78.

Mucha, B. B. 1980. A new species of yoke-toothed mastodont from the Pliocene of southwest USSR. In *Quaternary and Neogene Faunas and Floras of Moldavskaya SSR*, ed. I. Ya Yatsko, pp. 19–26. Kishiner: Shtiintsa (in Russian).

Nesin, V. A. 1996. An ancient fossil vole species (Rodentia, Cricetidae) from the Lower-Pontian of the South Ukraine. *Vestnik Zoologii* 3:74–75.(in Russian).

Nesin, V. A. and A. Nadachowski. 2001. Late Miocene and Pliocene small mammal faunas (Insetivora, Lagomorpha, Rodentia) of southeastern Europe. *Acta Zoologica Cracovvensia* 44(2):107–135.

Nesin, V. A. and G. Storch. 2004. Neogene Murinae of Ukraine (Mammalia, Rodentia). *Senckenbergiana Lethaea* 84(1–2):351–365.

Pevzner, M. A., V. N. Semenenko, and E. A. Vangengeim. 2003. Position of the Pontian of the Eastern Paratethys in the magnetochronological scale. *Stratigraphy and Geological Correlation* 11(5):482–491.

Pevzner, M. A., E. A. Vangengeim, V. I. Zhegallo, V. S. Zazhigin, and I. G. Liskun. 1982. Correlation of the upper Neogene sediments of Central Asia and Europe on the basis of paleomagnetic and biostratigraphic data. *International Geology Review* 25:1075–1085.

Popov, S. V., I. G. Shcherba, L. B. Ilyina, L. A. Nevesskaya, N. P. Paramonova, S. O. Khondkarian, and I. Magyar. 2006. Late Miocene to Pliocene palaeogeography of the Paratethys and its relation to the Mediterranean. *Palaeogeography, Palaeoclimatology, Palaeoecology* 238:91–106.

Rossina V. V., S. V. Kruskop, A. S. Tesakov, and V. V. Titov. 2006. First record of Late Miocene bat from European Russia. *Acta Zoologica Cracovensia* 49(1–2):125–133.

Semenov, Y. A. 2001. Stratigraphic distribution of the terrestrial carnivores in the Vallesian and Turolian of Ukraine. *Beiträge zür Paläontologie* 26:139–144.

Steenbrink J., F. J. Hilgen, W. Krijgsman, J. R. Wijbrans, and J. E. Meulenkamp. 2006. Late Miocene to Early Pliocene depositional history of the intramontane Florina–Ptolemais–Servia Basin, NW Greece: Interplay between orbital forcing and tectonics. *Palaeogeography, Palaeoclimatology, Palaeoecology* 238:151–178.

Steininger, F. F. 1999. The continental European Miocene. Chronostratigraphy, geochronology and biochronology of the Miocene "European Land Mammal Mega-Zones" (ELMMZ) and the Miocene "Mammal-Zones" (MN-Zones). In *Land Mammals of Europe*, ed. G. E. Rössner and K. Heissig, pp. 9–24. Munich: Dr. Friedrich Pfeil.

Storch, G. and T. Dahlmann. 1995. Murinae (Rodentia, Mammalia). In *The Vertebrate Locality Maramena (Macedonia, Greece) at the Turolian-Ruscinian Boundary (Neogene)*, ed. N. Schmidt-Kittler, pp. 121–132. *Münchner Geowissenschaftliche Abhandlungen (A)* 28.

Titov, V. 2008. Earliest *Paracamelus* of the Old World. In *33rd International Geological Congress Abstracts*, August 6–14, 2008, Oslo, p. 1173.

Titov, V. V. and V. V. Logvinenko. 2006. Early *Paracamelus* (Mammalia, Tylopoda) in the Eastern Europe. *Acta Zoologica Cracovensia* 49(1–2):163–178.

Titov, V .V., A. S. Tesakov, I. G. Danilov, G. A. Danukalova, E. N. Mashchenko, A. V. Panteleev, M. V. Sotnikova, and E. K. Sychevskaya. 2006. The first representative vertebrate fauna from the Late Miocene of southern European Russia. *Doklady Biological Sciences* 411:508–509.

Topachevsky, V. A., V. A. Nesin, and I. V. Topachevsky. 1998. Biozonal microtheriological scheme (stratigraphic distribution of small mammals, Insectivora, Lagomorpha, Rodentia) of the Neogene of the northern part of the Eastern Paratethys. *Vestnik Zoologii* 32(1–2):76–87 (in Russian).

Topachevsky, V.A., A. L. Tchepalyga, V. A. Nesin, L. I. Rekovets, and I. V. Topachevsky. 1994. Small mammal communities of the Pontian regional stage of Eastern Europe and their possible continental analogs. In *Voprosy Teriologii. Paleotheriology*, ed. L. P. Tatarinov, pp. 107–112. Moscow: Nauka (in Russian).

Trubikhin, V. M. 1989. Paleomagnetic data for the Pontian. In *Pontien-Pl1 (sensu F. Le Play, N.P. Barbot de Marny, N.I. Andrusov)*, ed. P. M. Stevanović, L. A. Nevesskaja et al., pp. 76–79. Serie Chronostratigraphie und Neostratotypen, Neogen der Westlichen ("Zentrale") Paratethys 8., Zagreb/Belgrade: JAZU and SANU.

Vangengeim, E. A. and A. S. Tesakov. 2008. Maeotian mammalian localities of eastern Paratethys: Magnetochronology and position in European continental scales. *Stratigraphy and Geological Correlation* 16(4):437–450.

Vangengeim, E. A., I. A. Vislobokova, A. Y. Godina, E. L. Dmitrieva, V. I. Zhegallo, M. V. Sotnikova, and P. A. Tleuberdina. 1993. On the age of mammalian fauna from the Karabulak formation of the Kalmakpai River (Zaisan depression, Eastern Ka-

zakhstan). *Stratigraphy and Geological Correlation* 1(2): 165–171.

Vislobokova, I. A., M. V. Sotnikova, and A. E. Dodonov. 2001. Late Miocene–Pliocene mammalian faunas of Russia and neighboring countries. *Bollettino Società Paleontologica Italiana* 40(2): 307–313.

Werdelin, L. and N. Solounias. 1991. The Hyaenidae: Taxonomy, systematics and evolution. *Fossils and Strata* 30:1–104.

Chapter 25

Recent Advances in Paleobiological Research of the Late Miocene Maragheh Fauna, Northwest Iran

MAJID MIRZAIE ATAABADI, RAYMOND L. BERNOR, DIMITRIS S. KOSTOPOULOS, DOMINIK WOLF, ZAHRA ORAK, GHOLAMREZA ZARE, HIDEO NAKAYA, MAHITO WATABE, AND MIKAEL FORTELIUS

The fossil localities of Maragheh are located in the eastern Azarbaijan province, northwest Iran, between 37°20'–37°30'N latitude and 46°10'–46°35' E longitude. The Maragheh fauna has long been considered one of the three most preeminent western Eurasian Late Miocene Pikermian faunas, along with those of Samos and Pikermi in Greece. As with Pikermi and Samos, Maragheh is a true "Lagerstätte" because of the shear abundance and diversity of its fauna. It is unique among the three classical Pikermian faunas in its clear layer-cake stratigraphy with several, laterally continuous volcanic ashes that are readily amenable to radioisotopic dating.

A Russian explorer, M. Khanikoff, has been credited with first finding the Maragheh site in 1840 and sending a small collection to Dorpat University (now University of Tartu, Estonia). The Maragheh fauna was initially studied in the latter half of the nineteenth century (Abich 1858; Brandt 1870; Grewingk 1881). These early works provided data on Maragheh's similarity to Pikermi. The Austrian paleontologist H. Pohlig was invited by a merchant from the nearby city of Tabriz to visit the locality in 1884, and it was Pohlig (1886) who made the first comprehensive collection and geological study of Maragheh. He explored extensively across the Maragheh Basin and would appear to have sampled fossils from nearly all, if not all, of the Maragheh sections. The Pohlig collection in the Naturhistorisches Museum, Vienna, is extraordinary as an early collection because much of it preserves locality information that facilitates an understanding of its stratigraphic provenance. Two other Austrian paleon-

tologists, A. Rodler and E. Kittl, visited Maragheh and made an extensive collection of fossils, which were later studied (Kittl 1887; Rodler 1890; Rodler and Weithofer, 1890; Schlesinger 1917). R. Damon, from the British Museum of Natural History, London, purchased a small collection of Maragheh fossils, which was briefly communicated by R. Lydekker in 1886. In 1897, the French paleontologist M. Boule secured permission to conduct a paleontological expedition to Maragheh in 1904. The 1904 French expedition to Maragheh was organized at a very grand scale for this time in paleontology. A group of French paleontologists assisted by 12 local laborers excavated a large sample of Maragheh fossils from Kingir, Kopran, Shol'avand, and Kermedjawa (de Mecquenem 1905, 1906, 1908, 1911, 1924–1925).

More than 50 years elapsed before other reported expeditions occurred at Maragheh. F. Takai (1958) of Tokyo University collected Maragheh fossils from Kerjabad. R. Savage of Bristol University also visited Maragheh in 1958 and collected fossils. H. Tobien (1968) from the Johannes-Gutenberg University, Mainz, made important excavations of the middle portion of the Maragheh sequence in the 1960s. During the 1970s, three scientific groups conducted research at Maragheh: a combined Dutch–German group led by B. Erdbrink (Erdbrink 1976a, 1976b, 1977, 1978, 1982, 1988; Erdbrink et al. 1976), a joint University of Kyoto–Geological Survey of Iran team led by T. Kamei (Kamei et al. 1977; Watabe 1990; Watabe and Nakaya 1991a, 1991b), and the Lake Rezaiyeh Expedition (LRE) led by B. Campbell

(Campbell et al. 1980). R. Bernor was a student charged with the study of vertebrate fauna for the LRE, which resulted in his Ph.D. (1978) and manuscripts on the fauna, biostratigraphy, and zoogeographic relationships of the fauna (Bernor 1986) as well as the systematics, biostratigraphy, and zoogeography of the hipparionine horses (Bernor, Woodburne, and Van Couvering 1980; Bernor 1985). An extensive review of the fauna with systematic, chronologic, and biogeographic comparisons to Pikermi and Samos was published by Bernor et al. (1996).

There are three important outcomes from the field work undertaken in the 1970s, including (1) collection of fossils with close regard to stratigraphic provenance, which has led to a biostratigraphy of the Maragheh fauna; (2) study of all collections to better understand the taxonomy and diversity of the mammalian fauna; and (3) application of a variety of geochronologic tools to secure well-resolved ages for the Maragheh section and its faunas.

After a 30-year cessation of excavation activities in the Maragheh Basin, Iran's Department of the Environment (DOE) and National Museum of Natural History (MMTT) started a new initiative and sponsored new excavations in the area, which resulted in the nomination of 10 km² of the Maragheh fossiliferous area as a national protected zone and the establishment of a field museum and research station in this area. The recent MMTT–University of Helsinki initiative, known as the International Sahand Paleobiology Expedition (INSPE), is currently in progress. This program has undertaken three field seasons between 2007 and 2009, discovering several new localities and numerous fossils. The program has further reinitiated studies of the mammalian fauna with the intention of bringing them into a contemporary taxonomic context for comparative paleoecological and paleobiogeographic studies.

GEOLOGY AND STRATIGRAPHY

The Maragheh Basin is bounded to the north by the northwest–southeast-trending Anatolian transform fault, also known as the Tabriz fault, to the west by the north–south-trending Urmiyeh fault and to the south by the northwest–southeast Mendelasar transform fault. Regionally, the Maragheh Basin and its associated transform faults are dominated by the Zagros crush zone to the south and west. Also, there are the Urmiyeh–Bazman (Urmiyeh–Dokhtar) volcanic belt in the northeast and the Sanandaj–Sirjan metamorphic belt (Mendelassar transform) in the southwest (figure 25.1). The

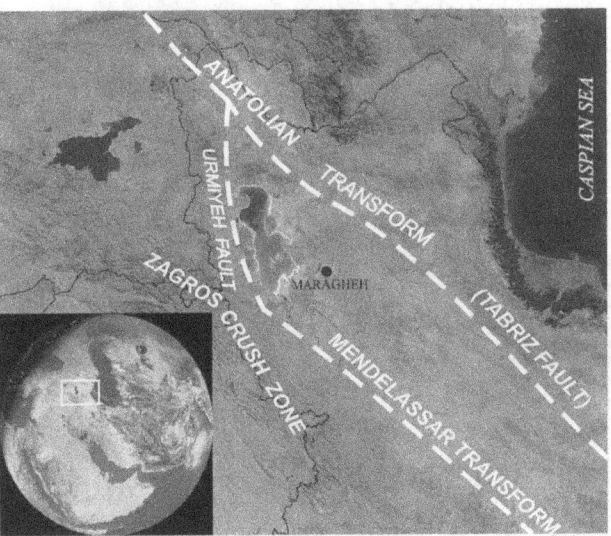

Figure 25.1 Geographic position and relationships of the Maragheh area to the major tectonic features. After Dewey et al. (1973) and Huber (1976) in northwest Iran.

Urmiyeh–Bazman volcanic belt with its northwest–southeast trend is believed to have resulted from the collision of the Arabian and Iranian plates (Davoudzadeh, Lammerer, and Weber-Diefenbach 1997). The Sanandaj–Sirjan metamorphic belt with a similar trend lies between the main Zagros thrust (crush zone) and the Urmiyeh–Bazman belt.

During the Paleogene, northwest Iran experienced a wide range of postcollisional arc volcanic activities. After this magmatism event, clastic, evaporite, and carbonate sediments were deposited during the late Paleogene and early Neogene (Lower Red and Qom formations). By the end of the Early Miocene, the last Tethyan seaway incursion regressed from this area resulting in the local carbonate deposition cycle (Aghanabati 2004). Consequently, at the beginning of the Neogene, this domain emerged above sea level and developed as incipient mountain ranges, basin troughs, and a topography resembling present conditions (Davoudzadeh, Lammerer, and Weber-Diefenbach 1997). The most significant deposits of this time are terrestrial sediments and evaporites known collectively as the Upper Red Formation. The remains of these deposits are not found in the Maragheh area but mostly occur north of the Tabriz fault and south of the Urmiyeh fault (see figure 25.1). It seems that these major faults in the area, which have been active since the Paleozoic (Aghanabati 2004), structurally controlled and prevented deposition of these units in the Maragheh Basin. Volcanic activity was reinitiated in the Late Miocene to Middle Pliocene interval in the Maragheh Basin and adjacent areas (Moin-Vaziri and Amin-Sobhani 1977).

The Late Miocene Maragheh stratigraphic sequence accumulated on the southern flank of the Mount Sahand volcanic massif. Mount Sahand is a large volcanic complex which covers an area of about 10,000 km² (Moin-Vaziri and Amin-Sobhani 1977) and, despite its circular outline, is not a single volcano. A series of distinct volcanic cones are arranged along an east–west trend collectively forming this enormous volcanic massif.

The Late Miocene deposits of the Maragheh Basin consist of a thick sequence of volcaniclastic continental strata with a basal pyroclastic unit. Kamei et al. (1977) named the entire 500–600-m-thick Late Miocene sequence the Maragheh Formation. They differentiated this formation into a lower fossiliferous member (160 m) and an upper nonfossiliferous unit forming the upland hills of Mt. Sahand. Campbell et al. (1980) restricted the Maragheh Formation to the lower 300 m volcaniclastic and fossiliferous series. They also referred to the basal pyroclastic unit as the Basal Tuff Formation. Hence, the fossil-bearing sequence of Maragheh Basin is confined to the lower 150 m of a 300-m-thick Maragheh Formation. The upper surface of the Maragheh Formation is erosional with a local Pliocene–Quaternary capping of heavily oxidized terrace gravels, pumice breccias, and boulder-ridden soils. These uppermost horizons are more than 350 m thick south of Sahand, but they can be as much as 1000 m in thickness in areas near the Anatolian transform (Bernor, Woodburne, and Van Couvering 1980).

THE MARAGHEH GROUP

We describe herein the sedimentary horizons of the Maragheh Group as they are expressed in the Maragheh Basin.

The Basal Tuff Formation represents a single air-fall unit of rhyolite tuff with local thickness of over 80 m. This unit is a uniform, unbedded, and structureless deposit of white, devitrified ash with randomly oriented crystals of mica and fresh fragments of feldspar and quartz. The unit represents a tremendous pyroclastic event with substantial outcrops south and northeast of the central fossiliferous area and has proven to be useful for long-range intrabasin correlations (Campbell et al. 1980; Bernor, Woodburne, and Van Couvering 1980; Bernor 1986).

The Maragheh Formation rests unconformably on the Basal Tuff Formation. It is eroded to a thickness of 300 m and consists of strata made up exclusively of detrital fragments of hornblende andesite and dacitic pumice and is interbedded at widely spaced intervals with layers of pumice-lapilli tuff. In general, the volcaniclastic beds are unlaminated and poorly sorted silty grits with lenses of andesite and pumice cobbles, in depositional units ranging in thickness from 1 m to 3 m. The top of each depositional unit is generally marked by a darker, weathered zone with root casts (Bernor, Woodburne, and Van Couvering 1980).

Maragheh Formation deposits are bound to the north by the Anatolian transform (Tabriz fault) (see figure 25.1). Between the northwest of Sahand massif and the city of Tabriz possible Miocene-Pliocene deposits are unlike the Maragheh Formation. These beds are composed of diatomites containing fish and mollusks and some lignites with plant remains. Hipparionine teeth and scarce mammalian bones are recorded from these deposits (Rieben 1934), whose nature is quite different from those of the Maragheh Formation. Recently, abundant mammalian fossils have been discovered in the areas north and northeast of Tabriz (Mirzaie Ataabadi, Zaree, and Orak 2011; Mirzaie Ataabadi, Mohammadalizadeh, Zhang 2011). Although this fossil material resembles that of the Maragheh Formation, their geology, sedimentology, and taphonomy differ. To the west, the Maragheh Formation is bounded by Lake Urmiyeh (or Urmiah), which is a shallow, hypersaline body of water formed in the Pleistocene by the activities of the Tabriz and Urmiyeh faults (Aghanabati 2004). To the southwest, the limit of Maragheh Formation is the Mendelassar ridge (see figure 25.1). This is uplifted lightly metamorphosed rocks known as the Sanandaj–Sirjan metamorphic belt.

Campbell et al. (1980) reported that some lithological horizons in the Maragheh Formation can be traced over wide areas allowing intrabasin correlations. The most distinctive unit for correlation is a diamictitic breccia named "Loose Chippings." This marker bed crops out best in the central portion of the study area and has been used for stratigraphic correlation of vertebrate localities in this area (figure 25.2). This bed is recorded as the "scoria bed" by Kamei et al. (1977) and as the "trachytic breccia" by Erdbrink et al. (1976).

Section C (see figure 25.2) represents the recent excavations in the Maragheh area by MMTT and INSPE teams. It is correlated to adjacent sections by a major pumice layer known as "Pumice Bed 2." Pumice Bed 2 is 5–7 m thick and has been widely traced in the study area. This bed was likely accumulated from a single

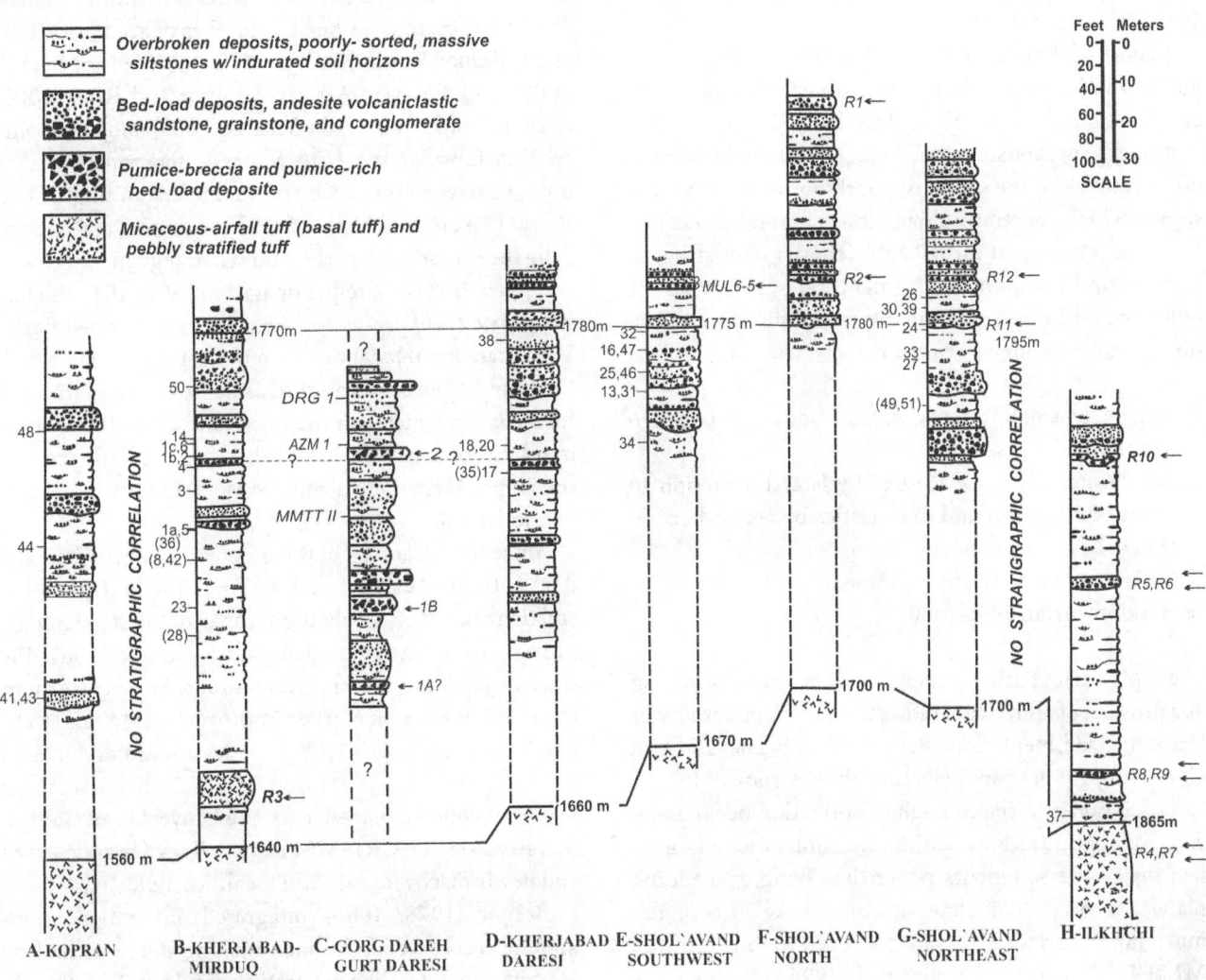

Figure 25.2 Lithology and stratigraphy of the Maragheh Formation, northwest Iran. (For location of sections A–H, see figure 25.3.) Sites A and H are correlated based on the basal tuff. Sites B and D–G are correlated by the "Loose Chippings" marker bed. Site C is correlated to nearby sections by "Pumice Bed 2" and corresponds to the recent excavations (MMTT II, DRG1, and AZM1) in Maragheh. The position of section C above the basal tuff and below the "Loose Chippings" is certain. However, the details of the base and top of the section are not recorded. The basal tuff is shown at the base of most sections, with topographic elevation. Numbers and letters to the left of each column are fossil localities. R1–R12 and MUL6-5 are sites from which radiometric age determinations were obtained (see also figure 25.3 and table 25.1). *1A*, *1B*, and *2* in site C refers to pumice beds.

large-scale flow event (T. Sakai, pers. comm.). The sections in the extreme northeast and southwest (A and H) are correlated based on the Basal Tuff Formation.

The Maragheh Formation seems to rest with a low angle regional unconformity on the Basal Tuff Formation. Based on the studies in the central fossiliferous area the regional dip of the Maragheh Formation is west-southwest with about 5 m/km inclination. Triangulation from the presently known exposures of the Basal Tuff Formation by the American team in the 1970s also indicates a consistent dip to the west-southwest with a general inclination of about 15 m/km. The differences

between the dips of these units suggest that the Basal Tuff draped over a west-sloping paleoslope/basin that was gradually filled by the Maragheh Formation. These successive beds prograded eastward as the base level rose (Campbell et al. 1980). This interpretation has been generally supported by radiometric (Campbell et al. 1980) and biostratigraphic evidence (Bernor 1978, 1986; Bernor, Tobien, and Van Couvering 1979; Bernor, Woodburne, and Van Couvering 1980). These data suggest that the beds and associated fossils farthest to the west (Kopran) are the oldest compared to the localities in the easternmost part (Ilkhchi), which are the youngest

(Bernor, Woodburne, and Van Couvering 1980; Bernor 1986).

Major sedimentary facies that have been distinguished in Maragheh Formation are pebble and cobble conglomerate, which make up less than 5% of the studied sections, gray sandstone and breccia facies, which make up about 25% of the sections, poorly sorted massive siltstones, which constitute about 70% of Maragheh Formation, and occasional air-fall tuff deposits consisting almost entirely of pumice fragments. It seems that the following sedimentary events are responsible for deposition of the Maragheh Formation:

- Erosion by small streams, which made a small disconformity at the base
- Deposition of coarse clastics by lateral accretion in point bar deposits and fine clastics by vertical accretion in overbank deposits
- Soil formation at the top of these units
- Random airfall deposition

These processes built the extensive Maragheh Formation as a product of alluviation rather than volcanic activity or lacustrine sedimentation (Bernor, Woodburne, and Van Couvering 1980; Campbell et al. 1980; Bernor 1986).

Fossil bones in the Maragheh Formation occur as localized concentrations within the unlaminated beds, floating in the sediments rather than lying on bedding planes. A single complete articulated skeleton of the mustelid *Promeles palaeattica* has been found from the MMTT 13 quarry (Bernor et al. 1996). Taphonomic studies of these fossil accumulations indicate autochthonous bone assemblages accumulated on overbank or floodplain deposits of fluvial systems. A large number of the bones are preserved with articulation of distal limb elements and early weathering stages. Pyroclastic events such as mudflows or ash falls were not directly responsible for the mortality of animals. On the other hand, biologic agents were the probable cause of death, as the bones were buried almost immediately or subaerially exposed only long enough to allow removal of some elements by scavengers (Morris 1997).

PALEONTOLOGY AND CHRONOLOGY

Bernor (1986) and Bernor et al. (1996) provided an account of the mammalian species reported from Maragheh. Since 1996, there have been a number of taxonomic revisions that affected the documentation of fossils at Maragheh and their comparisons with other penecon-

temporaneous Eurasian and African mammal faunas. Moreover, there have been a number of studies of Eurasian (Bernor, Kordos, and Rook 2003, 2005; Eronen et al. 2009) and Eurasian-African (Bernor and Rook 2008; Bernor, Rook, and Haile-Selassie 2009) biogeographic relationships for the Late Miocene interval that have brought new significance to our understanding of Old World (Pikermian) chronofaunas in general, and specifically the importance of the Maragheh sequence.

Figure 25.3 is a satellite image indicating the principal vertebrate fossil producing areas in the Maragheh Basin. Vertebrate fossil localities crop out across the Maragheh Basin and often are expressed as dense concentrations of fossils up to a meter in thickness and extending tens to hundreds of meters laterally. This is particularly true for the Upper Maragheh locality MMTT 13 near the village of Shol'avand.

Since the latter part of the nineteenth century, coincident with the Austrian exploration of the Maragheh Basin, there has been a growing archive for the stratigraphic provenance of the Maragheh fauna (Bernor 1986). The work by Japanese, Dutch, German, and American groups in the 1970s brought marked improvements to this stratigraphic record. Figure 25.2 provides a summary of eight stratigraphic sections, arrayed from west to east, of these principal collecting areas with the University of California, Riverside (UCR)–MMTT localities (Bernor 1986) and newly discovered INSPE localities indicated.

Bernor (1978, 1986) integrated all known stratigraphic records of fossil mammals to develop the first Maragheh mammalian biostratigraphy. He originally subdivided the Maragheh Formation into three units based on the stage of evolution of the *Hipparion* s.s. lineage: "Lower Maragheh" whose base was defined by the first occurrence of *Hipparion gettyi* at Kopran; "Middle Maragheh" by the first occurrence of *Hipparion prostylum*; and "Upper Maragheh" by the first occurrence of *Hipparion campbelli*. Recent studies of the Maragheh *Hipparion* samples housed by the Muséum National d'Histoire Naturelle, Paris (MNHN), and Howard University Laboratory of Evolutionary Biology suggest that *Hipparion prostylum*, originally defined based on skull morphology alone, may not occur at Maragheh. The MNHN sample originally referred to *Hipparion prostylum* (Woodburne and Bernor 1980; Bernor, Woodburne, and Van Couvering 1980; Bernor 1985) is not supported by the postcranial material: there are no metapodials or phalanges of *Hipparion prostylum* s.s (type species from Mount Luberon, France) in the MNHN sample. On the other hand, *Hipparion campbelli's* postcrania (Howard University Maragheh sample) are morphologically similar to

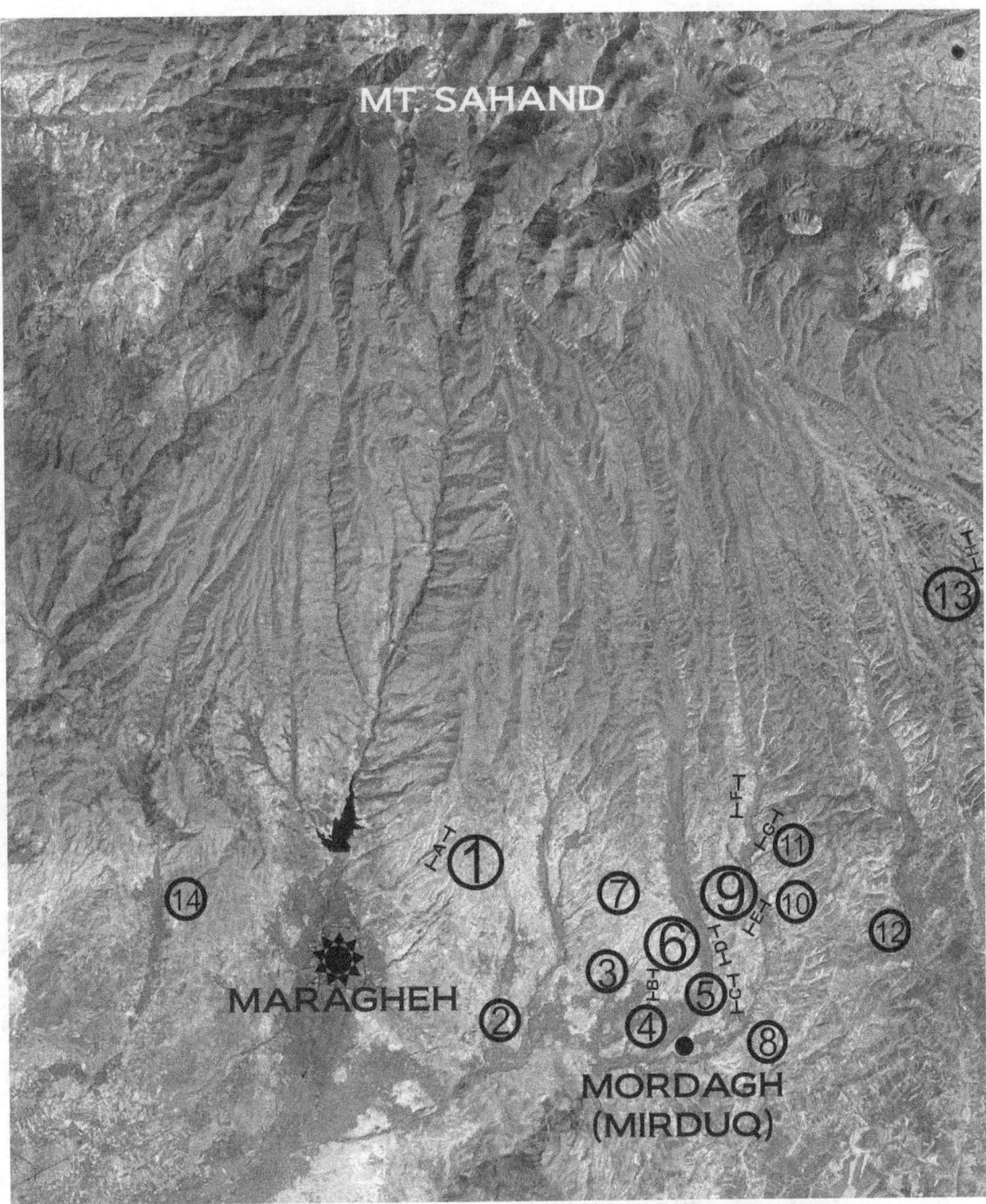

Figure 25.3 Geographic sites and fossil localities (UCR-MMTT, MMTT, and INSPE) of the Maragheh Basin, northwest Iran: (*1*) Kopran localities; (*2*) Varjoy localities; (*3*) Aliabad localities; (*4*) Mordagh (Mirduq, Mordaq) localities; (*5*) Dare Gorg (Gort Daresi) localities (including new MMTT and INSPE localities); (*6*) Karajabad (Kherjabad) localities; (*7*) Sumu Daresi locality; (*8*) E. Mordagh localities; (*9*) Shalilvand (Shol'avand) localities; (*10*) Ghartavol localities; (*11*) N. E. Shalilvand (Shol'avand) localities; (*12*) Khermejavand locality; (*13*) Ilkhchi localities; and (*14*) Ahagh (Ahaga), W. Maragheh localities. |—x—| correspond to the stratigraphic columns of figure 25.2.

Hipparion prostylum s.s. The postcrania in the collections of the MNHN and the Bayerische Staatssammlung für Paläontologie und Historische Geologie, Munich (BSP), are referable to *Hippotherium brachypus*; they co-occur with skulls that have a reduced preorbital fossa like

H. prostylum. This consistent co-occurrence in two distinct paleontologic collections suggests that it is possible that these skulls and postcrania may be of the same taxon, *Hippotherium brachypus.* We further find that the MNHN Maragheh *Hipparion* assemblage includes two

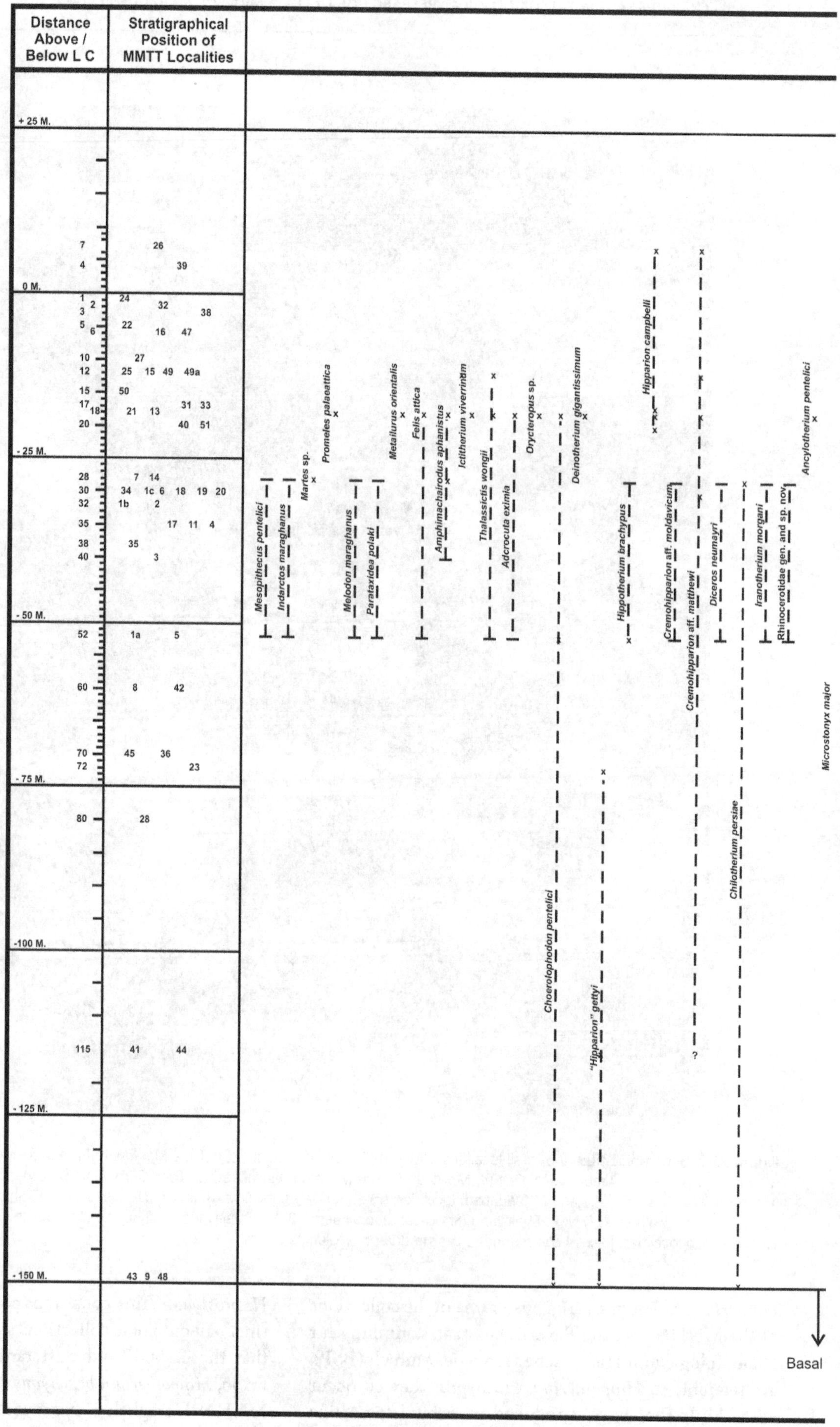

Distance Above / Below L C	Stratigraphical Position of MMTT Localities

+ 25 M.

0 M.

7 26
4 39

1 2 24 32 38
3 5 22 16 47
6
10 27
12 25 15 49 49a
15 50
17 21 13 31 33
18
20 40 51

- 25 M.

28 7 14
30 34 1c 6 18 19 20
32 1b 2
35 17 11 4
38 35
40 3

- 50 M.

52 1a 5

60 8 42

70 45 36
72 23

- 75 M.

80 28

-100 M.

115 41 44

- 125 M.

- 150 M.

43 9 48

Basal

Mesopithecus pentelici
Indarctos maraghanus
Martes sp.
Promeles palaeattica
Melodon maraghanus
Parataxidea polaki
Metailurus orientalis
Felis attica
Amphimachairodus aphanistus
Ictitherium viverrinum
Thalassictis wongii
Adcrocuta eximia
Orycteropus sp.
Deinotherium gigantissimum
Hipparion campbelli
Hippotherium brachypus
Cremohipparion aff. moldavicum
Cremohipparion aff. matthewi
Diceros neumayri
Iranotherium morgani
Rhinocerotidae gen. and sp. nov.
Ancylotherium pentelici
Microstonyx major
Choerolophodon pentelici
"Hipparion" gettyi
Chilotherium persiae

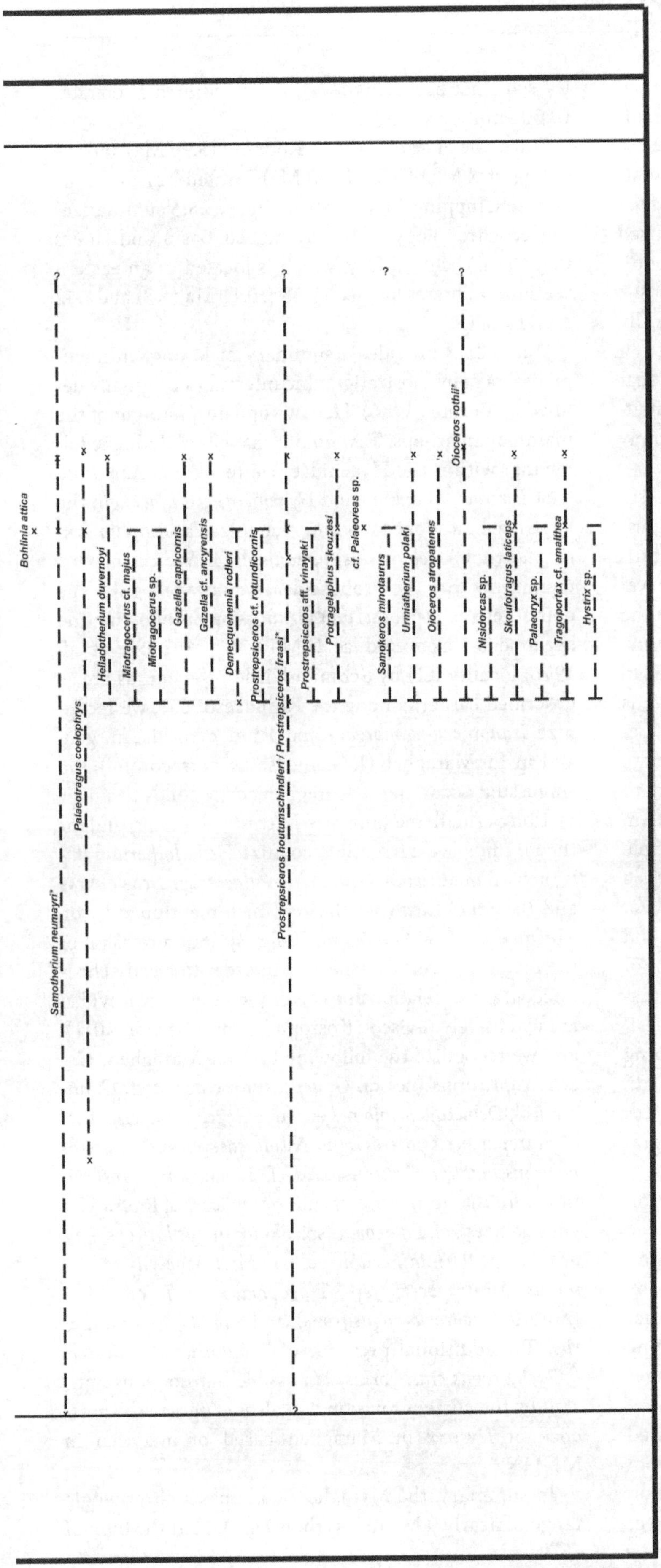

Bohlinia attica

Palaeotragus coelophrys

Samotherium neumayri*

Helladotherium duvernoyi

Miotragocerus cf. majus

Miotragocerus sp.

Gazella capricornis

Gazella cf. ancyrensis

Demecquenemia rodleri

Prostrepsiceros cf. rotundicornis

Prostrepsiceros houtumschindleri / Prostrepsiceros fraasi*

Prostrepsiceros aff. vinayaki

Protragelaphus skouzesi

cf. Palaeoreas sp.

Samokeros minotaurus

Urmiatherium polaki

Oioceros atropatenes

Oioceros rothii*

Nisidorcas sp.

Skoufotragus laticeps

Palaeoryx sp.

Tragoportax cf. amalthea

Hystrix sp.

Figure 25.4 Mammalian biostratigraphy of the Maragheh Formation, northwest Iran. The stratigraphic provenance of vertebrate localities is given above/below the "Loose Chippings" (LC) marker bed. Taxa with * are also recorded from "Upper Maragheh" Loc. 37.

distinct skull morphologies and two distinct postcranial morphologies. The skulls with a large and single preorbital fossa (POF) placed close to the orbit are believed to be associated with the elongate and slender metapodials and likely referable to *Cremohipparion* aff. *C. moldavicum*, whereas the skull with a single and highly reduced POF is likely associated with the short, stout metapodials referable to *Hippotherium brachypus*. In that the MNHN assemblage is believed to be stratigraphically derived mostly from the middle portion of the section ("Middle Maragheh"), we now need to recognize that the biostratigraphic subdivisions may not be attributable to the evolution of a continuous lineage, but simply as biozones.

Current evidence suggests the following horse sequence: *Hipparion gettyi* occurs at Kopran, the oldest set of localities in the Maragheh Basin with its stratigraphic range being from the –150 m to –75 m interval of the section ("–" and "+" refer to below and above the level of the "Loose Chippings"); *Hippotherium brachypus* and *Cremohipparion moldavicum* occur in the –52 m to –25 m interval of the section; *Hipparion campbelli* is first known to occur at the –20 m interval and is believed to be present at the +7 m interval of the section. Small horses that we believe to be best referred to *Cremohipparion?matthewi* occur from about the –115 m to +7 m interval of the section. There are likely multiple species of small *Cremohipparion* from Western Eurasia (i.e., *Cremohipparion matthewi*, *Cremohipparion nikosi*, *Cremohipparion minus*, and potentially others), and there is simply too little cranial and complete metapodial data to determine which of these occur at Maragheh. The Maragheh hipparion assemblage is numerically abundant and diverse in species and is undergoing extensive new systematic analysis by Mirzaie Ataabadi, Bernor, and Wolf (in progress). Figure 25.4 updates Bernor's (1986) and Bernor et al.'s (1996) biostratigraphy of the Maragheh mammal fauna.

Swisher (1996) provided new single crystal Argon ages for the Maragheh fauna that significantly revised Campbell et al.'s (1980) initial report on zircon fission track ages and K/Ar^{40} ages. These are summarized, by stratigraphic interval, in table 25.1, which shows that the ages are internally coherent with the oldest being stratigraphically succeeded by progressively younger ages.

Bernor et al. (1996) suggested that based on estimated sedimentation rates the Maragheh fauna ranges from about 9 Ma to 7.4 Ma. Table 25.2 represents a calculation of ages for localities in the western and central portion of

the sequence based on the estimated sedimentation rates (0.008 million yr/m).

Thus, the oldest locality is Kopran I (8.96 Ma) and the youngest is MMTT 26 (7.68 Ma), located 7 m above the "Loose Chippings" marker bed. Figure 25.5 summarizes the geochronology of the Maragheh Basin and shows that the Ilkhchi locality, which is located in an eastern section, calibrates between 7.58 ± 0.11 Ma (R8) and 7.42 ± 0.11 Ma (R10).

Table 25.3 provides a summary of Maragheh mammalian taxa by biostratigraphic interval as originally defined by Bernor (1986). Here we update a number of the mammalian groups. Taxonomic changes include the following: within the Hyaenidae, we recognize *Adcrocuta* (not *Percrocuta*) *eximia* and *Hyaenictitherium wongii*; the large machairodont cat is now recognized as *Amphimachairodus* (not *Machairodus*) *aphanistus* (L. Werdelin, pers. comm.); within the Proboscidea, we now recognize the deinothere as *Deinotherium gigantissimum* following the specimens discovered at MMTT31 (Erdbrink et al. 1976; locality K1) by Schmidt-Kittler; the equids are as described earlier; among the Rhinocerotidae, we recognize *Iranotherium morgani* and Rhinocerotidae n. gen. and sp. for Maragheh (I. Giaourtsakis, pers. comm.); the single suid occurring at Maragheh corresponds to a population of small/medium-size *Microstonyx major*; and for the giraffids, we currently recognize *Helladotherium duvernoyi*, *Samotherium neumayri*, *Palaeotragus coelophrys* and *Bohlinia attica* (not the specimen mentioned by de Mecquenem [1924–1925:pl. II, fig. 3]), but a revision of the material is certainly needed in order to clarify chronogeographic relationships. Additionally, the Bovidae are extensively revised (Kostopoulos and Bernor, 2011), and we recognize the following taxa for Maragheh: *Gazella capricornis* (not cf. *G. deperdita*), *Gazella* cf. *G. ancyrensis*, *Demecquenemia* n. gen. (not *Gazella*) *rodleri*, *Oioceros atropatenes*, *Oioceros rothii*, *Nisidorcas* sp., *Prostrepsiceros houtumschindleri*, *Prostrepsiceros* cf. *P. vinayaki*, *Prostrepsiceros fraasi*, *Prostrepsiceros* cf. *P. rotundicornis*, *Protragelaphus skouzesi*, cf. *Palaeoreas* sp., *Skoufotragus laticeps*, *Palaeoryx* sp., *Urmiatherium polaki*, *Miotragocerus* cf. *M. maius*, *Miotragocerus* sp., *Tragoportax* cf. *T. amalthea* (not *Miotragocerus rugosifrons*), and *Samokeros minotaurus*. The additional presence of *Skoufotragus schlosseri* (=*Pachytragus crassicornis*) is possible but not substantiated by the current revision. We also recognize the presence of *Hystrix* in Maragheh based on material in MNHN.

In summary, the Maragheh fauna has a chronologic range of nearly 9 Ma to less than 7.4 Ma, but the bulk of

Table 25.1

Summary of the Isotopic Age Determinations from the Maragheh and Basal Tuff Formations

		Ar-Ar			Zircon FT		K-Ar	
		Weighted Mean	Isochron Age	Sample	Sample		Sample	
+100	Korde-deh Pumice (Murdag Chai)							
+90 est.	Upper Pumice (Ilkchi)	7.420±0.107	7.536±0.111 (11 pt.)	R 10	7.2±0.4	R 10	7.5±0.4	R 10pl
+60 est.	Middle Pumice (Ilkchi)				7.3±0.5 R	R 5	7.7±0.5	R 5pl
							7.2±0.5	R 6pl
+15 est.	Lower Pumice (Ilkchi)	7.579±0.106	7.550±0.071 (8 pt.)	R 8	7.4±0.4	R 8	7.9±0.7	R 8pl
					7.0±0.4	R 8		
+15 est.	Village Pumice (Ilkchi)				5.2±0.3	R 9	5.5±1.0	R 9pl
					5.3±0.3	R 9		
+7	Layered Marker Pumice (Shollovend)	7.592±0.094	7.642±0.033 (8 pt.)	R 12	7.7±0.5	R 12	6.4±0.6	R 12pl
0	Loose Chippings Pumice (Shollovend)	7.748±0.136	7.787±0.139 (8 pt.)	R 11	7.8±0.4	R 11	7.4±0.3	R 11pl
−50	Gürt Dareseh Pumice (Murdag)							
−110	Ignimbritic Tuff (Murdag)	8.667±0.040	8.635±0.029 (9 pt.)	R 3	10.6±0.8	R 3	6.4±0.5	R 3pl
							8.8±0.5	R 3pl
							8.9±0.5	R 3pl
	Basal Tuff (western area)	10.391±0.024		R 4	11.2±0.6	R 4	10.1±0.5	R 4bi
							9.3±0.1	R 4an
	Basal Tuff (Ilkchi)	10.432±0.020		R 7	12.8±0.5	R 7	11.3±1.0	R 7bi
							9.3±0.1	R 7an

SOURCE: After Bernor (1986), Swisher (1996); Ar-Ar (Swisher 1996); Zircon FT and K-Ar (Campbell et al. 1980).

fossil material is from the middle and upper parts of the fossiliferous section ("Middle and Upper Maragheh"). The "Lower Maragheh" fauna is mostly composed of taxa with very long time distributions that cannot be used in fine time-resolution interregional comparisons (e.g., Kostopoulos and Bernor, 2011). For the best-known "Middle Maragheh" localities (1A, 1B, and 1C), where several groups have collected fossils, the current oldest age is for locality 1A, 8.16 Ma (see table 25.2 [estimated]), and the youngest localities ("Upper Maragheh") in the Shol'avand area (MMTT 26) date to 7.68 Ma (see table 25.2 [estimated]). To the east, the youngest localities of Ilkhchi (MMTT 37) would appear to be ca. 7.4 Ma.

The biochronological correlation of Maragheh with Samos and Pikermi, the classical mammal fossil localities of the Eurasian Late Miocene, is still open to discussion. Although this problem has been previously addressed (Bernor et al. 1996) and local biostratigraphy and geochronology of Maragheh and Samos have been greatly improved in the past years (Bernor et al. 1996; Swisher 1996; Kostopoulos, Sen, and Koufos 2003), the age of Pikermi remains badly resolved, being dated only indirectly.

According to the latest magnetostratigraphic correlation, the Samos fauna ranges from ca. 7.8 Ma to 6.7 Ma, but the core of the Samos fauna referred to as the Samos Dominant Faunal Assemblage (DMAS) is dated

Table 25.2

Locality Ages of UCR–MMTT Fossil Localities of the Maragheh Formation, Northwest Iran, Inferred from the Estimated Sedimentation Rates

UCR–MMTT Locality	Level (m) from Loose Chippings	District in Maragheh Area	Estimated Age (Ma)
26	7	Sholavand	7.68274
30	4	Sholavand	7.70728
39	4	Sholavand	7.70728
	0		**7.74***
24	−1	Sholavand	7.74818
32	−2	Sholavand West	7.75636
38	−3	Sholavand West	7.76454
22	−5	Sholavand East	7.7809
16	−6	Sholavand West	7.78908
47	−6	Sholavand West	7.78908
27	−10	Sholavand	7.8218
46	−12	Sholavand West	7.83816
25	−12	Sholavand West	7.83816
50	−15	Gort Daresi	7.8627
31	−17	Sholavand West	7.87906
33	−17	Sholavand	7.87906
13	−18	Sholavand West	7.88724
21	−18	Sholavand West	7.88724
51*	−20	Moghanjeq	7.9036
40	−20	Sholavand	7.9036
7*	−28	Sulu Dere	7.96904
52	−28	Sholavand West	7.96904
6	−29	Gort Daresi	7.97722
1C	−30	Gort Daresi	7.9854
18	−30	Sholavand West	7.9854
19	−30	Sholavand West	7.9854
20	−30	Sholavand West	7.9854
34	−30	Sholavand West	7.9854
1B	−32	Gort Daresi	8.00176
2	−32	Gort Daresi	8.00176
4	−35	Gort Daresi	8.0263
17	−35	Sholavand West	8.0263
35	−38	Sargezeh	8.05084
3	−40	Gort Daresi	8.0672
1A	−52	Gort Daresi	8.16536
5	−52	Gort Daresi	8.16536
8*	−60	Aliabad	8.2308
42*	−60	Aliabad	8.2308
36*	−70	Aliabad	8.3126
23	−72	Gort Daresi	8.32896
28	−80	Moghanjeq	8.3944
	−110		**8.64***
	−120	Kopran II	8.7216
	−150	Kopran I	8.967

NOTE: Ages with * are isotopic age determinations of zero level, which corresponds to "Loose Chippings," and −110 level, which corresponds to Mordaq (Murdag) Ignimbritic tuff (see also figure 25.5 and table 25.1). Localities with * have an estimated level with respect to "Loose Chippings."

between 7.2 Ma and 6.9 Ma (Kostopoulos, Sen, and Koufos 2003; Koufos, Kostopoulos, and Vlachou 2009). This chronology (figure 25.6) would make "Middle and Upper Maragheh" generally correlative with the lower fossil horizons (PMAS) at Samos (7.8–7.4 Ma), but the Maragheh localities, as currently understood, appear somewhat older than those producing the Samos Intermediary (IMAS) and Dominant Mammal Assemblages (7.4–6.9 Ma; Koufos, Kostopoulos, and Vlachou 2009).

However, this conclusion is not consistent with the evolutionary stage of several mammalian taxa and their combined presence in both Samos and Maragheh (i.e., *Melodon* [=*Parataxidea*] *maraghanus, Hyaenictitherium wongii, Adcrocuta eximia, Hippotherium brachypus,* "*Cremohipparion*" cf. *C. matthewi, Diceros neumayri, Skoufotragus laticeps, Miotragocerus* ex. gr. *valenciennesi, Gazella capricornis, Prostrepsiceros fraasi, Tragoportax, Palaeoryx, Protragelaphus, Palaeotragus, Samotherium*), which dates to the 7.4–6.9 Ma interval of Samos but appears to be older in "Middle and Upper Maragheh" intervals (for a detailed discussion, see Kostopoulos and Bernor 2011).

The Maragheh mammal association as it is shown in figure 25.4 is in part constructed from the old provenance data of the Vienna (NMW) and the Paris (MNHN) collections developed in Bernor (1986). However, the Tobien collection (BSP) does have stratigraphic provenance and was collected from the same part of the stratigraphic sequence where MMTT localities 1A–1C were collected (H. Tobien, pers. comm.).

One of us (RLB) correlates "Middle Maragheh," 8.1–7.9 Ma (estimated), with Pikermi based principally on the occurrence of *Hippotherium brachypus*, which is very similar to the Pikermi form, and *Cremohipparion moldavicum*, which is the sister taxon to *Cremohipparion mediterraneum*, also known from Pikermi (Koufos 1987). Another (DSK) holds that the apparent faunal relationships between Pikermi and Maragheh, also evidenced by several artiodactyl taxa, do not confirm such ages, as the

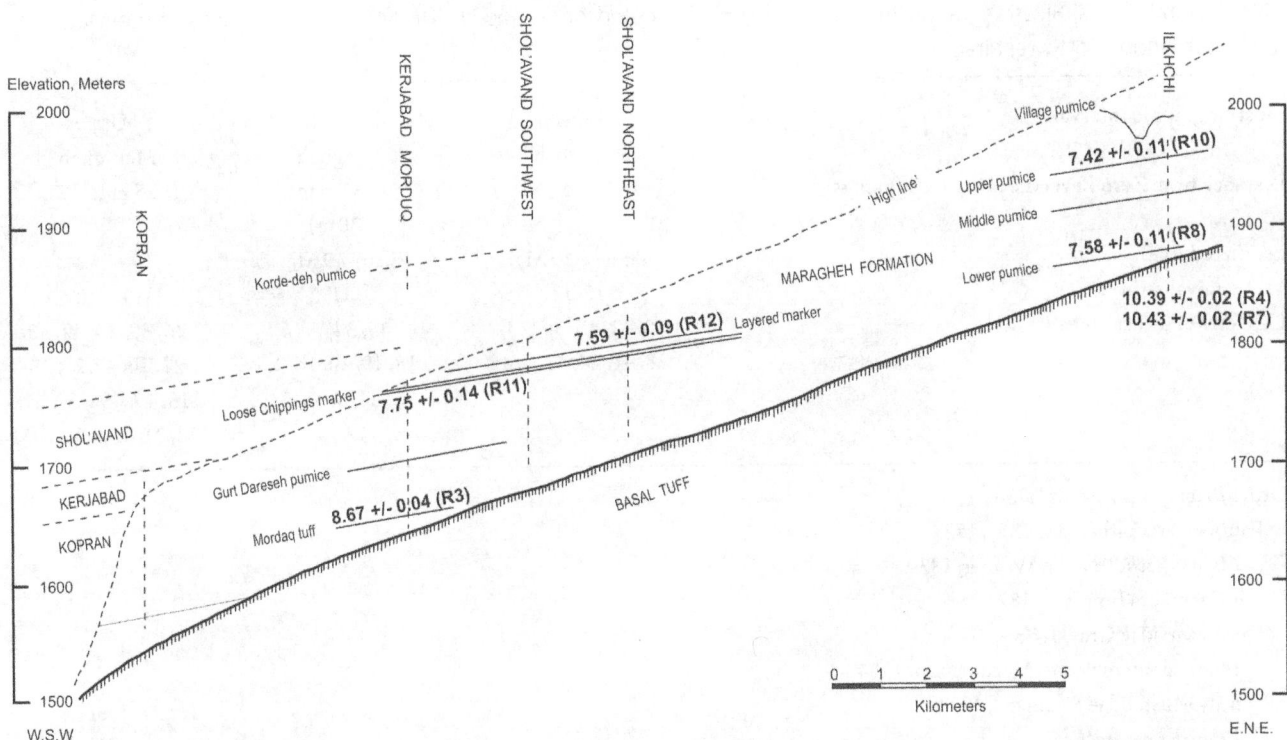

Figure 25.5 Stratigraphic array of isotopic age determinations from the Maragheh Formation and Basal Tuff. After Bernor (1986) and Swisher (1996).

mammal association of Pikermi appears to be more advanced than those from Vathylakkos-2, Prochoma, and Perivolaki (Greece), magnetostratigraphically dated between 7.4 Ma and 7.2 Ma (Koufos et al. 2006). In addition, *Hippotherium brachypus* and *Cremohipparion moldavicum* are also present in Akkaşdaği (Turkey; dated 7.1 Ma), while the type locality of *Cr. moldavicum* is Taraklija of the late Turolian.

Clearly, a conclusive correlation between the fauna of Maragheh, Samos, and Pikermi will have to await resolution of the apparent mismatch between the available geochronological and biochronological evidence. Without anticipating what the resolution will be, we note that the discrepancies can be explained by one or a combination of four possibilities:

- Mistakes in the radiometric or magnetostratigraphic dating
- Mismatches in the correlation of old localities with new stratigraphic evidence
- Major diachrony in the occurrence of species between these localities
- A mistaken attribution of ecomorphs to chronospecies

PALEOBIOGEOGRAPHY AND PALEOECOLOGY

We follow recent investigations on paleobiogeographic analysis by Bernor, Fortelius, and Rook (2001) and Fortelius et al. (1996) by undertaking genus-level faunal resemblance index (GFRI) studies using both the Simpson (1943) and the Dice (Sokal and Sneath 1963) indices. Dice FRI is highly recommended by Archer and Maples (1987) and Maples and Archer (1988) and is calculated as $2C/(A + B)$, where C is the number of shared taxa between two faunas and A and B are the total number of taxa present in fauna 1 and fauna 2, respectively. Simpson's FRI also has a long tradition of use (Bernor 1978; Flynn 1986; Bernor and Pavlakis 1987) and is calculated as C/smaller of (A or B).

Figure 25.7 illustrates the plot of GFRI in pairwise comparisons between "Middle and Upper Maragheh" and 10 other localities under consideration. Among the early Vallesian (MN 9) localities compared to Maragheh—the Western and Central European localities—all plot with an index value of less than 20% for both Dice and Simpson's GFRI. However, Sinap loc.12 (hominid zone), Turkey, has nine genera in common with Maragheh and a Dice GFRI

Table 25.3

Mammalian Species, Biostratigraphic Intervals and Their Estimated Ages, and UCR-MMTT Fossil Localities of the Maragheh Formation, Northwest Iran

Biostratigraphic Interval	"Lower Maragheh"	"Middle Maragheh"	"Upper Maragheh"
Distance from Zero Level (i.e., Loose Chippings)	(−150 m to −52 m)	(−52 m to −20 m)	(−25 m to +7 m)
Estimated Age	8.96–8.16 Ma.	8.16–7.9 Ma.	7.9–7.4 Ma.
MN-equivalent zone	10/11	11	11/12
UCR-MMTT fossil localities	8, 42, 45, 36, 23, 28, 41, 44, 43, 9, 48	7, 14, 34, 1c, 6, 18, 19, 20, 1b, 2, 17, 11, 4, 35, 3, 1A, 5	26, 39, 24, 32, 38, 22, 16, 47, 27, 25, 15, 49, 49a, 50, 31, 33, 21, 13, 37, 40, 51
Order Primates Linnaeus, 1758			
Family Cercopithecidae Gray, 1821			
Mesopithecus pentelici Wagner, 1839		X	
Order Carnivora Bowdich, 1821			
Family Ursidae Gray, 1825			
Indarctos maraghanus Mecquenum, 1924		X	
Family Mustelidae Swainson, 1835			
Promeles palaeattica Weithofer, 1888			X
Melodon maraghanus Kittl, 1887		X	
Parataxidea polaki Kittl, 1887		X	
Martes sp.		X	
Family Hyaenidae Gray, 1869			
Ictitherium viverrinum Roth and Wagner, 1854			X
Hyaenictitherium? wongii (Zdansky, 1924)		X	X
Adcrocuta eximia (Kaup, 1828)		X	X
Family Felidae Gray, 1821			
Metailurus orientalis Zdansky, 1924			X
Felis attica Wagner, 1857		X	X
Amphimachairodus aphanistus (Kaup, 1832)		X	X
Order Tubulidentata Huxley, 1872			
Family Orycteropodidae Bonaparte, 1850			
Orycteropus sp.			X
Order Proboscidea Illiger, 1811			
Family Gomphotheriidae Cabrera, 1929			
Choerolophodon pentelici Gaudry, 1862	X	X	X
Family Deinotheriidae Bonaparte, 1845			
Deinotherium gigantissimum Stefanescu, 1892			X
Order Perissodactyla Owen, 1848			
Family Equidae Gray, 1821			
"*Hipparion*" *gettyi* Bernor, 1985	X		
Hippotherium brachypus Hensel, 1862		X	
Hipparion campbelli Bernor, 1985			X
Cremohipparion aff. *C. moldavicum* Gromova, 1952		X	
Cremohipparion aff. *C. matthewi* Kormos, 1911	X	X	X
Family Chalicotheriidae Gill, 1872			
Ancylotherium pentelici (Gaudry, 1862)			X
Family Rhinocerotidae owen, 1845			
Diceros neumayri Mecquenem, 1905		X	

Biostratigraphic Interval	"Lower Maragheh"	"Middle Maragheh"	"Upper Maragheh"
Distance from Zero Level (i.e., Loose Chippings)	(−150m to −52m)	(−52m to −20m)	(−25m to +7m)
Estimated Age	8.96–8.16 Ma.	8.16–7.9 Ma.	7.9–7.4 Ma.
MN-equivalent zone	10/11	11	11/12
UCR-MMTT fossil localities	8, 42, 45, 36, 23, 28, 41, 44, 43, 9, 48	7, 14, 34, 1c, 6, 18, 19, 20, 1b, 2, 17, 11, 4, 35, 3, 1A, 5	26, 39, 24, 32, 38, 22, 16, 47, 27, 25, 15, 49, 49a, 50, 31, 33, 21, 13, 37, 40, 51
Chilotherium persiae Pohlig, 1887	X	X	
Iranotherium morgani (Mecquenem, 1908)		X	
Rhinocerotidae gen. and sp. nov.		X	
Order Artiodactyla Owen, 1848			
Family Suidae Gray, 1821			
Microstonyx major (Gervais, 1848)	X	X	X
Family Cervidae Gray, 1821			
Cervidae gen. and sp. indet.		X	
Family Giraffidae Gray, 1821			
Bohlinia attica (Gaudry and Lartet, 1856)		X	
Palaeotragus coelophrys Rodler and Weithofer, 1890	X	X	X
Samotherium neumayri Rodler and Weithofer, 1890	X	X	X
Helladotherium duvernoyi (Gaudry and Lartet, 1856)		X	X
Family Bovidae Gray, 1821			
Miotragocerus cf. *M. maius* (Meladze, 1967)		X	
Miotragocerus sp.		X	
Gazella capricornis (Wagner, 1848)		X	X
Gazella cf. *G. ancyrensis* Tekkaya, 1980		X	X
Demecquenemia rodleri (Pilgrim and Hopwood, 1928)		X	
Prostrepsiceros cf. *P. rotundicornis* Weithofer, 1888		X	
Prostrepsiceros houtumschindleri (Rodler and Weithofer, 1890)	X	X	
Prostrepsiceros fraasi (Andree, 1926)		X	
Prostrepsiceros cf. *P. vinayaki* (Pilgrim, 1939)		X	
Protragelaphus skouzesi Dames, 1883		X	X
Skoufotragus laticeps (Andree, 1926)		X	X
Tragoportax cf. *T. amalthea* (Roth and Wagner, 1854)		X	X
Urmiatherium polaki Rodler, 1889		X	X
Oioceros atropatenes Rodler and Weithofer, 1890		X	X
Oioceros rothii Wagner, 1857		X	X
Samokeros minotaurus Solounias, 1981		X	
Palaeoryx sp.		X	
Nisidorcas sp.		X	
cf. *Palaeoreas* sp.		X	
Order Rodentia Bowdich, 1821			
Family Hystricidae Burnett, 1830			
Hystrix sp.		X	

SOURCE: Modified after Bernor (1986).

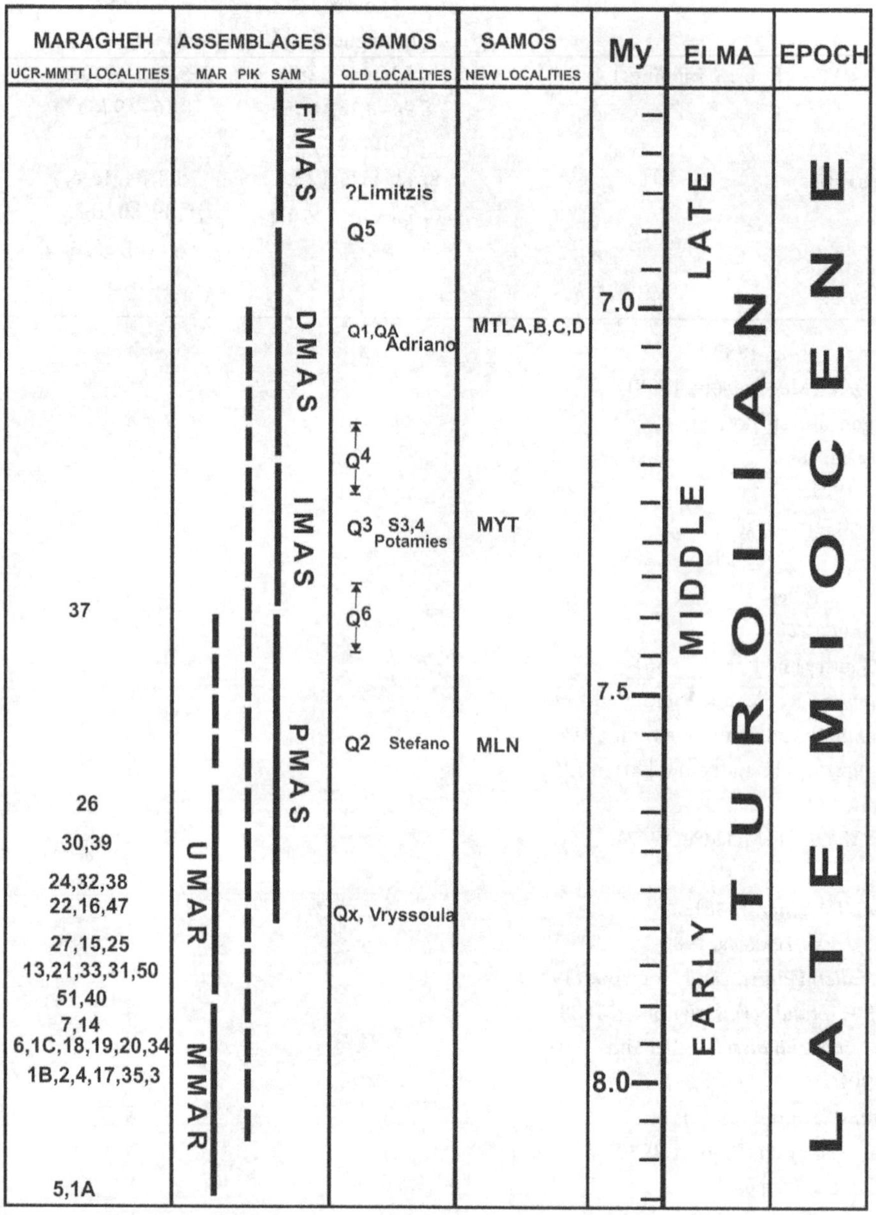

Figure 25.6 Chronostratigraphic position of Maragheh (MAR), Samos (SAM), and Pikermi (PIK) fossiliferous sites and faunal assemblages according to data presented in this chapter and Koufos, Kostopoulos, and Vlachou (2009).

of 30% and Simpson's GFRI of 47%, which is the highest among these early Vallesian localities. The shared genera between Sinap loc. 12 and Maragheh are *Chilotherium*, *Choerolophodon*, *Criotherium*, *Gazella*, *Hipparion*, *Orycteropus*, *Palaeoreas*, *Palaeotragus*, and *Tragoportax*.

We also used ungulate tooth crown height (Bernor, Fortelius, and Rook 2001) to show the contrast between the localities under consideration. The three-part subdivision of crown height includes the following: brachydont, whereby crown length of the upper second molar

(M2) is greater than its crown height; mesodont, whereby M2 crown length is roughly the same as crown height; and hypsodont, whereby M2 crown height is more than two times that of crown length.

All data have been downloaded from the NOW (Neogene of Old World) database on September 30, 2009 (Fortelius 2009; http://www.helsinki.fi /science/now/). Our analyses here consider only large mammals because small mammal records vary greatly across the faunas as a result of taphonomic and sampling bias.

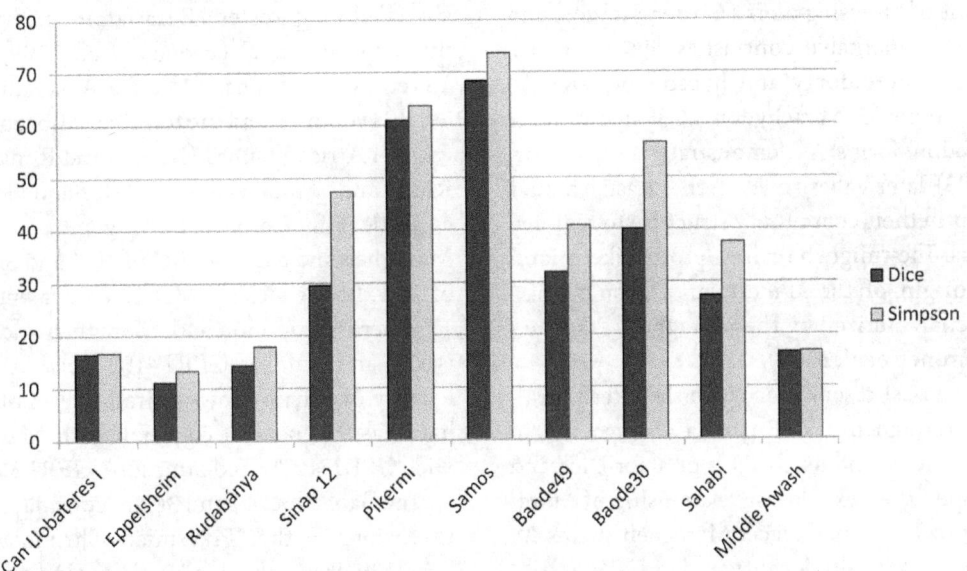

Figure 25.7 Genus-level Faunal Resemblance Indices (GFRI); pairwise comparison of Maragheh with localities under consideration. MN 9: Can Llobateres, Spain; Eppelsheim, Germany; Rudabanya, Hungary; Sinap loc. 12, Turkey. MN 12: Samos and Pikermi, Greece; Baode loc. 49, China. MN 13 equivalent: Baode loc. 30, China; Sahabi, Libya, and Middle Awash, Ethiopia.

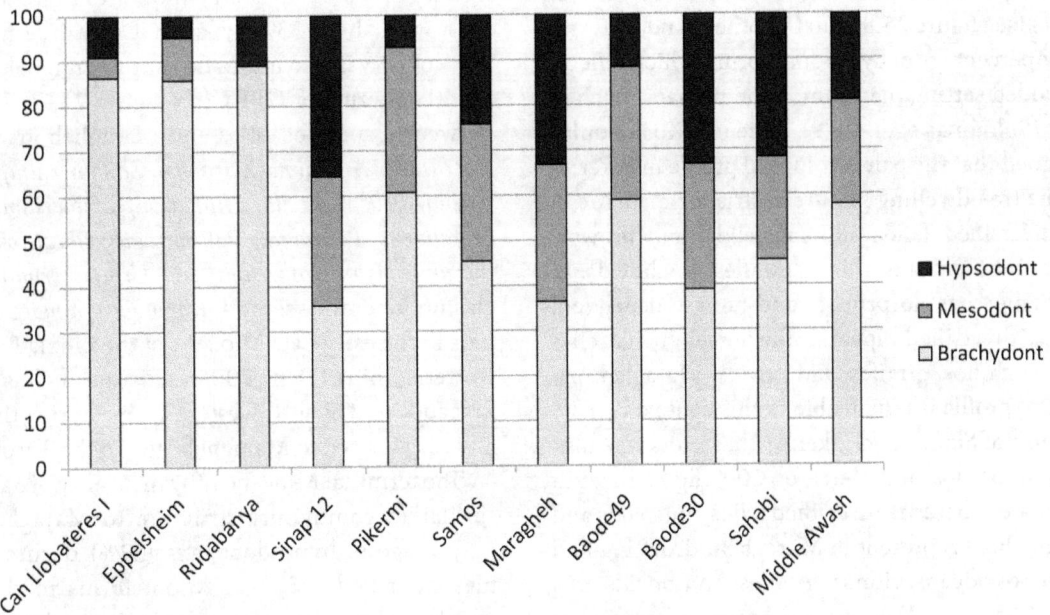

Figure 25.8 Crown height diagrams for Maragheh and Vallesian/Turolian localities under consideration. MN 9: Can Llobateres, Spain; Eppelsheim, Germany; Rudabanya, Hungary; Sinap loc. 12, Turkey. MN 12: Samos and Pikermi, Greece; Maragheh, Iran; Baode loc. 49, China. MN 13 equivalent: Baode loc. 30, China; Sahabi, Libya and Middle Awash, Ethiopia.

Analysis of crown height (figure 25.8) also demonstrates that Western and Central European (MN 9) localities of Rudabanya, Eppelsheim, and Can Llobateres have low incidence of mesodont and hypsodont forms (5–10% of mesodont and 5–10% hypsodont taxa) compared to more than 85% of brachydonts. This suggests that Central and Western Europe during the Vallesian was forested with warm climates and low seasonality. The hypsodonty in these faunas is only due to occurrence of the invasive species of hipparionine horses (*Hippotherium*) that show adaptation to closed environments with a significant amount of browsing diets (Bernor, Kordos, and Rook 2003). In remarkable contrast is Sinap loc. 12, showing stronger mesodonty and hypsodonty signals. Sinap loc. 12 exhibits 35 % brachydonts, 28% mesodonts, and 35% hypsodont forms. As demonstrated before (Fortelius et al. 2003), later Vallesian Western Eurasian faunas are very similar in their community structure to regional Turolian faunas. They might be called "proto-Pikermian," showing the origin of the "Pikermian Chronofauna," which is best characterized by the Maragheh–Pikermi–Samos triad (Eronen et al. 2009).

Maragheh's closest resemblance is to the Greek locality of Samos. Maragheh has 28 genera (22 species) in common with Samos, and its GFRI is 68% for Dice and 74% for Simpson. The next closest relationship of Maragheh is clearly to Pikermi, Greece. Maragheh shares 28 genera (20 species) with this locality, and its GFRI is 61% for Dice and 63% for Simpson. These classical localities of the "Pikermian Chronofauna" show stable community structures. Maragheh and Samos have crown-height percentages among its ungulate taxa as follows: about 45–35% of brachydont, 30% of mesodonts, and 25–35% of hypsodonts (see figure 25.8). On the other hand, Pikermi has a lower percentage of hypsodont forms, which reflects a more wooded setting than Samos and perhaps much of Maragheh (Solounias et al. 1999). Recently, Kostopoulos (2009) argued that the paleoecological profile in Pikermi includes the tree-dwelling semi-terrestrial primate *Mesopithecus*, diversified felids and mustelids, and browse-dependent proboscideans, rhinos, giraffes, and bovids. In contrast, Samos has no primate and browse-dependent taxa and has diversified hipparionine horses and gazelles with grazing rhinos, giraffes, and bovids. Maragheh paleoecological profile is remarkable in this context by having a mixture of Samos and Pikermi characters (see also Koufos, Kostopoulos, and Merceron 2009:fig.7). In addition to primates, there are diversified felids, mustelids and hipparionine horses present in Maragheh. Browsing and grazing proboscideans, rhinos, giraffes, and bovids also occur in Maragheh. Therefore, although Maragheh is somewhat more similar to Samos than it is to Pikermi (see figure 25.7), environmentally it indicates more wooded settings in its dominant grass/bushy vegetation. For this reason, as mentioned by Strömberg et al. (2007), an east to west climatic and vegetational gradient across the Greco-Iranian province is not clearly evident.

In the Middle Turolian, great intercontinental dispersion of large carnivores and ungulates of the "Pikermian chronofauna" evidently occurred (Kostopoulos 2009), and some of the core Pikermian genera extended their

geographic range so that a sizable number of genera are shared among western Eurasian members of the "Pikermian chronofauna" (Eronen et al. 2009), Chinese MN 13–equivalant faunas (Mirzaie Ataabadi et al., chapter 29, this volume), and the terminal Miocene northern and eastern African faunas (Bernor and Rook 2008, Bernor, Rook, and Haile-Selassie 2009). Baode loc. 30 (MN 13 equivalent in China) has 13 genera in common with Maragheh and a Dice GFRI of 40% and Simpson's GFRI of 56%. Baode loc. 49 (MN 12 equivalent in China) has 11 genera in common with Maragheh and its Dice GFRI is 32% and Simpson GFRI 41%. North African (Libyan) locality of Sahabi shows a similar level of similarity. Sahabi has 9 genera in common with Maragheh, and its Dice GFRI is 27% and Simpson's GFRI 37%.

The Baode loc. 49 and 30 are considered to be eastern extensions of the "Pikermian Chronofauna" (Mirzaie Ataabadi et al., chapter 29, this volume), having exhibited Pikermian-type community structures. The MN 12–equivalent locality of Baode 49 is similar to Pikermi, with about 12% hypsodont taxa, 30% mesodonts, and 59 % brachydonts. On the other hand, the MN 13–equivalent locality of Baode 30 is very similar to Maragheh and Samos, with about 33% hypsodont taxa (see figure 25.8). This locality is also more similar to Maragheh than locality 49 in terms of GFRI (see figure 25.7). Taxa shared between Baode localities and Maragheh are as follows: *Adcrocuta, Amphimachairodus, Chilotherium, Cremohipparion, Felis, Gazella, Hipparion, Ictitherium, Indarctos, Metailurus, Palaeoryx, Palaeotragus, Parataxidea, Samotherium, Hyaenictitherium,* and *Urmiatherium*. The large elasmothere rhinocerotid, *Iranotherium morgani*, also occurs in the early Late Miocene of the Linxia Basin, northwestern China (Deng 2005). This species has apparently first appeared in northwestern China during the Vallesian and immigrated to Maragheh later in the Turolian.

The terminal Miocene Libyan locality of Sahabi is also similar in community structure to Maragheh. A high percentage of hypsodont taxa (32%) occurs, with 23% mesodont and 45% brachydont forms in this locality, which is as similar to Maragheh, as are the Chinese Baode localities (see figure 25.7). Taxa shared between Sahabi and Maragheh are as follows: *Adcrocuta, Amphimachairodus, Ceratotherium* (*Diceros*), *Cremohipparion* and possibly *Hipparion, Gazella, Indarctos, Prostrepsiceros, Samotherium,* and *Tragoportax*.

The Late Miocene Middle Awash fauna (Ethiopia) is less than 20% similar to Maragheh in terms of both Dice and Simpson GFRIs. However, it has a somewhat similar community structure, but with fewer hypsodont and more brachydont taxa. These African faunas clearly had

"Pikermian" elements that were vicariant and evolved independently subsequent to an early–middle Turolian extension (Bernor and Rook 2008; Bernor, Rook, and Haile-Selassie 2009).

CONCLUSION

Apart from widely distributed "Pikermian Chronofauna" taxa (e.g., *Adcrocuta, Amphimachairodus, Deinotherium, Felis, Hipparion, Hyaenictitherium, Hystrix, Ictitherium, Indarctos, Metailurus, Palaeotragus*, and *Samotherium*), the giraffid *Bohlinia*, bovids such as *Demecquenemia rodleri*, the large *Palaeoryx*, and *Miotragocerus* cf. *M. maius*, and hipparionine horses such as *Cremohipparion moldavicum* indicate affinities of the Maragheh fauna to the Northern Black Sea region. Some taxa, such as *Prostrepsiceros* cf. *P. vinayaki*, show relationships with the western Asian (Arabia, Afghanistan, and the Indian Subcontinent) region, whereas *Hippotherium brachypus, Hipparion campbelli* (sister taxon to Samos *Hipparion dietrichi*), *Cremohipparion matthewi, Gazella* cf. *G. ancyrensis, Prostrepsiceros fraasi/houtumschindleri, Samokeros*, and *Skoufotragus* suggest affiliations with those from Anatolia and Samos. *Urmiatherium* and *Iranotherium* are taxa in common with China. Regardless of the absence of real-time resolution and the mismatch between some geochronological and biochronological data, it is evident that the Maragheh area was affected by biogeographically distinct Late Miocene areas, representing the crossroads of several Old World provinces.

ACKNOWLEDGMENTS

We appreciate the efforts of Dr. D. Najafi Hajipour and Dr. S. Montazami and their predecessors in Iran's Department of Environment for reinitiating and supporting new excavations in the Maragheh fossiliferous area. We also thank the Governor, the Mayor, and the local DOE office managers in Maragheh and Tabriz for their hospitality and assistance during INSPE fieldworks. MMA, MF, and fieldwork in Maragheh were partially supported by the Academy of Finland, RHOI project, and the Sasakawa Foundation. RLB wishes to acknowledge the National Science Foundation, including EAR-0125009 (grant to R. L. Bernor and M. O. Woodburne), BCS-0321893 (grant to F. C. Howell and T. D. White), and the Sedimentary Geology and Paleobiology Program (GEO: EAR: SEP) for supporting his research on this project. We thank Zhaoqun Zhang and the volume editors for their reviews and helpful comments. Z. Davoudifard extensively helped in preparation of several illustrations for this paper.

REFERENCES

Abich, H. W. 1858. Tremblement de terre observé à Tebriz en Septembre 1856, notices physiques et géographiques de M. Khanykof sur l'Azerbeidjan. *Bulletin de l'Académie (Imperiale) des Sciences de Saint Petersbourg, Classe Physico-Mathmatique.* 382 XVI-22 notes 27:339–352.

Aghanabati, A. 2004. *Geology of Iran.* Tehran: Geological Survey of Iran (in Persian).

Archer, A. W. and C. G. Maples. 1987. Monte Carlo simulation of selected binomial similarity coefficients (1): Effect of number of variables. *Palaios* 2:609–617.

Bernor, R. L. 1978. The mammalian systematics, biostratigraphy, and biochronology of Maragheh and its importance for understanding Late Miocene hominoid zoogeography and evolution. Ph.D. diss., University of California, Los Angeles.

Bernor, R. L. 1985. Systematic and evolutionary relationships of the hipparionine horses from Maragheh, Iran (Late Miocene, Turolian age). *Palaeovertebrata* 15:173–269.

Bernor, R. L. 1986. Mammalian biostratigraphy, geochronology, and zoogeographic relationships of the Late Miocene Maragheh fauna, Iran. *Journal of Vertebrate Paleontology* 6:76–95.

Bernor, R. L., M. Fortelius, and L. Rook. 2001. Evolutionary biogeography and paleoecology of the *Oreopithecus bambolii* Faunal Zone (Late Miocene, Tusco-Sardinian Province). *Bolletino della Società Paleontologica Italiana* 40(2):139–148.

Bernor, R. L., L. Kordos, and L. Rook. 2003. Recent advances on multidisciplinary research at Rudabánya, Late Miocene (MN9), Hungary: A compendium. *Palaeontographia Italica* 89:3–36.

Bernor, R. L., L. Kordos, and L. Rook. 2005. An introduction to the multidisciplinary research at Rudabánya. *Palaeontographia Italica* 90:9–10.

Bernor, R. L. and P. P. Pavlakis. 1987. Zoogeographic relationships of the Sahabi large mammal fauna. In *Neogene Paleontology and Geology of Sahabi*, ed. N. T. Boaz, A. El-Arnauti, A. W. Gaziry, J. De Heinzelin, and D. D. Boaz, 349–384. New York: Liss.

Bernor, R. L. and L. Rook. 2008. A current view of As-Sahabi large mammal biogeographic relationships, *Garyounis Scientific Bulletin*, Special Issue 5:285–292.

Bernor, R. L., L. Rook, and Y. Haile-Selassie. 2009. Paleobiogeography. In *Ardipithecus Kadabba: Late Miocene Evidence from the Middle Awash, Ethiopia*, ed. Y. Haile Selassie and G. Wolde Gabriel, 549–563. Berkeley: University of California Press.

Bernor, R. L., N. Solounias, C. C. Swisher III, and J. A. Van Couvering. 1996. The correlation of three classical "Pikermian" mammal faunas—Maragheh, Samos, and Pikermi—with the European MN unit system. In *The Evolution of Western Eurasian Neogene Mammal Faunas*, ed. R. L. Bernor, V. Fahlbusch, and H. W. Mittmann, 137–156. New York: Columbia University Press.

Bernor, R. L., H. Tobien, and J. A. Van Couvering. 1979. The mammalian biostratigraphy of Maragheh. *Annales Géologie Pays Hellenica* 1979(1):91–100.

Bernor, R. L., M. O. Woodburne, and J. A. Van Couvering. 1980. A contribution to the chronology of some Old World faunas based on hipparionine horses. *Geobios* 13(5):25–59.

Brandt, J. F. 1870. Ueber die von Herrn Magister Adolph Goebel auf Seiner Persischen Reise bie der Stadt Maragha in der Provinz Azerbeidjan gefundenen Säugethier-Reste. *Denkschrift des Naturforscher-Vereins zu Riga* 3:1–8.

Campbell, B. G., M. H. Amini, R. L. Bernor, W. Dickenson, W. Drake, R. Morris, J. A. Van Couvering, and J. A. H. Van Couvering. 1980. Maragheh: A classical Late Miocene vertebrate locality in northwestern Iran. *Nature* 287:837–841.

Davoudzadeh, M., B. Lammerer, and K. Weber-Diefenbach. 1997. Paleogeography, stratigraphy, and tectonics of the Tertiary of Iran. *Neues Jahrbuch für Geologie und Palaeontologie. Abhandlungen* 205:33–67.

de Mecquenem, R. 1905. Le gisement de vertèbres fossiles de Maragha. *Comptes Rendus Acadmie des Sciences de Paris* 14 1:284–286.

de Mecquenem, R. 1906. Les vertèbres fossiles de Maragha. *La Nature* 24:10.

de Mecquenem, R. 1908. Contribution à l'étude du gisement des vertèbres de Maragha et de ses environs. *Annales d'Histoire Naturelle* 1:27–79.

de Mecquenem, R. 1911. Contribution à l'étude du gisement des vertèbres de Maragha et de ses environs. *Annales d'Histoire Naturelle* 1:81–98.

de Mecquenem, R. 1924–1925. Contribution à l'étude des fossiles de Maragha. *Annales de Paléontologie* 13/14:135–160.

Deng, T. 2005. New discovery of *Iranotherium morgani* (Perissodactyla, Rhinocerotidae) from the Late Miocene of the Linxia Basin in Gansu, China and its sexual dimorphism. *Journal of Vertebrate Paleontology* 25:442–450.

Dewey, J. F., W. C. Pitman III, W. B. F. Ryan, and J. Bonnin. 1973. Plate tectonics and the evolution of the alpine system. *Geological Society of America Bulletin* 84(10):3137–3180.

Erdbrink, D. P. B. 1976a. A fossil Giraffine from the Maragheh region in N.W. Iran. *Bayerische Staatssammlung fur Paläontologie und Historische Geologie Mitteilungen* 16:29–40.

Erdbrink, D. P. B. 1976b. Early *Samotherium* and early *Oioceros* from an Uppermost Vindobonian fossiliferous pocket at Mordaq near Maragheh in N.W. Iran. *Bayerische Staatssammlung fur Paläontologie und Historische Geologie Mitteilungen* 16:41–52.

Erdbrink, D. P. B. 1977. On the distribution in time and space of three Giraffid genera with Turolian representatives at Maragheh in N.W. Iran. *Proceedings Koninklijke Nederlandse Akademie van Wetenschappen* B 80(5):337–355.

Erdbrink, D. P. B. 1978. Fossil Giraffidae from the Maragheh area, N.W. Iran. *Bayerische Staatssammlung fur Paläontologie und Historische Geologie Mitteilungen* 18:93–115.

Erdbrink, D. P. B. 1982. A fossil reduncine antelope from the locality K 2 east of Maragheh area, N.W. Iran. *Bayerische Staatssammlung für Paläontologie und Historische Geologie Mitteilungen* 22:103–112.

Erdbrink, D. P. B. 1988. *Protoryx* from three localities East of Maragheh, N.W. Iran. *Proceedings Koninklijke Nederlandse Akademie van Wetenschappen* B 91(2):101–159.

Erdbrink, D. P. B., H. N. A. Priem, E. H. Hebeda, C. Cup, P. O. Dankers, and S. A. P. L. Cloetingh. 1976. The bone bearing beds near Maragheh in N.W. Iran. *Proceedings Koninklijke Nederlandse Akademie van Wetenschappen* B 79(2):85–113.

Eronen, J. T., M. Mirzaie Ataabadi, A. Micheels, A. Karme, R. L. Bernor, and M. Fortelius. 2009. Distribution history and climatic controls of the Late Miocene Pikermian Chronofauna. *Proceedings of the National Academy of Sciences* 106(20):11867–11871.

Flynn, J. 1986. Faunal provinces and the Simpson coefficient. *Contributions to Geology, University of Wyoming Special Paper* 3:317–338.

Fortelius, M. (coordinator). 2009. *Neogene of the Old World Database of Fossil Mammals (NOW)*. University of Helsinki. http://www.helsinki.fi/science/now/.

Fortelius, M., J. Kappelman, S. Sen, and R. L. Bernor. 2003. *Geology and Paleontology of the Miocene Sinap Formation, Turkey*. New York: Columbia University Press.

Fortelius, M., L. Werdelin, P. Andrews, R. L. Bernor, A. Gentry, L. Humphrey, H. W. Mittmann, and S. Viranta. 1996. Provinciality, diversity, turnover, and paleoecology in land mammal faunas of the later Miocene of Western Eurasia. In *The Evolution of Western Eurasian Neogene Mammal Faunas*, ed. R. L. Bernor, V. Fahlbusch, and H. W. Mittmann, 414–448. New York: Columbia University Press.

Grewingk, C. 1881. Ueber fossile Saugethiere von Maragha in Persien. *Verhandiungen der K. K. Geologischen Reichsanstalt*, 1881:296.

Huber, H. (compiler). 1976. *Tectonic Map of Northwest Iran*. Tehran: National Iranian Oil Company, Exploration and Production.

Kamei, T., J. Ikeda, H. Ishida, S. Ishida, L. Onishi, H. Partodizar, S. Sasajima, and S. Nishimura. 1977. A general report of the geological and paleontological survey in Maragheh area, North-West Iran. *Memoirs of the Faculty of Sciences, Kyoto University* 43:131–164.

Kittl, E. 1887. Beiträge zur Kenntniss der fossilen Saugethiere von Maragha in Persien. 1 Carnivoren. *Annalen Naturhistorischen Museums Wien* 11:317–338.

Kostopoulos, D. S. 2009. The Pikermian Event: Temporal and spatial resolution of the Turolian large mammal fauna in SE Europe. *Palaeogeography, Palaeoclimatology, Palaeoecology* 274:82–95.

Kostopoulos, D. S. and R. L. B. Bernor. 2011. The Maragheh bovids (Mammalia, Artiodactyla): Systematic revision and biostratigraphic-zoogeographic interpretation. *Geodiversitas* 33(4):649–708.

Kostopoulos, D. S., S. Sen, and G. D. Koufos. 2003. Magnetostratigraphy and revised chronology of the Late Miocene mammal localities of Samos, Greece. *International Journal of Earth Sciences* 92:779–794.

Koufos, G. D. 1987. Study of the Pikermi hipparions. Part I: Generalities and taxonomy. *Bulletin du Muséum national d'Histoire naturelle Paris*, 4th ser., 9, sect. C, 2:197–252.

Koufos, G. D., D. S. Kostopoulos, and G. Merceron. 2009. Palaeoecology and Palaeobiogeography. In *The Late Miocene Mammal Faunas of the Mytilinii Basin, Samos Island, Greece: New Collection*. ed. G. D. Koufos and D. Nagel, pp. 409–428. *Beiträge zur Paläontologie* 31.

Koufos, G. D., D. S. Kostopoulos, and T. D. Vlachou. 2009. Biochronology. In *The Late Miocene Mammal Faunas of the Mytilinii Basin, Samos Island, Greece: New Collection*, ed. G. D. Koufos and D. Nagel, pp. 397–408. *Beiträge zur Paläontologie* 31.

Koufos, G. D., S. Sen, D. S. Kostopoulos, and T. D. Vlachou. 2006. The Late Miocene vertebrate locality of Perivolaki, Thessaly, Greece. 10. Chronology. *Palaeontographica Abteilung A* 276:185–200.

Lydekker, R. 1886. On the fossil Mammalia of Maragha in Northwest Persia. *Quarterly Journal Geological Society of London* 42:173–176.

Maples, C. G. and A. W. Archer. 1988. Monte Carlo simulation of selected binomial similarity coefficients (II): Effect of number of sparse data. *Palaios* 3:95–103.

M. Mirzaie Ataabadi, J. Mohammadalizadeh, Z. Zhang, M. Watabe, A. Kaakinen, and M. Fortelius. 2011. Late Miocene large mammals from Ivand (northwestern Iran). *Geodiversitas* 33(4):709–728.

M. Mirzaie Ataabadi, G. Zaree, and Z. Orak. 2011. Large mammals from the new Late Miocene fossil localities near Varzeghan, northwest Iran. *Vertebrata PalAsiatica* 49:311–321.

Moin-Vaziri, H. and E. Amin-Sobhani. 1977. *Volcanology and Volcanosedimentology of Sahand.* Tehran: Teacher Education University (in Persian).

Morris, R. S. 1997. The taphonomy and paleoecology of the Late Miocene terrestrial vertebrate locality near northwest Iran: A framework for paleoenvironmental analysis of Late Miocene hominoidea. Ph.D. diss., University of California, Los Angeles.

Pohlig, H. 1886. On the Pliocene of Maragha, Persia, and its resemblance to that of Pikermi in Greece; on fossil elephant remains of Caucasia and Persia; and on the resuits of a monograph of the fossil elephants of Germany and Italy. *Quarterly Journal Geological Society of London* 42:177–182.

Rieben, H. 1934. Contribution a la géologie de l'Azerbeidjan person. *Bulletin de la Société Neuchatel des Sciences naturelles* 59:19–144.

Rodler, A. 1890. Ueber *Urmiatherium polaki* n. gen., n. sp., einen neuen Sivatheriiden aus dem Knochenfeide von Maragha. *Denkschriften K. Akademie der Wissenschaften, Wien; Mathematisch-Naturwissenschaftlichen Klasse* 56:315–332.

Rodler, A. and K. A. Weithofer. 1890. Die Wiederkauer der Fauna von Maragha. *Denkschriften K. Akademie der Wissenschaften, Wien; Mathematisch Naturwissenschaftlichen Klasse* 62:753–772.

Schlesinger, G. 1917. Die Mastodonten des K. K. Naturhistorischen Hofmuseums. Morphoiogische-Phylogenetische Untersuchungen. *Denkschriften K. Akademie der Wissenschaften, Wien; Mathematische-Naturwissenschaftlichen Klasse* 1:1–230.

Simpson, G. G. 1943. Mammals and the nature of continents. *American Journal of Science* 241:1–31.

Sokal, R. R. and P. H. A. Sneath. 1963. *Principles of Numerical Taxonomy.* San Francisco: Freeman.

Solounias, N., J. M. Plavcan, J. Quade, L. Witmer. 1999. The Paleoecology of the Pikermian biome and the Savanna myth. In *The Evolution of Neogene Terrestrial Ecosystems in Europe,* ed. J. Agustí, L. Rook and P. Andrews, pp. 436–453. Cambridge: Cambridge University Press.

Strömberg, C. A. E., L. Werdelin, E. M. Friis, and G. Saraç. 2007. The spread of grass-dominated habitats in Turkey and surrounding areas during the Cenozoic: Phytolith evidence. *Palaeogeography, Palaeoclimatology, Palaeoecology* 250:18–49.

Swisher III, C. C. 1996. New 40Ar/39Ar dates and their contribution toward a revised chronology for the Late Miocene of Europe and West Asia. In *The Evolution of Western Eurasian Neogene Mammal Faunas,* ed. R. L. Bernor, V. Fahlbusch, and H. W. Mittmann, pp. 64–77. New York: Columbia University Press.

Takai, F. 1958. Vertebrate fossils from Maragha. *Institute of Oriental Culture, University of Tokyo* 26:7–11.

Tobien, H. 1968. Palaeontologische Ausgrabungen nach Jungtertiären Wirbeltieren auf der Insel Chios (Griechenland) und bei Maragheh (N.W. Iran). *Jahrbuch der Vereiningung "Freunde der Universität Mainz"* 1968:51–58.

Watabe, M. 1990. Fossil bovids (Artiodactyla, Mammalia) from Maragheh (Turolian, Late Miocene), Northwest Iran. *Annual Report of the Historical Museum of Hokkaido* 18:19–56.

Watabe, M. and H. Nakaya. 1991a. Phylogenetic significance of the postcranial skeletons of the hipparions from Maragheh (Late Miocene), NW Iran. *Memoirs of the Faculty of Sciences, Kyoto University* 56(1–2):11–53.

Watabe, M. and H. Nakaya. 1991b. Cranial skeletons of *Hipparion* (Perissodactyla, Mammalia) from Maragheh (Turolian, Late Miocene), NW Iran. *Memoirs of the Faculty of Sciences, Kyoto University* 56(1–2):55–125.

Woodburne, M. O. and R. L. Bernor. 1980. On superspecific groups of some Old World hipparionine horses. *Journal of Paleontology* 54(6):1319–1348.

Chapter 26

A Review of the Neogene Succession of the Muridae and Dipodidae from Anatolia, with Special Reference to Taxa Known from Asia and/or Europe

HANS DE BRUIJN, ENGIN ÜNAY, AND KEES HORDIJK

This overview of the Neogene Muridae and Dipodidae from Anatolia is an illustrated, updated, and emended version of an earlier biozonation (Ünay, de Bruijn, and Saraç 2003a). It aims at making the specialized scattered literature accessible for a larger group of geoscientists, to enhance long-distance correlations and detect migration patterns. In contrast to our earlier zonation, the associations presented here originate from single localities or from sites that are situated near to each other in the same formation. This procedure has the disadvantage that the content and composition of the assemblages is influenced by local biotopes and taphonomical conditions, but it has the advantage that mistakes due to incorrect correlation are excluded. The decision to restrict the information per time slice to one association makes these units a priori incomplete and unequal.

The localities selected for this review are dispersed over Anatolia (figure 26.1), which may add a geographical bias to the composition of the associations. However, the complete Mineral Research and Exploration Institute of Turkey (MTA) database (covering about 150 rodent assemblages) suggests that differences due to geographical position within Anatolia are of minor importance during the Neogene. With the exception of the assemblage from Altıntaş, which comes from a karst fissure filling, and the one from Iğdeli, which comes from fluviatile sediments, the rodent faunas studied originate from lignitic clay lenses in lacustrine deposits.

In order to somewhat limit the number of specimens figured, species that are known from Anatolia only are represented by one molar (preferably M1 or m1) per time slice (locality) on plates 26.1–26.4. Species from Asia and/or Europe that are either known or are expected to have been present are represented by a complete dentition (if available). Genera identified as immigrants into Anatolia are indicated by bold italics in figure 26.2 and were counted per time slice (figure 26.4). The proportion of species that Anatolia has in common with Europe (figure 26.3) and central Asia (figure 26.5) is calculated per time slice.

Species that are poorly represented will, as a rule, be identified to the generic level only. If a species is well represented, but not known to science, the genus name will be followed by "n. sp." Identifications of genera and species generally follow the literature, because descriptions and taxonomical revisions are beyond the scope of this chapter. This conservative approach may, unfortunately, disguise the real range of taxa in space and time. This phenomenon probably occurs in species of the, in our opinion oversplit, genera *Megacricetodon* and *Democricetodon*, while the identifications of the Plio-Pleistocene genera of the Cricetinae seem to suffer from the reverse, because these have not been studied in detail in recent years. If two generic names (separated by a slash) are entered in one column in figure 26.2, the sample contains morphotypes of both genera, so we cannot recognize them on the basis of the literature. Two species names

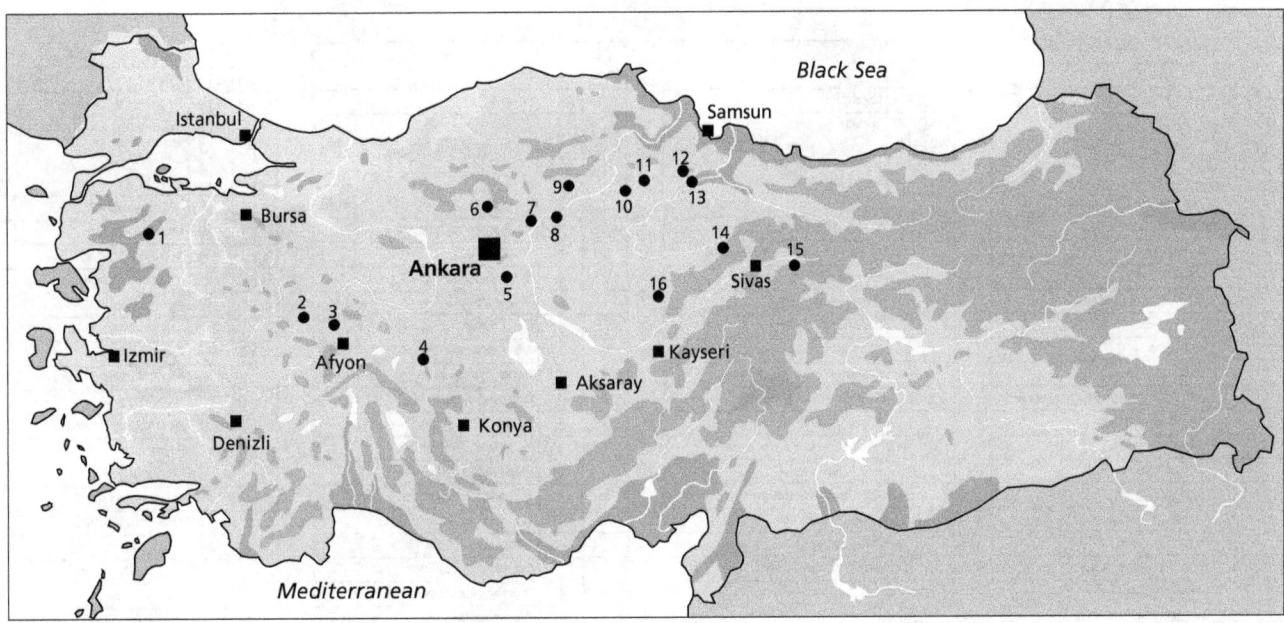

Figure 26.1 Sketch map of Turkey showing the approximate geographical position of the 16 localities that yielded the assemblages of Muridae and Dipodidae discussed. For locality names and coordinates, see appendix 5.

(separated by a slash) entered in the same column in figure 26.2 suggests that these species names may be synonyms.

The identification of a genus as immigrant is based on whether or not that genus has a (potential) ancestor in previous zones. This procedure has the effect that a number of genera that probably originated elsewhere are not recognized as immigrants. Examples of this phenomenon are the dipodid *Protalactaga minor* and the murine *Orientalomys*. The reference used for immigrants into unit A are the associations from the Late Oligocene localities Gözükızıllı and Yeniköy (Ünay, de Bruijn, and Saraç 2003b).

The geographical entities Anatolia, central Asia, and Europe are defined as Turkey east of the Dardanelles and by the borders between Asia and Europe. This means that the Greek islands off the Turkish coast are Europe, which has an impact on similarities in composition.

Generic and specific determinations will not be justified other than by figures. The taxa mentioned in figure 26.2 are listed alphabetically in appendices 1 and 2, accompanied by the taxon authorities and those references are gathered separately after the text references.

All figured cheek teeth in plates 26.1–26.4 are approximately ×15 and depicted as if they are from the left side. Taxa first appearing in a zone are figured on the left side

of the plates. This has the effect that genera that have a long stratigraphical range are successively pushed to the right side of the page.

THE EARLY MIOCENE RECORD

The Early Miocene assemblages of Muridae and Dipodidae are diverse in containing 5 to 12 species (see figure 26.2A; plate 26.4). The association of the Muridae from Kargı 2 (zone A) exclusively contains genera that are not known from the, presumably, Late Oligocene *Eucricetodon*-dominated rodent assemblages from Gozükızıllı and Yeniköy. Although the one and a half *Melissiodon* teeth from Kargı 2 are the only record of that genus from Anatolia so far, we have not listed it among the immigrants, because it is considered to be phylogenetically close to the genus *Edirnella*, which is known from the Late Eocene (Süngülü, Lesser Caucasus; de Bruijn, Ünay, and Yilmaz 2003) and the Oligocene (Kocayarma, Turkish Thrace; Ünay-Bayraktar 1989). A similar conservative assessment of the status of the poorly represented *Heterosminthus* from Kargı 2 is based on the occurrence of an unidentified dipodid in the Late Oligocene section of Yeniköy (Ünay and de Bruijn 1987).

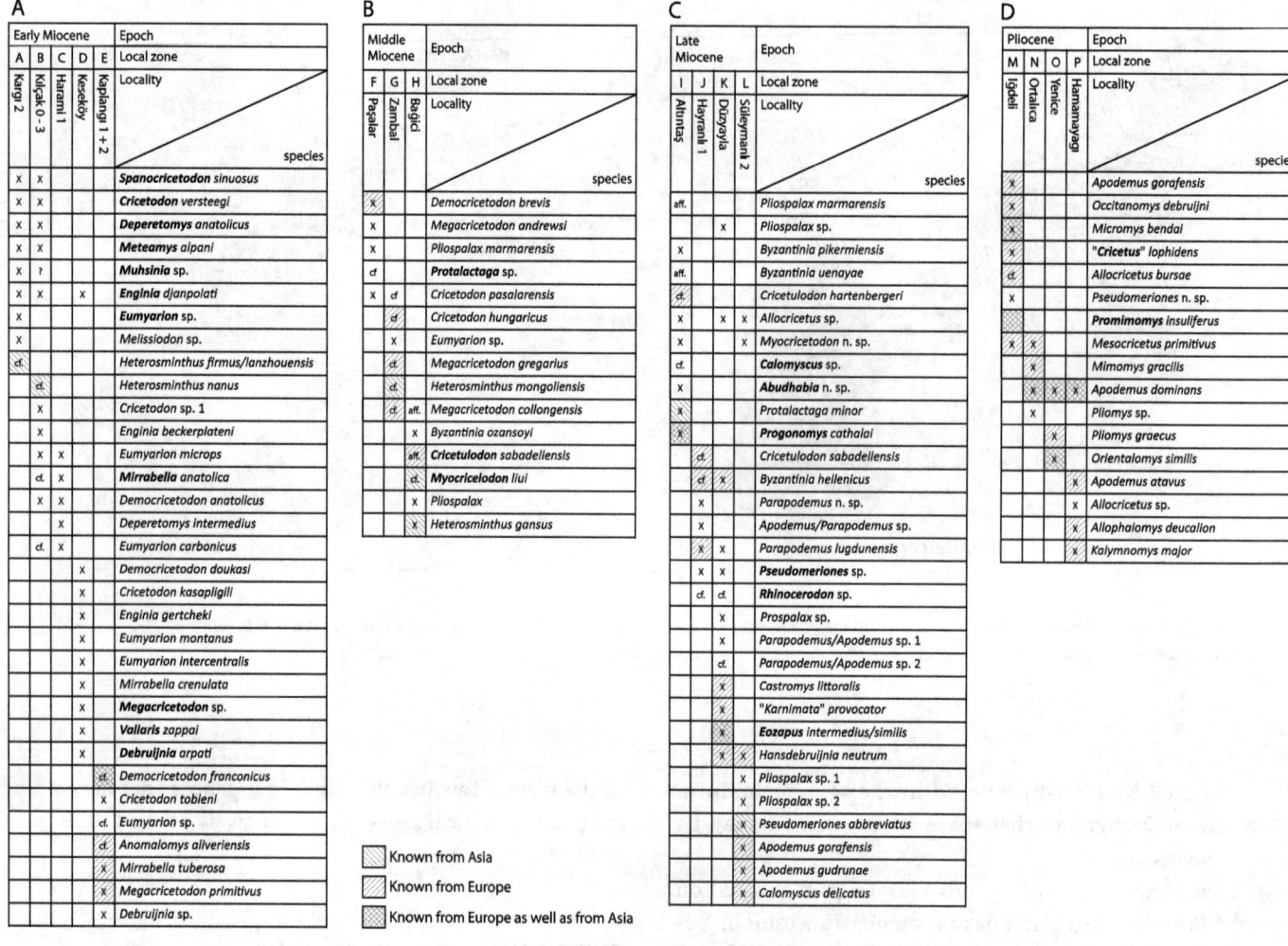

Figure 26.2 The identifications of the Muridae and Dipodidae from the 16 localities selected and the position of these localities relative to the informal biozonation of the Anatolian continental Neogene (Ünay et al. 2003a): (*A*) the Early Miocene record; (*B*) the Middle Miocene record; (*C*) the Late Miocene record; (*D*) the Pliocene record. Genera identified as immigrants are in bold italics. Occurrences outside Anatolia are indicated by different hatchings (see legend). x, species occurrence.

The Early Miocene succession of species of the genera *Enginia, Deperetomys, Eumyarion, Cricetodon, Spano/ Democricetodon,* and *Heterosminthus* suggests local evolutionary change. Unfortunately, the identification of the different species of *Heterosminthus* remains uncertain due to the poor material available. This applies in particular to the material from Kargı 2, which is entered as *H. firmus/lanzhouensis*—two species that are very similar and might be synonomous. *Muhsinia* sp. is listed in the locality Kılçak 0–3 by a question mark on the basis of the data presented by Theocharopoulos (2000). Because we could not find a single specimen in the Kılçak collection, its presence in this assemblage remains uncertain. Prominent new arrivals in Anatolia in the course of the Early Miocene are the aberrant cricetid

genus *Mirrabella* in zone B (not figured) and the spalacid *Debruijnia* in zone D.

THE MIDDLE MIOCENE RECORD

The Middle Miocene assemblages contain five or six species each and are thus less diverse than the Early Miocene ones (see figure 26.2B; plate 26.3). Since the collections from Paşalar and Bağiçi are large this relative paucity in taxa probably reflects the real situation. The association from Paşalar contains one tooth of an allactagine, which is the oldest record of a jerboa from Anatolia. The genus name *Pliospalax* is used here as including *Sinapospalax* Sarıca and Sen 2003.

Evolutionary change during the Middle Miocene is most notable in the Cricetodontinae and in the genus *Heterosminthus*, while the first representatives of the true Cricetinae and of the genus *Myocricetodon* appear in the upper part of this time-slice.

THE LATE MIOCENE RECORD

The association from the karst fissure deposit of Altıntaş (zone I) with nine murids and one dipodid is the most species rich of all assemblages in this time slice (see figure 26.2C; plate 26.2). It contains four immigrants: the murine *Progonomys cathalai*, a species of *Abudhabia* that is morphologically identical to *A. baheensis* Qiu, Zheng, and Zhang 2004 but considerably larger, the true cricetine *Allocricetus* and, if our assessment of the few small teeth figured on plate 26.2 (figures 70–73) is correct, a species of *Calomyscus*. Unfortunately, this material is limited, so the identification of what seems to be the oldest record of the genus anywhere remains uncertain for the time being.

Protalactaga minor (we do not recognize *Paralactaga* as generically different from *Protalactaga*) has not been counted as an immigrant, because of the presence of an undetermined allactagine in the assemblage of Paşalar.

The assemblage from Hayranlı 1 (zone J) contains the oldest record of *Pseudomeriones* and of *Rhinocerodon* in the Anatolian succession. The murines have become diverse in this level and are represented by teeth in three size groups. The medium-size group, identified as *Apodemus/Parapodemus* sp., represents a morphologically very heterogeneous species containing *Progonomys*, *Parapodemus*, and *Apodemus* morphotypes. Further diversification of the murines, which is possibly also due to immigration, is seen in the assemblage from Düzyayla. This locality contains a few teeth of *Eozapus*, which have been entered as *E. intermedius/similis*, species from Europe and Asia, respectively, that are very similar in size and morphology and may eventually prove to be synonymous.

The first certain *Calomyscus*, *C. delicatus*, occurs in Süleymanlı 2 (zone L) and is associated with, among others, two *Apodemus* species and *Hansdebruijnia neutrum*.

THE PLIOCENE RECORD

With the exception of the association from Iğdeli, the number of specimens in our Pliocene assemblages is limited (see figure 26.2D; plate 26.1), because these were collected for biostratigraphical purposes within the framework of mapping projects of the MTA general directorate. As a result, the diversity of these associations is limited and likely significantly undersampled.

In general, one can conclude that the diversity within the Murinae and Cricetinae remains high during this time slice, while the Arvicolinae rapidly diversify. Whether the increase of the number of the genera and species of voles is the result of in situ radiation or of immigration cannot be decided on the basis of the record available.

BIOSTRATIGRAPHICAL CORRELATION OF THE ASSEMBLAGES OF MURIDAE AND DIPODIDAE FROM ANATOLIA WITH THE EUROPEAN AND CHINESE SUCCESSIONS

The Early Miocene zones share *Heterosminthus firmus/lanzhouensis* in zone A and *H. nanus* in zone B with the East Asiatic record. The type localities of both former species have been assigned ages near the Oligocene/Miocene boundary, and that of *H. nanus* is considered Early Miocene (appendix 3). The overall composition of the assemblage from Kargı 2 (zone A) is similar to that from the central Anatolian locality Inkonak (de Bruijn et al. 1992), but the *Muhsinia* teeth are somewhat more primitive and the assemblage does not contain *Heterosminthus*. The locality Inkonak has been correlated to magnetic chron C6Br (Krijgsman et al. 1996), which has an age between 22.268 Ma and 22.564 Ma based on the ATNTS2004 (Lourens et al. 2004). Krijgsman also sampled the section containing the localities Kargı 1–3, but the magnetic signal of these samples appeared to be too weak to obtain dependable results. A second tie point to the paleomagnetic time scale is the locality Harami 1, which has been correlated with chron C6Bn.2n (Krijgsman et al. 1996). Based on the ATNTS2004, the age of this chron is between 21.992 Ma and 22.268 Ma. Judging by the difference between the rodent associations from Kargı 2 and Inkonak (see figure 26.2A) we think that the age difference of 500,000 yr as suggested by the paleomagnetic results is far too small.

The first fauna exchange between Anatolia and Europe during the Early Miocene is documented in zone E by murids that characterize MN 4: *Democricetodon franconicus*, *Anomalomys aliveriensis*, *Mirrabella tuberosa*, and *Megacricetodon primitivus*. All four species potentially have ancestors in Anatolia, so it is assumed that they immigrated into Europe from the east. This migration provides a solid basis for correlating zone E with MN4.

The Middle Miocene zones F to H share *Heterosminthus mongoliensis* (zone G) and *Myocricetodon liui* and

Heterosminthus gansus (zone H) with China. The type locality of *H. mongoliensis*, Ulan-Tologoi, is considered to be early Middle Miocene in age, which is more or less in line with our age estimate of zone G. However, the type localities of *M. liui* and *H. gansus* have both been assigned to the early Late Miocene (appendix 3), which is younger than our estimate for the Anatolian occurrences.

The European species recognized in the zones F to H are *Democricetodon brevis* (zone F), *Cricetodon hungaricus*, *Megacricetodon gregarius*, and *Megacricetodon collongensis* (zone G), and *Cricetulodon sabadellensis* (zone H). With the exception of *C. hungaricus*, these species seem to occur in Europe in either older assemblages (*M. collongensis*) or younger ones (*M. gregarius*, *D. brevis*, *C. sabadellensis*). This discrepancy may be partly due to erroneous identification of the species of the, in our opinion oversplit, genera *Democricetodon* and *Megacricetodon*. The presence of *Cricetulodon* in Anatolia before the "*Hipparion* event" is probably real, because that has been suggested on the basis of data from the Dardanelles area also (Ünay and de Bruijn 1984).

The biostratigraphical evidence for our correlation of the zones F to H with the European MN 5 to MN 7/8 and the Chinese late Shanwangian/early Tungurian are unfortunately weak, and radiometric and paleomagnetic data are lacking.

The Late Miocene associations (I–L) share *Progonomys cathalai* and *Protalactaga minor* (both zone I), *Eozapus intermedius/similis* (zone K), and *Pseudomeriones abbreviatus* (zone L) with China (appendices 3 and 4). We follow Wessels (2009a) in considering *Progonomys sinensis* a junior synonym of *P. cathalai*. The type localities of the first two species are considered to have an early Late Miocene and a late Late Miocene age, respectively (appendix 3). The type locality of *Eozapus intermedius* has been correlated with MN 10/11 (=early Late Miocene); those of *E. similis* and *P. abbreviatus* with the late Late Miocene.

The Late Miocene assemblages from Anatolia (zones I–L) contain many species that have originally been described from Europe. This similarity is, however, in part the result of the definition of the geographical entities. *Byzantinia pikermiensis*, *B. hellenicus*, "*Karnimata provocator*," and *Hansdebruijnia neutrum* are known from Greece only in Europe. If we exclude these species from the discussion, Anatolia and Europe share *Cricetulodon hartenbergeri*, *Progonomys cathalai* (zone I), *Parapodemus lugdunensis* (zone J), *Eozapus intermedius/similis*, and *Castromys littoralis* (zone K), and *Apodemus gorafensis* and *Calomyscus delicatus* (zone L) during the Late Miocene. This faunistic similarity provides a reasonably sound ba-

sis for the correlation of the Late Miocene Anatolian zones to the MN scheme (appendix 4). Most unexpected among the Murinae is the presence of *Castromys littoralis* in the locality Düzyayla, because this species is otherwise known from its type locality (Crevillente 17, Spain) only.

The composition of the Pliocene assemblages from Anatolia (zones M–P) is largely similar to the time equivalent European MN units. Some species, such as *Orientalomys similis* and *Micromys bendai*, have exceptionally large geographical ranges. The former was originally described from Odessa and later recognized in Ertemte (Nei Mongol) and Tourkobounia 1 (Greece), and the range of the latter has been extended by synonymizing *Micromys tedfordi* from China with *M. bendai* from Europe (Hordijk and de Bruijn 2009). The content of European species does not exceed 70% in any of the Pliocene localities (figure 26.3). This is probably caused by the way these percentages were calculated: taxa that could not be identified to species level have not been counted as European.

Correlation of the Anatolian zones to MN 14–17 on the basis of the evolutionary stages of the voles is fairly straightforward (Ünay and de Bruijn 1998), although it cannot be excluded that the assemblage of Hamamayağı might not correlate with MN 17, but rather with the early Biharian. Correlation to the Chinese mammal age Yushean is less apparent, because we are insufficiently familiar with the voles from that area.

DISTRIBUTION PATTERNS AND MIGRATIONS OF NEOGENE MURIDAE AND DIPODIDAE

Figure 26.4 gives the number of immigrating genera of the Muridae and Dipodidae in the localities selected. In interpreting this figure, it should be realized that the data presented are based on one assemblage per time slice only and that the number of specimens collected from the various localities is very different. In spite of the resulting bias this causes, it is of interest to look closer at the immigration pattern that shows four peaks. The oldest one is seen in Kargı 2, a locality that is tentatively correlated with the Oligocene–Miocene boundary interval and contains *Spanocricetodon*, *Cricetodon*, *Muhsinia*, *Enginia*, and *Eumyarion*. None of these genera have a (potential) ancestor in older assemblages (Ünay, de Bruijn, and Saraç 2003a), so we conclude that there was a major turnover in the Muridae faunas of Anatolia during the Oligocene–Miocene boundary transition. This is in sharp contrast to the situation in southwestern Europe where

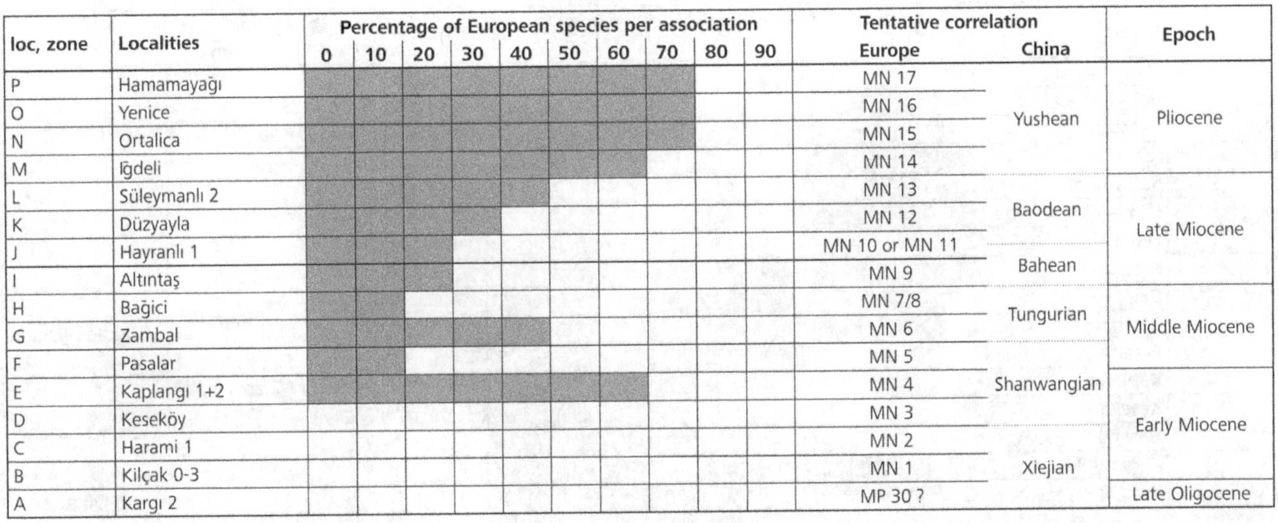

loc, zone	Localities	Percentage of European species per association										Tentative correlation		Epoch
		0	10	20	30	40	50	60	70	80	90	Europe	China	
P	Hamamayağı											MN 17	Yushean	Pliocene
O	Yenice											MN 16		
N	Ortalica											MN 15		
M	Iğdeli											MN 14		
L	Süleymanlı 2											MN 13	Baodean	Late Miocene
K	Düzyayla											MN 12		
J	Hayranlı 1											MN 10 or MN 11	Bahean	
I	Altıntaş											MN 9		
H	Bağici											MN 7/8	Tungurian	Middle Miocene
G	Zambal											MN 6		
F	Pasalar											MN 5		
E	Kaplangı 1+2											MN 4	Shanwangian	
D	Keseköy											MN 3		Early Miocene
C	Harami 1											MN 2		
B	Kılçak 0-3											MN 1	Xiejian	
A	Kargı 2											MP 30 ?		Late Oligocene

Figure 26.3 Graph showing the percentage of species shared by Anatolia and Europe by locality. Occurrences that could not be identified to the species level have been considered as different.

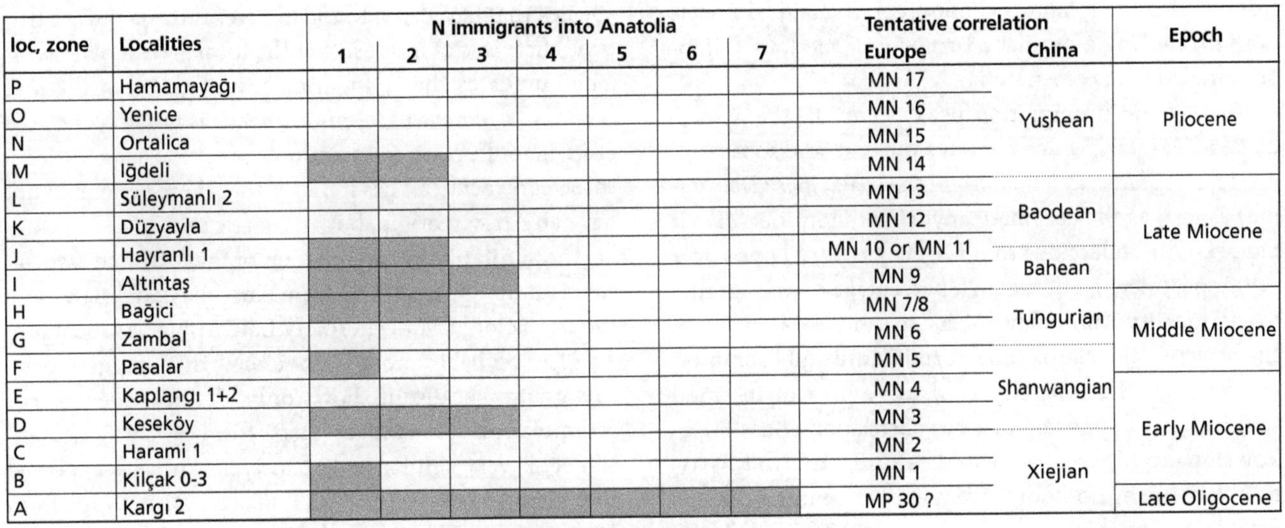

loc, zone	Localities	N immigrants into Anatolia							Tentative correlation		Epoch
		1	2	3	4	5	6	7	Europe	China	
P	Hamamayağı								MN 17	Yushean	Pliocene
O	Yenice								MN 16		
N	Ortalica								MN 15		
M	Iğdeli								MN 14		
L	Süleymanlı 2								MN 13	Baodean	Late Miocene
K	Düzyayla								MN 12		
J	Hayranlı 1								MN 10 or MN 11	Bahean	
I	Altıntaş								MN 9		
H	Bağici								MN 7/8	Tungurian	Middle Miocene
G	Zambal								MN 6		
F	Pasalar								MN 5		
E	Kaplangı 1+2								MN 4	Shanwangian	
D	Keseköy								MN 3		Early Miocene
C	Harami 1								MN 2		
B	Kılçak 0-3								MN 1	Xiejian	
A	Kargı 2								MP 30 ?		Late Oligocene

Figure 26.4 Graph showing the number of immigrant genera into Anatolia per locality ("zone") and a tentative correlation of the local zones to the European MN scheme and the Chinese mammal ages. All questionable entries have been omitted.

Late Oligocene–Earliest Miocene faunas are endemic and still contain the Oligocene murid genera *Eucricetodon*, *Pseudocricetodon*, and *Adelomyarion*.

Figures 26.2, 26.3, and 26.5 show that the murid genera dominating the Kargı 2 and Kılçak assemblages are known from neither Europe nor Asia during the Early Miocene. *Eumyarion*, *Mirrabella*, and *Deperetomys* reached Europe much later (de Bruijn and Saraç 1991, 1992; de Bruijn et al. 1992). Since neither *Eumyarion tremulus* (Lopatin 1996) from the Late Oligocene of Kazakhstan nor *Eumyarion*

kowalskii (Lindsay 1996) from the Late Oligocene of Pakistan belong in the genus *Eumyarion* (Wessels and de Bruijn 2001), *Democricetodon* s.l. seems to have been the only murid from Anatolia that conquered large parts of Eurasia during the Early Miocene (MN 4). The only species in the Kargı and Kılçak faunas that are also known from central Asia are the dipodids *Heterosminthus firmus/lanzhouensis* and *H. nanus* (see figure 26.2). The absence in Europe and Asia of the murid genera that characterize the Early Miocene Anatolian assemblages may

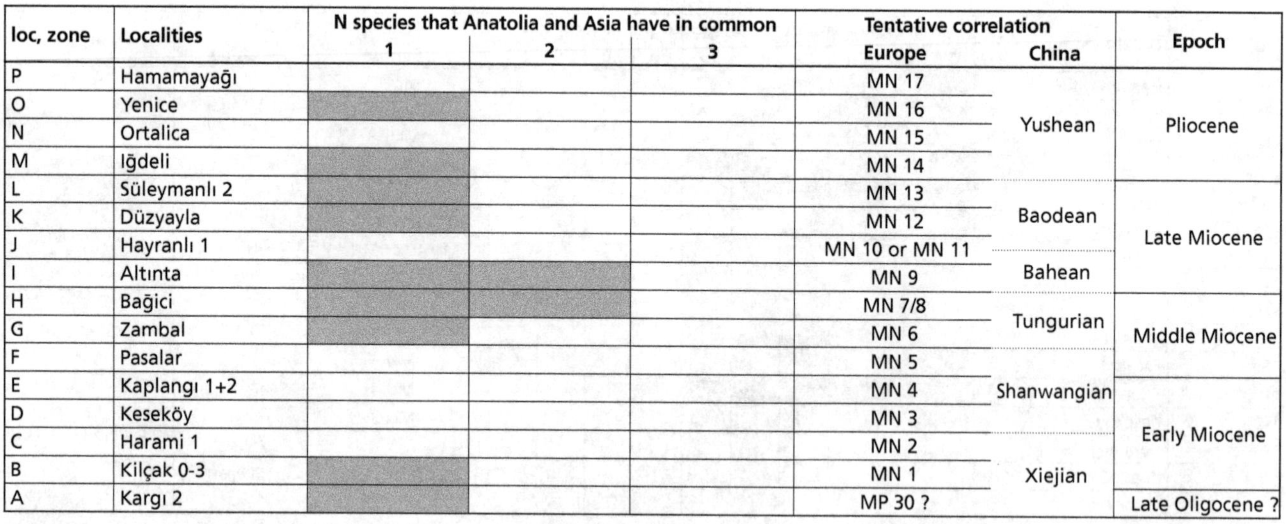

loc, zone	Localities	N species that Anatolia and Asia have in common			Tentative correlation		Epoch
		1	2	3	Europe	China	
P	Hamamayağı				MN 17		
O	Yenice				MN 16	Yushean	Pliocene
N	Ortalica				MN 15		
M	Iğdeli				MN 14		
L	Süleymanlı 2				MN 13	Baodean	Late Miocene
K	Düzyayla				MN 12		
J	Hayranlı 1				MN 10 or MN 11		
I	Altınta				MN 9	Bahean	
H	Bağici				MN 7/8	Tungurian	Middle Miocene
G	Zambal				MN 6		
F	Pasalar				MN 5		
E	Kaplangı 1+2				MN 4	Shanwangian	Early Miocene
D	Keseköy				MN 3		
C	Harami 1				MN 2		
B	Kilçak 0-3				MN 1	Xiejian	
A	Kargı 2				MP 30 ?		Late Oligocene ?

Figure 26.5 Graph showing the number of species per locality that are known from Anatolia as well as from central Asia.

well indicate that these genera arrived as immigrants from the Iranian block (Popov et al. 2004; Wessels 2009b), causing a dramatic turnover of the Late Oligocene murid fauna of Anatolia.

The second immigration peak occurs in the assemblage of Keseköy, which contains the first, and so far oldest, record of *Megacricetodon* and the spalacines (*Debruijnia*) as well as the enigmatic myocricetodontine *Vallaris zappai*. Our collection from Keseköy is very large (over 2000 specimens), but the high percentage of immigrants in this locality is probably not due to sample size, because the newcomers *Vallaris* and *Debruijnia* are quite common and may therefore be expected to be present in small samples also. Supposing that our correlation of the Keseköy fauna to MN 3 is correct, the immigrants in this assemblage are again known from neither central Asia nor Europe during this time slice (see figures 26.3 and 26.5). Since the younger fossil record of the spalacines suggests that that they had an eastern Mediterranean origin, it is against our expectations that there is no potential ancestor known from older levels in Anatolia.

The third immigration peak in our sequence occurs in the localities Bağiçi, Altıntaş, and Hayranlı, which are tentatively correlated to the late Middle Miocene/early Late Miocene (MN 7/8 to MN 10) interval. The older of these localities (Bağiçi) contains the first true *Myocricetodon* and *Cricetulodon* in Anatolia, the next locality (Altıntaş) contains the newcomers *Calomyscus*, *Abudhabia*, and *Progonomys*, and the third locality (Hayranlı) contains the first *Pseudomeriones* and *Rhinocerodon*. Figures 26.2, 26.3, and 26.5 show that, in contrast to earlier

periods of enhanced immigration, there are a number of species that have been found all over Eurasia during this time slice. It is of interest that the local arrival of *Cricetulodon* predates the "Hipparion" event and that the first record of *Calomyscus* in our sequence is much older than elsewhere. In contradistinction to previous suggestions, these early occurrences indicate that these genera do not have an African origin, but an Asiatic one.

The similarity in composition of the Muridae associations (at the species level) from Anatolia and Asia during the late Middle Miocene/early Late Miocene (see figure 26.3) is probably more pronounced than suggested by this graph, because it shows only identical species, not resemblances. *Abudhabia* n.sp. from Altıntaş, for instance, is very similar to *A. baheensis*, and it is expected that direct comparison of the *Rhinocerodon* material from Hayranlı and Düzyayla with the species described by Zazhigin (2003) will reveal greater similarity. Unfortunately, we were unable to express the similarity in composition of the Anatolian and Asiatic associations (see figure 26.3) in percentages as we did for Anatolia and Europe (see figure 26.5), because our knowledge of the Murinae from Asia in this time slice is not good enough to do so.

The fourth immigration peak is seen in the locality Iğdeli with the appearance of the first true vole, *Promimomys*, and the enigmatic "*Cricetus*" *lophidens*. By this time, the murine fauna of Anatolia had become very similar to the one of Europe, in particular that from Greece (see figure 26.5). The graph showing the number of species that Anatolia and Asia have in common suggests that the

Pliocene murid assemblages of these areas are rather different. However, this may in part be an artifact due to the lack of studies comparing the cricetines from Asia and Anatolia and to the relatively poor Pliocene rodent record from Anatolia (distribution of all sites given by figure 26.1 and appendix 26.5).

CONCLUSION

The Neogene record shows that the interaction of the Anatolian Dipodidae and Muridae, with Europe and Asia follows different patterns. The history of the Dipodidae of Anatolia is relatively simple, because all species identified from that area are known from central Asia and presumably originated there. Their occurrences in Anatolia through time are not correlated with the expected increase in similarity of the associated Muridae. The only dipodid that reached Europe during the Neogene is *Eozapus*. The occurrences of some of the Asiatic species in Anatolia provide only a weak basis for the correlation of the Anatolian biozones with the Chinese mammal ages.

The Neogene history of the Anatolian Muridae is quite complex. The latest Oligocene/earliest Miocene assemblages contain seven genera that are neither known from older levels nor from other areas. Our working hypothesis is that these genera are immigrants from the Iranian block (see paleogeographical map by Popov et al. 2004). Correlation of the Early Miocene zones A–D with the European MN scheme is therefore at best a sophisticated guess. However, zone E contains four murid species that are known from European faunas assigned to MN 4. The Anatolian associations F–H and the European Middle Miocene ones do not share any characteristic murid.

The first murid species that Asia and Anatolia have in common during the Neogene is *Myocricetodon liui* in zone H, a level that we tentatively correlate with MN 7/8 in Europe and the Late Tungurian of China. The entry of the murine *Progonomys cathalai* in zone I occurs with the European *Cricetulodon hartenbergeri*, the Asiatic dipodid *Protalactaga minor*, and a new species of *Abudhabia* that is similar to *A. baheensis*. This association allows the correlation of zone I with the Bahean of China and MN 9 in Europe. Similarly, the occurrence of *Micromys bendai/tedfordi* in zone M suggests a correlation of this zone with the early Yushean of China and MN 14-15 in Europe. The younger zones J–P show a gradually increasing content of European murids, indicating more fauna exchange with Anatolia during the Late Miocene and Pliocene. It is peculiar that the eastern Mediterranean and western Mediterranean Cricetodontini remained different during this period, suggesting the presence of a barrier (ecological?) between these areas.

ACKNOWLEDGMENTS

The incentive to review the Anatolian record of the Muridae and Dipodidae was given by the organization of the "Neogene Terrestrial Biostratigraphy and Chronology in Asia" symposium (Beijing, June 8–14, 2009). The second author (EÜ) expresses her gratitude for receiving financial support enabling her to participate in that workshop, and she thanks Nihal Çınar and Bahar Beşter (MTA) for their help with the PowerPoint presentation. We are grateful for the sustained support of the General Directorate of the Mineral Research and Exploration institute of Turkey (MTA) for our field program during the last 30 years. The first author (HdB) acknowledges the hospitality and the facilities of the Faculty of Geosciences of Utrecht University. Many geologists working with the MTA have contributed to make the collections that form the basis of this overview. In thanking Gerçek Saraç (MTA) for his never failing interest in the biostratigraphy of the continental basins of Turkey and for his quality as a friend and fossil hunter, we gratefully acknowledge all those colleagues who participated over the years.

The good council and discussion received from Gudrun Daxner-Höck (Natural History Museum, Vienna) on the identification of the *Heterosminthus* remains from Anatolia is very much appreciated. We thank Hans Meeldijk (Department of Cell Biology, Utrecht University) and Tilly Bouten (Department of Earth Sciences, Utrecht University) for making the larger part of the SEM pictures. The technical knowhow and help of Fred Trappenburg (GeoMedia, Utrecht University) in making the plates and figures has been an essential contribution.

REFERENCES

Text

de Bruijn, H., E. Ünay, L. W. van den Hoek Ostende, and Saraç. 1992. A new association of small mammals from the lowermost Miocene of Anatolia. *Geobios* 25(5):651–670.
de Bruijn, H., E. Ünay, and A. Yilmaz. 2003. A rodent assemblage from the Eo/Oligocene boundary interval near Süngülü, Lesser Caucasus, Turkey. In *En torno a Fósiles de Mamíferos: Datación, evolución y Paleoambiente*, ed. N. López-Martínez, P. Peláez-Campomanes,

and M. Hernández Fernández. *Coloquios de Paleontología* Volumen extraordinario 1:47–76.

Hordijk, K. and H. de Bruijn. 2009. The succession of rodent faunas from the Mio/Pliocene lacustrine deposits of the Florina-Ptolemais-Servia Basin (Greece). *Hellenic Journal of Geosciences* 44:21–103.

Krijgsman, W., C. E. Duermeijer, C. G. Langereis, H. de Bruijn, G. Saraç, and P. A. M. Andriessen. 1996. Magnetic polarity stratigraphy of the Late Oligocene to Middle Miocene mammal-bearing continental deposits in Central Anatolia (Turkey). *Newsletters on Stratigraphy* 34:13–29.

Lindsay, E. H. 1996. A new eumyarionine cricetid from Pakistan. *Acta Zoologica Cracoviensia* 39(1):279–288.

Lopatin, A. V. 1996. New Early Miocene Zapodidae (Rodentia, Mammalia) from the Aral Formation of the Altyn Schokysu locality (North Aral region). *Paleontological Journal* 33:182–191.

Lourens, L., F. Hilgen, N. J. Shackleton, J. Laskar, and D. Wilson. 2004. The Neogene Period. In *A Geologic Time Scale 2004*, ed. F. M. Gradstein, J. G. Ogg and A. G. Smith, pp. 409–440. Cambridge: Cambridge University Press.

Popov, S. V., F. Rögl, A. Y. Rozanov, F. F. Steiniger, I. G. Shcherba, and M. Kovac. 2004. Lithological-paleogeographic maps of Paratethys. *Courier Forschungsinstitut Senckenberg* 250:1–46.

Qiu, Z.-d., S.-h. Zheng, and Z.-q. Zhang. 2004. Gerbillids from the Late Miocene Bahe Formation, Lantian, Shaanxi. *Vertebrata PalAsiatica* 42(3):193–204.

Sarıca, N. and S. Sen. 2003. Spalacidae (Rodentia). In *Geology and Paleontology of the Miocene Sinap Formation, Turkey*, ed. M. Fortelius, J. Kappelman and R. L. Bernor, pp. 141–162. New York: Columbia University Press.

Theocharopoulos, C. D. 2000. Late Oligocene–Middle Miocene Democricetodon, Spanocricetodon and Karydomys n. gen. from the eastern Mediterranean area. *Gaia* 8:1–92, 11 plates.

Ünay, E. and H. de Bruijn. 1984. On some Neogene rodent assemblages from both sides of the Dardanelles, Turkey. *Newsletters on Stratigraphy* 13(3):119–132.

Ünay, E. and H. de Bruijn. 1987. Middle Oligocene to Early Miocene rodent assemblages from Turkey, a preliminary report. In *International Symposium on Mammaliam Biostratigraphy and Paleoecology of the European Paleogene*, ed. N. Schmidt-Kittler, pp. 203–210. *Münchner Geowissenschaftliche Abhandlugen*, Series A 10.

Ünay, E. and H. de Bruijn. 1998. Plio-Pleistocene rodents and lagomorphs from Anatolia. *Mededelingen Nederlands Instituut voor Toegepaste Geowetenschappen TNO* 60:431–466.

Ünay, E., H. de Bruijn, and G. Saraç. 2003a. A preliminary zonation of the continental Neogene of Anatolia based on rodents. In *Distribution and Migration of Tertiary Mammals in Eurasia*, ed. J. W. F. Reumer and W. Wessels, pp. 539–547. DEINSEA 10.

Ünay, E., H. de Bruijn, and G. Saraç. 2003b. The Oligocene rodent record of Anatolia: A review. In *Distribution and Migration of Tertiary Mammals in Eurasia*, ed. J. W. F. Reumer and W. Wessels, pp. 531–537. DEINSEA 10.

Ünay-Bayraktar, E. 1989. *Rodents from the Middle Oligocene of Turkish Thrace*. Utrecht Micropaleontological Bulletins, Special Publication.

Wessels, W. 2009a. *Progonomys* from the Kütahya area (Turkey). In *Miocene Rodent Evolution and Migration: Muroidea from Pakistan, Turkey and Northern Africa. Geologica Ultraiectina* 307:89–125.

Wessels, W. 2009b. Miocene rodents, faunal exchange and migration routes between Eurasia and Africa. In *Miocene Rodent Evolution and Migration: Muroidea from Pakistan, Turkey and Northern Africa. Geologica Ultraiectina* 307:229–252.

Wessels, W. and H. de Bruijn 2001. Rhizomyidae from the Lower Manchar Formation (Miocene, Pakistan). *Annals of Carnegie Museum* 70(2):143–168.

Zazhigin, V. S. 2003. New genus of Cricetodontinae (Rodentia, Cricetidae) from the Late Miocene of Kazakhstan. *Russian Journal of Theriology* 2(2):65–69.

Taxon Authorities

Aguilar, J.-P., L. D. Brandy, and L. Thaler. 1984. Les rongeurs de Salobrena (Sud de l'Espagne) et le problème de la migration Messinienne. *Paléobiologie Continentale* 14(2):3–17.

Argyropulo, A. I. and I. G. Pidoplichka. 1939. Recovery of a representative of Muridae (Glires, Mammalia) in Tertiary deposits of the USSR. *Comptes Rendus (Doklady) de l'Academie des Sciences de l'USSR* 23(2):209–212.

Bachmayer, F. and R. W. Wilson. 1970. Small mammals from the Kohfidisch fissures of Burgenland, Austria. *Annalen des Naturhistorischen Museum Wien* 83:351–386.

de Bruijn, H. 1976. Vallesian and Turolian rodents from Biotia, Attica and Rhodes (Greece). *Koninklijke Nederlandse Akademie van Wetenschappen Proceedings B* 79(5):361–384.

de Bruijn, H. 1993. Early Miocene rodent faunas from the eastern Mediterranean area. Part 3. The genera Deperetomys and Cricetodon with a discussion of the evolutionary history of the Cricetodontini. *Proceedings of the Koninklijke Nederlandse Akademie van Wetenschappen* 96(2):15–216.

de Bruijn, H., M. R. Dawson, and P. Mein. 1970. Upper Pliocene Rodentia, Lagomorpha and Insectivora (Mammalia) from the isle of Rhodes (Greece). Part 1–3. *Koninklijke Nederlandse Akademie van Wetenschappen Proceedings B* 73(5):535–584.

de Bruijn, H., and G. Saraç. 1991. Early Miocene rodent faunas from the eastern Mediterranean area. Part 1: The genus *Eumyarion*. *Koninklijke Nederlandse Akademie van Wetenschappen Proceedings B* 94(1):1–36.

de Bruijn, H., and G. Saraç. 1992. Early Miocene rodent faunas from the eastern Mediterranean area. Part 2: *Mirabella* (Paracricetodontinae, Muroidea). *Koninklijke Nederlandse Akademie van Wetenschappen Proceedings B* 95(1):25–40.

de Bruijn, H., E. Ünay, G. Saraç, and G. Klein Hofmeyer. 1987. An unusual new eucricetodontine from the Lower Miocene of the eastern Mediterranean *Koninklijke Nederlandse Akademie van Wetenschappen Proceedings B* 90(2):119–132.

de Bruijn, H., E. Ünay, L. W. van den Hoek Ostende, and G. Saraç. 1992. A new association of small mammals from the lowermost Lower Miocene of central Anatolia. *Geobios* 25(5):651–70.

de Bruijn, H., L. W. van den Hoek Ostende, and S. K. Donovan. 2007. *Mirrabella*, a new name for the genus *Mirabella* de Bruijn et al., 1987 (Mammalia); preoccupied by *Mirabella* Emeljanov, 1982 (Insecta). *Contributions to Zoology* 76(4):279–280.

de Bruijn, H. and A. J. van der Meulen. 1975. The Early Pleistocene rodents from Tourkobounia-1. *Koninklijke Nederlandse Akademie van Wetenschappen Proceedings B* 78(4):314–338.

de Bruijn, H. and W. von Koenigswald. 1994. Early Miocene rodent faunas from the eastern Mediterranean area. Part 5: The genus *Enginia* (Muridae), with a discussion of the structure of the incisor enamel. *Koninklijke Nederlandse Akademie van Wetenschappen Proceedings B* 97(4):381–405.

de Bruijn, H. and P. J. Whybrow. 1994. A Late Miocene rodent fauna from the Baynunah Formation, Emirate of Abudhabi, United Arab Emirates. *Koninklijke Nederlandse Akademie van Wetenschappen Proceedings B* 97(4):407–422.

Fahlbusch, V. 1964. Die cricetiden (Mammalia) der Oberen Süsswassermolasse Bayerns. *Abhandlungen der Bayerischen Akademie der Wissenschaften* 118:1–136.

Fahlbusch, V. 1966. Cricetidae (Rodentia, Mammalia) aus der mitteImiozänen Spaltenfüllung Erkertshofen bei Eichstätt. *Mitteilungen der Bayerischen Staatssammlung für Paläontologie und Historische Geologie* 6:109–131.

Fahlbusch, V. 1992. The Neogene mammalian faunas of Ertemte and Harr Obo in Inner Mongolia (Nei Mongol), China. Part 10, *Eozapus* (Rodentia). *Senckenbergiana Lethaea* 72:199–217.

Forsyth Major, C. I. 1902. Exhibition of, and remarks upon some jaws and teeth of Pliocene voles (*Mimomys* gen. nov.). *Proceedings of the Zoological Society of London* 1:102–107.

Freudenthal, M. 1963. Entwicklungsstufen der miozänen Cricetodontinae (Mammalia, Rodentia) Mittelspaniens und ihre stratigraphische Bedeutung. *Beaufortia* 10:51–157.

Freudenthal, M. 1967. On the mammalian fauna of the *Hipparion*-beds in the Calatayud-Teruel basin. Part 3: *Democricetodon* and *Rotundomys* (Rodentia). *Koninklijke Nederlandse Akademie van Wetenschappen Proceedings B* 70(3):298–315.

Freudenthal, M. 1970. A new *Ruscinomys* (Mammalia, Rodentia) from the Tertiary (Pikermian) of Samos, Greece. *American Museum Novitates* (2402):1–10.

Hartenberger, J. L. 1965. Les Cricetidae (Rodentia) de Can Llobateres (Neogene d'Espagne). *Bulletin de la Société Géologique de France* 7(7):487–498.

Jacobs, L. L. 1978. Fossil rodents (Rhizomyidae & Muridae) from the Neogene Siwalik deposits, Pakistan. *Museum of Northern Arizona Press* 52:1–95.

Kaup, J. 1829. Entwicklungsgeschichte und natürlich System der europäische Tierwelt 1:154.

Kordos, L. 1986. Upper Miocene hamsters (Cricetidae, Mammalia) of Hasznos and Szentendre: A taxonomic and stratigraphic study. *Magyar Állami Földtani Intézet Évi Jelentéze* 1984: 523–553 (in Hungarian).

Kormos, T. 1930. Neue Wühlmäuse aus dem Oberpliozän von Püspökfürdö. *Neues Jahrbuch für Mineralogie, Geologie und Paläontologie. Beilageband B* 69:323–346.

Kormos, T. 1932. Neue Pliozäne Nagetiere aus der Moldau. *Paläontologische Zeitung* 14:193–200.

Kowalski, K. 1958. An early Pleistocene fauna of small mammals from the Kadzielna Hill in Kielce (Poland). *Acta Paleontologica Polonica* 1:1–47.

Kretzoi, M. 1955. *Promimomys cor* n.g. n.sp. ein altertümlicher arvicolide aus dem Ungarischen Unterpliozän. *Acta Geologica Academiae Hungaricae* 3:89–94.

Kretzoi, M. 1959. Insectivoren, Nagetiere und Lagomorphen der jüngstpliozännen Fauna von Csanóta im Villanyer Gebirge (Südungarn). *Vertebrata Hungarica* 11:155–193.

Kuss, S. E. and G. Storch. 1978. Eine Säugetierfauna (Mammalia: Artiodactyla, Rodentia) des älteren Pleistozäns von der Insel Kalymnos (Dodekanes, Griechenland). *Neues Jahrbuch für Geologie und Paläontologie. Monatshefte* 1978:206–227.

Lartet, E. 1851. Notice sur la colline de Sansan. *Portes, Auch* Volume 1(A) 46pp.

Lavocat, R. 1952. Sur une faune de mammifères Mioènes découverte à Beni-Mellal (Atlas Marocain). *Comtes Rendus de l' Ácademie de Sciences de Paris* 235:189–191.

Li, C.-k. 1977. A new cricetodont rodent of Fangshan, Nanking. *Vertebrata PalAsiatica* 15(1):67–75.

Martín Suáres, E. and M. Freudenthal. 1994. *Castromys*, a new genus of Muridae (Rodentia) from the Late Miocene of Spain. *Scripta Geologica* 106:11–33.

Méhely, L. V. 1908. Species generis *Spalax*. Die arten der Blindmäuse in systematischer und phylogenetischer Beziehung. *Mathematische und naturwissenschaftliche Berichte aus Ungarn* 28:1–390.

Méhely, L. V. 1914. Fibrinae Hungariae. *Annales Historico-Naturalis Musei Nationalis Hungarici* 12:157–243.

Mein, P. 1958. Les mammifères de la faune sidérolitique de Vieux Collonges. *Nouvelles Archives du Muséum d'Histoire Naturelle de Lyon* 5:1–122.

Mein, P. and M. Freudenthal. 1971. Les cricetidae de Vieux-Collonges. *Scripta Geologica* Part I, 5:1–51.

Michaux, J. J. 1969. Muridae (Rodentia) du Pliocène supérieur d'Espagne et du midi de la France. *Palaeovertebrata* 3:1–25.

Milne Edwards, A. 1867. *Recherches pour servir à l'histoire des Mammifères* 71:375.

Nehring, A. 1898a. Die gruppe der *Mesocricetus* arten. *Archiv für Naturgeschichte* 1:373–392.

Nehring, A. 1898b. Über *Dolomys* nov. gen. *Zoologischer Anzeiger* 21:13–16.

Peláez Campomanes, P., and R. Daams. 2002. Middle Miocene rodents from Pasalar, Anatolia, Turkey. *Acta Palaeontologica Polonica* 47:125–132.

Preble, E. A. 1899. Revision of the jumping mice of the genus *Zapus*. *North American Fauna* 15:6–39.

Qiu, Z.d., S.-h. Zheng, and Z.-q. Zhang. 2004a. Gerbillids from the Late Miocene Bahe Formation, Lantian, Shaanxi. *Vertebrata PalAsiatica* 42:193–204.

Qiu, Z.-d., S.-h. Zheng, and Z.-q. Zhang. 2004b. Murids from the Late Miocene Bahe Formation, Lantian, Shaanxi. *Vertebrata PalAsiatica* 42(1):67–76.

Ruiz Bustos, A., C. Sese, C. Dabrio, J. A. Penna, and J. Padial. 1984. Micromammíferos del nuevo yacimiento del Plioceno inferior de Gorafe A (Depressíon de Guadix-Baza, Granada). *Estudios Geologicos* 40:231–241.

Rummel, M. 1998. Die Criceiden aus dem Mittel- und Obermiozän der Türkei. *Documenta Naturae* 123:1–300.

Saríca, N., and S. Sen. 2003. Spalacidae (Rodentia). In *Geology and Paleontology of the Miocene Sinap Formation, Turkey*, ed. M. Fortelius, J. Kappelman, and S. Sen, pp. 141–162. New York: Columbia University Press.

Schaub, S. 1920. *Melissiodon* nov. gen., ein bisher übersehener Oligozäne Muride. *Senckenbergiana* 2:43–47.

Schaub, S. 1925. Die hamsterartigen Nagetiere des Tertiärs und ihre lebende Verwandten. *Abhandlungen der Schweizerischen Palaeontologischen Gesellschaft* 45:1–114.

Schaub, S. 1930a. Fossile Sicistinae. *Eclogae Geologicae Helvetiae* 23(2):616–637.

Schaub, S. 1930b. Quartäre und jungtertiäre Hamster. *Abhandlungen Sweizerischen Palaeontologisches Gesellschaft* 49:1–48.

Schaub, S. 1934. Über einige fossile Simplicidentaten aus China und der Mongolei. *Abhandlungen der Schweizerischen Palaeontologischen Gesellschaft* 54:1–40.

Schaub, S. 1938. Tertiäre und Quartäre Murinae. *Ahandlungen der Schweizeriischen Palaeontologischen Gesellschaft* 61:1–39.

Schaub, S. and H. Zapfe. 1953. Die Fauna der miozänen Spaltenfüllung von Neudorf an der March. Simplicidentata. *Sitzungsberichte der Österreichschen Akademie der Wissenschaften, Mathematisch-Naturwissenschaftliche Klasse* 162(1):1–114.

Sen, S., J. J. Jaeger, M. Dalfes, J. M. Mazin, and H. Bocherens. 1989. Découverte d'une faune de petits mammifères Pliocènes en Anatolie occidentale. *Comptes Rendus de l'Academie des Sciences, Série 2: mécanique, physique, chimi, sciences de l'univers, sciences de la terre* 309(2):1729–1734.

Storch, G. and T. Dahlmann. 1995. Muridae (Rodentia, Mammalia). In *The Vertebrate Locality Maramena (Macedonia, Greece) at the Turolian-Ruscinian Boundary (Neogene)*, ed. N. Schmidt-Kittler, pp. 121–132. *Münchner Geowissenschaftliche Abhandlugen*, series A 28.

Teilhard de Chardin, P. 1926. Description des mammiféres tertiaires de Chine et de Mongolie. *Annales de Paléontologie* 15:1–51.

Thaler, L. 1966. Les rongeurs fossiles du Bas-Languedoc dans leurs rapports avec l'histoire des faunes et la stratigraphie d'Europe. *Mémoires du Muséum National d'Histoire Naturelle, Serie C* 17:1–195.

Theocharopoulos, C. D. 2000. Late Oligocene-Middle Miocene *Democricetodon, Spanocricetodon* and *Karydomys* n. gen. from the eastern Mediterranean area. *Gaia* 8:1–92.

Thomas, O. 1905. On a collection of mammals from Persia and Armenia presented to the British Museum by Col. A. C. Bailward, *Proceedings of the Zoological Society*, London 2:519–527.

Tobien, H. 1978. New species of Cricetodontini (Rodentia, Mammalia) from the Miocene of Turkey. *Mainzer Geowissenschaftliche Mitteilungen* 6:209–219.

Ünay, E. 1978. *Pliospalax primitivus* n. sp. (Rodentia, Mammalia) and *Anomalomys gaudryi* Gaillard from the *Anchitherium* fauna of Sariçay. *Bulletin of the Geological Society of Turkey* 21:121–128.

Ünay, E. 1980. The Cricetodontini (Rodentia) from the Bayraktepe section (Çanakkale, Turkey). *Koninklijke Nederlandse Akademie van Wetenschappen Proceedings B* 83(4):399–418.

Ünay, E. 1990. A new species of *Pliospalax* (Rodentia, Mammalia) from the Middle Miocene of Paşalar, Turkey. *Journal of Human Evolution* 19(4/5):4–53.

Ünay, E. 1996. On fossil Spalacidae (rodentia). In *The Evolution of Western Eurasian Neogene Mammal Faunas*, ed. R. L. Bernor, V. Fahlbusch, and H. W. Mittmann, pp. 246–252. New York: Columbia University Press.

van de Weerd, A. 1976. *Rodent Faunas of the Mio-Pliocene Continental Sediments of the Teruel-Alfambra Region, Spain*. Utrecht Micropaleontological Bulletins Special Publication 2:5–181.

van de Weerd, A. 1979. Early Ruscinian rodents and lagomorphs (Mammalia) from the lignites of Ptolemais (Macedonia, Greece). *Koninklijke Akademie van Wetenschappen, Proceedings B* 82(2):127–170.

Wang, B.-y. and Z.-x. Qiu. 2000. Dipodidae (Rodentia, Mammalia) from the lower member of the Xianshuihe Formation in Lanzhou Basin, China. *Vertebrata PalAsiatica* 38:12–35.

Wessels, W., C. D. Theocharopoulos, H. de Bruijn, and E. Ünay. 2001. Myocricetodontinae and Megacricetodontinae (Rodentia) from the lower Miocene of NW Anatolia. In *Papers in Paleontology Honoring Prof. Dr. Oldřich Fejfar*, ed. I. Horáček and J. Míkovský, pp. 371–388. *Lynx* 32.

Wu, W.-y. and L. J. Flynn. 1992. New murid rodents from the Late Cenozoic of the Yushe Basin, Shaanxi. *VertebrataPalAsiatica* 30(1):17–38.

Young, C. C. 1927. Fossile Nagetiere aus Nord-China. *Paleontologia Sinica* 3(3):1–82.

Zazhigin, V. S. 2003. New genus of Cricetodontinae (Rodentia, Cricetidae) from the Late Miocene of Kazakhstan. *Russian Journal of Theriology* 2(2):65–69.

Zazhigin, V. S. and A. V. Lopatin. 2000. The history of the Dipodidae (Rodentia, Mammalia) in the Miocene of Asia: *Heterosminthus* (Lophocricetinae). *Paleontological Journal* 34(3): 319–332.

Zeng, S. 1982. Middle Pliocene micromammals from the Tianzhu loc. 80007 (Gansu province). *Vertebrata PalAsiatica* 20(2): 138–147.

APPENDIX 1

Alphabetical List of the Genera and Species of Muridae (with Taxon Authorities) Identified in the Localities Selected

Genus	Species	Genus	Species
Abudhabia de Bruijn and Whybrow 1994		*Cricetulodon* Hartenberger 1965	
Abudhabia	n. sp.	*Cricetulodon*	*hartenbergeri* Freudenthal 1967
Allocricetus Schaub 1930		*Cricetulodon*	*sabadellensis* Hartenberger 1965
Allocricetus	*bursae* Schaub 1930	*Cricetulus* Milne Edwards 1867	
Allophaiomys Kormos 1930		*Cricetulus*	sp.
Allophaiomys	*deucalion* Kretzoi 1959	*"Cricetus"* in: de Bruijn et al. 1970	
Apodemus Kaup 1829		*"Cricetus"*	*lophidens* de Bruijn et al. 1970
Apodemus	*dominans* Kretzoi 1959	*Debruijnia* Ünay 1996	
Apodemus	*gorafensis* Ruiz Bustos et al. 1984	*Debruijnia*	*arpati* Ünay 1996
Apodemus	*gudrunae* van de Weerd 1976	*Democricetodon* Fahlbusch 1964	
Byzantinia de Bruijn 1976			
Byzantinia	*hellenicus* (Freudenthal 1970)	*Democricetodon*	*anatolicus* Theocharopoulos 2000
Byzantinia	*ozansoyi* Ünay 1980	*Democricetodon*	*brevis* Schaub 1925
Byzantinia	*pikermiensis* de Bruijn 1976	*Democricetodon*	*doukasi* Theocharopoulos 2000
Byzantinia	*uenayae* Rummel 1998	*Democricetodon*	*franconicus* Fahlbusch 1966
Calomyscus Thomas 1905		*Deperetomys* Mein and Freudenthal 1971	
Calomyscus	*delicatus* Aguilar et al. 1984		
Castromys Martín Suárez and Freudenthal 1994		*Deperetomys*	*anatolicus* de Bruijn et al. 1973
		Deperetomys	*intermedius* de Bruijn et al. 1987
Castromys	*littoralis* Martín Suárez and Freudenthal 1994	*Enginia* de Bruijn and von Koenigswald 1994	
Cricetodon Lartet 1851		*Enginia*	*beckerplateni* de Bruijn and von Koenigswald 1994
Cricetodon	*hungaricus* (Kordos 1986)		
Cricetodon	*kasapligili* de Bruijn et al. 1993		
Cricetodon	*pasalarensis* (Tobien 1978)	*Enginia*	*djanpolati* de Bruijn and von Koenigswald 1994
Cricetodon	sp. 1 in de Bruijn et al. 1993		
Cricetodon	*versteegi* de Bruijn et al. 1993		

(continued)

Genus	Species	Genus	Species
Enginia	*gertcheki* de Bruijn and von Koenigswald 1994	*Mirrabella*	*tuberosa* (de Bruijn et al. 1987)
Eumyarion Thaler 1966		*Muhsinia* de Bruijn et al. 1992	
Eumyarion	*carbonicus* de Bruijn and Saraç 1991	*Muhsinia*	n. sp.
		Myocricetodon Lavocat 1952	
Eumyarion	*intercentralis* de Bruijn and Saraç 1991	*Myocricetodon*	*liui* Qiu et al. 2004
		Occitanomys Michaux 1969	
Eumyarion	*microps* de Bruijn and Saraç 1991	*Occitanomys*	*debruijni* (Sen et al. 1989)
Eumyarion	*montanus* de Bruijn and Saraç 1991	*Orientalomys* de Bruijn and van der Meulen 1975	
Eumyarion	*weinfurteri* Schaub and Zapfe 1953	*Orientalomys*	*similis* (Argyropulo and Pidoplichka 1939)
Hansdebruijnia Storch and Dahlmann 1995		*Parapodemus* Schaub 1938	
Hansdebruijnia	*neutrum* (de Bruijn 1976)	*Parapodemus*	*lugdunensis* Schaub 1938
Kalymnomys Kuss and Storch 1978		*Pliomys* von Méhely 1914	
Kalymnomys	*major* Kuss and Storch 1978	*Pliomys*	*graecus* de Bruijn and van der Meulen 1975
Karnimata Jacobs 1978		*Pliospalax* Kormos 1932	
"*Karnimata*"	*provocator* (de Bruijn 1976)	*Pliospalax*	*marmarensis* Ünay 1990
Megacricetodon Fahlbusch 1964		*Pliospalax*	*primitivus* Ünay 1978
Megacricetodon	*andrewsi* Peláez Campomanes and Daams 2002	*Pliospalax*	*tourkobouniensis* de Bruijn and van der Meulen 1975
Megacricetodon	*collongensis* (Mein 1958)	*Progonomys* Schaub 1920	
Megacricetodon	*gregarius* Schaub 1925	*Progonomys*	*cathalai* Schaub 1938
Megacricetodon	*primitivus* (Freudenthal 1963)	*Progonomys*	*sinensis* Qiu, Zheng, and Zhang 2004
Melissiodon Schaub 1920		*Promimomys* Kretzoi 1955	
Melissiodon	sp.	*Promimomys*	*insuliferus* Kowalski 1958
Mesocricetus Nehring 1898		*Prospalax* von Méhely 1908	
Mesocricetus	*primitivus* de Bruijn et al. 1970	*Prospalax*	sp.
Meteamys de Bruijn et al. 1992		*Pseudomeriones* Schaub 1934	
Meteamys	*alpani* de Bruijn et al. 1992	*Pseudomeriones*	*abbreviatus* (Teilhard de Chardin 1926)
Micromys Dehne 1841			
Micromys	*bendai* van de Weerd 1979	*Rhinocerodon* Zazhigin 2003	
Micromys	*tedfordi* Wu and Flynn 1992	*Rhinocerodon*	sp.
Mimomys Forsyth Major 1902		*Spanocricetodon* Li 1977	
Mimomys	*gracilis* (Kretzoi 1959)	*Spanocricetodon*	*sinuosus* Theocharopoulos 2000
Mirrabella de Bruijn et al. 2007		*Vallaris* Wessels et al. 2001	
Mirrabella	*anatolica* (de Bruijn and Saraç 1992)	*Vallaris*	*zappai* Wessels et al. 2001
Mirrabella	*crenulata* (de Bruijn and Saraç 1992)		

APPENDIX 2

Alphabetical List of the Genera and Species of Dipodidae (with Taxon Authorities) Identified in the Localities Selected

Genus	Species
Eozapus Preble 1899	
Eozapus	*intermedius* Bachmayer and Wilson 1970
Eozapus	*similis* Fahlbusch 1992
Heterosminthus Schaub 1930	
Heterosminthus	*gansus* Zeng 1982
Heterosminthus	*mongoliensis* Zazhigin and Lopatin 2000
Heterosminthus	*nanus* Zazhigin and Lopatin 2000
Heterosminthus	*lanzhouensis* Wang and Qiu 2000
Heterosminthus	*firmus* Zazhigin and Lopatin 2000
Protalactaga Young 1927	
Protalactaga	*minor* Zheng 1982

APPENDIX 3

Alphabetical List of the Asiatic Species Recognized in Anatolia and Their Type Localities and Ages

Genus and Species	Type Locality	Mammal Age	Epoch
Abudhabia baheensis	Shaanxi loc. 12	Bahean	early Late Miocene
Eozapus similis	Ertemte 2	Baodean	late Late Miocene
Heterosminthus firmus	Ayaguz		~Oligo/Miocene transition
Heterosminthus gansus	Tianzhu loc. 80007	Baodean	late Late Miocene
Heterosminthus lanzhouensis	Shangxigou GL9601		~Oligo/Miocene transition
Heterosminthus mongoliensis	Ulan-Tologoi		early Middle Miocene
Heterosminthus nanus	Batpakunde		Early Miocene
Myocricetodon liui	Shaanxi loc. 12	Bahean	early Late Miocene
Orientalomys similis	Odessa		late Late Miocene
Progonomys sinensis	Lantian loc. 19	Bahean	early Late Miocene
Protalactaga minor	Tianzhu loc. 80007	Baodean	late Late Miocene
Pseudomeriones abbreviatus	King Yan Fou	Baodean	late Late Miocene

APPENDIX 4

Alphabetical List of the European Species Recognized in Anatolia and Their Type Localities and Ages

Genus and Species	Type Locality	MN Zone	Mammal Age	Epoch
Allocricetus bursae	Csarnota	MN 15	Ruscinian	Late Pliocene
Allophaiomys deucalion	Villany 5		Biharian	Early Pleistocene
Anomalomys aliveriensis	Aliveri	MN 4	Orleanian	Early Miocene
Apodemus gorafensis	Gorafe	MN 13	Turolian	Late Miocene
Apodemus gudrunae	Valdecebro 3	MN 13	Turolian	Late Miocene
Byzantinia hellenicus	Samos 4	MN 11	Turolian	Late Miocene
Byzantinia pikermiensis	Pikermi 4	MN 12	Turolian	Late Miocene
Calomyscus delicatus	Salobrena	MN 13	Turolian	Late Miocene
Castromys littoralis	Crevillente 17	MN 12	Turolian	Late Miocene
Cricetodon hungaricus	Hasznos	MN 6	Astaracian	Middle Miocene
Cricetulodon hartenbergeri	Pedregueras 2c	MN 9	Vallesian	Late Miocene
Cricetulodon sabadellensis	Can Llobateres	MN 9	Vallesian	Late Miocene
"Cricetus" lophidens	Maritsa	MN 13/14	Late Turolian/Early Ruscinian	L. Miocene/E. Pliocene
Democricetodon brevis	La Grive	MN 7/8	Astaracian	Middle Miocene
Democricetodon franconicus	Erkertshofen	MN 4/5	Orleanian	Early Miocene
Eozapus intermedius	Kohfidisch	MN 10/11	Late Vallesian/Early Turolian	Late Miocene
Hansdebruijnia neutrum	Pikermi 4	MN 12	Turolian	Late Miocene
Kalymnomys major	Kalymnos		Biharian	Early Pleistocene
"Karnimata" provocator	Pikermi 4	MN 12	Turolian	Late Miocene
Megacricetodon collongensis	Vieux Collonges	MN 4/5	Orleanian	Early Miocene
Megacricetodon gregarius	La Grive	MN 7/8	Astaracian	Middle Miocene
Megacricetodon primitivus	Valtorres	MN 5	Orleanian	Early Miocene
Mesocricetus primitivus	Maritsa	MN 13/14	Late Turolian/Early Ruscinian	L. Miocene/E. Pliocene
Micromys bendai	Ptolemais	MN 14	Ruscinian	Early Pliocene
Mimomys gracilis	Csarnota	MN 15	Ruscinian	Pliocene
Mirrabella tuberosa	Aliveri	MN 4	Orleanian	Early Miocene
Occitanomys debruijni	Maritsa	MN 13/14	Late Turolian/Early Ruscinian	L. Miocene/E. Pliocene
Parapodemus lugdunensis	Mollon	MN 11	Turolian	Late Miocene
Pliomys graecus	Tourkobounia 1	MN 16	Villanyian	Late Pliocene
Progonomys cathalai	Montredon	MN 10	Vallesian	Late Miocene
Promimomys insuliferus	Podlesice	MN 14	Ruscinian	Early Pliocene

APPENDIX 5

Locality Names and Coordinates Shown in Figure 26.1

Number	Locality	Coordinates	
1	Paşalar	39°58'4.0"N	26°17'50.5"E
2	Kaplangı	38°48'18.1"N	29°50'11.3"E
3	Altıntaş	39°33'97"N	30°71'56.8"E
4	Harami 1	38°27'27.2"N	31°49'30.8"E
5	Bağiçi	39°31'56.8"N	32°54'42.0"E
6	Keseköy	40°39'51.6"N	32°40'51.5"E
7	Kılcak 0-3	40°12'52.4"N	33°24'20.5"E
8	Süleymanlı 2	40°30'54.4"N	33°37'27.6"E
9	Ortalıca	41°3'2.1"N	34°14'56.0"E
10	Kargı 1	40°52'41.1"N	34°52'56.4"E
11	Zambal	41°2'59.8"N	34°59'3.0"E
12	Hamamayağı	40°58'25.3"N	35°47'9.3"E
13	Yenice	40°59'20.0"N	35°48'15.7"E
14	Hayranlı 1	39°44'52.9"N	36°48'51.7"E
15	Düzyayla	39°55'33.4"N	37°18'56.5"E
16	Iğdeli	39°11'25.4"N	35°52'38.4"E

Chapter 27

Late Miocene Fossils from the Baynunah Formation, United Arab Emirates

Summary of a Decade of New Work

FAYSAL BIBI, ANDREW HILL, MARK BEECH, AND WALID YASIN

The region of Al Gharbia (previously known as the Western Region) comprises much of the area of Abu Dhabi Emirate west of the city of Abu Dhabi and bears the only known late Miocene terrestrial fossil biota from the entire Arabian Peninsula. Driving along the Abu Dhabi-As Sila' highway, which runs parallel to the Gulf coast and connects the United Arab Emirates to Saudi Arabia, one encounters a landscape of low dunes to the south, and sabkha (supratidal salt flats) and the sea to the north. Interspersed on both sides of the highway are low-lying jebels (hills), at most rising about 60 m above the surrounding terrain. These jebels are capped by a resistant gypsum-anhydrite-chert bed that produces characteristic table-top forms, or mesas. The sediments of these flat-topped jebels have been formally described as the Baynunah Formation (Whybrow 1989) and are composed mainly of reddish fine- to medium-grained sands of dominantly fluvial origin, along with brown and green silts, alternating sand-carbonate sequences, gypsum, and fine intraformational conglomerates. Several horizons within the Baynunah Formation bear fossils, either body parts or traces, of vertebrate, invertebrate, and plant taxa (figures 27.1 and 27.2).

From the sites of Rumaitha in the east to Jebel Barakah in the west, and from the sites of Shuwaihat in the north to Jaw Al Dibsa in the south, the area across which the Baynunah Formation is exposed forms a long east–west trending quadrangle measuring 180 × 45 km, covering 8100 km² (figure 27.3). Interspersed within this area are over two dozen documented sites from which fossil

remains have been collected over three decades. These fossils include remains of hippopotamus, giraffes, crocodiles, rodents, turtles, catfish, ratites, machairodont felid, and hyaenids, as well as bivalves, gastropods, and fossil wood. The Baynunah Formation records a time when a perennial river supported a rich ecosystem in what is now a hyperarid part of the world.

By way of biochronology, the Baynunah fossil fauna is estimated to be between 8 Ma and 6 Ma (Whybrow and Hill 1999). No other terrestrial fossil sites of late Miocene age are known from the remainder of the Arabian Peninsula. The Baynunah fauna, then, represents the sole sample available to chart the biotic continuity between late Miocene Arabia and neighboring contemporaneous fossil sites in Asia (e.g., Siwaliks, Maragheh), the Mediterranean (Pikermi, Samos), and Africa (Sahabi, Toros-Menalla, Lothagam, Tugen Hills).

Since the publication of a monographic treatment of the Baynunah fossils (Whybrow and Hill 1999), renewed fieldwork activities have brought new light to elements of the Baynunah fossil fauna. This chapter summarizes the latest knowledge on the fossil biota of the late Miocene Baynunah Formation.

HISTORY OF EXPLORATION

The first indications of the presence of vertebrate fossils from Al Gharbia came with the explorations of oil geologists working in the 1940s (Hill, Whybrow, and Yasin

Figure 27.1 Excavating lower jaws of a proboscidean at the site of Hamra 3-1. To the south, in the background, are the upper beds of the Baynunah Formation. Taken on December 17, 2007.

1999). The first publication of fossil remains from Al Gharbia came with Glennie and Evamy's (1968) description of fossil root casts and mention of proboscidean tooth remains from Jebel Barakah. Peter Whybrow (Natural History Museum, London [N.H.M.]) made fossil discoveries in Al Gharbia in 1979 and 1981, and described additional fossil root casts that he interpreted as mangroves (Whybrow and McClure 1981). In 1983, an archaeological survey of Al Gharbia by the Abu Dhabi Department of Antiquities and Tourism including Yasin and a team of German archaeologists (Vogt et al. 1989) resulted in the first significant collection of fossils from a number of sites in Al Gharbia. Upon invitation of the Department of Antiquities, Hill in 1984 visited Abu Dhabi to evaluate the recently collected fossils. In 1986, collaborative work began between Whybrow and Hill, and in 1989 a joint Yale–N.H.M. expedition was initiated to study the Al Gharbia fossil deposits. Extensive work on the Baynunah Formation, up to 1995, was given monographic treatment in Fossil Vertebrates of Arabia (Whybrow and Hill 1999). Beginning in 2002, the Abu Dhabi Islands Archaeological Survey, including Beech, undertook work on the Baynunah deposits. This included the discovery and salvage

of important new sites (Beech and Higgs 2005), including sites preserving footprint trackways (Higgs et al. 2003). In 2003, Bibi was invited by the Abu Dhabi Public Works Department to further explore the Baynunah deposits. Two visits to Al Gharbia in that year with small teams resulted in the discovery of new sites and some 200 new fossil specimens, including the ratite *Diamantornis laini* (Bibi et al. 2006). In December 2006, Bibi and Hill joined Beech and Yasin under the invitation of the newly established Abu Dhabi Authority for Culture and Heritage (A.D.A.C.H.) for a brief survey of the Al Gharbia sites. Since the completion of this chapter, A.D.A.C.H. has become the Abu Dhabi Tourism and Culture Authority (A.D.T.C.A.). December 2007 saw the inception of the current Yale–A.D.A.C.H. project, with annual fieldwork expeditions to the Baynunah Formation.

FOSSIL FAUNA OF THE BAYNUNAH FORMATION

An up-to-date listing of the entire Baynunah fossil fauna is given in table 27.1. Fieldwork since 1995 (Whybrow and

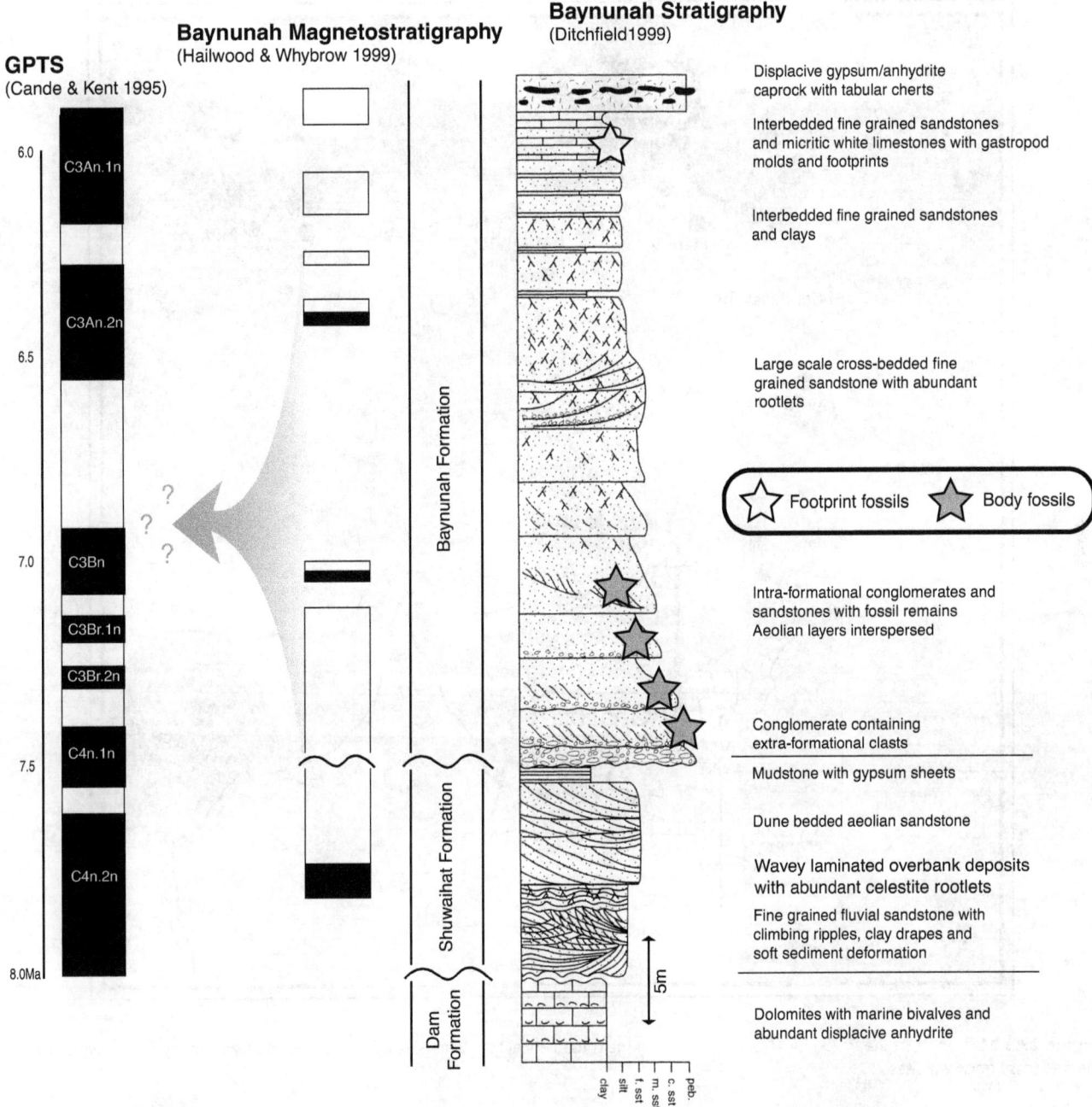

Figure 27.2 Stratigraphy and magnetostratigraphy of the Baynunah Formation. Body fossils (of vertebrates, invertebrates, and plants) are recovered from multiple horizons in the lower part of the Baynunah, while footprint fossils are recorded from carbonates correlating to the upper portion. Stratigraphic column reproduced from Ditchfield (1999:fig. 7.2), and paleomagnetic data from Hailwood and Whybrow (1999:fig. 8.5). The high frequency of polarity reversals in the period between 8 Ma and 6 Ma (Cande and Kent 1995) means any correlation of the Baynunah with the GPTS is equivocal, but the presence of at least four reversals in the Baynunah suggests a duration of 300,000 yr or more for this formation.

Hill 1999) has added the following taxa to the Baynunah fossil faunal list: two species of snake, one of which may be a colubrid; a sawfish (Pristidae); eggshells of two different ratites, *Diamantornis laini* and an aepyornithid-like form (Bibi et al. 2006); an anhinga (Stewart and Beech 2006); the giraffid *Palaeotragus* cf. *germaini*; and a sciurid rodent (Kraatz, Bibi, and Hill 2009). These come in addition to significant new material found of already-recorded taxa that will increase knowledge of the recorded forms and result in further taxonomic resolution. Among these

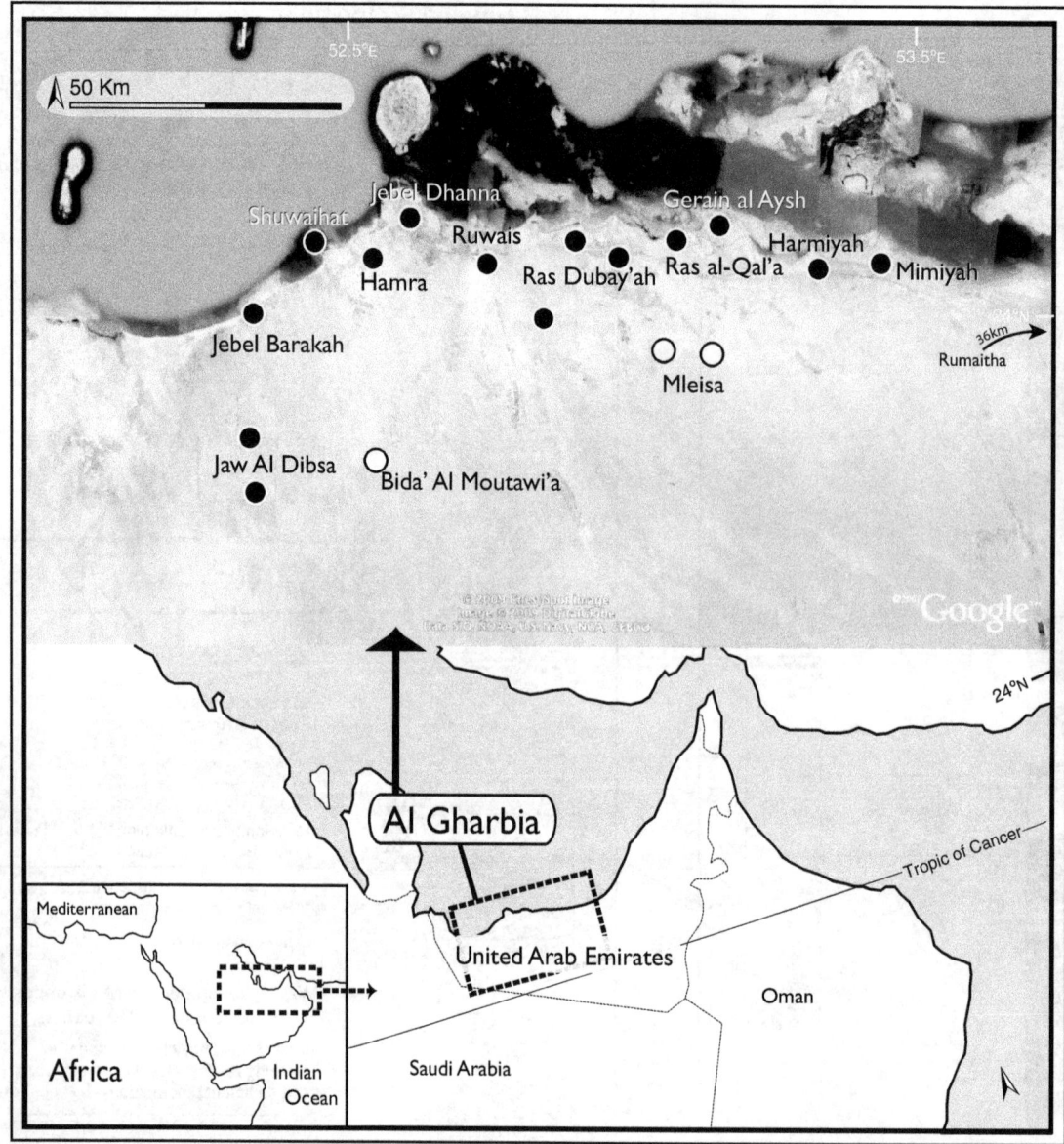

Figure 27.3 Map and satellite image of Al Gharbia showing main fossil localities. Black circles denote localities with body fossils; white circles denote fossil trackway sites.

are ostracods and foraminifera from the carbonates of the upper Baynunah; additional remains of the thryonomyid (cane rat) that suggest assignment to *Paraulacodus* and additional remains of the endemic gerbil *Abudhabia baynunensis*; a ratite synsacrum larger than that of modern *Struthio* that might be associated with either fossil eggshell type; new mandibular remains of *Hipparion* that will help further characterize *H. abudhabiense* (originally diagnosed on a partial mandible); relatively abundant remains of fossil proboscideans, including several mandibles, tusks and tusk fragments, a cranium, and postcrania attributable to

Stegotetrabelodon syrticus, and mandibles of the shovel-tusked proboscidean (*Amebelodon*/"*Mastodon*" *grandincisivus*); and a deciduous monkey premolar, only the second primate specimen to have come from the Baynunah Formation (Hill and Gundling 1999).

In addition, fossil trackway sites have been discovered (figure 27.4), adding a new dimension of study to the Baynunah fossil fauna (Higgs et al. 2003; Higgs, Gardener, and Beech 2005). Currently, at least three sites are known at which footprint remains of proboscideans and ungulates are preserved in carbonates (figure 27.3). These

Table 27.1

Baynunah Faunal List

Taxon			Lower Nawata	Toros Menalla 266	Sahabi Mb. U	Africa	Greco-Afghan	Siwaliks Dhok Pathan	Remarks
Plantae	"Algae"		Gen. et sp. indet.						
	Leguminosae		?Acacia sp.						
Protista	Foraminifera		(in progress)						
Mollusca	Gastropoda	Buliminidae	?Subzebrinus or ?Pesudonapaeus						"Palaearctic and Oriental" (Mordan 1999)
	Bivalvia	Mutelidae	Mutela sp.				G		
		Unionidae	Leguminaia sp.					G	Iraq
Crustacea	Ostracoda	Cytherideidae	Cyprideis sp.						G
Pisces	Siluriformes	Clariidae	Clarias sp.	(G)	G	(G)	G		
		Bagridae	Bagrus shuwaiensis	(G)	G		G		
	Cypriniformes	Cyprinidae	Barbus sp.				G		
	Pristiformes	Pristidae	(in progress)						
Reptilia	Crocodilia	Crocodylidae	Crocodylus cf. niloticus	G/(S)	G/(S)	G/(S)			
			Crocodylus sp.						
			?Ikanogavialis				G/(S)		S. America, Australasia
		Gavialidae	Gavialidae gen. et sp. indet.						
	Testudines	Trionychidae	Trionyx sp.		G	G	G	G	
			cf. Mauremys sp.			(G)	(G)	(G)	
		Testudinidae	Geochelone (Centrochelys) aff. sulcata		G	G	G		Subgenus Centrochelys Libya, Tunisia, possibly Saudi Arabia (de Lapparent de Broin and van Dijk 1999)
Aves	Squamata	cf. Colubridae	Colubridae spp. (in progress)						
	Ratitae	Incertae sedis	Diamantornis laini	S			S		
			'Aepyornithid-type' eggshell	'G'			'G'		'G'
	Pelicaniformes	Anhingidae	Anhinga sp.	G		G	G		Today worldwide
	Ciconiiformes	Ardeidae	Gen. et sp. indet.						
Mammalia	Artiodactyla	Hippopotamidae	Archaeopotamus aff. lothagamensis	G/(S)		(G)	G/(S)		
		Bovidae	Pachyportax latidens						S
			Prostrepsiceros aff. libycus			G/(S)	G/(S)		
			Prostrepsiceros aff. vinayaki					(S)	
			Gazella aff. lydekkeri			G			(S)
			Tragoportax cyrenaicus	(S)		S	S		(S)

(continued)

Table 27.1 (continued)

Order	Family	Taxon	Toros Menalla — Lower Nawata	Toros Menalla — 266	Sahabi Mb. U	Africa	Greco-Afghan	Siwaliks Dhok Pathan	Remarks
	Giraffidae	cf. Neotragini (in progress)							
		Palaeotragus germaini (in progress)	S			S			
		? Bramatherium gen. et sp. indet.					(G)	(G)	
	Suidae	Nyanzachoerus syrticus	S	S	S	S			
		Propotamochoerus hysudricus						S	
Carnivora	Mustelidae	Plesiogulo praecocidens				G	G	G	Species from China (Barry 1999)
	Felidae	Machairodontinae gen. et sp. indet.							
	Hyaenidae	gen. et sp. indet. "very large"							
		gen. et sp. indet. "medium-size"							
Perissodactyla	Equidae	Hipparion abudhabiense		(S)		(S)			
		Hipparion sp.							
	Rhinoceratidae	gen. et sp. indet.							
Rodentia	Muridae	Abudhabia baynunensis			G	G	G	G	Genus in China also
		Myocricetodon (sp. nov.?)			G	G	G	G	Genus in China also
		Parapelomys cf. charkhensis					G	G	
		Dendromus aff. melanotus				G			Genus in Spain also
		Dendromus sp.							
	Dipodidae	Zapodinae gen. et sp. indet.							"Asiatic" (de Bruijn 1999)
	Thryonomyidae	cf. Paraulacodus (in progress)	(G/S)			(G/S)			Genus recorded from mid-Miocene Siwaliks (Flynn and Winkler 1994)
	Sciuridae	gen. et sp. indet.							
Insectivora	Soricidae	gen. et sp. indet.							
Primates	Cercopithecidae	gen. et sp. indet.							
Proboscidea	Deinotheriidae	gen. et sp. indet.							
	Elephantidae	Stegotetrabelodon syrticus	G		S	S	(S)		Species occurrence in Italy (Ferretti, Rook, and Torre 2003)
	Gomphotheriidae (Amebelodontidae)	cf. Amebelodon / ?"Mastodon" grandincisivus			(G)	(G)	(G)		

NOTE: Genera or species in common with other sites are denoted by G or S, parentheses denoting uncertainty (usually result of a cf. designation).

SOURCES: Agustí (2008); Badgley et al. (2008); Barry (1999); Bishop and Hill (1999); Boisserie (2005); de Bruijn (1999); de Lapparent de Broin and van Dijk (1999); Eisenmann and Whybrow (1999); Ferretti, Rook, and Torre (2003); Flynn and Winkler (1994); Forey and Young (1999); Gentry (1999); Geraads and Güleç (1999); Higgs et al. (2003); Hill and Gundling (1999); Jeffrey (1999); Mordan (1999); Rauhe et al. (1999); Sanders (2008); Sen (1998); Tassy (1999); Whybrow and Clements (1999); Whybrow et al. (1990).

Figure 27.4 One of at least 10 proboscidean trackways at the site of Mleisa 1.

same carbonates also preserve molds of gastropods proposed to be "?cerithid" (Ditchfield 1999).

PALEOENVIRONMENTS OF THE BAYNUNAH

Despite the presence of a perennial river system in Al Gharbia, the environment in this region was clearly arid during the late Miocene. Though arid environments are clearly recorded in the sediments of the Shuwaihat Formation underlying the Baynunah (Bristow 1999), there is no unambiguous sedimentological evidence for arid environments in the Baynunah Formation. Fine- to medium-grained quartz sands with frosted grains (D. Peppe, pers. comm.) and large-scale cross-beds are commonly observable. Gypsum-cemented root casts are present in many such layers, which Glennie and Evamy

(1968) interpreted as the remains of dune-stabilizing vegetation (contra Whybrow and McClure 1981).

Paleocurrent readings indicate Baynunah rivers derived from the northwest or west as outlined by Friend (1999). Topographically, the most likely western watershed and source that could provide a river system to the Al Gharbia area is the Wadi Sahba in Saudi Arabia (e.g., Whybrow and Hill 2002). A modern analogue for the Baynunah River can be sought in the Nile, which today flows its course through hyperarid parts of Sudan and Egypt. A late Miocene river system that is broadly analogous to the Baynunah one is the As Sahabi River, which took its source largely from the Tibesti Mountains in Chad and flowed north through a proto-Sahara to empty in the Mediterranean (Drake, El-Hawat, and Salem 2008).

Though no clear aeolian beds are present in the Baynunah, the underlying Shuwaihat Formation includes within it large aeolian dune beds (Bristow 1999). Much as Schuster et al. (2006) interpreted aeolian beds from the northern Chad Basin to indicate the onset of desertic conditions in the Sahara by around 7 Ma, the presence of aeolian deposits in the Shuwaihat Formation provides early evidence for the presence of desert conditions in the Arabian Peninsula going back to at least the late Miocene. The Shuwaihat Formation dune beds may push the evidence for desertification in Arabia even further back, if the age of this formation is accepted as being middle Miocene (15 ± 3Ma in Hailwood and Whybrow 1999).

The upper parts of the Baynunah Formation bear about a 15 m sequence of fine sands and clays alternating with carbonates. The carbonates are resistant to erosion and are extensively exposed several kilometers inland as deflated surfaces that are highly reflective and easily visible from satellite imagery (see figure 27.1). Within these large exposures are located three footprint sites, two of these recording proboscidean trackways (Higgs et al. 2003) and one site with an ungulate trackway. These same carbonates include ostracods and molds of gastropods resembling cerithids, suggesting a marine or brackish water origin. Shells of *Melanoides* reported from above the Mleisa 1 trackway carbonate by Higgs et al. (2003) are suspiciously well preserved. Despite years of fieldwork, no other occurrences of *Melanoides* have ever been found, neither at Mleisa 1 nor at any other site from the Baynunah Formation. Accordingly, we suspect the Mleisa 1 *Melanoides* to be surface finds not deriving from the Baynunah Formation itself.

The Baynunah deposits are exposed over a long area that is parallel to the coastline and that at its widest reaches at least 45 km (see figure 27.3). It is apparent that the original river system ranged over an even wider area in the course of its temporal span. A single fossil vertebra recovered in December 2007 is identifiable as a sawfish (Chondrichthyes: Pristidae). Living sawfish are known to inhabit shallow marine or estuarine environments, often swimming far up river. The discovery of sawfish remains among the Baynunah fauna begs further consideration of the proximity of the marine shoreline to the Baynunah fluvial system. It is conceivable that a proto-Gulf seaway may have been in existence, even though intermittently, in the area to the north or west of late Miocene Al Gharbia.

PALEOBIOGEOGRAPHY OF THE BAYNUNAH

The geographic location of the Baynunah Formation, at a locus between the Asian, European, and African continents, is reflected in its fossil fauna (see table 27.1), which comprises a unique mix of taxa not found together elsewhere. Taxa known from the Siwaliks, such as the bovid *Pachyportax latidens* and the suid *Propotamochoerus hysudricus*, are found with otherwise African taxa, including the ratite *Diamantornis laini*, the hippopotamid *Archaeopotamus*, and the giraffid *Palaeotragus* cf. *germaini*. Perhaps with the exception of the bivalve *Leguminaia*, all Baynunah fossil taxa are shared between either African or southern Asian (Siwaliks) faunas, with no taxa uniquely indicative of the Greco-Irano-Afghan zone (de Bonis et al. 1992). This is intriguing given the relative proximity of sites such as Injana (Brunet and Heintz 1983) and Maragheh (Bernor 1986). It appears that faunal exchange was very much a function of latitude, with environmental factors permitting biotic dispersal in east–west directions and restricting it in north–south directions. The Himalayan-Zagros-Tauride mountain range has been proposed as a dispersal barrier throughout the late Miocene between sites in the Siwaliks/Iraq and the Greco-Afghan biogeographic zone (Brunet and Heintz 1983; Brunet et al. 1984; Beden and Brunet 1986).

The Baynunah fauna includes a much larger number of taxa in common with African rather than Siwaliks assemblages (see table 27.1). Three African sites dating to between 8 Ma and 5 Ma and with extensive faunal lists are chosen for comparison with the Baynunah: Toros-Menalla (Chad; Vignaud et al. 2002), the Lower Member of the Nawata Formation at Lothagam (Kenya; Leakey and Harris 2003), and Sahabi (Libya; Boaz et al. 2008), but collections from the Mpesida Beds and Lukeino Formation of the Tugen Hills, Kenya, are also relevant to this

discussion (Hill et al. 1985; Hill 1999a, 1999b, 2002). Each of these three sites is situated within a different African region, and each shares taxa in common with the Baynunah fauna. Baynunah taxa with affinities to Sahabi include *Geochelone* (*Centrochelys*) aff. *sulcata*, *Myocricetodon*, and *Amebelodon*/"*Mastodon*" *grandincisivus*. Taxa in common between the Baynunah and the Lower Nawata include *Archaeopotamus* aff. *lothagamensis*, *Diamantornis laini*, and perhaps *Paraulacodus*.

CHRONOLOGY AND DURATION

No tuffs or other volcanics are present in the Shuwaihat or Baynunah formations. As a result, age estimates for the Baynunah so far derive exclusively from biochronological correlations. Previous age estimates based on biochronology had placed the Baynunah fauna at somewhere between 8 Ma and 6 Ma in age. More recently, *Archaeopotamus* (Boisserie 2005) and *Diamantornis laini* (Harris and Leakey 2003; Harrison and Msuya 2005; Bibi et al. 2006) have been determined to be present in the Baynunah fauna. These two taxa are also present in the Nawata Formation, specifically in the Lower Member but not the Upper Member. While the maximum age of the fossil fauna from the Lower Member is not precisely determined, the upper limit of this member is firmly established as at 6.5 Ma by the Marker Tuff (McDougall and Feibel 2003). Similarly, the suid *Propotamochoerus hysudricus* (Bishop and Hill 1999) ranges from 10.2 Ma to 6.8 Ma (Badgley et al. 2008). Assuming similar taxonomic age ranges applied in Arabia, these three taxa tentatively propose a minimum age limit of 6.5 Ma for the Baynunah.

Kingston (1999) sampled fossil enamel from the Baynunah for stable carbon isotopes and discovered a dominant C_4-feeding signal in teeth of *Stegotetrabelodon* sp., *Hipparion abudhabiense*, *Hipparion* sp., ?*Bramatherium* sp., *Tragoportax cyrenaicus*, and *Archaeopotamus* aff. *lothagamensis*. Analysis of paleosol carbonates indicated the presence of mixed C_3-C_4 habitats (but no C_4-dominated habitats [Kingston 1999]). In the Siwaliks and the Tugen Hills, the first enamel isotope values indicating a pure C_4 diet ($\delta^{18}C$ values > −2‰) do not appear until around 7 Ma (Cerling, Wang, and Quade 1993; Morgan, Kingston, and Marino 1994). The first C_4-dominated habitats (paleosols) do not appear in the Siwaliks until 7.37 Ma (Quade et al. 1989; Quade and Cerling 1995; Barry et al. 2002), though this is not apparent in the paleosols of the Tugen Hills sequence (Kingston, Marino, and Hill 1994). If the presence of a significant C_4 component in habitats and diets can be assumed to have had a similar chronology in the Arabian Peninsula, then the presence of a strong C_4 dietary signal in the Baynunah fauna might establish the maximum age of this assemblage as being around 7.4 Ma.

On the basis of the above evidence, then, we tentatively suggest a constrained age of between 7.5 Ma and 6.5 Ma for the Baynunah Formation. Continued fossil discoveries and analysis will help more precisely constrain the age of this fossil assemblage. Most promising among these is perhaps the taxonomic determination of the ostracods from the carbonates of the upper Baynunah.

How long the Baynunah river system was in existence is currently not known. The collective fossil fauna appears consistent with a single temporal window, but in reality a late Miocene fauna could sample a million years or more before biochronological inconsistencies are detected. The presence of up to four polarity reversals in the magnetostratigraphy of the Baynunah Formation (see figure 27.2; Hailwood and Whybrow 1999:fig. 8.5) indicates at least some geologically significant duration of time is represented. According to the geomagnetic polarity timescale (Cande and Kent 1995), 12 polarity reversals are present in the period between 8 Ma and 6 Ma, with magnetochron durations varying between 34 ka and 368 ka. This gives some indication that a span of over 300 ka might easily be accommodated by the Baynunah sediments, though without a direct chronological tiepoint, it is not possible to say more at the moment. Ongoing paleomagnetic work promises to build a more complete and precise local Baynunah magnetostratigraphy that will help better determine the age and duration of this set of deposits.

SITE DOCUMENTATION, MAINTENANCE, AND SALVAGE

Exploratory work since 1999 has significantly increased the recognized areal extent of the Baynunah fossil deposits. In particular, this has come with the discovery of fossils from a number of areas further inland than had previously been known. Primary among these are the sites of Jaw Al Dibsa (or Umm al-Ishtan), Bida' Al Mutawa'ah, and Mleisa. Jaw Al Dibsa comprises a collection of sites bearing remains of proboscideans, bovids, giraffids, fish, ratites eggshells, and fossil wood. The Bida' Al Mutawa'ah and Mleisa sites are fossil trackway sites where proboscidean and ungulate footprints have been discovered (Higgs et al. 2003). The discovery of these sites has in-

creased the potential for further fossil discoveries in the Al Gharbia region. Current annual fieldwork efforts continue to focus on the survey and documentation of fossiliferous exposures inland of the coastal sites.

Paleontological work in Al Gharbia is a race against the rapid rate of development characteristic of the United Arab Emirates. All the Baynunah fossil sites are in real danger of development activities that either destroy them or make them inaccessible to scientists. For example, sites on Shuwaihat and Jebel Dhanna from which important fossil specimens were discovered and described (Whybrow and Hill 1999) have been appropriated by military and oil refinery installations. The Baynunah Formation type section at Jebel Barakah has itself in recent years been greatly disturbed by earth-moving activities and is now in a military area. Access to coastal and inland Baynunah exposures is being continually restricted by military, municipality, and oil industry appropriation. At this rate, it will be barely a matter of a decade before almost all the fossil sites are compromised.

The Abu Dhabi Authority for Culture and Heritage is taking important measures toward site maintenance and awareness that in some cases have proved successful at protecting fossil sites. These include fencing-off certain sites such as the Mleisa trackways and the placement of signs informing of the proximity of sites. These are, however, effectively temporary measures, and without the enactment of legislation and enforcement the fossil sites of Al Gharbia continue to remain in real threat of damage. Paleontological work in the Al Gharbia region takes on the aspect of a salvage mission, whereby fossiliferous deposits are studied and documented in anticipation of their being lost in the near future.

CONCLUSION

Since 1999, fossil discoveries have continued to be made in the late Miocene Baynunah Formation in Al Gharbia, Abu Dhabi Emirates. Since 1996, annual fieldwork efforts were restarted as a collaboration between Yale University and the Abu Dhabi Authority for Culture and Heritage, headed by the four authors. All in all, renewed efforts have resulted in the collection of hundreds of new fossil specimens, containing among them new taxa not before known from the Baynunah, and better representation of known taxa that promises further refinement of the faunal list. This comes in addition to the discovery of new fossil sites much further inland than had previously been known, including sites preserving trackways of proboscideans and an ungulate. Analyses seeking pollen, phytoliths, and an improved local paleomagnetic stratigraphy are underway, as are sedimentological studies of the sands and carbonates of the Baynunah Formation. These, in conjunction with the information provided by the fossil remains, should help better determine the age, duration, and paleoenvironments of the Baynunah Formation.

ACKNOWLEDGMENTS

We would like to thank Xiaoming Wang and the organizers of the Asian Mammal Stratigraphy conference that took place at the IVPP in Beijing in 2009 for the opportunity to participate in the meeting and contribute to this volume. The Society of Vertebrate Paleontology generously supported costs of travel and attendance for F. Bibi. Paleontological work in Al Gharbia is supported by the Abu Dhabi Authority on Culture and Heritage (ADACH), the Revealing Hominid Origins Initiative (National Science Foundation, USA, grant no. 0321893), an NSF International Research Fellowship Award (grant no. OISE-0852975), Yale Peabody Museum, Yale University Provost's Office, Institute de Paléoprimatologie Paléontologie Humaine (IPHEP), Agence Nationale de la Recherche ANR-09-BLAN-0238. Additional support has come from the Abu Dhabi National Oil Company and the Abu Dhabi Public Works Department. We are also grateful to the following individuals for support: M. Al Neyadi, H. H. Sheikh Sultan bin Zayed Al Nahyan, J.-R. Boisserie, Jacques Gauthier, F. C. Howell, P. Vignaud, Elisabeth S. Vrba, T. White, and to the following for their contributions to fieldwork and study: B. Kraatz, N. Craig, D. Evans, M. Fox, A. Haidar, W. Joyce, S. Lokier, O. Otero, D. Peppe, M. Schuster, A. Attar, S. Majzoub, E. Moacdieh, K. Zreik. M. Fortelius, L. Flynn, A. Gentry, and one anonymous reviewer provided helpful reviews and editorial assistance.

REFERENCES

Agustí, J. 2008. New data on the rodent fauna from As Sahabi, Libya. *Garyounis Scientific Bulletin Special Issue* 5:139–143.

Badgley, C., J. C. Barry, M. E. Morgan, S. V. Nelson, A. K. Behrensmeyer, T. E. Cerling, and D. Pilbeam. 2008. Ecological changes in Miocene mammalian record show impact of prolonged climatic forcing. *Proceedings of the National Academy of Sciences* 105:12145–12149

Barry, J. C. 1999. Late Miocene Carnivora from the Emirate of Abu Dhabi, United Arab Emirates. In *Fossil Vertebrates of Arabia: With Emphasis on the Late Miocene Faunas, Geology, and Palaeoenviron-*

ments of the Emirate of Abu Dhabi, United Arab Emirates, ed. P. J. Whybrow and A. P. Hill, pp. 203–208. New Haven: Yale University Press.

Barry, J. C., M. E. Morgan, L. J. Flynn, D. Pilbeam, A. K. Behrensmeyer, S. M. Raza, I. A. Khan, C. Badgley, J. Hicks, and J. Kelley. 2002. Faunal and environmental change in the late Miocene Siwaliks of northern Pakistan. *Paleobiology* 28(S2):1–71.

Beden, M. and M. Brunet. 1986. Faunes de mammifères et paléobiogéographie des domaines indiens et péri-indiens au Néogène. *Sciences de la Terre* 47:61–87.

Beech, M. and W. Higgs. 2005. A new late Miocene fossil site in Ruwais, Western Region of Abu Dhabi, United Arab Emirates. In *Emirates Heritage*, ed. P. Hellyer and M. Ziolkowski, vol. 1 pp. 6–21. Abu Dhabi: Zayed Centre for Heritage and History.

Bernor, R. L. 1986. Mammalian biostratigraphy, geochronology, and zoogeographic relationships of the late Miocene Maragheh fauna, Iran. *Journal of Vertebrate Paleontology* 6(1):76–95.

Bibi, F., A. B. Shabel, B. P. Kraatz, and T. A. Stidham. 2006. New fossil ratite (Aves: Palaeognathae) eggshell discoveries from the Late Miocene Baynunah Formation of the United Arab Emirates, Arabian Peninsula. *Palaeontologia Electronica* 9(1); 2A:13 pp., 554KB; http://palaeo-electronica.org/paleo/2006_1/eggshell /issue1_06.htm.

Bishop, L. C. and A. Hill. 1999. Fossil Suidae from the Baynunah Formation, Emirate of Abu Dhabi, United Arab Emirates. In *Fossil Vertebrates of Arabia: With Emphasis on the Late Miocene Faunas, Geology, and Palaeoenvironments of the Emirate of Abu Dhabi, United Arab Emirates*, ed. P. J. Whybrow and A. Hill, pp. 254–270. New Haven: Yale University Press.

Boaz, N. T., A. El-Arnauti, P. Pavlakis, and M. J. Salem, eds. 2008. Circum-Mediterranean geology and biotic evolution during the Neogene Period: The perspective from Libya. *Garyounis Scientific Bulletin, Special Issue* 5.

Boisserie, J.-R. 2005. The phylogeny and taxonomy of Hippopotamidae (Mammalia: Artiodactyla): A review based on morphology and cladistic analysis. *Zoological Journal of the Linnean Society* 143:1–26.

Bristow, C. S. 1999. Aeolian and sabkha sediments in the Miocene Shuwaihat Formation, Emirate of Abu Dhabi, United Arab Emirates. In *Fossil Vertebrates of Arabia: With Emphasis on the Late Miocene Faunas, Geology, and Palaeoenvironments of the Emirate of Abu Dhabi, United Arab Emirates*, ed. P. J. Whybrow and A. Hill, pp. 50–60. New Haven: Yale University Press.

Brunet, M. and E. Heintz. 1983. Interprétation paléoécologique et relations biogéographiques de la faune de vértebrées du Miocène supérieur d'Injana, Irak. *Palaeogeography, Palaeoclimatology, Palaeoecology* 44:283–293.

Brunet, M., E. Heintz, and B. Battail. 1984. Molayan (Afghanistan) and the Khaur Siwaliks of Pakistan: An example of biogeographic isolation of late Miocene mammalian faunas. *Geologie en Mijnbouw* 63(1):31–38.

Cande, S. C. and D. V. Kent. 1995. Revised calibration of the geomagnetic polarity timescale for the Late Cretaceous and Cenozoic. *Journal of Geophysical Research-Solid Earth* 100(B4): 6093–6095.

Cerling, T. E., Y. Wang, and J. Quade. 1993. Expansion of C4 ecosystems as an indicator of global ecological change in the late Miocene. *Nature* 361(6410):344–345.

de Bonis, L., M. Brunet, E. Heintz, and S. Sen. 1992. La province greco-irano-afghane et la répartition des faunes mammaliennes au Miocène supérieur. *Paleontologia i Evolució* 24–25:103–112.

de Bruijn, H. 1999. A Late Miocene insectivore and rodent fauna from the Baynunah Formation, Emirate of Abu Dhabi, United Arab Emirates. In *Fossil Vertebrates of Arabia: With Emphasis on the Late Miocene Faunas, Geology, and Palaeoenvironments of the Emirate of Abu Dhabi, United Arab Emirates*, ed. P. J. Whybrow and A. Hill, pp. 186–197. New Haven: Yale University Press.

de Lapparent de Broin, F. and P. P. van Dijk. 1999. Chelonia from the late Miocene Baynunah Formation, Emirate of Abu Dhabi, United Arab Emirates: Paleogeographic implications. In *Fossil Vertebrates of Arabia: With Emphasis on the Late Miocene Faunas, Geology, and Palaeoenvironments of the Emirate of Abu Dhabi, United Arab Emirates*, ed. P. J. Whybrow and A. Hill, pp. 136–162. New Haven: Yale University Press.

Ditchfield, P. W. 1999. Diagenesis of the Baynunah, Shuwaihat, and Upper Dam Formation sediments exposed in the Western Region, Emirate of Abu Dhabi, United Arab Emirates. In *Fossil Vertebrates of Arabia: With Emphasis on the Late Miocene Faunas, Geology, and Palaeoenvironments of the Emirate of Abu Dhabi, United Arab Emirates*, ed. P. J. Whybrow and A. Hill, pp. 61–74. New Haven: Yale University Press.

Drake, N. A., A. S. El-Hawat, and M. J. Salem. 2008. The development, decline and demise of the As Sahabi River System over the last seven million years. *Garyounis Scientific Bulletin Special Issue* 5:95–109.

Eisenmann, V. and P. J. Whybrow. 1999. Hipparions from the late Miocene Baynunah Formation, Emirate of Abu Dhabi, United Arab Emirates. In *Fossil Vertebrates of Arabia: With Emphasis on the Late Miocene Faunas, Geology, and Palaeoenvironments of the Emirate of Abu Dhabi, United Arab Emirates*, ed. P. J. Whybrow and A. Hill, pp. 234–253. New Haven: Yale University Press.

Ferretti, M. P., L. Rook, and D. Torre. 2003. Stegotetrabelodon (Proboscidea, Elephantidae) from the late Miocene of southern Italy. *Journal of Vertebrate Paleontology* 23(3):659–666.

Flynn, L. J. and A. J. Winkler. 1994. Dispersalist implications of Paraulacodus indicus: A South Asian rodent of African affinity. *Historical Biology* 9:223–235.

Forey, P. L. and S. V. T. Young. 1999. Late Miocene fishes of the Emirate of Abu Dhabi, United Arab Emirates. In *Fossil Vertebrates of Arabia: With Emphasis on the Late Miocene Faunas, Geology, and Palaeoenvironments of the Emirate of Abu Dhabi, United Arab Emirates*, ed. P. J. Whybrow and A. Hill, pp. 120–135. New Haven: Yale University Press.

Friend, P. F. 1999. Rivers of the Lower Baynunah Formation. In *Fossil Vertebrates of Arabia: With Emphasis on the Late Miocene Faunas, Geology, and Palaeoenvironments of the Emirate of Abu Dhabi, United Arab Emirates*, ed. P. J. Whybrow and A. Hill, pp. 39–49. New Haven: Yale University Press.

Gentry, A. W. 1999. Fossil pecorans from the Baynunah Formation, Emirate of Abu Dhabi, United Arab Emirates. In *Fossil Vertebrates of Arabia: With Emphasis on the Late Miocene Faunas, Geology, and Palaeoenvironments of the Emirate of Abu Dhabi, United Arab Emirates*, ed. P. J. Whybrow and A. Hill, pp. 290–316. New Haven: Yale University Press.

Geraads, D. and E. Güleç. 1999. A Bramatherium skull (Giraffidae, Mammalia) from the late Miocene of Kavakdere (Central Tur-key). Biogeographic and phylogenetic implications. *Bulletin of the Mineral Research and Exploration Institute of Turkey* 121:51–56.

Glennie, K. W. and B. D. Evamy. 1968. Dikaka: Plants and plant-root structures associated with aeolian sand. *Palaeogeography, Palaeoclimatology, Palaeoecology* 4:77–87.

Google Earth, http://www.google.com/earth/index.html (accessed May 2009).

Hailwood, E. A. and P. J. Whybrow. 1999. Palaeomagnetic correlation and dating of the Baynunah and Shuwaihat Formations, Emirate of Abu Dhabi, United Arab Emirates. In *Fossil Vertebrates of Arabia: With Emphasis on the Late Miocene Faunas, Geology, and Palaeoenvironments of the Emirate of Abu Dhabi, United Arab Emirates*, ed. P. J. Whybrow and A. Hill, pp. 75–87. New Haven: Yale University Press.

Harris, J. M. and M. G. Leakey. 2003. Lothagam birds. In *Lothagam: the Dawn of Humanity in Eastern Africa*, ed. M. G. Leakey and J. M. Harris, pp. 161–166. New York: Columbia University Press.

Harrison, T. and C. P. Msuya. 2005. Fossil struthionid eggshells from Laetoli, Tanzania: Taxonomic and biostratigraphic significance. *Journal of African Earth Sciences* 41:303–315.

Higgs, W., A. Gardner, and M. Beech. 2005. A fossil proboscidean trackway at Mleisa, Western Region of Abu Dhabi, United Arab Emirates. In *Emirates Heritage*, vol. 1, *Proceedings of the 1st Annual Symposium on Recent Palaeontological and Archaeological Discoveries in the Emirates*, ed. P. Hellyer and M. Ziolkowski, pp. 21–27. Al Ain: Zayed Centre for Heritage and History.

Higgs, W., G. Kirkham, G. Evans, and D. Hull. 2003. A Late Miocene Proboscidean trackway from Mleisa, United Arab Emirates. *Tribulus* 13:3–8.

Hill, A. 1999a. Late Miocene sub-Saharan vertebrates, and their relation to the Baynunah fauna, Abu Dhabi, United Arab Emirates. In *Fossil Vertebrates of Arabia: With Emphasis on the Late Miocene Faunas, Geology, and Palaeoenvironments of the Emirate of Abu Dhabi, United Arab Emirates*, ed. P. J. Whybrow and A. Hill, pp. 420–429. New Haven: Yale University Press.

Hill, A. 1999b. The Baringo Basin, Kenya: From Bill Bishop to BPRP. In *Late Cenozoic Environments and Hominid Evolution: A Tribute to Bill Bishop*, ed. P. Andrews and P. Banham, pp. 85–97. London: Geological Society of London.

Hill, A. 2002. Paleoanthropological research in the Tugen Hills, Kenya. *Journal of Human Evolution* 42:1–10.

Hill, A., R. Drake, L. Tauxe, M. Monaghan, J. Barry, A. K. Behrensmeyer, G. Curtis, B. Fine Jacobs, L. Jacobs, N. M. Johnson, and D. Pilbeam. 1985. Neogene palaeontology and geochronology of the Baringo Basin, Kenya. *Journal of Human Evolution* 14:749–73.

Hill, A. and T. Gundling. 1999. A monkey (Primates; Cercopithecidae) from the late Miocene of Abu Dhabi, United Arab Emirates. In *Fossil Vertebrates of Arabia: With Emphasis on the Late Miocene Faunas, Geology, and Palaeoenvironments of the Emirate of Abu Dhabi, United Arab Emirates*, ed. P. J. Whybrow and A. Hill, pp. 198–202. New Haven: Yale University Press.

Hill, A., P. J. Whybrow, and W. Yasin. 1999. History of palaeontological research in the Western Region of the Emirate of Abu Dhabi, United Arab Emirates. In *Fossil Vertebrates of Arabia: With Emphasis on the Late Miocene Faunas, Geology, and Palaeoenvironments of the Emirate of Abu Dhabi, United Arab Emirates*, ed. P. J. Whybrow and A. Hill, pp. 15–23. New Haven: Yale University Press.

Jeffrey, P. A. 1999. Late Miocene swan mussels from the Baynunah Formation, Emirate of Abu Dhabi, United Arab Emirates. In *Fossil Vertebrates of Arabia: With Emphasis on the Late Miocene Faunas, Geology, and Palaeoenvironments of the Emirate of Abu Dhabi, United Arab Emirates*, ed. P. J. Whybrow and A. Hill, pp. 111–115. New Haven: Yale University Press.

Kingston, J. D. 1999. Isotopes and environments of the Baynunah Formation, Emirate of Abu Dhabi, United Arab Emirates. In *Fossil Vertebrates of Arabia: With Emphasis on the Late Miocene Faunas, Geology, and Palaeoenvironments of the Emirate of Abu Dhabi, United Arab Emirates*, ed. P. J. Whybrow and A. Hill, pp. 354–372. New Haven: Yale University Press.

Kingston, J. D., B. Marino and A. Hill. 1994. Isotopic evidence for Neogene hominid paleoenvironments in the Kenya Rift Valley. *Science* 264:955–959.

Kraatz, B. P., F. Bibi, and A. Hill. 2009. New rodents from the Late Miocene of the United Arab Emirates. *Journal of Vertebrate Paleontology* 29:129A.

Leakey, M. G. and J. M. Harris, eds. 2003. *Lothagam: The Dawn of Humanity in Eastern Africa*. New York: Columbia University Press.

McDougall, I. and C. S. Feibel. 2003. Numerical age control for the Miocene-Pliocene succession at Lothagam, a hominoid-bearing sequence in the northern Kenya Rift. In *Lothagam: The Dawn of Humanity in Eastern Africa*, ed. M. G. Leakey and J. M. Harris, pp. 43–64. New York: Columbia University Press.

Mordan, P. B. 1999. A terrestrial pulmonate gastropod from the late Miocene Baynunah Formation, Emirate of Abu Dhabi, United Arab Emirates. In *Fossil Vertebrates of Arabia: With Emphasis on the Late Miocene Faunas, Geology, and Palaeoenvironments of the Emirate of Abu Dhabi, United Arab Emirates*, ed. P. J. Whybrow and A. Hill, pp. 116–119. New Haven: Yale University Press.

Morgan, M. E., J. D. Kingston, and B. D. Marino. 1994. Carbon isotopic evidence for the emergence of C_4 plants in the Neogene from Pakistan and Kenya. *Nature* 367(6459):162–165.

Quade, J. and T. E. Cerling. 1995. Expansion of C-4 grasses in the late Miocene of northern Pakistan—Evidence from stable isotopes in paleosols. *Palaeogeography, Palaeoclimatology, Palaeoecology* 115(1–4):91–116.

Quade, J., T. E. Cerling, and J. R. Bowman. 1989. Development of Asian monsoon revealed by marked ecological shift during the latest Miocene in northern Pakistan. *Nature* 342(6246):163–166.

Rauhe, M., E. Frey, D. S. Pemberton, and T. Rossmann. 1999. Fossil crocodilians from the late Miocene Baynunah Formation of the Emirate of Abu Dhabi, United Arab Emirates: Osteology and palaeoecology. In *Fossil Vertebrates of Arabia: With Emphasis on the Late Miocene Faunas, Geology, and Palaeoenvironments of the Emirate of Abu Dhabi, United Arab Emirates*, ed. P. J. Whybrow and A. Hill, pp. 163–185. New Haven: Yale University Press.

Sanders, W. J. 2008. Review of fossil Proboscidea from the Late Miocene–Early Pliocene site of As Sahabi, Libya. *Garyounis Scientific Bulletin Special Issue* 5:241–256.

Schuster, M., P. Duringer, J. F. Ghienne, P. Vignaud, H. T. Mackaye, A. Likius, and M. Brunet. 2006. The age of the Sahara Desert. *Science* 311(5762):821–821.

Sen, S. 1998. The age of the Molayan mammal locality, Afghanistan. *Geobios* 31:385–391.

Stewart, J. R. and M. Beech. 2006. The Miocene birds of Abu Dhabi (United Arab Emirates) with a discussion of the age of modern species and genera. *Historical Biology* 18:103–113.

Tassy, P. 1999. Miocene elephantids (Mammalia) from the Emirate of Abu Dhabi, United Arab Emirates: Palaeobiogeographic implications. In *Fossil Vertebrates of Arabia: With Emphasis on the Late Miocene Faunas, Geology, and Palaeoenvironments of the Emirate of Abu Dhabi, United Arab Emirates*, ed. P. J. Whybrow and A. Hill, pp. 209–233. New Haven: Yale University Press.

Vignaud, P., P. Duringer, H. T. Mackaye, A. Likius, C. Blondel, J. R. Boisserie, L. de Bonis, V. Eisenmann, M. E. Etienne, D. Geraads, F. Guy, T. Lehmann, F. Lihoreau, N. Lopez-Martinez, C. Mourer-Chauviré, O. Otero, J. C. Rage, M. Schuster, L. Viriot, A. Zazzo, and M. Brunet. 2002. Geology and palaeontology of the Upper Miocene Toros-Menalla hominid locality, Chad. *Nature* 418(6894):152–155.

Vogt, B., W. Gockel, H. Hofbauer, and A. A. Al-Haj. 1989. The coastal survey in the Western Province of Abu Dhabi. *Archaeology in the United Arab Emirates* 5:49–60.

Whybrow, P. J. 1989. New stratotype; the Baynunah Formation (Late Miocene), United Arab Emirates: Lithology and palaeontology. *Newsletters on Stratigraphy* 21:1–9.

Whybrow, P. J. and D. Clements. 1999. Arabian Tertiary fauna, flora, and localities. In *Fossil Vertebrates of Arabia: With Emphasis on the Late Miocene Faunas, Geology, and Palaeoenvironments of the Emirate of Abu Dhabi, United Arab Emirates*, ed. P. J. Whybrow and A. Hill, pp. 460–473. New Haven: Yale University Press.

Whybrow, P. J., and A. Hill, eds. 1999. *Fossil Vertebrates of Arabia: With Emphasis on the Late Miocene Faunas, Geology, and Palaeoenvironments of the Emirate of Abu Dhabi, United Arab Emirates*. New Haven: Yale University Press.

Whybrow P. J. and A. Hill. 2002. Late Miocene fauna and environments of the Baynunah Formation, Emirate of Abu Dhabi, United Arab Emirates: The Mid-East "monsoon"? *Annales Géologiques des Pays Helléniques* 34: 353–362.

Whybrow, P. J., A. Hill, W. Yasin, and E. A. Hailwood. 1990. Late Miocene primate fauna, flora and initial palaeomagnetic data from the Emirate of Abu Dhabi, United Arab Emirates. *Journal of Human Evolution* 19:583–588.

Whybrow, P. J. and H. A. McClure. 1981. Fossil mangrove roots and palaeoenvironments of the Miocene of the eastern Arabian peninsula. *Palaeogeography, Palaeoclimatology, Palaeoecology* 32:213–225.

Chapter 28

Neogene Mammal Biostratigraphy and Chronology of Greece

GEORGE D. KOUFOS

The Neogene continental deposits of Greece are expansive, include several mammal fossiliferous sites, and provide useful information for their chronology and biostratigraphy. One of the first discovered Greek Neogene mammal localities is Pikermi (Attica, near Athens) found in 1835; its fauna is very rich and includes several new taxa, found subsequently in Eurasia and Africa. During the end of the nineteenth century, the mammal localities of Samos found by Forsyth Major yielded a great amount of fossils housed in various museums and institutes (Forsyth Major 1894). In 1993, a new series of excavations started in Samos, and the results are published in a separate volume of *Beiträge zur Paläontologie* (Koufos and Nagel 2009). The mammal localities of Axios Valley (Macedonia, Greece) were discovered during the beginning of the twentieth century (1915–1916) (Arambourg and Piveateu 1929). A new campaign of excavations in Axios Valley started at the beginning of the 1970s and continues now. The collection includes a rich fauna from known, stratigraphically controlled localities (Koufos 2006a), which can be used as a reference collection for the late Miocene of the Eastern Mediterranean.

During the last 40 years, several new excavations have been carried out in the Neogene of Greece. Several new Miocene fossiliferous sites (Maramena, Perivolaki, Nikiti, Kerassia, Antonios, Silata) have been found, and a great amount of fossils has been unearthed (Koufos, Kostopoulos, and Koliadimou 1991; Schmidt-Kittler 1995a; Koufos and Syrides 1997; Vasileiadou, Koufos, and Syrides 2003; Theodorou et al. 2003; Koufos 2006b,

2006c). The latest Pliocene mammal localities found in northern Greece (Volax, Dafnero, Sesklon) provided rich collections of fossils (Sickenberg 1968; Koufos et al. 1991b; Symeonidis 1992). This Villafranchian (Villanyian) collection is the best known in the eastern Mediterranean with certain stratigraphic control and age.

All new material, as well as the revision of the old collections that provided significant biochronological data, allowed the precise dating of the localities and the development of the biostratigraphy of the continental deposits of Greece. Moreover, other chronological methods (magnetostratigraphy, radiochronology) have been applied in several localities, providing data on their numerical ages. At the same time, the palaeoecological study of the fossil faunas and their comparison with the modern ones provided significant data for reconstructing the Neogene palaeoenvironment and its changes. References are numerous, and the present article is an effort to summarize them and to illustrate the biostratigraphy of the Neogene mammal localities of Greece. Extensive effort began in 1992, when the first faunal lists of the late Miocene were prepared for the Neogene Old World (NOW) database, and continues to the present with revision and the addition of new data (Koufos 2006a and ref. therein).

EARLY MIOCENE

The early Miocene mammal localities of the eastern Mediterranean are relatively few, and information from the

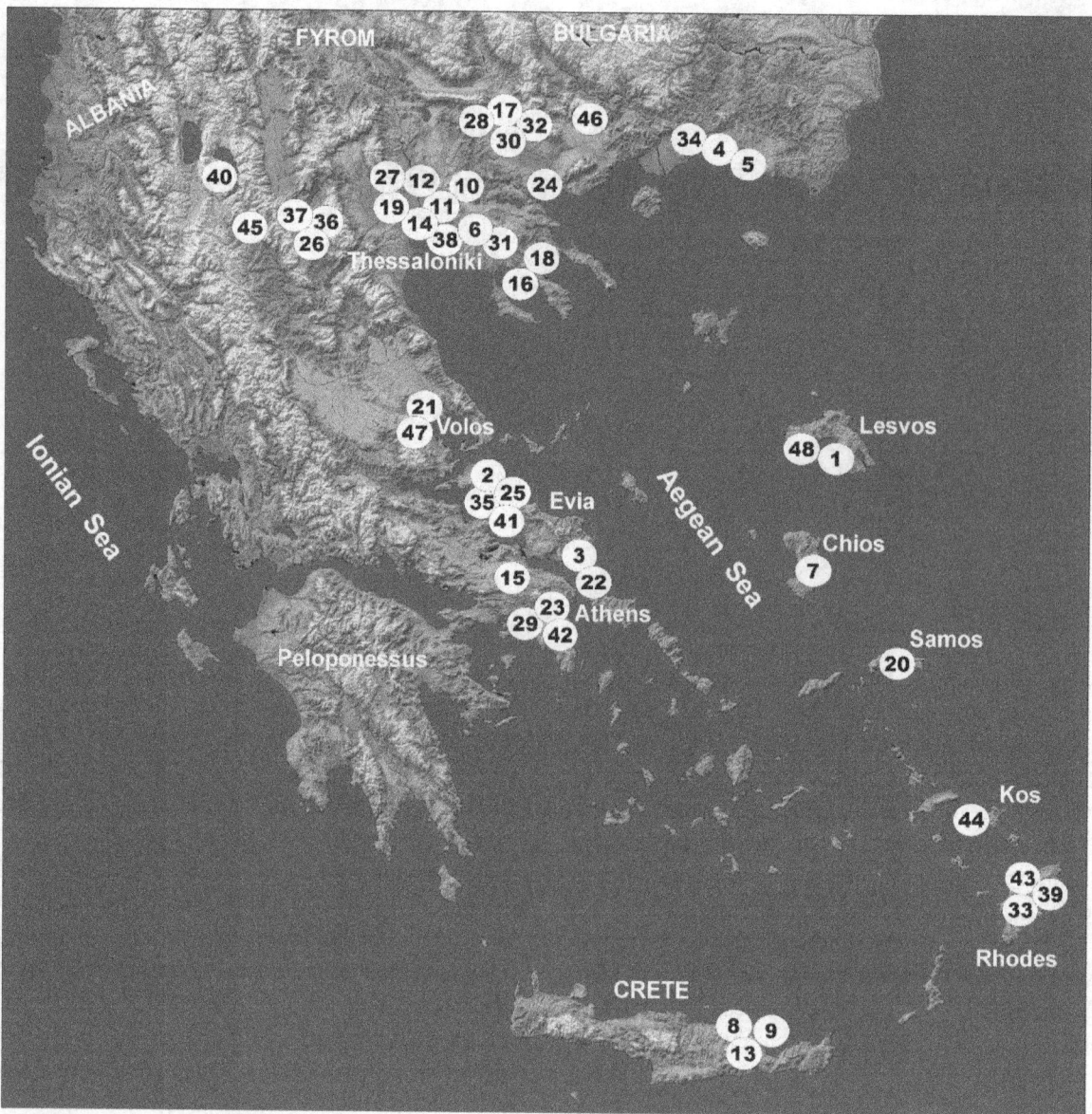

Figure 28.1 Map of Greece indicating the position of the main Neogene mammal localities of Greece: (*1*) Gavathas (GVT); (*2*) Kalimeriani (KLM); (*3*) Aliveri (ALI); (*4*) Karydia I+II (KAR); (*5*) Komotini (KOM); (*6*) Antonios (ANT); (*7*) Thymiana-A, B, C (THA, THB, THC); (*8*) Melambes (MLB); (*9*) Plakia (PLK); (*10*) Chrysavgi (CHR); (*11*) Pentalophos-1 (PNT); (*12*) Xirochori-1 (XIR); (*13*) Kastelios (KAS); (*14*) Ravin de la Pluie+Ravin des Zouaves-1 (RPl+RZ₁); (*15*) Biodrak (BDK); (*16*) Nikiti-1 (NKT); (*17*) Lefkon (LFK); (*18*) Nikiti-2 (NIK); (*19*) Ravin des Zouaves-5 (RZO), Vathylakkos localities (VLO, VTK, VAT), Prochoma-1 (PXM); (*20*) Samos-Mytilinii localities; (*21*) Perivolaki (PER); (*22*) Halmyropotamos (HAL); (*23*) Chomateres (CHO)-Pikermi (PIK); (*24*) Rema Marmara (REM); (*25*) Kerassia (KRS); (*26*) Lava-2 (LAV); (*27*) Dytiko localities (DTK, DIT, DKO); (*28*) Ano Metochi (MTH), Monasteri (MNS); (*29*) Pyrgos Vassilissis (PYV); (*30*) Maramena (MAR); (*31*) Silata (SLT); (*32*) Spilia-0-4 (SPO, SPL, SPI); (*33*) Marites (MRT); (*34*) Kessani-1, *2* (KES); (*35*) Limni-3 (LIM); (*36*) Kardia (KRD); (*37*) Ptolemais-1, *3* (PTO, PTL); (*38*) Megalo Emvolon (MEV); (*39*) Apolakkia (APK); (*40*) Kastoria-1 (KST); (*41*) Limni-6 (LMN); (*42*) Tourkovounia-1-5 (TRK, TKV); (*43*) Damatria (DAM); (*44*) Kardamena (KRM); (*45*) Dafnero (DFN); (*46*) Volax (VOL); (*47*) Sesklon (SES); (*48*) Vatera (VTR).

Balkan Peninsula is unknown or doubtful. In Greece, the early Miocene mammal faunas are limited. This is possibly due to lack of study and prospecting of the various areas in the eastern Mediterranean that have continental deposits.

The oldest known Miocene mammal locality of Greece is Gavathas (GVT) located on Lesvos Island (figure 28.1 [1]). The fossiliferous site is situated in the upper part of the Lacustrine Unit, consisting of alternating marly limestones and marls with thin lignitic intercalations; this unit is strongly silicified because of younger volcanic action in the area. The Lacustrine Unit is overlain by the Pyroclastic Unit, known in the area as the

"Sigri pyroclastics," consisting of ignimbrites, volcanic ashes, debris and mud flows, as well as glassy lavas (Koufos, Zouros, and Mourouzidou 2003 and ref. therein). Both lower tooth rows of a deinothere were found in Gavathas; they belong to the primitive *Prodeinotherium bavaricum* with similarities to the African early Miocene deinotheres (Koufos, Zouros, and Mourouzidou 2003). The first deinotheres and gomphotheres arrived in Eurasia by the "*Gomphotherium* land bridge" during early Orleanian, at ~19.0 Ma (Rögl 1999). The overlying Pyroclastic Unit has been dated radiometrically as younger than 18.4 ± 0.4 Ma (Pe-Piper and Piper 1993). Thus, the age of Gavathas locality must be older than 18.4 Ma, and taking in mind its stratigraphic position an age of ~19.0 Ma, corresponding to early Orleanian, MN 3 is quite reasonable for it (figure 28.2); this age fits quite well with the migration of the first deinotheres into Eurasia (Koufos, Zouros, and Mourouzidou 2003 and ref. therein).

The locality of Aliveri is situated on Evia Island in a lignitic pit, near the homonymous village (see figure 28.1 [3]). The fauna of Aliveri includes mainly micromammals; however, dental remains of two carnivores, *Euboictis* and *Palaeogale*, are recognized (Schmidt-Kittler 1983; Doukas 1986; de Bruijn et al. 1987; Hofmeijer and de Bruijn 1988). A complete list of the Aliveri fauna based on the known bibliography is given by Koufos (2006a). The species *Democricetodon franconicus*, *Megacricetodon primitivus*, and *Eumyarion weinfurteri* are the oldest representatives of the genera in western Europe, while the former two are slightly more primitive than those from the type localities (Hofmeijer and de Bruijn 1988). Both type localities (Erkertschofen and Valtorres) are MN 4 age (Mein 1990). Thus, Aliveri must be dated as middle Orleanian, MN 4 (see figure 28.2). This age was also proposed for Aliveri by Mein (1990) and de Bruijn et al. (1992).

There is a reference to the presence of the anthracothere *Brachyodus onoideus* in the Neogene of Evia Island. A mandibular fragment of this taxon was discovered near the village of Kalimeriani (see figure 28.1 [2]) and described by Melentis (1966). However, the exact locality and stratigraphic level of this specimen are unknown, and there is no information on accompanying fauna. The taxon is referred from the early to middle Orleanian (MN 3–4) of Europe (de Bruijn et al. 1992), and a similar age is possible for Kalimeriani. The area needs more prospecting for the relocation of the site, while the collecting of more material will provide more biochronological data. The locality of Kalimeriani is referred in figure 28.2 from MN 3 to MN 4 with a question mark.

Two early Miocene localities, Karydia I and II, are known from the area of Komotini, Thrace (see figure 28.1 [5]), which yielded a micromammalian fauna. The cricetids of this fauna have been studied extensively; the genera *Cricetodon* and *Anomalomys* are more derived than those of Aliveri, suggesting a younger age; thus, Karydia I+II are referred to late MN 4 (Theocharopoulos 2000). Another locality from the same area, named Komotini (KOM), yielded a faunule of small mammals. The few faunal data do not permit a definitive age assignment, but it should be similar to Karydia I+II (de Bruijn et al. 1992); thus, it is referred to MN 4 with a question mark (see figure 28.2).

MIDDLE MIOCENE

Although the middle Miocene mammal localities of Greece are more numerous than those of the early Miocene, they are still few, and the corresponding fauna is poorly known. The locality Antonios (ANT), Chalkidiki, Greece (see figure 28.1 [6]), was discovered in 1996; it is situated near Thessaloniki and includes a micro- and macromammalian fauna. The fossils are concentrated in the karstic fissures of the basement limestone and recovered during excavations for the construction of a road. Before that, the fissures were covered by the oldest sediments of the basin, which also partially filled them (Koufos and Syrides 1997). The initial determination of the collected fossils and their comparison with the known European material suggested an age from the upper part of MN 4 to the lower part of MN 5 (Koufos and Syrides 1997). The sanitheres of Antonios have smaller teeth than those from Thymiana, Chios Island, indicating their more primitive character (Koufos 2007). As the Thymiana fauna is dated to MN 5, at ~15.5 Ma, an older age at the beginning of MN 5 is quite possible for the Antonios fauna. The study of the micromammals indicated that they are more evolved and thus younger than those from Karydia I+II (MN 4). The stratigraphic range of the micromammalian taxa as well as their morphology and similarities suggest a correlation of Antonios with the MN 4/5 boundary interval (Vasileiadou and Koufos 2005), confirming the initial age determination.

One of the well-known middle Miocene fossiliferous sites of Greece is Thymiana, Chios Island (see figure 28.1 [7]). The locality was discovered in the 1940s, and several scientists excavated there (Paraskevaidis 1940; Melentis and Tobien 1967; Tobien 1980; Kondopoulou et al. 1993; Koufos, de Bonis, and Sen, 1995; de Bonis, Koufos, and Sen 1997a, 1997b, 1998; López-Antoñanzas, Sen, and Koufos 2004). The last excavations in Thymiana were carried out by a Helleno-French team in 1991–1993.

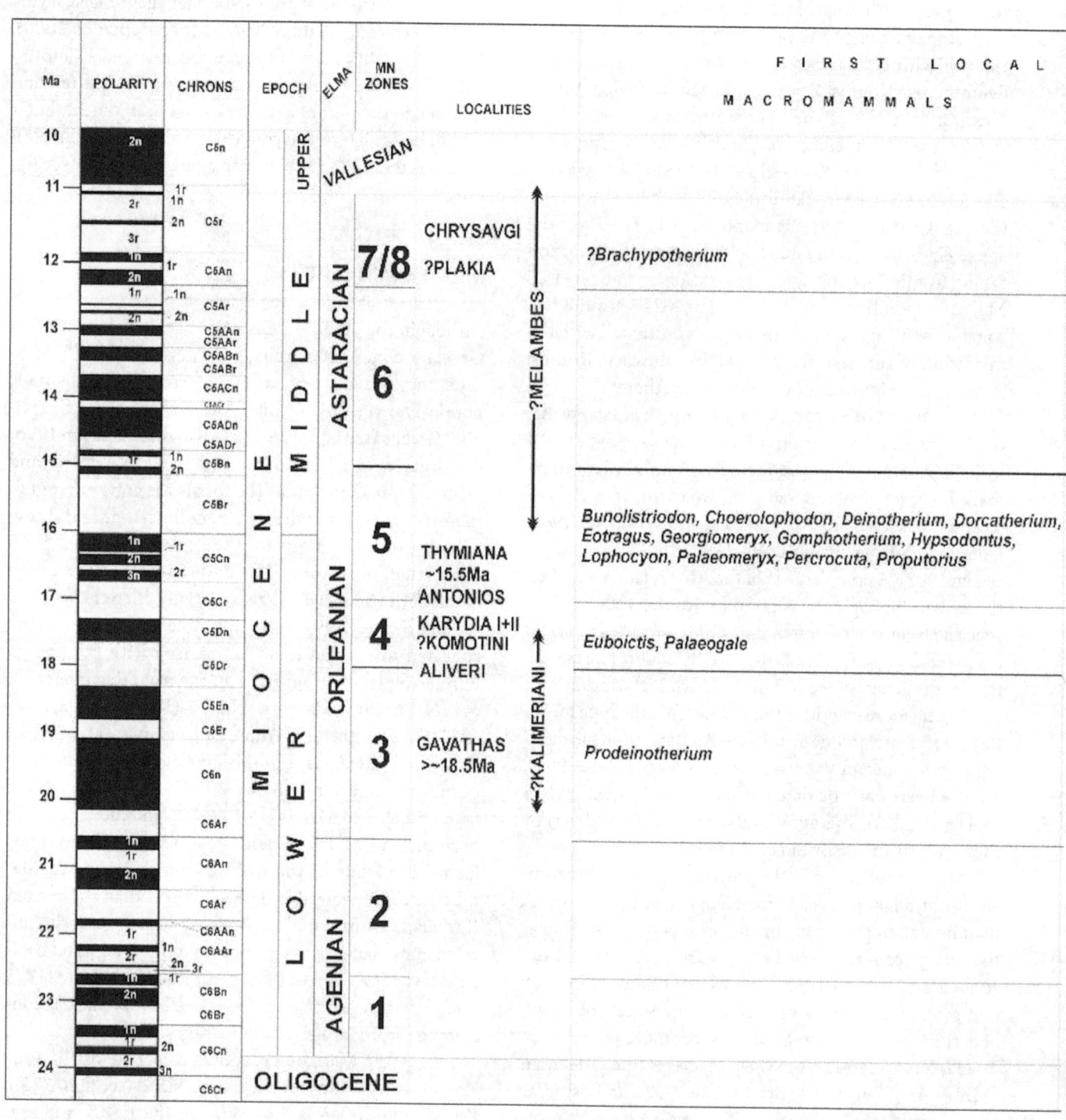

Figure 28.2 Biostratigraphy of the early to middle Miocene mammal localities of Greece with the first-appearing taxa. The geomagnetic time scale is from Cande and Kent (1995), and the MN zone boundary ages from Steininger (1999).

OCCURRENCE

MICROMAMMALS

Byzantinia, Cotimys   ← FLA *Byzantinia*

Alloptox, Myomimus, Peridyromys,   ← FLA *Choerolophodon*
Pliospalax, Prolagus, Sayimys,
Schizogalerix, Spermophillinus, Turkomys

Albertona, Aliveria, Anomalomys, Brasantoglis,   ← FLA *Eotragus, Proputorius,*
Cricetodon,Democricetodon, Desmanodon, Eumyarion,   *Sanitherium, Palaeomeryx*
Galerix,, Glirundinus,Glirulus,Glis, Heramys, Heterosorex,
Megacricetodon, Microdyromys, Miopetaurista, Mirabella,   ← FLA *Cricetodon, Megacrice-*
Myxomygale, Palaeosciurus, Plesiodimylus,   *todon, Democricetodon*
Pseudotheridomys, Tamias

← FLA *Prodeinotherium*

FLA= First Local Appearance

Three fossiliferous horizons have been recognized, Thymiana-A, B, C (THA, THB, and THC, respectively) The horizons THA and THC include small mammals, while THB only macromammals (Kondopoulou et al. 1993). In spite of extensive fieldwork, the Thymiana collection of large mammals is relatively poor because the fossiliferous deposits are extremely hard and it is difficult to collect the fossils. The Thymiana *Sanitherium schlagintweiti* is somewhat more evolved than that of Antonios (Koufos 2007), and as the latter fauna is dated to MN 4/5, the age of Thymiana must be MN 5. The rodent fauna of the Thymiana locality comes from two horizons, THA, a few meters below horizon THB, and THC, a few meters above; horizon THB contains the large mammals. The preliminary study of the Thymiana rodents indicates that the fauna includes several elements that are different from those of the typical European MN 4 faunas; it includes one ctenodactylid, five cricetids, and the ochotonid *Alloptox* (S. Sen, pers. comm.). The Thymiana ctenodactylids have been studied and belong to the species *Sayimys intermedius*; this taxon is known from Pakistan, the Arabian Peninsula, and Turkey (López-Antoñanzas, Sen, and Koufos 2004). In Turkey, it is known from the locality Paşalar, considered MN 5 or MN 6 (Begun, Güleç, and Geraads 2003; van der Made 2003, 2005). Although there are some affinities between the Thymiana and Paşalar faunas (de Bonis et al. 1977a; de Bonis, Koufos, and Sen 1997b), disagreement about the age of Paşalar does not allow certain age determination. The magnetostratigraphic study of the Thymiana section indicates that all mammal localities are included in a long reverse-polarity zone with alternations of short normal and reverse polarities (Kondopoulou et al. 1993; S. Sen, pers. comm.). The long reverse zone with the fossiliferous horizons can be correlated to the interval of chrons C58n.1n to C5C. This interval corresponds to the time interval from 16.0 Ma to 15.2 Ma according to Cande and Kent (1995), which correlates to the zone MN 5, as suggested by Steininger (1999). Considering all of this and the stratigraphic position of the fossiliferous sites, an age estimate of ~15.5 Ma is quite possible for the Thymiana fauna.

The locality Chryssavgi (CHR) is situated in Mygdonia Basin (Macedonia, Greece; see figure 28.1 [10]) and was rediscovered at the end of the 1980s. The CHR fauna includes mainly micromammals, but some dental remains of ?*Brachypotherium* are also referred from this locality (Koufos 2006a). The Chryssavgi fauna includes *Byzantinia bayraktepensis*, which is similar to the type material from the Turkish locality of Bayraktepe-1. The latter is dated to the late Astaracian, MN 7+8 (Ünay 1980), and a similar age is quite possible for the CHR

fauna. The genus *Megacricetodon* is absent in Bayraktepe-1 but present in the older Kalamis locality with a form intermediate between *Cricetodon* and *Byzantinia* (Ünay and de Bruijn 1984); thus, CHR is younger than Kalamis. In the younger locality of Yeni Eskihisar, *Desmanodon minor* is replaced by the more advanced *D. major*. The presence of *D. minor* in CHR suggests an age older than that of Yeni Eskihisar; thus, the CHR fauna dates to the late Astaracian, MN 7+8 (see figure 28.2). Its age is similar to that of the Turkish locality of Bayraktepe-1 and between the localities Kalamis and Yeni Eskihisar (Koliadimou and Koufos 1998).

The locality Melambes (MLB) is situated near Heraklion, Crete (see figure 28.1 [8]) has provided a limited fauna, including the remains of a hyracoid, cf. *Prohyrax hendeyi*, and a tragulid, *Dorcatherium naui* (Bonneau and Ginsburg 1974; van der Made 1996). Based on the tragulid, Bonneau and Ginsburg (1974) proposed a Vallesian age for Melambes. Later, the locality was referred to the late Astaracian, MN 7+8 (Benda and Meulenkamp 1990). Recently, van der Made (1996) used the hyracoid to suggest an early Astaracian (MN 6) age for Melambes. It is obvious that this age debate is due to poor material and that more fossils are necessary for a certain age determination; this is the reason that the site Melambes is given with a long stratigraphic range and question mark in figure 28.2.

LATE MIOCENE

Contrary to the small number of the early and middle Miocene mammal localities of Greece, those of late Miocene are numerous. The late Miocene fauna is rich and includes mainly macromammals; in fact, the occasional rarity of micromammals is due to lack of work done on them. Among the late Miocene mammal localities of Greece, some (Pikermi, Samos, Axios Valley) are very well known and their faunas are very important for biochronology. Moreover, several new localities have been discovered during the last 40 years, yielding very rich fauna and providing many useful data for biochronology. As the late Miocene localities are numerous, only the most important ones and those for which there are recent chronological data will be noted in the following.

Axios Valley

The late Miocene mammal localities of Axios Valley were known from the Arambourg collection housed in the

Museum National d'Histoire Naturelle of Paris (MNHN); it is reported as "Salonique" or "Saloniki" in the old biostratigraphic tables of the European mammal localities. The problem is that Arambourg's collection includes material from various localities and belonging to different stratigraphic horizons. This mixing does not lead to a certain age determination. On the other hand, the name "Salonique" or "Saloniki" has no meaning, as the fossiliferous sites are far from the city of Thessaloniki. Since 1972, a new series of excavations in the Axios Valley localities has yielded a great amount of fossils, which provided significant data about the biostratigraphy of late Miocene deposits. More than 20 fossiliferous sites have been discovered in Axios Valley, belonging to three different stratigraphic horizons.

Nea Messimvria Localities

This group includes the localities of Pentalophos-1 (PNT), Xirochori-1 (XIR), Ravin de la Pluie (RPl), and Ravin des Zouaves-1 (RZ1) (see figure 28.1 [11, 12, 14]), all situated in the Nea Messimvria Formation. The localities RPl and RZ1 possibly belong to the same stratigraphic horizon, as they are geographically close each other; RPl and XIR are well known because of the presence of the hominoid Ouranopithecus macedoniensis (de Bonis et al. 1974; de Bonis et al. 1990a). The RPl fauna is rich, providing certain data for its age. The fauna includes the large-size Hipparion primigenium, which is more evolved than the typical form from Eppelsheim (MN 9), suggesting younger age for RPl (Koufos 1986, 1990). The bovids are represented in the RPl fauna mainly by Mesembriacerus melentisi and Samotragus praecursor. The first taxon has more primitive features than the Turolian ovibovines, while the second one is smaller with more primitive characters than Samotragus crassicornis from the Turolian of Samos (Bouvrain and de Bonis 1984, 1985). The locality of S. crassicornis is unknown, and thus, although there are certain ages for the Samos localities, it cannot be used for a more precise age determination. The RPl Prostrepsiceros vallesiensis is smaller than the Turolian Prostrepsiceros with primitive morphology, indicating an older age (Bouvrain 1982). The well-known late Miocene hyaenid Adcrocuta eximia is represented in RPl by a subspecies with more primitive features than the typical Turolian one, suggesting an older age (de Bonis and Koufos 1981). The main RPl giraffid was referred to ?Decennatherium (Geraads 1979), but recently it was transferred to the new genus Palaeogiraffa by de Bonis and Bouvrain (2003). The genus is Vallesian and is represented by three species in the three Vallesian localities of

Axios Valley (PNT, XIR, RPl). The larger species, D. major, with the more evolved milk molars is present in RPl, indicating that it is younger than PNT and XIR (de Bonis and Bouvrain 2003). The rodent Progonomys cathalai was also recognized in RPl, confirming its Vallesian age. Based on all of the preceding, RPl can be dated to late Vallesian, MN 10. The magnetostratigraphic study indicated that RPl is included in a normal polarity zone, which can be correlated to chron C4Ar.1n (9.31–9.23 Ma according to Cande and Kent 1995); thus, an estimated age of ~9.3 Ma is quite possible for the RPl fauna (Sen et al. 2000), consistent with its late Vallesian (MN 10) age.

The XIR locality, situated close to RPl (~1.5 km), yielded few fossils. In spite of the poor fauna overall, the presence of the hominoid Ouranopithecus macedoniensis (de Bonis et al. 1990a; de Bonis and Koufos 1993), the giraffid Palaeogiraffa (de Bonis and Bouvrain 2003), and the bovid Samotragus praecursor (G. Bouvrain, pers. comm.) suggest strong similarities to the RPl fauna and indicate a similar late Vallesian age. The magnetostratigraphic record of the XIR section indicates that the fossiliferous site is situated in the base of a normal polarity zone correlated to chron C4Ar.2n, corresponding to a 9.64–9.58 Ma time interval (see figure 28.2); this interpretation suggests an estimated age of ~9.6 Ma for the XIR fauna (Sen et al. 2000).

The locality PNT is placed close to RPl and XIR, and all occur within the Nea Messimvria Formation. Although the PNT fauna is quite rich, it has a peculiar character as most of its faunal elements differ from those of the other two Vallesian localities. Koufos (2000a) studied the PNT hipparions, and based on them, as well as on the other available data, he suggested a late MN 9 or early MN 10 age. The presence of the giraffid Palaeogiraffa macedoniae in PNT and its more primitive features than Palaeogiraffa from XIR and RPl (de Bonis and Bouvrain 2003) indicates that PNT is older than XIR and RPl. The PNT aardvark is similar to Amphiorycteropus pottieri from the Vallesian levels of Sinap Tepe in Turkey (de Bonis et al. 1994), suggesting a similar age for the PNT fauna. A. pottieri was recognized in the localities of Sinap 108, 72, and 12, dated from ~10.2 Ma to 9.6 Ma (Fortelius, Nummela, and Sen 2003). The MN 9/10 boundary is estimated to 9.7 Ma according to Steininger (1999) or to 9.64 Ma according to Sen (1997); thus, A. pottieri is dated to the uppermost MN 9, and a similar age is possible for PNT. The bovids from PNT are different from those of RPl and RZ1. The Protoryx from Pentalophos is more derived than the Protoryx of the Beglia Formation, Tunisia (MN 8/9), but less than that of Pikermi and Samos

(Greece) *Protoryx* (Bouvrain 1997), suggesting a Vallesian age. Taking all this in mind, PNT is certainly Vallesian and older than RPl and XIR, the latter estimated at ~9.6 Ma (Sen et al. 2000). With the MN 9/10 boundary estimated at ~9.6 Ma, PNT would be correlated to the uppermost early Vallesian (MN 9).

Vathylakkos Localities

The better-known localities of this group are Ravin des Zouaves-5 (RZO), Vathylakkos-1, 2, 3 (VLO, VTK, VAT), and Prochoma-1 (PXM) (see figure 28.1 [19]); some other smaller fossil concentrations were also found in the Vathylakkos Formation, but the collected material is poor and difficult to attribute. The locality RZO yielded a rich fauna and provided good biochronological data. The presence of *Mesopithecus*, more exactly its largest form, *M. delsoni*, indicates early Turolian age (de Bonis et al. 1990b; Koufos 2009a, 2009b). The equids *Hipparion dietrichi* and *H. proboscideum* suggest an age from the early to the middle Turolian, too; both are known from the Turolian of Samos and Turkey (Koufos and Vlachou 2005; Vlachou and Koufos 2009). The RZO *Adcrocuta eximia* differs from that of RPl, but it resembles the typical form from Pikermi and other Turolian localities (Koufos 1980, 2000b; de Bonis and Koufos 1994). The RZO bovids include several taxa. *Palaeoreas zouavei* is more primitive than the typical *P. lindermayeri* from Pikermi, suggesting an older age (Bouvrain 1980). *Nisidorcas planicornis* is a Turolian element; more precisely, the RZO form has more primitive features than Vathylakkos and Perivolaki (MN 12) forms (Kostopoulos 2006), suggesting early Turolian age. The RZO suid *Propotamochoerus* cf. *hysudricus* is similar to that from an unknown locality of Samos (de Bonis and Bouvrain 1996), indicating Turolian age. Therefore, the RZO fauna can be dated to early Turolian, MN 11. The magnetostratigraphy of the RZO section indicates that the locality is included in a reverse polarity zone, which can be correlated to chron C4r.1r, corresponding to the time interval from 8.23 Ma to 8.07 Ma; this interpretation provides an estimated age of ~8.2 Ma for RZO, fitting well with the biochronological data (see figure 28.2).

The Vathylakkos (VLO, VTK, VAT) and Prochoma (PXM) localities (see figure 28.1 [19]) are close to each other and include similar fauna with Turolian characters. The equids include *Hipparion dietrichi*, known from the early to the middle Turolian of Greece and Turkey (Vlachou and Koufos 2002, 2006, 2009; Koufos and Vlachou 2005). The presence of *Nisidorcas* is another

evidence for a Turolian age; the Vathylakkos *Nisidorcas* belongs to the more evolved middle Turolian forms (Kostopoulos 2006). The hyaenid *Adcrocuta eximia* belongs to the Turolian subspecies of the taxon, suggesting a similar age (Koufos 2000b). The Vathylakkos and Prochoma *Microstonyx* belong to group A, dated to MN 11–12 (Kostopolous, Spassov, and Kovachev 2001). The Vathylakkos bovids include some taxa closer to early Turolian (MN 11) ones, but some other taxa are closer to the middle Turolian (MN 12; Bouvrain 2001); this suggests an age closer to the lower part of MN 12. The presence of *Parapodemus schaubi* in Vathylakkos (Koufos 2006a) suggests a middle Turolian (MN 12) age. The Vathylakkos *Mesopithecus* is smaller than *M. delsoni* from RZO, and its morphology is between this species and *M. pentelicus* from Pikermi; thus, it is younger than RZO and older than Pikermi (de Bonis et al. 1990b; Koufos et al. 2004; Koufos, Kostopoulos, and Merceron 2009; Koufos, Kostopoulos, and Vlachou 2009). The magnetostratigraphy of the PXM and Vathylakkos sections situates the PXM fossiliferous site in a reverse polarity zone (Kondopoulou et al. 1992; incorrectly referred to the normal polarity zone in the section by Sen et al. 2000). Corrected placement of PXM correlates to the lower reverse polarity zone of the Vathylakkos section. The normal polarity zone of the Vathylakkos section fits better with chron C3Br.2n (7.37–7.34 Ma), and the PXM reverse polarity zone with chron C3Br.3r (7.43–7.37 Ma). This interpretation agrees with the presence of *Parapodemus schaubi* and the "more evolved than PXM" character of the Vathylakkos fauna (Koufos et al. 2006). Thus, a corrected age of ~7.4 Ma and ~7.3 Ma for the PXM and Vathylakkos localities, respectively, is adopted.

Dytiko Localities

Three localities, Dytiko-1, 2, 3 (DTK, DIT, DKO), are close to one another (see figure 28.1 [27]), situated in the Dytiko Formation, and have quite similar fauna. The Dytiko fauna includes several mammal taxa that allow unambiguous biochronology. *Mesopithecus* from Dytiko differs from the typical *M. pentelicus* from Pikermi in being slightly smaller. Moreover, a small-size form similar to *M. monspessulanus* is present (de Bonis et al. 1990b; Koufos 2009a, 2009b), indicating that Dytiko is younger than Pikermi and belongs to the late Turolian, MN 13. The Dytiko bovids are also different from the Pikermi species. The genus *Protragelaphus* is represented by *P. theodori*, which is more derived than *P. skouzesi* in Pikermi (Bouvrain 1979), suggesting a younger age. The suid *Mi-*

crostonyx is larger than that of Pikermi, suggesting a younger age (de Bonis and Bouvrain 1996). Kostopolous, Spassov, and Kovachev (2001) consider the Dytiko *Microstonyx* large size and in group-C, which is younger than Pikermi of MN 12–13. The new genus *Dytikodorcas longicornis* is very similar to *Dytikodorcas libycus* (=*Prostepsiceros libycus*) from Sahabi; the latter is dated to the late Turolian, MN 13, and thus a similar age is quite possible for Dytiko. Among the Dytiko hipparions there is a very small form, like that known from Q5 of Samos; Q5 is dated at about 6.9–6.7 Ma (Koufos, Kostopoulos, and Merceron 2009), corresponding to MN 13. All of these data suggest a late Turolian (MN 13) age for Dytiko localities.

Mytilinii Basin, Samos

The mammal localities of Samos yielded numerous fossils, but most of the old collections are without locality indications, making impossible their correlation with the local stratigraphy. Even those collections (Forsyth Major's in Lausanne and London, as well as Brown's in New York) having locality information are difficult to correlate with the local stratigraphy (details in Kostopoulos et al. 2009). Thus, there is a significant problem in the dating of the localities, and several interpretations are known. The Samos fauna is considered either as a single homogeneous and isochronous fauna deposited in a short time span (Solounias 1981) or as belonging to two different stratigraphic horizons (Gentry 1971; Sondaar 1971). The radiochronology of the various volcanic deposits or tuffs provided some ages, which could not solve the problem. In 1993, a new campaign of fieldwork and excavations began in Samos by a team of palaeontologists led by the author. The main goals of this research were as follows:

- Relocation of the old fossiliferous sites and discovery of new ones
- Detailed stratigraphy with location of the fossiliferous sites
- Creation of new fossil collections from select localities with clear stratigraphic position
- Extended magnetostratigraphic study of the fossiliferous formation
- Study of new material together with revision of the old collections and comparison with other faunas from Eurasia to obtain biochronological data
- Determination of age and palaeoenvironment for all localities

The results of the first 12 years of field work is published (Koufos and Nagel 2009).

All known localities of Samos are situated in the Mytilinii Formation, a volcaniclastic series of sediments, representing a continental phase in the history of the Mytilinii Basin (Kostopoulos et al. 2009 and ref. therein). All the new localities, as well as the old ones, were correlated with the local stratigraphy; for the old localities, the correlation is based on personal field observations, bibliographic references, and study of the old collections. The study of the new material from Samos and its comparison with the old Samos collections, as well as with the other Eurasian faunas, indicates that four different faunal assemblages can be distinguished (Koufos, Kostopoulos, and Merceron 2009).

The oldest new locality is Mytilini-4 (MLN), situated at the base of the main fossiliferous beds of the Mytilinii Formation (Kostopoulos et al. 2009). The majority of the determined taxa suggest Turolian age. The MLN hipparions indicate similarities with the early/middle Turolian (MN 11–12) forms; their detailed comparison suggests an age close to the base of MN 12 (Vlachou and Koufos 2009). The MLN giraffids and bovids suggest rather an early Turolian age, although there are middle Turolian elements (Kostopoulos 2009a, 2009b). The magnetostratigraphy of the Mytilinii Formation has been studied extensively, using two long sections (Mylos, Dromos), plus short ones for better correlation (Kostopolous, Sen, and Koufos 2003). The lower part of the main fossiliferous bed, including MLN, can be correlated to chrons C3Br.2n–C4N.1n ranging from 7.65 Ma to 7.45 Ma and suggesting an estimated age of ~7.5 Ma for MLN. The old locality Q2 of Brown and "Stefano" label of Forsyth Major probably correspond to MLN and have a similar age (Kostopoulos et al. 2009; Koufos, Kostopoulos, and Merceron 2009).

The new locality, Mytilini-3 (MYT), is situated in the upper part of the main fossiliferous beds of the Mytilinii Formation (Kostopoulos et al. 2009). The hipparions from MYT are slightly different from those of MLN and indicate middle Turolian (MN 12) age (Vlachou and Koufos 2009). *Samotherium boissieri* known from MLN is replaced by the larger *S. major* in MYT, indicating younger age (Kostopoulos 2009a). The MYT bovids also include middle Turolian (MN 12) taxa, like *Gazella pilgrimi*, *Sporadotragus parvidens*, *Skoufotragus zemalisorum*, and *Palaeoryx* sp. (Kostopoulos 2009b). The association of "*Diceros*" *neumayri*, *Dihoplus pikermiensis*, and *Ancylotherium pentelicum* (Giaourtsakis 2009; Giaourtsakis and Koufos 2009) suggests also middle Turolian age. The part of the section, including MYT, is magnetostratigraphically correlated with chron C3Br.2r corresponding to the 7.43–7.37 Ma time interval; this interpretation suggests an estimated age of ~7.3 Ma for MYT, middle Turolian age. The locality MYT is probably identical to Q3 of Brown, "Potamies" label of Forsyth Major and S2-3 of Solounias; a similar age is quite possible for these localities.

The younger new locality is Mytilini-1 (MTL) situated in Adrianos ravine and in the upper part of the main fossiliferous beds. This locality includes several fossiliferous sites named MTLA, MTLB, MTLC, MTLD, and MTLE (Kostopoulos et al. 2009). Their fauna is quite rich and includes several taxa (Koufos, Kostopoulos, and Merceron 2009). The rodents *Pseudomeriones pythagorasi* and "*Karminata*" *provocator* indicate middle Turolian age (Vasileiadou and Sylvestrou 2009). The presence of *Choerolophodon pentelici* in MTL suggests Turolian age (Konidaris and Koufos 2009). The MTL hyaena belongs to the Turolian subspecies of *Adcrocuta eximia*, while the resemblance of *Machairodus giganteus* from MTL to that from Vathylakos and Halmyropotamos suggests middle Turolian age (Koufos 2009c). The rhinocerotid association "*Diceros*" *neumayri*, *Dihoplus pikermiensis*, and *Ancylotherium pentelicum* suggests middle Turolian (MN 12) age, too (Giaourtsakis 2009; Giaourtsakis and Koufos 2009). The hipparionine horse *H. brachypus* known from Pikermi, Akkaşdaği (Turkey), and Hadjidimovo (Bulgaria) also suggests middle Turolian age (Vlachou and Koufos 2009). The large-size *Microstonyx* of MTL indicates an age at the upper part of the middle Turolian (Sylvestrou and Kostopoulos 2009). The giraffid *Samotherium major* from MTL is larger than that from MYT (lower MN 12), indicating younger age (Kostopoulos 2009a). The bovids *Palaeoryx pallasi* and *Gazella capricornis* are typical middle Turolian taxa known from Pikermi, and their presence in MTL indicates a similar age (MN 12); more precisely, MTL must be considered in the upper part of MN 12. This part of the Mytilini Formation is correlated magnetostatigraphically with chron C3Br.1n, ranging from 7.17 Ma to 7.13 Ma, while an estimated age of ~7.1 Ma is quite possible for the fauna (Kostopoulos, Sen, and Koufos 2003; Koufos, Kostopoulos, and Merceron 2009); this age corresponds to the end of the middle Turolian (MN12) according to Sen (1997) or to late MN 12 according to Steininger (1999). The old localities "Adriano" of Forsyth Major, "Adrianos or Stefanidis ravine" of Melentis, and Q1 of Brown correspond to one of the MTL fossiliferous sites and have similar age.

The detailed stratigraphic correlation of the old fossiliferous sites with the local stratigraphy of the Mytilinii Basin indicates two other fossiliferous horizons. An old one situated in the base of the Mytilinii Formation and another one in the uppermost part of the formation (Kostopoulos et al. 2009; Koufos, Kostopoulos, and Merceron 2009). The localities Qx of Brown and "Vryssoula" of Forsyth Major and Skoufos are included in this horizon. Although we relocated these sites, we did not excavate to get new material. However, the magnetostratigraphic record indicates that this part of the Mytilinii Formation can be correlated to chron C4n.2n, corresponding to 8.0–7.6 Ma and indicating an early Turolian (MN 11) age (Kostopolous, Sen, and Koufos 2003; Koufos, Kostopoulos, and Merceron 2009). The uppermost part of the Mytilinii Formation includes the old localities Q5 of Brown and ?Limitzis; Q5 was relocated, but our efforts to find fossils were unsuccessful. However, this part of the Mytilinii Formation can be correlated to chrons C3Ar–C3Bn, corresponding to 7.1–6.5 Ma (Kostopolous, Sen, and Koufos 2003; Koufos, Kostopoulos, and Merceron 2009). Hence, Q5 is dated to MN 13 or to latest MN 12 according to Sen (1997) or to Steininger (1999), respectively (see figure 28.2).

Perivolaki, Thessaly

The locality of Perivolaki (PER) is situated in central Greece, Thessaly (see figure 28.1 [21]); it was discovered in 1996 by geologists from the Institute of Geological and Mineral Exploration of Greece (IGME). A team of palaeontologists, led by the author, excavated there from 1997 to 2004; the collected material has been studied extensively, and a magnetostratigraphic study of the section has been carried out. The study of the PER fauna, chronology, and palaeoecology are given in a separate volume of *Palaeontographica* (Koufos 2006b).

The fossiliferous site PER is located in the upper part of the lithostratigraphic unit B (Koufos et al. 1999; Sylvestrou and Koufos 2006) and includes quite a rich mammal fauna with a clear Turolian character. The occurrence of *Mesopithecus* is an evidence for Turolian age, and its resemblance to Vathylakkos *Mesopithecus* suggests similar middle Turolian (MN 12) age (Koufos 2006d). The PER *Adcrocuta eximia* resembles that from Pikermi and Vathylakkos, while *Plesiogulo crassa* from PER is similar to the Vathylakkos form (Koufos 2006e), indicating a middle Turolian age. The PER hipparions include the taxa *H. dietrichi*, *H. macedonicum*, and *H. mediterraneum*, suggesting also an early to middle Turolian age (Vlachou and Koufos 2006). The PER *Microstonyx* is metrically similar to that of Vathylakkos and also suggests middle Turolian age (Sylvestrou and Kostopoulos 2006). The coexistence of *Gazella pilgrimi*, *Nisidorcas planicornis*, and

Tragoportax rugosifrons gives a primitive feature to the PER fauna, indicating early Turolian (MN 11) age. However, the last two taxa seem to be more evolved than those of Vathylakos, suggesting a slightly younger age for PER (Kostopoulos 2006:162–163, 174). On the other hand, the PER *Prostrepsiceros* cf. *fraasi* resembles the middle Turolian forms of the genus, while cf. *Helladorcas* represents a descendant of *Protragelaphus*, indicating a MN 11 to early MN 12 age (Kostopoulos 2006). Hence, all the biochronological data from PER suggest correlation to the lower part of the middle Turolian (MN 12). A number of sections in the surrounding area of PER including the Miocene deposits have been sampled for magnetostratigraphy. The fossiliferous site is situated in a long reverse polarity zone, which can be correlated to chron C3Br.2r, corresponding to 7.3–7.1Ma (Koufos et al. 2006). Taking in mind that the PER fauna is slightly more derived than the Prochoma (~7.4 Ma) and Vathylakkos (~7.3 Ma) ones and less than Pikermi (~7.1 Ma), an age of ~7.2 Ma is possible for it.

Nikiti, Chalkidiki

Until the early 1990s, late Miocene faunas were almost unknown in the Chalkidiki Peninsula, although there were some sporadic references to specimens in the area. The localities of Nikiti (see figure 28.1 [18]; Koufos et al. 1991) are under excavation by the author. More than five fossiliferous sites have been found, two of them yielding a rich fauna. Both localities are situated in the Nikiti Formation. The sediments of the Nikiti Formation are coarse (gravels, pebbles, coarse sands), and only in the upper part are there some lenticular intercalations of sandy clays and finer sands, which are fossiliferous (Koufos et al. 1991). In such deposits, it is impossible to get samples for magnetostratigraphy, and thus the age determination of the Nikiti localities is based on biochronological and stratigraphical data.

The older locality is known as Nikiti 1 (NKT), well known because of the presence of the hominoid *Ouranopithecus macedoniensis* (Koufos 1993, 1995). The occurrence of this hominoid indicates a late Vallesian age. The NKT hipparions include the taxa *H. primigenium* and *H. macedonicum*; this association characterizes the late Vallesian of the Axios Valley (Koufos 1990, 2000c). The bovids from NKT include *Miotragocerus* cf. *pannoniae*, *Prostrepsiceros syridisi*, *Oioceros* cf. *atropatenes*, and ?*Gazella* sp. (Kostopoulos and Koufos 1997; Kostopoulos 2005). *M. pannoniae* is known from the Vallesian of Central Europe (Fortelius 2009). The NKT *Prostrepsiceros*, although

it resembles *P. houtumschindleri* from Maragheh (MN 11–12), has some differences that indicate an older age (Kostopoulos and Koufos 1997). The giraffids from NKT provided an age between late Vallesian–early Turolian (Kostopolous, Koliadimou, and Koufos 1996). The NKT fauna includes some late Vallesian taxa but also includes the early appearances of some Turolian ones; thus, a latest Vallesian age is quite possible for it. According to Steininger (1999), the Vallesian–Turolian boundary is dated at ~8.7 Ma; considering that RPl is dated at ~9.3 Ma and the NKT fauna is slightly more advanced than the RPl one, an age between 9.3 Ma and 8.7 Ma is quite possible for NKT.

The younger locality Nikiti 2 (NIK) is situated 20 m above NKT and yielded a completely different fauna with clear Turolian elements. Its stratigraphic position above the latest Vallesian locality NKT is a strong evidence for a younger, early Turolian age. The presence of *Mesopithecus* (Koufos 2009b), replacing the hominoid *Ouranopithecus*, suggests a Turolian age, too. The NIK hipparion association with *H. dietrichi*, *H. macedonicum*, and *H. proboscideum* suggests an early Turolian age (Vlachou and Koufos 2002). The NIK assemblage of bovids is more advanced than that of NKT, indicating an early Turolian (MN 11) age (Kostopoulos and Koufos 1999). The localities of Nikiti probably include the Vallesian–Turolian boundary in Greece, but the lack of magnetostratigraphy prevents its certain recognition. The NIK fauna is the oldest Turolian one for Greece, marking the beginning of this time in the area.

Pikermi

The classical site of Pikermi is situated in the well-known Pikermi ravine, across which there are several fossiliferous sites. All the old collections come from these sites but with uncertain provenance. It is sure that the material is mixed, and the available collections cannot provide clear biochronological data. Thus, the several analyses and comparisons of the faunal data suggest various ages for Pikermi.

However, based on the available data, an age determination of Pikermi will be attempted. The Pikermi faunal list used for the various comparisons is based on Gaudry's collection housed in MNHN; according to Gaudry (1862–1867), he excavated in one site of the ravine, and it is quite possible that its collection may be homogeneous. The Pikermi fauna is different from that of the middle Turolian Vathylakkos. The hipparions of Pikermi are different (*H. mediteraneum–H. brachypus* in Pikermi versus

H. dietrichi–*H. macedonicum* and possibly *H. probosci-deum* in Vathylakkos). The Pikermi bovids are also different; *Tragoportax amalthea* and *T. gaudryi* instead of *T. rugosifrons*, *Gazella capricornis* in place of *G. pilgrimi*, and the absence of *Nisidorcas planicornis*. The *Microstonyx* of Pikermi is larger than that of Vathylakkos, indicating younger age (de Bonis and Bouvrain 1996). The Pikermi *Microstonyx* belongs to the medium-size group-B, and it is younger than the small-size group-A in which Vathylakkos-Prochoma *Microstonyx* belongs (Kostopolous, Spassov, and Kovachev 2001). The Pikermi *Meso-pithecus* is also smaller and different from Vathylakkos *Mesopithecus* (de Bonis et al. 1990b; Koufos et al. 2004; Koufos 2009b), indicating a younger age than Vathylakkos. Finally, the presence of *Parapodemus gaudryi* in Pikermi versus *Parapodemus schaubi* in Vathylakkos indicates a younger age. Thus, the Pikermi fauna is younger than that of Vathylakkos, which means younger than ~7.3 Ma. The Pikermi fauna is similar to the fauna of the Samos locality Mytilinii-1 (MTL), dated at 7.1 Ma (Koufos, Kostopoulos, and Vlachou 2009), and thus a similar age is quite possible for it. The Perivolaki fauna is slightly different from that of Pikermi; its ~7.2 Ma age is consistent with a possible age of ~7.1 Ma for Pikermi (Koufos et al. 2006). One of the recently collected and studied Turkish faunas is that of Akkashdaği, which has similarities to that of Pikermi; taking in mind that the Akkashdaği age is ~7.0 Ma (Karadenizli et al. 2005), a similar age is quite possible for Pikermi. The PIK fauna matches those of DYTI and CHO (see figure 28.7b [sub-cluster B5]), indicating age similarities, too. All these data and correlations indicate that Pikermi could be dated to late middle Turolian (MN 12), according to Steininger (1999) or to the middle/late Turolian boundary according to Sen (1997), at ~7.1 Ma. However, a definitive age for Pikermi will come from new research in the area studing the stratigraphy, making new collections, and applying new methods for age determination (magnetostratigraphy, or radiochronology if possible).

Kerassia, Evia Island

The locality Kerassia (KER) is situated in the northern part of Evia Island (see figure 28.1 [25]). It was discovered in 1981 and excavated by H. de Bruijn and C. Doukas in 1982, the collected fossils are housed in Athens. A group of palaeontologists from the University of Athens, led by Prof. G. Theodorou, started a new campaign of excavations at Kerassia in 1992 and collected a significant number of fossils (Theodorou et al. 2003). According to them, the Kerassia fossiliferous sites belong to two fossiliferous horizons. The lower horizon includes the sites K2, K3, and K4, while the upper one the sites K1 and K6; faunal lists are given separately but seem similar.

The first reference to KER age is given in a brief description of a *Microstonyx* skull by van der Made and Moya-Sola (1989) suggesting a middle Turolian age. Later comparison of KER *Microstonyx* indicated that it is similar to that from Perivolaki, Vathylakkos, and Prochoma, belonging to group A (Kostopolous, Spassov, and Kovachev 2001). All these localities are dated to MN 12, 7.5–7.3 Ma, and thus a middle Turolian (MN 12) age is possible for Kerassia. The Kerassia *Adcrocuta eximia* belongs to the Turolian subspecies *A. e. eximia* and resembles that from Vathylakkos, Prochoma, Adrianos-Samos, and Pikermi (Roussiakis and Theodorou 2003); all these localities are dated to the middle Turolian, MN 12, and given the absence of *A. eximia* in the Greek late Turolian (MN 13) faunas, a middle Turolian age is quite possible for Kerassia. The felid *Metailurus parvulus* from KER is similar to that from the middle Turolian localities Chomateres, Pikermi, and Mytilinii-1-Samos (Roussiakis, Theodoro, Ilipoulos 2006; Koufos 2009e). The giraffid association *Palaeotragus rouenii*, *Helladotherium duvernoyi*, *Bohlinia attica*, and *Samotherium major* is referred from KER (Iliopoulos 2003). The previous three taxa have a long stratigraphic range in Greece, covering the Vallesian and Turolian. *Samotherium major* is a more derived form than *S. boissieri*. The latter is known from the Samos locality Mytilinii-4 (MLN), dated at the beginning of MN 12, and it is replaced in the younger Mytilinii-3 (MYT) by *S. major* (Kostopoulos 2009a); MLN is dated at ~7.5 Ma, while MYT at ~7.3 Ma. The age of Kerassia may be closer to that of MYT. Caution is wise, however, as the bovids and the equids from KER are still unstudied. Complete study and comparison of the fauna will also allow appraisal as to whether the two fossiliferous levels of Kerassia correspond to different faunas.

Other Late Miocene Localities

In addition to the previously mentioned late Miocene mammal localities, there are several others, distributed across the whole country. Some of them include a rich fauna allowing age determination, while others include old collections either from doubtful stratigraphic horizons or with poor or mixed material, causing several problems in age determination. These localities with their faunal lists are given by Koufos (2006a). However, it is important to note some of them as they are well known or they include some significant taxa.

A well-known late Miocene locality is Pyrgos Vassilissis (PYV), situated near Athens (see figure 28.1 [29]) because of the presence of a mandible from the hominoid *Graecopithecus freybergi*. Although the PYV fauna is referred as similar to that of Pikermi, it is doubtful that an age more precise than late Miocene can be provided (Koufos and de Bonis 2005 and ref. therein). The locality of Halmyropotamos (HAL) is situated on Evia Island (see figure 28.1 [22]); T. Skoufos collected some material at the beginning of the twentieth century, which was studied later by Melentis (1967). The fauna is quite similar to that of Pikermi (middle Turolian, MN 12), but it also has some minor differences, suggesting a slightly older age. However, as the HAL collection is an old one, it is quite possible that the material was mixed with that from other collections, and thus its age relative to other middle Turolian faunas is doubtful. A new locality named Chomateres (CHO) was discovered near Pikermi at the beginning of the 1970s. N. Symeonidis from the University of Athens and H. Zapfe from the Museum of Vienna excavated there and collected some material, which is unpublished. Although some information about the locality and the fauna was given, there is no extensive description and comparison of the fauna, except for the felid *Metailurus* and the cercopithecid *Mesopithecus* (Marinos and Symeonides 1974; Symeonidis 1978; Zapfe 1991). The CHO fauna is similar to that from Pikermi, and a middle Turolian (MN 12) age can be supposed. The CHO *Mesopithecus* belongs to *M. pentelicus*, but it retains some primitive features suggesting slightly older age (Koufos 2009a), which fits well with the earlier opinion of Mein (1990) and de Bruijn et al. (1992) that Chomateres is slightly older than Pikermi.

In Serres Basin (Macedonia, Greece), there are a number of late Miocene localities that mainly include small mammals. The locality Lefkon (LFK) includes a fauna with the Vallesian *Progonomys cathalai* and the Turolian *Parapodemus lugdunensis*, indicating a latest Vallesian age (de Bruijn and Van der Meulen 1979; de Bruijn et al. 1992). A poor micromammalian fauna has been collected from the locality Rema Marmara (REM; see figure 28.1 [24]). The presence of *Parapodemus gaudryi* suggests a middle Turolian age (MN 12), while the presence of a primitive *Micromys* suggests a younger age (MN 13); thus, an age at the middle/late Turolian boundary is quite possible for this locality (de Bruijn 1989; Koufos 2006a). A

couple of localities from the Serres Basin (Macedonia, Greece), named Ano Metochi 1, 2 (MTH) yielded a micromammalian fauna (see figure 28.1 [28]); the presence of *Pliopetaurista dehneli*, *Apodemus gudrunae*, and *Apodemus dominans*, as well as the absence of primitive arvicolids, suggest a latest Turolian (MN 13) age. Another locality, Monasteri (MNS), close to the Ano Metochi ones (see figure 28.1 [28]), yielded a similar fauna, indicating a late Turolian (MN 13) age (de Bruijn 1989). In western Macedonia, there is the locality of Lava (LAV) with a micrommalian fauna (see figure 28.1 [26]). The absence of middle Turolian (MN 12) elements, the presence of some Turolian/Ruscinian taxa, and the absence of taxa shared with Pikermi and Samos suggest an age younger than Pikermi, early MN 13 (de Bruijn et al. 1999).

Recently a Vallesian fauna has been reported from Crete and from the locality of Plakias. First dated to late Astaracian (MN 7+8), the locality Plakias or Plakia (PLK), near Heraklion (see figure 28.1 [9]), was discovered in the 1970s and included few remains of micromammals (de Bruijn and Meulenkamp 1972; Steininger, Bernor, and Fahlbusch 1990; van der Made 1996). The locality was rediscovered in 2011, and the study of a new collection of micromammals has provided more faunal data that allow its certain dating at the lower part of the early Vallesian, MN 9 (de Bruijn et al. 2012).

THE END OF THE MIOCENE IN GREECE

The end of the Miocene is characterized by the extinction of characteristic Miocene mammalian taxa (*Choerolophodon*, *Adcrocuta*, *Microstonyx*, *Helladotherium*, *Samotherium*, *Bohlinia*, *Palaeotragus*, *Prostrepsiceros*, *Palaeoreas*, *Oioceros*, *Protragelaphus*, *Tragoportax*, *Pliocervus*, *Ceratotherium* etc.) and appearance of new elements. This is quite clear from the analysis of the faunal data (Koufos 2006c, 2006f). Two Greek mammal localities mark this change, the localities of Maramena (MAR) and Silata (SLT) situated in Macedonia, Greece (see figure 28.1 [30 and 31]).

The locality of Maramena is situated in the Serres Basin and has yielded a rich micro- and macromammalian fauna. The detailed study of the locality and its fauna is given in a separate volume of *Münchner Geowissenshaftliche Abhandlungen* (Schmidt-Kittler 1995a). The MAR carnivores include the Miocene taxa *Promeles* and *Promephitis*, as well as *Lutra*, which is a Ruscinian element (Schmidt-Kittler 1995b). The MAR hipparions have more advanced hypsodonty and metapodial morphology than the Turolian forms, also indicating Ruscinian age (Sondaar and Eisenmann 1995). There are several bovids in MAR fauna that mainly are Miocene taxa (*Tragoportax gaudryi*, *Tragoportax* cf. *amalthea*, *Ouzocerus*). However, there is a caprini-like bovid (*Norbertia hellenica*), marking the early occurrence of the caprini, which were known from the end of Ruscinian (Köhler, Moya-Sola, and Morales 1995). The insectivores include both Turolian and Ruscinian species (Doukas et al. 1995). The majority of the MAR glirids are known in Ruscinian localities, indicating an age younger than Turolian (Daxner-Höck 1995). The murid *Apodemus gorafensis* is more advanced than the late Turolian *Apodemus gudrunae*, while it is less advanced than the early Ruscinian *Apodemus gorafensis* (Storch and Dalmann 1995); this indicates an age at the end of the Turolian or the beginning of the Ruscinian. The flying squirrels *Pliopetaurista dehneli* and *Miopetaurista thaleri* are known from the late Turolian to late Ruscinia (MN 13–15), while *Spermophilinus turolensis* is known from the early Vallesian (MN 9) to late Turolian (MN 13); this association suggests a late Turolian (MN 13) or younger age for MAR (de Bruijn 1995). The correlation of the available data suggests that MAR can be dated to Turolian/Ruscinian boundary (MN 13/14); however, its several large mammals typically of Miocene age seem to have more a Miocene rather than Pliocene character.

The other latest Turolian locality is Silata (SLT), situated in the Chalkidiki Peninsula (see figure 28.1 [31]). The SLT fauna consists mainly of micromammals, while few macromammalian remains have been found. The study of the SLT micromammals indicates that they are similar to the Maramena micromammalian assemblage. The taxa *Spermophilinus turolensis*, *Asoriculus gibberodon*, and *Pliopetaurista dehneli* are present in both localities, indicating an age at the Turolian/Ruscinian boundary (MN 13/14) for SLT. Two other faunal elements (*Amblycoptus jessiae* and *Deinsdorfia kerkhoffi*), common in SLT and MAR, indicate close age similarities (Vasileiadou, Koufos, and Syrides 2003). The last occurrence of *Spermophilinus turolensis* is reffered in MAR (de Bruijn 1995), and its presence in SLT fauna indicates a similar age. The SLT macromammals are few with fragmentary material and cannot provide certain biochronological data (Koufos 2006c). The SLT faunal assemblage includes some early Ruscinian taxa (*Erinaceus*, *Asoriculus gibberodon*, *Micromys paricioi*, *Mesocricetus primitivus*, *Pliopetaurista dehneli*), as well as Turolian ones (*Spermophilinus turolensis*, *Paramachairodus orientalis*, *Hipparion* cf. *mediterraneum*, *Mi-*

crostonyx major, *Helladotherium* or *Samotherium*), hence suggesting an age at the Turolian/Ruscinian boundary (MN13/14). The SLT and MAR faunas are compared with the latest Turolian fauna of Metochi (Serres Basin) and the Ruscinian fauna Maritses (Rhodes Island) by cluster analysis. The result is that the SLT and MAR faunas are very similar and both are closer to Metochi than to Maritses fauna; hence, the SLT and MAR faunas have more Miocene than Pliocene character (Koufos 2006c), marking the end of Miocene in Greece and Southeastern Europe. The transitional character of these faunal assemblages is also indicated by their separation as a cluster different from those of late Miocene and early Pliocene (see figure 28.7).

PLIOCENE

Although there are several Pliocene mammal localities in Greece, most of them include poor faunas and cannot provide certain biochronological data. The majority of the Pliocene faunas includes mainly micromammals. Pliocene large mammal faunas are generally rare in the eastern Mediterranean, and the available material is relatively poor. However, the late Pliocene (Villafranchian s. l.) localities with large mammals are relatively common, including quite rich faunas that were studied and provide good biochronological evidence. The biochronology and biostratigraphy of the Pliocene mammal localities was given earlier in various articles (Van der Meulen and Van Kolfschoten 1986; Koufos and Kostopoulos 1997; Koufos 2001, 2006a). A summary of these data is given in figure 28.3. However, it is important to note particular Pliocene localities, as they include new data and reveal important events for the local and eastern Mediterranean biostratigraphy.

The locality of Kessani (KES), discovered in 1996, is situated in the Xanthi-Komotini Basin (Thrace, Greece) (see figure 28.1 [34]) and includes a micromammalian fauna plus some large mammals. The continental deposits with the mammals overlies a series of marine deposits containing a typical molluscan fauna of Paratethyan type, dated to the late Turolian (Pontian) and indicating a younger age for the KES fauna (Syrides, Koliadimou, and Koufos 1997). The Kesani fauna was recently revised and according to the biochronological ranges of the identified taxa, the absence of characteristic early Pliocene rodents such as *Promimomys*, its faunal similarity with both the late Turolian and early Ruscinian faunas of the neighboring Ptolemais Basin, it is dated from the latest Turolian to the earliest Ruscinian (Vasileiadou et al. in press). Another interesting locality is Megalo Emvolon (MEV; see figure 28.1 [38]), one of the few known Pliocene localities of the eastern Mediterranean with large mammals (Arambourg and Piveteau 1929). More recently, visiting scientists found some fossils, and a small collection was made by the author; there is no fossiliferous horizon in M. Emvolon but sporadic small bone concentrations appearing by erosion. It is necessary to visit the section from time to time (usually after winter) to check for fossils. The fauna includes *Dolichopithecus ruscinensis*, which is similar to that from Perpignan, suggesting a late Ruscinian (MN 15) age; the presence of *Trischizolagus dumitrescuae* indicates also a late Ruscinian age. Thus, the Megalo Emvolon fauna is dated to late Ruscinian, MN 15 (see figure 28.3) (Koufos 2006a and ref. therein). Among the Ruscinian localities, it is important to refer Apolakkia (APK) on Rhodes Island (see figure 28.1 [39]). The Apolakkia fauna includes mainly small mammals, suggesting a late Ruscinian age (MN 15; see figure 28.3) and few large mammals. The presence of *Pachycrocuta pyrenaica* and the possible presence of *Hipparion crassum* confirm that age (Koufos and Kostopoulos 1997 and ref. therein). It is worth mentioning that the genus *Hipparion* makes its last appearance in Apolakkia. The locality Damatria (DAM) on Rhodes Island (see figure 28.1 [43]) is situated above Apollakia, and its micromammalian fauna suggests MN 16 age (Benda, Meulenkamp, and van de Weerd 1977; Van der Meulen and Van Kolfschoten 1986). It is interesting to note here that some remains of *Equus* and *Leptobos* were found in the Damatria Formation (Benda, Meulenkamp, and van de Weerd 1977). This Damatria record of *Equus* may correspond to the first appearance of the genus in Greece.

A number of localities (Dafnero [DFN], Volax [VOL], Sesklon [SES]) yielded a rich fauna of large mammals (see figure 28.1 [45–47]). The collections from Dafnero and Sesklon are new, and that from Volax originates from a single known site; thus, stratigraphic positions are certain (Sickenberg 1968; Koufos and Kostopoulos 1997; Athanassiou 1998). The study of these faunas indicates that they are similar, dating to lower MN 17 at ~ 2.5 Ma (Koufos and Kostopoulos 1997). Another interesting new late Villafranchian or Villanyian fauna is that of Vatera (VTR) on Lesvos Island (see figure 28.1 [48]), including the cercopithecid *Paradolichopithecus arvernensis*. Two fossiliferous sites Vatera-F and Vatera-DS yield a rich large mammal fauna suggesting latest Pliocene age ~ 2.0 Ma (de Vos et al. 2002; Lyras and Van der Geer 2007).

FAUNAL COMPOSITION OF THE GREEK NEOGENE MAMMAL LOCALITIES

The early and middle Miocene localities are few and their faunal composition is not representative (figure 28.4). Although the material is poor, most of the common mammal families are known in the middle Miocene. On the other hand, the late Miocene is richer in number of localities and families (figure 28.5). However, there are significant differences in the faunal composition of the various MN biozones depending on the number of localities and their richness. The biozone MN 9 is unknown in Greece; it is possible that the locality Pentalophos 1 (PNT) of Axios Valley belongs to this biozone. The biozone MN 10 was also unknown until the 1970s, when its first evidence was discovered in Axios Valley and Crete. The MN 10 taxomomic composition (see figure 28.5a) includes some micromammalian families; the cricetids, glirids, and murids dominate. The macromammals are represented by 14 families from which the hyaenids, giraffids, and bovids dominate the fauna with 40% of the determined species; an abundance of equids and rhinos is also present in the MN 10 fauna. The biozone MN 11 is relatively poor because of the few known mammal localities. It contains 11 families of large mammals, while the small mammals are absent (see figure 28.5b). A significant change from Vallesian to Turolian faunas is the replacement of the hominoids by the cercopithecids and the absence of the percrocutids. The hyaenids, giraffids, equids, and bovids continue to dominate MN 11. The MN 12 fauna is the richest of the Turolian, including 18 macromammalian families (see figure 28.5c). Among them, the hyaenids, felids, mustelids, equids, giraffids, and mainly bovids dominate; the bovids are abundant, representing 27% of the fauna (28 different species; table 28.1). It is worth mentioning the appearance of the tragulids and cervids in MN 12 (see figure 28.5c). The MN 13 fauna includes both small and large mammals. Among the small mammals the glirids, cricetids, and murids dominate (29%), while in the large mammals the hyaenids, mustelids, equids, giraffids, and bovids are the dominant families; bovids are represented by nine species corresponding to 13% of the fauna (see figure 28.5d).

The taxonomic composition of the Pliocene mammal biozones (figure 28.6) is limited, as the faunas are poor; most of the families are represented by a single taxon, and in many cases the determinations are doubtful. The MN 14 fauna is characterized by many murids (36%) and cricetids (14%); the known large mammals are very few

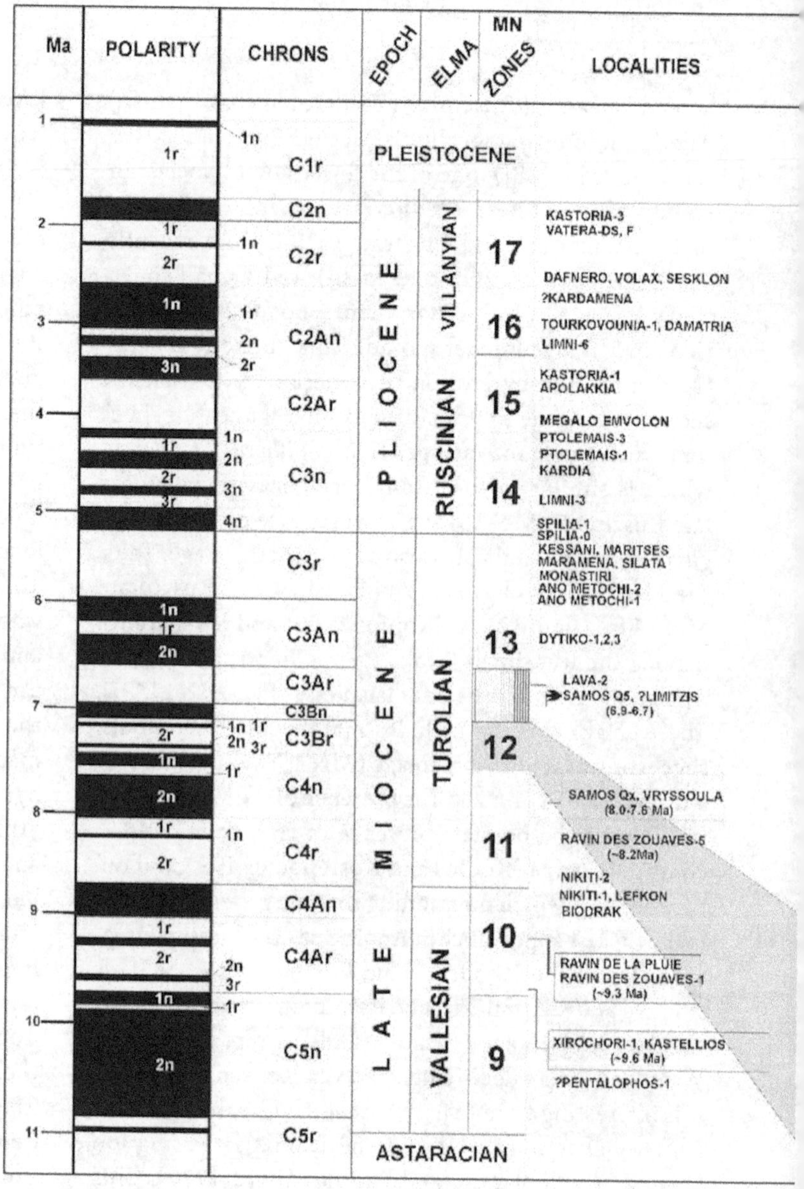

Figure 28.3 Biostratigraphy of the late Miocene mammal localities of Greece with first-appearing taxa. The geomagnetic time scale is from Cande and Kent (1995), and the MN zone boundary ages from Steininger (1999).

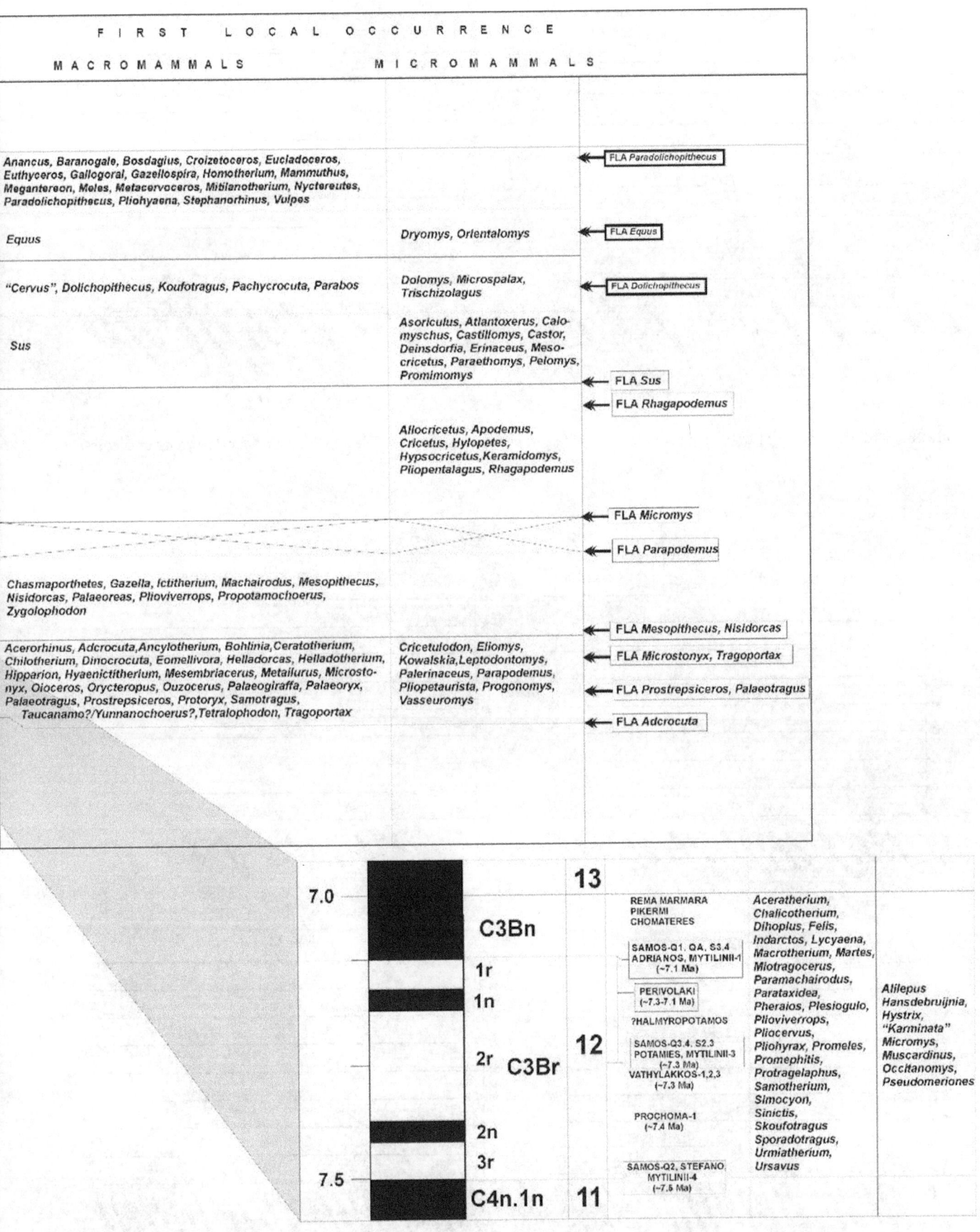

FIRST LOCAL OCCURRENCE

MACROMAMMALS   MICROMAMMALS

Anancus, Baranogale, Bosdagius, Croizetoceros, Eucladoceros,
Euthyceros, Gallogoral, Gazellospira, Homotherium, Mammuthus,
Megantereon, Meles, Metacervoceros, Mitilanotherium, Nyctereutes,
Paradolichopithecus, Pliohyaena, Stephanorhinus, Vulpes

FLA *Paradolichopithecus*

Equus   *Dryomys, Orientalomys*

FLA *Equus*

"Cervus", Dolichopithecus, Koufotragus, Pachycrocuta, Parabos   *Dolomys, Microspalax,
Trischizolagus*

FLA *Dolichopithecus*

Sus   *Asoriculus, Atlantoxerus, Calo-
myschus, Castillomys, Castor,
Deinsdorfia, Erinaceus, Mesocricetus, Paraethomys, Pelomys,
Promimomys*

FLA *Sus*

FLA *Rhagapodemus*

*Allocricetus, Apodemus,
Cricetus, Hylopetes,
Hypsocricetus,Keramidomys,
Pliopentalagus, Rhagapodemus*

FLA *Micromys*

FLA *Parapodemus*

Chasmaporthetes, Gazella, Ictitherium, Machairodus, Mesopithecus,
Nisidorcas, Palaeoreas, Plioviverrops, Propotamochoerus,
Zygolophodon

FLA *Mesopithecus, Nisidorcas*

Acerorhinus, Adcrocuta,Ancylotherium, Bohlinia,Ceratotherium,   *Cricetulodon, Eliomys,*   FLA *Microstonyx, Tragoportax*
Chilotherium, Dinocrocuta, Eomellivora, Helladorcas, Helladotherium,   *Kowalskia,Leptodontomys,*
Hipparion, Hyaenictitherium, Mesembriacerus, Metailurus, Microsto-   *Palerinaceus, Parapodemus,*   FLA *Prostrepsiceros, Palaeotragus*
nyx, Oioceros, Orycteropus, Ouzoceros, Palaeogiraffa, Palaeoryx,   *Pliopetaurista, Progonomys,*
Palaeotragus, Prostrepsiceros, Protoryx, Samotragus,   *Vasseuromys*   FLA *Adcrocuta*
    Taucanamo?/Yunnanochoerus?,Tetralophodon, Tragoportax

7.0	C3Bn	13	REMA MARMARA, PIKERMI, CHOMATERES	Aceratherium, Chalicotherium,
	1r		SAMOS-Q1, QA, S3.4 ADRIANOS, MYTILINII-1 (~7.1 Ma)	Dihoplus, Felis, Indarctos, Lycyaena, Macrotherium, Martes, Miotragocerus,
	1n		PERIVOLAKI (~7.3-7.1 Ma)	Paramachairodus, Parataxidea,
			?HALMYROPOTAMOS	Pheraios, Plesiogulo, Plioviverrops, Pliocervus,
	2r C3Br	12	SAMOS-Q3.4, S2.3 POTAMIES, MYTILINII-3 (~7.3 Ma) VATHYLAKKOS-1,2,3 (~7.3 Ma)	Pliohyrax, Promeles, Promephitis, Protragelaphus, Samotherium,
	2n		PROCHOMA-1 (~7.4 Ma)	Simocyon, Sinictis, Skoufotragus,
	3r			Sporadotragus, Urmiatherium,
7.5	C4n.1n	11	SAMOS-Q2, STEFANO, MYTILINII-4 (~7.5 Ma)	Ursavus

*Alilepus
Hansdebruijnia,
Hystrix,
"Karminata"
Micromys,
Muscardinus,
Occitanomys,
Pseudomeriones*

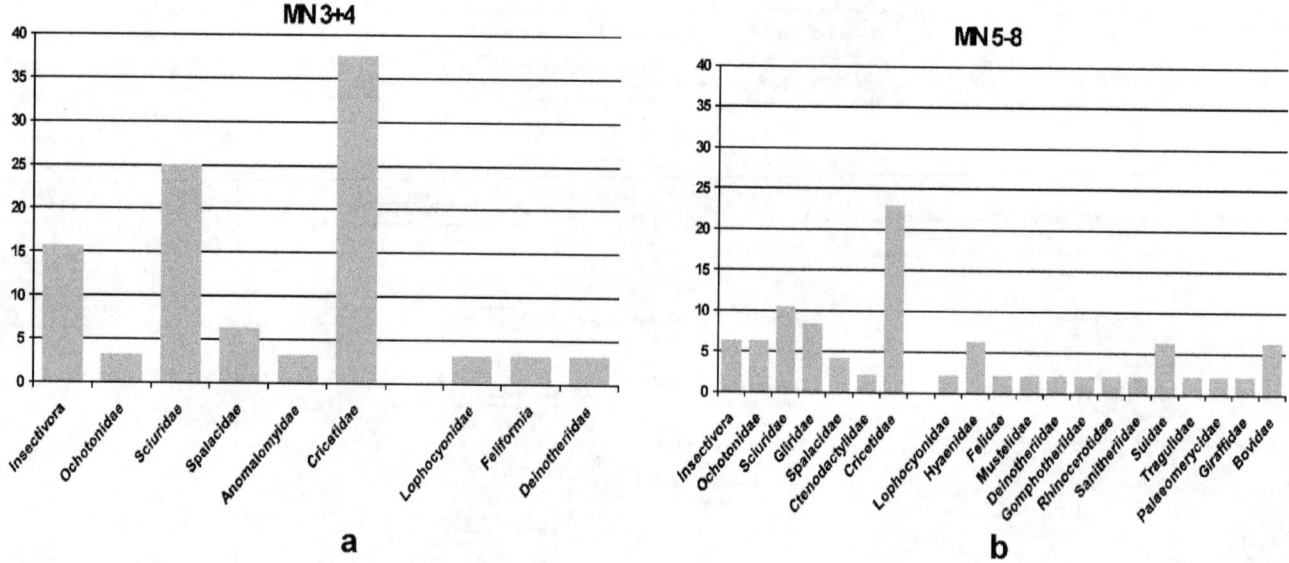

Figure 28.4 Taxonomic composition (% number of species per family) of the early (*a*) and middle (*b*) Miocene mammal biozones of Greece.

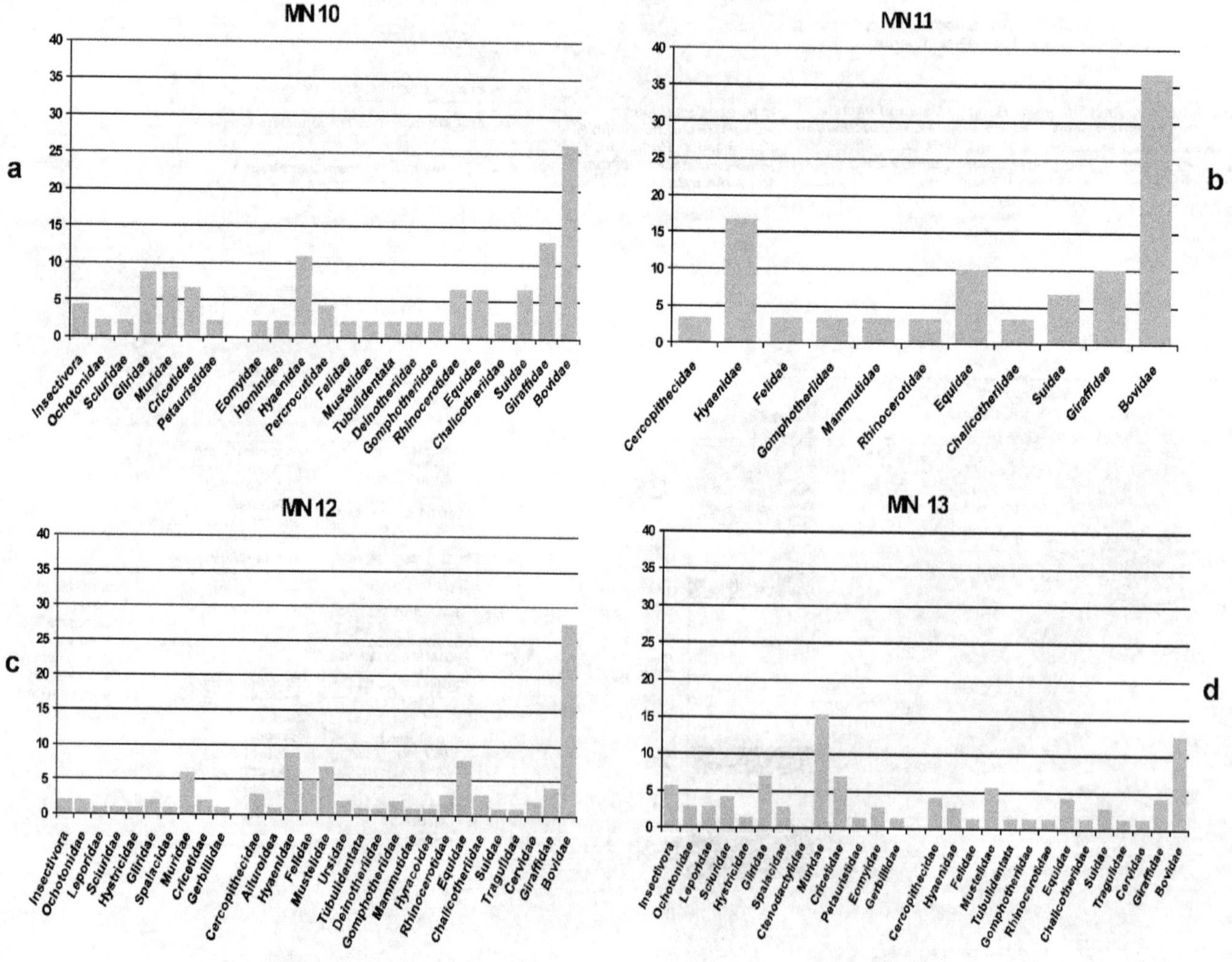

Figure 28.5 Taxonomic composition (% number of species per family) of the late Miocene mammal biozones of Greece.

Table 28.1

Matrix of Presence/Absence for 262 Mammal Species from Several Neogene Localities of Greece

Family	MN 3	MN 4	MN 5	MN 7+8	MN 10	MN 11	MN 12	MN 13	MN 14	MN 15	MN 16	MN 17
Insectivora	0	5	2	1	2	0	2	4	0	0	0	0
Ochotonidae	0	1	1	2	1	0	2	2	0	0	0	0
Leporidae	0	0	0	0	0	0	1	2	0	2	0	0
Sciuridae	0	8	2	3	1	0	1	3	2	0	1	0
Castoridae	0	0	0	0	0	0	0	0	1	1	0	0
Hystricidae	0	0	0	0	0	0	1	1	0	0	1	0
Gliridae	0	0	2	2	4	0	2	5	2	2	4	1
Spalacidae	0	2	2	0	0	0	1	2	1	1	1	0
Anomalomyidae	0	1	0	0	0	0	0	0	0	0	0	0
Ctenodactylidae	0	0	1	0	0	0	0	0	0	0	0	0
Muridae	0	0	0	0	4	0	6	11	10	6	5	1
Cricetidae	0	12	7	4	3	0	2	5	4	0	0	0
Petauristidae	0	0	0	0	1	0	0	1	1	0	0	0
Eomyidae	0	0	0	0	1	0	0	2	1	1	0	0
Gerbillidae	0	0	0	0	0	0	1	1	1	0	0	0
Arvicolidae	0	0	0	0	0	0	0	0	1	4	3	1
Cercopithecidae	0	0	0	0	0	1	3	3	0	1	0	1
Hominidae	0	0	0	0	1	0	0	0	0	0	0	0
Lophocyonidae	0	1	1	0	0	0	0	0	0	0	0	0
Feliformia	0	1	0	0	0	0	0	0	0	0	0	0
Ailuroidea	0	0	0	0	0	0	1	0	0	0	0	0
Canidae	0	0	0	0	0	0	0	0	0	1	0	3
Hyaenidae	0	0	3	0	5	5	9	2	0	1	0	2
Percrocutidae	0	0	0	0	2	0	0	0	0	0	0	0
Felidae	0	0	1	0	1	1	5	1	0	0	0	3
Mustelidae	0	0	1	0	1	0	7	4	0	0	0	2
Ursidae	0	0	0	0	0	0	2	0	0	0	0	2
Tubulidentata	0	0	0	0	1	0	1	1	0	0	0	0
Deinotheriidae	1	0	1	0	1	0	1	0	0	0	0	0
Gomphotheriidae	0	0	1	0	1	1	2	1	0	0	0	1
Mammutidae	0	0	0	0	0	1	1	0	0	0	0	0
Elephantidae	0	0	0	0	0	0	0	0	0	0	0	1
Hyracoidea	0	0	0	0	0	0	1	0	0	0	0	0
Rhinocerotidae	0	0	0	1	3	1	3	1	0	1	0	1
Equidae	0	0	0	0	3	3	8	3	1	1	1	1
Chalicotheriidae	0	0	0	0	1	1	3	1	0	0	0	0
Sanitheriidae	0	0	1	0	0	0	0	0	0	0	0	0
Suidae	0	0	2	1	3	2	1	2	1	1	0	0
Tragulidae	0	0	1	0	0	0	1	1	0	0	0	0
Cervidae	0	0	0	0	0	0	2	1	1	1	0	5
Palaeomerycidae	0	0	1	0	0	0	0	0	0	0	0	0
Giraffidae	0	0	1	0	6	3	4	3	0	0	0	2
Bovidae	0	0	3	0	12	11	28	9	1	3	1	9

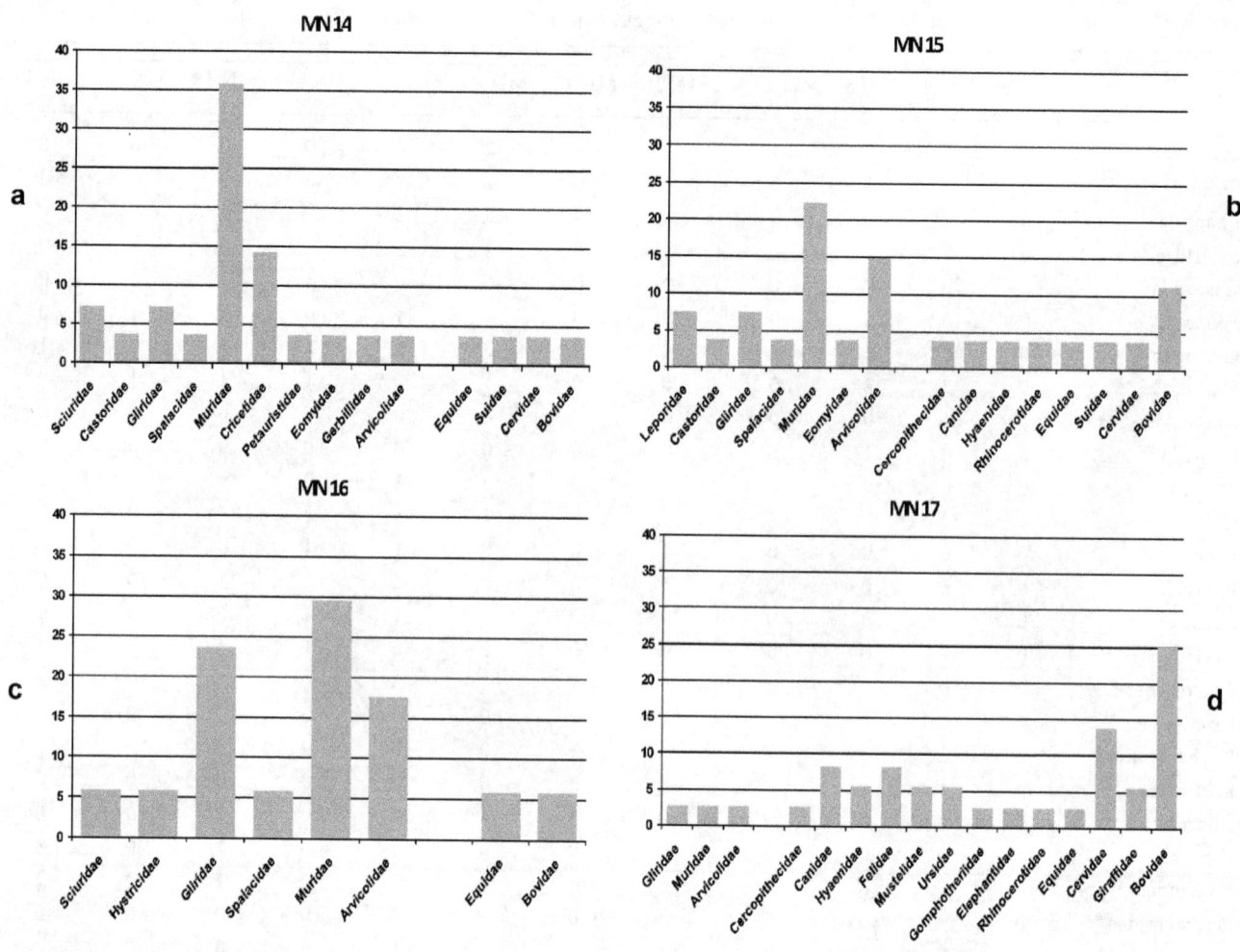

Figure 28.6 Taxonomic composition (% number of species per family) of the Pliocene mammal biozones of Greece.

(see figure 28.6a). This is artificial, as the available localities of large mammals are few and with poor material. The arvicolids make their appearance in MN 14, while the canids appeared in MN 15 (see figure 28.6b). The more representative fauna is that of MN 17, characterized by the dominance of the canids, felids, cervids, and bovids, the last two families representing 25% and 14% of the fauna, respectively (see figure 28.6d). The first elephants appear in MN 17.

The faunal relationships and age similarities of the various localities are analyzed using multivariate techniques of the PAST software (Hammer, Harper, and Ryan 2001). New localities or old ones from which there are new collections, as well as quite rich faunas, are included in the analysis. There is only one exception, the Pikermi fauna; the faunal list for Pikermi is based on Gaudry's collection, housed in MNHN. A presence/absence matrix of

262 species (the matrix is available on request) is analyzed by various methods, using different indices. The results of the varied methods are not very different, and those of two are given in figure 28.7. Resolved species and those referred as cf. or aff. are included in the matrix; those referred as sp. or with question mark are excluded.

The analysis by nonmetric MDS using the Raup–Crick index (see figure 28.7a) gives good separation of the faunas. Coordinate 1 distinguishes the late Miocene and Miocene/Pliocene faunas from the early–middle Miocene and Pliocene ones, while coordinate 2 separates the Miocene faunas from the Miocene/Pliocene and the Pliocene ones. The early Miocene faunas (ALI, KAR) are separated from the middle Miocene ones (THY, ANT); their separation seems to be clear, although the available data are few. The late Miocene faunas cluster in one group. Among this group, there is an age discrimination

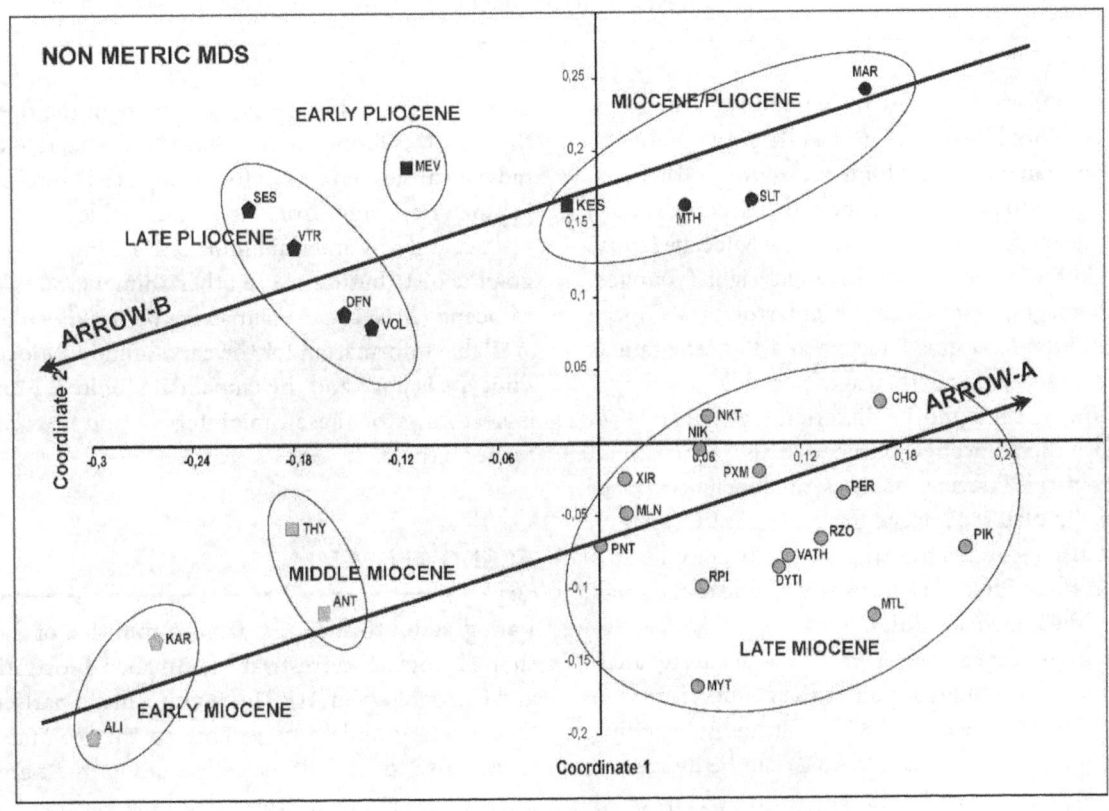

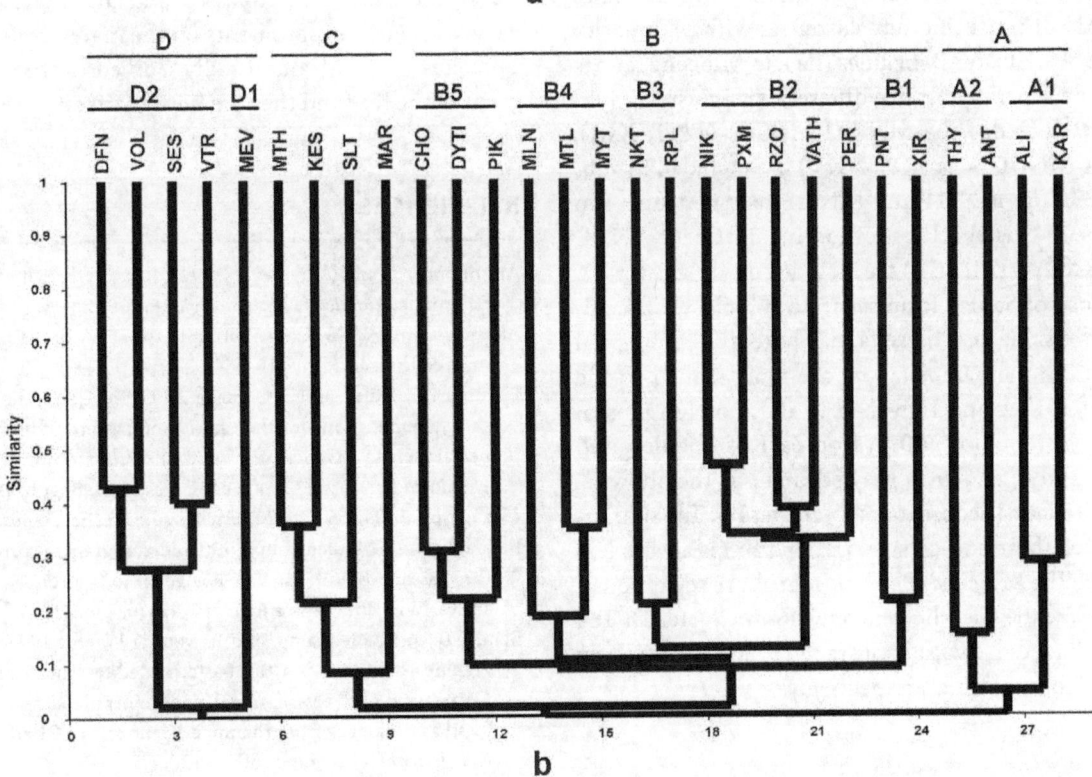

Figure 28.7 Taxonomic similarity of the Neogene mammal faunas of Greece: (*a*) nonmetric MDS analysis, using Raup-Crick's index; (*b*) cluster analysis, using Jaccard's index.

of the faunas from the Vallesian (left limit of the group) through early-middle Turolian (middle of the group) to the late Turolian ones (right limit of the group). This classification seems to extend throughout the Miocene faunas; there is a clear age discrimination of the Miocene faunas from the lower left (older) to the upper right (younger) end of the diagram (see figure 28.7a [arrow A]). On the other hand, the Miocene/Pliocene and Pliocene faunas are classified according to their age from older to younger ones, following the opposite direction (see figure 28.7a [arrow B]). The Miocene/Pliocene faunas are separated clearly from the Pliocene ones, as they include taxa of both ages. The latest Pliocene faunas (SES, VOL, DFN, VTR) are also separated from the early Pliocene one. This separation of the Pliocene faunas may be due to the small number of the known localities.

The analysis of the same data matrix with Cluster analysis using Jaccard's index separates the studied faunas in four main clusters (see figure 28.7b). Cluster A, including the early and middle Miocene faunas, can be divided in two sub-clusters. The early Miocene faunas (ALI, KAR) constitute sub-cluster A1 and the middle Miocene ones (THY, ANT) constitute sub-cluster A2 with a similarity less than 5%. Cluster B includes the late Miocene faunas and can be separated in five different sub-clusters. Sub-cluster B1 includes the MN 9/10 faunas (PNT, XIR), while the MN 10 ones (RPl, NKT) are included in sub-cluster B3. The MN 11 and MN 12 faunas define two different sub-clusters. Those of continental Greece (NIK, RZO, PXM, VATH, PER) are included in sub-cluster B2, while those of Samos form separate sub-cluster B4. Although the same age, these faunas have different faunal composition; the Samos faunas are more similar to the faunas of Turkey and Iran than to those of continental Greece (Kostopoulos 2009c; Koufos, Kostopoulos, and Vlachou 2009), and thus they separate in the analysis. The MN 13 faunas constitute sub-cluster B5. The similarity between these sub-clusters (B1 and B5) is always less than 10%. The Miocene/Pliocene faunas are separated in Cluster C, while the Pliocene ones form Cluster D. The latter is divided in two sub-clusters: D1 and D2 are early and late Pliocene faunas, respectively.

STRATIGRAPHIC DISTRIBUTION OF THE GREEK NEOGENE MAMMAL TAXA

The stratigraphic distribution of the known Neogene genera of Greece is given in the appendix. It is quite clear that there are many gaps in the stratigraphic distribution of the various taxa. This is more acute in the rodents, as their study is limited in the Greek Neogene; few and sporadic localities have been found and studied or the studied material comes from large mammal localities, which yield few if any micromammals. Looking to the stratigraphic distribution of the other animal groups, the late Miocene (MN 10–13) faunas are better known, but the available information for the early–middle Miocene and Pliocene is poor and the faunal data limited. More work is necessary for these time intervals and research must focus on them.

ACKNOWLEDGMENTS

I am grateful to the organizing committee of the workshop "Neogene Terrestrial Mammalian Biostratigraphy and Chronology in Asia" for inviting me to participate in this meeting and to contribute to this volume. Many thanks also to all Chinese colleagues for their great hospitality during the meeting. Thanks are also due to Prof. M. Woodburne, Dr. N. Spassov, and Dr. V. Titov for their review and useful comments on the manuscripts. Dr. M. Fortelius and R. Camp kindly worked for the improvement of the text and they are both thanked.

REFERENCES

Arambourg, C. and J. Piveteau, J. 1929. Les vertébrés du Pontien de Salonique. *Annales de Paléontologie* 18:59–138.

Athanassiou, A. 1998. Contribution to the study of the fossil mammals of Thessaly. *Gaia* 5:1–354.

Begun, D., E. Güleç, and D. Geraads. 2003. The Çandır hominoid locality: Implications for the timing and pattern of hominoid dispersal events. In *Geology and Vertebrate Paleontology of the Miocene Hominoid Locality of Çandır*, ed. E. Güleç, D. Begun, and D. Geraads, pp. 240–265. *CourierForschung-Institut Senckenberg* 240.

Benda, L. and J. Meulenkamp. 1990. Biostratigraphic correlations in the Eastern Mediterranean Neogene and chronostratigraphic scales. *Newsletter of Stratigraphy* 23:1–10.

Benda, L., J. Meulenkamp, and A. van de Weerd. 1977. Biostratigraphic correlations in the Eastern Mediterranean Neogene 3. Correlation between mammal, sporomorph and marine microfossil assemblages from the upper Cenozoic of Rhodes, Greece. *Newsletter of Stratigraphy* 6:117–130.

Bonneau, M. and L. Ginsburg. 1974. Découverte de *Dorcatherium puyhauberti* Arambourg et Piveteau (Mammalia) dans les facies continentaux de base de la molasse tertiaire de Crete (Grèce): Consequences stratigraphiques et tectoniques. *Compte Rendus Sommaire Société Géologique France* 16(1):11–12.

Bouvrain, G. 1979. Un nouveau genre de Bovidé de la fin du Miocène. *Bulletin de la Société Géologique de France* 21:507–511.

Bouvrain, G. 1980. Le genre *Palaeoreas* (Mammalia, Artiodactyla, Bovidae): Systematique, et extension geographique. *Paläontologische Zeitschrift* 54:55–65.

Bouvrain, G. 1982. Revision du genre *Prostrepsiceros*. *Paläontologische Zeitschrift* 56:113–124.

Bouvrain, G. 1997. Les bovidés du Miocène supérieur de Pentalophos (Macédoine, Grèce). *Münchener Geowissenschftische Abhandlungen* A34:5–22.

Bouvrain, G. 2001. Les Bovidés (Mammalia, Artiodactyla) des gisements du Miocène supérieur de Vathylakkos (Grèce du Nord). *Neues Jahrbuch für Geologie und Paläontologie Abhandungen* 220:225–244.

Bouvrain, G. and L. de Bonis. 1984. Le genre *Mesembriacerus* (Artiodactyla, Mammalia), un oviboviné du Vallésien (Miocène supérieur) de Macédoine (Grèce). *Palaeovertebrata* 14(4):201–223.

Bouvrain, G. and L. de Bonis. 1985. Le genre *Samotragus* (Artiodactyla, Bovidae) une antilope du Miocene Superieur de Grèce. *Annales de Paléontologie* 71(4):257–299.

Cande, S. C. and D. V. Kent. 1995. Revised calibration of the geomagnetic polarity time scale for the late Cretaceous and Cenozoic. *Journal of Geophysical Research* 100:6093–6095.

Daxner-Höck, G. 1995. Some glirids and Cricetids from Maramena and other late Miocene localities in northern Greece. In *The Vertebrate Locality Maramena (Macedonia, Greece) at the Turolian/Ruscinian Boundary (Neogene)*, ed. N. Schmidt-Kittler, pp. 103–120. *Münchner Geowissenschftische Abhandlungen* 28.

de Bonis, L. and G. Bouvrain. 1996. Suidae du Miocène supérieur de Grèce. *Bulletin du Muséum nationale d'Histoire naturelle Paris*, 4th ser. 18, C(1):107–132.

de Bonis, L. and G. Bouvrain. 2003. Nouveaux Giraffidae du Miocène supérieur de Macédoine (Grèce). *Advances in Vertebrate Paleontology* 3:5–16.

de Bonis, L., G. Bouvrain, D. Geraads, and G. D. Koufos. 1990a. New hominid skull material from the late Miocene of Macedonia in Northern Greece. *Nature* 345:712–714.

de Bonis, L., G. Bouvrain, D. Geraads, and G. D. Koufos. 1990b. New remains of *Mesopithecus* (Primates, Cercopithecidae) from the late Miocene of Macedonia with the description of a new species. *Journal of Vertebrate Paleontology* 10:473–483.

de Bonis, L., G. Bouvrain, D. Geraads, G. D. Koufos and S. Sen. 1994. The first aardwarks from the Miocene of Macedonia (Greece). *Neues Jarbüch für Geologie und Paläontologie Monatchhefte* 194:343–360.

de Bonis, L., G. Bouvrain, D. Geraads and J. K. Melentis. 1974. Première découverte d'un primate hominoïde dans le Miocène supérieur de Macédoine (Grèce). *Compte Rendus Academie Sciences Paris* D278:3063–3066.

de Bonis, L. and G. D. Koufos. 1981. A new hyaenid (Carnivora, Mammalia) in the Vallesian (late Miocene) of Northern Greece. *Scientific Annals Faculty Physics and Mathematics, University of Thessaloniki* 21:79–94.

de Bonis, L. and G. D. Koufos. 1993. The face and the mandible of *Ouranopithecus macedoniensis*. Description of new specimens and comparisons. *Journal of Human Evolution* 24:469–491.

de Bonis, L. and G. D. Koufos. 1994. Some hyaenids from the late Miocene of Macedonia (Greece) and the phylogeny of hunting hyaenas. *Münchener Geowissenschaftliche Abhandlungen* 26:81–96.

de Bonis, L., G. D. Koufos, and S. Sen. 1997a. A giraffid skull and mandible from the middle Miocene of the island of Chios (Aegean Sea, Greece). *Palaeontology* 40:121–133.

de Bonis, L., G. D. Koufos, and S. Sen. 1997b. The sanitheres (Mammalia, Suoidea) from the middle Miocene of Chios Island, Aegean Sea, Greece. *Revue Palaeobiologique* 16(1):259–270.

de Bonis, L., G. D. Koufos, and S. Sen. 1998. Ruminants (Bovidae and Tragulidae) from the middle Miocene (MN 5) of the island of Chios, Aegean Sea (Greece). *Neues Jahrbuch fur Geologie und Paläontologie Abhandlungen* 210:399–420.

de Bruijn, H. 1995. The Vertebrate locality Maramena (Macedonia, Greece) at the Turolian/Ruscinian boundary (Neogene). 8. Sciuridae, Petauristidae and Eomyidae (Rodentia, Mammalia). In *The Vertebrate Locality Maramena (Macedonia, Greece) at the Turolian/Ruscinian Boundary (Neogene)*, ed. N. Schmidt-Kittler, pp. 87–102. *Münchener Geowissenschftische Abhandlungen* 28.

de Bruijn, H. 1989. Smaller mammals from the Upper Miocene and lower Pliocene of the Strimon basin, Greece. Part 1. Rodentia and Lagomorpha. *Bolletino Società Paleontologica Italiana* 28:189–195.

de Bruijn, H., R. Daams, G. Daxner-Hock, V. Fahlbusch, L. Ginsburg, L. P. Mein, and J. Morales. 1992. Report of the RCMNS working group on fossil mammals, Reisensburg 1990. *Newsletter of Stratigraphy* 26(2/3):65–117.

de Bruijn, H., S. C. Doukas, L. W. van den Hoek Ostende, and J. W. Zachariasse. 2012. New finds of rodents and insectivores from the upper Miocene at Plakias (Crete, Greece). *Swiss Journal of Palaeontology* 131: 61–75.

de Bruijn, H. and J. Meulenkamp, J. 1972. Late Miocene rodents from the Pandanassa Formation (prov. Rethymnon), Crete, Greece. *Proceedings Koniklijke Nederlandse Akademie Wetenschappen* B78(4):314–338.

de Bruijn, H., G. Saraç, G., L. W. van den Ostende and S. Roussiakis 1999. The status of the genus name *Parapodemus* Schaub, 1938; new data bearing on an old controversy. *DEINSEA* 7:95–112.

de Bruijn, H., E. Ünay, G. Saraç and G. K. Hofmeijer. 1987. An unusual new eucricetodontine from the Lower Miocene of the Eastern Mediterranean. *Proceedings Koniklijke Nederlandse Akademie Wetenschappen* B90(2):119–132.

de Bruijn, H. and A. Van der Meulen. 1979. A review of the Neogene rodent succession in Greece. *Annales géologiques Pays helleniques*, non-series vol. 1:207–217.

de Vos, J., J. van der Made, A. Athanassiou, G. Lyras, P. Sondaar and M. Dermitzakis. 2002. Preliminary note on the late Pliocene fauna from Vatera (Lesvos, Greece). *Annales géologiques Pays helleniques* 39:37–69.

Doukas, K. 1986. The mammals from the lower Miocene of Aliveri (island of Evia, Greece). Part 5. The insectivores. *Proceedings Koniklijke Nederlandse Akademie Wetenschappen* B89:15–38.

Doukas, C., L. W. van den Hoek Ostende, C. Theocharopoulos, and J. F. Reumer. 1995. The Vertebrate locality Maramena (Macedonia, Greece) at the Turolian/Ruscinian boundary (Neogene). 5. Insectivora (Erinaceidae, Talpidae, Soricidae, Mammalia). In *The Vertebrate Locality Maramena (Macedonia, Greece) at the Turolian/Ruscinian Boundary (Neogene)*, ed. N. Schmidt-Kittler, pp. 43–64. *Münchener Geowissenschftische Abhandlungen* 28.

Forsyth Major, C. J. 1894. *Le gisement ossifère de Mytilinii et catalogue d'ossements fossiles recueillis à Mytilini, île de Samos, et déposés au Collège Galliard, à Lausanne.* Lausanne: Georges Bridel.

Fortelius, M. (coordinator). 2009. *Neogene of the Old World Database of Fossil Mammals (NOW).* University of Helsinki. http://www.helsinki.fi/science/now/.

Fortelius, M., S. Nummela, and S. Sen. 2003. Geology and vertebrate paleontology of the Miocene hominoid locality of Çandir: 9. Orycteropodidae (Tubulidentata). In *Geology and Vertebrate Paleontology of the Miocene Hominoid Locality of Çandır,* ed. E. Güleç, D. Begun, and D. Geraads, pp. 194–201. *CourierForschung-Institut Senckenberg* 240.

Gaudry, A. 1862–67. *Animaux fossils et géologie de l'Attique.* Paris: Savy.

Gentry, A. 1971. The earliest goats and other antelopes from the Samos *Hipparion* fauna. *Bulletin British Museum Natural History, Geology* 20:231–296.

Geraads, D. 1979. Les Giraffinae (Giraffidae, Mammalia) du Miocène supérieur de la région de Thessalonique (Grèce). *Bulletin Museum Nationale Histoire Naturelle Paris,* 4th ser., C1(4):377–389.

Giaourtsakis, J. 2009. Rhinocerotidae. In *The Late Miocene Mammal Faunas of the Mytilinii Basin, Samos Island, Greece: New Collection,* ed. G. D. Koufos and D. Nägel, pp. 157–187. *Beiträge zur Paläontologie* 31.

Giaourtsakis, J. and G. D. Koufos. 2009. Chalicotheriidae. In *The Late Miocene Mammal Faunas of the Mytilinii Basin, Samos Island, Greece: New Collection,* ed. G. D. Koufos and D. Nägel, pp. 189–205. *Beiträge zur Paläontologie* 31:.

Hammer, C., D. A. T. Harper, and P. D. Ryan. 2001. PAST: Paleontological Statistics Software package for education and data analysis. *Palaeontologia Electronica* 4(1); 4A: 9 pp., 178kb; http://palaeo-electronica.org/2001_1/past/issue1_01.htm.

Hofmeijer, G. K. and H. de Bruijn. 1988. The mammals from the Lower Miocene of Aliveri (island of Evia, Greece). *Proceedings Koniklijke Nederlandse Akademie Wetenschappen* B91(2):185–204.

Iliopoulogs, G. 2003. The Giraffidae (Mammalia, Artiodactyla) and the study of histology and chemistry of fossil mammal bone from the Late Miocene of Kerassia (Euboea Island, Greece). Ph.D. diss., University of Leicester.

Karadenizli, L., G. Seyitoglu, S. Sen, N. Arnaud, N. Kazanci, G. Saraç, and C. Alçiçek. 2005. Mammal bearing late Miocene tuffs of the Akkaşdağı region; distribution, age, petrographical and geochemical characteristics. In *Geology, Mammals and Environments at Akkaşdağı, Late Miocene of Central Anatolia,* ed. S. Sen, pp. 553–566. *Geodiversitas* 27(4).

Köhler, M., S. Moya-Sola, and J. Morales. 1995. Bovidae and Giraffidae (Artiodactyla, Mammalia). In *The Vertebrate Locality Maramena (Macedonia, Greece) at the Turolian/Ruscinian Boundary (Neogene),* ed. N. Schmidt-Kittler, pp. 167–180. *Münchener Geowissenschftische Abhandlungen* 28.

Koliadimou, K. and G. D. Koufos. 1998. Preliminary report about the Neogene/Quaternary micromammals of the Mygdonia Basin. *Romanian Journal of Stratigraphy* 78:75–82.

Kondopoulou, D., L. de Bonis, G. D. Koufos, and S. Sen. 1993. Palaeomagnetic and biostratigraphic data from the middle Miocene vertebrate locality of Thymiana (Chios Island, Greece). *Proceedings 2nd Congres Geophysical Society of Greece,* pp. 626–635.

Kondopoulou, D., S. Sen, G. D. Koufos, and L. de Bonis 1992. Magneto and biostratigraphy of the late Miocene mammalian locality of Prochoma (Macedonia, Greece). *Paleontologia i Evolució* 24/25:135–139.

Konidaris, G. and G. D. Koufos. 2009. Proboscidea. In *The Late Miocene Mammal Faunas of the Mytilinii Basin, Samos Island, Greece: New Collection,* ed. G. D. Koufos and D. Nägel, pp. 139–155. *Beiträge zur Paläontologie* 31.

Kostopoulos, D. S. 2005. The bovidae (Mammalia, Artiodactyla) from the late Miocene locality of Akkaşdağı, Turkey. *Géodiversitas* 27(4):735–791

Kostopoulos, D. S. 2006. Cervidae and Bovidae. In *The Late Miocene Vertebrate Locality of Perivolaki, Thessaly, Greece,* ed. G. D. Koufos, pp. 151–183. *Palaeontographica* Abt. A, 276(1–6).

Kostopoulos, D. S. 2009a. Giraffidae. In *The Late Miocene Mammal Faunas of the Mytilinii Basin, Samos Island, Greece: New Collection,* ed. G. D. Koufos and D. Nägel, pp. 299–343. *Beiträge zur Paläontologie* 31.

Kostopoulos, D. S. 2009b. Bovidae. In *The Late Miocene Mammal Faunas of the Mytilinii Basin, Samos Island, Greece: New Collection,* ed. G. D. Koufos and D. Nägel, pp. 345–389. *Beiträge zur Paläontologie* 31.

Kostopoulos, D. S. 2009c. The Pikermian Event: Temporal and spatial resolution of the Turolian large mammal fauna in SE Europe. *Palaeogeography, Palaeoclimatology, Palaeoecology* 274:82–95.

Kostopoulos, D. S., K. K. Koliadimou, and G. D. Koufos. 1996. The giraffids (Mammalia, Artiodactyla) from the late Miocene mammalian localities of Nikiti (Macedonia, Greece). *Palaeontographica* 239:61–88.

Kostopoulos, D. S. and G. D. Koufos. 1997. Late Miocene bovids (Mammalia, Artiodactyla) from the locality "Nikiti 1" (NKT), Macedonia, Greece. *Annales de Paléontologie* 81:251–300.

Kostopoulos, D. S. and G. D. Koufos. 1999. The bovidae (Mammalia, Artiodactyla) of the "Nikiti 2" (NIK) faunal assemblage (Chalkidiki peninsula, N. Greece). *Annales de Paléontologie* 85(3):193–218.

Kostopoulos, D. S., G. D. Koufos, G. D., I. A. Sylvestrou, G. D. Syrides, and E. Tsombachidou. 2009. Lithostratigraphy and fossiliferous sites. In *The Late Miocene Mammal Faunas of the Mytilinii Basin, Samos Island, Greece: New Collection,* ed. G. D. Koufos and D. Nägel, pp. 13–26. *Beiträge zur Paläontologie* 31.

Kostopoulos, D., S. Sen, and G. D. Koufos. 2003. Magnetostratigraphy and revised chronology of the late Miocene mammal localities of Samos, Greece. *International Journal of Earth Sciences* 92:779–794.

Kostopoulos, D., N. Spassov, and D. Kovachev. 2001. Contribution to the study of *Microstonyx*: Evidence from Bulgaria and the SE European populations. *Géodiversitas* 23(3):411–437.

Koufos, G. D. 1980. Palaeontological and stratigraphical study of the Neogene continental deposits of Axios Valley, Macedonia, Greece. Ph.D. diss., Aristotle University of Thessaloniki. *Scientific Annals Faculty of Physics-Mathematics, University Thessaloniki* 19(11):1–322 (in Greek with English summary).

Koufos, G. D. 1986. Study of the Vallesian hipparions of the lower Axios valley (Macedonia, Greece). *Geobios* 19:61–79.

Koufos, G. D. 1990. The hipparions of the lower Axios valley (Macedonia, Greece): Implications for the Neogene stratigraphy and the evolution of hipparions. In *European Neogene Mammal Chronology,* ed. E. Lindsay, V. Fahlbusch and P. Mein, pp. 321–338. New York: Plenum Press.

Koufos, G. D. 1993. A mandible of *Ouranopithecus macedoniensis* from a new late Miocene locality of Macedonia (Greece). *American Journal of Physical Anthropology* 91:225–234.

Koufos, G. D. 1995. The first female maxilla of the hominoid *Ourano-pithecus macedoniensis* from the late Miocene of Macedonia, Greece. *Journal of Human Evolution* 29:385–399.

Koufos, G. D. 2000a. New material of Vallesian hipparions (Mammalia, Perissodactyla) from the lower Axios valley, Macedonia, Greece. *Senckenbergiana Lethaea* 80:231–255.

Koufos, G. D. 2000b. Revision of the late Miocene carnivores from the lower Axios valley. *Münchener Geowissenschaften Abhandlungen* (A) 39:51–92.

Koufos, G. D. 2000c. The hipparions of the late Miocene locality "Nikiti 1," Chalkidiki, Macedonia, Greece. *Revue Palaeobiologique* 19(1):47–77.

Koufos, G. D. 2001. The Villafranchian mammalian faunas and biochronology of Greece. *Bolletino della Società Paleontologica Italiana* 40(2):217–223.

Koufos, G. D. 2006a. The Neogene mammal localities of Greece: faunas, chronology and biostratigraphy. *Hellenic Journal of Geosciences* 41:183–214.

Koufos, G. D. 2006b. The late Miocene Vertebrate locality of Perivolaki, Thessaly, Greece. *Palaeontographica* Abt. A 276:1–221.

Koufos, G. D. 2006c. The large mammals from the Miocene/Pliocene locality of Silata, Macedonia, Greece with implications about the Latest Miocene palaeoecology. *Beitrage zur Paläontologie* 30:293–313.

Koufos, G. D. 2006d. Primates. In *The Late Miocene Vertebrate Locality of Perivolaki, Thessaly, Greece*, ed. G. D. Koufos. *Palaeontographica* Abt. A 276:23–37.

Koufos, G. D. 2006e. Carnivora. In *The Late Miocene Vertebrate Locality of Perivolaki, Thessaly, Greece*, ed. G. D. Koufos. *Palaeontographica* Abt. A 276:39–74.

Koufos, G. D. 2006f. Palaeoecology and chronology of the Vallesian (late Miocene) in the Eastern Mediterranean region. *Palaeogeography, Palaeoclimatology, Palaeoecology* 234:127–145.

Koufos, G. D. 2007. Suoids (Mammalia, Artiodactyla) from the early/middle Miocene locality of Antonios (Macedonia, Greece). *Senckenbergiana Lethaea* 87(2):171–186.

Koufos, G. D. 2009a. The genus *Mesopithecus* (Primates, Cercopithecidae) in Greece. *Bolletino della Società Paleontologica Italica* 48(2):157–166.

Koufos G. D. 2009b. The Neogene cercopithecids (Mammalia, Primates) of Greece. *Géodiversitas* 31(4):817–850.

Koufos, G. D., 2009c. Carnivora. In *The Late Miocene Mammal Faunas of the Mytilinii Basin, Samos Island, Greece: New Collection*, ed. G. D. Koufos and D. Nägel, pp. 57–105. *Beiträge zur Paläontologie* 31.

Koufos, G. D. and L. de Bonis. 2005. The Late Miocene hominoids *Ouranopithecus* and *Graecopithecus*: Implications about their relationships and taxonomy. *Annales de Paléontologie* 91:227–240.

Koufos, G. D., L. de Bonis, D. S. Kostopoulos, L. Viriot, and T. D. Vlachou. 2004. New material of *Mesopithecus* (Primates, Cercopithecidae) from the Turolian locality of Vathylakkos 2, Macedonia, Greece. *Paläontologische Zeitschrift* 78(1):213–228.

Koufos, G. D., L. de Bonis and S. Sen. 1995. *Lophocyon paraskevaidisi* a new viverrid (Carnivora, Mammalia from the middle Miocene of Chios Island (Greece). *Geobios* 28(4):511–523.

Koufos, G. D. and D. S. Kostopoulos. 1997. Biochronology and succession of the Plio-Pleistocene macromammalian localities of Greece. In *Actes du Congrès Biochrom'97*, ed. J.-P. Aguilar, F. Leg-endre, and J. Michaux, pp. 619–634. Memoires et Traveaux E.P.H.E., Institut Montpellier, 21.

Koufos, G. D., D. S. Kostopoulos, and K. K. Koliadimou. 1991. Un nouveau gisement de mammiferes dans le Villafranchien de Macéédoine occidentale (Gréce). *Compte Rendus Academie Sciences Paris*, 2nd ser., 313:831–836.

Koufos, G. D., D. S. Kostopoulos, and G. Merceron. 2009. Palaeoecology–Palaeobiogeography. In *The Late Miocene Mammal Faunas of the Mytilinii Basin, Samos Island, Greece: New Collection*, ed. G. D. Koufos and D. Nägel, pp. 409–430. *Beiträge zur Paläontologie* 31.

Koufos, G. D, D. S. Kostopoulos, and T. D. Vlachou. 2009. Chronology. In *The Late Miocene Mammal Faunas of the Mytilinii Basin, Samos Island, Greece: New Collection*, ed. G. D. Koufos and D. Nägel, pp. 397–408. *Beiträge zur Paläontologie* 31.

Koufos, G. D., A. Koutsouveli, D. Galanakis, I. A. Sylvestrou, and T. D. Vlachou. 1999. A new late Miocene locality from Velestinon, Thessaly, Greece. Contribution to the biochronology of the neogene deposits. *Compte Rendus Academie Sciences Paris* 328:79–483.

Koufos, G. D. and D. Nagel, eds. 2009. *The Late Miocene Mammal Faunas of the Mytilinii Basin, Samos Island, Greece: New Collection. Beiträge zur Paläontologie* 31:1–430.

Koufos, G. D., S. Sen, D. S. Kostopoulos, I. A. Sylvestrou, and T. D. Vlachou. 2006. Chronology. In *The Late Miocene Vertebrate Locality of Perivolaki, Thessaly, Greece*, ed. G. D. Koufos, pp. 185–200. *Palaeontographica* 276.

Koufos G. D. and G. E. Syrides. 1997. A new mammalian locality from the early-middle Miocene of Macedonia, Greece. *Compte Rendus Academie Sciences Paris* 325:511–516.

Koufos, G. D., G. E. Syrides, K. K. Koliadimou, and D. S. Kostopoulos. 1991. Un nouveau gisement de Vertébrés avec hominoide dans le Miocene supérieur de Macédoine (Gréce). *Compte Rendus Academie Sciences* Paris, Serie II, 313:691–696.

Koufos, G. D. and T. D. Vlachou. 2005. Equidae (Mammalia, Perissodactyla) from the late Miocene of Akkaşdaği, Turkey. In *Geology, Mammals and Environments at Akkaşdaği, Late Miocene of Central Anatolia*, ed. S. Sen, pp. 633–705. *Géodiversitas* 27(4).

Koufos, G. D., N. Zouros, and O. Mourouzidou. 2003. *Prodeinotherium bavaricum* (Proboscidea, Mammalia) from Lesvos Island, Greece: The appearance of deinotheres in the Eastern Mediterranean. *Geobios* 36:305–315.

López-Antoñanzas, R., S. Sen, and G. D. Koufos. 2004. A ctenodactylid rodent (Mammalia: Rodentia) from the Middle Miocene of Chios Island (Greece). *Geobios* 38:113–126.

Lyras, G. A. and A. A. E. Van der Geer. 2007. The Late Pliocene vertebrate fauna of Vatera (Lesvos Island, Greece). *Cranium* 24(2):11–24.

Marinos, G. and N. Symeonidis. 1974. Neue Funde aus Pikermi, Attika und eine allgemeine geologische bersicht dieses paläontologischen Raumes. *Annales géologiques Pays hellenique* 26:1–27.

Mein, P. 1990. Updating of MN zones. In *European Neogene Mammal Chronology*, ed. E. Lindsay, V. Fahlbusch, and P. Mein, pp. 73–90. New York: Plenum Press.

Melentis, J. 1966. Der erste Nachweis von *Brachyodus onoideus* (Mammalia, Anthracotheriidae) aus Griechenland und die datierung der fundschichten. *Annales géologiques Pays helleniques* 17:221–235.

Melentis, J. K. 1967. Die pikermifauna von Halmyropotamos (S. Eüboa/Griechenland). *Proceedings Academy of Athens* 41:261–266.

Melentis, J. and Tobien, H. 1967. Paläontologische Ausgrabungen auf der Insel Chios (eine vorläufige Mitteilung). *Proceedings Academy of Athens* 42:147–152.

Paraskevaidis, E. 1940. Eine obermiocäne Fauna von Chios. *Neues Jahrbuch fur Geologie Mineralogie und Paläontologie* 83(B):363–442.

Pe-Piper, G. and J. W. Piper. 1993. Revised stratigraphy of the Miocene volcanic rocks of Lesbos, Greece. *Neues Jarbuch für Geologie und Paläontologie Monatshefte* (2):97–110.

Rögl, F. 1999. Circum-Mediterranean Miocene paleogeography. In *The Land Mammals of Europe*, ed. G. Rössner and K. Heissig, pp. 39–48. Munich: Dr. Friedrich Pfeil.

Roussiakis, S. and G. Theodorou, G. 2003. Carnivora from the late Miocene of Kerassia (Northern Euboea, Greece). DEINSEA 10:469–497.

Roussiakis, S., G. Theodorou, and G. Iliopoulos. 2006. An almost skeleton of *Metailurus parvulus* (Carnivora, Felidae) from the late Miocene of Kerassia (Northern Euboea, Greece). *Geobios* 39:563–584.

Schmidt-Kittler, N. 1983. The mammals from the lower Miocene of Aliveri (island of Evia, Greece). On a new species of *Sivanasua* PILGRIM, 1931 (Felidae, Carnivora) and the phylogenetic position of the genus. *Proceedings Koniklijke Nederlandse Akademie Wetenschappen* B86:301–318.

Schmidt-Kittler, N. 1995a. The Vertebrate locality Maramena (Macedonia, Greece) at the Turolian/ Ruscinian boundary (Neogene). *Münchner Geowissenschftische Abhandlungen* 28:1–180.

Schmidt-Kittler, N. 1995b. Carnivora, Mammalia. In *The Vertebrate Locality Maramena (Macedonia, Greece) at the Turolian/Ruscinian Boundary (Neogene)*, ed. N. Schmidt-Kittler, *Münchner Geowissenschftische Abhandlungen* 28:75–86.

Sen, S. 1997. Magnetostratigraphic calibration of the Neogene mammal chronology. *Palaeogeography, Palaeoclimatology, Palaeoecology* 133:181–204.

Sen, S., G. D. Koufos, D. Kondopoulou, and L. de Bonis. 2000. Magnetostratigraphy of the late Miocene continental deposits of the lower Axios valley, Macedonia, Greece. In *Mediterranean Neogene Cyclostratigraphy in Marine–Continental Deposits*, ed. G. D. Koufos and Ch. Ioakim. *Bulletin Geological Society of Greece*, Special Publication 9:197–206.

Sickenberg, O. 1968. Die pleistozanen knochbrekzien von Volax (Griech. Mazedonien). *Geolgische Jahrbuch* 85:33–54.

Solounias, N. 1981. The Turolian fauna from the island of Samos, Greece. *Contribution in Vertebrate Evolution* 6:1–232.

Sondaar, P. Y. 1971. The Samos *Hipparion*. *Proceedings Koniklijke Nederlandse Akademie Wetenschappen* B74:417–441.

Sondaar, P. Y. and V. Eisenmann. 1995. The hipparions (Equidae, Perissodactyla, Mammalia). In *The Vertebrate Locality Maramena (Macedonia, Greece) at the Turolian/Ruscinian Boundary (Neogene)*, ed. N. Schmidt-Kittler, pp. 137–142. *Münchener Geowissenschftische Abhandlungen* 28.

Steininger, F.F. 1999. Chronostratigraphy, geochronology and biochronology of the Miocene "European Land Mammal Mega-Zones" (ELMMZ) and the Miocene "Mammal-zones" (MN-Zones). In *The Land Mammals of Europe*, ed. G. Rössner and K. Heissig, pp. 9–24. Munich: Dr. Friedrich Pfeil.

Steininger, F. F., R. L. Bernor, and V. Fahlbusch. 1990. European Neogene marine/continental chronologic correlations. In *European Neogene Mammal Chronology*, ed. E. Lindsay, V. Fahlbusch, and P. Mein, pp. 15–46. New York: Plenum Press.

Storch, G. and T. Dalmann. 1995. Murinae (Rodentia, Mammalia). In *The Vertebrate Locality Maramena (Macedonia, Greece) at the Turolian/Ruscinian Boundary (Neogene)*, ed. N. Schmidt-Kittler, pp. 121–132. *Münchener Geowissenschftische Abhandlungen* 28.

Sylvestrou, I. and D. S. Kostopoulos. 2006. Suidae. In *The Late Miocene Vertebrate Locality of Perivolaki, Thessaly, Greece*, ed. G. D. Koufos, pp. 121–133. *Palaeontographica* A 276.

Sylvestrou, I. A. and D. S. Kostopoulos. 2009. Suidae. In *The Late Miocene Mammal Faunas of the Mytilinii Basin, Samos Island, Greece: New Collection*, ed. G. D. Koufos and D. Nägel, pp. 283–297. *Beiträge zur Paläontologie* 31.

Sylvestrou, I. A. and G. D. Koufos. 2006. Stratigraphy and Locality. In *The Late Miocene Vertebrate Locality of Perivolaki, Thessaly, Greece*, ed. G. D. Koufos, pp. 1–9. *Palaeontographica* 276.

Symeonidis, N., 1978. Ein schädel von *Metailurus parvulus* (Hensel) aus Pikermi (Attica, Griechenland). *Annales géologiques des Pays helleniques* 29:698–703.

Symeonidis, N. 1992. Fossil mammals of L. Pleistocene (Villafrachian) age from the basin of Sesklon (Volos). *Annales géologiques des Pays helleniques* 35:1–42.

Syrides, G. E., K. K. Koliadimou, and G. D. Koufos. 1997. New Neogene molluscan and mammalian sites from Thrace, Greece. *Compte Rendus Academie Sciences Paris*, Serie IIa, 324:427–433.

Theocharopoulos, K. 2000. Late Oligocene–Middle Miocene Democricetodon Spanocricetodon and Karydomys n. gen. from the eastern Mediterranean area, Ph.D. diss. *Gaia* 8:1–91.

Theodorou, G., A. Athanassiou, S. Roussiakis, and G. Iliopoulos. 2003. Preliminary remarks on the late Miocene herbivores of Kerassia (Northern Euboea, Greece). DEINSEA 10:519–530.

Tobien, H. 1980. A note on the skull and mandible of a new choerolophodonte mastodont (Proboscidea, Mammalia) from the middle Miocene of Chios (Aegean sea, Greece). In *Aspects of Vertebrate History: Essays in Honor of Edwin Harris Colbert*, ed. L. L. Jacobs, pp. 299–307. Flagstaff: Museum of Northern Arizona Press.

Ünay, E. 1980. The Cricedontini (Rodentia) from the Hayraktepe section (Canakale, Turkey). *Proceedings Koniklijke Nederlandse Akademie Wetenschappen* B84(2):217–238.

Ünay, E. and H. de Bruijn. 1984. On some Neogene rodent assemblages from both sides of the Dardanelles, Turkey. *Newsletter of Stratigraphy* 13(3):119–132.

van der Made, J. 1996. Pre-Pleistocene land mammals from Crete. In *Pleistocene and Holocene Fauna of Crete and Its First Settlers*, ed. D. S. Reese, pp. 69–79. *Monographs in World Archaeology* 28.

van der Made, J. 2003. Suoidea (pigs) from the Miocene hominoid locality Candir in Turkey. In *Geology and Vertebrate Paleontology of the Miocene Hominoid Locality of Çandır*, ed. E. Güleç, D. Begun, and D. Geraads, pp. 149–179. *CourierForschung-Institut Senckenberg* 240.

van der Made, J. 2005. Errata and reply to Guests Editor's notes. *Courier Forschung-Institut Senckenberg* 254:473–477.

van der Made, J. and S. Moyà-Solà. 1989. European Suinae (Artiodactyla) from the late Miocene onwards. *Bolletino della Società Paleontologia Italiana* 28(2–3):329–339.

van der Meulen, A. and T. van Kolfschoten. 1986. Review of the Late Turolian to Early Biharian mammal faunas from Greece and Turkey. *Memoires Società Geologica Italica* 31:201–211.

Vasileiadou, K. and G. D. Koufos. 2005. The micromammals from the Early/Middle Miocene locality of Antonios, Chalkidiki, Greece. *Annales de Paléontologie* 91:197–225.

Vasileiadou, K., G. D. Koufos, and G. E. Syrides. 2003. Silata, a new locality with micromammals from the Miocene/Pliocene boundary of the Chalkidiki peninsula, Macedonia, Greece. DEINSEA 10:549–562.

Vasileiadou, K. and I. A. Sylvestrou. 2009. Micromammals. In *The Late Miocene Mammal Faunas of the Mytilinii Basin, Samos Island, Greece: New Collection*, ed. G. D. Koufos and D. Nagel, pp. 37–55. *Beiträge zur Palaontologie* 31.

Vlachou, T. and G. D. Koufos. 2002. The hipparions (Mammalia, Perissodactyla) from the Turolian locality of Nikiti 2, Chalkidiki, Macedonia, Greece. *Annales de Paléontologie* 88:215–263.

Vlachou, T. and G. D. Koufos. 2006. Equidae. In *The Late Miocene Vertebrate Locality of Perivolaki, Thessaly, Greece*, ed. G. D. Koufos. *Palaeontographica* Abt. A 276:81–119.

Vlachou, T. and G. D. Koufos. 2009. Equidae. In *The Late Miocene Mammal Faunas of the Mytilinii Basin, Samos Island, Greece: New Collection*, ed. G. D. Koufos and D. Nägel, pp. 207–281. *Beiträge zur Paläontologie* 31.

Zapfe, H. 1991. *Mesopithecus pentelicus* Wagner aus dem Turolian von Pikermi bei Athen. Odontologie und osteologie. *Neue Denk-Schriften des Naturhistorisches Museum Wien* 5:13–203.

APPENDIX
Temporal Distributions (MN Zones 1–17) of the Various Neogene Taxa of Greece

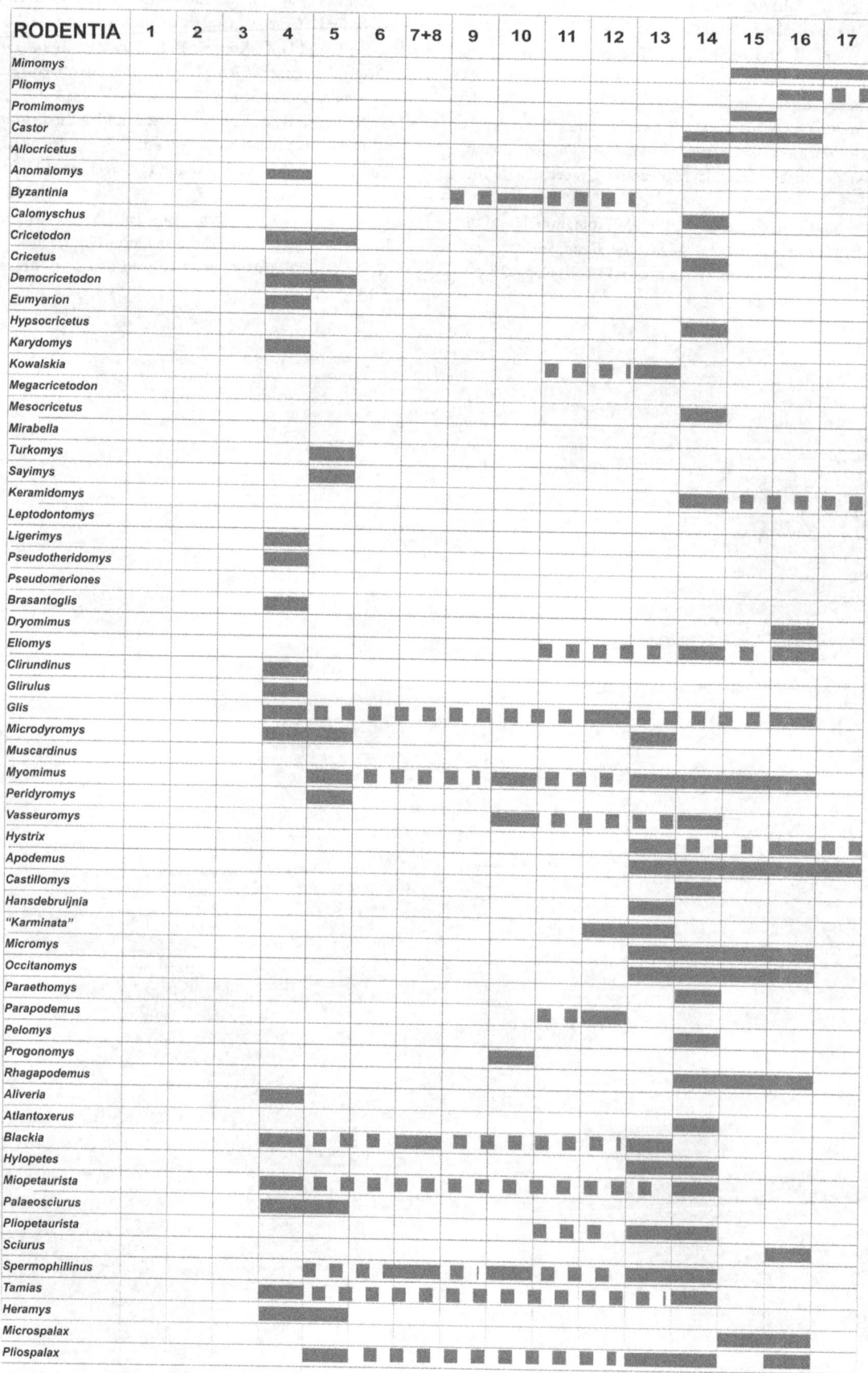

A

CARNIVORA	1	2	3	4	5	6	7+8	9	10	11	12	13	14	15	16	17
Bosdagius																▬
Indarctos											▬	▬				
Ursus																▬
Baranogale																▬
Eomellivora									▬							
Martes												▬				
Meles																▬
Palaeogale				▬												
Percrocuta					▬											
Plesiogulo											▬					
Promeles											▬					
Promephitis											▬	▬				
Proputorius																
Sinictis											▬					
Adcrocuta									▬	▬						
Chasmaporthetes										▬	▪	▬	▪	▪	▪	▬
Dinocrocuta									▬							
Hyaenictis											▬					
Hyaenictitherium									▬	▬	▬					
Ictitherium									▬	▬						
Lycyaena											▬					
Pliohyaena														▬	▪	▬
Plioviverrops										▬						
Protictitherium				▬	▬				▬	▪	▪	▬				
Thalassictis											▬					
Felis											▬					
Homotherium																▬
Machairodus										▬	▬					
Megantereon																▬
Metailurus											▬					
Paramachairodus												▬				
Pseudailurus					▬											
Nyctereutes														▬	▬	▬
Vulpes																▬

B

(continued)

ARTIODACTYLA	1	2	3	4	5	6	7+8	9	10	11	12	13	14	15	16	17
Brachyodus			▫	▫	▫											
Sanitherium					■											
Bunolistriodon					■											
Korynochoerus												■				
Listriodon					■											
Microstonyx									■	■	■	■				
Propotamochoerus										■						
Sus													■		▫	▫
Dorcatherium					■		▫	▫	▫	▫						
Croizetoceros																■
Eucladoceros																■
Metacervoceros																■
Pliocervus											■					
Bohlinia									■	■	■	■				
Georgiomeryx					■											
Helladotherium										■	■	■				
Miotragocerus										■	■	■				
Mitilanotherium																■
Palaeogiraffa									■							
Palaeomeryx					■											
Palaeotragus									■	■	■	■				
Samotherium										■	■	■				
Criotherium										■						
Eotragus					■											
Euthyceros					■											
Gallogoral																■
Gazella									■	■	■	■	■	■	■	■
Gazellospira																■
Helladorcas									■							
Hispanodorcas												■				
Hypsodontus					■											
Koufotragus														■		
Leprobos															■	
Mesembriacerus									■							
Nisidorcas										■						
Norbertia												■				
Oiceros									■	▫						
Ouzoceros									■				■			
Palaeoreas										■	■					
Palaeoryx									■	■	■	■				
Parabos														■		
Prostrepsiceros									■	■	■	■				
Protoryx									■	■						
Protragelaphus											■					
Samotragus									■							
Skoufotragus											■					
Sporadotragus											■					
Tethytragus					■											
Tragoportax									■	■	■	■				

C

	1	2	3	4	5	6	7+8	9	10	11	12	13	14	15	16	17
PROBOSCIDEA																
Anancus																▬
Choerolophodon					▬	▪	▪	▪	▬	▬	▬	▬				
Tetralophodon											▬					
Zygolophodon										▬	▬					
Deinotherium					▬	▪	▪	▪	▬	▬	▬	▬				
Prodeinotherium			▬													
Mammuthus																▬
PERISSODACTYLA																
Equus															▬	▬
Hipparion									▬	▬	▬	▬				
RHINOCEROTIDAE																
Aceratherium									▬	▪	▬					
Ceratotherium									▬	▬	▬	▬				
Dihoplus											▬					
Stephanorhinus																▬
CHALICOTHERIIDAE																
Ancylotherium									▬	▪	▬					
Chalicotherium											▬					
Macrotherium										▬	▪	▬				
TUBULIDENTATA																
Orycteropus									▬	▪	▪	▬				
HYRACOIDEA																
Pliohyrax												▬				
PRIMATES																
Dolichopithecus														▬		
Mesopithecus										▬	▬	▬				
Ouranopithecus									▬							
Paradolichopithecus														▬		

D

Part V

Zoogeography and Paleoecology

Chapter 29

Continental-Scale Patterns in Neogene Mammal Community Evolution and Biogeography

A Europe-Asia Perspective

MAJID MIRZAIE ATAABADI, LI-PING LIU, JUSSI T. ERONEN, RAYMOND L. BERNOR,
AND MIKAEL FORTELIUS

Spatial diachrony is a key question in stratigraphic correlation. If taxa appear at different times in different places, their first occurrences obviously will not represent useful time horizons. For terrestrial mammals, diachrony is closely intertwined with the question of faunal provinciality, including how provinces have themselves developed over time.

For Eurasia, the modern phase of research into diachrony and provinciality may be said to have begun with the 1975 Regional Committee on Mediterranean Neogene Stratigraphy meeting in Bratislava, Czechoslovakia. There, Pierre Mein produced his first iteration of his famous European Mammal Neogene (MN) zones (Mein 1975). This biochronologic system recognized reference faunas and struck correlations of European, North African, and West Asian vertebrate localities based on characteristic associations and first occurrences of taxa that migrated over great geographic distances in short intervals of geochronologic time. However, work in the Late Miocene of Maragheh, Iran (Bernor 1978, 1985, 1986), and Samos, Greece (Solounias 1981), developed biostratigraphic, biochronologic, and geochronologic work that did not support isochronic correlations of eastern Mediterranean/southwest Asian Turolian large mammal faunas with those of Europe (reviewed in Bernor, Solounias, et al. 1996 and Fortelius et al. 1996). This prompted Bernor and others (Bernor 1978; Bernor et al. 1979; Bernor 1984; Bernor and Pavlakis 1987) to propose several Eurasian and African biogeographic provinces that were distinct in their faunal character and ecology with resulting time-transgressive dispersion of successive chronofaunas (Olson 1952).

For a well-resolved answer to questions concerning temporal occurrence of fossil taxa, an independent chronology is necessary. That is to say, a chronology that is not based on the occurrence or evolution of the taxa concerned or on information that is likely to be correlated with those phenomena. In practice, this would mean a chronology that excludes biochronology, a constraint that still severely reduces the amount of data available for analysis. Here, we use a different approach. We accept all stratigraphic information of adequate quality as given in the NOW database (Fortelius 2009), including mammal biochronology, and investigate the spatial and temporal distribution of two attributes that reflect the regional development of faunas and environments: faunal resemblance and humidity estimated from mean hypsodonty. We ask whether within this biochronologically driven stratigraphic framework evidence still exists for diachronous development of faunas and environments in different regions. This is a conservative approach in that both faunal resemblance and hypsodonty depend on evolution over time, as does biochronology. If diachrony is nonetheless detected, the evidence for it must be regarded as strong.

Our two-pronged approach also allows us to ask whether the faunal resemblance patterns are related to regional differences in environmental conditions. We have previously shown that the mean "ordinated hypsodonty" (i.e., molar crown height assigned to hypsodonty classes; see the following section on methods) of large herbivorous mammals indicates major regional differences in the general trend of Neogene midlatitude aridification (Fortelius et al. 2002, 2003, 2006), and that these patterns are due to the genera and species with the highest occupancy (relative locality coverage; Jernvall and Fortelius 2002). Very recently, we addressed the dynamics of land mammal provinciality using faunal similarity to chart the origin, rise, acme, decline and extinction of the Pikermian chronofauna between 12 Ma and 5 Ma. We showed that the westward expansion of the Pikermian chronofauna from the late Middle Miocene to the medial Late Miocene correlates closely with the expansion of arid habitats as well as with the output of a paleoclimate modeling study (Eronen et al. 2009). We used Olson's concept of chronofauna in a quantitative sense for a faunal assemblage defined by taxonomic similarity, as described in the "Data and Methods" section.

Here, we turn our attention to the development of the Middle and Late Miocene mammal assemblages of East Asia, which we refer to as the Tunggurian and Baodean chronofaunas, after their most classic localities of occurrence (Tunggur-Moergen and Baode). Well known as the "pre-*Hipparion*" and "*Hipparion* faunas" of China, they respectively became the bases of the Tunggurian and Baodean land mammal ages in East Asia (Li, Wu, and Qiu 1984; Qiu and Qiu 1990, 1995).

The recently reviewed faunal composition of Tunggur-Moergen and Baode locality 49 mammal localities is shown in table 29.1. The traditional Tunggur "*Platybelodon* fauna" has recently been appointed to the Tunggur-Moergen fauna, to distinguish it from the underlying Tairum Nor fauna (Qiu, Wang, and Li 2006). The Tunggur-Moergen locality, the paleomagnetic correlation for which is between chrons C5Ar.3r and C5r.3r (13–11.8 Ma; Wang, Qiu, and Opdyke 2003), is reported as being correlative with MN-equivalent 7+8 (Qiu, Wu, and Qiu 1999). The Baode Red Clay fauna as a whole is now correlated to MN-equivalent 13 (Zhang and Liu 2005; Kaakinen et al., chapter 7, this volume). The paleomagnetic age for the Baode Formation in the type area is 7.23–5.34 Ma (Zhu et al. 2008), with an estimated age for locality 49 at 7.0 Ma (Kaakinen et al., chapter 7, this volume). In this chapter, we have selected Tunggur-Moergen and Baode locality 49 as the type localities for the two chronofaunas.

DATA AND METHODS

We downloaded the dataset from the NOW database on April 24, 2009 (see http://www.helsinki.fi/science/now). We used all large mammal data between 20.5 Ma and 3.4 Ma, encompassing the European Neogene mammal units MN 3 to MN 15.

The final dataset as analyzed for similarity indices comprised 358 localities and 516 taxa and is available upon request from the authors. We calculated different Genus-level Faunal Resemblance Indices (GFRIs) to Baode locality 49 and Tunggur-Moergen locality from the dataset using PAST (Hammer, Harper, and Ryan 2001; Hammer and Harper 2006; see http://folk.uio.no/ohammer/past/). We used Dice, Jaccard, Simpson, and Raup-Crick GFRIs to compare differences in the overall trends and between different GFRIs. Dice FRI is highly recommended by Archer and Maples (1987) and Maples and Archer (1988) and is calculated as $2C/(A + B)$, where C is the number of shared taxa between two faunas, and A and B are the total number of taxa in the compared faunas. We mapped the results only for Dice GFRI and have presented other GFRI results (which show similar patterns) in the appendix.

We applied the following criteria for undertaking these analyses: localities were excluded when they (1) could not be assigned to the temporal span of a single "MN-equivalent" unit, (2) had fewer than 7 taxa identified at the genus level, or (3) lacked geographic positional data (coordinates). We tested our method with different minimum numbers of taxa. We found that when we used fewer than 7 taxa the analytical noise increased, and when we limited the analysis to localities with 10 or more taxa, the number of available localities was low. Overall, we found the number of taxa used did not affect the spatial geographic patterns.

As provisional common coinage for Eurasian mammal biostratigraphy, we use ad hoc "MN-equivalents" (MNEQ). These are defined and computed from minimum and maximum age estimates for the locality ages as given, according to where the computed range midpoint falls in the correlation scheme in use in the NOW database (Steininger et al. 1996). In this study, we only included localities where the age span is less than or equal to that of the corresponding MN unit.

Olson (1952, 1958) developed the concept of the chronofauna for "a geographically restricted, natural assemblage of interacting animal populations that has maintained its basic structure over a geologically

Table 29.1

Updated Faunal Lists of Tunggur-Moergen and Baode Locality 49

Order	Tunggur-Moergen[a]	Baode Locality 49[b]
Carnivora	*Amphicyon tairumensis*	*Amphicyon* indet.
	?*Melodon* indet.	?*Melodon incertum*
	Metailurus mongoliensis	*Melodon majori*
	Plithocyon teilhardi	*Metailurus parvulus*
	Pseudarctos indet.	*Felis* indet.
	Gobicyon macrognathus	*Machairodus palanderi*
	Mionictis indet.	*Megantereon* indet.
	Martes indet.	*Adcrocuta eximia*
	Tungurictis spocki	*Hyaenictitherium hyaenoides*
	Percrocuta tungurensis	*Hyaenictitherium wongii*
	Sansanosmilus indet.	*Ictitherium viverrinum*
		?*Lycyaena dubia*
		Lutra aonychoides
		Plesiogulo brachygnathus
		Promephitis cf. *maeotica*
		Proputorius minimus
Proboscidea	*Zygolophodon gobiensis*	*Mammut borsoni*
	Platybelodon grangeri	*Mammut* indet.
Perissodactyla	*Anchitherium gobiense*	"*Hipparion*" *plocodus*
	Chalicotherium brevirostris	"*Hipparion*" *platyodus*
	Acerorhinus zernowi	"*Hipparion*" *hippidiodus*
	Hispanotherium tungurense	"*Hipparion*" *coelophyes*
	Rhinocerotidae indet.	"*Hipparion*" indet.
		Acerorhinus palaeosinensis
		Chilotherium indet.
		Stephanorhinus orientalis
Artiodactyla	*Listriodon splendens*	?*Dorcadoryx lagrelii*
	Kubanochoerus indet.	*Gazella dorcadoides*
	Palaeotragus tungurensis	*Gazella gaudryi*
	Stephanocemas thomsoni	*Gazella* indet.
	Lagomeryx triacuminatus	*Gazella* ?*paotehensis*
	Euprox grangeri	*Palaeoryx sinensis*
	Dicrocerus indet.	*Urmiatherium intermedium*
	Micromeryx indet.	*Cervavitus novorossiae*
	Turcoceros grangeri	*Procapreolus latifrons*
	Turcoceros noverca	*Honanotherium schlosseri*
		Palaeotragus microdon
		Palaeotragus indet.
		Propotamochoerus hyotherioides
		Chleuastochoerus stehlini

[a]Wang, Qiu, and Opdyke (2003).

[b]Kaakinen et al. (chapter 7, this volume).

significant period of time" (Olson 1952:181). We define a computational equivalent of this concept in terms of faunal resemblance: a chronofauna is a set of localities united by faunal similarity to an arbitrarily selected type or standard locality. In this exploratory study, we have not set numerical limits to chronofaunas—although that would be possible—but have simply mapped and described the rise and fall of these entities over a series of time steps based on correlation with the MN system.

For hypsodonty calculations, we used only large herbivorous mammals; all small mammals (orders Lagomorpha, Chiroptera, Rodentia, and Insectivora) as well as carnivores (orders Carnivora and Creodonta) were deleted. Three classes of hypsodonty are recorded in the NOW database: brachydont, mesodont, and hypsodont. The criteria for assigning species to these classes are ultimately up to the taxonomic coordinators of the NOW advisory board, but the rule of thumb is based on the ratio of height to length of the second molar (upper or lower). Brachydont teeth have a ratio of less than 0.8, mesodont have a ratio in the range 0.8–1.2, and hypsodont over 1.2. For this study, the hypsodonty classes were assigned the values 1, 2, and 3, respectively. This is a relatively conservative procedure, as the difference in crown height between a hypsodont and a brachydont species is usually more than 3:1. The mean hypsodonty value (see appendix) was calculated for each locality by averaging these ordinated scores (excluding localities with a single species).

All maps depicting GFRI and hypsodonty values were made in MapInfo Professional 8.5 using the color grid interpolation and inverse distance weighted (IDW) algorithm with the following settings: cell size 20 km, search radius 600 km, grid border 600 km, number of inflections 10, values rounded to 0.01 (0.1 for hypsodonty) decimal. The inflection values were manually set to range from 1 to 3 for hypsodonty, from 0 to 0.55 for Baode FRI, and 0.5 for Tunggur FRI maps. We used opacity of 25% for the color interpolation to show the base map below the interpolated values.

"Hipparion" here is applied to all the species of hipparionine horses that are not explicitly assigned to any one of the following genera: *Hippotherium*, *Cremohipparion*, *Cormohipparion*, *Plesiohipparion*, or *Proboscidipparion*. We have presented the shared taxa between some major localities and also listed the genera with the highest incidence of occurrence in some time intervals in order to show the taxa driving the enhanced similarity patterns.

RESULTS

The Tunggurian chronofauna begins to emerge in MNEQ 4 (plate 29.1*B*), when the overall similarity to the Tunggur-Moergen locality is still low in all of Eurasia, as it was during the preceding interval, MNEQ 3 (see plate 29.1*A*). The Sihong-Songlinzhuang fauna (see plate 29.1*B* [SS]), shows a higher similarity to Tunggur (Dice 0.28) than European localities. It shares six genera with the Tunggur-Moergen fauna: *Anchitherium*, *Dicroceros*, *Lagomeryx*, *Micromeryx*, *Palaeotragus*, and *Stephanocemas*—all brachydont ungulates primarily associated with closed environments. Tunggur genera with the highest incidence of occurrence in the whole data matrix for MNEQ 4 are *Anchitherium*, *Amphicyon*, and *Lagomeryx*. Another region of emerging similarity is southwestern Europe (for geographical regions, see figure 29.1). Córcoles (see plate 29.1*B* [CC]), Spain (Dice 0.23) and Bézian (see plate 29.1*B* [BZ]), France (Dice 0.23) share *Amphicyon*, *Chalicotherium*, *Anchitherium*, *Hispanotherium*, *Lagomeryx*, *Martes*, *Mionictis*, *Plithocyon* and *Stephanocemas* with the Tunggur-Moergen fauna. The humidity analysis from hypsodonty (see plate 29.1*I*) suggests that East Asia (see figure 29.1) was humid at this time (blue pattern of East Asia in plate 29.1*I*), with incipient aridification in western Europe (notice the green pattern of southwestern Europe in plate 29.1*I*). This indicates that the "*Hispanotherium* fauna," which inhabited the seasonal dry conditions of the lower Middle Miocene of the Iberian peninsula in Europe (Antunes 1979; Agusti and Anton 2002), had not yet extended to East Asia.

More Tunggurian elements appear in MNEQ 5 (see plate 29.1*C*); the genera with the highest incidence of occurrence are *Anchitherium*, *Amphicyon*, and *Lagomeryx*. The highest similarity to Tunggur is found for Esvres-Marine Faluns (see plate 29.1*C* [EMF]), France (Dice 0.38), and the second highest for Kalkaman Lake (see plate 29.1*C* [KL]), Kazakhstan (Dice 0.33). There are 13 Tunggur-Moergen genera at Esvres-Marine Faluns, a significant component of western origin and some immigrant taxa, including *Hispanotherium*, *Chalicotherium*, *Anchitherium*, *Lagomeryx*, *Dicrocerus*, *Euprox*, *Amphicyon*, *Martes*, *Mionictis*, *Plithocyon*, *Pseudarctos*, *Sansanosmilus*, and *Zygolophodon*. In China, the Shanwang (see plate 29.1*C* [SW]) fauna (Dice 0.18) retains *Lagomeryx* and *Stephanocemas* from Sihong-Songlinzhuang fauna, suggesting that the climate and environment remained stable with little change in humidity (see plate 29.1*J*).

The maximum extent of the Tunggurian chronofauna occurred in MNEQ 6 (see plate 29.1*D*). Hezheng-Laogou

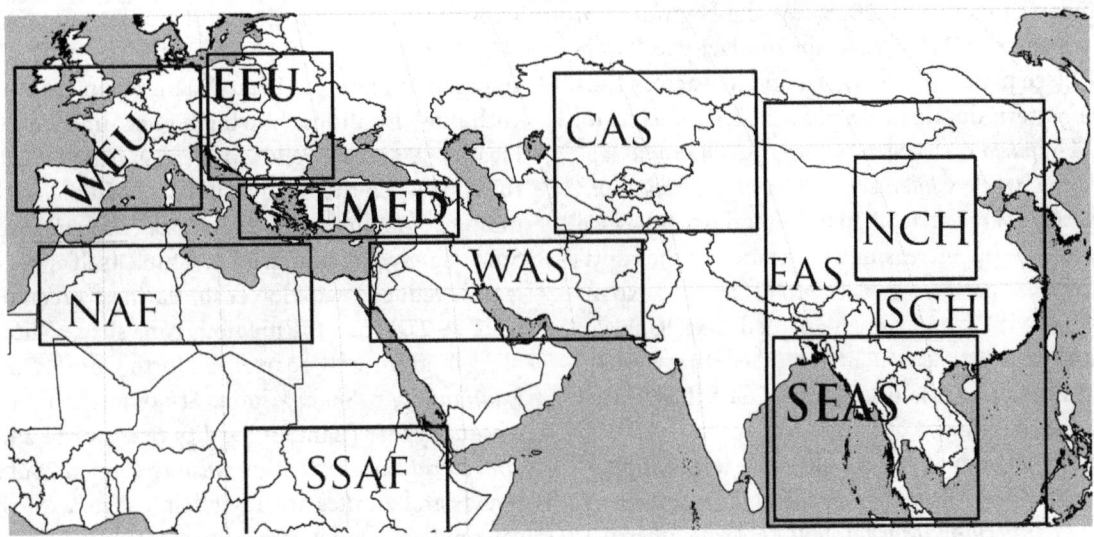

Figure 29.1 Geographical regions applied in this chapter. WEU = western Europe; EEU = eastern Europe; EMED = eastern Mediterranean; WAS = western Asia; CAS = central Asia; EAS = eastern Asia; SEAS = southeast Asia; NCH = North China; SCH = South China; NAF = North Africa; SSAF = Sub-Saharan Africa.

(see plate 29.1*D* [HL]) in China has the highest similarity to Tunggur (Dice 0.5), while high similarity is also seen for the locality Sansan (see plate 29.1*D* [S]), France (Dice 0.35), as well as Catakbagyaka (see plate 29.1*D* [CB]), Turkey (Dice 0.32), and Tongxin-Dingjiaergou (see plate 29.1*D* [TD]), China (Dice 0.32). The genera with the highest incidence of occurrence are *Anchitherium*, *Listriodon*, and *Amphicyon*. According to the hypsodonty proxy, humidity is uniform in Eurasia at this time with the exception of Ayibaligi Mevkii locality (see plate 29.1*K* [AM]) in Turkey, which shows dry conditions (red pattern).

During MNEQ 7+8, the low number of localities with high similarity to Tunggur indicates that the Tunggurian elements decline, except of course in Tunggur-Moergen itself (see plate 29.1*E* [TM]). The genera with the highest incidence of occurrence are *Listriodon*, followed by *Euprox*, *Micromeryx*, and *Chalicotherium*. At this time, the latitudinal climate gradient increased and arid belts appeared in the mid-latitudes of Asia (Flower and Kennett 1994; Liu, Eronen, and Fortelius 2009). In addition, eastern Mediterranean and eastern Europe (see figure 29.1) became more arid (notice the yellow patterns in plate 29.1*L*). The sparse data from southern East Asia (Chiang Muan [see plate 29.1*E* (CM)], Thailand) show significant difference from Tunggur-Moergen (Dice 0, no shared taxa). Although both areas have yielded primate-bearing

faunas, the northern Tunggur and southern Chiang Muan faunas are now separated by an arid belt (yellow pattern of East Asia in plate 29.1*L*) and have fully dissimilar faunas.

In MNEQ 9, the Tunggurian signal has essentially vanished from Eurasia (mostly blue patterns of no significant similarity in plate 29.1*F*), with some lingering similarity still seen in western Europe, including Spanish localities like Can Ponsic I (see plate 29.1*F* [CP]) (Dice 0.29) and Can Llobateres I (see plate 29.1*F* [CL]; Dice 0.23). The mid-latitude drying in Asia and the eastern Mediterranean area continued in MNEQ 9 (orange/red patterns in plate 29.1*M*) and coincided with a replacement of the Tunggurian chronofauna by a "*Hipparion* fauna" of Late Miocene aspect. In MNEQ 10, even the low similarity in western Europe disappears (dominance of blue patterns in plate 29.1*G*). The continued drying and cooling in this area (yellowish patterns in plate 29.1*N*) causes the extinction of the remaining Middle Miocene taxa in western Europe known as the Vallesian Crisis (Agustí and Moya-Sola 1990; Agustí and Anton 2002; Agustí, Sanz de Siria, and Garcés 2003) at the MN 9–10 boundary. In East Asia, continued drying of MNEQ 10 favored the expansion of the "*Hipparion* fauna" (plate 29.2*B*).

The Baodean chronofauna is the East Asian expression of the Pan-Eurasian "*Hipparion* fauna" (Kurtén 1952), corresponding to its western counterpart the Pikermian chronofauna (Eronen et al. 2009). At the beginning of the Late Miocene (MNEQ 9), genus similarity to Baode locality 49 (see plate 29.2E [B]) was still low across Eurasia (blue pattern dominant in plate 29.2A). *Adcrocuta*, *Gazella*, "*Hipparion*," *Ictitherium*, and *Machairodus* appear in China, while *Chilotherium*, *Gazella*, "*Hipparion*," and *Palaeotragus* appear at Sinap in Central Anatolia (see plate 29.2A [S]). The increasing aridification of the continent (orange/red patterns in plate 29.2H) favored expansion of open habitats and open-adapted taxa. *Gazella*, "*Hipparion*," and *Ictitherium* had reached the western margin of Eurasia at this time (Subsol de Sabadell [see plate 29.2A (SS)], Spain).

In MNEQ 10, more Baodean elements (e.g., "*Hipparion*," *Chilotherium*, *Gazella*, *Palaeotragus*, *Acerorhinus*, *Adcrocuta*, *Ictitherium*, *Machairodus*, *Hyaenictitherium*, and *Indarctos*) appear in Asia and in the eastern Mediterranean–West Asian region. Both North Chinese localities at this time show high similarity to Baode locality 49 (notice the orange pattern of northern China in plate 29.2B, which indicates Dice > 0.3), as do Grebeniki (see plate 29.2B [G]), Ukraine, and Poksheshty (see plate 29.2B [P]), Moldova (Dice > 0.35). The Baodean genera like *Adcrocuta*, *Cervavitus*, *Chilotherium*, *Gazella*, *Honanotherium*, "*Hipparion*," *Ictitherium*, *Machairodus*, *Metailurus*, *Palaeotragus*, *Plesiogulo*, and *Promephitis* appear widely in Eurasia, except for western Europe. The low similarity in western Europe may be related to its more humid climate (notice the blue/green pattern of western Europe in plate 29.2I) and the later expansion of the Pikermian chronofauna there (Eronen et al. 2009). The genera with the highest incidence of occurrence at this time across Eurasia are *Gazella*, "*Hipparion*," *Chilotherium*, and *Palaeotragus*.

The pattern in MNEQ 11 is similar to MNEQ 10 for the faunal similarity (see plate 29.2C), but for humidity, expansion of arid areas is observed (increase in yellow/orange patterns in plate 29.2J). The genera with the highest incidence of occurrence at this time are *Gazella*, "*Hipparion*," *Chilotherium*, *Palaeotragus*, and *Adcrocuta*. In addition to northern Chinese localities, Karacahasan (K), Kemiklitepe 1, 2 (KT), and Garkin (G) in Turkey, Thermopigi (T) in Greece, and Novaja Emetovka (NE) in Ukraine also show a high similarity (Dice > 0.3) to Baode locality 49 (see plate 29.2C). Western Europe still retains a remarkably low similarity to Baode locality 49 at this time (notice blue patterns of western Europe in plate 29.2C), and in addition, South China is dissimilar to North China (see figure 29.1 and plate 29.2C: notice light blue pattern in southern China compared to orange/red pattern in North China).

During the middle Turolian (MNEQ 12), the genera with the highest incidence of occurrence are *Gazella*, "*Hipparion*," *Palaeotragus*, and *Adcrocuta*. The Baodean signal in the Europe–West Asian region grows stronger, with more localities showing higher similarity to Baode locality 49 (green/yellow pattern in plate 29.2D). This is the time when the Pikermian chronofauna had its maximum geographic range in Eurasia (Eronen et al. 2009). Surprisingly, the similarity in China itself appears significantly reduced, with lower similarities (green pattern in plate 29.2D) than the previous time slice, although sampling is admittedly sparse. In North China, *Dorcadoryx*, *Lantiantragus*, *Shaanxispira*, *Schansitherium*, and *Dinocrocuta* appear (Lantian locality 6; see plate 29.2D [L]). *Dinocrocuta* is apparently a western element, but the large bovids and giraffes are all endemic genera. Such significant endemism in North China was previously found for species and genera of rodents and ungulates by Fortelius and Zhang (2006). In South China, some western elements do appear, and the similarity accordingly rises somewhat (green pattern in plate 29.2D). From MNEQ 11 to MNEQ 12, North China is among the driest areas in Eurasia (notice red patterns in plate 29.2J–K), and the endemic Chinese ruminant fauna seems to have evolved in response to these regionally harsh conditions (Fortelius and Zhang 2006).

During MNEQ 13, the genera with the highest incidence of occurrence are still "*Hipparion*," *Gazella*, and *Adcrocuta*. With the onset of red clay deposition at about 7 Ma (Sun et al. 1998; Qiang et al. 2001; Zhu et al. 2008; Kaakinen et al., chapter 7, this volume), approximately correlative with the beginning of MNEQ 13, the mammal record of North China improves dramatically (more North Chinese localities in plate 29.2E). The immigration of genera from adjoining areas results in a large area of high similarity across north central Asia and North China (notice red patterns of these areas in plate 29.2E). The change coincides with a marked increase in humidity (green patterns of plate 29.2L) in northern China, interpreted as a result of a strengthening Asian Monsoon by Fortelius et al. (2002). In contrast to northern China, northern central Asia is dry at this time (orange pattern of this region in plate 29.2L). Similarity in the western part of the continent, where conditions in contrast appear drier in this interval, is now reduced (compare orange patterns of western Europe in plate 29.2L with green patterns of same region in plate 29.2E).

As far as the poor record suggests, the Baodean chronofauna declines drastically and is effectively gone during MNEQ 14–15 (no reddish pattern present in North China and northern central Asia in plate 29.2F–G). This is despite little change in humidity of central Asia and China in MNEQ 14 (see yellow/red patterns of these regions in plate 29.2M).

DISCUSSION

It appears that the regionally differentiated history of environmental change is matched by patterns of diachrony in faunal similarity, even within a mainly biochronologic framework expected to minimize diachrony in species ranges. As a general finding, this is neither unexpected nor does it have new methodological implications for biostratigraphy. The details are more interesting and also potentially of greater practical concern.

The two cases that we have described here are quite different. The Tunggurian chronofauna appears to have originated in and extended from western Europe (yellow/orange/red patterns in plate 29.1B–E), where its early incarnation in the late Early/lower Middle Miocene is often referred to as the "Hispanotherium fauna." Its strongest expression in East Asia (orange/red patterns of this region in plate 29.1D–E) was at the very end of its history, in the terminal Middle Miocene, when it had already disappeared from other parts of its earlier range (see low similarity patterns of western Europe in plate 29.1D–E). In broad terms, this chronofauna seems to have first developed as a response to continental drying and increased seasonality (seasonal dryness) in Western Europe (notice green/orange patterns of this region in plate 29.1I–J), and to have followed the widening arid belt toward the east (red/yellow patterns in plate 29.1K–L). Its successive retreat from the western part of its range seems to reflect increased humidity in western Europe (notice blue patterns in plate 29.1K–L), which would have favored forest adapted taxa. It may also have been impacted by the increased aridity in the eastern Mediterranean (see red/yellow pattern of this region in plate 29.1K–L), which evidently drove the concurrent emergence of the Pikermian chronofauna there. Similarly, its ultimate extinction in East Asia (note the demise of yellow/red patterns in plate 29.1F–G and emergence of orange patterns in plate 29.2B) also seems to coincide with the expansion of the Pikermian-like Baodean chronofauna.

The protracted extension of the Tunggurian chronofauna across Eurasia (see change in location of yellow/orange patterns in plate 29.1C–D) implies a continuum of similar environments across its geographic range. Such a continuum is suggested by the remarkable uniformity of the Eurasian mammal fauna in MNEQ 6 (plate 29.1D, notice the yellow/red pattern). Whether this reflects the lingering biogeographic effects of the immediately preceding Mid-Miocene Climatic Optimum (climax time of Neogene warmth between 17 Ma and 15 Ma) or whether it is a direct result of the contemporary Mid-Miocene Cooling event between 14.8 Ma and 14.1 Ma (Flower and Kennett 1994; Zachos et al. 2001; Böhme 2003) is currently difficult to assess. In MNEQ 7+8, the overall similarity definitely declines with low number of localities with high similarity to the Tunggur locality (fewer yellow/red patterns in plate 29.1E) compared to the next older time slice (see plate 29.1D). This is coincidental with the incipient development of mid-latitude aridity (yellow patterns in plate 29.1L) as the cooling trend continues. In this perspective, the Tunggur-Moergen fauna (see plate 29.1E [TM]) appears as a last remnant of the Tunggurian chronofauna in a stable environment, which in the context of overall mid-latitude drying (Liu, Eronen, and Fortelius 2009) appears relatively humid (notice the light-blue pattern of Tunggur locality [TM] in plate 29.1L).

The history of the Baodean chronofauna is quite different. Although part of the Pan-Eurasian "Hipparion fauna" (Kurtén 1952), the Baodean chronofauna is distinct from the Pikermian chronofauna in its taxonomic content. The mature Northeast Asian Baodean chronofauna has a high proportion of immigrants compared with its precursors (Fortelius and Zhang 2006) and differs from the West Eurasian Pikermian chronofauna by relative diversity and abundance of mustelids, rhinocerotids, and cervids in combination with rarity of primates, giraffids, and mastodonts (Deng 2006). In part this may simply reflect its younger age, but growing evidence is emerging that the difference is also an expression of the relatively high endemism of this chronofauna, strongly dominated by ruminant and equid lineages that extend from the eastern Mediterranean and Balkans to China (Fortelius and Zhang 2006; Watabe 1992). In fact, vicariant East Asian radiations paralleling the West Asian (Pikermian) ones may be more common than has been recognized previously—for example, among the hipparionine horses (Bernor, Koufos, et al. 1996) and aceratherine rhinoceroses (Fortelius et al. 2003). Hipparionine horses in particular show long-distance extensions across this range and include species belonging to the genera Hipparion, Cremohipparion and later, Plesiohipparion. In the Late Miocene, geographic extension of Hipparion (especially Hipparion dietrichi and Hipparion campbelli) and Cremohipparion (Cr. moldavicum, Cr. mediterraneum, Cr. proboscideum, and Cr. matthewi) from western Eurasia to East Asia (Hipparion hippidiodus and Cr. licenti) would appear to have been followed by vicariance.

In contrast to the Tunggurian chronofauna, the Baodean chronofauna thus seems to have developed via a stage-of-endemism in a harsh environmental setting, followed by expansion when summer rainfall brought richer resources and immigrants to northern China in the latest Miocene. It peaked in its species diversity significantly later (MNEQ 13) than the Pikermian chronofauna (MNEQ 12), but on current evidence neither Pikermian (Eronen et al. 2009) nor Baodean chronofaunas continued into the Pliocene (vanished yellow/red pattern in plate 29.2F–G). The slower fall of the Baodean chronofauna in North China possibly benefited from little change in the humid conditions before 4 Ma as suggested by red clay lithology (Ding et al. 1999; Vandenberghe et al. 2004), fossil molluscs (Wu et al. 2006), pollen records (Wang et al. 2006), dust flux in the North Pacific (Rea, Snoeckx, and Joseph 1998), and isotopic evidence (Passey et al. 2008). An interesting contrast is the Yushe Basin (see plate 29.2F [Y]) hipparions, which included advanced members of the Late Miocene Pikermian and Baodean lineages (*Cremohipparion licenti*) as well as endemic Chinese lineages (*Proboscidipparion pater* and *Plesiohipparion houfenense*; Qiu, Huang, and Guo 1987).

Buildup of the similarity of mammal species across the known range of Pikermian chronofaunas, from China in the east, to Europe in the west, and to northern and Sub-Saharan Africa in the southwest (Bernor, Rook, and Haile-Selassie 2009) likely happened as a result of a large number of more or less independent range extensions, with diachrony in both appearances and extinctions of discrete mammalian lineages. The end-Miocene decline and base-Pliocene extinction of Pikermian chronofaunas was evidently a much more definite event, with ultimate extinction virtually instantaneous in a biochronological sense (Eronen et al. 2009). Chronofaunas, like individual species and genera, typically appear to have unimodal histories including buildup, climax, and decline. However, in contrast to the individual genera and species of these faunas, where rise and fall appear symmetrical (Jernvall and Fortelius 2004), the decline of the two chronofaunas investigated here appears significantly more abrupt than their gradual buildup.

Both the Tunggurian and the Baodean chronofaunas moved into East Asia from the west. Although the details are different, these chronofaunas apparently moved across the continent in response to changing climate (humidity). This appears to be the case of expansion as well as contraction. Despite the underlying assumption of taxonomic isochrony, the standard biochronologic framework does appear to be stable enough to reveal regionally diachronous change at the continental scale. This is strong evidence that taxon ranges are generally highly diachronous across Eurasia.

ACKNOWLEDGMENTS

We thank the Academy of Finland for project funding on long-term work in China (to MF). LPL is indebted to the support from the Major Basic Research Projects of MST of China (2006CB806400), the NSFC (40872018, 40672010), and State Key Laboratory of Paleobiology and Stratigraphy (Nanjing Institute of Geology and Paleontology, CAS, No.083114). JTE and LPL acknowledge funding support from Kone Foundation. RLB wishes to acknowledge funding from the National Science Foundation, including EAR0125009 and BCS-0321893 (Revealing Hominid Origins Initiative to F. Clark Howell and Tim White). We thank Anu Kaakinen, Zhaoqun Zhang, and Ben Passey for discussions. We are also grateful to Michael Woodburne and Nikos Solounias for their productive reviews and helpful comments.

REFERENCES

Agustí, J. and M. Antón. 2002. *Mammoths, Sabertooths, and Hominids.* New York: Columbia University Press.

Agustí, J. and S. Moya-Sola. 1990. Mammal extinctions in the Vallesian (Upper Miocene). *Lecture Notes in Earth Science* 30:425–432.

Agustí, J., A. Sanz de Siria, and M. Garcés. 2003. Explaining the end of the hominoid experiment in Europe. *Journal of Human Evolution* 45(2):145–153.

Antunes, M. T. 1979. *Hispanotherium* fauna in Iberian Middle Miocene, its importance and paleogeographical meaning. *Annales Geologiques Des Pays Helleniques*, non-series vol. 1:19–26.

Archer, A. W. and C. G. Maples. 1987. Monte Carlo simulation of selected binomial similarity coefficients (1): Effect of number of variables. *Palaios* 2:609–617.

Bernor, R. L. 1978. The mammalian systematics, biostratigraphy and biochronology of Maragheh and its importance for understanding Late Miocene hominoid zoogeography and evolution. Ph.D. diss., University of California, Los Angeles.

Bernor, R. L. 1984. A zoogeographic theater and a biochronologic play: The time/biofacies phenomena of Eurasian and African Miocene mammal provinces. *Paléobiologie Continentale* 14:121–142.

Bernor, R. L. 1985. Systematic and evolutionary relationships on the hipparionine horses from Maragheh, Iran (Late Miocene, Turolian age). *Palaeovertebrata* 15:173–269.

Bernor, R. L. 1986. Mammalian biostratigraphy, geochronology, and zoogeographic relationships of the Late Miocene Maragheh fauna, Iran. *Journal of Vertebrate Paleontology* 6:76–95.

Bernor, R. L., P. A. Andrews, N. Solounias, and J. A. van Couvering. 1979. The evolution of "Pontian" mammal faunas: Some zoogeographic, palaeoecological and chronostratigraphic considerations. *Annales Pays Helleneniques*, non-ser. vol. 1:81–89.

Bernor, R. L., G. D. Koufos, M. O. Woodburne, and M. Fortelius. 1996. The evolutionary history and biochronology of European and South West Asian Late Miocene and Pliocene Hipparionine horses. In *The Evolution of Western Eurasian Neogene Mammal Faunas*, ed. R. L. Bernor, V. Fahlbusch, and H. W. Mittmann, pp. 307–338. New York: Columbia University Press.

Bernor, R. L. and P. P. Pavlakis. 1987. Zoogeographic relationships of the Sahabi large mammal fauna. In *Neogene Paleontology and Geology of Sahabi*, ed. N. T. Boaz, A. El-Arnauti, A. W. Gaziry, J. De Heinzelin and D. D. Boaz, pp. 349–384. New York: Liss.

Bernor, R. L., L. Rook, and Y. Haile-Selassie. 2009. Paleobiogeography. In *Ardipithecus kadabba: Late Miocene Evidence from the Middle Awash, Ethiopia*, ed. Y. Haile-Selassie and G. Wolde Gabriel, pp. 549–563. Berkeley: University of California Press.

Bernor, R. L., N. Solounias, C. C. Swisher III, and J. A. Van Couvering. 1996. The correlation of three classical "Pikermian" mammal faunas—Maragheh, Samos and Pikermi—with the European MN unit system. In *The Evolution of Western Eurasian Neogene Mammal Faunas*, ed. R. L. Bernor, V. Fahlbusch, and H. W. Mittmann, pp. 137–156. New York: Columbia University Press.

Böhme, M. 2003. Miocene climatic optimum: Evidence from lower vertebrates of Central Europe. *Palaeogeography, Palaeoclimatology, Palaeoecology* 195: 389–401.

Deng, T. 2006. Paleoecological comparison between Late Miocene localities of China and Greece based on *Hipparion* faunas. *Geodiversitas* 28(3):499–516.

Ding, Z. L., S.-f. Xiong, J.-m. Sun, S.-l. Yang, Z.-y. Gu, and T.-s. Liu. 1999. Pedostratigraphy and paleomagnetism of a ~7.0 Ma eolian loess–red clay sequence at Lingtai, Loess Plateau, North-Central China and the implications for paleomonsoon evolution. *Palaeogeography, Palaeoclimatology, Palaeoecology* 152:49–66.

Eronen, J. T., M. Mirzaie Ataabadi, A. Micheels, A. Karme, R. L. Bernor, and M. Fortelius. 2009. Distribution history and climatic controls of the Late Miocene Pikermian chronofauna. *Proceedings of the National Academy of Sciences* 106(20):11867–11871.

Flower, B. P. and J. P. Kennett. 1994. The Middle Miocene climatic transition: East Antarctic ice sheet development, deep ocean circulation and global carbon cycling. *Palaeogeography, Palaeoclimatology, Palaeoecology* 108:537–555

Fortelius, M. (coordinator). 2009. *Neogene of the Old World Database of Fossil Mammals (NOW)*. University of Helsinki. http://www.helsinki.fi/science/now/.

Fortelius, M., J. T. Eronen, L. Liu, D. Pushkina, J. Rinne, A. Tesakov, I. Vislobokova, Z. Zhang, and L. Zhou. 2002. Fossil mammals resolve regional patterns of Eurasian climate change during 20 million years. *Evolutionary Ecology Research* 4:1005–1016.

Fortelius, M., J. T. Eronen, L. Liu, D. Pushkina, A. Tesakov, I. Vislobokova, and Z. Zhang. 2003. Continental-scale hypsodonty patterns, climatic palaeobiogeography, and dispersal of Eurasian Neogene large mammal herbivores. In *Distribution and Migration of Tertiary Mammals in Eurasia, a Volume in Honour of Hans de Bruijn*, ed. J. Reumer and W. Wessels, pp. 1–11. DEINSEA 10.

Fortelius, M., J. T. Eronen, L. Liu, D. Pushkina, A. Tesakov, I. Vislobokova, and Z. Zhang. 2006. Late Miocene and Pliocene large land mammals and climatic changes in Eurasia. *Palaeogeography, Palaeoclimatology, Palaeoecology* 238: 219–227.

Fortelius, M., L. Werdelin, P. Andrews, R. L. Bernor, A. Gentry, L. Humphrey, H. W. Mittmann, and S. Viranta. 1996. Provinciality, diversity, turnover, and paleoecology in land mammal faunas of the later Miocene of Western Eurasia. In *The Evolution of Western Eurasian Neogene Mammal Faunas*, ed. R. L. Bernor, V. Fahlbusch, and H. W. Mittmann, pp. 414–448. New York: Columbia University Press.

Fortelius, M. and Z. Zhang. 2006. An oasis in the desert? History of endemism and climate in the Late Neogene of North China. *Palaeontographica A* 277:131–141.

Hammer, Ø. and D. A. T. Harper. 2006. *Paleontological Data Analysis*. Oxford: Blackwell.

Hammer, Ø., D. A. T. Harper, and P. D. Ryan. 2001. PAST: Paleontological Statistics Software Package for Education and Data Analysis. *Palaeontologia Electronica* 4(1): 9 pp.; http://palaeo-electronica.org/2001_1/past/past.pdf.

Jernvall, J. and M. Fortelius. 2002. Common mammals drive the evolutionary increase of hypsodonty in the Neogene. *Nature* 417:538–540.

Jernvall, J. and M. Fortelius. 2004. Maintenance of trophic structure in fossil mammal communities: Site occupancy and taxon resilience. *American Naturalist* 164:614–624.

Kurtén, B. 1952. The Chinese *Hipparion* fauna. *Commentationes Biologicae Societatis Scientiarum Fennicae* 13:1–82.

Li, C.-k., W.-y. Wu, and Z.-d. Qiu. 1984. Chinese Neogene: Subdivision and correlation. *Vertebrata PalAsiatica* 22(3):163–178.

Liu, L., J. T. Eronen, and M. Fortelius. 2009. Significant mid-latitude aridity in the Middle Miocene of East Asia. *Palaeogeography, Palaeoclimatology, Palaeoecology* 279:201–206.

Maples, C. G. and A. W. Archer. 1988. Monte Carlo simulation of selected binomial similarity coefficients (II): Effect of number of sparse data. *Palaios* 3:95–103.

Mein, P. 1975. In *Report on Activity of the RCMNS Working Group (1971-1975)*, ed. J. Senes, pp. 78–81. Bratislava, Slovakia: I.U.G.S. Commission on Stratigraphy, Subcommission on Neogene Stratigraphy.

Olson, E. C. 1952. The evolution of Permian Vertebrate Chronofauna. *Evolution* 6:181–196.

Olson, E. C. 1958. Fauna of the Vale and Choza: A summary, review and integration of the geology and faunas. Fieldiana. *Geology* 10 (32):397–448.

Passey, B. H., L. K. Ayliffe, A. Kaakinen, Z.-q. Zhang, J. T. Eronen, Y.-m. Zhu, L.-p. Zhou, T. E. Cerling, and M. Fortelius. 2008. Strengthened East Asian summer monsoons during a period of high-latitude warmth? Isotopic evidence from Mio-Pliocene fossil mammals and soil carbonates from northern China. *Earth and Planetary Science Letters* 277:443–452.

Qiang, X.-k., Z.-x. Li, C. Powell, and H.-b. Zheng. 2001. Magnetostratigraphic record of the Late Miocene onset of the East Asian monsoon, and Pliocene uplift of northern Tibet. *Earth and Planetary Science Letters* 187:83–93.

Qiu, Z.-x., W.-i. Huang, and Z.-h. Guo. 1987. The Chinese hipparionine fossils. *Palaeontologia Sinica* 175(NS25):1–250.

Qiu, Z.-x. and Z.-d. Qiu. 1990. Neogene local mammalian faunas: Succession and ages. *Journal of Stratigraphy* 14(4):241–260.

Qiu, Z.-x. and Z.-d. Qiu. 1995. Chronological sequence and subdivision of Chinese Neogene mammalian faunas. *Palaeogeography, Palaeoclimatology, Palaeoecolology* 116:41–70.

Qiu, Z.-d., X. Wang, and Q. Li. 2006. Faunal succession and biochronology of the Miocene through Pliocene in Nei Mongol (Inner Mongolia). *Vertebrata PalAsiatica* 44 (2):164–181.

Qiu, Z.-x., W.-y. Wu, and Z.-d. Qiu. 1999. Miocene mammal faunal sequence of China: Palaeozoogeography and Eurasian relationships. In *Land Mammals of Europe*, ed. G. E. Rössner and K. Heissig, pp. 443–457. Munich: Dr. Friedrich Pfeil.

Rea, D. K., H. Snoeckx, and L. H. Joseph. 1998. Late Cenozoic eolian deposition in the North Pacific: Asian drying, Tibetan uplift, and cooling of the northern hermisphere. *Paleoceanography* 13:215–224.

Solounias, N. 1981. The Turolian Fauna from the Island of Samos, Greece. *Contributions to Vertebrate Evolution* 6:1–232.

Steininger, F. F., W. A. Berggren, D. V. Kent, R. L. Bernor, S. Sen, and J. Agustí. 1996. Circum-Mediterranean Neogene (Miocene-Pliocene) marine continental chronologic correlations of European mammal units. In *The Evolution of Western Eurasian Neogene Mammal Faunas*, ed. R. L. Bernor, V. Fahlbusch, and H. W. Mittmann, pp. 7–46. New York: Columbia University Press.

Sun, D., J. Shaw, Z. An, M. Cheng, and L. Yue. 1998. Magnetostratigraphy and paleoclimatic interpretation of a continuous 7.2. Ma Late Cenozoic eolian sediments from the Chinese Loess Plateau. *Geophysical Research Letters* 25:85–88

Vandenberghe, J., H.-y. Lu, D.-h. Sun, J. van Huissteden, and M. Konert. 2004. The Late Miocene and Pliocene climate in East Asia as recorded by grain size and magnetic susceptibility of the Red Clay deposits (Chinese Loess Plateau). *Palaeogeography, Palaeoclimatology, Palaeoecology* 204: 239–255.

Wang, L., H.-y. Lu, N.-q. Wu, J. Li, Y.-p. Pei, G.-b. Tong, and S.-z. Peng. 2006. Palynological evidence for Late Miocene–Pliocene vegetation evolution recorded in the red clay sequence of the central Chinese Loess Plateau and implication for palaeoenvironmental change. *Palaeogeography, Palaeoclimatology, Palaeoecology* 241:118–128.

Wang, X.-m., Z.-d. Qiu, and N. O. Opdyke. 2003. Litho-, bio-, and magnetostratigraphy and paleoenvironment of Tunggur Formation (Middle Miocene) in central Inner Mongolia, China. *American Museum Novitates* 3411:1–31.

Watabe, M. 1992. Phylogeny of Chinese hipparion (Perissodactylia, Mammalia): Their relationships with the western old world and North American hipparionines. *Paleontologia i Evolució* 24/25:155–174.

Wu, N.-q., Y.-p. Pei, H.-y. Lu, Z.-t. Guo, F.-j. Li, and T.-s. Liu. 2006. Marked ecological shifts during 6.2–2.4 Ma revealed by a terrestrial molluscan record from the Chinese Red Clay Formation and implication for Palaeoclimatic evolution. *Palaeogeography, Palaeoclimatology, Palaeoecology* 233:87–299.

Zachos, J., M. Pagani, L. Sloan, E. Thomas, and K. Billups. 2001. Trends, rhythms, and aberrations in global climate 65Ma to present. *Science* 292:686–693.

Zhang, Z. and L. Liu. 2005. The Late Neogene mammal biochronology in the Loess Plateau, China. *Annales de Paléontologie* 91:257–266.

Zhu, Y.-m., L.-p. Zhou, D.-w. Mo, A. Kaakinen, Z. Zhang, and M. Fortelius. 2008. A new magnetostratigraphic framework for late Neogene Hipparion Red Clay in the Eastern Loess Plateau of China. *Palaeogeography, Palaeoclimatology, Palaeoecology* 268:47–57.

APPENDIX

Complete List of the Localities Used in This Study with
Geographical Position (Coordinates), Age Correlation
(MN-equivalent), Genus-level Faunal Resemblance Indices
(Dice, Jaccard, Simpson, and Raup-Crick GFRI),
and Mean Hypsodonty Values

Locality	Age (MNEQ)	Latitude	Longitude	T/Dice	T/Jaccard	T/Simpson	T/Raup-Crick	B/Dice	B/Jaccard	B/Simpson	B/Raup-Crick	Mean Hypsodonty
Ad Dabtiyah	4	26.45	48.59	0	0	0	0.0975	0	0	0	0.0525	1.1
Aérotrain	4	48.05	1.85	0	0	0	0.115	0.054054	0.027778	0.125	0.39	1.5
Akcahisar 1	4	38.05	27.62					0	0	0	0.075	1
Armantes 1	4	41	-2	0.17647	0.096774	0.375	0.9375	0.037037	0.018868	0.04	0.0175	1.2222
Artenay	4	48.1	1.9	0.11765	0.0625	0.12	0.3025					1
Artesilla	4	41.18	-1.5	0.13953	0.075	0.17647	0.6125	0.043478	0.022222	0.058824	0.08	1.1111
Baggersee Freudenegg 2	4	48.33	10.01									1
Baggersee Freudenegg 3	4	48.33	10.01									1
Belchatow C	4	51.25	19.33									1
Bézian	4	44.2	0.9	0.23529	0.13333	0.24	0.8675	0.037037	0.018868	0.04	0.025	1.0714
Buñol	4	39.42	0.79	0.093023	0.04878	0.11765	0.2625	0	0	0	0.01	1.1
Can Canals	4	41.2	0.7	0.095238	0.05	0.125	0.3375	0	0	0	0.01	1.1538
Can Julia	4	41.41	1.83	0.057143	0.029412	0.11111	0.335	0.052632	0.027027	0.11111	0.32	1.125
Can Mas	4	41.2	0.7									1
Córcoles	4	40.48	-2.65	0.23529	0.13333	0.5	0.98	0	0	0	0.115	1
Dera Bugti 6	4	29.03	69.15	0.088889	0.046512	0.10526	0.3175	0.041667	0.021277	0.052632	0.0525	1
Eggingen–Mittelhart 3	4	48.36	9.87	0.044444	0.022727	0.052632	0.0625	0.041667	0.021277	0.052632	0.1025	1.1053
El Canyet	4	41.45	1.99	0.054054	0.027778	0.090909	0.2725	0	0	0	0.055	1
Els Casots	4	41.33	2.16	0.10256	0.054054	0.15385	0.5075	0.047619	0.02439	0.076923	0.1625	1
Erkertshofen 2	4	48.97	11.2	0.15385	0.083333	0.23077	0.8	0	0	0	0.04	1
La Romieu	4	44.2	0.9	0.17778	0.097561	0.21053	0.7675	0.041667	0.021277	0.052632	0.0525	1.0714
Langenau 1	4	48.48	10.09	0.12766	0.068182	0.14286	0.4475	0.04	0.020408	0.047619	0.0375	1.0667
Langenau 2	4	48.5	10.1									1
Les Cases de la Valenciana	4	41.2	0.7	0.058824	0.030303	0.125	0.455	0	0	0	0.085	1.1429
Lisboa V	4	38.44	-9.08									1
Mae Soi	4	19.33	97.83									1
Moli Calopa	4	41.33	2.08	0.10526	0.055556	0.16667	0.5525	0.097561	0.051282	0.16667	0.5	1.4167
Montreal-du-Gers	4	43.95	0.2	0.05	0.025641	0.071429	0.24	0	0	0	0.0425	1.6667
Munébrega 1	4	41.25	-2									1
Munébrega 2	4	41.25	-1.7									1
Munébrega 3	4	41.25	-1.7									1
Munébrega AB	4	41.25	-1.7									1
Oberdorf 3 (O3)	4	47.07	15.15									1
Oberdorf 4 (O4)	4	47.07	15.15									1
Pellecahus	4	43.86	0.53	0.20408	0.11364	0.21739	0.7625	0	0	0	0.0025	1.2857
Petersbuch 2	4	48.97	11.19	0.11765	0.0625	0.25	0.7025	0	0	0	0.1075	1
Quinta da Farinheira	4	38.71	-9.13									2
Quinta das Pedreiras	4	38.71	-9.13									1
Quinta Grande	4	38.93	-8.51									2

Locality												
Rubielos del Mora	4	40.18	-0.66									1
Sihong-qizhui	4	33.16	118.33									1
Sihong-shuanggou	4	33.25	118.2									1
Sihong-songlinzhuang	4	33.3	118.25	0.28571	0.16667	0.375	0.9925	0.088889	0.046512	0.125	0.28	1
Sihong-xiacaowan	4	33.45	109.1									1
Torralba 2	4	40.3	-2.28									1
Wintershof-West	4	48.9	11.16	0.047619	0.02439	0.0625	0.13	0.044444	0.022727	0.0625	0.1275	1
Aktau Mountain	5	44.4	79	0.11765	0.0625	0.25	0.7975	0.054054	0.027778	0.125	0.385	1.1429
Al-Sarrar	5	26.58	48.58	0.21053	0.11765	0.33333	0.94	0.04878	0.025	0.083333	0.205	1
Antonios (ANT)	5	40.3	23	0.055556	0.028571	0.1	0.3375	0.051282	0.026316	0.1	0.27	1
Arroyo del Olivar	5	40.38	-3.65	0.17391	0.095238	0.2	0.6675	0.040816	0.020833	0.05	0.0525	1.25
Baigneaux-en Beauce	5	48.1	2.15	0.26923	0.15556	0.26923	0.9375	0.036364	0.018519	0.038462	0.0125	1.1667
Beaugency-Tavers	5	47.78	1.63	0.29091	0.17021	0.30769	0.975	0.068966	0.035714	0.068966	0.075	1.3333
Belometchetskaja	5	44.6	42									1.3158
Candir (Loc. 3)	5	40.24	33.48									1.4286
Candir (Loc.1)	5	40.29	33.48									1.5
Chios	5	38.36	26.13									1.4286
Contres MN 5	5	47.41	1.43									1.0556
Crastes	5	43.73	0.73									1
Dzhilanchik (Uly-Dzhilanchik)	5	49.4	65	0.055556	0.028571	0.1	0.33	0.054054	0.027778	0.054054	0.0725	1.25
Edelbeuren-Maurerkopf	5	48.09	10.03	0.25532	0.14634	0.28571	0.9425	0.027778	0.014286	0.034483	0.0425	1
Edelbeuren-Schlachtberg	5	48.09	10.02	0.16667	0.090909	0.3	0.86	0	0	0	0.08	1
Eibiswald	5	46.68	15.24	0.058824	0.030303	0.125	0.405	0	0	0	0.3375	1.2
Engelswies	5	48.03	9.13	0.15385	0.083333	0.23077	0.75	0.10526	0.055556	0.22222	0.0375	1
Estación Imperial (Madrid)	5	40.4	-3.68	0.17143	0.09375	0.33333	0.855	0.028169	0.014286	0.034483	0.6575	1.75
Esvres - Marine Faluns	5	47.3	0.8	0.38235	0.23636	0.5	1	0	0	0	0.0025	1.1111
Faluns of Touraine & Anjou	5	45.5	4.5	0.11765	0.0625	0.25	0.74	0	0	0	0.0875	1.5
Georgensgmünd	5	49.19	11.01	0.11429	0.060606	0.22222	0.6975	0	0	0	0.095	1.25
Gisseltshausen	5	48.7	11.95									1
Grund	5	48.63	16.06									1
Guanghe-dalanggou	5	35.46	103.46									1
Guanghe-wangshijie	5	35.51	103.5									1.4
Göriach	5	47.55	15.3	0.25532	0.14634	0.28571	0.955	0	0	0	0.0025	1
Hambach 6C	5	50.9	6.45	0.14286	0.076923	0.1875	0.615	0	0	0	0.0075	1.1
Heggbach	5	48.15	9.89	0.16667	0.090909	0.3	0.865	0.051282	0.026316	0.1	0.255	1
Henares 2	5	37.6	-1.91									1
Häder	5	48.21	10.38	0.23529	0.13333	0.5	0.98	0	0	0	0.105	1.1111
Kalkaman lake	5	51.7	76.5	0.33333	0.2	0.6	1	0.051282	0.026316	0.1	0.295	1
Karagac 1	5	39.02	30.04	0.29268	0.17143	0.4	0.98	0.090909	0.047619	0.13333	0.3425	1.2727
La Hidroelectrica Madrid	5	40.25	-3.43	0.15789	0.085714	0.25	0.8	0	0	0	0.035	1.3333
La Retama	5	40.1	-2.7									

(continued)

Locality	Age (MNEQ)	Latitude	Longitude	T/Dice	T/Jaccard	T/Simpson	T/Raup-Crick	B/Dice	B/Jaccard	B/Simpson	B/Raup-Crick	Mean Hypsodonty
Langhian Sables Fauves	5	43.5	0.5									1.3333
Li Mae Long	5	17.71	98.9									1
Luçane	5	43.7	16.62									1.3333
Münzenberg (Leoben)	5	47.4	15.1									1
Mala Miliva	5	44.13	21.43									1
Manchar 1	5	26	68	0.17391	0.095238	0.2	0.695	0.040816	0.020833	0.05	0.055	1.1
Manchar 2	5	26	68	0.15	0.081081	0.21429	0.7325	0.046512	0.02381	0.071429	0.16	1
Manciet base	5	43.81	0.5									1
Montejo de la Vega	5	41.55	-3.65	0.11765	0.0625	0.25	0.77	0.054054	0.027778	0.125	0.3875	1
Moratines	5	40.5	-3.5	0.11765	0.0625	0.25	0.76	0	0	0	0.09	1.6
O'Donnell	5	40.44	-3.59									1.25
PAR-Penuelas (Madrid)	5	40.4	-2.31									1.6667
Paseo de la Esparanza (Madrid)	5	40.4	-3.68	0.11765	0.0625	0.25	0.75	0	0	0	0.125	1.4286
Paseo de la Esparanza 7 (Madri	5	40.4	-3.68									1.6667
Paseo de las Acacias (Madrid)	5	40.4	-2.31	0.16667	0.090909	0.3	0.875	0.10256	0.054054	0.2	0.5475	1.5
Pontlevoy	5	47.4	1.2	0.26415	0.15217	0.26923	0.9525	0.035714	0.018182	0.037037	0.0225	1.0833
Poudenas-Peyrecrechen	5	44.05	0.21	0.21622	0.12121	0.36364	0.93	0	0	0	0.0425	1.0909
Puente de Toledo	5	40.38	-3.7									1.4286
Puente de Vallecas	5	40.2	-3.5	0.29268	0.17143	0.4	0.985	0.045455	0.023256	0.066667	0.115	1
Réaup	5	44.08	0.2									1
Reisensburg	5	48.46	10.3									1
Rimbez - Lapeyrie base	5	44.03	0.03	0.15789	0.085714	0.25	0.805	0	0	0	0.0575	1
Rothenstein 1	5	51.11	7.68	0.10811	0.057143	0.18182	0.565	0.05	0.025641	0.090909	0.2275	1
San Isidro	5	40.4	-3.73									1.25
Sandelzhausen	5	48.62	11.8	0.26087	0.15	0.3	0.97	0.040816	0.020833	0.05	0.045	1.0714
Sant Mamet	5	41.49	2.04	0.15789	0.085714	0.25	0.8	0	0	0	0.0375	1
Savigné-sur-Lathan	5	47.45	0.31	0.14286	0.076923	0.1875	0.615	0	0	0	0.025	1.0556
Seegraben (Leoben)	5	47.4	15.1									1
Shanwang	5	32.01	116.49	0.18605	0.10256	0.23529	0.8375	0.043478	0.022222	0.058824	0.0725	1
Somosaguas-Sur	5	41.49	-2.04									1
Stallhofen	5	46.66	14.13									1
Tarazona	5	41.2	0.5									2
Thymiana B	5	38.31	26.13	0.057143	0.029412	0.11111	0.395	0	0	0	0.0725	1.5
Villafeliche 3	5	41.2	-1.5									1
Voggersberg	5	48.47	12.06									1
Vordersdorf b.Eibiswald	5	46.71	15.23									1
Zhongxiang-Xiaodian	5	31	112.5									1
Ziemetshausen 1b	5	48.3	10.5									1
Zinda Pir 3	5	30.1	70.1	0.057143	0.029412	0.11111	0.3775	0	0	0	0.085	1

Al Jadidah	6	25.5	49.65	0.058824	0.030303	0.125	0.4625	0	0	0	0.1025	1
Arroyo del Val	6	41.3	-1.5	0.17143	0.09375	0.33333	0.885	0	0	0	0.0675	1
Arroyo del Val VI	6	41.18	1.3									1.3333
Ayibaligi mevkii	6	38.52	26.62									3
Castelnau-d'Arbieu	6	43.88	0.7	0.051282	0.026316	0.076923	0.1775	0.047619	0.02439	0.076923	0.1975	1.2222
Catakbagyaka	6	37.12	28.17	0.32432	0.19355	0.54545	0.9975	0.05	0.025641	0.090909	0.2425	1
Devinská Nová Ves - Fissures	6	48.2	17.01	0.21739	0.12195	0.25	0.9025	0.040816	0.020833	0.05	0.065	1
Devinská Nová Ves - Sandhill	6	48.2	17	0.22642	0.12766	0.23077	0.8575	0.071429	0.037037	0.074074	0.0925	1.0833
Elgg	6	47.5	8.86									1
Gallenbach 2b	6	48.4	11.07									1
Haulies	6	43.56	0.66	0.11429	0.060606	0.22222	0.7375	0	0	0	0.08	1.3636
Hezheng-laogou	6	36.46	103.41	0.5	0.33333	0.71429	1	0.093023	0.04878	0.14286	0.38	1.2857
Inönü I (Sinap 24A)	6	40.53	32.64	0.26667	0.15385	0.31579	0.9725	0	0	0	0.015	1.2222
Jiulongkou	6	37	114	0.21622	0.12121	0.36364	0.9225	0.05	0.025641	0.090909	0.2075	1
Junggar-Botamoyin	6	46	89	0.26316	0.15152	0.41667	0.985	0.14634	0.078947	0.25	0.75	1
Junggar-botamoyindong	6	47	89									1
Junggar-chibaerwoyi	6	47	88									1
Junggar-Ganqikairixi	6	46	89									1
Junggar-Tieersihabahe	6	46.66	88.5	0.16216	0.088235	0.27273	0.835	0.05	0.025641	0.090909	0.2575	1
Kentyubek	6	50	65.66									1
Klein Hadersdorf	6	48.6	16.6									1.1429
La Barranca	6	40.75	-4.01									1.5
La Ciesma	6	41.2	0.5									1
Lantian-koujiacun	6	34.16	109.16									1
Laymont	6	43.41	0.98									1
Liet	6	50.6	2.35									1
Liuhe-lingyanshan	6	32	121									1
Lussan	6	43.63	0.73									1
Marciac	6	43.53	0.16									1
Miélan	6	43.43	0.31									1
Montesquiou-sur-L'Osse	6	43.58	0.33									1
Montpezat	6	43.4	0.96									1
Murero	6	41.16	-1.48									1
Póvoa de Satarém	6	38.85	-9.06									1
Paracuellos 3	6	41.3	-1.5	0.17647	0.096774	0.375	0.9375	0.054054	0.027778	0.125	0.3775	1.2
Paracuellos 5	6	40.5	-3.5	0.21053	0.11765	0.33333	0.93	0	0	0	0.0325	1.1111
Pasalar	6	37.43	30.06	0.29508	0.17308	0.34615	0.9475	0.09375	0.04918	0.10345	0.09	1.381
Prebreza	6	43.19	21.15									1.4
Qaidam-Olongbuluk	6	37	97.33									1.4
Rümikon	6	47.38	8.55									1
Riedern	6	47.71	8.3									1

(continued)

Locality	Age (MNEQ)	Latitude	Longitude	T/Dice	T/Jaccard	T/Simpson	T/Raup-Crick	B/Dice	B/Jaccard	B/Simpson	B/Raup-Crick	Mean Hypsodonty
Sansan	6	43.9	-0.5	0.35821	0.21818	0.46154	1	0.057143	0.029412	0.068966	0.0025	1
Simorre	6	44	0.3	0.29268	0.17143	0.4	0.9925	0.045455	0.023256	0.066667	0.155	1
Steinberg	6	48.82	10.62									1
Stätzling	6	48.4	10.96	0.31818	0.18919	0.38889	0.995	0	0	0	0.0175	1.05
Tüney	6	40.08	33.61									1
Tairum Nor	6	43.41	113.11	0.4	0.25	0.77778	1	0.052632	0.027027	0.11111	0.37	1.4
Thannhausen	6	48.3	11.5	0.15	0.081081	0.21429	0.6975	0	0	0	0.0125	1.125
Tongxin	6	36.98	105.9	0.28571	0.16667	0.55556	0.995	0	0	0	0.0925	1.3333
Tongxin-dingjiaergou	6	37	106	0.32432	0.19355	0.54545	1	0	0	0	0.0725	1.2857
Tongxin-gujiazhuang	6	37	106									1
Tongxin-jinzuizigou	6	37	106									1
Tongxin-maerzuizigou	6	37	106									1
Tongxin-shataigou	6	37	106									1
Tongxin-Yehuliquezi	6	37	106									1
Trimmelkam	6	48	12.9									1
Wiesholz	6	47.7	8.81									1
Alan (N.D. de Lorette)	7+8	43.21	0.95									1
Alan (Pompat)	7+8	43.21	0.95									1
Aleksandrodar	7+8	47.9	28.5									2
Anwil	7+8	47.37	7.83	0.22857	0.12903	0.44444	0.9575	0	0	0	0.0775	1
Atzgersdorf (WIEN)	7+8	48.1	16.3									1
Bachas	7+8	43.25	0.93									1
Beni Mellal	7+8	32.5	-6.5									1
Can Almirall	7+8	41.2	0.7									1
Can Feliu	7+8	41.5	2.13	0.17647	0.096774	0.375	0.935	0	0	0	0.1025	1
Can Mata 1	7+8	39.98	-5.8									1
Cassagnabère	7+8	43.23	0.8									1
Castell de Barberà	7+8	41.52	2.14	0.29787	0.175	0.33333	0.9875	0.12	0.06383	0.14286	0.4	1
Cerro del Otero	7+8	40.06	-1.01									1
Chiang Muan	7+8	18.93	100.23	0	0	0	0.135	0.054054	0.027778	0.125	0.3675	1.1111
Coca	7+8	41.21	-4.51									1
Coueilles (Baguent)	7+8	43.35	0.88									1
Dang Valley	7+8	28	82.33	0.057143	0.029412	0.11111	0.3875	0.052632	0.027027	0.11111	0.325	1
Dera Bugti W	7+8	29	69									1
Escanecrabe	7+8	43.28	0.75									1
Escobosa	7+8	41.2	0.7									1
Fangxian	7+8	32.1	110.7									1.8
Felsőtárkány 3/2 (Güdör-kert)	7+8	47.97	20.41									1
Gebeciler	7+8	38.77	30.75									1.6667

	7+8											
Hachan	7+8	43.28	0.45	0.28	0.16279	0.29167	0.9425	0.11321	0.06	0.125	0.275	1
Hostalets de Pierola Inferior	7+8	41.2	0.5	0.29412	0.17241	0.625	0.9975	0.054054	0.027778	0.125	0.34	1
Junggar-duolebulejin	7+8	47	88									1.57
Junggur_dingshanyanchi	7+8	46.42	87.48									1
Kaiyuan-Xiaolongtan	7+8	23.58	103.25									1
Kutsaj M	7+8	45.4	42.7									1.6667
La Cisterniga	7+8	41.61	-4.68	0.26316	0.15152	0.38462	0.895	0.075949	0.039474	0.10345	0.0125	1
La Grive St. Alban	7+8	44	0									1.1176
Laichingen	7+8	48.5	9.7									1
Lantian-gaopo-64004	7+8	34	109									1.6667
Lantian-gaopo-64008	7+8	34.1	109.3									1.5
Lanzhou-Quantougou	7+8	36.11	103.66	0.35294	0.21429	0.75	1	0.054054	0.027778	0.125	0.3625	1.6667
Lintong-lengshuigou	7+8	34.1	109.3									1.8
Lussan-Adeilhac	7+8	43.25	0.96									1
Mannersdorf	7+8	48	16.5									1.4286
Masquefa	7+8	41.2	0.7									1
Massenhausen	7+8	48.4	11.8	0.26316	0.15152	0.41667	0.985	0.04878	0.025	0.083333	0.2275	1
Minhe-lierbao	7+8	36.6	102.8									1.5
Minhe-nanhawangou	7+8	36.3	102.9									1.5
Minisu de Sus	7+8	46.28	22.01									1
Nombrevilla-2	7+8	41.11	-1.35									1
Opole 2	7+8	50.66	17.91									1
Péguilhan	7+8	43.31	0.7									1
Poudenas-Cayron	7+8	44.05	0.21	0.22857	0.12903	0.44444	0.9775	0	0	0	0.085	1
Przeworno 2	7+8	50.7	17.2									1
Saint-Gaudens (Les Pujaments)	7+8	46.11	0.28									1
Saint-Gaudens (Valentine)	7+8	43.1	0.71									1
Sant Quirze	7+8	41.52	2.08	0.28571	0.16667	0.30435	0.975	0.038462	0.019608	0.043478	0.0375	1.1667
Saricay	7+8	37.33	27.79									1
Serbian lake	7+8	43.83	21.42									1
Siziwangqi-Damiao	7+8	42	111.58	0.22727	0.12821	0.27778	0.9075	0.085106	0.044444	0.11111	0.205	1.4
Sofca	7+8	39.64	30.17									1
Solera	7+8	37.68	-3.35									1
Sopron	7+8	47.68	16.6									1
St. Gaudens	7+8	43.07	0.44									1.0667
St. Stephan im Lavanttal	7+8	46.8	14.9									1
Steinheim	7+8	48.7	10.05	0.22222	0.125	0.23077	0.82	0.035088	0.017857	0.035714	0.01	1
Türkenschanze (WIEN)	7+8	48.23	16.33									1
Toril 3	7+8	41.13	-1.38									1
Tunggur-Moergen	7+8	43	112	1	1	1	1	0.18182	0.1	0.19231	0.6425	1.3529
Xinan	7+8	34.66	112.18	1	1	1	1					1.6667

(continued)

Locality	Age (MNEQ)	Latitude	Longitude	T/Dice	T/Jaccard	T/Simpson	T/Raup-Crick	B/Dice	B/Jaccard	B/Simpson	B/Raup-Crick	Mean Hypsodonty
Yaylacilar	7+8	38.77	30.64									2
Yeni Eskihisar 1	7+8	37.31	28.07									2
Yeni Eskihisar 2	7+8	37.3	28.06									2
Yenieskihisar	7+8	37.31	28.04									2
Amuwusu	9	42.36	112.74									1
Arapli (Igdebaglar)	9	40.65	27.14									1
Atavaska	9	47	28.8	0	0	0	0.1025	0.10526	0.055556	0.22222	0.66	1.3333
Aveiras de Baixo	9	39.11	-8.86									1
Azambujeira inf.	9	39.26	-8.78									1
Ballestar	9	40.4	0.11	0.14634	0.078947	0.2	0.6675	0.13636	0.073171	0.2	0.5875	1.1667
Bermersheim	9	49.77	8.09									1
Bonnefont	9	43.25	0.35									1
Breitenfeld	9	47.16	15.5									1
Buzhor 1	9	46.56	28.16	0.05	0.025641	0.071429	0.1825	0.23256	0.13158	0.35714	0.9475	1.5714
Buzhor 2	9	44.56	28.16									1.5
Can Llobateres I	9	41.2	0.7	0.23188	0.13115	0.30769	0.755	0.16667	0.090909	0.2069	0.27	1.2381
Can Ponsic	9	41.2	0.7	0.21277	0.11905	0.2381	0.815	0.16	0.086957	0.19048	0.6275	1
Can Ponsic I	9	41.53	2.07	0.29091	0.17021	0.30769	0.9725	0.17241	0.09434	0.17241	0.5275	1.1667
Charmoille	9	47.43	7.21	0.1	0.052632	0.14286	0.4275	0.093023	0.04878	0.14286	0.38	1.2308
Creu Conill 20	9	41.67	2									1.4
Das	9	42.35	1.86									1
Dinotheriensande	9	49	8									1
Dongxiang-wangji	9	35.61	103.61									3
Doué-la-Fontaine	9	47.2	-0.3									1
El Firal	9	42.36	1.45	0.2	0.11111	0.28571	0.885	0.18605	0.10256	0.28571	0.8875	1.2308
Eppelsheim	9	49.72	8.97	0.14286	0.076923	0.15385	0.3625	0.16949	0.092593	0.17241	0.4975	1.1905
Esme Akçaköy	9	38.4	28.96	0.052632	0.027027	0.083333	0.225	0.04878	0.025	0.083333	0.23	1.2222
Esselborn	9	49.7	8.1	0.16667	0.090909	0.3	0.8775	0.051282	0.026316	0.1	0.235	1.0909
Estevar	9	0	0	0.11765	0.0625	0.25	0.755	0.21622	0.12121	0.5	0.9675	1.75
Esvres - Upper Faluns	9	47.3	0.8									1.3333
Gaiselberg	9	48.53	16.71									1.3333
Gau-Weinheim	9	49.84	8.04									1.1667
Grakali	9	41.93	44.3									2
Guonigou	9	35.83	103.75	0	0	0	0.12	0.10811	0.057143	0.25	0.755	2.5714
Hammerschmiede	9	47.9	10.61									1
Himberg	9	48.1	16.5									1
Hinterauerbach	9	48.4	12.01									1
Hostalets de Pierola Superior	9	41.2	0.5	0.2439	0.13889	0.33333	0.9475	0.090909	0.047619	0.13333	0.3425	1.2
Höwenegg	9	47.7	8.7	0.11765	0.0625	0.25	0.7925	0.10811	0.057143	0.25	0.735	1.4

Inzersdorf	9	48.13	16.33									2
Isakovo	9	47.22	28.43									1.4
Kalfa	9	46.54	29.23	0.086957	0.045455	0.1	0.2575	0.16327	0.088889	0.2	0.6025	1.1667
Laaerberg	9	48.16	16.4									1.6667
Lantian-12	9	34	109									2
Lapushna	9	46.9	28.5	0.17778	0.097561	0.21053	0.74	0.20833	0.11628	0.26316	0.8525	1
Los Valles de Fuentidueña	9	41	-4.2									1.2222
Mariathal	9	48.6	16.1									2
Masia de la Roma 2	9	40.5	-1									2
Melchingen	9	48.36	9.15									1.3333
Middle Sinap	9	40.56	32.7	0.14286	0.076923	0.1875	0.59	0.13333	0.071429	0.1875	0.615	1.375
Mistelbach	9	48.6	16.5									1
Molina de Aragon	9	40.85	-1.88									2
Nebelbergweg	9	47.38	7.63									1
Nikolsburg	9	48.8	16.63									1.5
Nombrevilla	9	41.07	-1.21	0.16667	0.090909	0.3	0.885	0	0	0	0.08	1.2
Nombrevilla-9	9	41.11	-1.35									1
Oberföhring	9	48.1	11.38									1
Oberhollabrunn	9	48.55	16.06									1
Oshin-II-5 upper	9	47	92.4	0	0	0	0.11	0.16216	0.088235	0.375	0.8925	2.3333
Priay II	9	46	5.3									1
Qaidam-Tuosu	9	37	97	0.15385	0.083333	0.23077	0.7725	0.19048	0.10526	0.30769	0.845	1.6
Rudabánya	9	48.71	20.63	0.14815	0.08	0.15385	0.4425	0.10526	0.055556	0.10714	0.195	1.2222
Salmendingen	9	48.35	9.1	0.15385	0.083333	0.23077	0.755	0.047619	0.02439	0.076923	0.175	1.3333
Santiga	9	41.2	0.5									1.2
Santiga (Sabadell)	9	41.55	2.1									1.2
Seu d'Urgel	9	42.3	1.6									1
Sevastopol (Sebastopol)	9	44.5	33.6	0.15789	0.085714	0.25	0.8075	0.2439	0.13889	0.41667	0.97	1.4444
Sinap 108	9	40.54	32.7									2.3333
Sinap 114	9	40.55	32.71									2
Sinap 14	9	40.53	32.62									2.5
Sinap 4	9	40.55	32.69									1.5
Sinap 64	9	40.56	32.7									2
Sinap 72	9	40.56	32.7	0.05824	0.030303	0.125	0.48	0.21622	0.12121	0.5	0.98	2
Sinap 88	9	40.56	32.7									3
Sinap 8A	9	40.54	32.7									2
Sinap 8B	9	40.54	32.7									2.5
Sinap 91	9	40.55	32.71									2
Sinap 94	9	40.55	32.69									1.5
Sop Mae Tham	9	16.75	99									1.4
Subsol de Sabadell	9	41.33	2.07	0.15	0.081081	0.21429	0.7	0.13953	0.075	0.21429	0.6525	1.3

(continued)

Locality	Age (MNEQ)	Latitude	Longitude	T/Dice	T/Jaccard	T/Simpson	T/Raup-Crick	B/Dice	B/Jaccard	B/Simpson	B/Raup-Crick	Mean Hypsodonty
Trie-sur-Baïse	9	43.33	0.36									1
Varnitsa	9	47.2	28.9	0.11429	0.060606	0.22222	0.6775	0.15789	0.085714	0.33333	0.86	1.5
Wartenberg	9	48.4	11.98	0.17647	0.096774	0.375	0.9275	0	0	0	0.11	1.1111
Westhofen	9	49.55	8.25									1
Wuzhong-ganhegou	9	37.99	106.2									1.6667
Vösendorf	9	48.1	16.31	0.057143	0.029412	0.11111	0.3925	0.10526	0.055556	0.22222	0.655	1.25
Zheltokamenka	9	45.26	33.45									1.3333
Zhongning-ganhegou	9	37.49	105.66									2.3333
Berislav	10	46.83	33.42									1.3333
Biru-Bulong	10	31.56	93.83									2.6667
Botamojnak	10	42.95	77.3	0.10811	0.057143	0.18182	0.6075	0.2	0.11111	0.36364	0.9575	2
Can Purull	10	41.55	1.97	0.17021	0.093023	0.19048	0.63	0.28	0.16279	0.33333	0.97	1.3333
Can Trullàs	10	41.55	1.96									1.5
Croix-Rousse	10	45.78	4.83									1
Eldari1	10	41.3	45.7	0.13953	0.075	0.17647	0.6025	0.30435	0.17949	0.41176	0.99	1.4286
Fugu-Laogaochuan-lamagou	10	39	111	0.097561	0.051282	0.13333	0.385	0.40909	0.25714	0.6	1	1.6
Fugu-Laogaochuan-wangdafuliang	10	39	111									1.6
Grebeniki	10	46.88	29.81	0.085106	0.044444	0.095238	0.2	0.36	0.21951	0.42857	1	1.7333
Grossulovo	10	47.35	30									1.75
Guanghe-houshancun	10	35.483333333333	103.5	0.04878	0.025	0.066667	0.1525	0.31818	0.18919	0.46667	0.9975	1.7
Guanghe-sigou	10	35.48	103.5	0.055556	0.028571	0.1	0.32	0.35897	0.21875	0.7	1	1.2
Gülpinar	10	39.53	26.11	0	0	0	0.0625	0.097561	0.051282	0.16667	0.4525	1.8571
Götzendorf	10	48	16.58	0.10811	0.057143	0.18182	0.5725	0.1	0.052632	0.18182	0.52	1
Hezheng-Dashengou	10	35.85	103.73	0.10714	0.056604	0.11538	0.145	0.40678	0.25532	0.41379	1	1.6111
Kohfidisch	10	47.16	16.35	0.1	0.052632	0.14286	0.4325	0.27907	0.16216	0.42857	0.995	1
La Cantera	10	40.06	-1.01									1.5
La Roma 2	10	40.66	-1.01	0.054054	0.027778	0.090909	0.2825	0.05	0.025641	0.090909	0.215	1.6
La Tarumba I	10	41.55	1.97	0.1	0.052632	0.14286	0.51	0.18605	0.10256	0.28571	0.8875	1.3
Masia del Barbo	10	40.4	-1.1									1.6
Masia del Barbo 2	10	40.66	0.83									1.6
Masia del Barbo 2B	10	40.66	-1.16									1.6
Montredon	10	44.4	3.36	0.1	0.052632	0.14286	0.3825	0.13953	0.075	0.21429	0.63	1.6
Novoukrainka	10	48.3	31.5									1.6667
Orignac	10	43.11	0.16									1.3333
Pentalophos 1 (PNT)	10	41	23	0.093023	0.04878	0.11765	0.375	0.17391	0.095238	0.23529	0.7	1.6364
Poksheshty	10	47.24	28.68	0.15385	0.083333	0.23077	0.79	0.38095	0.23529	0.61538	1	1.875
Ravin de la Pluie (RPL)	10	41	22.95	0.05	0.025641	0.071429	0.205	0.13953	0.075	0.21429	0.6225	1.6667
Ravin des Zouaves 1	10	41	22.95	0	0	0	0.115	0.054054	0.027778	0.125	0.395	2.4
Respopeny	10	47.45	28.36									1.625

Locality	N											
Sant Miquel de Taudell	10	41.55	1.98	0.2	0.11111	0.28571	0.905	0.18605	0.10256	0.28571	0.8475	1.3
Sherullah 9	10	34.3	69.4				0.255	0.20833	0.11628	0.26316	0.845	2
Sinap 1	10	40.54	32.69	0.088889	0.046512	0.10526	0.39	0.18182	0.1	0.26667	0.7375	1
Sinap 12	10	40.56	32.7	0.097561	0.051282	0.13333	0.7175	0.054054	0.027778	0.125	0.44	1.8462
Sinap 49	10	40.58	32.93	0.11765	0.0625	0.25						1.5882
Sinap 83	10	40.55	32.69	0.17021	0.093023	0.19048	0.655	0.04	0.020408	0.047619	0.05	1
Soblay	10	45.1	5.5	0.052632	0.027027	0.083333	0.25	0.2439	0.13889	0.41667	0.985	1.6
Stratzing	10	48.45	15.6	0.23256	0.13158	0.29412	0.9375	0.043478	0.022222	0.058824	0.0825	1.6667
Terrassa	10	41.56	2	0	0	0	0.095	0.15789	0.085714	0.33333	0.845	1.3333
Udabno1	10	41.31	45.23	0	0	0	0.0675	0.25	0.14286	0.45455	0.9825	1.4286
Wissberg	10	49.85	8.02	0.055556	0.028571	0.1	0.3675	0.10256	0.054054	0.2	0.57	1
Xirochori 1 (XIR)	10	41	23	0.047619	0.02439	0.0625	0.15	0.13333	0.071429	0.1875	0.5375	1.3333
Altan-Teli	11	47	92.6	0	0	0	0.0425	0.046512	0.02381	0.071429	0.16	2.2308
Baccinello V0	11	42.7	11.1	0.042553	0.021739	0.047619	0.0775	0.12	0.06383	0.14286	0.3875	2
Bala Yaylaköy	11	39.74	33.08	0.098361	0.051724	0.11538	0.06	0.125	0.066667	0.13793	0.155	1.4444
Corakyerler	11	40.6	33.63	0.11429	0.060606	0.22222	0.7225	0.26316	0.15152	0.55556	0.9925	1.9167
Crevillente 2	11	38.4	-0.5	0.090909	0.047619	0.11111	0.2775	0.34043	0.20513	0.44444	1	1.3
Csakvar	11	47.4	18.5	0.04878	0.025	0.066667	0.155	0.22727	0.12821	0.33333	0.95	1.7
Darayaspoon	11	38.66	69.86	0.054054	0.027778	0.090909	0.245	0.1	0.052632	0.18182	0.5075	1
Dorn Dürkheim 1	11	49.77	8.26	0	0	0.14286	0.0275	0.34783	0.21053	0.47059	1	1.25
Dzhuanaryk	11	42.2	75.6	0.1	0.052632	0.047619	0.4475	0.13953	0.075	0.21429	0.6325	1.75
Eminova	11	40.11	31.95	0.042553	0.021739	0.1	0.055	0.32	0.19048	0.38095	0.9875	1
Garkin	11	38.41	30.31	0.055556	0.028571	0	0.325	0.20513	0.11429	0.4	0.955	1.5
Injana	11	34.33	44.66	0	0	0.090909						1.6
Kalimanci 1	11	42.5	23.1	0.054054	0.02381	0	0.0125	0.2449	0.13953	0.3	0.9325	1.875
Karacahasan	11	39.86	33.24	0	0	0.058824						1.8889
Kayadibi	11	37.21	30.8	0.046512	0.026316	0.076923	0.0275	0.34783	0.21053	0.47059	1	1.4167
Kayadibi 3	11	37.59	32.26	0.1	0.052632	0.14286	0.4475	0.13953	0.075	0.21429	0.6325	1
Kemiklitepe 1.2	11	38.39	29.14	0.042553	0.021739	0.047619	0.055	0.32	0.19048	0.38095	0.9875	1.9286
Kemiklitepe D	11	37.5	29	0.055556	0.028571	0.1	0.325	0.20513	0.11429	0.4	0.955	2
Kocherinovo1	11	42	23.08									1.75
Kocherinovo 2	11	42.03	23.08									1.6667
Küçükçekmece	11	40.98	28.76	0	0	0	0.0125	0.2449	0.13953	0.3	0.9325	1.3
Lantian-30	11	34	109									2.6667
Lower Maragheh	11	37.41	46.35	0	0	0	0.0875	0.15789	0.085714	0.33333	0.8525	2
Nikiti 1 (NKT)	11	40	23.5	0.054054	0.027778	0.090909	0.2575	0.1	0.052632	0.18182	0.47	1.6
Nikiti 2 (NIK)	11	40	23.5	0	0	0	0.1075	0.10811	0.057143	0.25	0.7075	1.8571
Novaja Emetovka	11	46.65	30.6	0.046512	0.02381	0.058824	0.125	0.30435	0.17949	0.41176	0.97	1.7273
Peralejos D	11	40.48	-1.03									2
Piera	11	41.2	0.5	0.051282	0.026316	0.076923	0.2275	0.095238	0.05	0.15385	0.4025	1.3333
Puente Minero	11	40.36	-1.08	0.095238	0.05	0.125	0.385	0.13333	0.071429	0.1875	0.5825	1.625

(continued)

Locality	Age (MNEQ)	Latitude	Longitude	T/Dice	T/Jaccard	T/Simpson	T/Raup-Crick	B/Dice	B/Jaccard	B/Simpson	B/Raup-Crick	Mean Hypsodonty
Qaidam-shenggou	11	37.08	97.25	0.11111	0.058824	0.2	0.635	0.35897	0.21875	0.7	1	1.6667
Ravin des Zouaves 5	11	41	22.95	0.085106	0.044444	0.095238	0.185	0.28	0.16279	0.33333	0.97	1.625
Sabuncubaglari 1	11	41.68	26.59									1
Samos-Q6	11	37.75	26.97									2.2
Samos-Qx	11	37.75	26.97									2
Strumyani 1	11	42.83	23.08									2.25
Strumyani 2	11	42.83	23.1									1.8571
Taghar	11	34.35	69.41									1.75
Thermopigi	11	41.28	23.36	0.045455	0.023256	0.055556	0.1075	0.34043	0.20513	0.44444	0.9975	1.7273
Vivero de Pinos	11	40.33	-1.08									1.5
Yangjiashan	11	35.86	103.76	0.17778	0.097561	0.21053	0.8	0.5	0.33333	0.63158	1	1.25
Yuanmou-baozidong	11	25.66	101.83	0.20513	0.11429	0.30769	0.895	0.19048	0.10526	0.30769	0.8925	1.5
Yulafli (CY)	11	41.2	27.82	0.052632	0.027027	0.083333	0.2175	0.04878	0.025	0.083333	0.19	1.3
Özluce	12	37.24	28.52									2.5
Akgedik-Bayir	12	37.25	28.25	0	0	0	0.045	0.15	0.081081	0.27273	0.785	2.125
Akkasdagi	12	38.49	33.64	0.076923	0.04	0.076923	0.095	0.43636	0.27907	0.46154	1	2.0625
Aljezar B	12	40.33	-1.08									2
Baccinello V2	12	42.7	11.1									1.5
Baltavar	12	47	17	0.047619	0.02439	0.0625	0.18	0.13333	0.071429	0.1875	0.5275	1.5455
Baynunah	12	24.16	52.56	0.051282	0.026316	0.076923	0.2375	0.2381	0.13514	0.38462	0.9675	1.5
Belka	12	50.81	28.18	0.1	0.052632	0.14286	0.4475	0.27907	0.16216	0.42857	1	1.375
Casa del Acero	12	38.1	-0.5									1.6667
Casteani	12	42.5	11.15									1.4
Cerro de la Garita	12	40.38	-1.08	0.11321	0.06	0.11538	0.2425	0.25	0.14286	0.25926	0.9375	1.2308
Chimishlija (Cimislia)	12	46.5	28.8	0.081633	0.042553	0.086957	0.1575	0.30769	0.18182	0.34783	0.985	1.4737
Chobruchi (Tchobroutchi)	12	46.6	29.7	0.090909	0.047619	0.11111	0.3175	0.25532	0.14634	0.33333	0.925	1.7143
Chomateres	12	38.08	23.88	0.10526	0.055556	0.16667	0.535	0.097561	0.051282	0.16667	0.5	1.4
Çobanpinar (Sinap 42)	12	40.21	32.53	0	0	0	0.0375	0.18605	0.10256	0.28571	0.825	1.8
Concud	12	40.4	-1.13	0.086957	0.045455	0.1	0.21	0.16327	0.088889	0.2	0.615	1.2727
Concud Barranco	12	40.4	-1.01									1.5
Crevillente 15	12	38.25	-0.8									1.375
Crevillente 16	12	38.23	-0.8	0	0	0	0.1	0.054054	0.027778	0.125	0.3625	1.3333
Duzyayla	12	39.92	37.31	0	0	0	0.105	0.15789	0.085714	0.33333	0.87	1.4286
Ebic	12	38.8	35.65									1.5
Elekci	12	37.34	28.17									1.6667
Fiume Santo	12	40.8	8.3									1.6667
Gura-Galben	12	46.42	28.42	0	0	0	0.07	0.21053	0.11765	0.44444	0.9725	1.5556
Hadjidimovo-1	12	41.5	23.83	0.11111	0.058824	0.11538	0.1775	0.2807	0.16327	0.28571	0.9475	1.4762
Halmyropotamos (HAL)	12	38.5	24.2	0.043478	0.022222	0.05	0.065	0.20408	0.11364	0.25	0.805	1.6429

Locality												
Jilong	12	28.86	85.18									1.8
Kalimanci 2	12	42.5	23.1									1.7
Kalimanci 4	12	42.5	23.1									2.25
Kalimantsi-Pehtsata	12	42.5	23.1									1.3333
Kavakdere (Turolian)	12	40.24	32.56									1.75
Kemiklitepe A-B	12	37.5	29									2
Kerassia 1	12	38.5	23									1.6667
Kerassia 4	12	38.5	23									1.25
Kizilören	12	37.85	32.1									1.3333
Kromidovo 2	12	41.45	23.36									2
Lantian-6	12	34	109									2.3333
Las Pedrizas	12	36.83	-4.5									1.6667
Los Aljezares	12	37.75	-0.85									1.5
Los Mansuetos	12	40.35	-1.11	0.047619	0.02439	0.0625	0.135	0.13333	0.071429	0.1875	0.57	1.4545
Lufeng-shihuiba	12	25.3	102.4	0.04878	0.025	0.066667	0.1675	0.18182	0.1	0.26667	0.8175	1
Mahmutgazi	12	38.02	29.4	0.10526	0.055556	0.16667	0.4925	0.2439	0.13889	0.41667	0.97	1.6667
Masada del Valle 2	12	40.38	-0.43	0.057143	0.029412	0.11111	0.365	0.26316	0.15152	0.55556	0.9925	2
Middle Maragheh	12	37.4	46.36	0	0	0	0.075	0.15789	0.085714	0.33333	0.9	2
Molayan	12	34.31	69.41	0.11111	0.058824	0.2	0.6375	0.25641	0.14706	0.5	0.99	1.25
Montemassi	12	42.9	11	0.058824	0.030303	0.125	0.395	0.10811	0.057143	0.25	0.7225	2
Mt. Luberon	12	43.7	5.4	0.12	0.06383	0.125	0.315	0.33962	0.20455	0.375	0.99	1.75
Mytilinii 1A	12	37.75	26.96	0.13793	0.074074	0.15385	0.2725	0.22951	0.12963	0.24138	0.765	2.1111
Mytilinii 1B	12	37.75	26.96	0.054054	0.027778	0.090909	0.3075	0.2	0.11111	0.36364	0.9425	2
Mytilinii 1C	12	37.75	26.96	0.071429	0.037037	0.076923	0.0575	0.27119	0.15686	0.27586	0.915	1
Mytilinii 3	12	37.75	26.96	0.047619	0.02439	0.0625	0.13	0.31111	0.18421	0.4375	0.995	2.5
Mytilinii 4	12	37.75	26.96	0	0	0	0.1275	0.21622	0.12121	0.5	0.9575	2
Novo-Elizavetovka	12	47	31	0.05	0.025641	0.071429	0.2025	0.23256	0.13158	0.35714	0.9325	1.5
Ortok	12	42.4	75.6	0.054054	0.027778	0.090909	0.2825	0.2	0.11111	0.36364	0.9275	1.7143
Perivolaki	12	38.41	23.5	0.093023	0.04878	0.11765	0.3425	0.30435	0.17949	0.41176	0.9975	1.8333
Pikermi	12	38.01	23.99	0.057143	0.029412	0.11111	0.3675	0.26316	0.15152	0.55556	1	1.48
Pikermi-MNHN (PIK)	12	38.08	23.88	0.040816	0.020833	0.043478	0.045	0.26923	0.15556	0.30435	0.955	1.5
Pinaryaka	12	41.19	32.49	0.11111	0.058824	0.15385	0.08	0.34667	0.20968	0.44828	0.99	1.2
Prochoma	12	41	22.93	0	0	0	0.065	0.20513	0.11429	0.4	0.95	1.5455
Ravin X	12	38.4	23.48	0.04878	0.025	0.066667	0.1475	0.18182	0.1	0.26667	0.8075	1.5
Ribolla	12	43	11.1	0	0	0	0.04	0.18605	0.10256	0.28571	0.8575	2
Salihpasalar	12	37.25	28.27	0.052632	0.027027	0.083333	0.2375	0.2439	0.13889	0.41667	0.98	1.6667
Salihpasalar 1	12	37.25	28.25	0.055556	0.028571	0.1	0.3175	0.35897	0.21875	0.7	1	1.5714
Salihpasalar 2	12	37.25	28.26	0.055556	0.028571	0.1	0.2975	0.35897	0.21875	0.7	1	1.5
Samos	12	37.75	26.97	0.081633	0.042553	0.086957	0.175	0.15385	0.083333	0.17391	0.4725	1.7407
Samos (A-1)	12	37.75	26.97	0.072727	0.037736	0.076923	0.0975	0.34483	0.20833	0.34483	0.99	1.9231
Samos Main Bone Beds	12	37.5	26.87	0.055556	0.028571	0.076923	0.005	0.34667	0.20968	0.44828	0.9875	1.9697

(continued)

Locality	Age (MNEQ)	Latitude	Longitude	T/Dice	T/Jaccard	T/Simpson	T/Raup-Crick	B/Dice	B/Jaccard	B/Simpson	B/Raup-Crick	Mean Hypsodonty
Samos Old Mill Beds	12	37.5	26.87									2.5
Samos White Sands	12	37.5	26.87									2.2
Samos-Q5	12	37.75	26.97									1.875
Sandikli Kinik	12	38.56	30.13	0.097561	0.051282	0.13333	0.37	0.45455	0.29412	0.66667	1	2
Serefköy	12	37.36	28.23	0.052632	0.027027	0.083333	0.2525	0.19512	0.10811	0.33333	0.905	1.8
Serrazzano	12	43.6	10.8									2
Sinap 26	12	40.56	32.64	0.10811	0.057143	0.18182	0.565	0.2	0.11111	0.36364	0.91	1.9
Sinap 27	12	40.56	32.64									2.25
Sinap 33	12	40.55	32.64	0.055556	0.028571	0.1	0.3075	0.20513	0.11429	0.4	0.945	1.8
Sinap 70	12	40.55	32.64									2.5
Sor	12	39.27	67.65	0.054054	0.027778	0.090909	0.285	0.3	0.17647	0.54545	0.995	1.7143
Taraklia	12	46.5	29	0.075472	0.039216	0.076923	0.12	0.39286	0.24444	0.40741	1	1.7917
Tudorovo	12	46.26	30.03	0	0	0	0.105	0.27027	0.15625	0.625	0.9975	1.6667
Upper Maragheh	12	37.4	46.41	0.088889	0.046512	0.10526	0.2675	0.33333	0.2	0.42105	1	1.9091
Valdecebro 5	12	40.35	-1.03									2
Vathylakkos 1 (VLO)	12	41	23									1.6667
Vathylakkos 2 (VTK)	12	41	23	0	0	0	0.07	0.097561	0.051282	0.16667	0.5075	1.5556
Vathylakkos 3 (VAT)	12	41	23	0	0	0	0.02	0.2	0.11111	0.2381	0.76	1.4667
Almenara-Casablanca M	13	40	0									1
Amasya 2	13	37.65	28.48									2
Ananjev	13	47.7	30									1.5
Arenas del Rey	13	36.96	-3.9									2
Arquillo 1	13	40.4	-1.1	0.090909	0.047619	0.11111	0.285	0.29787	0.175	0.38889	0.9875	2
Baccinello V3	13	42.7	11.1	0.051282	0.026316	0.076923	0.21	0.28571	0.16667	0.46154	0.9875	1.6
Brisighella	13	44.5	11.6	0	0	0	0.0575	0.2439	0.13889	0.41667	0.985	1.9091
Bunker de Valdecebro	13	40.35	-1.05									3
Douaria	13	35	10									1.5
Dytiko 1 (DTK)	13	41	23	0	0	0	0.0275	0.13636	0.073171	0.2	0.63	1.8462
Dytiko 2 (DIT)	13	41	23	0.055556	0.028571	0.1	0.3225	0.15385	0.083333	0.3	0.8575	1.5
Dytiko 3 (DKO)	13	41	23	0	0	0	0.045	0.05	0.025641	0.090909	0.2	1.6667
El Arquillo 1	13	40.35	-1.05	0.086957	0.045455	0.1	0.21	0.28571	0.16667	0.35	0.9675	1.9333
Fugu-Laogaochuan-miaoliang	13	39	111	0.057143	0.029412	0.11111	0.3725	0.36842	0.22581	0.77778	1	1.3333
Gravitelli	13	38.2	15.5	0.11111	0.058824	0.2	0.65	0.30769	0.18182	0.6	0.9975	1.6667
Gusinyy perelyot	13	50.25	67.83	0.085106	0.044444	0.095238	0.2025	0.44	0.28205	0.52381	1	2.1
Hatvan	13	47.66	19.68	0.10526	0.055556	0.16667	0.5375	0.4878	0.32258	0.83333	1	1.25
Hsin-An-Loc.12	13	34.6	112.2	0.095238	0.05	0.125	0.385	0.44444	0.28571	0.625	1	1.2857
Kalmakpaj	13	47.3	85.3	0.057143	0.029412	0.11111	0.3525	0.26316	0.15152	0.55556	1	2.2
Karabastuz	13	50.3	79.16	0	0	0	0.1225	0.27027	0.15625	0.625	0.9925	2
Khirgis-Nur II-lower	13	49.2	93.4					0.27027	0.15625	0.625	0.995	2.3333

Khololochi Nor	13	45.4	100.9									2
Krivaja Balka	13	46.5	30.8									1
La Alberca	13	38	-0.6									2
La Gloria 5	13	40.38	-1.06	0.10811	0.057143	0.18182	0.615	0.55	0.37931	1		3
Lantian-42	13	34	109							1	1	1.571
Lantian-44	13	34	109									1
Las Casiones	13	40.43	-1.11	0.044444	0.022727	0.052632	0.0575	0.16667	0.090909	0.21053	0.7275	2
Librilla	13	38	-0.6	0	0	0	0.08	0.21053	0.11765	0.44444	0.9775	2.2
Milagros	13	41.35	-3.41	0.088889	0.046512	0.10526	0.265	0.125	0.066667	0.15789	0.45	2
Nurpur	13	32	77									1.4
Odessa 2	13	46.5	30.8									2.5
Olkhon (Sarayskaya. hor. 3)	13	53.1	107.2									1
Pao-Te-Lok.108	13	39	111	0.097561	0.051282	0.13333	0.3925	0.54545	0.375	0.8	1	1.75
Pao-Te-Lok.109	13	39	111	0.054054	0.027778	0.090909	0.2875	0.55	0.37931	1	1	1.5
Pao-Te-Lok.110	13	39	111	0.055556	0.028571	0.1	0.32	0.35897	0.21875	0.7	1	1.8
Pao-Te-Lok.30	13	39	111	0.081633	0.042553	0.086957	0.2025	0.53846	0.36842	0.6087	1	2
Pao-Te-Lok.31	13	39	111	0.095238	0.05	0.125	0.3575	0.35556	0.21622	0.5	0.9975	1.875
Pao-Te-Lok.43	13	39	111	0.045455	0.023256	0.055556	0.0975	0.51064	0.34286	0.66667	1	1.75
Pao-Te-Lok.44	13	39	111	0.1	0.052632	0.14286	0.4675	0.51163	0.34375	0.78571	1	1.7647
Pao-Te-Lok.49	13	39	111	0.18182	0.1	0.19231	0.6425	1	1	1	1	1.4375
Pao-Te-Lok.52	13	39	111	0	0	0	0.0975	0.36842	0.22581	0.77778	1	1.5
Pavlodar	13	52	77	0.081633	0.042553	0.086957	0.175	0.42308	0.26829	0.47826	1	2
Polgardi	13	47	18.3	0	0	0	0.0225	0.13636	0.073171	0.2	0.63	1.5
Qingyang-Lok.115	13	36	108	0.10526	0.055556	0.16667	0.505	0.53659	0.36667	0.91667	1	1.875
Qingyang-Lok.116	13	36	108	0.10256	0.054054	0.15385	0.4825	0.47619	0.3125	0.76923	1	1.8889
Rambla de Valdecebro 0	13	40.22	-1.05									3
Rambla de Valdecebro 3	13	40.22	-1.05									2.6667
Rambla de Valdecebro 6	13	40.22	-1.05									2.3333
Sahabi	13	30.08	20.75									1.5
Saint Arnaud	13	36.08	5.4									2.3333
Songshan-Loc.2	13	36.96	103.29	0.18605	0.10256	0.23529	0.8275	0.3913	0.24324	0.52941	1	1.4444
Songshan-Loc.3	13	36.96	103.27	0.15789	0.085714	0.25	0.805	0.43902	0.28125	0.75	1	1.4286
Taskinpasa 1	13	38.49	34.94									2
Tha Chang 2	13	15	102.33	0	0	0	0.08	0.052632	0.027027	0.11111	0.315	2
Valdecebro 3	13	40.35	-1.08									2.25
Venta del Moro	13	39.4	-0.4	0	0	0	0.01	0.23529	0.13333	0.27273	0.8825	1.8182
Villastar	13	40.01	-1.01									2.4
Wu-Hsiang-Loc.71	13	36.8	112.8									1
Wu-Hsiang-loc.78	13	36.8	112.9									1.4286
Wu-Hsiang-Lok.70	13	36.8	112.9									1.4286
Wu-Hsiang-Lok.73	13	39	111	0	0	0	0.1	0.42105	0.26667	0.88889	1	1.25

(continued)

Locality	Age (MNEQ)	Latitude	Longitude	T/Dice	T/Jaccard	T/Simpson	T/Raup-Crick	B/Dice	B/Jaccard	B/Simpson	B/Raup-Crick	Mean Hypsodonty
Yushe-hounao	13	37	112.9	0	0	0	0.0525	0.2439	0.13889	0.41667	0.99	1.5
Alcoy	14	38.6	-0.5	0	0	0	0.09	0.15385	0.083333	0.3	0.8275	2
Celleneuve	14	43.5	3.5									1.6667
Chono Hariah 1 lower	14	48.3	93									3
Dorkovo	14	42.03	24.13	0	0	0	0.12	0.10811	0.057143	0.25	0.695	1.5
Fonelas	14	37.25	-3.1	0	0	0	0.025	0.095238	0.05	0.15385	0.405	2
Gödöllö	14	47.6	19.36									1.5
Gaozhuang	14	38.28	113.11	0	0	0	0.065	0.04878	0.025	0.083333	0.2675	2
Gorafe 1	14	37.29	-3.02									1
Gorafe 4	14	37.29	-3.02									2
Guanghe-shilidun-LX200014	14	35.5	103.5	0.054054	0.027778	0.090909	0.285	0.3	0.17647	0.54545	0.995	1.86
Kessani 1.2	14	41	25	0.055556	0.028571	0.1	0.2725	0.051282	0.026316	0.1	0.3075	1
Khirgis-Nur II	14	49.2	93.4	0	0	0	0.0825	0.26316	0.15152	0.55556	1	2.5
Khirgis-Nur II-upper	14	49.2	93.4	0	0	0	0.04	0.23256	0.13158	0.35714	0.9525	2.4
Kosyakino	14	43.9	46.68									1.2857
La Gloria	14	40.23	-5.48									1.6667
La Gloria 4	14	40.38	-1.06	0	0	0	0.0925	0.15385	0.083333	0.3	0.81	2
Montpellier	14	43.6	3.89	0	0	0	0.0125	0.24	0.13636	0.28571	0.8875	1.4615
Olkhon (Sarayskaya. hor. 5)	14	53.1	107.2									2
Orrios	14	40.35	-1									2
Peralejos	14	40.29	-1.02									1
Peralejos E	14	40.48	-1.03									1
Saint Laurent des Arbres	14	43.93	4.8									1.6667
Trévoux	14	45.93	4.76									1.4

Locality											
Vendargues	14	43.65	3.96								2
Villeneuve de la Raho	14	42.63	2.91								1.5
Yushe-Gaozhuang	14	37	113	0	0	0.0525	0.047619	0.02439	0.076923	0.1675	2
Anvers 1	15	51.21	4.41								1
Apolakkia	15	36.06	27.8	0	0	0.0225	0.13953	0.075	0.21429	0.69	1.6667
Çalta 2	15	40.23	32.54	0	0	0.03	0.13953	0.075	0.21429	0.625	2
Capeni	15	46.01	25.55	0	0						1
Ciuperceni 2	15	44.19	25.22	0	0	0.045	0.04878	0.025	0.083333	0.1925	1
Csarnota 2	15	45.89	18.21	0	0	0.0475	0.1	0.052632	0.18182	0.485	1.2
Ivanovce	15	48.2	17.9	0	0	0.0525	0.15	0.081081	0.27273	0.7375	1.2
La Calera	15	39.31	-5.15	0	0	0.0825	0.15	0.081081	0.27273	0.775	2.1667
Layna	15	41.05	-2.19	0	0	0.0675	0.097561	0.051282	0.16667	0.4225	1.6667
Malushteni	15	46.13	28.08	0	0						2
Megalo Emvolon (MEV)	15	40.3	22.45	0	0	0.0775	0.15385	0.083333	0.3	0.81	1.4
Mianxian-yangjiawan	15	33.16	106.7	0	0	0.0825	0.15385	0.083333	0.3	0.8475	1.1667
Muselievo	15	43.3	24.4	0	0	0.035	0.13333	0.071429	0.1875	0.53	1.6667
Odessa Catacombs	15	46.5	30.8	0	0	0.01	0.15385	0.083333	0.17391	0.48	1.5
Perpignan	15	42.68	2.88								1.6667
Sugas-Bai	15	45.5	25.5								1
Val di Pugna	15	43.3	11.5								1
Varghis	15	46.13	25.53	0.090909	0	0.2825	0.15	0.081081	0.27273	0.81	1
Weze 1	15	50.08	18.78	0.054054	0	0.07	0.05	0.025641	0.090909	0.245	1
Wölfersheim	15	50.4	8.85	0.027778	0						1.25

NOTE: T = similarity with Tunggur-Moergen; B = similarity with Baode locality 49. Localities with blank GFRI values had fewer than seven distinctive genera and were not used in similarity analysis.

Chapter 30

Intercontinental Dispersals of Sicistine Rodents (Sicistinae, Dipodidae, Rodentia) Between Eurasia and North America

YURI KIMURA

The early definitive sicistine, *Allosminthus*, first appeared in the late Late Eocene of Qujing, Yunnan Province, China (Wang 1985). Sicistines became abundant and diverse during the Oligocene and the Miocene of Asia, with recent accounts listing nine genera in the Oligocene to early Miocene (*Tatalsminthus, Shamosminthus, Gobiosminthus, Parasminthus, Plesiosminthus, Litodonomys, Sinodonomys, Omoiosicista,* and *Arabosminthus*) and five genera in the Miocene (*Sicista, Heterosminthus, Lophocricetus, Lophosminthus,* and *Sibirosminthus*) (Holden and Musser 2005; Kimura 2010). The fossil record of the Sicistinae illustrates their paleogeographic ranges in both the Old and New World. Nevertheless, of all living and fossil sicistines, only *Plesiosminthus*- and *Sicista*-like mice are known from the Great Plains of North America in Early Oligocene to Late Miocene and early Pleistocene sediments (Korth 1994; Martin 1994). The *Plesiosminthus*-like genus *Schaubeumys* is common in the late Oligocene to middle Miocene, contemporaneous with or younger than the European *Plesiosminthus* record (e.g., Wilson 1960; Engesser 1979; Korth 1994; Martin 1994). The *Sicista*-like genera *Macrognathomys* and *Miosicista* are found in the Middle to Late Miocene, older than the previously known Asian *Sicista* record (e.g., Hall 1930; Green 1977; Korth 1993). The similarities between *Schaubeumys* and *Plesiosminthus* were explained either by the presence of plesiomorphic characters, resulting from early separation with little morphological modification (e.g., Galbreath 1953; Black 1958; Engesser 1979) or by phylogenetically close relations (e.g., Wilson 1960;

Green 1977, 1992; Hugueney and Vianey-Liaud 1980). North American *Macrognathomys* and *Miosicista* were described without verifying generic characters distinct from Eurasian *Sicista*. The purpose of this study is to refine the phylogenetic relationships of *Plesiosminthus* and *Sicista* in order to clarify the hypothesis of dispersal of *Plesiosminthus* from Eurasia to North America and to test the hypothesis that *Sicista* dispersed from Asia, not from the Great Plains of North America.

The Sicistinae is a subfamily of the superfamily Dipodoidea, which includes jerboas, jumping mice, and birch mice (Holden and Musser 2005). The Dipodoidea is osteologically characterized by enlarged infraorbital foramen to transmit the medial masseter muscle, a small foramen ventromedial to the infraorbital foramen for nerve and blood vessels, the zygomatic plate narrow and undeveloped, the vertical process of the jugal contacting or approaching the lacrimal, and the angular process of mandible not deflected outward and frequently with a perforation (Ellerman 1940; Klingener 1984). Many paleontological studies traditionally recognized two families, Zapodidae (including Sicistinae and Zapodinae) and Dipodidae (e.g., Simpson 1945; Black 1958; Hugueney 1969; Green 1977; Wang 1985; Korth 1994; Martin 1994). However, osteological and myological characters are gradational among dipodoids (e.g., Vinogradov 1930; Klingener 1964, 1984), and more recent phylogenetic systematic studies have placed Zapodinae (jumping mice) and Sicistinae (birch mice) as two subfamilies of Dipodidae (e.g., Stein 1990; McKenna and Bell 1997),

Table 30. 1

Genera of Sicistinae

Family Dipodidae Fischer von Waldheim 1817

    Subfamily Sicistinae Allen 1901

        Allosminthus Wang 1985

        Arabosminthus Whybrow et al. 1982

        Gobiosminthus Huang 1992

        Heterosminthus Schaub 1930

        Litodonomys Wang and Qiu 2000

        Lophocricetus Schlosser 1924

        Lophosminthus Zazhigin and Lopatin 2002

        Megasminthus Klingener 1966

        Omoiosicista Kimura 2010

        Parasminthus Bohlin 1946

        Plesiosminthus Viret 1926

        Shamosminthus Huang 1992

        Sibirosminthus Zazhigin and Lopatin 2002

        Sicista Gray 1827

        Sinodonomys Kimura 2010

        Sinosminthus Wang 1985

        Tatalsminthus Daxner-Höck 2001

        Tyrannomys Martin 1989

which is followed here. Table 30.1 lists the genera attributed to Sicistinae up to now.

For this study, *Plesiosminthus geringensis* (Martin 1974) was considered a synonym of *P. clivosus*, following Korth (1980), although specimens referred to *P. geringensis* have longer m1 in comparison to those of *P. clivosus* (Martin 1974; Storer 2002). Storer's (2002) specimens of *P. "geringensis"* may represent two species, because his limited sample shows extreme size variation. The lower first molars are 0.3 mm longer than M1 on average. The difference of 0.3 mm is larger than observed in any species of *Plesiosminthus*. For example, m1 is longer than M1 by only 0.05 mm in *P. schaubi* (Hugueney and Vianey-Liaud 1980) and *P. moralesi* (Alvarez Sierra, Daams, and Lacomba Andueza 1996); the maximum observed difference is 0.13 mm in *P. barsboldi* (Daxner-Höck and Wu 2003). *Plesiosminthus galbreathi* (Wilson 1960) and *Sicista vinogradovi* (Topachevskii 1965) were considered synonymous with *P. grangeri* and *S. praeloriger*, respectively, following Green (1977) and Kowalski (1979).

Institutional Abbreviations. CM, Carnegie Museum of Natural History, Pittsburgh, Pennsylvania; **SDSM**, South Dakota School of Mines and Technology, Museum of Geology, Rapid City, South Dakota; **UCMP**, Museum of Paleontology, University of California, Berkeley; **UNSM**, University of Nebraska State Museum, Lincoln,

Nebraska; **USNM**, Smithsonian Institution, National Museum of Natural History, Washington, DC; **V**, specimen prefix of **IVPP**, Institute of Vertebrate Paleontology and Paleoanthropology, Beijing, China. This study follows Wang (1985) for dental terminology as in figure 30.1. Dental formula for the Sicistinae: 1.0.1.3/1.0.0.3. Upper cheek teeth are shown in large caps; lower cheek teeth in lowercase.

PREVIOUS HYPOTHESES OF PHYLOGENETIC RELATIONSHIPS OF *PLESIOSMINTHUS* AND *SICISTA*

North American species of *Plesiosminthus* were referred to either *Schaubeumys* (e.g., Black 1958; Engesser 1979; Korth 1980, 1994) or *Plesiosminthus* (e.g., Wilson 1960; Green 1977, 1992; Hugueney and Vianey-Liaud 1980; Flynn 2008). Although there are no consistent differences to distinguish these genera, several dental variations of *Schaubeumys* are not present in *Plesiosminthus*: ectolophid connecting to metalophid, distinct anterocone, protoloph connecting to endoloph near mesoloph, and curved metaloph (Wilson 1960; Engesser 1979; Korth 1980). Considering the above variations, the small species *P. clivosus* appears to belong to the Eurasian genus *Plesiosminthus*, and three large species (*S. grangeri, S. sabrae*, and *S. cartomylos*) have been referred to *Schaubeumys*.

In analyses that predated cladistic phylogenetic analysis, Schaub (1930) and Hugueney and Vianey-Liaud (1980) concluded that *P. promyarion–P. myarion* form an evolutionary lineage, to the exclusion of the more derived *P. schaubi*. This hypothesis was widely accepted (e.g., Engesser 1987; Alvarez Sierra, Daams, and Lacomba Andueza 1996; Comte 2000). The *P. promyarion–P. myarion* lineage shares primitive characters such as strong posterior arm of protoconid on m2 and m3, frequent occurrence of double protoloph on M2, and unreduced m3 (Schaub 1930; Hugueney and Vianey-Liaud 1980). The *P. moralesi–P. schaubi* lineage has simple morphological structures, characterized by a high ratio of length for M1 to M2, reduced m3, and absence of the primitive characters of the *P. promyarion–P. myarion* lineage (Schaub 1930; Hugueney and Vianey-Liaud 1980; Alvarez Sierra, Daams, and Lacomba Andueza 1996). Other Eurasian species, notably Asian *P. asiaticus, P. tereskentensis*, and *P. barsboldi*, are difficult to place relative to the evolutionary tree of Hugueney and Vianey-Liaud (1980).

Kowalski (1979) proposed two evolutionary lineages within *Sicista*. *Sicista bagajevi–S. praeloriger*–modern *S. betulina* have complicated dental structures, whereas *S.*

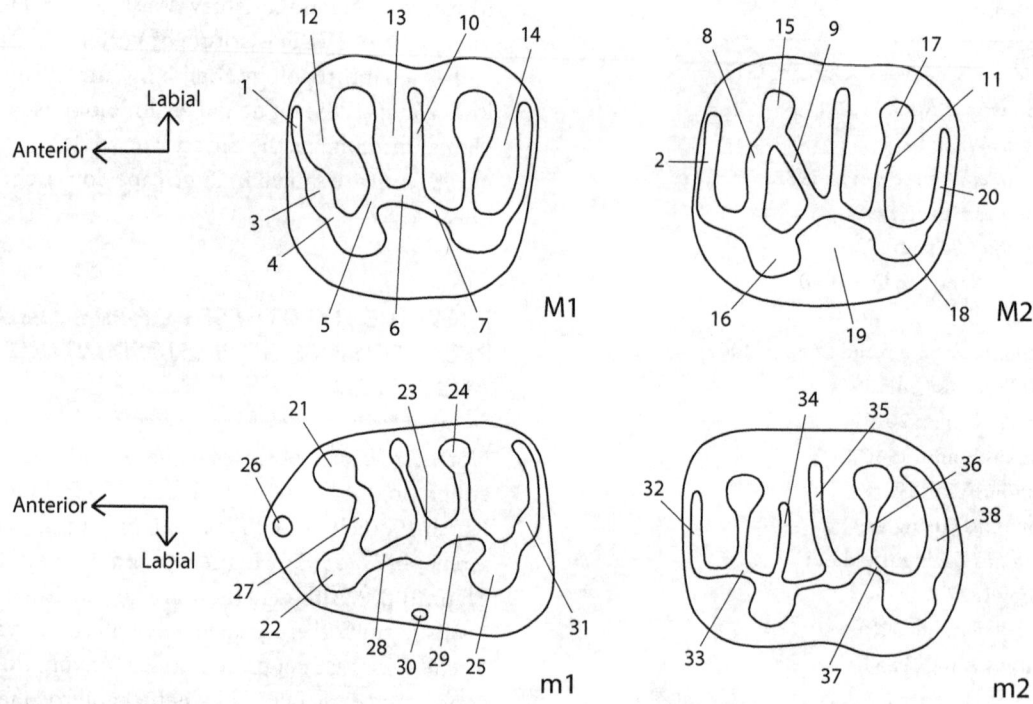

Figure 30.1 Dental terminology used in this chapter: (*1*) anteroloph I; (*2*) anteroloph II; (*3*) anterocone; (*4*) anterior arm of protocone; (*5*) posterior arm of protocone; (*6*) anterior endoloph; (*7*) posterior endoloph; (*8*) protoloph I; (*9*) protoloph II; (*10*) mesoloph; (*11*) metaloph; (*12*) anterior fossette; (*13*) middle fossette; (*14*) posterior fossette; (*15*) paracone; (*16*) protocone; (*17*) metacone; (*18*) hypocone; (*19*) endosinus; (*20*) posteroloph; (*21*) metaconid; (*22*) protoconid; (*23*) mesoconid; (*24*) entoconid; (*25*) hypoconid; (*26*) anteroconid; (*27*) metalophid; (*28*) anterior ectolophid; (*29*) posterior ectolophid; (*30*) ectostylid; (*31*) hypoconulid; (*32*) anterolophid; (*33*) anterior arm of protoconid; (*34*) posterior arm of protoconid; (*35*) mesolophid; (*36*) arm of entoconid; (*37*) anterior arm of hypoconid; (*38*) posterolophid.

pliocaenica–modern *S. subtilis* have simplified dental structures. Both lineages occurred in the Pliocene (Kowalski 1979). Green (1977) and Korth (1994) suggested *Plesiosminthus clivosus* was sister to *Macrognathomys*. Green (1977) suggested without explanation that Asian *Sicista bagajevi* belonged to *Macrognathomys*.

PLESIOSMINTHUS AND *SICISTA* FOSSILS FROM ASIA

Plesiosminthus is common in the late Oligocene to earliest Miocene (MP 28 to MN 1) of France, Switzerland, Spain, and Germany (e.g., Hugueney and Vianey-Liaud 1980; Engesser 1987; Ziegler 1994; Alvarez Sierra, Daams, and Lacomba Andueza 1996). In Asia, fossils of *Plesiosminthus* were recently discovered, including a classic European species, *P. promyarion* (Daxner-Höck and Wu 2003). Three species of *Plesiosminthus* were described from Xinjiang Province (China), Mongolia, and Kazakhstan (table 30.2). *Sicista* fossils are frequently recovered from late Miocene to Recent deposits within their mod-

ern range (e.g., Savinov 1970; Erbajeva 1976; Kowalski 1979, 2001; Qiu and Storch 2000), which is restricted to Central Asia and Europe between 40°N and 60°N, from Lake Baikal to Norway in the north, and to Romania and Austria in the south (Holden and Musser 2005). Three fossil species of *Sicista* are known from East Asia (see table 30.2). Although *Sicista* specimens are known from several Chinese faunas, these specimens are usually not identified at the specific level, because of the difficulty of identifying species in *Sicista*, which is the only extant genus of the Sicistinae. Some of its species have only been recognized karyologically. In Nei Mongol (Inner Mongolia), five faunas produce *Sicista* fossils.

Among the Nei Mongol faunas, the Gashunyinadege Fauna includes both *Plesiosminthus* (originally listed as *Parasminthus*; see Qiu and Wang 1999) and *Sicista*. The Gashunyinadege Fauna is currently considered Shanwangian in age and correlates roughly with European Neogene Mammal (MN) Faunal Zone 4 or 5 (Qiu and Wang 1999). Qiu, Wang, and Li (2006) estimated the age of the Gashunyinadege Fauna to be approximately ~17 Ma by reanalyzing faunal compositions. Therefore, *Sicista* sp.

Table 30.2

Fossils of *Plesiosminthus* and *Sicista* from Asia

Genus	Species	Locality	Approximate Age	Reference
Plesiosminthus	*P. promyarion*	Tavan Ovoony Deng, Valley of Lakes, Central Mongolia	Late Oligocene	Daxner-Höck and Wu (2003)
	P. asiaticus	Junggar Basin, Xinjiang, China	25.2 Ma	Daxner-Höck and Wu (2003); Ye, Meng, and Wu (2003); Zhang, Yue, and Wang (2007)
	P. barsboldi	Unkhltseng, Valley of Lakes, Central Mongolia; Gashunyinadege, central Nei Mongol, China	Oligocene/Miocene transition; 17 Ma	Daxner-Höck and Wu (2003); Qiu, Wang, and Li (2006) for age of the Gashunyinadege Fauna; this study for the Gashunyinadege sample
	P. tereskentensis	Altynshokysu, North Aral region, Kazakhstan	Early Miocene	Lopatin (1999)
	Plesiosminthus sp. 1	Junggar Basin, Xinjiang, China	23.9 Ma	Daxner-Höck and Wu (2003); Ye, Meng, and Wu (2003); Zhang, Yue, and Wang (2007)
	Plesiosminthus sp. 2	Junggar Basin, Xinjiang, China	23.9 Ma	Daxner-Höck and Wu (2003); Ye, Meng, and Wu (2003); Zhang, Yue, and Wang (2007)
	Plesiosminthus sp.	Gashunyinadege, central Nei Mongol	17 Ma	Kimura, this study (2010)
Sicista	*S. wangi*	Bilike, central Nei Mongol	Early Pliocene (4.5 Ma)	Qiu and Storch (2000); Qiu, Wang, and Li (2006) for age
	S. bagajevi	Gusinyi Perelet, Kazakhstan	Pliocene	Savinov (1970)
	S. pliocaenica	Chikoi River Valley, Western Transbaikalia	early Pleistocene	Erbajeva (1976)
	Sicista sp.	Gashunyinadege, Nei Mongol	17 Ma	Kimura, this study (2010, 2011)
	Sicista sp.	Shala, central Nei Mongol	Late Miocene (8.5 Ma)	Qiu and Wang (1999); Qiu, Wang, and Li (2006) for age
	Sicista sp.	Yushe Basin, Shanxi, China	late Late Miocene (6 Ma)	Flynn, Tedford, and Qiu (1991)
	Sicista sp.	Ertemte, central Nei Mongol; Harr Obo, central Nei Mongol	late Late Miocene (6 Ma); earliest Pliocene (5 Ma)	Fahlbusch, Qiu, and Storch (1983); Qiu, Wang, and Li (2006) for age
	Sicista sp.	Bilike, central Nei Mongol	Early Pliocene (4.5 Ma)	Qiu and Storch (2000); Qiu, Wang, and Li (2006) for age
	Sicista sp.	Gaotege, central Nei Mongol	late Early Pliocene (4 Ma)	Li (2006)

from this locality represents the earliest record of the genus. These sicistine specimens bridge both the chronologic and biogeographic gap between sicistine rodents in Europe and North America.

CLADISTICAL ANALYSIS

A cladistic analysis was performed in PAUP* 4.0b10 (Swofford 2002), in order to test a biogeographic hypothesis regarding the dispersal of *Plesiosminthus* and *Sicista*. Thirty-three dental characters (appendix 1) were scored (appendix 2) for 29 terminal taxa, including unnamed Nei Mongol taxa. All characters were equally weighted and unordered. The data set was obtained by observation of museum collections (appendix 3) and from the literature (Bohlin 1946; Galbreath 1953; Hugueney 1969; Savinov 1970; Shotwell 1968, 1970; Chaline 1972; Green 1977; Engesser 1979, 1987; Kowalski 1979; Hugueney and Vianey-Liaud 1980; Ziegler 1994; Alvarez Sierra et al. 1996; Lopatin 1999; Comte 2000; Daxner-Höck and Wu 2003). *Allosminthus ernos* and *Heosminthus primiveris* were used as outgroups. *Allosminthus* is well documented as the most primitive sicistine (McKenna and Bell 1997; Wang 1985; Wang and Qiu 2000). *Heosminthus* was chosen as outgroup because this genus showed sister relationship to *Plesiosminthus* plus other Asian dipodids in the cladistic analysis of Wang and Qiu (2000), which did not include *Sicista*. Phylogenetic relationship between *Heosminthus* and *Sicista* is important in considering intercontinental dispersals of sicistines. *Plesiosminthus tereskentensis* (Lopatin 1999), *Schaubeumys cartomylos* (Korth 1987), *Sicista pliocaenica* (Erbajeva 1976), and *Macrognathomys nanus* (Hall 1930) were excluded from the analysis because they are incompletely known and could not be adequately scored. According to Flynn (2008), North American *Miosicista* and *Tyrannomys* form a monophyletic group with *Plesiosminthus*, and this group is sister to *Macrognathomys*. Nevertheless, *Miosicista* and *Tyrannomys* were excluded from the analysis due to the incompletely known dentition. Of 13 modern *Sicista* species (Holden and Musser 2005), the most common species (*Sicista betulina* and *S. subtilis*) and Chinese birch mice (*S. concolor*) were chosen for the analysis. *Sicista* species are identified mainly based on karyological studies (Sokolov, Kovalskaya, and Bashevic 1987; Baskevich 1996) and morphology of male reproductive organs (e.g., Sokolov, Bashevic, and Kovalskaya, 1986; Shenbrot 1992). Although morphologically identified *Sicista betulina* and *Sicista subtilis* have been found to include two karyological species, respectively (Baskev-

ich 1996), morphologically identified species were considered in this study because interspecific variation among the morphologically similar species is unknown.

Results

A heuristic search generated 2114 equally most parsimonious trees (tree length = 102, C.I. = 0.54, R.I. = 0.82, R.C. = 0.44). The resultant trees are summarized in figures 30.2 and 30.3. The strict consensus tree (see figure 30.2) shows *Sicista* and *Macrognathomys* forming a monophyletic group united by four unambiguous synapomorphies of lower teeth: anteroconid connecting to metalophid (character 17), protoconid directed transversely (character 18), longitudinal ridge present (character 23), and entoconid connecting to hypoconid (character 30). "*Macrognathomys*" *gemmacollis* is a basal sister to other *Sicista* species, including *Sicista* sp. from Gashunyinadege and "*Macrognathomys*" sp. (Shotwell 1970) of the Bartlett Mountain Local Fauna. The clade exclusive of "*Macrognathomys*" *gemmacollis* is supported by three synapomorphic features: protoloph attaching to center of protocone (character 9), arm of entoconid extending posterolabially (character 22), and metaconid in line with anteroconid (character 26). In the 50% majority-rule consensus tree (see figure 30.3), *Sicista* sp. from Gashunyinadege is a successive sister taxon to the remaining *Sicista* species. *Sicista wangi*, *S. praeloriger*, and *S. bagajevi* are successive sister taxa to a monophyletic group comprised of *Sicista* sp. from Bilike and *S. betulina*. This clade accommodates the *S. bagajevi–S. praeloriger*–modern *S. betulina* lineage of Kowalski (1979), characterized by complicated tooth morphology.

In contrast to the strong support of *Sicista*, the clade (*Heosminthus* (*Plesiosminthus* ("*Schaubeumys*" (*Megasminthus*)))) is weakly supported by one unambiguous synapomorphy (character 2) with a low bootstrap value (17%). However, the topology of the clade is generally congruent with previous work:

1. *Heosminthus* is a sequential genus to *Plesiosminthus* (Wang 1985; Wang and Qiu 2000).
2. *P. myarion* is more derived than *P. clivosus* (Wilson 1960).
3. *P. conjunctus* is closest to *P. promyarion* (Ziegler 1994).
4. *Megasminthus* is more derived than *P. grangeri* (Klingener 1966; Green 1977).
5. *P. moralesi* is more basal than *P. schaubi* (Alvarez Sierra, Daams, and Lacomba Andueza 1996).

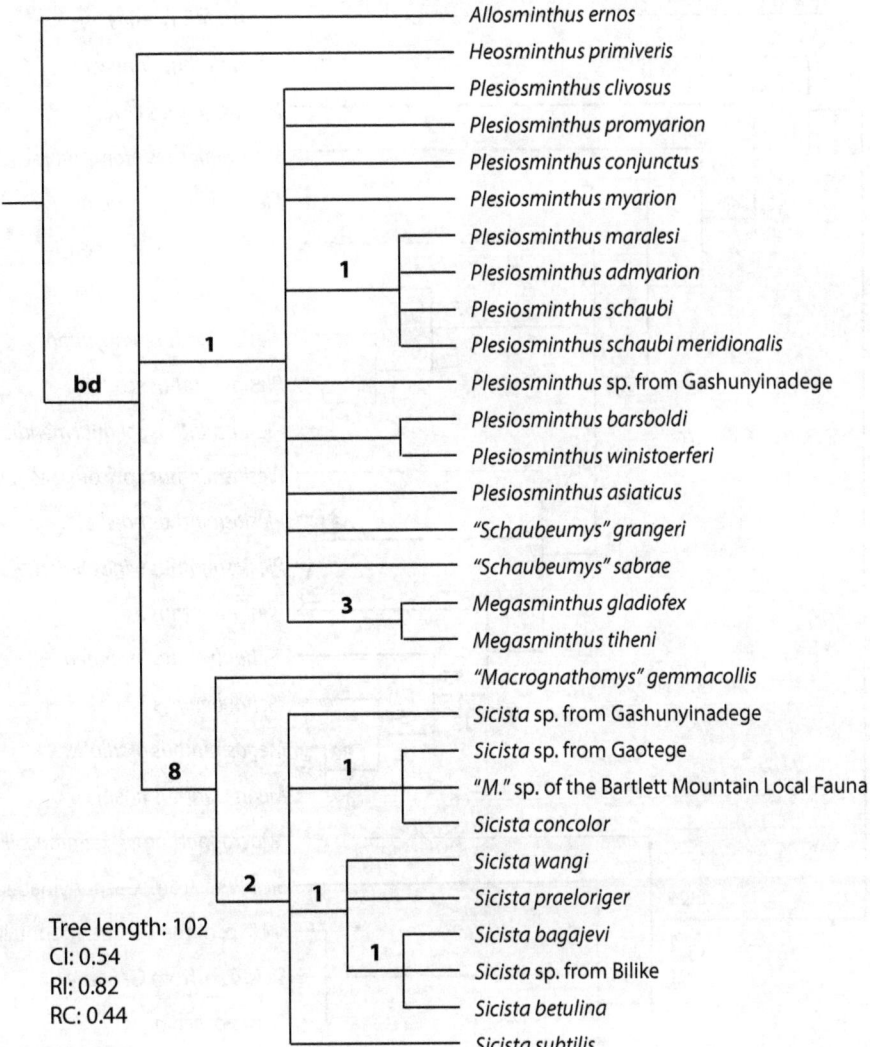

Figure 30.2 Strict consensus tree of *Plesiosminthus*, *Megasminthus*, and *Sicista*, resulting from 2114 most parsimonious trees. Tree length = 102; consistency index = 0.54; retention index = 0.82; rescaled consistency index = 0.44bd, Bremer decay value.

6. *P. schaubi* and *P. schaubi meridionalis* are sister taxa (Comte 2000).

7. The clade including *P. winistoerferi* is separate from both the *P. myarion* and *P. schaubi* clades (Engesser 1987).

Statistical support of the clade (*Plesiosminthus* ("*Schaubeumys*" (*Megasminthus*))) is low, with a bootstrap value of 28%. The two synapomorphic features of this clade are a long anteroloph II of M1 (character 6) and the presence of the anterocone (character 7). *Plesiosminthus clivosus*, *P. promyarion*, and *P. conjunctus* are successive sisters to clade A (remaining *Plesiosminthus* ("*Schaubeumys*" (*Megasminthus*))). The monophyletic group of (*P. conjunctus*, clade A) is characterized by one synapomorphy,

a complete protoloph II on M2 (character 15). *Plesiosminthus conjunctus* has a complete protoloph II in 80% of the specimens (9/11), whereas this structure is absent or incomplete in *P. clivosus* and *P. promyarion*. *Plesiosminthus admyarion* was recovered in the clade (*P. moralesi* (*P. admyarion* (*P. shaubi*, *P. schaubi meridionalis*))), in contrast to Comte (2000), who found *P. admyarion* was closer to *P. myarion* than to *P. schaubi* based on morphologically unreduced m3 with the mesolophid and the arm of the entoconid present (64/95: 67%) and the presence of the posterior arm of the protoconid on m2 (23/128: 18%). Monophyly of *P. schaubi* and *P. schaubi meridionalis* is supported by the absence of the posterior arm of the protoconid and the mesolophid on m3 (character 32), resulting

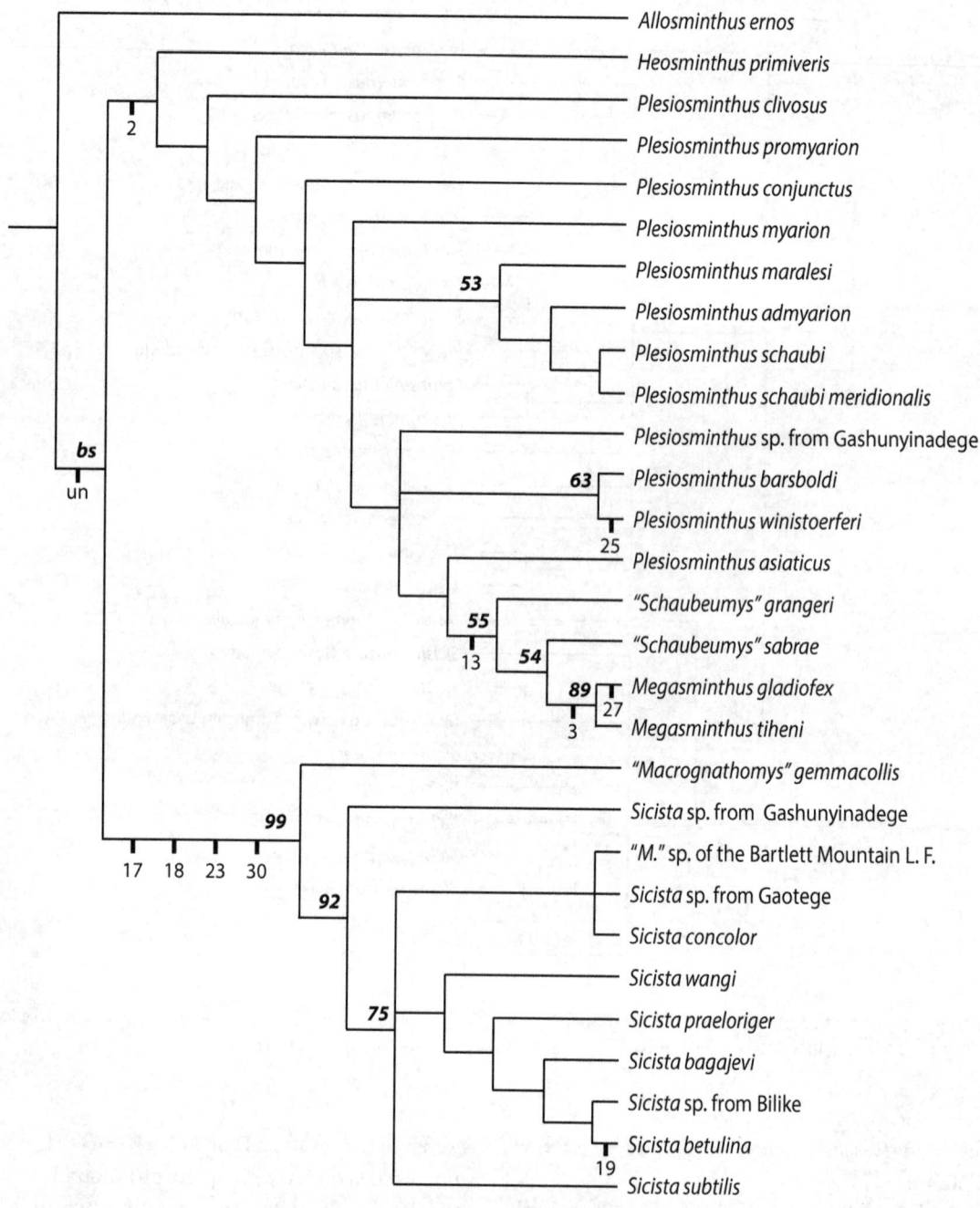

Figure 30.3 Fifty percent majority-rule consensus tree of *Plesiosminthus*, *Megasminthus*, and *Sicista*, resulting from 2114 most parsimonious trees. bs = bootstrap value; un = unambiguous character.

in more simplified morphology of m3 than in *P. admyarion*. *Plesiosminthus barsboldi* and *P. winistoerferi* have the arm of the entoconid of m1 extending posterolabially (character 22). Character 29 (arm of the entoconid of m2 extends posterolabially) is shared by *Plesiosminthus* sp. from Gashunyinadege and *P. winistoerferi*. This character is homoplasic with *Sicista*. *Plesiosminthus asiaticus* is a basal

sister taxon to the North American clade ("*Schaubeumys*" (*Megasminthus*)). The North American clade has a synapomorphy of curved metaloph (character 13). One synapomorphy of endosinus open to the anterior fossette (character 3) supported a monophyly of *Megasminthus*. The metaconid connecting to protoconid (character 27) is an autapomorphic feature for *Megasminthus gladiofex*.

DISCUSSION

North American Synonyms of *Plesiosminthus* and *Sicista*

The resultant trees (see figures 30.2 and 30.3) falsify the previously proposed hypotheses of Green (1977) and Korth (1994) that "*Macrognathomys*" is descended from North American *Plesiosminthus clivosus* and is ancestral to *Sicista*. North American "*Schaubeumys*" is synonymous with *Plesiosminthus* because *P. grangeri* and *P. sabrae* do not form a monophyletic group. Tooth morphology of Asian *P. barsboldi* and North American *P. grangeri* is illustrated in figure 30.4. In contrast to the nonmonophyly of "*Schaubeumys*," *Megasminthus* is strongly supported and should be retained as a valid genus. Consequently, *Plesiosminthus* sensu stricto is cladistically paraphyletic. However, this study retains the generic name as *Plesiosminthus* sensu lato because the characters examined in this study are limited to dental morphology.

The genus *Megasminthus* is conventionally included in the subfamily Zapodinae, separate from the sicistine *Plesiosminthus* (e.g., Klingener 1966; Green 1977; Korth 1980; McKenna and Bell 1997), considering its intermediate dental morphology between *Sicista* and modern zapodines. However, *Megasminthus* lacks two osteological synapomorphies of the Zapodinae: the small foramen medial to the infraorbital foramen is dorsally closed, and two venous foramina are absent in the posterior part of the palate (Klingener 1966). The only dental character shared by *Megasminthus* and modern zapodines is the endosinus opening to the anterior fossette on M1 (Klingener 1966; character 3 in appendices 1 and 2). Based on the phylogenetic results of this study and the available osteological characters, *Megasminthus* is considered to belong to the Sicistinae instead of the Zapodinae.

Comte (2000) considered the sample of *Plesiosminthus* from Venelles, France, to be distinct at the subspecific level from *P. schaubi* from Coderet and named it *P. schaubi meridionalis*. A long mesolophid is present on m2 in all 59 specimens of *P. schaubi*, whereas the mesolophid is reduced in 77% (20 out of 26 specimens) of *P. schaubi meridionalis* (Comte 2000:tables 26–27). The paracone is always connected to the anterior arm of the protocone in *P. schaubi*, whereas the protoloph of *P. schaubi meridionalis* connects to the posterior arm of the protocone (Comte 2000:table 25). These diagnostic features of *P. schaubi meridionalis* are less variable within a species of *Plesiosminthus*. Thus, *P. meridionalis* could be recognized as a distinct species.

North American "*Macrognathomys*" is a junior synonym of *Sicista*. *Sicista* sp. of the Bartlett Mountain Local Fauna from the Late Miocene (~7.5 Ma) of Oregon (Shotwell 1970) forms a clade with Chinese birch mice,

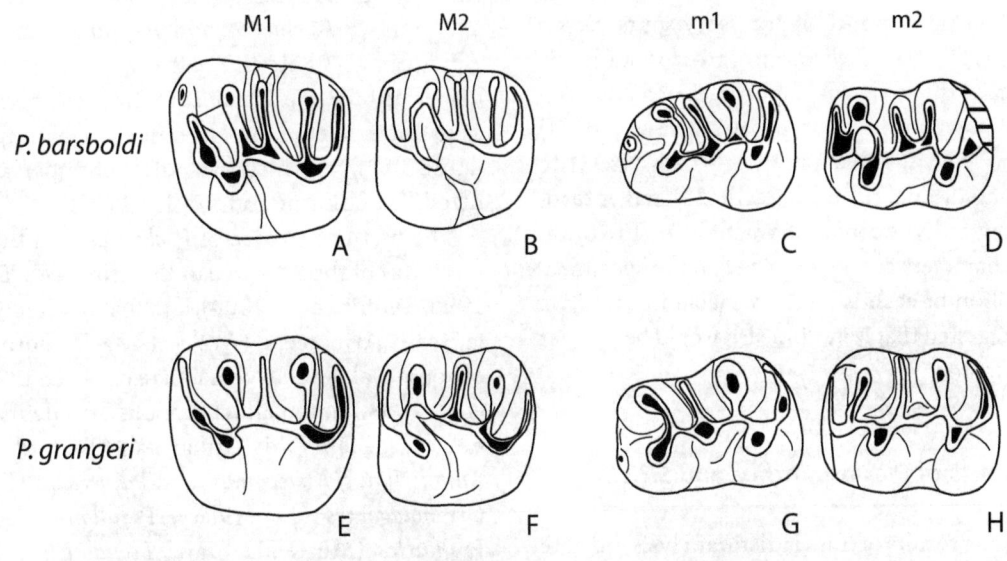

Figure 30.4 Comparison of *Plesiosminthus barsboldi* from Gashunyinadege and *Plesiosminthus grangeri* from the early Hemingfordian Rosebud Formation of South Dakota: (*A*) right M1 (V15889.1, reversed); (*B*) left M2 (V15888.1); (*C*) left m1 (V15889.7); (*D*) left m2 (V15889.12); (*E*) right M1 (SDSM 7954, reversed); (*F*) left M2 (SDSM 7954); (*G*) left m1 (SDSM 67150); (*H*) right m2 (SDSM 67144, reversed. Scale bar = 1 mm.

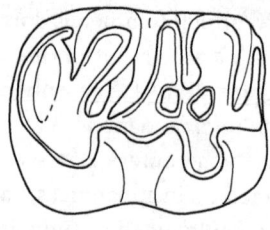

Figure 30.5 Right m2 of *Sicista gemmacollis* from the Valentine Formation of South Dakota (holotype, SDSM 8396), corrected from Green (1977). Scale bar = 1 mm.

Sicista sp. from Gaotege, and *Sicista concolor*. Shotwell (1970) originally described specimens from both the Bartlett Mountain Local Fauna and Black Butte Local Fauna (~10 Ma) of Oregon as "*Macrognathomys*" cf. *nanus*, and he was followed by Green (1977). However, Shotwell's (1970) specimens are substantially smaller than *Sicista nanus* (UCMP 29634) and have an accessory spur from the hypoconid on m1 and unreduced m3, as in *S. gemmacollis*. From the illustration in Shotwell (1970), the m2 from the Black Butte Local Fauna is indistinguishable from that of the holotype *S. gemmacollis* (figure 30.5). Therefore, the three lower teeth from the Black Butte Local Fauna should be assigned to *S. gemmacollis*. In addition, two "*Macrognathomys*" specimens from the Quartz Basin (Shotwell 1968) almost certainly belong to *S. gemmacollis*. Nevertheless, the specimens of the Bartlett Mountain Local Fauna (Shotwell 1970) seem different from *S. gemmacollis*, judging from the illustration. Based on large size and the absence of accessory spurs, *Sicista nanus*, as well as Siberian *S. pliocaenica*, are close to *S. subtilis*. Considering the four synapomorphies of *Sicista*, *Miosicista* (only known by the holotype jaw [UNSM 45424]) can be judged from the illustration of Korth (1993) to possess the longitudinal ridge (character 23) and entoconid connecting to hypoconid (character 30). The other two derived characters are not observed in the specimen, but the condition of each is highly variable in primitive *Sicista*. It is suggested that *Miosicista* also would be a junior synonym of *Sicista*.

Dispersal Events of *Plesiosminthus* and *Sicista*

The phylogenetic results of this cladistic analysis indicate that *Plesiosminthus* dispersed from Asia to North America at least two times (figures 30.6 and 30.7), in contrast to the previous hypothesis that an intercontinental dispersal of *Plesiosminthus* happened once in the early Oli-

gocene or earlier (e.g., Wilson 1968; Engesser 1979; Hugueney and Vianey-Liaud 1980; Korth 1994; Tedford et al. 1987; Tedford et al. 2004). The migrations were achieved by sisters of *P. clivosus* and *P. grangeri*, and *P. meridionalis*. Due to low density of fossil occurrences, *P. clivosus* and the otherwise somewhat later *P. grangeri* could be simultaneous immigrants. Two outgroups (*Allosminthus* and *Heosminthus*) are known only from Eurasia. *Allosminthus* is endemic in East Asia (Wang 1985), and *Heosminthus* is more widely distributed in East Asia (Daxner-Höck 2001) and Anatolia (de Bruijn et al. 2003). Based on the fossil evidence, Asian taxa dispersed to Europe and North America, although the most basal species, *Plesiosminthus clivosus* and *Sicista gemmacollis*, are found only in North America. The trans-Beringia route was used for intercontinental dispersals of sicistines, based on the phylogenetic results and geographic distribution of *P. promyarion* and *P. asiaticus*, as was the case for many dispersal events of land mammals after the early Eocene when the North Atlantic Thulean route was no longer available (Tedford et al. 1987; Tedford et al. 2004; Webb and Opdyke 1995; Woodburne and Swisher 1995).

The dispersal timing of *P. clivosus* is not consistent with any major dispersal event of land mammals (see figure 30.7). In fact, *Plesiosminthus clivosus* is the only allochthonous species that defines the beginning of the Arikareean in Tedford et al. (2004). On the other hand, the dispersal timing of *P. grangeri* could correspond to a second-order episode, defined as an event of five to eight immigrants to North America (see figure 30.7). Based on the first occurrences of the species, it is estimated that ancestors of *P. clivosus* and *P. grangeri* reached North America by ~30 Ma and ~28 Ma, respectively. Timing of dispersal of *P. meridionalis* to North America is uncertain, but it likely took place in the Hemingfordian (see figure 30.7), for which two pulses of dispersals are recognized (Woodburne and Swisher 1995).

The first occurrence of *P. clivosus* is in the beginning (~30 Ma) of the Arikareean (Martin 1974; Tedford et al. 1996; Tedford et al. 2004), and that of *P. grangeri* is in the late early Arikareean (Ar2, ~28 Ma; Tedford et al. 2004) of South Dakota and Nebraska (e.g., Green 1977; Carrasco et al. 2005). Seven allochthonous taxa define the beginning of the late Early Arikareean (Tedford et al. 2004): *Amphechinus*, *Metechinus*, and *Parvericius* (Erinaceidae), *Gripholagomys* (Leporidae), *Parallomys* (Allomyidae), *Promartes* (Mustelidae), and *Pseudotheridomys* (Eomyidae). Thus, *Plesiosminthus grangeri* is the eighth species to define this phase.

Dispersal rates of living small mammals are estimated from ~3 km/yr to ~10 km/yr in mountainous terrain of

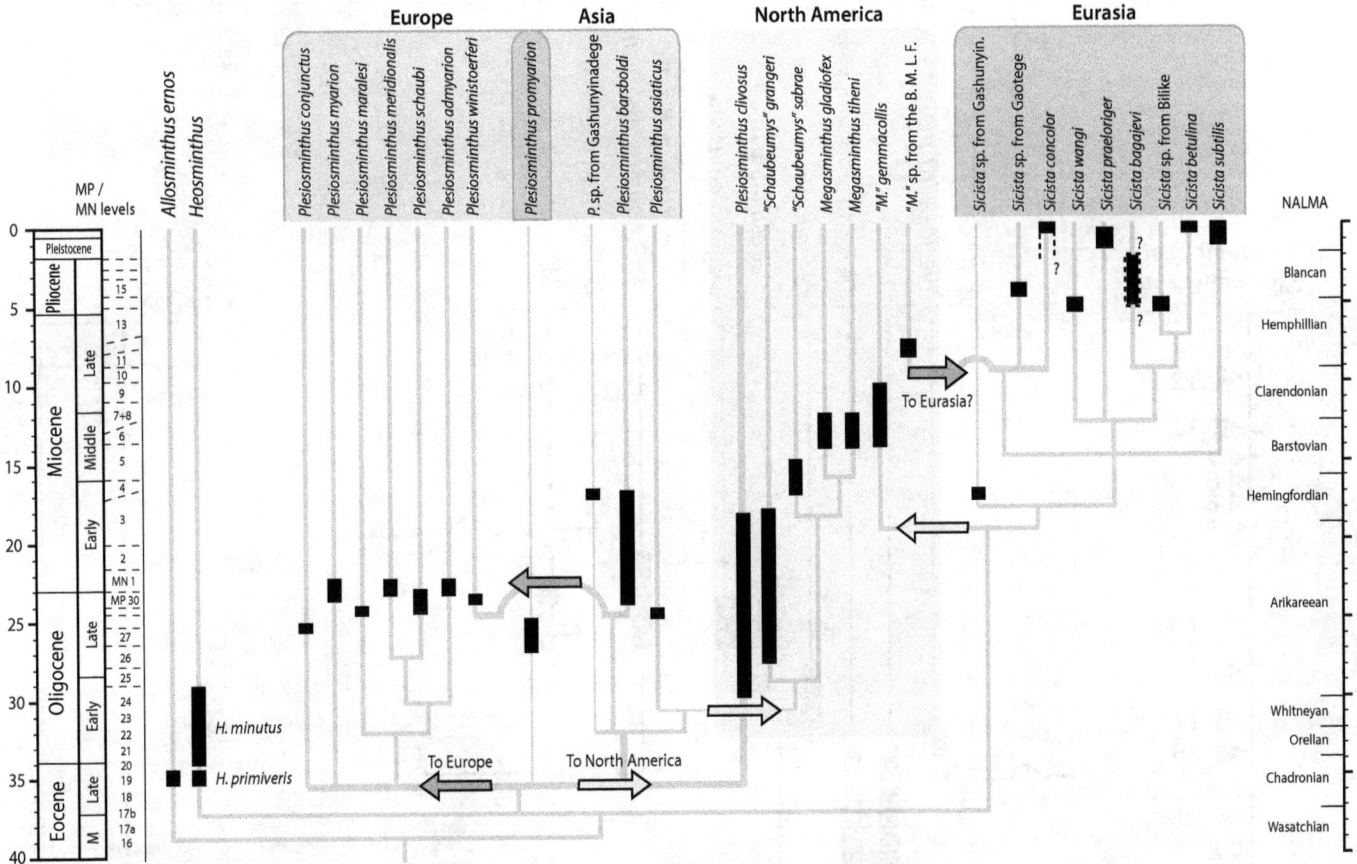

Figure 30.6 Chronological ranges and phylogenetic relationships of *Plesiosminthus*, *Megasminthus*, and *Sicista*, based on the 50% majority-rule consensus tree. Nodes supported by low bootstrap (<25%) values are collapsed. Arrows indicate the possible direction of dispersal events but not dispersal timing. MP reference levels and MN zones follow Legendre and Lévêque (1977) and Schlunegger et al. (1996), and Agustí et al. (2001), respectively.

the Andes (Jaksic et al. 2002). *Plesiosminthus* and *Sicista* may have migrated at a much slower pace under a presumably severe climate that filtered land mammals in Beringia. Modern *Sicista* hibernates about six months a year in their short lifetime of up to three years (Nowak 1999). The distance traveled by *Plesiosminthus* and *Sicista* in dispersing to North America approximated 20,000 km. Even in the most conservative scenario, they could have reached the northern Great Plains in less than 1 m.y.

Faunal exchanges of land mammals frequently took place between Europe and Asia during the Neogene (van der Made 1997; Mein 1999; Qiu and Li 2003) due to the lack of geographic barriers (Rögl 1999). In the Chinese Neogene, 35% (44/126) of rodents are congeneric with European faunas, whereas only ~10% (12/126) are found in North American faunas (Qiu and Li 2003).

Asian sicistines, including *P. promyarion* (Daxner-Höck and Wu 2003), dispersed twice to Europe (see figure 30.6).

Dispersals of sicistines to Europe were not related to the Grande Coupure (Stehlin 1909), a major Asian dispersal episode at 33.5 Ma (Hooker et al. 2004), because the first European appearance of *P. promyarion* is ~6 m.y. later in MP 26 (~27 Ma; Schlunegger et al. 1996) at Mas de Pauffiè, France (BiochroM'97 1997) and Mümliswil-Hardberg, Switzerland (Engesser and Mayo 1987). After *P. promyarion* arrived in Europe, European *Plesiosminthus* species evolved independently of Asian species except for *P. winistoerferi*. *Plesiosminthus winistoerferi* first occurs in MP 30 (~23.5 Ma; Schlunegger et al. 1996) and probably has a separate Asian origin from other European *Plesiosminthus* based on this phylogeny (see figure 30.6).

The phylogenetic results of this study indicate that *S. gemmacollis* dispersed from Asia to North America in the Hemingfordian dispersal pulse because it split from Asian *Sicista* before Gashunyinadege *Sicista* sp. evolved

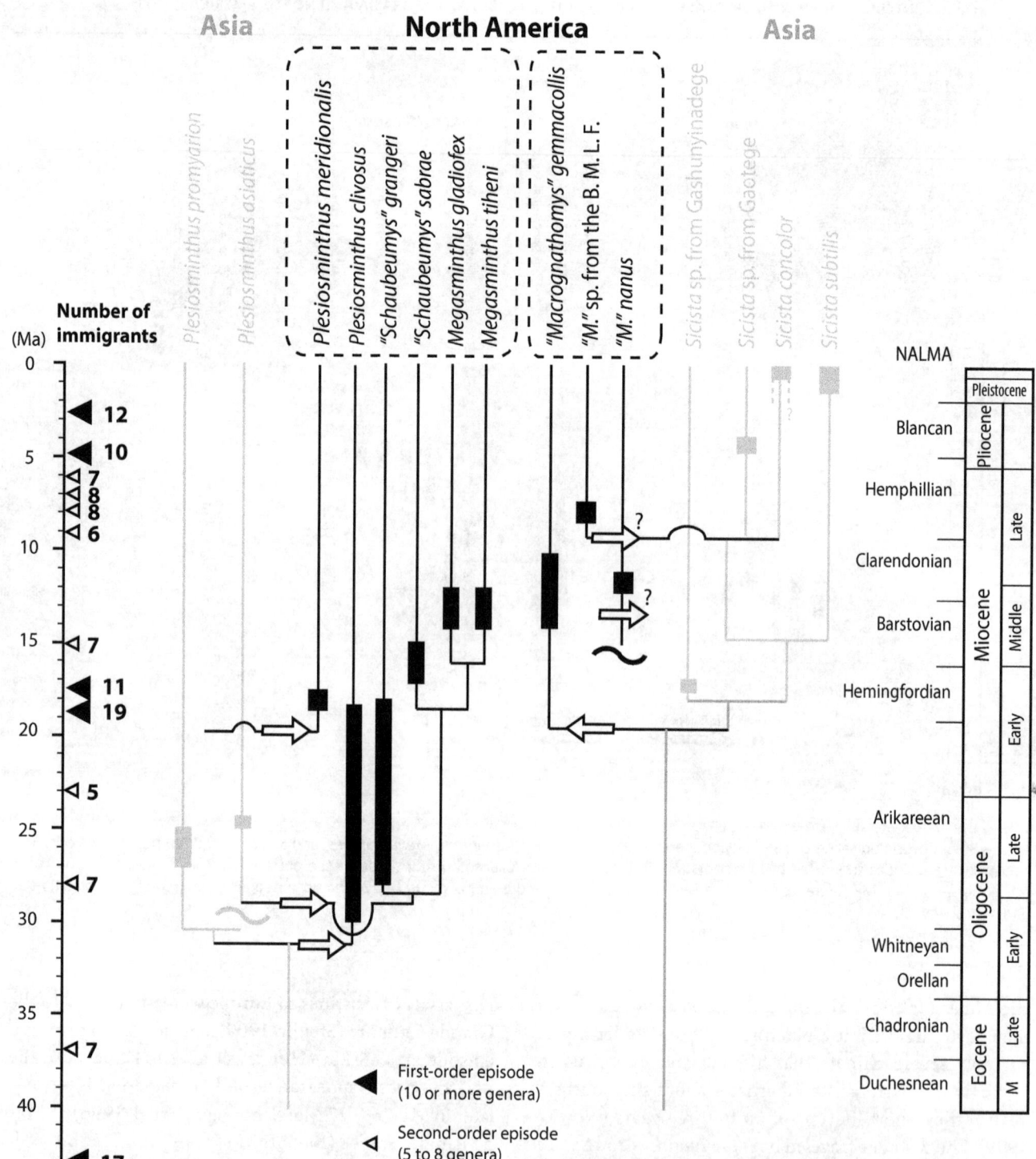

Figure 30.7 Chronological ranges of North American sicistine species and timing of dispersal episodes of mammals into North America. Open arrows indicate the dispersal direction and timing. First-order episodes (with 10 or more genera) and second-order episodes (with 5 to 8 genera) are indicated by solid triangles and open triangles, respectively. Modified from Webb and Opdyke (1995). The number of immigrants is taken from Woodburne and Swisher (1995) for the Duchesnean to Whitneyan; Tedford et al. (2004) for the Arikareean to Hemphillian; and Webb and Opdyke (1995) for the Blancan. North American land mammal age and geological time scale follow Woodburne (2004) and Gradstein, Ogg, and Smith (2004), respectively.

(see figure 30.7). Other small mammalian participants in these dispersal events include *Eomys* (Eomyidae), *Euroxenomys* (Castoridae), *Petauristodon* (Sciuridae), and two soricids (Tedford et al. 2004). Nevertheless, derived *Sicista* may be descended from North American immigrants. North American *Sicista* species are older than Asian species except *Sicista* sp. from Gashunyinadege (see figure 30.7). Within the clade that contains Chinese birch mice, *Sicista* sp. from the Bartlett Mountain Local Fauna is ~3.5 m.y. older than its sister *Sicista* sp. from Gaotege. *Sicista nanus*, showing morphological resemblance to *S. subtilis*, is ~10 m.y. older than the first appearance of modern *Sicista subtilis* (~1.5 Ma; Kowalski 2001).

Geographic Ranges of North American Sicistines

North American sicistines were distributed in the Great Plains and northern Rocky Mountains from the beginning of the Arikareean (30 Ma) to Hemphillian (~7.5 Ma) and in the Pleistocene (e.g., Carrasco et al. 2005). The northernmost occurrences are from southern Saskatchewan (Storer 1975, 2002). *Plesiosminthus clivosus* is from the Arikareean Kealey Springs Local Fauna (Storer 2002) and from the Hemingfordian Topham Local Fauna (Skwara 1988). *Megasminthus tiheni* from the Wood Mountain Fauna is contemporaneous with the Barstovian taxa of Nebraska (Storer 1975). The southernmost occurrence is recorded by "*Schaubeumys*"-type *Plesiosminthus* sp. in southern California (Raschke 1984; Reynolds 1991). In eastern North America, a single M2 (USNM 475851) of *Plesiosminthus* sp. was discovered in the Pollack Farm Local Fauna (18 Ma) of Delaware (Emry and Eshelman 1998). This specimen possesses a diagnostic feature of *P. meridionalis* (the paracone is connected to the posterior arm of the protocone). Apparently a Hemingfordian relic of *P. meridionalis*, this specimen displays a minor difference from *P. meridionalis* of the type locality in the absence of the endostyle.

CONCLUSION

Phylogenetic analyses utilizing Asian specimens reveal that North American *Schaubeumys* and *Macrognathomys* are synonymous with *Plesiosminthus* and *Sicista*, respectively, and that *Plesiosminthus* and *Sicista* dispersed from Asia to North America. *Megasminthus* belongs to the Sicistinae, rather than the Zapodinae. *Plesiosminthus* is paraphyletic if *Megasminthus* is maintained. The

most basal species of *Plesiosminthus* and *Sicista* (*P. clivosus* and *S. gemmacollis*) are found in North America. *Plesiosminthus admyarion* is closer to *P. schaubi* than to *P. myarion*. *Plesiosminthus* sp. from Gashunyinadege and *P. winistoerferi* possess the arm of the entoconid of m2 extending posterolabially, which is a homoplasy shared with *Sicista*. *Plesiosminthus meridionalis* is a distinct species rather than a subspecies of *Plesiosminthus schaubi*, based on two diagnostic features of *P. meridionalis* that are not observed in *P. schaubi* (short mesolophid on lower teeth and presence of protoloph II of M2). A monophyletic group of *Sicista* species, including North American taxa, is united by four unambiguous synapomorphies of the lower teeth: anteroconid connecting to metalophid, protoconid directed transversely, longitudinal ridge present, entoconid connecting to hypoconid.

The trans-Beringia route allowed intercontinental dispersals of sicistines, based on the phylogenetic results and geographic distribution of *P. promyarion* and *P. asiaticus*, as it did for many dispersal events during the Neogene. Intercontinental dispersal of *Plesiosminthus* took place perhaps three times via the Bering Land Bridge. Sisters of *P. clivosus* and *P. grangeri* reached North America by ~ 30 Ma and ~ 28 Ma, respectively. *Plesiosminthus grangeri* was the eighth member of a second-order immigration episode in the early Arikareean. The third event is represented by a single specimen of 18 million-year-old *P. meridionalis* recovered from eastern North America.

Asian species migrated to Europe twice. Dispersals of sicistines to Europe were not related to the Grande Coupure, as the first European appearance of *P. promyarion* is ~6 m.y. later at ~27 Ma (MP 26). A primitive sister taxon of *P. winistoerferi* also migrated westward. Dispersal timing of this event is uncertain.

Timing and direction of dispersals of *Sicista* are more uncertain than those of *Plesiosminthus*. Phylogenetic analyses indicate that an ancestor of *S. gemacollis* dispersed to North America, probably during the Hemingfordian, along with *P. meridionalis*. Asian-derived *Sicista* may be descended from North American species because occurrences of North American species are older than Asian species except for the oldest Asian *Sicista* from Gashunyinadege, Nei Mongol, China.

ACKNOWLEDGMENTS

This paper is a portion of a master's thesis. I would first and foremost like to thank Dr. Zhuding Qiu for exceptional opportunities and the academic support he has given me, Dr. Wenyu Wu for translating Ziegler (1994) into English and allowing me to access undescribed specimens of *Litodonomys*, and Dr. Banyue Wang for valuable discussion of early dipodids. I greatly thank Drs. Louis Jacobs, Yukimitsu Tomida, Dale Winkler, and Alisa Winkler for valuable comments on this manuscript, and Dr. Lawrence Flynn for providing us with his manuscript for "Evolution of Tertiary Mammals of North America." I sincerely appreciate Drs. Xiaoming Wang and Tao Deng for teaching me their field technique, and Drs. Qiang Li and Yingqi Zhang for providing me with tips to identify small mammals. This manuscript was greatly improved by careful comments and advice of an anonymous referee and Dr. Lawrence Flynn.

Financial support was provided by the Institute for the Study of Earth and Man, Southern Methodist University, the Geological Society of America, and Mikasa Museum, Hokkaido, Japan.

REFERENCES

Agustí, J., L. Cabrera, M. Garcés, W. Krijgsman, O. Oms, and J. M. Parés. 2001. A calibrated mammal scale for the Neogene of Western Europe. *Earth Science Reviews* 52:247–260.

Allen, J. A. 1901. Note on the names of a few South American mammals. *Proceedings of the Biological Society of Washington* 14:183–185.

Alvarez Sierra, M. A., R. Daams, and J. I. Lacomba Andueza. 1996. The rodents from the Upper Oligocene of Sayatón 1, Madrid Basin (Guadalajara, Spain). *Proceedings of the Koninklijke Nederlandse Akademie van Wetenschappen* 99:1–23.

Baskevich, M. I. 1996. On morphologically similar species in the genus *Sicista* (Rodentia, Dipodoidea). *Bonner Zoologische Beiträge* 46:133–140.

BiochroM'97. 1997. Synthèses et tableaux de corrélations. In *Actes du Congrès BiochroM'97*, ed. J.-P. Aguilar, S. Legendre, and J. Michaux, pp. 769–805. Mémoires et Travaux E.P.H.E., Institut de Montpellier 21.

Black, C. C. 1958. A new sicistine rodent from the Miocene of Wyoming. *Breviora* 86:1–7.

Bohlin, B. 1946. The fossil mammals from the Tertiary deposit of Taben-buluk, western Kansu, Part II: Simplicidentata, Carnivora, Artiodactyla, Perissodactyla, and Primates. *Palaeontologia Sinica*, n.s. C, no. 8B, Sino-Swedish Expedition Publication 28:1–259.

Carrasco, M. A., B. P. Kraatz, E. B. Davis, and A. D. Barnosky. 2005. *Miocene Mammal Mapping Project (MIOMAP)*. University of California Museum of Paleontology, http://www.ucmp.berkeley.edu/miomap/. Accessed February 26, 2009.

Chaline, J. 1972. *Les rongeurs du Pléistocène moyen et supérieur de France. Cahiers de Paléontologie*. Paris: Centre National de la Recherche Scientifique.

Comte, B. 2000. Rythme et modalités de l'évolution chez les rongeurs à la fin de l'Oligocène: Leurs relations avec les changements de l'environnement. *Palaeovertebrata* 29:83–360.

Daxner-Höck, G. 2001. New zapodids (Rodentia) from Oligocene-Miocene deposits in Mongolia, Part 1. *Senckenbergiana Lethaea* 81:359–389.

Daxner-Höck, G. and W.-y. Wu. 2003. *Plesiosminthus* (Zapodidae, Mammalia) from China and Mongolia: Migrations to Europe. In *Distribution and Migration of Tertiary Mammals in Eurasia: A Volume in Honour of Hans de Bruijn*, ed. J. W. F. Reumer and W. Wessels, pp. 127–151. DEINSEA 10.

de Bruijn, H., E. Ünay, G. Saraç, and A. Yïlmaz. 2003. A rodent assemblage from the Eo/Oligocene boundary interval near Süngülü, Lesser Caucasus, Turkey. *Coloquios de Paleontología* Extra 1:47–76.

Ellerman, J. R. 1940. *The Families and Genera of Living Rodents*. Vol. 1, *Rodents Other Than Muridae*. London: British Museum (Natural History).

Emry, R. J. and R. E. Eshelman. 1998. The early Hemingfordian (early Miocene) Pollack Farm local fauna: First Tertiary land mammals described from Delaware. In *Geology and Paleontology of the Lower Miocene Pollack Farm Fossil Site*, ed. R. N. Benson, pp. 153–173. *Delaware Geological Survey Special Publication* 21.

Engesser, B. 1979. Relationships of some insectivores and rodents from the Miocene of North America and Europe. *Bulletin of Carnegie Museum of Natural History* 14:1–68.

Engesser, B. 1987. New Eomyidae, Dipodidae, and Cricetidae (Rodentia, Mammalia) of the Lower Freshwater Molasse of Switzerland and Savoy. *Eclogae Geologicae Helvetiae* 80:943–994.

Engesser, B. and N. A. Mayo. 1987. A biozonation of the Lower Freshwater Molasse (Oligocene and Agenian) of Switzerland and Savoy on the basis of fossil mammals. *Münchner Geowissenschaftliche Abhandlungen* 10:67–84.

Erbajeva, M. A. 1976. Fossiliferous bunodont rodents of the Transbaikal area. *Geologiya i Geofizika* 17:144–149.

Fahlbusch, V., Z.-d. Qiu, and G. Storch. 1983. Neogene mammalian faunas of Ertemte and Harr Obo in Nei Monggol, China. Report on field work in 1980 and preliminary results. *Scientia Sinica*, ser. B, 26:205–224.

Fischer von Waldheim, G. 1817. Adversaria Zoologica. *Mémoires de la Société Impériale des Naturalistes de Moscou* 5:357–472.

Flynn, L. J. 2008. Dipodidae. In *Evolution of Tertiary Mammals of North America*, ed. C. M. Janis, G. F. Gunnell, and M. D. Uhen, vol. 2, pp. 406–414. New York: Cambridge University Press.

Flynn, L. J., R. H. Tedford, and Z.-x. Qiu. 1991. Enrichment and stability in the Pliocene mammalian fauna of North China. *Paleobiology* 17:246–265.

Galbreath, E. C. 1953. A contribution to the Tertiary geology and paleontology of northeastern Colorado. *University of Kansas Paleontological Contributions, Vertebrata* 4:1–120.

Gradstein, F., J. Ogg, and A. Smith. 2004. *A Geologic Time Scale 2004*. New York: Cambridge University Press.

Gray, J. E. 1827. Synopsis of the species of the Class Mammalia, as arranged with reference to their organization, by Cuvier, and other naturalists, with specific characters, synonyma, etc. In *The Animal Kingdom Arranged in Conformity with Its Organization, by the Baron Cuvier, with Additional Descriptions of All the Species*

Hitherto Named, and of Many Not Before Noticed, ed. E. Griffith, C. H. Smith, and E. Pidgeon, vol., 5, pp. 207–273. London: G. B. Whittaker.

Green, M. 1977. Neogene Zapodidae (Mammalia, Rodentia) from South Dakota. *Journal of Paleontology* 51:996–1015.

Green, M. 1992. Comments on North American fossil Zapodidae (Rodentia: Mammalia) with reference to *Megasminthus, Plesiosminthus,* and *Schaubeumys. Occasional Papers of the Museum of Natural History, the University of Kansas* 148:1–11.

Hall, E. R. 1930. Rodents and lagomorphs from the later Tertiary of Fish Lake Valley, Nevada. *Bulletin of the Department of Geological Sciences, University of California Publications* 19:295–312.

Holden, M. E. and G. G. Musser. 2005. Family Dipodidae. In *Mammal Species of the World: A Taxonomic and Geographic Reference,* ed. D. E. Wilson and D. M. Reeder, pp. 871–893. Baltimore: Johns Hopkins University Press.

Hooker, J. J., M. E. Collinson, and N. P. Sille. 2004. Eocene-Oligocene mammalian faunal turnover in the Hampshire Basin, UK: Calibration to the global time scale and the major cooling event. *Journal of Geological Society, London* 161:161–172.

Huang, H.-s. 1992. Zapodidae (Rodentia, Mammalia) from the middle Oligocene of Ulantatal, Nei Mongol. *Vertebrata PalAsiatica* 30:249–286.

Hugueney, M. 1969. Les rongeurs (Mammalia) de l'Oligocène supérieur de Coderet-Bransat (Allier). *Documents du Laboratoire de Géologie de la Faculté des Sciences de Lyon* 34:1–227.

Hugueney, M. and M. Vianey-Liaud. 1980. Les Dipodidae (Mammalia, Rodentia) d'Europe occidentale au Paléogène et au Néogène inférieur: Origine et évolution. *Palaeovertebrata Mémoire jubilaire en homage à R. Lavocat,* pp. 303–342.

Jaksic, F. M., J. A. Iriarte, J. E. Jimènez, and D. R. Martínez. 2002. Invaders without frontiers: Cross-border invasions of exotic mammals. *Biological Invasions* 4:157–173.

Kimura, Y. 2010. New material of dipodid rodents (Dipodidae, Rodentia) from the Early Miocene of Gashunyinadege, Nei Mongol, China. *Journal of Vertebrate Paleontology* 30:1860–1873.

Kimura, Y. 2011. The earliest record of birch mice from the Early Miocene Nei Mongol, China. *Naturwissenschaften* 98:87–95.

Klingener, D. 1964. The comparative myology of four dipodoid rodents (genera *Zapus, Napaeozapus, Sicista,* and *Jaculus*). *Miscellaneous Publications, Museum of Zoology, University of Michigan* 124:1–100.

Klingener, D. 1966. Dipodid rodents from the Valentine Formation of Nebraska. *Occasional Papers of the Museum of Zoology, University of Michigan* 644:1–9.

Klingener, D. 1984. Gliroid and dipodoid rodents. In *Orders and Families of Recent Mammals of the World,* ed. S. Anderson and J. K. Jones Jr, pp. 381–388. New York: Wiley.

Korth, W. W. 1980. Cricetid and zapodid rodents from the Valentine Formation of Knox County, Nebraska. *Annals of Carnegie Museum* 49:307–322.

Korth, W. W. 1987. New rodents (Mammalia) from the late Barstovian (Miocene) Valentine Formation, Nebraska. *Journal of Paleontology* 61:1058–1064.

Korth, W. W. 1993. *Miosicista angulus,* a new sicistine rodent (Zapodidae, Rodentia) from the Barstovian (Miocene) of Nebraska. *Transactions of the Nebraska Academy of Sciences* 20:97–101.

Korth, W. W. 1994. *The Tertiary Record of Rodents in North America.* New York: Plenum Press.

Kowalski, K. 1979. Fossil Zapodidae (Rodentia, Mammalia) from the Pliocene and Quaternary of Poland. *Acta Zoologica Cracoviensia* 23:199–212.

Kowalski, K. 2001. Pleistocene rodents of Europe. *Folia Quaternaria* 72:1–389.

Legendre, S. and F. Lévêque. 1997. Étalonnage de l'échelle biochronologique mammalienne du Paléogène d'Europe occidentale: vers une intégration à l'échelle globale. In *Actes du Congrès BiochroM'97,* ed. J.-P. Aguilar, S. Legendre, and J. Michaux, pp. 769–805. Mémoires et Travaux E.P.H.E., Institut de Montpellier 21.

Li, Q. 2006. Pliocene rodents from the Gaotege Fauna, Nei Mongol (Inner Mongolia). Ph.D. diss., Graduate School of the Chinese Academy of Sciences, Beijing, China (in Chinese).

Lopatin, A. V. 1999. New Early Miocene Zapodidae (Rodentia, Mammalia) from the Aral Formation of the Altynshokysu Locality (North Aral Region). *Paleontological Journal* 33:429–438 (in Russian in *Paleontologicheskii Zhurnal* 1999:93–102).

López-Antoñanzas, R. and S. Sen. 2006. New Saudi Arabian Miocene jumping mouse (Zapodidae): Systematics and phylogeny. *Journal of Vertebrate Paleontology* 26:170–181.

Martin, J. E. 1976. Small mammals from the Miocene Batesland Formation of South Dakota. *Contributions to Geology, the University of Wyoming* 14(2):69–98.

Martin, L. D. 1974. New rodents from the lower Miocene Gering Formation of western Nebraska. *Occasional Papers of the Museum of Natural History, the University of Kansas* 32:1–12.

Martin, R. A. 1989. Early Pleistocene zapodid rodents from the Java local fauna of north-central South Dakota. *Journal of Vertebrate Paleontology* 9:101–109.

Martin, R. A. 1994. A preliminary review of dental evolution and paleogeography in the zapodid rodents, with emphasis on Pliocene and Pleistocene taxa. In *Rodent and Lagomorph Families of Asian Origins and Diversification,* ed. Y. Tomida, C.-K. Li, and T. Setoguchi, pp. 99–113. National Science Museum Monographs 8. Tokyo: National Science Museum.

McKenna, M. C. and S. K. Bell. 1997. *Classification of Mammals: Above the Species Level.* New York: Columbia University Press.

Mein, P. 1999. European Miocene mammal biochronology. In *The Miocene Land Mammals of Europe,* ed. G. E. Rössner and K. Heissig, pp. 25–38. Munich: Dr. Friedrich Pfeil.

Nowak, R. M. 1999. *Walker's Mammals of the World.* 6th ed. Baltimore: Johns Hopkins University Press.

Qiu, Z.-d. and C.-k. Li. 2003. Rodents from the Chinese Neogene: Biogeographic relationships with Europe and North America. *Bulletin of the American Museum of Natural History* 279:586–602.

Qiu, Z.-d. and G. Storch. 2000. The early Pliocene micromammalian fauna of Bilike, Inner Mongolia, China (Mammalia: Lipotyphla, Chiroptera, Rodentia, Lagomorpha). *Senckenbergiana Lethaea* 80:173–229.

Qiu, Z.-d. and X.-m. Wang. 1999. Small mammal faunas and their ages in Miocene of central Nei Mongol (Inner Mongolia). *Vertebrata PalAsiatica* 37:120–139.

Qiu, Z.-d., X.-m. Wang, and Q. Li. 2006. Faunal succession and biochronology of the Miocene through Pliocene in Nei Mongol (Inner Mongolia). *Vertebrata PalAsiatica* 44:164–181.

Raschke, R. E. 1984. Stratigraphy and paleontology of the Upper Oso Dam area, Orange County, California. M.S. thesis, California State University, Long Beach.

Reynolds, R. E. 1991. Biostratigraphic relationships of Tertiary small vertebrates from Cajon Valley, San Bernardino County, California. *SBVMA Quarterly* 38:54–59.

Rögl, F. 1999. Mediterranean and Paratethys: Facts and hypotheses of an Oligocene to Miocene paleogeography (short overview). *Geologica Carpathica* 50:339–349.

Savinov, P. R. 1970. Jerboas (Dipodidae, Rodentia) from the Neogene of Kazakhstan. *Byulleten' Moskoskovo Obshchestva Ispytatelei Prirody, Otdel Biologicheskii* 1970:91–134 (in Russian).

Schaub, V. S. 1930. Fossile Sicistinae. *Eclogae geologicae Helvetiae* 23:616–637.

Schlosser, M. 1924. Tertiary vertebrates from Mongolia. *Palaeontologia Sinica*, ser. C, 1:1–133.

Schlunegger F., D. W. Burgank, A. Matter, B. Engesser, C. Mödden. 1996. Magnetostratigraphic calibration of the Oligocene to Middle Miocene (30–15 Ma) mammal biozones and depositional sequences of the Swiss Molasse Basin. *Eclogae geologicae Helvetiae* 89:753–788.

Shenbrot, G. I. 1992. Cladistic approach to analysis of phylogenetic relationships among Dipodoidea (Rodentia; Dipodoidea). *Archive of the Zoological Museum of Moscow State University* 29:176–200 (in Russian).

Shotwell, J. A. 1968. Miocene mammals of southeast Oregon. *Bulletin of the Museum of Natural History, University of Oregon* 14:1–67.

Shotwell, J. A. 1970. Pliocene mammals of southeast Oregon and adjacent Idaho. *Bulletin of the Museum of Natural History, University of Oregon* 17:1–103.

Simpson, G. G. 1945. The principles of classification and a classification of mammals. *Bulletin of the American Museum of Natural History* 85:1–350.

Skwara, T. 1988. Mammals of the Topham Local Fauna: Early Miocene (Hemingfordian), Cypress Hills Formation, Saskatchewan. *Saskatchewan Museum of Natural History, Natural History Contributions* 9:1–169.

Sokolov, V. E., M. I. Baskevič, and Y. M. Kovalskaya. 1986. *Sicista kazbegica* sp. n. (Rodentia, Dipodidae) from the basin of the Terek River upper reaches. *Zoologicheskii Zhurnal* 65:949–952 (in Russian).

Sokolov, V. E., Y. M. Kovalskaya, and M. I. Baskevič. 1987. Review of karyological research and the problems of systematics in the genus *Sicista* (Zapodidae, Rodentia, Mammalia). *Folia Zoologica* 86:35–44.

Stehlin, H. G. 1909. Rémarques sur les faunules de mammifères des couches éocènes et oligocènes de Bassin de Paris. *Bulletin de la Société Géologique de France* 9:488–520.

Stein, B. R. 1990. Limb myology and phylogenetic relationships in the superfamily Dipodoidea (birch mice, jumping mice, and jerboas). *Zeitschrift für Zoologische Systematik und Evolutionsforschung* 28:299–314.

Storer, J. E. 1975. Tertiary mammals of Saskatchewan; Part III, the Miocene fauna. *Life Sciences Contributions, Royal Ontario Museum* 103:1–134.

Storer, J. E. 2002. Small mammals of the Kealey Springs Local Fauna (early Arikareean; late Oligocene) of Saskatchewan. *Paludicola* 3:105–133.

Swofford, D. L. 2002. *PAUP*: Phylogenetic Analysis Using Parsimony (*and Other Methods)*, Version 4. Sunderland: Sinauer Associates.

Tedford, R. H., L. B. Albright III, A. D. Barnosky, I. Ferrusquia-Villagranca, R. M. Hunt Jr., J. E. Storer, C. C. Swisher III, M. R. Voorhies, S. D. Webb, and D. P. Whistler. 2004. Mammalian biochronology of the Arikareean through Hemphillian interval (Late Oligocene through early Pliocene epochs). In *Late Cretaceous and Cenozoic Mammals of North America*, ed. M. O. Woodburne, pp. 169–231. New York: Columbia University Press.

Tedford, R. H., T. Galusha, M. F. Skinner, B. E. Taylor, R. W. Fields, J. R. Macdonald, J. M. Rensberger, S. D. Webb, and D. P. Whistler. 1987. Faunal succession and biochronology of the Arikareean through Hemphillian interval (late Oligocene through earliest Pliocene Epochs), North America. In *Cenozoic Mammals of North America: Geochronology and Biostratigraphy*, ed. M. O. Woodburne, pp. 153–210. Berkeley: University of California Press.

Tedford, R. H., J. B. Swinehart, C. C. Swisher III, D. R. Prothero, S. A. King, and T. E. Tierney. 1996. The Whitneyan–Arikareean transition in the High Plains. In *The Terrestrial Eocene–Oligocene Transition in North America*, ed. D. R. Prothero and R. J. Emry, pp. 312–334. New York: Cambridge University Press.

Topachevskii, V. A. 1965. *Nasekomoyadnye i gryzuny nogaiskoi pozdnepliotsenovoi fauny*. Kiev: Naukova Dumka (in Russian).

van der Made, J. 1997. Intercontinental dispersal events, eustatic sea level and Early and Middle Miocene stratigraphy. In *Actes du Congrès BiochroM'97*, ed. J.-P. Aguilar, S. Legendre, and J. Michaux, pp. 769–805. Mémoires et Travaux E.P.H.E., Institut de Montpellier 21.

Vinogradov, B. S. 1930. On the classification of Dipodidae (Rodentia). I. Cranial and dental characters. *Izvestiya Akademii Nauk SSSR* 1930:331–350.

Viret, J. 1926. Nouvelles observations relatives à la faune de rongeurs de Saint-Gérand-le-Puy. *Comptes Rendus de l'Académie des Sciences, Paris* 183:71–72.

Wang, B.-y. 1985. Zapodidae (Rodentia, Mammalia) from the Lower Oligocene of Qujing, Yunnan, China. *Mainzer geowissenschaftliche Mitteilungen* 14:345–367.

Wang, B.-y. and Z.-x. Qiu. 2000. Dipodidae (Rodential, Mammalia) from the lower member of Xianshuihe Formation in Lanzhou Basin, Gansu, China. *Vertebrata PalAsiatica* 38:10–35.

Webb, S. D. and N. D. Opdyke. 1995. Global climatic influence on Cenozoic land mammal faunas. In *Effects of Past Global Change on Life: Studies in Geophysics*, ed. J. Kennett and S. Stanley, pp. 184–208. Washington, D. C.: National Academy Press.

Whybrow, P. J., M. E. Collinson, R. Daam, A. W. Gentry, and H. A. McClure. 1982. Geology, fauna (Bovidae, Rodentia) and flora from the early Miocene of eastern Saudi Arabia. *Tertiary Research* 4:105–120.

Wilson, R. W. 1960. Early Miocene rodents and insectivores from northeastern Colorado. *University of Kansas Paleontological Contributions, Vertebrata* 7:1–92.

Wilson, R. W. 1968. Insectivores, rodents and intercontinental correlation of the Miocene. *23th International Geological Congress* 10:19–25.

Woodburne, M. O. 2004. *Late Cretaceous and Cenozoic Mammals of North America*. New York: Columbia University Press.

Woodburne, M. O. and C. C. Swisher III. 1995. Land mammal high-resolution geochronology, intercontinental overland dispersals, sea level, climate, and variance. In *Geochronology, Time Scales,* and *Global Stratigraphic Correlations: A Unified Framework for a Historical Geology,* ed. W. A. Berggren, D. V. Kent, M. P. Aubry, and J. Hardenbol, pp. 335–364. Society for Sedimentary Geology Special Publication 54. Tulsa: Society for Sedimentary Geology.

Ye, J., J. Meng, and W.-y. Wu. 2003. Oligocene/Miocene beds and faunas from Tieersihabahe in the northern Junggar Basin of Xinjiang. *Bulletin of the American Museum of Natural History* 279:568–585.

Zazhigin, V. S. and A. V. Lopatin. 2002. The history of the Dipodoidea (Rodentia, Mammalia) in the Miocene of Asia: 6. Lophodont Lophocricetinae. *Paleontological Journal* 36:385–394 (translated from *Paleontologicheskii Zhurnal* 2002:62–71).

Zhang, R., L.-p. Yue, and J.-q. Wang. 2007. Magnetostratigraphic dating of mammalian fossils in Junggar Basin, northwest China. *Chinese Science Bulletin* 52:1526–1531.

Ziegler, R. 1994. Rodentia (Mammalia) aus den oberoligozänen Spaltenfüllungen Herrlingen 8 und Herrlingen 9 bei Ulm (Baden–Württemberg). *Stuttgarter Beiträge zur Naturkunde,* ser. B (Geologie und Paläontologie) 196:1–81.

APPENDIX 1

Description of Characters and Character States Employed in the Phylogenetic Analysis of *Plesiosminthus*, *Megasminthus*, and *Sicista*

UPPER DENTITION

1. Longitudinal groove [WQ]: absent (0); present (1)
2. Shape of endosinus: symmetrical (0); anteriorly oblique (1)
3. Endosinus: closed (0); open to the anterior fossette (1)
4. Posteroloph: continuous with the hypocone (0); discontinuous (1)

UPPER M1

5. Enamel lake of M1: absent (0); one or two (1); three or more (2)
6. Anteroloph [= anterior arm of protocone, WQ]: absent or weak (0); short, incomplete (1); long, complete (2)
7. Anterocone: absent (0); present as a swollen part of the anteroloph (1); distinct from the anteroloph (2)
8. Protoloph I [modified from LS]: located on the labial side of the anterior arm of the protocone (0); connecting to the anterocone or, if the anterocone is absent, located on the lingual side of the anterior arm of the protocone (1); absent (2)
9. Protoloph II: absent (0); connecting to the center of the protocone (= protoloph) (1); attaching to the posterior arm of the protocone (2); connecting to the mesocone (3)

10. Middle fossette: open (0); closed by a strong longitudinal ridge (1)
11. Mesoloph located: closer to the hypocone (0); at the middle of the endoloph (1)
12. Endostyle: absent (0); present (1)
13. Metaloph [modified from WQ and LS]: straight, connecting to the center of the metacone (0); curved, connecting to the posterolabial corner of the metacone (1); metaloph II (2)
14. Metacone: rounded or less compressed (0); anteroposteriorly compressed, forming a loph with the metaloph (1)

UPPER M2

15. Protoloph II [modified from WQ and LS]: absent (0); if present, incomplete (1); completely present, connecting to the posterior arm of the protocone (2); completely present, connecting to the mesocone (3)
16. Posterior concavity behind hypocone [modified from WQ and LS]: absent (0); present (1)

LOWER M1

17. Anteroconid [modified from LS]: isolated from the metaconid and protoconid (0); located on the base of

NOTE: Characters taken or modified from Wang and Qiu (2000) and López-Antoñanzas and Sen (2006) are indicated as WQ and LS in square brackets, respectively. All characters are equally unordered and unweighted.

the metaconid (1); connecting to the metalophid (2); double anteroconid (3)

18. Protoconid directed: posterolabially (0); transversely (1)

19. Anterior ectolophid connecting to [modified from WQ]: protoconid (0); absent or no connection (1); metalophid (2)

20. Ectostylid: absent (0); present (1)

21. Mesolophid [modified from WQ]: absent or short (0); present as pseudomesolophid (1); long, weaker and lower than the ecotolophid (2); long, as high as the ectolophid (3)

22. Arm of entoconid extending: transversely or anterolabially (0); posterolabially (1)

23. Longitudinal ridge between posterolophid and hypoconid: absent (0); present (1)

24. Hypoconulid: absent (0); present (1)

25. Accessory spur into posterior fossette: absent (0); present, extending from the hypoconulid (1); present, extending from the hypoconid (2)

LOWER M2

26. Location of metaconid: opposite or slightly anterior to the protoconid (0); in line with the anteroconid (1)

27. Metaconid connecting to [modified from WQ]: anterior arm of the protoconid (0); anteroconid (1); protoconid (2)

28. Posterior arm of protoconid: complete, connecting to the metaconid or incomplete, extending to the base of posterior wall of the metaconid (0); connecting to the mesostylid (=psuedomesolophid, LS) (1); connecting to the mesolophid (2); absent (3)

29. Arm of entoconid extending: transversely or anterolabially (0); posterolabially (1)

30. Entoconid connecting to: anterior arm of the hypoconid (0); hypoconid (1)

31. Mesolophid [WQ and LS]: absent or weak (0); short (1); long (2)

LOWER M3

32. Posterior arm of protoconid or mesolophid: either one present (0); both present (1); both absent (2)

33. Arm of entoconid extending [modified from WQ]: transversely or anterolabially (0); posterolabially (1); absent (2)

APPENDIX 2

Character Matrix for Phylogenetic Analysis of *Plesiosminthus, Megasminthus,* and *Sicista*

	5	10	15	20	25	30	
Allosminthus ernos	00000	00000	00000	00000	0A0A0	00000	00B
Heosminthus primiveris	01000	10220	10011	00000	00010	00000	200
Plesiosminthus clivosus	?1000	21220	10011	00000	30000	00CA0	20A
Plesiosminthus promyarion	11010	D1220	10011	0000A	300A0	00000	200
Plesiosminthus conjunctus	11010	21220	10012	000A0	30000	00000	200
Plesiosminthus myarion	?1010	21220	10012	00000	30010	00000	200
Plesiosminthus schaubi	11010	21220	11000	01000	30010	00300	220
Plesiosminthus meridionalis	?1010	11220	11B12	01001	C0010	00300	120
Plesiosminthus moralesi	?1010	20220	11010	000A0	30010	00300	D02
Plesiosminthus admyarion	?1010	D12D0	11010	03001	3001B	00300	200
Plesiosminthus sp. from Gashunyinadege	?1000	21220	10012	0000A	3A0A0	00210	20A
Plesiosminthus barsboldi	11000	21220	10012	0A000	30010	01000	211
Plesiosminthus winistoerferi	11000	21220	10012	000A0	31001	01110	21A
Plesiosminthus asiaticus	11000	22220	10002	000A0	30010	00000	200
"Schaubeumys" grangeri	11000	212F0	10102	00010	30010	01100	00A
"Schaubeumys" sabrae	?1000	22230	10100	01010	30010	01300	200
Megasminthus gladiofex	?1100	22230	10103	00021	30010	03300	200
Megasminthus tiheni	?1110	22230	10103	00021	30010	01300	200
"Macrognathomys" gemmacollis	?000A	D222?	1001?	?2120	F0102	00311	201
Sicista sp. from Gashunyinadege	00000	12210	10011	12120	31102	1B111	2A2
Sicista sp. from Gaotege	00001	12211	A0010	12120	1110B	12E11	201
"Macrognathomys" sp. from the Bartlette Mountain Local Fauna	?0001	1221?	10010	1212A	21102	12311	201
Sicista wangi	00000	12101	10010	12121	2110B	12311	202
Sicista praeloriger	00001	02101	10010	12120	2110B	AB311	202
Sicista bagajevi	00002	12101	10011	12120	21100	12311	201
Sicista sp. from Bilike	00001	12101	01011	12120	21102	1B311	20D
Sicista betulina	00002	02101	00011	12130	2110B	1B111	002
Sicista concolor	00001	12211	10012	12120	21102	1B311	201
Sicista subtilis	00000	12211	10010	12120	21100	12311	222

NOTE: Characters are listed in appendix 1. Missing or unknown data are coded as ?, A, B, C, D, E, and F are designated for polymorphic character states, 0/1, 0/2, 0/3, 1/2, 1/3, and 2/3, respectively.

APPENDIX 3

Museum Specimens Examined for This Study

Species	Museum	Catalog Number	Reference
Plesiosminthus galbreathi	CM	36543 (cast of KU10250)	
Plesiosminthus sabrae	CM	13527, 13529, 13943-13948, 13950, 14774, 14967, 149678, 15854, 15927, 159278, 18809–10	Engesser (1979)
Plesiosminthus schaubi	CM	10932	
Allosminthus ernos	IVPP	V7632.1–41, V7633	Wang (1985)
Heosminthus primiveris	IVPP	V7630.1–30	Wang (1985)
Plesiosminthus barsboldi	IVPP	V15886 to V15889	Kimura, this study (2010)
Plesiosminthus sp. from Gashunyinadege	IVPP	V15882 to V15885	Kimura, this study (2010)
Sicista sp. from Gashunyinadege	IVPP	V15896 to V15899	Kimura, this study (2010, 2011)
Sicista sp. from Bilike	IVPP	V11905.1–718	Qiu and Storch (2000)
Sicista sp. from Gaotege	IVPP	Catalog number not given yet	Li (2006)
Sicista wangi	IVPP	V11930, V11931.1–130	Qiu and Storch (2000)
Megasminthus gladiofex	SDSM	7971, 7997-8009	Green (1977)
Plesiosminthus clivosus	SDSM	7815–19, 7824, 7826, 7827, 7850	Martin (1976); Green (1977)
Plesiosminthus grangeri	SDSM	7952, 7954, 7956, 7958, 8117, 67144, 67148, 67150, 67153, 67154	Green (1977)
Sicista betulina	USNM	257391, 254986	
Sicista concolor concolor	USNM	173796–7, 173800, 173806, 200398	
Sicista concolor flava	USNM	173790, 173792, 173794, 173795	
Sicista concolor leathemi	USNM	354392, 354393, 354386, 354388	
Sicista subtilis nordmanni	USNM	122117 to 122119	

NOTE: CM = Carnegie Museum of Natural History, Pittsburgh, Pennsylvania; IVPP = Institute of Vertebrate Paleontology and Paleoanthropology, Beijing, China; SDSM = South Dakota School of Mines and Technology, Museum of Geology, Rapid City; USNM = Smithsonian Institution, National Museum of Natural History, Washington, D.C.

Chapter 31

Paleodietary Comparisons of Ungulates Between the Late Miocene of China, and Pikermi and Samos in Greece

NIKOS SOLOUNIAS, GINA M. SEMPREBON, MATTHEW C. MIHLBACHLER, AND FLORENT RIVALS

Extant ungulates occupy a great diversity of Recent habitats such as forests, woodlands, savannas, grasslands, steppe, deserts, steep mountain slopes, and arctic tundra. They also exhibit a diverse array of morphological adaptations that are suited to these varied ecological conditions. Many modern ungulates, particularly ruminants, are descended from lineages that adaptively radiated during the Middle and Late Miocene, a time of dramatic climate cooling during which open grassy and more arid habitats appeared and began to spread. The C_4 grasslands had a sudden rise and spread to ecological dominance 8–3 Ma (Edwards et al. 2010). The new Late Miocene open habitats intermingled with more archaic closed habitats and have now largely replaced them in many parts of continents. All of the early mammalian studies (1900–1970s) documenting the opening of habitats are based on the appearance of new taxa of fossil ungulates. Isotopic and tooth microwear and mesowear approaches did not exist then. Paleobotanical studies have also documented patterns regarding these changes (Dorofeyev 1966; Ortega 1979; Mai 1995). Recently, paleobotanical and isotopic studies have become more focused and have shown the complexity of these paleoecological events (Quade, Solounias, and Cerling 1994; Retallack 2001; Strömberg et al. 2007; Bond 2008; Zachos, Dickens, and Zeebe 2008). It is undisputed that such habitats did open somewhat during the later Miocene, with an onset at around 8 Ma. These new open habitats represented untapped ecological conditions within which many ungulate groups expanded, diversified, and adapted (Jacobs

et al. 1999; Janis, Damuth, and Theodor 2000; Bond 2008; Janis 2008). In addition, the evolution of hypsodonty has recently been used as proxy for the degree of aridity or tree cover in various extinct habitats (Fortelius et al. 2003; Eronen et al. 2010a, 2010b). Overall these paleoecological events need more research.

The paleobotanical record of the Late Miocene (11–6 Ma) indicates the persistence of forests and woodlands along with the spreading of grasslands (Axelrod 1975; Mai 1995; Retallack 2001; Bruch et al. 2006; Bruch, Uhl, and Mosbrugger 2007). In other words, habitats during this time interval formed a widespread grassland/savanna/woodland continuum over much of the northern continents. The ecological changes that occurred during the Late Miocene and the complex ecological relationships that ungulates had with these evolving habitats, make the study of ungulate paleobiology both challenging and interesting.

During the end of the Miocene and through the Pliocene, the climate of the peri-Mediterranean Sea changed from a warm wet subtropical to a drier and cooler temperate climate (Dorofeyev 1966; Axelrod 1975; Fortelius et al. 2003; Strömberg et al. 2007; Eronen et al. 2010a, 2010b; Solounias, Rivals, and Semprebon 2010). The Pikermi, Samos, and Maragheh faunas fall at a critical time in the cooling trend of this climatic change and probably were temperate faunas. Considered as a fauna, or chronofauna, the late Miocene of China (hereafter denoted LMC) samples more time depth, but parts of it were near the time of Pikermi and Samos. The modern

East African savanna has been used as the primary model for the late Miocene ungulate-rich ecosystems of the northern continents, within which hypsodonty in many ungulates evolved (Osborn 1910; Matthew 1926; Abel 1927; Kurtén 1952, 1971; de Bonis et al. 1992; Coppens 1994). In this model, the woodland component of the savanna was not considered. An emphasis on open grassy savanna plains is apparent in murals and other reconstructions of the time. Other ecologies such as the woodlands of India and of Europe were not considered sufficiently as alternative model biomes for the Late Miocene. Kurtén (1952) represented the ideas and paradigms of many, and he proposed the presence of a vast open and arid region (spanning seven time zones) extending from Greece to China. The periphery of that steppe was hypothesized to have been surrounded by more forested habitats. In Kurtén's study, particularly referencing his model map, there were three types of faunas: a vast steppe fauna, a peripheral forest fauna, and a transitional fauna bridging the steppe and forest habitats. Pikermi in Greece and Hsin-An-Hsien and Wu-Hsiang-Hsien in the Shanxi Province of China were proposed to be transitional localities, and Samos was part of the steppe (Kurtén 1952).

It has been proposed that many of these late Miocene faunas represent a geographically widespread extinct biome, the Pikermian Biome, within which the dominant habitats were closed woodlands (Solounias et al. 1999). The physiognomy of the Pikermian Biome was different from the vast expanses of open savanna and grassland in present-day East Africa. In reviewing the paleobotanical record, Solounias and colleagues (Solounias et al. 1999; Solounias, Rivals, and Semprebon 2010) found that the woodland aspect of the savanna is more similar in plant physiognomy to the later Miocene. The Pikermian Biome was widespread during the late Miocene from Greece through Bulgaria, Ukraine, Turkey, Southern Russia, Iran, and east all the way to China (Deng 2006). Taxa from this biome have also been found in southern France, Spain, and North Africa (Thenius 1959). The cohesion of a single Pikermian Biome is most strongly supported by the faunal similarities of the Late Miocene faunas from China to Greece, Spain, and North Africa (Bernor, Koufos et al. 1996; Solounais et al. 1999; Solounais, Rivals, and Semprebon 2010). The biome is characterized by *Indarctos*, marchairodontids, proboscideans, and chalicotheres. It also includes many species of hyaenids, rhinos, equids, and giraffids unlike those of the African savanna. The bovids are also numerous. Advanced Miocene hypsodont bovids have small masseter muscles; their masticatory morphology resembles extant Tragelaphini (Solounias, Moelleken, and Plavcan 1995). In many cases,

the very same species are found in late Miocene of China (LMC) and at Pikermi and Samos.

The modern Savanna Biome includes open grassy plains, low-density tree cover woodlands, and densely treed woodlands (Beerling and Osborne 2005; Judith Harris, pers. comm. 2010). Herbivore dietary interpretations by Solounias, Rivals, and Semprebon (2010) based on analyses of ungulate tooth microwear suggest that Pikermi and Samos were most likely a woodland mosaic with more extensive tree cover with fewer and/or smaller patches of grassy open habitat in comparison to the African open savanna. Pikermian Biome habitats provided a diversity of opportunities for species that depended on browsing as well as species that grazed. Savanna C_4 grasses appear to have been subdominant to woody plants and to C_3 grasses that would have grown in shaded areas of the woodland, glades, and margins of water. The ungulate components of the LMC Pikermi and Samos faunas were more species rich and more diverse in diet than the ungulates observed in modern African forests, woodlands, or savannas, yet dietarily most similar to the ungulates found in the woodlands of India (based on tooth microwear; unpublished data). It is unlikely that the Pikermi and Samos ungulates inhabited dense forests, because there is no microwear evidence for concentrated fruit browsing. Conversely, widespread grasslands are unlikely because many mixed feeders are present as well as browsers. Koufos, Kostopoulos, and Merceron (2009) agreed in an additional study of mesowear and microwear for Samos and found an environment of "open bushland with a thick grassy herbaceous layer."

In this study, we use tooth mesowear analysis to compare the paleodietary patterns of the classic Late Miocene faunas of Pikermi and Samos from Greece to those of China. A new scale is used for evaluating mesowear, a technique that uses the sharpness of molar cusp apices as a proxy for dietary abrasion. Details of this mesowear scale can be found in Mihlbachler et al. (2011). Animals with high-abrasion diets, particularly grazers, have high incidences of rounded and blunted cusp apices with low relief, while animals with low-abrasion diets, particularly folivorous browsers, have higher incidences of sharpened cusp apices with high relief. Specifically, this study evaluates similarity in mesowear and in hypsodonty between Pikermi, Samos, and the LMC, which record key samples of species of the Pikermian Biome. It tests the integrity of the biome and how the Pikermian Biome compares to the African savanna—the only biome today that rivals these Late Miocene faunas in ungulate diversity. These comparisons are based on mesowear and hypsodonty.

GEOLOGICAL AND BIOSTRATIGRAPHIC SETTING

The Pikermi localities are stratigraphically and geographically very confined and best approximate a fauna. The localities are concentrated in a small area: 0.2 km² with a sedimentary thickness of 40 m. The sediments are relatively horizontal and all samples come from the exposures of a small ravine (Megalo Rema; Gaudry 1862–1867; Abel 1927). There are no volcanic deposits at Pikermi, and though the radiometric age is not known, Pikermi is considered to be of Turolian age.

The Samos samples are mixed from six primary localities. Older collectors did not record locations during their excavations, which resulted in the mixing of fossils from different stratigraphic levels. The bone beds are concentrated on the upper part of the Main Bone Bed Member of the Mytilini Formation. The geographic area is small (2 km²) (Solounias 1981; Weidmann et al. 1984; Koufos and Nagel 2009). The estimated overall time span for these bone beds is approximately 800,000 yr for the core Samos faunas (our data). An average radiometric age of 7.2 Ma is given for the upper part of these sediments (the Main Bone Beds, Turolian; present study). Fossils were derived from small depression troughs where some bone beds may have taken thousands of years to form. Thus, unlike Pikermi, Samos represents composite faunas.

The data from the LMC are preliminary. The samples are mixed from numerous localities and time intervals and definitely do not represent a single fauna. The time depth is at least 4 million yr and the geographic area is very large. The LMC, as represented by our samples, is a proxy for a chronofauna.

RELATIVE AGE OF PIKERMI BASED ON THE SAMOS SPECIES

Evidence for the relative ages of Pikermi versus Samos comes from the diversity of eight groups of herbivores (for general fauna lists, see Solounias 1981; Bernor, Solounias, et al. 1996; and Koufos and Nagel 2009). A few species listed here are useful in the relative dating of Pikermi, but for some we have no mesowear data. Not all of these species were used in the dietary interpretations that follow.

1. One species of *Aceratherium* at Pikermi versus four *Chilotherium* species at Samos (the Samos taxon cluster is more derived).

2. Two species of hipparionine at Pikermi (*Cremohipparion mediterraneum* and *Hippotherium brachypus*) versus six at Samos (*Cremohipparion proboscideum, Cremohipparion matthewi, Cremohipparion nikosi, Hippotherium* sp., *Hipparion dietrichi,* and "*Hipparion*" sp. (R. L. Bernor, pers. comm. 2010).

The numerous horses at Samos follow a "punctuated" event in the Subparatethyan Province recorded in the latest Vallesian levels of Sinap, Turkey (Bernor et al. 2003; Scott et al. 2003), in which several superspecific clades originated across the geographic extent of the Pikermian Biome (Eronen et al. 2009). Thus, Samos is younger than this punctuated event. The Pikermi species *Hippotherium brachypus* is more primitive than, and could have evolved into, the more derived Samos form of *Hippotherium* sp. (Bernor, Koufos, et al. 1996). The same is plausible for Pikermi *Cremohipparion mediterraneum* and the more derived form *Cremohipparion proboscideum* (Bernor, pers. comm. 2010). The paucity of grazing species such as equids at Pikermi suggests that the habitat was more closed and that Pikermi may be older than Samos (Solounias, Rivals, and Semprebon 2010).

3. *Tragoportax rugosifrons* and *T. curvicornis* from Samos can be derived from the more primitive *T. amalthea* found at Pikermi (Solounias 1981). The presence of *Tragoportax amalthea* at both localities, however, does suggest that they were close in age.

4. *Protragelaphus skouzesi* is found both at Pikermi and Samos, but *Prostrepsiceros houtumschindleri,* a probable descendant of *Protragelaphus,* is found only on Samos.

5. As with equids, there are more species of *Gazella* at Samos, which suggests a diversification from the single species found at Pikermi.

6. *Oioceros rothii* (Pikermi) is fairly primitive and could have evolved into a more derived Samos species (*Oioceros wegneri, Samotragus crassicornis,* or *Prosinotragus kuhlmanni*).

7. *Palaeoreas* is found at both Samos and Pikermi. The taxon is common at Pikermi but rare at Samos. *Criotherium argalioides* and *Parurmiatherium rugosifrons* (a rare species whose dentition is not known and hence not included in the mesowear tables) most likely descended from *Palaeoreas* and are also found to occur at Samos but not at Pikermi. In addition, at Samos, *Palaeoreas* occurs only at the Stefana and Tholorema locality of Q6. These localities are within the Old Mill Beds (estimated at 8.2 Ma). This observation substantiates that *Palaeoreas* could have been ancestral to *Criotherium* and *Parurmiatherium* which are found in the younger horizons (Main Bone Beds).

8. *Protoryx carolinae* from Pikermi is more primitive and probably ancestral to the two *Pachytragus* species

found at Samos (*P. laticeps* and *P. crassicornis*; Solounias 1981).

In conclusion, although Pikermi is not dated radiometrically, these patterns of closely related species with descendents occurring at Samos but not at Pikermi suggest that Pikermi was slightly older than Samos. The age of Pikermi may be estimated at about 8 Ma.

MATERIALS AND METHODS

Mesowear data for the Late Miocene species from China were obtained primarily from specimens in the collections at the American Museum of Natural History. Additional specimens of giraffids and *Hezhengia bohlini* were obtained at the Hezheng Paleozoological History Museum. We did not include specimens from the Late Miocene–Pliocene of China housed in the Uppsala collection. For Pikermi and Samos, original specimens and casts of teeth were used from the following institutions: American Museum of Natural History, New York; British Museum of Natural History, London; Carnegie Museum, Pittsburgh; Hessischer Landesmuseum, Darmstadt; The National Museum of Natural History (Smithsonian), The Natural History Museum, London; Naturhistorisches Museum, Basel; Naturhistorisches Museum, Vienna; Musée Géologique, Lausanne; Muséum National d'Histoire Naturelle, Paris; Museum of Comparative Zoology, and Peabody Museum at Harvard, Cambridge, Massachusetts; Yale Peabody Museum, New Haven, Connecticut; Museum of Palaeontology and Geology, National and Kapodistrian University of Athens; Geologisch-Paläontologisches Institut, Münster; Paläontologisches Institut der Universität Wien, Vienna; Sammlung für Paläontologie und Historische Geologie, Munich; Senckenberg Forschungsinstitut und Naturmuseum, Frankfurt; and Württenemberische Naturalen-Sammlung, Stuttgart.

The extant mesowear data were collected at the American Museum of Natural History, Natural History Museum, London, Carnegie Museum, Museum of Comparative Zoology and Peabody Museum of Harvard, National Museums of Kenya, National Museum of Natural History, and Yale Peabody Museum.

The African Savanna Biome includes woodlands and grasslands. Several species found in the African savanna also inhabit the African woodlands—an overlap. Open grasslands are also included in the savanna biome. The African forest species are distinct from the other ecologies. Table 31.1 summarizes the species, number of speci-

mens, mesowear average for each species, and the hypsodonty category.

Mesowear is a measure of the relative amount of abrasive and attritional wear on teeth as reflected by the shapes of the worn cusp apices and results in an apical shape formed by years of feeding. Because mesowear relies on a macroscopic amount of dental wear, it reflects diet over a much longer time period than microwear and is not susceptible to the "last supper effect" (Fortelius and Solounias 2000).

Originally, mesowear had the following subdivisions: high apices or low = occlusal relief; sharp round or blunt = cusp shape (Fortelius and Solounias 2000). The new scale of mesowear used in this study is composed of seven states (0–6). Observation of many dentitions indicates that there are very few to no teeth that have high and blunt cusps. Thus, this category could be excluded. The same is true with low and sharp cups, which may occur in certain individuals of zebra but are again extremely limited and can be excluded. Thus, the old scale is reduced basically to high relief or low, with high subdivided as sharp or round apices, and low apices subdivided into round or blunt. The new scale can be summarized as follows: high and sharp becomes 0 or 1 (depending on the degree of sharpness); in like fashion, high and rounded becomes 2 or 3 and low and rounded becomes 4 or 5, while low and blunt equals 6. Each number is blunter than the previous (0 is the sharpest and 6 the bluntest). The new scale was built on real *Equus* molars exhibiting differential mesowear. These teeth were used as a basis for assigning fossil teeth to mesowear stages. These seven cusp shapes were chosen to represent standardized intervals along a mesowear continuum, ranging from sharp to blunt as explained previously. Cusp apices of the seven selected *Equus* molars were cast onto an epoxy stage and used as a sort of mesowear "ruler," such that a fossil tooth is compared to the seven reference cusps and assigned a score ranging from 0 (sharpest) to 6 (bluntest). Figure 31.1 shows the mesowear scale and its application. The mesowear strip is three dimensional, and this greatly facilitates apical cusp comparisons.

This technique is described more fully by Mihlbachler et al. (2011) and differs somewhat from the original formulation of the mesowear method (Fortelius and Solounias 2000) in which teeth were assigned to two categories, relative sharpness and relief, without any clearly demarcated boundaries between mesowear stages. Cusp sharpness and degree of relief are probably not independent variables. For example, blunt cusps by definition have no relief. In this newer system, mesowear is treated as a univariate variable reflecting a continuum from

Table 31.1

Extant and Extinct Species Studied Using Mesowear and Hypsodonty

Species	N	Mesowear Score	Hypsodonty	Locality
Aepyceros melampus	17	1.2941	3	Savanna
Alcelaphus buselaphus	76	3.0789	3	Savanna
Antidorcas marsupialis	26	0.6923	3	Savanna
Ceratotherium simum	24	4.6666	3	Savanna
Connochaetes taurinus	52	3	3	Savanna
Damaliscus lunatus	5	4	3	Savanna
Diceros bicornis	34	0.11764	2	Savanna
Equus burchelli	121	4.67768	3	Savanna
Gazella granti	17	1.29411	3	Savanna
Gazella thomsoni	146	1.20547	3	Savanna
Giraffa camelopardalis	61	0.68852	1	Savanna
Hippotragus equinus	26	2.23076	3	Savanna
Hippotragus niger	20	2.6	3	Savanna
Kobus ellipsiprymnus	22	2.0909	3	Savanna
Litocranius walleri	69	1.42028	1	Savanna
Ourebia ourebi	128	1.6875	3	Savanna
Redunca redunca	77	2.12987	3	Savanna
Sigmocerus lichtenstanii	17	2.58823	3	Savanna
Syncerus caffer aequinoctialis	31	2	3	Savanna
Taurotragus oryx	21	0.66667	2	Savanna
Tragelaphus imberbis	31	0.8387	1	Savanna
Tragelaphus strepsiceros	7	2	1	Savanna
Boocercus eurycerus	27	1.1851	1	African forest
Cephalophus dorsalis	28	2	1	African forest
Cephalophus natalensis	6	2	1	African forest
Cephalophus niger	31	1.5483	1	African forest
Hyaemoschus aquaticus cottoni	18	1.66666	1	African forest
Okapia johnstoni	8	0.25	1	African forest
Tragelaphus scriptus	47	1.02127	1	African forest
Bohlinia attica	6	2.4	1	Samos
Ceratotherium pachygnathus	13	2.92308	3	Samos
Criotherium argalioides	18	1.77777	3	Samos
Dicerorhinus schleiermacheri	1	1	2	Samos
Gazella gaudryi-capricornis	26	1.2	2	Samos
Gazella larger	1	0	2	Samos
Gazella sp. very small	1	0	2	Samos
Gazella U skull	1	0	2	Samos
Gazella V skull	1	1	2	Samos
Graecoryx valenciennesi	1	1.4	1	Samos
Hippotherium dietrichi	11	2.6	3	Samos
Hippotherium giganteum	1	0	3	Samos
Hippotherium matthewi	6	3.3	3	Samos
Hippotherium primigenium	14	2.5	3	Samos
Hippotherium proboscideum	19	3.33333	3	Samos
Oioceros wegneri	3	2	2	Samos
Pachytragus crassicornis	23	1.65217	2	Samos
Pachytragus laticeps	49	1.26531	2	Samos

Species	N	Mesowear Score	Hypsodonty	Locality
Palaeoryx pallasi	14	1.57143	2	Samos
Palaeotragus coelophrys	2	1	1	Samos
Palaeotragus rouenii	13	1.5	1	Samos
Prostrepciceros houtumschinderi	4	0.75	2	Samos
Protragelaphus skouzesi	1	1	2	Samos
Pseudotragus capricornis	1	2	2	Samos
Samokeros minotaurus Rongia	3	1	2	Samos
Samotherium sensu lato	1	2	2	Samos
Samotherium Munich	1	4	2	Samos
Samotherium sansu lato	9	2.88888	2	Samos
Sporadotragus parvidens	13	0.69231	2	Samos
Tragoportax amalthea	21	1.38095	2	Samos
Tragoportax amalthea *Graecoryx*-like	1	0	2	Samos
Tragoportax amalthea flat front horn	1	2	2	Samos
Tragoportax amalthea weavy front horn	1	0	2	Samos
Tragoportax curvicornis	1	2	2	Samos
Tragoportax rugosifrons	17	1.47059	2	Samos
Gazella Munich	4	1.5	2	Pikermi
Graecoryx valenciennesi	4	1.4	1	Pikermi
Helladotherium duvernoyi	2	1	1	Pikermi
Hippotherium primigenium	14	2.5	3	Pikermi
Oioceros rothii	5	1.33333	2	Pikermi
Palaeoreas lindermayeri	24	1.29167	2	Pikermi
Palaeoryx pallasi	2	3	2	Pikermi
Palaeotragus rouenii	16	0.875	1	Pikermi
Tragoportax amalthea	2	2	2	Pikermi
Bohlinia sp.	2	2.5	1	Miocene of China
Chilotherium harbereri	13	2.33333	2	Miocene of China
Dorcadoryx triquetricornis bent braincase	1	2	2	Miocene of China
early *Bos*-like	1	3	3	Miocene of China
Gazella ?dorcadoides	4	0.5	2	Miocene of China
Gazella hypsodont	1	1	3	Miocene of China
Gazella large one fossa deep	2	1	2	Miocene of China
Gazella-like but more hypso	1	1.5	2	Miocene of China
Gazella-like Samos ones but hypso	1	1.5	2	Miocene of China
Gazella paotehensis 3 in a block	3	1.66666	2	Miocene of China
Gazella smaller than Samos	4	0.75	2	Miocene of China
Gazella type *gaudryi-capricornis*	4	1.26923	2	Miocene of China
Gazella very very large	7	2.14286	3	Miocene of China
Hezhengia bohlini Hez	10	3.4	3	Miocene of China
Hippotherium dermatorhinum	3	3.5	3	Miocene of China
Hippotherium fossatum	3	2	3	Miocene of China
Hippotherium kreugergi	2	3	3	Miocene of China
Hippotherium placodus	2	3	3	Miocene of China
Hippotherium tylodus	3	4	3	Miocene of China
Honanotherium schlosseri Hez	6	3.2	2	Miocene of China
Honanotherium sp. (*schlosseri*)	5	1.8	2	Miocene of China
Pachytragus largrelli small brain bent	3	0	2	Miocene of China

(continued)

Table 31.1 (*continued*)

Species	N	Mesowear Score	Hypsodonty	Locality
Palaeoryx sinensis	13	1.46154	2	Miocene of China
Palaeotragus very large	1	2	1	Miocene of China
Palaeotragus coelophrys Hez	13	2.6	1	Miocene of China
Palaeotragus microdon-rouenii	8	2	1	Miocene of China
Palaeotragus rouenii	1	2	1	Miocene of China
Palaeotragus sp. (*coelophrys*)	18	1.72222	1	Miocene of China
Palaoeoreas lindermayeri	1	2	2	Miocene of China
Plesiaddax depereti	5	1	3	Miocene of China
Samotherium mini species	1	1	2	Miocene of China
Samotherium boissieri Hez	7	3.2	2	Miocene of China
Samotherium neumayri Hez	2	1	2	Miocene of China
Samotherium sinense Hez	1	2.5	2	Miocene of China
Samotherium sp. (*sinense*)	2	4	2	Miocene of China
Schansitherium tafeli Hez	6	2.3	2	Miocene of China
Sinoryx bombifrons	3	1.66666	3	Miocene of China
Sinotragus wimani	7	2	3	Miocene of China
Urmiatherium intermedium	14	1.92857	3	Miocene of China

NOTE: Mesowear scale is from 0 to 6 (sharpest to most blunt); crown height is 1 = brachydont 2 = mesodont, 3 = hypsodont.

sharp cusps with high relief to blunt cusps with zero relief, similar to an approach adopted by Mihlbachler and Solounias (2006). Table 31.1 lists the mesowear data.

Crown-height categories (brachydont, mesodont, hypsodont) for the extant species are primarily from Janis (1988). Those for the fossil species were extracted from the NOW database of Eurasian Neogene fossil mammals (Fortelius 2009), available online (http://www.helsinki.fi/science/now). The dataset was downloaded on May 23, 2009. Crown-height categories were not available for two species, but in those cases we concordantly used measurements collected on museum specimens for *Palaeoryx pallasi* and *Samokeros minotaurus*, which are mesodont (hypsodonty index = 0.84 and 0.82, respectively). Some species were assigned a hypsodonty category by visual matching to other known species and without measuring tooth heights.

Statistical tests comparing mesowear data between the faunas were performed on SPSS v14.0. For analyses involving two faunas (e.g., African forest fauna vs. African savanna fauna) we used the nonparametric Mann–Whitney U test. We used Kruskal–Wallis tests when three or more faunas were included in the comparison.

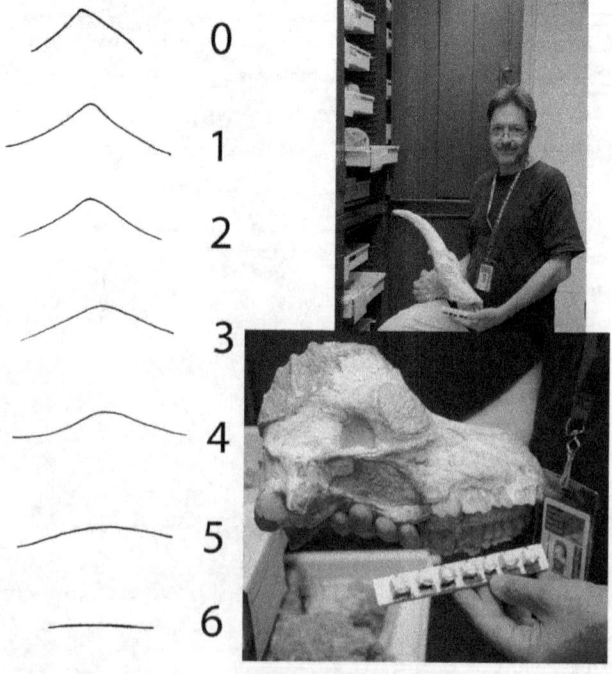

Figure 31.1 The new mesowear scale, a plastic strip with selected teeth in a sequence from sharpest to most blunt. NS holding the strip at the American Museum with bovids *Palaeoryx* and *Urmiatherium* from China.

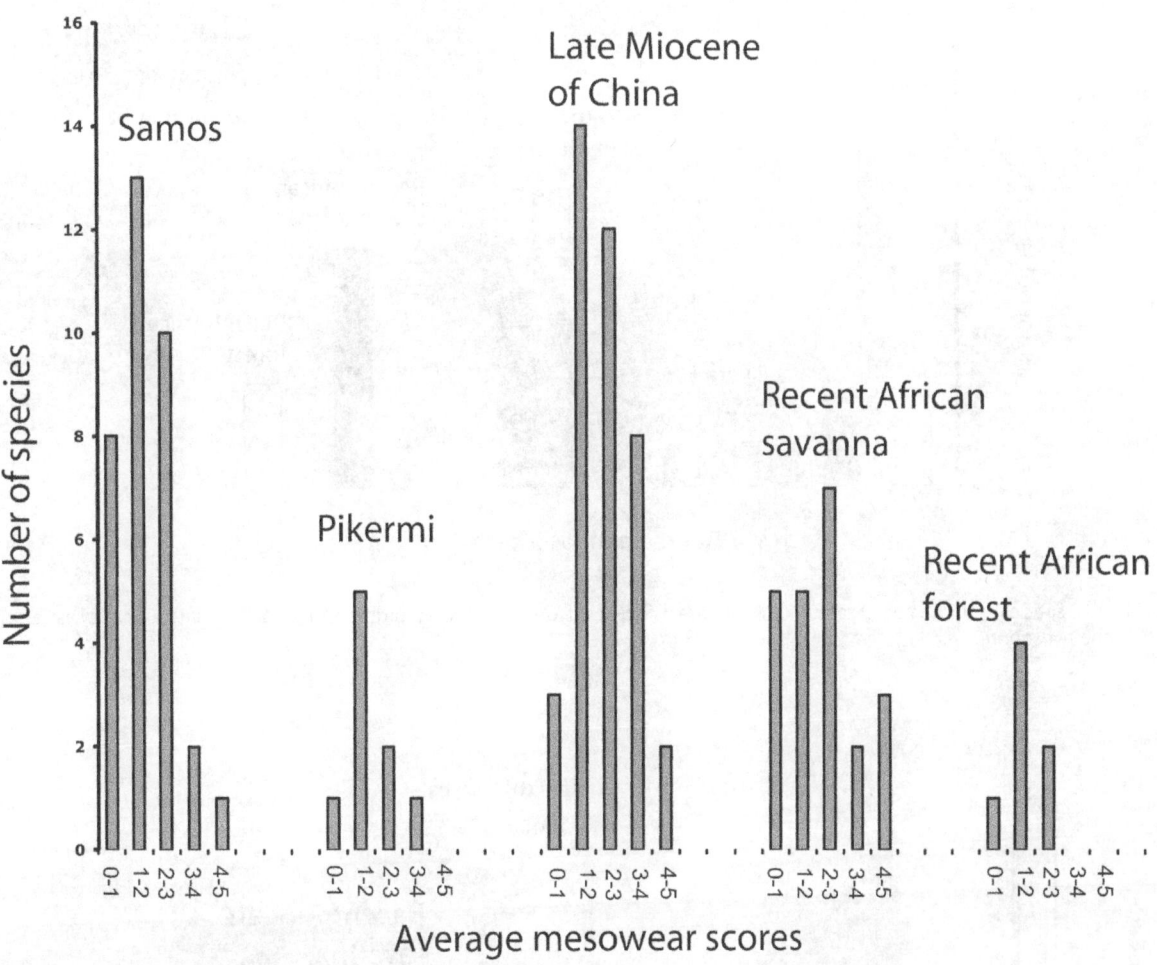

Figure 31.2 All species from Samos, Pikermi, and the Late Miocene of China plotted along with the extant African rainforest species and the African woodland–savanna continuum. The numbers are the mesowear scores.

RESULTS

The first hypothesis is that the distribution of mesowear scores between the three fossil faunas (Samos, Pikermi, and the late Miocene of China; LMC) is the same. The frequency distributions of the average mesowear scores (per species) are plotted in figure 31.2. The overall distributions of the mesowear scores among the three fossil faunas are similar. We were unable to statistically detect significant differences between the average mesowear scores for the three fossil faunas ($P = 0.075$). However, the small number of species in the Pikermi fauna significantly reduced the power of the test. Significant differences were found between the average mesowear scores of the two larger faunas, Samos and China ($p = 0.031$), with the LMC having overall higher mesowear scores

(i.e., blunter cusps with lower relief), suggesting slightly more abrasive diets overall.

The second hypothesis is that the distributions of mesowear scores in each of the individual fossil areas do not differ from the African savanna. The average mesowear score distribution of the African savanna fauna is intermediate between that of Samos and the LMC, and significant differences between the extant fauna and the two fossil faunas were not found ($p = 0.158$; $p = 0.798$).

The third hypothesis is that the levels of hypsodonty between the three fossil areas are the same. In comparing these areas, the average mesowear score distributions contrast with the distribution of crown height types. Figure 31.3 shows the histograms of the distributions of crown-height types among the three areas. Although the mesowear scores of the African savanna fauna were found

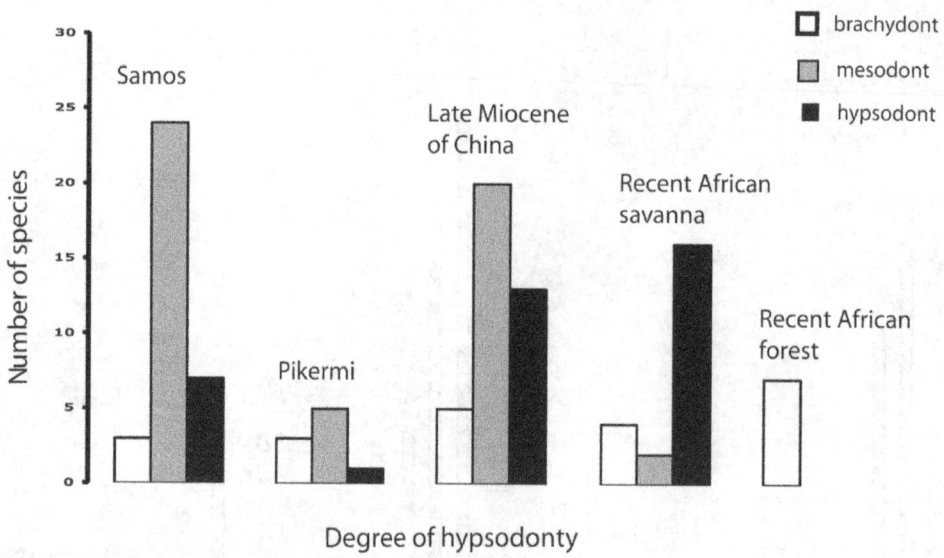

Figure 31.3 All species from Samos, Pikermi, and the Late Miocene of China sorted according to hypsodonty into three levels: brachydont, mesodont, and hypsodont.

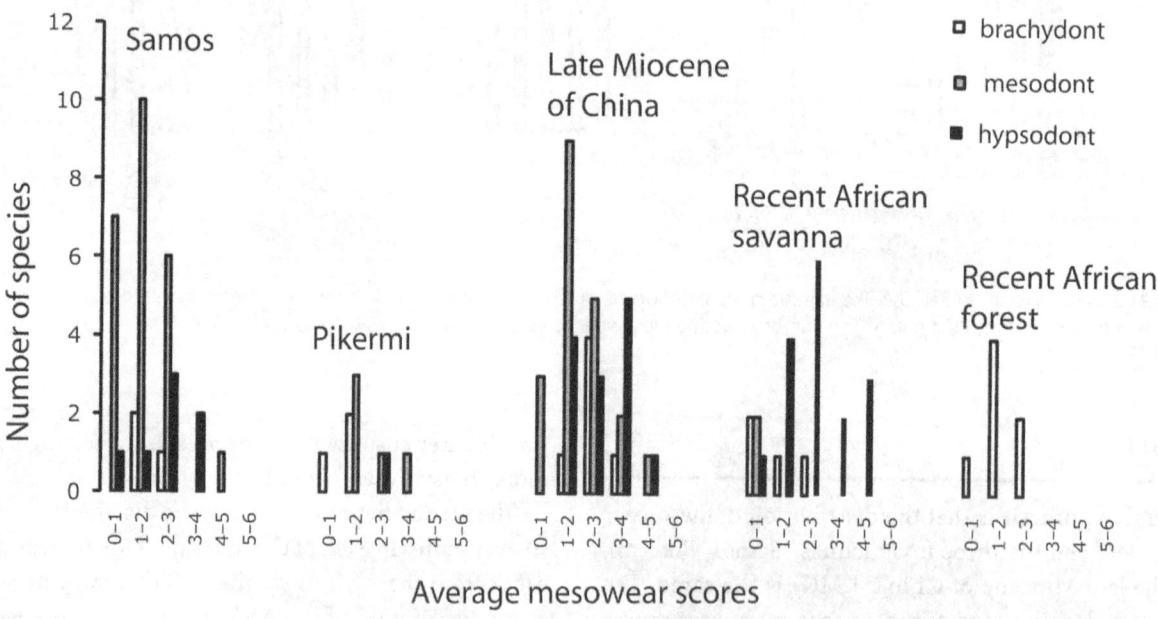

Figure 31.4 All species from Samos, Pikermi, and the Late Miocene of China plotted along with the extant African rainforest species and the African woodland–savanna continuum. The numbers are the mesowear scores. The hypsodonty is subdivided into three levels: brachydont, mesodont, and hypsodont.

to be intermediate with respect to the fossil areas, the extant African savanna fauna is markedly more hypsodont overall in comparison to all of the studied Miocene areas. (Samos: $P = 0.04$, Pikermi: $P = 0.018$, LMC: $P = 0.027$). All three fossil faunas are dominated by mesodont taxa, while the African Savanna is dominated by hypsodont taxa. LMC appears to have slightly more hypsodont taxa in comparison to Samos and Pikermi, although the difference is not significant (Kruskal–Wallis test $P = 0.231$). Elimination of the small Pikermi fauna does not yield a significant change in the statistical result ($P = 0.466$).

The fourth hypothesis is that the levels of hypsodonty within the three fossil areas are not different from the modern African savanna. The fifth hypothesis is that there is no correlation between hypsodonty and mesowear scores within any of the three fossil areas. Figure 31.4 plots the distribution of average mesowear scores per crown-height type. If hypsodonty is a reliable proxy for dietary abrasion, higher-crowned species are expected to have higher mesowear scores. Statistically significant differences were found in the average mesowear scores between the crown-height types in the African savanna fauna ($P = 0.013$) and Samos fauna ($P = 0.049$). Exclusion of the few brachydont species from the Samos fauna increases the significance ($P = 0.019$). A weaker relationship between crown-height type and mesowear was found in the LMC ($P = 0.112$). When brachydont taxa are excluded from the Chinese areas, the test is still insignificant ($P = 0.068$). These results generally suggest that crown height type and mesowear are weakly related, with hypsodont taxa having blunter cusps overall than brachydont and mesodont taxa. However, a relationship between mesowear and crown-height type is questionable among brachydont and mesodont taxa. Furthermore, among the three fossil faunas, although the mesowear scores for hypsodont taxa are higher on average than those of mesodont taxa, mesodont species show a greater range of mesowear values, in two instances (Samos and LMC) completely overlap the mesowear ranges of the brachydont and hypsodont taxa, and in two instances (Samos, Pikermi) include species with lower (blunter) mesowear scores than any of the hypsodont taxa.

MESOWEAR PATTERNS OF EXTINCT SPECIES

Various patterns have been observed in the mesowear scores and hypsodonty of extinct species (table 31.2). The Miocene species distribution in terms of mesowear and hypsodonty is different from the African savanna. *Hippotherium* is not like the zebra (*Equus burchelli*). The Miocene *Hippotherium* species are spread throughout the mesowear score spectrum which is suggestive of a diversity of diets. On the contrary, zebras are grazers. Similarly, the mesowear of *Ceratotherium pachygnathus* is sharper than that of the modern species *Ceratotherium simum*. The other two genera of extinct rhinocerotids are *Chilotherium* and *Dicerorhinus*. *Chilotherium* has notably blunter apices than *Dicerorhinus*.

There are many species of Giraffidae, and unlike today, they appear to be diverse in mesowear and were probably similar to bovids in their feeding habits. Giraf-

fids are dietarily diverse, as has been shown in previous microwear research (Solounias et al. 2000). Giraffids are spread throughout the mesowear scores and differ from *Giraffa camelopardalis*, which is a committed browser. *Samotherium* sensu lato was broadly adapted dietarily. *Samotherium* spp. are interesting in having the bluntest apices in the samples of both Greece and China. *Palaeotragus* differs from *Samotherium* in having sharper apices. Overall, its diet is less abrasive than that of *Samotherium*.

Honanotherium is notably different from *Bohlinia* in mesowear. This result reinforces that the two genera are distinct. The *Bohlinia* mesowear is very blunt, unlike the modern giraffe (*Giraffa*), but they are closely related. *Helladotherium* has sharp apices and strongly differs from the Pleistocene African *Sivatherium* (both genera are Sivatheriinae).

The systematic and hypsodonty relationship of bovid species to mesowear is complex. Most Miocene bovid species have sharper apices than *Samotherium*, *Honanotherium*, and most *Hippotherium*. They are clearly different in mesowear scores from the majority of *Hippotherium*. In the modern African savanna, grazing bovids have the same mesowear as the zebra. The only bovids that have extreme bluntness of 3 are *Palaeoryx pallasi* and *Hezhengia*. *Palaeoryx pallasi* and *Hezhengia*, however, are not more hypsodont than other bovids (e.g., *Pseudotragus*, *Palaeoreas*, and *Pachytragus*). In the modern African savanna, grazer bovids are more hypsodont than many of the mixed feeders. Mesodont bovid taxa such as *Tragoportax*, *Prostrepsiceros*, *Sporadotragus*, and *Gazella* have sharper scores than the brachydont taxon *Graecoryx*. There is a single hypsodont species (*Pachytragus largelli*) with sharp apices. In the modern African savanna, there are no hypsodont bovids with sharp apices. During the late Miocene, we find taxa similar in hypsodonty that differ in mesowear scores. Examples are *Sporadotragus* and *Pseudotragus*, *Prostrepsiceros* and *Protragelaphus*, *Tragoportax* and *Samokeros*, *Pachytragus* and *Sinotragus*, *Criotherium* and *Sinotragus*, and *Urmiatherium* and *Sinotragus*. Similar or conspecific taxa with the same hypsodonty can differ in mesowear scores. Examples are *Gazella dorcadoides* and *Gazella gaudryi*, *Tragoportax amalthea* from Samos and *Tragoportax rugosifrons* from Samos, *Oioceros rothii* and *Oioceros wegneri*, *Pachytragus largrelli* and *Pachytragus laticeps*, *Hippotherium giganteum* and *Hippotherium primigenium*, and *Samotherium neumayri* and *Samotherium boissieri*.

Mesodont species of *Tragoportax*, *Oioceros*, *Palaeoryx*, *Palaeoreas*, and *Gazella* have blunter apices than *Prostrepsiceros*, *Sporadotragus*, and other *Tragoportax* and *Gazella*

Table 31.2

Tooth Mesowear and Hypsodonty for the Pikermian Biome Taxa

	0	1
HYPSODONT	*Pachytragus largelli* CH *Gazella* hypsodont CH **Hippotherium giganteum S**	**Pachytragus laticeps S** **Pachytragus crassicornis S** **Criotherium argalioides S** Plesiaddax depereti CH Sinoryx bombifrons CH Urmiatherium intermedium CH Gazella-like but more hypsodont CH
MESODONT	**Helladotherium duvernoyi S P** **Tragoportax, Graecoryx-like S** **Tragoportax amalthea weavy S** **Prostrepsiceros houtumschindleri S** **Sporadotragus parvidens S** **Gazella U skull S** *Gazella* sp. very small CH *Gazella? dorcadoides* CH	Samotherium mini species CH Samotherium neumayri CH Honanotherium sp. CH **Palaoeoryx pallasi S** **Tragoportax rugosifrons S** **Tragoportax amalthea S** **Samokeros minotaurus S** **Oioceros rothii P** **Gazella gaudryi P S** **Gazella U skull S** **Gazella capricornis P** **Gazella Munich sample P** Gazella-like Samos but hypsodont CH Gazella more hypsodont CH Gazella gaudryi-capricornis CH Gazella large one fossa deep CH Gazella larger CH Gazella smaller than Samos CH Gazella paotehensis CH **Protragelaphus skouzesi P S** **Palaeoreas lindermayeri P** **Dicerorhinus schleiermacheri S**
BRACHYDONT	**Palaeotragus rouenii P**	**Palaeotragus rouenii S** **Palaoeotragus coelophrys S** Palaoeotragus coelophrys CH **Graecoryx valeciennesi P S**

NOTE: P (Pikermi), S (Samos), and CH (Chinese Miocene).

2	3	4
Sinotragus wimani CH	early *Bos* CH	
	Hippotherium tylodus CH	
	Hippotherium dermatorhinum CH	
	Hippotherium fossatum CH	
Ceratotherium pachygnathus P S	*Hippotherium kreugergi* CH	
	Hippotherium placodus CH	
Hippotherium dietrichi S	**Hipotherium proboscideum** S	
Hippotherium primigenium P S	**Hippotherium matthewi** S	

2	3	4
Samotherium sensu lato S	*Honanotherium schlosseri* CH	
Samotherium major S		**Samotherium Munich** S
Samotherium boissieri S	*Samotherium boissieri* CH	*Samotherium sinense* CH
Samotherium very large CH		
Samotherium large CH		
Samotherium sinense CH		
Schansitherium tafeli CH		
Palaeoryx sinensis CH	**Palaeoryx pallasi** P	
Tragoportax amalthea P		
Tragoportax curvicornis S		
Pseudotragus capricornis S		
Samodorcas kuhlmanii S		
Oioceros wegneri S		
Pseudotragus capricornis S		
Dorcadoryx triquerticornis CH	*Hezhengia bohlini* CH	
Gazella gaudryi-capricornis S		
Gazella very large CH		
Palaeoreas lindermayeri CH		
Chilotherium harbereri CH		

2	3	4
Bohlinia attica P S		
Bohlinia sp. CH		
Palaeotragus sp. large CH		
Palaoeotragus coelophrys CH		
Palaeotragus micorodon-rouenii CH		
Palaeotragus rouenii CH		

species, which have similar crown height. The most hypsodont bovids are *Pachytragus*, *Criotherium*, *Sinotragus*, *Plesiaddax*, *Sinoryx*, and *Urmiatherium*. These, however, do not have the same mesowear scores.

DISCUSSION

Mesowear, Microwear, and the New Method

Dental wear of herbivorous ungulates is primarily caused by tooth-on-tooth attrition and abrasion from the plants ingested, soil, sand, dust, and other contaminants. These occlusal interactions produce microscopic scars (pits and scratches), frequently described as microwear. As the tooth wears, dental microwear features have a high turnover rate, and microwear reflects what an animal ate in the short term, probably within hours to weeks, depending on the specific dental wear rate (Solounias, Fortelius, and Freeman 1994). In contrast, macroscopic aspects of cusp shape (mesowear) involve substantially more dental wear and therefore reflect diet over a much longer time interval and are not perceptibly subject to daily dietary fluctuation but rather reflect overall diet over the course of months to years (Fortelius and Solounias 2000). Mesowear and microwear reflect aspects of diet on different temporal and spatial scales. Incongruence between these results may not mean that one of these methods is misleading. Rather, incongruence between microwear and mesowear suggests differences between the proximate diet at the time of death and the average diet over the course of seasons (years). Microwear ought to be biased toward the habitats proximate to the location and season of the death of an individual, whereas mesowear ought to be biased toward the overall average diet of an individual. A methodology needs to be developed for an effective comparison of mesowear and microwear on the same samples.

The new mesowear scale of seven states of wear in a gradation from 0 to 6 is simple and easy to use. Each apex represents a particular state but also somewhat sharper states. For example, a score of 3 represents the apical morphology of 3 and all apices that are in between 3 and 2. The state 0 represents quite a sharp tooth but also includes sharper teeth. The fact that we used horse teeth to construct the scale is not relevant. It can be used with any type of selenodont tooth, but we must admit that it is easier to use with perissodactyls than with artiodactyls. We find the plastic strip with the selected apices to be the preferable way to score teeth. The fact that the aspect of high or low relief and relative cusp bluntness have been

combined into one variable is simpler and gives more resolution. The apices of the scale are three dimensional and this enables a more precise comparison with the unknown in question.

Paleoecological Approaches

Overall three other approaches show a similar pattern for the Pikermian Biome:

1. An isotopically C_3-dominated ecosystem was found in various areas of the Miocene of Greece and Turkey, including Pikermi and Samos, which represents mostly woodland (Quade, Solounias, and Cerling 1994; Quade et al. 1995; Cerling et al. 1997).

2. A recent dental microwear study of the Samos and Pikermi faunas suggests that the Samos and Pikermi faunas occupied woodland habitats with limited areas of open savannas (Solounias, Rivals, and Semprebon 2010). Overall, the ungulates were mixed feeders, suggesting seasonality or a complex woodland environment with some grasses. Many giraffids, bovids, and hippotheres were mixed feeders. The grazers were primarily giraffids and hippotheres. Most of the browsers were Antilopini. The discovery of seasonality from microwear is novel, and in some ways better than paleobotanical inference. Most Miocene plant taxa span such wide latitude that it is difficult to decipher seasonality.

3. Paleobotanical data from various localities during the Late Miocene of the Mediterranean are also in agreement with C_3 vegetation and woodland environments (Axelrod 1975; for a summary, see Solounias, Rivals, and Semprebon 2010). Under the woodland scenario, the presence of C_3 grasses in the woodland clearings, meadows, and water margins was possible. The presence of C_3 grassland needs to be proven but could not occur under present-day carbon dioxide conditions. The Late Miocene is also marked with the onset of C_4 grasses (Edwards et al. 2010), which were limited and did not necessarily grow in the areas encompassed under this investigation.

The differences of the overall mesowear scores between the species studied from Greece and China are minor, suggesting that the distribution and range of diets of the ungulates are similar in terms of abrasion (see figure 31.2). The mesowear scores of the Chinese fauna (LMC) are slightly higher in comparison to the Greek faunas, possibly suggesting slightly more open or arid conditions. This small difference could be due to sampling, and a broader study of fossils from China may change this result. The prevailing signal from mesowear suggests paleoecological uniformity throughout a vast area (eight time zones, Spain to China) during the Late Miocene. This result, which is congruent with the concept of a single Pikermian Biome, is not dissimilar to what a basic comparison of faunas would suggest; as discussed in this chapter, many ungulate species unify the Pikermian faunas from Spain to China (see table 31.1).

The smaller number of species for the Pikermi fauna is consistent with the suggestion that it is older (about 8 Ma instead of 7.2 Ma), because it might have had fewer taxa in the newly developing biome. By "developing biome," we mean that the Pikermi fauna samples species from the onset of the Pikermian Biome. Richer faunas derive from slightly younger faunas such as Samos, the core of Maragheh and the main Shanxi and Gansu faunas. We hypothesize that from 8 Ma to 7.2 Ma, there was a major adaptive radiation of ungulate species. Pikermi, however, is fundamentally similar in mesowear scores to Samos and China.

The differences between the overall mesowear scores between the three fossil samples and the modern African savanna are minor, suggesting overall similar levels of abrasion (see figures 31.2 and 31.3). This similarity in mesowear is limited and needs to be evaluated by comparisons to other modern faunas such as those from the Indian and European forests. Currently, such data are not available. The modern African forest is notably different from the Pikermian Biome.

Individual Species

Preliminary comparisons solely of the average mesowear scores can be conveniently subdivided into three: the bluntest apices with score averages 3 and 4, an intermediate range with averages 1 and 2, and the sharpest apices scoring near 0.

At Samos and the LMC, the bluntest dentitions (mesowear score averages of about 3 and 4) are from the *Hippotherium* species, the bovid *Hezhengia*, and the giraffids *Honanotherium* and *Samotherium*. The modern African savanna with comparable mesowear score averages includes the alcelaphine bovids (*Alcelaphus* and *Connochaetes*), the grazing rhinocerotid *Ceratotherium*, and the zebra *Equus burchelli*. Clearly, the two clusters of taxa are derived from different kinds of species; similar mesowear is found in such different species. The hippotheres differ from zebras but are both Equidae. The Alcelaphini, as a bovid tribe, are grazing hypsodont species, but there is no such grazing tribe during the Miocene. Sys-

tematic mesowear comparisons need to be developed further.

The intermediate range in the late Miocene includes several hippotheres, giraffids, and rhinocerotids. In addition, there are numerous *Gazella* species and many other bovids. The giraffids are numerous. The rhinocerotid *Ceratotherium pachygnathus* (an archaic form of *Ceratotherium*) and *Dicerorhinus* fall in this range. In the modern African savanna, this range is filled primarily by *Gazella*, the gazelle-like *Aepyceros*, and various bovids (but of different genera than the Miocene such as *Hippotragus* species and *Syncerus*). *Syncerus* is in the tribe Bovini, which, like the Alcelaphini, includes many more species than in the Miocene past. During the Late Miocene, *Samokeros* is a Bovini precursor, and in China a more *Bos*-like species was present (not yet described). In the two clusters, taxa are again derived primarily from different systematic groups. Modern *Gazella* has the same mesowear with the extinct species of *Gazella*. In the middle range, hippotheres are particularly abundant. In contrast, the modern zebra falls in the group with the bluntest apices. Equids from the intermediate range of mesowear scores are absent in the modern African savanna.

In the low range of mesowear scores (0–1) are several *Gazella* species and other bovids, hippotheres, and giraffids. The giraffids are of the more archaic type (*Palaeotragus rouenii* and *Helladotherium*). Early Caprini (*Pachytragus* and *Sporadotragus*) have mesowear similar to Antilopini. The *Gazella* species are present, but they also have a wide range of mesowear scores. In the modern African savanna, this mesowear range (0–1) is occupied by *Tragelaphus* species, the giraffe, *Diceros*, and the Antilopini *Antidorcas*.

The findings show that similar or different mesowear—and presumably diets—can occur between species that are systematically close. For example, the giraffid *Samotherium* strongly differs in mesowear from the recent giraffe. A mesodont *Samotherium* can have mesowear similar to a *Hippotherium*, which is significantly more hypsodont. The mesowear of Miocene *Ceratotherium* is different from the present day *C. simum*.

Overall Picture and Comparisons

In contrast to the similar mesowear score distributions, the hypsodonty of the modern African savanna is radically greater. In this case, the difference is clearly influenced phylogenetically. The notable hypsodonty difference is attributed to the presence of several extant species of the tribes Alcelaphini, Reduncini, Hippotragini, as well as *Syncerus* (Bovini) and the rhinoceros *Ceratotherium simum*. During the Miocene, none of these highly hypsodont taxa were present. In contrast, the pool of available species in the Miocene of Eurasia was mostly mesodont Bovidae (archaic Antilopini, Caprini, and Boselaphini). Thus, the recent African savanna species are more hypsodont, but it is interesting that their mesowear is very similar to that of the Miocene localities.

Despite having relatively fewer hypsodont species, Pikermian Biome paleodiets (in terms of mesowear) appear to have been similar to those of the modern African savanna, that is to say, they exhibit a range of mesowear consistent with a similar mixture of browsers, mixed feeders, and grazers. Hypsodonty is a morphological characteristic that has evolved within many lineages of herbivorous mammals, including ungulates, rodents, and other extinct groups. Yet we know of no instance where hypsodonty has been secondarily lost. It has been suggested (Feranec 2003; Mihlbachler and Solounias 2006; MacFadden 2008) that while hypsodonty might be an adaptation to high-abrasion diets such as grazing, it does not constrain a species to an abrasive diet. Rather, it expands niche breath by increasing the amount of abrasion a species can tolerate. Our data support this hypothesis to some degree. For example, *Hippotherium* spans from 0 to 3 in mesowear scores, although the hypsodonty of these taxa is about the same. Similarly, *Samotherium* spans from 2 to 4 in mesowear values, *Pachytragus* from 0 to 1, and *Gazella* from 0 to 2. Similarly, the mesowear scores for these fossil localities and the recent African savanna overlap considerably. Despite a weak correlation, hypsodonty is a poor predictor of mesowear using the present data. Given the lack of phylogenetic evidence for crown-height character reversal, the prevalence of hypsodonty is expected to increase through time, as long as episodes of selection for hypsodonty repeatedly occur. However, the increasing frequency of hypsodonty does not require a prevailing overall dietary or associated climatic trend such as opening habitat. Even if climate from time to time reverses to cooler wetter phases, hypsodonty levels will not record this. This suggests that relative hypsodonty levels may be informative of paleoecology within a narrow window of time, among ecologically linked regions (such as Samos and China in the Miocene), that can potentially draw from a similar taxonomic pool. For instance, the Chinese Miocene fauna is slightly more hypsodont overall in comparison to that of Samos. This small difference is proportional to the small difference in average mesowear scores. The LMC is both slightly more hypsodont with slightly higher (blunter) mesowear scores. These differences are congruent and proportional. In contrast, the modern African savanna fauna and the Miocene faunas have similar mesowear levels, but the modern African savanna is disproportionately hypsodont in comparison. We suggest the high degree of overall hypsodonty is less a function of recent climate and more a consequence of 8 myr of cumulative evolution for hypsodonty in numerous ungulate lineages. Comparisons of hypsodonty levels of taxonomic groups or faunas from widely different time periods are complicated by the additional phylogenetic inertia characterizing the younger taxa and faunas.

In conclusion, we find, as in previous studies (Fortelius and Solounias 2000; Mihlbachler and Solounias 2006; Rivals, Mihlbachler, and Solounias 2007), that mesowear is a useful tool for comparisons of groups of taxa. The ecological cohesion of the Pikermian Biome is supported by mesowear. Combining the results of this mesowear analysis with an earlier microwear analysis suggests a rich mosaic with woodland but including a limited area of open savanna. We also find the hypsodonty of this biome to be distributed in harmony with the mesowear and with evolutionary levels for hypsodonty.

ACKNOWLEDGMENTS

We thank the committee of the Neogene Terrestrial Mammalian Biostratigraphy and Chronology Symposium in China. We also thank the curators of natural history museums in Paris, London, Vienna, Lausanne, Basel, Bern, New York, Frankfurt, Munich, Münster, and Stuttgart. This study was supported by funds from three previous National Science Foundation grants (one specifically for Samos, NSF dissertation improvement grant EAR 76-00515); the others for microwear of ruminants and equids, BSR 8605172 1985–86 and IBN 9628263 1995–96, respectively. Tooth mesowear was recorded in 2009 on specimens covered by these grants. We thank Judith Harris for inspiration and numerous discussions and collaborations. We also thank Louis Abatis, Peter Andrews, Athanassios Athanassiou, Raymond Bernor, Shao-kun Chen, Eric Delson, Vera Eisenmann, Burkart Engesser, Mikael Fortelius, Jens Franzen, Henry Galiano, Alan Gentry, Ursula Göhlich, Judith Harris, Elmar Heizmann, Kurt Heissig, Thomas Keiser, Ursula Menkveld, Doris Nagel, Malcolm McKenna, Clemens Oekentorp, David Pilbeam, Karl Rausen, Peter Robinson, Socrates

Roussiakis, George Schaller, Shan-qin Chen, Fritz Steininger, Pascal Tassy, Richard Tedford, George Theodorou, Herbert Thomas, John Van Couvering, Alan Walker, Xiaoming Wang, Marc Weidmann, and Eileen Westwig.

REFERENCES

Abel, O. 1927. *Lebensbilder aus der Tierwelt der Vorzeit*. 2nd ed. Stuttgart: Jena Verlag Gustav Fisher.

Axelrod, I. D. 1975. Evolution and biogeography of the Madrean Tethyan sclerophyll vegetation. *Annals of the Missouri Botanical Garden* 62:280–334.

Beerling, D. J. and C. P. Osborne. 2005. The origin of the savanna biome. *Global Change in Biology* 12:2023–2031.

Bernor, R. L., G. D. Koufos, M. O. Woodburne, and M. Fortelius. 1996. The evolutionary history and biochronology of European and Southwest Asian late Miocene and Pliocene hipparionine horses. In *The Evolution of the Western Eurasian Neogene Mammal Faunas*, ed. R. L. Bernor, V. Fahlbusch, and H.-W. Mittmann, pp. 307–347. New York: Columbia University Press.

Bernor, R. L., R. S. Scott, M. Fortelius, J. Kappelman, and S. Sen. 2003. Equidae (Perissodactyla). In *The Geology and Paleontology of the Miocene Sinap Formation, Turkey*, ed. M. Fortelius, J. Kappelman, S. Sen, and R. L. Bernor, pp. 220–281. New York: Columbia University Press.

Bernor, R. L., N. Solounias, C. C. Swisher III, and J. A. Van Couvering. 1996. The correlation of the classical "Pikermian" mammal faunas—Maragheh, Samos, and Pikermi—with the European MN unit system. In *The Evolution of the Western Eurasian Neogene Mammal Faunas*, ed. R. L. Bernor, V. Fahlbusch, and H.-W. Mittmann, pp. 137–154. New York: Columbia University Press.

Bond, W. J. 2008. What limits trees in C_4 grasslands and savannas? *Annual Review of Ecology and Systematics* 39:641–659.

Bruch, A. A., D. Uhl, and V. Mosbrugger. 2007. Miocene climate in Europe—patterns and evolution: A first synthesis of NECLIME. *Palaeogeography, Palaeoclimatology, Palaeoecology* 253:1–7.

Bruch, A. A., T. Utescher, V. Mosburugger, I. Gabrielyan, and D. A. Ivanov. 2006. Late Miocene climate in the circum-Alpine realm—a quantitative analysis of terrestrial paleofloras. *Palaeogeography, Palaeoclimatology, Palaeoecology* 238:270–280.

Cerling, T. E., J. M. Harris, B. J. MacFadden, M. G. Leakey, J. Quade, V. Eisenmann, and J. R. Ehleringer. 1997. Global change through the Miocene/Pliocene boundary. *Nature* 389:153–158.

Coppens, Y. 1994. The east side story: The origin of humankind. *Scientific American* 270:88–95.

de Bonis, L., G. Bouvrain, D. Geraads, and G. Koufos. 1992. Diversity and paleoecology of Greek late Miocene mammalian faunas. *Palaeogeography, Palaeoclimatology, Palaeoecology* 91:99–121.

Deng, T. 2006. Paleoecological comparison between late Miocene localities of China and Greece based on *Hipparion* faunas. *Geodiversitas* 28:499–516.

Dorofeyev, P. I. 1966. Flora of the *Hipparion* epoch. *International Geological Reviews* 8:1109–1117.

Edwards, E. J., C. P. Osborne, C. A. E. Strömberg, and S. T. Smith, and C_4 Grasses Consortium. 2010. The origins of C_4 grasslands: Integrating evolutionary and ecosystem science. *Science* 328: 587–591.

Eronen, J. T., M. M. Ataabadi, A. Micheels, A. Karme, R. L. Bernor, and M. Fortelius. 2009. Distribution history and climatic controls of the Late Miocene Pikermian chronofauna. *Proceedings of the National Academy of Sciences* 106:11867–11871.

Eronen, J. T., K. Puolamäki, L. Liu, K. Lintulaakso, J. Damuth, C. Janis, and M. Fortelius. 2010a. Precipitation and large herbivorous mammals II: Application to fossil data. *Evolutionary Ecology Research* 12:235–248.

Eronen, J. T., K. Puolamäki, L. Liu, K. Lintulaakso, J. Damuth, C. Janis, and M. Fortelius. 2010b. Precipitation and large herbivorous mammals I: Estimates from present-day communities. *Evolutionary Ecology Research* 12:217–233.

Feranec, R. S. 2003. Stable isotopes, hypsodonty, and the paleodiet of *Hemiauchenia* (Mammalia: Camelidae): A morphological specialization creating ecological generalization. *Paleobiology* 29:230–242.

Fortelius, M. (coordinator). 2009. Neogene of the Old World Database of Fossil Mammals (NOW). University of Helsinki, http://www.helsinki.fi/science/now/.

Fortelius, M., J. Eronen, L.-p. Liu, D. Pushkina, A. Tesakov, I. Vislobokova, and Z.-q. Zhang. 2003. Continental-scale hypsodonty patterns, climatic paleobiogeography and dispersal of Eurasian Neogene large mammal herbivores. DEINSEA 10:1–11.

Fortelius, M. and N. Solounias. 2000. Functional characterization of ungulate molars using the abrasion-attrition wear gradient: A new method for reconstructing paleodiets. *American Museum Novitates* 3301:1–36.

Gaudry, A. 1862–1867. *Animaux fossiles et géologie de l'Attique*. Paris: Martinet.

Jacobs, B. F., J. D. Kingston, and L. L. Jacobs. 1999. The origin of grass-dominated ecosystems. *Annals of the Missouri Botanical Garden* 86:590–643.

Janis, M. C. 1988. An estimation of tooth volume and hypsodonty indices in ungulate mammals and the correlation of these factors with dietary preferences. In *Teeth Revisited: Proceedings of the VII International Symposium on Dental Morphology*, ed. D. E. Russel, J. P. Santorio, and D. Signogneu-Russel, pp. 367–387. Paris: Muséum National de Histoire Naturelle.

Janis, C. M. 2008. An evolutionary history of browsing and grazing ungulates. In *The Ecology of Browsing and Grazing*, ed. I. J. Gordon and H. H. T. Prins, pp. 21–45. New York: Springer.

Janis, C. M., J. Damuth, and J. M. Theodor. 2000. Miocene ungulates and terrestrial primary productivity: Where have all the browsers gone? *Proceedings of the National Academy of Sciences* 97:7899–7904.

Koufos, G. D., D. Kostopoulos, and G. Merceron 2009. Palaeoecology–Palaeogeography. In *The Late Miocene Mammal Faunas of Samos*, ed. G. D. Koufos and D. Nagel, pp. 409–427. *Beiträge zur Paläontologie* 31.

Koufos, G. D. and D. Nagel, eds. 2009. *The Late Miocene Mammal Faunas of Samos*. *Beiträge zur Paläontologie*, 31.

Kurtén, B. 1952. The Chinese *Hipparion* fauna. *Commentationes Biologicae Societatis Scientiarum Fennicae* 13:1–82.

Kurtén, B. 1971. *The Age of Mammals.* New York: Columbia University Press.

MacFadden, B. J. 2008. Geographic variation in diets of ancient populations of 5-million-year-old (early Pliocene) horses from southern North America. *Palaeogeography, Palaeoclimatology, Palaeoecology* 266:83–94

Mai, D. H. 1995. *Tertiäre Vegetationsgeschichte Europas, Methoden und Ergebnisse.* Jena: Gustav Fischer Verlag.

Matthew, D. W. 1926. The evolution of the horse. *Quarterly Review of Biology* 1:139–185.

Mihlbachler, M. C., F. Rivals, N. Solounias, and G. Semprebon. 2011. Dietary change and evolution of horses in North America. *Science* 331:1178–1181.

Mihlbachler, M. C. and N. Solounias. 2006. Coevolution of tooth crown height and diet in oreodonts (Merycoidontidae, Artiodactyla) examined with phylogenetically independent contrasts. *Journal of Mammalian Evolution* 13:11–36.

Orgeta, M. 1979. Erste Ergebnisse einer palynologischen Unterschuchung der Lignite von Pikermi/Attica. *Annales Géologiques des pays Helléniques* 2:909–921.

Osborn, H. F. 1910. *The Age of Mammals.* New York: Macmillan.

Quade, J., T. E. Cerling, P. Andrews, and B. Alpagut. 1995. Paleodietary reconstruction of Miocene faunas from Paşalar, Turkey, using stable carbon and oxygen isotopes of fossil tooth enamel. *Journal of Human Evolution* 28:373–384.

Quade, J., N. Solounias, and T. E. Cerling. 1994. Stable isotopic evidence from paleosol carbonates and fossil teeth in Greece for forest or woodlands over the past 11 Ma. *Palaeogeography, Palaeoclimatology, Palaeoecology* 108:41–53.

Retallack, G. J. 2001. Cenozoic expansion of grasslands and climatic cooling. *Journal of Geology* 109:407–426.

Rivals F., M. C. Mihlbachler, and N. Solounias. 2007. Effect of ontogenetic-age distribution in fossil samples on the interpretation of ungulate paleodiets using the mesowear method. *Journal of Vertebrate Paleontology* 27:763–767.

Scott, R. S., M. Fortelius, K. Huttunen, and M. Armour-Chelu. 2003. Abundance of "Hipparion." In *Geology and Paleontology of the Miocene Sinap Formation, Turkey,* ed. M. Fortelius, J. Kappelman, S. Sen, and R. L. Bernor, pp. 380–397. New York: Columbia University Press,

Solounias, N. 1981. The Turolian fauna from the island of Samos, Greece: With special emphasis on the hyaenids and the bovids. *Contributions to Vertebrate Evolution* 6:1–232.

Solounias, N., M. Fortelius, and P. Freeman. 1994. Molar wear rates in ruminants: A new approach. *Annales Zoologici Fennici* 31:219–227.

Solounias, N., W. S. McGraw, L.-A. C. Hayek, and L. Werdelin. 2000. The paleodiet of the Giraffidae. In *Antelopes, Deer, and Relatives,* ed. E. S. Vrba and G. B. Schaller, pp. 84–95. New Haven: Yale University Press.

Solounias, N., S. M. C. Moelleken, and J. M. Plavcan. 1995. Predicting the diet of extinct bovids using masseteric morphology. *Journal of Vertebrate Paleontology* 15:795–805.

Solounias, N., M. Plavcan, J. Quade, and L. Witmer. 1999. The Pikermian Biome and the savanna myth. In *Evolution of the Neogene Terrestrial Ecosystems in Europe,* ed. J. Agustí, P. Andrews, and L. Rook, pp. 427–444. Cambridge: Cambridge University Press.

Solounias, N., F. Rivals, and G. Semprebon. 2010. Dietary interpretation and paleoecology of herbivores from Pikermi and Samos (late Miocene of Greece). *Paleobiology* 36:113–136.

Strömberg, C. A. E., L. Werdelin, E. M. Friis, and G. Saraç. 2007. The spread of grass-dominated habitats in Turkey and surrounding areas during the Cenozoic: Phytolith evidence. *Palaeogeography, Palaeoclimatology, Palaeoecology* 250:18–49.

Thenius, E. 1959. *Tertiär.* Stuttgart: Ferdinand Enke Verlag.

Weidmann, M., N. Solounias, R. E. Drake, and G. H. Curtis. 1984. Neogene stratigraphy of the Mytilinii Basin, Samos Island, Greece. *Geobios* 17:477–490.

Zachos, J. C., G. R. Dickens, and R. E. Zeebe. 2008. An early Cenozoic perspective on greenhouse warming and carbon-cycle dynamics. *Nature* 451:279–283.

CONTRIBUTORS

Chinese names are reversed to facilitate citations.

NADEZHDA ALEXEEVA
Geological Institute
Siberian Branch
Russian Academy of Sciences
Ulan-Ude 670047
Russia
ochotona@mail.ru

PIERRE-OLIVIER ANTOINE
Equipe de Paléontologie
Institut des Sciences de l'Évolution
Université Montpellier 2
F-34095 Montpellier Cedex 5
France

Géosciences-Environnement Toulouse
Université de Toulouse
F-31400 Toulouse
France

Mission Paléontologique Franco-Balouche
F-34270 Le Triadou
France
pierre-olivier.antoine@univ-montp2.fr

MAJID MIRZAIE ATAABADI
Faculty of Science
University of Zanjan
45371-38791 Zanjan
Iran
majid_mirzaie@yahoo.com

Department of Geosciences and Geography
University of Helsinki
FI-00014 Helsinki
Finland
majid.mirzaie@helsinki.fi

DEMCHIG BADAMGARAV
Paleontological Center
Mongolian Academy of Sciences
Ulaanbaatar 210644
Mongolia
dbadamgarav@yahoo.com

CATHERINE E. BADGLEY
Museum of Paleontology
University of Michigan
Ann Arbor, Mich. 48109
cbadgley@umich.edu

JOHN C. BARRY
Department of Human Evolutionary Biology
Harvard University
Cambridge, Mass. 02138
jcbarry@fas.harvard.edu

MARK BEECH
Historic Environment Department
Abu Dhabi Tourism and Culture Authority (ADTCA)
P.O. Box 2380
Abu Dhabi
United Arab Emirates
mark.beech@adach.ae

ANNA K. BEHRENSMEYER
Department of Paleobiology
Smithsonian Institution
Museum of Natural History
Washington, D.C. 20013
behrensa@si.edu

RAYMOND L. BERNOR
Sedimentary Geology and Paleobiology Program
National Science Foundation
Arlington, Va. 22230

College of Medicine
Department of Anatomy
Laboratory of Evolutionary Biology
Howard University, Washington D.C. 20059
rbernor@comcast.net

SHUN-DONG BI
Department of Biology
Indiana University of Pennsylvania
Indiana, Pa. 15705
shundong.bi@iup.edu

FAYSAL BIBI
Department of Vertebrate Paleontology
American Museum of Natural History
Central Park West at 79th Street
New York, NY 10024
fbibi@amnh.org

MADELAINE BÖHME
Section Paleontology
University of Munich
80333 Munich
Germany
m.boehme@lrz.uni-muenchen.de

BAO-QUAN CAI
Department of History
Xiamen University
361005 Xiamen
Fujian Province
China
caibq@163.com

YAOWALAK CHAIMANEE
Department of Mineral Resources
Paleontology Section
Bangkok 10400
Thailand
yaowalak@dmr.go.th

OLIVIER CHAVASSEAU
Université de Poitiers
F-86022 Poitiers
France
olivier.chavasseau@univ-montp2.fr

I. U. CHEEMA
Department of Environmental Sciences
Arid Agriculture University
Rawalpindi
Pakistan
iucheema1962@yahoo.com

PAULINE COSTER
Université de Poitiers
F-86022 Poitiers
France
pauline.coster@univ-poitiers.fr

J.-Y. CROCHET
Mission Paléontologique Franco-Balouche
F-34270 Le Triadou
France

GUDRUN DAXNER-HÖCK
Geological-Paleontological Department
Natural History Museum
1010 Vienna
Austria
gudrun.hoeck@nhm-wien.ac.at

HANS DE BRUIJN
Institute of Earth Sciences Utrecht
Utrecht University
3584 CD Utrecht
Netherlands
hdbruijn@geo.uu.nl

TAO DENG
Institute of Vertebrate Paleontology and Paleoanthropology
Chinese Academy of Sciences
100044 Beijing
China
dengtao@ivpp.ac.cn

WEI DONG
Institute of Vertebrate Paleontology and Paleoanthropology
Chinese Academy of Sciences
100044 Beijing
China
dongwei@ivpp.ac.cn

EDOUARD-GEORGES EMONET
Université de Poitiers
F-86022 Poitiers
France

MARGARITA ERBAJEVA
Geological Institute
Siberian Branch
Russian Academy of Sciences
Ulan-Ude 670047
Russia
erbajeva@gin.bscnet.ru

JUSSI T. ERONEN
Department of Geosciences and Geography and
Helsinki Institute for Information Technology
Department of Computer Science
University of Helsinki
FI-00014 Helsinki
Finland
jussi.t.eronen@helsinki.fi

LAWRENCE J. FLYNN
Peabody Museum of Archaeology and Ethnology and
Department of Human Evolutionary Biology
Harvard University
Cambridge, Mass. 02138
ljflynn@fas.harvard.edu

MIKAEL FORTELIUS
Institute of Biotechnology and
Department of Geosciences and Geography
University of Helsinki
FI-00014 Helsinki
Finland

Institute of Vertebrate Paleontology and Paleoanthropology
Chinese Academy of Sciences
100044 Beijing
China
mikael.fortelius@helsinki.fi

AKIRA FUKUCHI
Hanshin Consultants
Satsumasendai
Kagoshima
afkc@goo.jp

URSULA BETTINA GÖHLICH
Geological-Paleontological Department
Natural History Museum
1010 Vienna
Austria
ursula.goehlich@nhm-wien.ac.at

WULF A. GOSE
Jackson School of Geosciences
University of Texas
Austin, Tex. 78712

ANDREW HILL
Department of Anthropology
Yale University
New Haven, Conn. 06520
andrew.hill@yale.edu

KEES HORDIJK
Institute of Earth Sciences Utrecht
Utrecht University
3584 CD Utrecht
Netherlands
hordijk@geo.uu.nl

SU-KUAN HOU
Institute of Vertebrate Paleontology and Paleoanthropology
Chinese Academy of Sciences
100044 Beijing
China
kk338@126.com

JEAN-JACQUES JAEGER
Université de Poitiers
F-86022 Poitiers
France
jean-jacques.jaeger@univ-poitiers.fr

ANU KAAKINEN
Department of Geosciences and Geography
University of Helsinki
FI-00014 Helsinki
Finland
anu.kaakinen@helsinki.fi

YOSHINARI KAWAMURA
Aichi University of Education
448–8542 Kariya
Aichi Prefecture
Japan
yskawamr@auecc.aichi-edu.ac.jp

AUNG AUNG KHYAW
Department of Archaeology
National Museum and Library
Mandalay
Myanmar

YURI KIMURA
Department of Earth Sciences
Southern Methodist University
Dallas, Tex. 75205
yuri.mammut@gmail.com

ANNETTE KOSSLER
Institut für Geologische Wissenschaften
Freie Universität Berlin
FR Paläontologie
12249 Berlin
Germany
kossler@zedat.fu-berlin.de

DIMITRIS S. KOSTOPOULOS
Geological Department
Museum of Geology and Paleontology
Aristotle University of Thessaloniki
GR-54124 Thessaloniki
Greece.

GEORGE D. KOUFOS
Laboratory of Geology and Palaeontology
Department of Geology
Aristotle University of Thessaloniki
GR-54124 Thessaloniki
Greece
koufos@geo.auth.gr

CHUAN-KUI LI
Institute of Vertebrate Paleontology and Paleoanthropology
Chinese Academy of Sciences
100044 Beijing
China
li.chuankui@pa.ivpp.ac.cn

QIANG LI
Institute of Vertebrate Paleontology and Paleoanthropology
Chinese Academy of Sciences
100044 Beijing
China
liqiang@ivpp.ac.cn

JOSEPH C. LIDDICOAT
Department of Environmental Science
Barnard College
Columbia University
New York, N.Y. 10027
jcl31@columbia.edu

EVERETT H. LINDSAY
Department of Geosciences
University of Arizona 85721
Tucson, Ariz.
ehlind@cox.net

LI-PING LIU
Institute of Vertebrate Paleontology and Paleoanthropology
Chinese Academy of Sciences
100044 Beijing
China
liuliping@ivpp.ac.cn

Department of Geosciences and Geography
University of Helsinki
FI-00014 Helsinki
Finland

JUHA PEKKA LUNKKA
Department of Geology
University of Oulu
FI-90014 Oulu
Finland
juha.pekka.lunkka@oulu.fi

LAURENT MARIVAUX
Mission Paléontologique Franco-Balouche
F-34270 Le Triadou
France

Equipe de Paléontologie
Institut des Sciences de l'Évolution
Université Montpellier 2
F-34095 Montpellier Cedex 5
France
Laurent.Marivaux@univ-montp2.fr

JIN MENG
Department of Paleontology
American Museum of Natural History
New York, N.Y. 10024
jmeng@amnh.org

GREGOIRE MÉTAIS
Mission Paléontologique Franco-Balouche
F-34270 Le Triadou
France

Centre de Recherche sur la Paléobiodiversité et les
    Paléoenvironnements
Laboratoire de Paléontologie
Muséum National d'Histoire Naturelle
F-75005 Paris
France
akssyria@hotmail.com

MATTHEW C. MIHLBACHLER
Department of Anatomy
New York College of Osteopathic Medicine
Old Westbury, N.Y. 11568

Department of Paleontology
American Museum of Natural History
New York, N.Y. 10024
mmihlbac@nyit.edu

KAZUNORI MIYATA
Fukui Prefectural Dinosaur Museum
911–8601 Katsuyama
Fukui Prefecture
Japan
k-miyata@dinosaur.pref.fukui.jp

MICHÈLE E. MORGAN
Department of Human Evolutionary Biology
Harvard University
Cambridge, Mass. 02138
memorgan@fas.harvard.edu

RYOHEI NAKAGAWA
Mie Prefectural Museum
514–0006 Tsu
Mie Prefecture
Japan
nakagr00@pref.mie.jp

HIDEO NAKAYA
Department of Earth and Environmental Sciences
Faculty of Science
Kagoshima University
890–0065 Kagoshima
Kagoshima Prefecture
Japan
nakaya@sci.kagoshima-u.ac.jp

XI-JUN NI
Institute of Vertebrate Paleontology and Paleoanthropology
Chinese Academy of Sciences
100044 Beijing
China
nixj@amnh.org

NEIL D. OPDYKE
Department of Geology
University of Florida
Gainesville, Fla. 32611
drno@ufl.edu

ZAHRA ORAK
Department of Environment
Muze Meli Tarikh Tabiei (National Museum of Natural History)
Tehran
Iran

MAEVA J. ORLIAC
Equipe de Paléontologie
Institut des Sciences de l'Évolution
Université Montpellier 2
F-34095 Montpellier Cedex 5
France
maeva.orliac@univ-montp2.fr

BENJAMIN H. PASSEY
Department of Earth and Planetary Sciences
Johns Hopkins University
Baltimore, Md. 21218
bhpassey@jhu.edu

RAJEEV PATNAIK
CAS in Geology
Panjab University
Chandigarh
India
rajeevpatnaik@gmail.com

HANNELE PELTONEN
Department of Geosciences and Geography
University of Helsinki
FI-00014 Helsinki
Finland
hk.peltonen@gmail.com

LAURI J. PESONEN
Department of Physics
Laboratory for Solid Earth Geophysics
University of Helsinki
FI-00014 Helsinki
Finland
lauri.pesonen@helsinki.fi

DAVID PILBEAM
Department of Human Evolutionary Biology
Harvard University
Cambridge, Mass. 02138
pilbeam@fas.harvard.edu

GUO-QIN QI
Institute of Vertebrate Paleontology and Paleoanthropology
Chinese Academy of Sciences
100044 Beijing
China
qiguoqin@ivpp.ac.cn

ZHAN-XIANG QIU
Institute of Vertebrate Paleontology and Paleoanthropology
Chinese Academy of Sciences
100044 Beijing
China
qiuzhanxiang@ivpp.ac.cn

ZHU-DING QIU
Institute of Vertebrate Paleontology and Paleoanthropology
Chinese Academy of Sciences
100044 Beijing
China
qiuzhuding@ivpp.ac.cn

ABDUL RAHIM RAJPAR
Earth Sciences Division
Pakistan Museum of Natural History
44000 Islamabad
Pakistan
rajpar55@yahoo.com

S. MAHMOOD RAZA
Higher Education Commission
Sector H-9
Islamabad
Pakistan
smraza@hec.gov.pk

FLORENT RIVALS
Institut Català de Paleoecologia Humana i Evolució Social (IPHES)
Campus Sescelades URV (Edifici W3)
43007 Tarragona
Spain
frivals@iphes.cat

G. ROOHI
Earth Sciences Division
Pakistan Museum of Natural History
44000 Islamabad
Pakistan
roohighazala@yahoo.com

MANA RUGBUMRUNG
Department of Mineral Resources
Paleontology Section
Bangkok 10400
Thailand

HARUO SAEGUSA
Museum of Nature and Human Activities
669–1546 Sanda
Hyogo Prefecture
Japan
saegusa@hitohaku.jp

GINA M. SEMPREBON
Bay Path College
Longmeadow, Mass. 01106
gsempreb@baypath.edu

SEVKET SEN
Centre de Recherches sur Paléobiodiversité et
    Paléoenvironments
UMR7207 du CNRS
Muséum National d'Histoire Naturelle
75005 Paris
France
sen@mnhn.fr

AUNG NAING SOE
Department of Geology
Dagon University
Yangon
Myanmar

NIKOS SOLOUNIAS
Department of Anatomy
New York College of Osteopathic Medicine
Old Westbury, N.Y. 11568

Department of Paleontology
American Museum of Natural History
New York, N.Y. 10024
nsolouni@nyit.edu

GARY T. TAKEUCHI
Page Museum at the La Brea Tar Pits
5801 Wilshire Blvd.
Los Angeles, CA 90036
gtakeuch@tarpits.org

HIROYUKI TARUNO
Osaka Museum of Natural History
546–0034 Osaka
Osaka Prefecture
Japan
taruno@mus-nh.city.osaka.jp

RICHARD H. TEDFORD†
Department of Paleontology
American Museum of Natural History
New York, N.Y. 10024
†1929–2011

ALEXEY S. TESAKOV
Geological Institute
Russian Academy of Sciences
Moscow
Russia
tesak@ginras.ru

VADIM V. TITOV
Institute of Arid Zones
Southern Scientific Center
Russian Academy of Sciences
Rostov-on-Don
Russia
vvtitov@yandex.ru

YUKIMITSU TOMIDA
National Museum of Nature and Science
305-0005 Tsukuba
Ibaraki Prefecture
Japan
y-tomida@kahaku.go.jp

ZHIJIE J. TSENG
Department of Vertebrate Paleontology
Natural History Museum of Los Angeles County
Los Angeles, Calif. 90007

Program of Integrative Evolutionary Biology
University of Southern California
Los Angeles, Calif. 90007
jack.tseng@usc.edu

SOE THURA TUN
Myanmar Geosciences Society
Hlaing University Campus
Yangon
Myanmar

ENGIN ÜNAY
Mineral Research and Exploration Institute of Turkey
06800 Çankaya, Ankara
Turkey
unayengin@yahoo.com.tr

ELEONORA VANGENGEIM[†]
Geological Institute
Russian Academy of Sciences
Moscow
Russia
[†]1930–2012

BAN-YUE WANG
Institute of Vertebrate Paleontology and Paleoanthropology
Chinese Academy of Sciences
100044 Beijing
China
wangbanyue@pa.ivpp.ac.cn

XIAOMING WANG
Department of Vertebrate Paleontology
Natural History Museum of Los Angeles County
Los Angeles, Calif. 90007

Institute of Vertebrate Paleontology and Paleoanthropology
Chinese Academy of Sciences
100044 Beijing
China
xwang@nhm.org

MAHITO WATABE
Hayashibara Museum of Natural Sciences
700–0907 Okayama
Okayama Prefecture
Japan
moldavicum@pa2.so-net.ne.jp

JEAN-LOUP WELCOMME
Mission Paléontologique Franco-Balouche
F-34270 Le Triadou
France
jl@welcomme.org

WILMA WESSELS
Faculty of Geosciences
Department of Earth Sciences
Utrecht University
2584 CD Utrecht
Netherlands

DOMINIK WOLF
Forschungsinstitut Senckenberg
ROCEEH
Senckenberganlage 25
60325 Frankfurt am Main
Germany
d_wolf_palaeo@yahoo.de

MICHAEL O. WOODBURNE
Department of Geology
Museum of Northern Arizona
Flagstaff, Ariz. 86001
mikew@npgcable.com

WEN-YU WU
Institute of Vertebrate Paleontology and Paleoanthropology
Chinese Academy of Sciences
100044 Beijing
China
wuwy@yahoo.com

GUANG-PU XIE
Gansu Provincial Museum
730050 Lanzhou
Gansu Province
China
xgp2008.ok@163.com

WALID YASIN
Historic Environment Department
Abu Dhabi Tourism and Culture Authority (ADTCA)
Al Ain National Museum
Al Ain
United Arab Emirates
walid.ismail@adach.ae OR w.yasin@adach.ae

JIE YE
Institute of Vertebrate Paleontology and Paleoanthropology
Chinese Academy of Sciences
100044 Beijing
China
jieye@hotmail.com

GHOLAMREZA ZARE
Department of Environment
Muze Meli Tarikh Tabiei (National Museum of Natural History)
Tehran
Iran

ZHAO-QUN ZHANG
Institute of Vertebrate Paleontology and Paleoanthropology
Chinese Academy of Sciences
100044 Beijing
China
zhangzhaoqun@ivpp.ac.cn

SHAO-HUA ZHENG
Institute of Vertebrate Paleontology and Paleoanthropology
Chinese Academy of Sciences
100044 Beijing
China
zhengshaohua@ivpp.ac.cn

TAXONOMIC INDEX

Note: Page numbers followed by *f* and *t* indicate figures and tables, respectively. For variations in spellings of generic names by different authors in different chapters, we followed M. C. McKenna and S. K. Bell, *Classification of Mammals Above the Species Level* (New York: Columbia University Press, 1997), for the name with priority to be listed first and followed by the variant in parentheses, such as *Dicrocerus* (=*Dicroceros*). For species names, we did not attempt to reconcile differences and simply listed them as equal alternatives separated by a "/," (e.g., *Ancylotherium pentelici/pentelicum*). Nor did we attempt to standardize taxonomic opinions of binomials, and certain species may be listed under different genera by different authors, (e.g., *Cormohipparion antelopinum*, *Hipparion antelopinum*, and *Hippotherium antelopinum*).

GENERAL INDEX